Lecture Notes on Data Engineering and Communications Technologies **170**

The aim of the book series is to present cutting edge engineering approaches to data technologies and communications. It will publish latest advances on the engineering task of building and deploying distributed, scalable and reliable data infrastructures and communication systems.

The series will have a prominent applied focus on data technologies and communications with aim to promote the bridging from fundamental research on data science and networking to data engineering and communications that lead to industry products, business knowledge and standardisation.

Indexed by SCOPUS, INSPEC, EI Compendex.

All books published in the series are submitted for consideration in Web of Science.

Jemal H. Abawajy · Zheng Xu ·
Mohammed Atiquzzaman · Xiaolu Zhang
Editors

Tenth International Conference on Applications and Techniques in Cyber Intelligence (ICATCI 2022)

Volume 1

Editors
Jemal H. Abawajy
Faculty of Science, Engineering and Built Environment
Deakin University
Geelong, VIC, Australia

Mohammed Atiquzzaman
School of Computer Science
University of Oklahoma
Norman, OK, USA

Zheng Xu
School of Computer Engineering and Sciences
Shanghai Polytechnic University
Shanghai, China

Xiaolu Zhang
Department of Information Systems and Cyber Security
The University of Texas at San Antonio
San Antonio, TX, USA

ISSN 2367-4512 ISSN 2367-4520 (electronic)
Lecture Notes on Data Engineering and Communications Technologies
ISBN 978-3-031-29096-1 ISBN 978-3-031-29097-8 (eBook)
https://doi.org/10.1007/978-3-031-29097-8

This Springer imprint is published by the registered company Springer Nature Switzerland AG
The registered company address is: Gewerbestrasse 11, 6330 Cham, Switzerland

Foreword

The 10th International Conference on Applications and Techniques in Cyber Intelligence (ATCI 2022), building on the previous successes online conference (2021 and 2020 due to COVID-19), Huainan, China (2019), Shanghai, China (2018), Ningbo, China (2017), Guangzhou, China (2016), Dallas, USA (2015), Beijing, China (2014), and Sydney, Australia (2013), is proud to be in the 10th consecutive conference year. ATCI 2021 has moved online due to COVID-19.

The purpose of ATCI 2022 is to provide a forum for presentation and discussion of innovative theory, methodology and applied ideas, cutting-edge research results, and novel techniques, methods, and applications on all aspects of cyber and electronics security and intelligence. The conference establishes an international forum and aims to bring recent advances in the ever-expanding cybersecurity area including its fundamentals, algorithmic developments, and applications.

Each paper was reviewed by at least two independent experts. The conference would not have been a reality without the contributions of the authors. We sincerely thank all the authors for their valuable contributions. We would like to express our appreciation to all members of the program committee for their valuable efforts in the review process that helped us to guarantee the highest quality of the selected papers for the conference.

We would like to express our thanks to the strong support of the general chairs, publication chairs, organizing chairs, program committee members, and all volunteers.

Our special thanks are due also to the editors of Springer Thomas Ditzinger and Suresh Dharmalingam for their assistance throughout the publication process.

Jemal Abawajy
Zheng Xu
Mohammed Atiquzzaman
Xiaolu Zhang

Organization

General Chairs

Hui Zhang	Tsinghua University, China
Liang Wang	Chinese Academy of Sciences, China

Online Conference Organizing Chairs

Xianchao Wang	Fuyang Normal University, China
Shibing Wang	Fuyang Normal University, China

Program Chairs

Jemal Abawajy	Deakin University, Australia
Zheng Xu	Shanghai Polytechnic University, China
Mohammed Atiquzzaman	University of Oklahoma, USA
Xiaolu Zhang	The University of Texas at San Antonio, USA

Publication Chairs

Mazin Yousif	T-Systems International, USA
Vijayan Sugumaran	Oakland University, USA

Publicity Chairs

Kewei Sha	University of Houston, USA
Neil. Y. Yen	University of Aizu, Japan
Shunxiang Zhang	Anhui University of Science and Technology, China

Program Committee Members

William Bradley Glisson	Sam Houston State University, USA
George Grispos	University of Nebraska at Omaha, USA
V. Vijayakumar	VIT Chennai, India
Aniello Castiglione	Universit di Salerno, Italy
Florin Pop	University Politehnica of Bucharest, Romania
Neil Yen	University of Aizu, Japan
Xianchao Wang	Fuyang Normal University & Tech., China
Feng Wang	Fuyang Normal University & Tech., China
Jia Zhao	Fuyang Normal University & Tech., China
Xiuyou Wang	Fuyang Normal University & Tech., China
Gang Sun	Fuyang Normal University & Tech., China
Ya Wang	Fuyang Normal University & Tech., China
Bo Han	Fuyang Normal University & Tech., China
Xiuming Chen	Fuyang Normal University & Tech., China
Xiangfeng Luo	Shanghai Univ., China
Xiao Wei	Shainghai Univ., China
Huan Du	Shanghai Univ., China
Zhiguo Yan	Fudan University, China
Abdulbasit Darem	Northern Boarder University, Saudi Arabia
Hairulnizam Mahdin	Universiti Tun Hussein Onn, Malaysia
Anil Kumar K. M	JSS Science & Technology University, Mysore, Karnataka, India
Haruna Chiroma	Abubakar Tafawa Balewa University, Bauchi, Nigeria
Yong Ge	University of North Carolina at Charlotte, USA
Yi Liu	Tsinghua University, China
Foluso Ladeinde	SUNU, Korea
Kuien Liu	Pivotal Inc., USA
Feng Lu	Institute of Geographic Science and Natural Resources Research, Chinese Academy of Sciences, China
Ricardo J. Soares Magalhaes	University of Queensland, Australia
Alan Murray	Drexel University, USA
Yasuhide Okuyama	University of Kitakyushu, Japan
Wei Xu	Renmin University of China, China
Chaowei Phil Yang	George Mason University, USA
Hengshu Zhu	Baidu Inc., China
Morshed Chowdhury	Deakin University, Australia
Elfizar	University of Riau, Indonesia
Rohaya Latip	Universiti Putra, Malaysia

The 10th International Conference on Applications and Techniques in Cyber Intelligence (ATCI2022)

18 June 2022
ATCI 2022 has moved online due to COVID-19.
http://atci.com.cn/

Program Book

Conference Program at a Glance

Saturday, June 18, 2022, Tencent Meeting Online Link		
10:00–10:10	Opening ceremony by conference PC Chair	Tencent Meeting
10:10–10:50	Keynote 1: Kim-Kwang Raymond Choo	Tencent Meeting
10:50–11:30	Keynote 2: Jemal Abawajy	Tencent Meeting
Saturday, June 18, 2022, Tencent Meeting Online Link		
13:00–18:00	Session 1	Tencent Meeting
	Session 2	Tencent Meeting
	Session 3	Tencent Meeting
	Session 4	Tencent Meeting
	Session 5	Tencent Meeting
	Session 6	Tencent Meeting
	Session 7	Tencent Meeting

Please download Tencent Meeting.
We will send an online conference link to your email.
https://meeting.tencent.com/sg/en/index.html

ATCI 2022 Keynotes

Kim-Kwang Raymond Choo holds a Ph.D. in information technology from Queensland University of Technology, Australia. Prior to starting his Cloud Technology Endowed Professorship at UTSA, Dr. Choo spent five years working for the University of South Australia and five years working for the Australian Government Australian Institute of Criminology. He was also a visiting scholar at INTERPOL Global Complex for Innovation between October 2015 and February 2016 and a visiting Fulbright scholar at Rutgers University School of Criminal Justice and Palo Alto Research Center (formerly Xerox PARC) in 2009.

In April 2017, he was appointed an honorary commander, 502nd Air Base Wing, Joint Base San Antonio-Fort Sam Houston, USA. He is also a fellow of the Australian Computer Society, a senior member of IEEE, and co-chair of IEEE Multimedia Communications Technical Committee's Digital Rights Management for Multimedia Interest Group.

Jemal Abawajy is a faculty member at Deakin University and has published more than 100 articles in refereed journals and conferences as well as a number of technical reports. He is on the editorial board of several international journals and edited several international journals and conference proceedings. He has also been a member of the organizing committee for over 60 international conferences and workshops serving in various capacities including best paper award chair, general co-chair, publication chair, vice-chair, and program committee. He is actively involved in funded research in building secure, efficient, and reliable infrastructures for large-scale distributed systems. Toward this vision, he is working in several areas including: pervasive and networked systems (mobile, wireless network, sensor networks, grid, cluster, and P2P), e-science and e-business technologies and applications, and performance analysis and evaluation.

Contents

Cyber Intelligence for Business and Management Innovations

TOPSIS Model Establishment in the Context of Internet Finance

Ruonan Gu[1], Yuanyuan Deng[1](✉), Di Zhao[1], Lei Cai[1], and Mujeeb Ur Rehman[2]

[1] School of Sciences, Shenyang Jianzhu University, Shenyang, Liaoning, China
dyy_0124@163.com

[2] Huazhong University of Technology, Wuhan, Hubei, China

Abstract. With the rapid development of Internet finance in China, the loan scale of Chinese financial institutions is also increasing year by year. China's financial institutions have a wide range of service objects and users, and a large amount of money. If they cannot accurately identify users with high risk of default, the healthy development of market economy will be affected. Therefore, establishing an accurate credit risk assessment model of Internet finance to reduce non-performing loans is conducive to the healthy development of China's market economy and social harmony and stability. In this paper, TOPSIS method is adopted to select 7 evaluation indicators (corporate structure, tax rate, etc.) to quantify, standardize and standardize the indicators successively. We build best-plan, worst-plan and scoring formulas. Finally, the enterprise credit risk assessment model is established to measure the ability of enterprises to repay loans, so as to assess the credit risk of enterprises. In this paper, the data of 10 companies as an example, the application of the established model, through MATLAB to solve, and draw the results into a graph. Finally, SPSS was used for systematic clustering of enterprises, and combined with the scores, the 10 smes were divided into low risk, medium risk and high risk categories. Banks can make credit decisions according to the final scores and clustering results of each enterprise.

Keywords: Internet Finance · TOPSIS Method · MATLAB Decomposition · Credit Decision

1 Introduction

Corporate loans are one of the main sources of income for banks. At the same time, Banks and companies have formed a mutually beneficial pattern. The enterprise obtains sufficient funds to carry out project activities by borrowing from the bank, which banks get profit from. However, there is problem when banks grant loans to small and medium-sized enterprises. This is because those enterprises are often small-scale and lack hypothecated asset, which lead to the result that they fail to repay their loans if the companies don't operate well. Therefore, it is essential to study the problem of bank credit decision-making for small and medium enterprises [1, 2].

Before making a credit decision, we must first assess the credit risk of the company. The traditional risk assessment models generally adopt the analytic hierarchy process

J. H. Abawajy et al. (Eds.): ICATCI 2022, LNDECT 170, pp. 3–11, 2023.
https://doi.org/10.1007/978-3-031-29097-8_1

(AHP) [3–5], which constructs a judgment matrix for the indicators that may cause the company to be unable to repay loans, obtains the weight of each indicator, and finally calculates the corporate score. Determine whether to grant loans to enterprises. But this model has some weaknesses: 1. Traditional models generally use expert evaluation and scoring methods when constructing the judgment matrix. The disadvantage of this method is that it is highly subjective, and seem to lack objectivity. 2. At the same time, the traditional evaluation model has serious polarization in the index selection process. Too few indexes will affect the accuracy of the final judgment result. Too many indexes will increase the complexity of calculation in the process of applying the model and may also cause confusion in the evaluation scale. 3. The traditional evaluation model only considers whether the company meets the criteria for granting loans, but does not consider how to make decisions to maximize benefits when funds are limited.

According to references [6], We have established an evaluation index system to measure the credit risk of enterprises [7, 8], which consist of seven indicators including the company's credit rating, default record, company's net income, and the number of effective invoices for sales. Then, the TOPSIS method is adopted to establish an evaluation model that is based on the relevant data of the existing 10 companies to evaluate the company's ability to repay loans, thereby predicting the companies' credit risk. After that, we introduced weights and made a progress based on the original model. In order to maintain the robustness of the model, this paper adopts three calculation methods to calculate the weights and then gained the average value [9]. Finally, a corresponding credit risk assessment model is established that is applied to evaluate 10 companies' credit risk.

2 TOPSIS Model Establishment

2.1 Credit risk Assessment Indicators

Relevant information shows that credit risk in a broad sense refers to the uncertainty of bank earnings. The profitability, operating ability and reputation of banks all reflect the bank's ability to repay loans to a certain extent, and these factors will all have an impact on bank credit decisions [10, 11]. This article takes the relevant data of 10 companies as an example (The data used in this paper are available at http://www.mcm.edu.cn/), and selects 7 indicators such as reputation rating, whether there is a default record, company net income, company structure, tax rate, number of effective transactions, and the proportion of invalid exchanges to assess the credit risk of these companies Evaluation. The Risk Evaluation Index System is shown in Fig. 1.

Among them, credit rating and default record are indicators to measure the company's creditworthiness, and the company's net income and tax rate are used to measure the company's profitability. The formula for calculating net income is Net income = total output price tax-total input price tax.

The tax rate is calculated by the ratio of value-added tax to the current taxable sales income. The tax rate of companies with weak profitability will be lower. The calculation formula for this indicator is tax rate = current value-added tax payable / current taxable sales income Value-added tax = output tax-input tax.

The number of transactions, the proportion of obsolete exchanges, and the company structure can reflect the company's operational capabilities and stability to a certain extent. The proportion of voided transactions refers to the ratio of voided transactions to the total number of transactions [12–14]. Companies with large structures are generally more stable, and the risk of banks granting loans to them is relatively small.

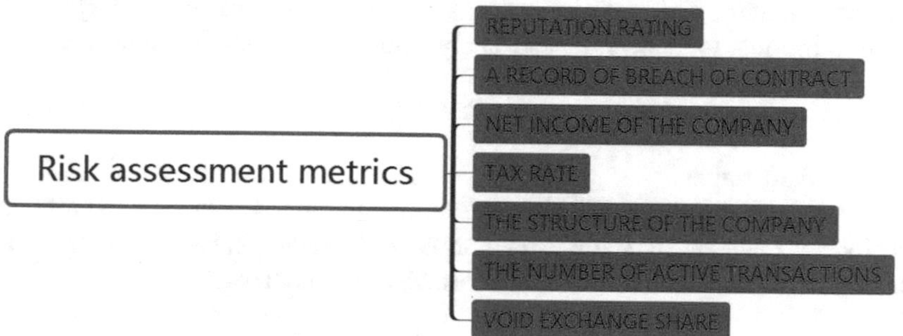

Fig. 1. Risk assessment metrics

2.2 Index Preprocessing

Step 1: the quantification of indicators

Some of the indicators need to be quantified, such as default records and reputation ratings [15]. The company's reputation ratings are divided into three categories: A, B, and C, and default records are divided into yes or no. The specific quantitative indicators are shown in Table 1 and Table 2.

Table 1. Reputation rating metrics

Reputation rating	Quantify metrics
A	3
B	2
C	1

Table 2. Quantitative indicators of default records

Whether to default	Quantify metrics
Yes	0
No	1

Step 2: Positive processing of indicators

Among the selected indicators, the larger the indicator, the stronger the company's ability to repay the loan, and the higher the corresponding score, such as the company's net income; some indicators are the opposite. The bigger the better the index becomes the very large index, the smaller the better the index is called the extremely small index. We need to perform indicator conversion on the indicators, and the indicators are processed in a positive direction, that is, all indicators are converted into extremely large ones. The calculation formula for converting extremely small to extremely large indicators is:

$$\max -x \tag{1}$$

Step 3: Standardization

Different indicators have different dimensions. In order to eliminate the influence of dimensions, we need to standardize the normalized indicators. If there are n evaluation objects and m evaluation indicators, the normalization matrix is:

$$X = \begin{bmatrix} x_{11} & x_{12} & \cdots & x_{1m} \\ x_{21} & x_{22} & \cdots & x_{2m} \\ \vdots & \vdots & \cdots & \vdots \\ x_{n1} & x_{n2} & \cdots & x_{nm} \end{bmatrix} \tag{2}$$

Denote the normalized matrix as Z, and the elements in Z are:

$$z_{ij} = x_{ij} / \sqrt{\sum_{i=1}^{n} x_{ij}^2} \tag{3}$$

2.3 Corporate Loan Repayment Ability Score

We define the optimal solution $Z^+ = (Z_1^+, Z_2^+, \cdots, Z_m^+)$:

$$= (\max\{z_{11}, z_{21}, \cdots, z_{n1}\}, \max\{z_{12}, z_{22}, \cdots, z_{n2}\}, \cdots, \max\{z_{1m}, z_{2m}, \cdots, z_{nm}\}) \tag{4}$$

Then define the worst solution $Z^- = (Z_1^-, Z_2^-, \cdots, Z_m^-)$:

$$= (\min\{z_{11}, z_{21}, \cdots, z_{n1}\}, \min\{z_{12}, z_{22}, \cdots, z_{n2}\}, \cdots, \min\{z_{1m}, z_{2m}, \cdots, z_{nm}\}) \tag{5}$$

The distance between the i-th evaluation object and the optimal solution is:

$$D_i^+ = \sqrt{\sum_{j=1}^{m} (Z_j^+ - z_{ij})^2} \tag{6}$$

The distance between the i-th evaluation object and the worst solution:

$$D_i^- = \sqrt{\sum_{j=1}^{m} (Z_j^- - z_{ij})^2} \tag{7}$$

Then, we can calculate the unnormalized score of the i-th evaluation object:

$$S_i = \frac{D_i^-}{D_i^+ + D_i^-} \tag{8}$$

Finally, according to the scores of each company, the company's loan repayment ability is evaluated. The higher the score, the stronger the loan repayment ability and the lower the credit risk.

3 Application of the Model

We take the relevant data of the companies we have (part of the data are shown in Table 3) as an example, and use the established credit risk assessment model to score these companies.

Table 3. Some data on enterprise-related metrics

Company number	Reputation rating	Whether to default	Void transactions
E1	3	1	0.03
E2	2	1	0.1
E3	1	1	0.11
E4	1	1	0.01
E5	2	1	0.11
E6	1	1	0.02
E7	1	1	0.74
E8	2	1	0.08
E9	3	1	0.04
E10	1	1	0.05

Then the matrix is processed by normalization and normalization. We use MATLAB to solve the problem and get the scores of each company which is shown in Fig. 2.

From the Fig. 2, we can see that the enterprise numbered E9 (hereinafter referred to as E9) has the highest score, while the enterprise numbered E1 (hereinafter referred to as E9) has the lowest score. It shows that E9 has strong loan repayment ability and low credit risk, while E1 is the opposite.

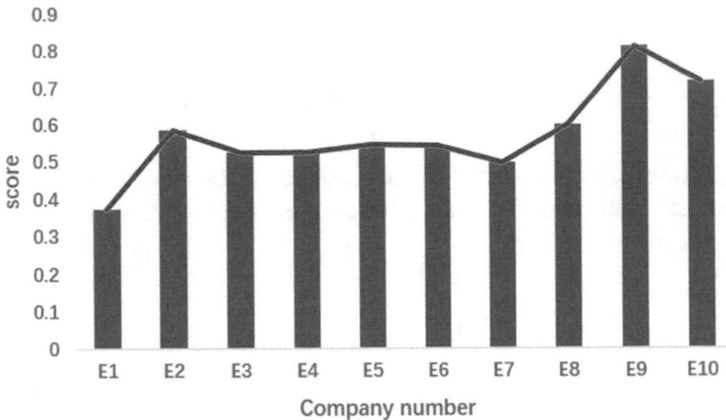

Fig. 2. Corporate loan repayment ability score

4 Introduce Weight

Taking into account that different indicators have different degrees of influence on credit risk, this article introduces indicator weights. Here, this article uses the analytic hierarchy process to set the weights. In order to maintain the robustness of the model, we use arithmetic average method, geometric average method, and eigenvalue three methods are used to calculate the weight of each indicator, and finally the average of the weights of the three methods is selected as the final weight of each indicator. The calculation results of the weights are shown in the Table 4.

Table 4. The result of the calculation of the weight

	Arithmetic average	Geometric averaging	The feature value method	Average
Tax rate	0.3369	0.3417	0.3367	0.34
Net income of the company	0.1832	0.1855	0.1833	0.18
Reputation rating	0.0691	0.0674	0.0683	0.07
Whether to default	0.0356	0.0345	0.0354	0.04
Void exchange share	0.11	0.1077	0.1112	0.11
The number of active transactions	0.1688	0.1696	0.1689	0.17
The structure of the company	0.0964	0.0936	0.0961	0.1

The weighted scoring results of each company are shown in the Fig. 3.

It can be seen that the company's score has not changed much after the introduction of weights, indicating that the original model is reasonable.

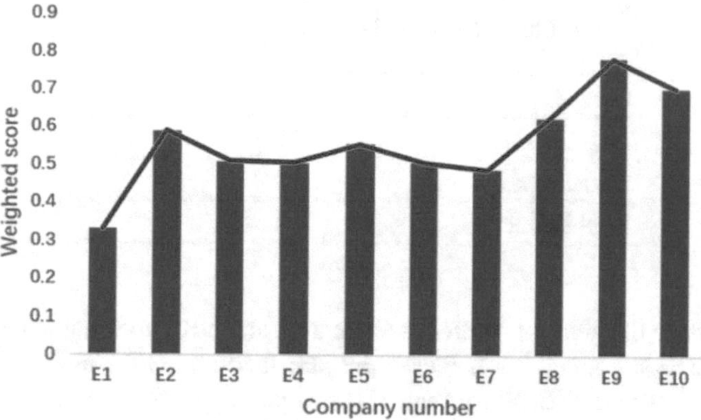

Fig. 3. Enterprise-weighted loan repayment capacity score

5 Cluster Analysis

Then we use SPSS software to perform cluster analysis on the scoring results, and the clustering results are shown in Fig. 4:

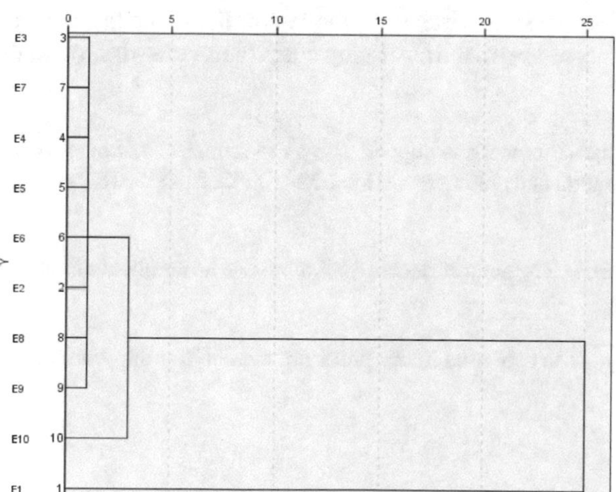

Fig. 4. Clustered genealogy

Finally, we divide companies into three categories: low-risk, medium-risk, and high-risk. The details are shown in the Table 5:

Table 5. The classification of the enterprise

Category	Number
Low-risk class	2
Moderate risk class	7
High-risk class	1

You will have the greatest control over the appearance of your figures if you are able to prepare electronic image files. Please prepare the image files in PostScript (PS) or Encapsulated PostScript (EPS) formats. Use a separate file for each image. File names should be of the form "Fig. 1.ps" or "Fig. 2.eps".

6 Conclusion

This paper adopts the TOPSIS method, selects 7 evaluation indicators of company structure, tax rate, company income, etc., and establishes a credit risk evaluation model of the company to measure the company's ability to repay loans, thereby evaluating the company's credit risk.

Finally, we use SPSS clustering to divide small and medium-sized enterprises into three categories according to the clustering coefficient: low-risk, medium-risk, and high-risk.

Acknowledgments. This work is supported by the t fund of national undergraduate training program for innovation and entrepreneurship (D202006262135250782).

Conflicts of Interest. The authors declare that there are no conflicts of interest.

Data Availability. The data used in this paper are available at http://www.mcm.edu.cn/.

References

1. Shaista, W., Nabila, N.: Debt financing decisions of SMES in emerging markets: empirical evidence from Malaysia. Int. J. Bank Mark. **37**(1), 258–277 (2019)
2. Assef, F., Steiner, M.T., Steiner Neto, P.J., de Barros Franco, D.G.: Classification algorithms in financial application: credit risk analysis on legal entities. IEEE Latin Am. Trans. **17**(10), 1733–1740 (2019)
3. Shukla, U.P., Nanda, S.J.: Designing of a risk assessment model for issuing credit card using parallel social spider algorithm. Appl. Artif. Intell. **33**(1–4), 191–207 (2019)
4. Lappas, P.Z., Yannacopoulos, A.N.: A machine learning approach combining expert knowledge with genetic algorithms in feature selection for credit risk assessment. Appl. Soft Comput. **107**(29), 107391 (2021)
5. Masmoudi, K., Abid, L., Masmoudi, A.: Credit risk modeling using Bayesian Network with a latent variable. Expert Syst. Appl. **127**, 157–166 (2019)

6. Liang, L., Jiang, R.Y., Liang, Y.: A Multicriteria decision method with uncertain information in financial credit loan decision-making. Appl. Mech. Mater. **52–54**, 1868–1872 (2011)
7. Gordy, M.B.: A comparative anatomy of credit risk models. J. Bank. Financ. **24**(1–2), 119–149 (2000)
8. Gustafson, C.R., Pederson, G.D., Gloy, B.A.: Credit risk assessment. Agric. Financ Rev. **65**(July), 201–217 (2005)
9. Twala, B.: Multiple classifier application to credit risk assessment. Expert Syst. Appl. **37**(4), 3326–3336 (2010)
10. Gourieroux, C.: Affine models for credit risk analysis. J. Financ. Economet. **4**(3), 494–530 (2006)
11. Marinakis, Y., Marinaki, M., Doumpos, M., Matsatsinis, N., Zopounidis, C.: Optimization of nearest neighbor classifiers via metaheuristic algorithms for credit risk assessment. J. Glob. Optim. **42**(2), 279–293 (2008). https://doi.org/10.1007/s10898-007-9242-1
12. Doumpos, M., Zopounidis, C.: Model combination for credit risk assessment: a stacked generalization approach. Ann. Oper. Res. **151**(apr.), 289–306 (2007)
13. Bekhet, H.A., Eletter, S.F.K.: Credit risk assessment model for Jordanian commercial banks: neural scoring approach. Rev. Dev. Financ. **4**(1), 20–28 (2014)
14. Hilscher, J., Wilson, M.I.: Credit ratings and credit risk: is one measure enough. Manage. Ence, **63**(10) (2013)
15. Ferreira, F.A.F., Santos, S.P., Dias, V.M.C.: An Ahp-based approach to credit risk evaluation of mortgage loans. Int. J. Strateg. Prop. Manag. **18**(1), 38–55 (2014)

An Analysis of Internet Financial Risk Prevention Strategies from the Perspective of Network Security

Jianwei Ma[1], Baoping Zhang[1(✉)], and Asia Ullah[2]

[1] School of Economics and Management, Lanzhou University of Technology, Lanzhou 730050, Gansu, China
zbp18822034512@126.com

[2] University of Balochistan, Quetta, Pakistan

Abstract. With the rapid development of the Internet in recent years, its regulatory issues have become increasingly prominent and have attracted great attention from all walks of life. Among them, issues related to network security have become a research hotspot for many scholars. Based on the various Internet financial network trust issues, security and legal issues facing the current Internet development process, this article studies the problems, influencing factors and risk prevention of Internet financial companies from the perspective of Internet financial users in my country. Finally starts from "Improving the internal control system", "Establishing a risk early warning system and evaluation mechanism", "Focusing on talent cultivation", and putting forward relevant suggestions and countermeasures.

Keywords: Cyberspace Security · Cloud Computing · Internet Financial Risks

1 Introduction

1.1 Background Description

The birth of the Internet has benefited from the integration and development of computer software and hardware technology and communication technology. With the recent rapid development of social economy, the Internet has brought a profound impact to the entire society. In particular, the application of Internet finance in all walks of life has caused companies to face a reshuffle of their original market position and competitive landscape. In such a rapidly changing environment, companies are faced with many opportunities while facing many opportunities. With numerous challenges. As a direct participant in micro-level enterprises and economic activities, what choice should be made in the face of this current situation? Financing is a necessary link in the process of enterprise development. Internet finance has undoubtedly become a necessary means to solve corporate financing difficulties. It has become an indispensable way for today's enterprises to respond to economic globalization and participate in competition. It is essential for the development of enterprises. When the virtual digital world and the physical world become an unstoppable trend, under the application of third-party payment, P2P lending mode, crowdfunding mode, and small and micro financial mode represented by Alibaba Microfinance Group, the network Space security is no longer a simple technical issue.

J. H. Abawajy et al. (Eds.): ICATCI 2022, LNDECT 170, pp. 12–19, 2023.
https://doi.org/10.1007/978-3-031-29097-8_2

1.2 Literature Review

It is pointed out that the application of financial technology and digital technology has reduced the starting point for residents to obtain financial services, the number of people enjoying financial services has increased, and the marginal transaction cost has been reduced (Hanno Beck, 2001). Although Internet financial intermediary platforms can reduce financing costs, the problem of the failure of both parties in the transaction can easily cause borrowers to face huge selection risks and disrupt the order of the lending market (Freedman 2008). Herbert (2008) pointed out that banking financial technology risks can be divided into five categories, in which the definition, measurement methods and identification methods of Internet financial operational risks are analyzed in detail. Classers et al. (2002) believe that digital finance will have a huge impact on the structure of the financial industry, and the prevention of digital financial risks needs to focus on security and soundness, public policy and other aspects at the same time [1]. Internet technology is an important way to extend the boundaries of financial business. However, the steady development of digital finance requires vigilance against multiple risks such as technology, law, and management, and a sound digital financial supervision, risk control and other systems (Oliver Gajda, 2013). Since the 1980s, with the help of database classification technology and blockchain technology, Western developed countries led by the United States have deeply embedded Internet technology in various fields such as transportation, information, energy, and finance. Popular vocabulary such as "Internet traffic", "Internet communication", "Internet energy" and "Internet finance" [2]. This phenomenon is called the "inversion principle" by Rochet and Tirole (2004), that is, if one party of the platform market is charged a higher fee, the other party of the platform will be charged a lower fee. Everett (2015) studied the impact of online borrowers' information asymmetry on adverse selection, moral hazard and deferred repayment [3].Most empirical studies believe that the development of Internet finance has increased the level of risk-taking by banks [4], WANG believes that the combination of supervised and unsupervised machine learning algorithms can accurately assess the credit risk of Internet finance and improve the supervision efficiency of regulators. From the perspective of behavior, some scholars use evolutionary games to explore the evolutionary equilibrium strategy of Internet financial supervision behavior [5].

2 The Status Quo of Internet Finance Development and Financial Risks

2.1 Status Quo of Internet Finance Development

At present, my country's Internet financial business is mainly divided into five categories: "digital payment, online financing, digital wealth management, new digital insurance, and digital credit investigation". The security of online payment is the most direct manifestation of online payment risk exposure. Risk events Promote the risk exposure of online payment and evolve into an online payment security issue. Under certain conditions, not all online payment risk factors are manifested as online payment security issues [6]. Only those risk factors that accompany risk events that receive social attention will be transformed into security issues that all parties to online payment are concerned about.

Compared with traditional financing activities, through the use of high-tech technology, Internet finance allows both parties to complete various transactions without leaving home, which greatly reduces transaction costs and improves transaction convenience [7]. With the development, the network trust problem of Internet financial transactions has gradually been exposed. For example, although digital wealth management platforms require a license granted by the state to operate, this only means that the state allows such business operations, but due to the scale of assets and risks of digital wealth management platforms The management level is uneven, and at the same time there is a lack of national credit as a redemption guarantee. Some financial products are not fully contracted by insurance companies, which have high credit risks and so on. In addition, it also includes liquidity risk, market risk, operational risk, long-tail risk, etc. [8].

2.2 Internet Financial Risks

Since Internet financial risks are the interweaving of Internet and traditional financial risks, the consequences caused by their risks also have different characteristics. The former pays more attention to user interaction, entertainment, and cross-border, while the latter tends to be rigorous and steady. Internet product innovation is guided by the ultimate user experience, and the product design of traditional financial institutions is oriented towards safety and stability. For example, In 2015, American scholar Melanie Swan first proposed the concept of blockchain finance, Any block in the blockchain records all the information of the previous block, and any information can be traced back to each block of the blockchain, so any transaction can be traced back, and any node is retained. The complete blockchain, and cannot be tampered with through timestamps [9].

Technical risk factors. The current various technical risks of Internet finance itself are mainly caused by the technical security of the network information system and the huge data capacity. For example, Internet finance companies actively or passively lead to the leakage of customer information, the entire system collapses due to imperfect technology, Purposeful hacking attacks, etc., technical factors are fatal to enterprises and customers. Once these risks occur, they will directly lead to the loss of a large amount of funds in the financial system, the leakage of customer personal information, and the blow of the Internet finance industry. Is devastating [10].

Regulatory risks mainly refer to financial risks caused by a vacuum or inadequate supervision of the financial supervision of my country's Internet market. Generally speaking, they are financial problems that are "unmanageable" and "unmanageable", and "unmanageable" are financial laws and regulations. The problem of "uncontrollable" is the invalidity of financial supervision and law enforcement methods. At present, my country's economic development is shifting from high-speed development to high-quality development. The problems left over from the high-speed development stage cannot be dealt with quickly. The current legal system related to Internet finance in my country is not perfect, and people have many problems. Based on his review of the Internet lending regulatory system, he proposed to strengthen the inclusiveness and democracy of the regulatory system to promote industry innovation and achieve financial inclusion. Based on a microscopic perspective, most scholars have explored the impact of big data, blockchain and other technologies on Internet financial supervision. The formulation and implementation of relevant policies still need a long way to go.

Taking P2P as an example, its online lending refers to the establishment of direct lending relationships between the supply and demand parties of funds in a specific network environment, information interaction through the network, the establishment of electronic contracts involving the amount of funds, deadlines and interest rates, and the protection of electronic signature technology Its legal effect. It also satisfies the interests of both borrowers and lenders, building bridges and providing services for both parties [11] (Fig. 1).

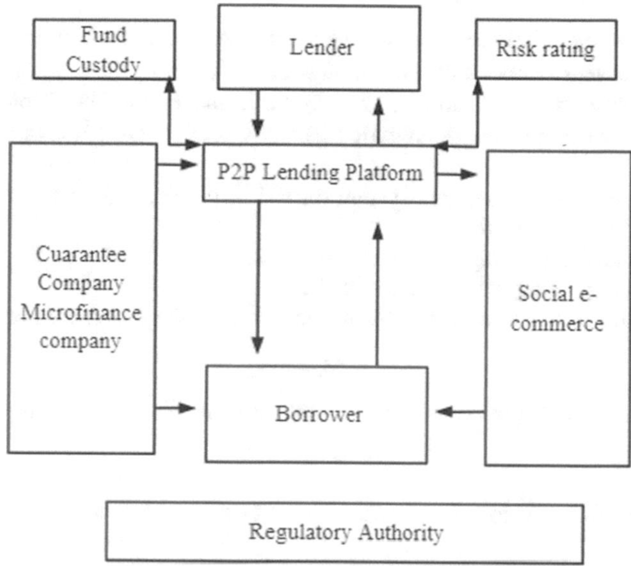

Fig. 1. P2P Major Participants in Lending

As a small financing method, self-employed individuals, small and micro enterprises and even personal fund transfers are its main customers. Most platforms have low qualifications and high borrowing interest. The lack of strict examination and approval has led to occasional defaults by borrowers. This has caused difficulties in platform operation and even bankruptcy. Through consulting related literature, expert consultation, questionnaire surveys and other methods, it is found that the current influencing factors affecting Internet financial risks mainly include "Internet financial product differentiation, platform qualifications, government supervision, the degree of perfection of relevant laws, the nature of the institution's legal person, and the registered capital of the platform. Interest rates, etc." Take the average annual interest rate of Internet financial institutions as an example. The higher the average annual interest rate of the institution, the greater the financial risk [12] (Fig. 2).

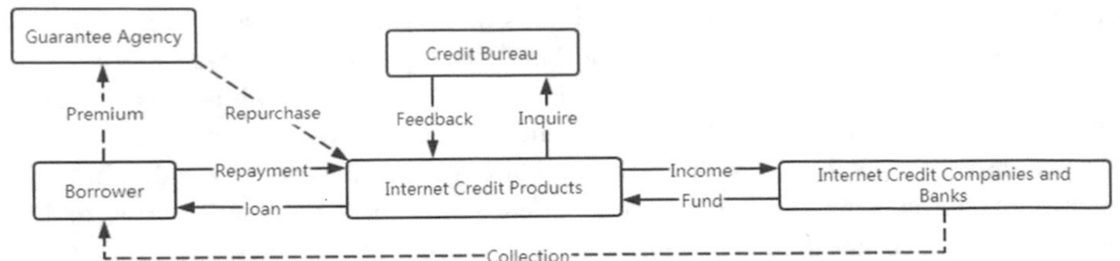

Fig. 2. Internet Microfinance Product Operation Model

2.3 Influencing Factors

At present, many scholars' research on the influence factors of Internet financial risks is more qualitative research, and seldom pays attention from the empirical aspect. The analytic hierarchy process is more applied to the influence factors of Internet financial risks. The application process of analytic hierarchy process (AHP) is as follows.

(1) Choosing an evaluation index system for risk influencing factors.
(2) Construct a judgment matrix.
(3) Calculate the index weight and consistency test.
(4) Single-level ranking and consistency check.
The first step is to calculate the row product of the judgment matrix Mi.

$$M_i = A_{ij} \tag{1}$$

The second step is to open the Nth power of the row product W_i.

$$W_i = \sqrt[n]{M_i} \tag{2}$$

The third step is to calculate the feature vector, which is the weight W_i'.

$$W_i' = \frac{W_i}{\sum_{i=1}^{n} W_i} \tag{3}$$

The fourth step is to calculate the maximum eigenvalue λmax.

$$\lambda_{max} = \sum_{i=1}^{n} \frac{[AW]i}{nw_i} \tag{4}$$

The fifth step is to query the RI value of the corresponding level of the judgment matrix and calculate the CI value.

$$CI = \frac{\lambda_{max} - n}{n - 1} \tag{5}$$

The sixth step is to calculate the consistency ratio CR.

$$CR = {}^{CI}/_{RI} \tag{6}$$

When the random consistency ratio $CR \leq 0.10$ or $CI = 0.00$, it means that the structure of the matrix single sorting meets the requirements of consistency, otherwise the judgment matrix should be revised.
(5) Comprehensive evaluation of Internet financial risk level.

3 Case Analysis

3.1 Case Introduction

Ant Financial Services Group (hereinafter referred to as "Ant Financial Services") was established in October 2014. As early as March 7, 2013, Alibaba Group decided to prepare for the establishment of Alibaba Small and Micro Financial Services Group based on the Alipay account system and integrate the Alibaba system. All financial innovations for small and micro businesses and individual consumers. After 19 months of preparatory work, Ant Financial was established. In recent years, the proportion of Ant Financial's main business income is shown in the table below (Table 1).

Table 1. Proportion of Ant Group's Main Business Income

Project	2017 years	2018 years	2019 years	2020 years
	Proportion of amount	Proportion of amount	Proportion of amount	Proportion of amount
Digital payment and merchant services	54.88%	51.75%	43.03%	35.86%
Digital Fintech Platform	44.33%	47.38%	56.20%	63.39%
Innovative business and others	0.79%	0.87%	0.77%	0.75%
Total	100%	100%	100%	100%

As can be seen, it has a complete business structure and complete functions. It uses "Alipay" to have the settlement function of a commercial bank, uses "Alipay" to have the deposit function of a commercial bank, and uses "Ant Micro Loan" to have the loan function of a commercial bank. Obviously, Ant Ant Financial has become a representative and model in the field of Internet finance in China.

3.2 Operating Basis

Ant Financial's products such as Alipay, Yu'ebao, and micro-loan services have been loved by the general public once they came out, but their development should be attributed to the development of computers and the Internet, which stems from big data, cloud computing and credit investigation Wait for the support of the underlying platform.

Cloud computing based on big data can provide powerful storage and computing capabilities. It not only has the ability to quickly process a large amount of data and information, but also has a powerful effect in cooperation with big data. Through the cooperation of many apps and Alipay, it has a large amount of original information, including not only the seller's business data, but also personal bank statements, utility payment records, credit card repayment records, funds transactions with others, and occupations. Related information, etc. Through these raw data and the perfect combination of big data and cloud computing, Ant Financial's position in my country's Internet finance is unshakable.

One of the in-depth purposes of Ant Financial for big data is to control everyone's credit status through the analysis of each data behavior. However, in the Internet finance era, sharing data is more important than analyzing data, and the more sharing, the greater the value created. Therefore, Ant Financial is still working hard to build an Internet credit investigation system.

4 Internet Financial Risk Prevention Strategy

4.1 Improve the Internal Control System

A feasible and effective internal control system can provide a reliable guarantee for companies to avoid financing risks. It needs to gradually break the barriers of traditional thinking, introduce professional managers, and improve the level of financial management. Financial management departments at all levels should strengthen communication, clarify the responsibilities and powers of financial personnel, establish a management structure where the chief financial officer supervises the overall situation, financial personnel at all levels perform their duties, and the external audit department conducts real-time supervision, and implements financial management work. In every business link, do a good job of matching powers and responsibilities to maximize corporate value.

4.2 Establish a Risk Early Warning System and Evaluation Mechanism

Facing the rapidly changing internal and external environment, Internet finance companies need to establish and improve financing risk early warning systems and evaluation mechanisms. An effective risk early warning system can enable companies to timely discover credit risks, business management risks and policy risks that are beyond their risk controllable range. The financing evaluation mechanism is to determine the intensity of their risks and determine whether they are within the controllable range, and then formulate Corresponding measures such as risk avoidance, loss control, risk transfer, risk retention, etc., are eliminated when they have not occurred or are in the bud. "Everything will be established in advance, and if not, the risk warning system must be updated in real time according to changes in the internal and external environment, so that the results will be accurate. In addition, the evaluation results of the risk system can also be used as the basis for performance appraisal of managers and employees to form a top-down risk management system with full participation.

4.3 Focus on Talent Cultivation

The construction of a risk management and control talent team is essential to the sustainable development of enterprises. Internet finance companies need to improve their awareness and ability to manage and control financing risks while emancipating the minds of senior managers. For those who cannot meet their development needs due to lack of personnel, the staffing should be increased in time, a set of perfect, capable of meeting demand and attractive talent introduction mechanism should be established within the enterprise, and the recruitment and recruitment of investment and financing departments should be increased. Cultivate strength. For the selection of talents, we must strictly control the entrance, and expand the truly capable talents to relevant departments. At the same time, it is also necessary to conduct regular job training for all financial staff to encourage the most basic level to understand their development status and enhance management awareness and management. Responsibility to help companies to carry out effective capital management. In addition, a sound organizational system and staffing only meet the rigid requirements of risk management. The final implementation of the work requires relevant management methods to promote self-learning and continuous improvement in accordance with the market environment and regulatory requirements.

References

1. Classers, S., Thomas, G., Daniela, K.: Electronic finance: reshaping the financial landscape around the world. J. Financ. Serv. Res. **1**, 29–61 (2020)
2. Wright, A.: Decentralized Blockchain Technology and the Rise of Lex Cryptographia. Soc. Sci. Electron. Publishing **15**(1), 11–12 (2018–09)
3. Everett, C.R.: Group membership, relationship banking and loan defaultrisk: the case of online social lending. Bank. Finance Rev. **7**(2), 15–54 (2015)
4. Wang, R., Liu, J., Luo, H.: Fintech development and bank risk taking in China. Eur. J. Finance **27**(4–5), 397–418 (2021)
5. Wang, B., Ning, L.J., Kong, Y.: Integration of unsupervised and supervised machine learning algorithms for credit risk assessment. Expert Syst. Appl. **128**, 201–315 (2019)
6. Allen, F.: E-Finance: an introduction. J. Financ. Serv. Res. **22**(2), 5–27 (2002)
7. Bertino, E., Bettini, C., Ferrari, E., Somarati, P.: An access control model supporting periodicity constraints and temporal reasoning. ACM Trans. Database Syst. **23**(3), 231–285 (2000)
8. Berger, S., Gleisner, F.: Emergence of Financial Intermediaries on Electronic Markets: The Case of Online P2P Lending. In: Working Paper, University of Frankfurt (2006)
9. Kamble, S.S., Gunasekaran, A., Sharma, R.: Modeling the blockchain enabled traceability in agriculture supply chain. Int. J. Inform. Manage. **52**, 101967 (2020)
10. Kumar, V.L., Nataraian, S., Keerthana, L.: Credit risk analysis in peer-to-peer lending system. Bus. Horiz. **1**, 28–36 (2016)
11. Yu, T., Shen, W.: Funds sharing regulation in the context of the sharing economy: understanding the logic of chin' s P2P lending regulation. Comput. Law Secur. Rev. **35**(1), 42–58 (2019)
12. Stephenson, P.: Applying forensic techniques to information system risk management first steps. Comput. Fraud Secur. **2003**(12), 17–19 (2003)
13. Dhillon, G., Torkzadeh, G.: Value-focused assessment of information system security in organizations. Inf. Syst. J. **16**(3), 293–314 (2006)

Prediction of Transaction Risk in Financial Market Based on Neural Network Model

Qiyang Li(✉)

Opto-Electronics Information Science and Engineering, College of Optical and Electronic Technology, China Jiliang University, Hangzhou, Zhejiang, China
liqiyang20000118@163.com

Abstract. In recent years, China's economic development has grown rapidly. As an important part of China's national economy, the financial market has become more and more important for risk avoidance and prediction in the transaction process in the financial market. China's per capita income level has been continuously improved, and the national policies have been adjusted for high interest rate sectors such as real estate and the Internet, The prediction of transaction risk in financial market is becoming more and more important for investors in the market. With the continuous development of computer technology, neural network is widely used in the financial market. Through the neural network under computer technology, the prediction model of the financial market is built, and the data of the financial market are collected and analyzed in real time, so as to provide sufficient data for the study of real social information and people's work. The experimental results show that the risk aversion efficiency of financial market transaction risk prediction based on neural network model is improved by 7.6%. Combined with computer intelligence, it fully displays the complexity of financial time series prediction model in wavelet clustering model and phase space reconstruction prediction model. For the limitations of manually analyzing the situation and one sidedness of the market, the risk points of historical market transactions in big data are sorted and reminded, the neural network algorithm is updated and iterated in an innovative way, and the financial risk prediction model is constructed by using the wavelet clustering prediction method.

Keywords: Neural Network · Financial Market · Transaction Risk · Modeling

1 Introduction

Participants in the financial market always hope to predict the future return and volatility of assets, and on this basis, reasonably allocate their own assets and effectively disperse and manage the risk volatility of assets. The research on volatility is of great theoretical significance in the financial field [1, 2]. Compared with the traditional prediction method based on historical data, the results show that the prediction model based on network information has better stability, closer to the actual transaction volume, and has certain application value [3, 4].

J. H. Abawajy et al. (Eds.): ICATCI 2022, LNDECT 170, pp. 20–28, 2023.
https://doi.org/10.1007/978-3-031-29097-8_3

Economic globalization and capital internationalization have strengthened the economic ties between countries and regions, making the interaction between different global financial markets unprecedented. For example, the "subprime mortgage crisis" in the United States in 2008 directly caused heavy damage to China, the European Union, Japan and even the whole global economy. With the continuous improvement of China's financial market system, It is undoubtedly of great practical significance to further discuss the relevant relations between the current markets [5, 6]. However, the existing literature has not comprehensively compared and studied multiple markets in China's current financial market system, and lacks a comprehensive analysis of the whole market system. Especially with the listing and trading of China's stock index futures, the stock index futures are compared with other markets such as funds, treasury bonds there are few literatures on the unification of foreign exchange and currency markets [7, 8].

The innovation of this paper is to put forward the transaction risk prediction of financial market based on neural network model. For the research on risk prediction of financial crisis, chaos model and data mining technology are mainly used abroad. The following is a brief analysis of the application of these two technologies. Chaos is a common phenomenon in nature. In order to better explore the laws of nature, chaos theory has become a common research theory, and chaos research methods are also very common. The unconditional coverage test method and economic index method are used to conduct a posteriori analysis on the VR prediction value. The bootstrap method is used to predict the stock price and rise and fall trend. Based on the above background, the research and analysis of the trend of the securities market has always been a hot topic, and the research and analysis of securities have attracted more and more attention of practitioners and researchers.

2 Neural Network in Financial Market Analysis

2.1 Neural Network Technology

With the deepening of research, a large number of analysis methods and auxiliary decision-making algorithms are available. We use intelligent algorithms and analysis methods to analyze the transaction risks in the financial market. More and more people review and screen the information on the network, and people get more information more quickly through the Internet [9, 10]. Arioua l uses the knowledge of Finance and econometrics, and uses spa to test the statistical significance of the results when using the loss function method. In the research of tail risk measurement, AR and es are used to measure the risk of China's Shanghai and Shenzhen 300 stock index futures market.

With the deepening of research, Krishnan B's view has fundamentally changed. More and more scholars have studied the definition and essence of risk. Risk is the uncertainty of the future; Risk is the possibility of loss. Risk is the degree of possible loss. The working principle of the neural network is that the total input of the neurons in the first hidden layer is obtained by the inner product operation between the input samples and the initial weight connected to the first hidden layer, and the output is obtained after the action of the nonlinear activation function. This output is used as the input of the next layer and calculated in turn until the output layer obtains the output of the network, which has the following forms.

$$P(h/v) = \prod_{j=1}^{N} P(h_j/v)$$

2.2 Risk Application Analysis of Financial Market

The securities market plays an important role in social and economic life. At present, the Western securities market has been relatively perfect after hundreds of years of development. After the founding of new China, China's securities market began to develop continuously in China. Especially after the reform and opening up, China's securities market developed rapidly and made remarkable achievements. The market risk is analyzed by neural network technology: that is, the stock index futures change with the change of the underlying stock index. Liquidity risk: that is, the risk to investors caused by the failure to realize the contract in time at an appropriate price. Cash flow risk: that is, the risk suffered by investors when they are unable to meet the capital requirements of normal transactions in time at the position building stage or when the performance trading margin required to maintain their position at the position holding stage is insufficient.

3 Risk Prediction and Analysis of Neural Network

Through the neural network system, looking back on the development and changes of reform and opening up, especially since the 1990s, China's financial market has gradually become a market system structure with typical characteristics such as sound legal system, deepening function, high degree of openness and strong resource allocation. Securities prediction refers to base on the information of securities itself (historical price, transaction quantity, etc.) Predict the future trend of securities by combining relevant company information and policies. Stock index futures are emerging gradually, which is of great significance to the capital market and its participants. Its primary role is to avoid risks. China's stock market implements short restriction, while the stock index futures market implements multi short two-way trading mechanism, which will help investors hedge system risks, alleviate the sharp rise and fall of unilateral market and stabilize the market. It is conducive to the formation of a reasonable market price and the maturity of the market.

4 Financial Environment and Risk Estimation and Analysis

4.1 Financial Market Risk Analysis

Due to the differences in the environment and the constraints of laws and regulations, the same kind of goods publicly traded in different markets often show different price trends, which undoubtedly increases the difficulty in the management of international financial activities to a great extent. Whether network information can be used to predict the financial market is the primary problem we need to solve. The specific results are shown in Fig. 1.

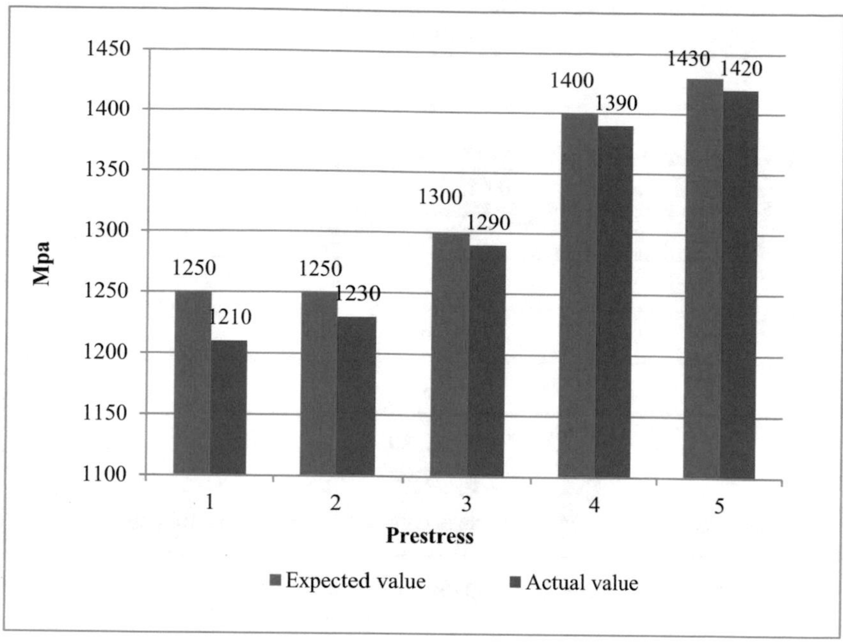

Fig. 1. The optimal capacity allocation

4.2 Market Risk Prediction Modeling

After empirical determination of the optimal parameters of the model, it is extended to continuous prediction, and the accuracy of continuous prediction in the short term is acceptable. Finally, the fitting model of time series is used, and the prediction error is smaller. By selecting the appropriate number of neurons in the input layer, the error can be controlled in a small range, indicating that the fitting model is optimal in the univariate prediction model (in Fig. 2).

Firstly, the obtained data are used to construct an undirected graph based on graph theory, and then the features in the undirected graph are counted, mainly including 11 features such as the number of edges and points of the graph, the average distribution of points, the maximum diameter and the number of published people. The representative feature data (maximum price and trading volume) in financial information are selected to calculate the correlation with the statistical features respectively, Research to test whether the network information is related to the financial market information: prediction is the ultimate goal of the research. The optimal fitting model is determined, the univariate dynamic prediction model is established, and the decision function and state function are added to make investment decision (in Table 1).

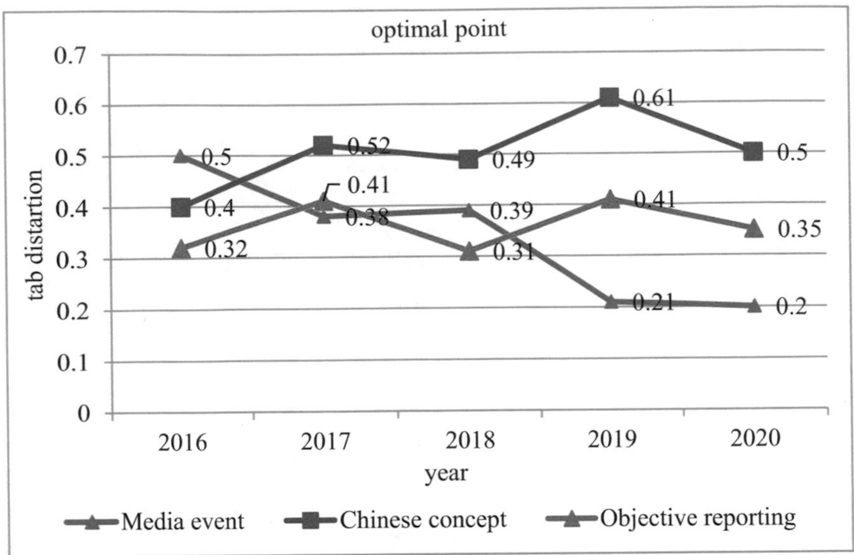

Fig. 2. Comparison of running time

Table 1. Statistical table of sample library

years	Main business income	Revenue growth rate	Total profit	Profit growth rate
2018	19967.1	9.4	1563.7	5.9
2019	21655.9	8.5	1662.1	6.3
2020	23595.8	9.4	1792	7.8

5 Conclusions

Although this paper has great deficiencies in the transaction risk prediction of the financial market based on the neural network model, due to the one-sided data that may be collected, with the continuous improvement of China's financial market system, it is undoubtedly of great practical significance to further discuss the relevant relations between the current markets. All walks of life conduct their own research and form a variety of prediction methods. Moreover, they are limited to the prediction of their own business departments. Commercial banks have their own prediction systems, and securities institutions also have their own prediction systems, which are incompatible with each other and have great limitations. However, the existing literature has not made a comprehensive comparative study of multiple markets in China's current financial market system, and lacks a comprehensive analysis of the whole market system. Especially with the listing and trading of China's stock index futures, there is less literature on the unified study of stock index futures and other markets such as funds, treasury bonds, foreign exchange and money.

6 BP Principle and Experimental Results

Let the three-layer BP network be as shown in Fig. 3. The input layer has M nodes, the output layer has L nodes, and the hidden layer has only one layer with N nodes.

There is only one layer with N nodes. In general, N > M > L. Let the input of the neural nodes in the input layer be ai (i = 1,2,...,M); the output of the nodes in the hidden layer is aj (j = 1,2,...,N); the output of the neural node in the output layer are yk (k = 1,2,...,L); the output vector of the neural network is ym the desired network output vector is yp.

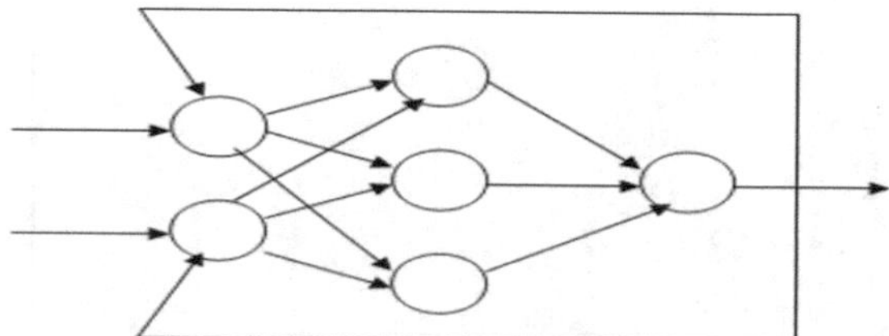

Fig. 3. The neural mode

The input of the i-th node of the input layer is

$$net_i = \sum_{i=1}^{M} x_i + \theta_i$$

where xi (1,2,...,M); is the input to the neural network, and ϑi is the threshold value of the i-th node. The corresponding output is

$$a_i = f(net_i) = \frac{1}{1 + \exp(-net_i)} = \frac{1}{1 + \exp\left(-\sum_{i=1}^{M} x_i - \theta_i\right)}$$

In BP network learning, the learning of nonlinear properties is mainly done by the hidden layer and the output layer. The general order.

$$a_i = x_i$$

The input to the jth node of the hidden layer is

$$net_j = \sum_{j=1}^{N} \omega_{ij} a_i + \theta_j$$

where ω_{ij}, θ_{ij} are the weights of the hidden layer and the threshold value of the jth node, respectively. The corresponding output is

$$a_j = f(net_j) = \frac{1}{1 + \exp(-net_j)} = \frac{1}{1 + \exp\left(-\sum_{j=1}^{N} \omega_{ij} a_i - \theta_j\right)}$$

The input to the kth node of the output layer is:

$$net_k = \sum_{k=1}^{l} \omega_{jk} a_j + \theta_k$$

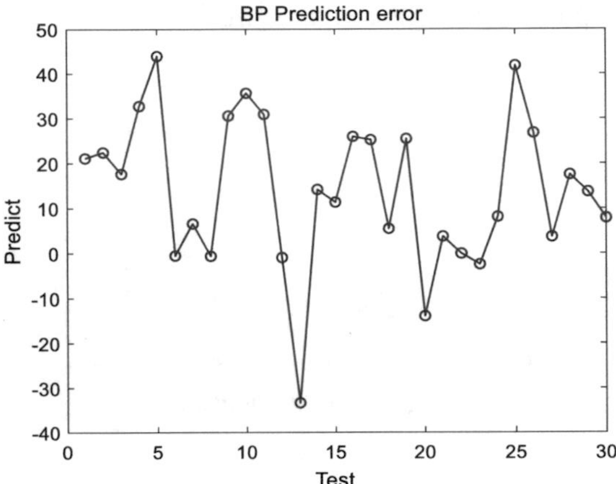

Fig. 4. BP Prediction error

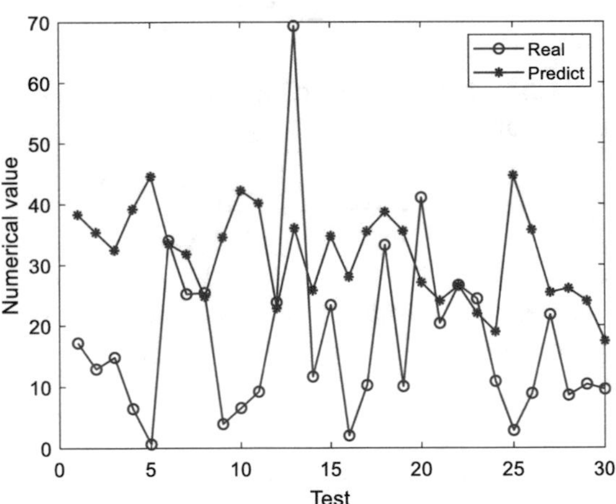

Fig. 5. The comparison of real and predict

Where ω_{jk}, θ_k are the weights and the kth node and threshold of the output layer, respectively. The corresponding outputs are

$$y_k = f(net_k) = \frac{1}{1 + \exp(-net_k)} = \frac{1}{1 + \exp(-net_k)}$$
$$= \frac{1}{1 + \exp\left(-\sum_{k=1}^{l} \omega_{jk} a_j - \theta_k\right)}$$

Predicted results (in Figs. 4, 5, and 6).

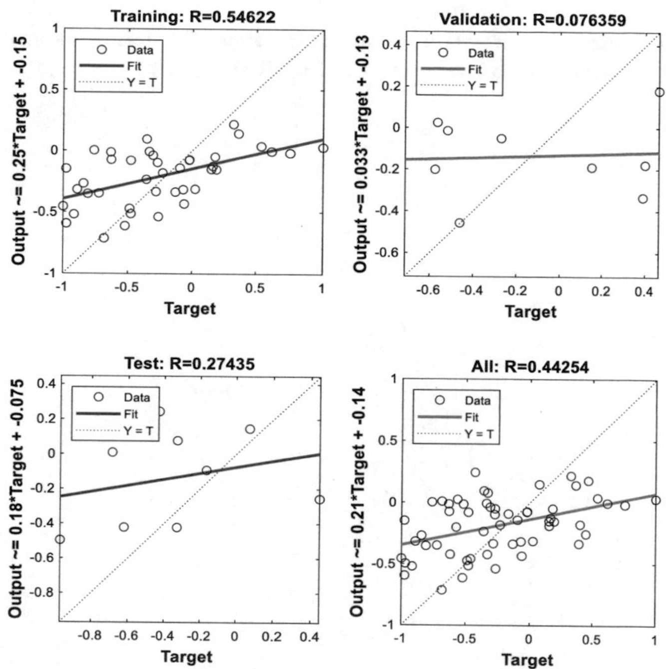

Fig. 6. The comparison of validation and target

References

1. Kwon, H., Lee, D., et al.: Dynamic forecasts of bankruptcy with recurrent neural network model. J. Intel. Inform. Syst. **23**(3), 139–153 (2017)
2. Hassabis, D., Kumaran, D., Summerfield, C., et al.: Neuroscience-inspired artificial intelligence. Neuron **95**(2), 245–258 (2017)
3. Ribeiro, G.T., Santos, A., Mariani, V.C., et al.: Novel hybrid model based on echo state neural network applied to the prediction of stock price return volatility. Expert Syst. Appl. **184**(1), 115490 (2021)
4. Bukhari, A.H., Raja, M.A.Z., Sulaiman, M., Islam, S., Kumam, M.S.P.: Fractional Neuro-sequential ARFIMA-LSTM for financial market forecasting. IEEE Access **8**, 71326–71338 (2020)
5. Camerer, C.F.: Artificial intelligence and behavioral economics. NBER Chapters **24**(18), 867–871 (2018)
6. Hassabis, D.: Artificial intelligence: chess match of the century. Nature **544**(7651), 413–414 (2017)
7. Kalaiselvi, K., Velusamy, K., Gomathi, C.: Financial prediction using back propagation neural networks with opposition based learning. J. Phys: Conf. Ser. **1142**(1), 012008 (2018)
8. Aryal, S., Nadarajah, D., Rupasinghe, P.L., et al.: Comparative analysis of deep learning models for multi-step prediction of financial time series. J. Comput. Sci. **16**(10), 1401–1416 (2020)

9. Rout, A.K., Dash, P.K., Dash, R., et al.: Forecasting financial time series using a low complexity recurrent neural network and evolutionary learning approach. J. King Saud Univ. – Comput. Inform. Sci. **29**(4), 536–552 (2017)
10. Kotoky, N., Dutta, A., et al.: Hybrid testing for evaluation of seismic performance of highway bridge with pier made of HyFRC. Structures **20**, 848–865 (2019)

Intelligent Business Model Data Analysis Based on Particle Swarm Optimization

Yanni Wu[1(✉)] and Rajit Ragab[2]

[1] School of Economics, Shanghai University, Shanghai, China
13262668187@163.com
[2] Vellore Institute of Technology, Vellore, India

Abstract. This paper analyzes business data based on particle swarm optimization algorithm (PSOA). Firstly, this paper introduces the development status of business data, then expounds the impact of business data on economy, and then studies PSOA. At the same time, the framework of business data analysis process based on the algorithm is studied, and the clustering effect of the algorithm is simulated. Finally, the experimental results show that with the continuous increase of the number of iterations, the fitness of the function decreases and finally approaches zero, indicating that the distance between the business data and the cluster center is decreasing. When the number of iterations is about 500, the Euclidean distance of fitness measurement function tends to zero, and the data clustering effect is better. This may indicate that PSOA can meet the needs of business data analysis.

Keywords: Particle Swarm Optimization Algorithm · Data Analysis · Business Model

1 Introduction

In the context of global economic integration, the financial industry is also facing more and more fierce and complex competition. Traditionally, the management of financial enterprises is mainly through banks and other financial institutions to achieve their own development [1, 2]. However, with the continuous change, development and progress of the times, people have higher and more comprehensive requirements for the financial industry, which is not only limited to simply recording data, analyzing the relationship between things and processing information, but also hope to obtain more effective, accurate and measurable decision-making methods and more scientific and systematic models and means. So as to provide customers with better service quality and efficiency [3, 4].

Many scholars have carried out relevant research on PSOA. The domestic literature on data mining and decision tree algorithm mainly studies the methods based on particle swarm optimization (NMP), object-oriented and kernel function. Some scholars put forward the kernel function as the core idea, and the combination of paradigm and heuristic. Other scholars pointed out two basic modes based on particle swarm optimization strategy, joint evolutionary search strategy and collaborative optimization in

J. H. Abawajy et al. (Eds.): ICATCI 2022, LNDECT 170, pp. 29–35, 2023.
https://doi.org/10.1007/978-3-031-29097-8_4

their articles [5, 6]. Another scholar proposed a clustering analysis method based on the training process. This method needs to process a large number of repetitive sample sets to obtain classification results before reaching a conclusion, so as to judge whether different types have similar attribute relations or large degree of differentiation. At the same time, a certain amount of prediction errors will be added to the training data [7, 8]. The above research has laid a research foundation for this paper.

In the field of economy and finance, data analysis is an important research direction, while the traditional clustering algorithm does not have real-time and accuracy due to time constraints. The model is established based on the random renewal principle, and the dynamic programming theory is used for iterative calculation. The optimal solution is obtained as the experimental result, and its feasibility and effectiveness are verified by simulation test. Finally, it is applied in the field of economy and finance.

2 Discussion on Economic and Financial Data Based on Particle Swarm Optimization Algorithm

2.1 Development Status of Economic and Financial Data Analysis

More and more new types of information service demand also began to appear [9, 10]. For example, e-commerce, online education and other fields have had a certain impact on the traditional financial industry. In the analysis of economic and financial data, we mainly use the traditional clustering algorithm, which combines the historical database. The development of economic and financial data analysis is a long and complex process. It requires us to collect, sort and classify these information, and apply it to the decision tree generation model. This method has many disadvantages. Firstly, it will lead to a large number of problems such as repeatability and redundancy. Secondly, because everyone focuses on the selection of the same index or variable by users at different levels, it may lead to differential treatment and decision deviation. Finally, economic and financial data analysis needs strong computing power. If the algorithm is too complex, it cannot meet the needs of practical application [11, 12].

2.2 Impact of Financial Data on Economy

The analysis of financial data mainly refers to the research on some phenomena generated in economic operation, so as to predict the future development trend and make decisions. In economic data analysis, we generally use financial indicators to reflect social value, and these indicators are abstract and general. Economic development is inseparable from finance, and the financial industry is the most important, dynamic and potential stock in the national economy. The pace of economic system reform has accelerated. Under the influence of national macro-control policies, the loan business of commercial banks has been impacted. There are links between the financial sector and the financial industry. They interact and cooperate with each other to form an organic system, and various industries or enterprises will be impacted to varying degrees, resulting in their development and changes. The analysis of financial data is to find the laws existing in these economic activities and find out their development trend. In the market economy, information transmission channels between enterprises and individuals are very convenient.

The analysis of financial data is mainly to study the economic impact in the process of financial industry development. In this field, people often pay attention to macro policies, which leads to the impact of many industries to varying degrees. The most important is the fluctuation brought about by the interest rate and the correlation between the price indexes of the stock market and the bond market to a certain extent, while the change in the exchange rate may also be caused by the types of financial instruments, trading volume and other reasons. Therefore, financial data have a significant impact on the economy.

2.3 Particle Swarm Optimization Algorithm

Particle swarm optimization algorithm is an optimization algorithm based on swarm intelligence under the influence of long-term observation of bird swarm and fish eye search activities. Through cooperation and competition, each bird forms a huge common force to quickly obtain food and avoid threats. The algorithm is an optimized search algorithm based on population intelligence. It not only maintains the overall population-based search strategy, but also introduces a practical "variable speed" model to realize the algorithm. Particle swarm optimization algorithm allows particles with dual functions of learning and memory to learn continuously, and dynamically adjust their search strategy according to the situation. Moreover, there are no specific operations such as crossover, selection and mutation of cluster members in the calculation process of the algorithm. Only a particle with mass and no volume can simulate birds in a flock of birds. The optimal solution of the problem can be: as a "particle field" to be found, each particle is limited to a region by a fitness function, and the particle has two attributes of position and velocity at the same time. Firstly, particles can be allocated as needed. In the iterative optimization cycle of the algorithm, each particle independently looks for the optimal predictive solution space in the search space according to certain rules, and takes it as the current individual extreme value. Then, the sharing mechanism is used to share the information with other particles in the population as a whole. Each particle adjusts its speed and position according to the direction of the current predicted optimal global solution, and finally converges to the current global position of the predicted optimal solution space. Particle swarm optimization has small parameters in practice and is easy to implement and master. At the same time, based on the biological and social background, it is helpful to solve more advanced nonlinear problems with excellent performance. This method is applied to optimize the performance of financial data analysis. Because the algorithm has the characteristics of strong overall search ability, fast convergence speed and few set parameters, it has attracted extensive attention in academic circles in recent years. Many improved algorithms have also been proposed to increase the ability to jump to and improve local extremes.

Mathematical description of PSO: the dimension of the search space is n, the population is m, and the population X = {X1,..., Xi,..., Xm}, where the position Xi of the ith particle = (Xi1, Xi2,..., Xin) T and the velocity VI = (Vi1, vi2,..., Vin) t. The individual extremum of particle I is pi = (Pi1, Pi2,..., pin) T, and the global extremum of the whole is Pg = (Pg1, PG2,..., Pgn) T. Xi particles update their velocity and position according to Eqs. (1) and (2).

$$v_{id}(k+1) = v_{id}(k) + c_1 rand\big(p_{id}(k) - x_{id}(k)\big) + c_2 rand\Big(p_{gd}(k) - x_{id}(k)\Big) \quad (1)$$

$$x_{id}(k+1) = x_{id}(k) + v_{id}(k+1) \quad (2)$$

where I = 1,2,..., m, M is the total number of particles representing m solutions in the group. N is the dimension of the particle search space. Xif (k) is the d-dimensional component of the k-th generation particle I. Pgd (k) is the best point for the k-generation population to find the component of the d-Dimension itself. C1 and C2 are generally a constant set 2. Appropriate C1 and C2 can accelerate the convergence of the algorithm and make it more difficult for particles to fall into local optimization.

3 Experiment

3.1 Financial Data Analysis Process Framework Based on Particle Swarm Optimization Algorithm

Financial data analysis based on PSOA mainly includes the following processes (shown in Fig. 1): (1) objective function. An objective function is determined according to the initial value of the system. In this process, the global search strategy is used. (2) The setting methods of decision variables, feedback factors and related parameters involve input and output. The specific steps are as follows: firstly, we need to select the operation to get the corresponding result set and establish the model, then complete it by solving the optimal solution, and finally calculate the financial index. The final results can be used to guide practical work.

3.2 Experimental Test Steps of Clustering Effect of Algorithm Framework

In this experiment, we mainly study the economic and financial data analysis based on PSOA, and classify and screen the original sample set through clustering. (1) Data preprocessing. Before clustering, we need to classify the samples to select the appropriate features, and then store these information in the database. First, make sure that each sample belongs to a certain attribute. Then judge whether it is a candidate identification according to the obtained single marking symbol. (2) Algorithm flow chart. Firstly, the data dimension parameters such as time period length and dimension required by each step are defined. Secondly, they are sorted after clustering and the results are obtained. Finally, a global optimal solution can be calculated through the formula.

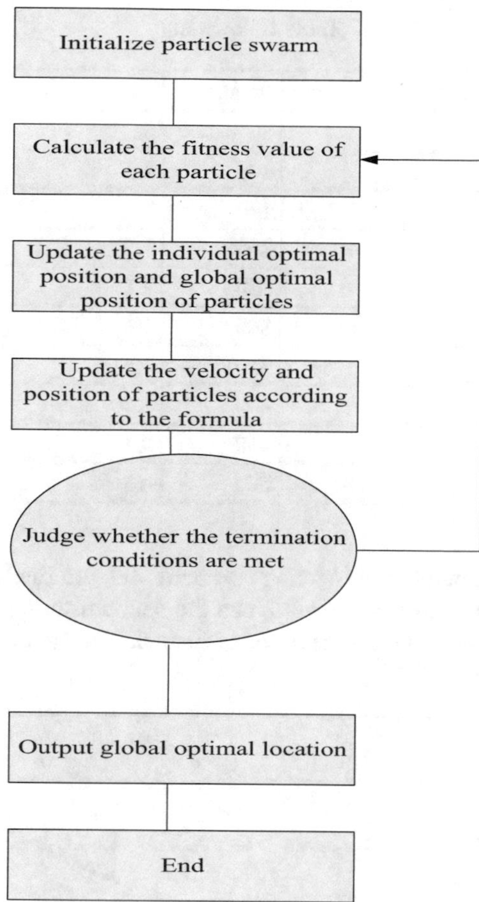

Fig. 1. Particle group co-optimization algorithm

4 Discussion

4.1 Test and Analysis of Clustering Effect of Particle Swarm Optimization Algorithm

Table 1 shows the clustering effect analysis data of financial data based on PSOA.

In this paper, we cluster the financial data and get the results. The algorithm can combine economic indicators with other aspects of performance. Particle swarm optimization algorithm can find a better solution and calculate the optimal value. At the same time, it can use different types of functions to deal with different problems. Finally, the use of these objective functions and the relationship between different parameters will also produce corresponding effects. The analysis results are shown in Fig. 2. We can clearly see from the comparison diagram of the cluster that with the continuous increase of the number of iterations, the fitness of the function decreases and finally approaches zero, indicating that the distance between the financial data and the cluster center is

Table 1. Test data

Financial data sample	Iterations	Function fitness
1	50	5
2	100	4.5
3	150	3
4	200	2.5
5	250	2.1
6	300	1.8
7	350	1.5
8	400	1.2
9	450	0.5
10	500	0.3

decreasing. When the number of iterations is about 500, the Euclidean distance of fitness measurement function tends to zero, and the data clustering effect is better. This may indicate that PSOA can meet the needs of financial data analysis.

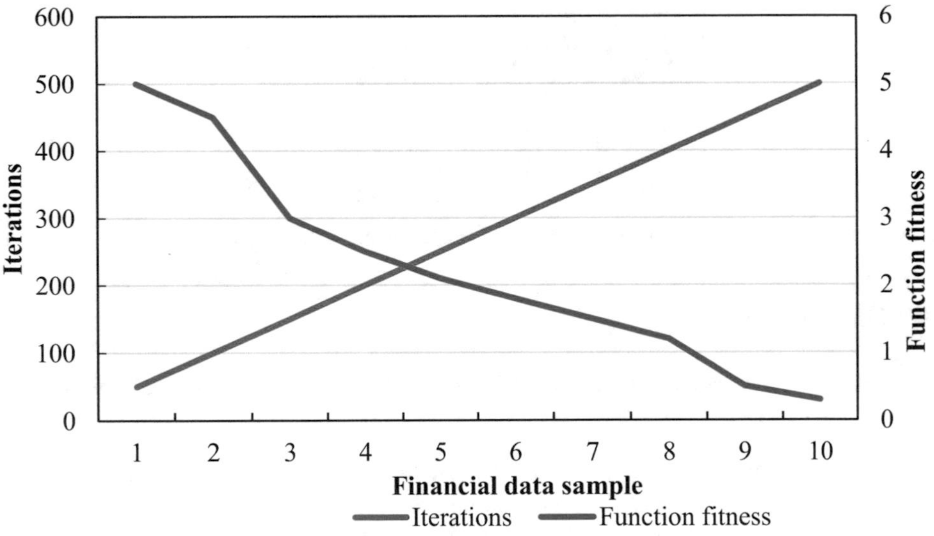

Fig. 2. Effect comparison of the particle group cluster clustering algorithm

5 Conclusion

Under the background of today's era, economic and financial development has also ushered in a golden age. All industries and fields need funds for investment. Economic and financial data analysis is a comprehensive evaluation of an enterprise's financial status and operating results, which provides an important basis for investors to make investment decisions. In this paper, the fuzzy clustering method is optimized based on PSOA, so as to effectively solve the problems and defects existing in the development of traditional economic industries.

References

1. Zhao, X.G., Zhang, Z.-Q., Xie, Y.-M., Meng, J.: Economic-environmental dispatch of microgrid based on improved quantum particle swarm optimization. Energy **195**, 117014.1-117014.15 (2020)
2. Jahed Armaghani, D., Shoib, R.S.N.S.B.R., Faizi, K., Rashid, A.S.A.: Developing a hybrid PSO–ANN model for estimating the ultimate bearing capacity of rock-socketed piles. Neural Comput. Appl. **28**(2), 391–405 (2015). https://doi.org/10.1007/s00521-015-2072-z
3. Fei, X., Tian, G.: Research on data mining algorithm based on neural network and particle swarm optimization. J. Intell. Fuzzy Syst. **35**(3), 2921–2926 (2018)
4. Jain, N.K., Nangia, U., Jain, J.: Economic Load Dispatch Using Adaptive Social Acceleration Constant Based Particle Swarm Optimization. J. Inst. Eng. (India), Ser. B. **9**(5), 431–439 (2018)
5. Abbas, G., Gu, J., Farooq, U., Raza, A.: Solution of an economic dispatch problem through particle swarm optimization: a detailed survey - part I. IEEE Access **5**(1), 15105–15141 (2017)
6. Ashtiani, M.N., Toopshekan, A., Astaraei, F.R., Yousefi, H., Maleki, A.: Techno-economic analysis of a grid-connected pv/battery system using the teaching-learning-based optimization algorithm. Sol. Energy **203**, 69–82 (2020). https://doi.org/10.1016/j.solener.2020.04.007
7. Papadakis, S., Markaki, M.: An in depth economic restructuring framework by using particle swarm optimization. J. Clean. Prod. **215**, 329–342 (2019)
8. Mohammadian, M., Lorestani, A., Ardehali, M.M.: Optimization of single and multi-areas economic dispatch problems based on evolutionary particle swarm optimization algorithm. Energy **161**, 710–724 (2018)
9. Jain, N.K., Nangia, U., Jain, J.: Economic load dispatch using adaptive social acceleration constant based particle swarm optimization. J. Instit. Eng. (India): Ser. B **99**(5), 431–439 (2018)
10. Nejadghanbar, N., Ghazizadeh-Ahsaee, M.: Particle swarm optimization algorithm-based fault location using asynchronous data recorded at both sides of transmission line. Revue Roumaine des Sciences Techniques. Serie Electrotechnique et Energetique **62**(2), 148–153 (2017)
11. Yang, J., Liu, J.: Influence maximization-cost minimization in social networks based on a multiobjective discrete particle swarm optimization algorithm. IEEE Access **6**, 2320–2329 (2018)
12. Wu, C., Yang, F., Wu, Y., Han, R.: Prediction of crime tendency of high-risk personnel using c5.0 decision tree empowered by particle swarm optimization. Math. Biosci. Eng. **16**(5), 4135–4150 (2019)

Multi-dimensional Data Perception and Intelligent Analysis Framework Design of Management Information System

Lei Pan[1(✉)] and Hany Abdullah[2]

[1] Tianjin Bohai Vocational Technology College, Tianjin, China
jxpanlei@163.com
[2] University of Sulaimani, Sulaymaniyah, Iraq

Abstract. The management information system can efficiently manage all kinds of information in the database, and with technical support, it can realize the function of intelligently analyzing data to help managers improve work efficiency. Therefore, this article designs the network framework of the management information system based on the B/S structure, sets the field length of the system database, and finds that the system's CPU occupancy rate is 43% when testing the system operation function, the system response time is 2.1s, and the system is stable. And the scalability is 96% and 90%, respectively. These test values are within the theoretical range, indicating that the actual operation of the system is feasible. Since the system can be used for personnel information management, the system's perception strength and recognition rate of personal identity characteristics are tested from the three dimensions of face information, voice information, and fingerprint information. It is found that the system is more sensitive to face information and its perception strength and the recognition rate reached 98% and 95% respectively, realizing the intelligent analysis of identity information.

Keywords: Management Information System · B/S Structure · Perceived Strength · Intelligent Analysis

1 Introduction

With the development of the information age, life is flooded with diversified information. How to manage this information in an orderly manner has become an urgent problem to be solved. The development of management information systems can intelligently manage tedious information. Therefore, this article uses advanced technology to design a management information system to help people quickly analyze and process information and data is of great significance.

Many scholars at home and abroad have conducted in-depth research on the multi-dimensional data perception and intelligent analysis framework design of management information systems, and have achieved good research results. For example, in some areas, government agencies use computers or Internet technology to manage a small amount of information. Because the management methods and management software

J. H. Abawajy et al. (Eds.): ICATCI 2022, LNDECT 170, pp. 36–44, 2023.
https://doi.org/10.1007/978-3-031-29097-8_5

used are tailored to their actual work needs, many functions cannot be used. For other related modules, no related data exchange and interaction occurred. This management software is still the same as the traditional closed management method, without any improvement in the workflow [1]. The management information system of some enterprises in our country has transformed from traditional management mode to information management mode, which makes the management of enterprise personnel and business more refined, and the system has added intelligent identification functions. The identity of enterprise personnel can be realized by comparing the information in the database. Information matching, but the disadvantage of the system is that the development technology is not mature enough [2]. Although the research on multi-dimensional data perception of management information systems is progressing smoothly, system development is quite difficult, and professional technical personnel are required to solve system development problems.

This article first explains the conditions for the development of management information systems, that is, to develop on the basis of the B/S structure and three frameworks, and then analyze the feasibility of the system from the three levels of technology, economy, and operation, and then based on the non-functional requirements of the system Based on the system development structure and framework, the basic structure and network structure of the management information system are designed, and the system operation performance is tested. Finally, the perception strength and recognition rate of the system in personnel information management are analyzed.

2 Management Information System Design Requirements

2.1 System Development Structure and Framework

2.1.1 B/S Structure

The B/S structure is a typical three-layer structure. Its presentation layer is the server used for browsing, and there is no possibility of business development in this layer [3]; on the functional layer, the Web App server is responsible for business development, login page management, and management of job requests from client browsers. And execute the corresponding program according to the task query instructions [4]; the data layer has data processing functions and is mainly connected to the database server software. Its task is to receive various data operation requests from the user in the client server through the WebApp. The database server is required Inquire about data, modify and calculate statistics, and transmit the data processing result to the Web application server.

2.1.2 Struts Framework

Struts is an open source subroutine within Software, which effectively separates business logic, performance logic and management logic.

2.1.3 Spring Framework

Spring is also an open source development framework. Spring is a new framework designed to reduce the complexity of developers using the J2EE platform to develop systems. Its core work is the IOC (Inversion of Control) container technology based on Java Bean properties, passing object dependencies to Spring for control, and providing developers with a declarative integrated MVC WEB AOP transaction management framework. It also helps loosen connect and simplify the deployment process [5].

2.1.4 Hibernate Framework

The Hibermate framework is one of the best ORM frameworks so far. It uses ORM mechanism to intelligently reflect the relationship between database Java objects, easily integrate JDBC, and avoid overly complex functions [6].

2.2 System Feasibility Analysis

2.2.1 Technical Feasibility

According to market demand analysis and growth, the management system adopts ASP.NET which is highly adaptable, stable and efficient in management. The ease of use and maintainability of Microsoft are the reasons for choosing it as the database. At the same time, Microsoft has strong server support, and the management system can be deployed faster by using advanced tools [7, 8]. Therefore, the use of these technologies promotes rapid design, simplifies system development procedures and final implementation processes, and can be used without any technical troubles.

2.2.2 Economic Feasibility

The management information system based on B/S is a good platform for notifying managers. The system development technology is very popular, the market price is also very cheap, and it is economically feasible.

2.2.3 Operational Feasibility

The operating interface of the system is very user-friendly, the screen content is clear and easy to understand, and the system is extremely convenient to use. Managers do not need to pass systematic computer training, nor do they need to be proficient in a lot of computer knowledge. They only need a short-term contact to master the use of the management system and improve the level of management improvement [9].

2.3 Analysis of Non-Functional Requirements of the System

2.3.1 Availability

The system must fully protect and utilize existing information resources, integrate different resources on the basis of existing information resources, increase new actual needs, and determine the steps and scale of system construction and transformation.

2.3.2 Manageability and Maintainability

Several objects are mainly involved in the operation of the system: system users and system administrators. For system users, the system should have a humanized operation page. Creating a management platform is very important for system administrators. Having a good management platform can reduce the overall maintenance cost of the system and reduce the difficulty of system maintenance and management [10, 11].

2.3.3 Scalability

The system should have good scalability. The system architecture should be able to increase the necessary data interfaces so that it can support the information requirements of various other related business management systems. The system should also have the ability to smoothly upgrade to meet the needs of future scale expansion of the system. All system functions can not only meet the needs of the current work, but also meet the needs of the foreseeable future work modules, such as the expansion of functional applications and the expansion of work services. Ensure that when the future system has new technologies that can be upgraded, the existing resources can be protected and fully utilized.

2.4 Information Extraction and Matching

Information can usually be regarded as a node in the coordinate system, and information can be matched according to the coordinate information by establishing exclusive coordinates for the information node [12].

$$p = \begin{bmatrix} x \\ y \\ z \end{bmatrix} = \begin{bmatrix} r\cos\phi\cos\beta \\ r\cos\phi\cos\beta \\ r\sin\phi \end{bmatrix} \tag{1}$$

$$s = \begin{bmatrix} r \\ \beta \\ \phi \end{bmatrix} = \begin{bmatrix} \sqrt{x^2 + y^2 + z^2} \\ \arctan 2(y, x) \\ \arctan 2(z, \sqrt{x^2 + y^2}) \end{bmatrix} \tag{2}$$

Among them, x, y, z are the abscissa, ordinate, and vertical coordinates of the information node, r is the distance of the point, β is the azimuth angle of the information point, ϕ is the elevation angle of the information point, and p and s are the coordinates of different levels.

3 Management Information System Design

3.1 Overall System Architecture

(1) Presentation layer: This interface mainly interacts through JSP. When a user query is received, the data sent by the system to the query will be displayed on the JSP page at the user's request. The user transfers the form data to the Action, and returns the corresponding JSP interface to the user after completing the operation. The Struts framework contains a library of configuration files used in the JSP interface.

(2) Control layer: It is mainly composed of a system controller, which can interrupt user requests and call business methods at the business logic level. After receiving the corresponding results, it controls the interface of different screen levels. The control layer is implemented in the system through special Actions, and the management of the control layer is also transferred to the Struts architecture, and complete management of page queries is realized through XML files.
(3) Business logic layer: It can reduce the dependency between layers. This feature is called free coordination. This is conducive to the further expansion of the system, that is, when the operation of the system changes, only the corresponding interface needs to be changed, without the need to change the operating services of the entire system.
(4) Data persistence layer: Use the Hibermate framework to implement the data persistence layer. The data persistence layer is mainly responsible for the interaction

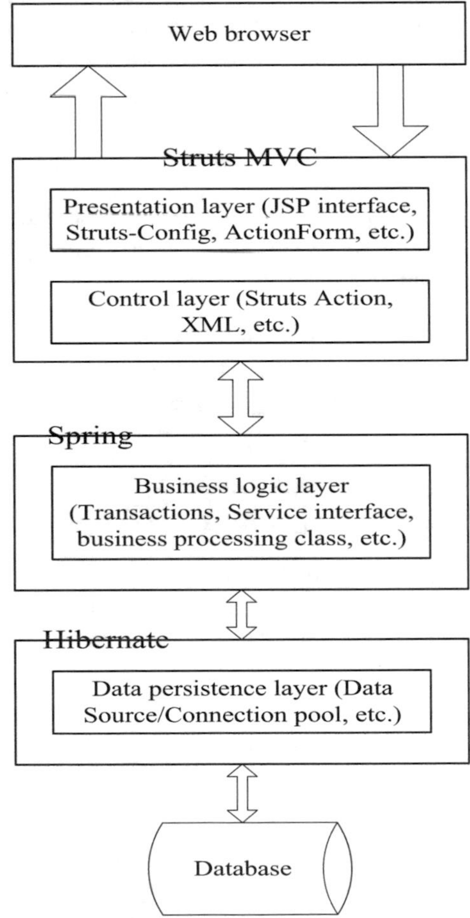

Fig. 1. Management information system architecture

with the database, combined with the business logic level, to achieve certain system functions. This layer can also be designed as an interface to provide interfaces with different performance for different databases. When the business logic layer has data that needs to be queried, the corresponding data function can be executed according to the so-called method or parameter type.

The architecture of the system is shown in Fig. 1.

3.2 System Network Framework Design Based on B/S

The management information system adopts the B/S network design mode. The system is composed of three parts, which facilitates the distributed deployment and centralized management of the system. This mode determines that a web browser can be used to access the system from any place and perform the system Operation, access to the system through the domain name, when the Web personnel information management system and the system database are configured on the same server, the system's multi-dimensional data collection function supports the retrieval of portrait information, calendar punching, and trajectory analysis, and supports peers Analysis, feature collision analysis, support portrait deployment control and early warning management. The network structure that the system runs is shown as in Fig. 2.

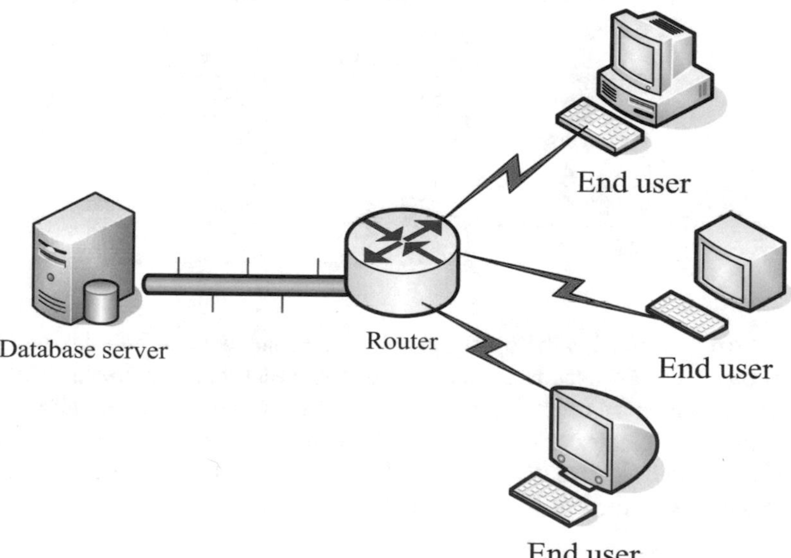

Fig. 2. Network framework of management information system based on B/S

3.3 Database Design

As shown in Table 1 is the information data collection of the management information system before the user login, in which the field lengths of the user ID, login name, user

Table 1. Database field settings

	Field length	Allow spaces
User id	20	No
Log-in name	15	No
Login password	25	No
User's real name	7	No
User gender	2	No

login password, user real name, and user gender in the database are set to 20, 15, 25, 7 and 2, and these information do not allow spaces.

4 System Application

4.1 System Operation Test

Table 2. System running test results

	Theoretical value	Test value
CPU usage	≤55%	43%
System response time	≤3.5s	2.1 s
Stability	≥92%	96%
Scalability	≥85%	90%

As shown in Table 2 is the operation test of the management information system. The test content is CPU occupancy rate, system response time, operational stability and scalability. Compare the test value with the theoretical value. If the test value is within the rotation range, it indicates that the system is operating reliably. According to the data in the table, it can be known that these four test values are all exceeding the theoretical value, which verifies the feasibility of system operation.

4.2 The Realization of the Information Display and Query Function of the System

There are two display modes for system information display, one is multi-level viewing and the other is horizontal display. The multi-level information display function starts from the root node of the information sub-relation tree and displays system information level by level. Users can click to enter the next level of information and view the subject content. This page displays the current content path. Users can go directly to the top information through the pop-up menu. Horizontal display is to display all system information in the same list. The user can view the information path through the path ID column.

The query of system information is divided into two types: basic query and advanced query. Simple query in the drop-down menu "Show", you can choose to display all information, only display business requirements, and only display technical requirements. Enter information keywords in the text box to receive system information queries. Users can click the "Advanced" link in the main query panel to expand the query panel. In advanced query, users can search system information through the contents of multiple fields, and each field can define multiple query terms.

4.3 System Multi-dimensional Data Perception and Intelligent Analysis

The management information system designed in this paper can be applied to personnel information management. As shown in Fig. 3, the management information system retrieves the identity information, and extracts the identity features from the three dimensions of face comparison, voice comparison, and fingerprint comparison. From the data in the figure, it can be seen that the perceived strength of face comparison and The recognition rate is the largest, the voice comparison is in the center, and the fingerprint comparison has the smallest perceptual strength and recognition rate, indicating that the system is most sensitive to facial features, so that the correct rate of identifying a person's identity can be the highest.

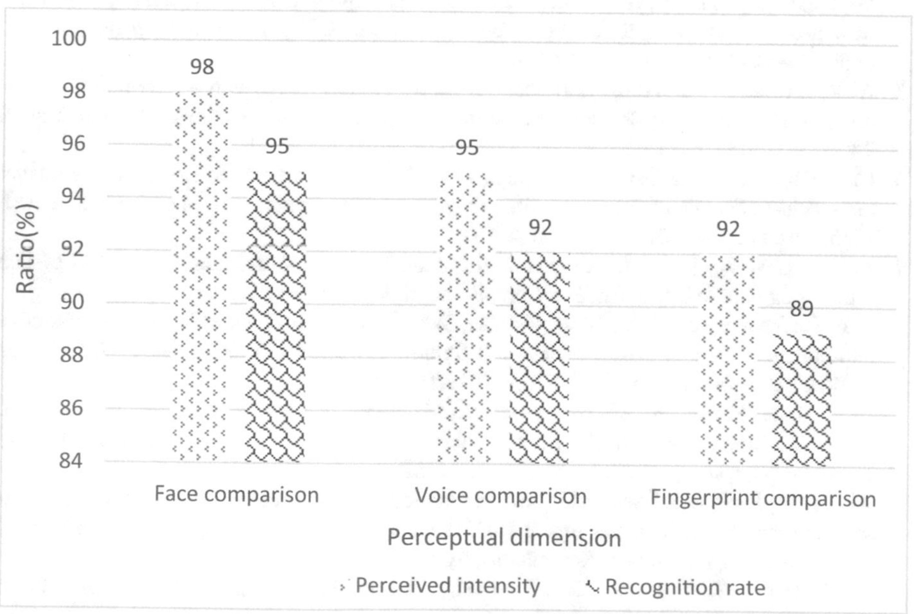

Fig. 3. Analysis of system perception strength and recognition rate

5 Conclusion

This paper designs a management information system based on the B/S structure and software framework according to the system development conditions and requirements. The most common use of the system is personnel identity information management. After the system is successfully designed, the information display and query functions and system operation are tested. After verifying that the system can operate normally, the multi-dimensional perception and recognition capabilities of the system are analyzed, and the intelligent application of the management information system in identity recognition is realized.

References

1. Xu, P., Jinli, W., Xiaolong, Z., et al.: Multi-dimensional data quality evaluation method for intelligent distribution network. Zhongguo Dianji Gongcheng Xuebao/Proc. Chin. Soc. Electr. Eng. **38**(5), 1375–1384 (2018)
2. Liu, Y., Wang, X., Boudreau, G., Sediq, A.B., Abou-Zeid, H.: A multi-dimensional intelligent multiple access technique for 5G beyond and 6G wireless networks. IEEE Trans. Wireless Commun. **20**(2), 1308–1320 (2021)
3. Liu, X., Cao, Z., Gu, N., Nahavandi, S., Zhou, C., Tan, M.: Intelligent line segment perception with cortex-like mechanisms. IEEE Trans. Syst. Man & Cybern. Syst. **45**(12), 1522–1534 (2017)
4. Lu, D., Asian, S., Ertek, G., Sevinc, M.: Mind the perception gap: an integrative performance management framework for service supply chains. Int. J. Phys. Distrib. Logistics Manage. **49**(1), 33–51 (2019)
5. Al-Refaie, A., Al-Alaween, W., Diabat, A., Li, M.-H.: Solving dynamic systems with multi-responses by integrating desirability function and data envelopment analysis. J. Intell. Manuf. **28**(2), 387–403 (2014)
6. Chagnon, M., Morsy-Osman, M., Plant, D.V.: Multi-dimensional formats and transceiver architectures for direct detection with analysis on inter-polarization phase modulation. J. Lightwave Technol. **35**(4), 885–892 (2017)
7. Baciu, G., Wang, Y., Li, C.: Cognitive visual analytics of multi-dimensional cloud system monitoring data. Int. J. Softw. Sci. Comput. Intell. **9**(1), 20–34 (2017)
8. Cao, B., Zhang, L., Li, Y., Feng, D., Cao, W.: Intelligent offloading in multi-access edge computing: a state-of-the-art review and framework. IEEE Commun. Mag. **57**(3), 56–62 (2019). https://doi.org/10.1109/MCOM.2019.1800608
9. Falahati, M., Karimi, A., Mohammadfam, I., Mazloumi, A., Reza Khanteymoori, A., Yaseri, M.: Multi-dimensional model for determining the leading performance indicators of safety management systems. Work **67**(4), 959–969 (2020)
10. Kim, D.C., Kim, W.K., Seo, Y.K.: An examination of multi-dimensional constructs of resistance to supply chain management (SCM) change for a small and medium sized food production company. J. Soc. Korea Indus. Syst. Eng. **42**(3), 206–216 (2019)
11. Yu, J., Deng, W., Yao, G.F., Cao, X.: Design of intelligent integrated power management system based on multi source information fusion. J. Comput. Theor. Nanosci. **14**(3), 1473–1477 (2017)
12. Wei, Y., et al.: Computational design of three-dimensional multi-constituent material microstructure sets with prescribed statistical constituent and geometric attributes. Multiscale Sci. Eng. **2**(1), 7–19 (2020)

Resource Transaction Management System Based on PageRank Algorithm

Jie Liu(✉)

Weifang Municipal Party School of CPC, Weifang, Shandong, China
ljwfdx@163.com

Abstract. With the rapid development of Internet technology, the application of information technology in life is also increasing. Among the resource transaction management systems of traditional cooperatives, it is mainly manifested as a traditional and restricted closed regional transaction management system. Due to information technology with the continuous opening up, the backward resource transaction management will be replaced by an advanced and open public resource transaction management system. This paper proposes a resource transaction management system based on PageRank algorithm, which can integrate various businesses to the greatest extent by integrating information expansion technology and modern electronic technology into the resource transaction management system and daily business operations. The main research content of this paper is the application of PageRank algorithm to the resource transaction management system, which provides a feasible solution for the resource transaction management system. The resource transaction management system of PageRank algorithm can solve such problems very well. The final results of the study show that when the total number of returned results for keywords is 346, the total number of selected results is 50, and the number of relevant results is 44, the precision is 88%. From this, it can be seen that the precision rate of the resource transaction management system query ranking and comparison statistics under the PageRank algorithm is relatively high.

Keywords: Internet Technology · Resource Trading · Public Resources · Resource Sharing

1 Introduction

As social public resources enter the Internet era, the efficiency of public resource allocation is insufficient, and there are also some problems in the resource transaction management system, such as non-sharing of resources, unorganized systems, imperfect systems, and incomplete supervision. In view of the above-mentioned problems in resource transactions, starting from improving the efficiency of resource transaction management and establishing a shared, open and efficient resource transaction management system, a brand-new mode of electronic transaction of public resources relying on the network

J. H. Abawajy et al. (Eds.): ICATCI 2022, LNDECT 170, pp. 45–52, 2023.
https://doi.org/10.1007/978-3-031-29097-8_6

has become the development trend and current Hotspot [1]. The use of PageRank algorithm combined with resource transaction management provides a feasible solution to solve the above problems.

In recent years, many researchers have studied the resource transaction management system based on PageRank algorithm, and achieved good results. For example, Karla Hahn believes that my country's current resource trading mechanism needs to be improved. As the scale of resource trading increases year by year, it is the most important to improve the informatization level of resource trading services to meet the needs of the current information age [2]. Mahboob V A believes that with the continuous development and progress of my country's economy and society, as well as the continuous improvement of information network technology, the resource transaction management system has entered the information age, and an efficient and safe resource transaction system can be embodied [3]. At present, scholars at home and abroad have carried out a lot of research on the resource transaction management system. These previous theoretical and experimental results provide a theoretical basis for the research in this paper.

Based on the theoretical basis of PageRank algorithm, combined with the analysis of resource transaction management system, this paper provides an information-based development method for resource transaction, improves the overall public resource management system with resource transaction management, and promotes the implementation of relevant national policies. PageRank algorithm It is used to calculate the set of relevant data in the data set and feed it back to the user, providing an efficient and fast sharing service function.

2 Related Theoretical Overview and Research

2.1 Introduction to Algorithms Related to Resource Transaction Management System

(1) The basic principle of PageRank algorithm
From the very beginning, the PageRank algorithm has played a great role in the search section of web pages. In actual operation, the links from page A to page B are called forward links to page A through the links between web pages and pages. A link to the back of Page B, and a forward link from Page A to Page B is considered a vote for Page B on Page A. Generally speaking, the more votes a page gets, the more weight PageRank will consider on the page [4, 5]. PageRank is mainly based on the random walk model. In the random walk model, after a user has entered and viewed a web page, clicks a link to it to enter another web page, and the user continues to click from one web page to another. Fits a convergent probability distribution. Its likelihood can also be considered an influencer on a web page.

(2) The basic principle of binary search algorithm
The binary search algorithm, also known as the halving search algorithm, can quickly and efficiently find the location of the data to be searched in order. The complexity of the current binary search algorithm is O(logn), which is faster than other common search algorithms. The binary search algorithm can be implemented

recursively or iteratively. The algorithm describes the iterative-based binary search algorithm and the recursion-based binary search algorithm, which is also a basic part of the K-PRSCAN algorithm [6]. The distance between two objects represents the probability that the two objects belong to a cluster. The smaller the distance between two objects, the more likely the two objects belong to the same object. The probability that two objects belong to the same cluster is measured by the similarity between the two objects.

2.2 Management Requirements of Resource Trading System

(1) Business requirements of resource transaction management system
To enable the resource transaction management system to meet the needs and design goals of various businesses, the system must provide system login and system homepage, login registration, notification, industrial and commercial registration, appointment site, expert selection, bid opening, bid evaluation, announcement, contract signing, withdrawal and withdrawal fees, inspection statistics, etc. Each link provides management, and the resource transaction management system includes many users, including staff in each link, bidding agents, and supervisors. Managers, etc., people review and enter the elements of each project. The system provides operation management functions, which mainly realize project registration, expert selection, bid opening, bid evaluation, fee refund, contract signing, etc. [7, 8]. Manage access and through this system, staff can standardize and scientific bidding process, effectively improve work efficiency and the incidence of violations of laws and regulations.

After each institution goes to the administrative supervision department for project review, bring the registration form approved by the competent department. Enter the entrance registration window to register, generate a transaction number, publish an announcement, and enter the registration information line publishing settings. Potential bidders or intended assignees can register and use this module again until registration is complete. Reserve the venue before the opening of the bid, sign the company before the opening of the bid, and list all the registered and used companies' payment information to log in and enter the login information [9]. When it is time to open the bid, the bid will be opened and recorded with the company's accompaniment information. List all expert information and log in before bid evaluation begins. After the evaluation, the public announcement of the winning bid will be combined with the online public announcement of the winning company.

(2) Resource Trading System Management Performance Requirements
The resource transaction management system based on the Pagerank algorithm mainly uses the B/S mode and the reorganization mode of J2EE technology. The remote service function of the system, the system provides an XML-based business data interface, follows the J2EE specification, and can provide remote business services and remote data services [10, 11]; on the basis of ensuring security, it supports the exchange of e-government information, and various The office network of business cooperation units is interconnected to realize information sharing and business collaboration, and achieve effective development. Improve the efficiency

of public resource transaction management. In general, the following points are required:

Reliability: When abnormal data is input, the system can provide friendly prompts and guidance for normal data input. The data is normal, making the system unusable, ensuring normal data input, no abnormal data output, and ensuring the logical flow of data, companies that register but do not pay fees cannot check in during the check-in session [12].
Efficiency: AJAX is widely used in the system for asynchronous server interaction, so that data can be exchanged between pages and loaded on demand, reducing server load. Every time a request is submitted, there is no need to refresh the entire page and improve responsiveness.
Continuity: From the selection of the platform system to the design of the application system, the continuity of the system is fully considered. Adopt rapid maintenance system platform design to ensure simple operation and low maintenance cost.

3 Experiment and Research

3.1 Experimental Method

(1) K-PRSCAN algorithm of resource transaction management system
DB-Index (DaviesBouldinIndex) and ARI (AdjustedRandIndex) are used to evaluate the pros and cons of K-PRSCAN and other clustering algorithms. In general, DB-Index is suitable for evaluating spherical clusters, as can be seen from its calculation formula. The smaller the value of DB-Index, the better the clustering results. Given two clusters Ci and Cj, let:

$$D_{\mathrm{Ci,\,Cj}} = \frac{\sigma \mathrm{C_i} + \sigma \mathrm{C_j}}{\mathrm{d_{Ci,\,Cj}}} \tag{1}$$

Among them, DCi refers to the average distance between each object in the cluster Ci and the cluster center, and dCi,Cj refers to the distance between the Ci cluster center and the Cj cluster center. Assuming that there are k clusters in the dataset, then DB-Index is defined as follows:

$$\mathrm{DB\text{ - }Index} = \frac{1}{\mathrm{k}} \sum_{\mathrm{i=1}}^{\mathrm{k}} \mathrm{maxi} \neq \mathrm{j}\left(D_{\mathrm{ci,\,cj}}\right) \tag{2}$$

Maxi is expressed as the maximum value of DB-Index. Generally speaking, the larger the value of DB-Index, the better the clustering effect of the corresponding algorithm, and vice versa, the less ideal clustering effect of the algorithm. The maximum value of DB-Index is 1, which indicates that the clustering result of the algorithm is the same as the original clustering result.

3.2 Experimental Requirements

In this experiment, the resource transaction management system is combined according to the PageRank algorithm. In the implementation of the K-PRSCAN algorithm in this experiment, the impact factor of the object is obtained by the PageRank algorithm after 100 iterations, that is, the threshold e is not considered as the end condition, and only the Set the parameter max_iter to 100. The damping factor of the PageRank algorithm is set to 0.85. The calculation programs of DB-Index and ARI are also implemented using Python 2.7.

4 Analysis and Discussion

4.1 Query Ranking and Comparison Statistics of Resource Transaction Management System

In this experiment, 5 keywords will be selected for retrieval query, and each time the web pages in the first 5 pages of the query results will be selected for analysis and comparison. The query result comparison statistics table is shown in the following table.

Table 1. Resource transaction management system query ranking comparison statistics table

query value	return knot total fruit	Total number of selected results	Number of relevant results	Precision(%)
Keyword one	315	50	45	90
Keyword two	346	50	44	88
Keyword three	193	50	43	86
Keyword four	234	50	46	92

It can be seen from the data in Table 1 and Fig. 1 that, in the resource transaction management system query ranking comparison statistics table, it can be seen that when the total number of returned results for keywords is 315, the total number of selected results is 50, and the number of related results is 45, the precision rate is 90%. When the total number of returned results for keywords is 346, the total number of selected results is 50, and the number of related results is 44, the precision rate is 88%. From this, it can be seen that the precision rate of the resource transaction management system under the PageRank algorithm is relatively high in the query ranking and comparison statistics.

4.2 Analysis of Clustering Results on Pagerank Algorithm Dataset

Each clustering algorithm will be run 5 times for each data set, and the following data will be recorded: the number of clusters in the clustering result, the data value of DB-Index, the data value of ARI, the average running time, and the result As shown below.

As shown in Fig. 2, it can be seen from the experimental data that the DB-Index value of the K-medoids algorithm is 0.6977, the ARI value is 0.7634, and the average

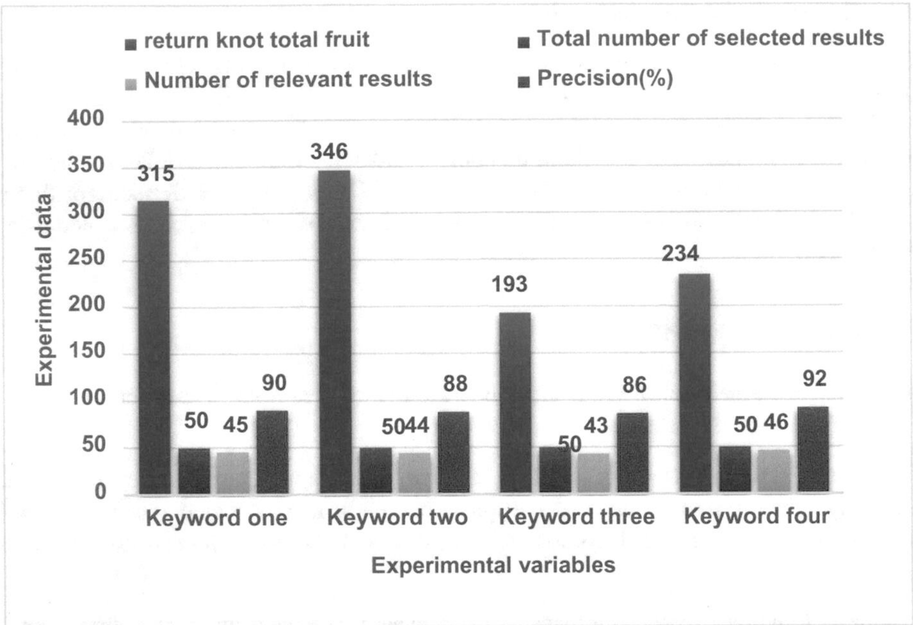

Fig. 1. Query ranking comparison statistics of resource transaction management system

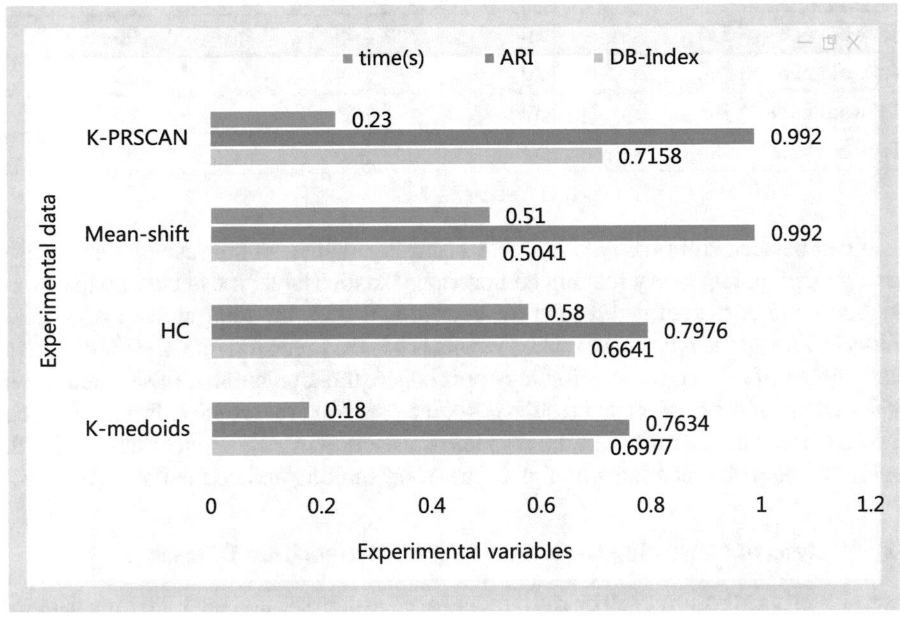

Fig. 2. Analysis diagram of clustering results on the PageRank algorithm dataset

running time is 0.18 s, while the DB-Index value of the K-PRSCAN algorithm is 0.992, the ARI value is 0.7158, and the running time average is 0.23. At the same time, the DB-Index value of the Mean-shift algorithm is 0.992, the ARI value is 0.5041, and the average running time is 0.51. The DB-Index value of the HC algorithm is 0.7976, the ARI value is 0.6641, and the average running time is 0.58. K-PRSCAN algorithm gets correct clustering results.

5 Conclusions

In this paper, the PageRank algorithm is used in combination with the resource transaction management system for analysis, and relevant experimental simulation analysis is carried out. Under the PageRank algorithm, this system is to collect and share the resource transaction data information through the resource transaction center to realize the resource transaction information. Centralized release to promote the formation of resource transaction bidding mechanism; through information monitoring of resource transaction process, realize dynamic supervision of resource transaction process and standardize resource transaction behavior; through statistical analysis of resource transaction information, provide reference for resource transaction and macro decision-making in accordance with. Accelerate the establishment of a unified and standardized resource trading market, continuously improve the efficiency and quality of resource allocation, and promote the continuous improvement of the socialist market economic system. This is a policy actively advocated by the current government. It is also an effective way to regulate government procurement and solve corruption problems. One of the measures, therefore, the establishment of a resource transaction management system conforms to the government's work orientation and the trend of social and economic development. At the same time, ensure cost control in the process, minimize resource consumption, and ensure quality, so that government departments can obtain maximum benefits, reduce resource waste, and reduce social costs.

References

1. Djungu, S.-J.A.O., Manneback, P., Congo, K.R.D.: SpeedSiteRank: PageRank algorithm distributed in websites. Int. J. Comput. Sci. Issues **17**(2), 13–18 (2020)
2. Hahn, K.L.: SERU (shared electronic resource understanding): opening up new possibilities for electronic resource transactions. D-Lib Mag. **13**(11/12), 11 (2007)
3. Mahboob, V.A., Jalali, M., Jahan, M.V., Barekati, P.: Swallow: resource and tag recommender system based on heat diffusion algorithm in social annotation systems. Comput. Intell. **33**(1), 99–118 (2017)
4. Strech, K.R.: Insurance on demand transaction management system. US (2013)
5. Ahmad, F.: A framework of transaction management in distributed database system environment. Int. J. Sci. Res. Sci. Eng. Technol. **3**, 35–53 (2014)
6. Tripathi, K.P.: Role of management information system (MIS) in human resource. Int. J. Comput. Sci. Technol. **2**(1), 58–62 (2012)
7. Prusak, R., Waszkielewicz, W., Kulawik, A.: Modifcation and improvement of human resource management system in metallurgical enterprise. Metalurgija **46**(2), 129–133 (2007)

8. Henriksen, I.E.: System and method involving resource description framework distributed database management system and/or related aspects (2015)
9. Ghosh, R.A.: Ranking and selecting entities based on calculated reputation or influence scores. US (2014)
10. Dlamini, T., Vilakati, S.: LSTM-based traffic load balancing and resource allocation for an edge system. Wireless Commun. Mob. Comput. **2020**, 1–15 (2020)
11. Belgacem, A., Beghdad-Bey, K., Nacer, H.: Dynamic resource allocation method based on symbiotic organism search algorithm in cloud computing. IEEE Trans. Cloud Comput. **10**, 1714–1725 (2020)
12. Strumberger, I., Bacanin, N., Tuba, M., Tuba, E.: Resource scheduling in cloud computing based on a hybridized whale optimization algorithm. Appl. Sci. **9**(22), 4893 (2019)

Research on Intelligent Student Management Evaluation Information System Based on Data Mining Algorithm

Xiao Wei(✉)

Chongqing Medical and Pharmaceutical College, Chongqing, China
wxcq331@163.com

Abstract. The evaluation of students' comprehensive information management is of great significance in the school's management practice, so the practical effect value of strengthening students' comprehensive information management is significant. From the current specific analysis, to improve the specific effect of students' comprehensive information management, on the one hand, advanced technology must be used, and on the other hand, information materials need to be mined and classified. The information will be more abundant. In short, in the practice of student information management evaluation, mining the deep connection of information is of great help to the formulation of management strategies, and to achieve the purpose of information mining, data mining technology must be applied. This paper research on the intelligent student management evaluation information system based on data mining algorithm.

Keywords: Data mining · Mining algorithm · Intelligence · Information Management · Management Evaluation

1 Introduction

School management is a work that must be strengthened in the orderly development of school teaching activities, especially at this stage, each school is actively carrying out modernization development, and the improvement of management in modernization development is an important content, so school management was placed in an important position [1]. As far as the specific analysis of school management is concerned, to improve the efficiency and quality of management, it is very necessary to build an information management system. On one hand, the construction of an information management system requires the introduction of advanced technology, and on the other hand, it needs to import rich data. Use technology to analyze data relationships, so that the formulation of student management strategies can be more in line with student practice. In short, the specific construction of an intelligent student management system has significant practical significance for school management, so the practical value of using data mining technology to serve the system construction [1].

J. H. Abawajy et al. (Eds.): ICATCI 2022, LNDECT 170, pp. 53–61, 2023.
https://doi.org/10.1007/978-3-031-29097-8_7

2 The Significance of Intelligent Student Management Evaluation Information

2.1 Realize the Dynamic Opening of the Classroom

The smart classroom is essentially a dynamic and open system. With the help of emerging information technologies such as cloud computing and mobile Internet, various smart terminals such as smart phones and wearable computing devices are used to make the classroom system transcend the time and space limitations and enable dynamic information exchange, making the pre-class, in-class, and after-class integrated, and the single and closed classroom teaching develops to a diversified open teaching [2].

2.2 Improve Classroom Interaction

Using intelligent mobile learning tools, the communication and exchanges between teachers and students, and between students and students are more 3D, and real-time communication can be carried out without barriers, which greatly improves classroom interaction ability and teaching efficiency [3].

2.3 Promote Collaborative Inquiry Learning

According to the needs of meaning construction, relying on the learning environment constructed by the information platform, learning methods such as group consultation, discussion and cooperative inquiry can be adopted in smart classrooms [2]. At the same time, teachers can also conduct real-time digital evaluation on group cooperation through the platform, and guide and help study group discussions and cooperative exploration.

2.4 Facilitate Personalized Learning

Big data makes it possible to implement individualized teaching and teaching, and truly realize the transition from group education to individual education. Learning analysis based on big data can accurately grasp the status of each learner's knowledge, realize the personalized learning ability assessment, make teachers' awareness of each student clearer, and formulate teaching in a targeted manner plans and tutoring strategies, push personalized learning materials, and conduct individualized "micro-class" assignments and tutoring after class, truly realizing student-centered, "one-to-one" personalized teaching.

2.5 Implement Guided Teaching

In the new classroom teaching mode, teachers are no longer the imparters and instillers of knowledge, but the guides and helpers of students' learning, and they always take an important guiding role in the whole teaching process. Before class, through situation construction and question stimulation, teachers guide students to become interested in the preview content, and actively check materials, use their brains, and discuss and research the preview materials and tests pushed by teachers; in class, through interactive communication, teachers guide students to explain The process of recognizing the preview content, expressing their own opinions, and guiding students to discover new problems and conduct discussions; after class, by arranging personalized homework and tutoring, teachers guide students to form an overall mastery of knowledge and a deeper understanding [4].

2.6 Improve Classroom Teaching Tact

Classroom teaching is ever-changing, and even the best teaching presets cannot foresee all the situations that may arise in the classroom. In smart classroom teaching, teachers are more required to have a strong ability to adapt to changes. According to new situations and new problems that may arise at any time in the teaching process, use the student information support provided by the smart classroom informatization platform, based on dynamic learning evaluation analysis and real-time learning [3].

3 Data Mining Algorithms

3.1 The Concept of Data Mining

The data mining technology plays a key role in it. Combined with the current development situation, data mining has been in the field of artificial intelligence and database. Knowledge [5, 6], to solve the problems caused by huge data sets, using data mining technology can analyze a lot of data and find information that is conducive to social development.

Data mining is a product that combines knowledge of multiple disciplines. It includes many disciplines. These disciplines are now the research hotspots of various departments, mainly including artificial intelligence, machine learning, pattern recognition, statistics, visualization technology, etc. Data mining can be explained as a process that automatically helps us analyze data and obtain potential knowledge from the data, to help decision makers make reasonable and correct decisions [5].

3.2 The Process of Data Mining

The data mining process mainly includes the following aspects: data selection or construction, preprocessing data, data transformation, modeling data, and analysis and evaluation of results [6].

1) Data set selection or construction: A large amount of data is composed of different data sources. To solve a data mining problem, the first step is to select or construct a data set and select appropriate data according to the purpose of the task and your own needs [7].
2) Data preprocessing: Data preprocessing is a key step in data mining. The types and formats of data we collect are diverse, and not all data sources meet our experimental needs, resulting in errors and inaccuracies in the results. Preprocessing is required.

Data preprocessing includes data cleaning, data integration, and standardized formatting of data. This process, in layman's terms, is to transform "dirty data" into "clean data" for our use.

3) Data conversion: Data conversion is to convert the processed data into features. These features should be prepared as much as possible to describe the data and make the data mining algorithm optimal.
4) Data mining: The core of data mining is pattern discovery [7], that is, using data mining algorithms and data mining tools to analyze the converted data and dig out the results we expect and want.
5) Result analysis and evaluation: When the results are obtained according to the data mining algorithm, analyze the obtained results and theoretical results, and draw corresponding conclusions.

In summary, the process of data mining is shown in Fig. 1:

Fig. 1. Data mining process

3.3 Clustering Algorithms for Data Mining Algorithms

According to the idea of clustering:

1) Partition-based clustering:
K-means, k-medoids (representing a sample point in each category), CLARANS. k-means is to minimize the value of the following expression:

$$V = \sum_{i=1}^{k} \sum_{x_j \in S_i} (x_j - \mu_i)^2$$

Advantages of k-means algorithm:

2) The k-means algorithm is a classic algorithm for solving clustering problems, and the algorithm is simple and fast.
3) The algorithm is relatively scalable and efficient for processing large datasets because its complexity is about O(nkt), where n is the number of all objects, k is the number of clusters, and t is the iteration number of times. Usually, $k < < n$. This algorithm usually converges locally.
4) The algorithm tries to find the k partitions that minimize the value of the squared error function. The clustering effect is better when the clusters are dense, spherical, or clump-like [8].

4 Intelligent Management Evaluation Information System Design of Data Mining Algorithm

The student information management system can be used for student information management, query, update and maintenance in schools and other institutions. It is easy to use, easy to use, and has a clear graphical interface. The software is written in java language, uses SQLServer2005 database as the background database to store information, and uses SQL statements to complete the query, input, modification, deletion and so on of student information. Use JDBC driver to realize the connection between front-end Java and back-end SQL database [8]. The Java language is highly cross-platform and can be used under windows, linux, ubuntu and other systems. It is convenient and simple, and has good security, as shown in Fig. 2. The SQLServer2005 database is efficient and safe, and the combination of the two can make use of their respective advantages.

The approximate functions implemented by the system:

1) User login interface. This interface can select the user's identity, "Administrator, Student". Different identities have different operation interfaces and functional permissions. Enter the correct ID number and password to log in.
2) Admin management interface has the highest authority. Allows adding, modifying, deleting, and querying student information [6–8].
3) Student login interface. Students can only query their own information.
4) The logged-in user information is stored in the "Administrator Login Table" and "Student Login Table" in the SQL database. If the user information does not exist in these two tables, there will be no right to log in to the management system. The security of this management system is guaranteed [9].

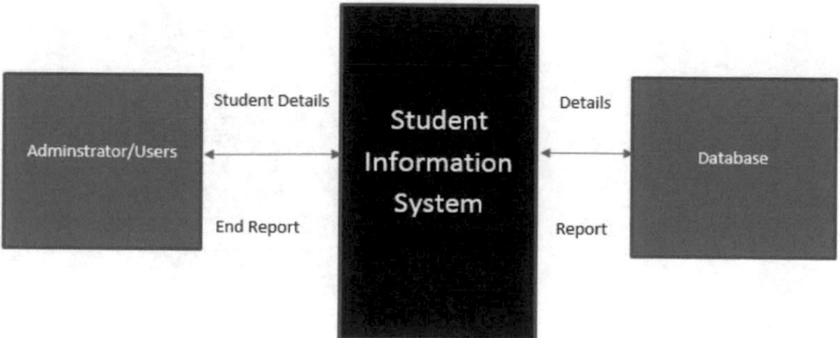

Fig. 2. Main functions of the system

4.1 Analysis of the Basic Requirements of the System

Login, user interface requirements: concise, easy to understand, easy to use, friendly user interface.

Security and confidentiality requirements: Only by logging in to the system with a username and password can information management be performed.

4.2 System Outline Design

The fundamental goal of the design phase is to determine how the required system should be implemented specifically after the design work in this phase, an accurate description of the target system should be obtained, so that the description can be directly translated into a certain system in the coding phase. A program written in a programming language [8–10]. The advantage of the program flow chart is that it is intuitive and clear, and it is easy to learn and master in Fig. 3, 4 and 5.

4.3 Basic Function Settings

Add, modify, delete, query and other operations on student information. The detailed design of its system E-R is shown in Fig. 6.

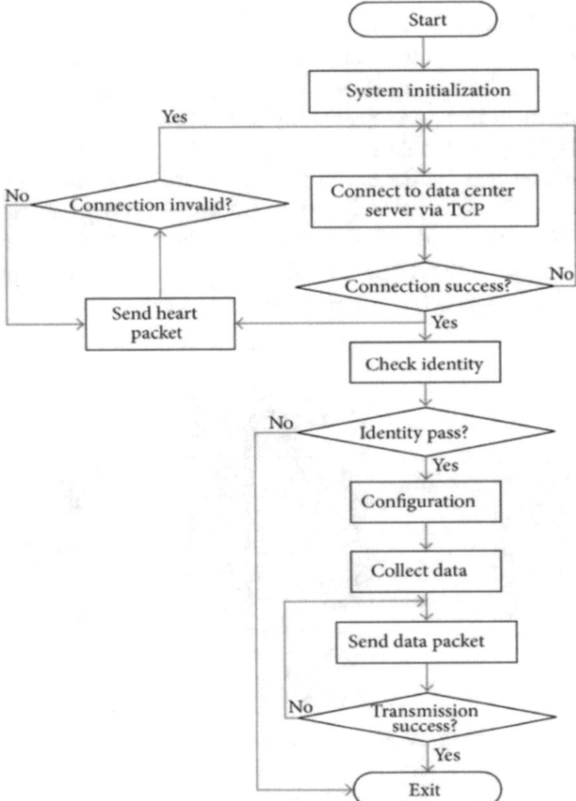

Fig. 3. Main flow chart of the system

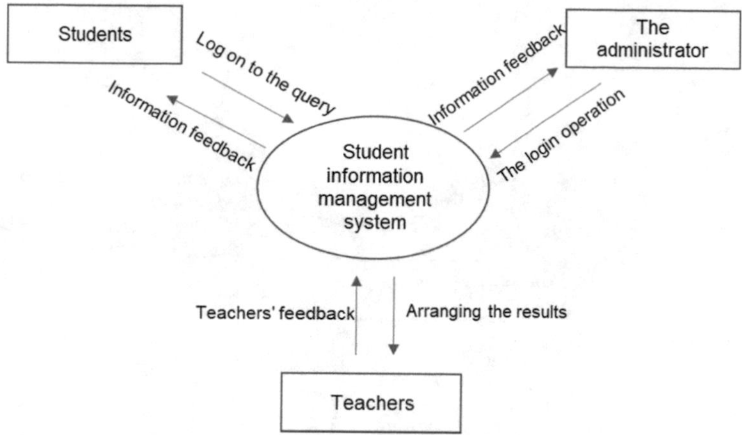

Fig. 4. Top-level data flow diagram of student management system

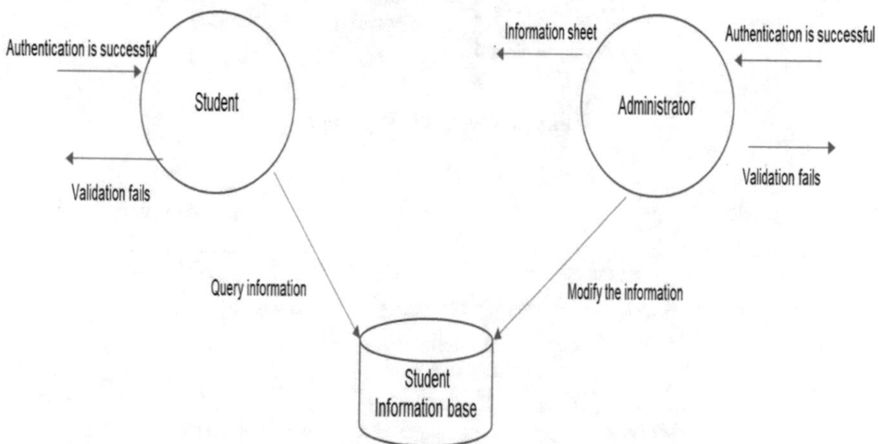

Fig. 5. Data flow diagram of the first layer of the student management system

4.4 System Function Module Division

After analyzing the needs of the system, the student information management system is mainly divided into two parts: student information query and student information management, including all aspects of student information [10]. The specific composition is shown in Fig. 7.

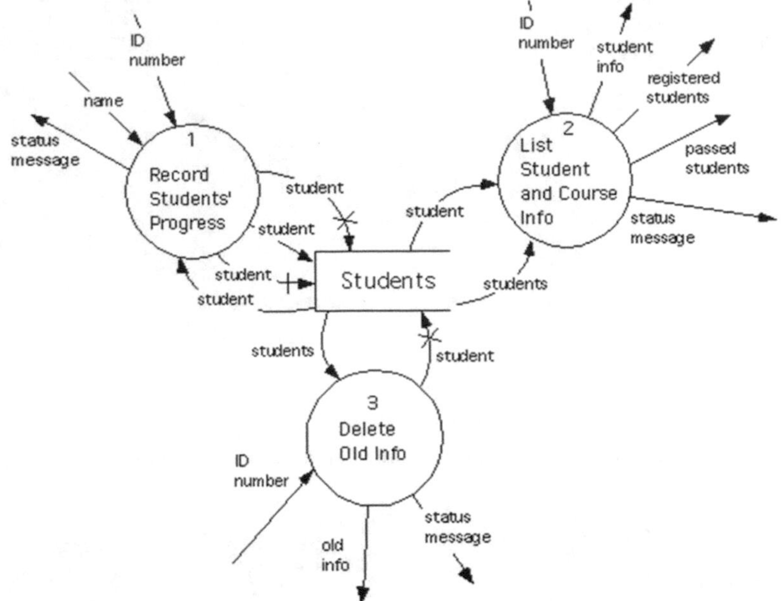

Fig. 6. Overall E-R design

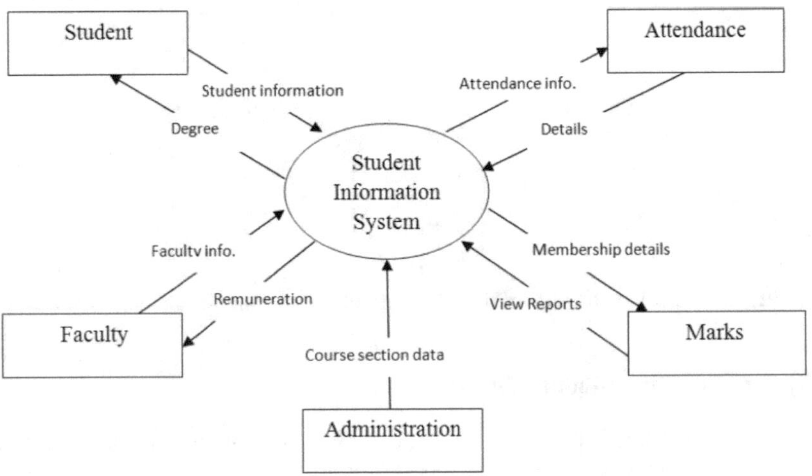

Fig. 7. Information type management

5 Conclusion

To sum up, in the construction of the school's information management system, in order to ensure that system has a more prominent management effect, it is necessary to use advanced technology. As a superior data analysis and processing method, data mining technology takes a prominent role in the connection construction of information data and the logical connection analysis of data. Therefore, based on the application analysis of data mining technology, students' intelligent management evaluation Information system, which is of great help for system utilization.

References

1. Hu, Q.-C., Liu, M.-C., Hu, J.: Design and implementation of database management system rules for information network. J. Comput. Appl. Softw. **33**(1), 16–20 (2016)
2. Feng, G.-Q., Cui, D.-L., Zhang, Y.-J., Zhou, P.: Hybrid artificial neural network-genetic algorithm for multi-objective optimization design of complex products from a sample perspective. Comput. Integr. Manuf. Syst. **22**(6), 1403–1414 (2016)
3. Agrawal, D.C., Hou, H.-Y., Cheng, T.-M.: The evaluation of competency-based diagnosis system and curriculum improvement of information management. Int. J. Inform. Commun. Technol. Educ. **17**(2), 87–102 (2021)
4. Chen, S., Tan, D.: Cognitive mechanism modeling and recognition optimization algorithm based on SA-ANN. Acta Electron. Sin. **46**(8), 211–216 (2018)
5. Pryimak, V., Kasian, S., Holubnyk, O.: Qualitative information in evaluation of the management efficiency of enterprises marketing activities. In: 2021 11th International Conference on Advanced Computer Information Technologies (ACIT), Deggendorf, Germany, pp. 428–432 (2021)
6. Zhang, H., Jia, Z., Chen, L., Guo, Y.: Hyperspectral image unmixing algorithm based on ANN end-member estimation. J. Comput. Appl. **37**(4), 1221–1225 (2020)
7. Bermeo-Andrade, H., González-Bañales, D.: Appropriation intention of a farm management information system through usability evaluation with PLS-SEM analysis. In: Yang, X.-S., Sherratt, S., Dey, N., Joshi, A. (eds.) Proceedings of Sixth International Congress on Information and Communication Technology. LNNS, vol. 236, pp. 633–641. Springer, Singapore (2022). https://doi.org/10.1007/978-981-16-2380-6_56
8. Xu, J.: Research on the application of ANN algorithm in the construction of intelligent student management evaluation information system. Adhesive **14**(3), 88–91 (2021)
9. Liu, H.: Student organization management evaluation of BAS-BPNN from the perspective of PDCA Cycle Management. Appl. Micro Comput. **12**(9), 60–62 (2020)
10. Kebede, M., Adeba, E., Chego, M.: Evaluation of quality and use of health management information system in primary health care units of east Wollega zone, Oromia regional state, Ethiopia. BMC Med. Inform. Decis. Mak. **20**(1), 107–110 (2020)

Purchasing Management System Based on Genetic Algorithm

Xiaoqian Zhou(✉)

Chongqing College of Architecture and Technology, Chongqing, China
xiaoqian_zhou@yeah.net

Abstract. With the popularization of Internet technology, more and more information is transmitted through the Internet. Since the traditional procurement mode is only horizontal integration, each node of the supply chain forms a network structure, and this mode of supply has poor flexibility. In order to improve this traditional procurement mode, Internet technology and procurement enterprises are combined to form a new procurement mode to reduce procurement costs, improve procurement efficiency, and promote enterprise informatization transformation. Based on this, this paper builds a purchasing management system based on genetic algorithm. The system is based on J2EE and uses B/S architecture technology. The entire application system is developed using Java language. After testing and verifying the system function, the system has been successfully developed.

Keywords: Genetic Algorithm · Purchasing Management System · B/S Architecture · Java Language

1 Introduction

The establishment of a procurement management system is one of the most important parts of the entire enterprise information management and control system, bearing the center of corporate strategy. The establishment of a procurement management system is of great significance to effectively manage the procurement process, improve the level of business management, and at the same time make the information in all aspects of procurement transparent, so as to prompt suppliers to improve the quality of supply, deliver goods in time, and arrive in time.

At present, the research on the design of purchasing management system based on genetic algorithm has been deeply discussed at home and abroad. For example, foreign companies implement a business management model, and material procurement personnel can get rid of the mechanical procurement model and invest more time in business procurement analysis, such as controlling the current procurement market, verifying the demand for procurement, strictly controlling procurement costs and strengthening risk identification aspect identification. At the same time, it emphasizes the collaborative work of multiple departments, unites procurement technicians, production personnel, optimizes various functions in the procurement process and negotiates bulk purchases,

J. H. Abawajy et al. (Eds.): ICATCI 2022, LNDECT 170, pp. 62–70, 2023.
https://doi.org/10.1007/978-3-031-29097-8_8

and establishes long-term cooperative partnerships with suppliers with stable cooperation advantages [1, 2]. Some scientists introduce genetic algorithm into the development of purchasing management system, use SQL Server to store the data involved in purchasing business, and perform automated customer data processing and warehouse management, which can effectively improve purchasing efficiency and purchasing level [3]. Although the design results of the procurement management system based on genetic algorithm are more remarkable, the application of the promotion system can really improve the procurement efficiency.

This paper first introduces the technology used in the development of procurement management, and then builds the adaptive function model of genetic algorithm, uses JAVA language development and B/S structure to design the Web data structure and functional modules of the procurement management system based on genetic algorithm, the system uses Java EE The development technology uses a relatively mature MVC model design, and the database used in the background is implemented by free and open source MySQL. Finally, the operation test, performance test, and login implementation of the system are carried out to verify that the system can be put into use normally and ensure normal operation.

2 Related Technologies and Genetic Algorithms

2.1 System Development Related Technologies

(1) Overview of MySQL
MySQL has significant advantages such as the ability to perform a variety of functions, a flexible and easy-to-use system interface, and a simple and practical structure. Because of these characteristics, MySQL excels with system applications. In particular, the combination of MySQL with PHP/PERL and Apache can have richer results and meet larger processing requirements [4].

(2) J2EE technology
J2EE is a specification guide. Different from the traditional technical architecture, J2EE has been widely used in database linking, login, query, transaction processing and so on. It can allow built-in development of new multiple servers based on integration customization to meet future business development needs. At the same time, the emergence of J2EE architecture allows developers to develop applications more easily and flexibly without worrying about database access issues, thereby improving development efficiency [5, 6].

(3) JDBC
JDBC (Java Data Base Connectivity) is a technical specification for running databases in Java, basically a type of specification specially designed to implement the JDBC interface. By running such a specification, Java and JDBC can be easily activated to understand the purpose of reusing a particular application in the application database. The JDBC jars we see usually refer to these standard classes [7]. With the help of JDBC, developers can create relatively independent software code through pure Java API, and directly access the database with the code of the program, thereby improving the writing efficiency of the system [8].

The JDBC API hides some details of the various databases from the developer. Because JDBC is written in Java, the JDBC API is platform-independent when used to connect to, read, write, and search databases. Developers can use JDBC to connect to the database, and only need to consider whether the database provides a JDBC driver, regardless of the characteristics of the database [9, 10].

2.2 Overview of Genetic Algorithms

The modeling of biological evolution by genetic algorithm can be divided into three stages. First, it is necessary to assume a simulation object, select a group that will work, select the existence rules of the strongest in nature, as the background environment, and then reorganize. Finally, there is mutation, which corresponds to the mutation in nature. Through the cycle of constantly generating new abnormal genes and the continuous evolution, selection and recombination of genes, it corresponds to the evolution process of the biological world in the genetic algorithm [11, 12].

Genetic algorithm also has some problems, such as its complexity. When the value range is large or the value range is uncertain, there are many parameters, which will easily reduce the convergence speed, and the required solution time will be very long. When it is harsh, it is possible that a feasible solution cannot be obtained [13, 14].

Adaptive Genetic Algorithm:

(1) Simple mapping, namely fitness function

$$\mathrm{f(x)} = \mathrm{f_a(x)} \tag{1}$$

$$f(x) = -f_a(x) \tag{2}$$

(2) The fitness of this form can be adopted for the problem of knowing the bounds

The question sought is the maximum value:

$$f(x) = \begin{cases} f_a(x) - C_{min}, f_a(x) > C_{min} \\ 0,\ \text{else} \end{cases} \tag{3}$$

The question asked is the minimum value:

$$f(x) = \begin{cases} C_{max} - f_a(x), f_a(x) < C_{max} \\ 0,\ \text{else} \end{cases} \tag{4}$$

Among them, f_a is the objective function, f(x) is the fitness value, and f_a and C_{min} are constants.

2.3 Analysis of System Design Requirements

(1) Personnel management needs
Personnel management in the system is mainly the management of personnel information market. Purchasing personnel are the executors of the specific work

of the purchasing department. Purchasing personnel should be familiar with the current product market, according to the implementation of product procurement procedures, and at the same time understand the progress of product procurement.

(2) Budget and planning management needs
Purchasing work should be planned in advance according to the company's production plan or sales level, especially the quantity and price of materials needed within a period of time should be estimated to form a good business procurement budget. Make a budget in the business procurement plan to save business costs and avoid unnecessary waste. The business budget and procurement plan should be formulated together with the business plan, material requirements and material business resources. In addition, corporate procurement budgets and plans must also be dynamically managed according to market changes. Therefore, enterprises need to grasp and study the market situation of the materials needed by the enterprise on the basis of understanding the complete market research, and adjust the plan and budget in real time according to the actual situation, so as to improve the procurement efficiency.

(3) Economic feasibility needs
At present, the purchasing department of the enterprise has all realized computer informatization, and established a network at the same time. Purchasing managers can exchange information through the network. On this basis, from an economic point of view, secondary development can be carried out, and the relative value of development is low.

(4) Social feasibility needs
In today's society, the competition of enterprises is fierce. In order to improve the efficiency of purchasing raw materials, the use of the network of information systems can facilitate the efforts of enterprises to strengthen purchasing management. In summary, the system development is feasible.

3 Design of Purchasing Management System

3.1 System Development Environment

This article uses the popular Intellij IDEA development tools for system development. The whole system is developed in JAVA language. Based on the security and openness of JAVA, the security of the system is improved to a certain extent. The use of mature, stable and open JAVA components further reduces the risk of development and maintenance, and improves the system scale and scalability. In order to meet the collaborative development needs of many developers, the system uses SVN as a branch management tool to record any changes made by developers, which is convenient for monitoring and updating the development code. Another obvious advantage of SVN is that when a conflict occurs when multiple developers modify a line of code at the same time, it is convenient for developers to resolve the conflict in time.

3.2 System Architecture Design

Provide a standard WEB SERVICE interface for the existing business system of the enterprise, reduce the connection with external systems, provide rich secondary development interfaces, and support the use of simple dynamic scripting language for operation specifications. According to the actual situation, the system structure adopted by the procurement management system is shown in Fig. 1.

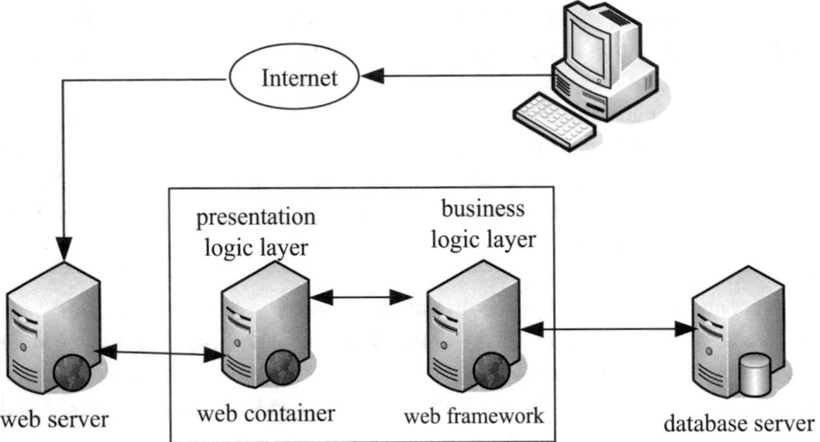

Fig. 1. System B/S structure

3.3 System Function Design

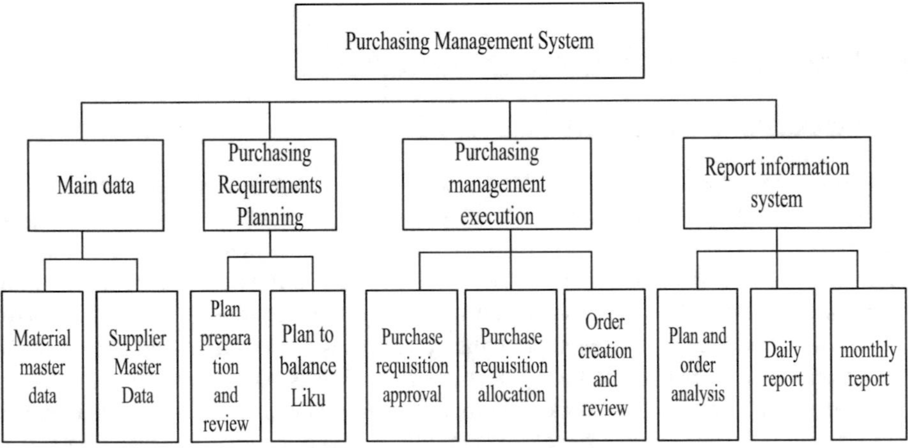

Fig. 2. System function modules

Figure 2 shows the four major modules of the procurement management system. The functions of each module are described below.

(1) Master data: The data management of the procurement management system consists of two parts, one is the maintenance of material master data, and the other is the maintenance of supplier master data. Users in the system can create, modify and display all kinds of key data, and can also view the change history of key data in the system. At the same time, the system also has the functions of freezing, marking and deleting basic data.
(2) Procurement requirements planning: For procurement planning, there is an option to create specific types of designs, scheduling times, project submitters and recipients, providing users with one-to-one line input and systematic checks on key data elements, while modifying the design, delete, commit and delete, allowing reviewers to modify it. It supports the calculation of purchasing requirements with reference to the existing available inventory and the required quantity, completes the balance of the inventory, and allows manual adjustment of requirements and multi-level review.
(3) Purchasing management execution: In the system, the purchaser can generate a purchase requisition and create a purchase order according to the "purchase demand plan" library; the purchase order can be manually entered. According to the material classification control principle and supply control principle, the system can determine the corresponding hierarchical approval process by combining conditions such as supply type, supply scale, internal or external suppliers, etc. After the purchase order is submitted for approval, it will go through the corresponding approval process, and the confirmer can choose to approve or reject it. During the approval process, when the user changes or confirms the purchase order, they can see the detailed information of the document, the previous approval status and the approval opinion.
(4) Report information system: The system provides inquiries of various plan types, and the query conditions include reporting unit, planning period, etc.; the supply unit can summarize the requirements according to the planning period; you can view the order through the order execution table the execution results, historical orders and their execution status can be tracked and analyzed; the daily report of material invoicing and the monthly report of material invoicing can check the purchase and sales of a certain material in a certain interval.

4 System Implementation

4.1 System Test

(1) Run the test

As shown in Table 1, test the system's CPU usage, memory usage, system IO usage (that is, the disk transfer speed, and the waiting time should be as small as possible), and the command feedback time (from when the user logs in to the system and initiates an command to when the command is received and the command is returned to the system) The average time spent), etc., the maximum CPU usage percentage should not exceed 60%, the physical memory should not exceed 45%, the IO usage should not exceed 53%, the instruction response time should be less than 1200 ms, the actual tested CPU usage

Table 1. Run Test Results

	Theoretical value	Test value
CPU usage	< 60%	46%
Memory usage	< 45%	32%
System IO usage	< 53%	39%
Command feedback time	≤ 1200 ms	300 ms-800 ms

It is 46%, the memory usage rate is 32%, the IO usage rate is 39%, and the feedback time is 300 ms-800 ms. These four test values are all within the theoretical range, so the test is successful.

(2) Performance test

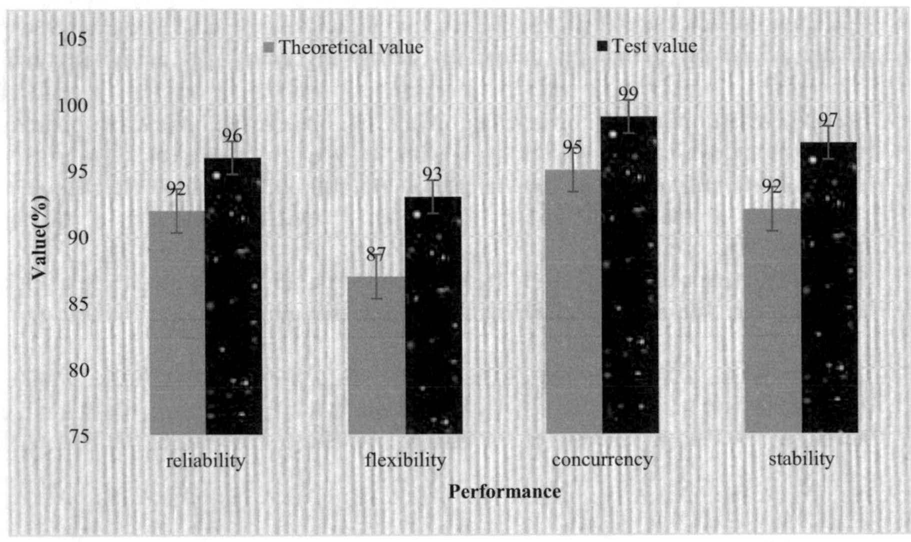

Fig. 3. Performance test results

The performance test results of the reliability, flexibility, concurrency, stability, etc. of the system are shown in Fig. 3. System reliability: Authenticate all user connections to prevent unauthorized user connections and malicious attacks. System data must be able to withstand intersection and SQL attacks to prevent system data corruption. At the same time, system data should be backed up regularly, and system data recovery should be supported when necessary. System flexibility: In order to meet the needs of business development, the system must be able to support extended or extended functions to facilitate system upgrades. Furthermore, considering the portability of the system, the system has been thoroughly checked in terms of deployment architecture, deployment language and server selection. Once the system is moved to the platform, there are no data errors or app development failures. System concurrency: can support multiple users to log in at the same time. System stability: The system can still perform operations when the network is unstable. According to the test results, the test values of reliability, flexibility, concurrency, and stability are 96%, 93%, 99%, and 97%, respectively, which are higher than the theoretical setting values of system performance, so the system performance test is successful.

4.2 System Login Implementation

The key to creating a login module is to correctly verify the legal identity of the user. Users must first register an account and set a password in the system. After being recognized and saved by the system, the user can use the account and password to enter the system and operate the system when logging in to the system again. If it does not belong to an account recognized by the system, the system will prevent the user from logging in.

5 Conclusion

The traditional procurement model is not only time-consuming, but also unable to manage the procurement process. This paper builds a procurement management system based on genetic algorithm, and combines MySQL, J2EE and other related development technologies and B/S structure to build a system system, and describes in detail the role of each functional module of the system. It can be seen from the system test results that the stability of the system operation is guaranteed, and the procurement management system designed in this paper can achieve an efficient management level.

References

1. Mustafa, A., Heppenstall, A., Omrani, H., et al.: Modelling built-up expansion and densification with multinomial logistic regression, cellular automata and genetic algorithm. Computers, Environment and Urban Systems **67**(jan.), 147–156 (2018)
2. Arroyo, P., Carrete, L.: Motivational drivers for the adoption of green energy: The case of purchasing photovoltaic systems. Manag. Res. Rev. **42**(5), 542–567 (2019)
3. Stritch, J.M., Darnall, N., Hsueh, L., et al.: Green technology firms and sustainable public purchasing. IEEE Eng. Manage. Rev. **46**(1), 128–131 (2018)
4. Rezapour, A., Hakimzadeh, S.M., Panahi, S., et al.: Designing a pragmatic model for strategic purchasing of health services in health insurance companies: The feasible pivot of strategic purchasing in a developing country. Clinical governance **24**(1), 42–55 (2019)
5. Johnsen, T.E.: Purchasing and supply management in an industrial marketing perspective. Industrial Marketing Management **69**(FEB.), 91–97 (2018)
6. Bartnik, R.: Aligning purchasing portfolio management with sourcing negotiation styles. Manag. Decis. **56**(11), 2341–2356 (2018)
7. Karttunen, E.: Purchasing and supply management skills revisited: an extensive literature review. Benchmarking An International Journal **25**(3), 00 (2018)

8. Knight, L., Tate, W.L.: Quality and integrity in purchasing and supply management research. J. Purch. Supply Manag. **24**(3), 177–182 (2018)
9. Savastano, M., Bellini, F., D'Ascenzo, F., et al.: Technology adoption for the integration of online–offline purchasing: Omnichannel strategies in the retail environment. Int. J. Retail & Distribution Manage. **47**(5), 474–492 (2019)
10. Lamenza, A., Fontainha, T.C., Leiras, A.: Purchasing strategies for relief items in humanitarian operations. J. Humani. Logi. Supply Chain Manage. **9**(2), 151–171 (2019)
11. Backstrand, J., Suurmond, R., Raaij, E.V., et al.: Purchasing process models: Inspiration for teaching purchasing and supply management. J. Purcha. Supply Manage. **25**(5), 100577.1–100577.11 (2019)
12. Foo, M.Y., Kanapathy, K., Zailani, S., et al.: Green purchasing capabilities, practices and institutional pressure. Manage. Environ. Quality An Int. J. **30**(5), 1171–1189 (2019)
13. Jaeaeskelaeinen, A.: Comparison of performance measurement in different purchasing and supply management practices. Int. J. Product. Perform. Manag. **67**(8), 00 (2018)
14. Bals, L., Schulze, H., Kelly, S., et al.: Purchasing and supply management (PSM) competencies: Current and future requirements. J. Purcha. Supply Manage. **25**(5), 100572.1–100572.15 (2019)

Credit Risk Prediction Model of Digital Rural Finance Based on PSO-SVM

Juan Huang(✉)

Wuhan University of Engineering Science, Wuhan, Hubei Province, China
hjwuhan@iaec.cn

Abstract. With the help of the Internet, the use of digital technology to develop the financial industry can give full play to the advantages of the network to practice the rural financial credit risk prediction model, farmers can obtain more financial products and services in a convenient and fast way, and can enhance remote areas. Financial availability of residents. The purpose of this paper is to study the credit risk prediction model of digital rural finance based on PSO-SVM. The support vector machine is selected as the basis for establishing the personal credit evaluation model, but the accuracy of the support vector machine classification model mainly depends on the selection of kernel function parameters. The algorithm optimizes the parameters of the support vector machine. The model before and after optimization is programmed in MATLAB, and the model before and after optimization is further verified by combining real Internet financial data. The comparative experimental results show that the support vector machine model optimized by particle swarm optimization has better evaluation effect than the unoptimized support vector machine. Significant improvement, can effectively reduce the second type of misjudgment rate.

Keywords: Digital Finance · Rural Finance · Credit Risk · Prediction Model

1 Introduction

Rural financial development has always been at a disadvantage. In the process of development, the shortage of funds in the rural financial market and the difficulty in obtaining loans are the main problems. These problems have affected the development of "agriculture, rural areas and farmers" [1]. From the perspective of rural economic development, the impact of rural finance is more important, and the backward development of rural finance has a negative impact on the rural economy [2]. As rural finance develops at a slower pace and can generate less real benefits, banks are more willing to target urban development. In response to this situation, the state has made a series of efforts to promote the development of the rural economy and issued a series of policies to support the rural economy [3]. From the three aspects of capital, taxation and interest rate, it has promoted the stable development of rural credit cooperatives, so that rural credit cooperatives have quickly achieved good development results [4].

Credit risk prediction has gradually become a research hotspot, and Correia M strengthens the analysis of credit risk through detailed analysis of basic information.

J. H. Abawajy et al. (Eds.): ICATCI 2022, LNDECT 170, pp. 71–78, 2023.
https://doi.org/10.1007/978-3-031-29097-8_9

Fundamental measures of volatility include (i) historical volatility in profitability, margins, turnover, operating income growth, and sales growth; (ii) the spread of analysts' forecasts for future earnings; (iii) the quartiles of the profitability distribution Quantile regression prediction for a range of numbers. As a test case for the advantages of volatility forecasting, the improved ability to predict future excess credit returns is documented, especially when using fundamental measures of volatility [5]. Charoontham K researches on different economic determinants (i.e. maximum credit risk investment constraint, maximum credit risk investment constraint, opportunity cost and opacity of the credit derivatives market for Thai banks). Digital credit default swaps are an optimal derivative contract that can be Sends a credible signal when the bank is constrained by the maximum investment constraint. In addition, the bank's profit is reduced because the cost of the optimal derivatives contract is higher when the bank is subject to positive opportunity costs and opaque credit derivatives markets [6] it is of practical significance to study the credit risk prediction model of digital rural finance.

This paper studies the rural finance in my country, highlights the important role of the digital rural finance credit risk prediction model in the development of rural finance, synthesizes the current research and application of credit evaluation at home and abroad, and analyzes the relevant domestic credit evaluation. This paper analyzes the principle and idea of SVM in depth, and proposes the PSO-SVM model for the problems existing in its application in the field of credit risk assessment. The model is verified by experiments and the results are analyzed.

2 Research on Credit Risk Prediction Model of Digital Rural Finance Based on PSO-SVM

2.1 Support Vector Machines

SVM is different from previous intelligent algorithms that simulate biometric data. It is based on the VC dimension theory of statistical learning and the principle of structural risk minimization. The best choice can get the best promotion ability [7]. Since the popularization of support vector machines, it has been implemented in many fields. Compared with other algorithms, SVM has many advantages, the main advantages are as follows: SVM has great advantages in small sample problems, it can get the optimal solution in the case of limited sample information currently held, not the optimal solution. The solution must be obtained when the amount of sample data tends to infinity, and the process of SVM solving eventually becomes a quadratic problem [8]. From the perspective of business research, the result of solving the quadratic problem is that the overall optimal solution does not have the "local edge" problem in the neural network solution; the kernel function maps the original problem to the high-level space, and non-new types of problems are in the original space. Transform the problem into a linear problem in a large space, and construct a linear separation surface in the large space. The complexity of the high-dimensional space algorithm has nothing to do with the sample size of the low-dimensional space, avoiding the "curse of dimensionality" [9].

2.2 Parameter Optimization Method Based on PSO Algorithm

Although both GA and PSO algorithms are heuristic algorithms, compared with GA, PSO has fewer adjustable parameters and stronger global search ability, so its optimization accuracy is often higher than GA, and the algorithm convergence speed is also faster than GA [10]. Based on this, this paper will introduce PSO to select the penalty parameters and kernel function parameters of the SVM intelligent prediction model to improve the prediction performance of the model. When using PSO to find the optimal penalty parameter C and kernel function parameter of SVM, it is necessary to give its initial value first. At the same time, in PSO, the penalty parameter C and the kernel function parameter are also called particles. Different from other optimization models, PSO gives each particle two different initial values, namely position x and velocity v. Specifically, the position x is the value of each particle itself, which is the same as the other optimization models, and the velocity v represents the displacement of each particle iteratively in the search space, which is different from the other optimization models where [11, 12].

However, if the particles updated at each iteration are optimal, an evaluation criterion must be set, also known as PSO suitability. In this paper, the accuracy of cross-validation (CV) is chosen as the criterion for determining the fitness of repeated particle updates. If the calculated CV accuracy is higher, it means that the fitness value of the particle is higher, and the x position of the particle is also higher, and the optimal position can be made, that is, the value at this time is the optimal value of the particle.

In order to find the optimal particle, it is necessary to find the optimal particle through the following types of iterative information, given the initial particle position x and velocity v. The formula is as follows:

$$v_{ij}^{t+1} = wv_{ij}^{t} + c_1\gamma_1\left(p_{ij}^{t} - x_{ij}^{t}\right) + c_2\gamma_2\left(g_{ij}^{t} - x_{ij}^{t}\right) \tag{1}$$

$$x_{ij}^{t+1} = x_{ij}^{t} + v_{ij}^{t+1} \tag{2}$$

In the above formula, i = 1, 2,..., k; j = 1, 2,..., d; t is the number of iterations of the PSO algorithm; Pij represents the optimal value of the particle itself under the current refinement, That is, the optimal value of each individual, gij represents the optimal value of the particle at each refinement, that is, the optimal value of the group, and xij represents the i-th particle located in the j-th particle. Exponential variable. In order to control the repeated process of each particle within a given span, this paper gives each particle a specific position and velocity, namely [–xmax, xmax], [–vmax, vmax], so that it can be repeatedly updated and controllable. c1 and c2 represent the acceleration factor, where c1 controls the speed at which the particle moves to the optimal value side, and c2 controls the speed at which the particle moves to the optimal value side r1 and r2 are randomly distributed numbers in the center [0.1]. w represents the inertia density, and its role is to balance the optimal performance of the PSO for the best global value and the best local value.

3 Investigation and Research on Credit Risk Prediction Model of Digital Rural Finance Based on PSO-SVM

3.1 Research Tools

This article uses the lissvm toolbox to complete the programming and experimentation of the classification model under the R2017b version of MARLAB. In order to more objectively compare the effect of the model before and after optimization, a unified hardware environment is used for implementation. The specific information is as follows:

Processor: Intel-core-i5-3210M
Running memory: 4 G
Main frequency: 2.5 G
System Type: 64-bit

3.2 Datasets

The datasets used in this article are from credit data and credit data in the UCI machine learning database. These two datasets are real personal data provided by customers in villages M and B when they apply for credit cards, and they are classic datasets commonly used in the field of credit evaluation.

Among them, the files included in the M rural data set are: user data files, digital data files and data description files, a total of 1000 credit data records, including 700 "good" samples and 300 "bad" samples, each sample contains 20 an attribute.

Rural credit data set B has a total of 690 pieces of credit data, consisting of 307 "good" samples and 383 "bad" samples, each of which contains 10 features. For privacy protection, the dataset does not give the specific meaning of all the features, only the first 9 features are also the expression of customer personal information, and the 10th feature is the category of the real performance of customer credit.

3.3 Data Preprocessing

The lack of credit data brings certain difficulties to credit evaluation, and it is necessary to take a reasonable approach to solve the missing data. How to deal with shortage:

(1) Case elimination method: delete all data with incomplete characteristic data. This method is suitable for cases where the total number of samples is large and the percentage of missing data is small.
(2) Average attribution method: When the exclusion method affects the integrity of the data, the attribution method can be considered for data processing. First, the characteristics of variables can be divided into arithmetic and non-arithmetic. For missing values in numeric formulas, use the mean value of the attribute to fill; for missing values in non-numeric formulas, use the attribute function to fill in.

3.4 Test Criteria for the Model

Calculation of the first type of false positive rate:

$$error\text{I} = \frac{FN}{TP + FN} \times 100\% \tag{3}$$

The calculation of the second type of false positive rate:

$$error\text{II} = \frac{FP}{TN + FP} \times 100\% \tag{4}$$

TP, correct and affirmative: predict the non-compliance customers as non-compliance customers;

TN, true negative: predict the defaulting customer as the defaulting customer;

FP, false positive: predicting a defaulting customer as a non-compliant customer (the second type of misjudgment);

FN, false negative: predicting non-compliance customers as defaulting customers (type 1 misjudgment).

4 Analysis and Research of Digital Rural Financial Credit Risk Prediction Model Based on PSO-SVM

4.1 Construction of PSO-SVM Credit Evaluation Model

The general understanding of the PSO-SVM Internet Credit Assessment online personal credit rating model is as follows: collect data in advance, examine the data points using the first component analysis method, and determine the final input model. After training, model evaluation and model evaluation are carried out, and then the parameters are optimized by the swarm particle algorithm; finally, the model before and after optimization is tested with test samples to confirm the probability of the model. According to the model construction tips, combined with the actual personal credit rating process, the steps of implementing PSO-SVM for personal credit assessment in this paper are shown in Fig. 1.

4.2 Comparative Analysis of Model Results

This section uses the test set data to test the performance of the SVM model and the PSO-SVM model. The comparative experimental results are summarized in Table 1:

After the above comparative analysis, the following conclusions can be drawn:

In terms of general model specification, both SVM and PSO-SVM have good predictive performance. The SVM model after PSO optimized kernel performance parameters improved by an average of 11% compared to the standard automated machine, as shown in Fig. 2.

Comparing the values of the first-type false positive rate and the second-type false positive rate of the standard SVM model and the PSO-SVM model, it can be seen that the SVM-optimized PSO model has both the first-type false positive rate. Alarm rate and

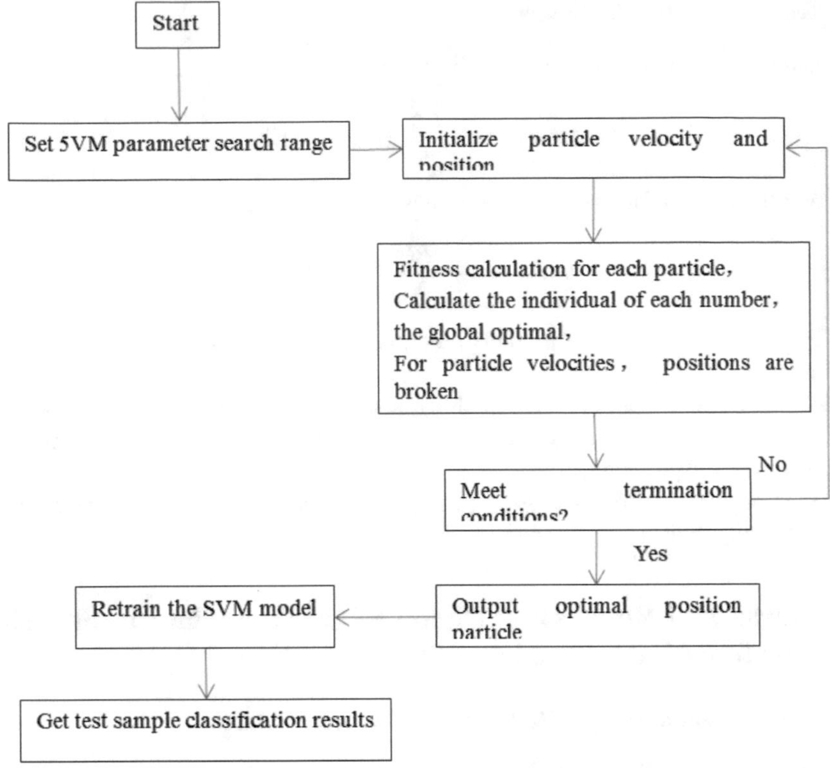

Fig. 1. PSO-SVM model implementation flow chart

Table 1. Comparison of the classification prediction results of the two models

Evaluation indicators	SVM	PSO-SVM	Change value
Type 1 false positive rate	12%	5%	–7%
Type II false positive rate	15%	6%	–9%
Overall classification accuracy	85%	96%	11%
Running time (s)	124	12	–112

Type II false alarm rate. The reduction is significant. When the inequality rate for the first category is higher, it means that the model has a stronger credit rating for borrowers. However, when Type II inflation is high, it can result in huge losses for credit institutions.

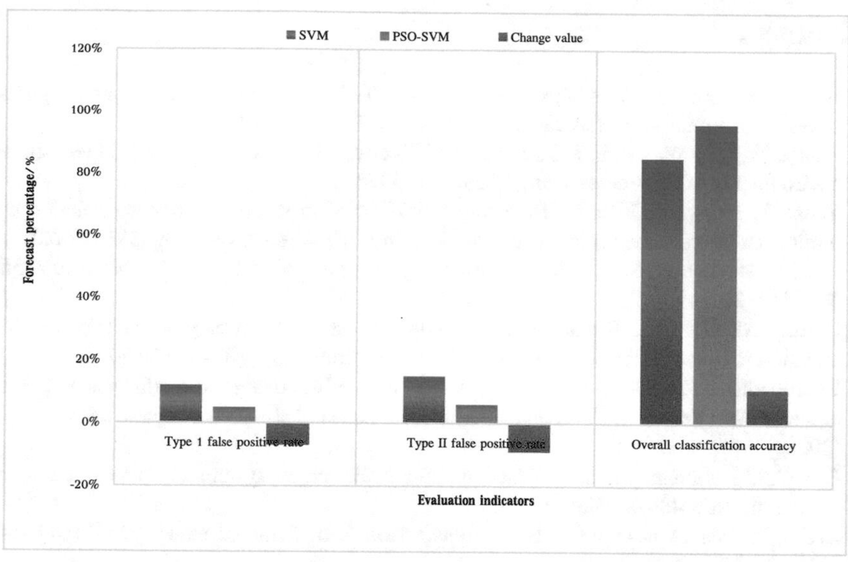

Fig. 2. Comparison of the classification prediction results of the two models

The second type of error analysis of the SVM model performed in this study was as high as 15%, and the second type of error analysis of the optimized model was also as high as 6%. Due to the small number of default models in the data samples, the SVM fails to learn the default sample features sufficiently, resulting in a high error rate of the model when checking the default sample components. However, in order to respect the original data structure, there is no intervention in the sample data structure. In terms of performance, optimizing the model is not only a fairly immediate improvement, but also a significant improvement in performance.

The above conclusions show that the method of optimizing SVM parameters using PSO in this study is effective, and PSO can improve the overall prediction results and performance of the SVM model.

5 Conclusions

Our country's economy is developing rapidly, and rural finance is constantly advancing in the development. My country's rural economy is still in the stage of exploration and reform, and a financial theoretical system with my country's regional characteristics has been formed in the development. Taking rural credit forecasting as a research guide, this paper demonstrates intelligent research (SVM) technology, which is based on the rural financial credit risk forecasting system (SVM) for early warning, and uses it to conduct research. Early warning of loan problems. An example of rural lending is an early warning example based on the PSO-SVM, which is an excellent tool to use the credit control unit to deal with the impact of the loan. Regulators can use the PSO-SVM risk assessment framework built into this document to better and more accurately prepare for the impact of future lending on rural finance, and develop and implement relevant policies to handle credit, thereby enhancing market control and prevention.

References

1. Ashour, S., Hao, Q.: Do analysts really anchor? Evidence from credit risk and suppressed negative information. J. Bank. Finance **98**(Jan), 183–197 (2019)
2. Zhang, W., Lu, W., Chen, R.S., et al.: An Effective digital system for intelligent financial environments. IEEE Access **7**(99), 155965–155976 (2019)
3. Yoon, B., Jeong, Y., Kim, S.: Detecting a risk signal in stock investment through opinion mining and graph-based semi-supervised learning. IEEE Access **8**, 2169–3536 (2020)
4. Nisar, T.M., Yeung, M.: Attribution modeling in digital advertising. J. Advert. Res. **58**(4), 399–413 (2018)
5. Correia, M., Kang, J., Richardson, S.: Credit market investors may not be paying enough attention to fundamental asset volatility. Rev. Acc. Stud. **23**(1), 37–94 (2018)
6. Charoontham, K., Kanchanapoom, K.: Credit derivatives design to facilitate loan purchase agreements in the secondary loan market in Thailand. J. Asia Business Stud. **14**(5), 561–580 (2020)
7. Lee, C.S.: Datafication, dataveillance, and the social credit system as China's new normal. Online Inf. Rev. **43**(6), 952–970 (2019)
8. Butler, T.: What's next in the digital transformation of financial industry? IT Professional **22**(1), 29–33 (2020)
9. Ge, Q., Guo, C., Jiang, H., et al.: Industrial power load forecasting method based on reinforcement learning and PSO-LSSVM. IEEE Trans. Cybernet. 52(2), 1112–1124 (2022)
10. Priyadarshi, N., Padmanaban, S., Holm-Nielsen, J.B., et al.: An experimental estimation of hybrid ANFIS–PSO-based MPPT for PV grid integration under fluctuating sun irradiance. IEEE Syst. J. **14**(1), 1218–1229 (2020)
11. El-Soud, M., Zyout, I., Hosny, K.M., et al.: Fusion of orthogonal moment features for mammographic mass detection and diagnosis. IEEE Access **8**, 129911–129923 (2020)
12. Mohamed, M.A., Hajjiah, A., Alnowibet, K.A., et al.: A secured advanced management architecture in peer-to-peer energy trading for multi-microgrid in the stochastic environment. IEEE Access **9**, 92083–92100 (2021)

Cache High Availability Intelligent Stall Management System Based on Redis Sentinel Mechanism Architecture

Qingjie Wang, Lijie Xiao, and Juan Xiao(✉)

School of Computer and Artificial Intelligence, XiangNan University, Chenzhou, Hunan, China
keyanxj@163.com

Abstract. Stall economy refers to an economic form formed by setting up stalls to obtain sources of income. It is one of the most primitive and viable commercial activities of mankind. It prospers the economic market and makes up for a blank spot in people's shopping. Its unique advantages, in the current context, can alleviate the employment pressure to a certain extent. However, because stall economy is a kind of marginal economy in the city. The reason why it is not elegant is that it affects the appearance and environment of the city. In view of the negative impact of the stall economy, many users, messy market and difficult management, an intelligent stall management system is designed, which is managed and supervised by government staff in the background and managed by stall households independently. The redis sentinel mechanism architecture is adopted to solve the problems of many users and strong real-time data concurrency, so as to provide urban managers, stall owners Consumers designed and developed an intelligent stall economic management system based on redis sentinel mechanism cache high availability architecture.

Keywords: Redis Database · Sentinel Mechanism · Cache Is Highly Available · Intelligent Stall Management System

1 Introduction

On May 27, 2020, the central civilization office made it clear that the evaluation indicators of national civilized cities in 2020 do not include road occupation, road market and mobile vendors. The promulgation of this policy and the positive response of all provinces to this initiative stimulated the emergence of a large number of stall vendors. Stall economy has the "three low" characteristics of "low threshold", "low risk" and "low consumption", which makes its development have unique advantages. The stall economy has solved the problem of employment difficulties to a certain extent. Because of the epidemic, the unemployed can achieve flexible employment by setting up stalls. At the same time, it also has a negative impact, which is reflected in the adverse impact of the road occupation operation of the stall on the traffic and environment. The stall economy eliminates more consumer demand after meeting the consumer's demand for cheap products, which damages the interests of regular stores to a certain extent, the

J. H. Abawajy et al. (Eds.): ICATCI 2022, LNDECT 170, pp. 79–86, 2023.
https://doi.org/10.1007/978-3-031-29097-8_10

site competition between vendors, and the growing contradiction between vendors and urban management, this series of problems are urgent for us to solve.

2 Research on Key Technologies

2.1 Redis Database

Traditional relational databases (Oracle, DB2, mysql, Microsoft SQL Server) have poor data type expression ability and poor complex query function [1]. They will cause downtime in the face of highly concurrent data queries. Once they are down, they are not easy to repair.

Redis is an open source, advanced, memory based, data persistent key value storage system. It is characterized by high reading and writing efficiency, rich data types and persistent storage, so it is especially suitable for high concurrent application scenarios. Redis can implement a variety of sorting methods. At the same time, it caches all data in memory to improve performance [2]. In addition, redis supports rich data types, mainly including string, list, set, sortedset and hash. All data types can call quite a number of methods, such as push/pop, add/remove, intersection, union and difference. As shown in the following Fig. 1, redis internally uses a redisobject object to represent all keys and values. Where type represents a value object, which data type it is. Encoding is the encoding storage method of different data types in redis.

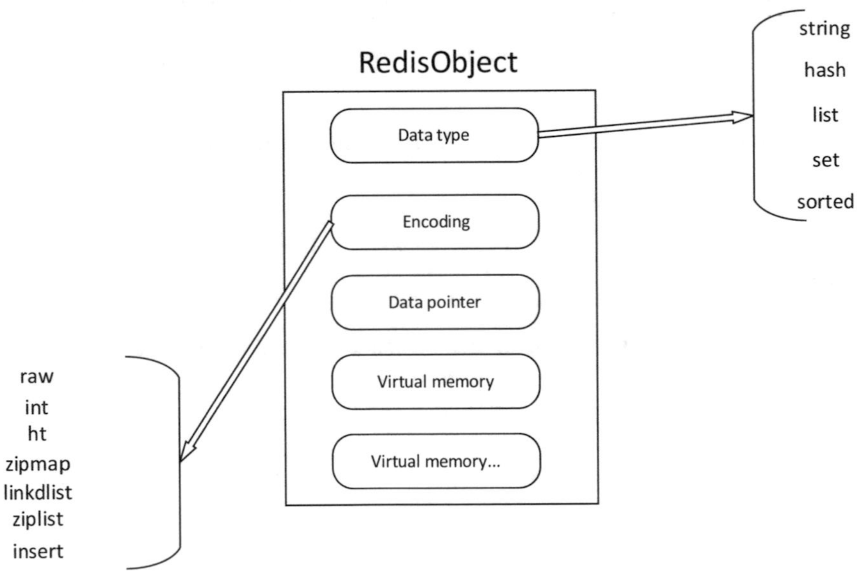

Fig. 1. Redis Object

2.2 Sentinel Mechanism

Sentinel mechanism is a special mechanism. Let's take a look at the working process of sentinel mode. First, redis provides sentinel commands. Sentinel is a process that can run independently. Its principle is that sentinel sends commands and waits for the response of the redis server to monitor multiple running redis instances. Then send the command to the master server and return it to the slave server through redis. When sentinel detects that the master server is down, it will automatically switch the slave server to the master server, and then notify other slave servers to modify the configuration file through the publish subscribe mode, while allowing them to switch hosts. The working principle is shown in the Fig. 2 below.

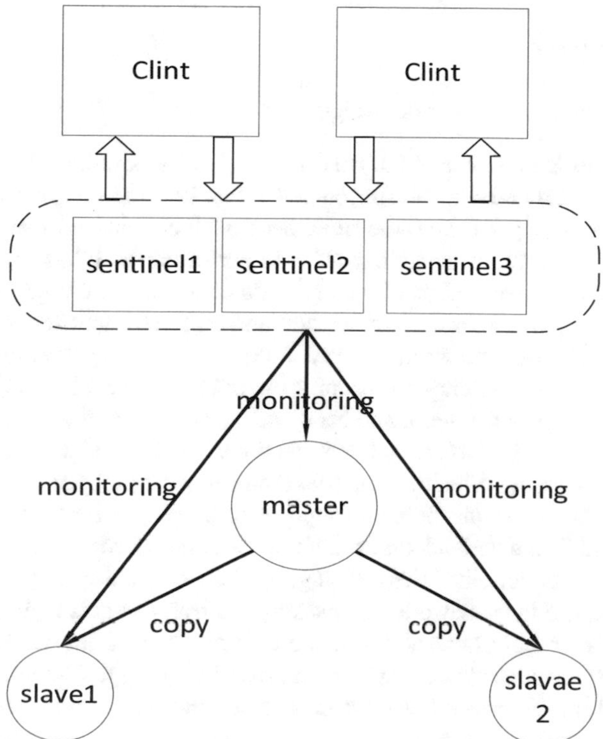

Fig. 2. Schematic diagram of redis sentry mechanism

2.3 SPRING

Spring is an open source framework to solve the complex problems of enterprise application development [3]. It uses basic JavaBeans to do things that could only be done by EJBs before.

Advantages of spring: 1 Lightweight; 2. Non intrusive: generally, objects in spring applications do not depend on their specific classes.

Inversion of control: spring uses inversion of control technology to realize loose coupling between modules. When the inversion of control technology is applied, other objects that the object depends on will be passed in passively, rather than creating or finding dependent objects by the object itself.

Aspect oriented:spring provides rich direction oriented programming support for realizing high cohesion within the module, and realizes cohesive development by separating the business logic of the application from the system level services (such as auditing and transaction management).

At present, spring MVC integrated into spring web flow is an upgrade of spring framework [7]. It separates the roles of controller, model object, scheduler and handler object, which makes them easier for users to customize.

3 System Design

3.1 Overall System Framework Design

The system adopts the architecture of front and rear end separation. Through this structure, the front and rear end codes are decoupled to facilitate their respective development and maintenance [4, 5]. At the same time, because the front and rear ends are separated, the later system can be better expanded in performance. It has established a good foundation for the transformation into large-scale distributed architecture, elastic computing architecture, micro service architecture and multi terminal services (a variety of clients, such as PC, MAC, browser, Android, IOS, etc.), which is also the reason why the separation of front and rear ends has become the trend of Internet project development.

The system adopts a front-end and back-end separation architecture, and the front-end mainly displays the interface and presents data. The latter side focuses on business logic and data processing. The front end uses Android technology, uses app for interface display, and the back end runs based on application server, such as Tomcat (see Fig. 3).

The back-end framework adopts springboot, which is accompanied by spring 4.0 0 was born. Springboot inherits the excellent gene of the original spring framework. It can quickly build the spring framework. The background framework is logically divided into an application layer that provides business interfaces to the front end and a data layer that provides data services to the application layer. From top to bottom, the application layer is divided into interface layer, business layer and basic service layer, which are respectively responsible for interface mapping, business logic entities and providing various basic services [6]. The application layer runs in the tomat server. The data layer is divided into access layer and storage layer. The access layer encapsulates the access interfaces to different entities, while the storage layer provides the access layer with the real operation interface to the database. The data layer adopts MySQL + redis and uses redis as cache to solve the problem of overload of a large number of users accessing the database.

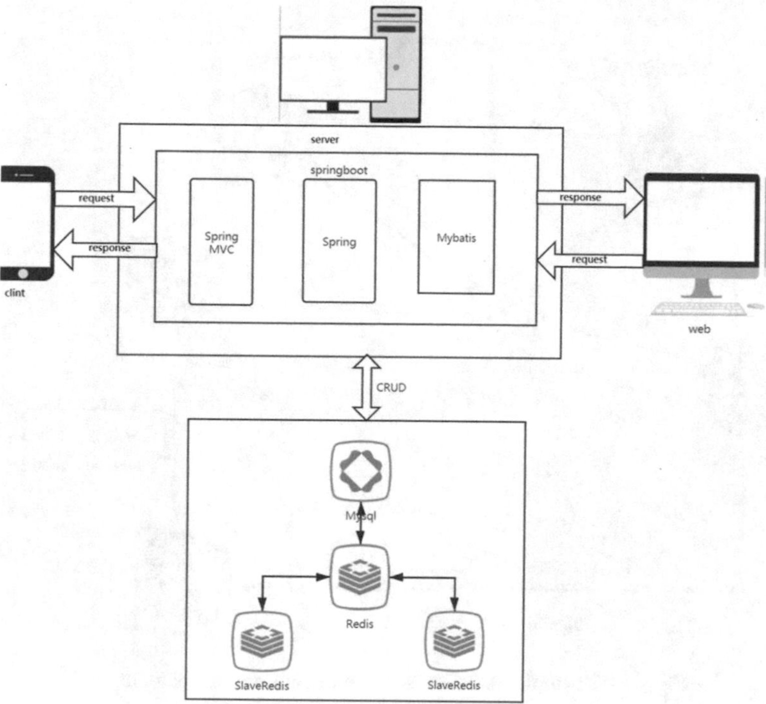

Fig. 3. Overall system framework

3.2 System Database Design

The database structure of the system is the combination of redis + MYSQL, and the implementation scheme of sentinel mechanism is shown in Fig. 4.

3.3 Synchronize MySQL Data to Redis

Set the cache time in the redis database. When the cache time of the data expires, it will be automatically released and re queried in the database [11]. However, in this case, the data we put in the cache does not require high data consistency before it can be put into the cache [10]. Use MySQL based on binlog_ udf_ Redis synchronizes the data in the database to redis. Automatically refresh redis through MySQL Synchronization.

3.4 Synchronize Redis to MySQL

When reading data, check it in redis first. If not, check it in the database, write it to redis at the same time, and set the expiration time. When saving data, you should analyze the specific situation [8, 9]. You can choose to insert it into the database and redis at the same time (if it is stored in redis, you'd better set the expiration time), or you can choose to insert it directly into the database without considering some problems.

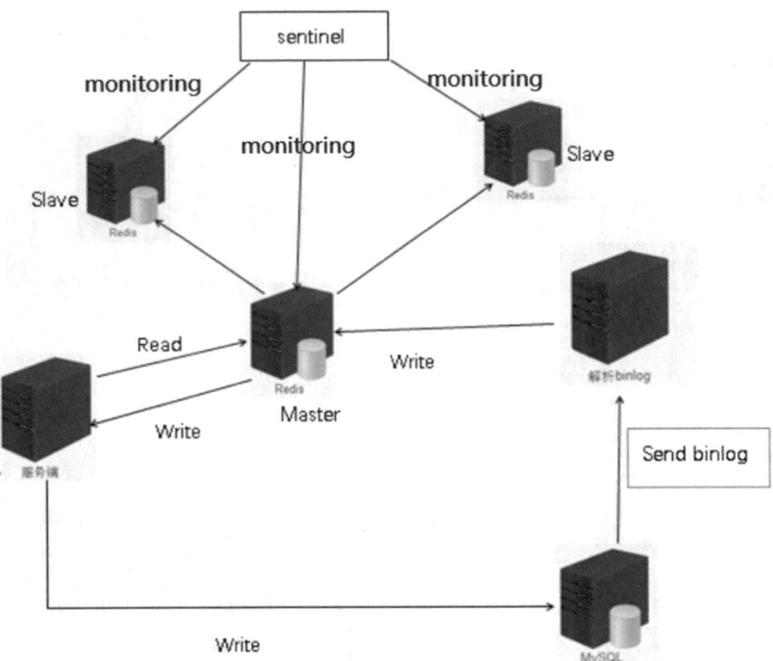

Fig. 4. The implementation scheme of sentinel mechanism

Table 1. Mysql database table design of the system

Table	Attribute 1	Attribute 2	Attribute 3	Attribute 4	Attribute 5
Audit information	Account (nchar(30))	ID number (nchar(30))	Name (nchar(30))	Administrator account nchar(30)	Time (datetime)
Comment	Vendor account (nchar(30))	Comment content (nchar(30))	Commentator (nchar(30))	Time (time)	
Complaint	Complained account (nchar(30))	Complaint content (nchar(30))	Complainant's account (nchar(30))	Management account (nchar(30))	Time (datetime)
Stall owner information table	Vendor name (nchar(30))	Password (nchar(30))	ID number (nchar(30))	ID picture (mediumblob)	Account (nchar(30))
Stall owner information table	Telephone (nchar(30))	Sex (nchar(30))	Age (nchar(30))	Product Photo 1(mediumblob)	Product Photo 2(mediumblo)
Booth Information Sheet	Vendor number (nchar(5))	State of the booth (int)	Booth longitude(double)	Stall latitude (double)	
Stall owner rental form	Vendor ID (nchar(5))	Vendor number (nchar(5))	Start clock in time (datetime)	End clock in time (datetime)	

The MySQL database table is shown in Table 1:

The data information required by the system includes audit information, user information, comment information, complaint information, punch in information, booth information, etc.

The table attribute structure composed of specific information is shown in Table 1:

4 System Implementation

4.1 System Implementation Process

The users of the whole system are divided into three categories – manager, user and stall. After opening the app, the stall owner can view the stalls existing in the current location and the location of the stalls. After selecting the stalls, determine the stall setting time, and enter the approval status after submitting the application. When the stall starts and ends, it is necessary to upload the health status of the current stall and upload it to the server for review. The administrator reviews the stall setting application and daily clock in submitted by stall users through app, and processes unqualified stall owners. In addition, the administrator plans the stalls, plans the places where stalls are often set up, and uploads them to the server for stall owners to choose. According to the user's current longitude and latitude, the system automatically pushes the booth only 500 m away from the user. In addition, users can set their own search conditions and select distance and type. The user can click the beach user to navigate the route.

4.2 System Execution Process

When the user queries the stall master data and enters the search criteria in APP, the data will be sent to the Tomcat server first, and Tomcat will request the Redis database to check whether the data exists in the current cache database. If the data does not exist in the current database, Tomcat will request the mysql database data. The queried data is returned to Tomcat, which sends it to APP for users to view.

5 Conclusion

In the early stage, they also adopted the mode of policy intervention, and urban management officers managed the areas in charge, but this consumed human resources. In order to improve the efficiency of the government and reduce the consumption of human resources. Using information means, design and develop intelligent stall management system standardized management of stall economy, including stall management, stall environmental management, stall owner management and so on. It provides information service for the benign cycle and sustainable development of the land sharing economy and the coordinated development of the land sharing economy and urban civilization. There are also some problems in the system. The intelligentization is not enough when the existing users intelligentize and recommend the booths, and the algorithm needs to be improved in the future.

Acknowledgments. 2020 innovation and entrepreneurship training program for college students in Hunan Province(NO.3688).

References

1. Belalem, G., Matallah, H., Bouamrane, K.: Evaluation of NoSQL Databases: MongoDB, Cassandra, HBase, Redis, Couchbase, OrientDB. Int. J. Softw. Sci. Comput. Intell. **12**(4), 71–91 (2020)
2. Alba, P.-Z., Clemente, R., Juan, S., Rubén, G.T., Víctor, V.-Y.: Ideal and predictable hit ratio for matrix transposition in data caches. Mathematics **8**(2), 184 (2020)
3. Palotás, B., Takács, P., Fink, J.: Simulation of a SCSC plate with a spring framework model including the effects of inelastic slip. Stahlbau **90**(6), 462 (2021)
4. Zhang, T.J., et al.: iBTune: individualized buffer tuning for large-scale cloud databases. Proc. VLDB Endowment **12**(10), 1221–1234 (2019)
5. Shen, Z., Lee, P.P.C., Shu, J., Guo, W.: Correlation-aware stripe organization for efficient writes in erasure-coded storage: algorithms and evaluation. IEEE Trans. Parallel Distrib. Syst. **30**(7), 1552–1564 (2019)
6. Pokorny, J.: NoSQL databases: a step to database scalability in web environment. Int. J. Web Inf. Syst. **9**(1), 69–82 (2013)
7. Fangmei, N., Xuemei, H., Jinjuan, M.: Application of spring boot integrated Redis cache technology in enterprise one-card system. Electr. Technol. Softw. Eng. **24**(2), 133–134 (2019)
8. Antas, J., Silva, R.R., Bernardino, J.: Assessment of SQL and NoSQL systems to store and mine COVID-19 data. Computers **11**(2), 29 (2022)
9. de Oliveira, V.F., Pessoa, M.A.O., Junqueira, F., Miyagi, P.E.: SQL and NoSQL Databases in the Context of Industry 4.0. Machines **10**(1), 20 (2021)
10. Riahi, F.: Oliver Schulte Model-based exception mining for object-relational data. Data Min. Knowl. Disc. **34**(3), 681–722 (2020)
11. Satoshi, G., Mariko, T.: Yadohisa Hiroshi Clustering for time-varying relational count data. Comput. Stat. Data Anal. **156**, 107123 (2021)

The Application of BIM Technology in the Construction of Project Construction Management Platform

Chunhua Song[1](✉) and Manoj Kautish[2]

[1] Shandong Huayu University of Technology, Dezhou, Shandong, China
sch15275712255@126.com
[2] LBEF Campus, Kathmandu, Nepal

Abstract. The construction industry is one of the pillar industries of my country's economic development . BIM technology plays a very important role in improving the level of project construction management. At present, my country lacks the overall planning of the project construction management system in the construction of the project construction management platform. System isolation, lack of scientific and reasonable standard specification system and lack of unified platform management model, these problems have a certain impact on the cost, construction period, quality and other aspects of the project. How to improve the level of project construction management is one of the issues we need to pay attention to at present. We need to formulate a scientific and reasonable BIM project construction management system, introduce BIM technology into the project construction management platform to achieve visual management, and use BIM5D model data for resource optimization management. In order to improve the project construction management mode, and improve the construction management level.

Keywords: BIM Technology · Project Construction · Management Platform

With the rapid development of my country's economic strength, the process of urbanization has gradually accelerated. Especially since the beginning of the 21st century, my country's construction industry has achieved unprecedented development, and the competition in the construction industry has become increasingly fierce. In the traditional construction process, resource consumption is relatively serious, and the level of information-based construction is relatively low. The level of construction management and technology can no longer meet the needs of the stable development of the current social construction industry. The introduction of BIM technology in the construction of construction management platforms for construction projects has become an inevitable trend in the development of the construction industry. BIM technology can carry out scientific and reasonable management and control of project construction, such as achieving reasonable management and control of project cost, project progress, construction quality and safety, and improving the project. Construction management efficiency.

J. H. Abawajy et al. (Eds.): ICATCI 2022, LNDECT 170, pp. 87–95, 2023.
https://doi.org/10.1007/978-3-031-29097-8_11

1 An Overview of BIM Technology

1.1 Introduction of BIM Technology

The application of BIM technology in the project construction management platform has effectively improved the information management level of the project project. BIM technology uses information technology to establish a three-dimensional building model, which includes the geometric structure of the building, the size information of components, professional attributes, space status and other information [1]. The construction management departments of the project can obtain construction information through the BIM building information model, ensuring that Consistency, accuracy and efficiency of information transmission, BIM technology provides a project data exchange and sharing platform for relevant departments of project engineering. The functional structure of BIM model is shown in Fig. 1 [2].

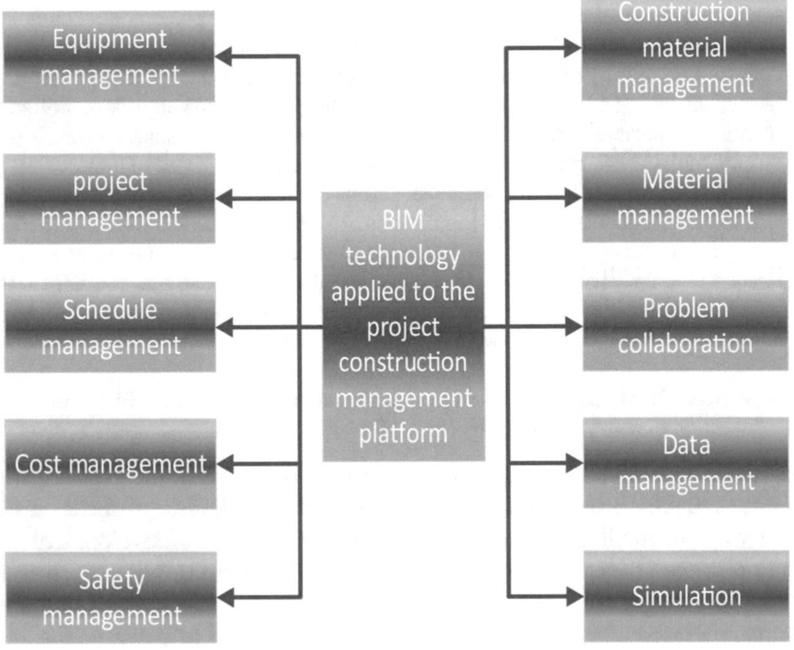

Fig. 1. Functional structure of BIM technology model

1.2 Features and Advantages of BIM Technology

(1) Visibility

Traditional architectural design generally adopts the method of graphic design drawings plus text. Such two-dimensional drawings are difficult to show the detailed information of the building. BIM technology has visibility [3]. The detailed dimensional information is displayed, the plane drawings are converted into three-dimensional models, and the construction content and process of the project can be visualized. BIM visualization makes it easier for managers to familiarize and manage the overall project [4].

(2) Coordination

The design of some construction projects is relatively complex, and BIM technology can be used to find out the problems that are not easy to be found in complex projects, and solve and optimize the problems reasonably [5]. In the process of project construction, it may happen that the design drawings and construction plans cannot meet the construction needs well, or the management personnel, technicians and construction personnel cannot communicate in a timely and efficient manner in terms of project information, which leads to Project information is asymmetric, and BIM technology can solve these problems well [4]. Through BIM technology, project related personnel can grasp and transmit the construction situation of the project in time, and can make the design drawings and construction plans meet the needs of construction, and improve the communication efficiency of each link of the project. BIM technology can properly solve the problem before construction, and also help to improve the cooperation ability and coordination ability between various departments [5].

(3) Information relevance

The BIM building information model stores the information of all stages of the project design. The logical relationship of the engineering construction organization is relatively complete [8]. The BIM building information model has integrity and relevance. When the technicians adjust the parameters of the construction project, other related parameters will also be Automatic update, improve work efficiency. Project construction managers can manage the construction process more efficiently by using the relevance of the BIM information model to avoid conflicting data generated during the construction process [9].

2 Problems Existing in the Construction of the Current Project Construction Management Platform

2.1 Lack of Overall Planning for the Project Construction Management System

Project construction management plays an important role in the smooth development of the project. At present, my country's construction industry lacks systematic overall planning in construction management, which may lead to waste of raw materials during the construction process, which will increase the construction cost virtually, and the construction period may be shortened [10]. Delayed, and even affect the quality of the project, increasing the risk of hidden dangers. At the same time, during the construction of the project, the staff will inevitably encounter some unexpected situations, including human factors and non-human factors [11]. These uncertain factors lack overall planning guidance and emergency measures, which hinder the normal development of the project, so it is reasonable to The advanced project construction management system planning is the basic guarantee to ensure the smooth implementation of the project project, which can effectively reduce or avoid the possible risks and losses of the project project [12].

2.2 Lack of a Unified Platform Management Model, and the Construction Quality of the Project Construction Management Platform is Low

At present, on the project construction management platform, there are great differences in the management modes of various departments, lack of a unified management mode, and the construction quality of the project construction management platform is low [13]. In addition, during the construction process of the project construction management platform, the management mechanisms of various majors are not unified, and the staff are not guided by standardized construction during the actual construction process, resulting in the overall management of the project project in a state of confusion [7]. It is difficult for departments to coordinate and unify, and the update and transmission of project construction information lags behind, which affects the project construction management platform to play its due role [14].

3 Application Research of BIM Technology in the Construction of Project Construction Management Platform

3.1 Develop a Scientific and Reasonable BIM Project Construction Management System

In terms of project construction management, the construction unit can improve the efficiency of project construction and ensure the quality of the building by formulating a sound BIM project construction management system. First, it is necessary to clarify the task assignment and post setting of the construction management personnel of the BIM project. For example, the personnel allocation of the construction enterprise should be fully integrated with the engineering project, and the balance of all aspects of the project should be fully considered, from experts, engineers, financial personnel, The administrative staff and other aspects comprehensively optimize the post work and clarify the setting of post responsibilities, so as to improve the construction management level of the BIM project. Second, clarify the management standards and specifications of project projects. In the construction management of BIM projects, construction companies should clarify the work content and management standards of managers at all levels according to the actual construction situation, and refine and decompose complex or medium and long-term work to improve Work efficiency, which is also conducive to ensuring the construction quality of the project (see Fig. 2).

3.2 Realize Visual Management During Construction

Technicians build a 3D model of the project through BIM technology, and upload the 3D model data to the construction management sharing platform. Relevant units can check

Fig. 2. Professional model of project construction management platform

the construction situation on the platform, such as construction progress, resource consumption, construction simulation, and construction site pictures. Or video, and design drawings can also be integrated into the BIM building information model. By comparing the construction site, managers can better view the construction status of the project and solve problems that are not easily found in site management or construction management. The platform centralizes the image data. Storage, other departments can also watch the construction situation in real time through the project construction management platform, and managers can achieve visual management during the construction process through BIM technology (see Fig. 3) [15].

Fig. 3. Project general contract management system

3.3 Improve the Space Utilization Rate of the Construction Site and the Utilization Rate of Engineering Materials

The management personnel give corresponding solutions according to the problems found, and use BIM technology to simulate the scheme. Each department obtains the relevant information of the rectification measures through the construction management platform, which can be directly applied to the construction site after multi-party review, which greatly improves the construction efficiency, the 3D model reviewed by professionals can be exported to form a video, which can also be used as construction site safety training. The BIM construction site layout procedure is shown in Fig. 4. Through BIM technology, managers can reduce ineffective processes, formulate a reasonable and standardized material placement area, and determine the best construction site layout plan. When obtaining the required construction materials, the staff can easily Find it quickly and improve work efficiency. In addition, in the BIM design process, the modeling technology should be used skillfully to design the model as a whole. For example, in the design process of wells and water collection corridors, the following formulas should be fully utilized for scientific modeling. Modeling of seepage should use Darcy's law of seepage and Furyo's formula:

$$\nu = \kappa \frac{\mathrm{H}}{l} = KJ \tag{1}$$

$$\nu = u = KJ \tag{2}$$

$$J = -\frac{dH}{dS} \tag{3}$$

Another example is the modeling of water collection corridors, which can be substituted into the following equation:

$$Q = A\nu = -bzk\frac{dz}{dS} \tag{4}$$

In short, modeling and design of project engineering plays an important role in shortening the construction period and ensuring the accuracy of the construction process.

3.4 Resource Optimization Management Using BIM5D Model Data

Based on the BIM5D construction project construction management platform, managers can have a clearer understanding of the specific conditions of project construction. BIM5D software can simulate the construction process and help managers have an in-depth understanding of the project's capital usage trends. First, import the plan prepared by Project into the Zebra schedule software, and then use the Zebra schedule software to draw the project progress bar chart. Finally, the construction schedule table is imported into the Glodon BIM5D software for project construction simulation. For example, in the process of project construction, according to the actual materials used in the construction, the management personnel establish a BIM5D model data center of the project project to optimize the resource management of the project construction,

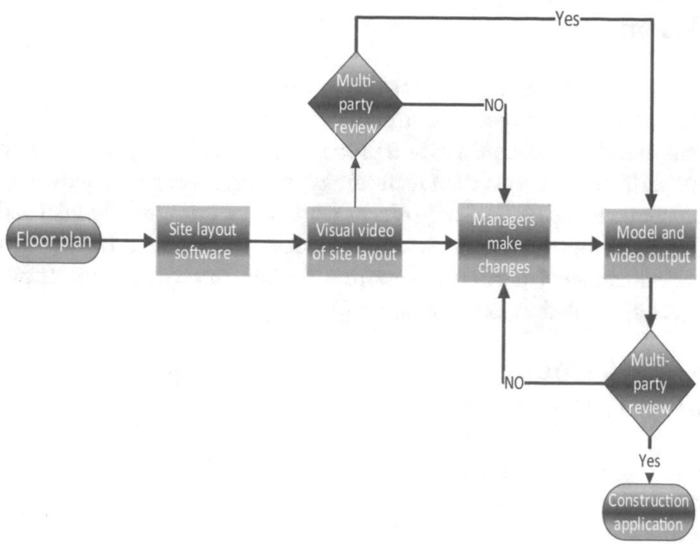

Fig. 4. Flow chart of BIM site layout design

and optimize the resources of some complicated and difficult-to-manage processes of the project project. It is planned to allocate a reasonable number of personnel or equipment required for each stage of construction to ensure that the construction needs of each stage are met. Through the BIM5D model, managers can grasp the use of funds in the project construction process and the time nodes of project construction, and if there is a deviation between the actual construction progress and the planned progress, the managers can analyze the cause of the deviation according to the model and formulate corrective measures for the deviation, so that The management of the project project is more efficient [15]. The management personnel can also determine a reasonable construction plan according to the cost plan analysis, and then decide to formulate the optimal construction plan through comparison. The BIM5D resource optimization management process is shown in Fig. 5 below.

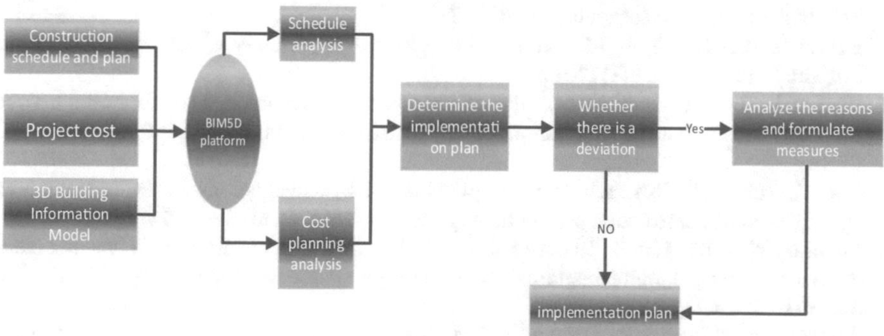

Fig. 5. Flow chart of BIM resource optimization management

4 Conclusion

BIM technology can effectively correct the deviation between the actual construction stage and the plan in the project construction management, and can solve some complex problems in the project construction process. By establishing a project construction management platform based on BIM technology, managers can more easily achieve reasonable management and control of project costs, project progress, construction quality and safety, and the information exchange between various departments in the project construction management platform is more efficient, improving the efficiency of the project. Project construction management efficiency.

Acknowledgment. R & D Center of BIM Technology and Application in Engineering (2019-02) of Shandong Huayu University of Technology.

References

1. Shaqour, E.N.: The role of implementing BIM applications in enhancing project management knowledge areas in Egypt. Ain Shams Eng. J. **13**(1) (2022)
2. Cabauatan, R.R., Uy, C., Manalo, R.A., de Castro, B.: Factors affecting intention to use blended learning approach in the tertiary level: a quantitative approach. Higher Educ. Future **8**(2) (2021)
3. Vignali, V., Acerra, E.M., Lantieri, C., Di Vincenzo, F., Piacentini, G., Pancaldi, S.: Building information Modelling (BIM) application for an existing road infrastructure. Autom. Constr. **128** (2021)
4. Manzoor, B., Othman, I., Gardezi, S.S.S., Harirchian, E.: Strategies for adopting building information modeling (BIM) in sustainable building projects—a case of Malaysia. Buildings **11**(6) (2021)
5. MuleroPalencia, S., ÁlvarezDíaz, S., AndrésChicote, M.: Machine learning for the improvement of deep renovation building projects using as-built BIM models. Sustainability **13**(12) (2021)
6. Subekti, H., Purnomo, A.R., Susilo, H., Ibrohim, Suwono, H.: Analysis of preservice science teacher information literacy towards research skills. J. Phys. Conf. Ser. **1006**(1) (2018)
7. Ines Nordhaus Dipl.-Ing.,Margarita Pelych Dipl.-Ing.,Burkhard Pott Dipl.-Ing.,Dr. Ing. Jens Tandler,Werner Breinig Dipl.-Ing.. BIM im konstruktiven Ingenieurbau – ein Erfahrungsbericht der DEGES. Bautechnik, 2020,97(2)
8. Bensalah, M., Elouadi, A., Mharzi, H.: Overview: the opportunity of BIM in railway. Smart Sustain. Built Environ. **8**(2) (2019)
9. Warren, R.: Low levels of teacher information literacy awareness and collaboration between librarians and teachers in information literacy instruction. Evid. Based Libr. Inf. Pract. **13**(3) (2018)
10. Liu, Y., Zub, A.T.: Research on the application of BIM technology in the efficiency and effectiveness of construction project management. J. Phys. Conf. Ser. **1574**(1) (2020)
11. Sumarni, W., Sudarmin, S., Kadarwati, S.: Creative skill improvement of the teacher candidates in designing learning programs through a project-based blended learning. J. Phys. Conf. Ser. **1918**(3) (2021)
12. Armando, N., Almeida, R., Fernandes, J.M., Silva, J.S., Boavida, F.: End-to-end experimentation of a 5G vertical within the scope of blended learning. Discov. Internet Things **1**(1) (2021)

13. Ershadi, M., Jefferies, M., Davis, P., Mojtahedi, M.: A building information modelling (BIM) approach to the systematic management of construction projects. IOP Confer. Ser. Mater. Sci. Eng. **1165**(1) (2021)
14. Qi, X., Liu, Y., Zhang, Y., Cao, S., Zhu, W., Tang, J.: Research on BIM model identification system for substation project management. J. Phys. Conf. Ser. **1486**(4) (2020)
15. Hu, Y., Pei, W.P., Gu, K., Jia, D.Y., Tao, Q.L.: BIM-based solution of management optimization and inspection during construction of a Stadium. J. Phys. Conf. Ser. **2044**(1) (2021)

Innovation of Financing Mode for Small and Micro Enterprises on Digital Finance Technology Alorithm

Tiantian Xia[1], Hui Wang[1,2], Yashi Che[2](✉), Yinan Gao[1,2], and Josep Maria[3]

[1] Jiangxi Regional Development Research Institute, Nanchang 330000, Jiangxi, China
[2] School of Finance and Economics, Jiangxi University of Technology, Nanchang 330000, Jiangxi, China
kamui2022@163.com
[3] Univ Pompeu Fabra, Tecnocampus Mataro Maresme, C Ernest Lluch 31, 08300 Mataro, Spain

Abstract. Digital banking has the endowment features of accurate management ministrant, pension network credit volume and intellect hazard management, which can comply with the demands of financial innovation for the high-quality blossom of small and micro-sized businesses. From the spectacular of digital technology infiltration influence, the bumf argument the handle flow and key mastery opinion design of render concatenation finance invention manner in the application of small and micro business, and aims at the risk management, technology application and jurisprudence constraints that may exist in the course of numeral finance serving small and micro commerce at the present stage. It puts forward some counterplan and recommendation, such as controlling the source of risks and avoiding risks, speeding up the blaze new trails of digital technology empowerment and perfecting the law to standardize the development.

Keywords: Digital Finance · Small and Micro Business · Precise Service · Credit Enhancement · Intelligent Risk Control

Digital banking is the outcome of render concatenation banking hugging numerical engineering [1]. As reveal in diagram 1, by mean of the large datum, cloud computing, webset of somethings, and the application of the technology such as block chain, digital financial present a complex network structure, the participation main body of supply chain finance platform have unlimited extension, nodal enterprises of the supply chain upstream and downstream operation structure is exist no more restricted to heritage chain organizations, formed the energy coupled cluster web organized system [2]. Government, financial institutions, logistics and other upholder of supply chain banking activities also reflect the criss-cross interests, and extend more service canals through the terrace space [3]. In a complicated network system, the function of the supply chain financial platform plays a neural hub, to master the operation of the supply chain business procedure, cash flow, interflow of goods and information flow of coordinated, therefore, have an intimate knowledge of the work of supply chain system key and pain points, in a duly mode for the lack of funds in the terrace provides a variety of services, including financial, make

J. H. Abawajy et al. (Eds.): ICATCI 2022, LNDECT 170, pp. 96–104, 2023.
https://doi.org/10.1007/978-3-031-29097-8_12

the enterprise focusing on the pullilatet of kernel contend, It is beneficial to improve the overall competitive advantage of supply chain [4] (Table 1).

Table 1. Model type

	Type
Model 1	Digital Finance Prepayment Mode
Model 2	Digital Finance Inventory Pledge Financing Model
Model 3	Digital Finance Accounts Receivable Model
Model 4	Digital Finance Intellectual Property Pledge Financing Model

1 Advanced Adhibition of Numerica Banking in Small and Micro Endeavor

By mean of the adhibition and penetration of numerical engineeringy, great changes have taken place in the functional design of supply chain finance such as value estimation of actual acres or rational possession pledges, accredite appraisalof players and venture governance of fund regression [5]. Consequently, counted on the donation traits of numerical banking [6], This paper discusses the reformatory engineer of the manipulation treat and curx spot exposure of the supply chain finance mode applied by small and micro enterprises in different scenarios [7].

1.1 Digital Finance Prepayment Model

Digital Finance prepaid account receivable financing mode mainly occurs in the small micro enterprise procurement procedures [8], emphasis on science and technology financing enterprises commitment to repurchase agreement with upper reaches providers, and usage the platform to specify founded storehouse receipt of confirmation of logistics enterprises request for mortgage from monetary organizations, ultimately to science and technology enterprise later marketing income for financing source about repayment [9]. As shown in Fig. 1, the specific business process is as follows:

Technology-based financing companies symbol procurement and sales protocols with upper reaches providers, and request about prepaid fundraising internet consequently; The terrace snaps the manufacturing and trading scenes of science -oriented financing companies and upper reaches providers by large information science, withdraw, consolidate and dissect the affairs imformation under the assistant of cloud calculating, and produce loan assessment presentation and suppliers them to monetary organizations; Financial institutions conduct multi - dimension assessment under the basement of the offsite loan status of small and micro companies; The terrace signed buyback and character guaranteed deals with upper reaches providers and warehouse monitor protocols with rears companies; The terrace notifies upper reaches suppliers to provider cargos to the warehouse of the specified rear companies. The specified rear companies gathers the materials of goods passed LoT science and produce video storehouse receipt, which is shared with monetary organizations at the moment; Technology-based financing enterprises commit the cash savings for picking up cargo, and monetary organizations release indicates to specified rear companies on internet to unleash the right to pick up goods of corresponding amount to technology-based financing enterprises.

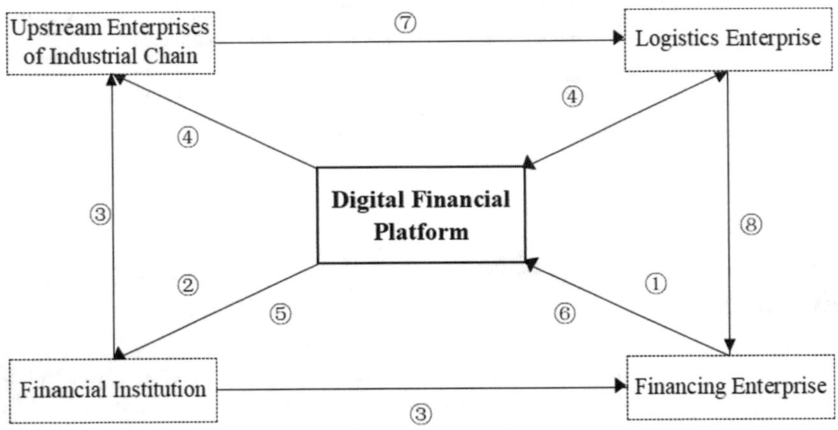

Fig. 1. Digital Finance Prepayment Mode

Efficiency Calculation Formula:

$$Y = \beta_0 + \beta_1 X + \beta_2 C + \varepsilon \tag{1}$$

Y is Finacial Institution Efficiency, X is Digital Finance Prepayment (Finacial Enterprise, Logistics Enterprise, Upstream Enterprise of Industrial Chain), C is Control Variable, ε is Random Error.

1.2 Digital Finance Inventory Pledge Financing Model

Digital Finance inventory impawn financing pattern usually occurs in the pledge of small micro enterprise production inventory, required by the Internet of things, such as cloud computing technology for small micro enterprise movables pledge to implement dynamic monitoring and evaluation, by the small micro enterprise inventory as the pledge to invite a loan from banking hierarchy system, and with its ulterior inventory sales inflows as source of repayment [10]. As shown in Fig. 2, the specific business process is as follows:

Technology-based financing endeavor ulterior online stock-sheet certification loan to tender enslave banking hathpace by excellence of its stock-sheet; The hathpace carry in the stock-sheet cognition of mechanics -based financing corporations in thescheme with the facilitate of normal interfaces via the webset of somethings electronic brand mechanics; Apply crucial information, webset of somethings and other numerical skills to realize real-time stock-sheet significance estimate and issue evaluation certificates, which will be yielded to fiscal hierarchy system by the hathpace; Fiscal hierarchy system make multi-dimensional judgement on the offline confidence status of mechanics -based financing corporations and make credit granting decisions; The platform signs pledge contracts with technology-based financing corporations and demands them to dispose of their inventories to appointed same-city logistics enterprises; Theoffice support enterprise accepts the stock-sheet and incessantly supervises the stock-sheet capacity to regulate real-time information for ffiscal hierarchy system to make loans; The goods income of the technology financing enterprise shall be mechanically and strictly averted to the monetary institution by the consideration and disposal localise of the hathpace, and the pledge contract shall become invalid after the loan repayment is completed.

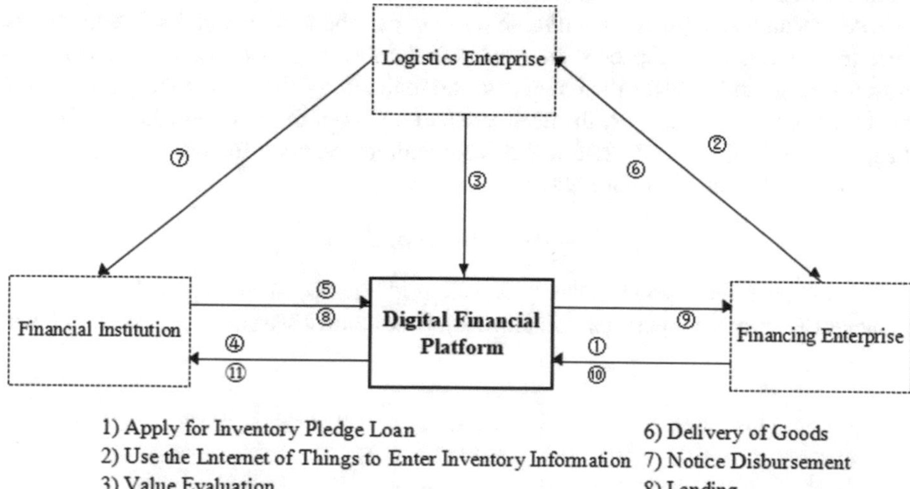

Fig. 2. Digital Finance Inventory Pledge Financing Model

Efficiency Calculation Formula:

$$Y = \beta_0 + \beta_1 X + \beta_2 C + \varepsilon \tag{2}$$

Y is Finacial Institution Efficiency, X is Digital Finance Inventory Pledge Financing (Finacial Enterprise, Logistics Enterprise), C is Control Variable, ε is Random Error.

1.3 Digital Finance Accounts Receivable Model

Accounts receivable financial affairs pattern of numerical Finance always appear in ales link of small and micro enterprises. Technology financing enterprises of upwards suppliers deeds with accounts receivable as certification, raise funds for the next round of production, and take customers' future payment as repayment source [11]. As shown in Fig. 3, the specific business process is as follows: Small and micro corporations symbol sales deeds with lower occupants and deeds webset with notes receivable as commitment; The hathpace makes use of blockchain and webset of somethings mechanics to overally inspect and score the operating conditions, liquidation competence and faith criterion of upwards and lower commitment, and advocate the estimate rept to Fiscal hierarchy system; Fiscal hierarchy system conduct multi-spatial synthetic appraisal based on internet appraisal records and offline loan investigation reports of upwards and lower companies; Fiscal hierarchy system regulate loan to mechanics-based financing companies, and the punishment and transactions mediate of the hathpace allocates money to mechanics-based financingcompanies; Enterprises lower of the tender restriction shall dispose of the sales income as liquidation authorship to the pay the money and resolution core of the hathpace, and the hathpace shall reimburse the financial institutions with headmaster and loving, and transfer the balance to the technology financing enterprises; If the fiscal hierarchy system acquire the president and loving of this banking, the certification engagement of opines acceptable will mechanicallycome noneffective.

Efficiency Calculation Formula:

$$Y = \beta_0 + \beta_1 X + \beta_2 C + \varepsilon \tag{3}$$

Y is Finacial Institution Efficiency, X is Digital Finance Inventory Pledge Financing (Finacial Enterprise, Downstream Enterprise), C is Control Variable, ε is Random Error.

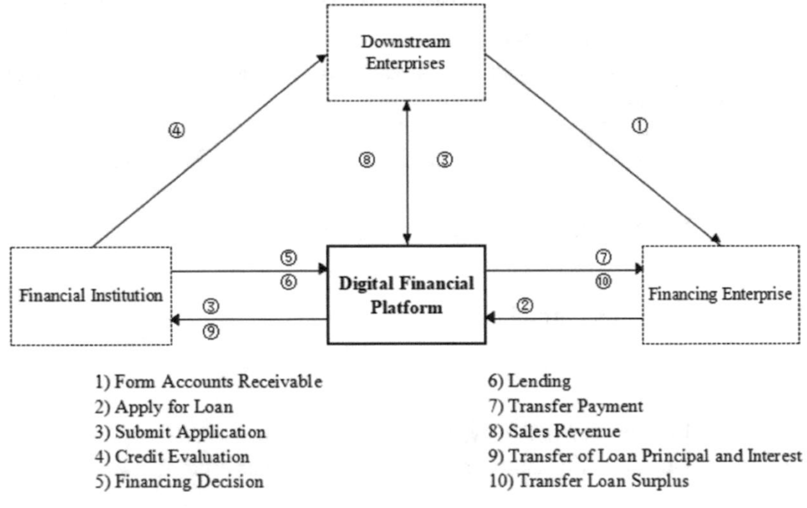

Fig. 3. Digital Finance Accounts Receivable Model

1.4 Digital Finance Intellectual Property Pledge Financing Model

Digital intellectual property pledge financing mode for Yu Xiaowei enterprise and supply chain partnership between the transfer of intellectual property rights trading scenario, required to exploit the information sharing module, digital technology intangible assets financing enterprises use understanding performance to the alienee career future earnings as the guarantee loans from financial institutions, The transfer fee of wise capability jurisdiction payable by the alienee shall be the source of repayment [12]. As shown in Fig. 4, the specific business process is as follows: Technology-based financing careers book in their understanding ownership in concerned presidency sectors, and the hathpace introduces understanding estate information mastered by involved presidency sectors (such as science and technology management departments and understanding presidency commerce centers) across normal contacts Relying on digital technology, the platform builds an feedback communion component of discarnate capitals to connect technology-based financing enterprises with transferee enterprises and sign agreements on the use or transfer of intellectual property rights; Technology financing presidency maintain online adhibition for clever capability commitment credit; The hathpace invests numerical mechanics to curtain and estimate the intellectual property cognition, evaluate the adhibition significance, prospect income, manipulation ranking, exploitation feasibility and science reconnoitre of the smart nature of the mechanics -oriented financing companies, and analyze the operation and profit status of the transferee enterprises before and after the adhibition of clever performance; Fiscal hierarchy system make credit determination depends on the assessment sequelae yielded by the flatform and the agreement affixed by the mechanics supplier and the offline loan ranking of the mechanics financing corporations and the operating execution and loan ranking of the grantee corporations; The grantee company uses discarnate capitals such as patents and brand to organize production, and the negotiate agreed in the agreement shall be the source of debt liquidation of the banking company.

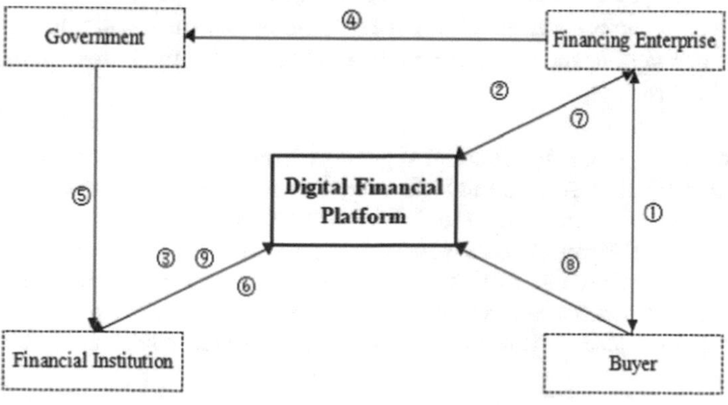

Fig. 4. Digital Finance Intellectual Property Pledge Financing Model

Efficiency Calculation Formula:

$$Y = \beta_0 + \beta_1 X + \beta_2 C + \varepsilon \quad (4)$$

Y is Finacial Institution Efficiency, X is Digital Finance Intellectual Property Pledge Financing (Finacial Enterprise, Buyer, Government), C is Control Variable, ε is Random Error.

2 Countermeasures and Suggestions for Small and Micro Enterprises to Apply Digital Finance

Precision management services, intelligent network to improve credit and risk control three endowment characteristics can promote the different modes of supply chain finance for small micro enterprise high quality service, however, the current number of financial development in our country is still in its infancy [13], in the small micro enterprise financing service practice there are some constraints, restricted the impact of banking services. To begin with, the operation of the provide catenary financial risk including outside surroundings risk, risk and danger of provide catenary enterprise provide catenary network, small micro enterprise the pledge of high technical characteristics increased the risk of the external environment and enterprise itself, at the same time digital financial dynamics of complex network structure and conductivity characteristics is the supply chain from the enterprise financial risk source end extending to the platform network end [14]; Second, the application of digital technology is not fully brought financial supply chain technology bottleneck, chain blocks, such as the Internet of things technology application in the balloon of provide catenary banking is not full-fledged, from the industrial, financial and other services the separation of different data system even more difficult for big data play to utility, lack of digital technology application become barriers to financial innovation of the supply chain [15]. Ultimately, the legitimate presupposes concerned with digital banking is not sound, and the now available legislation and rules are distant from catch sight of the legitimate needs of correlative commerce evolution. So as to far away the efficacious services of numerical banking for small and micro public institution, it is much significance to program rational tactics for the issues existing in the evolution course [16, 17].

Acknowledgment. We appreciate the financial support of Jiangxi Social Science Foundation Project (21GL51, 22ZXQH16); Science and technology research project of Jiangxi Provincial Department of Education (GJJ202013, GJJ212017); Jiangxi University Humanities and social sciences research project (GL21226); Collaborative education project of industry university cooperation of the Ministry of Education (202102013004, 202102348008); University excellence research project of Chinese Academy of Social Sciences (2021-KYLX02–02); Special project of Jiangxi Institute of regional development, Jiangxi University of science and technology(QYFZ2101, QYFZ2104).

References

1. Tiantian, X., Zhenduo, Z., Huan, X., Jing, X., Wentong, J.: The curvilinear relationship between job control and voice: role of emotional resistance to change and supervisor

developmental feedback. SAGE Open (2021). https://doi.org/10.1177/21582440211027960. April

2. Raya, J.M., Vargas, C.: How to become a cashless economy and what are the determinants of eliminating cash. J. Appl. Econ. **25**(2), 543–562 (2022). https://doi.org/10.1080/15140326.2022.2052000
3. Ng, S.-H., Zhuang, Z.: Exploring herding behavior in an innovative-oriented stock market: evidence from ChiNext. J. Appl. Econ. **25**(1), 523–542 (2022). https://doi.org/10.1080/15140326.2022.2050992
4. Mulaga, A.N., Kamndaya, M.S., Masangwi, S.J.: Spatial disparities in impoverishing effects of out-of-pocket health payments in Malawi. Global Health Action **15**(6), 13–15 (2022). https://doi.org/10.1080/16549716.2022.2047465
5. Lee, S.T., Yang, E.B.: Factors affecting social accountability of medical schools in the Korean context: exploratory factor and multiple regression analyses. Medical Education Online **27**(4), 1–6 (2022). https://doi.org/10.1080/10872981.2022.2054049
6. Qiu, T., Ma, X., Luo, B., Choy, S.T.B., He, Q.: Land defragmentation in China: does rental transaction inside acquaintance networks matter?. Journal of Applied Economics **25**(9), 259–278 (2022). https://doi.org/10.1080/15140326.2022.2043720
7. Jiang, M., Qi, J., Zhang, Z.: Under the same roof? the green belt and road initiative and firms' heterogeneous responses. J. Appl. Econ. **25**(7), 315–337 (2022). https://doi.org/10.1080/15140326.2022.2036566
8. Nester, M.S., Hawkins, S.L., Brand, B.L.: Barriers to accessing and continuing mental health treatment among individuals with dissociative symptoms. European Journal of Psychotraumatology **13**(1), 5–12 (2022). https://doi.org/10.1080/20008198.2022.2031594
9. Zembe-Mkabile, W., Sanders, D.: 'I know what I should be feeding my child': foodways of primary caregivers of Child Support Grant recipients in South Africa. Global Health Action **15**(12), 123–126 (2022). https://doi.org/10.1080/16549716.2021.2014045
10. Wildberger, J., Wenzel, K., Fishman, M.: Assessing clinical impacts and attitudes related to COVID-19 among residential substance use disorder patients. Substance Abuse **43**(4), 756–762 (2022). https://doi.org/10.1080/08897077.2021.2010249
11. Wang, Z., Wang, X., Xu, Y., Cheng, Q.: Are green IPOs priced differently? Evidence from China. Research in International Business and Finance **61**(8), 13–19 (2022). https://doi.org/10.1016/j.ribaf.2022.101628
12. Castro, C.M.: A curiosa história do Real Collegio dos Nobres. Ensaio: Avaliaçãoe Políticas Públicasem Educação **30**(8), 550–565 (2022). https://doi.org/10.1590/s0104-40362022003003764
13. Kim, H.M., Turesson, H., Laskowski, M., Bahreini, A.F.: Permissionless and permissioned, technology-focused and business needs-driven: understanding the hybrid opportunity in blockchain through a case study of insolar. IEEE Trans. Eng. Manage. **69**(10), 776–791 (2022). https://doi.org/10.1109/TEM.2020.3003565
14. Núñez, M.L., Cáceres Ruiz Díaz, M., Correa, E.S.: Características de los proyectos financiados por el Consejo Nacionalde Cienciay Tecnologíade Paraguay. Convocatorias 2013–2015. Poblacióny Desarrollo **28**(12), 55–67 (2022). https://doi.org/10.18004/pdfce/2076-054x/2022.028.54.055
15. Gulosino, C., Maxwell, P.: A comprehensive framework for evaluating shelby county school district's voluntary preschool program: the challenges of equity, choice, efficiency, and social cohesion. Urban Education **57**(5), 779–813 (2022). https://doi.org/10.1177/0042085918801885

16. Pérez, O.: La influencia de la innovación educativa utilizando las metodologías ABP en la cultura institucional de los posgrados de tres universidades paraguayas. Academo (Asunción) **9**(7), 23–37 (2022). https://doi.org/10.30545/academo.2022.ene-jun.3
17. Upham, P., Sovacool, B., Monyei, C.: Digital bricolage: Infrastructuring lower carbon digital space via Nordic datacentre development. Political Geography **96**(6) (2022). https://doi.org/10.1016/j.polgeo.2022.102617

Coal Energy Management Estimation System Based on Artificial Intelligence Algorithm

Yi Zheng(✉)

School of Management, Tianjin University of Technology, Tianjin, China
zyjust98@163.com

Abstract. Carbon energy management has become a factor restricting the progress of informatization of coal enterprises. How to improve the safety, reliability and availability of coal energy management system and reduce energy cost has become a key issue. The purpose of this paper is to study the coal energy management estimation system based on artificial intelligence algorithm. On the basis of artificial intelligence algorithm, this paper studies the digital management of coal yard, the intelligent, informatization and hierarchical interactive management of the whole process of large fuel. According to the characteristics of thermal power plant enterprises, the comprehensive evaluation index system of the coal yard intelligent management system is determined. Aiming at the special working environment of coal mines, this paper deeply analyzes the principles of current sensor technology, electronic technology and communication technology, as well as the relevant specifications and design requirements of the coal intelligent monitoring system, and designs a low-cost and cost-effective coal energy management estimation system. Experimental research shows that the number of logical disk reads for unoptimized query statements in this system has reached more than 1700 times, while the optimized query statement queries are only read 154 times, which fully reflects the superior performance of the system and greatly improves query performance.

Keywords: Artificial Intelligence Algorithm · Coal Energy Management · Data Collection · Comprehensive Evaluation

1 Introduction

The demand for automation, informatization and even intelligence of coal yard and coal handling system is very strong in domestic thermal power plants. Large thermal power plants have a high degree of automation and high level of informatization, but the construction of coal yard informatization is lagging behind, and the efficiency of the comprehensive data management system is not high. This has become a shortcoming of coal energy management systems for coal enterprises. The management estimation intelligent system is of great significance to the development of my country's coal enterprises [1, 2].

In the research of coal energy management estimation system based on artificial intelligence algorithm, many scholars have studied it and achieved good results. For

J. H. Abawajy et al. (Eds.): ICATCI 2022, LNDECT 170, pp. 105–112, 2023.
https://doi.org/10.1007/978-3-031-29097-8_13

example, N Liu designed a fuel management and coal quality improvement system in the power plant, which recognizes the data sharing between the local power plant network and has stable and reliable performance [3]. Jo B W chose the SAP ERP platform as the software platform of the coal energy optimization information management system, and can see the integration of the basic application system [4]. It can be seen that the research on coal energy management estimation system based on artificial intelligence algorithm is of great significance.

This work mainly uses artificial intelligence algorithms to comprehensively calculate the overall control system of the coal plant, and deeply examines the establishment of the comprehensive index system and comprehensive evaluation model. It comprehensively and effectively demonstrates the basic value level of the artificial intelligence algorithm intelligent control system, helping thermal power plants to maximize production efficiency and operational benefits. This paper establishes a practical, automatic, simple and efficient coal energy intelligent system evaluation system, which will play a leading role in the balance and quality assurance of coal factory construction.

2 Design and Research of Coal Energy Management Estimation System Based on Artificial Intelligence Algorithm

2.1 Coal Distributed Intelligent Quality Management System

(1) Download the weight in a custom time period
According to the company's requirements, the quality inspection department needs to take samples from each mine on the day when the weighbridge room is off work, and test the quality of the samples to reflect the coal quality of the mine sent to the shipping station on the day. It takes several hours to test the data of each component of the sample. The weighbridge room usually gets off work around 5:00 pm. If a test is performed, the quality inspector needs to work overtime, so the quality test needs to be carried out in different time periods. At present, they take a sample for testing before 3:00 pm, and wait until the next day for testing after 3:00 pm, which avoids the need for laboratory personnel to stay up late to work. Based on this need, the quality inspection system provides an interface for downloading weights within a custom time period. Through the user's choice, the weight information of the scale house within a period of time is downloaded to the corresponding table of the quality inspection database.

(2) Design of custom formula library with parameters
The custom formula library is the formula expression defined by the user according to the actual needs, so that the excess data can be flexibly and conveniently deweighted according to the set formula. The implementation steps of the custom formula library are divided into three parts: parameter conversion, formula expression analysis, and formula verification.

(3) Automatic weight deduction design
The automatic weight deduction of test items mainly uses the test item maintenance table, the test item critical value table, and the custom formula library table. The basic idea is that each test item in a test item maintenance table corresponds to

multiple test item critical values, and each critical value corresponds to a custom formula. Among them, the detection items, the threshold value of the detection items, and the formula library are all maintainable. In this way, it becomes flexible and convenient to deduct the excess weight of the inspection items in the coal quality management.

2.2 Intelligent Detection Technology and Principle of Coal Content

(1) Full water detection
The average moisture content is the external humidity and the internal humidity. In order to facilitate the work and ensure the accuracy of the measurement results, it can sometimes be carried out in two steps, that is, the first degree of external humidity, and then the internal humidity is measured. There are two basic types of water: free water and crystallized water. The coal industry essentially considers free water, which can be removed by infrared heating and measure energy. Carbon with a moisture content of not more than 5% is low-moisture carbon, while carbon with a moisture content of more than 15% is high-moisture carbon [5, 6].

(2) Nitrogen content detection
The traditional determination of nitrogen content in carbon is the Kjeldahl method, which involves long analysis times, the use of multiple chemical reagents, and the analysis of only one sample at a time, resulting in much lower yields. Nitrogen content in carbon can also be determined by chemiluminescence, which performs small-volume, high-speed analysis of samples without the need for chemical reagents, allowing repeated analysis of samples, which is hundreds of times higher than traditional methods..

(3) Sensor technology
The method of accepting sensors based on measured values is to select sensors based on the detection requirements of mass objects. The sensor is mainly used for intelligent detection of carbon content to monitor the environmental status of carbon and the content of various carbon components. Among them, the discovery of carbon ecosystems mainly uses gas sensors, carbon monoxide sensors, wind speed sensors and light detectors. The determination of the content of each carbon component generally adopts physical and chemical methods, combined with electronic detection technology, such as infrared sensors can be used to determine the sulfur content.

(5) Temperature sensor
Radiation thermometry is only suitable for temperatures limited to dark skin (which absorbs all light). Generally, the surface of the object needs to be adjusted to measure temperature accurately. In automatic measurement and control, the actual temperature of the medium can be obtained after the medium temperature is corrected.

2.3 Comprehensive Evaluation Index System of Intelligent Coal Yard Management System

(1) System function

The thermal power plant intelligent coal yard management system is a comprehensive fuel management solution for operators and management level personnel. The core functions include intelligent coal blending management, quality inspection management, cost management and efficiency supervision, digital coal yard, and fuel work management., to effectively control and optimize the allocation of the entire fuel business process and information resources, and improve the efficiency of fuel management by improving the level of management informatization data integration and application.

(2) System performance
The efficiency and fault-tolerant function of the intelligent coal yard management system are the system performance that can be directly felt by users, and the evaluation method of system efficiency is also relatively standardized and general. System turnaround time, response speed, and throughput are related to and reflected in other system performance indicators, and are not included in the evaluation index system. The fault-tolerant function of the system should be complete and perfect, and the system should be able to operate stably, reliably and safely for a long time [7, 8]. Its design principles are "mature and advanced, safe and reliable, open and variable, integrated and manageable, practical and convenient", and meet the requirements of usability, correctness, adaptability, maintainability, and completeness of documents.

(3) Economic benefits
The intelligent coal yard management system can improve the fuel management efficiency of thermal power plants, reasonably mix multiple coal types, improve the fuel cost of thermal power plants, improve the safety of power generation, and effectively control the emissions of sulfur dioxide, nitrogen oxides and dust, bringing huge economic benefits. Benefits and social benefits.

(4) System application
The actual application effect achieved after the completion of the intelligent coal yard management system is also called the efficiency index. The index system selects two indicators: user satisfaction and system function application degree [9, 10]. The application degree of system functions reflects the utilization degree of the intelligent coal yard management system; the degree of coordination between each functional module and equipment program control [11, 12]; the user satisfaction directly reflects the practical convenience, humanization and user experience of the system.

2.4 Calculation Method of Index Weight Based on Artificial Intelligence Algorithm

In order to check the consistency of the judgment matrix, the negative average CI of the remaining eigenroots except the largest eigenroot of the judgment matrix is introduced as an index to measure the deviation consistency of the judgment matrix.

$$CI = (\lambda max - n)/(n - 1) \tag{1}$$

The smaller the CI value, the better the consistency of the judgment matrix.

The random consistency ratio is:

$$CR = CI/RI \tag{2}$$

When CR < 0.10, it is considered that the relative weights have satisfactory consistency, otherwise, the judgment matrix needs to be adjusted to make it have satisfactory consistency.

3 Research on Experimental Design of Coal Energy Management Estimation System Based on Artificial Intelligence Algorithm

3.1 Preprocessing of Report Generation

The summary and generation of statistical reports is one of the main daily tasks of the quality management system. The generation of reports requires efficient and fast response capabilities, and the performance of report generation also directly affects the overall performance of the system, especially when generating reports for large tables, most of the queries used in report generation are with Group By subordinates. Sum, Avg, Max, Min and other summary queries of sentences require a lot of system resources.

Therefore, when designing the system, the reports that need to be used frequently and have a large amount of data can be classified according to statistical categories, and a summary result summary table can be generated in the system. Make statistical calculations.

3.2 Preprocessing of the Query

The user mainly feeds back the running results through the processing program in the foreground. When the user queries through different conditions in the foreground, the performance of the database system will also have different effects. The larger the amount of data queried, the longer the response time. When the user only enters the ending year and month in the foreground, but not the starting year and month, the response time is completely different from inputting the starting year and month and the ending year and month at the same time, so the reasonable design layout and increase in the foreground Query constraints have a great effect on improving system performance.

4 Experimental Analysis of Coal Energy Management Estimation System Based on Artificial Intelligence Algorithm

4.1 Optimization of Join Queries

Join operation is one of the commonly used operations in database query. Since join operation often generates large temporary tables, especially the join operation of multiple tables, SQL query statements will generate a huge amount of calculation and consume a lot of money under unoptimized conditions. Memory, resulting in strain on system resources. This paper takes the query optimization of the master-slave table of incoming

Table 1. Join query performance comparison

Type	Unoptimized query	Optimized query statement
CPU time (ms)	106	71
Logical Disk Reads	1863	1809
Query response time (ms)	1254	970

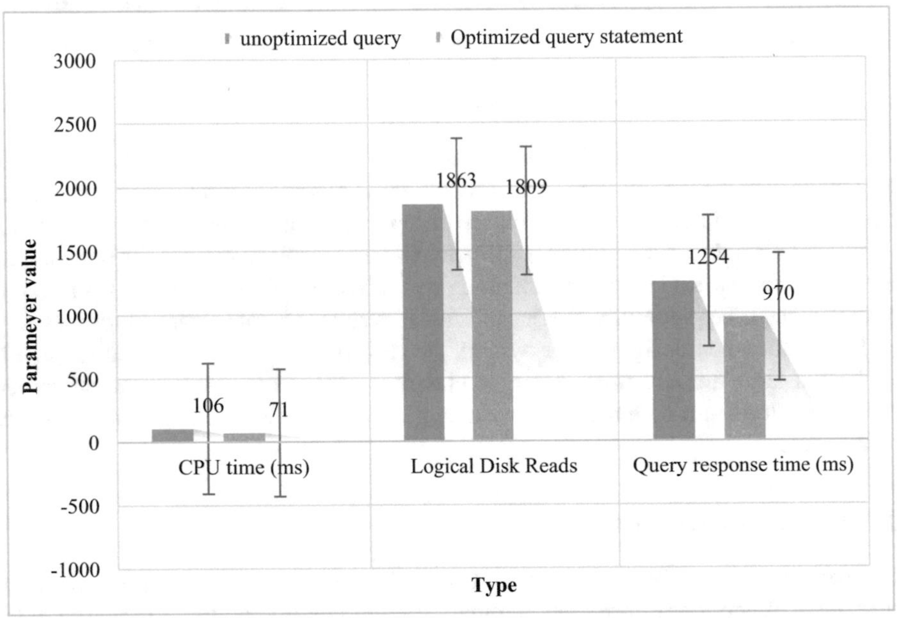

Fig. 1. Join query performance comparison

quality information in the quality management system as an example, and studies the comparison of query efficiency. The experimental results are shown in Table 1.

As can be seen from Fig. 1, the application of stored procedures can effectively avoid frequent data exchange between the client and the server, and make full use of the superior performance of the server, greatly improving the query performance.

4.2 Adjustment and Analysis of Inefficient SQL Statements

During performance analysis, the two proxy server locations resulted in a small amount of SQL message execution. One case is the incorrect use of query query results. There is also a situation where the order of the application conditions is disordered, resulting in the application being forced to stop. This article will use the index to process the query after optimization. The query performance comparison between the two is shown in Table 2.

Table 2. Nested query performance comparison

Type	Unoptimized query	Optimized query statement
CPU time (ms)	472	253
Disk reads	1742	154
Check the corresponding time	725	519

As can be seen from Fig. 2, the number of logical disk reads for the unoptimized query statement has reached more than 1700 times, while the optimized query statement query is only read 154 times. The optimized query statement greatly improves the query performance.

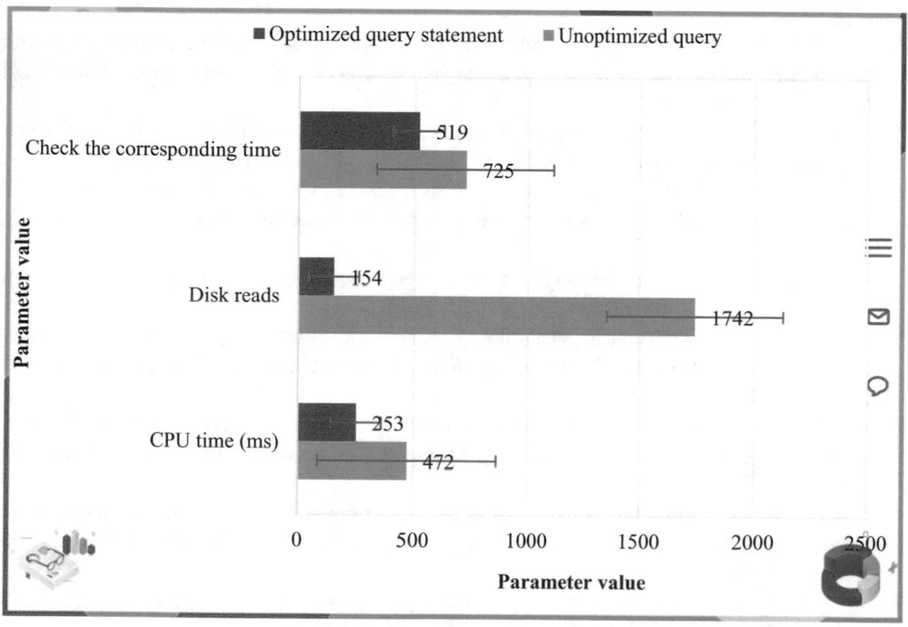

Fig. 2. Nested query performance comparison

5 Conclusions

This paper carefully reviews and analyzes the relevant specifications and design requirements of coal detection systems, designs a low-cost and efficient carbon detection system, and evaluates data processing. Due to the difficult operating conditions of carbon detection systems, small filters or other filters can be used during sampling and signal transmission to increase the system's ability to resist environmental disturbances. When collecting and transmitting data, it is possible to make full use of current speed sensor network and Internet of Things technologies to obtain more wireless data transmission and management technologies.

References

1. Rathor, S., Saxena, D.: Decentralized energy management system for LV microgrid using stochastic dynamic programming with game theory approach under stochastic environment. IEEE Trans. Indu. Appli. (99), 1–1 (2021)
2. Zineb, B., Abdelghani, C.: Mohammadia School of Engineers EMI, Mohammed V University, Ibn sina Avenue B.P. Agdal Rabat, Morocco. Development of a readiness for change maturity model: an energy management system implementation case study. Production Engineering Archives **28**(1), 93–109 (2022)
3. Liu, N.: Application of Intelligent Means and Advanced Production Technology in Coal Mine Production. IOP Conference Series: Earth and Environmental Science **300**(2), 022112 (2019). (5pp)
4. Jo, B.W., Khan, R., Javaid, O.: Arduino-based intelligent gases monitoring and information sharing Internet-of-Things system for underground coal mines. J. Ambi. Intell. Smart Envi. **11**(2), 183–194 (2019)
5. Bijańska, J., Wodarski, K.: Model of process management system in enterprises of the hard coal mining industry. Manage. Sys. Prod. Eng. **28**(2), 112–120 (2020)
6. Shi, F., Cao, H., Wang, C., et al.: A system dynamics model for ecological environmental management in coal mining areas in China. Int. J. Environ. Res. Pub. Health **17**(6), 2115 (2020)
7. Nie, Y.: Integrated management system for coal mine locomotive transportation. Electr. Sci. Technol. Appl. **7**(4), 109 (2021)
8. Velikosel'Skiy, A.V.: Scientific, methodological and institutional support for the development of the management system of a coal company. News of the Ural State Mining University **1**(2), 200–209 (2020)
9. Amalina, N.N., Larasati, H.E.: The implementation of contractor safety management system to prevent work accidents at coal mining company. The Indonesian J. Occupat. Safety and Health **9**(3), 338 (2020)
10. Kurniawan, R., Feinnudin, A.: Assessing the implementation of the energy management system in the first ISO 50001 building in Indonesia. Indonesian Journal of Energy **4**(2), 129–139 (2021)
11. Yue, M., Wang, X.: Research on control strategy of ship energy management system based on hybrid GA and PSO. Int. Core J. Eng. **6**(5), 185–193 (2020)
12. Liu, M., Zhang, X., Ma, Y., et al.: Thermo-economic analyses on a new conceptual system of waste heat recovery integrated with an S-CO2 cycle for coal-fired power plants. Energy Conversion and Management **161**(APR.), 243–253 (2018)

Establishment of Enterprise Management Information System Based on Big Data Information Technology

Yuan Song[1(✉)] and Asia Ullah[2]

[1] School of Economics and Management, Shanghai University of Political Science and Law, Shanghai, China
songyuan@shupl.edu.cn
[2] University of Balochistan, Balochistan, Pakistan

Abstract. We are now stepping forward from the traditional industrial society to a modern informational one as information and technology progress. Informatization is becoming an inevitable trend to promote economic and social reform especially for current China. Informatization has fundamentally changed the way that business develops. China's medium and micro enterprises are shifting from mainly relying on capital and human resource investment to technological progress and improving overall quality of employees. It is studied that optimization path of enterprise management information system with big data in this article. Firstly, literature review is conducted through methods of questionnaire survey, normative research and empirical research. Then, system analysis is conducted to develop system model, combining qualitative analysis and quantitative analysis together. Optimization programme of enterprise management information system is proposed as conclusion of the article in the end. Empirical research shows that 48% of businesses often use information technology in their management information systems; 39% of businesses use it occasionally; and 13% of companies never use it. Enterprises should strengthen the application of big data in management information systems, acquire real-time information management and decision-making, finally improve the efficiency of enterprise management information.

Keywords: Big Data · Management Information System · Information Technology

1 Introduction

Information science and technology has significantly influenced global economy, thus completely changed traditional mode of China's enterprise management and economic structure in the evolution of modern human society [1, 2]. The information environment in which the company is located is no longer just an environment that belongs to the material economy in the past, but an information economy environment that organically combines the technological development, organizational structure, manufacturing, and

J. H. Abawajy et al. (Eds.): ICATCI 2022, LNDECT 170, pp. 113–120, 2023.
https://doi.org/10.1007/978-3-031-29097-8_14

corporate governance through the Internet [3, 4]. Nowadays, enterprises develop by providing customers and consumers good experience of enterprise information management through innovation of business model, evolutionary innovation of product technology, optimization and integration of human resources, data analysis and sharing [5, 6]. Enterprises grow and benefit from transformation of management driven by informatization, changing of management philosophy, management mode and management efficiency as well [7, 8].

With further application of enterprise management information systems, the amount of data and the number of users have increased rapidly, thus performance problems of the system have become more and more prominent. Taking information system operation and asset management system as examples, performance weaknesses are comprehensively analyzed from four aspects of terminal, network, hardware, software platform and application system [9]. After performance optimization measures are implemented, system users get better using experience, also, it reduces maintenance operation burden for maintenance personnel.

On the basis of clarifying status quo and deficiencies of enterprise management information system and its implementation, enterprise needs, the necessity and appropriateness of optimizing enterprise management information system is analyzed in the article. Next, enterprise management information system optimization plan is proposed. In the end, operation and maintenance, security measures and evaluation of enterprise management information system are elaborated. The research goal of this article is to realize real-time data acquisition and analysis during the process of enterprise information management, thereby beneficially supporting management and decision-making of managers [10].

2 Optimization Path of Enterprise Management Information System with Big Data

2.1 Research Method

2.1.1 Questionnaire Survey

Questionnaire survey is used in the article to designedly and systematically investigate the use of commodity trade system. Data are collected to analyze the use of the system and to compare with the original one, to analyze the shortcomings of the system.

2.1.2 Normative Research

On the basis of careful literature review, analyzing research application status quo and future development, the status and characteristics of enterprise management information, three elements of management information system is analyzed, then management information system model is built. At the same time, system design and implementation could be achieved by analyzing goals, demands and functions of management information system with big data technology.

2.1.3 Empirical Research

It is disclosed in the article that what the current enterprise informatization development should be like, what problems still exist based on lots of empirical research. Then normative research is combined with empirical research to elaborate what the reasonable management information system should be, how to deal with the problems enterprises currently have.

2.2 Problems of Enterprise Management Information System in Enterprise L

2.2.1 Insufficient Information Mining Causing Failure of Information Management Functions

Enterprise L only uses the commodity trading system as a reminder of the inventory balance of production system. This cannot bring its function into play if it is only used for product sales approval process and dealing with sales processing, with lots of information not fully utilized. It is suggested to add models, make full use of system data, therefore increasing use efficiency and contribution of the system.

2.2.2 Processing Speed of Product Sales Affected by Tedious Operation

Commodity transaction sub-information system must be split into four columns before it is imported to sales management information table. In addition, after transaction data list is imported to the system, serial number of commodity transaction would no longer be displayed, so in sales management system, it can only be re-ordered according to the date of sales or the amount of money. In this case, it is difficult for managers to check the data, thereby affecting operation of enterprise internal management.

2.2.3 Unreasonable Setting of System Permissions Leading to Increasing Risks

The commodity trading subsystem provides setting function of transaction data approval route. Currently, the permission setting for route maintenance is only in the hands of the head office and branch sales administrators. Therefore, it is relatively complicated for regional sales agents to conduct commodity transactions. System commodity transaction authority is messy, with no uniformity, without combining market sales control and risk control. Poor system commodity transaction troll and large system loopholes are caused because there is no simple and easy-to-operate commodity transaction authority control setting.

2.3 Optimization Design of Enterprise Management Information System of Enterprise L

2.3.1 Customer-Centric Principle

The optimization of the management information system must be customer-centric, maximizing service value to customers, re-designing system to improve customer satisfaction, and responding to customer needs as quickly as possible. The information center must understand the needs of customers multi-dimensional, provide strong support and guarantee for enterprise L.

2.3.2 Principle of Corporate Strategic Development

The successful implementation of management information system optimization requires L company to formulate a set of detailed and feasible marketing strategy plans. It also needs continuous learning of management science, engineering technology, operation research and humane studies. Establishment of various standards such as the organizational structure and corporate culture of enterprise L, the establishment of a management information system is not only a technical thing, but a strategic planning system project for the future market development.

2.3.3 Principle of Enterprise Product Innovation System First

Continuous strengthening analysis, acquisition, sorting out and perceiving of market information, then feedback to decision makers, providing powerful help for them is required for optimizing management information system. Using the products developed by enterprise L, staying at the forefront of the market brings huge benefits to enterprise L.

2.3.4 Principle of Organic Combination of Management and Technology

The focus of the management information system is the integration of management and information technology. In the process of system optimization, the principle of organic integration of management and technology must always be maintained. Both of them must be reasonably applied to the process of system design and application. Only when the two are jointly, value of system optimization can be truly brought into play.

2.4 Optimization Algorithm of Enterprise Management Information System Based on Big Data

As the data stream continues to enter, the fit of the decision tree model is getting higher and higher. At this time, the accuracy of the decision tree classification should become more and more stable. The classification accuracy in the sliding window is P_{real} and the most ideal classification. The difference between accuracy P_{optimal} is also getting smaller and smaller. We define $\Delta P = P_{optimal} - P_{real}$ to detect the error between the optimal classification accuracy and the real-time classification accuracy in the sliding window. Suppose we give a confidence δ for the accuracy error, the confidence interval h can be obtained by the inequality in Hoeffding Theorem 1:

$$P(\overline{X} - E[\overline{X}] \geq h) \leq e^{-2nh^2} \tag{1}$$

Suppose $\overline{X}$ is the estimated accuracy rate, and $E[\overline{X}]$ is the true accuracy rate, then the meaning of the above inequality is the estimated accuracy rate and the true set as the estimated accuracy rate, and the probability that the deviation between the accuracy rates exceeds h is not greater than e^{-2nh^2}. According to symmetry, there are:

$$P(-\overline{X} + E[\overline{X}] \geq h) \leq e^{-2nh^2} \tag{2}$$

That is, for a given confidence δ, within a confidence interval with a width of 2h for the expected accuracy rate $E[\overline{X}]$, there is

$$\delta \leq 2e^{-2nh^2} \quad (3)$$

3 Experimental Research on Optimization Path of Enterprise Management Information System with Big Data

3.1 Research Purpose

Use and dealing with information is becoming more and more vital with expanding of enterprise scale and its organizational structure. Therefore, management information system is supposed to provide sufficient and convenient inquiry access with enterprises. Establishing a solid management information system can significantly improve production efficiency, strengthen information management, and improve the economic performance of enterprises. Traditional information management has many shortcomings, such as low efficiency, poor confidentiality, difficulty in finding, updating and maintaining information. It is very necessary to manage data through information, which will bring a lot of convenience to the company because it is easy to find, safe and easy to use, confidential, large storage capacity, and low cost. These can improve management efficiency.

3.2 Demand Analysis of Database

As the focus of system development, database requirements analysis is essentially to understand what data users need, what data will be used, and what kind of connections exist between these data. Only after obtaining these data can we formulate accurate and complete system requirements. Inaccurate or incomplete requirements analysis will cause the designed database to fail to meet user needs. If the accuracy and completeness of the requirements analysis are not enough, it is impossible to design a database that meets the needs of users. Enterprises are mainly composed of different businesses. The operation of enterprises is actually the operation of the business, and the operation process of the business is the core of the enterprise management system. Therefore, database must be designed in accordance with the business process to achieve business coordination and improve the rapid response capability of enterprises.

4 Experimental Analysis of Optimization Path of Enterprise Management Information System with Big Data

The most important thing in the innovation of enterprise management is to make full use of computer information technology to realize the construction of enterprise management informatization. So first of all, it is necessary for the managers and employees to understand information technology. Therefore, this article does some related surveys on the level of enterprise personnel's understanding of information technology, as shown in Table 1:

Table 1. Knowledge of information technology

	Managers	Staffs
Know very well	30%	26%
General know	45%	43%
Don't know	25%	31%

According to Fig. 1(below), there are 56% of enterprise personnel who know information technology very well. Generally, those who know information technology account for a relatively large proportion, 88%, and those who do not know information technology account for 56%. These data reflect that both managers and employees do not have a high level of understanding of information technology. If you want a company to go further, you must follow the trend of the times, update yourself in a timely manner and update the company, and strive to stay behind.

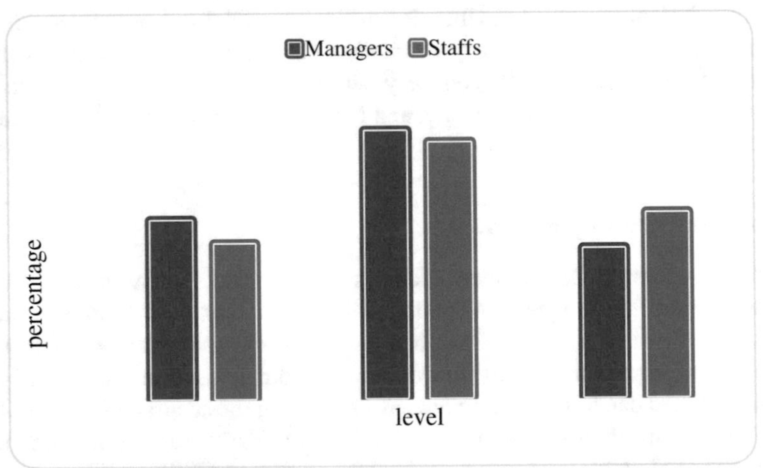

Fig. 1. Knowledge of information technology

In addition to surveys of corporate personnel's understanding of information technology, this article also randomly surveys several companies to explore how often they use information technology in management information systems. The results of the survey are shown in Fig. 2:

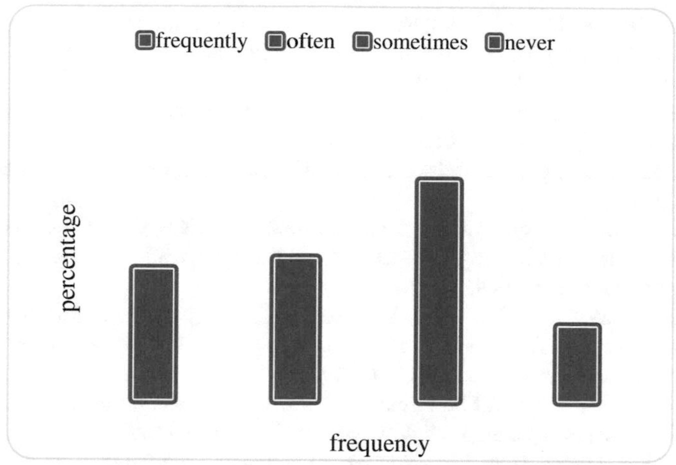

Fig. 2. Frequency of using information technology

It can be seen from Fig. 2 that 48% of companies often use information technology in management information systems; 39% of companies that use it occasionally; and 13% of companies that never use it. Enterprises should further deepen the application of big data ideas in management information systems, realize real-time information management and decision-making, and improve the efficiency of enterprise management information.

5 Conclusion

Enterprise information management is the management of the implementation process of enterprise information. This is not only a technical problem, but also a kind of management change. It is a major change to the original production organization method, information flow, and management mechanism of enterprises. Therefore, it is vital to develop management information system through professional management.

In recent years, enterprise L has made great progress in the construction of management information system, but with the continuous changes in business development, with the updating of product styles, the market environment has been continuously expanded, while traditional management cannot respond to the rapid development of today's information, marketing, data statistics, service innovation and other aspects. Enterprise L cannot keep up with the pace of business development, therefore restricting L's business development. So, it becomes vital for enterprise L to analyze and search for the shortcomings when making use of enterprise management information system, solving existing problems, and continuously optimizing management information system to promote business transformation and realize its strategic development.

References

1. Bartoněk, D., Bureš, J., Opatřilová, I.: The optimization of global navigation satellite systems measurement by the geographic information systems support. J. Comput. Theor. Nanosci. **13**(11), 8345–8354 (2016)

2. Borisova, V.V., Demkina, O.V., Mikhailova, A.V., Zieliński, R.: The enterprise management system: evaluating the use of information technology and information systems. Polish J. Manage. Stud. **20**(1), 103–118 (2019)
3. Schwaninger, M.: Organizing for sustainability: a cybernetic concept for sustainable renewal. Kybernetes **44**(6/7), 935–954 (2016)
4. Dong, Q., Li, B., Leung, H.: Understanding the API usage in Java. Inf. Softw. Technol. **73**(73), 81–100 (2016)
5. Feng, X., Chen, L.: Average shortest path optimization method of the course-scheduling system in universities based on geographic information system information. J. Comput. Theor. Nanosci. **13**(12), 10486–10491 (2016)
6. Galuszka, A., Krystek, J., Swierniak, A., Lungoci, C., Grzejszczak, T.: Information management in passenger traffic supporting system design as a multi-criteria discrete optimization task. Arch. Control Sci. **27**(2), 113–116 (2017)
7. Xu, W., Nie, Y.: RETRACTED: research on the legal protection of personal information security in the era of big data. J. Phys.: Conf. Ser. **1648**(3), 032067 (2020)
8. Utomo, I.G.W., Darma, G.S.: Measuring optimization of digital military programs: an innovation of information and communication systems in industrial digitalization 4.0. Int. Res. J. Eng., IT Sci. Res. **6**(2), 39–46 (2020). https://doi.org/10.21744/irjeis.v6n2.862
9. Guha, S., Kumar, S.: Emergence of big data research in operations management, information systems, and healthcare: past contributions and future roadmap. Prod. Oper. Manag. **27**(9), 1724–1735 (2018)
10. Golovina, T.A., Avdeeva, I.L., Parakhina, L.V.: Prospects for the development of information technologies in the modern management system. Central Russian J. Soc. Sci. **15**(1), 242–254 (2020)

Application Analysis of Intelligent Information Technology in Transaction Management System

Lin Xue(✉)

Haojing College of Shaanxi University of Science and Technology, Xi'an, Shaanxi, China
XL2022ZC@126.com

Abstract. With the rapid development of modern science and technology and the progress of society, our country has entered a new information age. In corporate office activities, office automation technology emerges as the times require. Office automation integrates advanced technologies such as intelligent information technology and network technology to accelerate information processing and effectively improve office efficiency and quality. The purpose of this paper is to study the application and analysis of intelligent information technology in transaction management system. On the basis of analyzing the problems existing in transaction management system, introduction of related technologies and analysis of system requirements, a transaction management system based on intelligent information technology is designed. The performance of the system was tested. The test results show that the system can still operate smoothly after 50 accounts are entered at the same time, and the success rate is 100%. The response time of the system login is below 800ms, and the response time of the other three modules is below 1000ms.

Keywords: Intelligent Information Technology · Transaction Management · Office Automation · System Design

1 Introduction

In recent years, due to the rapid development of world network information technology and economic globalization, the scientific and technological level of government office automation is also constantly improving, and the impact on national social policies and economic and social development is also increasing. Especially in today's information construction society, its significance and influence have been highly valued by government departments [1, 2]. Office automation information technology is a comprehensive interdisciplinary subject based on modern technologies such as computer science, information, behavioral science, and communication technology. The main purpose of developing this technology is to improve the comprehensive business level of enterprises and assist management decision-making [3, 4].

In recent years, many experts and scholars have been in-depth research on enterprise affairs management, and have made good research results. Some researchers believe that there are five basic elements of transaction management architecture: presentation layer,

J. H. Abawajy et al. (Eds.): ICATCI 2022, LNDECT 170, pp. 121–129, 2023.
https://doi.org/10.1007/978-3-031-29097-8_15

base layer, data access layer, big data resource layer and Web business operation layer. The author also points out that the bottom layer of presentation is the basic service interface provided by the system to users, while the basic layer is the core of the enterprise management system, and the Web service function layer provides the basic application functions required by users. In addition, the author also emphasizes that the rapid development of WAP devices has brought great flexibility to the development of the system [5, 6]. Regarding the function of the office system, some scholars have also emphasized the necessity of circularly managing and designing various functional departments within the company according to the principle to exchange information [7, 8]. Some R & D personnel emphasize the rational use of Oracle database system for transaction management design. Because the Oracle database system has great flexibility and flexibility, it can work freely in a cluster environment, and rational use of the Oracle database system can significantly reduce management costs and improve management system performance. Based on the advantages of using the Oracle database system in terms of cluster technology, availability, business intelligence and stability, the Web transaction management supported by the Oracle database is highly maintainable and efficient [9, 10]. Clarified with other developers the focus of messaging and acquisition quality in transaction management applications. By integrating functions according to the characteristics of transaction management of each unit, the optimal message transmission design is carried out, the upper and lower message links are seamlessly connected, and messages are transmitted objectively and actively without personal attitude [11, 12]. From the perspective of future development, the modern office system based on office automation technology will develop in the direction of multimedia, networking, simplification, integration, collaboration and intelligence, and then develop into a huge industry integrated system.

On the basis of consulting a large number of relevant references, combined with the problems existing in the transaction management system, the introduction of related technologies and the analysis of system requirements, this paper designs an enterprise transaction management system. The system mainly includes four functional modules, namely system login module, system management module, document management module and conference management module. Finally, the performance of these four modules is tested to verify whether they meet the requirements of this paper.

2 Application Analysis of Intelligent Information Technology in Transaction Management System

2.1 Problems Existing in the Transaction Management System

(1) Safety aspects
Security issues are very important. Security has been a very important factor since the advent of computers. In the Internet age, virus outbreaks have destroyed many important records and caused a lot of losses to businesses. The nature of many high-value business information can also be targeted by espionage hackers and theft of Internet link information. The main precautions we can take include installing antivirus software, adding hardware firewalls and using encrypted information technology.

(2) The rules and regulations are not perfect
Today, many companies have very inadequate office automation systems. For the new workflows just released, many people still don't understand how to control and use them. People with traditional manual methods are so uncomfortable changing the way they work that some people resist and question office automation. Therefore, it is necessary to establish appropriate management rules and regulations, properly publicize ideas, change concepts, integrate knowledge, and implement effective education and learning.
(3) Support and attention from relevant personnel
Some enterprise IT leaders do not support, there is a misunderstanding of the office automation system, the leaders do not pay attention to it, cannot check the official documents in time, and even do not turn on the computer for a long time. Leader support is an important aspect of the implementation of the office automation system, because the office documents have been stagnant for a long time. Seriously affect the practicability of the office automation system.

2.2 Introduction of Related Technologies

(1) Three-tier Architecture

(1) Presentation layer
The presentation layer is the application program interface that the user sees and is responsible for communicating with the user and interacting with the application operation. Export and display user data using a graphical interface by recording the information and data entered by the user. It's complicated and not easy to learn. When changing the interface, simply change the control screen to view the data, no other levels will be affected. User requests are sent to the logic layer through a web browser and the response data is displayed to the user.
(2) Logic layer
The logical layer is also called the middle layer, which corresponds to the middle layer. Receives presentation tier user requests, sends requests to data tier servers, receives data server-level data, and sends them to presentation tier users. In the middle tier, there are user-mode applications and the ability to log data processing, many of which are developed using visualization tools.
(3) Data layer
The data layer, also known as the database layer, uses a large relational database system to read and write data. Most of the data is in the SQL language, so you can quickly retrieve and modify large amounts of data. After the server accepts the data request, it applies the relevant requirements to the database, and then returns the result to the server.

Each layer in the three-layer B/S structure module is relatively independent, which reduces the development complexity, reduces the burden on the client, and reduces the configuration cost. The browser is easy to operate and has a friendly interface, allowing users to work. Simple and easy to use, effectively isolate data, prevent unauthorized users from reading and writing, increase the security of your data system, and improve the level of business management.

(2) MVC Framework Pattern

(1) Model
The model is the main body of the application, it implements the processing work plan, and represents the business data and logical relationships. When data changes, views are notified and models can provide multiple views, increasing the reusability of the application. Controllers can also access services and provide them. When the controller sends a user request to the model, the model returns the corresponding data representation to the controller. The model remains the same and contains the basic part of the application logic, which is used in the troubleshooting process.

(2) View
A view is the interface with which the user interacts. It does not process business data, but only receives and displays the data received by the user. The viewer can also send status query commands to the model, receive business update events from the model for data synchronization, but cannot change the model. Views can display model data in different ways. Sometimes there are many opinions. The displayed view is controlled by the controller, which sends the user's needs to the controller.

(3) Controller
When the controller receives a user action process, it finds the corresponding model element to handle the event and display the result of the data returned from the view. The user needs to change the state of the model. This can be achieved through a controller. The controller selects the appropriate model, processes and modifies the data, and tells the viewer to view the updated data.

MVC clearly divides the application into three levels. The code is fresher, the program is easier to maintain, and the difficulty of program design and development is reduced, the reusability is improved, and the division of labor among developers is more logical software design technology. Development provides a good foundation for management capabilities.

2.3 System Requirements Analysis

With the rapid development of global informatization and networking, rich network information has become an important part of people's life, work and research. The company handles managing and scheduling many employees, configuring and assigning various tasks, and planning and executing tasks to complete a range of transactions. Existing manual recording methods are not only inefficient, but also prone to errors. Normal management of the company's work. Create user-friendly, systematic, accurate, secure, and well-structured information for timely and accurate transaction processing, including personnel flow inquiries, job completion statistics, job scheduling, and execution plan feedback. A standardized management system makes management projects systematic, standardized and procedural, avoids randomness in management, improves the accuracy and speed of information processing, makes management scientific and modern, and meets the needs of scientific and technological development. With the development and expansion of the company, the company has higher and higher requirements for the accuracy and automation of employee management, and this management system has been developed.

3 Experiment

3.1 System Login Module

To log in to the system designed in this article, you need to enter the user name, password, and verification code for authentication, and enter the correct login name to log in to the system interface, which is the default operation. If the input is incorrect, return to the system login interface and log in again. The technologies involved in the authentication process include database connection, database query, verification code generation and transmission. System security measures include: the password field adopts MD5 encryption and decryption and anti-SQl injection technology. The MD5 encryption and decryption algorithm is given by:

$$F(x, y, z) = (x\&y)/((\sim x)\&z) \tag{1}$$

$$G(x, y, z) = (x\&Z)/(y\&(\sim z)) \tag{2}$$

where & and ~ represent AND and NOT operations, respectively.

Figure 1 is a flow chart of the system login module.

3.2 System Management Module

This part belongs to the system administrator authority function and is responsible for maintaining the basic functions of the entire system. Some managed objects include users, groups, roles, user permissions, and so on. In addition, the system management module also includes system data backup and system announcements.

User management is the processing of adding, deleting, modifying and checking user information. User and group role management are roles and groups that maintain system settings. Rights management includes corresponding rights assignment. Create multiple match relationships with each user for multiple users and permissions. Administrators should also back up system data regularly to protect the security of system data. System announcements are displayed on the main interface of the system, and administrators can update and maintain them in time according to the specific content.

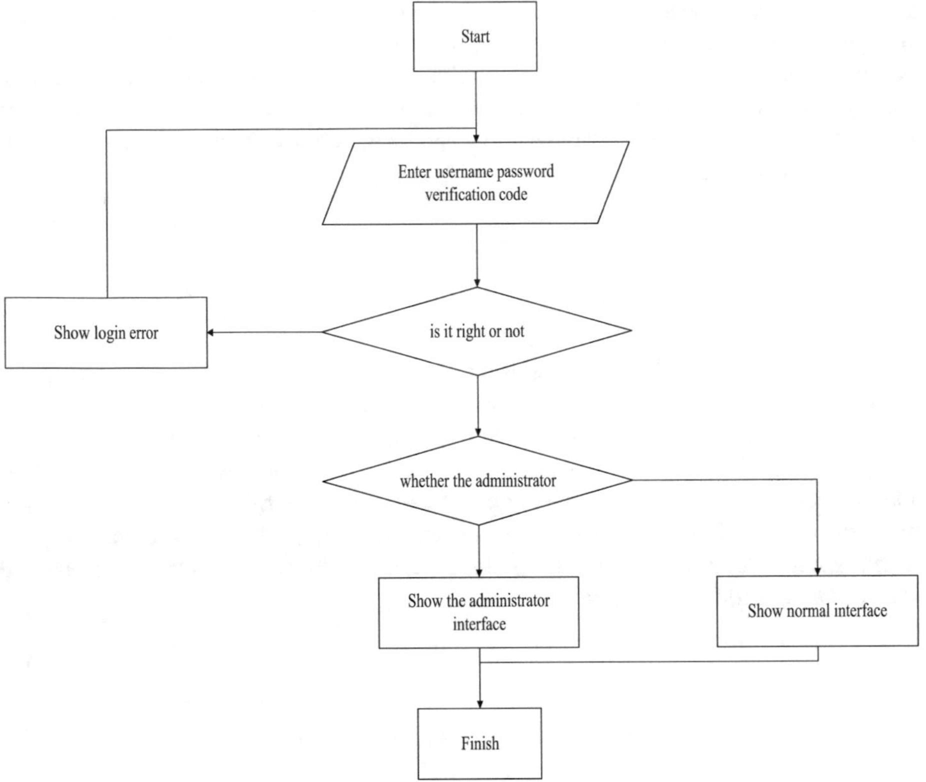

Fig. 1. System login flow chart

3.3 Document Management Module

During the normal operation of the company, there will be a lot of official documents. During the entire official document operation process from the official document writing to the official document distribution to the offices of various departments, a lot of processing affairs need to be done on the official documents. Send and receive process monitoring.

The official document operation process is as follows: the company manager decides to start with an official document. First, the official document is written by a special office staff. After approval by the leader, it is sent to the offices of various departments, and then distributed and circulated according to specific affairs, and finally the official document is archived.

3.4 Conference Management Module

The meeting management module manages the meeting list, meeting application, meeting review, weekly list, etc.

In the meeting management process, the meeting requester first sends the meeting request to the company. It should include all the details such as meeting topic, time, number of people, location, etc. Reviewers must verify the completeness and accuracy of the conference request, and if there are any issues, return the conference request to the applicant and document the issue. If the review is passed, it will enter the meeting review, and if the review is passed, it will enter the meeting process, such as notifying the participants, convening a meeting, etc. If not, meeting candidates will be notified and the application process completed.

4 Discussion

The tool used in the performance test of the networked office automation system designed in this paper is HPloadRunner9.0.

As can be seen from Table 1 and Fig. 2, through the performance test of the system, the system can still operate smoothly after 50 accounts are entered at the same time, and the success rate is 100%. The response time of the system login is less than 800ms, and the response time of the other three modules is less than 1000ms. The performance of the system meets the requirements of this paper.

Table 1. Concurrent operation test table

concurrent number	system login	System Management	Document management	Conference management
10	0.240	0.357	0.523	0.426
20	0.348	0.435	0.641	0.434
30	0.521	0.711	0.725	0.582
40	0.617	0.718	0.833	0.634
50	0.722	0.845	0.914	0.813

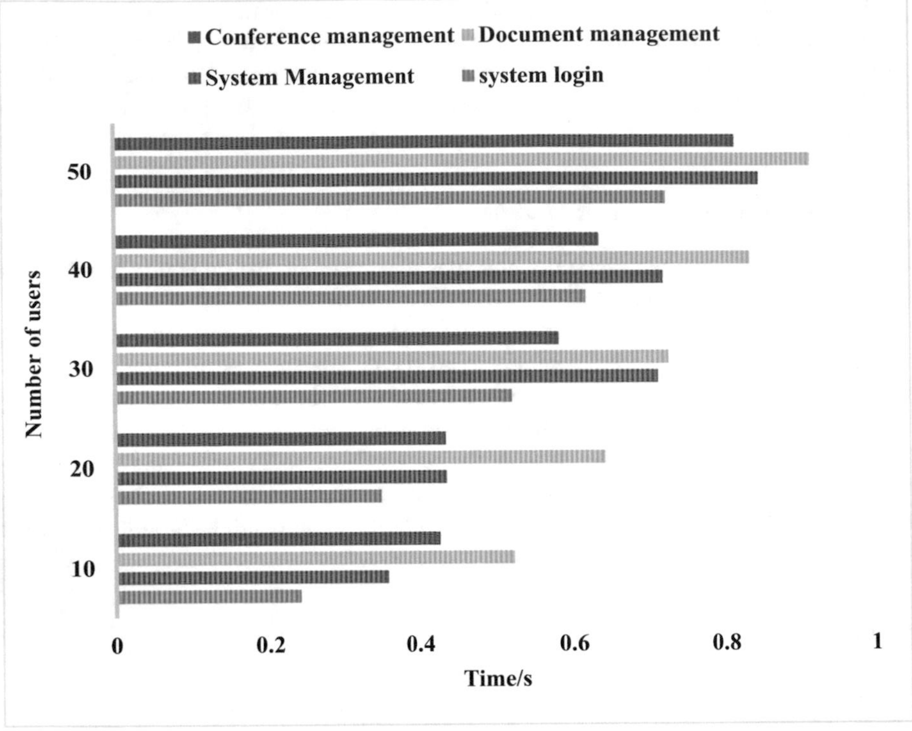

Fig. 2. Concurrent operation test chart

5 Conclusions

The rapid development of the Internet and intelligent information technology has changed people's business management model and lifestyle. The continuous development of the global economic environment, the continuous innovation of enterprise management ideas and methods, and the continuous development of computer network technology are all promoting the development and reform of enterprise management information systems.

References

1. Kim, J.J.: HBase based multi-row transaction management techniques. J. Eng. Appl. Sci. **14**(12), 4102–4108 (2019)
2. Lehdonvirta, V., Kassi, O., Hjorth, I., et al.: The global platform economy: a new offshoring institution enabling emerging-economy microproviders. J. Manag. **45**(2), 567–599 (2019)
3. Fekete, A.D.: Chiller: contention-centric transaction execution and data partitioning for modern networks (Technical Perspective). ACM SIGMOD Rec. **50**(1), 14 (2021)
4. Kurylo, M., Klochko, A., Volchenko, N., et al.: The use of biometric technologies for bank transaction security management against the background of the international experience: evidence from Ukraine. Banks and Bank Systems **16**(2), 47–58 (2021)

5. Yin, H., Yin, Y.L., Wang, D., et al.: Bibliometrics and visualization of research on trust in construction projects. Management Procurement and Law **174**(2), 1–12 (2020)
6. Jang, K.J.: Research on product management and transaction system development based on BlockChain. The e-Business Studies **21**(4), 57–70 (2020)
7. Rizal, A., Dewanti, P., Apriliani, I.M., et al.: Bumpy road to understanding transaction cost and fisheries co-management. Scientific World Journal **124**(2), 252–263 (2019)
8. Zhou, Z., Zeng, X., Xiao, H., et al.: Multiperiod portfolio optimization for asset-liability management with quadratic transaction costs. J. indu. Manage. Optimiz. **15**(3), 1493–1515 (2019)
9. Yuan, Y., Chu, Z., Lai, F., et al.: The impact of transaction attributes on logistics outsourcing success: a moderated mediation model. Int. J. Produc. Eco. **219**(Jan.), 54–65 (2020)
10. Gomes, L., Vale, Z.A., Corchado, J.M.: Multi-agent microgrid management system for single-board computers: a case study on peer-to-peer energy trading. IEEE Access **99**, 1 (2020)
11. Arvidsson, N., Jonsson, S., Snickare, L.K.: The transaction-relationship paradox effects of bank operational capabilities on bank performance throughout a shift from relationship lending to transaction. Manag. Financ. **45**(9), 1253–1271 (2019)
12. Qua-Enoo, A.A., Bervell, B., Nyagorme, P., et al.: Information technology integration perception on ghanaian distance higher education: a comparative analysis. Educational Research **20**(2), 1–31 (2021)

Design of Management Cloud Information System for Residential Buildings Based on High-Performance Computing and Face Recognition Technology

Na Zou and Xiaowei Wang(✉)

Jilin Engineering Normal University, Changchun, Jilin, China
iamwjzn@163.com

Abstract. With the rapid development of electronic information technology, social informatization has become an inevitable trend. This paper studies the management cloud information system of residential buildings based on high-performance computing and face recognition technology, and understands the relevant theories of the management cloud information system of residential buildings on the basis of literature data, and then discusses the management cloud information system based on high-performance computing and face recognition. The technical residential building management information system is designed, and the designed system is tested. The test results show that as the number of users increases, the concurrent processing time also increases, but the increasing trend is relatively slow. It can be seen that the concurrent processing performance of the system is good.

Keywords: Face Recognition · Residential Building · Management System · Cloud Information

1 Inductions

With the rise of Internet technology and the popularization of the Internet in people's daily lives [1, 2], people's life patterns, work patterns, and learning patterns have undergone earth-shaking changes, which have not only changed the conventional way of thinking, but also brought people in order to experience the rapid speed, the development of the Internet has brought people a convenient lifestyle [3, 4]. The Internet of Things technology is a very important part of the new generation of information technology era [5, 6]. It connects people and people, people and things, which is different from the traditional Internet. It connects people and things through communication technology, allowing people to log in and share information anytime, anywhere. It is an important stage of the development of the "information age" [7, 8].

The original version of this chapter was revised: the author's affiliation has been changed to "Jilin Engineering Normal University, Changchun, Jilin, China". The correction to this chapter is available at https://doi.org/10.1007/978-3-031-29097-8_122

J. H. Abawajy et al. (Eds.): ICATCI 2022, LNDECT 170, pp. 130–137, 2023.
https://doi.org/10.1007/978-3-031-29097-8_16

In view of the research of face recognition technology, some researchers have proposed a method for recognition using the features of points and bands of different brightness in the face image. Location information refers to the facial protrusions (nose and cheekbones) collected due to the short distance from the camera. Because the distance between the face and the camera is relatively short, the brightness of the picture is higher, but at the same time, the sunken part of the face (eyes) is also received, and the brightness of this part of the picture is also lower. Since the film information is matched by the contours of the face, mouth, eyes, and eyebrows into two triangular geometric distributions, the film signal is added on the base of these two triangles [9]. Therefore, some researchers have proposed to use the gray-scale spatial scale of facial images to describe the texture characteristics of human faces. This solution can determine the face structure in low-resolution facial images through the system. The color of low-resolution images is enhanced first, and then the face structure signal is obtained from a gray matrix composed of an inequality that can understand the face area, and a structured face signal model is used. In face matching recognition, the unequal intersection between the compared person and the typical face represents the closeness of the two, which can be used as a criterion for evaluating the degree of recognition crisis [10]. Some researchers believe that the face recognition method is mainly to measure the face portrait through spatial geometry, color and motion model, and then segment the face portrait through the low-rate encoding method. According to different subdivisions, there are different colors, and then the pre-defined three-dimensional geometric model is used to find the approximate division between the image to be compared and the standard image, and finally the motion characteristics of each part are used for face measurement [11]. In summary, there are more researches on face recognition technology, but relatively few researches on the application of enterprise information systems.

This paper conducts research on the management cloud information system of residential buildings with high-performance computing and face recognition technology, and analyzes the feasibility of the management cloud information system of residential buildings and the theory of face recognition technology on the basis of literature data. The management information system of residential buildings based on performance calculation and face recognition technology is designed, and the designed system is tested, and relevant conclusions are drawn through the test results.

2 Research on the Management Cloud Information System of Residential Buildings

2.1 Feasibility Analysis of Management Cloud Information System for Residential Buildings

(1) Economic feasibility. Due to the rapid development of computer technology and the widespread use of computer technology in enterprise management, it is very necessary to use computers for enterprise personnel management [12]. The 21st century will be an era full of competition and challenges. In this century, high performance, systematization, standardization and automation have become synonymous with modern business. However, because of the complicated data processing

methods, business managers must manage these data more efficiently. At the same time, due to the progress of the current information age, business managers also particularly need a kind of management. Therefore, like the complex information processing methods in the past, traditional manual management the method does not require a lot of personnel or material resources. The traditional machine management method also requires a large number of personnel, so the production cost is high, the error rate is also high, and the personnel flow and training expenses for new entrants are not small. A complete community management system is continuously available and can be continuously updated to meet the time and work needs of managers. For long-term use, the cost of the development system is calculated on average with the annual cost of traditional management methods. Compared with this, the price/performance ratio is possible. Therefore, the economic feasibility of developing this system is very high.

(2) Technical feasibility. The SQL database system developed by Microsoft used in the system uses its own grammar and programming language to write related function codes, that is, the program network and the protocol are separated. There are also several verification methods, including security verification, user role verification and login verification, and WPF in Visual Studio developed by Microsoft is developed. It provides a unified programming model, language and framework, and completely separates the work of interface designers and developers. At the same time, it provides a new multimedia interactive graphical user interface.

(3) Convenient functions. In this community management system, engineers skilled in development and operating systems provide user teams with specialized system training before using the system. The training content should include the adequacy of the various functions of the system and the brief maintenance of the system. In the process of program development and design, the intuitive interface and the text description of the controls allow users to have a complete understanding of the functions. In today's popularization of computers, this system is suitable for microcomputers, and most people can use and manage it, so the operation of this system should be regarded as a coordinated and simple manual operation. Therefore, the system can be run.

2.2 Face Recognition Technology

In recent years, face recognition has been a research hotspot in many fields, including pattern recognition, computer graphics, and artificial intelligence. Compared with other biological characteristics, human faces have the following advantages:

(1) Hide collection method

It is suitable for identifying and identifying criminal suspects in large public places, but biometric technologies such as fingerprints and iris cannot be used.

(2) Do not touch the collection device

The collection process does not require close contact between the user and the collection equipment, is inconspicuous, and is susceptible to most users.

(3) Strong interaction and strong follow-up monitoring ability

Face recognition can be used in combination with human-computer interaction to improve the reliability of the system. At the same time, monitoring and analysis can be performed by recording photos of people at the time of the event.

(4) Low cost

Although low-end cameras are cheap, they can already meet the accuracy requirements of face recognition. As the performance of image and image processing equipment improves, the accuracy of face recognition is also improving. However, in practical applications, equipment with requirements for backgrounds, expressions, postures, etc., is still very expensive. Different algorithms have their own advantages and disadvantages in different scenarios. The parameter selection and combination of multiple methods also have a significant impact on face recognition. How to choose the right face recognition algorithm and the right data to enhance the complete product experience in a specific application framework.

2.3 Face Recognition Algorithm

The invariant rotation of the feature point in the SIFT algorithm is achieved by using the pixel information around the feature point to artificially determine the fixed reference direction of the feature point. The SIFT method uses the image gradient method to obtain the pixel gradient and direction distribution characteristics near the feature point. The formula for calculating the order factor and direction is:

$$m(x, y) = \sqrt{(L(x+1, y) - L(x+1, y))^2 + (L(x, y+1) - L(x, y+1))^2} \tag{1}$$

$$\theta(x, y) = tan^{-1} \frac{L(x, y+1) - L(x, y-1)}{L(x+1, y) - L(x-1, y)} \tag{2}$$

Among them, L is the pixel value of the scale image where the feature point is located, the m gradient is the modulus value, and the θ gradient direction.

3 Design of Management Cloud Information System for Residential Buildings Based on High-Performance Computing and Face Recognition Technology

3.1 System Design Analysis

According to the functional analysis of the system, the framework of the system designed in this paper is shown in Fig. 1:

3.2 System Module Design

(1) Household management

The content of resident management is the management of various basic information of all local residents. Only by clearly understanding the information of the locals can the safety of the area be ensured. The administrator can modify, delete, add, and query basic family information, but users can only query the family information.

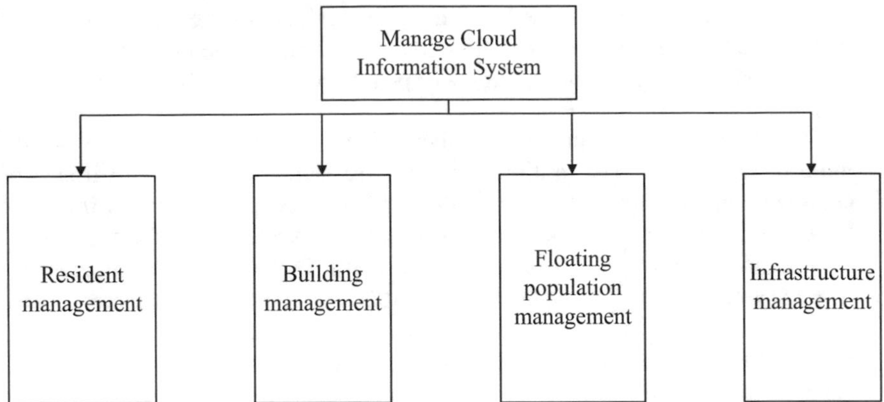

Fig. 1. Framework of the system

(2) Building management

The administrator of the cloud information system for the management of residential buildings can modify, cancel, add, and query the basic information of the building. For individual users, the management cloud information system of residential buildings can also query the information functions in the community.

(3) Management of Floating Population

The administrator of the management cloud information system of residential buildings needs to conduct a detailed survey of each floating population, and store the basic information of each floating population in the management system, including name, age, gender and contact information. In this way, the administrator can query, add and modify external demographic information at any time. With the emergence of the management cloud information system for residential buildings, the efficiency of community-based mobile management has been significantly improved.

(4) Infrastructure management

The infrastructure around the community sets important standards for the quality of life of residents. According to statistics, residents in areas with infrastructure are much happier than residents in areas without infrastructure. Therefore, it is necessary to add an infrastructure management information system to the regional information management system, so that residents can keep track of the surrounding area infrastructure.

3.3 Realization of Face Recognition

In the face recognition system of this article, this function uses OpenCV computer vision library technology as the technical basis for image recognition. The terminal camera captures and collects facial features, and uses important information from Vision Collector to capture and preview the captured images. Surface inspection is used to compare and track the collected images, and use cloud technology for face recognition and cloud recognition.

(1) Image acquisition process

The system mainly uses lines to collect face images from the camera computer terminal in real time, and captures the settings of existing cameras with certain characteristics. Then, put the image into line capture, insert each frame of the image captured by the camera into the buffer, filter the histogram, zoom and resize the image to fit the user's face. Hal's separation appeared on his face, finally revealing the interface.

(2) The pre-drawing process

Face recognition algorithms are highly sensitive to light, face angle and size, color complexity, face position and representation. Precise processing of the captured image on each panel.

(3) Face View program

The "Data/Horcascades/Path" installation folder provides some face-to-face partitions for computer users. The user only needs to add it to the program through CvLoad. Then see the square areas with object attributes in the image on a specific target, present these areas as a square box, and record the final result of the line in the variable.

4 System Test

The system built in this article is mainly the management cloud information system for residential buildings, and the operating logic of the system is relatively simple for the system itself. Black box testing (functional testing) can complete the testing of the management cloud information system of residential buildings. The black box test is a functional level test of the community information system. The specific method is to analyze the investment and cost of the operation business and verify the operation of the management cloud information system of the Accuracy residential building.

Simultaneous testing process: On the basis of the basic testing process, log in with multiple accounts, perform online management and form submission, and test various functions of the management cloud information system of residential buildings. The management cloud information system of residential buildings will be processed at the same time, and test processing capacity. The experimental results are shown in Table 1.

Table 1. System test results

	Online management/s	Submit form/s
100	2	1
150	3	2
200	3	2
250	3	3

It can be seen from Fig. 2 that the response time of the system in this paper increases with the increase of the number of users, but the increasing trend is relatively slow. It can be seen that the system has a good ability to handle concurrent events.

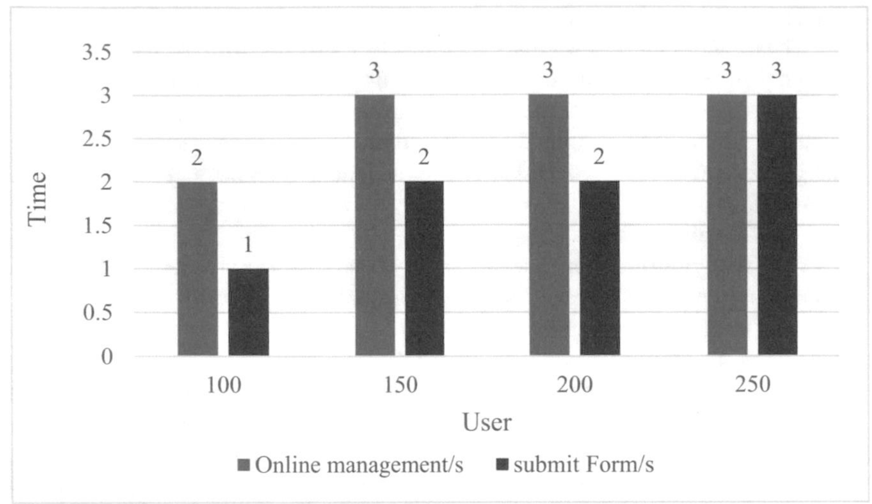

Fig. 2. System test results

5 Conclusions

This paper studies the residential building management cloud information system based on high-performance computing and face recognition technology. After understanding the relevant theories, the residential building management cloud information system based on high-performance computing and face recognition technology is designed, and the designed system is tested. The test results show that the system has a good ability to handle concurrent events.

Acknowledgements. Research on the construction of residential culture and ecology in Northeast China under the background of rural revitalization strategy. The subject number is JJKH20210197SK.

References

1. Ozdemir, D., Ugur, M.E.: A model proposal on the determination of student attendance in distance education with face recognition technology. Turk. Online J. Dist. Educ. **22**(1), 19–32 (2021)
2. Gentzel, M.: Biased face recognition technology used by government: a problem for liberal democracy. Philos. Technol. **34**(4), 1639–1663 (2021). https://doi.org/10.1007/s13347-021-00478-z
3. Bist, A.S., Febriani, W., Lukita, C., et al.: Design of face recognition AttendX for recording student attendance data based on artificial intelligence technology. Solid State Technol. **63**(2s), 4505–4518 (2020)
4. Delač, K., Grgić, M., Grgić, S., et al.: Image compression effects in face recognition systems. Mar. Policy **34**(3), 690–698 (2017)

5. Matsumura, H., Taketomi, T., Kato, H.: Impact of facial contour compensation on self-recognition in face-swapping technology. Multimed. Tools Appl. **80**(5), 7727–7748 (2020). https://doi.org/10.1007/s11042-020-09866-7
6. Eid, A.: Combined classifiers for invariant face recognition. Pattern Anal. Appl. **3**(4), 289–302 (2016)
7. Srinivasan, S., Mandal, S.K., Kumar, L., et al.: A structured protective cohesive health care information system using security and storage mechanism in cloud. Int. J. Eng. Trends Technol. **69**(3), 29–33 (2021)
8. Kumar, K.: Text query based summarized event searching interface system using deep learning over cloud. Multimed. Tools Appl. **80**(7), 11079–11094 (2021). https://doi.org/10.1007/s11042-020-10157-4
9. Palermo, R., Rossion, B., et al.: Do people have insight into their face recognition abilities? Q. J. Exp. Psychol. **70**(2), 218–233 (2017)
10. Carlucci, D., Renna, P., Materi, S., et al.: Intelligent decision-making model based on minority game for resource allocation in cloud manufacturing. Manag. Decis. (2020, ahead-of-print)
11. Pradhan, S., Tripathy, S.: FRAC: a flexible resource allocation for vehicular cloud system. IET Intell. Transp. Syst. **14**(14), 2141–2150 (2020)
12. Poux, F., Billen, R., Kasprzyk, J.P., et al.: A built heritage information system based on point cloud data: HIS-PC. Int. J. Geo-Inf. **9**(10), 588 (2020)

Development of Student Management System Based on WeChat Public Platform

He Sun(✉)

Kunming University of Science and Technology Oxbridge College, Kunming 650106, Yunnan, China
sunhe_associateprofessor@trymeimage.com

Abstract. With the rapid development of mobile Internet and the rapid popularization of smart phones, WeChat has an increasing share in universities and mobile phones. The push function of WeChat public platform has also made it an efficient channel for many colleges and universities (CAU) to release new information. The purpose of this paper is to study the student management system based on the WeChat public platform. The basic data interface designed in this paper and the shared, free and open resources available on the WeChat public platform, such as the powerful speech recognition interface provided by the WeChat public platform, use the Sina cloud computing platform as the data server, and use the MYSQL database to store data. This paper implements a student management WeChat subscription service system based on the WeChat public platform. This system has changed the status quo of untimely information transmission in the past, and it is convenient for students to inquire about personal attendance, discipline violation, conduct score information, course arrangement and course grades, etc. It also facilitates teachers to follow up and deal with class matters in a timely manner, effectively improving Improve the learning and work efficiency of college teachers and students. Experimental studies have shown that 40% of students use the platform 16–24 times a month, with a moderate frequency of use. It can be seen from the frequency of use that the platform has been recognized by most students in the application.

Keywords: WeChat Public Platform · Student Management · Information Query · Management System

1 Introduction

Student management is an important part of school management. At present, the vast majority of secondary vocational schools have a relatively complete student management system, but the management mode is still in the traditional non-network management mode [1, 2]. In this paper, the actual situation of student management in CAU, the combination of student management system and WeChat public platform can not only provide a convenient communication platform for students and administrators, but also a high-speed information transmission channel, which can provide student administrators, class teachers, students and even students. Parents can provide more convenient query

J. H. Abawajy et al. (Eds.): ICATCI 2022, LNDECT 170, pp. 138–145, 2023.
https://doi.org/10.1007/978-3-031-29097-8_17

services, grasp the situation of students in school, solve the problem that information cannot reach the audience in time, and reduce the workload of administrators and improve efficiency.

In the study of student management in CAU, many scholars have conducted research on it and achieved good results. For example, Razak F proposed to immediately change management strategies, strengthen teacher-student communication, give students independence, and improve student management plans [3]. Djuraevna A Z analyzed the problems existing in the management of students in CAU, and proposed solutions such as "implementing the 'convergence management' of Chinese and foreign students" and "actively carrying out the second classroom teaching" [4]. Therefore, it is of great significance to study the student management system based on the WeChat public platform.

By analyzing the shortcomings of the traditional student management system, this paper creates a student information management system based on a three-tier system model, builds a cloud server, and develops and designs the WeChat public platform. This article accesses the SQL Server database through JDBC, and completes the operations of adding, modifying and deleting the database. The system is easy to operate, has a friendly interface, greatly improves the efficiency and quality, and has good security, practicability and operability.

2 Research on the Design of Student Management System Based on WeChat Public Platform

2.1 WeChat Public Platform Has the Characteristics of Managing College Students

(1) Privacy, Diversity and Personalization of Disseminated Content

In terms of content distribution, since the WeChat public platform is a closed-loop communication method, the content distributed to users in this way is private. At the same time, the public WeChat platform supports the transmission of text, pictures and audio. To users and other multimedia information, which allows platforms to distribute content with different characteristics; in addition, popular content also has personal characteristics because public platforms can promote personal information to users by linking to third-party services [5, 6]. The privacy, diversity and personalization of the content distributed on the WeChat platform not only reflects respect for students' privacy, but also promotes harmony among students, and to a certain extent meets the needs of many schools for student management.

(2) Immediacy and accuracy of dissemination effects

In terms of communication ability, the WeChat public platform is instant, that is to say, as long as WeChat users log in, they can instantly download and view the reports pushed by the public, which is different from the information browsing function of Weibo; In many exchanges, the information from the platform will be passed on to every user, and the information growth rate is almost 100%. In addition, everyone supports the service platform. Forward message packets to users to ensure the accuracy of information dissemination [7, 8]. The consistency and accuracy of

WeChat communication can ensure that schools can meet time and community constraints and manage students in a timely and accurate manner.

(3) Younger and more educated objects of communication

From the perspective of mobile devices, WeChat, a product of the Internet development era, is easily accepted and promoted by students and office workers. WeChat provides a space for college students to use and spread in schools. Meanwhile, a study found that WeChat has become the most popular social networking tool among students in China.

(4) Free and simple operation and use

In terms of platform services and public use, WeChat is completely free, and the operation of the public WeChat platform does not require additional fees. The WeChat official account is simple to operate and can be used by anyone with a basic understanding of smartphone software. Likewise, the functions of a formal WeChat account do not require sophisticated techniques and can be effectively understood through training and practice.

The free and easy-to-use interface provides great flexibility for operating and using WeChat Work Accounts. At present, CAU need a certain amount of funds to manage students, and free platform services save the management costs of CAU [9, 10]. Second, public WeChat accounts are easy to use and available in multiple languages, providing flexibility and usability for students. At the same time, after certain training and counseling, student leaders can manage their own work skills.

2.2 Research on the Requirements of the Administrator's Main Interface

(1) Grade Management Module

One of the effective means to understand the actual learning situation of students is grades. In the grade management module, it is possible to add, query, revise and delete students' grade information. In this system, only the teacher has the operation authority of this module, and other users can only query.

(2) Student user operation process

After successfully logging into the system, students can perform two operations, one is user information management, and the other is result query. The former is to modify personal information and passwords. There are two ways to query results: query by student number and query by course number.

2.3 Overall System Design

(1) Realization of message management function

The head teacher user can receive the messages posted by the student users on the mobile terminal through the background management system. The messages are sorted according to the time of posting. The head teacher user can selectively reply. The reply content can only be seen by the student user who posted the message, and the class teacher user has the right to delete any message.

(2) Realization of information release function

This function means that the head teacher user publishes some notices of the school to the "notification bar" of the WeChat public platform through the background management system for students to view. This function is mainly used for class teacher users to release information and push notifications for students during weekends, holidays, winter and summer vacations and other periods when students are not in school. For example, the head teacher can push information such as tuition fees to be paid for the next school year, the start date of the school, and the specific time of the report to the student users during the summer vacation.

(3) Student Information Query Test

When the head teacher user inquires the personal basic information of the student user, he needs to enter the correct student number or name of the student user, otherwise the relevant information cannot be inquired. When inquiring about student users' activity registration information or scholarship application information, they need to enter the correct student number or name, or the registration project or application project, and then the relevant information can be inquired [11, 12]. The following is the specific operation process for querying personal information, mainly to test whether the function is running normally.

2.4 Evaluation Index System Based on AHP

When calculating the weights of each indicator, this paper chooses the AHP method, which can unify quantitative analysis and qualitative analysis, transform qualitative into quantitative, and ensure the final data results are more accurate. In:

$$CI = \frac{\lambda \max - n}{n - 1} \tag{1}$$

$$CR = \frac{CI}{RI} \tag{2}$$

where CI is the consistency test, RI is the query value, and CR is the variance test.

3 Experimental Research on Student Management System Based on WeChat Public Platform

3.1 Experimental Subjects

This paper selects the students of a higher vocational college as the object of practical application of the platform, and selects 100 students as the users of the platform, and the student managers are the corresponding student managers of the class, involving 3 student managers. Through the interviews and questionnaires conducted by the application objects, feedbacks are collected, and corresponding evaluations are made to the higher vocational student management platform.

3.2 Analysis of Student Feedback

The design of the questionnaire is carried out from the two modules of the platform's usage and the platform's functional section. Through the platform application object students filling out the questionnaire, the feedback of the platform application is collected, and the analysis is carried out to obtain the evaluation of the use of the platform. In this paper, a questionnaire survey was conducted on 100 selected students, and a total of 200 questionnaires were recovered, with a recovery rate of 100%.

4 Experimental Research Analysis of Student Management System Based on WeChat Public Platform

4.1 Analysis of Platform Usage Times

This paper studies the number and situation of student platform use, and the specific data is shown in Table 1.

Table 1. Analysis of student platform usage times

Frequency	Male	Female
Daily use	14.7	13.1
4–6 times a week	38.8	42.6
4–7 times a month	39.2	38.4
Less than 4 times a month	7.3	5.9

As shown in Fig. 1, among the majority of male students, 40% of the students use the platform 16–24 times a month, and the frequency of use is moderate. From the frequency of use, we can know that the platform has been recognized by most students in the application, and students will use the platform.

Among them, the frequency of use of the education and management modules has occupied the majority, indicating that students use the functions of these two modules the most and have more demand for them. Regarding the convenience and ease of operation of the platform, students believe that it is convenient and that there is no difficulty in operation. In addition, in the use of major modules or sections, students believe that the three sections of mental health education, class affairs, and real-time notification are the most popular, and they also think that the three sections of safety education, rules and regulations, and contact me should be further improved. At the same time, most students believe that the platform has brought them convenience and progress, which has a positive impact. At the same time, they also believe that the update speed of the platform information is slow and needs to be improved.

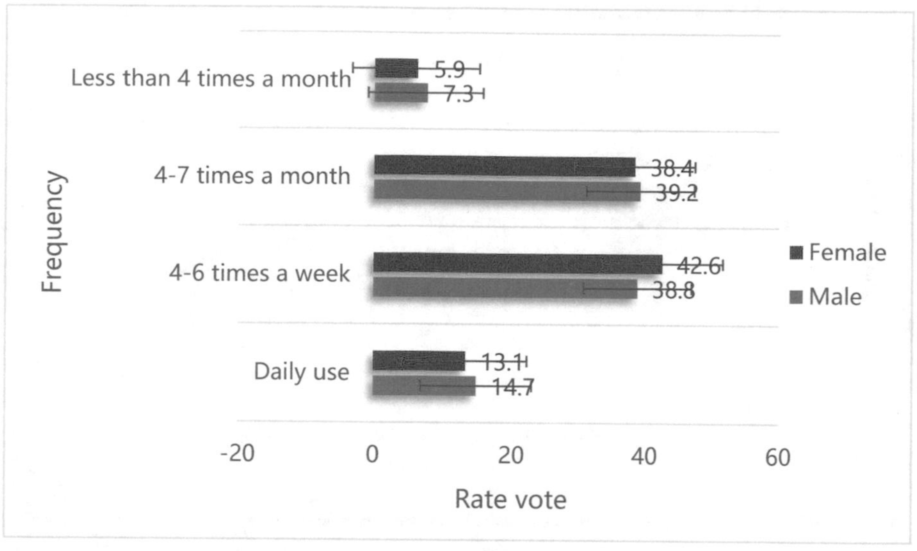

Fig. 1. Analysis of student platform usage times

4.2 Analysis of the Usage of Student Management Platform

The main data are shown in Table 2.

As shown in Fig. 2, it can be seen from the analysis results of the interviews and questionnaires of the application objects that the application of the platform can basically meet the needs of student managers and students for functions and content, and bring positive effects to their daily life, indicating that these The functions and content can still be used, but some content still needs to be enriched and improved, and the speed of updating the content still needs to be further improved.

Under the strong support of the WeChat platform, the university student management platform has obtained many useful functions. These functions make student management more convenient and make management more targeted. Personalized education, management and service based on student characteristics. The combination of the platform and the traditional management model forms a "two-dimensional integration" management

Table 2. The use of plates in the management module

Options	Male	Female
Rules and regulations	36.2	37.4
Dormitory management	14.7	15.5
Campus Activities	13.2	11.8
Reward and Punishment Notice	24.6	23.0
Class work	11.5	12.3

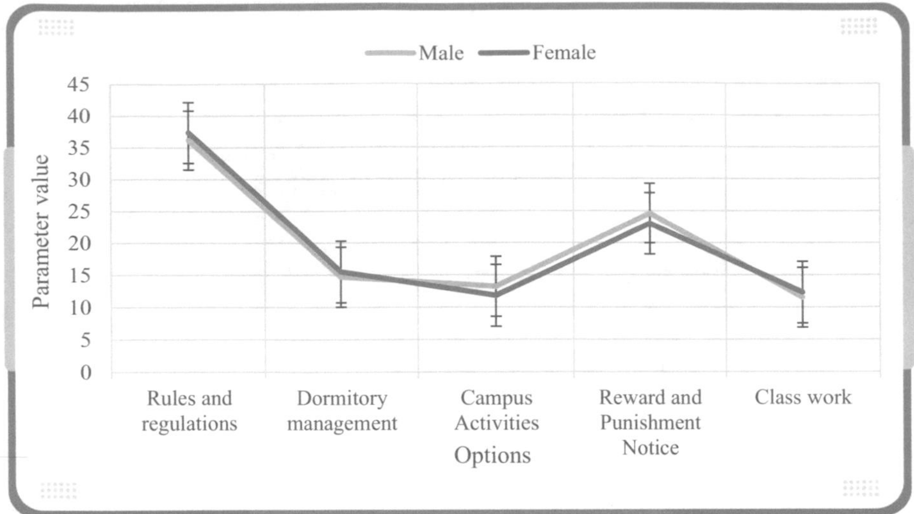

Fig. 2. The use of plates in the management module

model, which complements the advantages and disadvantages, making student management more dynamic, and making individual student managers handier in management. When students use the platform, they fully demonstrate the ubiquitous and ubiquitous advantages of education, management and services, obtain convenient, effective and timely help that cannot be obtained from the traditional student management model, and obtain the application of the platform from the questionnaire feedback.

5 Conclusions

Starting from the actual needs of CAU, using the principles of software engineering, this paper first analyzes the needs of the system designed in this paper, and then conducts overall design and detailed design of the entire system, and finally successfully develops a student management system. It provides great convenience for CAU to manage student information. In the design and development, the system adopts the B/S three-tier structure technology, and organizes the code through the separation of business logic, data and interface. By optimizing the existing management system, a web and mobile terminal management system with faster response speed, more efficient query and real-time query function is realized.

References

1. Luna, A.C., et al.: Mixed-integer-linear-programming-based energy management system for hybrid PV-wind-battery microgrids: modeling, design, and experimental verification. IEEE Trans. Power Electron. **32**(4), 2769–2783 (2017)
2. Alameri, I.A., Radchenko, G.: Development of student information management system based on cloud computing platform. J. Appl. Comput. Sci. Math. **11**(2), 9–14 (2017)

3. Razak, F., Mokhtar, A.E., Rahman, A.A., et al.: Investigating the linkage between service quality and satisfaction in context of student management system: an evidence from Malaysia. J. Phys. Conf. Ser. **1793**(1), 012033 (2021)
4. Djuraevna, A.Z.: Formation of student management activities in the higher education management system. Psychology (Savannah, Ga.) **58**(2), 1494–1499 (2021)
5. Jasmis, J., Aziz, A.A., Jono, M., et al.: An analysis model for an integrated student activities management system for higher education during RMO/CMCO/PASCA COVID-19 period in Malaysia. Proc. Comput. Sci. **179**(2), 798–803 (2021)
6. Alkhateeb, M.A., Abdalla, R.A.: Factors influencing student satisfaction towards using learning management system Moodle. Int. J. Commun. Technol. Educ.: Off. Publ. Inf. Resour. Manag. Assoc. **17**(1), 138–153 (2021)
7. Omodan, B.I.: A decolonial strategy to reconstruct student-management relationships in a university system. Acad. J. Interdiscip. Stud. **10**(2), 10 (2021)
8. Chandra, T.: Evaluation of student satisfaction in using the learning management system for online learning at XYZ University. Turk. J. Comput. Math. Educ. (TURCOMAT) **12**(6), 2810–2816 (2021)
9. Martellucci, L., Krishna, K.K.: Analysis of air-cooling battery thermal management system for formula student car. J. Transp. Technol. **11**(3), 436–454 (2021)
10. Jamaluddin, J., Bundu, P., Anshari, A., et al.: Course and training models online application services (LAO-KURSUS) and learning management system (LMS) to improve student competencies. Univ. J. Educ. Res. **9**(6), 1258–1273 (2021)
11. Thamrin, R.M., Andriani, R.: Design web-based registration and data management of student thesis information system. SISFOTENIKA **11**(1), 101 (2021)
12. Duevi, I.: Student satisfaction vs student achievements – should quality management system in higher education aim at student satisfaction or student achievements? Poslovna izvrsnost - Bus. Excell. **14**(2), 51–67 (2020)

Application of Artificial Intelligence Technology in Human Resource Delicacy Management System

Shimin An and Ying Chen(✉)

Department of Economics and Management, Lanzhou University of Technology, Lanzhou 730050, Gansu, China
y13454553011@126.com

Abstract. The rapid development of digital economy has created a good economic and technological environment for the development of artificial intelligence (AI). Under the influence of AI technology, traditional human resource management (HRM) is facing certain opportunities and challenges. Therefore, after clarifying the AI technology, this paper combined it with the delicacy management of human resources to form a complete delicacy management system of human resources, mainly including six major systems such as Intelligent Decision Support System. Specifically, the application foundation of AIHRM system is established from three levels of artificial intelligence application, and the application framework was constructed. In order to promote the reform of the working mode of human resource management and create sustainable competitive advantages for enterprise human resource management, it is expected to play a reference role for the follow-up research in the field of AI and HRM.

Keywords: Artificial Intelligence · Delicacy Management · Human Resource Management System

1 Introduction

In the era of mobile Internet, the maturity of representative technologies such as big data, cloud computing and Internet of Things makes artificial intelligence take on the technical characteristics of machine learning, neural network, intelligent computing and man-machine cooperation. Different from the characteristics of previous new technologies, AI has emerged more and more intelligent production methods and products to satisfy the personalized needs of users. At the same time, it has changed people's lifestyle and learning style, and gradually penetrated into the workplace, bringing opportunities and challenges to the traditional human resource management model. According to McKinsey's 2020 global *AI Survey Report*, 48% of respondents said that their organizations use at least one form of AI technology, and a considerable number of large companies will use a full range of technologies in the future.

J. H. Abawajy et al. (Eds.): ICATCI 2022, LNDECT 170, pp. 146–154, 2023.
https://doi.org/10.1007/978-3-031-29097-8_18

Artificial intelligence was first proposed at the Dartmouth society in 1956. It is an interdisciplinary subject integrating the theoretical basis of computer science, mathematics, physiology, philosophy and so on. Its connotation has been derived into a series of theories, methods, technologies and application systems that can simulate, expand and extend human intelligence [1]. Ai Rui defines AI as the combination of AI industry and AI technology. AI industry mainly includes various value systems such as the application of technical algorithms, and AI technology is the part that uses machines to help human beings replace or even surpass human beings to realize the functions of cognitive recognition, analysis and decision-making.

Delicacy management is an enterprise management concept originated from Japan. It clarifies and concretizes the management responsibility to achieve "accuracy, detail, depth and standardization". Its essence is to find key problems in the whole process of management, grasp weak links, refine each part, make targeted corrections and deal with them in time, with the main goal of minimizing management costs, improving work efficiency and maximizing enterprise value. The earliest delicacy management of human resources was based on the six basic modules of human resources management, that is, to achieve "precision, accuracy, detail and strictness" in human resources planning, training and development, recruitment and allocation, performance management, benefit management and labor relations management. Subsequently, based on the characteristics of the digital economy era, many scholars applied AI modules to human resource management to make its functions intelligent, efficiency accurate and practical.

Human resource management system refers to the cross system and process of human resource management and information technology. At present, AI and other technical means have been applied to the major functional modules of human resource management, such as AI recruitment [2], employee training of intelligent design [3], performance evaluation of fuzzy artificial neural algorithm [4], salary management based on algorithm control [5], business practice applications such as intelligent algorithm predicting turnover rate [6] and employee relationship measurement based on natural language processing and machine learning algorithm [7] are in the ascendant. In addition, during the epidemic period, the remote work of enterprise employees further strengthened the technology application ability of enterprises, and AI technology optimized the management effect.

2 Application of Artificial Intelligence Technology in Human Resource Delicacy Management System

2.1 Application Basis

First of all, AI is different from the prior theory of traditional management research. After forming assumptions, data analysis is carried out to obtain the results, but big data is obtained first, and then logical derivation is carried out accordingly. Human resource analysis is to collect, integrate and analyze data first, so as to lay data support for the follow-up development of human resource management, which also changes the logic of management decision-making.

Secondly, for HRM researchers, the impact of AI mainly appears in two aspects: one is the impact of AI technology on human resource management activities (AI on HRM), such as how the introduction of AI technology reshapes the organization's recruitment activities and efficiency; The second is the impact of the combination of AI technology and HRM activities on HRM activities themselves and their management objects (people, employees, interpersonal relations, man-machine cooperation, etc.) (AI + HRM).

In addition, the basic support, platform architecture and technology of artificial intelligence application lay a theoretical foundation for the proposal of the conceptual model of "AIHRM" (Artificial Intelligence Human Resource Management) [8]. Specifically, artificial intelligence technology and human resource management function modules are combined to form six "AIHRM" systems: intelligent decision-making help system, intelligent evaluation system, human-computer interaction system, intelligent training system, consulting system and intelligent incentive system [9]. Based on this, this paper explores the application framework of AI technology in human resources delicacy management system.

2.2 Application Framework

The application of AI technology in the delicacy management of human resources needs a mass of data as the application basis. In the past, the data involved in human resources were institutional data in the human resources information system, mainly including basic data such as resume, gender, age, education level and family situation. The introduction of big data technology makes the data content mastered by the organization more refined. Some scholars divided human resources big data into three basic types: physiological big data, behavioral big data and relational big data. The main sources are internal data and external data, as shown in Table 1.

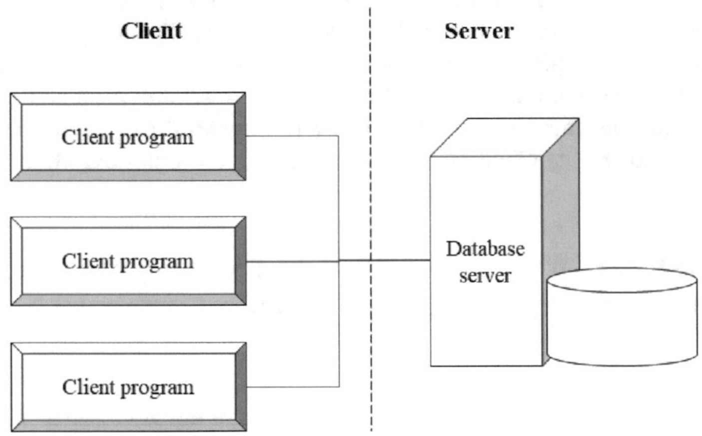

Fig. 1. C/S system architecture

Among them, the internal data of the organization can adopt the management system under the C/S (client and server) structure, as shown in Fig. 1. The C/S structure is to reasonably allocate the operation functions of the management system to specific clients and servers, which is more suitable for the use of Intranet in enterprises. Although it has certain limitations, it has strong confidentiality. It can be used to detect and count the structured data of the company's employees, which is conducive to the establishment of talent information base and facilitate intelligent decision-making. In addition, the management system under B/S architecture can connect with the external server through WWW browser to realize the user's working interface. Specifically, the application scope of the management system based on B/S architecture is not only applicable to the LAN inside the enterprise, but also applicable to the WAN outside the enterprise. Therefore, it can not only ensure the confidentiality needs of enterprises' designated computers, but also meet the unrestricted office under the Internet, realize flexible office, and meet the needs of enterprises' global management.

The core technology at the basic support level of AI can be applied to human resource planning, recruitment and allocation, training and development, performance management, salary and welfare management, etc. At the level of platform architecture, the core technologies include high-tech network construction and deep learning platform, which can be applied to employee growth training, performance management and so on. At the technical level, the core technologies include intelligent search, robot technology, biometric technology and natural language processing technology, which can be applied to HRM six modules. The specific application of AI technology in the fine management of human resources will be reflected in the "AIHRM" system, as shown in Fig. 2.

Intelligent Decision Support System (IDSS): In the digital era, many enterprise managers expect to introduce AI algorithm to realize fine management system in human resource management. This function not only has fast operation speed, high degree of automation and accurate analysis to improve decision-making efficiency, but also can change the management decision-making idea from "limited rationality" to "extreme rationality", optimize the enterprise decision-making environment and help human resource management make better decisions. Many forward-looking managers prefer to use big data-driven AI based on computer deep learning algorithm to integrate internal and external information of the organization and assist managers in making decisions on complex and key issues [10]. The new generation of decision support system combines decision support system (DSS) and professional system (ES) to form intelligent decision support system (IDSS). Among them, DSS can provide several types of memory assistance to support the use of representation and operation, such as database, database view and so on. DSS can also control assistive tools, which are designed to help decision makers use representation, operation and memory to integrate the decision-making process according to their personal style, skills and knowledge. For example, LinkedIn intelligently sorts the structured data and unstructured data of excellent talents through artificial intelligence recommendation algorithm. On this basis, enterprises match the recruitment needs with the personal information of job seekers to realize intelligent recruitment decision-making; Google uses artificial intelligence technology based on natural language processing algorithm to predict whether employees have turnover intention by analyzing the specific text information in employees' emails. After understanding the basic situation of employees, Google takes preventive measures in advance to decide whether to retain them and reduce the brain drain rate of enterprises; BP neural network algorithm can simulate the neural system of human brain [11]. By integrating multiple neural network nodes, it can continuously learn and train the input enterprise

Table 1. Three levels of artificial intelligence

HR big data		Primary coverage	Implementation mode
Classification	Physiological big data	Detect physiological status and health level	Other analysis technologies such as wearable devices
	Behavioral big data	Educational behavior big data	Data records such as purchase and borrowing of online educational materials
		Job search behavior big data	Browse recruitment information online, deliver resumes, etc.
		Work behavior big data	Use of office software and hardware equipment, frequency of using seminar room, etc.
		……	……
	Relational big data	Online interactive behavior	The contact behavior between members of an organization, including telephone, e-mail and other communication devices
		Offline interactive behavior	Communication and cooperation within the team, non workplace interaction, etc.
Source	Organization internal data	Basic data: resume and other basic information	The company's smart phones (APP, etc.), wearable devices, office Internet of things and sensor settings, internal communication system, digital office system, ERP system, etc. are mainly related to the integration and sharing of cross departmental data within the company

(*continued*)

Table 1. (*continued*)

HR big data		Primary coverage	Implementation mode
		Project data: training, league construction, labor union and other activities and project participation records	
		Performance data: behavioral big data such as performance evaluation	
	Organize external data	Mining social network data of candidates using artificial intelligence algorithm	Carry out online recruitment, e-commerce platform, social networking, etc.

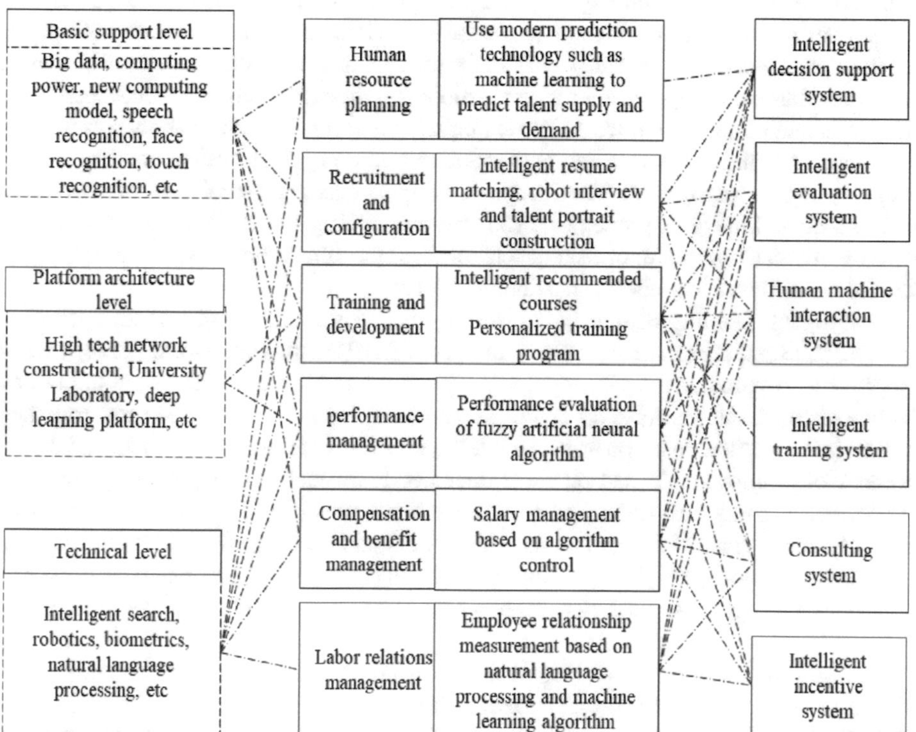

Fig. 2. Application framework of artificial intelligence technology in delicacy management of human resources

big data until the emergence of minimum error, and then design and construct Intelligent Decision Support System to form a fair and objective salary evaluation system [9].

Intelligent Evaluation System: Intelligent evaluation is mainly applied to the delicacy management of human resources. For example, the recruitment robot Mya can rank the performance of applicants, and the recruitment robot Hirevue can evaluate the EQ and personality traits of applicants by analyzing their voice, intonation, wording and facial expressions [12]. Many large enterprises have introduced recruitment robots to reduce the workload of HR and focus on complex and strategic issues [2]. In addition, because artificial neural networks deal with the nature of potential classification and prediction tasks, they are obviously suitable for turnover prediction tasks. The function of artificial neural network is to predict employees' turnover intention, which can specifically understand which specific employees may leave in the future, and reveal the unknown reasons affecting employees' turnover through analysis.

Human Machine Interaction System: According to the relevant research of human-computer interaction, in the process of technology application, the interaction between machine and human forms two human-computer cooperation modes: Intelligence enhancement and automation [13]. Intelligence enhancement refers to that the machine expands the intelligent boundary of human beings and completes the work together with human beings; Intelligent substitution means that machines replace human intelligence and complete work independently [14]. In the process of delicacy management of human resources, human-computer interaction is more intelligent to help HR make intelligent decisions and other system work. For example, recruitment robots, robots recommend personalized training courses and chat robots based on Q & A system can use natural language processing to understand employees' questions and search the corresponding answers in the knowledge base [15]; The artificial intelligence recruitment competition held by SourceCon, the robot Brilent screened out the appropriate resume in only 3.2 s, 28,124 times faster than the top headhunting team.

Intelligent Training System: By changing the way of training, providing flexible training time and personalized training methods, AIHRM maximizes the willingness of employees to voluntarily train, so as to promote the quality of enterprise training and build a channel and platform for enterprises and employees to learn and grow together. The training method based on artificial intelligence aims at the career growth and development of employees. Based on the career needs of each employee, it changes the traditional learning and training methods, such as converting written documents into more dynamic visual forms, and customizes personalized learning plans and training programs for employees [16]. In addition, research shows that robots can push appropriate training programs according to employees' ability status and future career interests [3].

Consulting System: Human resources service enterprises, driven by technology, take high-quality service as the core, reduce customer enterprise management costs, provide important decision-making consultation, establish their own brand and reputation of human resources service enterprises, and enhance market competitiveness. In the delicacy management of human resources, it is mainly used in the process of personnel recruitment. The amount of resume, candidate application and job consulting data generated in this process increases exponentially, forming a large amount of candidate information. The traditional manual resume screening or simple keyword screening method of intermediary recruitment website cannot effectively deal with this massive data. For example, employees can get help more easily at any time and place through intelligent consulting service platforms such as chat robots, which greatly reduces the cumbersome tasks of HR practitioners in dealing with daily affairs.

Intelligent Incentive System: With the diversification of employees in the workplace, enterprises gradually realize the need to formulate different evaluation mechanisms and incentive policies for different types of employees. Intelligent incentive system is mostly used for incentive salary, performance incentive and training and learning incentive. For example, using big data to understand the basic situation of training objects is conducive to the determination of incentive objectives and incentive methods, so as to improve the pertinence and accuracy of incentives. In addition, Club Med found that based on the mining of a large amount of data generated by employees in the workplace after entry, personalized incentive measures can be formulated to improve employees' job satisfaction.

3 Conclusions

To sum up, the enterprise will also face various challenges and opportunities. Human resource managers must realize that the rise of AI is unstoppable. Enterprises can only actively accept and embrace the challenges and opportunities brought by emerging technologies, help employees establish a sense of competition and change, quickly adapt to the changes brought by AI, flexibly use AI technology, effectively build management system and realize the delicacy management of human resources.

In addition, In the process of delicacy management of human resources, AI technology needs to gather and analyze a mass of data related to prediction, which is closely related to people's privacy and possible ethical and moral problems. For example, the analysis of e-mail text, surveillance video in office, routine clock in and clock out, and app use data on mobile phones and computers may be the most direct manifestation of privacy infringement [17]. Common ethical problems include data leakage, abuse and improper treatment of employee data by enterprises [18]. Therefore, in the future research, enterprises should establish clear moral and ethical boundaries, reduce or avoid various moral problems caused by tracking and monitoring employees' psychology, behavior and other data through risk management, increase information security system and improve the delicacy management system of human resources.

References

1. Taddy, M.: The Technological Elements of Artificial Intelligence. NBER Working Paper (2018)
2. Allal-Chérif, O., Aránega, A.Y., Sánchez, R.C.: Intelligent recruitment: how to identify, select, and retain talents from around the world using artificial Intelligence. Technol. Forecast. Social Change **169**, 120822 (2021)

3. Malik, A., Budhwar, P., Patel, C., Srikanth, N.R.: May the bots be with you! Delivering HR cost-effectiveness and individualised employee experiences in an MNE. Int. J. Hum. Resour. Manag. (in press)
4. Huang, L.-C., Huang, K.-S., Huang, H.-P., Jaw, B.-S.: Applying fuzzy neural network in human resource selection system. In: IEEE Annual Meeting of the Fuzzy Information, Processing NAFIPS 2004, pp. 169–174 (2004)
5. Duggan, J., Sherman, U., Carbery, R., McDonnell, A.: Algorithmic management and app-work in the Gig economy: a research agenda for employment relations and HRM. Hum. Resour. Manag. J. **30**(1), 114–132 (2020)
6. Rosenbaum, E.: IBM Artificial Intelligence Can Predict with 95% Accuracy Which Workers Are About to Quit Their Jobs (2019). https://www.cnbc.com/2019/04/03/ibm-ai-can-predict-with-95-percent-accuracy-which-employees-will-quit.html. Accessed 10 Oct 2021
7. Garg, R., Kiwelekar, A.W., Netak, L.D., Ghodake, A.: i-Pulse: a NLP based novel approach for employee engagement in logistics organization. Int. J. Inf. Manag. Data Insights **1**(1), 100011 (2021)
8. Brooks, R.A.: Intelligence without representation. Artif. Intell. **47**(1–3), 139–159 (1991)
9. Jia, Q., Guo, Y., Li, R., Li, Y., Chen, Y.: A conceptual artificial intelligence application framework in human resource management. In: ICEB 2018 Proceedings (2018)
10. Daugherty, P.R., Wilson, H.J.: Book highlight—The self ware factory floor: AI in production, supply chain, and distribution. Global Bus. Organiz. Excell. **38**(1), 53–60 (2018)
11. Richard, M.D., Lippmann, R.P.: Neural network classifiers estimate Bayesian a posteriori probabilities. Neural Comput. **3**(4), 461–483 (1991)
12. Hmoud, B., Laszlo, V.: Will artificial intelligence take over human resources recruitment and selection. Netw. Intell. Stud. **VII**(13), 21–30 (2019)
13. Jarrahi, M.H.: Artificial intelligence and the future of work: human-AI symbiosis in organizational decision making. Bus. Horiz. **61**(4), 577–586 (2018)
14. Raisch, S., Krakowski, S.: Artificial intelligence and management: the automation-augmentation paradox. Acad. Manag. Rev. **46**(1), 192–210 (2021)
15. Majumder, S., Mondal, A.: Are Chatbots really useful for human resource management? Int. J. Speech Technol. **24**(4), 969–977 (2021)
16. Premnath, E., Chully, A.A.: Artificial intelligence in human resource management: a qualitative study in the Indian context. J. Xi'an Univ. Archit. Technol. **XI**(XII), 1193–1205(2020).
17. Tursunbayeva, A., Di Lauro, S., Pagliari, C.: People analytics—A scoping review of conceptual boundaries and value propositions. Int. J. Inf. Manag. **43**, 224–247 (2018)
18. Calvard, T.S., Jeske, D.: Developing human resource data risk management in the age of big data. Int. J. Inf. Manag. **43**, 159–164 (2018)

Application of Data Mining Technology (DMT) in Human Resources Assessment Management System

He Ma(✉) and Meina Chen

Department of Human Resource Management, Dalian Neusoft University of Information, Dalian, Liaoning, China
mahe@neusoft.edu.cn

Abstract. With the tide of globalization and informatization surging all over the world, China's human resource management began the process of informatization. The informatization of human resource management has promoted the development of modern enterprise system and human resource management mode, and brought new changes to China's human resource management. The progress of society, the development of information technology and the gradual establishment of modern enterprise system in China have put forward higher requirements for human resource management system. Many early human resource management systems have been difficult to meet the human resource needs of modern enterprise management. Human resource data is a huge and cumbersome data set, which contains rich potential knowledge. Therefore, based on DMT and multiple linear regression analysis algorithm, this paper investigates and analyzes the evaluation of assessment management and salary treatment of different departments of company A and B. The survey results show that in the HR assessment and management system, the average salary is 7500, of which the proportion of people with a salary of 8000–10000 yuan is up to 27.6%; The evaluation of the company's assessment management generally tends to be excellent.

Keywords: Data Mining Technology · Human Resource Assessment · Management System · Application Research

1 Introduction

Human resource is an important application field of enterprise survival. With the continuous development of enterprises, a large number of data related to HR will be generated within enterprises. Mining the value of these data and creating benefits for them has always been a problem ignored by enterprises. DMT can make full use of the enterprise's own data, vigorously improve the enterprise's development and operation level, and expand the business, so that the enterprise can better improve itself, serve the society and serve users. It helps the decision-makers of enterprises to take corresponding measures to determine the HR of enterprises, so as to continuously improve the business efficiency of enterprises and finally realize the optimal allocation of resources.

J. H. Abawajy et al. (Eds.): ICATCI 2022, LNDECT 170, pp. 155–162, 2023.
https://doi.org/10.1007/978-3-031-29097-8_19

The application of DMT in HR assessment management system has been studied by many scholars at home and abroad: Tzanova S studies the complex relationship between continuous physical reality and digital representation. By understanding the synergy between constructiveness, provability and computability (in the form of system research program), it can break the gap between the digital world of logic programming and the complex multi-dimensional scientific world with rich sensor experience. Explaining the digital and real world is a very unique text, which goes beyond the common cliché of describing digital as a global phenomenon [1]. The system proposed by Papineni S designs a model by considering different employee parameters and using random forest algorithm. It helps the human resources department to retain employees by identifying gaps, and helps the organization run smoothly with a good employee retention rate. The combination of human resources and data science helps to improve the productivity, collaboration and well-being of employees in the organization. It also helps to develop strategies that affect employee performance in terms of external and social factors [2]. Leiter M simulation is used for different allocation strategies, and two different indicators are used to evaluate the human resource allocation in different projects. Three types of companies are studied. For each type, virtual companies are created, and several scenarios of collaborators, projects and tasks are simulated to evaluate the staffing process. Research results: the research shows that for different simulations, the evaluation may adopt different allocation strategies and indicators, and there is no golden rule of staffing in an organization with multiple projects and multiple skill collaborators [3].

Because the traditional data mining methods are often only to obtain the surface information of these data. In order to master their exact internal attributes and connections, we need to change our ideas, study and use the data mining theory, and apply the DMT and demand forecasting technology to the field of human resource management, more simply, it is to analyze the situation of human resource management on the basis of objective data analysis, so that human resource data information can be used reasonably. Specifically, it is to give full play to the important value of big data in human resource management, sort out and analyze relevant human data, find the data basis for the specific implementation of human resource management, conduct deeper data mining and analysis, and scientifically customize human resource management strategies according to the analysis and prediction results of data. It can be said that after applying DMT to the field of human resource management, the value of data will be reflected in all aspects of human resource management such as analysis and prediction [3, 4].

2 Core Concept

2.1 Data Mining Objects

Data has become diversified and complicated with the changes of the information age. The information databases storing different data types provide the original data source for data mining. These data mining objects have their own mining schemes.

(1) In transactional databases, the specific technologies and algorithms of data mining have been relatively mature, and some general patterns can be directly applied. Now

people pay attention to how to improve the mining efficiency and Multi Strategy mining patterns.

(2) In relational database, because the entity relation model has been very mature, SQL query language and visualization tools provide a strong basis for data mining. Combined with the mining objectives, people can use association rules, clustering and other algorithms to explore new knowledge or models. For the data mining of relational database, its application value is quite significant. Simple multi-dimensional knowledge mining, multi table mining, multi-layer data mining and constraint data mining all belong to the problems faced by relational database mining.

(3) In the data warehouse, multidimensional data or multidimensional data cube model are generated around a certain topic, which will be related to many data sources. All preprocessed data warehouses provide an excellent data foundation for data mining with their auxiliary tools. At the same time, multi-dimensional operators can enable mining to complete data analysis efficiently and quickly. Online analytical processing (OLAP) has the same goal as data mining, but the focus is different. The former is to explore knowledge patterns, and the latter is to seek relevant methods of data aggregation. They are all based on deep analysis and advanced processing of data warehouse.

(4) Among the new databases, new mainly refers to those object-oriented or object relational databases. Their emergence puts forward new heights and requirements for data mining, which has become an inevitable trend [5].

In a word, for different data sources, we need to choose different data mining objects and mining schemes according to the data characteristics.

2.2 Assessment Method of Human Resource Management

2.2.1 Assessment Method

Choosing an appropriate performance evaluation method is one of the prerequisites for effective performance evaluation. Performance evaluation must correspond to certain evaluation methods. The determination of evaluation methods depends on the characteristics of employees and team work [6, 7].

① Key performance indicator method key performance indicator method, usually referred to as KPI - a performance evaluation method widely used in modern enterprises. The key indicator method refers to the performance evaluation of the organization, which is mainly the evaluation of key performance indicators. Key performance indicators are the key success factors restricting the operation of the organization. They are generally defined from the four aspects of time, quantity, quality and cost. Then the evaluation of key performance indicators is the performance evaluation of the organization, so that the evaluation results can be obtained. Therefore, the key performance indicator method focuses on how to select key performance indicators, and make different choices according to different job types, so as to make performance evaluation more intuitively [8, 9]. In recent years, the key performance indicator method has been used more and more widely and has become one of the mainstream assessment methods.

② 360° assessment method, as the name suggests, is a method that can make a comprehensive assessment. The comprehensiveness of this method is mainly due to two reasons: first, the comprehensiveness of relevant personnel - it can assess all relevant personnel, including superiors, subordinates, related rating colleagues, customers or service objects, as well as the appraiser himself. As the subject of assessment, it is still necessary to evaluate the appraisee; Second, the questionnaire is designed according to the objectives of performance appraisal, and the above-mentioned relevant personnel are invited to fill in. The evaluation results also come from different levels. The evaluation information is multi angle and multi-directional, which can make a more effective and comprehensive evaluation. Generally speaking, 360° assessment method is more objective and comprehensive [10, 11].

2.2.2 Assessment Process

Determining the assessment objective is the premise of project HR performance evaluation. Before performance evaluation, we should clearly put forward the expected objectives of employees and teams, make them understand the direction of their efforts, compare their completion with the expected objectives, reflect the business ability of employees and teams, and feedback the results.

2.2.3 Assessment Model

Performance evaluation is a system that is constantly improved in communication. Performance evaluation connects individual performance with the project to evaluate individual performance. Therefore, performance evaluation connects employees, projects and assessment. In the process of project promotion, employee performance may be affected by internal factors; It may also be affected by external factors [12, 13].

3 Construction of HR Assessment Management System Model

In the multiple regression model, the proportion of the variation of independent variables affected by the variation of dependent variables is determined by using the judgment coefficient X2, which is the ratio of the sum of regression squares to the sum of overall squares. Its calculation is as follows (1).

$$X^2 = \frac{\sum_{i=1}^{a} (\hat{y}_i - y)^2}{\sum_{i=1}^{a} (y_i - y)^2}, \quad 0 \leq X \leq 1 \tag{1}$$

The closer x is to 1, the better the fitting degree of the model is. The closer x is to 0, the worse the fitting degree of the model is.

With the increase of the number of samples, the determination coefficient X will increase, which is prone to the illusion that the more independent variables, the better the fitting effect. In order to reduce the influence of the number of independent variables and sample size on the results of regression analysis, correction X2 is introduced, as shown in formula (2).

$$\text{Adjusted } X^2 = 1 - \frac{\sum_{i=1}^{a} (y_i - \hat{y}_i)^2/(a-q-1)}{\sum_{i=1}^{a} (\hat{y}_i - y)^2/(a-1)} \tag{2}$$

Correction x2 divides the sum of squares of residuals and the sum of squares of total deviations by their respective degrees of freedom to eliminate the influence of the number of variables on the goodness of fit, which can accurately reflect the goodness of fit. Q – predicted value of dependent variable; Q – number of independent variables.

4 Investigation and Analysis of HR Assessment Management

In this paper, the questionnaire survey is carried out in the form of network distribution. The distribution objects are companies a and B. the evaluation of company assessment management and salary treatment by different departments of companies a and B are investigated. According to the analysis of KPI scores of each department, only a few departments show normal distribution, and the trend chart of assessment results is shown in Table 1 and Fig. 1.

Table 1. Trend chart of assessment results

	excellent	good	commonly	pass
production	31.32%	54.43%	0.26%	9.81%
Material control	60.11%	0.13%	21.02%	19.93%
technology	0%	100%	0.12%	0.32%
Warranty	19.%56	66.47%	0.11%	11.45%
After sale	96.23%	18.48%	0%	0.41%
application	80.13%	21.08%	0.3%4	0.27%
General manager's Office	51.23%	52.67%	0.12%	0.16%
Finance	0%	100%	0%	0.13%

This paper investigates the salary of companies a and B, which are divided into three echelons: technology, operation and management, as shown in Fig. 2.

According to the survey results, the average salary of companies a and B is 7500, of which the proportion of people with a salary of 8000–10000 yuan is up to 27.6%.

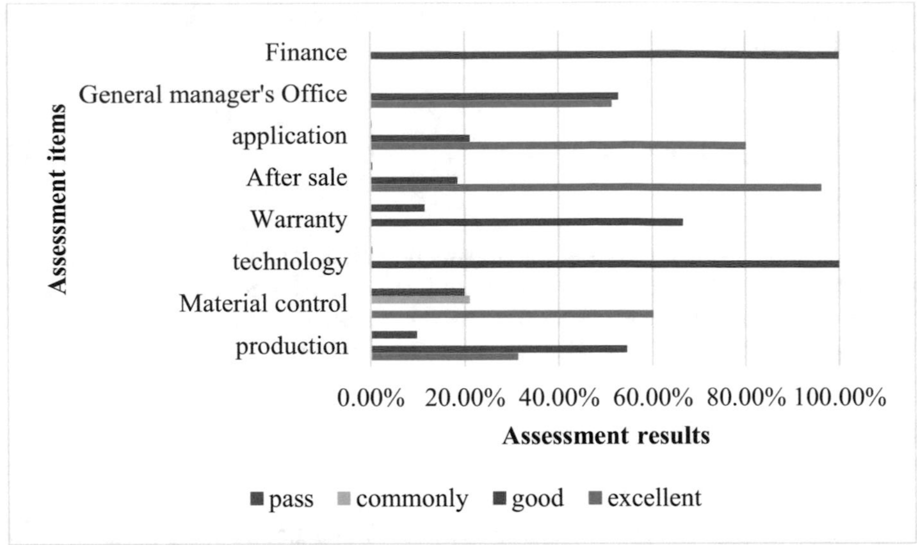

Fig. 1. Trend chart of assessment results

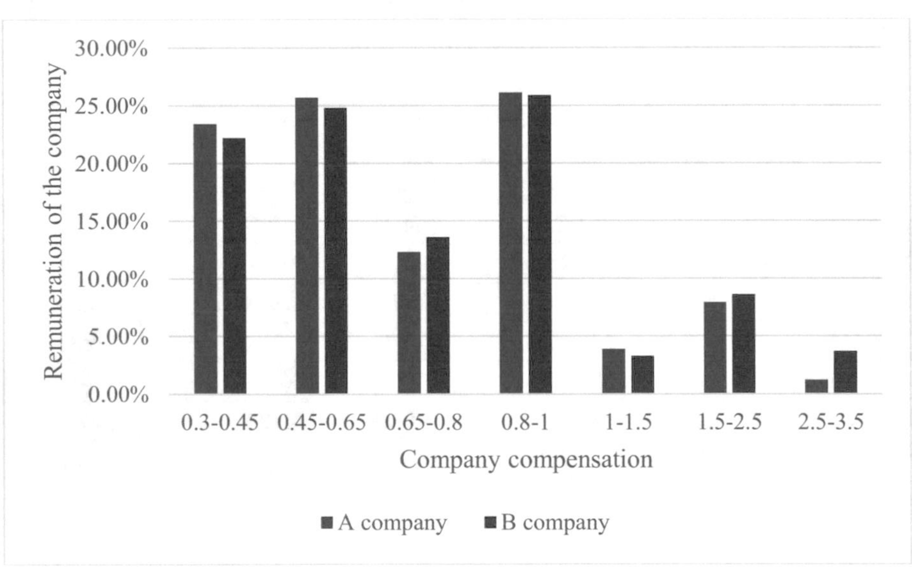

Fig. 2. Remuneration of the company

5 Conclusions

This paper has carried out a series of research on HR assessment, systematically expounded the basic theory of HR management, introduced the relevant data mining methods, constructed the HR assessment management system, and investigated the year-end performance assessment of employees in enterprises A and B. However, the research results of this chapter only put forward a new thinking for the major research topic of HR, and there are still many other problems worthy of in-depth discussion: the talent performance appraisal system is bound to move towards the talent performance comprehensive management mode in the future development, from the post personnel appraisal mode to the company's strategic performance management mode, The main characteristics of talent evaluation methods will change from subjective fuzzy qualitative to objective and accurate quantitative. The 21st century will be a new era of rapid development of economic informatization and a new era of information gathering and flooding. Using DMT to deeply explore the potential laws of a large number of talent data will create great social and economic benefits for social and economic development and people's life. The application of DMT in the field of HR will be more and more extensive and in-depth, which will contribute to the development of the field of HR management.

Acknowledgements. This work was supported by 2020 New Think Tank for Higher Education Institutions in Liaoning Province, 2021 Annual Planning Topic of China Private Education Association (project number: CANFZG21042), and Project of Dalian Academy of Social Sciences (project number: DLKX2021B10 and 2021DLSKY146).

References

1. Tzanova, S.: The digital and the real world: computational foundations of mathematics, science, technology, and philosophy. Comput. Rev. **60**(6), 239 (2019)
2. Shewchuk, J.P., Nussbaum, M.A., Kim, S., et al.: Simulation modeling and ergonomic assessment of complex multiworker physical processes. IEEE Trans. Hum.-Mach. Syst. **47**(6), 777–788 (2017)
3. Leite, M., Baptista, A.J., Ribeiro, A.M.R.: A trap of optimizing skills use when allocating human resources to a multiple projects environment. Team Perform. Manag. **23**(3–4), 110–123 (2017)
4. Ketmaneechairat, H.: Individual and collective graph mining: principles, algorithms and applications. J. Digit. Inf. Manag. **16**(1), 43 (2018)
5. Papineni, S., Reddy, A.M., Yarlagadda, S., et al.: An extensive analytical approach on human resources using random forest algorithm. Int. J. Eng. Trends Technol. **69**(5), 119–127 (2021)
6. Anshuman, A., Kunnath-Poovakka, A., Eldho, T.I.: Towards the use of conceptual models for water resource assessment in Indian tropical watersheds under monsoon-driven climatic conditions. Environ. Earth Sci. **78**(9), 1–15 (2019). https://doi.org/10.1007/s12665-019-8281-5
7. Graham, R.: Facing the crisis in human resources for eye health in sub-Saharan Africa. Community Eye Health **30**(100), 85–87 (2017)
8. Madera, J.M., Dawson, M., Guchait, P., et al.: Strategic human resources management research in hospitality and tourism: a review of current literature and suggestions for the future. Int. J. Contemp. Hosp. Manag. **29**(1), 48–67 (2017)
9. Mauro, A.D., Greco, M., Grimaldi, M., et al.: Human resources for Big Data professions: a systematic classification of job roles and required skill sets. Inf. Process. Manag. **54**(5), 807–817 (2017)

10. Razzaq, S., Aslam, U., Kashmir, A., et al.: The impact of human resources management practices on employee commitment: evidence from Pakistan telecom sector. Soc. Sci. Electron. Publ. **7**(77), 649–667 (2017). 2222-6990
11. Akbari, N., Ghaffari, A.: Verifying relationship of knowledge management initiatives and the empowerment of human resources. J. Knowl. Manag. **21**(5), 1120–1141 (2017)
12. Kadochnikov, S.M., Fedyunina, A.A.: The impact of financial and human resources on the export performance of Russian firms. Econ. Syst. **41**(1), 41–51 (2017)
13. Lee, E.B., Kim, J., Lee, S.G.: Predicting customer churn in mobile industry using data mining technology. Ind. Manag. Data Syst. **117**(1), 90–109 (2017)

Construction of Electric Energy Data and Carbon Emission Management Platform Under Computer Technology

Xu Yang[1](✉) and Saad Metawa[2]

[1] Shaoguan Power Supply Bureau of Guangdong Power Grid Co., Ltd., Shaoguan 512026, Guangdong, China
xxcmei99@163.com

[2] Mansoura University, Mansoura, Egypt

Abstract. So as to comply with the national energy revolution and "dual carbon" strategic goals, and actively respond to the overall requirements of establishing and improving green and low-carbon technology evaluation and trading system and scientific and technological innovation service platform, the research and development of intelligent "energy + dual carbon" service platform under the new situation policy has been carried out. Based on the regional smart energy service platform, relying on the application of energy big data, focusing on the whole-process management of carbon assets from the perspectives of the government, enterprises and the trading market, the construction of smart "energy + dual carbon" service platform will be carried out. By pooling massive energy information and combining with big data, artificial intelligence and other advanced technologies, the platform studies carbon accounting methods, calculation methods and prediction methods and forms a standard algorithm model to build carbon monitoring system, carbon assessment system and carbon emission prediction system, and build "dual carbon" energy digital intelligence products. It provides "carbon monitoring" for the government to assist carbon management, "carbon analysis" for enterprises to guide energy saving and carbon reduction, "carbon verification" for the trading market to optimize resource allocation, and promote the green and high-quality development of the social economy. This article studies a series of theories and knowledge about the construction of electric energy data and carbon emission management platform under computer technology, reveals the related content of the construction of electric energy data and carbon emission management platform, and carries on the construction of electric energy data and carbon emission management platform. By analyzing the actual effect of video image, the computer technology, the electric energy data and carbon management platform for the construction of the study, test results show that the electric energy data and carbon management platform for the construction of the modeling design in image recognition and differentiation of line identification, data reorganization and system performance of automatization of 81.02%, 91.03%, 92.15%, 97.05%.

Keywords: Computer Technology · Electric Energy Data · Carbon Emissions · Management Platform Construction

J. H. Abawajy et al. (Eds.): ICATCI 2022, LNDECT 170, pp. 163–171, 2023.
https://doi.org/10.1007/978-3-031-29097-8_20

1 Introduction

Based on the regional intelligent energy service platform, this paper conducts in-depth research on the module development and market operation of digital energy carbon innovative application. The multi-energy heterogeneous data collection and analysis method, energy efficiency optimization method based on "source-network-charge-storage" interactive scheduling, and carbon emission monitoring, carbon evaluation and carbon emission prediction system based on "energy-electric-carbon" algorithm model were constructed. As for the application of electric energy data and carbon emission management platform construction under computer technology, it is necessary to strengthen the construction and practical application capacity of electric energy data and carbon emission management platform under computer technology, so as to solve various problems in the construction of electric energy data and carbon emission management platform.

Many scholars at home and abroad have studied computer technology. In foreign studies, Mewda et al. proposed the application of technology based on computed tomography in these applications. Taking computer-aided planning and treatment of severe maxillofacial atrophy as an example, embedding and mesh transplantation were used to treat maxillofacial atrophy [1]. Kashihara K et al. mentioned building an intelligent computer-aided system to easily recover archaeological finds from fragments. Regional coding genetic algorithm (RCGA) is suitable for the localization of 3D restoration. The fitness function of RCGA was calculated according to the similarity between the target and the correction mode in the multi-view plane image. Simulation studies show that RCGA method, especially AKAZE technology, can effectively adjust the position of 3D debris automatically [2]. Kim PW et al. proposes an intelligent algorithm that provides teachers with information by measuring participation levels in real time. Internet of Things (IoT) algorithms for assessing student engagement levels are proposed to assess students' psychological states by measuring electrical skin activity (or skin electrical response). The proposed algorithm is innovative and allows teachers to provide feedback to students while monitoring them in real time. It provides a foundation for the application of the Internet of Things in teaching [3]. However, the construction of electric energy data and carbon emission management platform under computer technology is still in the initial stage, and there is still a certain gap compared with foreign systems.

In order to further improve the construction of electric energy data and carbon emission management platform based on computer technology in China, we must start from the following points: First, deepen the research and development of computer technology; Secondly, relevant data algorithms are optimized. Finally, strengthen exchange and study with foreign countries, improve the construction level of electric energy data and carbon emission management platform.

2 Research on the Construction and Application of Electric Energy Data and Carbon Emission Management Platform Based on Computer Technology

2.1 Computer Technology

Computer technology is a relatively new science and technology, and has become an essential and important component of People's Daily life [4, 5]. Computer technology is highly integrated and can integrate a variety of complex technologies and systems, such as electronic engineering, information engineering, ergonomics and other disciplines [6, 7]. Computer technology has completely overturned the passive situation of slow data processing and limited data calculation accuracy in the past. Computer technology has been applied to all fields of society. It can be said that computer technology is the core technology of building a digital society and the essential content of building a "smart city". The power of computer technology lies not only in the rapid processing and interactive sharing of data information, but more importantly in the centralized display of data information and the comprehensive utilization of the actual value of data information [8, 9].

In the development of automation, the application of computer technology is mainly reflected in the application of text and data processing technology in automation, such as Word, WPS and other computer editing and typesetting tools make editing processing, layout design, copy printing and image processing and other office activities efficiency; The application of the automation of data processing technology to make the office staff, Excel spreadsheets are concerned can be done directly on the computer table design, processing, and tabulation of all operations to achieve the form processing in each link of the automation in power and automation for the application of multimedia technology and image processing system, voice, fax, multi-functional multimedia workstations, integrated services The Digital Network has modernized the application of network technologies in office services and public automation, such as video conferencing, conducting face-to-face conversation through multimedia networks, and summarizing, printing and judging the feasibility of decisions through computers, allowing one person to participate in many at the same time [10, 11].

Computer application is also a high-quality representative product of the integration of communication technology and computer technology, which can realize massive processing, collection and utilization of data and information [12, 13]. The society in the new era has developed into a society dominated by big data, and the era of big data has quietly arrived, which is closely related to the vigorous development of computer technology. With the support of computer technology, the collection, storage and utilization of data information have been comprehensively strengthened, and the unified collection and sorting of scattered data can be better completed, without wasting any valuable data information, and the effective utilization rate of data information has reached a relatively high level [14, 15]. With the continuous development of computer technology, the storage capacity and processing capacity of data have been comprehensively enhanced, and the overall utilization rate is higher and the radiation range is wider, which can bring good opportunities for the construction and development of multiple industries in the society [16, 17]. Computer technology has also promoted the rapid development of artificial intelligence, which is also the concentrated embodiment of the good integration of communication technology and computer technology.

2.2 Construction of Electric Energy Data and Carbon Emission Management Platform

The development and application of smart "energy + double carbon" service platform complies with the national energy revolution and the "double carbon" strategic goal, and serves the major customer groups in the market, such as government agencies, enterprises, power grid companies, energy service providers and third-party institutions. It provides all-round carbon monitoring and measurement, multi-dimensional carbon analysis and assessment, comprehensive carbon asset management, personalized carbon account and emission reduction services, etc., to build an integrated intelligent control brain and resource center of "energy + dual carbon", and to help build a digital and dual carbon comprehensive service ecosystem.

(1) Platform architecture design
The overall architecture of the platform adopts mature IAAS, PAAS and SAAS cloud architectures. Based on the collection, identification and uploading of underlying data, it provides support for smart "energy + dual carbon" functional applications by means of data mining and data analysis. Through the three external service means of large-screen monitoring, Web application and mobile APP, Realize the 4-oriented service output, as shown in Fig. 1.

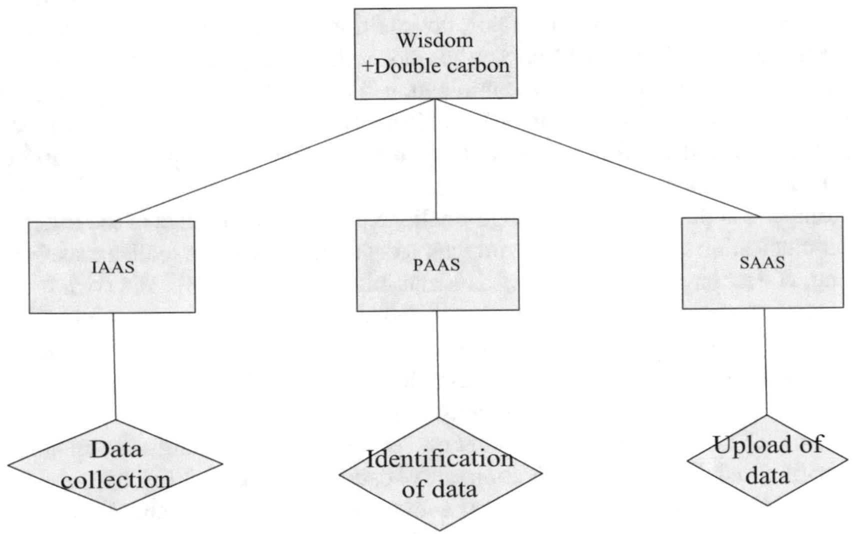

Fig. 1. Overall platform overall architecture flow chart

The platform can provide digital carbon applications such as carbon quota, regional carbon monitoring and analysis, carbon emission reporting and verification, carbon consulting, enterprise carbon asset management, enterprise carbon emission reduction, and carbon account.

Through carbon emission conversion and economic data integration to support scientific calculation, for regional, industrial and enterprise dimensions, carbon emission measurement, monitoring and carbon reduction decision-making and suggestions; Assist enterprises to carry out carbon asset management, carbon compliance tracking, automatic generation of carbon verification report, and carbon ecological services; The service system of carbon efficiency code, carbon account and green electricity points for users was established to quantify and count the value of energy conservation and emission reduction behaviors of enterprises, buildings and parks. Do a good job in supporting the carbon trading system under the quota distribution system, carry out market analysis of carbon market, match trading, and fully tap the value of enterprise carbon assets. As a whole, it will drive up and down the industrial and supply chains, jointly promote the transformation of energy and electricity from high carbon to low carbon, and from fossil energy to clean energy, and actively promote the realization of the "dual carbon" goal.

Based on the indoor structure of public buildings, a mathematical model of energy capacity is established. According to the objective function and constraint conditions, mixed integer linear or nonlinear programming method is used to calculate the comprehensive energy efficiency. Based on each branch energy source system, the optimization model is developed, and the total capacity is continuously optimized until the optimal and suboptimal solutions of each optimization subproblem are calculated. Through the implementation of comprehensive energy efficiency optimization calculation, fully coordinate the use of regional available new energy forms, support wind energy, solar energy, electric vehicles, etc., to achieve high efficiency of energy utilization. The energy efficiency optimization method based on "source-network-charge-storage" interactive scheduling has been formed, overcoming the problems of calculating comprehensive energy efficiency and energy management in the energy industry.

3 Research on the Construction and Application Effect of Electric Energy Data and Carbon Emission Management Platform Based on Computer Technology

3.1 Comparative Analysis

In this paper, comparative analysis method is adopted to collect data points of electrical energy data and construction of carbon emission management platform in established steps, classify and process each case to be tested, and construct platform data model. For the construction of electric energy data and carbon emission management platform, the system was modeled locally, and all the key points were collected and analyzed. In view of the different effects of the test, the electrical energy data and carbon emission management platform construction and non-algorithm two groups of data processing. Optimize the collected sample data.

3.2 Description and Analysis

In this paper, comparative analysis method is used to analyze and process the construction of electric energy data and carbon emission management platform, and data processing and processing of system samples. Through the construction of electric energy data and carbon emission management platform, the effect is analyzed.

3.3 Calculation Formula

$$C_1 = C_e + C_n \tag{1}$$

$$C_e = P * (1 - \beta) * \alpha \tag{2}$$

Represents energy carbon emissions, and represents indirect emissions from power use and carbon emissions from natural gas use. P represents regional power supply, represents the proportion of clean energy in the region, and represents power emission factor.

3.4 Details

Computer technology, the electric energy data and carbon management platform of all elements of the construction of the parse, the computer technology, the electric energy data and carbon management platform for the construction of the effect on the total integration, part of the resolution images, through this process, using the data model, to be blurred images, the best effect through the model design, And optimize the related parameters of electric energy data and carbon emission management platform.

4 Investigation and Research Analysis of Electric Energy Data and Construction of Carbon Emission Management Platform Based on Computer Technology

4.1 Effect Test

The test objects are divided into electrical energy data under computer technology and the construction of carbon emission management platform and non-algorithm effect are compared and analyzed. The model data of video automatic recognition system is established by processing the electric energy data and carbon emission management platform construction under computer technology. Test the two groups of data respectively, and sort out the improvement effect of the electric energy data and the construction of the carbon emission management platform under computer technology. Record the actual performance of the relevant data with the electric energy data and the construction of the carbon emission management platform: In the comparative analysis, computer technology under the electric energy data and carbon emissions management platform to promote the construction of the goal, effect testing, image analysis application effect is very obvious. Key techniques to enhance the application of image analysis have laid a solid foundation, with the results shown in Table 1 and Fig. 2.

Table 1. Manage platform data tables

	Carbon emission assistance	Value-added services	Digitization	Building cost
Number	132	301	412	652
Ratio	81.02%	91.03%	92.15%	97.05%
Percentage	91	93	94	98

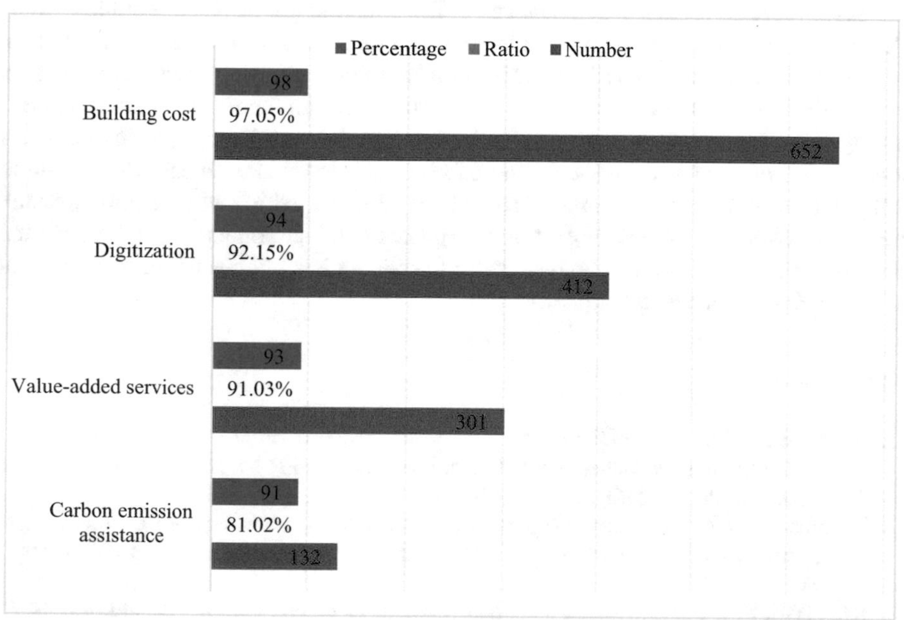

Fig. 2. Platform construction application data chart

The construction and application of electric energy data and carbon emission management platform based on computer technology is better than that of the traditional way. Excellent performance in carbon emission assistance, value-added services, digitalization and construction cost, etc., testing and analysis in model effect data processing and application, construction of electric energy data and carbon emission management platform based on computer technology is refined, scientific and efficient in electric energy and carbon emission management. In view of the electric energy and carbon emissions of the management platform, various special situations encountered in the testing process of the data model are analyzed from a more microscopic perspective to scientifically avoid various defects of the model. The construction of electric energy data and carbon emission management platform based on computer technology is conducive to the improvement of the construction of electric energy data and carbon emission management platform based on computer technology.

5 Conclusions

The construction and application of the smart "energy + dual carbon" service platform is fully in line with the development of national policies and the actual demand of the market. It is open to the government, enterprises and the carbon trading market, and has broad prospects. This platform can help the government to carry out carbon regulation, help enterprises to save energy and reduce emissions, help the accurate data verification of carbon trading market, and help the healthy development of social and ecological civilization. The future economic and social benefits are significant, and it has guidance and reference significance for the construction of other "double carbon" platforms. For carbon emission monitoring, the core basic function of the platform largely depends on the underlying collection and calculation of energy consumption data. In the future, on-site carbon emission monitoring equipment will be synchronously equipped to improve the scientific and accurate carbon emission monitoring data. The construction of electric energy data and carbon emission management platform under computer technology plays an important role in all fields, which will certainly promote the construction technology of electric energy data and carbon emission management platform under computer technology, and improve the level of electric energy data and carbon emission management platform.

References

1. Wong, M.E., Kau, C.H., Melville, J.C., et al.: Bone reconstruction planning using computer technology for surgical management of severe Maxillomandibular atrophy. Oral Maxillofac. Surg. Clin. North Am. **31**(3), 457–472 (2019)
2. Kashihara, K.: An intelligent computer assistance system for artifact restoration based on genetic algorithms with plane image features. Int. J. Comput. Intell. Appl. **16**(3), 1750021 (2017)
3. Kim, P.W.: Real-time bio-signal-processing of students based on an Intelligent algorithm for Internet of Things to assess engagement levels in a classroom. Future Gener. Comput. Syst. **86**(SEP.), 716–722 (2018)
4. Selin, C., Rawlings, K.C., Ridder-Vignone, K.D., et al.: Experiments in engagement: designing public engagement with science and technology for capacity building. Public Underst. Sci. **26**(6), 634 (2017)
5. Fonseca, B., Fernandes, E., Fonseca, M.: Collaboration in science and technology organizations of the public sector: a network perspective. Sci. Public Policy **44**(1), scw013 (2017)
6. Weber, S., Duarte, C.: Yield analysis for electrical circuit designs: many problems and some recent developments in electronic engineering. IEEE Solid-State Circuits Mag. **12**(1), 39–52 (2020)
7. Maheshwari, K.: A new sinusoidal quadrature oscillator for electronics engineering. Int. J. Electron. Inf. Eng. **10**(1), 45–50 (2019)
8. Maiti, M., Kayal, P.: Digitization: its impact on economic development & trade. Asian Econ. Financ. Rev. **7**(6), 541–549 (2017)
9. Stets, J.D., Corso, A.D., Nielsen, J.B., et al.: Scene reassembly after multimodal digitization and pipeline evaluation using photorealistic rendering. Appl. Opt. **56**(27), 7679–7690 (2017)
10. Ahmadjonova, K.: The use of multimedia technology in teaching foreign languages. Sci. Bull. Namangan State Univ. **1**(1), 37 (2019)

11. Kim, B., Chilamkurti, N., Dogra, D.P., et al.: Editorial note: emerging multimedia technology for multimedia-centric IoT. Multimed. Tools Appl. **77**(4), 4543 (2018)
12. Macedo, I.M.: Predicting the acceptance and use of information and communication technology by older adults: an empirical examination of the revised UTAUT2. Comput. Hum. Behav. **75**(oct.), 935–948 (2017)
13. Kosov, G.V., Stankevich, G.V., Gukasov, A.V., Romanko, L.V., Tekeeva, M.U.: Ethnopolitical process in the north Caucasus through the lens of North Caucasian online media. In: Popkova, E.G., Ostrovskaya, V.N. (eds.) ISC 2017. AISC, vol. 726, pp. 676–683. Springer, Cham (2019). https://doi.org/10.1007/978-3-319-90835-9_78
14. Xu, L., Jiang, C., Wang, J., et al.: Information security in big data: privacy and data mining. IEEE Access **2**(2), 1149–1176 (2017)
15. Kuang, L., Hao, F., Yang, L.T., et al.: A tensor-based approach for big data representation and dimensionality reduction. IEEE Trans. Emerg. Top. Comput. **2**(3), 280–291 (2017)
16. Kandasamy, N.K., Tseng, K.J., Boon-Hee, S.: A virtual storage capacity using demand response management to overcome intermittency of solar PV generation. IET Renew. Power Gener. **11**(14), 1741–1748 (2017)
17. Eicke, S., Seung, D., Egli, B., et al.: Increasing the carbohydrate storage capacity of plants by engineering a glycogen-like polymer pool in the cytosol. Metab. Eng. **40**(Complete), 23–32 (2017)

Application of Computer Electronic Information Technology in Engineering Project Management

Bing Liu(✉)

Intelligent Manufacturing and Automobile School, Chongqing Vocational College of Transportation, Chongqing 402247, China
1987liubin1987@163.com

Abstract. In recent years, the integration of information technology and civil engineering project management has never stopped. It can be said that information technology has penetrated into all stages and links of civil engineering project management, and can be seen in all aspects of civil engineering project management. This paper discusses the application of computer electronic information technology in engineering project management. In this paper, systems engineering theory, process management, information process, Based on technical knowledge such as historical science and integration process, combined with the uniqueness of technical production management, a set of overall research framework and conceptual model of engineering construction project integration and management has been established. The attrition rate of project management has been reduced by 0.7%, and the attrition rate of mechanical costs has been reduced by 1.5%. By informatizing project management, it can also reduce the difficulty of outsiders participating in project management, break industry boundaries, eliminate technical barriers, and allow management to return to its roots.

Keywords: Engineering Construction Project Management · Electronic Information Technology · Information Integration · Systems Engineering Theory

1 Introduction

The rapid development of modern information technology has had a profound impact on the way of life and work, and is gradually changing the pattern of the construction industry. The general application of modern information technology is not only the change of the whole construction industry system and mechanism, but also the change of its production structure mode, management concept, management method and management method. A revolution is taking place in the field of engineering project management, and the driving force of this revolution is modern information technology.

In the research on the application of computer electronic information technology in engineering project management, many scholars have studied it and achieved good results. For example, Li Y proposed to apply the concept of integrated computer system to engineering project management, and completed it through computer technology The integration and management of engineering information, the establishment

J. H. Abawajy et al. (Eds.): ICATCI 2022, LNDECT 170, pp. 172–180, 2023.
https://doi.org/10.1007/978-3-031-29097-8_21

of integrated project information management, thereby improving work efficiency [1]. Cajzek R and Klansek U of the United States believe that local area network, wide area network, telecommunication technology, business database and comprehensive project management system should be integrated and implemented in project management to provide a variety of project information services [2]. It can be seen that the importance of information construction for engineering project management lies.

Through research, this paper integrates and coordinates the cost target and time target of engineering construction projects, and uses many mathematical theoretical models, such as linear programming theory, fuzzy mathematical theory and so on. It is applied to the integration project of engineering construction goals, so that information construction becomes an important means to strengthen the execution of the system, so that information technology can really play its due role in improving the management level of engineering projects.

2 Engineering Construction Project Information Integration and Application

2.1 Information and Information Technology

(1) Standardization of engineering construction project information
Standardization is the basis of information integration, and standardization is the fundamental way and means to realize information integration. Due to the characteristics of different types of information and strong repetition of civil engineering projects, information standardization helps to unify, simplify, coordinate and streamline the relevant elements in the information. Information standardization can promote the reuse of previous information achievements and information resources, realize resource sharing, obtain greater benefits at lower costs, and achieve faster development in a shorter period of time [3]. Standardization is also an important part of science and technology, and advanced standards reflect advanced science and technology. Actively adopting relevant international standards is an important guarantee for the informatization, even process, and integration of participants in civil engineering projects, which can greatly improve the quality and impact of integration [4, 5].

(2) How information is transmitted in construction project management
In construction project management, information is first expressed as intention, that is, some ideas of the construction unit about the building to be built. How to express these ideas requires our design department to draw drawings according to the design intentions of the construction unit, and express them with text, graphics, symbols, numbers and other information through drawings [6, 7]. How does this information become a building? This involves the way we conduct project management and the method of project construction. The way of project management is to plan, organize, coordinate and control the project through what system and process and what organizational form we adopt. The method of project construction is to express various information in the drawings through what kind of construction technology, and express the information on the drawings through physical objects to form a building process.

(3) How to apply information technology in construction project management
First, we identified the tools and software we had at our disposal. Then, we have to understand the functions of these devices and the characteristics of the tools and software that can be used in construction project management, and we have to make choices among these devices and software [8, 9]. We also have to decide how we use these information technology and equipment according to the current handling of various aspects of construction project management. Due to the limited tools we currently use, the information technology we apply in the field of construction project management is also limited. We can study how information technology is applied in construction project management according to the process and method of information transmission in construction project management.

(4) Theoretical methods related to construction risk management of engineering projects
Traditional construction risks are always directed against three construction control objectives: time, cost, and quality. With the continuous development of many innovative technologies, people are more and more aware that the application of innovative technologies can solve construction control safety problems. Using this technology can realize real-time collection and monitoring of security information. These technologies include dynamic range systems (DGPS), cloud computing, sensors, remote technology and radio frequency identification (RFID). And the application of new technology can greatly reduce the risk of employee injury in high-risk areas. The flow of safety information has played a key role in the construction industry's transition from passive to active safety management. Active safety management requires smoother and more efficient flow of safety information. Basic construction safety management is the process of collecting, transmitting, storing, analyzing, displaying, and responding to safety information, but many innovative techniques can be used to manage the flow of safety information. Use radio frequency identification (RFID), laser detection, sensors and positioning systems (GPS) to obtain security-related information such as identification, location and environmental information. Wireless and broadband networks are used to send secure messages to the right people at the right time. Combine related analytical techniques (such as data mining, geographic information systems (GIS) and augmented reality) and optical techniques (such as 3D) to guide construction practices to prevent death or injury. Therefore, when studying construction safety risks, the loss is defined as an accident, and the possibility of a safety accident and the possible loss of personnel should be studied.

2.2 The Application of Informatization in Construction Project Management

(1) Scope of application of informatization in construction project management
Computers can be applied to the entire project design, technical design, construction, final registration and after-sales process in project management. Especially computer-based BIM design, computer-based precast concrete construction technology, and computer-based project management methods.

(2) The application of informatization in construction project management

In this paper, by integrating different services and applications in a common software platform so that their information can be shared, their services can meet the needs of construction management. BIM scheduling software and budget software with GPS can basically meet the needs of many services such as safety, improvement, quality, scheduling, budget settlement, project management costs, etc. [9]. In order to achieve open management and information management of housing. It can also make progress management simple and accurate.

2.3 The Application of Information Management in the Risk Control of Construction Personnel Behavior Safety

The construction personnel ID number (pid) is the unique identification of the construction personnel, which consists of 18 strings. The personnel number (vLicence) represents the participation responsibilities and authority of different participants on the construction site, and the personnel type (vType) represents the personnel. Define whether it belongs to the owner, the supervisor or the construction party. The date the record was generated when it was captured is denoted as (vDate) [10]. Each intelligent electronic monitoring device number expressed as (IEMDid) is the unique identifier of a certain intelligent electronic monitoring device, so the record of a person passing through an intelligent electronic monitoring device and being identified can be expressed as:

$$\mathrm{r} = (pid, vLicene, vType, vDate, IEMID) \tag{1}$$

Calculate the amount of records formed by the construction personnel during the time τ for each intelligent electronic monitoring device:

$$\mathrm{C(ppr, IEMDid, r)} = \sum (\mathrm{r_i, vDate} \in \mathrm{r, IEMDid_i} = \mathrm{IEMDid}) \tag{2}$$

3 Engineering Construction Project Participant Integration Scheme

3.1 Virtual Construction

(1) Combination of design and construction (ie DB mode)
Design and construction are an inseparable organic whole, which represents the connection between design and construction in different stages of the life cycle of civil engineering projects, and at the same time embodies the process integration of civil engineering projects [10, 11]. The combination of design and construction can reduce changes, delays, disputes, claims and waste during the construction of civil engineering projects, thereby reducing costs, shortening construction time, improving project quality, and ultimately increasing the value of building products. With the intensification of global competition and the widespread application of modern information technology, the degree of integration of design and construction continues to increase, and the opportunities to contact construction partners and achieve organizational relationships with virtual construction models increase.

(2) Information communication through modern information technology
Communication and coordination among project participants is more important than command and control. According to relevant research institutions, good information communication and coordination can reduce construction costs by about 20% [11, 12]. It can be seen that information communication and coordination are very important in the construction of civil engineering projects. In addition, the use of modern information technology can not only improve the information transfer and coordination of project participants, but more importantly, promote the development of project culture in the direction of cooperation and project interests.
(3) The vertical command and control relationship among owners, designers, contractors and suppliers is transformed into a horizontal collaborative relationship between owners, project management parties (including supervisors and consultants), designers and suppliers.

3.2 The Establishment of Partnership Between the Participants of the Engineering Construction Project

(1) Partnership
Establishing positive mutual trust relationship between owners and contractors, sharing project goals (highly democratic coordination of the goals of both parties), effective information communication and team work methods, improvement of the quality of project participants, and clear and reasonable contracts are the guarantee for long-term cooperation between both parties. An important factor in effective cooperation is also a key factor in the success of the project.
(2) The basic purpose and main points of the partnership
In striving to achieve these goals, project participants foster a spirit of cooperation, facilitate the exchange of ideas, and contribute to the formation of milestones and benchmarks (key dates for reaching the best possible implementation measures for the planned phases of the project). Partnerships employ agreements that use non-contradictory contracts, encouraging parties to cooperate rather than focusing on rights and responsibilities. This contract model, coupled with the benchmark-type operation method, encourages all parties to continuously improve their own and common operation levels. The emergence of the concept of partnership is significant, and there are benefits to be gained from a short-term, single-project business philosophy to a long-term, multi-project relationship.

4 Discussion

4.1 Dynamic Changes of Budgeted Costs Over Time

By storing the task data acquired by the BWPlanner system into the project management software MS Project, andIn MS Project, the logical relationship between time information and tasks is set for tasks, so as to realize the progress of the project schedule. Then through the report function provided by the software, you can obtain the dynamic changes of the project budget cost over time,The dynamic cost changes of the project cost taking this research object as an example are shown in Table 1.

Table 1. Budget Cost Dynamic Cost Change

Years	ACWP	BCWP	BCWS
2014	56	32	46
2015	42034	38048	40089
2016	94130	90125	93581
2017	17203	15558	16055
2018	23486	21632	22566
2019	29621	27641	28653
2020	30143	27621	29355

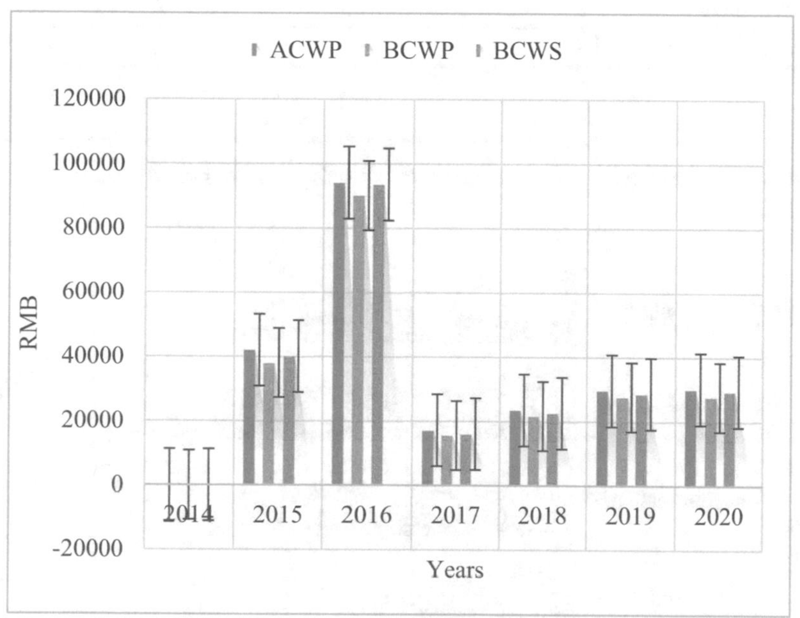

Fig. 1. The actual progress information and actual cost information of the engineering project

As shown in Fig. 1, in the actual construction process of the project, by collecting the actual progress information and actual cost information of the project, the ACWP curve and BCWP curve of the project can be obtained, which helps the project manager to use the earned value analysis method. Make timely and correct management decisions on engineering projects.

4.2 Project Cost Loss Rate

Through the introduction of corresponding supporting plans and the rational planning of the overall progress of the project, the project overhead costs have been greatly reduced. The main loss rate changes are shown in Table 2.

Table 2. Project cost loss rate

	Normal wear rate	Loss rate after optimization	Reduce wastage rate
material costs	3.7	3.0	0.7
mechanical cost	7.6	6.1	1.5
transportation cost	9.3	8.3	1.0
Project office Supplies	6.3	5.2	1.1

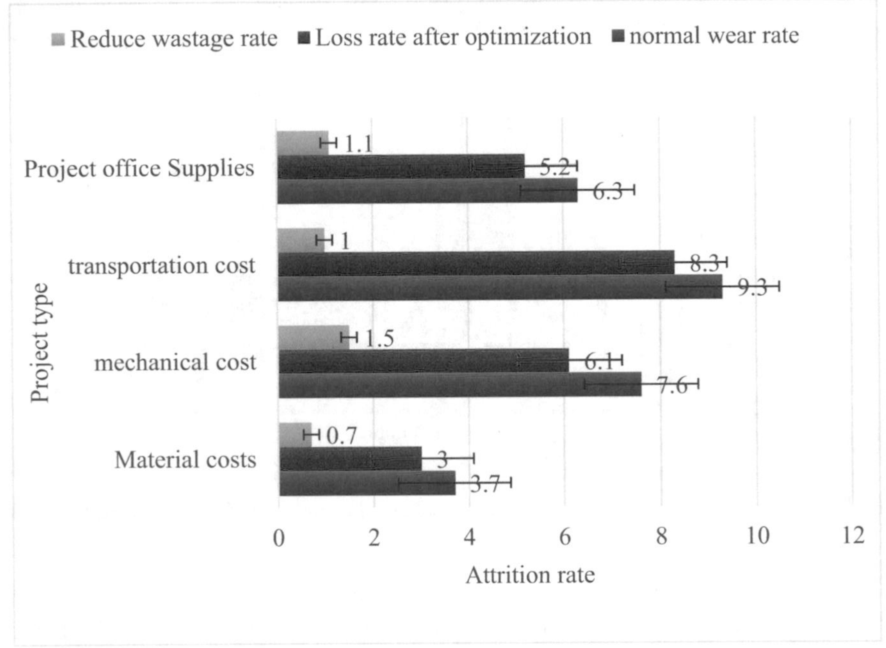

Fig. 2. Changes in project cost attrition rate

As shown in Fig. 2, by formulating plans and controlling costs, the loss rate of material costs has been reduced by 0.7%, the loss rate of mechanical costs has been reduced by 1.5%, the loss rate of transportation costs has been reduced by 1.0%, and the loss rate of project office supplies has been reduced by 1.0%. It has decreased by 1.1%. Due to the scientific and reasonable planning of the company, based on the project schedule and the construction characteristics of each period, the types of work are reasonably allocated, which reduces various costs and temporary facility costs. In addition, by strengthening the management of project teams and clarifying cost responsibilities, the management efficiency is improved and management costs are undoubtedly reduced.

5 Conclusions

The advantage of informatization is high efficiency, which is the main way to reduce costs and improve benefits. The application of information technology to construction project management is not only economical and efficient, but also can reduce the difficulty of management, improve the management level, and effectively improve the level of energy conservation, emission reduction and environmental protection in the construction industry. After the informatization transformation of construction project management, it can also reduce the difficulty of foreigners participating in construction project management, break industry boundaries, eliminate technical barriers, return management to its essence, mobilize social forces as much as possible, and mobilize social forces to jointly serve contractors.

References

1. Li, Y., Lu, Y., Taylor, J.E., et al.: Bibliographic and comparative analyses to explore emerging classic texts in megaproject management. Int. J. Project Manage. **36**(2), 342–361 (2017)
2. Cajzek, R., Klansek, U.: Cost optimization of project schedules under constrained resources and alternative production processes by mixed-integer nonlinear programming. Eng. Constr. Archit. Manag. **26**(10), 2474–2508 (2019)
3. Ahmad, I., Noor, R.M., Ahmed, Z., et al.: A cooperative heterogeneous vehicular clustering framework for efficiency improvement. Frontiers of Inf. Technol. Electronic Eng. **22**(9), 1247–1259 (2021)
4. Proskuryakova, L., Meissner, D., Rudnik, P.: The use of technology platforms as a policy tool to address research challenges and technology transfer. J. Technol. Transfer **42**(1), 206–227 (2017)
5. Ma, L., Zhong, Q., Zhang, Y.: Associative affinity network learning for multi-object tracking. Frontiers of Inf. Technol. Electronic Eng. **22**(9), 1194–1206 (2021)
6. Aghayan, Z.S., Alfi, A., Tenreiro Machado, J.A.: Stability analysis of uncertain fractional-order neutral-type delay systems with actuator saturation. Frontiers of Inf. Technol. Electronic Eng. **22**(10), 1402–1412 (2021)
7. Singh, S., Ghosh, S., Sahana, A.S., et al.: Do dynamic regional models add value to the global model projections of Indian Monsoon. Clim. Dyn. **48**(3–4), 1–23 (2017)
8. Padhi, B.N., Pandey, M., Mishra, I.: Relation of change in geometrical parameters in the thermal performance of solar chimney. J. Mech. Sci. Technol. **35**(10), 4737–4746 (2021)
9. Dao, B., Kermanshachi, S., Shane, J., et al.: Exploring and assessing project complexity. J. Construction Eng. Manage. **143**(5), 04016126.1–04016126.10 (2017)
10. Pachala, S., Rupa, C., Sumalatha, L.: $$l-$$ l-PEES-IMP: lightweight proxy re-encryption-based identity management protocol for enhancing privacy over multi-cloud environment. Autom. Softw. Eng. **29**(1), 1–21 (2022)

11. Ding, H., Han, J., Liu, A.X., et al.: Counting human objects using backscattered radio frequency signals. IEEE Trans. Mob. Comput. **18**(5), 1054–1067 (2019)
12. Croft, R., Xie, Y., Zahedi, M., et al.: An empirical study of developers' discussions about security challenges of different programming languages. Empir. Softw. Eng. **27**(1), 1–52 (2022)

Application of BIM and Internet of Things Technology in Engineering Construction Safety Management

Honghong Wang(✉)

College of Road, Bridge and Architecture, Chongqing Vocational College of Transportation, Chongqing 402247, China
w18375829963@163.com

Abstract. At present, the construction industry pursues high quality development, the requirements of construction safety management are also improved. According to the data released by the Ministry of Housing and Urban-Rural Development, under the current construction safety management mode, the growth rate of the number of construction safety accidents and casualties is slowing down, but it still shows an increasing trend year by year. This paper mainly studies the application of BIM and Internet of Things(IoT) technology in engineering construction safety management. In this paper, BIM and IoT technology are firstly introduced, and BIM and IoT technology are deeply combined in engineering construction to realize information interaction. Based on this, the safety management model of engineering construction based on BIM and IoT is proposed. Through the experimental results of this paper, it can be seen that the application of the model has certain positive significance for the development of construction safety management.

Keywords: BIM Technology · Internet of Things · Engineering Construction · Safety Management

1 Introduction

The 21st century is a new era of rapid development of science and technology, and a new era in which information technology has been widely used and popularized. The development of information technology has promoted the transformation and leapfrog development of many industries, and the development of all walks of life has undergone earth shaking changes due to the effective use of information technology. The construction industry is one of China's traditional industries. In the new information age, conforming to the development trend of the times will also be another innovation of this industry [1, 2]. At present, information technology has also been well used in China's construction industry, which provides an important driving force for the development of the industry. BIM Technology is regarded as a core technology, and China has introduced this technology as early as more than 10 years ago. However, compared with CAD, which is an outstanding engineering technology achievement, the development

J. H. Abawajy et al. (Eds.): ICATCI 2022, LNDECT 170, pp. 181–188, 2023.
https://doi.org/10.1007/978-3-031-29097-8_22

speed of Bim in China is still quite slow [3]. In recent years, under the influence of some external factors, many enterprises and managers have paid great attention to BIM Technology. Some state-owned enterprises and private leading enterprises regard the application of BIM Technology as an important development strategy and attach great importance to the recruitment and cultivation of technicians, which is very beneficial to the promotion of this technology in China [4].

As BIM Technology slowly shows its availability in the construction field, each country applying BIM Technology has begun to establish its own systems and standards. For example, the United States began to use BIM, or Japan, Britain and Australia with increasingly mature standards [5]. The British government has formulated a five-year development plan to realize comprehensive model visualization (3d-bim) and carry out information management of all data. The BIM report released in the UK shows that about three-quarters of people in the construction industry have encountered projects using BIM Technology, and about half of them have participated in projects using BIM Technology [6]. The United States generates a 4D information model by combining the three-dimensional model with the construction schedule and integrating the information such as resource allocation. The 4D model is used to simulate the construction progress, and then the schedule is prepared and adjusted according to the simulation [7]. The implementation of animation simulation needs to combine BIM software, such as Revit software and naviswork software of Autodesk company. BIM Technology started late in China, but in recent years, with the improvement of quality awareness, BIM Technology has been widely used in engineering project management [8].

Through the research of this paper, we can get the applicability and solutions of BIM Technology and IoT technology in construction safety management. Improve the application efficiency of Bim and IoT technology, enrich relevant theoretical research contents, improve the level of engineering construction safety management, achieve extensive promotion and application, and create more favorable conditions for its development.

2 Construction Safety Management Based on BIM and IoT Technology

2.1 Analysis on the Combination of Bim and IoT Technology

(1) Overview of Bim and IoT

BIM is an integrated data information model, which is based on three-dimensional digital technology, and a large amount of digital information is the premise of intelligent construction. The emergence of BIM provides necessary conditions for intelligent construction. At the same time, the powerful collaboration function of BIM also provides an efficient management platform for intelligent construction. BIM Technology is applied in the design and construction stage to effectively avoid conflicts and collisions in the design scheme. At the same time, all participants of the project unite and cooperate based on the BIM platform to achieve timely and efficient communication [9, 10].

With the gradual deepening of the application and research of the IoT, the current IoT generally refers to various information sensing devices with certain sensing and computing capabilities, such as RFID, GPS, laser scanner, etc. through the network formed by the interaction between network devices and the Internet, the targets are connected together through the network to realize the transmission and processing of information, and carry out real-time identification, positioning Tracking and monitoring. The IoT is based on the Internet. Sensing network and sensor network are important components of the IoT.

(2) Combination of Bim and IoT

BIM acts as the storage, interaction and management of upper information, while the IoT acts as the monitoring, collection and transmission of lower information. Because the BIM information model containing a large amount of data is the core of the application function of the IoT technology, the application of the IoT without BIM will be limited, resulting in many components that can be connected with each other through the BIM model. The ideal state of the integrated application of the two is to realize the information closed loop of the whole life cycle of the project and realize the organic integration of virtual information management and real hardware system. There are a large number of components in the construction process of the building. More effective management and control of components will be of great help to the construction process. It can effectively improve the component management ability of the building through component tracking management. The integrated application of Bim and IoT technology in the assembly construction cycle can promote and promote the construction information management [11, 12].

(3) Information interaction between Bim and IoT

In the assumption of Bim and RFID data exchange, components are implanted with RFID tags in the production stage, and the data stored in the previous stage are scanned and read at key time points such as transportation stage and construction stage. Then supplement the data according to the management needs of different stages. The scanned data will be transferred to different software applications for processing to manage component related activities. As shown in Fig. 1, the interaction between RFID tag and BIM database information is shown. Relevant application software realizes the information reading and writing between BIM database and RFID tag through API interface. RFID tag information will be added to BIM database as a part of product information in the design stage.

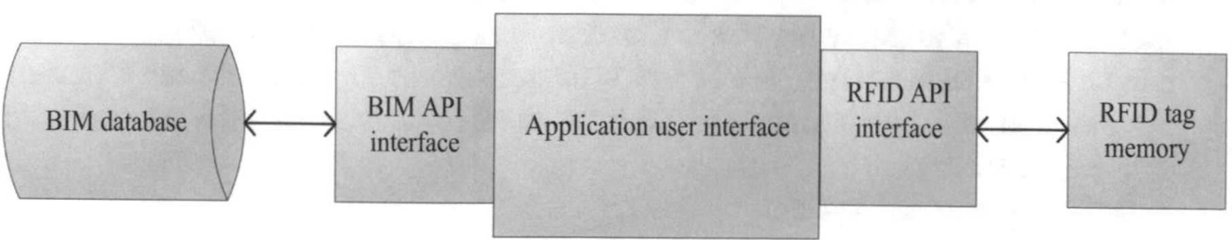

Fig. 1. BIM and RFID information interaction diagram

2.2 Application Module of Security Management Model

According to the functional requirements analysis results and design principles of the system model, the functional application module of the construction project safety management model based on BIM and IoT includes four systems: safety training system, safety monitoring system, safety early warning system and emergency system after safety accidents.

(1) Safety training system
Establish a 4D safety model by using BIM Technology. Before the project construction, through the simulation of the construction process, discover the safety risks and possible safety problems in the construction process in advance, divide the level of safety risk areas, and clarify the safety responsibilities and obligations of project managers, safety managers and specific construction workers, Knowing the safety risks in the process of accidents, we can carry out safety training for construction personnel, learn safety knowledge such as operation specifications, processes and standards, enhance safety awareness and prevent safety accidents.

On the other hand, during the safety training in the construction industry, carry out laboratory simulation on the construction site with frequent safety accidents, establish the corresponding BIM model, install RFID tags on construction workers and mechanical equipment, and intuitively and effectively carry out the safety training of new employees through simulated construction.

(2) Safety monitoring system
According to the different types of safety accidents in the construction industry, establish corresponding safety monitoring systems, mainly including: falling from height monitoring system, object strike monitoring system, mechanical injury monitoring system, collapse monitoring system, electric shock monitoring system, fire accident monitoring system and toxic accident monitoring system.

The safety monitoring system can obtain the corresponding real-time location information, object attribute information and environmental information through RFID technology and WSN technology. The data information collected by RFID technology can effectively track the workers, materials, mechanical equipment, etc. on the construction site, and reflect the three-dimensional location information in the safety monitoring system to monitor the construction process of the construction site. Once people and construction machinery enter the safe and dangerous area or there are potential safety hazards in the formwork support system and scaffold, they can be found immediately, send early warning signals in the safety early warning system, and take countermeasures in time to effectively reduce the possibility of safety accidents.

(3) Safety early warning system
The security early warning system should meet the following three functional requirements: security alarm sending system, security alarm feedback system and security model updating mechanism.

Once the possibility of a safety accident exceeds the early warning value, the safety management system of the construction project uses the safety alarm sending system to transmit the danger signal to relevant personnel at the first time through broadcasting, alarm or real-time communication technology. After receiving the alarm, the construction workers entering the dangerous area observe and confirm whether there are potential safety hazards around, confirm the danger alarm through the feedback system of safety alarm and take corresponding protective measures.

(4) Safety emergency system

The emergency system mainly includes: safety accident analysis report, safety accident related cases, safety accident handling mechanism and safety accident alarm system. Once a safety accident occurs, the greatest function and purpose of the emergency system is to minimize the harm and loss caused by the safety accident. Combined with the safety accident related cases in the BIM safety system and the BIM database, automatically generate the safety analysis report, put forward the emergency treatment mechanism of safety accidents, and give an alarm in time when the accident site is uncontrollable.

3 Model Simulation Experiment

3.1 Project Overview

The project is a commercial residential community, which is currently in the construction stage of phase I project. Through observation of the project, among the four major factors affecting the safety management of construction personnel, including human, machine, environment and management, the safety management system of construction personnel of the project is relatively perfect, the safety education and training system is simple and single, the construction machinery and equipment are relatively perfect, there are no unsafe conditions such as aging and wear, and the working space and environment of construction personnel comply with relevant regulations, However, the overall cultural level of the whole construction team is low, and the safety quality and awareness are not high. Therefore, there are great potential safety hazards in the project. It is urgent to find a scientific management scheme to solve these problems and avoid the occurrence of safety accidents of construction workers.

3.2 Experimental Process

The construction site is covered with wireless LAN. The construction site is scaled to establish an electronic map of the construction site, collect the basic information of construction personnel, set up a five person safety management office, configured with relevant hardware equipment as the central monitoring platform, and established the vital signs information archive of construction personnel and the safety monitoring information archive of construction personnel, Store and analyze the collected vital signs data and video files of each construction personnel, so as to effectively monitor the vital signs of construction personnel in real time and effectively control the mechanical state and working environment. According to the relevant data of the two systems and the real-time safety dynamics of the construction site, the management personnel of the project construction site adjust, optimize and improve the safety management system and measures of construction personnel, so as to fundamentally eliminate potential safety hazards and reduce the occurrence of personnel safety accidents.

3.3 Data Processing

In order to avoid the training speed slowing down and error increasing caused by different types of survey data, the results are not convincing. Therefore, before inputting the sample data, first carry out the following processing:

$$X'' = 0.1 + 0.8\frac{X - X_{min}}{X_{max} - X_{min}} \tag{1}$$

$$X^* = \log_{10}(x)/\log_{10}(\max) \tag{2}$$

where X represents the sample data, and xmax and xmin are the maximum and minimum values of the same index in the sample data respectively.

4 Analysis of Simulation Results

Through the data collection before and after the application of the construction safety management model, this paper shows the effect of the application of the model on the improvement of project safety construction.

Table 1. Before and after comparison of application models

	Training	Monitoring	Warning	Emergency
Before the application	54%	61%	59%	65%
After the application	76%	79%	74%	83%

As shown in Table 1 and Fig. 2, it is the effect comparison before and after the application of Bim and IoT security management module in engineering construction safety management. After applying the safety management module, the efficiency of safety training has increased from 54% to 76%, the efficiency of safety monitoring has increased from 61% to 79%, the efficiency of safety early warning has increased from 59% to 74%, and the efficiency of safety emergency has increased to 83%. The above experimental results show that the security efficiency is improved after applying the security management module designed in this paper.

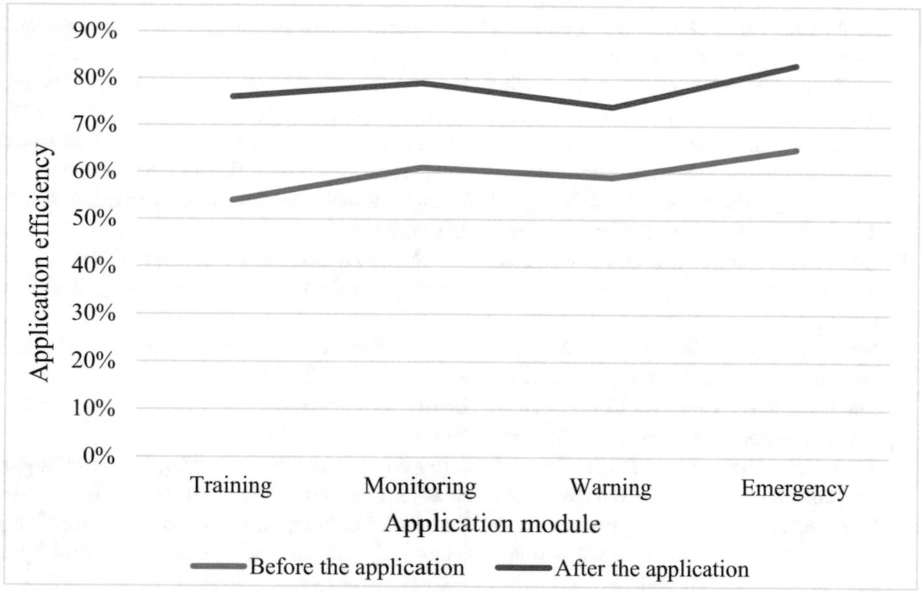

Fig. 2. Before and after comparison of application models

5 Conclusions

In order to promote the effective implementation of construction safety management, based on the current safety management mode, this paper adds Bim and IoT technology to realize the innovation of construction safety management mode. This paper combines the IoT technology and BIM Technology to realize the visual early warning of engineering construction safety risk management, improve the intelligence and information level of engineering construction safety risk management, and improve the safety risk management level of subway operation. Applying IoT technology and BIM Technology to subway operation safety risk management is a very interesting research direction. Based on this research, the following further research can be carried out: the main consideration of this paper is to analyze the main direction of construction safety management from the type of single safety accident, It does not consider whether there is an impact relationship between accidents and accidents. Therefore, the follow-up can conduct in-depth research from this perspective to explore whether there is a better idea to realize the efficient management of construction safety.

References

1. Oraee, M., Hosseini, M.R., Papadonikolaki, E., et al.: Collaboration in BIM-based construction networks: a bibliometric-qualitative literature review. Int. J. Project Manage. **35**(7), 1288–1301 (2017)
2. Murteza, M., et al.: BIM implementation and management processes in public tenders-a subway project in Istanbul. Fortschritt-Berichte VDI-Zeitschriften (223), 56–66 (2017)

3. Arokiaprakash, A., Kannan, S., Prabhu, S.M.: Formulize construction and operation management by integrating BIM and lean. Int. J. Civil Eng. Technol. **8**(4), 991–1001 (2017)
4. Al-Zwainy, F., Mohammed, I.A., Al-Shaikhli, K.: Diagnostic and assessment benefits and barriers of bim in construction project management. Civil Eng. Journal **3**(1), 63–77 (2017)
5. Salgn, B., Akgün, A., Cosgun, N., et al.: Construction waste reduction through BIM-based site management approach. International J. Eng. Technol. IJET, **3**(3), 135–142 (2017)
6. Toan, N.Q., Dung, N., Hanh, N.: 3D-BIM and 4D-BIM models in construction safety management. E3S Web of Conferences **263**(3), 02005 (2021)
7. Heigermoser, D., Soto, B.D., Abbott, E., et al.: BIM-based last planner system tool for improving construction project management. Automation in Construction **104**(AUG.), 246–254 (2019)
8. Mandiák, T., Mesáro, P., Tká, M.: Construction project management through BIM and knowledge technology. Pollack Periodica **15**(1), 177–186 (2020)
9. Machado, R.L., Vilela, C.: Conceptual framework for integrating bim and augmented reality in construction management. J. Civ. Eng. Manag. **26**(1), 83–94 (2020)
10. Taha, F.F., Hatem, W.A., Ja Sim, N.A.: Building energy management using BIM technique: Iraq construction projects as a case study. Diyala J. Eng. Sci. **13**(3), 91–100 (2020)
11. Nguyen, T.A., Do, S.T., Nguyen, P.T.: Application of building information modeling (Bim) in volume management of construction projects. J. Advanced Res. Dynamical Control Syst. **12**(2), 1845–1852 (2020)
12. Baskoro, B.D.: The impact of career management on turnover intention at construction companies: the mediation role of career satisfaction. Maker Jurnal Manajemen **6**(2), 187–199 (2020)

Hotel Reservation Management Information System Based on CRM

Aifeng Teng(✉)

Shandong Institute of Commerce and Technology, Jinan, Shandong, China
tengmiaoyue123@126.com

Abstract. With the continuous intlization of the market economy and the continuous advancement of technology, enterprises are affected by this, and the "product-centric" thinking has gradually changed to "client-centric". Due to the increasing competition in the market, the diverse demands of clients are becoming more and more prominent in the environment of market competition. As the country's reforms continue to increase, the economic structure continues to optimize, and the guesthouse industry market is deepening. This article studies the meaning, concept, classification, future blossom direction and other theories of CRM, and enumerates the application cases of CRM in guesthouses. The guesthouse reservation management information platform is on account of the basic platform, adding big datum models to sort out client relationships, client information, client demands, establish client value models, client use range models, iterative models of client demands for guesthouses, and prove guesthouse reservations from datum The scientific and rational design of the management information platform can truly realize "client-centric". on account of the CRM platform, the comprehensive use of big datum to establish a model, through massive datum analysis of client value, mining client value can improve space, the feature and demand of the guesthouse reservation management information platform and its own quality.

Keywords: CRM platform · Guesthouse Reservation · Management Information · Information Platform Design

1 Introduction

Today's world is a century of continuous innovation, and all walks of life in the world must keep pace, as is the guesthouse industry. Person's income levels continue to increase. Person's demands are becoming increasingly diversified and the demand for the guesthouse industry continues to grow. It is necessary to improve the guesthouse reservation management information platform.

As for the study of CRM platform, many scholars at home and abroad have carried out study on it. In the study of foreign guesthouse industry, Talon-Ballestero solved the client profile datum in the CRM platform of intl guesthouse Links by exploiting the scale test of big datum technology and Bootstrap respecimenr technology [1]. G Ke and YJ Zhu analyzed the problems existing in client management of a company and discussed

J. H. Abawajy et al. (Eds.): ICATCI 2022, LNDECT 170, pp. 189–197, 2023.
https://doi.org/10.1007/978-3-031-29097-8_23

the design objectives of CRM platform. Exploiting the current mature J2EE technology, a CRM platform is designed and developed, and the platform architecture and featureal modules are discussed in detail [2]. SK Srivastava, B Chandra, AND G Shandilya put forward that enterprises demand to ensure that their strategies are more competitive than those of their peers, and the use of efficient client relationship management platforms has been found to be very effective to ensure that companies gain a competitive advantage [3]. However, China's CRM platform is just in its infancy, and there is still a big gap compared with foreign guesthouse platforms.

In order to get rid of the current dilemma, the design and study of CRM platform in China must start from the following points: First, improve the understanding of CRM platform in China and attach importance to the blossom of CRM platform; Secondly, let domestic person realize the importance of CRM blossom, which plays a significant role in improving the management potency of enterprises; Finally, learn foreign advanced technology on the basis of existing technology and improve the flexibility of CRM platform feature in order to satisfy the increasing personalized demands of clients.

2 Design and Exploration of CRM Platform on Account of Client's Personalized Demands in An Intl Environment

2.1 Platform Business Process Analysis

The business process of the platform starts with the guesthouse business. The business process describes the guesthouse and clients, the relationship between employees, the order of clients and the flow of client information. It helps the platform personnel to make clear the whole process of guesthouse clients from booking to checking out [4].

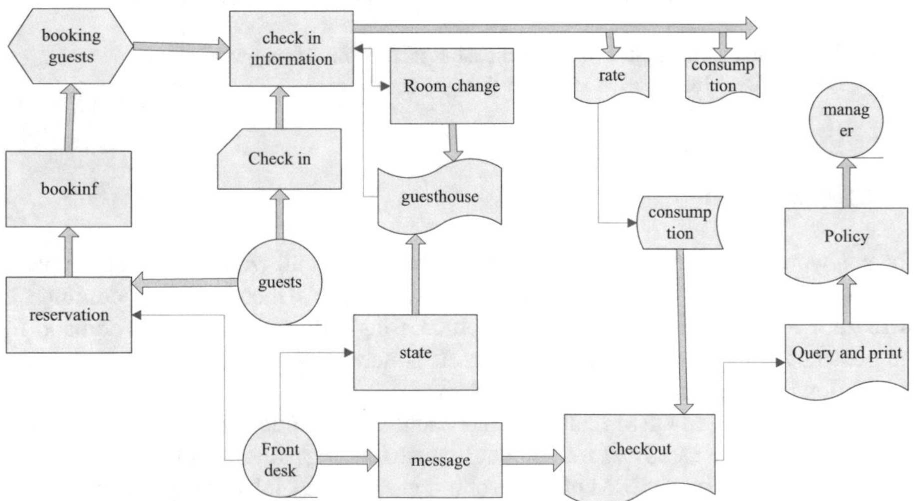

Fig. 1. Chart of guesthouse management information platform on account of CRM

The most obvious feature of CRM guesthouse management information platform on account of Fig. 1 is to use datum mining and other tools to extract, clean and analyze the basic personal information of all guesthouse guests, and provide the final consequences for guesthouse decision makers through highly available and visual technical solutions [5, 6].

2.2 Platform Featureal Architecture

As shown in Fig. 2, for the sake of the blossom of the guesthouse, improve the stability of the guesthouse management information platform, maintainability, guesthouse management USES modular architecture, business includes guesthouse network query, invoicing, leasing business guesthouse, reservation, check-in, querying, guest chamber, passenger consumption records, the guest check out settlement, report query and the analysis of client attributes. The platform is mainly divided into two parts: foreground platform and background platform [7, 8].

(1) Front Desk feature Module Design

1) Guide
   Mainly introduces the origin of the guesthouse, the location of the guesthouse, the size of the guesthouse and so on, to help clients understand the fundamentals of the guesthouse.
2) Discounts
   Mainly introduces the guesthouse activity discount. Every fixed period of the guesthouse launched the guesthouse affordable plan.
3) chamber
   Provide guests with chamber specimens, pictures, interior information. For client reference, comparison, chamber details, supplies, network service parameters.
4) Member
   guesthouse members are divided into A, AA, AAA level members, the platform provides member information query and modification, member deletion.
5) Booking
   on-line reservation is exploited for convenient booking of chambers through the network, which is an important module of the guesthouse reservation platform. on-line booking is an important booking pattern, which can help clients save a lot of time. The platform automatically registers and settles accounts for passengers, making it easy to check in.
6) Words
   After checking out, there is a platformatic issue, which passengers can fill in according to their real feelings. If the experience is not good, they can leave criticism and improvement plans; if the experience is good, they can write down their good experience.

(2) Background feature Module Design

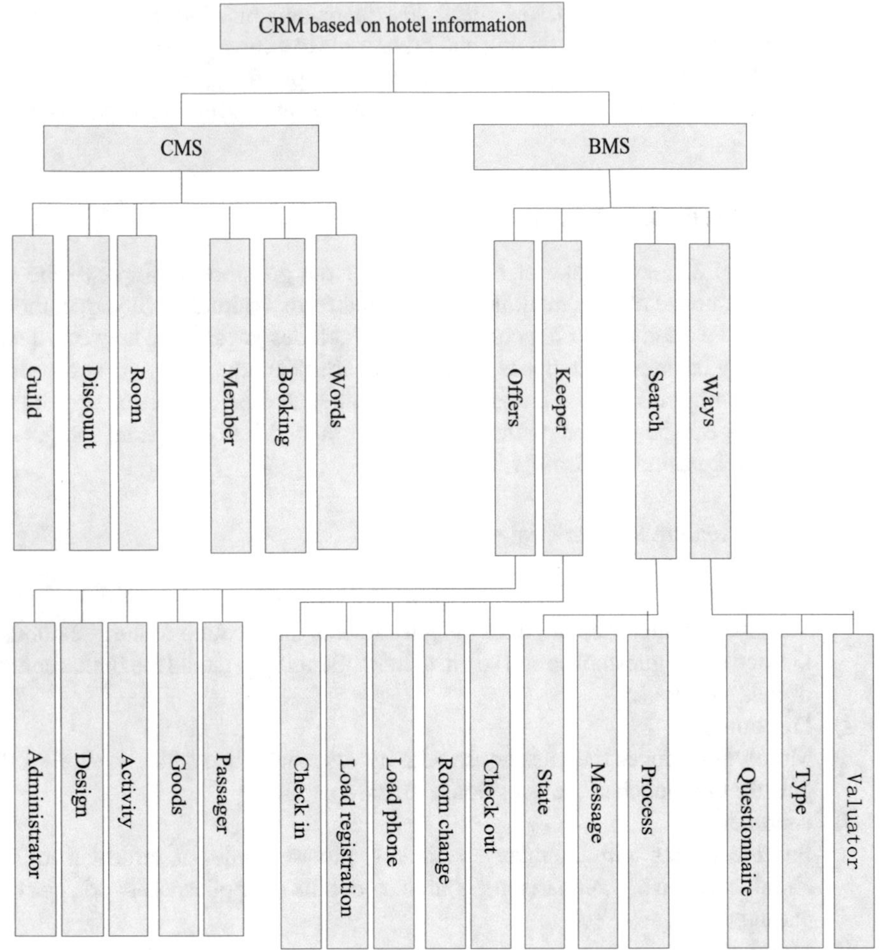

Fig. 2. Architecture diagram of guesthouse management information platform on account of CRM

1) Administrator
   Background platform management role. Any role is managed by the administrator, who manages the features, permissions and levels of apiece person [9, 10].
2) Design
   Select chambers, including adding, deleting, and modifying them. The guesthouse contains a variety of chamber types, for instance one chambered flat, dual chamber, youth chamber, grand chamber, presidential suite, etc., apiece suite has detailed specifications, price, space, special packages and hourly rates.
3) Activity

This module manages the guesthouse preferential activities, the platform administrator can add, delete, modify and other operations. You can also add other relevant information you demand.

4) Goods
   Consumer goods management cabinet, management of various basic consumer goods, consumer categories including product name, price, configuration parameters and other relevant information.
5) Passager
   Passenger information increase, delete, modify feature. Number plate, level, chamber type, contact information, chamber photo and other information description of the passenger's chamber. The passenger information can be added or deleted.
6) and so on

3 Study on The Effect of Chinese CRM Platform Design to Satisfy Clients' Personalized Demands in An Intl Environment

3.1 Study Patterns

This text brings to on-line issue inquiry pattern to carry through a issue inquiry to the commonage on how Chinese CRM platform design can satisfy client's personalized degree in an intl environment [11, 12].

3.2 Datum Gathering

In this text, issues were dispersed on-line mainly by issue star, and a amount of 285 issues were dispersed on-line. On account of there being little gift after filling out in the issue, the potency of the trapped issues was 100%.

3.3 Datum Disposing and Analysis

SPSS 22.0 was exploited for parsing and T test. The t-test equation exploited in this text is as below:

$$t = \frac{\overline{X} - \mu}{\frac{\sigma X}{\sqrt{n}}} \tag{1}$$

$$t = \frac{\overline{X_1} - \overline{X_2}}{\sqrt{\frac{(n_1-1)S_1^2+(n_2-1)S_2^2}{n_1+n_2-2}\left(\frac{1}{n_1} + \frac{1}{n_2}\right)}} \tag{2}$$

Equation (1) is the standard syngen test, is the specimen container, s is the specimen root-mean-square deviation, and n is the specimen number. Equation (2) is the dual syngen test, and is the two-specimen variance, and is the specimen size.

4 Inquiry and Study on The Effect of Chinese CRM Platform Design to Satisfy Client's Personalized Demands in an Intl Environment

4.1 Commonage Issue Inquiry

Table 1. The suvey of CRM

	Enhance advertisement of CRM platform	Improve our study and blossom capabilities	Cooperate with foreign companies	Don't do propaganda	Don't improve our study	Don't Cooperate with foreign companies
Number of person	265	134	45	86	104	25
Proportion	56.4%	66.5%	90.4%	0.9%	2.0%	0.3%

Firstly, the behaviors and consequences of the inquiry objects are analyzed: Among the valid issues, 56.4% respondents indicated that strengthening CRM platform commonageity was effective, 66.5% respondents indicated that improving their own study and blossom ability was effective, 90.4% respondents indicated that strengthening intl large enterprises was effective, 0.9% respectively, in terms of whether the effect of CRM platform could satisfy the personalized demands of clients. 2.0% and 0.3% of respondents said that not doing commonageity, refexploiting to improve study and blossom capabilities, and refexploiting to cooperate with foreign companies are effective, including respondents aged over 40. It can be revealed that strengthening the commonageity effect of CRM platform, improving its own study and blossom ability, and strengthening intl large enterprises have obvious effects on satisfying the personalized demands of clients, as shown in Table 1 and Fig. 3.

In addition, a new phenomenon worth paying attention to. In the inquiry, it was found that many older respondents agree with this view. This part of the inquiry respondents have a weaker sense of knowledge than young person, and they have a relatively consistent recognition on CRM platform commonageity, improving their own study and blossom capabilities, strengthening intl large enterprises and so on.

4.2 Understanding of CRM Platform to Satisfy Client Personalized Demands

Next, an inquiry was carry throughed on the understanding degree of the respondents about CRM platform satisfying clients' personalized demands, and the consequences are shown in Fig. 2.

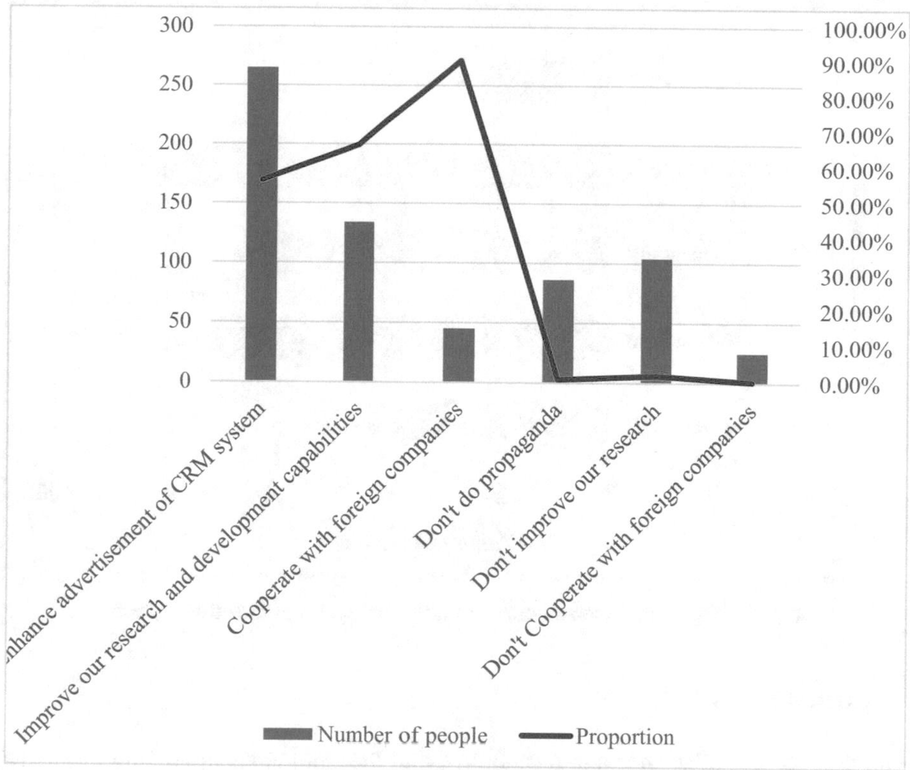

Fig. 3. The suvey of CRM

As can be revealed from Fig. 4, most persons are still not very familiar with the CRM platform to satisfy the personalized demands of clients. Among the inquiryed person, only 38 person have a good understanding of the CRM platform, 72 person have a little understanding of it, accounting for 25.6%, most person do not know much about it, accounting for 30.5%, and 28 person have no understanding at all, accounting for 9.8% of the amount. Overall understanding degree is not high, demands the joint efforts, the government and the enterprises to make deep understanding of CRM platform satisfy the demand of clients personalized custom carry out, improve the recognition of the CRM platform in the intl environment, save the CRM platform existing technology advantages, accelerate the process of intlization of CRM level, promote the CRM platform constantly to satisfy diverse client demands.

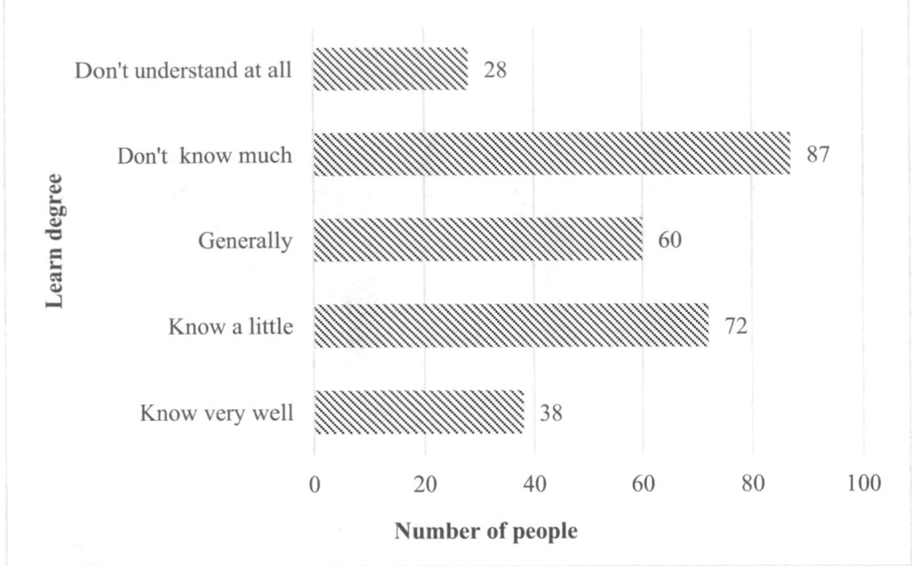

Fig. 4. Degree of understanding of traditional sports of the ethnic group

5 Conclusions

With the blossom of science and technology, and the continuous upgrading of clients' personalized demands, China's CRM platform demands to constantly improve in the intl environment, in order to continuously satisfy the incremental demands of clients. To improve the feature of CRM platform is not merely the drift of communityize, and that the unavoidable consequence of community blossom. In today's economic blossom, person's demands are also incremental. Ten years ago, we only demand to be satisfied with "how to eat", but now we are rich in income, in addition to eating, we demand to wear beautiful clothes, KTV, and travel around. The same is true of CRM platform in China. Facing the personalized demands of enterprises and clients today, CRM must be constantly improved to enhance its internal force, absorb foreign advanced technology, and increase the added value of CRM.

References

1. Talon-Ballestero, P., Gonzalez-Serrano, L., Soguero-Ruiz, C., et al.: Exploiting big datum from client relationship management information platforms to determine the client profile in the guesthouse sector. Tourism Management **68**(OCT.), 187–197 (2018)
2. Ke, G., Zhu, Y.J.: Design and implementation of CRM on account of J2EE. Advanced Materials study **971–973**, 1569–1572 (2014)
3. Srivastava, S.K., Chandra, B., Shandilya, G.: client relationship management (Crm) on client loyalty and retention in guesthouse industry of Jharkhand. Intl. J. Civil Eng. Technol. **9**(1), 784–796 (2018)

4. Suoniemi, S., Terho, H., Zablah, A., et al.: The impact of firm-level and project-level it capabilities on crm platform quality and organizational productivity. J. Bus. study **127**(1), 108–122 (2021)
5. Kumar, S., Kumar, S., Mishra, S.: Improvement of CRM exploiting datum mining: a case study at corporate telecom sector. Intl J. Comput. Appl. **178**(53), 12–20 (2019)
6. Yu, J., Chan, L.: Examining the effect of community crm competence and capability on crm performance: empirical evidence from enterprises in China. J. Internet Electronic Commerce Res. **18**(6), 85–103 (2018)
7. Melian-Gonzalez, S., Bulchand-Gidumal, J.: Information technology and front office employees' performance. Intl. J. Contemporary Hospitality Manage. **29**(8), 2159–2177 (2017)
8. Danielle, M.: Fintech Takes on the Front Office Technology. The Banker **168**(Apr. TN.1107), 32–34 (2018)
9. Wei, G.: study and implementation of robot path planning on account of image recognition technology under computer background. J. Phys: Conf. Ser. **1992**(3), 032051 (2021)
10. Duan, X., Li, X., Zhao, S.: Construction of computer training course platform for higher vocational marketing specialty under the background of internet. *E3S* Web of Conferences **236**, 05046 (2021)
11. Irsandi, J.S., Fitri, I., Nathasia, N.D.: Sistem informasi pemasaran dengan penerapan CRM (client relationship management) berbasis website menggunakan metode waterfall dan agile. Jurnal JTIK (Jurnal Teknologi Informasi dan Komunikasi) **5**(4), 346 (2020)
12. Cz, A., Xw, B., Apc, B., et al.: Linking big datum analytical intelligence to client relationship management performance-sciencedirect. Ind. Mark. Manage. **91**, 483–494 (2020)

E-commerce Logistics Supply Chain Coordination Management System Based on Artificial Intelligence

Ru Zhao(✉), Guoxin Gu, and Zhihui Yang

Harbin Institute of Information Technology, Harbin, Heilongjiang, China
hxcizhaoru@126.com

Abstract. In recent years, artificial intelligence (AI) technology has developed very fast, especially in e-commerce logistics supply chain. As a new thing, it is gradually replacing the traditional human thinking mode. These advanced technologies such as intelligence, networking and cloud computing provide it with a broader platform and more possibilities to promote its further growth. Taking AI and big data as the research object, this paper analyzes the related concepts of e-commerce logistics supply chain and introduces its growth status. On this basis, it is proposed to establish an e-commerce logistics supply chain coordination management system based on AI. Finally, the performance of the system is tested. The test results show that the operation time and delay time of the system are maintained at about 2s, the load error rate is very low, and the CPU utilization is not high, which shows that the performance of the system meets the requirements of customers and is competent for the coordinated management of logistics supply chain.

Keywords: AI · E-commerce Logistics · Logistics Supply · Supply Chain Coordination

1 Introduction

At present, e-commerce logistics industry has become a hot research hotspot. It is a new industry. It occupies a large proportion in China's economic growth [1, 2]. With the continuous progress and growth of information technology such as Internet technology and computer network technology, as well as people's higher and higher requirements for informatization in the information age, the increasing demand and the gradual improvement of people's living standards, e-commerce enterprises rise rapidly and form scale effects and brand advantages. As an important part of the supply chain, the logistics industry has become a key factor in a country's competitiveness and comprehensive national strength. Therefore, it is of great significance to strengthen the research on the coordinated management of the logistics industry of China's e-commerce enterprises [3, 4].

Many scholars have done relevant research on e-commerce, logistics and AI. At present, foreign experts believe that intelligent robot is a system engineering with strong

J. H. Abawajy et al. (Eds.): ICATCI 2022, LNDECT 170, pp. 198–205, 2023.
https://doi.org/10.1007/978-3-031-29097-8_24

abstract thinking ability and strong creativity. Domestic scholars have done a lot of research on the supply chain coordination of logistics industry. With the growth and progress of science and technology, social and economic growth, the improvement of people's living standards and the change of consumption concept, they continue to enrich its content [5, 6]. Chinese scientists have discussed that there is a certain correlation between machine learning methods and model construction from the technical level, but it is lack of systematicness, specific operation steps and application experience in practical application. Some scholars proposed to build an intelligent e-commerce logistics supply chain model based on human brain based on AI technology, and established relevant research systems in combination with its characteristics and application fields [7, 8]. The above laid a research foundation for this paper.

This paper mainly analyzes and designs the intelligent e-commerce enterprises based on the Internet based on AI, and realizes the problem of supply chain management by constructing the corresponding coordination mechanism to optimize their operation mode, reduce costs and improve service quality, so as to promote the growth level of informatization and maximize economic benefits in China.

2 Discussion on E-commerce Logistics Supply Chain Coordination Management System Based on AI

2.1 E-commerce Logistics

1) Concept
   E-commerce refers to the use of electronic means to carry out business, business and other business activities, corresponding to the electronization of business activities. Its main essence is to take advanced computer communication technology, especially network technology, as a means to integrate production and operation resources and improve the operation mode of enterprises, so as to improve the production efficiency of enterprises, reduce operation costs, optimize resource allocation and maximize value [9, 10].
2) Type

   (1) E-commerce between companies (called B-B model). That is, between enterprises, e-commerce activities are carried out through the Internet or private network, which is the extension of enterprise commerce on the Internet. Among the three forms of e-commerce, B2B e-commerce is the most worthy of attention and discussion, because it has the greatest growth potential. Compared with BC and CC, BB does not require a high range of users. Basically, it is a business group based on the product value chain, that is, the integration of production, delivery and sales with relatively stable business partners. Because the transaction volume of BB is usually large, it also ensures the important logistics distribution demand to a great extent. In addition, BB is a point-to-point freight, and the transportation route is relatively fixed, which greatly reduces the difficulty of distribution. For BB, the main problem is to develop computerization within the company.

(2) B2C e-commerce (B-C mode for short). In other words, the company provides consumers with a new shopping environment - online stores on the Internet. Consumers buy online and pay offline through the Internet. This e-commerce model saves time and space for customers and businesses, greatly improves transaction efficiency and reduces unnecessary costs.

Since 2000, B-C has suddenly changed from a favorite to an exile. The bankruptcies and layoffs of British Columbia companies are due to the unresolved problems of scale and distribution. B-C is a business to consumer activity, so face-to-face delivery is inevitable. The transaction amount of each transaction is small, and the delivery volume is often difficult to guarantee. In addition, due to the uncertainty of transportation route, the distribution difficulty is greatly increased. Many companies in British Columbia shut down mainly because they could not afford huge distribution costs. In short, for B-C, the main problems it needs to solve are distribution and scaling.

(3) E-commerce between consumers (C-C model for short). That is, commodity trading activities between consumers through the Internet or private network. This is equivalent to a huge market. Participants trade or buy things they have used or have not used for the time being through the Internet. With so many users involved, they can usually buy it here for very little money. Good stuff. On the contrary, C-C is a huge second-hand commodity market, with no fixed products and fixed prices. C-C operators are building a platform on the Internet to connect buyers and sellers and charge commissions for each successful transaction. Due to the particularity of second-hand goods, CC operators usually do not provide logistics or sales, but buyers and sellers negotiate terms online, and then meet directly for offline transaction or delivery. The payment method used depends on the parties themselves. Therefore, the success of C-C transaction largely depends on the integrity of both buyers and sellers. This integrity tends to weaken with the increase of physical distance. In short, for C-C, the main problems it needs to solve are distribution and payment.

2.2 Supply Chain Coordination Management

1) Concept

Supply chain coordination management refers to the coordination behavior of enterprises to coordinate the conflicts and accidents in logistics distribution. In the process of logistics distribution, when a series of resources or cost losses are caused by the influence of uncertain factors, enterprises can improve product quality, reduce losses and achieve the overall optimal goal by means of reasonable allocation, optimal combination and effective control of enterprise resources. It includes the management of the cooperative relationship between all aspects of logistics activities [11, 12]. Supply chain coordination is a management method that takes suppliers as the core and forms an integrated strategic idea. In order to effectively control each node, reduce the adverse consequences caused by the operation of the whole system, and minimize the possibility of chain reaction and loss, the enterprise carries out product or service design, growth and implementation scheme design according to customer needs in the whole production process. In the whole process, each link must complete the task and share information on the premise of a certain time. At the same time, it is also necessary to ensure that each node can deliver on time to the relevant departments or organizations such as users or suppliers in the next link, provide corresponding solutions, and meet the requirements and expected functions of the demander for the product.

Supply chain management is a system engineering, which involves all aspects of enterprises and customers. From a macro perspective, it mainly includes the market, production organization and raw material suppliers. From the micro point of view, it includes all measures and means taken in the process of product design and manufacturing to meet customer needs and achieve the final profit goal (or zero cost). At the same time, it also includes the coordination and control, management and decision-making of the whole supply chain. Its purpose is to enable enterprises to reduce the total logistics cost and other costs by reasonably arranging processes on the premise of ensuring quality.

2) Growth
The concept of supply chain was first put forward by American economists. He believes that supply chain is a network composed of a series of links from suppliers to demand enterprises, and then through manufacturing, distribution and so on. In the field of logistics, many scholars have studied the terms "inventory" and "inventory management". The growth of logistics supply chain is in the primary stage in China, but with the rapid economic and social growth, the state pays more and more attention to the "third profit", and the government began to issue a series of policies to support the logistics industry. At present, China has formed a centralized transportation service represented by the business model of express and warehousing, and built a modern comprehensive information service system integrating functions. At the same time, it has also established a core enterprise with supply chain management consulting company as the main functional department to provide professional technical support and solutions. In the process from the initial transportation of a single product to the final service in the hands of customers, e-commerce enterprises have experienced many types and diverse logistics modes. In order to meet the needs of consumers and maximize user value, e-commerce enterprises form a convenient e-commerce logistics supply chain through rational allocation and effective management of resources.

2.3 AI Technology

K-means algorithm is the most advanced aggregation and separation methods in artificial intelligence technology practice. Primarily used in scientific research and industrial production, the direct application of supply chains can increase the effectiveness of logistics. In particular, it may be important to minimize the value of the K-exponent. The basic idea is to exclude the Kay from the data for each element, first show the combined dynamics, and then calculate the distance between the other elements and the combined dynamics center. Then repeat the average of the next group. If the first two adjustments are not made, the process can be repeated. If there are significant changes, then the new cluster from the data heap is conservative in nature.

$$E = \sum_{j=1}^{K} \sum_{x_i} \|x_i - m_j\|^2 \tag{1}$$

In the next iteration, the K-means algorithm formulates elements to correct errors by comparing different elements and then completing the next iteration analysis by using or a center. The process continues until one of the three conditions is met. First, the core of the group does not change. The other group has no elements. Third, there is minimal difference. The formula is:

$$\overline{x}_i = \frac{1}{n} \sum_{x \in C_i} x \tag{2}$$

The K-means algorithm can measure the matching degree of two elements by distance. Therefore, this is a distance-based algorithm, in which the similarity between two elements will be higher. To express the shift in direction, it combines planarization and recognition patterns. It is like every margin of error it assumes. If we decide that there are different factions formed by the two K computations then we choose the lowest algorithm.

Experiment

E-commerce Logistics Supply Chain Coordination Management Framework

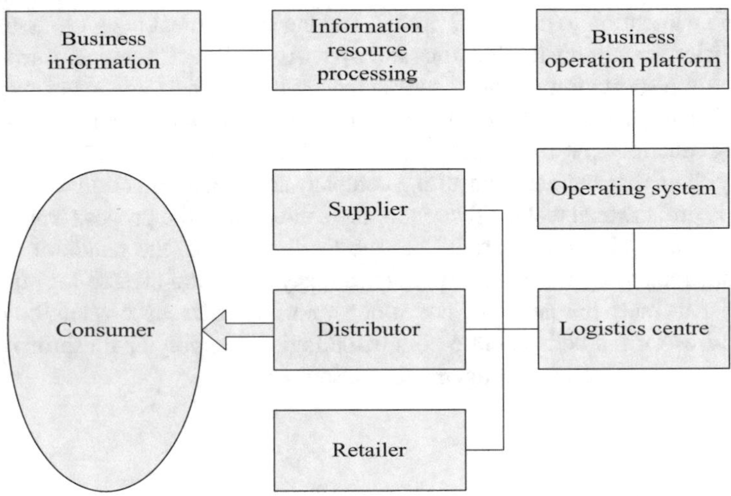

Fig. 1. Logistics and supply chain management process

Figure 1 is the flow chart of e-commerce logistics supply chain coordination management. The coordinated management of e-commerce logistics supply chain is a complex and huge system. It includes the integration of information, resources and knowledge in the whole process. In its process, it can improve the overall operation efficiency of the enterprise by establishing a coordination mechanism. It includes suppliers, distributors and retailers. Suppliers share information with distributors and retailers to reduce procurement costs and realize the optimization and integration of all aspects involved in the whole process. There are two parts in this function: one is the logistics center. The second is the seller (manufacturer). First, in order processing, it is necessary to feed back the goods submitted by the consumer to the manufacturer. This process is called "receiving". Then, the distributor will send it to retailers and suppliers to summarize and classify the purchased product information. When purchasing goods, the warehouse keeper needs to store the required goods on the storage equipment. When the goods arrive at the destination, he needs to inform the supplier in time to transport the goods to the designated place. If the owner does not receive the order, he will arrange the distribution personnel to reissue and return the products according to the actual situation, and re warehousing to ensure the rationality of the inventory, and track and record it for preparation in the logistics center.

2.4 System Test Process

In the process of system testing, it mainly analyzes user needs and e-commerce supply chain in detail, and formulates corresponding solutions according to the problems raised by customers. (1) Confirm whether the function module can operate normally. (2) Determine whether the software system can achieve the expected objectives, including whether the data processing process is correct and the interface design. When the business logic is consistent with the database operation, it is necessary to complete relevant work, and ensure that the collected data is complete, accurate and effective input and output information, otherwise it cannot meet the requirements.

3 Discussion

3.1 System Performance Test and Analysis

Table 1 shows the performance test data of e-commerce logistics supply chain coordination management system.

Table 1. System performance test data

Number of tests	Operate time (s)	Delay time (s)	Load error rate(%)	CPU availability(%)
1	1	1	2	1
2	2	1	3	2
3	1	1	2	2
4	3	2	3	1
5	1	1	1	2

Figure 2 shows the test of the operation time, delay time, load error rate and CPU utilization of the system. It can be seen from the figure that the operation time and delay time of the system are maintained at about 2s, the load error rate is very low, and the

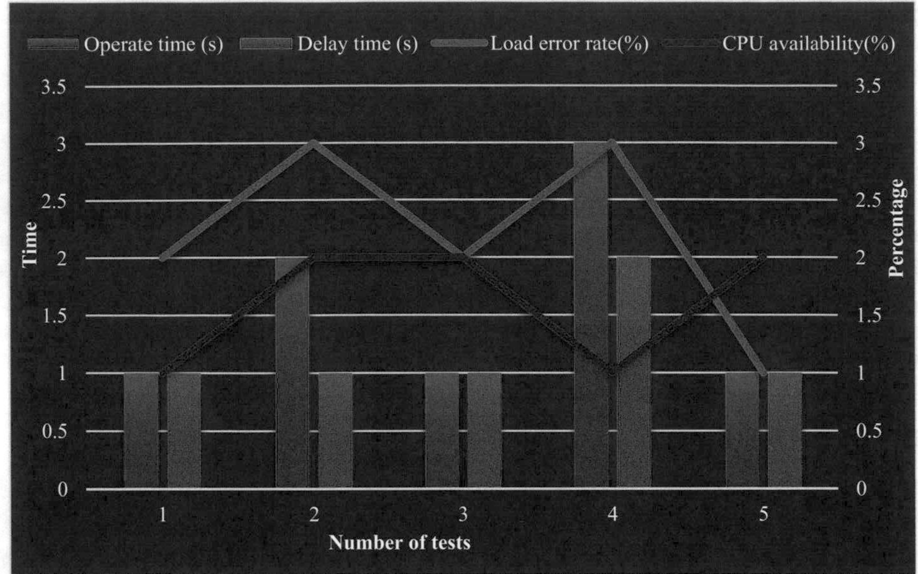

Fig. 2. System performance test comparison

CPU utilization is not high, which shows that the performance of the system meets the customer's requirements and is competent for the coordinated management of logistics supply chain.

4 Conclusion

This paper mainly studies the coordination management of e-commerce logistics supply chain. Firstly, it analyzes the growth status of e-commerce at home and abroad, and introduces its characteristics and application fields. Secondly, according to the actual needs, an intelligent supplier management system which meets the needs of enterprises and has a certain feasible solution is designed. Then, aiming at the shortcomings of traditional manual operation, such as low efficiency, slow data update and high cost, a new system model based on AI optimization method is proposed to reduce the probability of problems in supply chain coordination management, so as to improve the overall operation efficiency.

References

1. Cui, H.: Intelligent coordination distribution of the whole supply chain based on the internet of things. Complexity **2021**(1), 1–12 (2021)
2. Choy, K.L., Ho, G.T.S., Lee, C.K.H.: A RFID-based storage assignment system for enhancing the efficiency of order picking. J. Intell. Manuf. **28**(1), 111–129 (2014)
3. Liu, L.: Construction of e-commerce management performance model based on AI technology. Revista de la Facultad de Ingenieria **32**(12), 389–396 (2017)

4. Zhang, C., Ren, M.: Customer service robot model based on e-commerce dual-channel channel supply coordination and compensation strategy in the perspective of big data. Int. J. System Assurance Engineering and Manage. 1–11 (2021)
5. Liu, N., Ye, N.: Research on composition of social credibility index based on AI model. Wirel. Commun. Mob. Comput. **2020**, 1–6 (2020)
6. Preil, D., Krapp, M.: AI-based inventory management: a Monte Carlo tree search approach. Annals of Operations Res. 1–25 (2021)
7. Lopatin, A.: Use of rough set theory and neural networks methods in supply chain management. Modern Economics **22**(1), 44–49 (2020)
8. Yu, Z., Chen, Y.: Research on the optimization of after-sales parts supply chain management based on supplier management —taking SAIC general motors after-sales parts as an example. Open Access Library Journal **05**(11), 1–23 (2018)
9. Wang, S.: Research on the application of AI in sports meeting management system. Revista de la Facultad de Ingenieria **32**(16), 344–350 (2017)
10. Yi, X., Wu, J.: Research on safety management of construction engineering personnel under "Big Data + AI." Open Journal of Business and Manage. **08**(3), 1059–1075 (2020)
11. Arpa, M.D., Yolta, A., Onay, E., et al.: New therapeutic system based on hydrogels for vaginal candidiasis management: formulation-characterization, antibacterial activity, vaginal irritation and direct contact test. Pharmaceutical growth and Technol. **25**(10), 1238–1248 (2020)
12. Jakovljevic, P.J.: The future of AI & supply chain management. Supply Chain Brain **23**(1), 11 (2019)

Design and Development of Human Resource Information Management System Based on Big Data Technology

Jiaying Li[1,2](✉)

[1] City University of Malaysia, Kuala Lumpur, Malaysia
yingbao1030@163.com

[2] School of Management, Guangdong University of Science and Technology, Dongguan City, Guangdong Province, China

Abstract. The advent of the era of big data(BD) poses new challenges to human resource(HR) management in our country. How to make good use of massive unstructured and unstructured information to provide enterprises with valuable and effective services has become an urgent problem to be solved. For this reason, it is beneficial to study the HR information management system(MS) in this article, and the purpose is to solve the problem of inefficiency. This article mainly uses the experimental method, statistical method and comparative method to carry on the technical elaboration to the system, as well as the analysis on the functional module. Experimental data shows that the accuracy of the system designed in this paper is guaranteed, above 85%, and further improvement is needed in terms of time-consuming.

Keywords: Big Data Technology · Human Resources · Information Management · System Design

1 Introduction

The traditional HRMS mainly uses manual methods to complete the collation and statistical analysis of employee information and related personnel data. In the information age, the amount of data is rising sharply, the processing speed is accelerating, and the types of data are diverse, all of which need to be solved by modern technology. The design and development of HR information MS based on BD technology is necessary in the era of high demand for talents.

There are many researches on the design and development of HR information MS based on BD technology. For example, some scholars have pointed out that corporate HR management is very important for companies in the new era to improve their competitive advantages, and the application of information technology in HR management can help improve the management level [1, 2]. Some scholars also use the BD audit management platform to promote effective management planning and control of labor deployment [3,

J. H. Abawajy et al. (Eds.): ICATCI 2022, LNDECT 170, pp. 206–214, 2023.
https://doi.org/10.1007/978-3-031-29097-8_25

4]. Others use information technology to analyze and explore the long-term collected data, which can effectively realize the integration of data analysis and data resources [5, 6]. Therefore, this article also intends to use BD technology to conduct research on HR information system, in order to achieve more far-reaching progress based on the research results of predecessors.

This article first studies BD technology, and briefly describes several of the more common technologies. Secondly, it explains the HR information MS. Then from the principles, goals, development platform and database design, the system is designed. Finally, the designed system is tested through experiments and the results are obtained.

2 Design and Development of Human Resource Information MS Based on Big Data Technology

2.1 Big Data Technology

(1) Net

.NET Framework is the core and foundation of Microsoft.NET development. It can provide a suitable development environment for creating and running SQL Server database services. The object responsible for storing data in the.NET Framework platform is ADO.NET. ADO.NET is an updated version of Active Data Objects 2.6 (ADO). Using ADO.NET can more easily create distributed data sharing in the.NET Framework [7, 8].

(2) JSP technology

Java is a software development language with absolutely superior performance. Its biggest feature is simplicity, object appearance, portability and dynamics, which is especially suitable for developing application systems on the network. After continuous development in recent years, there have been many successful cases of using Java language development system to learn and learn from [9, 10].

(3) Tomcat application server

Tomcat is a large open source code server, and software developers can try it for free. It is more suitable for situations where the number of users accessing the system is not very large. Especially if there is not much software development experience like students, beginners can benefit from it. Use Tomcat to quickly respond to HTML access requests [11, 12].

Data mining is a comprehensive interdisciplinary science. Naive Bayes algorithm can classify personal information and improve the data processing ability of the system [13]. The basic probability formula that it needs to use:

$$P = p(x') = 1 - p(x) \tag{1}$$

If x and y are mutually exclusive events (disjoint events):

$$P(x \vee y) = p(x \cup y) = p(x) \times p(y|x) \tag{2}$$

Among them, p is the probability, and $P(x \vee y) = p(x) + p(y) - p(x \cap y)$.

2.2 Human Resources Information MS

The HRMS is a network-based, database-based, using various management techniques, through the analysis and processing of personnel data to achieve effective management of corporate employee information. In terms of personnel information, we must first establish job descriptions and determine what each person should do. Then assign the appropriate personnel to complete the task according to the job description requirements. Finally, the specific work can be started after the approval of the HRs manager. The content to be submitted will not become effective and officially released until the time limit is reviewed and signed by the supervisor. The HR MS is based on the network, which can realize the sharing of data and information resources and improve the efficiency of the enterprise in information processing. Use computer technology and database storage technology to perform unified operations on employees. The system adopts B/S structure mode to realize data access and query functions, user authority division and control methods.

The HRMS is based on BD technology, which manages the company's employee information, and at the same time provides more auxiliary functions for decision makers. With the acceleration of my country's economic and social development and the process of informatization, In modern society, people are paying more and more attention to how to better adapt to the new situation and meet challenges after improving their own quality and ability level has become one of the urgent problems to be solved. HR MS is composed of enterprise personnel MS and employee information MS. Among them, it includes basic information of personnel, departments, positions and other basic data.

The development of personnel MS is the transformation of HR management model to BD, realize information sharing, improve work efficiency, and reduce labor costs. Establish a HRs information system platform. Use network, database and other technologies to scientifically and effectively classify and organize employees of the company. At the same time, various information materials can be collected and analyzed through the Internet to meet the communication needs and individual needs of different departments; in addition, various types of personnel MSs can be integrated and used to achieve efficient office purposes. Realize the integration of the system with other business processes. HR information MS is an important part of enterprise management. Through scientific and effective analysis and utilization of employees, it can provide managers with timely and accurate decision-making basis.

The amount of data in the HR information MS is huge, and there are a lot of unstructured and semi-structured processing problems in the use process. Due to the wide application range of large database technology, its storage capacity is low. The database of this system is mainly adopted by the traditional enterprise employee information system; and with the progress and development of the times, the massive data is excavated, analyzed and utilized. The function of the HRMS is not perfect. At present, most companies use manual methods for management. The database is a large and complex data warehouse and processing center with a large amount of data and error-prone. The database itself also needs a certain amount of time to complete the storage process and maintenance work; at the same time, due to different software development environments, its performance and usage methods are different. There are some problems in the system due to other reasons, such as insufficient memory capacity or high occupancy rate of some files.

With the continuous and in-depth development of my country's socialist market economy, the competition for talents has become increasingly fierce, and HR management has attracted more and more attention. The era of BD has arrived. Introducing scientific, reasonable and effective methods in business management to improve staff quality and work efficiency is one of the important problems faced by modern business managers. The development of HR information MS is an important measure for the development of enterprise management information and scientific development in the era of BD.

2.3 System Design

(1) Design principles

In order to ensure the overall needs of actual users, the following principles should be followed when designing the system:

The main purpose of the system is to be practical, that is, to maximize the usefulness of the system in actual work. It is necessary to fully consider various issues that may affect the work, such as business level, management link processing, etc. Satisfaction management should be considered as the first element. At the same time, the interface design needs to be beautiful, but not cumbersome. It is easy for users to use, and the function keys are clear. On the basis of ensuring practicability, it is necessary to ensure advanced technology and mature software development and design, and all aspects must be considered. The system is required to be object-oriented, have a visual interface, support network environment work, and network computing mode to be in line with the latest technology, and as far as possible to walk on the cutting edge of the same type of software. System development is not the end, as business requirements continue to increase. It is very important to add new functions to the system in a timely manner. Therefore, there are also requirements for the scalability of the system. And in order to ensure the stability of the system and enable it to complete the work stably, the system must be maintainable. Only in this way can the system be effectively used in the enterprise. System data operations, etc., all include the company's confidential documents, etc., which cannot be leaked. Therefore, the safety and reliability of the system is very important. The system is required to have a strict enforcement system for user login. There must be no loopholes in the design of the certification process. There are also clear regulations and designs for the permissions of various user levels. In this way, the security of the system is increased, and the enterprise can use the system more at ease. The development of all system software, etc. provides source programs for expansion requirements. The design must comply with the industry and international standards.

(2) Design goals

Use the most advanced computer technology to develop a HRMS and provide network functions. And develop and design on the basis of accuracy, convenience and speed. So that the leadership can quickly and effectively understand company information, and provide convenient operations for HR management. Makes the work side simple and easy to operate, but effective. Realize computerization of internal management and realize information sharing. Improve the shortcomings of current human management through modern office methods.

(3) System structure and development platform

In this system, the daily changes and management functions of company information, employee information, etc. will be completed on the C/S structure based on the active database. The query function of employee information, company information and salary will be based on the B/S structure of the active database. The combination of C/S and B/S becomes the basic structure of the system. The C/S structure is a software system architecture. Its advantages are increased system online processing capabilities, enhanced system scalability, enhanced system security, the ability to utilize various resources, and improved application development productivity. Its disadvantages are the difficulty of system transplantation, high development and maintenance costs, and more complicated use. The full name of the B/S structure is Browser/Server, which is a 3-tier structure improved based on the C/S because of the increasing rise of network technology, and people feel that the C/S structure is insufficient.

(4) System database design

The core of the digital transformation of HRMS is the degree of rationalization of the database MS, which is the core of the system. The database cannot carry out effective data collection and management, so the processing and utilization of information is impossible to talk about. Therefore, building an efficient database MS is a prerequisite for the realization of a HRMS. The system database adopts the SQL Server 2000 database system, which can well realize the authentication of the database server. The design of the table structure is the basis for the development of database management software. A reasonable table structure can intuitively reflect the information contained in the data and the problems in the program.

(5) System architecture design

The HR MS is only a part of the power plant MS, and its development is based on the unified framework provided by the power plant MS (MIS system). The system is divided into seven subsystems, namely personnel information management, insurance management, education and training, attendance management, contract management, occupational health management, and system parameter design. The details are shown in Fig. 1:

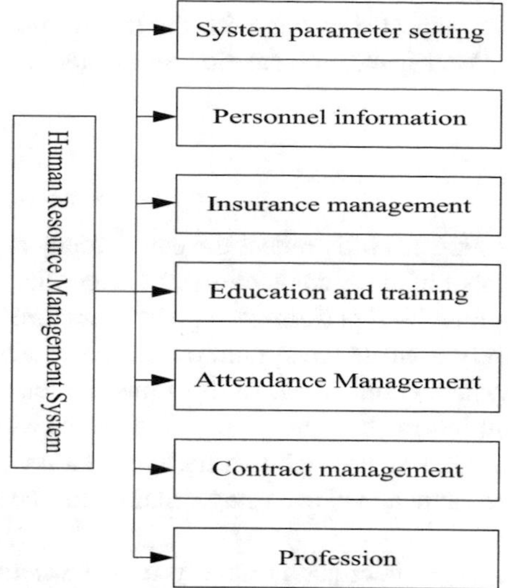

Fig. 1. Structure of the Personnel Information MS

Personnel information management includes high-level information, employee information, team information, and department information, among which the more important is employee information. Attendance management includes attendance records, overtime applications, attendance analysis, and attendance summary. This module is related to the payment of salaries by the Finance Department. Education and training management includes pre-job training of employees, training applications, and different types of training. Occupational health management, including occupational disease information, occupational health, occupational safety, and labor insurance. The system parameter setting includes the parameters of different modules.

3 System Implementation Test

3.1 Design Environment

Generally speaking, a software development environment (SDE) is a set of software, which is based on the existing hardware and software provided by the customer to assist in the development and maintenance of system software and software. It consists of software development tools and installation environment.

The development environment created by this system is said to use SQLServer2008 as the database design language, JAVA language to develop the front-end operating system platform, use JSP technology, support web mode and dynamic operation, use Java script scripts to obtain sound application server using Tomcat7.

3.2 Test Content

The core content of the system test process is the rational design of the test plan. The test plan mainly involves the following content, namely, the various functions of the test system, etc., compare the expected results with the test results, and find errors in the network system in time. Modify data information and system programs.

3.3 Test Method

(1) Performance test

Concurrency testing can fully reflect the concurrency capability, which mainly depends on the number of users, connection efficiency, etc., and its ultimate goal is undoubtedly to obtain related performance parameters through system access, and then comprehensively evaluate the system on this basis. The entire process of the user entering the system is simulated. At this time, the selected sample is 150, and the user will submit information every 10 s. The server will respond immediately after submitting the request, and will exit the system after 4 min. The tester needs to record in detail the situation of each stage, mainly for the server's response time.

(2) Function test

The test of the login interface is mainly to test whether the user name and password can be correctly prompted. What the system module cannot ignore is the query function. Among them, attention should be paid to attendance, salary, training, etc. Staff should pay attention to safety testing and make appropriate adjustments to the system process to avoid irretrievable results due to personal negligence.

4 Analysis of Test Results

4.1 Performance Test Results

This text carries on the performance test to the designed system, mainly analyzes from its duration, delay time and accuracy. Among them, this article selects the number of users in the first four minutes, as shown in Table 1:

Table 1. System Performance Test Results

	Duration	Delay time	Accuracy
150	280	0.1	0.9
300	290	0.15	0.89
450	300	0.2	0.86
600	310	0.3	0.85

As shown in Fig. 2, we can see that when the number of users increases, the duration of the system increases accordingly, and the delay time gradually increases. The accuracy of the system dropped from 90% to 85%.

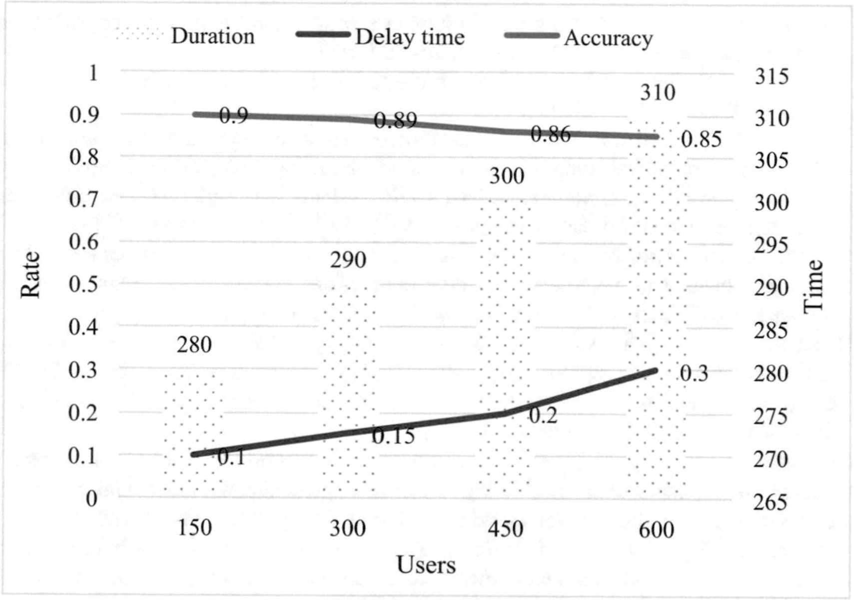

Fig. 2. System Performance Test Results

5 Conclusion

In recent years, HR management has occupied an important position in the operation of modern enterprises, and BD technology provides a good solution to these problems. This system realizes the BD analysis of HR management, and collects and organizes the basic information of a large number of employees and the requirements of relevant positions when collecting statistics on personnel information in the personnel department. The development of HR information MS is mainly to solve some problems in current management. The system designed in this paper can meet the basic requirements, but there is still a problem of insufficient data, and the system's data processing capabilities need to be strengthened.

References

1. Selvaraj, J.: Design and development of web module based biomedical database management system for voluntary health services multispecialty hospital. Shanghai Ligong Daxue Xuebao/Journal of University of Shanghai for Science Technol. **22**(12), 1379–1387 (2020)
2. Kuzneco, V.S., Kaschenko, M.: Robotization in modern conditions of human resource management. Actual Directions of Scientific Researches of the XXI Century Theory and Practice **8**(2), 90–94 (2020)
3. Miller, R.E., Dunn, P.: Teaching case: mihotel: applicant processing system design case. J. Inf. Syst. Educ. **29**(1), 21–24 (2018)
4. L'Ecuyer, F., Raymond, L., Fabi, B., et al.: Strategic alignment of IT and human resources management in manufacturing SMEs: empirical test of a mediation model. Empl. Relat. **41**(5), 830–850 (2019)

5. Bragina, D., Molodchik, N.: Big data in human resource management. Manage. Personnel and Intellectual Resources in Russia **9**(3), 76–80 (2020)
6. Nguyen, D.T.N., Teo, S.T.T.: HR orientations and HR department effectiveness in Vietnam. Pers. Rev. **47**(5), 1048–1066 (2018)
7. Waychal, P., Mohanty, R.P., Verma, A.: Leading indicators of innovation as a competence for individuals: an empirical study. J. Advances in Manage. Res. **8**(2), 301–322 (2017)
8. Nieves, J., Osorio, J.: Commitment-based HR systems and organizational outcomes in services. International Journal of Manpower **38**(3), IJM-09–2015–0144 (2017)
9. Turulja, L., Bajgoric, N.: Information technology, knowledge management and human resource management: investigating mutual interactions towards better organizational performance. VINE Journal of Information and Knowledge MSs **48**(3) (2018)
10. Hedberg, L.M., Luchak, A.A.: Founder attachment style and its effects on socioemotional wealth objectives and hr system design. Hum. Resour. Manag. Rev. **28**(1), 33–45 (2017)
11. Rana, G., Sharma, R.: Emerging human resource management practices in industry 4.0. Strategic HR Review **18**(4), 176–181 (2019)
12. El-Kassar, A.-N., Singh, S.K., et al.: Green innovation and organizational performance: the influence of big data and the moderating role of management commitment and hr practices - sciencedirect. Technological Forecasting and Social Change **144**, 483–498 (2019)
13. Ramaprasad, B.S., Prabhu, N., Lakshminarayanan, S., et al.: Human resource management practices and organizational commitment: research methods, issues and future directions (2001–2016). Ind. Commer. Train. **49**(6), 277–287 (2017)

Application of Computer Image Transformation Technology in Fashion Design

Baojuan Yang(✉)

Department of Construction Management and Real Estate, Chongqing Jianzhu College, Chongqing 400072, China
12232@163.com

Abstract. In order to play the role of computer image transformation technology, optimize the fashion design process, improve the quality of design, this paper will carry out relevant research. This paper first discusses the basic concept of computer image transformation technology, then introduces the application advantages of this technology in clothing style design, and finally puts forward the application strategy of technology in clothing style design combined with a case. Through the study of this paper, computer image transformation technology has a high application value in clothing style design, the adoption of relevant strategies can play a role in technology, make clothing style design process more concise, and design quality is guaranteed and improved.

Keywords: Fashion design · Computer image transformation technology · Clothing

1 Introduction

Clothing is a necessity in people's life, modern people like to beautify their image and decorate their life through various clothing, so they have a high pursuit of clothing style, especially the new style of clothing, which makes clothing style designers need to constantly innovate in the design work. However, in the past, the process of clothing style design was cumbersome, so the design quality could not be guaranteed, often fluctuated, and the design of finished products could not fully meet the personalized needs of people, indicating that the difficulty of clothing style design is very high, and the innovation of clothing style cannot be realized quickly. From this perspective, related areas that can use technology to design, with the help of the technology to design process is simplified, the quality will be guaranteed and ascend, after the idea has received recognition design personnel, so the designers began to seek feasible technology, Computer image transformation technology is in such a background into the fashion design.

With the help of the computer image transformation technique, design did achieve the above goal, makes the clothing design innovation difficulty is reduced, update the iteration speed faster, but because of the combination of the technology and design time is shorter, so there is still much garment designers fail to master the technology application strategy, it shows that the computer image transformation technology needs

J. H. Abawajy et al. (Eds.): ICATCI 2022, LNDECT 170, pp. 215–223, 2023.
https://doi.org/10.1007/978-3-031-29097-8_26

to be further popularized in fashion design. How to use this technology to do a good job in fashion design is an urgent problem to be solved. From this point of view, clothing style designers need to understand the basic concept and application advantages of computer image transformation technology, and master the application strategy of technology in the design, so it is necessary to carry out relevant research.

2 Research Background

Modern people have higher requirements for the quality of life. They begin to follow their own needs and pursue various aesthetic feelings in life. The relevant behaviors are mainly reflected in people's requirements for clothing. For example, many people in modern times pursue fresh aesthetic feeling, so they often choose some clothes with fashionable elements, but some people also pursue classical aesthetic feeling. In terms of clothing selection, they tend to choose clothes with classical elements, and even some people pursue the aesthetic feeling of the alternation of ancient and modern times, hoping to have both classical and fashionable elements in their clothes. It can be seen from this that modern people have various demands for clothing, which also provides a huge design space for the field of clothing design. In theory, clothing designers can give play to their own expertise and imagination, and design a variety of clothing works for workers to choose from, but the current situation seems to have backfired. According to a research survey, most of the clothes designed by fashion designers are very similar, which means that there is a widespread phenomenon of design plagiarism among designers. Another data reality is that less than 10% of the clothing design results can be favored by consumers in the market, and the remaining 90% can not be purchased by consumers at the first time, which means that these clothing design results do not meet people's aesthetic needs.

In the face of this situation, previous studies believed that the reason was that clothing designers had inspiration exhaustion problems, that is, after a long period of design work, art practitioners in any art field would face inspiration exhaustion problems, but experienced art practitioners would use their own methods to find inspiration, such as going out for travel, which could open their minds and get inspiration, Therefore, inspiration exhaustion is not the root cause of this situation. After realizing this, modern research has further analyzed the problem and confirmed that inspiration exhaustion is only the superficial cause of the current situation. The fundamental reason is that clothing designers do not have much time to seek inspiration, so they are in a state of inspiration exhaustion for a long time, which leads to the current situation and can not be solved for a long time. According to this achievement, people think that it is a good solution to provide clothing designers with enough time to find inspiration, but this method has the defect of insufficient realization under the restriction of enterprise rules and regulations, so people turn their attention to computer image transformation technology.

Computer image transformation technology is a kind of image processing technology. Different from other image processing technologies, this technology can directly change the image, generally including translation, rotation, scaling, shearing, projection, etc. Therefore, this technology is considered to be a good image design software and is now widely used in other image art design fields. Therefore, clothing designers can also try to use this technology to design clothing images. This technology can provide more visual inspiration for clothing designers, which cannot be realized in real environments. For example, people cannot achieve image scaling in real environments, but computer image transformation technology can do this, so clothing designers can design clothing in their virtual environments, Get more novel inspiration.

Computer image transformation technology is highly operable in garment design, that is, the same type of image processing technology or design software can enable designers to adjust the design image from different angles in garment design, but the results after each adjustment are difficult to overlap, which is prone to low quality design problems, because these technologies or software fail to take into account the logic between functions when designing image processing functions, However, the computer image transformation technology takes this into account, and the image adjustment results obtained by its translation, rotation, scaling, stagger tangent, projection and other transformation functions can be perfectly superimposed, which has multiple transformation functions, indicating that this technology is highly operable and provides many new design angles for garment designers.

3 Basic Concepts and Advantages of Computer Image Transformation Technology

3.1 Basic Concepts

Computer image transformation technology is a kind of image processing technology, mainly use the computer to analyze the image, and then adjust the defects in the image, or change the image according to other special needs, and finally achieve the result. The technology in essence belongs to the digital technology, that is, the technical system can only process the digital image, this image is composed of a variety of two-dimensional array, array is composed of a variety of pixels, each pixel has a specific gray value, image transformation is the technical means of adjusting these gray values. In principle, because a large number of two-dimensional array digital image connotation, after all array combination will form a large image array, so directly within the array spatial domain processing will be very tedious, need to go through a lot of calculation, but you can image transformation technology in Fourier transform and Walsh transform, discrete cosine transform, the principle of Indirect processing of the image space domain to transform the space domain, and then processing of the transform domain can quickly achieve the goal, and the processing effect is more advantageous than direct processing. It can be seen from the above that there are many basic principles of computer image transformation technology [1, 2]. Fourier transform is more commonly used at present, and its basic process is shown in Fig. 1.

Combined with Fig. 1, the Fourier transform domain is in the original space based on image transform processing, makes space domain, transform domain, and then in the transform domain based on pixel spectrum transformation, the formation of image transformation curve, come to an end when the curve is a goal in image processing, if the image processing results are not satisfied in the process, The parameters of transform domain can be changed by modifying the value to ensure the final quality of image

Fig. 1. Fourier transform

processing. Fourier transform of the advantage is that the process is convenient, easy to operate, but it is not perfect, on the contrary the Fourier transform is a very serious defect, is easy to generalize this kind of defect in general will only in the image under the condition of large scale two dimensional array, but many modern processed image two-dimensional array level has reached the level of mass, So the application value of the Fourier transform is weakened to some extent, presented some new transform principle, related areas such as wavelet transform, it has a strong ability of localization, basically will not face any image processing results of generalization problems, so the wavelet transform and Fourier change is as widely used two kinds of computer principle of image transformation technique. Figure 2 shows the wavelet transform [3–6].

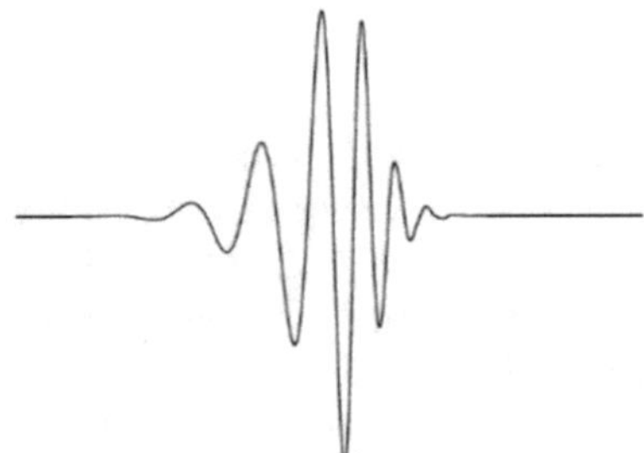

Fig. 2. Wavelet transform

According to the Fig. 2 shows that wavelet transform is based on the value of a constant, according to the standard parameter (under different conditions of different standard parameters, it failed to figure in the annotation) wave pattern transform, transform initial amplitude is small, but will soon reach the peak, peak and duration is shorter, also turn to level off, the complete transformation processing.

3.2 Advantages of the Application of Computer Image Transformation Technology in Clothing Style Design

Optimization of Two-Dimensional Plane Viewing Angle

The garment style design itself is to process the plane image, so the relevant work is completed in the two-dimensional plane perspective, such as designers need to adjust the brightness and specifications of the garment style image from this perspective, but the previous design work is more dependent on manual work, and manual work ability is limited. At work often need through their own subjective conjecture to adjust, this leads to design work often repeated, such as a designer in a suit design, first on the original observation, found that the shoulder manuscript, brightness is too high, so you need to cut, but specific to how much lower is unknown, so the designers made many attempts to It can be seen that the whole process is often repeated, and if there are more adjustments to be made, the repetition frequency will be higher. This phenomenon shows that under the previous two-dimensional perspective, clothing style design is cumbersome, inefficient and difficult, especially in the innovative design of clothing style, this phenomenon will be more prominent [8–10].

And computer image transformation technology can effectively optimize the two-dimensional graphic, makes the design efficiency and quality, namely the technology to the original plane clothing style images into digital images, the image of each pixel gray value of two dimensional array is can be controlled, so that the entire design process parameterization improvement, First of all, the designer can choose a two dimensional array, total cost of the grey value of the array observation, understand its brightness, secondly if design personnel are not satisfied with the brightness, you can choose the corresponding pixel gray value, and then adjust grey value parameters directly, in this way can be more convenient and fast to find the most reasonable brightness, complete relevant adjustment, And this way can also support designers to carry out various creative attempts in clothing style innovation. Figure 3 is the clothing style design process optimized by computer image transformation technology from a two-dimensional plane perspective.

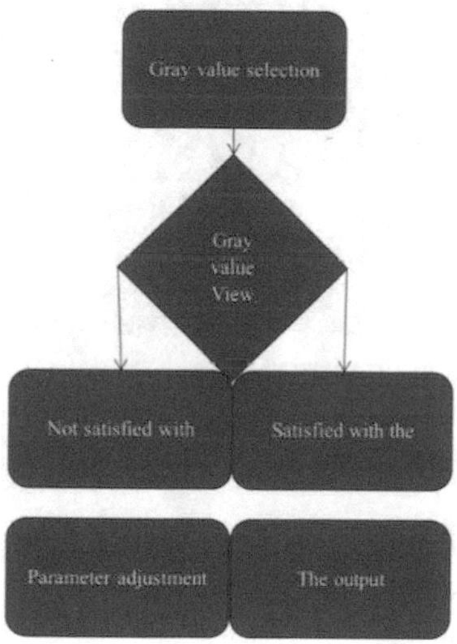

Fig. 3. Garment style design process optimized from two-dimensional plane perspective by computer image transformation technology

Interference Handling

There is no denying the fact that any technology in 2 d images into the process of digital images will be disturbed, lead to local noise in image, the defect makes the digital image cannot be used for the design of costume style, the reason is that 2 d image into a digital image is mainly done by cameras, in this process is very susceptible to interference, However, such interference can not be effectively dealt with by previous technical means, so the fashion design is often not good results. For example, a clothing style designer found in his work that there was obvious noise interference in the digital image, and the waist and cuff of the image were very blurred. In this case, clothing design could not be carried out, and the designer did not have a good method to deal with it.

Aiming at this point, although computer image transformation technology in the transformation process will be disturbed, but the function of image processing, the interference can be processing, key lies in the technology of compression processing functions, namely after the interference of the noise image can produce noise value, its initial state is more dispersed, so it's not easy to handle, But if these noise values can be concentrated, it can be effectively dealt with, and the compression processing function can play this role, all the noise values can be compressed to form an abnormal pixel point, and then the designer can delete the abnormal pixel point, and then fill the normal pixel point, the operation process is not complicated, can be completed manually. Figure 4 shows the basic flow of the compression processing function.

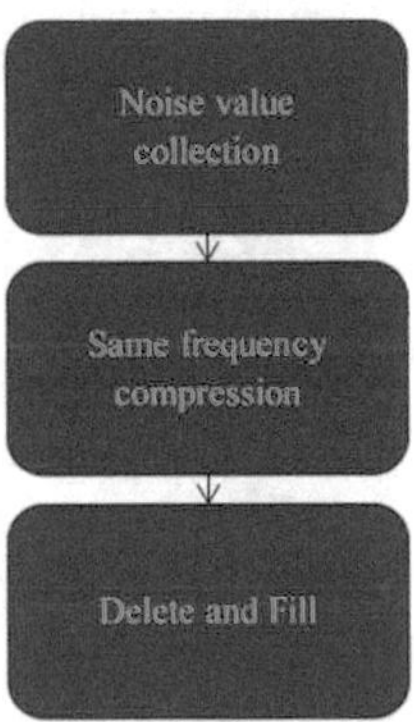

Fig. 4. Basic flow of compression processing function

4 The Application Strategy of Computer Image Transformation Technology in Clothing Style Design

4.1 Innovative Application of Style

Because modern higher request to the clothing style, and very like the new design, so design personnel in the work to try to innovate, and innovative attempt in general can be divided into two dimensions, one is the innovation to the appearance of clothing styles, this dimension of work can be finished using conventional processing software, The other is to innovate the brightness and color matching of clothing styles. The work on this dimension cannot be completed by conventional processing software, but needs the help of computer image transformation technology. From this point of view, designers should be familiar with the operation mode of computer image transformation technology. Usually, two strategies can be adopted to innovate the brightness and color collocation of clothing styles: First, brightness for dress design, designers can create multiple in computer image transformation technology system of the same clothing style template, and then according to oneself idea to adjust the brightness of the each template, after adjustment for each template can be compared, intuitive choice brightness is the best template, if all template will not be able to satisfy the designer, Then designers can continue to change the brightness of a template by adjusting parameters, so that they can finally find the best brightness and complete the purpose of innovation; Second, according to the colour collocation of the clothing style, because different colour in the chromaticity and color differences, so the designer can according to the color classification principle, the clothing style of the template 2 d array of digital image classification, and then according to their own ideas for each of the two dimensional array color chromaticity, color adjustment, For example, using uniform parameters to unify the tone of a two-dimensional array, so that innovation can be carried out quickly. Figure 5 Operation flow of color tone adjustment.

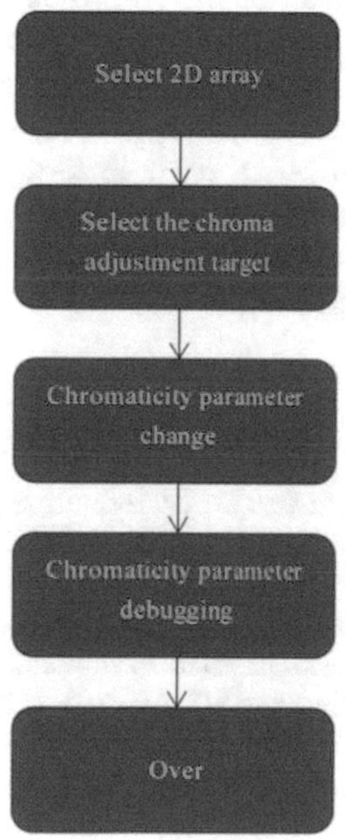

Fig. 5. Operation flow of color tone adjustment

4.2 Application of Compression Function

Although in general computer image compression transformation technology is mainly used for processing the noise, but in some special cases it can also be used to design itself, namely in the design of costume style, design staff occasionally need to combine some design elements, but because the design element is smaller, so very difficult, sometimes to solve the problem through the artificial, But by compression can help artificial point is reached, the designer directly add design elements on the digital image template, then use the compression function select multiple design elements of the two dimensional array, or pixels, restart compression to combine selected target, finally after amplification combined with two-dimensional array, or pixels, Adjust the style, brightness and other aspects of the multi-user combination relationship. It can be seen that computer image transformation technology can help designers to complete some design operations that cannot be completed manually, which improves the quality of clothing style design and is guaranteed. Therefore, this technology has a high application value in clothing style and is worth promoting.

5 Conclusion

In conclusion, design the importance in the modern social environment, is related to the quality of people's lives, such as garment designers need to do a good job of design, but because of the limitation of realistic factors, so many modern garment design work quality, efficiency and security, makes the design results did not meet the requirements of the present, that transformation is needed to design work. From this perspective, because the computer image transformation technique can provide effective help for design work, solve the problem of the past, so design transformation can apply for direction in computer image transformation technique, design personnel is necessary to study the technology of concrete application way, so as to give full play to the technology, promote transformation of design work.

References

1. Lee, U.H., Bi, J., Patel, R., et al.: Image transformation and CNNs: a strategy for encoding human locomotor intent for autonomous wearable robots. IEEE Robot. Autom. Lett. **5**, 5440–5447 (2020)
2. Esnaola, M.L., Bengoa, N.A., Araujo, B.S., et al.: Machine learning for video action recognition: a computer vision approach. In: 2018 14th International Conference on Signal-Image Technology & Internet-Based Systems(SITIS). IEEE (2019)
3. Bhuvaneswari, S., Subashini, T.S., Saravanan, M., et al.: A new approach for automatic scene text detection. Int. J. Comput. Inform. Technol. **11**(1), 81–88 (2019)
4. Ramasamy, R., Arumugam, V.: Robust image watermarking using fractional Krawtchouk transform with optimization. J. Ambient. Intell. Humaniz. Comput. **12**(7), 7121–7132 (2020). https://doi.org/10.1007/s12652-020-02379-z
5. Galteri, L., Seidenari, L., Uricchio, T., et al.: Preserving low-quality video through deep learning. IOP Conf. Ser. Mater. Sci. Eng. **949**, 012068 (2020)
6. Mian Qaisar, S.: Isolated speech recognition and its transformation in visual signs. J. Electric. Eng. Technol. **14**(2), 955–964 (2019). https://doi.org/10.1007/s42835-018-00071-z
7. Chen, B., Pan, B.: Calibrating mirror-assisted multi-view digital image correlation system using a speckled planar object. Measure. Sci. Technol. **32**(3), 034008(12pp) (2021)

8. Geninatti, S.R., Boemo, E.I.: Real-time reconfigurable processor to detect similarities in compressed video using generalized though transformation. IEEE Trans. Circ. Syst. Video Technol. **99**, 1 (2019)
9. Singh, G., Banga, V.K., Yingthawornsuk, T.: Inverse kinematics solution of Programmable Universal Machine for Assembly (PUMA) Robot. In: 2019 15th International Conference on Signal-Image Technology & Internet-Based Systems (SITIS) (2019)
10. Tajudin, A.S., Isa, I.S., Soh, Z., et al.: A new technique of flow voids segmentation on MRI image for cerebrovascular disease. In: 2018 IEEE 8th International Conference on System Engineering and Technology (ICSET). IEEE (2019)

Application of Data Mining Technology in Business Administration Data

Wei Li(✉)

Zhejiang Institute of Economics and Trade, Hangzhou 310018, China
yixiuge0124@163.com

Abstract. In order to play the role of data mining technology to do a good job in business management, this paper will carry out relevant research. This paper firstly introduces the basic concept of data mining technology, then analyzes the current situation of business administration, and puts forward the importance and difficulties of data application in management. Finally, the paper constructs the data application system of business administration centering on data mining technology. Through research, business management data application system based on data mining technology has good practical value, with the help of the system can play a technical role, so that business management in the support of data, can improve the quality of management.

Keywords: Business administration · Data application · Data mining technology

1 Introduction

The nature of business management is very special, involving all levels of social industry, enterprises and so on. It aims to ensure the sustainable and healthy development of the market and solve all kinds of industrial and commercial problems and violations. Therefore, business management is very important. But our country business management has long been some quality problems, including prominent management is not timely, slow speed of management decisions or inaccurate, etc., the existence of these problems is not conducive to healthy development of the market, will also help up unhealthy ethos, therefore, how to solve these problems, improving the quality of business management has been the subject of much attention. This background, the data mining technology is put forward for industrial and commercial management brings revelation, which rely on the technology of industrial and commercial administration, can rely on real-time data and the data have reveal essence of things and the role of internal problems, so according to the result of data mining can have a clear understanding of the situation, then can make accurate decisions quickly, therefore, the most important problem at present is how to apply data mining technology in business management data, which needs to carry out relevant research.

J. H. Abawajy et al. (Eds.): ICATCI 2022, LNDECT 170, pp. 224–232, 2023.
https://doi.org/10.1007/978-3-031-29097-8_27

2 Data Mining Technology Function and Value Analysis

Data mining technology is a technical tool to implement the data mining process. The so-called data mining is to conduct in-depth analysis on the basis of surface data information to obtain internal deep hidden information. For example, in the enterprise management data, the data shows that the enterprise's income is 10000 yuan and the cost is 30000 yuan. Through comparative analysis, it can be seen that the actual income of the enterprise is negative 20000 yuan, of which negative 20000 yuan is the income Hidden information behind cost data. It can be seen from this that data mining itself is not difficult, and it can be completed through simple comparative analysis. However, in actual enterprise management data applications, people face a huge amount of data and complex data relationships, which leads to a significant increase in the difficulty of data mining. Simple comparative analysis methods are not fully applicable, and many mining tasks need to rely on various algorithms, Data mining technology was born under this background.

The core idea of data mining technology is to regard the data target as a whole, thus setting a range, and then within this range, take any data as the base node, directly and indirectly connect it with other data to obtain the relationship model of all data on the basis of this data, and then replace the data base node, repeating the process until all data become the primary base node location. In this way, multiple relational models can be obtained. At this time, the data mining technology will carry out mining in two stages. The first stage is to analyze the relational model to obtain the hidden information behind the relevant data of the relational model. The second stage is to combine any relational models together to obtain the hidden information behind the relational model data group through mining. The first stage of the two stages must be implemented, In the second stage, you can choose not to execute, but if it is executed, it means that the mining results are more in-depth, which can better reveal the nature of the things represented by the data.

From this point of view, data mining technology has a high application value in the application of enterprise management data. The reason lies in three aspects: first, the amount of enterprise management data is huge, the type is complex, it is difficult to handle manually, and it can not be deeply mined by other technical means, but data mining technology can give consideration to both processing efficiency and result quality, which is basically the only choice in the application of modern enterprise management data; Secondly, enterprise management data plays a role in business administration, which has very strict requirements for the accuracy of results. If errors occur, it is likely to infringe on the rights and interests of enterprises, or worsen the non-standard behavior of enterprises. In the face of this requirement, the Data Mining Technology Section ensures that the accuracy of results reaches the standard; Third, data mining technology can help people deeply analyze data and understand the nature of the problem, so it can strengthen the strength of relevant management work based on enterprise management data and better supervise enterprises.

3 Basic Concepts of Data Mining Technology

Data mining technology is data processing technology in essence. The whole operation process is divided into five steps, as shown in Fig. 1.

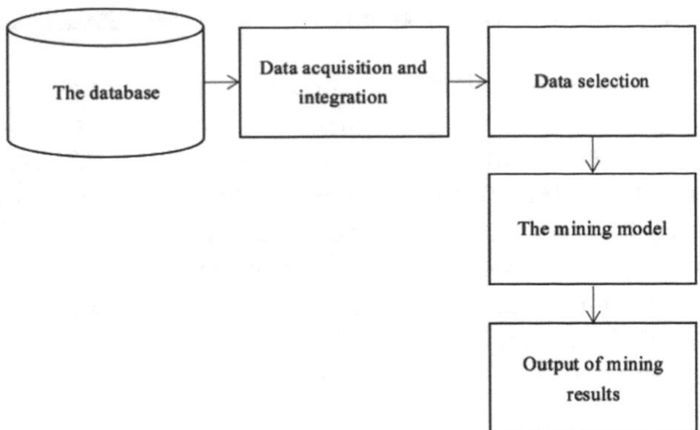

Fig. 1. Data mining technology operation flow

Combined with Fig. 1, the operation of data mining technology firstly obtains initial data through data collection, and the initial data will be stored in the database, thus forming the data source. Secondly, relevant data in the data source will be accepted by the processing system of data mining technology, and then conduct data cleaning and integration operations. Including data cleaning is to remove the initial data that exist in the low quality data, such as duplicate data, incorrect data, makes the initial data become more pure, and pure data will be classified according to characteristics such as keywords, conducted in system integration, respectively, after the completion of the system according to the actual needs to choose the classification of several data mining, mining rely mainly on relevant model, finally, mining results can be obtained on the basis of the model, and the results will be received to the user's computer and other devices after visual processing, so that users can make decisions after consulting [1–3].

You can see that the processes of data mining technology itself is not complicated, so in the practical application of this technology has the advantages of convenient operation, at the same time the technology is only responsible for the analysis of the data mining, don't reject any type of data, so this technology has a wide range of application, and the reasonable mining model, the technology of the output result is very accurate, the accuracy of the method is much higher than that of manual mining analysis. In addition, compared with the manual mode, the biggest advantage of data mining technology is that it can in a short period of time to deal with large scale, the relationship between the complexity of the data, at least among the industrial and commercial management data using the performance is very good, the human also has a good ability of data processing, but also has ability limitation, when large scale and complex relationship between data, artificial it must cost a lot of time for data analysis, and cannot guarantee the result, and the industrial and commercial management data is a large scale, the complexity of the relationship between characteristics, so the artificial cannot be efficient to analyze all the data, the results tend to be partial, and not accurate enough, but with the help of the data mining technology to solve all these problems [4, 5]. Therefore, data mining technology should be widely used in the application of business management data [6]. However, it is necessary to pay attention to data integrity in the application. In order to ensure the integrity of data, corresponding safeguard measures should be taken. The technical measures are shown in Fig. 2.

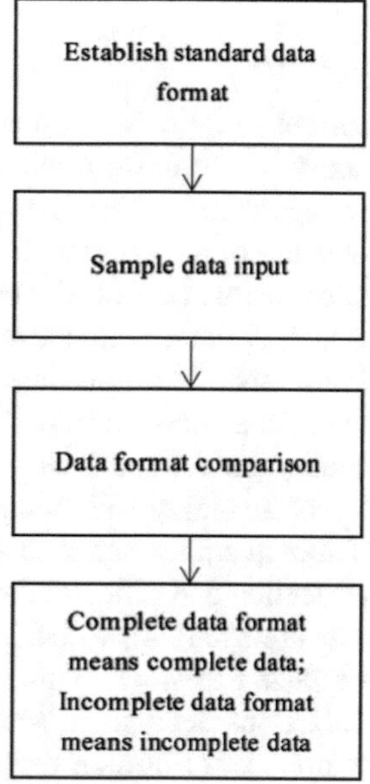

Fig. 2. Data Integrity Guarantee Measures

4 Status Quo of Business Administration and Importance and Difficulty of Data Application in Management

4.1 Overview of the Current Situation

Industrial and commercial management of the current work of the organization in a certain area as an example, through the observation and statistics results, the organization management jurisdiction there are approximately 120000 commercial organizations, each organization output at all relevant business data, so the administration for industry and commerce affairs is busy, and in order to cope with the busy work, the group has been expanded to all the year round, today in the organization at the grass-roots level workers have thousands, but even so still can't effectively cope with the busy work, the reason is that the organization's work mode to manual mode, still work rely on artificial, so the artificial ability, work efficiency limit is not enough, although thousands of by hand, but work in the face of hundreds of millions of data, there is a huge gap on the order of magnitude. Under this condition, many quality problems occurred in the work of the organization. For example, the organization failed to detect the irregular behavior of a local merchant in a commercial activity in time. As a result, the behavior was discovered nearly a year later, and the market was already damaged by the irregular behavior of the merchant. Thus, there are many problems in the current situation of business administration, which is also a challenge faced by many business management organizations.

4.2 Importance of Data Application

In fact, many problems in the current situation of business administration are not new. People have been aware of the existence of these problems for a long time, but people have not found a good solution for a long time. This phenomenon has attracted the attention of relevant fields, so the root cause of the problem is obtained through research and analysis. The business management status quo of all sorts of problems are caused by basic application because of the lack of data, which also puts forward the assumption: if in industrial and commercial administration can timely access to relevant data, and to deal with the data, before you make a decision to perform the work, can avoid work efficiency is low, lack of work quality and related issues. Put forward the assumption of the soon got the recognition of people, makes the industrial and commercial management organizations realize itself should strengthen data applications, two key elements which are highly efficient data collection and efficient data analysis, at the same time to guarantee the data analysis results are accurate, so business management organization can largely rely on artificial is reduced to work, This not only reverses the situation that business management organizations need to constantly expand their staff to deal with work needs, but also saves the cost of human resources. It can be seen that data application is very important for business management organizations [7–9].

4.3 Data Application Difficulties

Although data application is very important, but in fact many modern business management organizations have not yet effectively applied data in the work, the reason is that there are some difficulties in data application, the data application effect will be greatly reduced on the basis of the difficulties have not been solved, but also may mislead the final decision. At present, difficulties in data application are mainly reflected in two aspects: First, industrial and commercial management organizations in the face of the management of large scale data, status quo of the middleweight breakthrough is the norm, and the number of the hour in growth, such a large magnitude data lead to industrial and commercial management organizations were by no means ordinary or artificial technology to process the data, if as required by the efficiency of data processing, can capture only part of the data, unable to process all data, on the contrary, if you want to process all data, then the work efficiency can not be guaranteed, will be greatly delayed; Second, the management of the industrial and commercial management organizations face has very complex relation between data, from the angle of the quality of data processing, processing results, the relationship between the need to clearly show all data in the relationship between the high complexity of lead to human logic cannot be processed with traditional technology, this problem also caused difficulties to industrial and commercial management data applications [10]. Table 1 shows the mean magnitude and complexity assessment results of management data of a certain business management organization.

Table 1. Average magnitude and complexity assessment results of management data of a certain business management organization

Project	Content
Manage the magnitude mean of data	150 million
Complexity assessment Results	Very high

5 The Construction Scheme of Business Management Data Application System Based on Data Mining Technology

5.1 Reasons for Technical Selection

In the face of the two major difficulties of business and management data applications of artificial or unable to cope with the traditional technologies, but data mining technology is different from the first two, the efficiency of the technical characteristics of processing and efficiency analysis, that is, data mining technology is under the intelligent logic operation, the core of the logic for the neural network (see Fig. 3), according to the characteristics of the neural network data mining technology can quickly handle scale huge data, according to the statistics can be performed within hours of hundreds of millions of data processing, at the same time, intelligent logic is a kind of approximate human logic model, which makes the data mining system to human recognition, thinking, analysis of data, but also because it is a technology, there are no human limitations, so you can tease through the intricacies of the data and come up with the results. Thus, in the face of two difficulties, data mining technology is the best choice, is one of the few technical tools that can meet the application of business management data, so we should choose this technology to apply business management data.

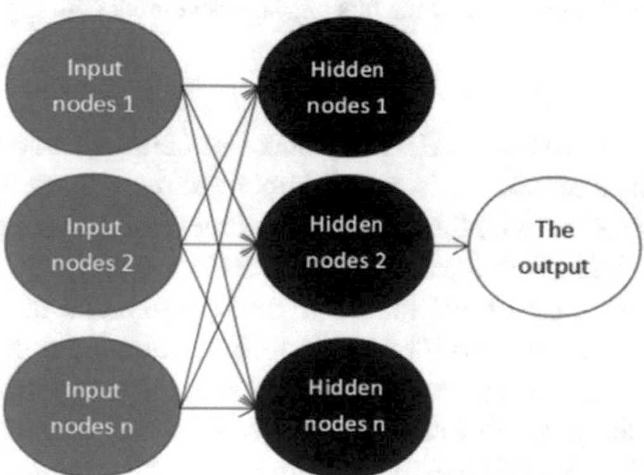

Fig. 3. Neural network (blue is the input layer, black is the hidden layer, and white is the output layer)

Figure 3 is a typical feedforward neural networks, the neural network model is composed of input layer, hidden layer and output layer, the input layer is composed of a number of input nodes, each input node represents an initial data, hidden layer is composed of a number of hidden nodes, each node represents a preset implied data, in business and management data system general specification requirements for data, data specification requirements such as the number of human resources in a business organization and the size of the organization. This basis, the input nodes of input layer will enter the hidden layer, and the hidden nodes interact within the hidden layer, the interaction of the relationship between the two types of nodes is determined by the rules, major in business and management data for comparison rules, association rules is compared to the initial data input node is consistent with the hidden node data, if it is consistent, it means that the initial data conforms to the specification; on the contrary, there is some error. The results can be obtained in this way, and the results will be visualized in the output layer for manual judgment.

5.2 System Construction Scheme

Figure 4 is the overall framework of the business management data application system.

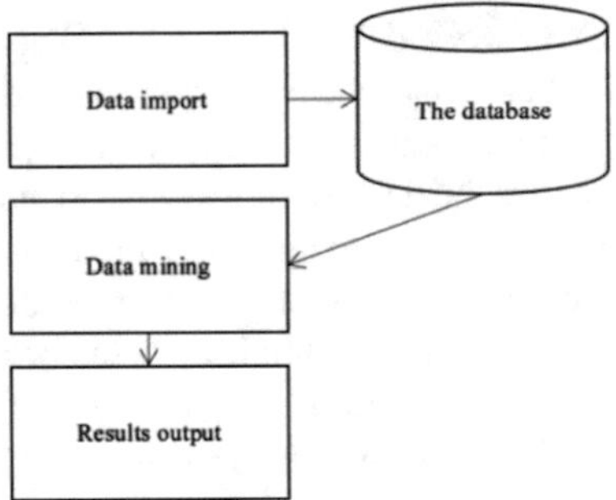

Fig. 4. The overall framework of the business management data application system

Data Import. Data import is a link in the process of data collection. In the past, it was mainly responsible by manual, that is, manual needs to record data and then import data into the database. However, with the significant increase of data magnitude, manual can no longer be responsible for this work, so it needs to use technical means to solve the problem. From this perspective, the industrial and commercial management organizations can use the current popularity of the mobile terminal and the Internet of things technology development business management system, through the system to automatically collect data, such as alcohol and tobacco products sell the situation in view of the retail merchant, ask its every time before selling to scan electronic product code, so you can get the data of business, these data will be automatically imported into the database through the management system.

Database. Because of the large amount of data, ordinary databases cannot meet the needs of industrial and commercial data storage and cannot form initial data sources. Therefore, business management organizations are advised to choose cloud databases, which meet the needs because of their unlimited capacity. It should be noted that although cloud database can meet the huge data storage requirements, its use will bring some security risks, so security protection should be done in database applications.

Data Mining. Refer to Fig. 1 construction corresponding data mining system, but in order to ensure system targeted for the work of business management, industrial and commercial management organizations to do a good job in the system configuration, such as the need to input relevant business management standard in the system, construct the data management system of the hidden layer of neural network in this way can system for industrial and commercial management data mining analysis, accurate results are obtained.

Result Output. Because data mining analysis results in the final show human need, so in terms of the output should consider the results of artificial reading some requirements, which is artificial generally want to be able to read as a result, at the same time, the results show the form to accord with human cognition, not is, in the form of electronic code results output, with which shall be connected to the artificial handheld terminal, computer terminal, etc., and the transcoding tool is used to obtain the results consistent with artificial cognition.

6 Conclusion

To sum up, there are many problems in the current situation of business management. These problems make business management organizations realize the importance of data application in work. Therefore, organizations should actively demand relevant technical means to apply data, and data mining technology is a good choice. With the help of this technology, business management organizations can timely understand the market situation, identify problems can quickly make accurate decisions, so as to maximize the quality of work.

References

1. Raut, A.R., Khandait, S.P.: Review on data mining techniques in wireless sensor networks. In: IEEE Sponsored 2nd International Conference on Electronics And Communication System (ICECS 2015). IEEE (2020)
2. Widyati, R., Ashari, A., Afiahayati.: A review of using data mining and machine learning for predicting drug loading modeling in solid lipid nanoparticles containing curcumin. J. Phys. Conf. Ser. **1918**(4), 042015 (2021)
3. Javier, L.Z., Torralbo, J., Cristobal, R.: Early prediction of student learning performance through data mining: a systematic review. Psicothema **202133**(3), 456–465 (2021)
4. Fan, J., Zhang, M., Sharma, A., et al.: Data mining applications in university information management system development. J. Intell. Syst. **31**(1), 207–220 (2022)
5. Shirono, T., Niizeki, T., Iwamoto, H., et al.: Therapeutic outcomes and prognostic factors of unresectable intrahepatic cholangiocarcinoma: a data mining analysis. J. Clin. Med. **10**(5), 987 (2021)

6. Liao, S.H., Widowati, R., Puttong, P.: Data mining analytics investigate Facebook Live stream users' behaviors and business models: The evidence from Thailand. Entertain. Comput. **41**(1), 100478 (2022)
7. Rodr Guezherrera, A., Reyesandrade, J., Rubioescudero, C.: Rationale for timing of follow-up visits to assess gluten-free diet in celiac disease patients based on data mining. Nutrients **13**(2), 357 (2021)
8. Linlin, L.I.: The intrusion data mining method for distributed network based on fuzzy kernel clustering algorithm. Int. J. Auton. Adapt. Commun. Syst. **14**(4), 1(2021)
9. Yang, C.Y., Chen, I., Ogata, H.: Toward precision education: educational data mining and learning analytics for identifying students' learning patterns with ebook systems. Educ. Technol. Soc. **24**(1), 1176–3647 (2021)
10. Avdagic-Golub, E., Begovic, M., Causevic, S., et al.: Profiling contact center customers for optimization of call routing using data mining techniques. In: 2021 20th International Symposium INFOTEH-JAHORINA(INFOTEH) (2021)

Cyber Intelligence for Network and Cloud Technologies

Application of Digital Media Technology in Computer Design Art Creation

Xian Du[1(✉)] and Rashid Gul[2]

[1] Shanghai Institute of Visual Arts, Shanghai, China
duxian@siva.edu.cn
[2] University of Peshawar, Peshawar, Pakistan

Abstract. With the development of today's technology, digital media technology (DMT) will be involved in many current fields. DMT is also changing computer design art creation. The use of digital media art will better perform computer design work, create higher-quality works, and better convey the author's artistic creation views. This article studies the application of DMT in computer design art creation, and aims to understand the intersection of DMT and computer design art creation, and to better play the role of DMT in creative work. This article mainly uses statistical analysis and experimental research methods to combine the theoretical analysis results with the actual situation of computer design art creation. It also conducts investigation and data analysis on the role of DMT in computer design creation. The results show that 81.5% of students prefer to use DMT in computer design and artistic creation, and they use DMT more frequently. It is necessary to study the application of DMT in computer design art creation.

Keywords: Digital Media Technology · Computer Design · Artistic Creation · Practical Application

1 Introduction

In the 21st century, digital new media art has become a relatively broad design technology, which has been applied in various industries. Traditional sculpture, cultural relics, literature and art can no longer increase the amount of information that the audience can receive, nor can they satisfy the audience's vision. Therefore, digital art design requires practitioners to achieve the best combination, and combine traditional media with new media organizations to achieve the best combination.

There are many research results in the application of DMT in computer design art creation. For example, RosliH proposed that since the existing exhibition does not provide all-DMT, most museum visitors do not have a high degree of experience with the real digital experience [1]. Some scholars said that my country's modern art creation is inseparable from DMT, and the application of DMT has had a great impact on art design and creation [2]. Some scholars believes that DMT is gradually being applied to various fields, and it has also brought some changes to the development of modern design art. The continuous development and improvement of DMT. The influence of

J. H. Abawajy et al. (Eds.): ICATCI 2022, LNDECT 170, pp. 235–242, 2023.
https://doi.org/10.1007/978-3-031-29097-8_28

DMT on modern design art helps to better apply DMT according to the needs of modern design art [3]. Therefore, this topic is aimed at the research on the application of DMT in computer design and artistic creation, which will be of great research value in related fields in the future.

The research content of this article has these aspects: expound the classification and expression of digital media art, analyze DMT, find out the intersection of computer design art creation and digital media art, and explore the existence of DMT in computer design art creation. Strengths and weaknesses, and investigate and analyze the effect of DMT on computer design art creation, what kind of development and change it brings to computer design art creation, and analyze the relationship between the two and finally get the result.

2 The Use of DMT in Computer Design Art Creation

2.1 Classification and Expression of Digital Media Art

2.1.1 Classification of DMT

The birth of digital media art stems from the advent of the electronic age and the rapid development of computers. "Digital media art works can be divided into two forms of artistic expression, static and dynamic; from the perspective of computer creation software, it can be divided into two creative methods, linear and interactive [4, 5].

2.1.2 The Expression of Digital Media Art

Digital media art uses computers as a creative or artistic method to digitize traditional media. However, compared with traditional media, digital media has more unique characteristics, such as interactivity. The field of interactive product design includes interactive design, information visualization design, UI design, game design, web design and multimedia design. This field pays more attention to user experience and user needs, and puts the user's interactive experience in the first place [6, 7].

2.2 The Intersection of Computer Design Art Creation and Digital Media Art

Modern computer design uses digital technology to create works, efficiently completes the design work, and enriches the form of computer design works to a greater extent. Digital media art has also become a pioneer in computer design art creation, bringing the entire industry Here comes a huge opportunity and creativity [8, 9].

2.2.1 The Intersection

In terms of Internet technology, under the background of network broadband speed, various new Internet concepts go hand in hand. New media network concepts such as big data, cloud, cloud storage, and cloud computing also provide a broader platform. The application of digital design to art design is comprehensive. The development provides technical support and creative sources for digital art. Art design uses feasible digital technology to enrich itself. Therefore, digital media art the use of art and design is also manifested at the media level [10, 11].

Looking at the history of art, every new concept of art is inseparable from the change of the media. Nowadays, the relationship between digital media art and media is obvious. Digital media art is composed of computer science, art and communication. The expression of modern digital media art is also inseparable from the expression. It can be the digitization of traditional media. It can be the performance of new media [12].

2.2.2 Intersection at the Media Level

The development of science provides technical support and creative sources for digital art. Art design uses feasible digital technology to enrich itself. Therefore, the use of digital media art and art design is also reflected in the media level. The digital technology environment will inevitably lead to the cultural form of the times. Affected, the art museum has an undoubtedly role in improving the quality of the people. Not only virtual reality technology has made great progress in art design, but other modern digital media can also be expressed in art design.

2.2.3 Intersection at the Application Level

Virtual and reality have become two important methods of art design. The application can be divided into online and offline. Online mainly relies on electronic terminals and digital network media, so that the audience can see the artistic effect close to Realistic effects, even surpassing special effects that cannot be achieved in reality; digital art design applications have huge development potential online and offline, and practitioners must also have relevant software and hardware application capabilities.

2.3 Advantages and Disadvantages of DMT in Computer Design Art Creation

2.3.1 Advantages of DMT in Computer Design Art Creation

Compared with traditional public art forms, DMT has the following advantages for public art works that incorporate DMT:

Digital technology itself has extremely powerful functions and advantages. After a high degree of development and popularization, it has profoundly changed and supported the way of life of our time. It is ubiquitous in all aspects, and it can be said that it penetrates into the global capillaries of social life. It is precisely because of this inherent characteristic advantage that the application of digital technology is inherently popular in various fields, fusion and versatility. The Internet has greatly digitized human society, and any behavior implemented through digital means will work without communication barriers.

Due to digital information processing and dissemination technology, digital media-based art in public spaces also has this characteristic.

2.3.2 The Disadvantages of DMT in Computer Design Art Creation

Based on digital technology, electricity will inevitably be used, and the demand for electricity is also an important constraint. Whether the power system and energy conservation and environmental protection can be solved in the public environment have become issues that have to be considered in artistic creation.

The application of digital technology in computer design art creation is limited by the maturity of the technology itself. For some newer technologies such as VR and AR, although the application prospects in public art creation are very broad, even it can become a new direction of expression, but its restricted conditions such as the use environment and the systematic cooperation with other technologies are not perfect. Therefore, this idealized application is only a concept at present.

2.4 Application of Video Compression Algorithm in Digital Media

The basic idea of the DCT transform is to describe the image signal with transform coefficients in the frequency domain. The process is to first divide the entire image into macro blocks, and after the transformation, the correlation between the transformation coefficients is greatly reduced. Only a few low-frequency components are allocated a fixed number of bits and transmitted, and the receiving end undergoes an inverse discrete cosine transform (IDCT) to return to the spatial domain to restore the original signal 1IS1. Use the following formula to calculate the 8 × 8 macroblock DCT transform:

$$G(\mathrm{m}, n) = \frac{1}{4}K(\mathrm{m})K(n)L\sum_{t=0}^{6}\sum_{a=0}^{8} d(s,f)\cos\frac{(4s+2)i\Gamma}{16}\cos\frac{(4f+2)n\Gamma}{16} \tag{1}$$

The two-dimensional inverse DCT transform (IDCT) formula is

$$g(\mathrm{s}, n) = \frac{1}{4}\sum_{\mathrm{m}=0}^{6}\sum_{n=0}^{6} K(\mathrm{m})K(n)D(m,n)\cos\frac{(4s+2)m\Gamma}{16}\cos\frac{(4f+2)n\Gamma}{16} \tag{2}$$

3 Experimental Investigation and Research on the Effect of DMT on Computer Design Art Creation

3.1 Questionnaire Design Process

The subjects were 100 students from the experimental class and the control class.

(1) When preparing the preliminary questionnaire, the number of questions should be minimized to avoid fatigue.

Distribute the questionnaire. The questionnaire is distributed online and on-site, and friends are invited to help friends and students around. A total of 100 questionnaires were distributed, including 100 completed questionnaires and 100 completed questionnaires.

The results of the questionnaire were analyzed, including students' attitudes towards DMT, frequency of use, etc.

3.2 Questionnaire Content

The design of the questionnaire is mainly divided into three parts:

The first part is a survey of the differences in the attitudes of the experimental class and the control class towards the use of DMT in computer design art creation;

The second part is the use of digital image processing, computer vision, video technology, computer graphics and other DMT information collection in artistic creation;

The third part is to organize the data collected from the questionnaire to understand students' attitudes and frequency of using DMT in computer art creation.

4 Experimental Investigation and Analysis of the Effect of DMT on Computer Design Art Creation

4.1 The Difference Analysis of the Experimental Class and the Control Class on the Use of DMT in Computer Design Art Creation

This article investigates and researches on the attitudes of the students in the experimental class and the control class to the use of DMT and the role of DMT in computer design art creation. The experimental results are shown in Table 1.

Table 1. Differences in students' attitudes and preferences towards DMT

Project	Test group (M ± SD)	Control group
Average value	4.34 ± 0.264	3.34 ± 0.632
Like very much	61.2	38.3
Like	20.3	37.6
General	8.3	12.4
Dislike	5.7	9.6
Dislike very much	4.3	4.1

It can be seen from Fig. 1 that after teaching, the experimental class used DMT to choose "very like" and "like", accounting for 81.5%, holding a "general" attitude accounted for 8.3%, holding "dislike" and "Very disliked" was 5.7% and 4.3% respectively. 75.9% of the students in the control class chose "very like" and "like" for using DMT, and 12.4%, 9.6%, and 4.1% of the attitudes of "average", "dislike" and "very dislike" were respectively. There are obvious differences in the survey results. After teaching, the experimental class students' attitudes and preferences for using DMT have been significantly improved.

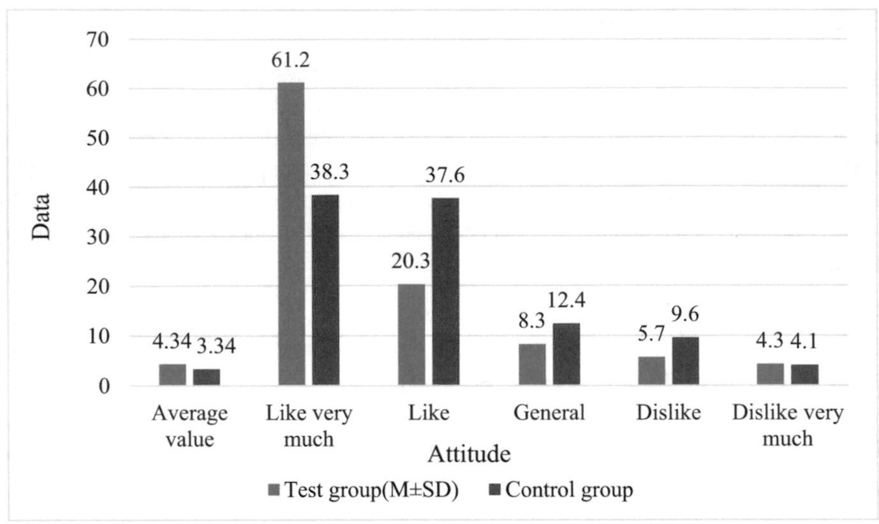

Fig. 1. Differences in students' attitudes and preferences towards DMT

4.2 Analysis of the Application of DMT in Artistic Creation

This paper investigates the frequency of using digital image processing, computer vision, video technology and computer graphics in artistic creation by 100 students in the experimental class and the control class. The experimental results are shown in Table 2.

Table 2. Differences in application of DMT in artistic creation

Project	Frequent	Occasional	Seldom
Digital image processing	16	7	2
Computer vision	18	6	1
Video technology	19	4	2
Computer Graphics	12	10	3

It can be seen from Fig. 2 that the number of people who frequently and occasionally use digital image processing, computer vision, video technology, and computer graphics in artistic creation account for 23, 24, 23, and 22, respectively, and more in artistic creation. The number of people who rarely use digital image processing, computer vision, video technology, and computer graphics who use DMT are 2, 1, 2 and 3, respectively. It can be seen that in the process of artistic creation, DMT is used more frequently.

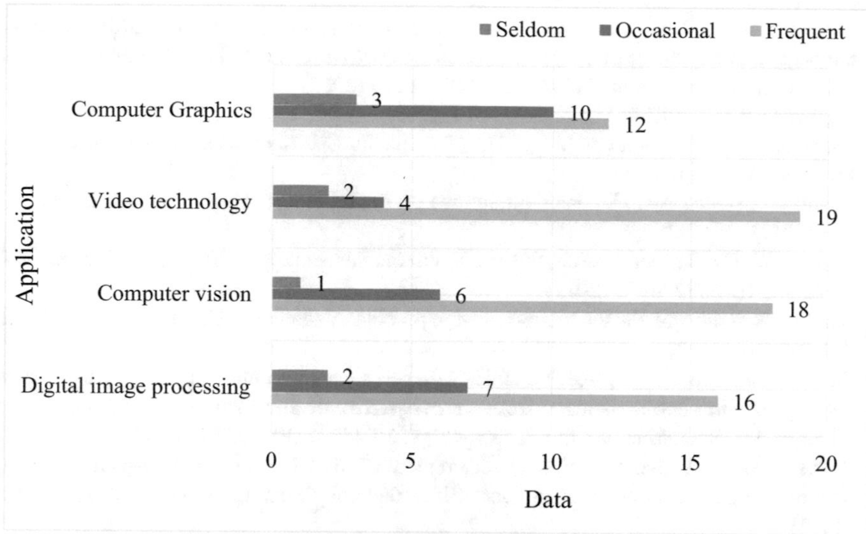

Fig. 2. Differences in application of DMT in artistic creation

5 Conclusions

In today's era, DMT has penetrated into various industries, especially in the computer field. Although in our country, digital media started late in our country, DMT has great room for improvement in the technology of computer design art creation. In today's and future society, the research and application of computer design art creation also need to be explored and developed. This article explores the intersection of computer design art creation and digital media art. It also introduces the advantages and disadvantages of DMT in computer design art creation, and conducts an experimental investigation on the role of DMT in computer design art creation. From the investigation and analysis, it can be seen that students' attitudes and preferences towards the use of DMT have increased significantly and that in the process of artistic creation, DMT is used more frequently. Therefore, the research on the application of DMT in computer design art creation has great practical value.

Acknowledgment. 2021 Shanghai Institute of Visual Arts through the construction of major—Major Capacity building of Advanced commercial photography Designing and creation.

References

1. Rosli, H., Kamaruddin, N.: Visitor experience's on DMT for the museum exhibition in Malaysia: a preliminary findings. Int. J. Sci. Res. **7**(2), 245–248 (2020)
2. Rizk, M., Baghdadi, A., Jezequel, M., et al.: No-instruction-set-computer design experience of flexible and efficient architectures for digital communication applications: two case studies on mimo turbo detection and Universal Turbo Demapping. Design Autom. Embedded Syst. **25**(1), 1–42 (2021)

3. Lühmann, A.V., Wabnitz, H., Sander, T., et al.: M3BA: a mobile, modular, multimodal biosignal acquisition architecture for miniaturized EEG-NIRS-based hybrid BCI and monitoring. IEEE Trans. Biomed. Eng. **64**(Issue 6), 1199–1210 (2017)
4. Bagherian Khosroshahy, M., Abdoli, A., Rahmani, A.M.: Design and power analysis of an ultra-high speed fault-tolerant full-adder cell in quantum-dot cellular automata. Int. J. Theoret. Phys. **61**(2) (2022)
5. Licata, J.I., Baker, A.E.: Updated guidelines on digital media use by children. Jaapa **30**(4), 1–3 (2017)
6. Thanavathi, C.: Teachers' perception on digital media technology. Turkish J. Comput. Math. Educ. **12**(10), 6972–6975 (2021)
7. Shade, L.R.: Civic media: technology, design practice. Canadian J. Commun. **44**(2), 289–291 (2019)
8. Shoai, A., Ladeveze, L.N., Acevedo, L.E.: Combining dialogue and technology: digital media and organization-public dialogue from the perspective of multinationals in Spain and Latin America. Revista Latina de Comunicacion Soc. **2020**(77), 309–327 (2020)
9. Romele, A.: Entangled in digital media: review of digital media: human-technology connection, by Stacey O'Neal Irwin. Soc. Phil. Technol. Quar. Electron. J. **21**(1), 106–113 (2017)
10. Nuenen, T.V., Scarles, C.: Advancements in technology and digital media in tourism. Tourist Stud. **21**(3), 146879762199041 (2021)
11. Coleman, S.: The digital difference: media technology and the theory of communication effects. J. Commun. **67**(6), E7–E8 (2017)
12. Schiller, D.: The filters through which we live: representing the emotional self in digital media. Int. J. New Media Technol. Arts **12**(1), 1–17 (2017)

Using Computer 3D Drawing Technology to Design Traditional Village Spatial Structure and Protection and Utilization Value Evaluation

Ying Xiong[1(✉)] and Feng Peng[2]

[1] Zhejiang International Studies University, Hangzhou, Zhejiang, China
ly13114613831@163.com

[2] Power China Huadong Engineering Corporation, Hangzhou, Zhejiang, China

Abstract. Traditional villages embody the essence of our country's agricultural civilization, local folk customs and architectural features, condensing the historical memory of the Chinese nation, and showing the cultural connotation of our country's harmonious coexistence with nature. However, the reality is not optimistic. For example, the unreasonable spatial structure design of traditional villages, the destruction of characteristic buildings, the imminent disappearance of traditional villages, etc., which make the protection and development of traditional villages face severe challenges. At the same time, the rapid development of modern information technology and network communication has also brought more convenience and comfort to people's lives. With the advancement and breakthrough of computer technology and the increasing popularity of mobile terminal equipment in the Internet era, more and more different types of two-dimensional drawing tools and other digital three-dimensional data processing software have emerged. Among them, 3D graphics technology is a new type of computer graphics processing technology, which has powerful functions and high integration characteristics, and has been widely used in graphics design, image production, virtual reality and other fields. This article mainly adopts the questionnaire survey method to study the design of traditional village spatial structure by computer 3D mapping technology and its protection and utilization value evaluation. It is intended to use computer 3D mapping technology to rationally plan and design the traditional village spatial structure to better protect and Utilize the unique resources of traditional villages. According to the survey results, quite a few of the villagers interviewed said that the spatial structure of the village was designed reasonably, but the design of certain areas still needs to be improved.

Keywords: Traditional Villages · Spatial Structure · Protection and Utilization · 3D Mapping Technology

1 Introduction

With the passage of time and the rapid progress of urbanization, many traditional villages that have not been discovered and protected have disappeared. In this context, the Chinese government and all walks of life are paying more and more attention to the

J. H. Abawajy et al. (Eds.): ICATCI 2022, LNDECT 170, pp. 243–250, 2023.
https://doi.org/10.1007/978-3-031-29097-8_29

protection and development of traditional villages. Traditional villages have witnessed the development of the Chinese nation. They are precious material and spiritual wealth in Chinese civilization. Modular planning management and uniform spatial structure design have also caused a large number of traditional villages to lose their cultural characteristics. Therefore, how to protect and make good use of traditional villages, reduce the destruction of traditional villages, and delay their decline are also problems that need to be solved urgently in our country. At present, computer information technology is developing rapidly, and computer 3D drawing technology is also widely used in many fields, such as graphic design, image production, and virtual reality. In view of this, this article studies how computer 3D mapping technology can innovate and transform the design of traditional village spatial structure, so as to better protect and utilize the advantageous resources of traditional villages, so as to cope with the existing problems of traditional villages.

At present, there are relatively rich research results on the design of traditional village spatial structure and its protection and utilization value evaluation. For example, Some scholars pointed out that in order to refine and protect the spatial structure of traditional villages and the characteristics of house courtyards, restoration measures and utilization methods are also needed [1]. Some scholars believes that the purpose of protecting and using traditional villages is to ensure the sustainable development of traditional villages [2]. Some scholars proposed that an effective understanding and mastering of the environmental quality and the characteristics of public activities in the public space of traditional villages can better optimize the public space [3]. Therefore, this article has reference significance and practical value for the future research on the application of computer 3D mapping technology in spatial structure design and the protection and utilization of traditional villages.

This article mainly includes these aspects: it explains the traditional village and its related concepts, what is the traditional village spatial structure and value evaluation; it also introduces the application of computer 3D mapping technology in the traditional village spatial structure design, and the image hash generation algorithm in In the application of traditional village spatial structure design; finally, the selected villages are selected to investigate whether their spatial structure design is reasonable and whether the protection and utilization are appropriate, and finally the analysis results are obtained.

2 Computer 3D Mapping Technology for the Design of Traditional Village Spatial Structure and Protection and Utilization Value Evaluation

2.1 Traditional Villages and Related Concepts

2.1.1 Traditional Village

Traditional villages are the main place of production and inheritance of Chinese history and culture, which is conducive to better inheriting traditional culture and highlighting the traditional cultural characteristics of the village. Through the changes of the times, it still retains the traditional architectural style, historical culture and folk customs, and has not been influenced by modern planning concepts. It has a relatively complete architectural complex with regional colors, which contains intangible historical culture and place spirit. These villages represent the historical development context of a certain region and a certain culture, and as carriers of residents' activities and spiritual culture, they contain a lot of traditional cultural information [4, 5].

Roughly speaking, our country has started to protect traditional villages since the beginning of this century, and we have seen actual results. For example, the living environment has been improved, but there are still some problems that have not been resolved, and some new problems have also emerged. There are fewer and fewer traditional villages, and the serious situation has caused the traditional villages to eventually decline and disappear. Natural decline is also one of the main reasons for the decrease in the number of traditional villages, such as fire and flood damage. There are also some traditional buildings under the influence of natural and man-made destruction, aging and collapse are serious, and they lack the necessary maintenance.

2.1.2 Spatial Structure of Traditional Villages

The spatial structure of traditional villages refers to the scale and form, function and land use layout of village settlements, landscape composition, social organization and cultural characteristics, and spatial evolution mechanism, etc., which can reflect the local social economy and culture to a certain extent.

Traditional village space, its spatial connotation includes three levels of artificial material, natural ecology and spiritual culture. Among them, the artificial material space element is the organic combination of the various components of the building and other material space levels, which constitute the spatial field of the village life and production; the natural ecological space mainly includes the natural environment's topography, water system, geology and other resources, which is the bearing the survival and development of the village. The foundation of natural space, spiritual space is a space maintained by local culture, humanistic spirit and blood ties in the process of people's utilization and transformation of material space in traditional villages. Its non-material cultural connotation is reflected in its material carrying space [6, 7].

The form and structure of the internal space of a traditional village are determined by the basic elements of its material space, such as streets and lanes, open spaces in squares, buildings and their courtyards. The internal space elements of traditional villages are the main research objects of space activation design. The research in this article emphasizes the general laws and distinctive features of the internal space of traditional villages at the material form space level of nodes, streets, courtyards and buildings.

2.1.3 Valuation

To evaluate the value of traditional villages, we must follow these principles. The first is to follow the principles of authenticity and uniqueness. The authenticity of traditional villages is the basis for evaluating their value. We should not only pay attention to the universal value of traditional villages, but also their unique value. The second is to follow the principle of completeness and richness. The value evaluation of traditional villages is a coherent process. While maintaining its integrity in this process, the richness of content and methods should also be adhered to, so as to cover all aspects of the value of traditional villages. The third is to follow the principles of sustainability and operability. Traditional villages are constantly changing over time. Value evaluation should adhere to sustainability and a process of dynamic change. At the same time, early warning and monitoring should be strengthened to prevent damage to value or damage. Factors that may cause threats need to be discovered in time and corresponding measures should be taken. The content of the comprehensive investigation also needs to have a certain correspondence with the content of the value assessment, the assessment report and the further protection plan, so as to enhance the operability of its value assessment.

Assessing the protection and utilization value of traditional villages, the environmental value of the village is the most intuitive and important traditional village value. Generally speaking, the evaluation of the value of traditional villages mainly focuses on these aspects: first, it mainly evaluates the regional traffic, natural geography, site selection environment, and traditional features of the village, whether the historical environmental information is relatively completely preserved, and reflects the wisdom and skill of adapting to local conditions. The second is to see whether the public space and landscape elements have well preserved the spatial pattern of traditional villages, and historical traces can be found from large to small scales. The third is to see whether the traditional architectural style and features can be retained, which can reflect the local cultural characteristics. After fully realizing the value of the three aspects of the village environment, it is possible to more clearly protect and use its value characteristics and condense the regional culture and the spirit of the place. The traditional life value of the crowd is the most easily overlooked but the most vivid traditional village value [8, 9].

The value evaluation of traditional villages is a close combination of qualitative analysis and quantitative analysis. Only when they complement each other can the value evaluation of traditional villages be more reasonable, systematic and comprehensive. At present, most of the value evaluation of traditional villages stays at the stage of qualitative analysis. Although there are some cases combined with quantitative analysis, they have not been widely used. The combination of qualitative analysis and quantitative analysis is a scientific and reasonable way. The content of the value assessment should be consistent with the comprehensive survey of traditional villages, and it should be more referential and operable in the process of further protection and utilization.

2.2 Application of Computer 3d Drawing Technology in the Design of Traditional Village Spatial Structure

Computer 3D graphics technology is a new method of using three-dimensional images in the field of computer technology. It can digitize, process and visualize image information. In the production process of 3D graphics, various information needs to be processed and analyzed, and the application of computer software technology is also relatively extensive. First, make a virtual building model, establish a relationship network system between the space in the building and the three-dimensional simulation environment, and then complete the creation, projection and conversion of the virtual reality building model. Then use computer-aided design methods to realize the modularization of architectural design functions, the standardization of processes, and the automation of design results generation [10].

It can model traditional village buildings and indoor spaces through two-dimensional planes or three-dimensional entities, so as to achieve reasonable division of different functional areas, landscape elements and environments in traditional villages, and can generate virtual models on a computer network platform. The 3D model can automatically create, modify and perfect the internal structural components and building facade structure of the entire village according to user needs; at the same time, it can also design the building structure and layout according to the actual situation of the traditional village.

2.3 Application of Image Hash Generation Algorithm in the Design of Traditional Village Spatial Structure

The image hash generation algorithm has a certain role in computer 3D mapping technology. It can low-pass filter the original image of the overall architectural space of the traditional village to obtain the low-frequency part of the image, and then extract the histogram of the low-frequency part of the image, according to certain rules divide the pixels into several groups, and use the ratio of the number of pixels in each pixel group to generate a hash value. By modeling the two-dimensional plane or three-dimensional entity of the spatial structure, a virtual model is generated under the computer network platform, and finally a reasonable design of the spatial structure of the traditional village is realized.

In order to make the hash algorithm sensitive to local tampering, we quantify and compare the wavelet decomposition coefficients of different groups to describe the local characteristics of the image, thereby generating another part of the image hash value, so that the final hash value of the image is determined by These two parts are composed of hash values. For preprocessing, regardless of the color information of the image, the color image is first converted into a grayscale image. The formula for the low frequency part of the original image U is expressed as.

$$U_{low}(z, q) = J(z, q, \rho) * U(z, q) \tag{1}$$

Extract the histogram N, get the histogram shape of the image, and increase the security of the hash algorithm.

$$O(K) = \{o_u | u = 1, \ldots, K, 0 \leq o_u \leq 263\} \tag{2}$$

Then use O(K) to extract a histogram composed of K gray levels from N.

$$N_K = \{n_k(j_u) | u = 1, \ldots, K\} \tag{3}$$

Among them, K is the number of gray levels of the histogram, and $n_K(j_u)$ is the number of pixels whose pixel value is j_u. The image hash generation diagram is shown in Fig. 1.

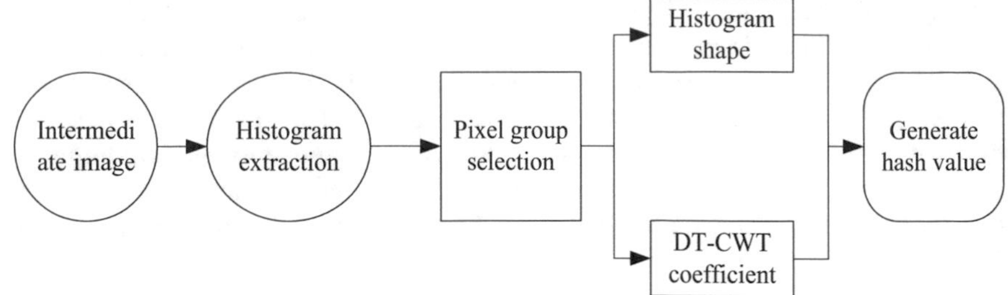

Fig. 1. Block diagram of the image hash

3 Investigation and Research on Traditional Village Spatial Structure Design and Protection and Utilization Value Evaluation

3.1 Questionnaire Design Process

This questionnaire selects a traditional village in G city for investigation, and the selected objects are the permanent residents of the village, with a total of 90 people. Through the issuance of paper questionnaires, collect and sort out the information filled in by the survey subjects, and then analyze them and draw the conclusions of the questionnaire.

(1) The preliminary preparation of the questionnaire takes into account that the interviewees may come from different ages and their literacy levels, and the difficulty and number of questions are designed to be as simple and understandable as possible to avoid fatigue or incomprehension of the interviewees.
(2) The questionnaire is distributed by means of inquiring and filling out on-site questionnaires. A total of 90 questionnaires were distributed and 90 valid questionnaires were recovered. The questionnaire recovery rate was 100%.
(3) Questionnaire analysis the collected questionnaire information is sorted out and the required information data is obtained. Analyze the results of the questionnaire. The analysis results mainly include the opinions of local residents of different ages on the spatial structure design of the village. Some of the results obtained from the questionnaire are as follows.

3.2 Questionnaire Survey Content

A questionnaire survey was conducted among 90 interviewees who selected the village on whether the spatial structure of the village was reasonable and whether the protection and utilization were appropriate; then the information collected in the questionnaire of these 90 interviewees was sorted out for the purpose of this. The analysis of the view on whether the spatial structure of the village is reasonable.

4 Investigation and Analysis of Traditional Village Spatial Structure Design and Protection and Utilization Value Evaluation

In this questionnaire survey, local villagers of different ages were interviewed to investigate whether the spatial structure of the village is reasonable and whether it is properly protected and used. The survey results are shown in Table 1.

Table 1. Views of villagers of different ages on the design of village spatial structure

Project	Teenagers	Middle-aged people	Senior people
Very reasonable	6	7	8
Reasonable	8	9	10
General	11	10	9
Unreasonable	5	4	3

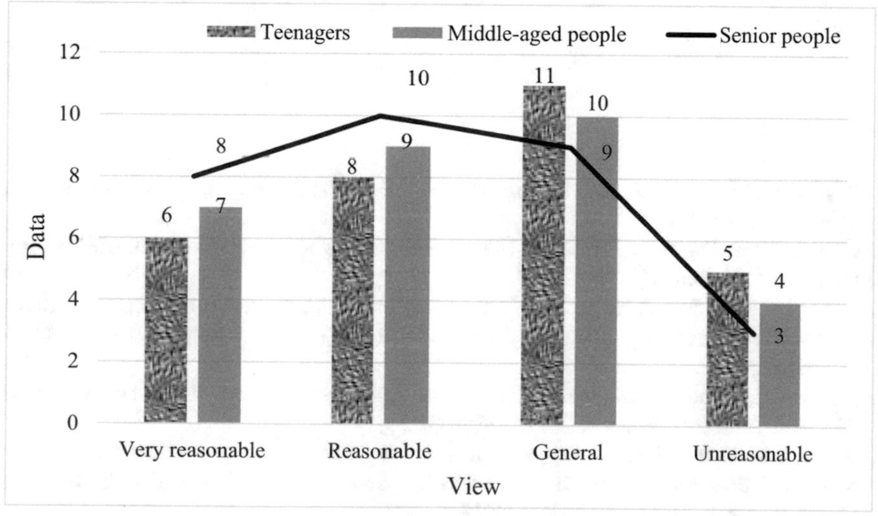

Fig. 2. Views of villagers of different ages on the design of village spatial structure

It can be seen from Fig. 2 that among the interviewed villagers, 14, 16, and 18 of the young, middle-aged, and old people respectively indicated that it was reasonable, which was more than half of the interviewees, but there were still 12 people who indicated that the design was unreasonable. It can be seen that quite a few of the interviewed villagers said that the spatial structure of the village is designed reasonably, but the design of certain places still needs to be improved.

5 Conclusion

Traditional villages, as our country's precious cultural heritage and treasures of non-renewable resources, contain the essence of agricultural civilization and Chinese excellent traditional culture. Our country is also paying more and more attention to the protection of traditional villages, re-planning the spatial structure of traditional villages rationally, establishing a relatively complete value evaluation process for traditional villages, evaluating the usable value, and making traditional villages better protected through value evaluation. At the same time, computer 3D graphics technology is a new technology that uses three-dimensional space visualization, and it has been widely used in many fields. The application of 3D mapping technology in the spatial structure design of traditional villages can better meet people's needs for traditional village environment and housing through rational utilization and transformation of existing resources and environmental conditions. At the same time, this has played a certain role in protection and utilization, to a certain extent, meets the requirements of the sustainable development of traditional villages, and also brings social benefits and other values.

Acknowledgment. Key projects of Humanities and Social Sciences in universities of Zhejiang Province.

References

1. Patrignani, M.: A split & push approach to 3D orthogonal drawing. J. Graph Algorith. Appl. **4**(3), 87–101 (2000)
2. Cho, S., Baek, D., Baek, S.Y., et al.: 3D volume drawing on a potter's wheel. IEEE Comput. Graphics Appl. **34**(3), 50–58 (2014)
3. Garvey, G.P.: Life drawing and 3D figure modeling with MAYA: developing alternatives to photo-realistic modeling. Leonardo **35**(3), 303–310 (2002). https://doi.org/10.1162/002409402760105325
4. Bernard, P., Mendez, J.D.: Drawing in 3D: using 3D printer pens to draw chemical models. Biochem. Mol. Biol. Educ. **48**(3), 253–258 (2020)
5. Sousa, L.R., Duarte, L.C., Coltro, W.K.T.: Instrument-free fabrication of microfluidic paper-based analytical devices through 3D pen drawing. Sens. Actuat. B Chem. **312**, 128018 (2020). https://doi.org/10.1016/j.snb.2020.128018
6. Borrero, M., Stroth, L.R.: A proposal for the standardized reporting of error and paradata regarding Structure from Motion (SfM) 3D models used in recording and consolidating archaeological architecture. Adv. Archaeol. Pract. **8**(4), 376–388 (2020)
7. Uchida, H., Ando, K.: Action program for small-scale riverbank protection in a Bangladesh village. Japan Agricul. Res. Quar. Jarq **45**(2), 137–143 (2012)
8. Maccone, C.: Moon farside protection, moon village and PAC (Protected Antipode Circle). Acta Astronaut. **154**, 233–237 (2019)
9. Laura, P.: Old village cemeteries from the perspective of heritage studies and heritage protection. Stud. Lithuania's History **2012**(29), 13 (2012)
10. Martínez, C., Encinas-Sanz, F., Serna, J., et al.: On the parametric characterization of the transversal spatial structure of laser pulses. Optics Commun. **139**(4), 299–305 (2020)

Statistical Analysis of Cloud Data in Internet-Based Product Design

Guang Yuan[1(✉)] and Rashid Gul[2]

[1] Beijing Shangpin Technology Development Co. Ltd., Beijing, China
Yuanguang2021@126.com
[2] University of Peshawar, Peshawar, Pakistan

Abstract. Internet-based product design platform theory is a key theoretical method in mass customization, which can meet diverse customer needs with low-cost and high-efficiency services and reduce the cost of enterprise product development. The purpose of this article is to study the statistical analysis of cloud data for Internet-based product design. This article studies the structure of the product design platform model for Internet cloud services, proposes a digital description technology collection of product platform models for cloud services, and proposes cloud-oriented services the integrated expression model of product platform resource information and its data structure. This article combines cloud computing technology and cloud services provided by Amazon, as well as applications such as HBase database, to study and implement a new massive data analysis system. Experimental research shows that the performance of the model proposed in this paper is significantly better than that of the MapReduce model by 20%, and as the amount of data increases, its advantages become more prominent based on the functional testing of each part of the system.

Keywords: Product Design · Cloud Computing · Data Analysis System · Performance Modeling

1 Introduction

In today's society, various data sources will generate huge amounts of data. Data is highly prosperous, science and technology are also highly developed, people have more and more methods and channels to obtain information and data, and their ability to obtain data is getting stronger and stronger [1, 2]. Nowadays, how to obtain data faster and more is no longer the main problem that people face; and how people can easily and effectively filter out the data that suits them from more and more data content has become an urgent need for people.

In the research on the statistical analysis of cloud data in Internet-based product design, many scholars have conducted research on it and achieved good results. For example, Spiegel R mentioned that the establishment of agricultural information platform construction should focus on agricultural information at the county and grassroots

J. H. Abawajy et al. (Eds.): ICATCI 2022, LNDECT 170, pp. 251–259, 2023.
https://doi.org/10.1007/978-3-031-29097-8_30

levels. Service organization construction and information management [3]. Feng B organized and analyzed the development status of cloud computing agricultural product information cloud data, and determined that cloud computing is the general direction of the future development of the information industry. Just like this, how to build a user-satisfied interface in data visualization design is the current design research that is lacking [4]. Therefore, the statistical analysis and research on cloud data of Internet-based product design is very necessary in today's society.

This paper analyzes the cloud data of cloud computing, integrates the design links in the product transaction process, and designs the model of cloud computing product transaction. When predicting the processing time, this paper analyzes the task allocation algorithm and the scheduling strategy of the cloud node, and proposes a mechanism for dividing the number of parallel tasks on the cloud node. Finally, the time expenditure prediction scheme proposed in this paper is verified through the benchmark test of big data.

2 Research on the Statistical Method of Cloud Data in Product Design Based on the Internet

2.1 The Data Structure of the Product Platform Resource Information Integration Expression Model

In order to effectively manage the resource data in the product platform design process in the cloud computing environment, according to the aforementioned product platform resource information integration expression model, an object-oriented method is used to define the product based on the analysis of the respective characteristics of the object model and the relational model. Platform class, process class, task class, resource class and view class, use the structural hierarchy of the class to express product platform information [5, 6].

2.1.1 Product Platform

The product platform mainly contains the relevant attribute variable information of the overall description information of the product platform design, which mainly includes four variables: attribute name, attribute type, platform usage mode, and industry description. The name is a character string that expresses the product platform name identification; the type is a character string that expresses the product platform category identification; the usage method describes the authorization method for the external delivery of product platform design services: the industry description expresses the industry information involved in the product platform.

2.1.2 Process Category

The process category describes the relevant attribute variables of the description information of the product platform development process, mainly including: process description, process characteristics, delivery and settlement methods and other variables [7, 8]. This class is the base class of the development process, by which the development view can be defined and mapped to the various stages of the product platform development process.

2.1.3 Task Category

Describes the task description information during the downward task decomposition of each development process of the product platform, which mainly includes variables such as task description, task information, and delivery and settlement methods. Inheriting this class can decompose development tasks into subcategories such as analysis tasks, planning tasks, design tasks, evaluation tasks, etc. The analysis tasks are oriented to the process of product platform requirements analysis and function transformation; the planning tasks are oriented to the product platform planning process; design The task is oriented to the product platform design and process design process; the evaluation task is oriented to the product platform evaluation and optimization process [9]. Face the development process of the characteristic product platform for each subtask and map to the specified view.

2.1.4 Resources

Describes a unified description of the distributed resources of the product platform, and its base class includes variables such as resource name, resource type, provider, and purchase time. Inheriting this class can decompose resources into subclasses such as design resources, information resources, product family instance resources, and knowledge resources [10, 11]. And these subclasses expand the description of static attribute information and dynamic attribute information. The inherent information of static attribute information resources, such as parameters, size, performance description and other information; dynamic attribute information is the dynamic information of the resource running in the cloud computing environment, Such as busy and idle status, running status, load status and other information. The resource category provides basic data for the product platform design service operation.

2.1.5 View Class

According to different product platform development process stages, the product platform development process decomposes and constructs information views for different stages and different personnel to support the different needs of the platform design process and resource invocation at different stages of the product platform development process in the cloud computing environment.

2.2 Research on Data Analysis System Based on Cloud Computing Platform

2.2.1 Implementation of the DB Manager Module

The main job of this module is the operation of the local database. Including the creation of tables in the database and its results.

Structure; operations such as adding, deleting, modifying, and checking data files in the database.

There are two tables stored in the local database of the system: one is for storing session information, and the other is for storing specific event information. The Session table contains the session id as the primary key and the start and end time of the session;

the Event table contains the id as the primary key, event information, and the data type of the status. The status is used to mark the Whether the record has been sent, its value is 0 means it has not been sent, and its value is 1 means the record has been sent [12].

2.2.2 Implementation of Network Status Check and Wifi Admin Modules

Before each http request is sent, this article will check the network status. For users of mobile devices such as mobile phones, this article will check whether their current network link is Wifi. For users of mobile devices, this article will only be sent when these devices have a Wifi connection.

2.2.3 Deployment and Implementation of Receiving Server

The receiving server is deployed in the Amazon EC2 environment, and the server is an EC2 instance created in Amazon. So at the beginning of this section, the creation of Amazon EC2 instances will be mentioned.

(1) Creation of Amazon EC2 instance
    Before using Amazon Cloud services, you need to create an AWS account. The account needs to be tied to the visa credit card to pay for the fees incurred by the user in the process of using Amazon cloud services. After the AWS account is created, use the account to log in to the Amazon Web Services page. After logging in successfully, you will enter the AWS Management Console page. In this page, choose to switch to the EC2 page.
(2) Design and implementation of EventCollector
    As mentioned in the previous content, the client of this system will send Http request in Http Post mode. The EventCollector in this article is a Java Servlet. The EventCollector class inherits the HttpServlet class and overrides the doGet and doPost methods.

EventCollector will accept requests sent by Http Post and Get, and save them in a local file. These local files will then be uploaded to the designated location on S3 by the Uploader program.

2.3 Queuing System of Cloud Computing Center

2.3.1 G/M/n Queuing System

In this paper, A(t) represents the cumulative distribution function (CDF) of the time interval, a(t) represents its probability density function (Probability Density Function, PDF), and the Laplace transform of the time interval (Laplace- Stieltjes Transform, LST) can be expressed as:

$$\mathrm{A}^*(s) = \int_0^{\infty} e^{-st} a(t) dt \tag{1}$$

LST is an important conversion function in queuing theory, which is mainly used to calculate PDF under CDF.

2.3.2 MlG/n/L Queuing System

In this queuing system, task arrival obeys Poisson distribution and the arrival rate is AAA, and customers (subtasks) obey the batch arrival mode. Therefore, this article assumes that the time interval of tasks obeys the exponential distribution C, and the CDF is C(t) = P(C < t), the PDF is BBB, and the LST of the interval time can be expressed as:

$$C^*(s) = \int_0^\infty e^{-st}c(t)dt \tag{2}$$

2.4 Service Technology of Product Platform Design Process

2.4.1 Cloud Service-Oriented Product Platform Design Process Description Model

The description model of the product platform design process for cloud services is to describe the product platform design and development activities in the cloud computing environment. It is the decomposition and combination of services at each stage of the product platform design process, and the call, configuration and integration of the product platform resources by the services of each stage The process can reflect the enterprise's application of product platform design services in the cloud service environment to realize the configuration design of product families and the manufacturing and generation of diversified products. It includes a variety of product platform services, such as demand analysis services, functional analysis services, and modular design. Services, configuration design services, product family customization services, product process services, manufacturing process services, and supply chain services.

2.4.2 Cloud Service Description of Product Platform Design Process

The product platform design service application SOA method defines the product platform design business as a sequence of steps from the perspective of business operations and processes: it is caused by the product platform requirements, and then the service information of the entire process of product platform design is transformed, and finally a service-based The combined product platform design business process output. SOA regards the business activities or processes performed in the product platform design process as services, and the cloud computing application system as a collection of services. The product platform design business process is a set of logically related task service combinations that realize the business results of a given product platform design, and is the optimal scheduling and interaction of product platform resources and design processes.

3 Experimental Research on Cloud Data Statistics Method for Product Design Based on the Internet

3.1 The Purpose of the Experiment

This paper proposes a hybrid parallel computing model with predictable performance. It uses the mainstream MapReduce as the benchmark model and integrates the iterative idea of the BSP (Bulk Synchronous Parallel Computing) model to reconstruct the basic components of the two models.

3.2 Experimental Process

This paper uses the WordCount application to verify the hybrid parallel computing model proposed in this paper and its corresponding data copy optimization scheme. First, a small sample benchmark test was run to obtain the relevant parameters of the model, and then the data copy optimization scheme was compared with HDFS. Finally, the scalability of the model was verified and compared with the MapReduce model under the Hadoop platform.

4 Statistical Research and Analysis of Cloud Data for Product Design Based on the Internet

4.1 Performance Prediction and Data Localization Analysis

This article applies for three nodes from the simulated cloud computing center built in the laboratory. The main configurations are as follows: Node1: 4-cores, 8G memory; Node2: 2-cores, 4G memory, Node3: 2-core, 2G memory. At the same time, this paper deploys the parallel computing model proposed in this paper to these three nodes, and the basic operating parameters of the model on the three nodes are shown in Table 1.

Table 1. Basic parameters of the model

	Pm (MB/S)	G (KB/S)	Inner-BS (ms)
Node1	24.26	163	34.27
Node2	14.22	241.63	53.50
Node3	10.32	210.99	40.17

As shown in Fig. 1, the time cost of the application can be predicted. There is an error between the predicted time and the actual time, mainly due to the volatility of CPU, network bandwidth and other performance factors. This article is selected after running multiple sets of data the average value of the parameters has caused a certain error, but the error range is still within the acceptable range.

4.2 Scalability and Practicality

This paper sets up four sets of experiments to compare the performance of the MapReduce model under Hadoop with the performance of the model proposed in this paper. The results are shown in Table 2.

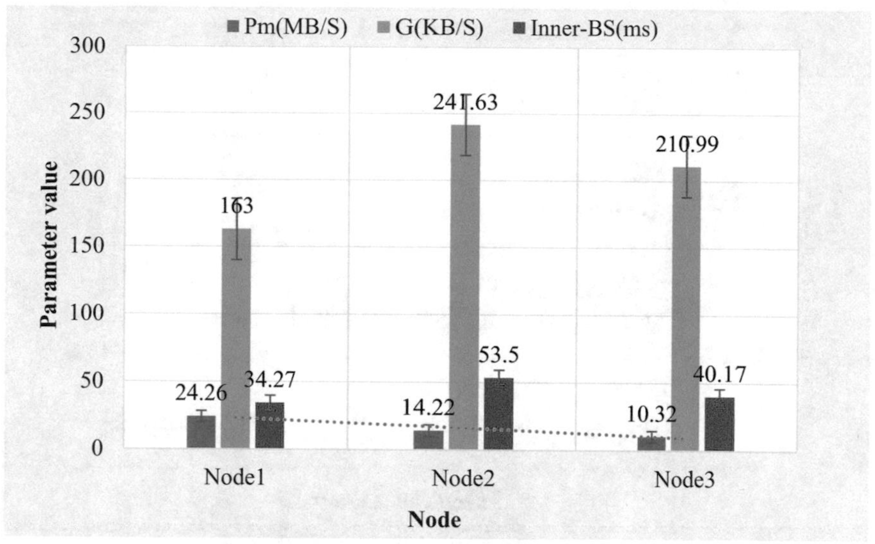

Fig. 1. Basic parameters of the model

Table 2. Performance comparison between hybrid model and MapReduce

Amount of Data (MB)	Synthetic Model	Hadoop
20 MB	5263	8881
40 MB	7361	1492
80 MB	7769	18861
120 MB	10072	23017

As shown in Fig. 2, the performance of the model proposed in this paper is significantly better than that of the MapReduce model by 20%, and as the amount of data increases, its advantages become more prominent based on the functional testing of each part of the system, the overall testing of the system, and The results of the above-mentioned stress and performance tests show that the system can fully meet the requirements of the enterprise for data analysis systems in terms of function and business requirements, and has stronger processing capabilities in terms of performance. Therefore, the system can replace the traditional data analysis system.

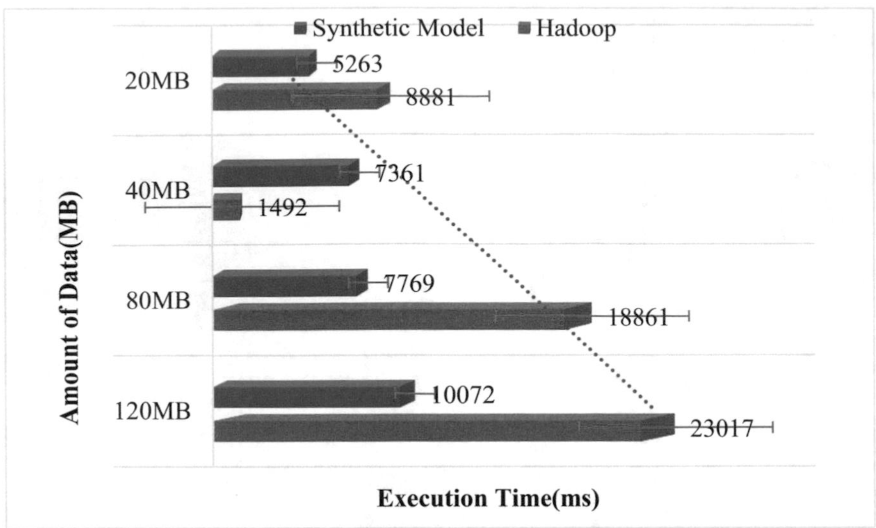

Fig. 2. Hybrid model and MapReduce performance comparison

5 Conclusions

This paper proposes a parallel computing model with predictable performance. It takes MapReduce as the benchmark model, highly integrates the design ideas of the BSP model, realizes the dynamic reconstruction of the model unit components, and provides a new way to shield the diversification of parallel computing models. At the same time, its performance prediction function provides a theoretical basis for users and cloud providers to apply for and manage resources.

References

1. Ad, A., Am, B.: Cloud attenuation statistics from radiometric measurements over a tropical location Kolkata, India. Adv. Space Res. **67**(1), 290–297 (2021)
2. Xiang, H., Zhang, T., Li, Z.: Big data cloud platform server load balancing algorithm based on improved chaotic partition algorithm. J. Phys. Conf. Ser. **1982**(1), 012116 (2021). https://doi.org/10.1088/1742-6596/1982/1/012116
3. Spiegel, R.: Big data and cloud computing have moved into design. Des. News **71**(10), 38–39 (2016)
4. Feng, B., Ma, X., Cheng, G., et al.: An efficient protocol with bidirectional verification for storage security in cloud computing. IEEE Access **4**(99), 7899–7911 (2017)
5. Prema, A., Renugadevi, R., Devi, R.R.: Data mining in cloud storage data with matlab statistics & machine learning toolbox. J. Inform. Comput. Sci. **10**(5), 685 (2020)
6. Anoruo, C.M.: Modelling and analysis of aerosol and cloud-precipitable water inter-hemispheric interactions of aerosol-satellite data using ground observation. Aerosol Sci. Eng. **4**(4), 331–350 (2020). https://doi.org/10.1007/s41810-020-00078-y
7. Kumar, J., Singh, A.K.: Performance assessment of time series forecasting models for cloud datacenter networks' workload prediction. Wireless Pers. Commun. **116**(6), 1–21 (2021)

8. Lyras, N.K., Efrem, C.N., Kourogiorgas, C.I., et al.: Medium earth orbit optical satellite communication networks: ground terminals selection optimization based on the cloud-free line-of-sight statistics. Int. J. Satell. Commun. Network. **37**(4), 370–384 (2019)
9. Wu, G., Yun, X., Wang, Y., et al.: A sketching approach for obtaining real-time statistics over data streams in cloud. IEEE Trans. Cloud Comput. **99**, 1 (2020)
10. Mochamad, R.P., et al.: The updated statistics of binary star clusters in the Large Magellanic Cloud. J. Phys. Conf. Ser. **1127**(1), 12053 (2019)
11. Dadashazar, H., Crosbie, E., Majdi, M.S., et al.: Stratocumulus cloud clearings: statistics from satellites, reanalysis models, and airborne measurements. Atmos. Chem. Phys. **20**(8), 4637–4665 (2020)
12. Ibrahim, M.M., Danbala, A.A., Ismail, M.: Towards attaining reliable and efficient green cloud computing using micro-smart grids to power internet data center. J. Comput. Commun. **07**(7), 195–205 (2019)

On the Application of Computer Digital Media Resources in Painting Design

Junjing Chen[1(✉)] and Fawaz Almulihi[2]

[1] Nanchang Institute of Technology, Nanchang, Jiangxi, China
jxchenjunjing@163.com
[2] Taif University, Taif, Saudi Arabia

Abstract. With the all-round popularization of the Internet and the development of science and technology, digital media (DM) painting has become an indispensable ppainting of our lives, and DM painting is still constantly updated, making the field of painting design always changing. Painting design is constantly innovating and optimizing with the change of the times. Computer DM resources have a wide range of application prospects in painting design. It makes painting design more diversified, integrated and technological, which not only brings us opportunities, It also brings challenges. Nowadays, we are in the society of the modern information technology era. The development of DM has made paintingists' expressions of painting more diverse. The main research content of this paper is the application of computer DM resources in painting design. By combining the application of computer DM in painting design, there are more diversified choices in the field of painting design. The final results of the research show that the creation time of the experimental class is significantly shorter than that of the control class. For example, the poster creation control class takes 18 class hours, while the experimental class only needs 4 class hours. The efficiency of painting design creation is greatly improved, and the computer digital model is used. Digital graphics drawing software technology for media resources has increased the time of creation and design, and the creation efficiency has been significantly improved.

Keywords: Digital Media · Painting Design · Information Technology · Media Resources

1 Introduction

DM has played many positive roles in all areas of human life, and human communication activities are gradually adapting to the requirements of reform and transformation. Traditional painting design has undergone several changes, and under the guidance of various emerging technologies, has undergone various optimizations, transformations and integrations. In such an environment that is easy to innovate and change, natural computer media resources are more important than traditional painting. In terms of design, it is a powerful push and change [1]. The use of DM resources in painting design, on the one hand, allows human senses to fully enjoy painting, on the other hand, it can increase

J. H. Abawajy et al. (Eds.): ICATCI 2022, LNDECT 170, pp. 260–267, 2023.
https://doi.org/10.1007/978-3-031-29097-8_31

the amount of information dissemination and enhance the dissemination effect. Technology and painting design, painting design provides a broader space for the company's development.

In recent years, many researchers have conducted research on the application of computer DM resources in painting design and have achieved good results. For example, Rossi believes that DM has played a vital role in the change of people's lifestyles. Through DM, people can better perceive the world and enjoy the things on the screen only on DM platforms, because DM resources. It is very meaningful for research [2]. Amelio A believes that painting design is always by our side. Traditional painting design is designed by paintingists using paper and pen on the drawing board. Today, computer technology can design more beautiful visual sensory painting, using painting design and computer DM. Together, it can produce unexpected effects [3]. At present, a lot of research has been done on the application of computer DM resources in painting design. These previous theories and experimental results provide a theoretical basis for the research of this paintingicle.

This paintingicle analyzes the construction of computer DM resources and combines the relevant characteristics and application analysis of computer DM resources [4]. Finally, it applies to paintingistic design. By studying the application of DM painting in painting design, the research field of DM painting has been expanded, the form of expression of computer media painting has been enriched, and the interaction between more people and painting design has been explored. Design painting for others.

2 The Concept and Development of Computer Digital Media

2.1 The Concept of Computer Digital Media

2.1.1 The Main Body of Computerdigital Media

The main body of DM is information technology and digital technology, the theoretical source is mass communication, and the guiding method is contemporary painting. It is a comprehensive science that applies communication and information technology to culture, business, painting, education, and management industries. In DM technology, they are integrated in a whole new way. Show it down. DM applications have a wide range of industries, such as the film and television industry, publishing industry, game industry, news and entertainment industry, etc., as well as network companies, television stations and other units [5]. Specifically, it includes comics, online games, mobile games, digital TV and movies, digital publishing and education, and mobile TV. According to a research report released by the country, China's DM revenue surpassed the global movie entertainment industry for the first time in 2008. The media industry plays a very important role in promoting the development of the country's cultural industry. The computer DM makes painting creation more efficient, more diversified and integrated in form, and has better visual expressive effects.

2.1.2 Development of Digital Computer Media

At present, communication network technology and bandwidth are gradually popularized, and the DM industry is an emerging industry that integrates DM, cultural industries, and network technology, and is developing rapidly in countries around the world. The rapid growth of this industry is inseparable from the continuous development and support of DM technology. The technology fully integrates digital information processing technology, digital communication technology, network technology and computer technology, and is an interdisciplinary and technical research field. DM is currently the most direct way of communication [6, 7]. It has gradually become an important channel for the public, government and enterprises to understand information and communicate with each other, and is gradually being implemented in various industries. After a long period of development, DM has gradually changed from text graphics to audio, and the traffic has become larger and larger. This has increased the bandwidth requirements of information networks to varying degrees, and the development of information networks has also promoted the rapid operation of DM. To the creative process. This is also the main method of distinguishing DM painting from other painting.

2.1.3 Prospects of Digital Computer Media

Judging from the current development situation, the prospect of DM is very broad, and it is very likely to become the pillar of the national economy. The state has increased its support for DM technology research and development, and has made considerable progress. Currently, the country lacks computer and digital technology. With the rapid development of the computer and DM industries, the demand for professional and technical personnel continues to increase. Talents with rich theoretical foundations, professional paintingistic achievements and professional practical capabilities are very popular in the talent market [8]. With the gradual expansion of the DM industry. The demand for advertising, animation, performances, and games will increase, and the value of DM will increase.

2.2 The Characteristics of the Application of Computer DM in Painting Design

2.2.1 Advantages of Computer Painting Design

Computer painting design focuses on the application of design methods. Compared with the complex techniques of traditional painting design, the computer painting design process is more convenient, and the modern aesthetics is stronger, especially in terms of the proportion of characters and the proportion of building structure. Through specific data. Computer painting design can repeatedly modify the work, and the modification of certain contents will not affect other ppaintings, which is unmatched by traditional painting design. In addition, there are abundant sources of computer painting design materials, which can be easily obtained through Internet resources. However, obtaining traditional painting design materials requires designers to refer to a lot of materials [9, 10]. The collection method is very complicated, and sometimes it may not be possible to design the project well. In the development of online games, for the requirements of image processing, if the performance of the material does not meet the requirements, then the designed game scene will not reach the best. In addition, when the paintingist paints, he has a relatively high requirement on the resolution of the drawing table to meet the requirements of the image.

2.2.2 The Concept of Computer Painting Design

The concept of computer painting design is to imitate the skills and techniques of traditional painting creation on the basis of traditional painting creation. From the perspective of the development of painting design, the development of computer painting still has a very broad prospect, but computer painting can still be traditional painting. The technique is the basic. However, the two are in a single opposite relationship. The root of computer painting design is inseparable from traditional painting. The computer painting drawing board is constantly improving on the basis of traditional painting design, especially in graphic design, sampling from the introduction of computer painting equipment is more in line with the real brushes, the work design efficiency is high, and the basic theory and aesthetics of traditional painting design are made up for Consciousness is not available in computer painting design. Therefore, the development of computer painting design has not had an impact on the development of traditional painting design. On the contrary, the two can develop together in a complementary manner [11, 12]. Computer painting design retains the core content of traditional painting design. The connection is closer, and the vitality of painting design is stronger.

3 Research on the Preparation of Computer Digital Media Painting Design Experiment

3.1 Experimental Method

The Sigmoid function is a common sigmoid function in digital computer models, also known as the sigmoid growth curve. In information science, because it is a single-increasing inverse function, the Sigmoid function is often used as a threshold function of neural networks to map variables between 0 and 1.

3.1.1 Sigmoid Function Algorithm

$$y = \alpha + \beta \ln(x + x_0) \tag{1}$$

where x, y represent the x-axis and y-axis of the image coordinate axis, α and β are independent variable digital models, k is 1 to k mathematical variables, polynomial function, it is a constant and independent variable x through a finite number of multiplication and The result of addition. Obviously, when k = 1, it is a linear function, when k = 2, it is a quadratic function

$$y = \beta_0 + \beta_1 x + \beta_2 x^2 + \beta_k x^k \tag{2}$$

3.2 Experimental Data Collection

This paintingicle mainly uses the data survey method to conduct an experimental survey of 200 students in the computer painting design course of a college of painting. The experimental objects are divided into two groups, which are divided into an experimental group and a control group. Perform painting design through the same painting creation topic, compare the time required for creation in the case of computer DM and creation under the traditional mode, analyze the data through the feedback of the investigator to the computer media resource library, and analyze the data by data model, Draw a more representative conclusion.

4 Analysis of Experimental Results of Computer Digital Media Painting Design

4.1 Comparative Analysis of the Creation Time of Painting Projects

Through comparative experimental analysis of five subjects: animation subject creation, packaging subject design, pattern subject creation, poster subject creation and advertising subject creation, the computer DM-assisted painting design application is the painting design experimental group in the school, and contrasts with it is the control group under the influence of the traditional painting design model. The analysis results are shown in Table 1.

Table 1. Comparison and analysis of the creation time of painting projects

Subject	Control class (lesson hour)	Experimental class (lesson hour)
Animation character creation	22	8
Packaging project design	26	7
Pattern project creation	27	6
Poster project creation	18	4
Advertising project creation	25	11

As shown in Fig. 1, by comparing the project creation time of the students in the control class and the students in the experimental class, it can be seen that the creation time of the experimental class is significantly shorter than that of the control class. It takes 4 class hours, and the efficiency of painting design is greatly improved. Compared with the traditional painting design mode, the computer-assisted painting teaching mode has two creative advantages in terms of data retrieval and digital technology application support in the creation of painting topics. The use of computer digital model media resources for digital graphics drawing software technology has increased the time of creation and design, and the creation efficiency has been significantly improved.

4.2 Analysis of Preferences for Computer Media Resources

This paper conducts a questionnaire survey and analysis after the completion of the project creation of 200 students who ppaintingicipated in the practice, and collects suggestions on the impact of DM resources on painting design. They are divided into four categories: useful, generally useful, not very useful, and not easy to use. Statistics are performed on each degree, and the analysis results are shown in Fig. 2.

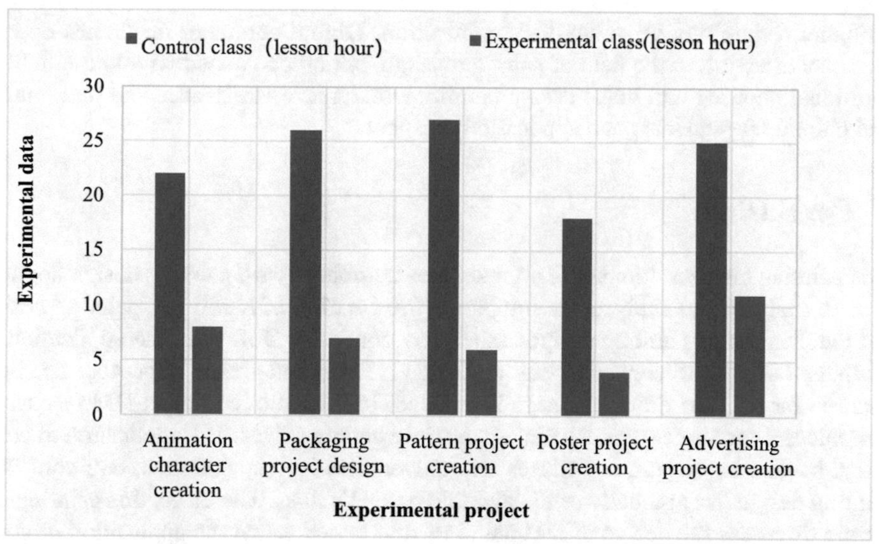

Fig. 1. Comparison and analysis of painting project creation time

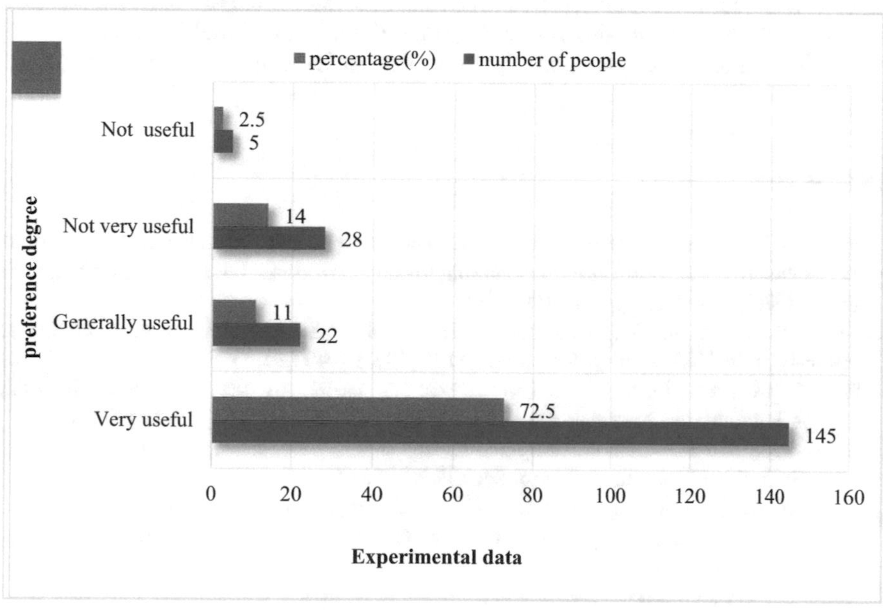

Fig. 2. Analysis of preferences for computer media resources

As shown in Fig. 2, 145 people find computer DM resources very helpful, but only 5 people find it useless. For the design of computer DM resources painting, it is an organic combination of computer and painting, and it is also technology and painting. The combination. In the application of design paintingworks, it is very common to use computer technology as an auxiliary application. Digital computer media has opened up a new direction in the field of painting design, which can penetrate into the field of traditional painting and make design paintingworks more easily accepted by humans. DM painting design has good application prospects.

5 Conclusions

This paintingicle uses computer DM resources combined with painting design applications to analyze, and analyzes by comparing the creation of traditional painting design and the creation of painting design assisted by computer DM. With the application of computer DM, the efficiency of painting design is faster and greatly improved. The time required for painting drawing design is reduced. In the era of computer DM resources, computer technology is gradually integrated into people's lives. The application of computers in painting design in colleges and universities is very common, and computer painting design has gradually penetrated into people's lives. Therefore, this paintingicle mainly discusses the design of DM painting, and believes that the application of computer DM resources in painting design has profound research significance. DM resources have a stimulating effect on painting design, and their diversity has changed the form of painting design to a certain extent, making it gradually transformed from the original traditional, unified, and flat form to a spatial, dynamic, and multidimensional form. Enrich and enhance the connotation of painting design. Design experience from different angles.

References

1. Sugiarto, E., Kurniawati, D.W., Febriani, M., et al.: Computer-based art in folklore illustration: development of mixed media digital painting in education context. IOP Conf. Ser. Mater. Sci. Eng. **1098**(3), 032017 (7pp) (2021)
2. Rossi, T.C.: Femininity and its imagens in digital media: questions to think gender and visuality in the XXI century. Tempo Soc, **2017**(29), 12–49 (2017)
3. Ferrari, M.C., Amelio, A., Nardelli, G.M., et al.: Assessment on the application of facilitated transport membranes in cement plants for CO2 capture. Energies **14**, 13–25 (2021)
4. Rossi, T.C., Rossi, T.C.: Femininity and its imagens in DM: questions to think gender and visuality in the XXI century. Tempo. Soc. **29**, 41–45 (2017)
5. Singh, A., Rabbani, A., Regenauer-Lieb, K., et al.: Computer vision and unsupervised machine learning for pore-scale structural analysis of fractured porous media. Adv. Water Resour. **147**(1), 12–46 (2020)
6. Boldyrev, Y.Y., Glukhov, V.V., Stelmashonok, E.V., et al.: Computer engineering as the knowledge-intensive core of digital Production: origin and development. IOP Conf. Ser. Earth Environ. Sci. **816**(1), 012006 (8pp) (2021)
7. Rao, P.S., Yedukondalu, K.: Hardware implementation of digital image skeletonization algorithm using FPGA for computer vision applications. J. Visual Commun. Image Represent. **59**(Feb.), 140–149 (2019)
8. Linke, J., Mcdermid, T.G.J.: spatially consistent landcover maps for reliable landscape monitoring: an object-based disturbance-inventory approach. Landscape Ecol. **24**(2), 157–170 (20190

9. Zolkin, A.L., Matvienko, E.V., Suchkov, D.K., et al.: Digital development of agrarian production - institutional approach. IOP Conf. Ser. Earth Environ. Sci. **988**(4), 042070 (5pp) (2022)
10. Butcher, A.R., Corfe, I.J.: Geo-inspired science, engineering, construction, painting and design. Geol. Today **37**, 58–74 (2021)
11. Lanza: Brushstroke cinema: the concept of matte painting in the work of peter ellenshaw and albert whitlock. Film History **32**(2), 80 (2020). https://doi.org/10.2979/filmhistory.32.2.04
12. Bekker, T., Barendregt, W., Skovbjerg, H.M., et al.: Editorial special issue on assumptions about the concept of childhood and the roles of children in design. Int. J. Child – Comput. Interact. **19**(MAR.), 89–92 (2019)

Motion Trajectory Control of Compound Forming Contour Accuracy Based on Internet of Things Technology

Wenliang Mao(✉)

Gansu Institute of Mechanical and Electronic Technology, Tianshui 741001, Gansu, China
mao.wen.liang@163.com

Abstract. High-complexity components have been widely used, and the performance and precision requirements of their manufacturing have become higher and higher. The additive-subtractive composite manufacturing technology is considered to be a very promising technical method because it has the advantages of both additive manufacturing and subtractive manufacturing. However, the current additive-subtractive manufacturing still faces the trajectory control and subtractive process Many problems, such as process trajectory control, have severely restricted the contour accuracy of compound forming by increasing or decreasing the material. The purpose of this paper is to study the motion trajectory control of the contour accuracy of the composite forming of increasing and decreasing materials based on the Internet of Things technology. Through a certain practice method of adopting a single-pass multi-layer cladding on the blade as the contour research and verification, the practical results show that when multi-layer cladding is performed on the blade with five-axis spiral trajectory and dynamic cutter axis vector, the contour accuracy of the blade far exceeds the contour accuracy that can be achieved by the traditional method of the adjacent layer of cutter axis vector by about 3 times.

Keywords: Internet of Things · Increase and Decrease Material Manufacturing · Motion Trajectory Control · Tool Axis Vector

1 Introduction

With the development and widespread use of various aerospace, biomedical and other cutting-edge technologies and emerging industries, the performance and accuracy requirements for its design and manufacturing have become more stringent [1, 2]. Since we are in the traditional manufacturing process, such as casting, machining, etc., there are universal problems such as difficult forming and processing, long manufacturing cycle, etc., especially for high-performance and difficult-to-process alloy materials, and highly functional gradient functions. Materials, internal complex structures, etc. are manufactured, and such problems are more prominent [3, 4]. Additive manufacturing is easy to realize the manufacture of functional structures, gradient materials and internal

J. H. Abawajy et al. (Eds.): ICATCI 2022, LNDECT 170, pp. 268–275, 2023.
https://doi.org/10.1007/978-3-031-29097-8_32

structures, and has the characteristics of rapid manufacturing and green manufacturing. It has become a hot spot in advanced manufacturing technology research [5, 6].

The composite manufacturing technology of adding and reducing materials is currently a hot research topic in the manufacturing industry. Many scholars and scientific research institutions at home and abroad have conducted extensive research on composite manufacturing technology of adding and reducing materials [7, 8]. These studies provide a good theoretical and experimental basis for understanding the multi-physical coupling process, multi-task art collaboration, and intelligent control of additive manufacturing [9, 10]. However, because laser additive manufacturing is a collaborative control of multiple physical quantities, including additive process parameters, material performance parameters, subtractive process parameters, collaborative process parameters, etc., it is necessary to composite manufacturing with high dimensional accuracy, contour accuracy, defect-free, and complete performance. Controlled laser additive and subtractive material manufacturing parts are still facing great challenges.

Based on the study of the optimization control of the motion trajectory, the laser additive process and the mechanism of improving the accuracy of the contour between the cladding layers, this paper has conducted research and verification through a certain experimental method of single-pass multi-layer cladding of the blade.

2 Research on the Motion Trajectory Control of the Contour Accuracy of the Composite Forming of Increasing and Decreasing Materials Based on the Internet of Things Technology

2.1 Optimal Control of Motion Trajectory

(1) Laser additive tool axis vector control

The traditional laser deposition method is to use parallel layering, and the knife axis vector is fixed. As the curvature of the part changes, an additional step effect is added between layers. In the layer-by-layer printing, errors such as fluid gravity and surface tension will accumulate during the forming process. Under the action, the occurrence of defects is aggravated. When the deposition height reaches a certain level, the edge collapse will occur at the part boundary, which will affect the final forming height and forming accuracy of the formed part. The proposed spiral trajectory and dynamic tool axis vector can reduce the step effect and the on and off frequency of the laser during the cladding process, thereby improving the contour accuracy of the cladding part.

(2) Non-linear error and tool wear error control

Because of the problems of looseness, pores, non-fusion, low density, etc., the additive forming parts have low forming accuracy and poor surface quality, which requires milling and cutting. Due to the complex structure of five-axis simultaneous machining, there are nonlinear errors; in addition, the tool wears quickly in cutting difficult-to-machine materials, and the contact position of the tool and the workpiece surface changes after wear. The existence of non-linear error and tool wear error causes the actual motion trajectory of the tool to deviate from the ideal interpolation trajectory, and finally leads to the reduction of the contour accuracy of the formed part. The proposed tool axis vector plane interpolation algorithm and tool 3D error compensation algorithm can correct the actual tool motion trajectory, which is close to the theoretical trajectory to the greatest extent, thereby improving the contour accuracy of the formed part.

(3) Constant surface knife contact speed control

Practice shows that the stability of the tool trajectory is an important factor that affects the contour accuracy of the formed part. In the additive process, the stable scanning speed can obtain high-quality deposition profile; in the subtractive process, the stable surface feed rate can not only obtain high surface roughness, but also greatly reduce the cutting force and tool wear.

2.2 Laser Additive Process

As the tool combines translational cloud motion and rotary motion, the resulting motion trajectory is extremely complex. To analyze and solve complex motion trajectories, precise solving methods and solving equations are required, and the solving process is cumbersome and difficult to implement. Combined with the relative motion relationship between the tool and the workpiece, the motion mode of the laser five-axis additive machine tool has the following three modes:

(1) The cladding nozzle only performs translational movement (double turntable five-axis structure). That is, the tool (cladding nozzle) does not perform rotational movement, the tool axis is always parallel to the Z axis, and the rotation of the workpiece is performed by the rotary axis. In this structure, the coordinate system of the machine tool is fixed, the workpiece is installed on a rotating worktable, and the rotation of the rotating table drives the workpiece to rotate relative to the tool. This structure is used for composite manufacturing of small workpieces. When the main shaft is replaced with a cladding nozzle, the powder beam sprayed by the nozzle will not change the focal position of the laser beam converging with the influence of gravity; at the same time, the molten pool is not prone to collapse.

(2) The cladding nozzle only performs rotary motion (double-rotor five-axis structure). Both rotary motions are executed by the main shaft. This structure is used for composite manufacturing of large workpieces. When the spindle is replaced with a cladding nozzle, the powder beam sprayed by the nozzle will change the focal position of the laser beam under the influence of gravity; at the same time, the molten pool is prone to collapse.

(3) The cladding head both translates and rotates (swing head turntable five-axis structure). The motion trajectory of the tool is produced by the combination of translation and rotation. This type of structure machine tool is often used for the composite manufacturing of medium-slender and complex curved parts, such as blade aviation parts.

2.3 Mechanism Analysis of Improving the Profile Accuracy Between Cladding Layers

In summary, the contour accuracy of the final formed part is greatly affected, especially for thin-walled parts with variable curvature. The five-axis spiral dynamic tool axis vector method proposedcan avoid these problems. The main mechanisms for improving the profile of the cladding part are as follows:

(1) The five-axis spiral dynamic tool axis vector is based on the five-axis spiral subtraction method. During the cladding process, the deposited layer spirals. Ascends, the tool axis vector between the nozzle and the cladding position changes dynamically with the surface curvature of the theoretical model. There is a step effect;
(2) In the actual cladding process, under the action of various factors such as shielding gas disturbance, fluid gravity and surface tension, when the deposition height reaches a certain level, due to the superposition of inter-layer errors, the moving molten pool is at the boundary where the curvature of the part changes greatly. It also produces edge collapse, which is much smaller than the parallel layering method;
(3) The spiral track does not have interlayer cladding nozzles intermittently lifting, so the laser does not need to be turned on and off frequently, which reduces the problem of poor overlap at the junction of the starting point and the end point of the cladding.

2.4 Vector Change Algorithm

The tool axis vector is converted into the rotation angle of the machine tool rotation axis in the post-processing. According to the structural characteristics of the machine tool, the rotation axis has the actual rotation range. In order to ensure the correct transformation of the tool axis vector, first determine the relationship between the rotation angle and the tool axis vector according to the machine motion model, and establish the change equation of the tool axis vector.

The table rotates relative to the tool, and the rotation angle is clockwise as the positive direction. First, the tool vector u is rotated clockwise around the Z axis to the coordinate plane (−Y) (+Z), and the angle C is calculated. Secondly, the tool axis vector u rotates clockwise around the X axis to calculate the angle A. When the tool rotates or swings relative to the workpiece, the tool axis vector transformation is completed. In the above process, the angle and motion coordinates can be expressed as:

$$A = \pm \arccos(u_z) \tag{1}$$

$$C = \begin{cases} \pi/2 + \arctan(u_y/u_x) & u_x > 0, u_y \geq 0 \\ \pi/2 & u_x > 0, u_y = 0 \\ 3\pi/2 + \arctan(u_y/u_x) & u_x < 0, u_y \geq 0 \\ 3\pi/2 & u_x < 0, u_y = 0 \\ 2\pi + \arctan(u_y/u_x) & u_x < 0, u_y < 0 \\ -\arctan(u_y/u_x) & u_x > 0, u_y < 0 \end{cases} \tag{2}$$

$$\begin{cases} X = x_c \cos C + y_c \sin C \\ Y = -x_c \cos A + y_c \cos C \cos A + (z_c + d) \sin A \\ Z = x_c \sin C \sin A - y_c \cos C \sin A + (z_c + d) \cos A \end{cases} \tag{3}$$

3 Experiment

3.1 Trajectory Virtual Simulation

Use the blade local tool path to simulate the trajectory and the tool axis vector to verify whether the transformation of the tool axis vector is correct. The simulation model is constructed through the MATLAB software platform, and the blade tool position source file is read in. Since the blade profile is formed by sweeping multiple sections along the axial direction, the direction of the knife axis vector is perpendicular to the surface contact point during material reduction, spiraling upward along the axial direction and dynamically changing with the normal direction of the section line of the blade surface; when adding material direction of the tool axis vector is parallel to the surface contact point, spiraling upward along the axis, and dynamically changing along the tangent direction of the surface. According to the simulation results of the local profile trajectory of the blade, the motion trajectory of the subtractive material and the direction of the tool axis vector are the same as the programmed path trajectory: the blade motion trajectory after vector transformation remains unchanged, but the direction of the tool axis vector follows the direction of the spiral trajectory. It is turned by 90°, which is consistent with the envisioned additive tool axis vector. Therefore, it is verified that the vector transformation algorithm processes the blade trajectory path correctly.

3.2 Experimental Design

316L stainless steel (022Cr17Ni12Mo2) powder with a diameter of 60–125 μm is used as the cladding material, and 316L stainless steel forgings are used as the substrate material, with a size of φ150 mm × 10 mm. The chemical composition of the powder is shown in Table 1.

Table 1. Stainless steel (316L) chemical composition

Element	Si	Cr	Ni	Mn	Mo	C	O	Fe
Standard	≤1.0	16.0–18.0	10.0–14.0	≤2.0	2.0–3.0	≤0.03	≤0.04	The remaining
Measured	0.59	17.62	12.43	1.45	2.53	0.022	0.027	The remaining

In order to obtain the best cladding results, the selected laser process parameters are shown in Table 2.

Table 2. Laser cladding process parameters

Laser power (W)	Scan speed (mm/min)	Powder flow rate (g/min)	Lift (mm)
900	600	9	0.5

The NC program with correct simulation is transmitted to the composite manufacturing center of additive and subtractive materials for laser cladding experiment verification. In order to obtain a comparative effect, two kinds of tool axis vector control methods are used for cladding experiments, one is the five-axis spiral dynamic tool axis vector method, and the other is the method of obtaining the tool axis vector by using the points of the adjacent layer.

4 Discussion

In order to quantitatively compare the differences between the two cladding methods, the blades obtained under the two conditions were tested. Because the blade is a special structural part, its blade profile is fitted by multiple cross-section distortions, and the accuracy of its profile profile is usually evaluated by the error of the cross-sectional measurement data. In order to compare the contour accuracy difference of the two cladding results, a three-coordinate measuring instrument was used to select the position of the larger twisted section in the theoretical model, and the section dimensions of the blades cladding in the two ways were measured respectively, and compared with the theoretical dimensions. The results are shown in Fig. 1 and Fig. 2. The detailed data is shown in Table 3. With the current cladding process parameters, the wall thickness of the cladding layer is 3 mm, and the center distance from the scanning track line is ±1.5 mm. The measured values all correspond to the envelope surface of the corresponding section.

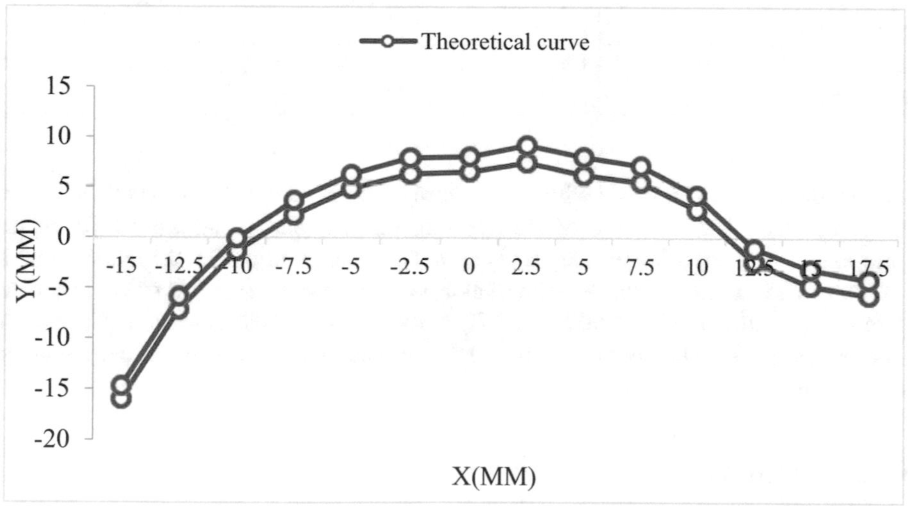

Fig. 1. Comparison between theory and adjacent tool axis vector curve

Figure 1, 2 and Table 3 show that the blade section curve of the spiral cutter shaft vector cladding has a uniform margin compared with the theoretical section curve. The maximum profile deviation value is 1.57 mm, and the minimum deviation value is 1.45 mm, which is consistent with the theoretical section data. The degree is high,

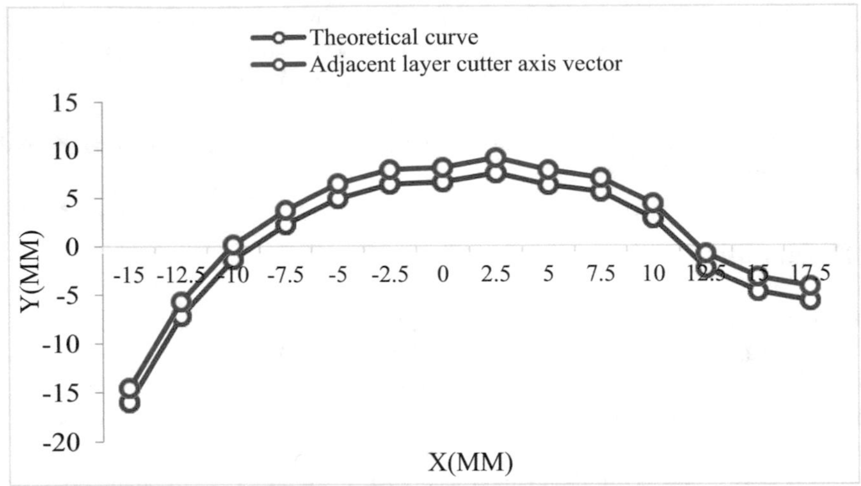

Fig. 2. Comparison of theoretical and dynamic spiral cutter axis vector curves

Table 3. Comparison of measurement data of two kinds of cladding results (mm)

Method	Theoretical deviation	Maximum contour deviation	Minimum contour deviation	Contour deviation value
Spiral knife shaft vector	1.50	1.57	1.45	0.12
Adjacent tool axis vector	1.50	1.82	1.28	0.54

and the tolerance is controlled within +0.1 mm. Compared with the theoretical section curve, the blade section curve of the cladding blade under the adjacent layer of the knife axis vector has a larger margin change. The maximum profile deviation value is 1.82 mm, and the minimum deviation value is 1.28 mm. In order to offset or rotate, there is a deviation of about ±0.32 mm. The maximum contour deviation of the blade obtained by the two cladding methods is 0.25 mm, and the minimum contour deviation is 0.27 mm.

5 Conclusions

This article analyzes the existing laser trajectory path through the principle of laser additive material and finds that the existing laser scanning path is based on three-axis linkage or approximately five-axis processing. It cannot effectively solve the step effect produced by the cladding process of complex thin-walled and variable-curvature parts; at the same time, due to the collapse of the molten pool, the deposition growth height is also limited. Based on the above problems, a new five-axis spiral dynamic tool axis vector control method is proposed. Through the five-axis milling/cladding vector transformation algorithm, the bottleneck problem is solved based on the post-processing technology, and the cladding contour accuracy of complex thin-walled parts is improved.

References

1. Kullberg, A., Bjrklund, C., Brkovic, I., et al.: Effects of learning addition and subtraction in preschool by making the first ten numbers and their relations visible with finger patterns. Educ. Stud. Math. **103**(3), 157–172 (2020)
2. Yuliwijayanti, A., Santoso, Madjdi, A.H.: Manipulative media technology for addition and subtraction of integers in elementary schools. J. Phys.: Conf. Ser. **1823**(1), 012096 (2021)
3. Rajpoot, N.C., Govindarajan, V.S.: Paper chromatographic determination of total capsaicinoids in capsicums and their oleoresins with precision, reproducibility, and validation through correlation with pungency in scoville units. J. Assoc. Off. Anal. Chem. **64**(2), 311–318 (1981)
4. Efstathiou, C., Pekmestzi, K., Moshopoulos, N.: On the diminished-1 modulo 2n+1 addition and subtraction. J. Circuits Syst. Comput. **29**(5), 2030005 (2020)
5. Barghash, S., El-Razeq, S.A., Elmansi, H., Elmorsy, M.A., Belal, F.: Intermolecular interactions of saxagliptin and vildagliptin with human serum albumin. J. Appl. Spectrosc. **88**(6), 1266–1275 (2022). https://doi.org/10.1007/s10812-022-01308-6
6. Fayol, M., Thevenot, C.: The use of procedural knowledge in simple addition and subtraction problems. Cognition **123**(3), 392–403 (2012)
7. Henklain, M., Carmo, J.: Stimulus equivalence and increase of correct responses in addition and subtraction problems. Paidéia (Ribeirão Preto) **23**(56), 349–358 (2013)
8. Lee, D.H.: An analysis on the word problems of the addition and subtraction in mathematics text books and its students' responses. Sch. Math. **11**(3), 479–496 (2009)
9. Gilmore, C.K., Spelke, E.S.: Children's understanding of the relationship between addition and subtraction. Cognition **107**(3), 932–945 (2008)
10. Singh, P.: An analysis of word problems in school mathematics texts: operation of addition and subtraction. J. Sci. Math. Educ. Southeast Asia **29**(1), 41–61 (2006)

Computer Network Technology in Big Data Distributed Cloud Storage Platform

Yuelin Xu[1(✉)], Yuran Wang[1], Baiyang Wan[2], and Ahthasham Ullah Khan[3]

[1] Department of Computational Mathematics and Cybernetics, Mathematics and Applied Mathematics, Shenzhen MSU-BIT University, Shenzhen 518172, Guangdong, China
1120190060@smbu.edu.cn

[2] Faculty of Economics, Shenzhen MSU-BIT University, Shenzhen 518172, Guangdong, China

[3] Institute of IT and Computer Science, Afghanistan, Kabul, Afghanistan

Abstract. In the era of big data, distributed cloud storage platforms are widely used to store massive amounts of data and information. Compared with traditional storage methods, distributed cloud storage platforms can provide flexible storage space to ensure the security and speed of platform operating data. This article uses computer network technology to design a distributed cloud storage platform, test platform file upload and download time and user login platform success rate to verify the reliability of platform operation, and realize the feasibility of user login function module operation, I hope this article distributed cloud storage platform can help people process data efficiently and realize the safe preservation of data and information.

Keywords: Big data · Computer network technology · Distributed cloud storage platform · Data processing

1 Introduction

Before the emergence of cloud storage services, users needed to upgrade and expand their personal storage systems in order to solve the storage needs of large amounts of data to obtain greater storage capacity and performance [1]. However, this approach will lead to increased storage costs for users. In addition, this kind of solution can only solve the current storage problem, and treat the symptoms rather than the root cause. After the emergence of cloud storage technology, users can easily solve the storage problem of large amounts of data by using cloud storage services. Users do not need to care about the physical equipment of the cloud storage service at all, and only need to pay a much lower cloud storage service fee than upgrading the storage device themselves to use the massive cloud storage service. Users only need to use the Internet to be able to access and store personal data anytime and anywhere.

Many scholars have conducted in-depth discussions on the design and implementation of computer network technology in the big data distributed cloud storage platform, and have obtained good research results [2]. For example, a big data cloud storage platform developed by a scholar can integrate users' NAS devices with cloud storage, and

J. H. Abawajy et al. (Eds.): ICATCI 2022, LNDECT 170, pp. 276–284, 2023.
https://doi.org/10.1007/978-3-031-29097-8_33

automatically manage users' storage space. At the same time, the platform can protect users' files through the snapshot mechanism and inherent redundancy capabilities of cloud storage. Through the management interface provided by the platform, users can customize file management and protection methods according to their own needs. At the same time, the platform provides an access interface consistent with traditional NAS devices to facilitate user data access. A company's project design is based on a cloud data center that converts a custom cloud storage interface into a special application of a traditional storage connection protocol, and uses distributed storage technology for the company's complete design and cloud platform business maintenance to provide customers with efficient and comprehensive, convenient services, convenient for users to use the cloud storage service platform more simply [3]. Although many researchers have designed big data distributed cloud storage platforms, the security of data storage needs to be strengthened.

This article first introduces the distributed storage technology, analyzes the platform requirements, and then designs the big data distributed cloud storage platform functional modules based on the cloud storage platform structure using computer network technology, and finally verifies the platform usability by testing platform performance and user login module processes.

2 Related Concepts of Distributed Cloud Storage Platform

2.1 Distributed Storage Technology

Distributed storage technology is to dynamically connect storage disks scattered in different networks through the network to create a virtual storage pool that users can access at any time [4].

This article mainly uses distributed cloud storage platform to manage storage files. Since there is almost no direct correlation between unstructured data objects, such as text information, image information, they are usually Blob data objects, so a distributed file system is used to store and manage data [5]. For specific implementation details, grouping data by data block is a common method for editing distributed file systems, and data storage processing operations are usually manifested in the underlying data block operations [6].

2.2 Platform Demand Analysis

(1) Client requirements
Every successful product (even a client APP with cloud storage as the core) requires a registration and login interface. Here we require these two interfaces to be as simple as possible. The use of virtual hard drives is very similar to our use of hard drives on personal PCs (such as C drive, D drive). It can copy files to, for example, "My Network Disk", or by pulling files to folders, uploading local files and moving between files in cloud disks [7].
(2) File encryption requirements
Since the adjustment and development of the Internet, personal data information has grown exponentially, and the Internet is generally used as a medium to store

its data [8]. Accompanying this is the security of its data, causing a lot of trouble. With the increase in data volume and data value of contemporary users and the rapid development of cloud storage technology, personal user data has suffered unprecedented security threats. If cyber hackers successfully steal a large amount of personal or corporate data, the negative impact will be unprecedented, and the consequences will be incalculable. So data security is very important.

From the perspective of application security, we can use security policies to encrypt data information to improve the security of the big data cloud storage system [9]. According to the actual situation, improving the security of the big data cloud storage platform can start from the following two aspects: one is to encrypt files locally through transparent encryption technology; the other is to encrypt files again during network transmission. The common encryption method is SSL, which improves the overall security of user data [10, 11].

(3) Storage and monitoring requirements

Data storage is a main function of the server, and the specific requirements are described in the following two aspects.

Configuration of the number of data copies: server nodes in the cluster are down from time to time, so backing up multiple copies to different nodes is a solution. At the same time, for different user levels, such as ordinary users, VIP users, and SVIP users, we will provide different levels of security storage levels, that is, provide different and user data storage copies [12].

Data remote storage configuration: In order to ensure that the security of user data stored on the big data cloud storage platform is sufficiently reliable, copies of user data can be stored on different nodes. After the big data cloud storage platform is properly built, it will be provided with corresponding monitoring functions, which can monitor the files stored in the system and their synchronization status in real time [13]. Specifically, the following content needs to be monitored: First, the file synchronization status, in an intuitive and concise way to show the user who is currently synchronizing data and the user's specific synchronization file, and can monitor the synchronization status of the specified file in real time; the second is metadata Node operating conditions, such as memory, CPU usage and thread count; the third is file data changes [14].

2.3 LNC Algorithm

The LNC algorithm comprehensively considers the document size, the popularity of the document, and the value of the document cached document. It is an algorithm suitable for the cache replacement of the document-based cache system.

The LNC algorithm points out: the access heat of a file is based on two factors, the access history and access trend of the file, which reflects the frequency and trend of the file being accessed over a period of time. For any file p, the calculation formula of its access heat H is:

$$H = 1/(T + AAT) \tag{1}$$

Among them, T is the elapsed time since the file p was accessed last time, and AAT is the average access interval time of the file.

$$AAT_k = aT_k + (1 - a)AAT_{k-1} \tag{2}$$

Among them, T_k represents the time interval between the kth access to the file and the k-1th access. AAT_{k-1} is the calculated average access time interval after the k-1th access to the file. AAT_k represents the average access time interval obtained after the k-th access to the file, and a is an impact factor.

3 Platform Design

3.1 Structural Design of Cloud Storage Platform

The core of cloud storage combines storage devices and application software to form a special form of architectural service. In the eyes of users, cloud storage is not a specific device, but a combination of multiple storage devices and servers distributed in different regions.

Cloud storage is very different from traditional storage devices. It is a more complex system that integrates storage devices, server clusters, authority management, access networks and other parts through the use of certain strategies to help users store data and Access business. Its structural model has four layers, namely the access area, application interface layer, basic management layer and storage layer.

3.2 Overall Design of Platform Function Modules

(1) Client module

The client is a module provided to users registered on the cloud data storage platform, and mainly has the functions of registration and login and file filtering drive. Registration and login means that platform users need to fill in the registration information provided by the platform when registering an account. When logging in, users enter the user name and login password according to the information during registration, and then enter the encryption key to open the hard disk, and then they can use big data distributed cloud storage Service function now. The file filter driver is the client-side secondary encryption, that is, the file filter encryption.

(2) Safe and reliable transmission module

This module determines the period for users to use the cloud storage platform, because it is related to the safety and reliability of user file transfer. This part should include two functions. The first is to use the HTTPS network transmission protocol to enhance the security of data transmission. The second is that the number of threads in the request processing system in non-blocking IO transmission is limited. If we use traditional blocking IO to transfer files through multiple threads, it will definitely cause the thread source utilization of the request processing system is extremely low. The IO type of the process is at the operating system level, so in order to solve the performance problems caused by blocking IO during the file transfer process, it was decided to use non-blocking IO to transfer files with a single thread, which is faster and better than multi-program file transfer.

(3) Server-side module

The server side is the service providing module of the entire platform. As mentioned earlier, it consists of two parts: one is the storage cluster, and the other is request processing. The cloud storage cluster is mainly responsible for massive data storage. The request from the client is responded to by the request processing system.

Figure 1 is the architecture of the above three functional modules in the big data distributed cloud storage platform.

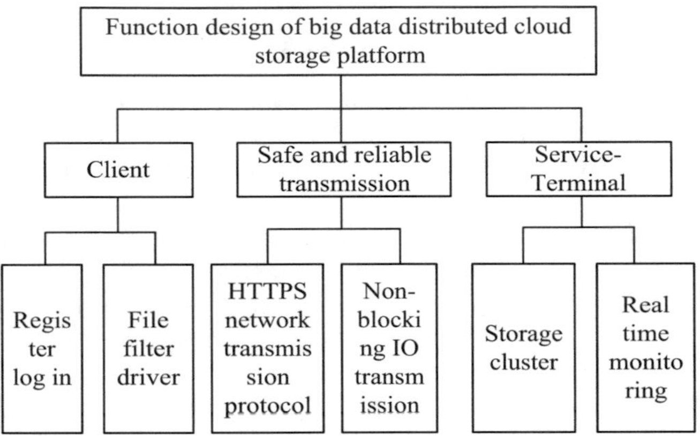

Fig. 1. Functional module design of big data distributed cloud storage platform

4 Platform Implementation

4.1 Platform Test

(1) Time test for uploading and downloading user files

The big data distributed cloud storage platform designed in this article is realized based on computer network technology. Our server side includes the data processing backend and cloud storage cluster, which together constitute the service port of the big data cloud storage platform. From the theoretical calculation of transmission, the actual theoretical value can reach 6.25 Mb/s. In order to test and compare the actual file upload and download time, this paper uses these 7 files with file sizes ranging from 1 to 7G to test the platform. The results are shown in Fig. 2. It can be seen from the data in the figure that the upload and download speeds of this big data cloud storage platform are basically the same. As the size of the file data gradually increases, the time required for each gigabyte of data becomes larger and larger. Analyzing the reasons for the above phenomenon, we can see that when the amount of file data is larger, the theoretically required upload or download time is longer, so the possibility of network failures in the process of file transmission

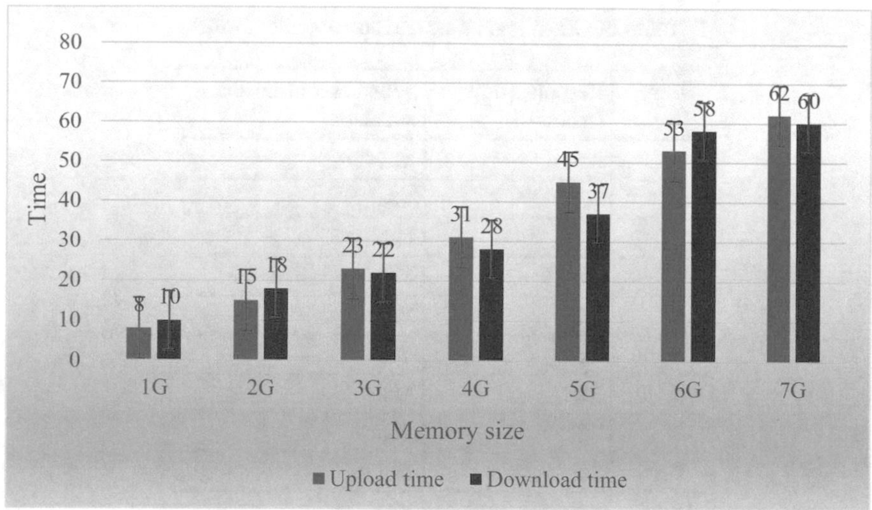

Fig. 2. Comparison result of file upload and download time

on the network will also increase linearly. Such as current network congestion, network signal instability, data frame loss and other network reasons. The file data transmission module uses multi-threading and resumable transmission technology. When a network failure causes the current data block divided by the resumable transmission technology to be interrupted in the network, the file transmission module will use the resumable transmission. The technology retransmits the data block. This leads to the above phenomenon. The larger the file, the longer the required transmission time will be than the theoretical value. On the other hand, we use multiple threads in the file data transmission module at the same time, so the file data transmission speed and data block retransmission rate are greatly improved.

(2) Responsive test of user login
The response test includes performance tests such as the response of user file synchronization and cloud storage system login. Only by reaching a certain performance index can our big data cloud storage platform be applied in practice. We first test the system login response time of the platform. In theory, the login response time should not exceed 3.5 s, and the login success rate should be above 80%. In the test, we used 12 different users to perform the login test. The specific test results are shown in Table 1 below.

It can be seen from the data in Table 1 that in these 12 tests, only two tests have a login time exceeding 3.5 s. According to the data, the login success rate is 83.33%, which exceeds the theoretical value, explain that the login response performance is in line with the expected demand.

Table 1. Client login time and success rate results

User	Response time (seconds)	Whether it meets the standard
1	2.5	Yes
2	3.2	Yes
3	3.7	No
4	2.8	Yes
5	2.1	Yes
6	2.6	Yes
7	3.0	Yes
8	2.7	Yes
9	2.8	Yes
10	2.6	Yes
11	4.1	No
12	3.3	Yes

4.2 Platform Login Function Realization

Users can register directly through the registration interface. When the user initiates a login request, it starts to block, waiting for the information returned by the request processing system. If the SSM request processing system finds no such user, it will randomly generate a verification code and ask the client to jump to the registration interface. In this interface, the user needs to fill in personal information, as well as a verification code. As for the key, the user does not need to worry about it, it will be randomly generated and stored by the request processing system. After the user clicks the registration request button, the login password will be encrypted first, and then the registration request will be sent to the request processing system. If the request processing system returns the login success information, just execute the steps in the preceding schematic diagram directly.With the above analysis of the process of client registration and login, we have a clear understanding of the entire process of client registration and login. On the premise of ensuring the completion of this part of the function, we must also fully consider the coupling of some login modules with other functional modules of the system. Based on the above analysis, the login process is shown in Fig. 3.

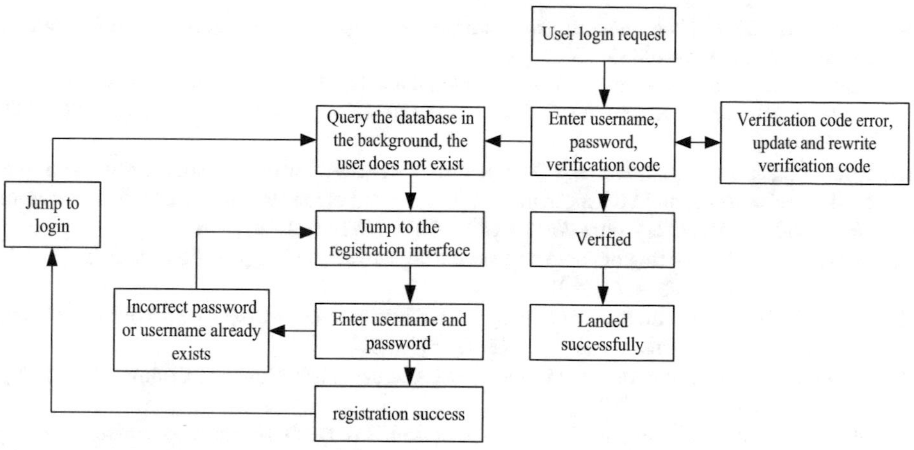

Fig. 3. Registration and login process

5 Conclusion

Cloud storage has many advantages, including low cost, high scalability, high security and reliability, and convenient use. Because of this, cloud storage technology has an unpredictable application prospect in the field of data storage. This article discusses the analysis of platform requirements and the design and implementation of functional modules, and tests the use of a distributed cloud storage platform to upload and download files and the response time of the platform, which confirms the usability of the platform.

References

1. Khan, S., Shakil, K.A., Alam, M.: Cloud based big data analytics: a survey of current research and future directions. Comput. Sci. **03**(5), 107–117 (2018)
2. Shorfuzzaman, M., Masud, M.: Leveraging a multi-objective approach to data replication in cloud computing environment to support big data applications. Int. J. Adv. Comput. Sci. Appl. **10**(3), 418–429 (2019)
3. Sharma, E.: A framework of big data as service platform for access control & privacy protection using blockchain network. Turkish J. Comput. Math. Educ. (TURCOMAT) **12**(11), 476–485 (2021)
4. Shah, S., Wu, W., Lu, Q., et al.: AmoebaNet: an SDN-enabled network service for big data science. J. Netw. Comput. Appl. **119**(OCT.), 70–82 (2018)
5. Barenji, A.V., Huang, G.Q.: Toward a blockchain cloud manufacturing system as a peer to peer distributed network platform. Robot. Comput.-Integr. Manuf. **54**(DEC.), 133–144 (2018)
6. Al Omar, A., Bhuiyan, M.Z.A., Basu, A., et al.: Privacy-friendly platform for healthcare data in cloud based on blockchain environment. Future Gener. Comput. Syst. **95C**(JUN.), 511–521 (2019)
7. Zegzhda, D.P., Moskvin, D.A., Myasnikov, A.V.: Assurance of cyber resistance of the distributed data storage systems using the blockchain technology. Autom. Control. Comput. Sci. **52**(8), 1111–1116 (2018). https://doi.org/10.3103/S0146411618080400

8. Ariffin, M.: Data leakage detection in cloud computing platform. Int. J. Adv. Trends Comput. Sci. Eng. **8**(1.3), 400–408 (2019)
9. Kumar, M.: Scalable malware detection system using big data and distributed machine learning approach. Soft. Comput. **26**(8), 3987–4003 (2021). https://doi.org/10.1007/s00500-021-06492-9
10. Patel, Y.S., Jaiswal, R., Misra, R.: Deep learning-based multivariate resource utilization prediction for hotspots and coldspots mitigation in green cloud data centers. J. Supercomput. **78**(4), 5806–5855 (2021). https://doi.org/10.1007/s11227-021-04107-6
11. Wilkinson, L.: Visualizing big data outliers through distributed aggregation. IEEE Trans. Vis. Comput. Graph. **24**, 256–266 (2018)
12. Singh, A., Garg, S., Kaur, K., et al.: Fuzzy-folded bloom filter-as-a-service for big data storage in the cloud. IEEE Trans. Industr. Inf. **15**(4), 2338–2348 (2019)
13. Marchiori, M.: Learning the way to the cloud: big data park. Concurr. Comput. Pract. Exp. **31**(2), e4234.1–e4234.17 (2019)
14. Mittal, S., Rakesh, N., Matam, R., et al.: An optimal storage and repair mechanism for group repair code in a distributed storage environment. Intell. Decis. Technol. **12**(4), 1–11 (2018)

Homomorphic Encryption Technology Based on Computer Cloud Computing

Minghu Tang(✉)

Qinghai Minzu University, Xining 810007, Qinghai, China
qh1580971@163.com

Abstract. With more and more attention paid to the privacy protection of data, how to ensure the security of cloud platform when storing data is also a current research hotspot. At the same time, the rapid expansion of the amount of data has also caused people to study whether the ciphertext can be directly manipulated, and then put forward homomorphic encryption. The purpose of this paper is to study the application of homomorphic encryption technology based on computer cloud computing. A cloud data sharing scheme with access control is proposed, which is not limited to homomorphic encryption to encrypt data, but uses the homomorphism of homomorphic encryption technology to complete access control. Choose to compare the existing FHE scheme and RLWE scheme. From the results of the experiment, the related data expansion rate is 200 times. Now the efficiency of RLWE is close to the commercial standard. In some simple and small data it can be applied in the environment, and the use of RLWE to provide cloud computing security is no longer a matter of paper.

Keywords: Computer cloud computing · Homomorphic encryption · Encryption technology · Technology application

1 Introduction

From the perspective of advanced technology, cloud computing has made great progress in recent years, but the security problem of cloud computing has not been solved, hindering the development of cloud computing technology [1]. Cloud security mainly focuses on monitoring, using encryption algorithms, standard security technologies, information security, etc., the most important of which is information security [2]. As long as all user data is stored in the cloud, the cloud can perform functions such as storage, data processing, cables, etc. The cloud contains all information about personal data. Once cloud service providers are unreliable or compromised, users' personal data will face unexpected threats [3].

In the field of cryptography, cryptography is called "the holy grail of cryptography", which attracts a large number of scientists [4]. The mature key program (MKFHE) required by some scholars is based on a simple, efficient, and efficient NTRU system with ciphertext and keys. This is an important step in post-quantum cryptography. A non-NTRU MKFHE type switch is required. This counteracts day-to-day attacks during

J. H. Abawajy et al. (Eds.): ICATCI 2022, LNDECT 170, pp. 285–292, 2023.
https://doi.org/10.1007/978-3-031-29097-8_34

homomorphic testing and reduces errors. According to the main points of RLWE and DSPR, the factors affecting the homomorphic error of the LTV12 program are analyzed in detail [5]. Some scholars proposed to protect environmental rights based on homomorphic secrets. It effectively addresses inaccuracies in confidentiality, confidentiality of business information, partnerships and monitoring of environmental rights activities. First, use blockchain and smart contracts to implement statistics based on homomorphic encryption, and then develop algorithms including pause generation, homomorphic chain encryption and decryption, and smart contracts. It's done. Since smart contracts work well in the blockchain, the blockchain solves the authentication protocol based on fully homomorphic encryption, verifies the actions of other transformers, and provides real-time [6]. The implementation of homomorphic technology solves the problem of information security in accounting and promotes the development of accounting, so the research on homomorphic technology is very important [7].

This book briefly introduces the development history and research background of homomorphic encryption technology, analyzes the main technical issues and key aspects of current homomorphic encryption technology. In order to apply homomorphic encryption in cloud computing in a targeted manner, several representative encryption algorithms of homomorphic encryption are summarized, and the development status of homomorphic encryption can be peeked from these algorithms and technologies. Then combined with the practical application, the nature of homomorphic encryption is analyzed, and finally the method of using homomorphic encryption in cloud computing is summarized.

2 Research on the Application of Homomorphic Encryption Technology Based on Computer Cloud Computing

2.1 Security Overview of Cloud Computing

Cloud computing is a new type of computing and business model, which is gradually developed with the continuous development. Its most basic concept is to use cloud computing clusters to automatically It is divided into many smaller subtasks, and then through a series of internal operations such as analysis and calculation through its internal multi-server, and finally returns the calculation result to the user [8]. Using the related technologies of cloud computing, the network server only needs a few seconds to process the huge data set, achieving high-intensity and high-efficiency network service capabilities [9]. Cloud service providers must ensure that software and hardware failures in their platforms will not affect the services provided, must ensure that they can provide reliable services to their users at any time, and must ensure that users are provided with stable and reliable services [11, 12].

2.2 Existing Fully Homomorphic Encryption Schemes

The main construction idea of the entire fully homomorphic encryption technology has gone through two stages:

The first stage is mainly to realize the transformation from a partial homomorphic encryption scheme to a fully homomorphic encryption scheme by means of re-encryption. Since some of the current homomorphic encryption schemes are not bootstrapped, the bootstrapping process has become the core of the research work at this stage. The main achievement at this stage is to successfully reduce the degree of decryption polynomial by introducing SSSP (in fact, other relevant parts of the scheme are also transformed, and other security assumptions are introduced), so as to realize the bootstrapping of the homomorphic encryption scheme, and finally use heavy Encryption implements enhanced gates.

The main work of this stage is to learn from the existing schemes, and then realize the bootstrapping of the scheme through clever construction. However, the cost of re-encryption is enormous. In order to adapt to the NAND gate circuit, it is necessary to convert the ciphertext and key into bits before re-encryption, which results in massive fragmentation operations that need to be processed. In addition, the introduction of SSSP to reduce the decryption polynomial requires a fragmentation (in order to decrypt correctly, the data needs to maintain the same precision), which makes the number of fragments to be calculated extremely large.

In addition, the security assumption of SSSP itself is weak, and there are many known attack methods. After the introduction of SSSP in the scheme, in order to ensure security, it is necessary to set SSSP to a large set, which further reduces the efficiency.

At this stage, the partial homomorphic encryption scheme itself has weak computing power, and can only use the enhanced circuit to realize the full homomorphic encryption scheme, which makes the performance very low. The main representatives of the first stage are based on ideal lattice scheme and based on integer scheme, the main technology is SSSP boot and re-encryption. Although the efficiency of the scheme at this stage is low, it points out the idea of fully homomorphic encryption for the first time, which has been used for reference by later researchers.

The second stage is mainly to design a more detailed hierarchical fully homomorphic scheme. The focus of this stage is to build an effective noise control method. The hierarchical fully homomorphic encryption scheme is a weakened fully homomorphic encryption scheme. It does not pursue a pure fully homomorphic encryption scheme, but constructs a homomorphic encryption scheme that can support a certain circuit depth. The biggest result of this process is the modulo exchange technique. Modulo switching technology only adds negligible time overhead to the scheme to enhance the support ability of homomorphic encryption scheme for homomorphic multiplication, in addition, it is possible to implement hierarchical and fully isomorphic encryption scheme through certain configurations. The main representatives of this stage are based on LWE and based on RLWE.

3 Investigation and Research on the Application of Homomorphic Encryption Technology Based on Computer Cloud Computing

3.1 Experimental Environment Settings

The operating environment is unified as Intel Corei73770 (3.4 GHz/L38M), 8G RAM.

The FHE scheme adopts the installation of NTL function library on DevC++ under Linux system and uses g++ to compile and run.

The modified RLWE scheme code is implemented by installing NTL algorithm library on vc++ 6.0 under windowsxp system.

3.2 RLWE-Based Scheme

The operation on the polynomial ring is based on the RLWE scheme. Here we first define the basic method of the polynomial operation. Because the degrees of polynomials in the scheme are equal, we only consider the case of polynomials of degree n − 1. Solving polynomial multiplication with FFT algorithm needs to go through three processes: (1) Convert coefficient representation to point value representation, this process is called evaluation. (2) The point value notation calculates the multiplication. (3) The point value representation is converted into the coefficient representation, and this process is called interpolation.

The encryption and decryption processes are relatively simple. For plaintext $m \in R_2^d$, define $r \leftarrow R_2^N$, define $m \leftarrow (m, 0) \in R_{qL}^2$, and calculate ciphertext as shown in Eq. 1.

$$c = m + A_L^T r \in R_{qL}^2 \tag{1}$$

Decryption can be performed at each layer or only at the last time as needed. That is, for the ciphertext $c \in R_q^2$, calculate the plaintext / as shown in Eq. 2.

$$m = [[\langle c, s_j \rangle]_{qj}]_2 \in R_q^2 \tag{2}$$

Generally speaking, the depth of the decryption operation is determined in a specific scenario, that is, L is determined (note that for ciphertext, the subscripts are arranged in reverse order). Therefore, in general, the last sj = s0, there is no need to generate a key of excess depth L.

3.3 Introduction to Verification Data

Cloud education is chosen as the experimental background of this paper. Many systems are included in conventional cloud education systems, using grade system data as authentication data. In order to make the test more realistic, the concept of fully homomorphic encryption is used to find the average size of a specific high school student, assuming 100 credits, the total number of students is 1000. In the experimental program, each student's grade needs 10 bits to represent, in order to avoid overflow, a 10-bit buffer is needed to store the sum of the student's grades. This process requires a total of 1999 additions and 1 multiplication.

4 Investigation and Analysis of the Application of Homomorphic Encryption Technology Based on Computer Cloud Computing

4.1 File Encryption Module

The flow chart of selecting a local file for encryption is shown in Fig. 1. First, select the file to be encrypted and uploaded locally. After selecting the file, the system uploads the file to HDFS, and then selects the parallel or serial encryption method. Two types of homomorphic encryption algorithms are used in the client side of the client, one is the RLWE algorithm in this paper, and the other is the FHE encryption algorithm. At the same time, when the encryption method of parallel encryption is selected, the number of encrypted nodes can be selected so as to clearly observe the comparison of encryption time under different number of nodes and serial encryption.

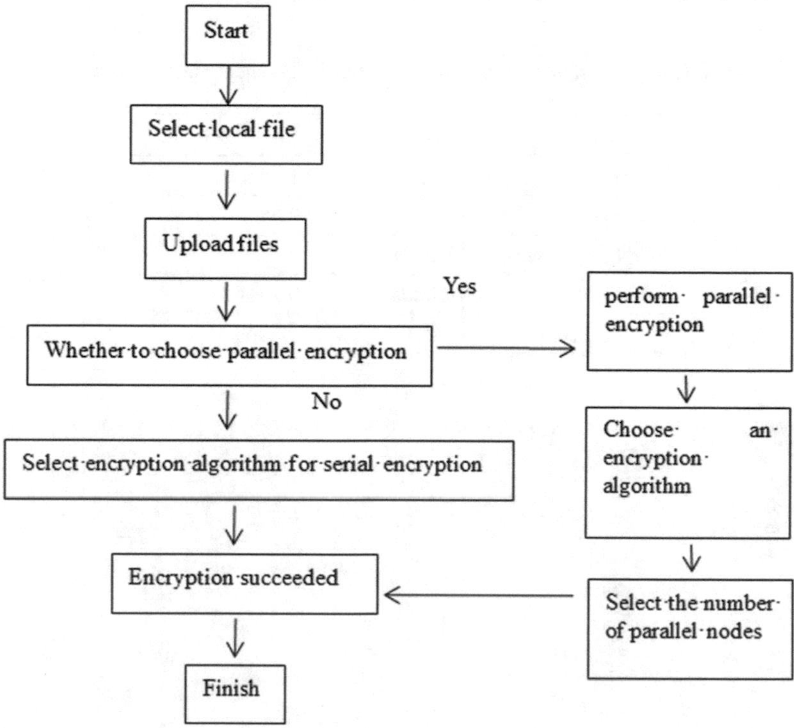

Fig. 1. Encryption flowchart

Click the browse button in the file encryption function sub-module to upload and encrypt the operation. First, the Select Local File dialog box will pop up, and start to select the file to be encrypted and uploaded. After completing the selection, serial and parallel selection and the selection of different homomorphic encryption algorithms, after the selection is complete, click the encryption button to encrypt. The files uploaded to the cloud server are in the form of ciphertext that has been homomorphically encrypted. When users want to access the server through the network to share files anytime and anywhere, they can download them through the download function in the browse module. Path dialog box, select the file path to be saved, click OK to decrypt, thus verifying that the client can complete the correct encryption and decryption functions.

4.2 Scheme Comparison

The calculation time and storage space of the FHE scheme can be calculated that its running time is about 350 h, and the ciphertext size after plaintext encryption is 0.1M, that is, the expansion rate before and after encryption is 100,000 times. The ciphertext space expands amazingly. The FHE solution is not suitable for big data environments such as education cloud. The time required for each fully homomorphic operation of the RLWE scheme and the FHE scheme is shown in Table 1:

Table 1. Comparison of running time between FHE scheme and RLWE scheme

Operate	FHE scheme	RLWE scheme
Encryption	1.22 s	1.38 s
Decrypt	0.15 s	0.45 s
Ciphertext refresh	18.1 s	0
Addition	0.5 ms	0.1 ms
Multiplication	1.2 ms	0.6 ms

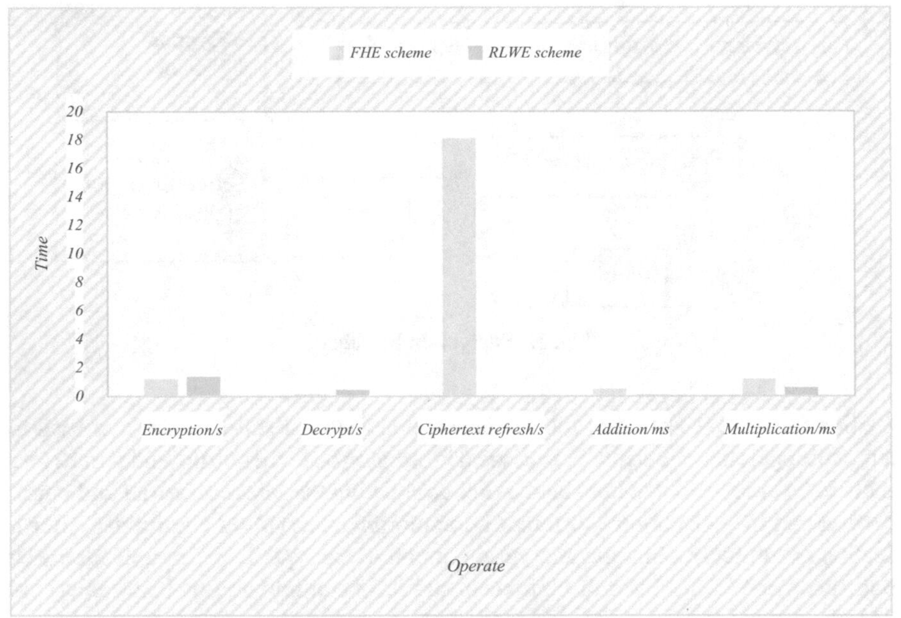

Fig. 2. Comparison of running time between FHE scheme and RLWE scheme

From the above comparison, it can be seen that the running time of the encryption and decryption, homomorphic addition and homomorphic multiplication of the RLWE scheme is slightly larger than that of the FHE scheme, but the RLWE scheme no longer needs the step of ciphertext refresh, which saves about 80% of the running time, At the same time, the RLWE scheme also uses SIMD technology to pack multiple plaintexts into a single ciphertext. These two techniques are explained in detail, which greatly saves the running time, as shown in Fig. 2. In the experiments of the RLWE scheme, the correlation data expansion rate is 200 times. From the experimental data, it can be seen that the RLWE scheme is operation independent. The time or ciphertext expansion rate is very close to the actual requirement.

5 Conclusions

With the development of cloud computing and the Internet, it is inevitable that more and more information needs to be handed over to the cloud (or server) for processing. Enterprises also need to use the computing power of the cloud to complete massive data computing tasks that could not be completed before. These are all scenarios for third-party data processing. Using fully homomorphic encryption technology, we can hand over data to a third party for processing while ensuring privacy. In order to solve the problem of information security in cloud computing, this paper analyzes the encryption technology, analyzes the structure of homomorphic coding technology research, and chooses the RLWE scheme as a guide. Finally, the application of homomorphic encryption technology in cloud computing is demonstrated.

References

1. Dogan, D.C., Altindis, H.: Storage and communication security in cloud computing using a homomorphic encryption scheme based Weil pairing. Elektronika ir Elektrotechnika **26**(1), 78–83 (2020)
2. Reis, D., Takeshita, J., Jung, T., et al.: Computing-in-memory for performance and energy-efficient homomorphic encryption. IEEE Trans. Very Large Scale Integr. (VLSI) Syst. **28**, 2300–2313 (2020)
3. Sharma, I.: Fully homomorphic encryption scheme with symmetric keys. Comput. Sci. **2011**(4), 1–4 (2013)
4. Doroz, Y., Ozturk, E., Sunar, B.: Accelerating fully homomorphic encryption in hardware. IEEE Trans. Comput. **64**(6), 1509–1521 (2015)
5. Kanan, K., Sharma, N.: A review of homomorphic encryption scheme analysis of cloud computing. J. Interdiscip. Cycle Res. **XIII**(VII), 458–467 (2021)
6. Boucenna, F., Nouali, O., Kechid, S., Tahar Kechadi, M.: Secure inverted index based search over encrypted cloud data with user access rights management. J. Comput. Sci. Technol. **34**(1), 133–154 (2019). https://doi.org/10.1007/s11390-019-1903-2
7. El-Yahyaoui, A., Kettani, E.: An efficient fully homomorphic encryption scheme. Int. J. Netw. Secur. **21**(1), 91–99 (2019)
8. Martins, P., Sousa, L.: A methodical FHE-based cloud computing model. Future Gener. Comput. Syst. **95**(JUN.), 639–648 (2019)
9. Hayward, R., Chiang, C.C.: Parallelizing fully homomorphic encryption for a cloud environment. J. Appl. Res. Technol. **13**(2), 245–252 (2015)
10. Breuer, P.T., Bowen, J.P.: Fully encrypted high-speed microprocessor architecture: the secret computer in simulation. Int. J. Crit. Comput.-Based Syst. **9**(1–2), 26–55 (2019)

11. Yagoub, M.A., Kazar, O., Beggas, M.: A multi-agent system approach based on cryptographic algorithm for securing communications and protecting stored data in the cloud-computing environment. Int. J. Inf. Comput. Secur. **11**(4/5), 413 (2019)
12. Awadallah, R., Samsudin, A., Teh, J.S., et al.: An integrated architecture for maintaining security in cloud computing based on blockchain. IEEE Access **9**, 69513–69526 (2021)

BIM Technology Combined with the Internet of Things Model Office Space Information Design Research

Yi Fu and Jinkuan Yang(✉)

School of Design and Art, Shenyang Jianzhu University, Shenyang, Liaoning, China
1106245859@qq.com

Abstract. In order to improve the informatization and industrialization in the transformation and upgrading of traditional office-in office space to informatization and Internet of Things indoor office space industry, and promote the integrated development of the two modernizations. This paper discusses the feasibility research based on BIM and Internet of Things technology in office space design, connects BIM and Internet of Things technology closely with office space design elements, and serves as the feedback, expresses the broad prospect of BIM and Internet of Things technology in office space design, greatly promotes the information development of office space design, and promotes the deep integration of traditional office space design and information data.

Keywords: Office space · BIM technology · The Internet of Things

1 BIM Technology Application Overview of BIM Technology Combined with the Internet of Things

The development speed of modern science and technology is getting faster and faster, the update speed of computer software and computing processing ability are more powerful, making the means of model information processing is also more and more important in the field of space design [1]. The typical regulation schemes can be exemplified as follows:

1.1 Software Overview of the BIM Technical Class

The development of science and technology makes the current design enterprises also more enough to rely on information design, engineering design and construction design in the early stage of project design and cooperation between various professional types. This working mode plays the purpose of shortening the time period, reduce the waste of resources and improve the economic efficiency.

At present, there are many software that can realize and have certain construction information model processing technology, mainly on the Revit and Rhino3D of American AutoDesk enterprises. Based based on this model information, China has launched a

J. H. Abawajy et al. (Eds.): ICATCI 2022, LNDECT 170, pp. 293–300, 2023.
https://doi.org/10.1007/978-3-031-29097-8_35

series of software such as Luban design and Guanglinda design. 1 AutoDesk enterprise Revit series software is outstanding in computational volume, and is relatively mature in visualization operations such as 3D display, model processing and real-time preview. There is also a strong industry recognition in deepening design and multi-software collaborative task processing.

In terms of 3D rendering and rendering performance, Enscape and other software (represented by GPU rendering) has shown strong vitality and operational space. In the near future, combined with the information flow of the Internet of Things can bring a new design experience and project display mode to the design of office space.

1.2 Application Prospect of the Internet of Things in Office Space Design

On the basis of combining the aforementioned research, after taking full consideration of China's national conditions and the current policies, the overall construction principle of office space is proposed, that is, overall planning, taking into account the overall, gradual implementation, procedure norms, demonstration guide, step by step [2]. In the construction process of office space, safe and efficient, applicable, economical and reasonable, so that the overall system has scalable, certain openness and flexibility. In office space, various infrastructure should be built simultaneously with each subsystem to prevent information island problems caused by differences in data types of application subsystems and stacking. Constantly increase the integration of new hardware and software subsystems, so that the management system is constantly improved. Application of advanced Internet of Things technology, to realize the non-inductive, convenient and efficient intelligent application, as well as dining, shopping, traffic, physical examination and other efficient user experience. Through the Internet of Things, cloud computing, artificial intelligence and other technologies to build an office space management platform of "cloud workbench aggregation and sharing application" for office space management. In order to realize the intelligent management, it is necessary to build the management cockpit, video artificial intelligence analysis, intelligent scenes, etc., so as to build the intelligent management brain in the office space (in Fig. 1).

1.3 The Combination of BIM Technology and Office Space Marketization

The issue that BIM technology is the future trend in the global architectural design industry has been basically affirmed, and the application of BIM technology still has great potential for development. Changes to the architectural design-related industries are disruptive and revolutionary [3]. The popularization and extensive application of building information model processing technology has a profound impact on the change of the design industry and is of great significance. According to the existing influence, the impact in the construction industry will surpass the current impact of electronic computers on the design industry. At present, there are two main important factors hindering the design: ① designer's repeated labor and disorderly personnel management of the obstacles of ② software coordination ability and information comprehensive processing ability. The main reason for this phenomenon is the great communication barriers of information and data interaction between many BIM (building information model processing technology) software, caused by the incompatibility of many software protocols

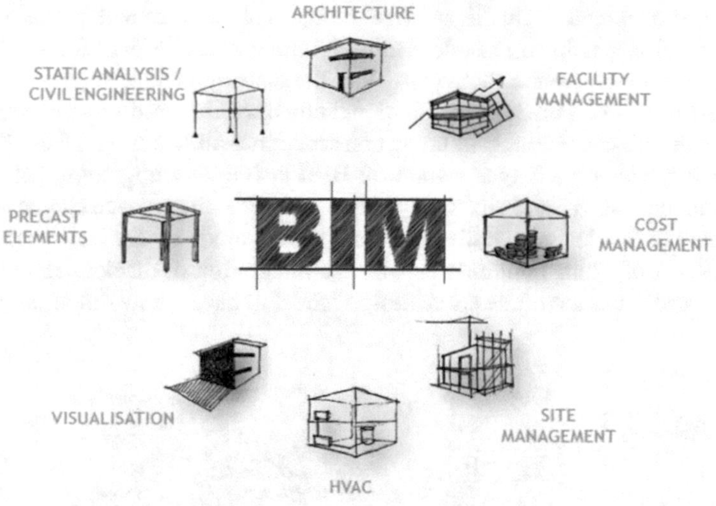

Fig. 1. Main types of BIM

and engineering file formats. It is exciting that the BIM data of the two driving endpoints in the current BIM software environment has had a very good development trend, and many software companies and design practitioners have made very gratifying optimization results. The core of the big data platform of the Internet of things includes two parts: distributed data storage and distributed computing. In order to make the office intelligent decision system adapt to the complex changing environment, the platform adopts BP neural network algorithm to realize the expected system functions and complete the output and iterative evolution of the network. The output of the neural network is:

$$y_j = f(Net_{in} - \theta_j) \tag{1}$$

In the expression, θ_j is the threshold of neurons; Net_{in} is the net of the i-th neuron. Enter, whose expression is:

$$Net_{in} = \sum\nolimits_{i=1}^{n} \omega_i \cdot x_i \tag{2}$$

In the expression, x_i is the input of neuron, which is the key of accident early warning model variables; ω_i is the connection weight to adjust the proportion of each input.

The data evolution of the Internet of things is assisted by BP algorithm. Relying on the chain derivation method in calculus, the error is back propagated to correct the weight of each connected neuron. Finally, a big data mining and analysis system that can be updated in real time and evolve continuously is realized.

1.4 Organic Combination of BIM Technology and the Internet of Things

For BIM, the combination with the Internet of Things can provide spatial positioning for all kinds of intelligent electromechanical equipment in the building. The spatial positioning of all kinds of intelligent electromechanical equipment in the building in the BIM model can help to provide a more intuitive analysis means for all kinds of maintenance and maintenance activities. With the development of smart cities, the use of "BIM + Internet of Things" to build digital cities will increasingly need to embrace BIM to obtain massive amounts of urban construction facilities model data. From BIM to CIM, it will become a bigger market for BIM technology upgrading [4]. All kinds of intelligent mechanical and electrical equipment in the office space can be selected at the beginning of the design and divided into different mode types. The combination of BIM + Internet of Things not only enables the integration of disciplines to show new vitality, but also gives the office space design model to have a new direction (in Fig. 2).

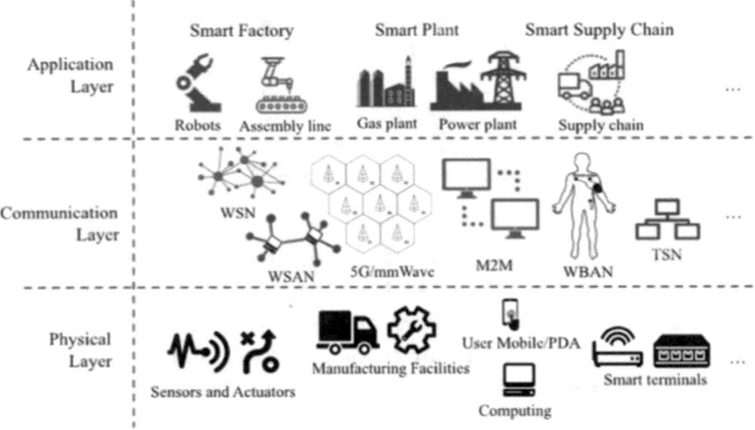

Fig. 2. Internet of things communication concept

2 Effective Application of BIM Technology Combined with the Internet of Things in Office Space Design

The application of BIM technology to building energy-saving design requires construction engineering designers to conduct a site survey of the surrounding environment of the building area, mainly including natural factors such as climate and geological characteristics. After the information processing of these factors, input the analysis software for the corresponding simulation building [5].

2.1 Field Analysis and Design

Therefore, it is necessary to conduct a comprehensive inspection of the factors affecting the energy saving of the construction project, and combined with the characteristics of the architectural design and the final quality of the building in the same environment, to provide a reasonable design scheme for the construction project, to scientifically investigate the hidden dangers in the construction structure design, and to provide a guarantee for the construction safety.

2.2 Drawing Design of the Office Space

The BIM technology can simulate the indoor office space, form a 3 D reference model, and realize the effective design optimization of the Internet of Things equipment through environmental simulation [6]. In addition, the use of BIM technology in the indoor office drawing design, which can generate a virtual 3 D space, so that designers can observe the drawings more intuitively, and reasonably judge the indoor office space layout. Three-dimensional drawings combined with the point layout of the mechanical and electrical equipment of the Internet of Things, so that the office space can intuitively reflect the multiple levels of the space, clear and concise three-dimensional drawings, and the convenient role of the Internet of Things equipment at the beginning of the design.

Make the data more accurate, so as to ensure the scientific and reasonable office space design. Using BIM technology for office space design has the advantages of visualization [7]. The designer can set the BIM technology to make a perfect design, and the construction personnel can better determine the construction target combined with the drawings to reduce the uncertainty in the construction process.

2.3 Lighting Analysis of the Office Environment

Good office space lighting design is a very important part of improving the modern interior office space design. Good lighting performance can not only improve the comfort of the indoor office environment, but also can affect people's use feelings in the psychological and emotional expression. Traditional interior office lighting design requires integrated environmental information of the building location to determine the sunrise side. However, this design method is cumbersome and designers have to provide reliable information through a lot of calculations. By using BIM technology to simulate the lighting conditions of the indoor office environment, designers can not only quickly draw up to a reasonable design scheme, but also conduct a comparative analysis of the indoor office lighting in combination with different seasons to find the best lighting design scheme.

2.4 Analysis of Electromechanical Equipment of Office Environment

The biggest advantage of BIM technology is that the virtual three-dimensional space can be built and adjusted according to the design needs. The contradiction between the wiring of mechanical and electrical equipment and various cables of indoor office space is existing in almost every office space. Through BIM technology processing, the design practitioners can not only optimize the arrangement of all kinds of cables, but also guide the optimal energy consumption and the optimal space location through data analysis. Combining the data of the Internet of Things equipment into the BIM three-bit software, the design practitioners can adjust the spatial position of each equipment through the software processing way, optimize the layout scheme of the mechanical and electrical equipment of the indoor office, and reduce the information interference of the wireless frequency bands and wired frequency bands (in Fig. 3) [8].

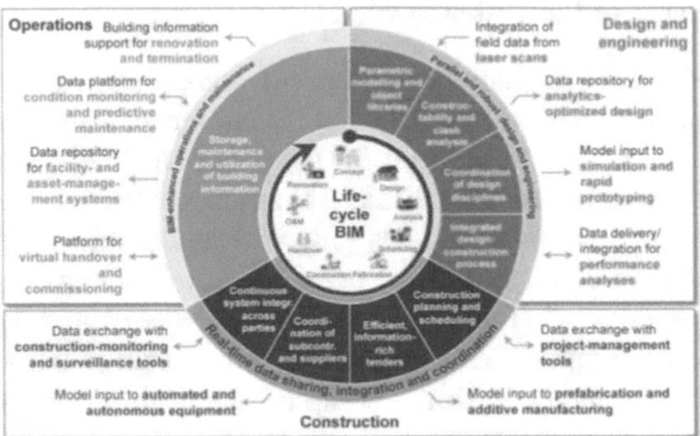

Fig. 3. Life cycle BIM

3 BIM Technology Combines the Advantages of the Internet of Things in Office Space Design

Traditional office space design is often limited to the factors inherent in the location of the project. For example: site area, construction factors, electrical factors layout, etc. The adoption of BIM technology can help design practitioners to better analyze and handle important design information such as personnel action trajectory information, spatial layout relationship, and comprehensive layout of lighting, building structure and electrical appliances in the early stage of the design. These comprehensive factors are often overlooked and very important information in office design.

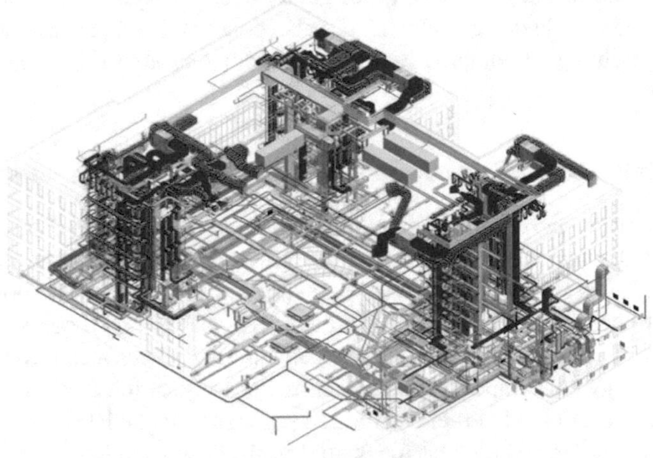

Fig. 4. Information processing of office space model

In the application of the office space of the Internet of Things Smart Park, it involves a variety of scenarios, namely, office space employees, visitors, managers and other scenes. The system integrates intelligent meetings, non-inductive attendance, mobile office, etc., and builds an intelligent staff management platform to realize the intelligent office space scene [9]. Visitor management has realized automation and intelligence, and integrates visitor application, park navigation and intelligent exhibition hall. Managers can intelligently manage the office space through various kinds of subsystems to realize the scene intelligence. Using the cloud work platform, aggregation, integration and sharing service application, it realizes the integration of office space application, and provides a multi-end platform of mobile office, efficient communication and collaboration. Operation management adopts 3 D modeling, including office space model customization service, building model customization service, indoor structure model customization service, equipment point digit authorization and services (in Fig. 4) [10].

4 Conclusion

To sum up, BIM technology, combined with the application effect of the Internet of Things, can promote the construction of intelligent office environment. According to the above design process, it can be seen that the Internet of Things technology presents a good application effect in the office system, and can support most of the online and mobile operation of the office. At the same time, the technology also gives various advantages, such as system convenience and security, etc. Therefore, the technology has a high feasibility in the design of the office system. The introduction of BIM technology combined with the Internet of Things will bring revolutionary changes to the office space design and even the whole industrial chain of the entire construction industry. This technology can integrate the whole life cycle of the whole industrial chain of the construction industry from the early planning of the project to the final demolition, and carry out information sharing, planning, management, operation and maintenance in the whole life cycle. While improving the efficiency in the whole life cycle, it effectively saves the resources and reduces the unnecessary waste of human resources, material resources and resources. From the perspective of office space design and even the development of the construction industry, the comprehensive design application of BIM technology combined with the Internet of Things can diversify and efficiently strengthen the processing ability of office space design, and better adapt to the future development direction of office space design. According to the above design process, it can be seen that the Internet of Things technology presents a good application effect in the office system, and can support most of the online and mobile operation of the office. At the same time, the technology also gives various advantages, such as system convenience and security, etc. Therefore, the technology has a high feasibility in the design of the office system.

References

1. Meyer, T., Brunn, A., Stilla, U.: Accuracy investigation on image-based change detection for BIM compliant indoor model. ISPRS Ann. Photogramm. Remote Sens. Spat. Inf. Sci. **2021**(4), 105–112 (2021)
2. Ryan, L., et al.: Enhancing a building information model for an existing building with data from a sustainable facility management database. Sustainability **13**(13), 7014 (2021)

3. Danylo, S., Tatjana, V.: Bibliometric analysis of building information modeling, geographic information systems and web environment integration. Autom. Constr. **128**, 103757 (2021)
4. O'Grady Timothy, M., et al.: Circular economy and virtual reality in advanced BIM-based prefabricated construction. Energies **14**(13), 4065 (2021)
5. GhaffarianHoseini, A., et al.: ND BIM-integrated knowledge-based building management: Inspecting post-construction energy efficiency. Autom. Constr. **97**, 13–28 (2019)
6. Okakpu, A., et al.: Exploring the environmental influence on BIM adoption for refurbishment project using structural equation modelling. Archit. Eng. Des. Manag. **16**(1), 41–57 (2020)
7. Seghier, T.E., et al.: Building envelope thermal performance assessment using visual programming and BIM, based on ETTV requirement of Green Mark and GreenRE. Jurnal Alam Bina **4**(3), 227–235 (2017)
8. Marzouk, M., Enaba, M.: Analyzing project data in BIM with descriptive analytics to improve project performance. Built Environ. Proj. Asset Manag. **9**(4), 476–488 (2019)
9. Fan, S.-L., et al.: Latent provisions for Building Information Modeling (BIM) contracts: a social network analysis approach. KSCE J. Civ. Eng. **23**(4), 1427–1435 (2019)
10. Chan, D.W.M., Olawumi, T.O., Ho, A.M.L.: Perceived benefits of and barriers to Building Information Modelling (BIM) implementation in construction: The case of Hong Kong. J. Build. Eng. **25**, 100764 (2019)

Big Data Automatic Classification Processing System Based on Cloud Computing

Jun Li[1](✉) and Ankit Singh[2]

[1] Jiangxi University of Software Professional Technology, Nanchang, Jiangxi, China
jxlijunmail@163.com

[2] Jawaharlal Nehru University, New Delhi, India

Abstract. How to quickly, accurately and fully obtain the information we need from social media, and how to effectively organize and manage this information are major challenges facing the current information technology space. Considering the needs of experimental research and practical application, this paper makes a detailed modular design of the automatic text classification processing system based on semantic Chinese. In terms of the impact of keyword weights on specific categories, the original algorithm is improved by introducing two parameters: the ratio of keywords to all keywords contained in the current category, and the average number of samples in this category compared to the current category. Number of samples. According to the principle of combining category concepts, the weights between keywords and categories are combined and calculated, and the highest weighted sorting number is obtained as the optimal sorting number of the text to be sorted. The final test results show that the improved algorithm has certain advantages compared with other algorithms in terms of recall rate, precision rate, F1 value, macro average and micro average. The macro-average precision, recall and FI of the improved algorithm are 4–5 percentage points higher than those of the other two algorithms.

Keywords: Cloud computing big data · Automatic classification · Classification processing · System design

1 Introduction

With the emergence of age information, people are filled with information every day, and how to find valuable information from the media quickly and regularly has become a problem that people are currently facing [1]. Therefore, various classification and filtering techniques have emerged [2]. Artificial intelligence-based automated voice design and programming can effectively and automatically organize a large amount of information, helping people find the information they need quickly and accurately in large and complex information [3]. Traditional text classification is to classify text into one or more predetermined categories, such as text content. The introductory text is hand-picked by experts and requires a high level of local knowledge and a high level of expertise [4].

With the development of technology, statistical methods and machine learning methods have been introduced into automatic text classification, and have achieved fruitful

J. H. Abawajy et al. (Eds.): ICATCI 2022, LNDECT 170, pp. 301–308, 2023.
https://doi.org/10.1007/978-3-031-29097-8_36

results [5]. Borwankar R. discusses a novel optical nondestructive inspection and automatic classification technique using custom image processing techniques. The initial database contains 100 parts with different reflectivity and surface conditions. The transform algorithm relies on Discrete Wavelet Transform (DWT) and Rotated Wavelet Transform (RWT) for feature extraction, while K-Nearest Neighbors (KNN) classifier is used for PUT classification. The maximum accurate classification efficiencies achieved by DWT and RWT are over 80% and 93%, respectively [6]. The best combination of Sevik U's successful classification of images into skin, burn, and background regions was found to be the fuzzy c-means algorithm for the segmentation part, and the multi-layer feedforward artificial neural network part trained by the back-propagation algorithm for the classification. With an F-score of 74.28% for classifying captured images without protocol, the proposed scheme manages to achieve similar results [7]. It is of practical significance to study the automatic classification and processing system of big data based on cloud computing [8].

This work designs and implements an automatic text classification processing system that can automatically classify batch unstructured text data, and perform accurate information analysis, rapid placement of information and secondary utilization of information from the batch information. Based on the industry's advanced cloud computing technology, the system performs classification feature training and large-scale particle extraction. Using the distributed solutions of traditional sorting algorithms and feature derivation algorithms, a sorting system is built on a distributed platform to sort texts in a distributed manner. The introduction of cloud computing solves the bottleneck of massive data processing in the traditional sorting process, and provides support for the extremely high requirements for computers and storage in the large-scale word processing process.

2 Research on the Design of Big Data Automatic Classification Processing System Based on Cloud Computing

2.1 Cloud Computing and Hadoop Platform

Cloud computing uses distributed storage for data storage and redundant storage, which generally has the characteristics of high reliability, high concurrent performance, high transmission rate, high scalability, automatic fault tolerance and easy management [9]. Google's distributed file system was designed by Google to store large amounts of search data [10]. Based on the principle of hardware failure, the block-based file is split, one write and one read, providing separate file upload and write functions, using master-slave with excellent performance, scalability and fault tolerance, it is open source Hadoop file system [11].

HDFS is designed as a distributed file system that runs on general-purpose hardware [12]. Fast detection and fast automatic recovery are the main architectural goals of HDFS. Therefore, HDFS not only has the characteristics of a general distributed file system, but also has its own design considerations in meeting the reliability of big data processing, and has made many assumptions, compromises and optimizations in its design ideas.

2.2 Analysis of Module Functional Requirements

2.2.1 Data Preprocessing

There are mainly two types of data sources provided by users, one is tabular data, and the other is relational database export data. In view of this data situation, a unified source data storage format is designed, data entry is completed, and preliminary data persistence is completed. After the initial persistence of the source data is completed, the results are stored in the database. Finally, the Chinese data preprocessing is performed on the cleaned data, including Chinese text segmentation, feature value extraction, and feature value calculation. These are all for subsequent text analysis algorithms. And visual display to provide data basis. The time of the user's source data is uncertain. In order to ensure certain real-time requirements, it is necessary to design a regular polling scheme. When new data is incoming, the data processing task is started.

2.2.2 Thematic Event Mining

It mainly mines the main topics that callers focus on within a fixed time span, so as to provide decision-making reference for analysts. Topic event mining requires certain real-time requirements, so it is necessary to improve the processing speed through distributed computing.

2.2.3 Automatic Text Classification

Users currently use manual annotation to classify and label incoming calls. In order to reduce labor costs, a solution for automatic text classification is provided. When a new incoming call comes in, the operator completes the content record of the incoming call, automatically classifies the incoming call, and lists the possibility of it, so as to provide reference for subsequent manual detection. At the same time, with the growth of time, the training set is continuously expanded, and the classification model is continuously updated in an iterative manner to improve the accuracy and recall of the model.

2.2.4 Web Visualization Display

As the most intuitive display of the analysis results, and also the content that users directly contact, the overall interface should be concise and intuitive, and the analysis results of the above parts should be displayed quickly and concisely. Each module corresponds to a sub-page, and each page provides multi-dimensional condition selection, and provides visual chart display after the dimension is determined.

3 Investigation and Research on the Design of Big Data Automatic Classification Processing System Based on Cloud Computing

3.1 Construction of the System Operating Platform

System platform: 5 personal computers, each with a CPU clocked at 2.0 GHz and 2G memory. Software platform operating system: 5 computers are connected to run a distributed cluster by transmitting Hadoop. The text classification program runs on Hadoop, and calls 5 computers to achieve distributed distribution through Hadoop. Planning development environment: lunar development platform, using maven's project management tool. Put the corresponding maven plugin on moon for related development.

3.2 System Architecture

The system consists of four main functional modules: training, testing, performance evaluation, and single text instant classification. The task of the training module is to obtain word-based category vectors and semantic concept-based category vectors during the training process. After the training process is completed, a semantic-based Chinese text classifier is obtained; the task of the test module is to use the text in the test set. The classifier performs classification and counts the classification results; the task of the performance evaluation module is to evaluate the performance of the classifier according to the classification parameters obtained by the test module; the instant classification of a single text is to use semantic-based methods for a new text. Chinese text classifier for classification.

3.3 Classification Performance Evaluation

The macro average is to first calculate the evaluation indicators of each category, namely recall, Accuracy, Fallout, Accuracy and Error, and then the average of all categories; the micro average is to first calculate the sum of A, B, C, D for all categories and then the above value Both are calculated using the sum of these values. The difference between average and average is that average gives each file the same weight, while average gives each category the same weight.

Text classification is essentially a mapping process, and the performance evaluation of text classification can reflect the accuracy of the classification system mapping. Commonly used evaluation indicators include recall, precision and F1 value. These indicators are related to the following values:

Recall rate:

$$recall = \frac{\text{Classified Extensive Text}}{\text{The amount of text the class should have}} = \frac{A}{A + C} \quad (1)$$

Accuracy:

$$precision = \frac{\text{Classified Extensive Text}}{\text{The actual number of texts assigned to this class}} = \frac{A}{A + B} \quad (2)$$

F1 value:

$$F1 = \frac{2 \times precision \times recall}{precision + recall} \quad (3)$$

It can be seen from the above definition that the recall rate is the probability that a document d should belong to category c, and the classifier also assigns it to category c; the precision rate is the correctness of document d being assigned to category c by the classifier probability of classification. Recall and precision reflect two different aspects of classification quality, both of which must be considered comprehensively, and the F1 value is a balance indicator that combines the two indicators.

4 Analysis and Research on the Design of Big Data Automatic Classification Processing System Based on Cloud Computing

4.1 System Modules

Training module: Insert the training text set, perform word preprocessing, feature selection, attribute weighting and other functions in the training text set, and create a word-based category vector for each category in the training text set. Then, semantic mapping is performed on the word-based vector space to obtain category vectors based on semantic concepts. Therefore, through the training module, the system can obtain the following data: word-based category vectors and semantic concept-based category vectors. The activity diagram of this module is shown in Fig. 1.

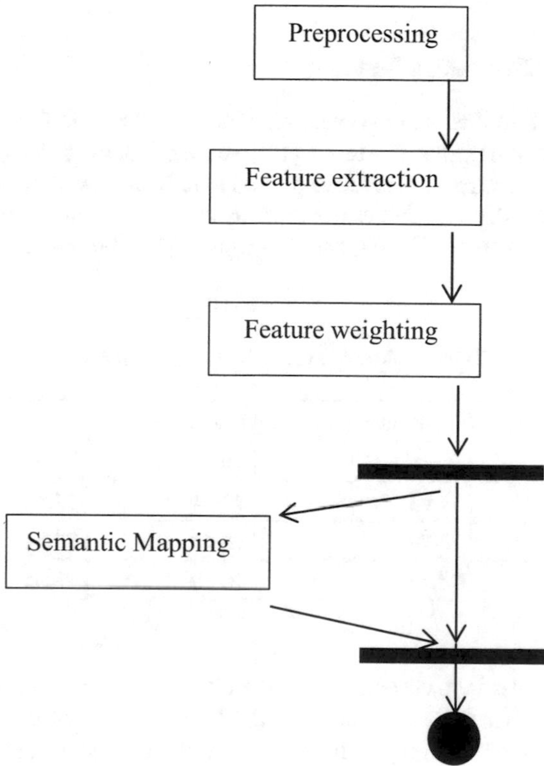

Fig. 1. Activity diagram of the training module

After the training process is completed, we obtain a Chinese text classifier based on the semantic concept vector space model.

Test module: Through the training module, we obtained a semantic-based Chinese text classifier. After preprocessing the test text, the feature vector of the text needs to be semantically transformed to realize the mapping from words to concepts. The reason is that if they are not semantically transformed, there will be a majority word mismatch error when comparing similarity. In the testing module, after classifying the texts in the test set, we will count the classification results. After the testing process is over, we will get 3 classification parameters. These 3 classification parameters include: the number of test documents belonging to this category, the number of texts that the classifier discriminates as this class, and the number of texts that belong to this class and are judged as this class. These parameters will be used in the performance evaluation module.

Performance evaluation module: For the performance evaluation test of the text scoring system, there are widely used performance evaluation indicators in the world, including the two key indicators of recall and accuracy.

Single-text instant classification module: The text ranking process generally includes six steps: acquiring datasets, creating text representation models, text training, classifier testing, classifier performance evaluation, and text classification.

4.2 Performance Evaluation Test

The test set selected in this paper comes from the database, and the data in the test set does not participate in training. The test set has a total of 13,592 pieces of data, including titles, keywords, and abstracts. Use the improved feature item weight calculation method to automatically classify and index the data in the test set, and compare it with the classification results of the LLR and MI algorithms. The obtained classification results are shown in Table 1:

Table 1. Macro average and micro average

	Micro average	Macro average		
Algorithm	P = R = F1	Precision	Recall	F1
LLR	71.23	73.44	72.32	73.24
MI	70.78	72.46	73.54	73.35
Improved algorithm	78.75	76.79	78.45	82.34

From the above test data, we can see that the classification results using the improved algorithm are better than the test results of the LLR and MI algorithms, whether from the test indicators of a single category or from the overall macro-average and micro-average results. The macro-average precision, recall and FI value of the improved algorithm are 4–5% points higher than the other two algorithms, and the micro-average is 7–8% points higher, as shown in Fig. 2. Since the test data are all class F, the degree of distinction between classes is relatively small, and there are many feature words shared between subclasses. If the data at the first-level discipline level is classified and tested, the categories are relatively coarse, and the differences between the various fields. The boundaries are clearer, and the test results will be better than the current ones. In addition, the uneven distribution of sample data is also one of the reasons for the large difference in classification results between subclasses.

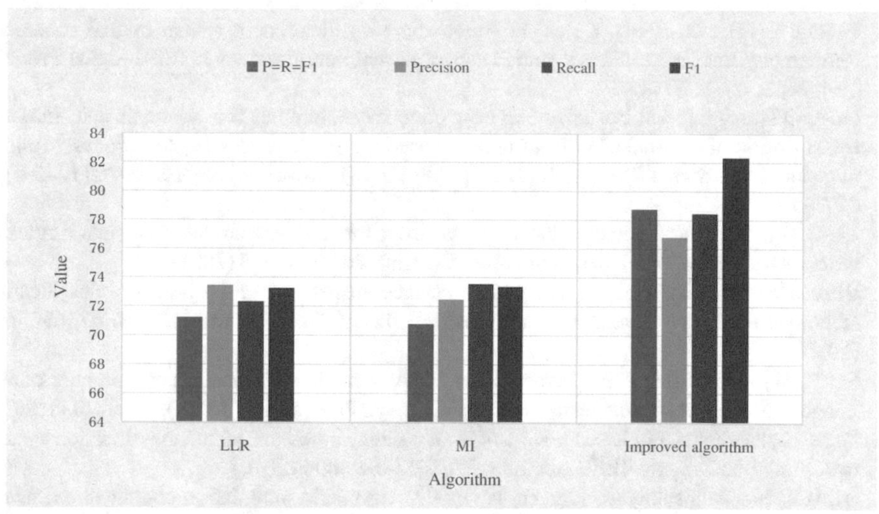

Fig. 2. Macro average and micro average

5 Conclusions

Human society has entered the information age. In people's daily activities, people regularly collect information, analyze it, and use it to judge their own behavior. To a certain extent, the amount of information has become an important factor that determines and limits the development of human society. This paper analyzes the shortcomings of traditional text classification algorithms, and at the same time combines the characteristics of unstructured texts, adopts a distributed solution to the text classification algorithm, and uses a distributed computing framework for classification and editing on a distributed platform. The system uses the supercomputer and storage capacity of cloud computing to solve the problems of large amount of text data and difficult to export features. Finally, the system creates a relatively complete text classification evaluation system, evaluates the classification results of the entire system, and analyzes the classification results accordingly, and proposes solutions to related problems to further optimize the system.

Acknowledgment. Jiangxi Province Educational Science Planning Project (No. 19YB266).

References

1. Demir, F., Sobahi, N., Siuly, S., et al.: Exploring deep learning features for automatic classification of human emotion using EEG rhythms. IEEE Sens. J. **21**, 14923–14930 (2021)
2. Mihandoost, S., Azimzadeh, E.: Introducing an efficient statistical model for automatic modulation classification. J. Signal Process. Syst. **92**(1), 123–134 (2019). https://doi.org/10.1007/s11265-019-01458-5

3. Huang, L., He, X., Fang, L., et al.: Automatic classification of retinal optical coherence tomography images with layer guided convolutional neural network. IEEE Signal Process. Lett. **26**, 1026–1030 (2019)
4. Hussain Hassan, N.M., Nashat, A.A.: New effective techniques for automatic detection and classification of external olive fruits defects based on image processing techniques. Multidimension. Syst. Signal Process. **30**(2), 571–589 (2018). https://doi.org/10.1007/s11045-018-0573-5
5. Li, C., Li, W.: Automatic classification algorithm for multisearch data association rules in wireless networks. Wirel. Commun. Mob. Comput. **2021**(1), 1–9 (2021)
6. Borwankar, R., Ludwig, R.: An optical surface inspection and automatic classification technique using the rotated wavelet transform. IEEE Trans. Instrum. Meas. **67**, 690–697 (2018)
7. Sevik, U., Karakullukcu, E., Berber, T., et al.: Automatic classification of skin burn colour images using texture-based feature extraction. Image Process. IET **13**(11), 2018–2028 (2019)
8. Eken, S., Menhour, H., Koksal, K.: DoCA: a content-based automatic classification system over digital documents. IEEE Access **7**(99), 97996–98004 (2019)
9. Alharbi, M., Alhuseini, A., Ragheb, A., et al.: Automatic modulation classification: investigation for millimeter wave over fiber channels. IEEE Photonics Technol. Lett. **31**(13), 1092–1095 (2019)
10. Baghdadi, Y., Bourree, A., Robert, A., et al.: Automatic classification of free-text medical causes from death certificates for reactive mortality surveillance in France. Int. J. Med. Inform. **131**(Nov.), 103915.1–103915.8 (2019)
11. Luque, A., Romero-Lemos, J., Carrasco, A., et al.: Non-sequential automatic classification of anuran sounds for the estimation of climate-change indicators. Expert Syst. Appl. **95**(APR.), 248–260 (2018)
12. Ponchard, C., Demolin, D., Hassid, S.: Automatic classification of French stops consonants. J. Acoust. Soc. Am. **146**(4), 3084–3085 (2019)

Discrete Artificial Bee Colony Algorithm for Prefabricated Building Resource Scheduling Based on Neural Network

Wei Wang[1], Xiaoying Wang[1], Xianhui Man[1](✉), and Xiang Lu[2]

[1] Chongqing College of Architecture and Technology, Chongqing, China
yeahmxh@163.com

[2] Zhongshe Engineering Consulting (Chongqing) Co., Ltd., Chongqing, China

Abstract. As the development focus of building industrialization, prefabricated building can shorten the construction cycle, save construction resources and protect the environment. In recent years, it has been vigorously promoted and developed. Different from the traditional construction methods, the key of prefabricated component construction is the construction control of prefabricated components. This paper analyzes the production process of prefabricated building resource scheduling, discusses the optimization of resource scheduling, summarizes the NN and discrete ABCA, and briefly introduces the structure of NN and the cost prediction and deficiency of NN. At the same time, the development status of prefabricated buildings is studied. The results show that the new prefabricated building area in China increases with the growth of years, and the new prefabricated building area in China will reach 47.943000 square meters in 2021.

Keywords: Neural Network · Prefabricated Building Resource Scheduling · Discrete Artificial Bee Colony Algorithm · Resource Scheduling Optimization

1 Introduction

The emergence of intelligent algorithm has greatly enriched the optimization technology, so that many optimization problems with high complexity and large scale can be better solved, and has been proved to be a highly stable optimization algorithm. Its extended discrete domain DABC algorithm has been applied to solve the problem of spectrum allocation and achieved good results. However, it still has the disadvantages of slow convergence and easy precocity.

With the continuous progress of science and technology, many experts have studied discrete artificial bee colonies. For example, Guo K and Zhang Q first analyzed the characteristics of LRP in RL. Secondly, the mathematical model of the problem is established. Then, a new greedy adjusted discrete ABCA is proposed [1]. Li x, Ma s proposed a new decomposition based multi-objective discrete ABCA modabc/D to solve the sequence dependent establishment time multi-objective replacement flow shop scheduling problem. The goal is to minimize completion time and total flow time [2]. Chen g, sun P, Zhang J established a repair model considering connection cost constraints and

J. H. Abawajy et al. (Eds.): ICATCI 2022, LNDECT 170, pp. 309–317, 2023.
https://doi.org/10.1007/978-3-031-29097-8_37

network connectivity. A model solving method based on discrete ABCA is designed [3]. Although the research results of discrete artificial bee colony are quite abundant, the research of discrete ABCA for prefabricated building resource scheduling based on NN is still insufficient.

In order to study the discrete ABCA for prefabricated building resource scheduling based on NN, this paper studies the prefabricated building resource scheduling and the discrete ABCA based on NN, and finds the resource-based goal. The results show that the NN is conducive to the scheduling of prefabricated component resources.

2 Method

2.1 Prefabricated Building Resource Scheduling

(1) Production process analysis of prefabricated building resource scheduling

In the process of factory production, the prefabricated components required for prefabricated houses are first restricted by factory time resources, human resources (renewable resources) and capital resources (nonrenewable resources); Second, the production of prefabricated components is similar to but different from that of ordinary factory components. At present, most prefabricated components do not have unified standard modules, and many are non-standard components. Orders need to be placed on site and customized by the factory. There is some uncertainty in this link. Therefore, this link has also become a restrictive factor affecting the production of prefabricated components. The design of prefabricated components needs to meet the needs of various disciplines at the same time. Finally, drawings with practical significance are given for production reference [4]. This requires that the information of each discipline of the prefabrication construction project be integrated, displayed on the drawings and fed back to the prefabrication factory. The prefabricated component factory shall communicate and coordinate with the general contractor according to specific needs and drawing requirements to finally determine the production of prefabricated components. In order to consider the impact of uncertainties, Two methods can be used to deal with the uncertainty of the problem: active scheduling and reactive power scheduling [5]. Active scheduling is a scheduling activity that is carried out before the scheduling task of the project occurs, by counting the occurrence probability of similar tasks in the past, predicting the future occurrence, considering uncertain factors and reducing the impact of these factors on the scheduling. Baseline scheduling can be used to solve and estimate the actual start time of the project scheduling process.

(2) Resource scheduling optimization problem

Generally speaking, the research purpose of resource constrained project scheduling problem is to find the scheduling scheme with the shortest construction period (or the lowest cost). A complete engineering project is composed of many works, which have a close relationship. The determination of scheduling scheme must meet the tight pre tight post constraint relationship between various operations of the project and the upper limit constraint of project resources. In this problem, there are many execution modes for the work in the project [6]. During scheduling, work execution modes can be combined and selected. Each execution mode consumes different resources and time. Compared with the first simulated test scheduling problem, the multi-mode resource constrained project scheduling problem needs to select all execution modes of the project and determine the start time of all jobs under resource constraints. Resource balance refers to the unified allocation of various resources used in the process of project implementation, so as to avoid the complete or overload use of one resource in a period of time, and rarely use or even idle these resources in another period of time. Unlike the resource constrained project scheduling problem, which seeks the shortest duration, the resource balance problem takes the project duration as the constraint and aims to minimize the fluctuation of resource use or the maximum consumption level of resources [7]. This kind of problem is also a strong NP hard problem. The imbalance of resource utilization is one of the important reasons for the low efficiency and rising cost of resource utilization.

2.2 Neural Network Discrete Artificial Bee Colony Algorithm

(1) Concept of discrete artificial bee colony algorithm

Discrete ABCA (ABC) is another excellent intelligent optimization algorithm after particle swarm optimization algorithm and ant colony algorithm. According to different division of labor, there are three types of bees. They perform their respective duties, exchange and share the quality information of honey sources, so as to find the highest quality honey sources. Once it was put forward, it attracted extensive attention at home and abroad. Especially in discrete optimization problems, the existing discrete ABCA (DABC) has the defect of redundant calculation. In ABC algorithm, each food source corresponds to the possible solution of the problem. There is a one-to-one correspondence between food sources and employed bees. The catcher evaluates each food source, and roughly selects the food source with high honey source quality as the object of neighborhood development according to the honey source quality of the food source location. In the whole search process, after limit development, a reconnaissance bee will appear in the food source where the quality of honey source cannot be improved. The Scout bee generates a new food source location to replace the old food source location according to a certain strategy, which helps the algorithm jump out of the local optimization. This effect is particularly important in the later stage of algorithm convergence [8].

(2) Concept of neural network

Since the NN was proposed, many different versions have been derived, and the NN algorithm is also a widely used algorithm. At present, after years of research and development, NN has formed a complete theoretical system and rigorous learning mechanism. Firstly, MP model is proposed to simulate the structure of neurons in the brain. Subsequently, a large number of experts and scholars have conducted rich research on MP model. Later, the academic community formed a unified NN model. In short, intelligent optimization algorithms have a common feature in solving global problems. They simulate natural processes, and the relationship between algorithms is also very close. If they are organically combined, they will learn from each other and perform better.

(3) Neural network cost prediction and its deficiency

The traditional cost is predicted by multiple regression analysis. Because there are many factors and indicators affecting the cost, and the data are highly dispersed, it is difficult to predict accurately. BP NN realizes cost prediction by establishing the mapping relationship between factor index and output target. The prediction process is as follows: select the input layer factor index and output layer target to be predicted, select the training sample set and prediction sample set, set the number of hidden layers and prediction nodes of the prediction model, select the activation function and loss function, select the optimization algorithm, and set the learning rate, iteration times and other parameters of the NN model training sample prediction, The prediction results of the prediction sample set are obtained. BP NN can analyze the influence index data affecting project cost and better solve the fitting problem between nonlinear data. NN are easy to fall into local minima [9]. BP NN can realize the complex nonlinear relationship between input nodes and output results. Its essence is an optimization model to solve practical problems. There may be multiple local minima in the actual problem itself, and it is difficult for BP NN to judge which local extreme point is the global extreme point in the process of data training and prediction. Therefore, when adjusting the network weight, it is likely to callback according to the local extreme point. Therefore, the final result is the local minimum rather than the global optimal solution, this leads to inaccurate network results [4].

(4) Structure of neural network

There is no connection between the hidden layers of the network, but a connection is formed between two adjacent layers. If the input layer has one node and the output layer has b nodes, the NN actually constitutes the Euclidean space mapping relationship from a to B. It can represent any nonlinear relationship from the input layer to the output layer [10]. In the process of network forward propagation, the network input layer information is known, processed from the input layer to the middle layer, and then transmitted to the output layer. If the difference between the actual output and the expected output of the output layer exceeds a certain range, the error signal will return in the opposite propagation direction along the original connection.

2.3 Resource Based Objectives

The objective function of the resource balance problem can be summarized as the following expression (1):

$$\min \sum_{K=1}^{K} \sum_{t=1}^{T} f_k(u_{k1}) \tag{1}$$

In Eq. (1), $f_k(u_{k1})$ is a function of the percentage of resource utilization. The fluctuation of resource use in the whole project cycle is studied. If the value of $f_k(u_{k1})$ is small, it means that the resource use is more balanced [11].

For general finite discrete d-dimensional optimization problems, n-valued finite discrete sets can be defined, as shown in Eq. (2):

$$S = \{s_1, s_2, ...s_n\} \tag{2}$$

s_i is the element in the discrete set where the discrete optimization problem is located, and its value is defined according to the specific problem.

According to the above definition, the general formula for generating the initial solution can be expressed as Eq. (3):

$$X_i = RAND(D; S, P(S)) \tag{3}$$

where $P(S) = \left\{P(S_1), P(S_2), ..., P(S_n)| \sum_{i=1}^{n} P(S_i) = 1\right\}$, $p(S_i)$ represents the probability of selecting S_i to solve one-dimensional parameter values.

3 Experience

3.1 Object Extraction

Experimental environment: HP PC, windows 764 bit operating system, processor: amd-phenom (TM) II × 4B97Processor1. 20 GHz, installed memory (RAM): 2.00 gb, simulation environment: matlabr2012a. In this experiment, 1000 data records in KDDCUP data set were randomly selected. There are three categories: normal, Neptune and Smurf. There are 10, 15 and 25 records in each category. The optimized BP algorithm and standard BP network structure are 11–11-2. Select half of the data as the training set and the other half for testing. The activation function of middle layer neurons is S-type function, and that of output layer neurons is simple linear function. The model parameters of ABCA are set as follows: the initial bee colony size is 10 and the number of iterations is 15. Half of the data is selected as the training set to form the network structure. The model parameters of the ABCA are set as follows: the initial bee colony size is 15 and the number of iterations is 150.

3.2 Experimental Analysis

First, the resource variance value will be recalculated each time the operation task is adjusted, which increases the computational complexity. It cannot be well applied to complex projects. In addition, if multiple running tasks need to be adjusted in the same period, the adjustment priority is unknown, which has the same defect as the peak shaving and valley filling method. Compared with the peak cutting and valley filling method and the difference method, the optimization method has integrity and specific application rules. Although the optimization goal is very clear, it still has the disadvantages of multiple adjustments and a large number of calculations. Similarly, in the relatively complex multi project resource allocation research, the amount of solving tasks is also large.

4 Discussion

4.1 Development Status of Prefabricated Buildings

Since the last century, many houses and buildings have been damaged after the two world wars. People are eager to rebuild their homes. With the rapid development of urbanization, the research on prefabricated buildings has sprung up at home and abroad. With the development of these years, the types of prefabricated buildings in China are constantly enriched, the technology is constantly mature, and the degree of standardization is becoming higher and higher. According to the data released by the National Bureau of statistics, the changes in the area of newly-built prefabricated buildings in China are shown in Table 1.

Table 1. Change of newly built prefabricated building area in China from 2019 to 2021

Particular year	The measure of area (10000 m^2)
2019	3362.6
2020	4623.6
2021	4794.3

It can be seen from the above that the newly built fabricated building area in China in 2019 is 33.626 million square meters; In 2020, the newly built prefabricated building area in China will be 46.236 million square meters; The newly built prefabricated building area in China in 2021 is 47.943 million square meters. The specific presentation results are shown in Fig. 1.

As can be seen from the above figure, the newly built prefabricated building area in China increases with the growth of years. It is not difficult to see that China's prefabricated construction industry is facing unprecedented opportunities and challenges. Driven by the state, prefabricated buildings in China show a vigorous development trend. Prefabricated component manufacturing enterprises should give full play to their subjective initiative, overcome technical problems, improve production efficiency and quality, and help China complete the transformation from a large manufacturing country to a powerful manufacturing country.

4.2 Comparison Experiment of Detection Accuracy

We selected 99 data records in two groups of KDDCUP data sets. The model parameters of artificial bee colony are the same as those in the experiment. The comparison of detection accuracy between BP NN and improved ABCA optimized BP NN (bp-abc) is shown in Table 2.

It can be seen from the above that the detection accuracy of BP NN data 1 is 75.6%, and the detection accuracy of BP NN data 2 is 66.7%; The detection accuracy of BP NN data 1 optimized by the improved ABCA is 83.5%, and the detection accuracy of BP

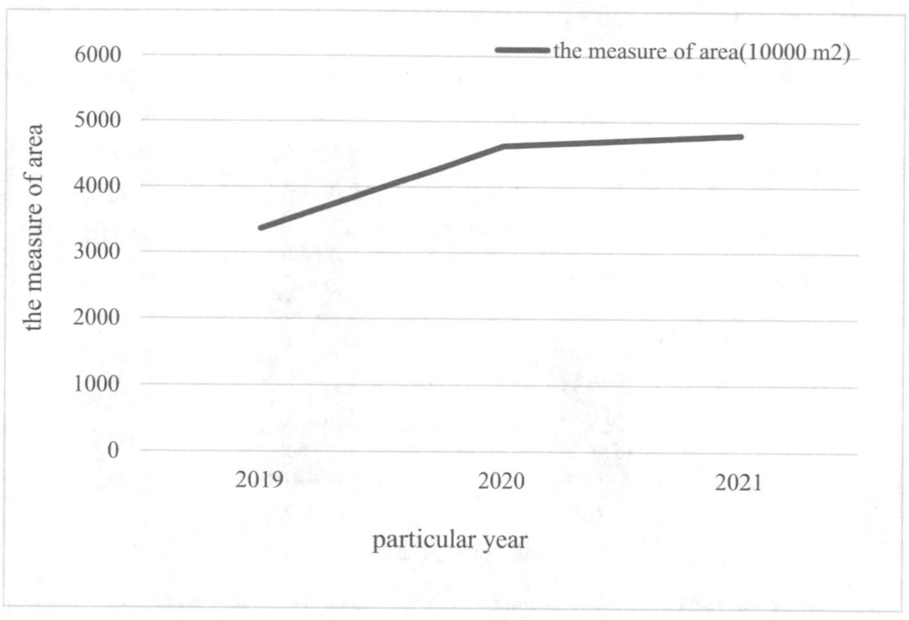

Fig. 1. Change of newly built prefabricated building area in China from 2019 to 2021

Table 2. Comparison of detection accuracy of three groups of data

Network model	Data 1 detection accuracy	Data 2 detection accuracy
BP	75.6%	66.7%
BP-ABC	83.5%	89.7%

NN data 2 optimized by the improved ABCA is 89.7%. The specific presentation results are shown in Fig. 2.

It can be seen from the above that the BP NN optimized by artificial bee colony has higher accuracy in intrusion detection. Because the BP NN will stop when it meets certain error requirements, it is possible that this error is not the global minimum, while the BP NN optimized by artificial bee colony will find a lower error to replace the error that may have met the requirements within the maximum number of iterations, so the latter will have a higher accuracy.

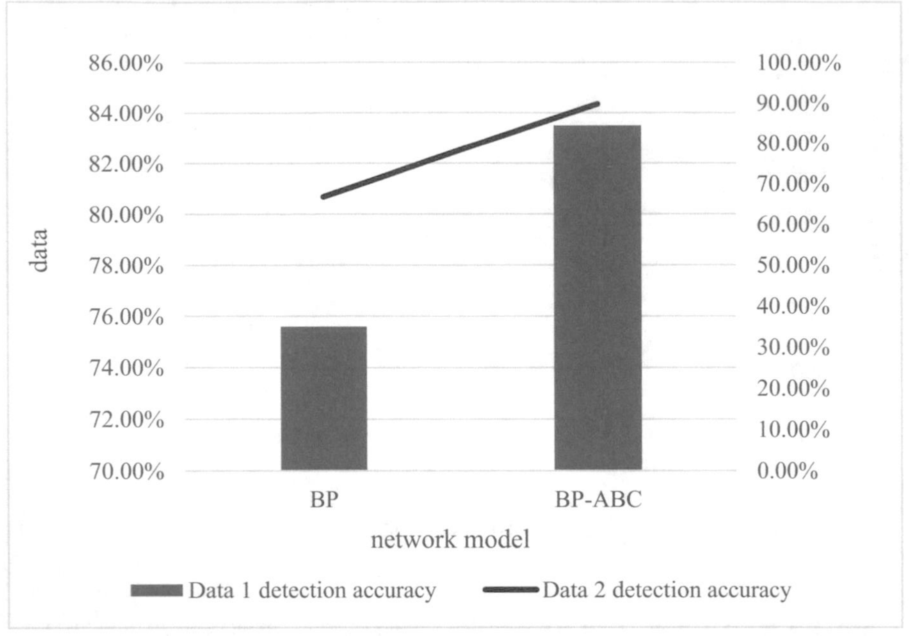

Fig. 2. Comparison of detection accuracy of three groups of data

5 Conclusion

As a new architectural form, prefabricated building is the trend of social development in China. This kind of building has the advantages of fast construction speed, green and environmental protection. The scheduling problem of prefabrication construction project is related to the cost and time efficiency of the project. An effective scheduling method for prefabricated component construction project can save construction period and cost and speed up the progress. In this paper, the detection accuracy of BP NN and improved ABCA optimized BP NN (BP ABC) is tested. The results show that the artificial bee colony Optimized BP NN has high accuracy in intrusion detection.

References

1. Guo, K., Zhang, Q.: A discrete artificial bee colony algorithm for the reverse logistics location and routing problem. Int. J. Inf. Technol. Decis. Mak. **16**(2), 1–19 (2017)
2. Li, X., Ma, S.: Multiobjective discrete artificial bee colony algorithm for multiobjective permutation flow shop scheduling problem with sequence dependent setup times. IEEE Trans. Eng. Manag. **2**, 1–17 (2017)
3. Chen, G., Sun, P., Zhang, J.: Repair strategy of military communication network based on discrete artificial bee colony algorithm. IEEE Access **99**, 1 (2020)
4. Zou, W., Pan, Q., Tasgetiren, M.F.: An effective discrete artificial bee colony algorithm for scheduling an automatic-guided-vehicle in a linear manufacturing workshop. IEEE Access **99**, 1 (2020)

5. Gomathi, B., Krishnasamy, K., Balaji, B.S.: Epsilon-fuzzy dominance sort-based composite discrete artificial bee colony optimisation for multi-objective cloud task scheduling problem. Int. J. Bus. Intell. Data Mining **13**(1–3), 247–266 (2018)
6. Oda, E.S., Abdelsalam, A., et al.: Optimal DGs allocation in distribution networks using modified flower pollination algorithm. In: Proc Int Mid East P, 2017(-), pp. 1424–1429 (2017)
7. Jia, Y., Du, J, Zhang, W.: [Lecture Notes in Electrical Engineering] Proceedings of 2017 Chinese Intelligent Systems Conference Volume 460||A Hybrid Discrete Artificial Bee Colony Algorithm for Multi-objective Blocking Lot-Streaming Flow Shop Scheduling Problem (2018). https://doi.org/10.1007/978-981-10-6499-9(Chapter57):593-602
8. Ozmen, O., Batbat, T., Ozen, T., et al.: Optimum assembly sequence planning system using discrete artificial bee colony algorithm. Math. Prob. Eng. pt. **4**, 1–14 (2018)
9. Wang, K., Li, X., Gao, L., et al.: A discrete artificial bee colony algorithm for multi-objective disassembly line balancing of end-of-life products. IEEE Trans. Cybern. **99**, 1–12 (2021)
10. Long, M., Wang, L.L.: S-box design based on discrete chaotic map and improved artificial bee colony algorithm. IEEE Access **99**, 1 (2021)
11. Jing, S., Lian, X., Chen, H., Zou, T., Ma, L.: Optimal layout and deployment for RFID system using a novel hybrid artificial bee colony optimizer based on bee life-cycle model. Soft. Comput. **21**(14), 4055–4083 (2016). https://doi.org/10.1007/s00500-016-2056-7

Cloud Computing Scheduling Algorithm Based on QoS Constraints

Chunping Wang[1(✉)] and Mohammed K. Kumar[2]

[1] Shanghai Technical Institute of Electronics & Information, Shanghai, China
wcp0617@126.com
[2] GLA University, Uttar Pradesh, India

Abstract. Task scheduling and resource allocation are two important core technologies in cloud computing. The business capabilities of cloud computing mainly focus on the services brought to end users. Depending on the virtualization technology it adopts, resource allocation will be parallelized with task scheduling differently than before. Since cloud computing is user-centric, service-oriented, and commercialized, the main workflow programming algorithms today are QoS-based programming algorithms, many of which are based on programming strategies in the original grid environment, but due to the cloud environment due to the unique characteristics of workflow, the original programming strategy may have problems in execution efficiency. This paper studies the QoS-constrained cloud computing scheduling algorithm, understands the relevant theoretical knowledge of cloud computing scheduling algorithms on the basis of literature, and then designs a QoS-constrained cloud computing scheduling algorithm, and tests the designed algorithm, through the test results, it is concluded that the algorithm in this paper can make the users in the system get better service quality assurance, and improve the user's service satisfaction as a whole.

Keywords: QoS Constraints · Cloud Computing · Scheduling Algorithm · Scheduling Strategy

1 Inductions

Due to the increase in the number of Internet users, how to allocate tasks accurately and effectively to improve network resource utilization and performance benefits [1, 2] has become one of the key issues in cloud computing technology research [3, 4]. At the same time, with the increasingly obvious cloud computing capabilities and the increasingly diverse business requirements of users, cloud service providers must pay more and more attention to application requirements, such as security when tasks are completed [5, 6]. Therefore, in order to maximize the service requirements of application-centric applications, improving application satisfaction is also a key research topic of current cloud computing technology [7, 8].

In the study of cloud computing software, some researchers claim that cloud computing refers to designing applications in a specific cloud environment according to

J. H. Abawajy et al. (Eds.): ICATCI 2022, LNDECT 170, pp. 318–325, 2023.
https://doi.org/10.1007/978-3-031-29097-8_38

local area network applications and the needs of different users. The perfect solution to cloud computing software problems is directly related to cloud computing performance, cloud computing stability, user satisfaction, and cloud computing providers' operating costs [9, 10]. Some researchers have proposed that the main purpose of cloud computing is to create business models. With the development of cloud computing technology, cloud computing technology has gradually changed from ordinary "computer" to "ordinary user". Therefore, how to create a work plan policy on the cloud compilation graph according to the different QoS requirements of different users is a problem. So far, heuristics like genetic algorithms have slowly been used as tools for planning work in cloud computing environments, but not as high-resolution and predictable solutions. Therefore, different QoS requirements (execution time, bandwidth, cost, reliability, etc.) of all active applications need to be considered when considering the calculation results, which can improve resource utilization and satisfy users' needs [11, 12]. To sum up, there are many research achievements on cloud computing scheduling, but the efficiency of cloud computing scheduling algorithm needs further research.

This paper studies the QoS-constrained cloud computing scheduling algorithm, analyzes the cloud computing task scheduling goals and QoS constraints on the basis of the literature, and then designs the QoS-constrained cloud computing scheduling algorithm, and tests the designed algorithm, and draw relevant research conclusions from the test results.

2 Research on Cloud Computing Scheduling Algorithm

2.1 The Goal of Cloud Computing Task Scheduling

In the cloud service system, the cloud programming center is a resource composed of multiple virtual machine execution nodes. In order to execute all m independent transfers sent by multiple users, it is mapped to a group according to resource usage and job information. That is, the cloud the essence of computing task scheduling is the process of allocating tasks to appropriate service execution nodes for processing according to the scheduling rules of the cloud service system. Through this process, task scheduling in the cloud environment not only meets the basic service requirements of user tasks, but also optimizes the scheduling efficiency of the entire cloud system. The research on task scheduling in cloud service systems mainly focuses on the following general indicators:

2.1.1 Best Time

The time period is the time from the arrival of the work to the completion of the execution, that is, the sum of the work transmission, waiting time, and execution, which can also be understood as the total time. The shorter the total task completion time, the faster the user needs are satisfied. For the same batch of user tasks with different requirements, the cloud system adopts different programming strategies to obtain different task scheduling results. The essence of task scheduling is to find a specific implementation scheme with the optimal scheduling result. This allows the user to minimize the scheduling time of all tasks and satisfy the service results in both parts of the service. Therefore, the optimal time period can be used as an important indicator to measure the effectiveness of cloud computing work planning strategies.

2.1.2 Quality of Service

Quality of service is often used as one of the key metrics for evaluating and standardizing cloud services. This is usually seen from two perspectives. One is the user's QoS requirements, and the other is the user's QoS requirements for system response time, security, and reliability. As a service provider, the job scheduling process considers system QoS requirements such as transmission bandwidth, storage capacity, and computing power. This increases system resource usage and reduces cloud system programming costs. This factor is a factor that users and service providers need to consider in order to obtain high-quality services, and it is also a factor that needs to be considered in the long-term development of cloud computing.

2.1.3 Economic Principles

There are countless service execution nodes distributed in the cloud service system. Due to differences in node performance, some nodes may not be used efficiently and service provider costs may be lost. Cloud computing is a way of doing business that requires attention to the impact of economic factors. There is also a need to maximize the profits of cloud service providers for cloud computing when allocating optimal resources to users of cloud computing. Therefore, the economy is in a position that cannot be ignored in the study of the development status of cloud computing.

2.1.4 Load Balancing

The essence of implementing a load balancing strategy in a cloud computing service system is to prevent some nodes from being overloaded and some resource nodes being inactive. Therefore, the programming efficiency of the entire system and users affects user satisfaction. Due to the huge difference in performance of service nodes in cloud computing systems and the huge number of user requests, how to improve task scheduling efficiency while maintaining system load balance is a daunting task.

2.2 Application of QoS Constraints

QoS-oriented grid resource management requires that all work submitted by users meets QoS constraints and must provide a QoS description when all resources are released. Based on these constraints and the QoS interpretation, the programming algorithm looks for a programming scheme that optimizes its global benefit. Finally, the programmer schedules the execution of tasks on the selected resources according to the scheduling scheme.Measurable QoS constraints for tasks include:

(1) Time limit

Limits on tasks related to time-related functions, such as total finish time, start time, and final finish time. Some real-time application tasks may have high time constraints. In this work, the researchers looked at the time to complete the work, but without loss of generality. The corresponding QoS of a resource is described as its computing power, from which the expected execution time of tasks within the resource can be derived.

(2) Reliability limitations

Resource failures can lead to long-running failures. Rerunning tasks repeatedly consumes resources and degrades system performance, but you can reduce the impact of job failures by assigning each task an acceptable reliability constraint and scheduling tasks to that requirement. The QoS constraints for such jobs mainly consist of their required minimum probability of success. The QoS corresponding to the resource is described as the failure rate per unit time, from which the probability of failure of the task executed on the resource within a specific time can be calculated.

(3) Security restrictions

Users may require different levels of security at work. Related research suggests a classification of security services, such as reliability, confidentiality, and integrity. Data confidentiality and integrity are two important data security measures. Among them, data confidentiality means that the data of a specific job is not illegally obtained by other processes, and data integrity means that the job data is not lost or error-free after transmission.

(4) Relative importance indicates the relative importance of the work. High-priority tasks need to be executed early, and if the benefits are the same, high-priority resources are used first.

3 Cloud Computing Scheduling Algorithm Based on QoS Constraints

3.1 Description of User Requirements and Resource Nodes

To select appropriate tasks to assign to idle resources, it is necessary to consider the user's quality service requirements for resources (ie, work resource requirements) and the resource characteristics of node resource workers. The first is called job demand characteristics, such as processor, memory, hard disk, bandwidth and other job requirements. The latter are called node resource characteristics, such as memory remaining, disk remaining, and CPU utilization on workers. Both can be represented by a set of vectors collectively called eigenvectors. Next, this article will introduce the following two types of vectors.

Job demand characteristics: Although QoS is a parameter derived from the network performance mechanism, in cloud computing technology, QoS services refer to various business attributes necessary to describe the work sent to users. In this paper, we will choose CPU requirements, memory requirements, network requirements, hard disk requirements, and values as QoS parameters, or the required attribute variables. This set of variables can be described by a one-dimensional vector.

$$\mathrm{Re}q = \{req_1, req_2, req_3, req_5\} \tag{1}$$

Reqp, req3, req4, and reqs represent CPU requirements, memory requirements, network requirements, hard disk requirements, and service prices, respectively. The values of these variables can be defined by the following two methods. One is estimated by learning historical information and analyzing the actual operation status of the current task, and the other is pre-determined when the task is performed in practical applications. The latter simpler approach is used in this work. For simple operations, the value range of the variable is [1, 10]. Here one is the minimum requirement of the resource, and ten is the maximum of the resource. In addition, since the algorithm in this paper refers to a large number of QoS parameters, it is also necessary to prevent the overload of resources. This parameter is also user-defined, which is the actual memory and hard disk usage required by the user. The higher the parameter, the higher the requirement.

At the same time, this setting is weighted using a weight vector to reflect the user's setting for each attribute variable.

$$k = \{k_1, k_2, k_3, k_4, k_5\} \tag{2}$$

Each weight variable of k represents the user's attention to this QoS parameter. At the same time, tasks can also be ranked by this vector. For this paper, what needs to be done is to select a weight parameter and assign a task type, as shown in Fig. 1. In practical use, we can design a more detailed task mapping scheme as required.

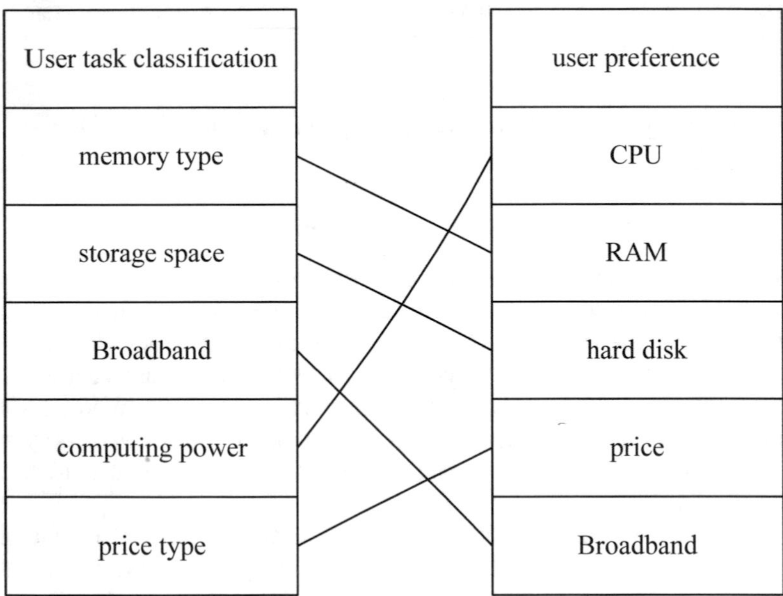

Fig. 1. Schematic diagram of task mapping scheme

3.2 Resource Scheduling Environment

In the cloud resource pool, resource scheduling is performed as follows: the responsible node first publishes information, other nodes evaluate local resources after receiving the work information, and finally the responsible node returns the information. Appropriately programmed nodes based on key metrics.

The evaluation system first needs to analyze the characteristics of each task and select resource nodes that satisfy hard QoS and constraints. This is the first step in evaluating resources, that is, the selected node must be able to provide a minimum level of resources for the job. For example, the memory attribute of a task is 512 MB. This means that the nodes required to execute the task must meet the memory resource requirements of more than 512 MB. This constraint is the minimum requirement for selecting resource nodes, such QoS requirement is QoS performance, mainly considering labor resource requirement, for example, job processors, memory, storage, and bandwidth.

Since cloud computing users' restrictions on service quality vary with system requirements, the QoS evaluation of user requirements must also be based on user weights. One is to consider meeting the requirements of hard indicators. If the requirements are not met, the found node cannot be used as a candidate node. Once resource nodes meet strict QoS constraints, cloud computing needs to consider users' QoS requirements. For example, the user wants the programming cost to be as low as possible and the completion time as short as possible, and then the user can set these two metrics themselves. The cloud computing system evaluates available resources according to user needs and system load, and selects appropriate scheduling nodes for tasks.

3.3 Algorithm Design

According to various settings of the QoS attributes of each task sent by the user, a priority calculation function based on QoS constraints is designed. The user must select the QoS preference level for the job when submitting the job, and pass the QoS feature preference value as a parameter when submitting the job. Priority calculation based on QoS constraints Priority is the two most important characteristics in QoS - time and cost.

The QoS-constrained multi-priority task scheduling algorithm sets the time fraction and cost fraction to 1, 2, 3, 4, 5, 6, 7, 8, 9, and 10, respectively, for a total of 10 levels. The time level 1–10 represents the increase in the urgency of the user's execution time, 1 represents the lowest level, 10 represents the highest level, and the cost level 1–10 represents the increase of the user's current labor price order, 1 is the lowest cost, and 10 is the highest fee. In cloud computing systems, users are charged according to the specific terms of each service provided. Therefore, when a user submits a job, the functional time and cost of the job should be prioritized, depending on the actual status of the job that is running this time. If the cloud computing system wants to prioritize the user's work, choose a higher time attribute value, and if the user only charges a lower cost value, choose a lower cost attribute value. The Job Client sends time and cost information to the JobTracker master node.

4 Algorithm Test

During the experiments, Hadoop was chosen to run a word count program. Using three different task scheduling algorithms: QoS-MP scheduler, FIFO scheduler and capacity scheduler, the same 5 sets of statistical word count experiments were performed on the same user of the cloud computing system, and the execution time was compared for each task group, and the experimental results are shown in Table 1.

Table 1. Algorithm performance test results

	FIFO Scheduler	Capacity Scheduler	QoS-MP Scheduler
1	6	13	18
2	4	8	12
3	8	16	28
4	6	14	21
5	5	10	15

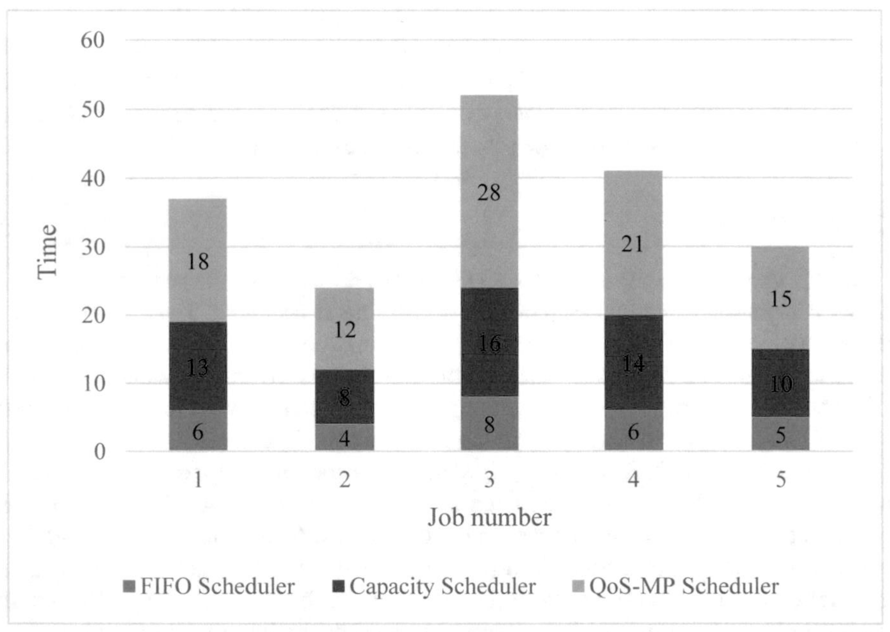

Fig. 2. Algorithm performance test results

As can be seen in Fig. 2, a graph comparing the execution time with the average execution time per task group shows that the performance of the QoS-MP scheduler optimizes as the number of tasks executed increases. The performance of the capacity scheduler is relatively stable as the number of tasks increases, but the average execution time of the cloud computing system using the QoS-MP scheduler is shorter, and the system performance is optimized, so users get better service quality assurance on the system, improve customer service satisfaction.

5 Conclusions

This paper studies the QoS-constrained cloud computing scheduling algorithm. After analyzing the relevant theoretical knowledge, the QoS-constrained cloud computing scheduling algorithm is designed, and the designed algorithm is tested. The test results are obtained, using QoS- The average job execution time in the cloud computing system of MP Scheduler will be smaller, and the performance of the system will be optimal, which can enable users in the system to obtain better service quality assurance, and overall improve user service satisfaction.

References

1. Kaur, A., Sharma, S.: An analysis of task scheduling in cloud computing using evolutionary and swarm-based algorithms. Int. J. Comput. Appl. **89**(2), 11–18 (2018)
2. Hamed, A.Y., Alkinani, M.H.: Task scheduling optimization in cloud computing based on genetic algorithms. Comput. Mater. Continua **69**(3), 3289–3301 (2021)
3. Meshkati, J., Safi-Esfahani, F.: Energy-aware resource utilization based on particle swarm optimization and artificial bee colony algorithms in cloud computing. J. Supercomput. **75**(5), 2455–2496 (2018). https://doi.org/10.1007/s11227-018-2626-9
4. Varshney, S., Sarvpal, S., et al.: A survey on resource scheduling algorithms in cloud computing. Int. J. Appl. Eng. Res. **13**(9 Pt.3), 6839–6845 (2018)
5. Panda, S.K., Pande, S.K., Das, S.: Task partitioning scheduling algorithms for heterogeneous multi-cloud environment. Arab. J. Sci. Eng. **43**(2), 913–933 (2017). https://doi.org/10.1007/s13369-017-2798-2
6. Bosmans, S., Maricaux, G., Schueren, F., et al.: Cost-aware hybrid cloud scheduling of parameter sweep calculations using predictive algorithms. Int. J. Grid Util. Comput. **10**(1), 63–75 (2019)
7. Srichandan, S., Kumar, T.A., Bibhudatta, S.: Task scheduling for cloud computing using multi-objective hybrid bacteria foraging algorithm - ScienceDirect. Future Comput. Inform. J. **3**(2), 210–230 (2018)
8. Samee, N., Ahmed, S.S., Seoud, R.: Metaheuristic algorithms for independent task scheduling in symmetric and asymmetric cloud computing environment. J. Comput. Sci. **15**(4), 594–611 (2019)
9. Kaur, D., Sharma, T.: Scheduling algorithms in cloud computing. Int. J. Comput. Appl. **178**(9), 16–21 (2019)
10. Umesh, A.S., Kumar, P., Patel, C.: Performance improvement of cloud computing data centers using energy efficient task scheduling algorithms. SSRN Electron. J. **4**(8), 633–636 (2018)
11. Geng, X., Yu, L., Bao, J., et al.: A task scheduling algorithm based on priority list and task duplication in cloud computing environment. Web Intell. Agent Syst. **17**(2), 121–129 (2019)
12. Sreenu, K., Malempati, S.: MFGMTS: epsilon constraint-based modified fractional grey wolf optimizer for multi-objective task scheduling in cloud computing. IETE J. Res. **65**(2), 201–215 (2019)

Implementation of Intelligent Mobile Translation Software for Computer Application

Yiqun Liang[1(✉)], Lingyi Yin[1], and Edris Zeinali[2]

[1] Haojing College of Shaanxi University of Science & Technology, Xi'an, Shaanxi, China
lyq19832022@126.com

[2] Islamic Azad University, Tehran, Iran

Abstract. With the expansion of global economic integration and exchanges and cooperation, the demand for Chinese-to-English translation in China's domestic and foreign translation markets is also increasing, and translation software has emerged as the times require. This paper studies the intelligent mobile translation software for computer applications, understands the relevant knowledge theory of intelligent mobile translation software on the basis of literature, and then designs the intelligent mobile translation software for computer applications. The test results show that different translation models have an impact on the translation. The paraphrase corpus translated by RNN is better than the model trained by using the corpus obtained by SMT.

Keywords: Computer Application · Translation Software · Software Design · Intelligent Mobility

1 Inductions

Due to the rapid development and globalization of information technology, there is an increasing demand for translation, especially technical translation of various engineering documents and technical manuals [1, 2]. However, the time required for translators to translate these documents has not decreased [3, 4]. In this regard, translators need to adapt to these changes in order to adapt to the new needs of the times [5, 6] and make themselves more competitive, and the use of computer-assisted translation software can greatly improve the quality and efficiency of translation work [7, 8].

To study translation software, some researchers have developed a Preliminary Research Model (CATSAM) based on Machine Acceptance of Machine-Assisted Translation Software. The model incorporates six elements of UTAUT theory: performance expectations, effort expectations, social influence, convenience, behavioral intent, and actual use. However, UTAUT theory has limitations in explaining the relationship between intentional and actual behavior. In order to further understand the transition between the translator's intention to use translation software and the application of user behavior, this paper presents the intended behavior and actual behavior in the preliminary model through theoretical analysis and questionnaire survey, and examines six possible effects: frequency of accepting similar tasks, frequency of accepting urgent tasks,

J. H. Abawajy et al. (Eds.): ICATCI 2022, LNDECT 170, pp. 326–333, 2023.
https://doi.org/10.1007/978-3-031-29097-8_39

frequency of accepting large-scale tasks, frequency of accepting collaborative tasks, frequency of accepting informative texts, and frequency of accepting tasks to be completed in terms of required terms [9, 10]. Some researchers pointed out that the extensive application of computer-aided translation technology in various industries over the years has fully confirmed the outstanding contribution of parallel corpora in the application field of MTI. At the same time, due to the further development of computer technology and electronic information technology, a large number of computer-aided translation software have been produced. From their names, it can be determined that they only have the function of assisting translation. The core meaning is to improve the effect and quality of translation. In a broad sense, the main function of CAT software is to assist teachers to carry out translation courses and to serve professional translators. Strictly speaking, it refers to machine or auxiliary translation tools professionally used to improve the effect and quality of translation, such as project management tools, alignment tools, terminology management tools, translation memory systems, etc. [11, 12]. To sum up, there are many researches on machine translation at present, but most of them focus on machine translation in parallel corpora, and there are few researches on intelligent mobile translation software.

This paper studies the intelligent mobile translation software for computer applications, analyzes the machine translation process and the functional requirements of intelligent mobile translation software on the basis of literature, and then analyzes the intelligent mobile translation software for computer applications.

2 Research on Intelligent Mobile Translation Software

2.1 Translation Process

So far, the training data of machine translation systems contains a large number of elements that are translated together, and these sentences are organized in the original language. Since bilingual data is limited, it cannot contain all the original language sentences, so if a new sentence needs to be translated, the system must first convert the existing bilingual sentences into smaller parts and save these translation elements in the original language as new translation elements. The topics of these parts may consist of complete sentences before the end of the language.

When constructing each bilingual system and obtaining bilingual components, it should try best to make the final source language and target language logically conform to the bilingual components, especially if the source language contains a large number of vocabulary, some of them must be word. In compiling bilingual sentences, the task of defining lexical-level coherence is called lexical coherence, and its interpretation also proves that the basic unity between the source and target languages of bilingual data is inseparable.

Once the basic grammar is established at the lexical level, bilingual data can be extracted from this database to remove basic units from translation models. The sensitivity of the bilingual part can be high or low. The smaller the bilingual section, the more coherent the sentences. When translating into group translation, that is, using word pairs as the basic unit of translation, adopt word-based translation principles or other translation principles or customize them to view the translation rules of different machine translation models. For example, if a program splits and retrieves a language grammar during translation, the program will receive a translation grammar based on that language grammar. If the content synchronization rules of translation grammar rules and possible synchronization rules cannot be translated, after closing a large number of translation rules, a possible model of translation rules must be developed. Commonly used methods include basic numerical estimation based on multiple possibilities, EM methods, and differential training methods.

In addition to the translation principles themselves, the main translation process has two main elements to measure the translation process, its function is to test the ease of translation design and use, the selection of the best translation candidates for all translation candidates, and the language target.

2.2 Analysis of Translation Software Requirements

2.2.1 Real-Time Translation

The interaction mode of traditional machine translation systems is that the text is input with language suggestions, and the system extracts the text suggested by the target language. This translation method will reduce real-time performance, especially in this era of mobile Internet, mobile phones have become mainstream translation systems, and it takes a long time to translate text to mobile phones, but real-time translation systems use voice or image recognition to input sentences, which Significantly reduces text entry time and improves real-time translation performance.

2.2.2 Translation Interaction

There is no interactivity in the text-to-text translation in traditional translation systems. In a real-world communication scenario, the counterparty may not need to manually enter text on an input device or read the mobile translation of each notification. The real-time translation system uses voice-over-picture recognition input and voice-over-picture synthesis output, which is more interactive.

2.2.3 Instant Translation

When translating with a traditional translation system, you need to input characters that the user can write, and the user must master the spelling of the input language. For example, if a user wants to hear an unknown word in their life and hear the result of the translation, it is difficult to answer such a question with a traditional translation system, but a real-time translation system can translate directly in real time. A real-time translation system is a more direct translation system that is closer to the actual communication scenario.

3 Intelligent Mobile Translation Software Design for Computer Applications

3.1 System Architecture

The design of the whole system adopts C/S structure and adopts thin client function. With the rapid development of electronic information technology and the continuous improvement of hardware technology, portable devices represented by smart phones have excellent computing power, storage capacity and battery power capacity. But compared to desktops such as desktops, laptops are resource-constrained in all aspects of their own resources, so thin clients are used when designing clients. When researchers try to move editing functions to the server side, the client side has the functions of taking pictures, editing simple images, sending requests and pictures, and receiving and displaying the results. After each translation, the client automatically logs out from the server and reconnects until the next translation request, which reduces client power consumption because the client does not have to be connected to the server all the time. Figure 1 shows the overall structure of the system.

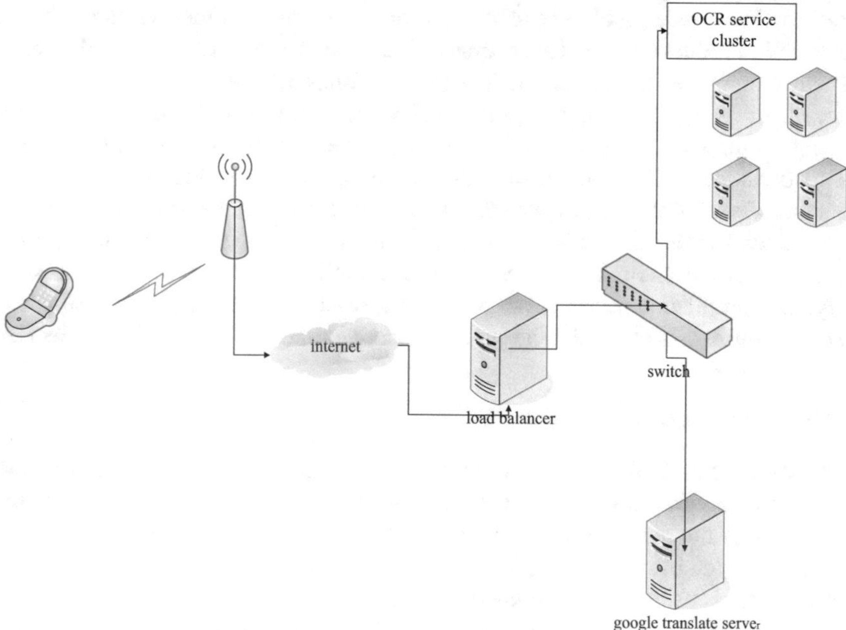

Fig. 1. System architecture diagram

As shown in Fig. 1, the system distributes high-load workloads on OpenNebula cloud-based support servers, including an OCR editor, embedded Google SaaS for web translation, and an electronic translation system for word processing based on an OCR image recognition engine, but the application is computer-intensive and consumes a lot of system resources when running. In addition, since OCR is the bottleneck of all system performance, multiple OCR processors are required, and the balancer is also required to be responsible for the distribution of OCR tasks, so the server side consists of three parts: load balancing, OCR processor, and conversion processor.

The mobile client uses the wireless communication network or connects with the mobile Internet to establish contact with the server. The wireless communication network may also use 4G or WIFI. In OpenNebula's cloud-based server, virtual machines running the load distribution server, the OCR processing server, and the translation server work together to construct a virtual local area network to enable an increase in the number of internal interactions between the different components of the server. In addition, the load balancing server and the translation server also have an Ethernet connection, and all Ethernet ports are assigned to a public network address. The balance server accepts the client's application through this address, connects to the Google Translate service through the public Ethernet network interface with the translation server, and feeds back the compilation result to the server. Since the number of OCR processing servers changes dynamically with application access requests, the settings for translation and load balancing servers also change dynamically with workload conditions.

The data collected by the client is based on the photos taken by the camera of the mobile phone or the photos originally stored on the mobile phone. The collected data is processed as needed and sent to the server based on the OpenNebula cloud. For the translation request, the collected image data is sent to the server, and the relevant parameters are passed to the request. The main settings are the original language of the client and the type of the target language. After the server receives the client's request, it distributes the OCR image scanning job to the OCR processing server through the OCR load balancer. The number of OCR processing servers depends on the number of users accessed. The text message is first sent to the server for translation, and then the translation server calls Google to translate the text message just obtained, and then Google returns the compilation result to the compilation server through translation. Finally, the compilation server feeds back the translated target language results to the server for online translation. The above is the main process of the entire architecture.

3.2 Neural Network Translation Model

(1) Position vector. RNN models require position vectors. This is because RNN models can receive information about the position of words in a sequence by computing the state of the recurrent hidden layers, while CNNs cannot. If it added a position vector, it need to specify the position of the word in the input sequence. Therefore, this vector is needed in this transformation model, which first transforms the input sequence $x = (x_1, ..., x_m)$ into a word vector represented by $w = (w_1, ..., w_m)$, and then adds the word position vector $p = (p_1, ..., p_m)$ to the model. This represents the absolute position of the word in the input sequence, so the input to the model is represented using a combination of word vectors and position vectors. It is expressed as follows.

$$e = (x_1 + p_1, ..., x_m + p_m) \tag{1}$$

(2) Convergence structure. A CNN system has two parts, an encoder and a decoder, both of which are computed from vector information through a convergent structure. The output of the first layer of the decoder is denoted as $h^l = (h_1^l, ..., h_n^l)$, and the output of the first layer of the decoder is denoted as $z^l = (z_1^l, ..., z_m^l)$. Each level consists of a convolution and a nonlinear layer. The convolution kernel mainly includes the convolution array W and the array array bias b.

(3) Multi-step attention mechanism. Each layer of the decoder is an independent attention mechanism. c_i^l represents the attention of the first layer of the encoder. This is derived from the weighted sum of output z_j^u and input word e_j at the end of the current encoding layer. The calculation method is as follows.

$$c_i^l = \sum\nolimits_{j=1}^{m} a_{ij}^l (z_j^u + e_j) \tag{2}$$

4 Software Testing

Table 1. Software performance test results

	Test 1	Test 2
SMT	30.12	30.34
SMT RNNPara SMT	29.5	29.3
RNN RNNPara SMT	29.3	29.4
SMT CNNPara SMT	29.5	29.6
CNN CNNPara SMT	29.7	29.2

In the experiments conducted in this work, the training body of the translation model is derived from parallel bodies in English and Chinese. The training set consists of 5234 pairs of English-Chinese parallel sentences, and the development set consists of 1200 pairs of English-Chinese parallel sentences. The test set consists of 1200 Chinese and English sentences. In this article, we will use the body based on three different translation models to train different neural networks, obtain different translation models, and select the SMT translation result as the basis. Table 1 shows the translation error rate results obtained by the model.

It can be seen from Fig. 2 that different translation models have an impact on the translation: for example, under the same conditions as the basic model when translating the same translation, the paraphrase corpus translated by RNN is better than the model trained using the corpus obtained by SMT.

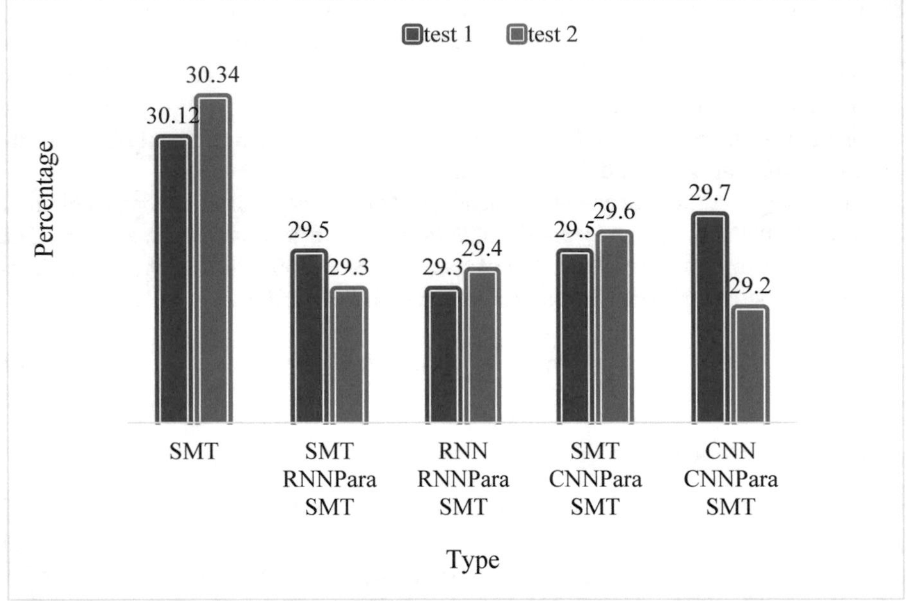

Fig. 2. Software performance test results

5 Conclusions

This paper studies the intelligent mobile translation software for computer applications. After analyzing the relevant theoretical knowledge, the intelligent mobile translation software for computer applications is designed, and the translation model of the software application is tested. Under the same conditions as the basic model of the same translation, the paraphrase corpus translated by RNN is better than the model trained by the corpus obtained by SMT.

References

1. Wang, X.: Building a parallel corpus for English translation teaching based on computer-aided translation software. Comput.-Aid. Des. Appl. **18**(S4), 175–185 (2021)
2. Farooque, M.: A comparative analysis of translation software using data science approach to Arabic statements. Int. J. Comput. Syst. Softw. Eng. **2**(3), 13–19 (2021)
3. Richa, E., Borde, E., Pautet, L.: Translation of ATL to AGT and application to a code generator for simulink. Softw. Syst. Model. **18**(1), 321–344 (2017). https://doi.org/10.1007/s10270-017-0607-8
4. Markoborodov, A.A., Skob, T.S., Ova, J.A., Volkanov, D.Y.: An approach to the translation of software-defined network switch flow table into network processing unit assembly language. In: Proceedings of the Institute for System Programming of RAS, **32**(3), 79–89 (2020)
5. Jiang, H.: Coastal atmospheric climate and artificial intelligence English translation based on remote sensing images. Arab. J. Geosci. **14**(6), 1–13 (2021). https://doi.org/10.1007/s12517-021-06713-3

6. Lv, M.: Agricultural climate change and multilingual GIS database translation system based on embedded database and artificial intelligence. Arab. J. Geosci. **14**(11), 1–20 (2021). https://doi.org/10.1007/s12517-021-07336-4
7. Xiao, Y., Keung, J., Bennin, K,E., et al.: Machine translation-based bug localization technique for bridging lexical gap. Inf. Softw. Technol. **99**(JUL.), 58–61 (2018)
8. Tabatabaei, O., Fatahipour, M., Sarab, M.M.: The impact of electronic vs. human observer feedback on improving teaching of translation skills to Iranian EFL students. Iran. J. Appl. Linguist. **22**(1), 154–196 (2021)
9. Wu, L., Wu, L.: Research on business English translation framework based on speech recognition and wireless communication. Mob. Inf. Syst. **2021**(4), 1–11 (2021)
10. Li, B.: Study on the intelligent selection model of fuzzy semantic optimal solution in the process of translation using english corpus. Wirel. Commun. Mob. Comput. **2020**(5), 1–7 (2020)
11. Pavlov, A.V., Polinov, A.A., Spirin, N.A., Onorin, O.P., Lavrov, V.V., Gurin, I.A.: Decision-making support in blast-furnace operation. Steel Transl. **49**(3), 185–193 (2019). https://doi.org/10.3103/S0967091219030082
12. Muravev, Y.: Teaching legal english translation by the case method in Russian-English language pair. Human. Soc. Sci. Rev. **8**(4), 961–971 (2020)

Probing into the Driving Mechanism of Computer Science and Technology to the Internet of Things

Xuehua Peng(✉)

College of Marxism, Sichuan University, Chengdu, Sichuan, China
pengxuehua2022@126.com

Abstract. The application of computer science and technology in the Internet of Things (IoT) has promoted the rapid development and wide application of the IoT in various fields. Useing a variety of methods such as positive method, graphic method and so on,this paper briefly introduces the computer science and technology exploited in building the IoT and demonstrates its role in promoting the IoT through IoT applications in real case scenarios.The development foundation of the IoT is more solid and the IoT can be more organized and promising. To investigate the driven mechanism of the IoT, this paper explores how science helps in building the IoT access system, in the following three aspects: design principles of the IoT access system, overall framework of the system and functions of the system. The ultimate goal is to make the IoT grow faster,and make life more and more convenient for modern people.

Keywords: Computer Science and Technology · The Internet of Things · IoT Access System

1 Introduction

Information transmission in the Internet is artificial [1] Owing to the rapid development and progress of advanced technology, mankind is no longer limited to using the network for information transmission whether in life or at work today. Instead, people gradually begin to realize connections between physical objects, so the Internet of Things (IoT) emerged. (IoT) is the general trend of the development of Internet. Over the past ten years since its concept was proposed, it integrates advanced technologies in the fields of sensing technology, communication technology, data technology and so on, aiming to realize the interconnection of all Things and provide technical support for building an intelligent and automated production and living environment for human beings. With the continuous development of Internet of Things technology, in recent years all walks of life have begun to use Internet of Things technology to carry out "intelligent" reform according to their actual characteristics.

J. H. Abawajy et al. (Eds.): ICATCI 2022, LNDECT 170, pp. 334–341, 2023.
https://doi.org/10.1007/978-3-031-29097-8_40

Research organizations around the world have similar but different divisions of IOT architectures. The more common IOT architectures are divided into three layers: perception layer, network layer and application layer. Although the Internet of Things platforms at home and abroad are constantly refining their own architectural systems with their own characteristics according to their own technical focus, it can still be found that most of them follow the three-layer basic division vertically. Foreign research on the construction of Internet of Things platform started early.The United States, for example, focuses on the advanced manufacturing system based on the Industrial Internet of Things due to its early urbanization and mature automation system.In recent years, three domestic operators have accelerated the progress of the Internet of Things field and actively promoted the construction of the Internet of Things platform and the improvement of related supporting hardware and software facilities.

Nowadays, people's requirements and expectations for the Internet of Things are also increasing. The existing Internet of Things platform and technology cannot meet people's production and life needs, and relevant research shows a certain lag. In IoT applications, both the connection and data communication among IoT devices require the support of computer science technology. The interplay and integration of the IoT, big data, machine learning, and artificial intelligence constituent an ecosystem to deal with huge industrial devices and data streams, which can resolve problems like mobility, storage, computational capacity etc. for real case scenarios [2].

2 The Key Technologies and Structure of the IoT

The key technologies of the IoT lie in computer science and technology. The IoT, literally, refers to connectivity between objects, which is different from the Internet. It does not completely correspond to information connection like the Internet, and it cannot be completely epitomized by virtual signals and data. Instead, the IoT means connection among physical objects in reality, which involves many technologies such as perception, information transmission, and information processing. Based on the current recognition of the IoT among the academia, the IoT can be defined by the following formula[1]:

$$\text{IoT} = (\text{cloud computing} + \text{sensor network}) \times \text{AI} \tag{1}$$

Within this formula:

$$\text{cloud computing} = (\text{big data} + \text{algorithm} + \text{devices}) \times \text{services} \tag{2}$$

$$\text{AI} = (\text{intelligent computing} \times \text{intelligent applications}) \text{ intelligent ecosystem} \tag{3}$$

From the definition, the key to realizing wide application of the IoT lies in the supports of cloud computing technology, sensor technology and artificial intelligence technology. Furthermore, the three kinds of technologies are based on computer science and technology in realizing software development, program design, information transmission,

[1] Defining formular of AI was issued by Baidu at the Fourth ABC SUMMIT 2019 Baidu Cloud Intelligence Summit.

numerical analytics and processing. Therefore, computer science and technology is the key to the IoT. It is precisely because of the support of intelligent sensing and identification technology, computational technology and communication technology in applied mathematics that the IoT can be widely applied in the Internet and real life in recent years.

The structure of the IoT is reliant on computer science and technology and is composed of three layers. The application of the IoT is based on computer programs, codes, and the Internet in computer science, and supported by processing devices and Internet platform, which brings convenience to people. The realization of the IoT's functions relies on six major elements, i.e. identification, sensing, communication, computation, services, and semantics [3]. Generally, the IoT's structure includes three layers: sensor layer, network layer and application layer.Structure of the IoT system is shown in Fig. 1.

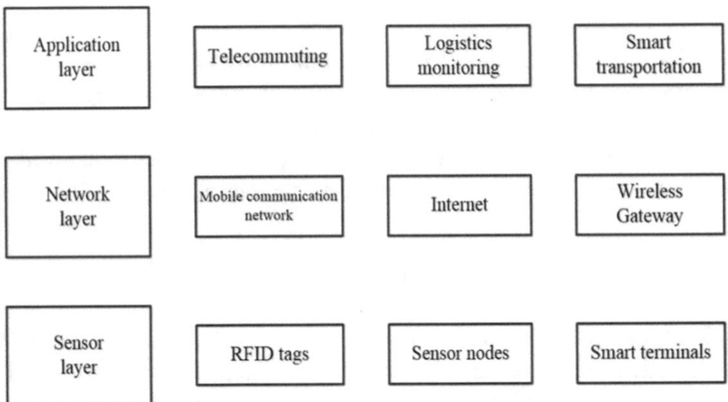

Fig. 1. Structure of the IoT system

Within the structure, sensor layer is the key to the industrialization of the IoT. The sensor layer can fully collect personal information, such as identity, travel, shopping and so on, and thus has a comprehensive and in-depth perception of the information [4]. The network layer, mainly for the transmission and processing of information, processes data acquired by the sensor layer and channel it to the application layer [5]. For instance, in 2019, Moreno, et al. proposed a decision-triggered protocol of data acquisition and transmission, which collects signals of terminal devices to the fusion center for decision-making and feed information back to executive devices. This new protocol greatly reduces the amount of information and further improves the efficiency of information exchange [6]. On this basis, Yu et al. proposed an updateable data sharing access control scheme, integrating attribute based encryption (ABE) algorithm and chameleon hash technology [7]. Moreover, Ren et al. combined ABE and blockchain technology to achieve cross-domain data access, token application and update. The application layer corresponds to the specific applications of the IoT, which is involved in many field such as telemedicine, logistics monitoring, and transportation network information [8].

3 The Role of Computer Science and Technology in Promoting the IoT

3.1 Computer Science and Technology Laying a Foundation for the Development of the IoT

Among the three structural layers of the IoT, the network layer is the pivotal and the most mature one. Compared with the sensor layer and the application layer, the network layer is more flexible and more convenient to apply in the IoT, for computer science can exploit well-established infrastructures like the Internet, Digital Video Broadcast (DVB) broadband and mobile communications network for IoT application. In fact, the IoT is not equal to a network environment, but is similar to a large-scale online store, which operates through the network and realizes the connection between the network and commodities. The IoT can display real structures of commodities through 3D pictures, so that people can have an in-depth understanding of each commodity, including its nature and structure. In this way, people can purchase inside their houses more conveniently. Thus the development and application of the IoT will directly reform people's way of life. In other words, computer science and technology will effectively empower connection between various physical objects with the network, realizing the integration of the network and commodities and laying a solid foundation for the development of the IoT.

3.2 Computer Science and Technology Propelling the Development of the IoT

The IoT will be better refined and more mature on the basis of more advanced computer science. At present, the IoT has penetrated into human's life and work in many aspects, such as smart furniture, telemedicine, online teaching, etc. Supported by the platform of the IoT, products can be presented to users in a more direct and visual way, providing customers with more choices and merchants more opportunities [9]. Computer science and technology provides technical support for the wide application of the IoT in real life with cutting-edge science and technology. The development of the Internet is built on the basis of computer science and technology, while the IoT requires more broad and direct connection between the Internet and physical objects.

3.3 Computer Science and Technology Making the Development of the IoT Well-Organized

Currently, computer science is already ripe and advanced, compared with IoT, a new type of technology. The development of the IoT develops fast to a large extent, which assists in solving problems emerged in the IoT operation.For instance, problems like how to fully present the goods on the network and how to optimize the operation of the IoT will resort to computer science and technology for a solution. Therefore, computer science and technology can make the internal development of the IoT more organized. In recent years, fields like home furnishing, education, transportation, and medical care have been gradually heading for intelligentization, intelligent technology has also been applied to the IoT. For example, the current IoT has been widely used in the medical industry. The IoT in the medical system also consists in three layers: sensor layer, network layer and application layer, as shown in Fig. 2 [10].

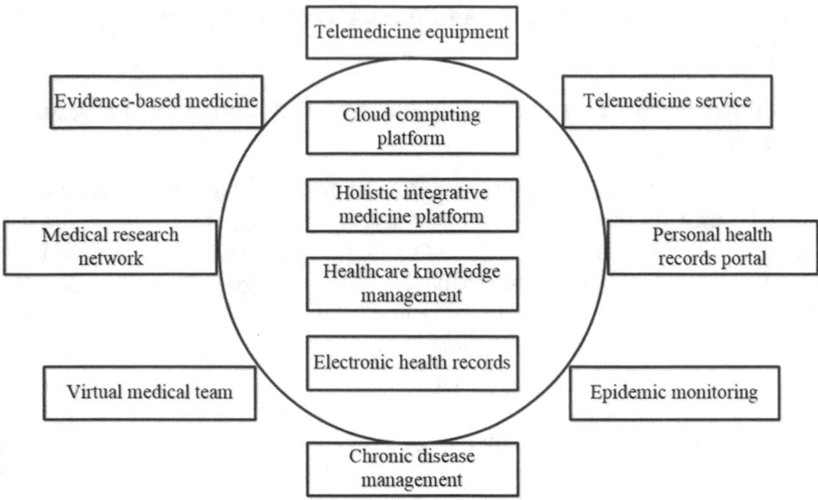

Fig. 2. Model of the IoT in medical system

3.4 Computer Science and Technology Vitalizing the IoT

The IoT has been developing fleetly and applied extensively. Enabled by the IoT, everyone can establish connections with physical objects by electronic tags and can position certain objects. In addition, with the IoT applied, people can manage and control various machines, equipment and personnel through a central computer in a centralized way. People can also control office facilities and home equipment with remote sensing and avoid the occurrence of theft. This function is similar to the automatic control system, which can also search for locations of items. The IoT can also aggregate small data it acquired into big data, providing basic data for the redesign of urban roads, the distribution of hospitals, the layout of schools and so on. Besides, it can provide reliable help for the prevention and control of natural disasters, crimes and epidemics. At present, the IoT has been widely applied. Under the support and promotion of computer science and technology in the future, the IoT will definitely set off a new wave of scientific and technological revolution.

4 Building an IoT Access System with Computer Science and Technology

IoT access system is a crucial part of the IoT. Some problems emerged in the development process of the IoT can be solved eventually. For example, science can encapsulate and parse data to ensure security during data transmission. Design of the IoT access system mainly adopts computer communication technology and computer application protocol as technical supports.

4.1 Principles of the IoT Access System Design

(1) The IoT access system's design should enable effective management of the equipment resources of the system. The system should be able to control and manage the equipment according to actual situations of the IoT application, so as to realize deletion and addition of equipment within the system in a dynamic manner.
(2) The system's design is to provide complete data services. The system needs to meet the actual needs of data application at the upper application layer of the system and realize data acquisition, encapsulation and transmission.
(3) The system's design should be user-oriented. Thus the system can be in line with real needs and better serve consumers.
(4) The design of the system should pursue diversity. It should include encapsulation of different functions of the system into different categories, so that one single category does not have to carry too many functions, which improves code utilization and the performance of the entire IoT system.
(5) The system's design should include expandability of the system. The interface design of the system should be abstract and general, so as to meet myriad needs of the system in real case scenarios and improve the expandability of the system in terms of hardware and software.

4.2 Design of the Framework of the IoT Access System

The overall framework of the IoT access system is shown in Fig. 3 [11].

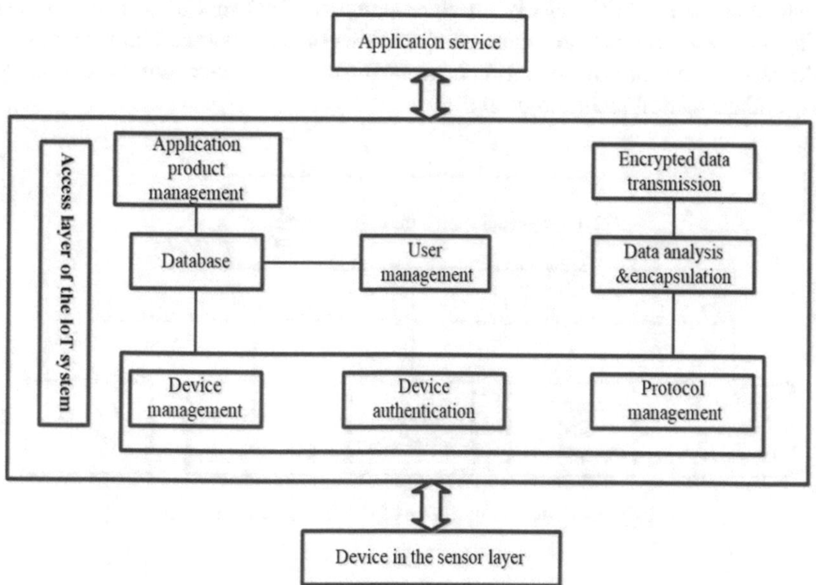

Fig. 3. Design of the framework of the IoT access system

In the overall framework of the system, the sensors or "devices in the sensor layer" at the bottom mainly acquire data in the environment, which is a key part of data acquisition in the system. The "access layer of the IoT system" provides data services to the "application service" on the top. The access layer stores and manages the data from the sensor layer through computer database technology, and then parses and encapsulates the acquired raw data through the IoT protocol. The data is managed in a unified format, and then the data is encrypted by computer encryption methods, so as to ensure security and accuracy of data transmission. The "application service" mainly provides services for the application system. Data transmitted to the "application service" will be parsed and processed for application.

4.3 Design of the Functions of the IoT Access System

In the function modules of the access system, the application product management module mainly manages the corresponding information of the application products connected to the IoT through computer science and technology. The system must create the information of the application products according to certain requirements. When an application product is successfully created, the product code and product key will be automatically generated. The data parsing and encapsulation module mainly encapsulates the collected raw data in a certain format on the device end, and then parses the data again in the upper layer. The parsed and encapsulated data becomes visual data, and then will be re-encapsulated adhering to a certain format and transmitted to the upper layer. The device management module is for managing devices and realizing data transmission between the upper and lower layer devices in the interaction process. The user management module majorly provides windows for users to operate the system, enabling users to maintain and manage operations of the system. The system security module ensures security of the entire network through data encryption technology and identity authentication technology in Fig. 4.

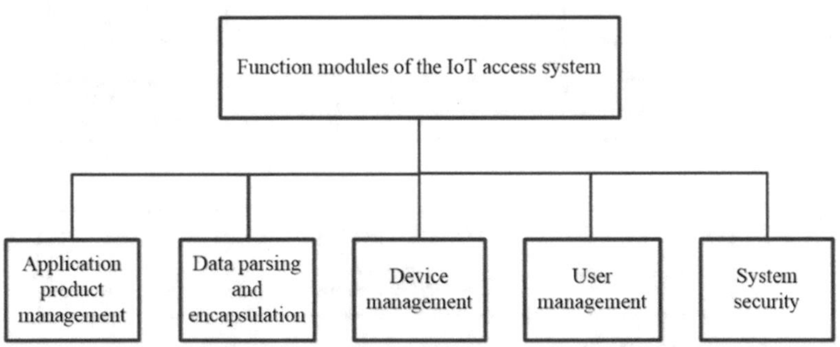

Fig. 4. Function modules of the IoT access system

5 Conclusions

Generally, the IoT has penetrated into many aspects of modern lives. Today, industries like logistics, medical education, economic business, intelligent home, etc., cannot operate so well without the IoT. The IoT is a milestone of the emergence of a new generation of information technology as well as a part and parcel of modern information technology. IoT access system not only solves technical problems of the IoT access equipment, but also improves the security of the entire IoT system. The science in the IoT gives rise to a speedily developing IoT system. The IoT has been establishing people-to-people and even people-to-objects communication that crosses regions and space. In the future, the IoT will continue to play a role in people's daily life, and will be more functional under the support of computer science and technology.

References

1. Denning, P.J.: The science in computer science. Commun. ACM **56**(5), 35–38 (2013)
2. Christian, R.: Towards IT peace research: challenges at the intersection of peace and conflict research and computer science. Sicherheit Frieden Secur. Peace, **38**(1), 10–16 (2020)
3. Naseer, Q.K., Sundus, Q., Najam, U.M., Gwanggil, J.: A secure data parallel processing based embedded system for internet of things computer vision using field programmable gate array devices. Int. J. Circuit Theory Appl. **49**(5), 1450–1469 (2021)
4. Raj, A., Raj, A., Ahmad, I.: Smart attendance monitoring system with computer vision using IOT. J. Mob. Multimedia,**17**(1–3), 115–126 (2021)
5. Kuzlu, M., Fair, C.,Guler, O.: Role of artificial intelligence in the internet of things (IoT) cybersecurity. Discover Internet Things, **1**(1), 1–14 (2021)
6. Moreno, C., González, A., Olazagoitia, J.L., et al.: The acquisition rate and soundness of a low-cost data acquisition system (LC-DAQ) for high frequency applications. Sensors **20**(2), 524 (2020)
7. Darrell, B.N., Maurice, D.E., Ashley, C.D., Sharon, B.L., Calvin, N., Delores, S.: Improving the quality of the Internet of Things instruction in technology management, cybersecurity, and computer science. Int. J. Inf. Commun. Technol. Educ. **16**(2), 59–70 (2020)
8. Robert, R., Lyle, P., Son, T.B., Richard, H., Meha, K.: An IoT smart rodent bait station system utilizing computer vision. Sensors **20**(17), 4670–4670 (2020)
9. Sahni, S., Mittal, A., Kidwai, F., Tiwari, A., Khandelwal, K.: Insurance fraud identification using computer vision and IoT: a study of field fires. Procedia Comput. Sci. **173**(C), 56–63 (2020)
10. Shakila, B., Saleh, A.A., Merlin, M.R.: A secured smart automation system for computer labs in engineering colleges using the Internet of Things. Comput. Appl. Eng. Educ. **29**(2), 339–349 (2020)
11. Bagaria, A., Khemani, H.: Mobile application based home automation with security advancements utilizing IoT and computer vision. Int. J. Innov. Technol. Explor. Eng. **9**(3), 3282–3288 (2020)

Personalized Internet Celebrity Check-In Recommendation System Based on Hybrid Recommendation Algorithm

Shutao Han[1], Qian Liu[1(✉)], Mingkun Xiao[1], Fadong You[1], Xuchu Li[1], and Junmin Kim[2]

[1] Liaoning Institute of Science and Technology, Benxi, Liaoning, China
lnkjxy_lq@163.com

[2] Tech University of Korea, Seoul, Republic of Korea

Abstract. Personalized Internet Celebrity Check-in Recommendation System Based on Hybrid Algorithm, aims at providing users with super quality, real and effective Internet celebrity check-in information. The system can offer users with personalized online celebrity check-in recommendation services based on disparate user characteristics, integrate lots of information that including essential information of check-in places, travel strategy, user experience and evaluation, along with corresponding travel notes, strategies and other recommended contents of check-in places. Through the hybrid recommendation algorithm, the system will complete user's personalized and precise recommendation in view of the user's feedback information, which will greatly improve travel efficiency of user's and the satisfaction of playing progress of the Internet celebrity check-in.

Keywords: Mixed Algorithm · Internet Celebrity Check-in · Recommended System

1 Introduction

With the continuous improvement of living standards, high-quality and popular things have gradually become the pursuit of the public, and the "Internet celebrity check-in place" spawned in the new context has also become a new choice for people to explore. On the one hand, consumers need to spend time deciding which check-in place to go, learning about the check-in place information and making travel plans. On the other hand, network information is complex and all-encompassing, making it difficult for consumers to quickly and accurately obtain interesting, high quality, authentic and reliable content. The influence of these two factors makes it increasingly difficult to meet the growing individual demand of the masses.

Through consulting relevant materials and surveys, it is found that currently the vast majority recommendation systems mainly recommend scenic spots that are relatively familiar to the public, which lacks attractiveness and freshness to the public. In addition, the recommended content is generally single and cannot provide users with unique and ingenious personalized recommendation services.

J. H. Abawajy et al. (Eds.): ICATCI 2022, LNDECT 170, pp. 342–349, 2023.
https://doi.org/10.1007/978-3-031-29097-8_41

Given the above phenomena and issues, this paper raises a personalized online celebrity check-in recommendation system based on a hybrid algorithm of collaborative filtering algorithm and content recommendation algorithm. This system will supply users with various travel routes and play routes, rich choices and unique travel experience through data analysis, data mining plus user's feedback [1].

2 System Introduction and Features

2.1 System Introduction

In recent years, online celebrity check-in places have become more and more popular among the public and personalized check-in places have gradually become the general voice of consumers. From this, we conceived the "Personalized Internet Celebrity Check-in Recommendation System Based on Hybrid Algorithm", based upon a hybrid algorithm due to both collaborative filtering and content recommendation to provide users with accurate online celebrity check-in recommendation services, pay attention to users' personalized needs and fully improve the travel efficiency and play satisfaction.

2.2 System Functions

The functions of this system mainly include recommendation of online celebrity check-in places, user sharing, check-in place strategy and travel recommendation, travel route plannings and personalized recommendation and other features. Among them, the user sharing and personalized recommendation functions need to log in. The detailed system functions are shown in Fig. 1.

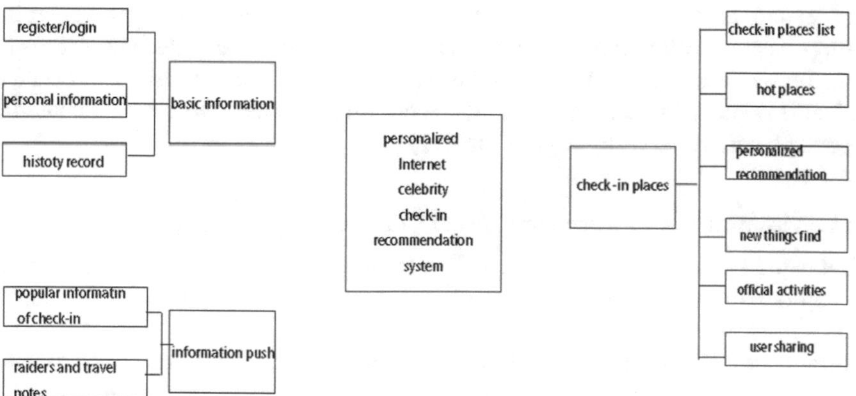

Fig. 1. The function of the system

3 System Function Design

The functional structure of this recommendation system is divided into front end and back end. The front end is broadly divided into user registration and website browsing,and the back-end is mainly divided into administrator and user. The administrator mainly maintains two aspects of data: basic data including online celebrity check-in information, popular recommendations, personalized recommendations, and user commentary strategies, as well as business data that content including registered user list, online celebrity check-in place list, article list in the system. Users mainly perform operations such as publishing comment guides, setting personal recommendation preferences, modifying registration information and obtaining personalized recommendation content.

4 Related Algorithms

4.1 Collaborative Filtering Algorithm

4.1.1 Concise Overview of Collaborative Filtering Algorithm

This system mainly uses item-based collaborative filtering. The main thought of item-based collaborative filtering is to recommend items similar to items that users like before. On the one hand, item-based collaborative filtering is more conforming to the actual situation that the change in the number of online celebrity check-in places in this system is significantly smaller than the change in the number of users. On the other hand, recommand reason that it's similar to the online celebrity check-in place that user has been to before, which is easy to convince.

And the item-based collaborative filtering algorithm is more inclined to maintain users' personal interests and are more suitable for meeting user's personalized needs. Its main features are that it is suitable for rich long-tail items (relatively non-popular items that represent a small number of user's personalized needs) and areas where users have strong personalized needs. As long as the user works on an item, it can recommend else products related to this to him, but cannot recommend new items to users without offline renewing the item similarity degree Table [2].

4.1.2 Advantages and Disadvantages of Collaborative Filtering

Advantages: (1) Only user behavior information is required, and does not depend on other information of users and items. (2) It is easy to implement distributed and can process massive data.

Disadvantages: (1) Problem of cold start: The collaborative filtering algorithm depends on user behavior to make recommendations.It is difficult to reflect value when the system is newly launched or when the user scale is small [3]. (2) Sparsity problem: The user base in the product is large and the number of items is big and the general user only operates a small amount of items. At this time, the user's operation matrix is very sparse, and the items calculated based on the behavior matrix that is too sparse the similarity is often not accurate enough,which will impact the precision of the recommendation results.

4.1.3 Implementation of Collaborative Filtering Algorithm

The item-based collaborative filtering algorithm mainly recommends the most similar items that have been manipulated by the user to the users.In this recommendation system, the similarity of check-in places is obtained by the cosine similarity. Because the non-zero value is much less than the overall value, so the data of a single user is sparse, and the cosine similarity will ignore the non-zero value.By calculating the cosine similarity of the angle between two n (n > 0) dimensional vectors, the similarity of check-in places is predicted. The value range of the cosine similarity is [−1,1], and the magnitude of the angle is inversely proportional to the cosine similarity. The obtained cosine value is close to 1, indicating that the two punching places are very similar; −1 indicates that they are completely opposite [4]. The formula for calculating cosine similarity is

$$sim(x_i, x_j) = \cos(x_i, x_j) = \frac{x_i \cdot y_i}{\|x_i\| * \|y_i\|} \tag{1}$$

Among them, x_i and x_j are two vectors used to represent the check-in place, the "·" sign represents the quantity product, $\|y_i\|$ represents the modulus of the vector y and its calculation formula is

$$\sqrt{\sum\nolimits_{i=1}^{n} y_i^2} \tag{2}$$

When using cosine similarity to calculate the similarity of two items, because of the phenomenon of "score inflation" (users get higher scores than they should, the overall average score increases or those with lower scores in the past get higher scores now), the meaning of all ratings needs to be subtracted from the user ratings, called the modified cosine similarity, also known as the Pearson correlation coefficient [5]. Given two vectors A = $(a_1, a_2, a_3, \ldots, a_n)$, B = $(b_1, b_2, b_3, \ldots, b_n)$, the calculation formula of the Pearson correlation coefficient of the two is:

$$\text{Pearson\_r} = \frac{\sum_{i=1}^{n}(a_i - \overline{a}) \times \left(b_i - \overline{b}\right)}{\sqrt{\sum_{i=1}^{n}(a_i - \overline{a})^2} \times \sqrt{\sum_{i=1}^{n}\left(b_i - \overline{b}\right)^2}}$$
$$\text{among it } \overline{a} = \frac{1}{n}\sum_{i=1}^{n} a_i, \overline{b} = \frac{1}{n}\sum_{i=1}^{n} b_i \tag{3}$$

However, there are cases where this formula requires multiple scans of the data and the following approximate formula is used to calculate [6]:

$$\text{Perason\_r} = \frac{\sum_{i=1}^{n} a_i b_i - \frac{\sum_{i=1}^{n} a_i \sum_{i=1}^{n} b_i}{n}}{\sqrt{\sum_{i=1}^{n} a_i^2 - \frac{\left(\sum_{i=1}^{n} a_i\right)^2}{n}} \sqrt{\sum_{i=1}^{n} b_i^2 - \frac{\left(\sum_{i=1}^{n} b_i\right)^2}{n}}} \tag{4}$$

That means the correlation coefficient of the two check-in place vectors a and b is defined as the rate of the covariance between the two variates and the product of the standard deviations of the two variables. The value of the Pearson correlation coefficient is between -1.0 and 1.0, below zero means a negative correlation and above means it is positive.The approached the correlation coefficient is to 1.0 or -1.0, the better the correlation and closer to zero the weaker the correlation. The idea of the Pearson correlation coefficient is filling in missing dimensions of the vector with zero and subtract the average value of the vector dimension with all dimensions.

Process of calculating the similarity of items with the algorithm is shown at Fig. 2 below [7].

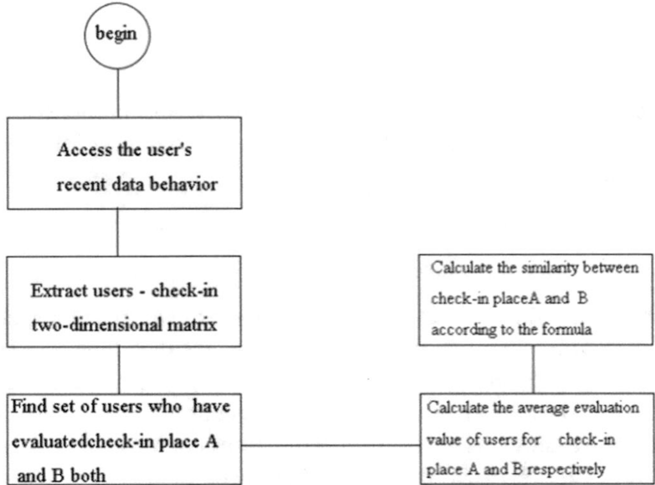

Fig. 2. The algorithm to calculate the similarity of items

And the overall flow of the algorithm is shown at Fig. 3 below:

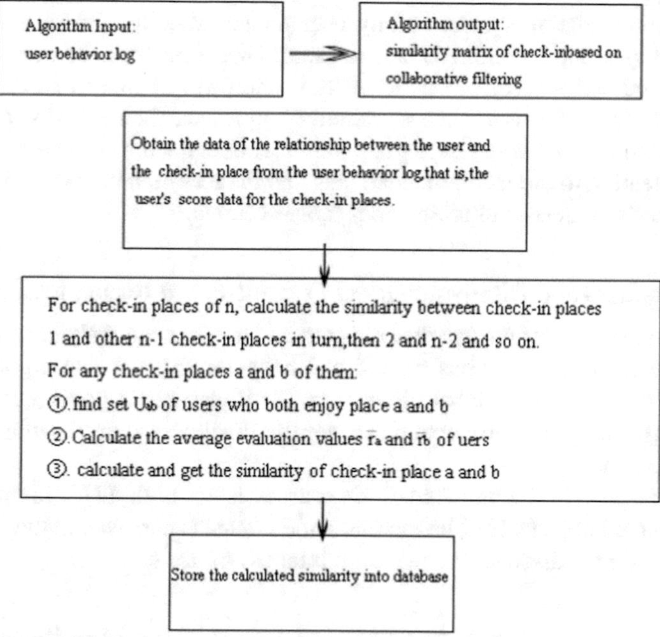

Fig. 3. The overall flow of the algorithm

After we get the similarity of the check-in place, we can calculate the interest of user u in check-in place a according to the following calculation formula:

$$P_{ua} = \sum\nolimits_{i \in N(u) \cap S(a,k)} w_{ai} r_{ui} \tag{5}$$

Here N(u) is a items set that the user likes, S(a,k) is the set of k check-in places that are most similar to the check-in place a, w_{ai} is the similarity between the check-in places a,i and r_{ui} is user u's interest in check-in place i (if it is an implicit feedback data set,user u has acted on check-in place i, then can make r_{ui}=1) this interest indicates that the more similar the item is to the user's historical interest item, taller ranking in the the user's recommendation list more probably it will get.

4.2 Content-Based Recommendation Algorithm

4.2.1 Brief Overview of Content-Based Recommendation Algorithm

The content-based recommendation algorithm is obtaining preferences of user according to the user's historical behavior and recommend items similar to the preferences for users. The content-based recommendation algorithm builds an algorithm model based on item-related information, user-related information and the user's action behavior on items. Generally, the following three steps are required: Firstly, construct user feature representation based on user information and user operation behavior. Secondly, construct the feature representation of the item depends on information of the item. Thirdly, feature representation of users and items, recommend items for users [8].

The recommendation system mainly uses content recommendation algorithms to recommend topics and recommend tags to users. Theme recommendation refers to recommending the user's favorite tags to users in the form of topics, so that each topic is an interest tag of the user, such as parent-child travel, theme parks, red culture in this recommendation system. Users can obtain recommended information with the tags by paying attention to the recommended tags and the recommendation system can also obtain the user's interests and preferences more accurately.

4.2.2 Advantages and Disadvantages of Content-Based Recommendation

Advantages: (1) It is easy to solve the cold start and the user can make recommendations based on the content if user has operational behavior [3]. (2) The algorithm can be recommended based on the label dimension. (3) It also has a good recommendation effect for niche areas. (4) Products that meet the timeliness requirements of the rapid growth of the item.

Disadvantages: (1) Recommendation accuracy is not high. (2) It is difficult to distribute long-tailed objects. (3) The recommended range is narrow and the novelty is not strong. (4) Unable to discover the potential interests of users.

5 Implementation of Hybrid Recommendation Algorithm

Each recommendation algorithm has its own unique advantages and disadvantages and applicable situations and a mixture of multiple recommendation algorithms should be adopted according to the actual situation to achieve a better recommendation effect [9]. This system uses a branch-mixing method to perform recommendation services, and related strategies are shown in Fig. 4 below:

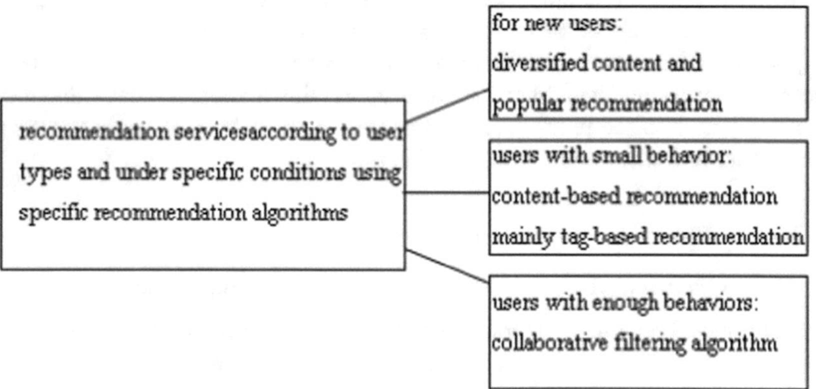

Fig. 4. Branch-mixing recommendation strategy [10]

6 Conclusions

The personalized Internet celebrity check-in recommendation system in view of the hybrid recommendation algorithm proposed in this paper mainly provides users with personalized recommendation through hybrid recommendation algorithm combined with user feedback and user preferences, which meets the personalized needs of users.

This recommender system basically realizes the expected basic functions, but the recommender system still has some deficiencies. For example, in the recommendation mechanism for users, there is still a cold start problem. The functional coverage of the system is relatively simple, and the types of Internet celebrity check-in locations cannot fully satisfy the diverse needs of users.

Acknowledgements. This work was financially supported by Liaoning Institute of Science and Technology 2022 Innovation and Entrepreneurship Training Program for College Students, Internet celebrity 'check-in'-exploring a cultural journey.

References

1. Dong, Z., Leng, C., Zheng, H.: Employment service system based on hybrid recommendation algorithm. In: BDCPS2020:Big Data Analytics for Cyber-Physical System in Smart City, pp. 368–375 (2021)
2. Bhosale, N.S., Pande, S.S.: A survery on recommendation system for big data applications. Data Mining Konwl. Eng. **7**(1), 42–44 (2015)
3. Abdullah, N.A., Rasheed, R.A., Nasir, M.H., Rahman, M.M.: Eliciting auxiliart information for cold start user recommendation: a survey. Appl. Sci. **11**(20), 9608 (2021)
4. Lekakos, G., Caravelas, P.: A hybrid approach for movie recommendation. Multimedia Tools Appl. **36**(1–2), 55–70 (2008)
5. Armstrong, R.A.: Should Pearson's correlation confficient be avoided?. Ophthalmic Physiol. Optics **39**, 16–327 (2019)
6. Balush, I., Cysotska, V., Albota, S.: Recommendation system development based on intelligent search, NLP and machine learning methods. Ceur Workshop Proc. **2917**, 584–617 (2021)
7. Moreno, A., Valls, A., Isern, D., Marin, L., Borràs, J.: SigTur/E-destination:ontology-based personalized recommendation of tourism and leisure activities. Eng. Appl. Artif. Intell. **26**, 643–644 (2013)
8. Sabitha, R., Vaishnavi, S., Karthik, S., Bhavadharini, R.M.: user interaction based recommender system using machine learning. Intell. Autom. Soft Comput. **31**(2), 037–1049 (2022)
9. Sharma, S., Rana, V., Malhotra, M.: Automatic recommendation system based on hybrid filtering algorithm. Educ. Inf. Technol. 1–16 (2021)
10. Mudita, Gupta, D.: A comprehensive study of recommender systems for the Internet of Things. J. Phys. Conf. Ser. 1969(1) (2021)

Design of Transmission Efficiency Model of Network Communication Signal Based on Information Fusion

Jue Li(✉)

School of Information Technology, Xingyi Normal University for Nationalities, Xingyi 562400, Guizhou, China
jkxjue@163.com

Abstract. With the rapid development of wireless communication and the Internet, the number of mobile devices is increasing day by day. As people adapt to mobile Internet life, people have higher and higher requirements for transmission quality and transmission speed. The purpose of this paper is to study the transmission efficiency model design of network communication signals based on information fusion. For dynamic restoration of network communication information and communication signal fusion based on information fusion, a complete transmission efficiency model is designed. It has been confirmed by experiments that the latest model is closer to the change of the transmission efficiency of the communication signal in the actual network compared with the traditional model, and the multi-component fusion of the communication signal in the network is realized.

Keywords: Information Fusion · Communication Signal · Transmission Efficiency · Model Design

1 Introduction

In today's world with developed information technology, computers are widely used in construction of all walks of life. People can use computer learning to communicate with each other and exchange information around the world through the network [1, 2]. However, the network is the carrier, and the real technology for exchanging information between people is network communication, so people have higher and higher requirements for communication quality [3, 4].

At present, some researchers are doing a lot of research on the problem of weakening the anti-jamming in the communication process, and have obtained some results. Some researchers first performed PCA transformation on communication data and reduced the impact on the entire communication process by sparse representation. This technique can quickly receive the communication signal processed by the interferometer, but it does not allow you to directly determine the accurate reading after the attenuation of the coherent superposition [5, 6]. For the first time, some designers use hard and soft threshold waves to divide the interference communication signal into multiple scales,

J. H. Abawajy et al. (Eds.): ICATCI 2022, LNDECT 170, pp. 350–358, 2023.
https://doi.org/10.1007/978-3-031-29097-8_42

automatically adjust the convergence degree of the particle coefficient, and give the interference attenuation threshold range used in the interference process. Although this technology can effectively reduce the interference of high-speed frequency conversion, there is a problem of obvious loss of high-frequency information in the communication signal [7, 8]. Some researchers also analyzed the interference communication signal from BEMD for the first time, and obtained a two-dimensional intrinsic coefficient function subband, and then estimated the interference change of each subband through the Gaussian mixture model, and used Damping to complete the damper. Although these methods can reasonably measure the comprehensive change of environmental noise and residual interference, if the existing method is used to reduce the interference, the total absolute value of the signal in the window cannot be measured, resulting in a large interference weakening error [9, 10]. Although some achievements have been made in network communication signal transmission at home and abroad, there are still many problems to be solved in practical applications, and a lot of research work to be done.

In this paper, on the basis of consulting a large number of related references, combined with the characteristics of network communication signals, the method of extracting characteristic parameters of network communication signals and the classification method of modulation identification, a new signal transmission efficiency model is designed, and the model is compared with the traditional model. For comparison, we can check that the model designed in this paper is efficient enough.

2 Design of Transmission Efficiency Model of Network Communication Signal Based on Information Fusion

2.1 Characteristics of Network Communication Signals

The so-called communication signal feature vector means that this set of vectors can essentially describe the characteristics of the signal, and different types of signals can be distinguished according to this set of vectors. Specifically, the attribute is a statistic on the original data, and the distribution of this statistic has obvious differences in different configuration modes [11, 12]. To determine the configuration mode of the communication signal, the derived attributes should not be sensitive to the signal-to-noise ratio and configuration parameters, but only to the signal modulation, which requires a large amount of calculation and needs to be avoided. Two types of statistics commonly used to describe communication signals include spectral characteristics and higher-order cumulant characteristics of communication signals.

2.1.1 Spectral Characteristics of the Signal

The spectral characteristics of a communication signal are statistics about the instantaneous amplitude, phase, frequency, and signal strength of the signal. Since the method of constructing a communication signal is closely related to amplitude information, phase information, frequency information, and signal strength information, a method can be found to exploit this information as a characteristic of the communication signal.

2.1.2 The Cumulant Characteristic of the Signal

The cumulative characteristic of the signal is the characteristic of the various torques of the signal. Typically, torques are calculated for the 2nd, 3rd, and 4th orders of the signal. There are two reasons. One is the high torque that can cancel the signal. This is because the high torque of Gaussian noise is 0. Second, for signals, the computation of higher-order torques is very complicated. Therefore, from these two perspectives, it is better to calculate the signal characteristics of the 2nd, 3rd and 4th order of the signal respectively.

2.2 Extraction Method of Network Communication Signal Characteristic Parameters

2.2.1 Feature Extraction Method Based on Instantaneous Parameters

When classifying modulated signals, the direct projection characteristics of instantaneous range interlayer, phase, frequency, etc. contained in the signal are relatively straightforward. Among the various signal assignment forms, MASK signal assignment information is included in the folder, MFSK signal assignment information is included in the frequency, and MPSK signal assignment information is included in the phase.

2.2.2 Feature Extraction Method Based on Wavelet Transform

The wavelet transformer has a good ability to study local details in both time and frequency. Like the processing method of digital signals, it can also better obtain the subtle features and instantaneous characteristics of information. Based on the characteristic parameters generated by the wavelet transform, it shows good characteristics in the environment of reducing the peak signal-to-noise ratio, and can also be fine-tuned well in the frequency-time domain.

2.2.3 Feature Extraction Method Based on Spectral Correlation

According to the characteristics of cyclic spectrum analysis, it is often used to identify and analyze digital data. Periodicity is very helpful in distinguishing digital communication signals from noise. This is also because digital communication signals are periodic, and noise is usually unstable, and its spectral correlation function is usually also periodic. The magnitude is zero if the spectrum is not zero, but more if the service spectrum is zero. Digital communications, on the other hand, are the exact opposite. The spectral correlation function includes the signal amplitude when the circulatory spectrum is non-zero and the signal amplitude when the circulatory spectrum is zero, the amplitudes of which are both zero. These features are helpful for reducing the influence of SNR on configuration identification results.

2.2.4 Feature Extraction Method Based on Constellation Diagram

Constellation diagrams are also a popular tool for analyzing digitally modulated information. Especially for MPSK and MQAM information, the constellation diagram can be a powerful feature parameter to distinguish these two information. The reconstruction of constellation information also often requires a priori signal information. The most efficient way to reconstruct a constellation is to regroup. However, the constellation method also requires prior knowledge of certain prior signals and appropriate estimation parameters. Therefore, when the feature extraction method based on constellation map is adopted, a perfect preprocessing process and reasonable fuzzy parameters are essential for constellation map reconstruction.

2.3 Classification Methods for Modulation Recognition

2.3.1 Pattern-Based Recognition Modulation

The most common digital communication signals are configuration information: instantaneous folder, signal frequency and phase. In theory, these three parameters can be used to identify how a digital signal is constructed. However, in practical applications, it is difficult to obtain this information in the presence of noise interference. Alternatively, it is difficult to accurately classify signal modulation patterns only by detecting this type of signal information. The main classification features used in determining configuration types using pattern recognition methods are: (1) Histogram features.Communication signals are mainly classified using signal amplitude, frequency and phase histograms, but the main problem of this method is that it uses a large number of bit features, which is necessary for classifier training and greatly increases the amount of computation. Decreasing the resolution of the histogram has a significant impact on the classification performance. (2) Statistical moment features.This method solves the problem that the feature dimension is too large when using histogram features to classify signal modulation functions. It proposes two efficient and relatively simple functions: the ratio of the peak amplitude value to the average amplitude value as a function mainly used to identify AM, SSB and CW signals. Another feature is the envelope and its rms ratio. It is mainly used to classify and identify DSB, SSB/FM and AM signals. (3)Transform Domain Features.In addition to using the temporal characteristics of the signal, it is also possible to modify the signal in other regions to obtain its characteristics. A feature of the transform domain is that it transforms a signal from the time domain to another property space to obtain features and use those properties to determine how the signal is configured.

Among the features of the transform region, the features widely used by scholars include the Fourier transform of the high-amplitude distribution function of the signal amplitude and the use of the position of the first zero point, and the use of the QAM signal constellation, decomposed into its damped wave communication signal receive function.

2.3.2 Maximum Likelihood Ratio Modulation Classification Method

The maximum likelihood ratio is another classical solution for classifying modulated signals, which mainly obtains classification statistics by correlating the probability functions of communication signals. Then, by using these statistics and comparing them to appropriate thresholds, the ability to tune pattern recognition is achieved. Classification by maximum likelihood ratio also requires prior knowledge of specific signal parameters, signal-to-noise ratio and distribution function format. Volatility and average. Generally speaking, the receiver works in a non-cooperative communication environment, so the constructed probability function has specific location parameters, and the usual processing method is to use the average probability classification criterion.

The advantage of the maximum likelihood ratio classification method is that it can use theoretical analysis to obtain a classification yield curve, which can theoretically ensure that the analysis results are optimal according to the Bayesian minimum cost principle. At the same time, for this reason, the maximum probability ratio analysis performance can also be used as the theoretical upper limit of efficiency to judge whether the selection of classification attributes is meaningful. However, this approach also has a number of drawbacks. The first is that using physical scale analysis requires a lot of prior knowledge. Then, the representation of the statistics is very complex due to the presence of positional parameters, which greatly reduces the actual performance. Simplifying the probability ratio function will miss a lot of analysis signals, which will affect the final analysis results. At the same time, the maximum likelihood ratio classification method is sensitive to parameter deviation, and the robustness is not strong. Any difference between the actual and estimated parameters can have a significant impact on the classification results.

3 Experiment

3.1 Constructing the Transmission Efficiency Sampling Matrix

To establish the transmission performance acquisition table, it is necessary to detect the reconstructed data signal through compression, and collect the original data of the dynamic data. For large numbers, synchronization of all dynamic data improves test quality for data transfer efficiency. Dilution of received information, and sparse basis sequence using discrete cosine transform:

$$\psi = [\psi_1, \psi_2, M, \psi_n]^T \tag{1}$$

where p stands for sparsity. The sampling matrix is constructed using formula (1):

$$H_{u,v} = \left(d_{u,v}\right)|h_{u,v}| * \psi \tag{2}$$

Through the operation of (2) in the formula, the factor data signal vector is further projected under the sampling matrix to obtain the signal vector, and the information vector also carries out the data signal acquisition at the same time.

3.2 Restore the Original Signal of Dynamic Data

As long as the sampling table of transmission performance is established, dynamic data can be collected iteratively step by step. Based on the transmission efficiency, the specific process of collecting dynamic data by sampling matrix is shown in Fig. 1.

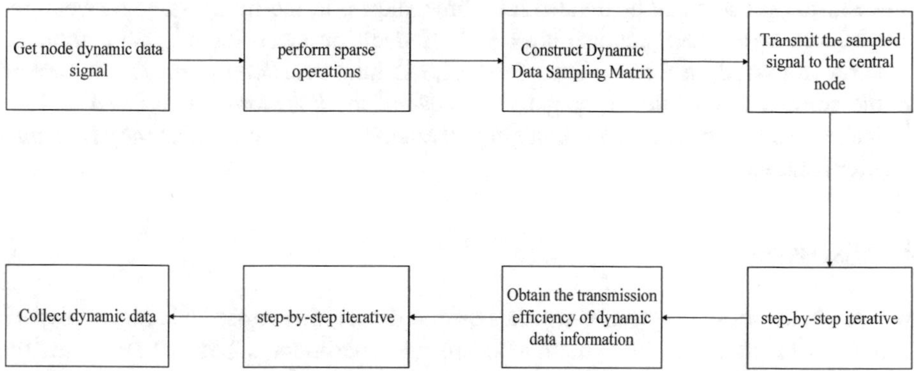

Fig. 1. Collect dynamic data flow chart

The transmission characteristic can directly restore the previous dynamic data signal after receiving the new dynamic data signal from the sampling matrix. Larger-scale data processing consolidation must be performed before the previous dynamic data signal can be recovered to continuously use dynamic data streaming applications. However, the assumed peak signal-to-noise ratio is also relatively small, so this reduces the data transfer rate. To sum up, noise removal of dynamic information and raw data processing have become a key part of the research process of big data streams. The development of driving communication signals relies on the unique way of generating dynamic data and information, and its industrial structure and product operation mode are constantly changing. Dynamic information is produced by a variety of complex network systems and interacts with these complex network systems to finally generate information. The dynamic data information has the functions of acquiring, storing and managing, processing and analyzing and sharing information. With the development of big data technology, more researchers have also joined the research on dynamic restoration of original data information. By selecting the current dynamic original information recovery method, and using the sampling matrix with the highest information transfer efficiency for recovery sampling, the multi-source communication signal fusion is completed.

3.3 Perform Multi-source Information Fusion

Multi-source information fusion refers to the process of obtaining reasonable estimates by correlating and integrating data from one or more sources. This is a more comprehensive approach to obtaining data from various sources using technical tools such as mathematics and computers. According to the difference in the degree of fusion, the entire fusion process can be divided into three stages: in the fusion stage between the data bottom layer, the functional layer and the decision layer, the original data needs to be preprocessed. In the management stage, the final decision information is formed by the combination of the convergence process of the decision-making layer and the decision of each source. It is generally highly customizable and requires very little data transfer bandwidth.

4 Discussion

In order to demonstrate the effectiveness of this model, this paper will conduct experimental verification. The configuration during the experiment is 3.55 GHz Pentium, the network communication signal buffer is 258 M, and the total transmission network communication signal volume is 100 G. The experiments in this paper are completed in the Windows 10 environment. The experiment compares the model in this paper with the traditional model, and the experimental results are shown in Table 1 and Fig. 2.

It can be seen from Table 1 and Fig. 2 that the transmission efficiency of the model in this paper is obviously closer to the actual transmission efficiency than that of the traditional model, so the network communication signal transmission efficiency model based on information fusion designed in this paper is effective.

Table 1. The results of the comparison between the model in this paper and the traditional model

number of experiments	actual transmission efficiency	The transmission efficiency of this model	Traditional Model Transfer Efficiency
1	168.24	168.21	141.22
2	157.25	156.84	120.14
3	191.25	191.10	152.46
4	185.21	174.25	141.25
5	144.25	143.25	102.14

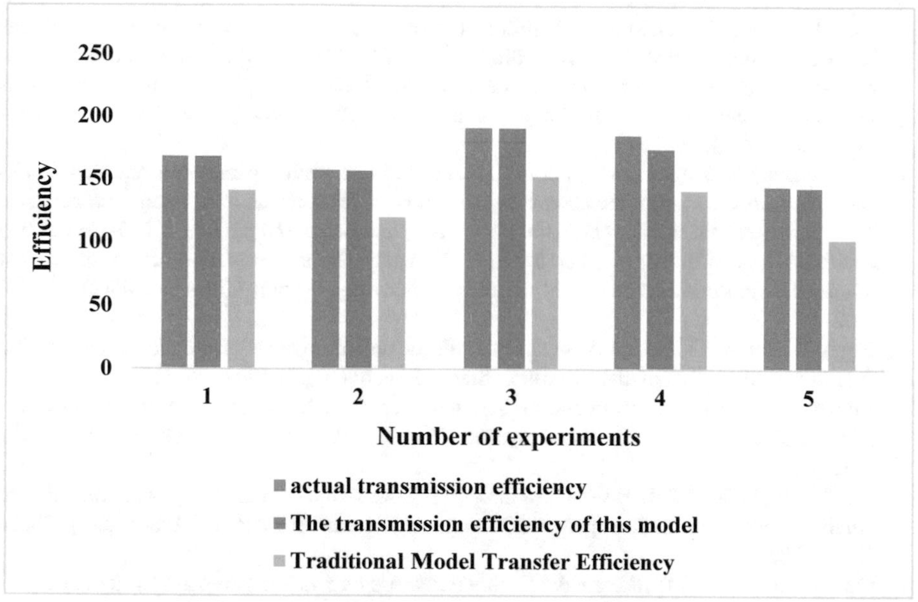

Fig. 2. The results of the comparison between the model in this paper and the traditional model

5 Conclusions

In this paper, through the research of network communication signal transmission efficiency modeling based on information fusion technology, in order to provide a complete efficiency model, the problem that the model cannot effectively integrate key data, and then to improve the traditional information continuity is poor, omissions. The research on the solutions to major problems has put forward the basis, in order to support the improvement and development of the communication signal transmission efficiency on the network in the future.

Acknowledgements. Fund Project: school level project of Xingyi NORMAL UNIVERSITY FOR NATIONALITIES (Project No.: 17xyjs27).

References

1. Chandra, B.S., et al.: Robust heartbeat detection from multimodal data via CNN-based generalizable information fusion. IEEE Transactions on Biomedical Eng. **66**(3), 710–717 (2019)
2. Koch, J.W.: On digital ethics for artificial intelligence and information fusion in the defense Domain. IEEE Aerosp. Electron. Syst. Mag. **36**(7), 94–111 (2021)
3. Bagwari, A., Bagwari, J., Tomar, G.S.: Smart sensor for the underwater communication signal. Wireless Pers. Commun. **116**(2), 1463–1480 (2021)
4. Dhayabaran, B., Raja, G.T., Magarini, M., et al.: Transmit signal shaping for molecular communication. IEEE Wireless Communication Letters **10**(7), 1459–1463 (2021)

5. Sarieddeen, H., Alouini, M.S., Al-Naffouri, T.Y.: An overview of signal processing techniques for terahertz communications. Proceedings of the IEEE **109**(10), 1628–1665 (2021)
6. Niezabitowska, A., Oleszkiewicz, A., Pieniak, M.: Does the frequency of using emoticons in computer-mediated communication signal creativity?. Creativity Theories – Research – Applications **6**(1), 66–76 (2019)
7. Ackermann, C., Beggiato, M., Bluhm, L.F., et al.: Deceleration parameters and their applicability as informal communication signal between pedestrians and automated vehicles. Transportation Research Part F Traffic Psychology and Behaviour **62**(APR.), 757–768 (2019)
8. Alzubi, A.A., Alarifi, A., Al-Maitah, M., et al.: Multi-sensor information fusion for internet of things assisted automated guided vehicles in smart city. Sustain. Cities Soc. **64**(1), 102539 (2021)
9. Paggi, H., Soriano, J., Lara, J.A., et al.: Towards the definition of an information quality metric for information fusion models. Comput. Electr. Eng. **89**(10), 106907 (2021)
10. Kumar, C., Kumar, G.: Performance assessment of hybrid optical amplifier for higher transmission efficiency with SD-WDM system. Wireless Pers. Commun. **116**(3), 2071–2082 (2020)
11. Abed, B.H., Majeed, J.H., Habeeb, N.A.: Optimization of bio-implantable power transmission efficiency based on input impedance. Computer Systems Science and Engineering **38**(1), 17–24 (2021)
12. Elbialy, S.E., Yousif, B., Samra, A.: Highly transmission efficiency of optoelectronic devices using active hybrid plasmonic coupler. Plasmonics **17**(1), 433–447 (2021)

Intelligent Recommendation System for Internet-Famous Sites Based on Web Crawler

Mingkun Xiao[1], Qian Liu[1(✉)], Shutao Han[1], Shuanggui Guo[1], and Ahthasham Ullah Khan[2]

[1] Liaoning Institute of Science and Technology, Benxi, Liaoning, China
lnkjxy_lq@163.com
[2] Institute of IT and Computer Science, Kabul, Afghanistan

Abstract. "Intelligent Recommendation System for Internet-famous Sites Based on Web Crawler" using Python as a development language, through Scrapy network crawler technology to capture the information data of "Internet-famous sites", using collaborative filtering algorithm to calculate user score similarity and user attribute similarity, recommend the most high-quality and most interesting information for users. The system can provide personalized recommendation service and decision support for travel planning according to the priority of keywords input by users, which saves users' browsing time and improves users' playing efficiency.

Keywords: Internet-famous Sites · Web Crawler · Collaborative Filtering · Recommendation System

1 Introduction

With the rapid development of we-media, "Internet-famous sites" have gradually set off a new wave of tourism, which not only provides new choices for people to travel, but also adds new impetus to the development of cultural and tourism industry. How to quickly pick out the favorite places in the mass of Internet-famous places has become the focus of users. In order to solve the problem of "information overload" [1] caused by massive data, this paper designed and implemented based on web crawler " Internet-famous sites" intelligent recommendation system, including " Internet-famous sites" management, recommendation, schedule planning, social interaction, so as to satisfy the user's personalized travel demand [2].

2 System Requirements Analysis

2.1 User Requirements Analysis

(1) Before punch in. You can easily find the introduction information of "Internet-famous sites" that meets your travel needs to screen your destination; Be able to

J. H. Abawajy et al. (Eds.): ICATCI 2022, LNDECT 170, pp. 359–366, 2023.
https://doi.org/10.1007/978-3-031-29097-8_43

make itinerary and strategy by yourself; Be able to understand the user evaluation information and other user information of all aspects of the destination.

(2) Punch in. Be able to share the local experience or pictures for everyone to see; Get real-time help during travel; Be able to understand the traffic arrival mode of each link of the journey.

(3) After punch in. Be able to easily organize your photos and feelings into travel notes for application analysis.

2.2 System Functional Requirements Analysis

(1) Provide system users with necessary information services of "Internet-famous sites". This demand is mainly for the selection of "Internet-famous sites", which serves people with tourism ideas, and provides users with corresponding tourism project information on the system platform, including online red, which is convenient for users to browse and determine their own punch in place.

(2) Provide system users with the navigation guidance function of "Internet-famous sites". This requirement is to allow users to simply and quickly complete the destination search, find any place they want to go, and understand their travel mode.

(3) Provide system users with the feedback function of "Internet-famous sites". This demand is mainly to collect user feedback behavior data and provide data support for personalized recommendation.

3 System Architecture

According to the demand analysis of the recommendation system of "Internet-famous sites", the overall system architecture can be divided into three parts: presentation layer, business logic layer and data access layer. The system architecture diagram is shown in Fig. 1.

(1) Presentation layer. This layer mainly provides different interactive pages for users. Ordinary users can directly feedback their behavior preferences after logging in; The administrator can calculate and manage the information generated by users and recommend for users based on the calculation results.

(2) Business logic layer. This layer is the core part of the system, which mainly contains user information management module, user operation module, "Internet-famous sites" recommendation module and "Internet-famous sites" trip planning module.

(3) Data layer. The data of this layer is collected from many aspects, including network resources, APP resources and information generated by users. According to the collected data, the system designs the user information table, Internet-famous sites information table, Internet-famous sites guide information table and Internet-famous Sites "rating information table respectively".

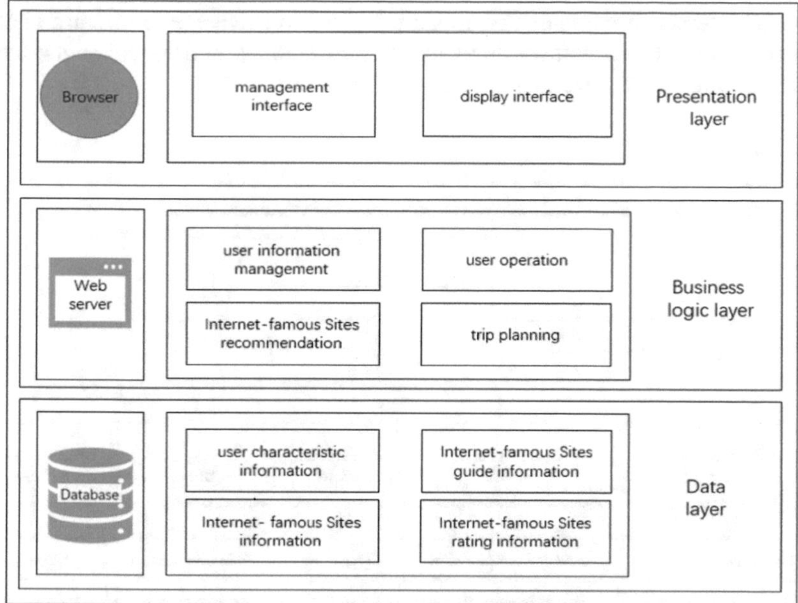

Fig. 1. System architecture diagram

4 System Design

4.1 Functional Structure Design

According to the design and analysis of the system architecture, the system is divided into Management side and client side. The management end includes an information management module; the client includes a recommendation module, a route planning module, a comment module, and a social interaction module. This paper is divided into five functional modules, including information management module, recommendation module, route planning module, comment module, social interaction module.

(1) Information management module. It is mainly divided into three major information management: users, administrators, and "Internet-famous Sites". Including registration and login, information acquisition and information editing.
(2) Recommended module. This module mainly includes popular recommendations and personalized recommendations. Among them, personalized recommendations include "Internet-famous sites" strategy recommendations, user favorite ranking recommendations, and route planning recommendations.
(3) Route planning module. The route planning module is one of the core modules of the system, including map display, route planning, route editing and other functions.
(4) Comment module. This module is an important data source for user modeling and system itinerary sharing. Including itinerary evaluation, score evolution, trip comment, etc. Itinerary evaluation means that users can rate and give feedback on the overall itinerary, which is used to feedback the user's satisfaction, so as to improve the system route planning and recommendation services.

(5) Social interactive module. This module is used to exchange information between users. Users can share travel strategies, text travel notes, and upload photos anytime and anywhere, let more users know. As shown in Fig. 2.

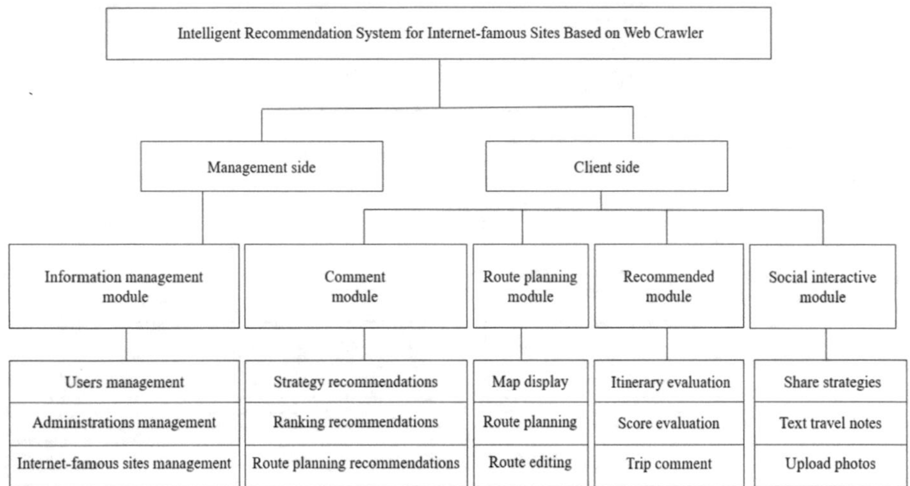

Fig. 2. System function structure

4.2 Database Design

This recommendation system uses MYSQL database, which mainly includes user information, Internet-famous sites information, guide information, scoring information, etc. The related tables are as Table1, Table2, Table3, Table4.

Table 1. User information

Field	Type	Description
user_id	Int	User number
user_name	Varchar	Username
user_pwd	Varchar	User password
user_sex	Varchar	User gender
user_age	Int	User age
user_destination	Varchar	Destination
user_cost	Int	User's cost

Table 2. Internet-famous sites information

Field	Type	Description
site_id	Int	Site number
site_name	Varchar	Site name
site_location	Varchar	Site address
site_type	Varchar	Site type
photo_link	Varchar	Photo's link

Table 3. Internet-famous sites guide information

Field	Type	Description
post_id	Int	Post number
post_title	Varchar	Post name
post_content	Varchar	Post content
post_comment	Varchar	Post comment
post_hits	Varchar	Post hits

Table 4. Internet-famous sites guide information

Field	Type	Description
user_id	Int	User number
site_id	Varchar	Site number
real_rate	Varchar	User Real ratings
predict_rate	Varchar	User Predict ratings

5 "Internet-Famous Sites" Recommendation System Implementation

5.1 "Internet-Famous Sites" Personalized Recommendation [3]

This module is a recommended display for different users. Firstly, clean the data resources that have been crawled and store them in the database. Secondly, feedback accurate data information to users. And then update the user-generated behavior information in real time. Finally, users are classified according to their behavior information. For ordinary users, the system will find the first few users with the highest similarity to the current user, and filter out the "Internet-famous sites" that the user has been to according to their behavioral habits, and find the "Internet-famous sites" with the highest similarity and recommend it to users. For those users who have not added any favorite places, the system will recommend those with high similarity of user attributes. For those users who have not rated the "Punch Place", the system will recommend those with high similarity in user ratings.

5.2 "Internet-Famous Sites" Route Planning

When the recommender system has won the trust of users, it is necessary to carry out route planning. This system uses the AutoNavi Map SDK to develop the map interface, and obtains all the routes in the AutoNavi map path planning [4]. Before completing this function, register as an AutoNavi map developer and create a key, and then perform route planning by referencing the SDK lib. After completing the route planning, you need to start the system navigation to ensure that the start point, end point, way point and related parameters are passed in.

5.3 "Internet-Famous Sites" Score Evaluation

This module is used to get user ratings. When a user inputs his own rating data and submits it, the user's rating data can be stored in the user rating information table in the database, which can be used as a reference index selected by other users and the main criterion for judging "Internet-famous sites".

6 Key Technologies for System Implementation

6.1 Web Crawler

Web crawler (also called web spider or web robot) is a program or script that can automatically capture information from the World Wide Web according to certain rules [5]. This system uses the Scrapy crawler framework [6]. The basic data comes from the "Internet celebrity check-in place" information on Mafengwo and Ctrip, such as the name, location, rating, type, photo, user comments, and comment clicks of the "Internet-famous sites". The page crawling process is as follows:

(1) Obtain the web page data through the Scrapy selector, and use the Xpath and CSS selectors to extract the name of the "Internet-famous Sites", the address of the image, etc. in the web page.
(2) After crawling all the pages on the address list, you can get the name, picture and relevant location of each "Internet-famous sites". At the same time, through the nested use of Xpath and CSS selectors, the data with specific tags in the web page can be obtained. After execution, the scores, types, user comments, and comment clicks of the "Internet-famous sites" can be obtained.
(3) Download and save the obtained relevant information and pictures into the corresponding fields of the database and locally.

6.2 Collaborative Filtering Algorithm

The main functions of collaborative filtering recommendation algorithms are prediction and recommendation [7]. The system uses the user collaborative filtering algorithm to calculate the similarity of user ratings and user attributes [8]. When two users have similar ratings for the same item, it indicates that their interests are similar [9]. In addition, we also need to solve the problems of new projects of the system and cold start of new users. This step is mainly calculation user attributes similarity. And user attributes determine the accuracy of recommended classifications, so as to provide users with personalized "Internet-famous sites".

6.3 Big Data Cluster Analysis

In order to make more accurate and detailed recommendations in massive data, we need to perform cluster analysis on users. After obtaining the similarity between items and users, in order to improve the scalability problem in the recommendation system, the users are clustered using a clustering algorithm [10]. Group users by applying K-means clustering algorithm, mean-shift clustering algorithm and t-SNE clustering algorithm. Then, by comparing the overall variance of the variable with the variance of the grouping variable, we can find the main variable that affects the grouping. Finally define labels for users and recommend suitable options.

7 Conclusions

"Intelligent Recommendation System for Internet-famous Sites Based on Web Crawler" can meet the personalized travel needs of users under the condition of saving time and effort. To a certain extent, the system can solve the problem of influencing users' destination decision caused by information overload of "Internet-famous sites", and it can also be used for reference for other system applications with greater commercial value. However, the system still has some challenges in terms of functional perfection. One is how to effectively embed map navigation into the system. Another aspect is the improvement of the accuracy of the recommendation results.

Acknowledgements. This work was financially supported by Liaoning Institute of Science and Technology 2022 Innovation and Entrepreneurship Training Program for College Students, Internet celebrity 'check-in'-exploring a cultural journey.

References

1. Zhang, X., Ding, X., Ma, L.: The influences of information overload and social overload on intention to switch in social media. Behaviour & Information Technol. **39**(2), 228–441 (2020)
2. Nitu, P., Coelho, J., Madiraju, P.: Improvising personalized travel recommendation system with recency effects. Big Data Mining and Analytics **4**(3), 139–154 (2021)
3. Fararni, K.A., Nafis, F., Aghoutane, B., Yahyaouy, A., Riffi, J., Sabri, A.: Hybrid recommender system for tourism based on big data and AI: A conceptual framework. Big Data Mining and Analytics **4**(1), 47–55 (2021)

4. Qadir, Z., Ullah, E., Munawar, H., S., Al-Turjman, E.: Addressing disasters in smart cities through UAVs path planning and 5G communications: a systematic review. Comput. Commun. **168**(15), 114–135 (2021)
5. Girardi, C., Ricca, F., Tonella, P.: Web crawlers compared. International J. Web Information Syst. **2**(2), 85–94 (2006)
6. Brejla, P., Gillbert, D.: An exploratory use of web content analysis to understand cruise tourism services. Int. J. Tour. Res. **16**(2), 157–168 (2014)
7. Konstan, J.A., Riedl, J.: Recommender systems: from algorithms to user experience. User Model. User-Adap. Inter. **22**(1–2), 101–123 (2012)
8. Bedi, P., Agarwal, S.K., Jindal, V.: MARST: multi-agent recommender system for e-tourism using reputation based collaborative filtering. Lecture Notes in Computer Science 8381, 189-201 (2014)
9. Singh, P.K., Pramanik, K.D., Choudhury, P.: An improved similarity calculation method for collaborative filtering- based recommendation, considering neighbor's liking and disliking of categorical attributes of items. J. Inf. Optim. Sci. **40**(2), 397–412 (2019)
10. Orhan, U., Hekim, M., Ozer, M.: EEG signals classification using the K-means clustering and a multilayer perceptron neural network model. Expert Syst. Appl. **38**(10), 13475–13481 (2011)

Application of Computer Network Technology in Vehicle System

Wenqiang Wang[1(✉)], Shanpeng Xia[1], Yongtao Nie[1], and Romany Viju[2]

[1] College of Automotive Engineering, Weifang Engineering Vocational College, Weifang 262500, Shandong, China
wwqzmr@163.com

[2] New Valley University, New Valley Governorate, Egypt

Abstract. A new concept of urban traffic is being formed, that is, vehicles run on the road network under the guidance of computer. Through the analysis of various characteristics of computer network, including electrical characteristics, communication principle, protocol types and security risks. Summarize the possible attack methods inside and outside the vehicle. Aiming at the corresponding threats, the intrusion detection model of vehicular computer network based on neural network is designed. Theoretically, the model should be able to resist injection attack, denial of service attack, replay attack and tamper attack. The results show that: compared with PCA-BP Neural network algorithm, Pb neural network algorithm takes 4 S, learning times is 500 times less than PCA-BP Neural network algorithm.

Keywords: Computer · Network Technology · Vehicle · Vehicle System

1 Introduction

With the continuous development of science and technology, many experts have studied the vehicle system. The energy feature image analysis method based on wavelet transform in UAV vehicle target detection and recognition system is introduced. A feature point extraction method based on wavelet transform is proposed. The importance of these points is confirmed by the energy estimation of their weights. For further image recognition, salient point descriptor based on wavelet transform is used. The simulation platform is used to simulate the UAV target detection and recognition system based on this method. This paper presents a new method of situational reasoning in V2V environment. Context and context are modeled based on context space, and Dempster Shafer combination rules are used for context reasoning. In the fusion mechanism based on discount rules, the reliability of each information source is increased. This method is applied to a context middleware framework to promote context and context reasoning and provide reliable support for collaborative applications in V2V environment. The implementation and experiment of the prototype system are discussed. In order to improve the performance of CAN bus, dynamic priority protocol is adopted. This can improve the

J. H. Abawajy et al. (Eds.): ICATCI 2022, LNDECT 170, pp. 367–374, 2023.
https://doi.org/10.1007/978-3-031-29097-8_44

real-time transmission of data, especially low priority data. Using CAN bus can reduce the complexity of vehicle connection. TFT display has many advantages, such as less heat and radiation, so that the image is very clear. The displayed image will be less dazzling / flickering, with higher quality, clearer everything, richer colors and clearer [1]. Some experts have studied the vehicle network management and communication technology, and put forward the concept of vehicle diagnosis system for emergency vehicle condition evaluation. A remote diagnosis system based on voice control is proposed. The system consists of three main parts. They are the module of the embedded global positioning system (gps-obd), voice control interface and vehicle monitoring server. At the same time, GPS on-board diagnosis module can be operated through voice control interface. Therefore, drivers can safely operate the proposed system through voice commands and keep their eyes on the road. The purpose of vehicle monitoring server is to provide real-time vehicle information retrieval service for authorized supervisors or maintenance personnel. Users can use a specific browser to retrieve real-time vehicle information, such as the date, time, speed and location of the monitored vehicle. The system can be used for automobile pollution detection, automobile fault alarm, remote diagnosis and road maintenance [2]. Some experts have studied the abnormal detection of CAN bus information in vehicles, and put forward a method to detect and track vehicles on expressways. This method is based on monocular computer vision and adopts two algorithms. The first method can locate the lane boundary in the image and get the three-dimensional shape of the road axis. The second algorithm uses the fixed lights embedded in the vehicle to detect, track and calculate the three-dimensional position in front of the vehicle. By combining the results of the two algorithms, a fusion step can let us know which vehicles are the most dangerous according to their location, speed and lane. The method has been implemented on velac and the whole system runs in real time. This paper presents a new method of situational reasoning in V2V environment. Context and context are modeled based on context space, and Dempster Shafer combination rules are used for context reasoning. In the fusion mechanism based on discount rules, the reliability of each information source is increased. This paper introduces the pedestrian safety measures, including strengthening infrastructure construction and vehicle passive safety characteristics. Then, the pedestrian detection active safety system based on vehicle and infrastructure sensors is systematically described. According to the type and configuration of sensors, as well as the video cues and classifiers used in the detection algorithm, pedestrian detection methods are classified. It is pointed out that collision avoidance requires not only pedestrian detection, but also pedestrian dynamics and behavior analysis to predict collision. In one embodiment, the processor is configured to execute application software for integrating audio data into the OBD system [3]. In order to study the application of computer network technology in vehicle system, the fuzzy inference engine is established through the research of computer network technology and vehicle system. The results show that the computer network technology is conducive to the application of vehicle system.

2 Method

2.1 Computer Network Technology

(1) Computer network structure

Computer network is a collection of computer systems with the functions of resource sharing and communication [4]. Automobile network is some control units or intelligent devices connected according to communication protocol. The sensor, actuator or interface control signal of the protocol controller is transmitted to the computer target system through the network [5]. Vehicular network is the interconnection structure of multiple LANs. Multiple LANs can be connected together through the gateway to form the Internet [6]. Gateway is a device that connects different networks and realizes different network protocol conversion [7]. The computer network service layer should complete the functions of system management, transceiver control service, asynchronous data transmission service, synchronous channel allocation and control service [8]. The system management completes the initialization, diagnosis and power management of the system [9]. The transceiver control completes the working state setting and monitoring functions of both sides, and controls both sides to complete the data transceiver operation required by the application layer [10].

2.2 The Concept of Computer Network

Computer network is a multilayer feedforward neural network trained by error back propagation algorithm, which is one of the most commonly used neural networks [11]. The basic idea of the algorithm is gradient descent algorithm, which uses the fastest gradient descent algorithm to adjust the weight of the network [12]. It can keep the error of backward return and associate it with the input at a certain time, so long-term memory network and short-term memory network are often used in the prediction and classification of time series. This is also the main reason why short-term memory network is superior to other methods such as Markov chain in the field of speech recognition and handwriting recognition.

(1) Concept of vehicle system

The vehicle position is determined by the on-board system according to the radio signal of the navigation satellite constellation, and the position provided by the navigation satellite constellation, the time tag provided by the navigation satellite constellation and the location determined by vehicle identification are sent to the traffic control center regularly. A central system for accurately locating at least one vehicle equipped with the system, which calculates the position of the current moment based on the vehicle position and speed vector and the time tag sent by the vehicle. Because the time tag provided by navigation satellite constellation can compensate for the fluctuation of transmission time between vehicle and traffic control center when the traffic control center receives the location, these systems can determine the position periodically and accurately.

(2) Vehicle television system

The vehicle television system, which only displays normal TV screen when the vehicle stops, is considered as a safety function. For the switching signal of the microcomputer image display, the image display condition of switching TV is switched. The information of setting parking brake, input information of automobile accessories and timer circuit timeout information are provided to the microcomputer. The computer determines the logical product of information and provides the image display switch signal. Only when the specific logic product is output, the normal image will be displayed on the TV. So the driver doesn't watch TV while driving, so he doesn't let the driver drive carelessly.

(3) On board network communication system

At present, the vehicle network communication system connects the bus of different protocol standards through gateway to realize the management of the vehicle network communication system. The vehicle network communication system mainly includes power transmission, chassis system, body system, sensing and execution system, entertainment multimedia system and line control operating system. High speed can bus and FlexRay bus can be used in power transmission and chassis system, with high real-time and reliability requirements, such as anti-lock braking system; low speed can bus can be used in body systems with low real-time requirements but more control units, such as window control. Because of the low speed can bus, the anti-interference ability of the system can be effectively improved and hardware cost can be reduced; LIN bus is mainly used for vehicle Some simple body sensing and execution systems on the vehicle; most buses can be used for entertainment and multimedia systems, which are most suitable for multimedia transmission interface with high transmission rate; FlexRay bus can also be used for line control operating systems with line control system, such as line control steering system.

2.3 Fuzzy Inference Engine

Fuzzy inference engine uses a certain method to derive the output of fuzzy controller from the input of sampling time and fuzzy control rules. The fuzzy inference engine consists of three parts: big premise, small premise and conclusion. The major premise is a number of multidimensional fuzzy conditional sentences, forming a rule base; the minor premise is a fuzzy judgment sentence, also known as fact. It is shown in formula (1):

$$R_{ij} = R_{ij}(A_i) \tag{1}$$

When the engine works, the change of speed is often caused by the load. When the speed and load change at the same time, it is necessary to determine who is the main factor causing the instability of the working condition. The calculation formula (2) is as follows:

$$\mu_{\mathrm{A \cap B}} = \mu_{\mathrm{A}}(\sigma)^{\wedge}\mu_{\mathrm{B}} \tag{2}$$

On the computer network bus, high-throughput data exchange is often carried out every few milliseconds, while low-frequency data transmission is carried out at low speed, and data exchange is carried out every tens of milliseconds or even hundreds of milliseconds. The calculation formula is shown in Eq. (3):

$$U = R_h(\frac{1}{d} \times B) \tag{3}$$

The eigenvector and eigenvalue of the covariance matrix Y (arranged from small to large) are calculated according to formula (3), as shown in formula (4):

$$\lambda(I - Y) \propto = 0 \tag{4}$$

3 Experience

3.1 Extraction of Experimental Objects

There are five main database access and operation objects in the database ado.Net Object model, including connections, commands, data readers, data adapters, and dataset objects. Among them, the connection object is mainly responsible for connecting to the database, the command object is mainly responsible for generating and executing SQL statements, and the datareader object is mainly responsible for reading the data in the database. ADO.NET It mainly provides two kinds of data providers, namely SQLServer.NETProvider as well as OLEDB.NET Providers.

3.2 Experimental Analysis

First of all, the field test generally requires a large number of vehicles and test personnel, and the field test process is expensive. The invention relates to a method and a device for using a plurality of generators to provide a power supply voltage for a load of an on-board electrical system of a vehicle. A regulator unit having at least one output stage transistor is assigned to each generator. The output transistor is controlled by PWM control signal. A control signal describing the pulse duty cycle of the PWM signal generated in or derived from one of the regulator units is also supplied to the regulator unit of the additional generator, where it is also used to control the output stage transistors. This procedure allows the load level of the generator to be agreed.

4 Discussion

4.1 Algorithm Performance Comparison

Under the same experimental data, PAC BP neural network algorithm and classical BP neural network algorithm can detect replay attack, injection attack and denial of service attack. Among them, the classical BP neural network and PCA-BP Neural network algorithm can improve the detection rate of tampering attack. As shown in Table 1.

It can be seen from the above that Pb neural network algorithm consumes 436 s and 1000 times of learning; PCA-BP Neural Network Algorithm consumes 432s and 1500 times of learning. The results are shown in Fig. 1.

It can be seen from the above that Pb neural network algorithm consumes 4 s more time than PCA-BP Neural network algorithm, and the learning times of Pb neural network algorithm is 500 times less than PCA-BP Neural network algorithm. The specific results are shown in Fig. 1.

Table 1. Replay attack detection comparison

Algorithm name	Algorithm time (s)	Learning times
Pb neural network	436	1000
PCA-BP Neural Network	432	1500

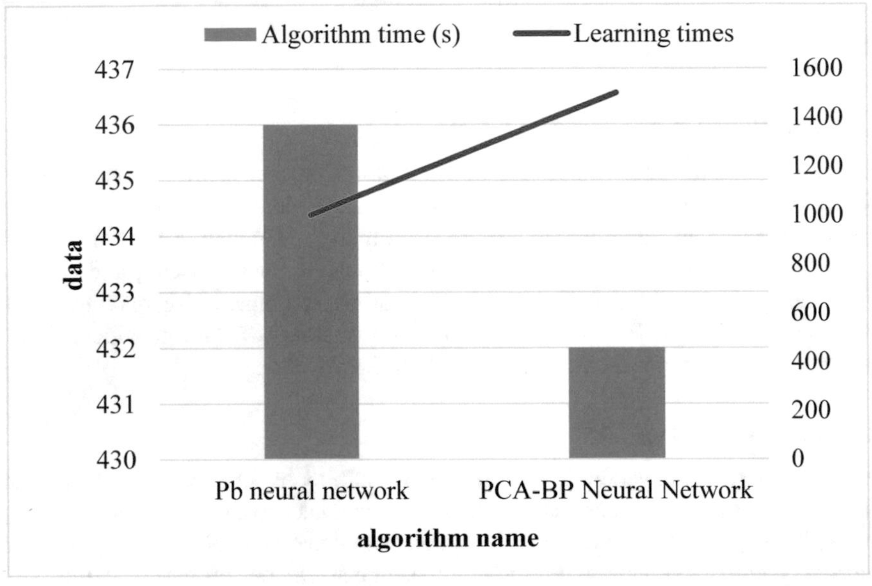

Fig. 1. Replay attack detection comparison

4.2 Experimental Results and Analysis

According to the different mobile speed of nodes, two different experimental scenarios of 6-22 m/s and 22-44 m/s are simulated, and the delay, packet arrival rate and overhead of different scenarios are compared. Then, by simulating different scenarios such as urban, highway and rural areas, the above three performance indicators of GSPR and TEAP under different scenarios are investigated are shown in Table 2.

Table 2. Message delay in different scenarios

	6-22 m\s	22-44 m\s
GSPR	3.63	1.48
TEAP	3.89	1.37

From the above, the performance index of GSPR protocol in 6-22 m\s scenario is 3.63, the performance index of TEAP is 3.89; the performance index of GSPR protocol in 22-44 m\s scenario is 1.48, and the performance index of TEAP protocol is 1.37. The specific results are shown in Fig. 2.

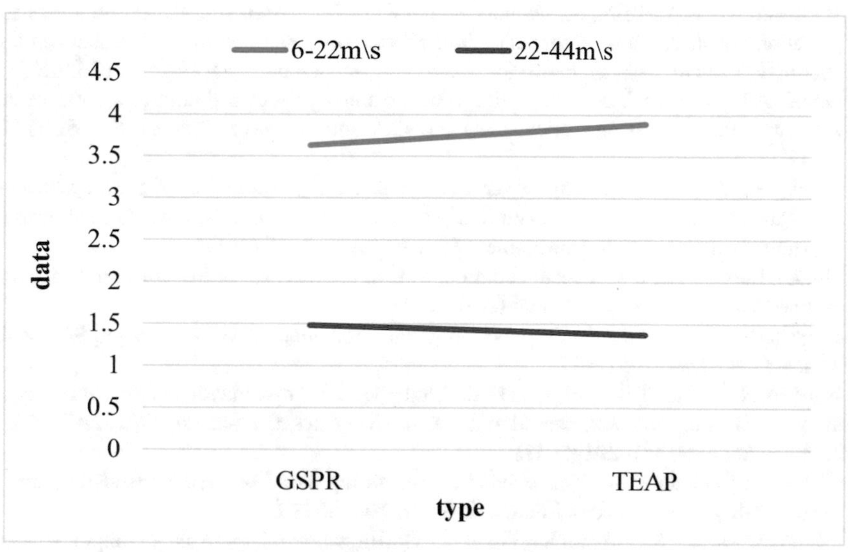

Fig. 2. Message delay in different scenarios

It can be seen from the above that the performance indexes of GSPR protocol and TEAP protocol in 6-22 m/s scenario are higher than those of GSPR protocol and TEAP protocol in 22-44 m/s scenario.

5 Conclusion

With the development of science and technology, navigation system has been widely used in the field of vehicle position prediction and life safety. As we all know, the integrated navigation system composed of GPS and INS can provide more accurate position prediction. The method of moment (MOM) is used to process the data collected by INS/GPS to extract the coordinate information and important information related to life safety. The system will be cost-effective and can be used as a black box for ships, cars, submarines and even small aircraft. This paper introduces four kinds of common automobile bus, their application in automobile network communication system, common network topology, three main topology types of FlexRay network, describes some advantages of FlexRay bus as a high-performance communication protocol, and the main content of FlexRay protocol specification, including FlexRay communication mode, encoding and decoding, frame format and clock synchronization.

References

1. Li, H., Han, D., Tang, M.: A privacy-preserving charging scheme for electric vehicles using blockchain and fog computing. IEEE Systems Journal **15**(3), 3189–3200 (2020)

2. Cecil, J., Albuhamood, S., Ramanathan, P., et al.: An Internet-of-Things (IoT) based cyber manufacturing framework for the assembly of microdevices. Int. J. Comput. Integr. Manuf. **4**, 1–11 (2019)
3. Hadiwardoyo, S.A., Patra, S., Calafate, C.T., Cano, J.-C., Manzoni, P.: An intelligent transportation system application for smartphones based on vehicle position advertising and route sharing in vehicular ad-hoc networks. J. Comput. Sci. Technol. **33**(2), 249–262 (2018)
4. Nejad, A.E., Romoozi, M.: Presenting a traffic management and control system in driver assistance form based on vehicular networks. Autom. Control. Comput. Sci. **51**(1), 1–12 (2017)
5. Singh, N., Tyagi, K.: Ranking of services for reliability estimation of SOA system using fuzzy multicriteria analysis with similarity-based approach. International Journal of System Assurance Engineering & Management **8**(jan.suppl.1), 1–10 (2017)
6. Hu, Q., Luo, F.: Review of secure communication approaches for in-vehicle network. Int. J. Automot. Technol. **19**(5), 879–894 (2018)
7. Bulysheva, L., Jones, J., Bi, Z.: A New approach for image databases design. Inf. Technol. Manage. **18**(2), 97–105 (2017)
8. Nguyen, N.T., Tran, T.N., Yin, S.H., et al.: Multi-objective optimization of improved magnetic abrasive finishing of multi-curved surfaces made of sus202 material. Int. J. Adv. Manuf. Technol. **88**(1–4), 381–391 (2017)
9. Cheng, J.: Evaluation of physical education teaching based on web embedded system and virtual reality. Microprocess. Microsyst. **5**(83), 103980 (2021)
10. Wang, Y., Moura, S.J., Advani, S.G., et al.: Optimization of powerplant component size on board a fuel cell/battery hybrid bus for fuel economy and system durability. Int. J. Hydrogen Energy **44**(33), 18283–18292 (2019)
11. Hong, C., Li, G.: Optimization design of network structure based on genetic algorithm. In: 2018 International Conference on Virtual Reality and Intelligent Systems (ICVRIS), 206–209 (2018)
12. Zhang, T., Ou, K., Jung, S., et al.: Dynamic analysis of a pem fuel cell hybrid system with an on-board dimethyl ether (DME) steam reformer (SR). Int. J. Hydrogen Energy **43**(29), 13521–13531 (2018)

Multi-regression Mathematical Model Prediction Based on Big Data Analysis and Cloud Computing

Yunqi Guo, Danhong Chen(✉), Yating Liang, and Yilin Wang

School of Economics and Management, Shenyang Aerospace University, Shenyang 110136, Liaoning, China
1507685405@qq.com

Abstract. Both the Internet of Things industry and the takeout industry are the hot industries in contemporary society. In the context of deepening supply-side structural reform and demand-side reform, the Internet of Things industry has entered. The third wave of development, along with cross-border integration, integrated innovation and large-scale development. The new stage. In the rapid development of the Internet of Things industry, the scale of the takeout industry will also follow Huge changes are analyzed and predicted by mathematical models and other ways. This model establishment of type algorithm can quickly provide effective scientific information, relevant enterprises can the development of its enterprise to make reasonable plans.

Keywords: The Internet of Things and the new Takeout Industry · Big Data · One-Yuan Linear Regression · Multi-Linear Regression

1 Introduction

Both the Internet of Things industry and the takeout industry are the hot industries in contemporary society.In the context of deepening supply-side structural reform and demand-side reform, the Internet of Things industry has entered the third wave of development, and has ushered in a new stage of cross-border integration, integrated innovation and large-scale development [1]. Under the rapid development of the Internet of Things industry, the scale of the takeout industry will also change greatly, which is analyzed and predicted through mathematical models and other ways. The establishment of this model algorithm can quickly provide effective scientific information, and the relevant enterprises can make reasonable plans for the development of their enterprises [2].

J. H. Abawajy et al. (Eds.): ICATCI 2022, LNDECT 170, pp. 375–384, 2023.
https://doi.org/10.1007/978-3-031-29097-8_45

2 Collection and Collation of Big Data

2.1 Selection of Big Data Indicators

The choice of index data is the basis of the research work. The Internet of Things industry and the takeout industry influence and penetrate each other, and the correlation is worth our further study [3]. There are many factors affecting the mutual influence of the Internet of Things industry and the takeout industry. The choice of index data should reflect the development of the Internet of Things industry and the takeout industry and the correlation of the feasibility of data collection and the achievability of calculation [4]. In order to objectively, comprehensively and measure the relevance of the Internet of Things industry and the takeout industry scientifically, adhere to the principle of scientific, practical and operability, and choose the following index data:

(1) The Internet of Things Industry Index Data

Metrics of the Internet of Things include (x1) (x2) (x3), which measure the market size of the overall Internet of Things, I OT hardware sales, and connections to represent the number of user users of the Internet of Things and the development of the industry [5].

(2) Takeout Industry Index Data

(y) Represents the scale of the takeout industry to represent the development achievements of the takeout industry. The scale of the takeout industry can intuitively display the development achievements of the takeout industry through its data [6].

3 Big Data Acquisition System

This paper collects relevant data indicators for the Internet of Things industry and the food delivery industry from 2015 to 2019,Results are presented in Table 1:

Table 1. 2015–2019 Iot of Things industry indicators data and takeout industry indicators data

	2015	2016	2017	2018	2019
X 1	7420	9220	10865	12600	14900
X 2	715	893	1188	1554	1914
X 3	6.39	9.04	15.35	22.56	31.25
y	491	1258	3053	4692	6536

Source: According to the public information of takeout professional members of China Communication Industry Association, Meituan Research Institute and China Hotel Association.

4 Classification Based on Univariate Regression Model

4.1 Ordinary Least-Squares Estimation

(1) Ordinary Least-Squares Regression Equation is Solved

According to the relevant literature,The market scale of the Internet of Things industry x1 has a relatively significant impact on the development of the new takeout industry y [7]. Through big data screening, we investigated the data of the market scale of the Internet of Things industry and the takeout industry scale from 2015 to 2019.

$$\widehat{\beta}_0 = y - \widehat{\beta}_1 x$$

$$\hat{\beta}_1 = \frac{\sum_{i=1}^{n}(x_i - x)(y_i - y)}{\sum_{i=1}^{n}(x_i - x)^2}$$

(In the formula, xi corresponds to the corresponding value of the market scale x1 of the Internet of Things industry, and yi corresponds to the corresponding value of the takeout industry scale).

Use the MathType Software, and the output results are presented in Table 2:

Table 2. Output of ordinary least squares regression on the scale of Internet of things industry and takeout industry

Coefficients					
model	Unstandardized coefficients		Standardization coefficients	t	Significance
	$\widehat{\beta}_0$	Standard error	$\hat{\beta}_1$		
The constant	−48483.680	6621.238	10.072	−9.316	0.000
The scale of the Internet of Things	11.083	0.221	0.995	45.163	0.000

Influential variable: the scale of the takeout industry

Regression coefficient significance test P = 7.4523E − 23, significantly below the significance level $\alpha = 0.05$, hence y with x1The linear regression was significant.

From the results in Table 2, $\widehat{\beta}_0 = -48483.680$, $\hat{\beta}_1 = 10.072$, and both P values were less than the significance level $\alpha = 0.05$, so both were highly significant.So the takeout industry y and the Internet of Things industry market scale is X11The established ordinary least-squares regression equation is:

$$\hat{y} = -48483.680 + 10.072x1$$

In sum up, it shows that for each unit of x1 of the Internet of Things industry, the corresponding takeout industry increases by an average of 10 72 units.

(2) Residues Analysis
Using SPSS software, the output results are shown in Fig. 1:

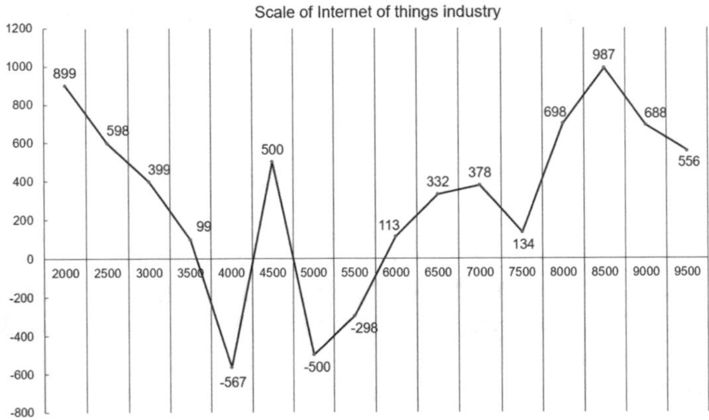

Fig. 1. Residual diagram of added value of market scale in different Internet of things industries

The mornay linear regression equation established from the above, [8] using the MathType software, calculated the value of the corresponding residual $e_i = y_i - \hat{y}_i$ for each term [9]. Take the residual as the vertical ordinate and the industrial scale of the Internet of Things as the abscissa [10].

Judging from the above residual map, the residual values are evenly scattered on both sides of the straight line y = 0, but there is some is ovariance, which can explain that the above ordinary least squares have certain limitations.

4.2 Build the Model by the Iterative Method

(1) Calculation of the Sample Correlation Coefficient
Using MathType software to calculate the correlation between the market size of the Internet of T industry and the output results are shown in Table 3:

Table 3. Output table of the correlation coefficient between Market size of Internet of Things industry and Scale of take-out industry

Correlation				
Spearman Rho			Industry market size	Scale of takeout industry
	Market size of the Internet of Things industry	Correlation coefficient	1.000	0.999**
		Sig. (Double tail)	0.0	0.000
		N	23	23
	Scale of takeout industry	Correlation coefficient	0.999**	1.000
		Sig.(Double tail)	0.000	0.0
		N	23	23

**.Significant correlation at level 0.01 (two-tailed).

From the output, a significant correlation between the two, the correlation coefficient $\rho = 0.999012$, very close to 1, so a special iterative method —— differential is adopted to eliminate residual autocorrelation.

(2) The Autocorrelation was Removed Using the Differential Method

Difference was calculated first $\Delta yt = yt - yt - 1$, $\Delta xt = xt - xt - 1$.

Δy least squares regression of Δx over origin was then done using MathType and the output results are shown in Table 4.

Varietals variable: the scale of the takeout industry after the difference; Linear regression over the origin.

Table 4. Results output table after establishing the different method regression model

Coefficients					
Model	Unstandardized coefficients		Standardization coefficients	t	Significance
	$\hat{\beta}_0$	Standard error	$\hat{\beta}_1$		
The industrial scale of the Internet of Things industry after the difference	0.179	0.048	0.549	2.078	0.004

According to the output, the model is significant, $\hat{\beta}_1 = 0.179$, so the difference method model is obtained as follows

$$\Delta y = 0.179\Delta x.$$

That is, $\hat{y}_t - \hat{y}_{t-1} = 0.179(x_t - x_{t-1})$

5 Analysis Based on Multiple Regression Mode

5.1 Ordinary Least-Squares Estimation of the Multiple Regression Model

(1) Establishment of the Regression Model

We are known from the relevant literature,The changes in the deliver y industry y are not only affected by the market size x1 of the Internet of Things industry, but also by x3. Therefore, by consulting relevant information, we can obtain relevant data on the market size of the Internet of Things industry and the hardware sales and number of the number of connections between 2015 and 2019.

Three illustrated variables, one illustrated variable, are determined by multiple linear regression of the regression coefficient formula [11].

$$\widehat{\beta} = \left(x^{T}x\right)^{-1}x^{T}y$$

Using MathType software, the output results are shown in Table 5:

Table 5. The output of the coefficient calculation ofmultivariate regression equation

Coefficients					
Model	Unstandardized coefficients		Standardization coefficients	t	Significance
	$\widehat{\beta}_0$	Standard error	$\hat{\beta}_1$		
The constant	0.008	0.038		0.305	0.788
Number of connections	1.000	0.000	0.398	452756.321	0.000
IOT hardware sales	1.000	0.000	0.510	398564.455	0.000

Influential variable: the scale of the takeout industry

As seen from the output, the significance P = of the constant variable is much greater than the significance level $\alpha = 0.05$, so the fitting effect of the regression equation is not ideal [12].

The correlation between the independent variables the output results are shown in Table 6:

Table 6. Independent variable correlation test output

Correlation					
Spearman Rho			Industry scale	Number of connections	Sales
	Industry scale	Correlation	1.000	0.98844	0.988768
		Sig (dual-tail)	0.000	0.000	0.000
		N	23	23	23
	Number of connections	Correlation	0.97785	1.000000	1.000000
		Sig (dual-tail)	0.000	0.000	0.000
		N	23	23	23
	IOT hardware sales	Correlation	0.991966	1.000000	1.000
		Sig(dual-tail)	0.000	0.000	0.000
		N	23	23	23

It can be seen from the output results that the market scale of the Internet of Things industry has a highly autocorrelation with IOT hardware sales, sales and number of connections, so it needs to do certain independent variable processing.

(2) Selected Independent Variables in the Stepwise Regression Method
Take αentry = 0.1, αremoval = 0.15, performed using MathType software and output results are shown in Table 7:

Table 7. The result of the stepwise regression

Coefficients					
model	Unstandardized coefficients		Standardization coefficients	t	Significance
	$\hat{\beta}_0$	Standard error	$\hat{\beta}_1$		
The constant	−89.528	1309.517		−0.076	0.857
IOT hardware sales	3.244	0.020	1.000	157.356	0.000
The constant	5422.358	638.456		7.883	0.000
IOT hardware sales	1.318	0.068	0.547	15.308	0.000
Number of connections	0.882	0.068	0.537	11.505	0.000

(*continued*)

Table 7. (*continued*)

Coefficients					
model	Unstandardized coefficients		Standardization coefficients	t	Significance
	$\hat{\beta}_0$	Standard error	$\hat{\beta}_1$		
The constant	0.009	0.043		0.312	0.744
IOT hardware sales	1.000	0.000	0.364	393918.723	0.000
Number of connections	1.000	0.000	0.398	472886.923	0.000
The Internet of Things market scale	1.000	0.000	0.089	148652.783	0.000

Influential variable: the scale of the takeout industry

From the output, the optimal subset model of stepwise regression is y and x2, x3, The model, namely, the market scale of the Internet of Things is excluded, and only the regression model of takeout industry scale and sales volume and connections is established, and the corresponding equation is:

$$\hat{y} = 5422.358 + 1.318x2 + 0.882x3$$

The above equations indicate that when the fixed connection number is x3unchanged, IOT hardware sales x2For each additional unit, the corresponding takeout industry scale has increased by 1.318 units on average. Also fixed sales x2invaristant, number of connections x3For each additional unit, the corresponding takeout industry scale has increased by an average of 0.882 units.

5.2 Residues Analysis

(1) The Residuals were Calculated and Drawn

Residual values were first calculated for each time period by SPSS software.

The diagram of the residual value and time using SPSS software is shown in Fig. 2:

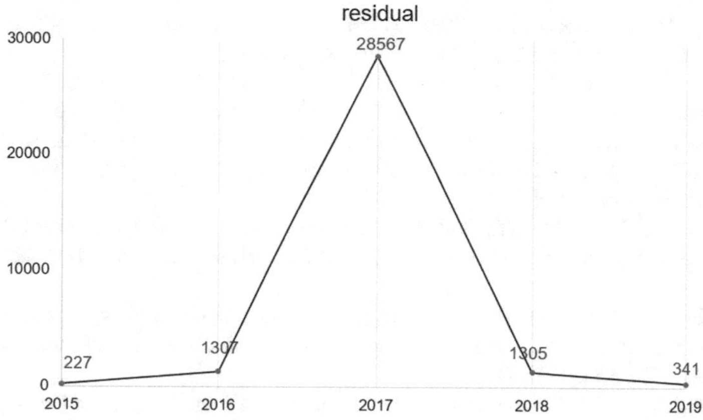

Fig. 2. Residuals graph of multiple linear regression

According to the image, except for the large residual value in 2017, the residual value was gentle in the remaining time period, and excluding the heteroscedasticity and autocorrelation phenomenon, the model is more reasonable.

6 Evaluation and Summary

In this paper, a regression model between the scale of the takeout industry and the market scale of the Internet of Things industry and the sales of the Internet and the number of connections, and the results are such as Eqs. (3–8) and (4–2).On the surface is two different regression models, but by the Internet of things industry market size and I OT hardware sales, connections have a strong correlation, that is, the Internet of things industry market size changes must have the corresponding IOT hardware sales and connection number changes, so essentially this article is through two different ways to establish two similar regression equations.This enables us to estimate the scale of China's takeout industry through the future development scale of the Internet of Things, so as to make some relevant policy adjustments.

Acknowledgments. This research was supported by the college students' innovative entrepreneurial training plan of Shenyang Aerospace University under grant X202110163199S. Danhong Chen is the corresponding author and instructor of this paper.

References

1. Bañuls, V.A., Salmeron, J.L.: Foresighting key areas in the Information Technology industry Technovation **28**(3), 22-35 (2007)
2. Beier, G., Niehoff, S., Ziems, T., et al.: Sustainability aspects of a digitalized industry – a comparative study from China and Germany. International Journal of Precision Engineering and Manufacturing-Green Technology **4**(2), 121–145 (2017)

3. Danilin, I.V.: The impact of the COVID crisis on the innovative potential of china's internet platforms. Her. Russ. Acad. Sci. **90**(6), 779–788 (2020)
4. Paiola, M., Agostini, L., Grandinetti, R., Nosella, A.: The Process of Business Model Innovation Driven by IoT: Exploring the Case of Incumbent SMEs **8**(1), 103–113 (2022)
5. Ammar, R., Samer, S.: Internet of Things from Hype to Reality - The Road to Digitization, Third Edition. Springer, ISBN 978–3–030–90157–8, pp. 232–261 (2022). https://doi.org/10.1007/978-3-319-99516-8
6. McKnight, L.W., Brooks, T.T.:Cloud to Edgeware - Wireless Grid Applications, Architecture and Security for the "Internet of Things". WorldScientific ISBN 9789814630801, 151–189 (2022)
7. Abdel-Basset, M., Mohamed, R., Elkomy, O.M.: Knapsack Cipher-based metaheuristic optimization algorithms for cryptanalysis in blockchain-enabled internet of things systems. Ad Hoc Netw. **128**, 102798 (2022)
8. Salim, S., Turnbull, B.P., Moustafa, N.: Data analytics of social media 3.0: Privacy protection perspectives for integrating social media and Internet of Things (SM-IoT) systems. Ad Hoc Networks **128**, 102786 (2022)
9. Mehdizadeh, N., Farzaneh, N.: An evidence theory based approach in detecting malicious controller in the multi-controller software-defined internet of things network. Ad Hoc Sens. Wirel. Networks **51**(4), 235–260 (2022)
10. Papanagnou, C.I.: Measuring and eliminating the bullwhip in closed loop supply chains using control theory and Internet of Things. Ann. Oper. Res. **310**(1), 153-170 (2022)
11. Aranda, J.A.S., dos Santos Costa, R., de Vargas, V.W., da Silva Pereira, P.R., Barbosa, J.L.V., Vianna, M.P.: Context-aware edge computing and internet of things in smart grids: a systematic mapping study. Comput. Electr. Eng. **99**, 107826 (2022)
12. Aly, S., Abdallah, A.S., Sherif, S., Atef, A., Adel, R., Hatem, M.: Toward Smart Internet of Things (IoT) for Apparel Retail Industry: Automatic Customer's Body Measurements and Size Recommendation System using Computer Vision Techniques, pp. 99–104 (2021)

Clothing Style Identification Technology Based on the Interactive Genetic Algorithm

Peipei Zhao[1(✉)], Dan Yu[1], Ning Yang[1], and Manoj Kautish[2]

[1] Art Department, Changchun Humanities and Sciences College, Changchun, Jilin, China
270783523@qq.com
[2] LBEF Campus, Kathmandu, Nepal

Abstract. With the rapid development of modern social economy, people's dress is more and more unique, distinctive and individualization. Successful clothing styles will always have obvious style characteristics. Clothing style can not only fully convey the concept of designers, but also express the emotional needs of consumers. In the traditional clothing customization system, the clothing design is only participated by designers, and the style is single. Facing many styles, users simply arranged and combined their styles repeatedly, and did not realize the innovative design of users. In order to solve such problems, the paper proposes a user-oriented method based on interactive evolutionary computing to design clothing and realize the innovative design of clothing to meet the customized needs of users and enhance the market competitiveness. Relevant research units and personnel at home and abroad for each module independent research has continued, for the cohesion between the module research or new problem, such as human body size data according to the automatic style of clothing screening, and from the human data according to style screening to clothing automatic import of fast sample design module are the key technology of clothing design development digital integration scheme. The paper used the literature research method, investigation method (100 valid questionnaires), interview method (24 experts) and other research methods, based on the interactive genetic algorithm, to study the clothing design identification technology, providing reliable and scientific technical support for the enterprise clothing design.

Keywords: Interactive Genetic Algorithm · Genetic Algorithm · Clothing Design · Personalized Design

1 Introduction

Today's social, economic and cultural level is gradually improving, and people's pursuit of enriching material life makes people need to more meet their personalized needs. Based on the interactive genetic algorithm, it integrates people's personalized needs with genetic algorithms into [1], and takes people's perceptual needs as the starting point and foothold. This model has become the dominant direction in the field of modern clothing design. The typical style is applied to the process of interactive genetic algorithm and assigned to the initial population as parameters. Through experiments,

J. H. Abawajy et al. (Eds.): ICATCI 2022, LNDECT 170, pp. 385–392, 2023.
https://doi.org/10.1007/978-3-031-29097-8_46

appropriate crossover and variant probability parameters are selected in order to accelerate the convergence speed of the algorithm and effectively relieve user fatigue [2]. It propose a user-oriented method based on interactive evolutionary computing to realize innovative clothing design, so as to meet user customized needs and enhance market competitiveness.

Interaction design is the design and definition of artifacts, environment, system behavior, and the appearance elements that convey this behavior, the concept of interaction design proposed by Western scholars. Some Chinese scholars have introduced the history and theoretical basis of interactive design, user-centered interactive design method and process, user research and usability evaluation method [3], and provided cutting-edge usability evaluation research and interactive design innovation research cases. In the interactive design of clothing, many scholars have studied the interactive genetic algorithms that provide the support of computer systems. It describes the application principle of genetic inheritance algorithm in clothing design: the clothing interaction design system provides some designed clothing for user evaluation. According to the user evaluation, the system will imitate the genetic inheritance of the designer clothing, including gene selection, crossover and mutation [4].

In the interactive design system, the interactive genetic algorithm can be used to complete the system design on the modular basis, fully considering the user's sensory evaluation and feedback to the computer system to gradually optimize the design. In the later display, 3 D software can be used to help present the design effect [5]. Ideal clothing interaction design system can provide many clothing design modules, color, fabric and other design elements. In use, the user will first choose the ideal style from the clothing style, and then determine the clothing to design, choose the different design elements (style, color, fabric) provided by the computer, the computer will combine the selected elements design, and export to the 3 D fitting software for simulation fitting, and then the results of the above design, combination until the ideal effect is achieved. Using the interactive design system, users can see the dress effect of different styles of clothing through computer 3 D simulation [6].

2 Proposed Method

2.1 The Concept of an Interactive Genetic Algorithm

Interactive genetic algorithms, belonging to interactive evolutionary computation, are a subset of interactive evolutionary techniques. Interactive genetic algorithm can also be known as human-computer interaction evolutionary optimization algorithm, which is a way that optimizes the design process to evolve in the optimal direction [7]. In the process of interaction evolution design, the communication bridge between users and computers is established through interaction to realize the interference and guidance of the evolution process. Through the participation of users, the algorithm does not rely on the fitness function, but evolves according to the user's interaction evaluation, which well expands the application field of genetic algorithm. Interactive genetic algorithm will human perceptual factors into the process of evolution, interactive genetic algorithm user evaluation is fitness value calculation, through human-computer interaction interface, the user according to their own preferences, the computer presented evolutionary individual phenotype evaluation, interactive interface will score value feedback to the genetic algorithm. The coding mode of the interactive genetic algorithm includes binary coding, real number coding, orderly serial coding, tree structure coding, etc. The required coding can be set according to the specific problems. Compared with other coding schemes, binary coding scheme is clear, simple, intuitive and widely used, applicable to solve large-scale optimization problems. Therefore, the coding method adopted in this paper is binary coding, that is, a binary string composed of numbers "0" and "1", this string is referred to as individual, l value is called individual chain length, l individual is called individual space [8].

Suppose that the value of chain length l is set to 17 depending on the encoding length, e. g:

$$S_1 = \{0, 1\}^{17} = \{00000000000000001\} \tag{1}$$

2.2 The Rationale of Genetic Algorithms

The population of the genetic algorithm is composed of a number of encoded individuals, among which each individual is composed of a gene sequence of the same length (the individual is the chromosome), and the complex gene sequence is not suitable for application. In order to simplify the problem of real number coding, binary coding often appears in practical problems. The ranking of chromosome gene sequences can characterize its own value, namely, the fitness function value [9]. According to its own value and the guidance of survival of the fittest, through selection, crossover and variation, and gradually make the population evolve towards the direction of the target solution space.

2.3 Genetic Process

Considering a to-be-optimized problem for maximizing the value of the objective function, it can generally be described as the following optimization model:

$$\max F(X) \tag{2}$$

$$s.t. X \in \Omega \tag{3}$$

In, X represents the decision variable; X Ω is a feasible solution for X satisfies the constraints.

2.4 Interaction Design

From websites to mobile devices, from industry to art, interactive design is already everywhere. Interactive design can be considered as a screen experience-based design. Nowadays, the common devices around us are mobile phones, tablets, e-books, copier panels, TV sets and various other devices that require people to input instructions to complete the task. The human-computer interaction model to the clothing interactive design system, the user is clothing design demand consumer or clothing designer, through input perceptual words to express the design module search and combination according to the requirements of the system, then present the design results to the user, the design process end [10], if satisfied, the user will feedback the design results, propose detailed requirements, and then continue the above cycle until the user is satisfied. The goal is to help users achieve the demand for personalized clothing, and to complete the initial part of the garment industry chain, namely, the garment design and the partial connection with the production.

3 Experiments

3.1 Experimental Content

Using the theory of interactive genetic algorithm and interactive design, in the clothing interactive design system, it can be accurately located, analyze the clothing design module, evaluate the clothing from the perspective of sensibility, collect and classify the perceptual words, and establishing the decision model can complete the functions of user interaction, user preference model acquisition and clothing recommendation.

3.2 Experimental Methods

The paper used literature research methods, survey methods, and interview methods. The literature research method is used as the basis to collect relevant theoretical knowledge. Fully collect relevant literature from domestic and foreign databases, and read the knowledge and discussion related to clothing interaction design. Including "genetic algorithm", "interactive design" and other basic concepts, related topics include clothing design element analysis, product family and database construction, batch customization, perceptual design, interactive design, molecularity, weight analysis method, clothing model, system design, genetic algorithm, etc. Using the survey method, 100 valid questionnaires were obtained, and the data were processed after interviews with 24 experts.

4 Discussion

4.1 Analysis of Clothing Style and User Emotional Needs

By investigating the configuration and weight of the user's perceptual needs, the garment part module, and the fuzzy relationship model between the two, and applying them to the system database, the interactive design system can be obtained. According to the idea of arrangement and combination, the various style characteristics in the module will be combined into a large number of clothing design results, which promotes the formation of clothing product families. The survey method of the paper combines the interview method and the questionnaire method. During the face-to-face interviews with the experts, the questionnaires (100 valid questionnaires) allowed the experts to score the displayed profile under six perceptual dimensions. The control variable method was used in the questionnaire. When examining certain style elements, other style elements remain unchanged. The variable style characteristics under the style elements were scored for each pair of perceptual words, and the corresponding relationship between variable style characteristics and perceptual words was obtained.Different consumers have different perceptual images of specific clothing, which is limited by various factors, such as consumer cultural background, growing regional culture, and understanding of fashion trend. Therefore, if you want to get comprehensive related data, a large number of samples need to be collected, which will inevitably lead to large differences in data results. However, experts and scholars with professional knowledge of clothing have more objective and accurate cognition of clothing. Therefore, the subjects of this article are experts with related knowledge of clothing. From clothing undergraduates, postgraduates, doctoral students, teachers and other groups, 24 experts were selected for survey.

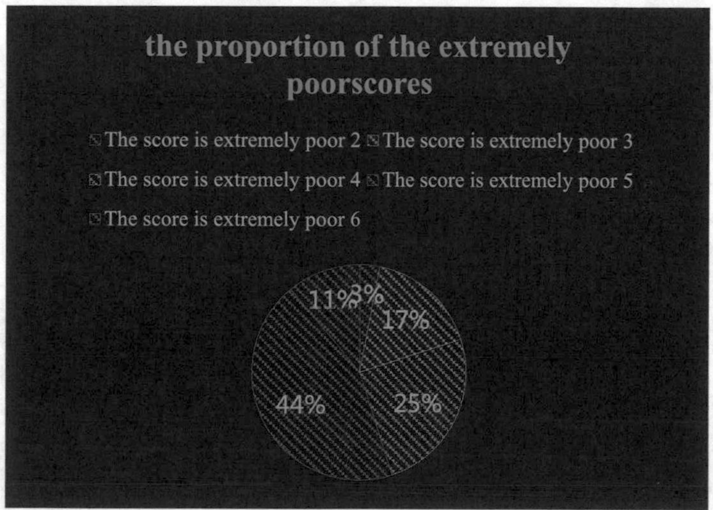

Fig. 1. The proportion of the extremely poor scores

After counting the original data, as shown in Fig. 1, the maximum extreme difference of 6 of 36; the maximum extreme difference of 5 was 44%, and 2 and 3 were only 20%. It can be seen that nearly half of the profiles and perceptual words are quite different from their subjective scores, and the scores are relatively scattered, so the correlation between these profiles and the corresponding dimension of the perceptual images is not particularly close, and the other half of the profiles give people more concentrated perceptual images. However, the widely different data does not mean that there is nothing to follow, because the vast majority of the 23 data in each group are still focused on several evaluation levels, so the data is p reprocessed to leave most of the evaluation results and eliminate the opinions of a few people, so as to get the more concentrated data.

4.2 Realize the Systematic Decision of Personalized Clothing Style

The design goal of the recommendation system is to recommend the best product according to customer needs, so the experiment mainly focuses on the recommended effect. The system starts with the user interaction module. The user first chooses the style of clothes he needs, and then the system presents the style of clothes in the clothing database to the user in a random order. Users began to score the clothes, how many clothes were determined by the user, the users scored one of the clothes, and recorded the chromosome coding and user scores. The system then recommends clothes for users according to the preference model, recommends clothes from all the selected styles of clothes in the database and scores them by the user. The system also requires users to score each dress, in order to count the user satisfaction, and users can also view all the attributes of the clothes. If the user chooses satisfied, the current preference model is saved into the user database. After the whole recommendation process, if the user chooses not satisfied, the system will update the preference model and interact again. The system recommended clothes for him again, and the recommended clothes for the user gave an average score of 82 points, and the user was basically satisfied. Famous users are tested separately below, and users first have a single interaction without updating the preference model. The results are shown in Fig. 2:

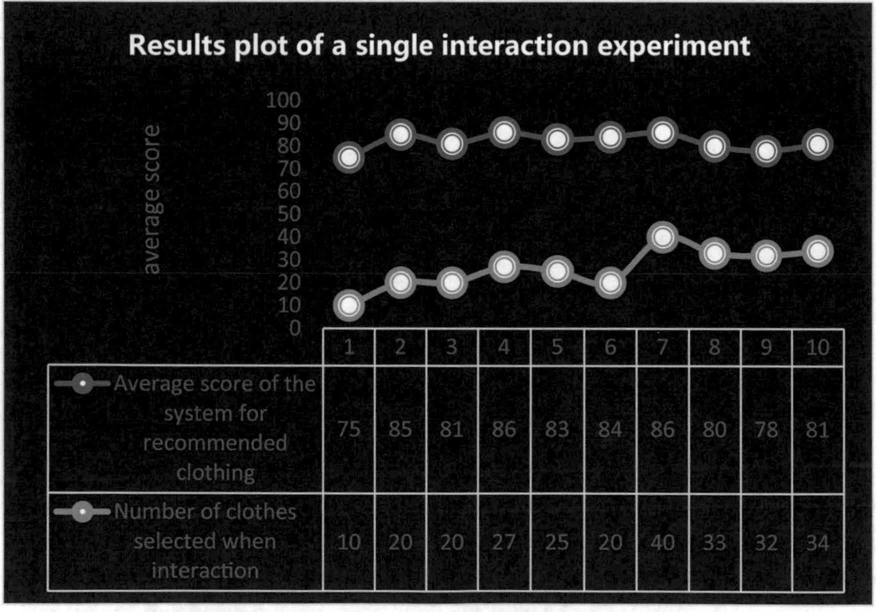

Fig. 2. Results plot of a single interaction experiment

The results show that the more preferences (selected style) in the system, the higher the score, the more information about the user, the greater the initial total group, the user can interact with the interaction system in the preference model, the experimental results are shown in Table 1:

Table 1. Results of testing for 5 users

User number	The number of interactions	Are you satisfied in the end	The average score for the system's recommended clothing
1	4	Unsatisfied	76.36
2	3	satisfied	90.54
3	1	satisfied	86.36
4	2	satisfied	82.4
5	4	satisfied	88.3

5 Conclusions

With the continuous prosperous development of the clothing industry and the more and more extensive application of intelligent recommendation technology, the application of intelligent recommendation technology to the clothing industry is another direction for the combination of the clothing industry and computer, and also an important direction to promote further breakthrough in the clothing industry. There are many intelligent algorithms in mathematics. After different algorithms are improved, they can be applied to clothing design. From different design angles, different algorithms can get different desired effects. More mathematical algorithms are being developed and applied in many fields of art design, and the beauty contained in the art of mathematics is constantly being explored and explored. Nowadays, with the development of the economy, the market competition among clothing enterprises is becoming more and fiercer. In order to make the enterprise in a dominant position, the traditional one-to-many clothing design method has fallen backward, and the contradiction between people's growing material and cultural needs is becoming more and more deepening. Therefore, genetic algorithms promote the integration of design theories in different fields during the interaction, providing scientific and effective technical support for the personalized development of enterprises.

Acknowledgements. Project Name: Construction of education mode of fashion performance specialty from the perspective of "three complete education". 2021 Jilin higher education scientific research project (JGJX2021D633).

References

1. Bartkowski, P., Zalewski, R., Chodkiewicz, P.: Parameter identification of Bouc-Wen model for vacuum packed particles based on genetic algorithm. Archives of Civil and Mechanical Eng. **19**(2), 322–333 (2019)

2. Santosa, S., Pramunendar, R.A., Prabowo, D.P., et al.: Wood types classification using back-propagation neural network based on genetic algorithm with gray level co-occurrence matrix for features extraction. IAENG International Journal of Computer Science **46**(2PT.141–263), 149–155 (2019)
3. Gaier, A., Asteroth, A., Mouret, J.B.: Data-efficient design exploration through surrogate-assisted illumination. Evol. Comput. **26**(3), 381–410 (2018)
4. Kumar, K.V., Sait, A.N.: Modelling and optimisation of machining parameters for composite pipes using artificial neural network and genetic algorithm. International Journal on Interactive Design and Manufacturing (IJIDeM) **11**(2), 435–443 (2014)
5. Tüzün, B., Saripinar, E.: Molecular docking and 4D-QSAR model of methanone derivatives by electron conformational-genetic algorithm method. Journal of the Iranian Chemical Society **17**(5), 985–1000 (2019)
6. Piemonti, A.D., Babbar-Sebens, M., Mukhopadhyay, S., et al.: Interactive genetic algorithm for user-centered design of distributed conservation practices in a watershed: an examination of user preferences in objective space and user behavior. Water Resour. Res. **53**(5), 4303–4326 (2017)
7. Varghese, B., Raimond, K., Lovesum, J.: A novel approach for automatic remodularization of software systems using extended ant colony optimization algorithm. Information and Software Technology **114**(OCT.), 107–120 (2019)
8. Sasaki, M., Laamrani, A., Yamashiro, M., et al.: Portfolio optimization by fuzzy interactive genetic algorithm. Journal of Advanced Management Science **6**(3), 124–131 (2018)
9. Ramasamy, V., Sidharthan, R.K., Kannan, R., et al.: Optimal tuning of model predictive controller weights using genetic algorithm with interactive decision tree for industrial cement kiln process. Processes **7**(12), 938 (2019)
10. Benabbou, N., Leroy, C., Lust, T.: An interactive regret-based genetic algorithm for solving multi-objective combinatorial optimization problems. Proceedings of the AAAI Conference on Artificial Intelligence **34**(3), 2335–2342 (2020)

Application of Cloud Computing in Computer Network Security Storage

Xuehua Peng[1,2](✉)

[1] College of Marxism, Sichuan University, Chengdu, Sichuan, China
pengxuehua2022@126.com

[2] School of Marxism, Southwest Petroleum University, Chengdu, Sichuan, China

Abstract. Network virus, hacking and other threats have been undermining data storage security in computer network. Thanks to some advanced technologies like cloud computing, it is possible to build a more secure storage system in the network. Therefore, this paper elaborates on the design of a secure storage system based on cloud computing. Firstly, this paper introduces key technologies of computer network security storage in cloud computing, including identity authentication, backup data, data encryption and key management. Then, resorting to methods like theoretical derivation, case analysis and process deduction, it further discusses the application of cloud computing technology in computer network security storage, in aspects of cloud architecture design, cloud computing services, node management model, load balancing mechanism and encrypted upload. The paper aims at finding a way to guarantee the storage security in the network, so as to meet the needs of users.

Keywords: Cloud Computing Technology · Computer Network · Security Storage · Storage System Design

1 Introduction

With the wide application of cloud computing technology, tremendous changes have been brought to traditional forms and ways of computing in computer network systems, which effectively relieved the pressure of computer data storage, improved the efficiency of using network data resources, and made data storage more secure and more convenient. Yet, since user resources are all stored in the cloud in the cloud computing environment, issues like user data security, privacy protection, remote storage of data, and stability, supervision and auditing of data have gradually become prominent [1].

Overseas, researches on the application of cloud computing technology in network storage started early, which endowed overseas countries and regions a solid research foundation, rich operation experience and a relatively mature business model. In the beginning, cloud computing is mainly developed and deployed by large companies like Microsoft, Google, Amazon, and IBM. At present, many developed and emerging information technology (IT) companies have taken cloud computing as an important branch in their business development. For example, Amazon is a well-known cloud

J. H. Abawajy et al. (Eds.): ICATCI 2022, LNDECT 170, pp. 393–401, 2023.
https://doi.org/10.1007/978-3-031-29097-8_47

service provider, and has issued its major cloud computing application—S3 (Amazon Simple Storage Service). S3 is mostly suitable for the storage of large files, which is able to store up to 5T of data. Furthermore, it provides relevant security mechanisms, where users can authorize who can get access to the data it stores.

In contrast, the development of cloud computing in China is relatively slow. So far, cities like Beijing, Shanghai, Shenzhen have become the first batch of cloud computing pilot cities, and many Chinese universities and research institutes have also stepped up their research on cloud computing. In March 2008, Tsinghua University became the first university in mainland China to participate in the Google cloud computing cooperation project. In July 2010, Beijing "Xiangyun Project" was officially launched. In August of the same year, "West Lake Cloud Computing" in Hangzhou, "Commercial Supercomputing Center" in Chengdu, also stated one after another. Alibaba Group officially launched the "Tianchi" platform in 2014. In year 2019 alone, the market size of China's cloud computing industry reached about 100 billion yuan. According to Gartner, an IT research and consulting company, from the perspective of investment, enterprises are steadily increase their investment in the cloud management service industry, from 18.59 billion US dollars in 2016 to 42.73 billion US dollars in 2020, an annual growth rate of about 20%.

At present, traditional encryption technology is widely used in cloud computing storage security, but simple application of only symmetric encryption algorithm or only asymmetric encryption algorithm no longer guarantees the security of data stored in the cloud. The former has a fast encryption speed, but not so safe, while the latter performs better in security but not in encryption efficiency. In this circumstance, according to the current situation of network storage security under the background of cloud computing, this paper proposes the application of cloud computing technology in computer network security storage, which provides technical support for the stable development of computer network. In general, this paper proposes a series of plans for cloud security storage system design like data encryption schemes that combine symmetric encryption algorithms and asymmetric encryption algorithms. The results show that the improved series of encryption methods based on cloud computing technologies is more efficient and highly secure, which can truly realize safe storage of data in the cloud computing environment.

2 Introduction on Cloud Computing

Cloud computing technology is a kind of Internet technology based on virtualization and distributed computing. Cloud computing platforms provide corresponding shared resources and services, mainly including software and hardware resources and various storage data, to fulfill various needs of different users. Cloud computing technology has many applications and functions. It employs data centers to store, process and share data. It enables access to computing resource pools through network and efficiently uses computing resources, stored data resources, application resources, IP addresses and other resources in the server, providing users with diverse types of services like computing and storage. There are three major categories of services involved in cloud computing, namely, Software as a Service (SaaS), Platform as a Service (PaaS) and Infrastructure as a Service (IaaS).

3 Key Technologies of Network Security Storage in Cloud Computing

3.1 Authentication Technology

Compared with traditional authentication methods, cloud computing authentication is undoubtedly faster, more effective and more secure. Yet at the same time, it has a relatively more complex and changeable resource structure. Main authentication technologies used in cloud computing are APP data call, dynamic password authentication and ID authentication authorization. First of all, take APP resource technology as an example. When a user calls data on this application, Web Service program interface will be used to access the data, but the program can only access the data when a third user completes identification. Another example is dynamic password authentication method. With this method, data can only be accessed through passwords plus authentication on hardware devices in a dynamic manner. In general, those technologies make cloud computing authentication more secure.

3.2 Backup Technology

Data backup is the current best way to avoid data loss. Storage in cloud computing data center carries two major functions, namely, recovery technology and disaster backup, which bring more security to the data center. Hard disks in computers can store loads of information, but their storage capacity is still limited. If the data in the hard disks is lost, it is extremely difficult to recover. Cloud computing technology can back up and integrate the data in the computer, so that whenever data loss occurs, the backup data in the cloud is still intact. With backup technology, cloud computing can avoid irreversible data loss and business loss like what happens in traditional storage and thereby ensure more secure information storage.

3.3 Data Encryption Technology

Encryption technology is the top priority of computer network security. Encryption technologies can be intricate and of myriad types, but it comes down to two general forms: symmetric encryption and asymmetric encryption. The most commonly used kind of symmetric encryption is Data Encryption Standard (DES). Each kind of data encryption that falls under symmetric encryption has a paired password key to protect the data [2]. The method of using two keys to encrypt three times is called encryption-decryption-encryption algorithm (EDE) and its formulae is as follows:

Encryption $C = E(K1, D(K2, E(K1, P))$
Decryption $P = D(K1, E(K2, D(K1, C))$

Decryption, the second step, is not due to requirements of cryptology. This is only for users who use triple DES to decrypt the data encrypted by single DES with this algorithm, which is due to the following formulae:

C = E (K1, D (K2, E (K1, P)) = E (K1, P)
P = D (K1, E (K2, D (K1, C)) = D (K1, C) [3]

In contrast, asymmetric encryption algorithm is completely different, for its keys are composed of two parts: a public key and a private key. The two are interspersed with each other, making encryption algorithm more complicated, and thereby bringing stronger security to data transmission and storage [4].

3.4 Key Management Technology

In traditional key maintenance and management of the data center, it is the enterprise administrator who controls the access to key data. Key management technology based on cloud computing can achieve key sharing through cloud computing service, so that users can access the stored data through shared key. Since data is encrypted, the key naturally requires careful management and protection. Cloud computing achieves effective access to data through key sharing of the sealed file in the cloud computing server.

4 Application of Cloud Computing in Computer Network Security Storage

4.1 Design of Cloud Architecture

First of all, cloud architecture strictly abides by the principle of proceeding from reality and is designed on the demands of the information storage system. General architecture of the computer network security storage system is shown in Fig. 1. The cloud architecture design should adhere to user's needs in reality [5]. The cloud architecture uses application software to store and access data. In designing modules of "login information" and "registration information", user's login and registration information is necessary and communications should be under HTTPS protocol. Meanwhile, function modules like login registration data certificate generation and system operation should be established in the system. A system database should be built as well. Users use function modules to issue corresponding operation instructions and input login and registration information, upload local data files and so on in the system operation interface, so as to introduce data into the system database.

The function design of computer network security storage system is mainly based on cloud computing technology. The system's function not only includes login and registration of users, but also covers effective transmission of different data under HTTPS protocol. When corresponding data is transmitted, the Web receives, decodes and encrypts the data and stores the encrypted data [6]. In terms of login page, there should be different layout on the page tailored for different requirements according to specific login process. After users enter the corresponding account and password set, the basic user operations in this mode can be enabled. The implementation code of the login function is as follows: [7].

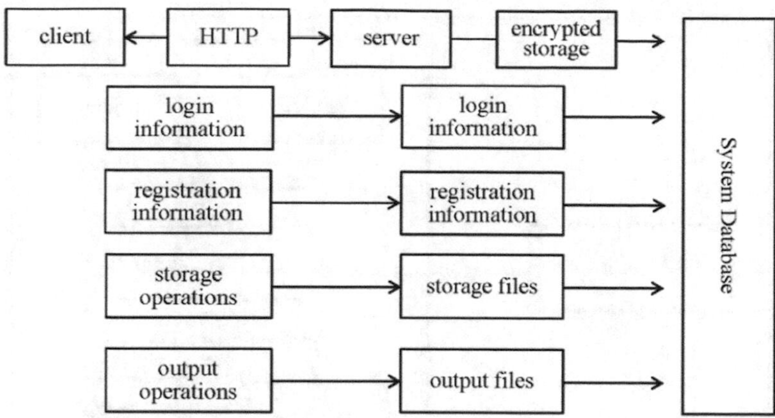

Fig. 1. General architecture of the computer network security storage system

```
String psnname=""; //User name
    if((String) session. Get Attribute ("validate-
code") !=null)// the system generating random string}
    if(validatedcode.equals(jym. To Lower Case ()))
    {message="wrong verification code";
    Char Encode.for ward (this.get Servlet Context( ),
Response,"/inc/tip.message="+Char Encode.gb2Un(i
message) + "&url=/");}
    else
```

Secondly, in order to endow the secure data storage system with the characteristics of cloud shrinking, a star topology configuration should be adopted and the system achieves dynamic allocation of operation goals based on the nodes in the cloud. User interface and the control center, the control center and nodes, and the nodes and the user interface will remain connected to each other, forming a closed loop system together. In this way, while operating, the control center of the system can send information such as processing request to each user, so that the control center and the client maintain effective communication. At the same time, the control center can also fully receive feedbacks from users.To sum up, Architecture of the cloud in computer network security storage system is shown in Fig. 2.

4.2 Design of Cloud Computing Service

Cloud computing service has high requirements on system security, because the cloud computing storage system has complete functions, extensive data sources, and highly valuable private user data, making it a target of external malicious attacks during operation [8]. Therefore, it is necessary to design the system and function modules of cloud service according to practical requirements. Relying on the superb data analysis and processing capabilities of cloud computing technology, the security risk assessment and

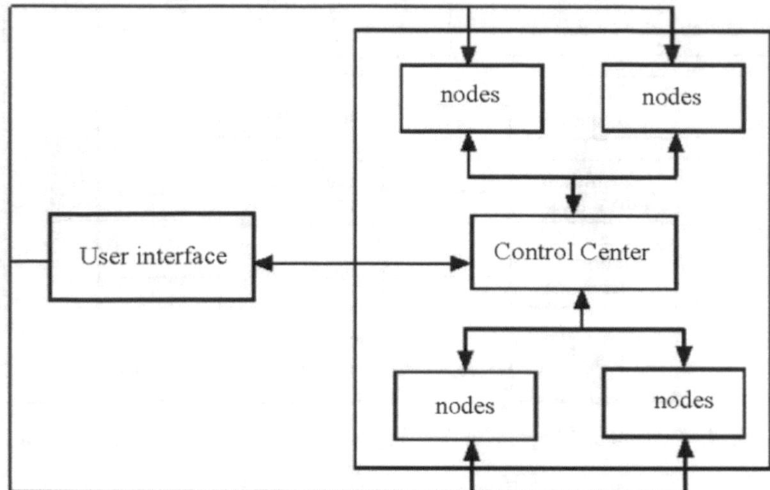

Fig. 2. Architecture of the cloud in computer network security storage system

attack inversion functions can be designed in the cloud service module. With such functions, the system can calculate the security risk coefficient of the stored data based on the statistical data information, predict potential network attacks, identify risk sources and replay the attack path when a user is attacked.

4.3 Design of Node Management Model

The node management model is also the key to continuous, stable and safe operation of the computer network security storage system. Structure of the node management appliance is shown in Fig. 3. The node management model includes: node allocation, initialization processing and many other aspects [9]. While the computer network security storage is in operation, the control center can conduct centralized and unified management of the data and resources in the computer network, and each node in the cloud can be effectively adjusted by using the corresponding algorithm. If there is a lack of stable running nodes in the cloud, it is difficult to process various requests, making it more difficult for the entire network system to run efficiently and stably. In order to solve this problem, in the design, the control center needs to be able to start a new node in time and quickly when there is a lack of running nodes [10]. Thanks to this design, the system can timely feedback various requests made by users, reasonably allocate the centralized power of nodes in the cloud, so as to better expand operation capacity of the cloud and improve the efficiency and quality of information and data processing of each node.

4.4 Design of Load Balancing Mechanism

In the operation process of computer network secure storage system based on cloud computing technology, there may be many obvious differences in the running status

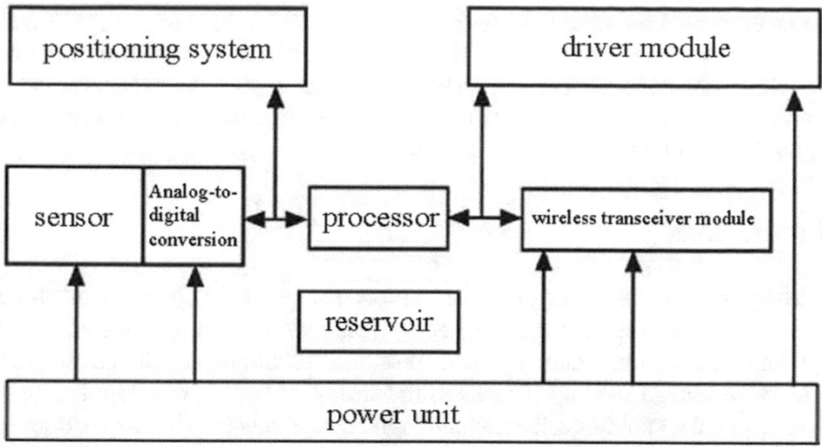

Fig. 3. Structure of the node management appliance

or performance of each node in the cloud. Therefore, it is necessary to establish an advanced and perfect load balancing mechanism to perform effective management of all nodes [11]. Nevertheless, the hardware performance of the computer network security storage system usually has a great impact on the stable operation of the node, which makes it inevitable to increase the operating load of the node and affect the operation efficiency of the whole system if there is too much information waiting to be processed by the same node. In order to solve this problem, the control center needs to process redundant information in time and allocate some data to other nodes to improve the overall efficiency of information processing.

The load of a node—NLi is determined by the estimated execution time of the assigned tasks on the node, and the total load TL is the average load of each node, namely:

$$TL = \frac{\sum_{i=1}^{n} NL_i}{n} \tag{1}$$

The total standard deviation BL is:

$$BL = \frac{\sqrt{\sum_{i=1}^{n} (NL_i - TL)^2}}{n} \tag{2}$$

The smaller the BL is, the better the load balancing performance will be.

4.5 Design of Encrypted File Upload

In traditional information storage systems, data is mainly transmitted in plaintext. When maliciously attacked by a third party, the traditional system in prone to suffer from data theft and tampering, which infringes the legitimate rights and interests of users, and even causes economic losses and negative social impacts. Therefore, it is imperative to design an encrypted file upload system in the cloud computing storage system. Based on practical needs of the system, the system can adopt different encryption technologies such as digital envelope encryption and PBE algorithm encryption to encrypt the file in transmission [12]. Then the recipient uses a password or corresponding key to decrypt the encrypted file to obtain the complete one.

5 Conclusions

In summary, although the emergence of computer network has brought much convenience to people's life and production, network security is still a prominent and serious issue, with viruses, hackers and system vulnerabilities continuously threatening to computer network storage security. Thanks to the steady and sound development of cloud computing, cloud computing related technologies can greatly help improve storage security. Against this backdrop, this paper discussed storage security technology in computer network based on cloud computing technology. Based on cloud computing, this paper proposed designs of a set of advanced secure storage system, which can better ensure the security of data storage in computer networks, and is worthy of wide application.

References

1. Mohan, A., Vamshikrishna, P.: Accounting and privacy preserving of data owner in cloud storage. Int. J. Innov. Technol. Explor. Eng. **10**(6), 14–17 (2021)
2. Aljumah, A., Ahanger, T.A.: Cyber security threats, challenges and defence mechanisms in cloud computing. IET Communications **14**(7), 1185–1191 (2020)
3. Stallings, W.: Cryptography and Network Security: Principles and Practice (5th Edition). Prentice Hall (2010)
4. Itani, W., Kayssi, A., Chehab, A.: Wireless Body Sensor Networks: Security, Privacy, and Energy Efficiency in the Era of Cloud Computing. Int. J. Reliable and Quality E – Healthcare **5**(2), 1–30 (2016)
5. Anakath, A.S., Kannadasan, R., Joseph, N.P., Boominathan, P., Sreekanth, G.R.: Insider attack detection using deep belief neural network in cloud computing. Computer Systems Science and Engineering **41**(2), 479–492 (2022)
6. Alimohammadi, K., Bayat, M., Javadi, H.H.S.: A secure key-aggregate authentication cryptosystem for data sharing in dynamic cloud storage. Multimedia Tools and Applications **79**(3), 2855–2872 (2020)
7. Reddy, G.N., Kumar, S.P.: RWWO: an effective strategy for workflow scheduling in cloud computing with predicted energy using Deep Maxout Network. Energy Harvesting and Systems **8**(2), 87–102 (2021)
8. Zulifqar, I., Anayat, S., Khara, I.: A review of data security challenges and their solutions in cloud computing. Int. J. Info. Eng. Electr. Bus. **13**(3), 30–38 (2021)
9. Chowdhury, M.: Time and energy-efficient hybrid job scheduling scheme for mobile cloud computing empowered wireless sensor networks. International Journal of Ad Hoc and Ubiquitous Computing **37**(1), 26–36 (2021)
10. Waziri, V., Adebayo, O., Danladi, H., Isah, A., Magaji, A., Abdullahi, M.: Network security in cloud computing with elliptic curve cryptography. Network and Communication Technologies **2**(2), 43 (2013)

11. Mishra, D., Kumar, V., Dharminder, D., Rana, S.: SFVCC: chaotic map-based security framework for vehicular cloud computing. IET Intel. Transport Syst. **14**(4), 241–249 (2020)
12. Zarezadeh, M., Mala, H., Ashouri-Talouki, M.: Multi-keyword ranked searchable encryption scheme with access control for cloud storage. Peer - to - Peer Networking and Applications **13**(6), 207–218 (2020)

Assessment Index System of Power Grid Projects Investment Risks Based on AHP-Improved Risk Matrix

Puji Yao[1], Weijun Wang[2], Liao Su[1], and Yicen Han[2](✉)

[1] Institute of Economics and Technology, State Grid Shaanxi Electric Power Company, Xi'an, Shaanxi, China

[2] Department of Economic Management, North China Electric Power University, Baoding, Hebei, China

ncepu_hyc@163.com

Abstract. In the context of the development of the energy Internet and the requirements of the power system reform, the pace of national unified power market construction accelerated, power grid investment also gradually began to change to a new investment model, power grid enterprises will face more complex investment risks. How to target the new situation of power grid investment management status quo, effective realization of power grid investment risk prevention and control and response, breakthrough power grid investment risk management constraints, become a comprehensive and in-depth promotion of power grid investment risk management is an important issue. This study fully combines the improved risk matrix with Delphi method and hierarchical analysis method, and establishes a power grid engineering investment risk index system consisting of 8 main indicators around power grid engineering economic risk, power grid engineering policy risk and power grid engineering environmental risk, quantifies and analyzes the probability of occurrence, influence, risk level, weight of each risk indicator and its acceptability, and then proposes countermeasures for power grid engineering investment risk mitigation, risk avoidance, risk acceptance and risk transfer under the new situation.

Keywords: Power Grid Projects Investment · Risk Matrix · Risk Identification · Risk Assessment

1 Introduction

Electricity construction project is an important industry closely related to China's national economy and people's livelihood, and it is also a public utility. Electricity has two parts: living materials and production materials, and sufficient, safe and stable supply of electricity is an important prerequisite for the healthy, stable and rapid development of our national economy. Power grid investment has a large capital scale, a long construction cycle, a wide social impact, high security and reliability requirements, energy security and national livelihood, etc., is the core key link of the power grid enterprises.

J. H. Abawajy et al. (Eds.): ICATCI 2022, LNDECT 170, pp. 402–411, 2023.
https://doi.org/10.1007/978-3-031-29097-8_48

Investment risk largely affects the power grid company's business development planning decisions, so risk analysis and risk management before investment decisions can make the power grid company's own level of competitiveness and management capabilities have greatly improved.

C.M. Tam et al. (2002), based on fuzzy mathematical theory, proposed the use of hierarchical analysis matrix in engineering projects to rank the importance of risk factors, get the risk points that need to be paid attention to in the project and analyze the corresponding measures [1]. Warszawski et al. (2008) collected a large amount of data by means of questionnaires and made a detailed study of risk management in transmission and substation projects, including risk perception, risk identification, cost management, etc., and derived corresponding risk counter measures [2]. Assili et al. (2008) analyzed the specific impact of investment on engineering projects in the power industry. Among them, the analysis was combined with the theory of engineering project risk management, and the impact of capital risk capacity on the effectiveness of the project was derived [3]. Boshan et al. (2011) studied the risk identification and control measures of electric power engineering projects, summarized the risk factors based on the theory of risk management, and proposed a risk response plan that maximizes risk avoidance [4]. Researchers have done a lot of research on power grid engineering investment risk identification, but less research has been done on quantitative assessment of power grid engineering investment risk. Therefore, Quantitatively assessing the possible risks in the investment of power grid enterprises has become an important issue that needs to be solved.

This study applies the improved risk matrix to structurally assess the risk of power grid engineering investment, in order to objectively and comprehensively identify the uncertainties and various potential risks affecting the achievement of power grid engineering investment goals, and effectively implement risk response measures to reduce the degree of risk to an acceptable level as far as possible, so as to provide a reference for the smooth development of China's power grid engineering investment and improve investment efficiency.

2 Research Methodology

2.1 Improved Risk Matrix

The improved risk matrix is based on the risk matrix analysis method to achieve the increase in the level of risk level three to five, the new risk acceptability assessment, and the new risk comprehensive evaluation to further enhance the scientific and operability of risk assessment and measurement. The main steps are as follows [5]: (1) calculating the risk probability and its ordinal value; (2) calculating the risk impact and its ordinal value; (3) determining the risk rank; (4) calculating the risk Broda value and ordinal value; (5) determining the risk acceptability.

The construction of power grid engineering investment risk assessment system will establish power grid engineering investment risk factors and weights, quantify and analyze the probability of occurrence, influence, risk level, weight and its acceptability of each risk indicator.

2.2 Delphi Method

In this study, a judging group of relevant professional experts in the field of grid investment risk was selected, and the grid investment risk indicators initially determined based on literature combing were revised after three rounds of anonymous expert correspondence or emails [6]. In the first round of survey, 20 questionnaires were distributed and 20 questionnaires were returned with a 100% return rate, 95% of which were valid; in the second and third rounds, 20 questionnaires were distributed with a 100% return rate, 90% of which were valid. The method of calculating the coordination coefficient of indicators was used to measure the degree of consistency of the experts' opinions on different indicators for testing.

2.3 Analytic Hierarchy Process

This study adopts the Analytic hierarchy process (AHP) to determine the weights of the grid investment risk index system. A panel of experts in the field of grid investment risk is invited to judge the importance of each index at each level by using the "1–9 scale method" through a questionnaire survey, and the judgment results are obtained to construct a two-comparison judgment matrix. At the criterion level, the survey respondents of economic benefit risk, social benefit risk, and environmental benefit risk made conclusions based on the relative importance of each evaluation index, and formed a comparison judgment matrix based on the results between the indexes. On the basis of constructing the judgment matrix of each evaluation index, the specific operation steps to determine the weight of each level of indicators are as follows [7]: (1) calculateing the value of relevant parameters; (2)doing the the CI consistency test of the judgment matrix, CI equal to 0 is the matrix with full consistency, CI < 0.1 is the matrix with basic consistency.

Based on the above calculation process, the CI of the primary evaluation index table is 0.0268 and the CI of the secondary evaluation index table is 0, 0.0550 and 0, respectively.

3 Grid Engineering Investment Risk Identification

The identified risks are classified according to the specific characteristics of grid construction projects of different project categories, deconstructed risks by objectives. Finally, according to the classification of investment risk to establish a grid construction project investment risk assessment index system, so as to make a comprehensive assessment of the index system and make a general assessment of the level of investment risk of the grid construction project [8].

3.1 Economic Risk of Power Grid Project

The economic risks of power grid project identified through the Delphi method are as follows:

(1) Operating Revenue Risk: Revenue from electricity sales is the main source of income for a power construction project investment, and small changes in the price of electricity can affect the return on investment of a power construction project. Increase or decreasing in demand from industrial users will have an impact on power grid infrastructure investment, increasing its infrastructure investment and operational risk.
(2) Cost overrun risk: The construction cycle of power grid construction projects is long, during construction, engineering raw materials are susceptible to price index fluctuations and changes in market supply and demand, resulting in increased procurement costs, while government industry policy adjustments, macro-control policy implementation, etc., may bring non-negligible impact on the cost of engineering projects.

3.2 Policy Risks of Power Grid Project

The policy risks of power grid project identified through the Delphi method are as follows:

(1) Power supply reliability risk: The reduction of average outage time after the implementation of the project does not meet expectations. Resulting in the initial investment target not being achieved. For high-voltage distribution networks (10 kV-220 kV); the reduction ratio of N-1 outage loss load can be equated to the reliability level improvement ratio. However, due to design and technical factors, the ratio does not meet expectations, which affects the safe and stable operation of the power system and causes investment risks.
(2) Grid security operation risk: The complexity of grid construction is high, and the voltage level of the grid is very high. After the grid construction is completed and enters the operation phase, there are many risk factors affecting the safe operation of the grid such as voltage fluctuation, current transfer, multiple DC landing, etc., which bring serious challenges to the grid security.
(3) Power quality risk: When the electricity from new energy projects is connected to the distribution network, it will bring certain risks to the power quality. New energy projects such as photovoltaic power supply need to convert DC power to AC power through inverters when they are connected to the grid, and this process is likely to produce harmonics, which will affect the power quality.
(4) Transmission power risk: After the completion of the project, due to the fierce competition in the power market, the transmission volume cannot reach the designed transmission volume and cannot meet the users' demand for electricity, and the economy of scale of the power industry is not reflected. The electricity price is reduced, resulting in difficulties in debt repayment.

3.3 Environment Risks of Power Grid Project

The environment risks of power grid project identified through the Delphi method are as follows:

(1) Clean energy consumption risk: The construction of the power grid project, does not meet the expected new energy grid-connected power, it will increase the use of primary fossil energy, increase the emission of carbon dioxide and other pollutants, resulting in investment risk.
(2) Electricity replacement power risk: The construction of power grid projects should meet the electricity demand of electric energy replacement, and reduce the consumption of fossil energy such as coal through electric energy replacement, thus reducing the emission of carbon dioxide, sulfur dioxide and other polluting gases and promoting emission reduction.

4 Risk Assessment System of Power Grid Engineering Investment Based on AHP-Improved Risk Matrix

Risk assessment is an important prerequisite for taking reasonable risk response measures, by comprehensively considering the source of risk, the possibility of occurrence and the severity of consequences with the risk level and its acceptability for comprehensive assessment. In the power grid engineering investment risk prevention, in order to improve the effectiveness of risk response measures, clarifying the risk level of each risk source is an important premise.

4.1 Power Grid Engineering Investment Risk Probability and Its Sequential Value

First should determine the likelihood P_i of risk occurrence in the mathematical expression of the risk matrix, for any risk indicators, combined with past grid engineering big data to calculate the probability of occurrence of the risk, the specific calculation formula is:

$$P_i = \frac{A_{Ri}}{A_i} \tag{1}$$

where, P_i is the approximate probability of occurrence of risk i in power grid investment projects; A_i is the total number of projects with possible risks obtained from the big data of past power grid projects; A_{Ri} is the number of projects with actual risk i among all projects with possible risk i. This paper uses the proportion of occurrence of risk based on big data to approximate the probability of actual occurrence of risk i in power grid investment projects. According to the 5 * 5 order matrix proposed by ISO 31000, the probability of grid investment risk is defined as five levels [9], and the corresponding values for each level are shown in the Table 1.

Table 1. Probability definition and Li values of the possible occurrence of grid investment risks

Serial number	Investment risk probability level	Probability intervals (%)	Li
1	Very unlikely probability P1	$0 \le P_i \le 10$	1
2	Unlikely probability P2	$10 < P_i \le 40$	2
3	Occasional probability P3	$40 < P_i \le 60$	3
4	Probable probability P4	$60 < P_i \le 90$	4
5	Extremely likely probability P5	$90 < P_i \le 100$	5

4.2 Risk Weights of Power Grid Engineering Investment

Calculating the risk factor weights of power grid engineering investment is more conducive to clarifying the risk level and the comprehensive level of risk. In this study, the analytic hierarchy process is used to determine the weights of the grid investment risk index system at each level and the total weights, and for different types of grid projects, the impact weights of investment risks are calculated as follows:

$$W_{Ri} = w_n \times w_{ni} \tag{2}$$

where W_{Ri} is the weight of risk i on the investment impact; w_n is the weight of the indicator at the target level to which risk i belongs; wni is the weight of risk i in its corresponding indicator at level n.

4.3 Power Grid Engineering Investment Risk Impact and Its Ordinal Value

For power grid investment risk assessment, the correspondence between the severity of risk index i for investment impact S_{Ri} and W_{Ri} is shown in the Table 2.

Table 2. Correspondence between the severity of investment impact s_{Ri} and W_{Ri}

Serial number	Investment impact severity level	The range of values of W_{Ri}	S_{Ri}
1	Negligible S1	$0 \le W_{Ri} \le 1/32$	1
2	Minor S2	$1/32 < W_{Ri} \le 1/16$	2
3	General S3	$1/16 < W_{Ri} \le 1/8$	3
4	Serious S4	$1/8 < W_{Ri} \le 1/4$	4
5	Critical S5	$1/4 < W_{Ri} \le 1$	5

The summary table of impact severity evaluation of investment risk indicators of power grid projects is shown in Table 3.

Table 3. Summary table of impact severity evaluation of investment risk indicators

Tier 1 indicators	W_n	Tier 2 indicators	W_{ni}	W_{Ri}	S_{Ri}
Economic risk A1	0.49	Operating Revenue Risk (A11)	0.67	0.33	5
		Cost overrun risk (A12)	0.33	0.16	4
Policy risks A2	0.31	Power supply reliability risk (A21)	0.30	0.09	3
		Grid security operation risk (A22)	0.12	0.04	2
		Power quality risk (A23)	0.17	0.05	2
		Transmission power risk (A24)	0.41	0.13	4
Environment risk A3	0.20	Clean energy consumption risk (A31)	0.33	0.07	3
		Electricity replacement power risk (A32)	0.67	0.13	4

4.4 Grid Engineering Investment Risk Level

The risk level of grid engineering investment is determined according to the product of the probability of occurrence of grid investment risk and the impact weight of grid risk, i.e., the formula for calculating the importance I_{Ri} of risk i is:

$$I_{Ri} = L_i \times S_{Ri} \tag{3}$$

The risk level is divided into 5 levels according to the value of the risk i significance IRi, as shown in the Table 4.

Table 4. Risk classification of power grid investment

Serial number	Investment impact severity level	The range of values of W_{Ri}
1	Significant Risk	$I_{Ri} \geq 21$
2	Greater risk	$16 \leq I_{Ri} \leq 20$
3	General Risk	$11 \leq I_{Ri} \leq 15$
4	Low Risk	$16 \leq I_{Ri} \leq 10$
5	Negligible risk	$I_{Ri} \leq 5$

4.5 Power Grid Engineering Investment Risk Borda Values and Ordinal Values

Based on the results of the likelihood and severity of consequences of the investment risk of power grid engineering, the value of Borda value and ordinal value of the investment risk of power grid engineering are calculated. The risk Borda value is an important indicator for assessing the risk level, and its calculation formula is:

$$B_i = (N - S_{Ri}) + (N - L_i) \tag{4}$$

where, N is the total number of risks, i is a particular risk, and k denotes a criterion. The Borda value is a ranking of the importance of the risks in terms of their impact and probability of occurrence. A Borda ordinal value bi of 0 indicates that the risk is the most important, bi of 1 indicates that there is another risk that is more important, and so on [10].

4.6 Acceptability of Investment Risk for Power Grid Projects

The acceptability of risk is an important basis for grid enterprises to take risk response measures. According to the level of acceptability of investment risk for grid enterprises, the risk acceptability is divided into five: negligible, more Acceptable, acceptable, undesirable and unacceptable, as shown in the Table 5.

Table 5. Summary table of investment risk assessment for power grid projects

Risk	S_{Ri}	L_i	Risk level	B_i	b_i	Risk acceptability	W_{Ri}
A11	5	5	Significant risk	6	0	Unacceptable	0.33
A12	4	4	Greater risk	8	1	Undesirable	0.16
A21	3	1	Negligible risk	12	6	Negligible	0.09
A22	2	2	Negligible risk	12	6	Negligible	0.04
A23	2	3	Low risk	11	4	More acceptable	0.05
A24	4	3	General risk	9	4	Acceptable	0.13
A31	3	5	General risk	8	1	Acceptable	0.07
A32	4	3	General risk	9	3	Acceptable	0.13

4.7 Selection of Investment Risk Response Measures for Power Grid Projects

Risk response is generally divided into four ways: risk reduction, risk transfer, risk avoidance and risk acceptance. The construction of two-dimensional moments of risk response (likelihood and severity of consequences) can provide a reference for the selection of reasonable risk response measures, and the mapping of investment risk response for power grid projects is specified in Fig. 1.

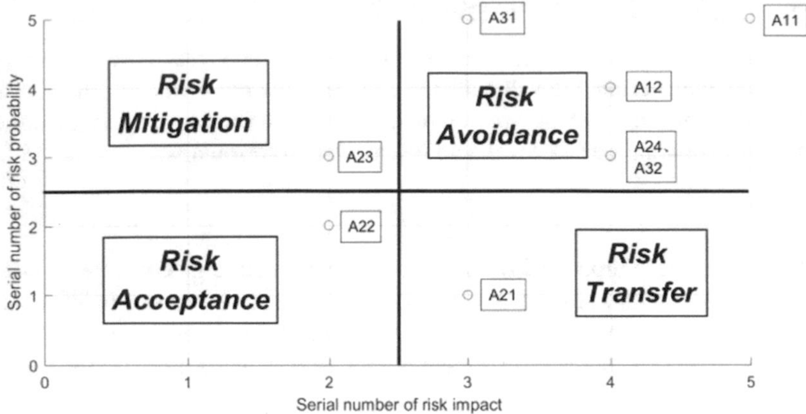

Fig. 1. Mapping of investment risk response for power grid projects

5 Conclusions

This study applies the AHP-improved risk matrix and to engineering project risk evaluation, constructs a power grid engineering investment risk index system, while adding risk acceptability assessment and new comprehensive risk evaluation, which has important guiding significance to improve the scientific and operability of power grid engineering investment risk assessment and measurement. This method is procedural, simple and easy to use, suitable for promotion in the power grid enterprises, as a powerful tool to assist investment decisions. In practice, care should be taken to develop more specific risk response plans for specific power grid projects based on the technical, financial and management levels of the power grid enterprises.

Acknowledgement. This research is supported by the project of State Grid Shaanxi Electric Power Company Institute of Economics and Technology (project number 5226JY200006).

References

1. Tam, C., Tong, T., Chiu, G., Fung, I.: Non-structural fuzzy decision support system for evaluation of construction safety management system. Int. J. Project Manage. **20**(4), 303–313 (2002)
2. Warszawski, A., Sacks, R.: Practical multifactor approach to evaluating risk of investment in engineering projects. J. Constr. Eng. Manag. **130**(3), 357–367 (2004)
3. Assili, M., HJDB, M., Ghazi, R.:An improved mechanism for capacity payment based on system dynamics modeling for investment planning in competitive electricity environment. Energy Policy **36**(10), 3703–3713 (2008)
4. Boshan, C., Rong, G.: Risk identification and control measures of construction project. Value Engineering (2011)
5. Murray, S., Grantham, K., Damle, S.: Development of a generic risk matrix to manage project risks. J. Indus. Sys. Eng. **5**(1), 35–51 (2011)

6. Kabir, G., Sumi, R.: Integrating fuzzy delphi method with artificial neural network for demand forecasting of power engineering company. Management Science Letters **2**(5), 1491–1504 (2012)
7. Mojavera, P., Khalilarya, S., Chitsaz, A., Assadi, M.: Multi-objective optimization of a power generation system based SOFC using Taguchi/AHP/TOPSIS triple method. Sustainable Energy Technologies and Assessments **38**(C), 100674 (2020)
8. Pennock, M., Haimes, Y.: Principles and guidelines for project risk management. Syst. Eng. **5**(2), 89–108 (2002)
9. Qazi, A., Shamayleh, A., El-Sayegh, S., Formaneck, S.: Prioritizing risks in sustainable construction projects using a risk matrix-based monte carlo simulation approach. Sustain. Cities Soc. **65**, 102576 (2021)
10. Koulinas, G., Xanthopoulos, A., Konstantinos, A., Dimitrios, E.: Risks ranking in a desalination plant construction project with a hybrid AHP, Risk Matrix, and simulation-based approach. Water Resour. Manage **35**, 3221–3233 (2021)

Intelligent Mental Pension System on Account of BP Nerve Network Algorithm

Xiuli Han[1(✉)] and Amar Jain[2]

[1] The Second Affiliated Hospital of Qiqihar Medical University, Qiqihar 161000, Heilongjiang, China
hxl18646632003@163.com

[2] Madhyanchal Professional University, Ratibad, India

Abstract. Artificial nerve network (Ann) algorithm has been developing rapidly for many years and has made great contributions to solving real model problems. BP nerve network algorithm has made great breakthrough in many economic fields and made great contribution to the design of intelligent mental pension system. Under the background of BP nerve network algorithm, it is a huge deal to strengthen the design of intelligent mental endowment system. The design of the intelligent mental endowment system is to solve the problem that the information of the decentralized management system of the traditional mental nursing home is isolated and the mental endowment service facilities cannot meet the needs of real-time nursing. The system unites the superiority of RFID technology and video linkage monitoring to monitor the condition of the elderly in real time, and children can know their parents' health status at any time. This article studies the concept and related knowledge on account of BP nerve network algorithm, explains a series of viewpoints and theories of intelligent mental pension system design on account of BP nerve network algorithm. In this text, the data test, on account of BP nerve network algorithm of intelligent mental pension system design research, test results show that the design of intelligent mental pension system on account of BP nerve network algorithm in system management efficiency, avoiding target deviation rate, model parameter optimization rate and adaptability reached 83.30%, 90.14%, etc. 92.85% and 98.28% .

Keywords: BP Nerve Network Algorithm · Intelligent Spirit · Mental Endowment · System Design

1 Introduction

Intelligent mental endowment is a widely known concept, involving the continuous development of the algorithm, BP nerve network algorithm processing a lot of content items, a lot of equipment, large intelligent systems, iot devices for interconnection. This system has made an impact on the process of supporting the elderly. Both the pension staff and the pension staff can make full use of this system, the use of many feature of the system, so that the pension problem of many abuses completely solved. On the view of algorithm, the importance of algorithm to the design of intelligent mental

J. H. Abawajy et al. (Eds.): ICATCI 2022, LNDECT 170, pp. 412–419, 2023.
https://doi.org/10.1007/978-3-031-29097-8_49

endowment system is highlighted. The design of intelligent mental endowment system on account of BP nerve network algorithm benefits from the superiority of the algorithm, provides solutions for the design of intelligent mental endowment system, especially makes an impact on the management and allocation of various resources. In the aspect of communication and learning of intelligent mental endowment system design on account of BP nerve network algorithm, it is necessary to strengthen the ability of intelligent mental endowment system design and application, and solve many problems in the realm of mental endowment.

Domestic and foreign scholar have studied the algorithm on account of BP nerve network. In foreign studies, Kuang Y Etc. Put forward a macroeconomic prediction model on account of modified multimedia assisted BP nerve network model and Ant colony optimization algorithm (ACO). Traditional methods should be enhanced by more research into the most advanced technologies. ACO was used to modify the BP model to make it more suitable for maintaining the characteristics of prediction probability [1]. Lv H. put forward a new method of appended momentum-elastic gradient descent. BP nerve network has better adaptability to learning speed. The algorithm is elevated to solve the problems of the traditional BP nerve network, for instance the selection of learning step, the size and direction of weight, and the learning rate is not easy to control. Experimental results show that the algorithm can predict martial arts competition routines and formulate scientific strategies while improving network scale and running time [2]. A scholar put forward an elevated SOC reckon tactics for lithium ion batteries on account of BP nerve network. Principal component analysis (PCA) and particle swarm optimization (PSO) are used to improve the accuracy and robustness of the algorithm. Principal component analysis (PCA) was used to select the most significant input features. PSO algorithm is used to determine the optimal value and learning rate of hidden layer neurons, because these parameters are the most critical factors to build the optimal BPNN model. The put forward model was tested and evaluated [3]. However, the design of intelligent mental pension system on account of BP nerve network algorithm is still in the initial stage, and there is still a certain gap compared with foreign systems.

To further improve the algorithm on account of BP nerve network in China, we must start from the following points: First, deepen the research on the algorithm system on account of BP nerve network; Secondly, the data model on account of BP nerve network algorithm is optimized. Finally, strengthen the communication and swap of foreign advanced technology, improve the technical level.

2 Exploration of Intelligent Mental Pension System on Account of BP Nerve Network Algorithm

2.1 On Account of BP Nerve Network Algorithm

BP nerve network algorithm is a classical algorithm, which solves many data models in application [4, 5]. BP network algorithm has been elevated and adjusted for many times [6, 7]. The algorithm deals with many practical problems on account of the characteristics of gradient information and can achieve very satisfactory results on many problems [8, 9].

As a branch of artificial intelligence network, BP nerve network is a scientific and popular algorithm in the realm of science. It can solve professional mathematical model problems and conduct data processing and model establishment for nonlinear problems. The algorithm includes other algorithms, for instance classification recognition, regression, compression, approximation and other related algorithms [10, 11]. This algorithm has been used and adopted in most fields of society, and has excellent fitness in various industries [12, 13]. The algorithm has strong application ability, simple model, easy to handle, and is very practical to deal with linear problems. The main theory of BP algorithm is the fastest descent algorithm on account of gradient. This model has the following defects, which must be clear: first, BP nerve network is prone to local minimum problem during data model sample training, so it is impossible to obtain the best solution set; Second, the convergence rate of the data model is low [14, 15].

(1) BP nerve network structure

BP nerve network consists of sevepochl important parts: input layer, hidden layer and output layer. Each part layer communicates and connects with each other, and is a single layer structure, so neurons can be set by themselves [14]. The hidden layer can be more than just a structure. The topology of a BP nerve network containing only a single hidden layer is shown in Fig. 1.

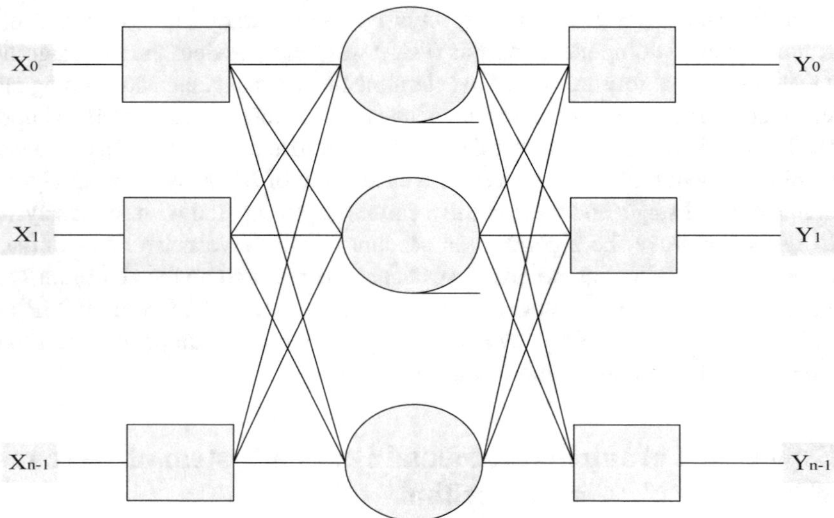

Fig. 1. Topology of BP nerve network

2.2 Design of Intelligent Mental Endowment System

The intelligent mental endowment system is a software system built on the network. Many scholars have defined the intelligent mental pension system. Some people think that "intelligent mental pension system is a collection of many independent computers, system users think the system is a big computer". In hardware, each computer runs independently, and physical and coherent resources in different places swap information through computer network technology. In terms of software, users think the whole system is a computer, and there is a dispersed executive system that manages computer resources by and large. You have to have both to call it a dispersed system. It is precisely because of the characteristics of intelligent mental pension system that it has a high degree of gathering strength and transparency.

The continuous deepening and intelligence of the pension system improves the pension level and means to the greatest extent, which is a huge innovation in the pension field. In the pension field, many high and new technologies, as well as computer network technology, are especially applied. Using these network technologies, the pension system can be upgraded to the intelligent field in the epoch of big data, which is a great progress in the pension field.

The feature of the five platforms of the system are mainly developed in establishing RFID positioning system, RFID middleware data fusion model, RFID data fault-tolepochnt control for service request, RFID-oriented video linkage system, information management system and special home system.

(1) Wear RFID wristband tags. To the pension personnel personal data identification, to all foster personnel issued wristband labels, and in the marking of old-age personnel information.
(2) Pension positioning and search. Pension workers should be taken care of in terms of safety to avoid accidents and problems, so special satellite tracking system should be set up for the positioning of personnel. In terms of management, relevant rules and regulations should be elevated to strengthen the life and entertainment management of old-age workers. Old-age workers can read and write information in the rest area, and send the reading and writing information to the intelligent system, and distribute the information to various supervisory departments and staff. After reading the information, the staff can timely know the position of the elderly and how to assist the elderly, and provide help for life, entertainment and daily living. The purpose of the intelligent system is to accurately track and manage the position of the elderly, and to deal with the elderly professionally.
(3) Emergency intelligent system for the elderly. This system is mainly to solve the elderly in the emergency to provide help, the staff according to the emergency intelligent system timely judge the location of the foster personnel, emergency dispatch related staff to the site. The elderly can easily and quickly press the intelligent system button when they encounter a critical situation. Emergency events can be quickly transmitted to the intelligent system center through the button. The staff can handle and manage the elderly according to the signal.
(4) Video system. The intelligent system will conduct video tracking for all positioned old-age workers. According to the video, it can accurately know when and where the old-age worker is, and what is found, and the old-age worker can be one-click informed.
(5) Information management system. Once the elderly are admitted to the nursing home, they begin to input their basic identity information and history.

Routine data for instance cases are stored daily update data until the elderly leave the nursing home.

(6) Expert system health management, SMS reminder. Every day, the elderly's physical information will be uploaded to the server. Doctors predict the physical condition of the elderly on account of their historical records and recent new physical data. The prediction results are sent to the relevant staff and children, so that the children do not have to worry about the elderly.

3 Research on the Design Effect of Intelligent Mental Endowment System on Account of BP Nerve Network Algorithm

3.1 Comparative Analysis

In this text, comparative analysis is used to test the effect of the design of the intelligent mental endowment system on account of BP nerve network algorithm, and the data model is analyzed according to the comparison of the actual effect. The model of empirical research is to carry out process data test according to the solution in the paper through the design of intelligent mental endowment system on account of BP nerve network algorithm, and to record and analyze the data of each test point. The method adopted this time is on account of BP nerve network algorithm and non-algorithm two groups of data processing. Stain all sorted data, add and remove data with relevant algorithms.

3.2 Overview

In this text, comparative analysis is used to design the intelligent mental endowment system on account of BP nerve network algorithm, and the data are recorded and processed. According to the complete process of the intelligent mental endowment system design effect analysis.

3.3 Design Formula

$$k = \frac{POS_C(D)}{|G|} \tag{1}$$

$$neg_R(X) = U - \overline{R}(X) \tag{2}$$

K refers to the degree of dependence of intelligent mental endowment on the system, D refers to the system input factor, G refers to the basic parameters of the system, U refers to the complete set, $\overline{R}(X)$ is the largest set in the set, and $neg_R(X)$ is the result set.

3.4 Calculation Principles

The influence factors of the design of the intelligent mental endowment system on account of BP nerve network algorithm are analyzed, the effect of the design of the intelligent mental endowment system on account of BP nerve network algorithm is scientifically decomposed, and the parameters of the design of the intelligent mental endowment system are processed and set.

4 Investigation and Research Analysis of Intelligent Mental Endowment System Design on Account of BP Nerve Network Algorithm

4.1 Test Effect

Test object can be divided into algorithm on account of BP nerve network intelligent spirit pension system design and the algorithm of intelligent spirit endowment system design effect comparison and analysis, respectively on two groups of testing data, compiled on account of the advantage of BP nerve network algorithm, on account of the BP nerve network algorithm of intelligent spirit endowment system design to effect the relevant data records: In the comparative analysis method, the design effect of the intelligent mental endowment system on account of BP nerve network algorithm can be on account of the design of the intelligent mental endowment system to improve the goal, the results show that the design effect of the intelligent mental endowment system on account of BP nerve network algorithm has a great advantage. The key problems of strengthening the design of intelligent mental endowment system on account of BP nerve network algorithm have been solved. The results are shown in Table 1 and Fig. 2.

Table 1. Mental endowment system data table

	Intelligent	Response performance	Integrated performance	Parameter tuning performance
Frequency	175	321	401	587
Ratio	83.30%	90.14%	92.85%	98.28%
Position	91	94	95	99

The design effect of the intelligent mental pension system on account of BP nerve network algorithm is better than that of the traditional way. In the system management efficiency, avoiding target deviation rate, model parameter optimization rate and adaptability to achieve high efficiency in the same field, in the model effect data processing and application of testing and analysis, intelligent mental pension system design performance in many aspects can be processed and analyzed. It focuses on all kinds of abilities required in the modeling process, including construction, mathematical, modeling, model verification and evaluation. The horizontal orientation focuses on the classification of modeling ability from the level and law of human cognitive development. The increase of research and development of the design and improvement of intelligent mental endowment system is a huge deal to the field of mental endowment.

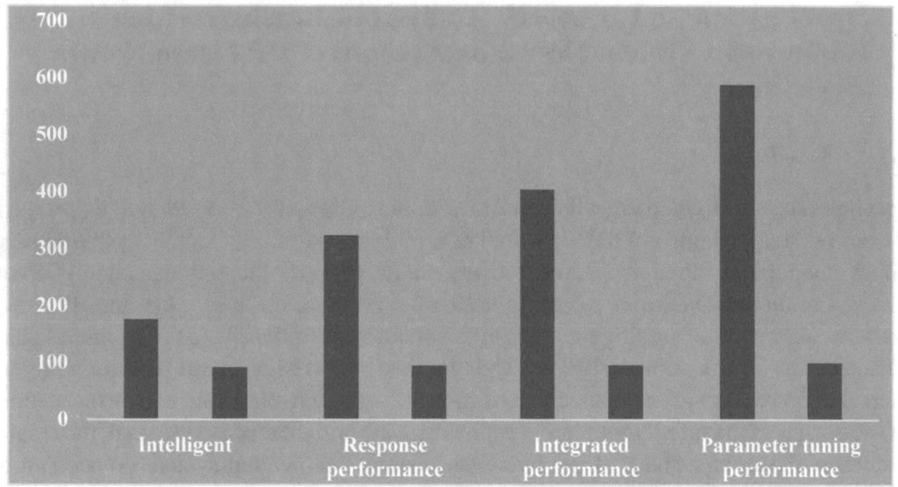

Fig. 2. Data distribution map of mental endowment system

5 Conclusions

Because of the self-organization and self-learning characteristics of nerve network, data fusion technology on account of BP is selected as the algorithm of data fusion model of dispersed pension system middleware. At the same time, the momentum term is introduced to improve the traditional BP algorithm. The elevated BP algorithm adjusts the weight along the avepochge direction of the error surface, and the weight changes to a reasonable direction in each itepochtion, which greatly enhances the stability of the algorithm. The design of intelligent mental endowment system on account of BP nerve network algorithm is established, and the reliability of the system is studied, which makes great technical support for the performance improvement of intelligent mental endowment system. Through the comprehensive analysis and research of the design process of the intelligent mental endowment system, not only the quality of the mental endowment system project is elevated, but also the resources of the mental endowment project are optimized.

Acknowledgements. Foundation: Periodical results of "Investigation and Research on the Demand and Service Status of mental Endowment for the Elderly in Heilongjiang Province", 2018 Philosophy and Social Science Research and Planning Project of Heilongjiang Province (Project No.: 18RKB069).

Positioning management of hospital mobile devices based on Internet of Things technology.

References

1. Kuang, Y., Singh, R., Singh, S., et al.: A novel macroeconomic forecasting model on account of revised multimedia assisted BP nerve network model and ACO. Multimedia Tools and Applications **76**(18), 18749–18770 (2017)

2. Lv, H.: Martial arts competitive decision-making algorithm on account of elevated BP nerve network. Journal of Healthcare Engineering **2021**(2), 1–8 (2021)
3. Lipu, M.H., Hannan, M.A., Hussain, A., et al.: Optimal BP nerve network algorithm for state of charge estimation of lithium-ion battery using PSO with PCA feature selection. Journal of Renewable & Sustainable Energy **9**(6), 064102 (2017)
4. Ahmad, F., Ahmad, T.: Content based image retrieval system based on deep convolution neural network model by integrating three-fold geometric augmentation. Optical Memory and Neural Networks **30**(3), 236–249 (2021). https://doi.org/10.3103/S1060992X21030061
5. Shakya, A., Biswas, M., Pal, M.: Evaluating the potential of pyramid-based fusion coupled with convolutional neural network for satellite image classification. Arab. J. Geosci. **15**(8), 1–22 (2022). https://doi.org/10.1007/s12517-022-09677-0
6. Yamaev, A., Chukalina, M.V., Nikolaev, D.P., et al.: Neural network for data preprocessing in computed tomography. Automation and Remote Control **82**(10), 1752–1762 (2021)
7. Kumi-Boateng, B., Peprah, M.S.: Modelling local geometric geoid using soft computing and classical techniques: a case study of the university of mines and technology (UMaT) local geodetic reference network. Int. J. Earth Sci. **2**(3), 166–177 (2020)
8. Belmahdi, B., Louzazni, M., Bouardi A.E.: Comparative optimization of global solar radiation forecasting using machine learning and time series models. Envir. Sci. Pollu. Res. **29**(10), 14871–14888 (2021)
9. Predicting carbon dioxide adsorption capacity on types 13X and 5A zeolites using artificial neural network modeling. International Nano Letters **12**(1), 107–112 (2021)
10. Xia, D., et al.: A parallel NAW-DBLSTM algorithm on Spark for traffic flow forecasting. Neural Computing and Applications **34**(2), 1557–1575 (2021)
11. Banerjee, A., Chatterjee, R.: Mapping of reservoir properties using model-based seismic inversion and neural network architecture in raniganj basin, India. J. Geol. Soc. India **98**(4), 479–486 (2022). https://doi.org/10.1007/s12594-022-2005-2
12. Ch, V., Rao, V.R., Dandumahanti, B.P.: Intelligent parking system. Int. J. Recent Technol. Eng. **8**(2s3), 550–550 (2020)
13. Pradhan, A.M.S., Kim, Y.-T., Shrestha, S., Huynh, T.-C., Nguyen, B.-P.: Application of deep neural network to capture groundwater potential zone in mountainous terrain, Nepal Himalaya. Environ. Sci. Pollut. Res. **28**(15), 18501–18517 (2020). https://doi.org/10.1007/s11356-020-10646-x
14. Mwakisisile, A.J., Larsson, T.: Analysis of a reform option for the tanzanian pension system. Tanzania Journal of Science **47**(1), 354–365 (2021)
15. Alpass, F.M.: Trajectories of material living standards, physical health and mental health under a universal pension. Br. J. Soc. Med. **74**(4), 362 (2020)

Image Encryption Algorithm Based on Internet of Things Technology

Hang Ruan[1(✉)] and Deepak Kumar Jain[2]

[1] Wuhan University of Science and Technology, Wuhan, China
1090440035@qq.com
[2] Chongqing University of Posts and Telecommunications, Chongqing, China

Abstract. Our production and way of life are increasingly affected by computer networks, and the problems of network security, data leakage, and information leakage are becoming increasingly prominent. With the rapid development of modern science and technology and network communication technology and the deepening of people's awareness of security awareness. Encrypting information is an important way to improve information security. The purpose of this paper is to study the image encryption algorithm based on the Internet of Things technology. This paper studies the image encryption algorithm based on the Internet of Things technology, uses the chaotic Duffing equation to sample and quantify the image, compares different image encryption algorithms to detect the encryption speed, and proposes an image encryption algorithm based on the Internet of Things technology, which makes image encryption more efficient and provides convenience for a broad audience. According to the speed comparison of the two algorithms for decrypting images of different sizes with 256 grayscales, it can be found that the substantial increase in encryption speed means that not only the Internet, but also the processing of digital images on smart devices is also enhanced, and the development of encryption applications can also be enhanced. Become reality.

Keywords: Internet of Things Technology · Image Encryption · Fast Encryption · Algorithm Research

1 Introduction

As a tool to transmit information, digital images can express information vividly, intuitively and vividly. With the rapid development of multimedia technology and Internet technology, the transmission of digital images on the Internet has become more and more frequent. Because image transmission is not only widely used in personal life, but also has a position that cannot be underestimated in social, political, military and other fields. At the same time, the security problems of digital images are also emerging. Therefore, how to protect image information from being stolen by illegal users in the process of image transmission has become a research hotspot and focus. Image encryption is a direct and effective way to protect the security of image information [1, 2].

Since image encryption technology has a very strong market demand, this technology has attracted a large number of researchers since the last century, and a large number

J. H. Abawajy et al. (Eds.): ICATCI 2022, LNDECT 170, pp. 420–428, 2023.
https://doi.org/10.1007/978-3-031-29097-8_50

of research results have emerged. Noura H applied chaos technology to digital image data information encryption for the first time, and proposed a chaos-based digital image data information encryption algorithm [3]. Boutros A proposed an image encryption scheme based on hyperchaotic and logistic systems. The scrambled matrix is generated by logistic chaos to scramble the image globally, and then the chaotic sequence is iteratively generated by the hyper-chaotic system to perform XOR operation with the plaintext image to diffuse the image [4]. From this point of view, the market demand for image encryption is strong.

This paper studies the image encryption algorithm based on the Internet of Things technology. This paper mainly analyzes several classical chaotic maps, and designs corresponding image encryption schemes by using the excellent characteristics of these chaotic maps. It also conducts preliminary research on image encryption preview technology, and proposes an image encryption scheme based on preview. And a solution based on chaotic tent mapping image encryption, which has better encryption effect, encryption speed and security.

2 Research on Image Encryption Algorithm Based on Internet of Things Technology

2.1 Image Digitization

That is, the image is stored in the digital format of the computer and processed by the computer. Generally, a computer is used to denoise, restore, segment, enhance, and extract features from images. Image digitization includes two processes: sampling and quantization: Sampling refers to discretizing the continuous spatial coordinates of an analog image into discrete spatial coordinates. If the sampling points are close enough, the sampled data is an accurate representation of the original image, and the image can be accurately reproduced by interpolation [5, 6]. According to the sampling theorem, the sampling frequency should not be lower than twice the highest frequency of the analog image, so as not to generate spectrum crossover during the sampling process and ensure no distortion during restoration; quantizing refers to the continuous The grayscale distribution is discretized into a discrete grayscale distribution.

2.2 Digital Image Processing

(1) Air domain method, transform domain method.

The air domain method treats the image as a two-dimensional function, and directly processes the pixels in the plane by the processing method of the function. The two-dimensional function is a set composed of pixel points; Then, various processing is performed on the coefficient array in the transform area according to the requirements. After the processing is completed, the coefficient array is inversely transformed and mapped back to the spatial domain through a series of operations. Various filtering operations, compressing data, extracting features, etc.belong to the transform domain method [7, 8].

(2) Digital image processing mainly includes:

Geometric processing, the physical changes of the image itself, such as scaling, rotation, and even distortion correction, panorama distortion and other operations are all geometric processing; arithmetic processing, performing arithmetic operations on the pixels of the image (addition, subtraction, multiplication, division, etc.), The operation method is intuitive and simple, which is very practical in medical applications;

Image enhancement, that is, highlighting the region of interest (ROI) in the image, enhancing effective information, weakening useless information, and facilitating processing or direct observation;

Image restoration, sharpening the blurred pixels, or removing the interfering pixels on the image, and restoring it to a more meaningful or distinguishable image for the human eye or computer, which is generally used in monitoring, image recognition and other occasions;

Image reconstruction refers to converting the data generated by the device into images that can be recognized by the human eye according to certain operating rules. A typical application of this processing is CT scan data imaging techniques;

The main purpose of image coding is to encode the image signal on the premise that the image can be restored according to its own characteristics of the input image signal. In the coding process, it is necessary to take into account the physiological and psychological characteristics of human beings, and coding cannot be done for the sake of coding;

Pattern recognition mainly includes fuzzy recognition method, syntactic structure recognition method, and statistical recognition method; respectively, it recognizes and processes images with the help of fuzzy mathematical theory, focuses on structure and primitive, and focuses on features [9, 10];

Image understanding, relying on image recognition, uses objective knowledge to promote computer understanding, imagination, thinking, and derivation, and further enables computers to understand image content,or show the state of understanding image content.

2.3 Image Compression

The reason why the image signal can be compressed is mainly due to the following two reasons: the image has redundant pixels, and there is a corresponding statistical relationship between the pixels, so it can be compressed; the premise of image compression is that it can be restored, that is, the compressed pixels It can be fully recovered during decoding; there are certain limitations in the human visual system. By reducing the accuracy, a certain objective distortion can be exchanged for data compression without being perceived by human subjective vision.

Generally, there are three ways to express the redundancy of digital images:

(1) Spatial redundancy, due to the value between adjacent pixels;
(2) Spectral redundancy, which is caused by the correlation of different color planes (such as RGB color images) or spectral bands;

(3) Temporal redundancy, due to the different number of frames in the sequence of images.

Due to the optical characteristics of the human eye, human beings have a certain upper limit on the resolution of images in terms of space, time, and color. It is not necessary to pursue the ultimate in the fineness of digital processing of images, but to meet the quality requirements. The excess part is meaningless to the human eye. Therefore, the purpose of image compression research is to reduce the space and bandwidth required for image storage and transmission by eliminating redundancy on the premise of satisfying human eye recognition.

Image compression is broadly classified into lossy and lossless. Lossless compression is reversible, the pixel reconstruction of each frame does not increase or decrease, the pixel value is equivalent to the original image, and no information is lost. It is the most ideal compression algorithm. Lossy compression will perform different degrees of loss on the image in the reconstructed image, and the compression process is irreversible.

Generally speaking, compression can be divided into three steps:

(1) Transformation: Transform the signal form, reduce the statistical correlation of the image signal itself, reduce its spatial redundancy, and represent the pixels in a computer-recognizable way;
(2) Quantification: analyze the application scenarios of different images, and determine the image accuracy under the premise of meeting the identification requirements of the corresponding occasions to reduce the image scale requirements;
(3) Encoding: Encode the image signal according to the needs of different scenarios.

Obviously, the quantization step will discard the image signal, so in the image compression process, transformation and encoding are reversible steps, while quantization is irreversible. The main difference between lossless compression and lossy compression is that lossy technology uses quantization to reduce the number of symbols output by encoding. The type and magnitude of quantization have a great impact on the bit rate, which directly affects the quality of the lossy scheme [11, 12]. According to the nature of the original input image, the image can be further divided into binary and continuous color images, and can also be divided into still images with spatial redundancy and image sequences with temporal redundancy, such as moving images. Many techniques used for still images can be used for image sequence compression, which is beyond the scope of this paper.

2.4 Chaos Duffing Oscillator

(1) Duffing equation

The Duffing equation (Duffing oscillator) is a nonlinear second-order differential equation used to model certain damped and driven oscillators as follows:

$$\ddot{x} + \delta\dot{x} + \alpha x + \beta x^3 = \varphi cos\omega t \quad (1)$$

where x = x(t) is the displacement at time t, $\dot{x}$ is the derivation of x with respect to time, ie velocity, and $\ddot{x}$ is the second derivative of x with time, ie acceleration, the parameters δ, α, β, γ and ω are Constant, Duffing's equation is an example of a dynamical system that exhibits chaotic behavior.

(2) Undamped Duffing equation:

According to the boundedness of the undamped natural oscillation solution, let γ = δ = 0, the natural undamped Duffing equation can be obtained:

$$\begin{aligned} & x(\ddot{x} + \alpha x + \beta x^3) = 0 \\ & \Rightarrow d|dt\left[1|2(\dot{x})^2 + 1|2\alpha x^2 + 1|4\beta x^4\right] = 0 \\ & \Rightarrow 1|2(\dot{x})^2 + 1|2\alpha x^2 + 1|4\beta x^4 = H \end{aligned} \tag{2}$$

where H is a constant, and the value of H is determined by the initial conditions x(0) and (0).

3 Experiment of Image Encryption Algorithm Based on Internet of Things Technology

3.1 Scrambling Algorithm

Scrambling encryption is to use certain rules to rearrange the position of the digital image matrix in space, where the pixel value ranges from 0 to 255, and the binary can also be represented by 8 bits; convert a pixel gray value into 8 bits. Binary, using a specific operation to change the position of a bit or a few bits in space can change one or some pixel gray values; at the same time, by converting from the time domain to the transform domain (frequency domain, DCT domain etc.), do some operations on the pixel value in the transform domain, change one or some pixel values in the frequency domain, and affect the entire pixel matrix in the time domain, which is also scrambling; the "position" of the pixel perform the·relevant operations.

3.2 Diffusion Algorithm

Although the image scrambling encryption algorithm can scramble the position of the original image pixels, it has certain shortcomings: usually it takes many iterations to achieve a noise-like effect; only shuffling the position of the image pixels will not affect the image. Statistical distribution of pixel values. Therefore, it is necessary to diffuse and improve the security of the image, so as to achieve the desired encryption effect. In the image encryption system, while keeping the position of the image pixels unchanged, the gray value information of any plaintext pixel is used to hide the pixel values in other ciphertext images as much as possible, so that the statistical characteristics of the plaintext are dissipated in the In the ciphertext, let each plaintext number influence the acquisition of multiple ciphertext numbers as much as possible.

4 Experimental Analysis of Image Encryption Algorithm Based on Internet of Things Technology

4.1 Image Encryption and Decryption Speed Comparison

Encryption speed is one of the important indicators to evaluate the performance of an image encryption system. To this end, the encryption speed of this scheme and other hyperchaos-based image encryption schemes was tested under the same environment on the same computer. The results are shown in Table 1:

Table 1. Image encryption and decryption speed comparison

Image size (pixels)	This algorithm	Quantum based image encryption
256 × 256 × 3	0.23–0.38	0.90–0.95
512 × 512 × 3	1.21–1.26	1.82–1.88
1024 × 1024 × 3	1.61–1.73	13.08–13.17
2048 × 2048 × 3	3.83–3.91	36.38–36.40

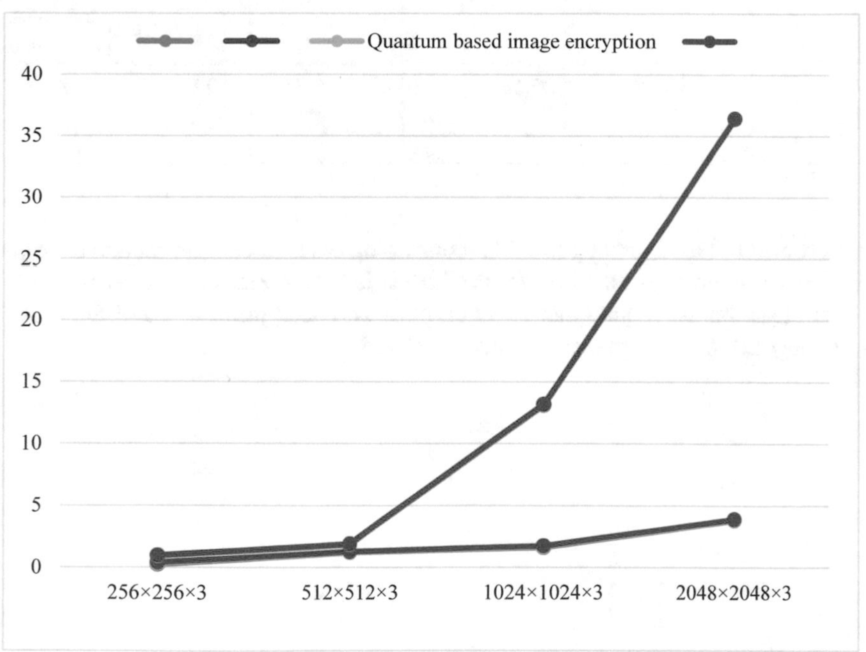

Fig. 1. Image encryption and decryption speed comparison

Specifically, as shown in Fig. 1, the speed of encryption and decryption of 256 grayscale images of different sizes using this algorithm and the non-parallel algorithm is compared. It must be found that the encryption speed of this algorithm has been greatly improved. The larger the image, the higher the speed. The larger the magnitude of, is even 10 times faster for 2048 × 2048 × 3 image processing.

4.2 Analysis of Encryption Speed

In addition to security, an encryption system also needs to consider the speed of encryption and decryption. The speed of encryption and decryption is also very important. The speed of encryption and decryption is too slow to have practical application significance. The time-consuming process of encrypting/decrypting digital images of different sizes is counted. As shown in the Table 2 below:

Table 2. Average time of image encryption and decryption for different algorithms (unit: seconds)

Image size (pixels)	This algorithm	Piecewise Nonlinear Chaos Map	Self-shrinking chaotic stream ciphers	Parallel Subimage Hyperchaotic Encryption	DNA-based new chaotic image encryption	Image Encryption Based on Hyperchaos and DNA Sequences
2048 × 2048 × 3	3.91	3.89	25.15	15	119.88	12.76

As shown in Fig. 2, among the time-consuming cases of encryption and decryption of the six different algorithms listed, the fastest is 3.89 s and the slowest is 119.88 s. After comparison, it is found that the encryption and decryption speed of the modified parallel algorithm has been significantly improved.

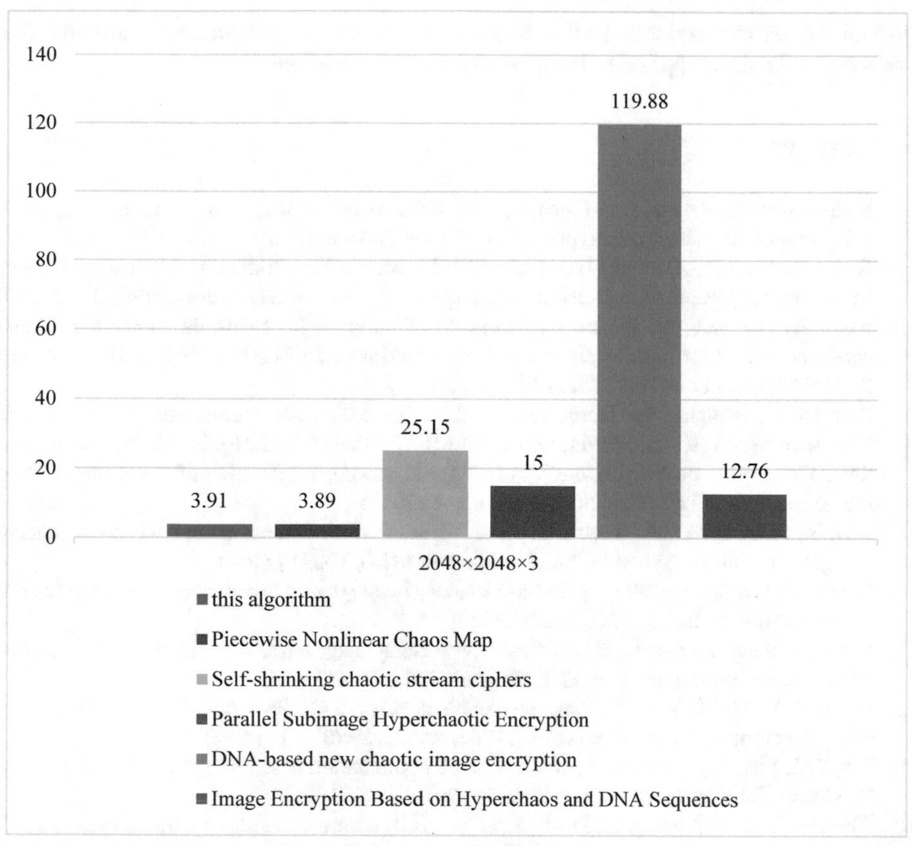

Fig. 2. Average time of image encryption and decryption for different algorithms

5 Conclusions

With the popularization of the Internet and the improvement of the performance of smart devices, people are more and more dependent on data, and the use of digital images has become more frequent, and at the same time, they are also concerned about privacy protection. The increasing value and abundance of digital image data makes image encryption algorithms a hot research focus. With the development of mobile device mobile phones, its performance is getting higher and higher, and the computing power is continuously improved, which will increase the processing of digital images of mobile devices such as mobile phones, so the encryption tools are real. It is wise to take a photo and save it encrypted with one click, even if your phone is lost or stolen, personal data such as photos are safe, which can better protect your personal photo data. Due to the characteristics of the image data itself, the traditional encryption methods have certain defects and cannot meet the needs of fast image encryption. Therefore, this paper focuses on the design and analysis of fast image encryption algorithms, and improves the efficiency on the premise of ensuring the reliability of image encryption as much as possible. After experimental testing and security analysis, these concepts are effective and safe, and can be applied to image encryption applications.

References

1. Muhammad, K., Hamza, R., Ahmad, J., et al.: Secure surveillance framework for IoT systems using probabilistic image encryption. IEEE Trans. Industr. Inform. 3679–3689 (2018)
2. Khan, J., Li, J.P., Ahamad, B., et al.: SMSH: secure surveillance mechanism on smart healthcare IoT system with probabilistic image encryption. IEEE Access **PP**(99), 1–1 (2020)
3. Noura, H., Chehab, A., Sleem, L., Noura, M., Couturier, R., Mansour, M.M.: One round cipher algorithm for multimedia IoT devices. Multimed. Tools Appl. **77**(14), 18383–18413 (2018). https://doi.org/10.1007/s11042-018-5660-y
4. Boutros, A., Hesham, S., Georgey, B., et al.: IEEE 2017 29th International Conference on Microelectronics (ICM) - Beirut, Lebanon (2017.12.10–2017.12.13). In: 2017 29th International Conference on Microelectronics (ICM) - Hardware acceleration of novel chaos-based image encryption for IoT applications, pp. 1–4 (2017)
5. Roy, S., Shrivastava, M., Rawat, U., et al.: IESCA: An efficient image encryption scheme using 2-D cellular automata. J. Inf. Secur. Appl. **61**(5), 102919 (2021)
6. Waqas, P., Byun, Y.: A blockchain-based secure image encryption scheme for the industrial internet of things. Entropy **22**(2), 175 (2020)
7. Shen, M., Deng, Y., Zhu, L., et al.: Privacy-preserving image retrieval for medical IoT systems: a blockchain-based approach. IEEE Network **33**(5), 27–33 (2019)
8. El-Latif, A, Abd-El-Atty, B., Venegas-Andraca, S.E., et al.: Providing End-to-end security using quantum walks in IoT networks. IEEE Access **PP**(99), 1–1 (2020)
9. Wu, T.Y., Fan, X., Wang, K.H., et al.: A DNA computation based image encryption scheme for cloud CCTV systems. IEEE Access **PP**(99), 1–1 (2019)
10. Dhawan, S., Chakraborty, C., Frnda, J., et al.: SSII: secured and high-quality steganography using intelligent hybrid optimization algorithms for IoT. IEEE Access **PP**(99):1–1 (2021)
11. Li, L., Wen, G., Wang, Z., et al.: Efficient and secure image communication system based on compressed sensing for IoT monitoring applications. IEEE Trans. Multimedia **22**(1), 82–95 (2020)
12. Thoms, G., Muresan, R., Al-Dweik, A.: Chaotic encryption algorithm with key controlled neural networks for intelligent transportation systems. IEEE Access,1–1 (2019)

Research on Cross-Border Electronic Evidence Network Forensics Cooperation and Response Mechanism

Lizhi Lin[1], Lijun Zhang[2(✉)], Fajian Xu[3], and Blessie Mae V. Gabriel[4]

[1] School of Big Data, Fuzhou University of International Studies and Trade, Fuzhou 350202, Fujian, China

[2] Fujian Ruiyan Judicial Expertise Institute, Fuzhou 350011, Fujian, China
zhanglijun6051@163.com

[3] Department of Computer and Information Security Management, Fujian Police College, Fuzhou 350007, Fujian, China

[4] Angeles University Foundation, Angeles City, Philippines

Abstract. Combating cybercrime is a difficult problem and challenge that all countries in the world need to face together. Countries should carry out research on application and cooperation mechanism in the legal evidence collection of electronic data; In view of the differences between countries and regions in terms of electronic forensics, forensics technology and software and hardware equipment standards, as well as differences in industry norms and legal norms, carry out research on electronic data laboratory certification and standardized management, cooperation system, third-party payment and virtual currency forensics, forensics confidentiality and cooperation in the field of certification in specific countries and regions Cooperation and response mechanism construction.

Keywords: Cross Border Forensics · Electronic Data · Ways of Obtaining Evidence · Cooperation Mechanism

New telecom network fraud cases emerge one after another, showing the characteristics of technology driven, industrial cooperation, cross-border implementation and so on. The black market industrial chain behind it has become a public hazard in telecommunications, Internet, finance and other industries. Globally, the implementation cost of telecom network fraud is low, the attack is difficult, the cost is high, the regional span is large, and the technical ability and organizational ability of fraudsters are becoming more and more sophisticated.

At present, the governance of cross-border Telecom fraud mainly focuses on the attack level, focusing on the direct attack on domestic and foreign fraud calls, online fraud platforms, underground banks and other gangs through cross-border police cooperation. Although the United Nations Convention against transnational organized crime, the Convention on cybercrime and the United Nations Convention against corruption provide an international legal basis for combating cross-border telecommunications fraud [1], due to the hiding places of telecommunications fraud gangs in African and Southeast Asian countries and regions, there is often a lack of real-time intelligence sharing

J. H. Abawajy et al. (Eds.): ICATCI 2022, LNDECT 170, pp. 429–437, 2023.
https://doi.org/10.1007/978-3-031-29097-8_51

mechanism, and even legal gaps, dislocation and even local protectionism. According to the barrel principle, countries and regions that lack effective control and strong legal response measures for cross-border Telecom fraud may still continue to be the hotbed and important source of Telecom fraud.

1 Legal Evidence Collection Approach and Application of Cross-Border Electronic Data

According to the rules for electronic data evidence collection in criminal cases issued, the measures and methods for public security organs to extract electronic data evidence include: detaining and sealing up the original storage medium; Extract electronic data on site; Extracting electronic data online through network; Freezing electronic data, accessing electronic data, etc. Taking the cross-border online gambling crime as an example, the electronic data evidence mainly includes: SMS, wechat and QQ chat records; Transfer records of payment tools such as WeChat, Alipay and bank cards. Domain name, website address, account number and password of gambling website; Online gambling betting amount, times, time, etc. These electronic data need to be obtained remotely through the network or investigated and collected by local personnel through various cross-border cooperation methods.

1.1 How to Legally Obtain Cross-Border Electronic Data in Civil Cases

According to Article 16 of several provisions of the Supreme People's Court on civil litigation evidence: "If the official documents and documentary evidence provided by the parties are formed outside the territory of the people's Republic of China, the evidence shall be certified by the notary organ of the country where they are located, or perform the certification procedures specified in the relevant treaties concluded between the people's Republic of China and the country where they are located. The evidence involving identity relationship formed outside the territory of the people's Republic of China shall be certified by the notary organ of the country where they are located and approved by the people's Republic of China The embassy or consulate of the people's Republic of China in that country shall certify or perform the certification procedures specified in the relevant treaties concluded between the people's Republic of China and the host country. If the evidence provided by the parties to the people's court is formed in Hong Kong, Macao and Taiwan, they shall go through relevant certification procedures. When a party provides documentary evidence or explanatory materials in a foreign language to the people's court, a Chinese translation shall be attached." If the parties want to use the chat records of overseas social software as evidence, they should notarize the identity of the chat subject and the content of the chat records in the notary organ of the country where they are located, and then go through the authentication formalities at the embassy and consulate of the people's Republic of China in that country; Finally, the notarized and certified materials and their Chinese translation can be submitted to the court as evidence.

1.2 How to Legally Obtain Cross-Border Electronic Data in Criminal Cases

Judging from the current implementation methods of international cross-border criminal electronic forensics, there are three main forms: first, it is carried out through the existing cross-border criminal judicial assistance mechanism; second, the investigation organ adopts remote login or technical means to enter the relevant system for unilateral acquisition [2], and third, the investigation organ is implemented through multinational enterprises or a third-party platform providing cross-border cloud computing services. Among them, it is necessary to follow the procedures of handling criminal judicial assistance cases through the Ministry of public security and handling criminal judicial assistance cases through diplomatic channels. Telecommunications network fraud is often a cross-border and cross regional crime, which determines that its prevention and combating must be based on multi-party cooperation, including cross-border cooperation and domestic cooperation.

1.3 Analysis and Application of Legal Evidence Collection Methods for Cross-Border Electronic Data

Telecommunications fraud has become an international problem. There is an urgent need for different countries and regions to work together to carry out cooperation and Application Research on legal evidence collection methods of cross-border electronic data. Through consultation and cooperation and signing relevant legal documents, countries actively build a cooperation model within the legal framework, give full play to the role of international organizations such as Interpol and relevant basic agreements, and carry out cross-border joint crackdown actions to encircle and suppress Telecom fraud. Strengthening cooperation and exchanges between different countries, regions and relevant international organizations and institutions to learn from each other is an important part of improving and improving China's ability to combat telecommunications fraud.

There have been some new changes in the development of cross-border criminal electronic evidence collection system in different countries and regions, and some new achievements have been made in technology application and system norms. Timely sorting out and summarizing this will help to accumulate and share the gains and losses of cross-border cooperation experience, and further improve the cooperation systems and common development of different countries and regions. We should also strengthen exchanges and cooperation in the legal evidence collection of cross-border electronic data, further deepen the cooperation mechanism in jointly combating the crime of new telecom network fraud and mutual legal assistance, and further carry out cross-border cooperation and research in electronic evidence collection, judicial expertise, evidence preservation, mutual recognition of evidence, information sharing, etc., Curb the rampant trend of cross-border cybercrime.

2 Research on the Differences of Electronic Data in Different Countries and Regions

The crime of telecommunication network fraud presents a high incidence trend, which seriously endangers the property safety of the state, the collective and the people. It has become a prominent crime with the fastest rise and the strongest response from the masses. Telecommunications fraud is usually domestic and foreign cooperation, with strong liquidity, poor evidence stability and high difficulty in case technical investigation. Telecommunications fraud originated in Taiwan, but the means of fraud are constantly updated, showing some new changes and trends. Combating telecommunications fraud has become a common problem and challenge for different countries and regions.

2.1 Comparison of Electronic Forensics Terms in Different Countries and Regions

Taking Taiwan Province as an example, due to the differences in the development process of technology and application between the mainland and Taiwan, we need to analyze and study the workflow and evidence application for rational use. There are many differences in some professional terms (as shown in Table 1).

Table 1. Electronic forensics terminology comparison

No.	Mainland electronic forensics terminology	Taiwan electronic forensics terminology
1	Electronic forensics	Digital forensics
2	Destruction of evidence	Eliminate syndrome
3	Forensics personnel	Identification personnel
4	Check	Examine
5	Executable program	Executive file

2.2 Differences Between Electronic Forensics Technology and Software and Hardware Equipment Standards

A series of standards for electronic forensics formulated by the international organization for Standardization (ISO) and the International Electrotechnical Commission (IEC). Information technology standards in different countries and regions basically refer to international standards and have certain commonality. There are slight differences in the level and classification of technology and software and hardware equipment standards. In terms of some professional terms, basic principles and software and hardware equipment involved in electronic data forensics and identification, different countries and regions can work together to fundamentally standardize the work of electronic data forensics and identification by formulating these basic standards, so as to lay a good foundation for case investigation, electronic data identification and mutual recognition of evidence.

There are great differences in the versions and quantities of international standards adopted by different countries and regions. In the past ten years, the electronic forensics technology in mainland China has developed rapidly, and a series of standard and perfect standard systems with Chinese characteristics and in line with the actual needs of the Chinese mainland have been formulated. Including the application technology, software and hardware equipment (especially for the special software technology, operating system, forensics object, instant messaging, network social software, etc.) according to our actual use, there are targeted standards and technical applications. Procedural standards include qualification standards for personnel involved in forensics, forensics equipment standards, environmental standards for networked equipment forensics, etc. Before carrying out the cooperation of electronic forensics, it is also necessary to carry out mutual authentication and benchmarking management for these links.

The standards in the identification stage are the main areas of concern of all countries. They are relatively standardized, including the recovery standards of electronic data, the search standards of classified data, and the standards of tools and methods, which are easy to integrate and recognize each other.

2.3 Differences in Industry Norms of Electronic Data Forensics

In terms of the industry norms of electronic data forensics, China is also at the forefront of the world. As of June 2021, various standards and technical specifications in the field of electronic data forensics in China include 5 national standards, 16 technical specifications for judicial expertise and 39 industrial standards for public safety (as shown in Table 2).

Table 2. Relevant industry standards

Industry standard	Standard number
General implementation specification for judicial authentication of electronic data	SF/Z JD0400001–2014
Technical specification for authenticity authentication of electronic documents	SF/Z JD0402004–2018
Technical specification for software function authentication	SF/Z JD0403004–2018
Operation specification for pseudo base station inspection	SF/Z JD0404001–2018

In this regard, different countries and regions need to strengthen exchanges and research, realize the transformation and mutual recognition of relevant technical specifications, and carry out corresponding localization transformation according to practical operation processes and legal norms.

2.4 Legal Norms

Between specific countries and regions, it is necessary to study the differences and transformation methods of different countries and regions in terms of evidence identification and operation norms such as personnel qualification, judicial expertise process and expertise report, so as to form a cooperative working mechanism in terms of mutual recognition and mutual verification and clue sharing.

3 Research on Cooperation Mechanism in Cross-Border Electronic Data Exploration, Evidence Collection and Identification

3.1 Cooperation and Research in Laboratory Certification and Standardized Management

The international organization for Standardization (IOS) and the International Electrotechnical Commission (IEC) officially issued the general requirements for the competence of testing and calibration laboratories (ISO/IEC17025:2017). CNCA issued the general requirements for inspection and testing institutions for the evaluation of qualification accreditation ability of inspection and testing institutions (RB/T 214-2017), which incorporates the contents of the new 17025 standard. China National Accreditation Commission for conformity assessment (CNAs) officially issued cnas-cl01:2018 accreditation criteria for the competence of testing and calibration laboratories. These laboratory certification and norms are directly related to the judicial procedure, evidence legitimacy, organization and personnel qualification of electronic data forensics and identification.

The certification of inspection and testing institutions in most European and American countries is also formulated with reference to the international standards of ISO/IEC, but there are many differences in institutions, personnel, site environment, equipment and facilities, management system and so on [3]. Further research is needed to realize the transformation and mutual recognition of management norms of inspection and testing institutions. The certification and standardized management system of inspection and testing laboratory is directly related to the process and results of electronic data forensics, clue research and judgment and judicial expertise. Therefore, we must strengthen research cooperation in this field, which is the basis of electronic data application and cooperation in different countries and regions.

3.2 Construction of Cross-Border Diversified Electronic Data Cooperation System

Combating cybercrime with cross regional characteristics highly depends on electronic evidence. However, in view of the criminal jurisdiction of different countries and regions and the restrictions on cross-border data flow, cross-border electronic evidence collection has become a difficulty for countries to combat cybercrime. We should reasonably learn from the cross-border electronic forensics modes that have been preliminarily formed in the international community, such as the mode of requesting criminal judicial assistance, the mode of signing multilateral or bilateral treaties, unilateral forensics by using technical means and unilateral forensics through legislation by network operators [4], and build a cross-border diversified cross-border electronic Forensics System on the basis of adhering to the principles of network sovereignty and legitimacy, Further simplify the criminal judicial assistance procedures of electronic evidence, avoid infringing on cyberspace sovereignty and personal privacy, and explore more efficient and efficient diversified cooperation mechanisms for cross-border electronic evidence collection.

3.3 Cooperation in Third-Party Payment and Virtual Currency Forensics in Different Countries and Regions

Virtual money laundering has the characteristics of strong concealment, fast cross-border circulation and diversified value exchange. Cyber crime and other illegal activities take virtual currency as the channel to carry out money laundering and capital circulation activities (as shown in Fig. 1). Moreover, the black industry chain of virtual currency money laundering has undergone continuous evolution, and gradually presents a large-scale and professional trend [5]. The judicial organs of various countries should explore effective governance countermeasures from the compound path of combining law with technology, so as to realize effective prosecution and crack down on crime.

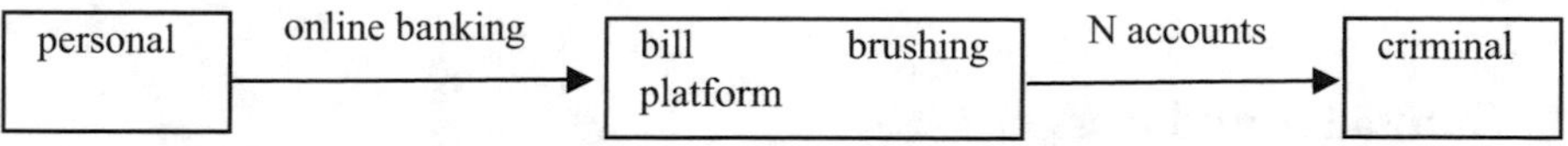

Fig. 1. Criminal capital circulation

Most of the cross regional and international cyber crimes are related to the third party, the fourth party payment and virtual currency. Therefore, in terms of electronic payment, countries should cooperate closely, establish a perfect consumer protection mechanism, improve user identity authentication and anti money laundering system [6], formulate industry norms for anti money laundering in the third-party payment industry, check the implementation and implementation of Payment institutions, and impose industry sanctions on Payment institutions that violate laws and Industry norms, After summarizing the relevant information, it shall be announced to the society regularly or irregularly, and urge Payment institutions and users to supervise themselves and maintain good social credit. Different countries and regions should sign cooperation agreements in the supervision of virtual currency, trading platform and payment platform to realize technology and information sharing. Take the illegal payment and settlement of virtual currency as the core of governance, adhere to the response mode of multi-party governance of technology, standardization and supervision, and comprehensively implement policies to realize the whole chain governance. Virtual currency trading platform, online payment platform and online social platform with payment and settlement function shall be included in the scope of obligation subjects, and risk management measures shall be established based on the risk status of money laundering.

3.4 Cooperation in the Field of Electronic Data Forensics, Confidentiality and Authentication

In order to meet the actual needs of case confidentiality and investigation, it is necessary for all countries to cooperate closely and establish a standardized system of electronic forensics security protection in line with their own work characteristics.

In some general operations of electronic forensics, such as data preservation, data recovery and other standards, we can learn from the standard achievements and certification technology of peers in specific countries; In the construction of electronic forensics confidentiality standard system, we need to fully consider the particularity of electronic data and confidentiality work. Starting from the characteristics of the generation, storage and transfer of electronic evidence, we should carry out normative research and the construction of confidentiality and authentication cooperation mechanism by categories, in order to form a scientific, reasonable and mutually recognized authentication security protection standard system [7].

3.5 Communication and Cooperation Mechanism and Personnel Training

Countries should also regularly carry out training, research, certification and interactive exchange meetings for electronic forensics personnel to form a reasonable and flexible long-term cooperation mechanism.

4 Summary and Prospect

Due to the differences of legal systems in different countries and regions, it is easy to lead to different rules for cross-border electronic evidence collection, which makes it more difficult for criminal judicial assistance. We should do a good job in domestic and overseas criminal judicial assistance on the basis of fully understanding, mastering and improving the technical norms of evidence collection [8]. Further establish and improve different applicable conditions to standardize unilateral cross-border electronic data forensics, so as to safeguard countries' network data sovereignty and national sovereignty; In the context of strengthening the free flow of global data, improve the data disclosure system [9]. On the other hand, on the basis of respecting sovereignty, we should learn from the compulsory disclosure mechanism, optimize the protection of civil rights and evidence collection procedures, and strengthen the legal obligations of network service providers [10], so as to balance the relationship between evidence collection efficiency and national sovereignty.

In short, all countries carry out research on electronic forensics technology for new telecommunications network fraud cases and other criminal acts, and improve the cooperation mechanism, which is conducive to improving the international level of combating cybercrime, jointly combating cross-border crime, ensuring the safety of people's property, improving the well-being of people all over the world, and safeguarding the stable development of countries and regions.

Acknowledgments. 1. 2021 Fujian young and middle-aged teacher education and scientific research project (Science and Technology): Research on information security of artificial intelligence equipment(JAT210523).

2. International law enforcement cooperation and exchange program of Fujian Police College: Regional security risk assessment based on temporal and spatial analysis of crime(GH202102).

3. 2021 college students' innovation and entrepreneurship training program (national level): Application of UAV in indoor field investigation.

References

1. Dipa, D., Ankita, C., John, W.: The anti-trafficking bill, 2018: does it fulfill India's commitment to the international community? J. Hum. Traffick. **7**(2), 224–237 (2021)
2. Tamara, L.: The role and position of the European public prosecutor in Croatian preliminary proceedings. Police and Security **28**(4), 490–516 (2019)
3. Rojszczak, M.: The uncertain future of data retention laws in the EU: Is a legislative reset possible?. Comput. Law Secur. Rev. **41** (2021)
4. Gooderham, P.N., Elter, F., Pedersen, T., et al.: The digital challenge for multinational mobile network operators. more marginalization or rejuvenation?. J. Int. Manag. **28**(4) (2022)
5. Zhamiyeva, R.M., Sultanbekova, G.B., Abzalbekova, M.T., et al.: The role of financial investigations in combating money laundering. Int. J. Electron. Secur. Digit. Forensics **14**(2), 188–198 (2022)
6. Nduka, B.O.E., Sechap, G.: Refocusing designated non-financial businesses and professions on the path of anti-money laundering and combating the financing of terrorism compliance. J. Money Laund. Control 24(4), 693–711 (2021)
7. Khalaf, O.I., Abdulsahib, G.M.: Optimized dynamic storage of data (ODSD) in IoT based on blockchain for wireless sensor networks. Peer - to - Peer Networking and Applications 1–16 (2021)
8. Maiorova, L.: Improving electronic evidence legal regulation in criminal proceedings. In: SHS Web of Conferences, vol. 134 (2022)
9. Schulman, D.J., Bateman, A.H., Greene, S.: Supply chains (Scope 3) toward sustainable food systems: an analysis of food and beverage processing corporate greenhouse gas emissions disclosure. Cleaner Production Letters **1** (2021)
10. Gijrath, S.J.H.: (Re-)defining software defined networks under the European electronic communications code. Comput. Law Secur. Rev. **40**, 105492 (2021)

A Review of Production Scheduling Research Based on Genetic Algorithm

Pengfei Xin(✉), Tianxing Sun, Jiaqi Wang, Ningbo Zhang, and Yawei Li

College of Mechanical and Electrical Engineering, Zhengzhou University of Light Industry, Zhengzhou 450002, Henan, China
1025929941@qq.com

Abstract. Production scheduling is one of the most important links in Intelligent manufacturing system, which has a significant impact on improving the production efficiency of enterprises. In the past decades, genetic algorithm (GA) has been extensively used in different production scheduling problems and has become the main method to solve scheduling problems. This paper makes a comprehensive literature review on the use of genetic algorithm to solve the production scheduling problem, comprehensively and systematically summarizes the workshop production scheduling such as JSP, FJSP, DFJSP, GSSP and FFJSP, and summarizes the research direction of GA in the future.

Keywords: Production Scheduling · Job Shop Scheduling · Flexible Job Shop Scheduling · Green Scheduling · Genetic Algorithm

1 Introduction

Along with the high-speed development of intelligent manufacturing, artificial intelligence, industry 4, industrial Internet plus, China made 2025, intelligent manufacturing system has become an important research content for enterprises to improve production efficiency and reduce production costs. As the key link of intelligent manufacturing system, production scheduling is not only an important process of production and manufacturing, but also the centralized allocation and sequencing of human, financial and material resources, which plays a key role in enterprise production and manufacturing. Enterprises solve the problems in production through the rational allocation of resources. Scheduling not only improves the utilization of resources, but also improves the production efficiency and reduces the production cost. Therefore, scheduling is widely used in intelligent manufacturing system. In recent years, in addition to the traditional job shop scheduling problem (JSP), flexible job shop scheduling problem (FJSP), dynamic flexible job shop scheduling problem (DFJSP), Green shop scheduling problem (GSSP), fuzzy flexible job shop scheduling problem (FFJSP) and other scheduling problems have attracted more and more attention. Figure 1 exhibits the different categories of production scheduling:

J. H. Abawajy et al. (Eds.): ICATCI 2022, LNDECT 170, pp. 438–445, 2023.
https://doi.org/10.1007/978-3-031-29097-8_52

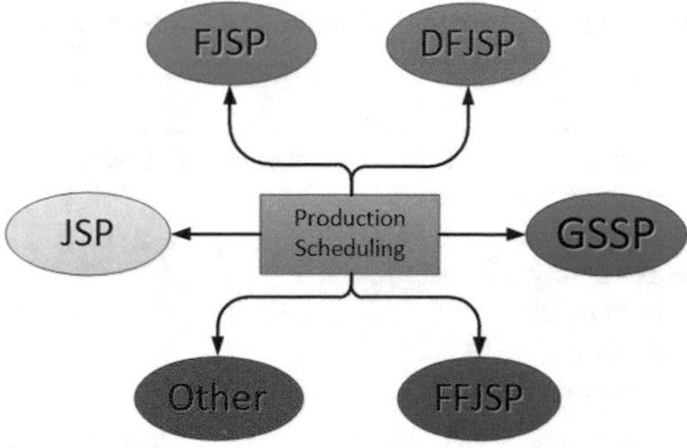

Fig. 1. Classification of production scheduling

In fact, the production scheduling problem is to reasonably allocate the machines and processing workpieces between different factories, so as to achieve the optimal allocation mode by the objectives of minimizing the completion time, minimizing the total machine load, minimizing carbon emission, minimizing the total delay and minimizing the penalty cost. Usually, the scheduling model is established based on the above objectives, and the optimal solution is obtained by using relevant algorithms. Different working hours of machines in different factories lead to various uncertain events among workers, machines and workpieces, such as abnormal production environment, machine failure, order change, operation delay, etc. As the scheduling problem becomes more and more complex, famous experts and scholars from home and abroad have done a lot of research and put forward many algorithms that can effectively solve the complex production scheduling problem. Most production scheduling problems are NP hard problems, so the traditional algorithms can not solve large-scale scheduling problems.

In recent years, swarm intelligence algorithm has been more and more favored by experts and scholars, and more and more swarm intelligence algorithms are used in scheduling problems. Swarm intelligence algorithms contain genetic algorithm (GA), particle swarm optimization (PSO), Simulated annealing algorithm (SAA), Ant colony algorithm (ACA) is mentioned to solve production scheduling problem. The advantages and shortcoming of the above algorithms are listed in the Table 1.

Genetic algorithm [1] was proposed by Professor Holland of the University of Michigan in 1975. Genetic algorithm supplies a general system theory for solving complex scheduling, and it doesn't rely on the particular field of the problem, and it has strong robustness. Therefore, genetic algorithm is mostly used in the fields of function combination optimization, intelligent control, image processing, path planning, production scheduling and so on. After decades of development, the genetic algorithm is used in various problems, which also proves the reliability of the algorithm. Most scholars have only reviewed the genetic algorithm is used in a single scheduling field, but the research on the application of genetic algorithm in various production scheduling problems has not been sorted out and summarized. Therefore, this paper comprehensively reviews and summarizes the application of genetic algorithm in JSP, FJSP, DFJSP, GSSP, FFJSP and other problems, analyzes the existing problems in this field, and discusses the future research direction.

Table 1. Advantages and shortcoming of the above algorithms

Algorithm name	GA	PSO	SAA	ACA
Advantage	1. Strong robustness 2. Strong global search ability 3. Easy to mix with other algorithms	1. Fast iteration speed 2. Few parameters	1. Efficient	1.Strong search ability
Shortcoming	1. Weak local search ability 2. Slow convergence	1. Weak convergence	1. The parameters are difficult to control 2. Easy to sink into local optimization	1. Slow convergence 2. Easy to sink into local optimization

2 GA Problem Description

GA is an efficient parallel global search algorithm formed by imitating biological evolution and heredity. It uses the elimination mechanism of survival of the fittest to solve the complex matter in the field of JSP. Each individual (i.e. chromosome) is compared to a feasible solution in the scheduling problem, and through the processing of coding and decoding of chromatids, Make each chromosome have the gene string of processing process or processing machine, and reasonably arrange all processes on the processing machine, so as to obtain the optimal scheduling scheme.

After chromosome encoding and decoding, GA starts iterative operation, and a new population will be generated after iteration. Each iteration operation includes population initialization, calculation of adaptive value, selection operation, crossover operation and mutation operation [5]. If the iteration conditions are met, a new generation of excellent population will be generated. Otherwise, continue the iteration until the iteration conditions are met. The basic iterative process of GA is shown in the Fig. 2.

GA has been developed and applied in JSP, FJSP, DFJSP, GSSP, FFJSP and other scheduling problems. Because genetic algorithm can provide the ability of global search and strong robustness, but genetic algorithm also has the disadvantages of easy premature and poor stability of solution, which leads to the singleness of genetic algorithm. This section will comprehensively describe the research progress of GA in the field of scheduling.

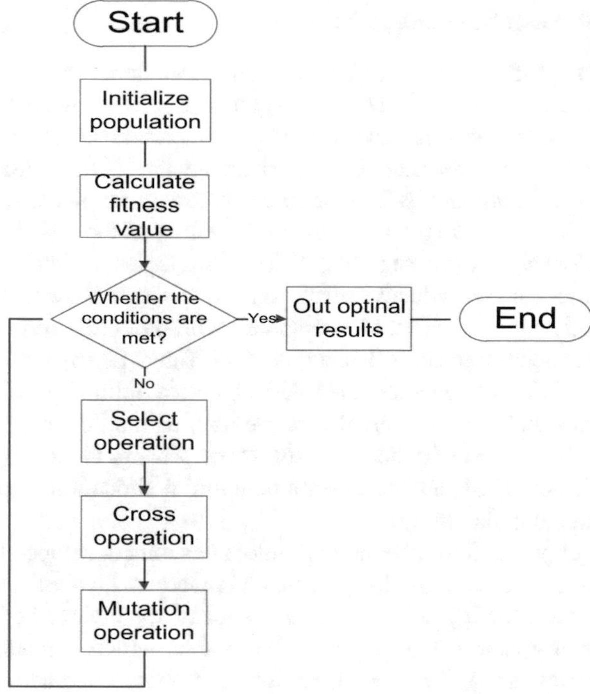

Fig. 2. The flow chart of GA

3 Research Progress of Genetic Algorithm in Production Scheduling

3.1 Job Shop Scheduling Problem

Job shop scheduling problem (JSP) [2] is one of the classical scheduling in traditional production scheduling. It is composed of m machines and n workpieces. This m machine is responsible for the specific processes of processing n workpieces. The processing time between different processes is determined, the processing sequence of each workpiece remains unchanged, and the processing process cannot be interrupted. GA has been widely used in JSP. The following mainly summarizes the research of GA in JSP.

JSP is actually a simplified model of many production scheduling, and the problem has important theoretical meaning and engineering significance. Essafi, I. [2] studied the job shop scheduling problem with delivery date. Aiming at minimizing the total delay, they combined genetic algorithm with local search algorithm, and set the longest path method by using disjunctive graph model. Experiments show the effectiveness of the algorithm. Ombuki, B. M. and Ventresca, M. [3] jointly proposed a strategy of hybrid GA and TS algorithm based on JSP, adopted genetic algorithm of genetic knowledge (gkGA) to solve JSP, and proposed mutation like operator of local search (LSGA), so as to better realize chromosome crossover. Rafsanjani, M. K. and Riyahi, M. [4] proposed a new crossover operator based on machine to solve the problem of premature convergence.

3.2 Flexible Job Shop Scheduling Problem

As an continuation of JSP problem, FJSP is a different NP hard problem, which is closer to the actual scheduling problem. Each process can be processed by different equipment, and the working time on each machine is strange. Compared with other algorithms, GA has strong robustness, so it has more obvious advantages in solving FJSP, but GA is still easy to sink into local optimal result. In order to explore more methods, many scholars have improved GA to improve the local search capability of the algorithm. Pezzella, F. et al. [5] proposed an algorithm integrating different strategies, which is used to generate the initial population and reproductive offspring. Experiments show that the integration strategy has significant advantages, and compared with other algorithms, it reflects the effectiveness of genetic algorithm. Huang, X. and Yang, L. [6] with SA algorithm, proposed a new hybrid GA to handle the multi-objective optimal problem considering transportation time, and used a external elite memory as the knowledge base to guide the GA to search the region with better performance. Zhang, G. et al. [7] proposed an effective GA to handle FJSP, and designed a new initial allocation method to enhance the quality of initial population.

The research of production scheduling problem has long developed from traditional job shop scheduling problem (JSP) to dynamic and practical. Dynamic flexible job shop scheduling problem (DFJSP) [8] belongs to a kind of dynamic scheduling problem. Its scheduling environment is complex and changeable, which is mainly disturbed by factors such as order randomness, machine failure, emergency order insertion and so on. DFJSP can solve the problem of dynamic event interference caused by machines, workpieces or resources in FJSP environment. For DFJSP, the research methods in the early years mainly include integer programming, heuristic rules and other methods, but in recent years, genetic algorithm has good efficiency in solving complex scheduling problems, which opens up a new idea for the research of dynamic scheduling problems. Rangsaritratsamee, R. and others [8] proposed an algorithm to solve DFJSP based on the double objective function considering efficiency and stability, and used genetic local search algorithm to generate scheduling at each rescheduling point. Fattahi, P. and Fallahi, A. [9] elaborated a meta heuristic algorithm by GA and established the multi-objective mathematical model of DFJSP, which solved the NP hard problem well. In order to solve the acceleration of rapid response in dynamic environment, Azadeh, A. et al. [10] proposed to integrate genetic algorithm and artificial neural network with computer simulation to explore the solution space in flexible flow shop. With the goal of minimizing delay, the output results of neural network are input into genetic algorithm. Experiments show the applicability of this method in obtaining the optimal solution.

3.3 Green Workshop Scheduling Problem

Green workshop scheduling problem (GSSP) is a scheduling problem for low-carbon manufacturing. As an extension of traditional workshop scheduling, GSSP is a hot issue in scheduling problem. Its advantages in energy consumption have brought high economic benefits to enterprises. It can not only save energy and protect the environment, but also reduce costs and improve efficiency. GSSP is a more complex scheduling problem, which is difficult to solve, but it can realize green manufacturing. Therefore, the research on this scheduling has more practical significance and application value. Seng, D. W. et al. [11] proposed an improved non dominated sorting genetic algorithm (NSGA-II) to solve the low-carbon scheduling of flexible job shop, taking the maximum completion time and low-carbon as the optimization objectives. The proposed low-carbon optimization algorithm has good global convergence. Sun, X. et al. [12] established a multi-objective FJSP mathematical equation based on minimum carbon emission, completion time and machine load, and proposed an improved NSGA-III combining multi-group coevolution and natural selection.

3.4 Fuzzy Flexible Job Shop Scheduling Problem

Fuzzy flexible job shop scheduling problem (FFJSP) is an uncertain FJSP, which extends the JSP through fuzzy working time or fuzzy priority constraints. The processing time of equipment in the workshop is usually set to a fixed value. This assumption is idealized and does not accord with the actual situation, because there are many uncertain interference factors in the actual workshop. However, many researchers use PSO and other intelligent algorithms to solve it. GA is also widely used in FFJSP. Lei, D. [13] proposed a decomposition integration GA for FJSP with fuzzy working time, and the algorithm aims to minimize the maximum fuzzy completion time. The algorithm solves the problem of FJSSP in fuzzy environment. Lei, D. [14] studied the link of fuzzy problem and flexible scheduling in job shop scheduling environment, and proposed an available co evolutionary GA (CGA) for minimizing fuzzy maximum completion time. The paper uses a new crossover operator, and the new selection operation are applied to the algorithm. In order to consider the FJSP with fuzzy delivery time, Shi, D. L. et al. [15] established the FJSP dynamic scheduling model with fuzzy delivery time, and solved the model by improving the immune genetic algorithm (IGA). The simulation verified that the established model and the improved algorithm have good practicability.

4 Conclusion and Prospect

From the analysis, GA is much more applied in solving JSP and FJSP problems, and rarely solves the current focus of scheduling research - low carbon scheduling and production logistics co scheduling. As stated previously, due to the combination characteristics of scheduling problems, hybrid GA is often used to meet the scheduling problem, mainly combining with its hybrid GA applications such as local search, DE, TS, SA and so on, and using genetic characteristics. The excellent features are retained and inherited, and the global search strategy is used to better search the optimal solution. As a multi population intelligent algorithm, the existing design of GA rarely considers the steps of Population Division, population evolution and communication between populations of multi population method, and the GA with global search strategy has not been paid attention to by researchers. The future research directions of GA in the following two aspects are discussed in detail below.

1) Low carbon scheduling. As an optimization problem that needs to optimize energy consumption objectives or constraints, low-carbon scheduling is not only the key

part of realizing carbon neutralization, but also the main way and method to improve the production efficiency of enterprises, and this problem strictly meets the requirements of contemporary environment and energy. At present, the research of GA in low-carbon scheduling has been fruitful. However, there are still some difficulties in its implementation in practical problems. Uncertainty is an inevitable factor in production scheduling. Although the current research has made great progress in the problem of uncertain scheduling, there is less research on the problem of uncertain low-carbon scheduling. It is necessary to combine uncertainty with low-carbon objectives or low-carbon constraints to obtain a more practical and interesting scheduling scheme. In short, the research on GA based low-carbon scheduling in deterministic or uncertain environment should be strengthened.

2) Collaborative scheduling of production logistics. As a scheduling more in line with the real-time workshop, production logistics collaborative scheduling not only has strong anti-interference, but also can improve customer satisfaction. For the current customized demand, many products have been customized, and production logistics collaborative scheduling can better meet customer requirements. Therefore, production logistics collaborative scheduling has more research potential. With the high-speed development of the Internet of things, collaborative scheduling of production logistics has also developed rapidly. In recent years, relevant scholars have done relevant research. At present, these problems have not been deeply studied, and GA is rarely used to meet these problems. In short, it is essential to pay more attention to the application of GA in production logistics collaborative scheduling.

References

1. Katoch, S., Chauhan, S.S., Kumar, V.: A review on genetic algorithm: past, present, and future. Multimed. Tools Appl. **80**(5), 8091–8126 (2020). https://doi.org/10.1007/s11042-020-10139-6
2. Essafi, I., Mati, Y., Dauzere-Peres, S.: A genetic local search algorithm for minimizing total weighted tardiness in the job-shop scheduling problem. Comput. Oper. Res. **35**(8), 2599–2616 (2006)
3. Ombuki, B.M., Ventresca, M.: Local search genetic algorithms for the job shop scheduling problem. Appl. Intell. **21**(1), 99–109 (2004)
4. Rafsanjani, M.K., Riyahi, M.: A new hybrid genetic algorithm for job shop scheduling problem. Int. J. Adv. Intell. Paradig. **16**(2), 157–171 (2020)
5. Pezzella, F., Morganti, G., Ciaschetti, G.: A genetic algorithm for the flexible job-shop scheduling problem. Comput. Oper. Res. **35**(10), 3202–3212 (2007)
6. Huang, X., Yang, L.: A hybrid genetic algorithm for multi-objective flexible job shop scheduling problem considering transportation time. Int. J. Intell. Comput. Cybern. **12**(2), 154–174 (2019)
7. Zhang, G., Gao, L., Shi, Y.: An effective genetic algorithm for the flexible job-shop scheduling problem. Expert Syst. Appl. **38**(4), 3563–3573 (2011)
8. Rangsaritratsamee, R., Ferrell, W.G., Jr., Kurz, M.B.: Dynamic rescheduling that simultaneously considers efficiency and stability. Comput. Ind. Eng. **46**(1), 1–15 (2004)
9. Fattahi, P., Fallahi, A.: Dynamic scheduling in flexible job shop systems by considering simultaneously efficiency and stability. CIRP J. Manuf. Sci. Technol. **2**(2), 114–123 (2010)

10. Azadeh, A., Goodarzi, A.H., Kolaee, M.H., Jebreili, S.: An efficient simulation-neural network-genetic algorithm for flexible flow shops with sequence-dependent setup times, job deterioration and learning effects. Comput. Netw. Commun. **31**(9), 5327–5341 (2019)
11. Seng, D.W., Li, J.W., Fang, X.J., Chen, J.: Low-carbon flexible job-shop scheduling based on improved nondominated sorting genetic algorithm-II. Int. J. Simul. Model. **17**(4), 712–723 (2018)
12. Sun, X., Wang, Y., Kang, H., Shen, Y., Chen, Q., Wang, D.: Modified multi-crossover operator NSGA-III for solving low carbon flexible job shop scheduling problem. Processes **9**(1), 62 (2021)
13. Lei, D.: A genetic algorithm for flexible job shop scheduling with fuzzy processing time. Int. J. Prod. Res. **48**(10), 2995–3013 (2010)
14. Lei, D.: Co-evolutionary genetic algorithm for fuzzy flexible job shop scheduling. Appl. Soft Comput. **12**(8), 2237–2245 (2012)
15. Shi, D.L., Zhang, B.B., Li, Y.: A multi-objective flexible job-shop scheduling model based on fuzzy theory and immune genetic algorithm. Int. J. Simul. Model. **19**(1), 123–133 (2020)

Design of Convolutional Neural Network Accelerator Based on RISC-V

Shaopeng Jin, Shuo Qi, Yilin Dai, and Yihu Hu(✉)

Electronic and Communication Engineering, Yanbian University, Yanji, Jilin, China
xuyh@ybu.edu.cn

Abstract. Nowadays, it is more and more difficult for general-purpose CPUs to meet the large-scale computing of neural networks, and research on domain-specific neural network accelerators is gradually emerging. Limited by the high cost of customization and lack of flexibility due to patents and licenses of x86 and ARM commercial architecture CPUs, this paper designs a RISC-V based convolutional neural network accelerators for volume and neural network hardware acceleration platforms to improve the arithmetic power of processors. Finally, verification is performed by writing test beach files under the vivado platform. The results show that for convolutional computation, the operation speed is effectively improved with the accelerators architecture compared to using only the basic instruction set.

Keywords: RISC-V · Convolutional Neural Network · Accelerator

1 Introduction

Since AlexNet won the ILSVRC competition in 2012, Convolutional Neural Network (CNN) started to attract attention and gradually applied to image classification and detection, face recognition and other fields [1–4]. However, with the rapid development of deep learning technology, CNN is improving the recognition accuracy while the network depth is deepening and the structure is becoming more complex, which leads to a dramatic increase in the computational load of the network [5, 6]. General-purpose processors are very inefficient and have a high energy overhead when dealing with large and complex convolutional computations. Therefore, it is necessary to design a hardware architecture specifically for accelerating convolutional computation.

RISC-V is an open source streamlined instruction set architecture proposed by the University of California, Berkeley (UCB) [7]. RISC-V consists of 3 basic instruction sets and 6 extended instruction sets. On the premise of implementing the basic instruction set, you can select the corresponding extended instruction set according to your own needs [8]. Compared to the x86 architecture, which is mainly used for desktop computers, and the ARM architecture, which is used for embedded devices, the RISC-V architecture has the advantage of being versatile and flexible, and the open source design approach avoids paying high licensing fees. Therefore, RISC-V has received a lot of attention from researchers in recent years and has achieved some development [9, 10].

J. H. Abawajy et al. (Eds.): ICATCI 2022, LNDECT 170, pp. 446–454, 2023.
https://doi.org/10.1007/978-3-031-29097-8_53

The calculation in CNN is mainly convolutional calculation, but also includes some other types of calculation [11]. Therefore, the idea of software and hardware co-design is usually adopted, that is, a special accelerator is designed to improve the efficiency of convolution calculation, and then a general-purpose processor is used to complete some calculations that occupy a small proportion in the network and do not need to be implemented in the hardware architecture. At present, the co-design of software and hardware based on convolution accelerator is mainly realized by building SoC system in FPGA platform, and most of the general-purpose processors used are based on ARM architecture. Some designs that combine RISC-V processors with convolutional accelerated computing have emerged, but overall the number is still small.

This paper extends the Hummingbird E203 kernel based on the open source RISC-V architecture with four custom CNN acceleration instructions and performs deep architectural fusion on the original RISC-V processor to achieve acceleration of convolutional neural networks.

2 Design of Convolutional Neural Network Accelerator Based on RISC-V

2.1 RISC-V Instruction Set Architecture

One of the most significant features of the RISC-V instruction set architecture is its scalability. The RISC-V instruction set is a combination of a basic instruction set and an extended instruction set. Both the base instruction set and the extended instruction set are composed of multiple modular instruction sets. Each modular instruction set is represented by an uppercase letter. For example, the letter I indicates the basic integer instruction set. The extended instruction set contains five categories, the letter M for the integer multiplication and division instruction set, the letter A for the atomic operation class, the letter F for the single-precision floating-point instruction set, the letter D for the double-precision floating-point instruction set, and the letter C for the compression instruction. The basic integer instruction set is the instruction set mandated by RISC-V, and the six extended module instruction sets are selected in different combinations according to the different areas of the processor application.

2.2 Overall System Architecture

The Hummingbird E203 processor is mainly used for embedded applications with low power consumption and small area. The design of this paper is based on the E203 processor core and connects the gas pedal to the E203 core through the NICE interface to achieve the function of passing the neural network instructions to the accelerator when the E203 core accepts them and returning the results to the E203 core after the accelerator has processed the instructions. The overall system architecture diagram is shown in Fig. 1.

The system is roughly divided into three parts: the E203 core, the NICE interface, and the coprocessor, etc. The main role of the E203 core is to receive instructions from the outside and to process other instructions except for the neural network instructions.

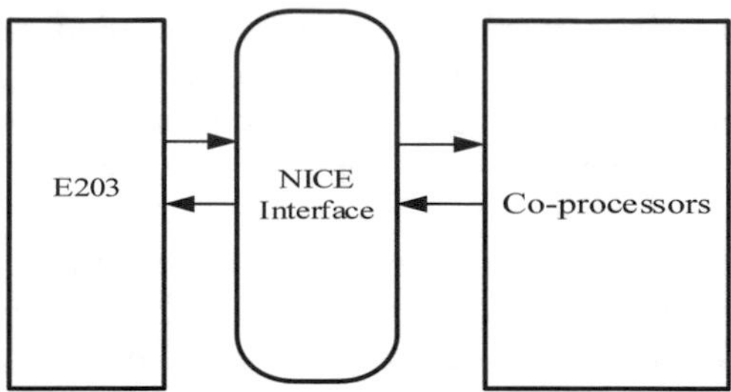

Fig. 1. System overall block diagram

It mainly consists of four types of signals: instruction request control signals, instruction feedback control signals, memory read/write request signals, and memory read/write feedback signals. The co-processor mainly completes the convolutional neural network related instructions and returns the results of the related instructions to the E203 core through the NICE interface.

2.3 NICE Interface

The E203 processor provides a NICE interface, which allows the integration of custom hardware co-processing units. Users can customize the corresponding co-processor computing units according to personal needs and the scalable Custom 0, 1, 2, and 3 instruction groups to reduce power consumption. Improve performance in specific areas. The opcodes of the Custom instruction are shown in Table 1.

Table 1. Extended instruction opcode

Custom	Inst [6:5]	Inst [4:2]	Inst [1:0]
Custom-0	00	010	11
Custom-1	01	010	11
Custom-2	10	110	11
Custom-3	11	110	11

The format of the extension instructions that can be recognized by the host processor and passed through the NICE interface is shown in Table 2.

Table 2. NICE instruction format

Command bits	[31:25]	[24:20]	[19:15]	[14]	[13]	[12]	[11:7]	[6:0]
Command bit name	Funct7	rs2	rs1	Xd	xs1	xs2	rd	opcode

Table 3 shows the detailed description of each field in the NICE directive.

Table 3. Description of each field of NICE command

field	Describe
opcode	Choose from 4 custom command groups
rd	target register index
xs2	If this bit is set, source register 2 will be read
xs1	If this bit is set, source register 1 will be read
xd rs1 rs2	If this bit is set, the target memory will be written to Source Register 1 Index Source Register 2 Index
funct	Coding different custom directives

The main function implemented in this design is convolutional operation. The extended instructions designed in this paper are shown in Table 4.

Table 4. Fouction description table of extension instruction and corresponding bit

Instruction	funct7	rs2	rs1	xd	xs1	xs2	rd
JSMP	0000000	Register Address 2	Register Address 1	0	1	1	-
FCONV	0000001	Register Address 2	Register Address 1	0	1	1	-
SCONV	0000010	Register Address 2	Register Address 1	0	1	0	-
JWLD	0000110	Type Configuration	Register Address 1	0	0	0	Pooling size

The instruction JSMP is responsible for loading the original image data required for the convolution operation into the co-processor. The instruction FCONV is responsible for loading the required convolution kernel data of the first convolution layer into the coprocessor. The instruction SCONV is similar to FVONV, except that SCONV loads the convolution kernel parameters of the second convolution layer into the coprocessor. The instruction JWLD is responsible for starting the coprocessor to complete the convolution operation.

2.4 Convolutional Neural Network Co-processor Hardware Architecture

Since convolutional neural networks involve a large amount of data, it is necessary to design the access memory to reduce the number of accesses and thus reduce the system power consumption. In this paper, we address this problem using an approach to improve data reuse. A pulsating array is a powerful data reuse method. A pulsating array consists of a large number of regular processing units, where data flows in each processing unit thus reusing data and reducing accesses, and the entire column can produce multiple computation results at the same time. Row fixed systolic array can flow data in 3 directions, which can greatly reuse data。Additive trees can also perform convolution calculations, the operations are performed by an integer column of parallel tree-like multipliers and adders. It can be easily computed in a pipelined manner. Inspired by the above research work, this paper designs a co-processor that can compute double layer convolution simultaneously as shown in the Fig. 2 to reduce the number of accesses and hardware resources, which can achieve the speedup function.

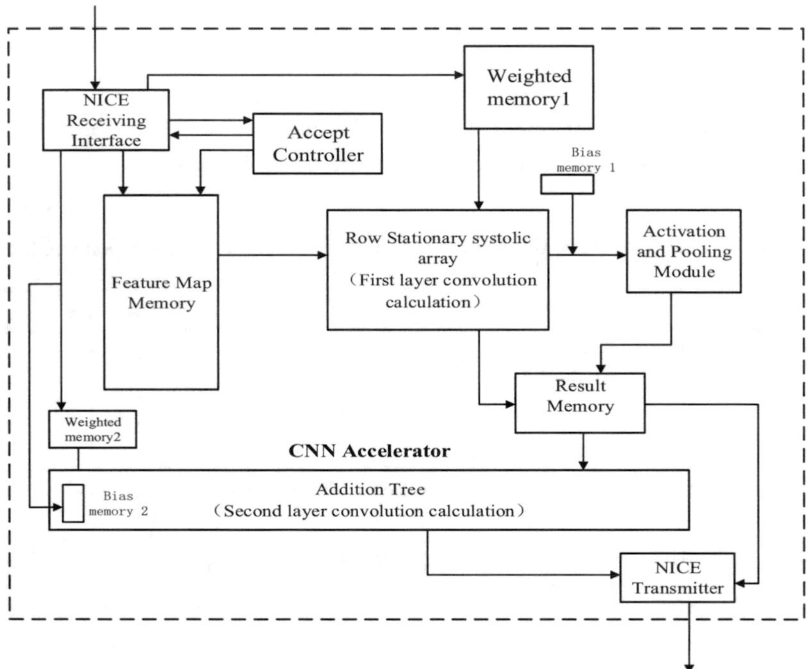

Fig. 2. Convolutional neural network co-processor internal overall module diagram

The NICE receiving interface is responsible for receiving instructions and data from the E203 processing. When the received instructions are the four coprocessor instructions defined in this paper, the coprocessor will start to work. When the received instruction is the co-processor instruction JSWP, the receiving controller will control the memory system of the E203 processor via the ICB bus of the NICE interface and carry the image data (feature map) to be processed stored in the E203 memory to the feature map memory of the co-processor. When the received instruction is the coprocessor instruction FCONV, the receiving controller will control the memory system of the E203 processor through the ICB bus of the NICE interface and load the convolution kernel 1 and bias 1 data stored in the E203 memory into the weight memory 1 and bias memory 1 of the coprocessor. When the received instruction is the coprocessor instruction SCONV, the receive controller will control the memory system of the E203 processor via the ICB bus of the NICE interface and load the convolution kernel 2 and bias 2 data stored in the E203 memory into the weight memory 2 and bias memory 2. When the received instruction is the coprocessor instruction JWLD, the receive controller will configure the line fixed pulsation array, the activation and pooling module result memory, the addition tree, the addition tree controller, and the NICE sender based on the received data. If it is a one-layer convolution calculation type, the receiving controller will load the data in the feature map memory and weight memory 1 into the row-fixed pulsed array for convolution calculation, and when the calculation is completed, the data output from the pulsed whole column will be added to the data in the bias memory 1 to produce the final convolution result and store the result into the result memory. When the storage is completed, the NICE generator will transfer the convolution results from the result memory to the memory of the E203 processor via the ICB bus. The whole process is a pipeline operation. In case of two-layer convolution operation type, the receiving controller will configure the activation and pooling module on top of the one-layer convolution operation. When the first layer of data is processed through the row fixed pulsation array, it is fed directly to the activation and pooling module. When the activation and pooling are completed, the data are written to the result memory, and then the additive tree controller will load these results and the data in the weight memory2 into the additive tree sequentially for the 2nd convolutional layer, and the NICE generator will return the computation results to the memory of the E203 processor synchronously. The whole computation process is also pipelined.

3 Simulation Verification and Results

3.1 System Function Simulation

The instruction JWLD is responsible for starting the coprocessor to complete the convolution operation and writing the result back to the memory of E203 processor. Before executing this instruction, the feature map, convolution kernel, and bias data to be processed should be moved to the coprocessor. To verify the instruction, a 4×4 feature map with a 3×3 convolution kernel is used for the convolution operation (resulting in 2×2), as shown in Table 5. The first few lines of code are used to load the feature map, convolution kernel, bias, and read the memory start address. The functional simulation waveform is shown in Fig. 3. When the instruction JSMP is executed (line 8 code), after some time, the access ICB bus of the NICE interface writes 4 convolutional results to the memory of the E203 processor, which is consistent with the theoretical number, and the instruction can be executed. This execution consumed a total of 24 clock cycles, which compares to the basic instruction architecture as shown in Table 6, and it can be seen that using the coprocessor architecture can greatly accelerate the convolutional computation compared to the basic instruction set architecture.

Table 5. Verification Code

1: 000000000100_00000_000_00001_0010011
2: 000000010000_00000_000_11111_0010011
3: 0000000_11111_00001_001_00001_0110011
4: 000000000100_00001_000_00001_0010011
5: 000000000001_00000_000_00010_0010011
6: 000000011100_00000_000_11111_0010011
7: 0000000_11111_00010_001_00010_0110011
8: 0000000_00010_00001_011_00000_0001011
9: 000000000011_00000_000_00011_0010011
10: 000000010000_00000_000_11110_0010011
11: 0000000_11110_00011_001_00011_0110011
12: 000000000011_00011_000_00011_0010011
13: 0000001_00010_00011_011_00000_0001011
14: 0000111_00000_00001_000_00000_0001011

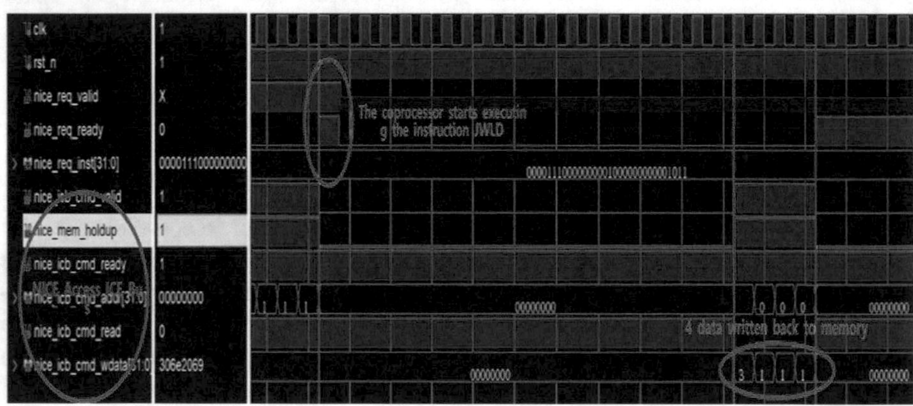

Fig. 3. Functional simulation waveform diagram

Table 6. Instruction Run Cycle

Instruction	RISC-V I&M	Accelerator command
Time (clock cycle)	13,267	24

3.2 Hardware Resource Consumption Analysis

Perform resource and consumption analysis of the overall system with the support of Vivado 2018.3 tools. The resource consumption of the main functional units is shown in Table 7. The hardware resources of the coprocessor account for a large proportion, and the LUT resources account for 95.2% of the overall system.

Table 7. Hardware resource consumption

Components	Slice LUTs	Block RAM Tile	DSPs	Slice Registers	F7 Muxes	F8 Muxes
General	71080	9	134	256,000	448	32
Co-processors	67644	7.5	134	N/A	192	0
E203	3,436	1.5	0	N/A	256	32

4 Conclusion

This paper implements a RISC-V based CNN accelerator. The accelerator is based on the Hummingbird E203 core, through the NICE interface, plus a coprocessor to achieve. The coprocessor is specialized to perform neural network operations, and its functions include convolution, activation, and pooling functions. In order to speed up the convolution operation, this paper uses systolic array, double-layer convolution, etc. for data multiplexing. The function of the coprocessor is verified on the vivado platform, and the results show that the architecture with the accelerator is 552.8 times faster than the architecture without the accelerator during the convolution calculation of the matrix, which greatly improves the computing power of the processor.

References

1. Poliyev, A.V., Korsun, O.N.: Speech recognition using convolutional neural networks on small training sets. IOP Conf. Ser.: Mater. Sci. Eng. **714**(1), 012024 (2020)
2. Li, X., Yang, Z., Wu, H.: Face detection based on receptive field enhanced multi-task cascaded convolutional neural networks. IEEE Access **8**, 174922–174930 (2020)
3. Sun, K., Li, Q., Li, D.: Face detection algorithm based on cascaded convolutional neural network. J. Nanjing Univ. Sci. Technol. **42**(1), 40–47 (2018)
4. Ren, S., He, K., Girshick, R., Sun, J.: Faster R-CNN: towards real-time object detection with region proposal networks. IEEE Trans. Pattern Anal. Mach. Intell. **39**(6), 1137–1149 (2017)
5. Qiao, Y., Shen, J., Xiao, T., Yang, Q., Wen, M., Zhang, C.: FPGA-accelerated deep convolutional neural networks for high throughput and energy efficiency. Concurrency & Comput. **29**(20), e3850.1-e3850.20 (2017)
6. Simonyan, K., Zisserman, A.: Very deep convolutional networks for large-scale image recognition. Comput. Sci. (2014)

7. Waterman, A., Lee, Y., Patterson, D.A., Asanović, K.: The RISC-V instruction set manual, volume i: user-level ISA. EECS Dept. **7**(9), 475 (2011)
8. He, Y.: Design and implementation of convolutional neural network accelerator based on RISCV. J. Phys.: Conf. Ser. **1871**(1), 012073 (2021)
9. Zhu, J., Wang, L., Liu, H., Tian, S., Deng, Q., Li, J.: An efficient task assignment framework to accelerate DPU-based convolutional neural network inference on FPGAS. IEEE Access **8**, 83224–83237 (2020)
10. Cavalcante, M., Schuiki, F., Zaruba, F., Schaffner, M., Benini, L.: Ara: a 1 ghz+ scalable and energy-efficient RISC-V vector processor with multi-precision floating point support in 22 nm FD-SOI. IEEE Trans. Very Large Scale Integr. Syst. **28**, 530–543 (2019)
11. Yoo, Y., Park, Y., Kim, I., Yi, K.: FPGA-based deep convolutional neural network accelerator design techniques for the handwritten number recognizer. Adv. Sci. Lett. **24**(3), 2152–2155 (2018)

Design and Implementation of English Teaching Analysis System Based on BP Neural Network Algorithm

Zhiyou Zhang[1,2](✉), Fangwen Chen[2], and Yue He[2]

[1] Department of Tourism, Sichuan Engineering Technical College, Deyang 610086, Sichuan, China
zzy@scetc.edu.cn

[2] Foreign Language Department, Sichuan Engineering Technical College, Deyang 610086, Sichuan, China

Abstract. With the development, my country's education level and the quality of students' learning have been rapidly improved. However, the teaching task is heavy, but the quality is not high and the efficiency is low, which seriously restricts the development of education. Under such circumstances, it is necessary to conduct teaching analysis, and effectively analyze various teaching data, which is conducive to discovering the internal connections, thereby promoting the efficient conduct of teaching decision-making and management. At the same time, the BP neural network (BPNN) is a self-adaptive, multi-input nonlinear system, which can obtain a method of approaching the optimal solution through learning and training. This article combines the BPNN algorithm to explore its role in the design and realization of the English teaching (ET) analysis system. The system has a higher degree of fit for teaching analysis, and the obtained analysis results have smaller errors. Correspondingly, it will also improve the efficiency of ET analysis.

Keywords: BP Neural Network · Teaching Analysis · Teaching Effect · Teaching Diagnosis

1 Introduction

BPNN algorithm is an algorithm for processing nonlinear functions. It can decompose complex problems into many easy to solve, easy to apply and strong robustness. In actual teaching analysis work, manual processing and analysis of large amounts of data is not ideal due to the large number of students and scattered teaching locations. Therefore, the use of BPNN can effectively gather and store these huge amounts of information on a small scale, which is of great help to the development of subsequent teaching analysis work.

Some scholars pointed out that the model based on the BPNN algorithm has good accuracy and generalization in the evaluation of the atmospheric particle index, which provides the possibility to scientifically and accurately evaluate and control the urban air

J. H. Abawajy et al. (Eds.): ICATCI 2022, LNDECT 170, pp. 455–461, 2023.
https://doi.org/10.1007/978-3-031-29097-8_54

fine particle index [1]. Some scholars believes that the BP network teaching quality evaluation method reduces the dimensions of multi-dimensional teaching quality evaluation indicators, and proves that the method is feasible. It solves the complex modeling problems of conventional evaluation methods, avoids subjective arbitrariness, and ensures Evaluate the validity of the results [2]. Some scholars proposed that in order to improve the scientific rationality of the comprehensive evaluation of the quality of pharmacy experimental training, an evaluation model based on improved BPNN was proposed [3]. This article explores the design of the ET analysis system based on the BPNN algorithm.

This article focuses on these aspects. First, BPNN and related studies were presented. In addition, the concepts of didactic analysis and related research were discussed. Finally, taking for example the ET analysis system, appropriate experimental results and conclusions from the analysis are obtained.

2 Related Theoretical Overview and Research

2.1 BPNN and Related Research

From the point of view of the application of neural networks, there are only a dozen kinds of researches at present, among which the most researched and most representative network is the BPNN.

As the neural network converges very slowly and the structure is difficult to determine, the scalability of the designed network cannot be guaranteed. The BPNN has these characteristics.

One is the capability of massively parallel processing. This ability is reflected in the distribution of interconnected neurons, and is superior to modern digital computers for serial processing of discrete symbols.

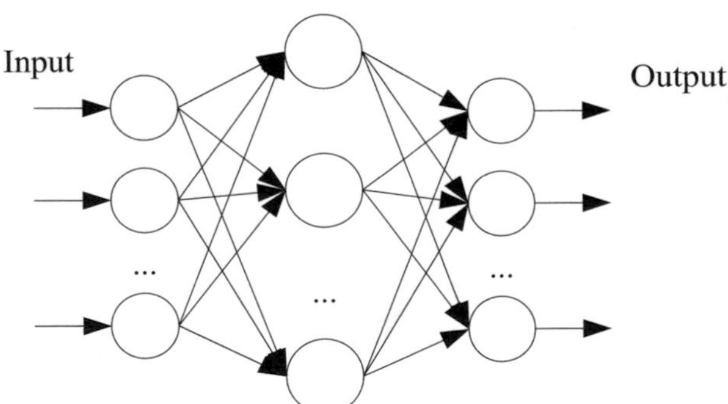

Fig. 1. Topological Structure of BPNN

The second is to have strong self-learning and self-adaptive capabilities. And the third is strong robustness and high fault tolerance with associative memory function [4, 5].

The neural network topology of a single hidden layer is shown in Fig. 1.

Generally, due to different network models, especially the number of layers and hidden nodes, the number is difficult to determine. If the number of single hidden layer nodes of BPNN is too small, the network may not be well trained. Too many nodes make the learning time too long, the error is also large, and may cause over-fitting to make the generalization ability of the network worse. Therefore, it is necessary to find the best single hidden layer node. The calculation method of the number of single hidden layer nodes is shown in formula (1) and (2).

$$g = \sqrt{a + b} + c \tag{1}$$

$$g = \log_2 a \tag{2}$$

Among them, g is the number of hidden layer nodes, a is the number of input neurons, b is the number of output neurons, and the value range of c is [1–10].

During the learning process, the input information will continue to increase as the output structure and parameters change, which makes it necessary to perform a lot of exercises to adjust the weights during neural network training. In a sample space, all neurons have their own unique processing capabilities. This processing method is to achieve the optimal solution optimization process of the entire system performance by calculating all the attribute weights on the data set corresponding to each layer of neurons, so that the algorithm convergence speed can be improved.

The algorithm of neural network is simple, easy to implement, and can be continuously improved according to specific conditions. It uses the input vector as the output variable to train the classifier parameters to determine the output weights, and then compares the various feature quantities on the input sample set with the corresponding weights, and optimizes and adjusts the results according to the obtained results [6, 7].

2.2 Analysis of ET and Related Research

Classroom teaching is the main channel for project dissemination and the first line to promote quality education. Teaching analysis is a means to improve teaching quality. The teaching process is the process of transferring information between teachers and learners, and also the process of classroom teaching behaviors developing and changing along the time axis.

Educational analysis includes clarifying the various components, elements and aspects of teaching by decomposing the classroom teaching or teaching process, so as to obtain an understanding and clear assessment of the teaching process. Specifically, teaching analysis is to analyze face-to-face teaching according to certain quality requirements. In this process, inspections are conducted according to the teaching rules, teaching goals and requirements, and the actual situation of the students. This is an objective decomposition of the complex teaching process, using modern educational technology, through the extraction and analysis of information in the face-to-face teaching process, the formation of analysis results, and provide a basis for objective evaluation of the teaching process.

Therefore, educational analysis should be based on educational theory and educational teaching rules, and data processing requires strict mathematical analysis methods and modern educational technology.

The basic theme of educational analysis is the teaching process. It is the process of describing, explaining, measuring and evaluating problems using certain analytical techniques in classroom practice. However, it is not enough to summarize the various elements and attributes of teaching based on the cognitive process. Although the analysis process will vary depending on the analysis perspective, the goal of promoting students' understanding in the learning process is always the same. The analyzed cognitive activities are accompanied by many other factors and have their own particularities [8, 9].

Educational analysis is a process influenced by many value-oriented factors. First of all, it is affected by the time when the educational development process takes place and the social environment. Different times and societies have different understandings and requirements for teaching, and the value orientation of teaching is inevitably restricted. Secondly, the analyst's background system has an impact on analysis teaching, and his knowledge system, social norm awareness, value system, etc. will all have an impact on analysis work.

Teaching analysis is based on the existing educational reality and proposes new ideas for optimizing education. In short, educational analysis is the process of learning about teaching activities. It is a cognitive activity that combines value, art, creation and production.

First, teaching analysis is an important way to diagnose classroom problems and provide classroom feedback information in order to adjust teaching strategies in a targeted manner.

Second, teaching analysis is the premise of teaching evaluation. The so-called educational evaluation mainly refers to the systematic process of objective measurement and scientific evaluation of educational activities and their effects through the collection of various measures and related data on the basis of certain objective standards.

Third, educational analysis is the need to objectively evaluate teaching. Only when the teaching evaluation information is specific and objective, can the results of the teaching evaluation be convincing, and only then can we find specific guidelines and targeted methods to improve teaching [10, 11].

In general, the analysis of teaching by observers of classroom teaching helps us to understand the teaching process objectively, to recognize the deficiencies in the teaching process, and to make adjustments and improvements.

2.3 Application of BPNN Algorithm in the Process of Constructing ET Analysis System

The BPNN method is a nonlinear analysis algorithm based on the learning process, which is mainly used in complex systems. This model is very robust and can improve the linear relationship between input and output to a certain extent.

With the increasing importance of urban and rural differences and individual personality differences of students, how to teach, correctly guide and cultivate high-quality, all-round and comprehensive talents according to the different characteristics of students has become a subject of continuous research and research. Therefore, the analysis of ET is also a subject of scientific research.

Among them, the classroom teaching evaluation system is an important part of teaching analysis. A complete classroom analysis grading system should include grading objects, grading topics, grading index system and grading standards. The subject of evaluation is the subject of evaluation, and the subject of evaluation is the subject of evaluation. The evaluation index system is a collection of all factors extracted according to its objectives and related principles and theories in the evaluation of teaching activities. It specifies the specific teaching activities to be restored.

The ET analysis system incorporates new curriculum concepts, creates a new evaluation standard for ET, and also provides a way for teachers to improve their professionalism. The use of self-reflection and diagnostic evaluation in classroom teaching practice has played a very good guiding role in the innovation of ET and the professional development of teachers.

As an effective platform for teachers to self-analyze classroom activities, the classroom teaching analysis system not only focuses on teacher teaching, but also focuses on learning. Therefore, considering the many advantages of the BPNN algorithm, this article applies the algorithm to the construction process of the ET analysis system, which will help reduce the design of the system, thereby promoting the scientific and efficient development of the ET analysis work [12, 13].

3 Experiment and Research

3.1 Experimental Environment

This experiment builds an ET analysis system. Since the education analysis system is an advanced database management system, in the case of a given data record, a complex analysis of the recorded data is required. Therefore, the feasibility and scalability of the database should be considered when designing the database system. The system decided to use SQL Server 2016 as the database development environment and Visual Studio 2015 as the integrated development tool. The hardware environment includes Pentium (R) + 512 M + 40 G PC. System operating environment: Windows XP.

3.2 Experimental Process

Four classes in a middle school in L city are selected for testing. The test items include teaching content, course structure, course management, teaching effect and so on.

4 Analysis and Discussion

Four classes of a middle school in L city are selected for testing. The test items include teaching content, course structure, course management, teaching effect and so on. The test results are shown in Table 1.

It can be seen from Fig. 2 that the degree of fit of the system to Class C ET analysis is the highest among these four classes, and the degree of fit of teaching content,

Table 1. ET Analysis Results

Project	Class A	Class B	Class C	Class D
Teaching content	0.91	0.95	0.9	0.92
Course structure	0.92	0.88	0.93	0.91
Course management	0.89	0.92	0.97	0.94
Teaching effect	0.95	0.94	0.96	0.89

course structure, course management, and teaching effect are 0.9, 0.93, 0.97, and 0.96, respectively. The system has a high degree of fit for teaching analysis, and the error of the analysis results obtained is also small, so that the efficiency of ET analysis will be improved accordingly.

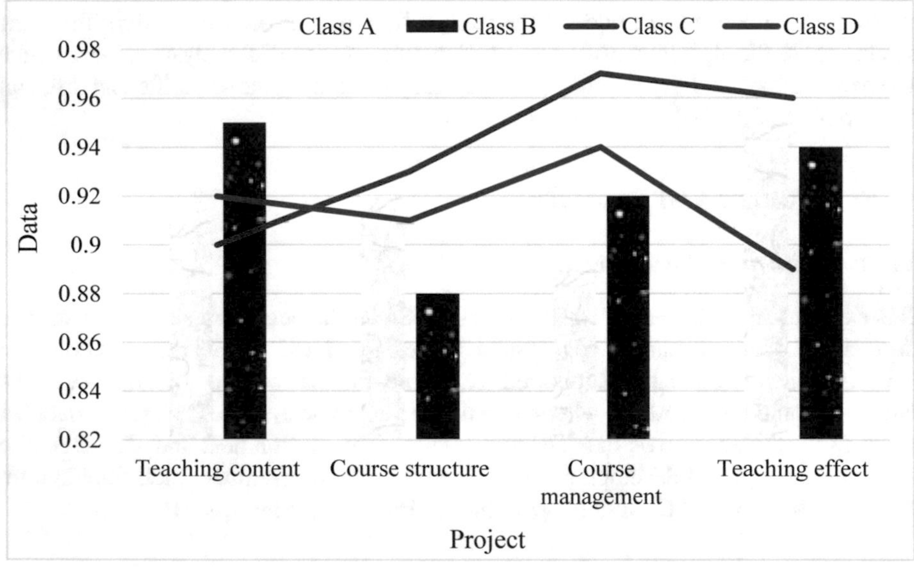

Fig. 2. ET Analysis Results of the System

5 Conclusion

Teaching analysis work has great practical significance for optimizing teaching work. Therefore, the development of teaching analysis work is very necessary. At the same time, the BPNN algorithm has strong learning ability and computing ability, and has been widely used in many fields. Therefore, applying the BPNN algorithm to the analysis of ET will improve the efficiency of the analysis of ET. Therefore, this article combines the BPNN algorithm to explore the design and implementation of the ET analysis system.

References

1. Chai, R., Tsourdos, A., Savvaris, A., Chai, S., Xia, Y., Chen, C.L.P.: Design and implementation of deep neural network-based control for automatic parking maneuver process. IEEE Trans. Neural Netw. Learn. Syst. **33**, 1400–1413 (2022)
2. Pandey, R.K., Lin, T.Y., Chao, C.P.: Design and implementation of a photoplethysmography acquisition system with an optimized artificial neural network for accurate blood pressure measurement. Microsyst. Technol. **27**(3), 1–23 (2021)
3. Nagata, F., Watanabe, K., Habib, M.K.: Design and implementation of convolutional neural network-based svm technique for manufacturing defect detection. Int. J. Mechatron. Autom. **8**(2), 1 (2021)
4. Wazir, A.S.B., et al.: Design and implementation of fast spoken foul language recognition with different end-to-end deep neural network architectures. Sensors **21**(3), 710 (2021)
5. Mohanraj, K.A.: Design and implementation of ann based scc with ggbs using auromix 400. Int. J. Recent Technol. Eng. **9**(2), 776 (2020)
6. Utomo, A.H., Gumilang, M.A., Ahmad, A.: Agricultural commodity sales recommendation system for farmers based on geographic information systems and price forecasting using probabilistic neural network algorithm. IOP Conf. Ser.: Earth Environ. Sci. **980**(1), 012061 (2022)
7. Fapi, C.B.N., Wira, P., Kamta, M., Tchakounté, H., Colicchio, B.: Simulation and dspace hardware implementation of an improved fractional short-circuit current mppt algorithm for photovoltaic system. Appl. Solar Energy **57**(2), 93–106 (2021)
8. Swetharani, K., Prasad, G.V.: Design and implementation of an efficient rose leaf disease detection and classification using convolutional neural network. Int. J. Image Min. **4**(1), 98 (2021)
9. Novita, R., Anzir, Q., Mustakim, Alhidayatillah, N., Suriani, J.: The Implementation of probabilistic neural network algorithm for classification of family hope program in Pekanbaru city. J. Phys.: Conf. Ser. **1783**(1), 012018 (2021)
10. Carvalho, C.B., Carvalho, E.P., Ravagnani, M.A.S.S.: Implementation of a neural network mpc for heat exchanger network temperature control. Braz. J. Chem. Eng. **37**(4), 729–744 (2020)
11. Mahmoud, M., Kumar, R.: A review on chatbot design and implementation techniques. Int. J. Eng. Technol. **7**(2), 2791 (2020)
12. Patel, S., Canoza, P., Salahuddin, S.: Logically synthesized and hardware-accelerated restricted boltzmann machines for combinatorial optimization and integer factorization. Nat. Electron. **5**(2), 92–101 (2022)
13. Morozov, A.Y., Abgaryan, K.K., Reviznikov, D.L.: Mathematical modeling of an analogue self-learning neural network based on memristive elements taking into account stochastic switching dynamics. Nanobiotechnol. Rep. **16**(6), 767–776 (2022)

Covering Device Structure Optimization Based on the Technology of Computer Data Analysis

Liyan Li, Xin Gong, Xuhao Lv, Wenchao He(✉), Ting Xu, Longkai Liang, Xuejun Li, and Liying Zhang

Institute of Technology, Changchun Humanities and Sciences College, Changchun, Jilin, China
xiao6666662021@126.com

Abstract. During the soil tank test, the uniform test scheme is generated by using the Design Expert software, and the test data are filled into the software. The data are subjected to multiple linear regression and binomial fitting through the analysis module of the software to obtain the mathematical model between the soil cover thickness, seed displacement and various factors, The variation of soil cover thickness and seed displacement with various factors are comprehensively analyzed in the computer software, and the optimal structure of soil cover is selected in the optimal combination module of the software.

Keywords: Design Expert · Computer Data Analysis · Structure Optimization

1 Soil Channel Test and Result Analysis

In the test process, the sliding knife opener is driven by the test bed to open the seed furrow, and the corn seeds are put into the seed furrow according to the pre-drawn position, and then the test bed drives the cover device to complete the soil covering. The speed of the car is maintained at about 1.2 m/s, and the data is manually measured and recorded [1–3].

2 Test Plan and Result Analysis for Measuring the Thickness of Overburden

In the soil tank test for measuring the thickness of the covering soil, the diameters of the small and large discs, the offset distance of the covering soil, and the opening angle of the disc are selected as the test factors. Each factor is selected at 3 levels. A uniform test scheme is used. Each group of tests measures ten corn seeds. Take the average value as the test result [4, 5].

Using Design Expert software to generate a uniform test plan, and fill in the test results in the table, as shown in Table 1. The maximum covering soil thickness is 60.763 mm, the minimum is 55.743 mm, and the average is 58.229 mm.

J. H. Abawajy et al. (Eds.): ICATCI 2022, LNDECT 170, pp. 462–469, 2023.
https://doi.org/10.1007/978-3-031-29097-8_55

Table 1. Test schedule and results

Test number	Diameter of small disc A (mm)	Diameter of large disc B (mm)	Overlying offset C (mm)	Disc opening angle D (°)	Cover thickness Y_1 (mm)
1	220.00	245.00	170.00	55.00	56.738
2	220.00	270.00	150.00	55.00	57.147
3	210.00	295.00	170.00	65.00	60.355
4	220.00	270.00	170.00	65.00	58.481
5	220.00	245.00	150.00	65.00	55.743
6	220.00	270.00	170.00	65.00	58.352
7	230.00	245.00	170.00	65.00	56.671
8	220.00	270.00	170.00	65.00	57.891
9	230.00	270.00	170.00	55.00	58.363
10	220.00	295.00	170.00	55.00	58.891
11	210.00	270.00	190.00	65.00	59.503
12	210.00	270.00	170.00	55.00	56.778
13	220.00	245.00	190.00	65.00	58.42
14	210.00	270.00	170.00	75.00	57.79
15	230.00	270.00	170.00	75.00	58.371
16	220.00	295.00	170.00	75.00	59.078
17	220.00	270.00	190.00	55.00	58.45
18	210.00	270.00	150.00	65.00	57.177
19	220.00	270.00	190.00	75.00	58.629
20	230.00	270.00	150.00	65.00	59.28
21	220.00	270.00	170.00	65.00	58.786
22	230.00	295.00	170.00	65.00	60.763
23	210.00	245.00	170.00	65.00	56.87
24	220.00	295.00	190.00	65.00	58.971
25	220.00	295.00	150.00	65.00	57.98
26	230.00	270.00	190.00	65.00	59.861
27	220.00	270.00	170.00	65.00	58.98
28	220.00	245.00	170.00	75.00	56.203
29	220.00	270.00	150.00	75.00	58.147
Average					58.229

The data is subjected to multiple linear regression and binomial fitting through the analysis module of the software, and the mathematical model between the thickness of the overlying soil and each factor is obtained as follows:

$$\begin{aligned} Y_1 = 58.5 + 0.4A + 1.28B + 0.7C + 0.15D + 0.15AB - 0.44AC - 0.25AD - \\ 0.42BC + 0.18BD - 0.21CD + 0.3A^2 - 0.34B^2 - 0.009625C^2 - 0.6D^2 \end{aligned} \quad (1)$$

Carrying out the significance analysis of the model, the significance P is 0.0024 < 0.05, and the correlation R is 0.8336, indicating that the regression equation is significant and reasonable, and the lack of fit is not significant, indicating that the test points can be described by the model. According to the data, the main factors (P < 0.05) that affect the thickness of the overburden are: B (diameter of the large disc), C (offset of the overburden), and D2 (square term of the opening angle of the disc) in descending order of the degree of influence [6–8].

Figure 1 is a single factor analysis diagram. Within a certain range, the thickness of the covering soil increases linearly with the increase of the diameter of the large disk and the offset distance of the covering soil. The former increases more obviously, and the opening angle of the disk is less than 65° The thickness of the overburden within a certain range shows an increasing trend, and the certain range with a disc opening angle greater than 65° shows a downward trend.

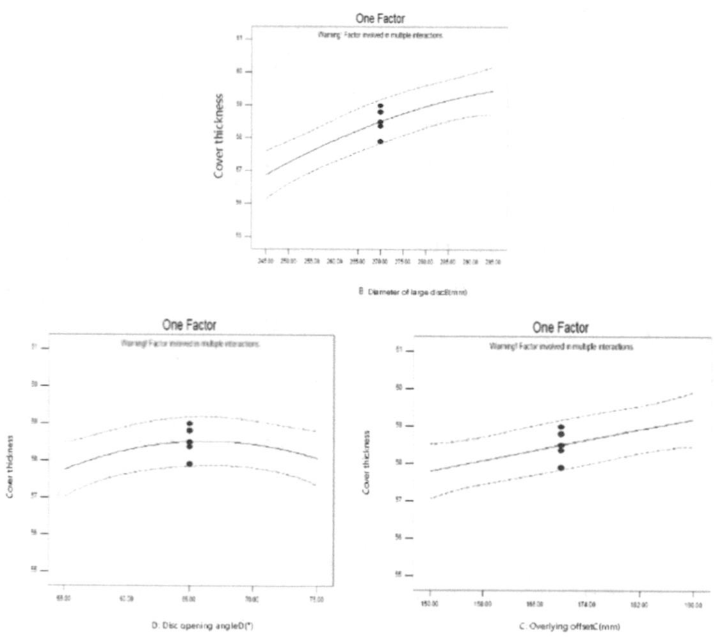

Fig. 1. Single factor data analysis

Figure 2 is the response surface of the overburden thickness to the diameter of the large and small overburden discs. It can be seen from the figure that under the condition that other factors remain unchanged, the overburden thickness increases with the increase of the diameter of the small disc and the large disc respectively, the latter trend is more obvious and the rising speed is faster, and the overburden thickness also increases with the increase of the interaction term of the diameter of the large and small discs.

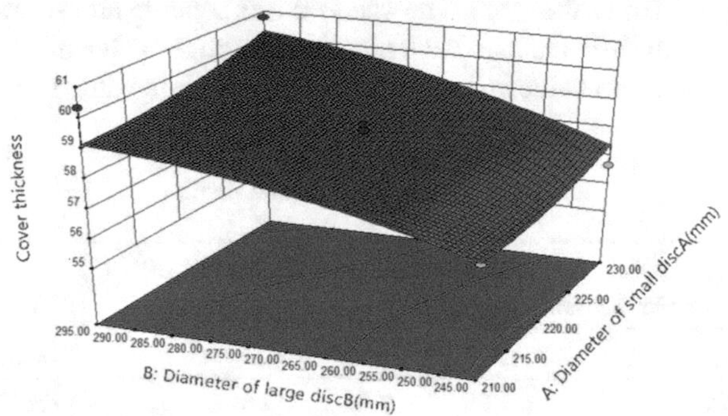

Fig. 2. The response of soil thickness by disc diameter

Figure 3 is the response surface of overburden thickness to overburden offset and disc opening angle. It can be seen that overburden thickness increases and decreases slowly with the increase of disc opening angle, and increases with the increase of overburden offset. The effect is obvious. With the increase of the interaction term, the overburden thickness increases and decreases slightly [9].

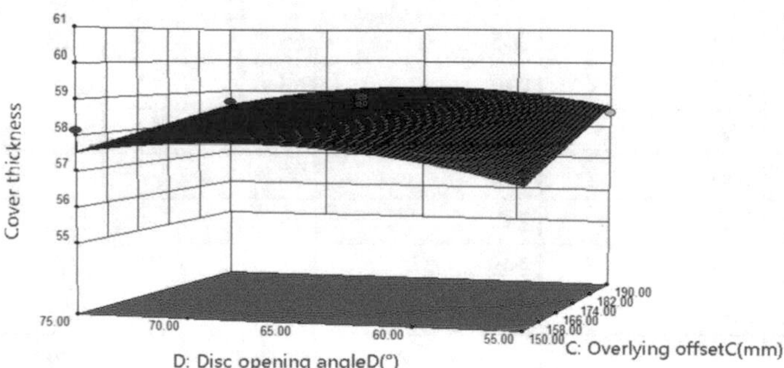

Fig. 3. The response of soil thickness by distance and open angle

3 Test Scheme and Result Analysis of Measuring Seed Displacement

The soil tank test for measuring the displacement of corn seeds selects the small disc diameter, large disc diameter and disc opening angle as the test factors, and each factor selects 3 levels. The test scheme is the same as the measurement method of soil cover thickness.

The uniform test scheme is generated by design expert software, and the test results of each group are filled in the table. The specific test scheme and results are shown in Table 2. It can be seen from the data table that the maximum value of seed displacement is 5.81 mm, the minimum value is 3.07 mm, and the average value is 4.402 mm.

Table 2. Test schedule and results

Test number	Small disc diameter A (mm)	Large disc diameter B (mm)	Disc angle C (°)	Seed displacement Y2 (mm)
1	210	270	55	3.07
2	210	295	65	3.45
3	210	270	75	3.63
4	220	270	65	4.45
5	230	245	65	4.96
6	220	270	65	4.3
7	220	270	65	4.45
8	220	270	65	4.3
9	220	295	75	4.95
10	220	295	55	4.49
11	210	245	65	3.15
12	220	270	65	4.45
13	230	270	75	5.23
14	220	245	55	4.09
15	220	245	75	4.52
16	230	295	65	5.81
17	230	270	55	5.55
Average				4.402

The mathematical model between seed displacement and various factors is obtained by multiple linear regression and binomial fitting of the data through the analysis module of the software, as follows:

$$Y_2 = 4.39 + 1.03A + 0.25B + 0.14C + 0.14AB - 0.22AC + 0.0075BC - 0.095A^2 + 0.047B^2 + 0.075C^2 \tag{2}$$

The significance of the model was analyzed. The significance P < 0.00001, the correlation R was 0.9902, and the degree of mismatch was not significant, indicating that the test points can be described by the model. According to the data, the main factors affecting the overburden thickness are: a (small disc diameter), B (large disc diameter), AC (interaction term between small disc diameter and disc angle), C (disc angle) and ab (interaction term between large and small disc diameter) [10].

Figure 4 is the single factor analysis diagram of small disc diameter, large disc diameter and disc opening angle of seed displacement. It can be seen from the diagram that the seed displacement increases with the increase of small disc diameter, large disc diameter and disc opening angle respectively, and the upward trend in the single factor analysis diagram of small disc diameter is more obvious.

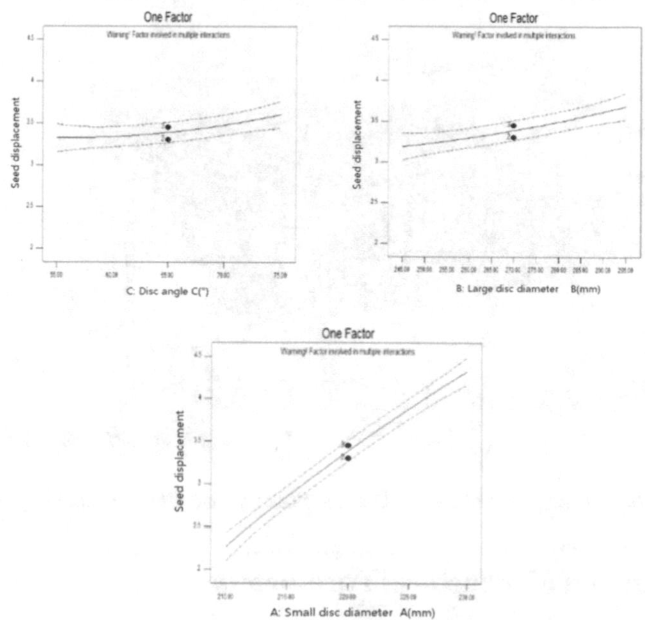

Fig. 4. Single factor data analysis

Figure 5 is the response surface of seed displacement to the diameter of large and small disks. It can be seen from the figure that the seed displacement increases with the increase of the diameter of large and small disks and the interaction term of the diameter of large and small disks. Figure 6 is the response surface of seed displacement to the diameter and opening angle of small disks. Under the joint action of the diameter and opening angle of small disks, The seed displacement showed an upward trend as a whole.

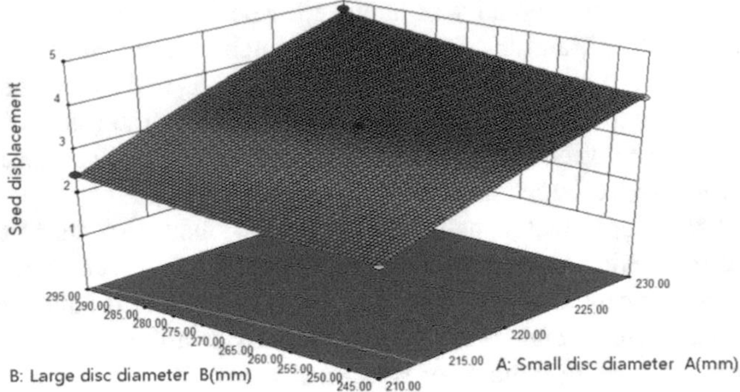

Fig. 5. The response of Seeds displacement by disc diameter

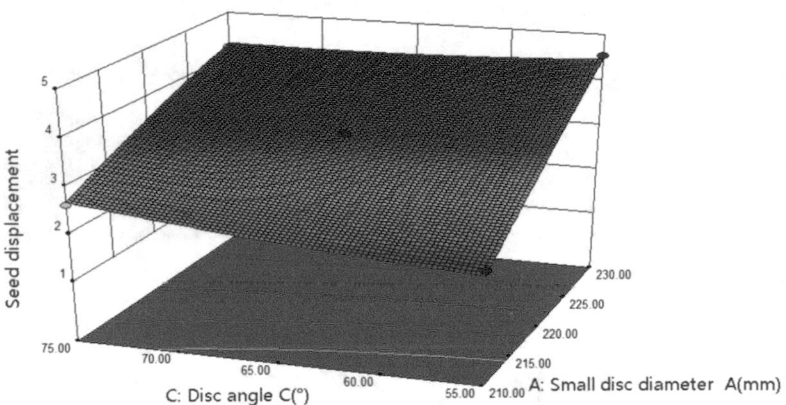

Fig. 6. The response of Seeds displacement by disc diameter opening angle

4 Optimization of Structural Parameters

Making design expert execute the experimental design document with corn seed displacement as the index, And enter the optimal combination (optimization) module, set the small disc diameter, large disc diameter and disc opening angle to the <in range> option, and set the test index seed displacement to the minimum <minimize> option. The system shows that when the small disc diameter is 210.4 mm, the large disc diameter is 271.63 mm and the disc opening angle is 55.76°, it is the recommended optimal structure, and the corresponding seed displacement is 3.05751 mm. Take the three values respectively 210 mm, 271 mm and 55° are input into the <target> option, which is the target value of the experimental design optimization combination module with the overburden thickness as the index. The system explicitly takes the first three as the target value. When the overburden offset is 187.66 mm, it is the recommended optimal structure, and the corresponding overburden thickness is about 58.58 mm, which is close to the average overburden thickness of 58.229 mm calculated above, indicating that the consistency of overburden thickness basically meets the requirements, At this time, the predicted value of seed displacement is 2.98 mm, which is close to the minimum value of 3.07 mm and the average value of 4.402 mm. There are 4 groups of optimal combination of seed displacement and 36 groups of optimal selection of soil cover thickness. At this time, the value of soil cover offset can be selected according to needs.

References

1. Mustafa, U., John, M., Fielke, C.S.: Defining the effect of sweep tillage tool cutting edge geometry on tillage forces using 3D discrete element modelling. Inform. Process. Agric. **2015**(2), 286–288 (2015)
2. Gabriel, K.P., Barrios, R.M., de Carvalho, A., Kwade, L., Tavares, M.: Contact parameter estimation for DEM simulation of iron ore pellet handling. Powder Technol. **2013**(1), 496–500 (2013)
3. Edgar, M., Burce, C., Kataoka, T., Okamoto, H., Shibata, Y.: Seeding depth regulation controlled by independent furrow openers for zero tillage systems. Eng. Agric., Env. Food **1**, 80–85 (2013)
4. Potyondy, D.O., Cundall, P.A.: A bonded-particle model for rock. Int. J. Rock Mech. Min. Sci. **8**, 100–132 (2004)
5. Cleary, P.W.: Large scale industrial DEM modelling. Eng. Comput. **2004**(2), 150–200 (2004)
6. Favier, J.F., Abbaspour-Fard, M.H., Kremmer, M., Raji, A.O.: Shape representation of axisymmetrical, non-spherical particles in discrete element simulation using multi-element model particles. Eng. Comput. **4**, 431–445 (1999)
7. Tahmish, F., Kumar, A.N.: Pseudomonas entomophila PE3 and Its exopolysaccharides as biostimulants for enhancing growth, yield and tolerance responses of sunflower under saline conditions. Microbiol. Res. **2021**(1), 244 (2021)
8. Wang, J., Feng, W., Zhang, H., Sun, J., Zhao, Q.: Experimental study on the effect of deep pine technology onwater and salt transport in soda saline land. Arab. J. Geosci. **7**, 560–567 (2021)
9. Javaheri, F., Esfandiarpour-Boroujeni, I., Farpoor, M.H., Holthusen, D., Stewart, R.D.: Counterions, smectite, and palygorskite increase microstructural stability of saline-sodic soils. Soil Tillage Res. **216**, 105258 (2021)
10. Ba, Y., Liu, J., Han, J., Zhang, X.: Application of vis-NIR spectroscopy for determination the content of organic matter in saline-alkali soils. Spectrochim. Acta Part A: Mol. Biomol. Spectrosc. **229**, 117863 (2020)

Anomaly Detection Method of Urban Rail Vehicle Timing Data Driven by KNN Algorithm

Jinsheng Chen(✉)

Guangzhou Vocational College of Technology and Business, Guangzhou 511442, China
CHENjinsheng002@163.com

Abstract. In order to detect the anomaly of sequential data of urban rail vehicles, this paper will focus on KNN algorithm, aiming to put forward an effective detection method based on the algorithm. Firstly, the concept of KNN algorithm is introduced, and the advantages of this algorithm in abnormal detection of vehicle timing data are discussed synchronously. Secondly, the flow of KNN algorithm is established. Finally, the system is designed for convenient detection. Through research, based on KNN algorithm, the system can detect abnormal data, and the detection effect is good, which can ensure the safety of vehicle operation.

Keywords: KNN algorithm · Urban rail vehicles · Timing data anomaly detection · Vehicle monitoring

1 Introduction

Along with the age and the development of economy, developing the urban rail vehicles, so railway vehicles has become a indispensable an important part of urban traffic, but the quantity of urban rail vehicles, some problems is made and bright, including the vehicle running safety problem, the problem has a sudden, the conventional method is difficult to prevent, once it happens, it may cause serious impacts, from affecting the economic operation of the city to threatening the life of passengers. Therefore, how to effectively prevent the occurrence of such problems has become the focus of people's attention. Based on this, puts forward some ideas of the abnormal data detection area, namely the vehicle running safety problem is not without warning, a matter of fact in a certain part of the vehicle before will be abnormal, thought its sudden, because people can't the first time found that the anomalies before, can not get the corresponding data information feedback, but under the technical support for all-weather monitoring of vehicles, by monitoring the operation of the first data is available, and if found abnormal data, means that the vehicle may be down at any time, and fault may cause safety problem, thus timely protection, at least reducing the incidence of security problems to a minimum.

Ideas put forward has been widely recognized, but also brought a new problem, namely the vehicle monitoring, how can get all the data, the problem is very complicated, the reason is that the vehicle causes a lot of security problems, involving vehicles of various mechanical systems, and each system contains a large number of parts, any

J. H. Abawajy et al. (Eds.): ICATCI 2022, LNDECT 170, pp. 470–478, 2023.
https://doi.org/10.1007/978-3-031-29097-8_56

abnormal parts could lead to security issues, at the same time, there are chain-type security problems, which involve more complex causes, so it is difficult to detect abnormal data. In this field have developed some effective methods in the early days, but these methods mostly on threshold theory, failure mechanism, so although have abnormal data detection ability, but the efficiency and accuracy method has defects, defect makes the previous approach in modern vehicle data magnitude increases increasingly, data logical relationship more complicated background no longer use, therefore, it is necessary to choose a new method to realize abnormal data detection, and KNN algorithm is a good choice. Therefore, in order to improve the anomaly detection of urban rail vehicle data based on this algorithm, it is necessary to carry out relevant research.

2 KNN Algorithm Concept and Application Advantages

2.1 Algorithm Concept

KNN algorithm is the most widely used machine learning algorithms, was first put forward by the Cover, hart, the function is to deal with the data, then according to the processing result of data classification, or to return to supervise, the characteristics of the algorithm is simple and intuitive, calculation process is programmed at the same time, does not involve parameters, therefore, no expression formula, just follow the standard template for programming. Two big functions of KNN algorithm show that the algorithm can be applied in the field, the corresponding algorithm of the application in different fields have the same way, the only difference is the final prediction decision link, namely if classified prediction, then adopts the majority voting method, the weighted majority voting method, but need to use the average value in regression prediction method, the weighted average method. Comparatively speaking, KNN algorithm is more suitable for classification prediction. In practical application, it can handle multiple classification problems at the same time and has strong data recognition ability, which enables the algorithm to classify rare events [1–3].

In terms of algorithm principle, KNN algorithm mainly relies on the idea of minority obeying the majority to make decisions and can effectively distinguish outliers in data. The so-called majority in the algorithm refers to the K values closest to the standard sample in the feature space of the exponential data. The category with the largest number of K values is the category of the sample point, while the minority will obey the majority and be divided into the same classification. It can be seen that the decision-making process of KNN algorithm mainly relies on three elements, namely K value, distance and decision rule. K value and distance have been discussed above, and the decision rule depends on the decision method used in the operation of the algorithm, namely majority voting method and weighted majority voting method mentioned above. At present, majority voting method is the most common classification decision rule in the application of KNN algorithm [4]. Its main content is: The majority classification of K points is regarded as the final classification attribute of the points to be measured, and the result can be obtained by selecting K value reasonably under this rule, and the rule for selecting K value is as follows: K is smaller, the smaller range forecast samples, the prediction error is smaller, but this will result in the prediction process becomes more complex, is likely to occur the phenomenon of excessive fitting, instead of K, the greater the sample prediction can lead to the results on a wide range error increases, but the prediction process is more simple, basic is not going to happen over fitting problem, According to this rule, the optimal result can be obtained by selecting a K value with relatively small error, low complexity and low incidence of over-fitting [5–7]. Figure 1 is the classification principle of KNN algorithm.

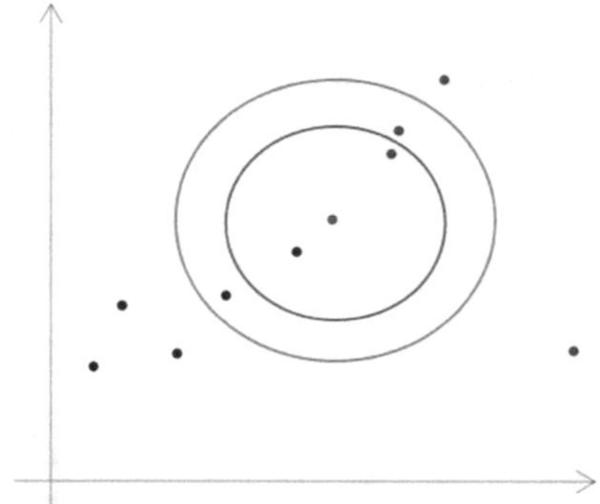

Fig. 1. Classification principle of KNN algorithm

With the development of the transportation field in China, urban rail vehicles are becoming more and more popular in China. Urban rail vehicles in some first tier cities have become the mainstream urban transportation tools. However, in the context of the constant popularity of urban rail vehicles, people find that the problem of abnormal time service data will occur in the practical application of urban rail vehicles, which has a great impact on the operation of urban rail vehicles, and even lead to safety accidents, so this problem should be paid attention to. At present, people's research on time service data anomalies of urban rail vehicles mainly focuses on the causes of anomalies. Through research, it is found that there are many causes of such problems, and it is very difficult to deal with them pertinently. Therefore, the most effective method at present is to detect time service data anomalies, and deal with them in a timely manner once they are detected. Although this way cannot avoid the occurrence of time service data anomalies, however, it can effectively solve the impact of this problem on the operation of urban rail vehicles.

That is, KNN algorithm is a simple and efficient algorithm, which has a wide range of applications, and can be applied to classification, regression, pattern recognition and other scenarios. The anomaly detection of urban vehicle timing data belongs to regression scenarios, so KNN algorithm can be used for actual detection work. In the practical application of KNN algorithm, we need to pay attention to two important parameters: sample weight and feature weight. Usually, we can use the feature weight algorithm based on SVM to confirm the specific weight. When the weight is clear, KNN algorithm can effectively improve the accuracy of the results compared with other algorithms. The reason is that the algorithm mainly constructs the feature space under specific conditions, and then classifies the samples according to the relationship between the k nearest samples in the feature space. At the same time, it shows that the sample characteristics are mathematical. Therefore, when the characteristics of the time service data are consistent with the feature space of the abnormal data, it represents that the group of time service data is abnormal. However, it should be noted that the KNN algorithm can only be used in the classification and decision making of a few adjacent samples, and the abnormal time service data of urban rail vehicles meets this requirement, that is, although the problem of abnormal time service data is not rare on the whole, the probability of occurrence of this problem is very small on an individual basis, which again proves that the KNN algorithm is suitable for the detection of abnormal time service data of urban rail vehicles.

2.2 Application Advantages

In the face of abnormal data detection problem, the commonly used methods are based on past theory and the birth of the threshold, failure mechanism, this method focuses on the functional, namely the previous method can found abnormal data from the data volume, and then analyze the data, to determine the source of the abnormal data, changes, etc., finally carries on the classification, but because of too much focus on the functional development, and the previous theoretical level is limited, processing efficiency is very low, so the traditional method in the practical application if there are any huge amounts of data, or data logic relationship between the phenomenon of complexity is too high, can't timely detection, but many modern abnormal data has large scale, in the field of data, the characteristics of the complex relationship between makes the previous method applicability is low [8, 9].

In contrast, KNN algorithm, the algorithm from the traditional formula, and adopt a kind of evaluation process to drive the machine learning, it can directly to evaluate data, so very simple evaluation is obtained as a result, this simple process makes it can process data quickly, that KNN algorithm even in the face of huge amounts of data can also results in a timely manner, at the same time, the thought of the minority obeying the majority also greatly weakens the data complexity, which not only does not affect

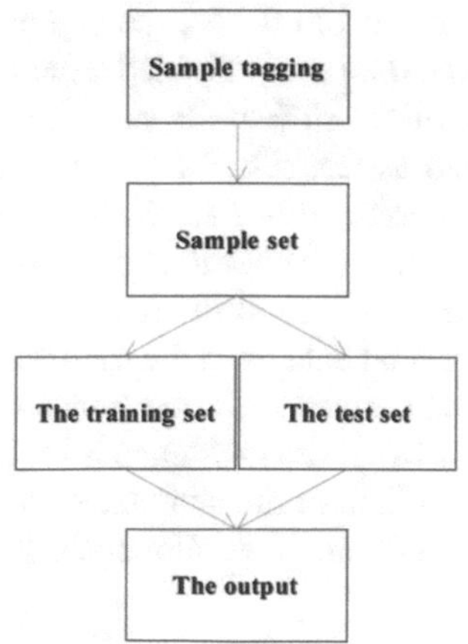

Fig. 2. Evaluation model of KNN algorithm

the detection efficiency, but also can get more accurate results. Figure 2 is the evaluation model of KNN algorithm.

2.3 Construction of Data Anomaly Detection Process of KNN Algorithm

Figure 3 shows the data anomaly detection flow of KNN algorithm.

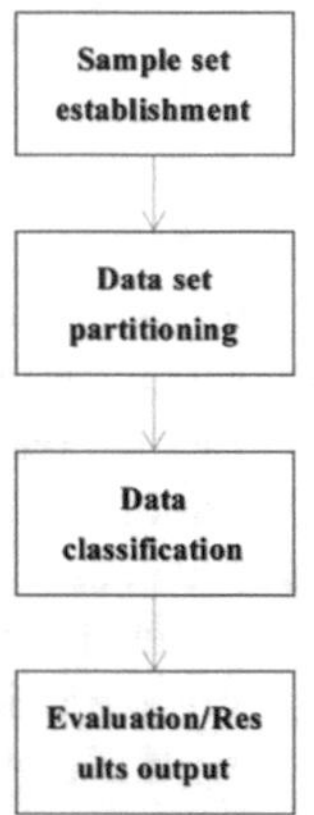

Fig. 3. Data anomaly detection flow of KNN algorithm

Combined with Fig. 3, because the abnormal detection of urban rail vehicle data is a typical data classification and evaluation problem, the detection process is established according to the majority voting rule of KNN algorithm, and the sample set is first established in the process: According to the standard data of railway vehicles all parts (normal data of standard data must be within the normal range of data range) based data evaluation, and then through real-time data acquisition device, the real-time data and standard data item by item comparison, known data to exceed bid, which set a standard data type for J, J class data exceeds bid for M, the eigenvalue i of the overall data is i = (M1, M2, …., Mn). Then the data set is divided. According to the minority follows the majority idea of KNN algorithm, 80% of the sample set is divided into training sets, and the remaining 20% is used as test sets. Majority voting rules in accordance with choice of K value, after the completion of data classification, the final assessment can get the results, the results can reveal whether the abnormal data, abnormal data belongs to classification, incidence, the incidence of them are the key information that can help people determine whether abnormal data from the transient fault, if the incidence is low, then the data exception may be derived from the transient fault, most of this situation does not need special attention, but if the incidence is on the rise, or has reached a high level, it shows that there is a real fault of the vehicle, according to the data classification can confirm the source of the fault, and then diagnosis and maintenance can eliminate the problem [10].

3 Construction of Abnormal Timing Data Detection System for Urban Rail Vehicles

3.1 System Framework

Figure 4 is the framework of urban rail vehicle timing data anomaly detection system. The system USES a three-tier design, physical layer, communication layer and terminal, respectively, which is mainly responsible for collecting the data, the physical communication layer is responsible for supporting the interaction between the communication with the terminal layer, the physical terminal layer is mainly responsible for data detection, and the result is given, the results will be sent via communication layer to the artificial handheld terminals, Corresponding instructions will also be generated and sent to the physical layer depending on the communication layer, in order to obtain more trust data information, or start the vehicle emergency system, to avoid the sudden occurrence of problems.

Fig. 4. Framework of urban rail vehicle timing data anomaly detection system

3.2 System Implementation Scheme

According to Fig. 4, the implementation scheme of the whole system is divided into three steps, and the specific contents of each step are as follows.

The physical. The physical layer is mainly composed of various sensors, sensor is the key equipment acquisition data information, to determine whether the data collected information is complete, accurate, so in order to ensure complete and accurate data information, need to carefully choose the sensor, here according to the mechanical structure of the urban rail vehicles, generally choose four types of sensors, they are mechanical kinetic energy sensors, temperature sensors, power sensors and speed sensors. The number of each type of sensor should be based on the specific situation of the vehicle, and there is no specific pattern. At the same time, the selected sensors should meet the performance requirements, otherwise it is prone to insufficient acquisition accuracy, real-time acquisition and other problems, affecting the quality of data information. Sensor selection, need to install the sensor in the specified location, the installation is complete every time want to test the sensor, confirm the sensor can collect data information, reasonable confirmation requires a bus technology and so on connect the sensor channels of communication and communication layer, ensure that the data collected information can be directly through the communication layer into the terminal. Because the data information collected by the sensor is electrical signal in the initial format, which cannot be read by the terminal device, it is necessary to install a transducer on the communication bus. The transducer's function is to convert the data information in the format of electrical signal into a digital format, which can be read by the terminal device. In addition, in addition to a number of sensors in the physical layer, there are also some control units, mainly responsible for the execution of relevant work according to the terminal layer instructions, each control unit can be achieved through mechanical programming and related mechanical equipment, the specific form is uncertain, but also need to be combined with the actual needs to set. Figure 5 shows the physical layer structure.

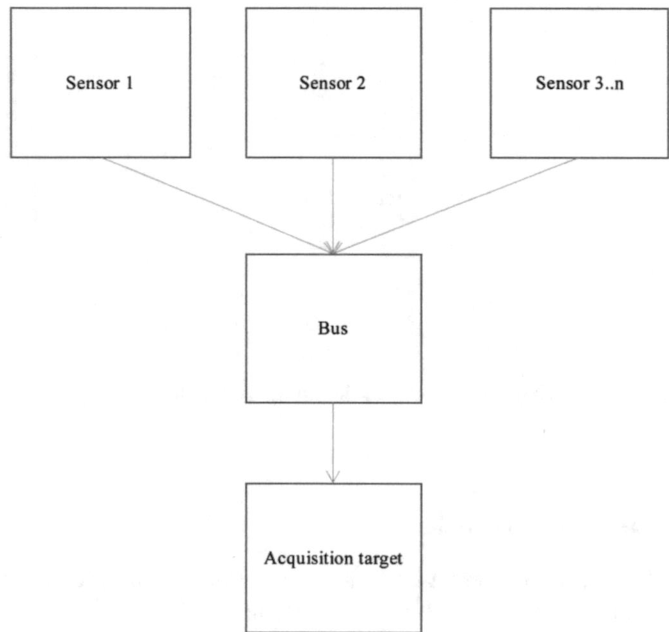

Fig. 5. Shows the physical layer structure

Communication layer. Communication layer's main function is to support the other two interactive communication, so the communication layer itself can also be divided into three components, respectively is signal transmitting interface, communication channels, signal receiving interface, including signal launch interface can be in sensor comes on the basis of the features of the signal emission, use an adapter to obtain transmitters, and will continue to send signals to the communication channels, The communication channel can be realized by ethernet, which has a wide range of communication protocols, a large communication capacity, and is very stable. It can operate in urban rail vehicles and is not susceptible to external interference. The communication receiving interface can adopt the same switching generator to promote the data and information to be sent to the terminal layer. Fig. 6 shows the communication layer structure.

Terminal layer. The terminal layer is where the KNN algorithm resides. It is loaded on the computer equipment of this layer, which can preprocess the data information received by the equipment. The purpose is to remove the data that does not meet the standard access conditions and duplicate data, ensure the data purity, and facilitate the data classification efficiency. At the same time, physical servers can be considered to build a data repository for storing pre-processed data and calculated data. People can query the results by scheduling related data in the database to determine whether there is an anomaly in the data. Fig. 7 shows the terminal layer structure.

4 Conclusion

To sum up, the operation of urban rail vehicles is related to passenger safety and urban economic development, so its operation quality and stability need to be guaranteed. In order to achieve this, KNN algorithm can be used to detect abnormal data, and the detection results can help people identify problems, find problems in time, and eliminate potential safety hazards as soon as possible.

Acknowledgments. Characteristic innovation project of university scientific research platform and scientific research projects of Guangdong Provincial Department of education in 2021: Research on Freight Express scheme of High-speed Railway Guangzhou south railway station (No.: 2021wtscx216).

References

1. Kim, H.I., Kim, H.J., Chang, J.W.: A secure kNN query processing algorithm using homomorphic encryption on outsourced database. Data Knowl. Eng. **123**, 101602 (2019)
2. Zardari, Z.A., He, J., Pathan, M.S., Qureshi, S.: Detection and prevention of jellyfish attacks using kNN algorithm and trusted routing scheme in MANET. Int. J. Netw. Secur. **23**(1), 77–87 (2021)
3. Yuan, C.: Deep learning of the SSL luminaire spectral power distribution under multiple degradation mechanisms by hybrid kNN algorithm. In: 2021 22nd International Conference on Thermal, Mechanical and Multi-Physics Simulation and Experiments in Microelectronics and Microsystems (EuroSimE) (2021)
4. Yaswanth, B.S., Murthy. V.: Smart safety and security solution for women using kNN algorithm and IoT. In: 2020 Third International Conference on Multimedia Processing, Communication & Information Technology (MPCIT) (2021)
5. Wang, H., Yu, Z., Guo, L.: Real-time online fault diagnosis of rolling bearings based on KNN algorithm. J. Phys.: Conf. Ser. **1486**, 032019 (2020)
6. Sal, B., et al.: Microwave dielectric property based classification of renal calculi: Application of a kNN algorithm. Comput. Biol. Med. **112**, 103366 (2019)

7. Zeng, J.-J., Fa-Mao, et al:. A design method of performance test case based on kNN algorithm. In: 2019 International Conference on Energy, Power, Environment and Computer Application (ICEPECA 2019) (2019)
8. Pavani, K.: Novel vehicle detection in real time road traffic density using haar cascade comparing with KNN algorithm based on accuracy and time mean speed. Revista Gestão Inovação e Tecnologias **11**(2), 897–910 (2021)
9. Saravanakumar, C., Geetha, M., Govindaraj, V., Mohan, P., Vijayakumar, K.: An efficient generic review framework for assessment of learners ability using KNN algorithm. In: Proceedings of the Fist International Conference on Advanced Scientific Innovation in Science, Engineering and Technology, ICASISET 2020, Chennai, India, 16–17 May 2020 (2021)
10. Qu, X., Dong, K., Zhao, J., Yu, Y.: An identification and location method for power quality disturbance sources in MMC converter based on KNN algorithm. In: 2021 4th International Conference on Energy, Electrical and Power Engineering (CEEPE), pp. 170–177. Chongqing, China (2021)

Intelligent Teaching Management System Based on Data Fusion Technology

Yonglin Zhao(✉)

Zhengzhou University of Science and Technology, Zhengzhou 450064, China
565815308@qq.com

Abstract. In order to construct the intelligent teaching management system of colleges and universities through data fusion technology, this paper will carry out relevant research. This paper first introduces the basic concept of data fusion technology, then puts forward the significance of the construction of intelligent teaching management system in colleges and universities, and finally analyzes the system construction scheme. Through the research of this paper, it is confirmed that data fusion technology is the basic support of intelligent teaching management system in colleges and universities, which can make all functions of the system more comprehensive, so the system can better serve the teaching work in colleges and universities.

Keywords: Data fusion technology · College teaching · Intelligent management system

1 Introduction

At present, in the face of the demand of different levels of the society put forward, education field to continuously seek to develop, but in the process of seeking development education groups are met with some difficulties, such as colleges and universities in their own development to improve the teaching of specific aim, requires teachers to fully understand the actual situation of each student, to carry out targeted teaching or instruction, makes the teacher to the students in the learning data acquisition and analysis, and teachers in data acquisition and analysis work soon encountered problem, namely each student learning contains huge magnitude, phyletic and various data, so the data collection and analysis work volume, high difficulty, is limited by their own capabilities, unable to effectively perform the work, this leads to the difficulty in improving the pertinence of teaching. From this perspective, the emergence of the problem that colleges and universities to rethink and realize the past teaching too dependent on manual mode, will be affected by artificial limitations, therefore, cannot achieve development goals, so in order to solve this problem, you need to introduce the relevant technology, this kind of circumstance smart technology got the high attention of university, the reason is that the technology of autonomous learning ability, able to work to a great extent, independently

J. H. Abawajy et al. (Eds.): ICATCI 2022, LNDECT 170, pp. 479–487, 2023.
https://doi.org/10.1007/978-3-031-29097-8_57

complete the data analysis, without human intervention, and also makes the artificial can according to the technical results of actual situation make accurate judgment and decision-making, thus improving the teaching pertinence, so for modern university hope that through intelligent technology to construct teaching intelligent management system, However, how to construct such a system has become a new problem. The reason is that intelligent technology only has the ability of data analysis, but not the ability of data collection, so the system cannot operate without data sources. At the same time, the operation of intelligent technology depends on intelligent logic. Therefore, how to input a logic to intelligent technology similar to the thinking of artificial in the analysis of scientific data is also an urgent problem to be solved. Presented to these questions, put the university began to look for other technical means, the system with other intelligent technology to college teaching management system support, and the data fusion technology can meet these requirements, therefore around how to based on data fusion technology to construct the teaching problem of intelligent management system, it is necessary to expand related research.

2 Basic Concepts of Data Fusion Technology

The intelligent teaching management system is a system tool focusing on intelligent technology and specially used for teaching management. With the help of intelligent technology, this tool can meet the needs of teaching management in many aspects, mainly including: first, it can analyze students' learning and help people supervise students' learning; Second, analyze the relevant information of students' learning, and give the main problems encountered in students' learning, so that managers can make targeted management decisions. The intelligent teaching management system is of great significance. The reason is that since the previous teaching management work, the artificial ability is limited, and many tasks can not be completed well. For example, to manage the overall students, you must first understand the learning situation of all students under different conditions. This work involves a lot of information, and it also requires all-weather uninterrupted information collection, which is impossible to do manually, the use of intelligent teaching management system can solve this problem, that is, as a system tool, it has the common feature of all technical systems, that is, it can operate around the clock, and obtain information efficiently. Therefore, with the help of this system, the limitations of artificial ability will no longer have a restrictive impact on teaching management. At the same time, the intelligent teaching management system can also replace manual analysis of information, and the results are more accurate and complete. Figure 1 is the basic template of intelligent teaching management system.

Data fusion technology is a kind of all data collection, then the data according to the preset rules of good fusion, to form the corresponding relationship between the various data, according to the relationship will be able to know the current situation of the data represented by the event and possible future development direction and the possibility of events in any direction development, for example, in teaching, student work related data type, data fusion and data access to know students often refer to a particular type of data, so the student work in the data and the data related to the title of the accuracy is high, the

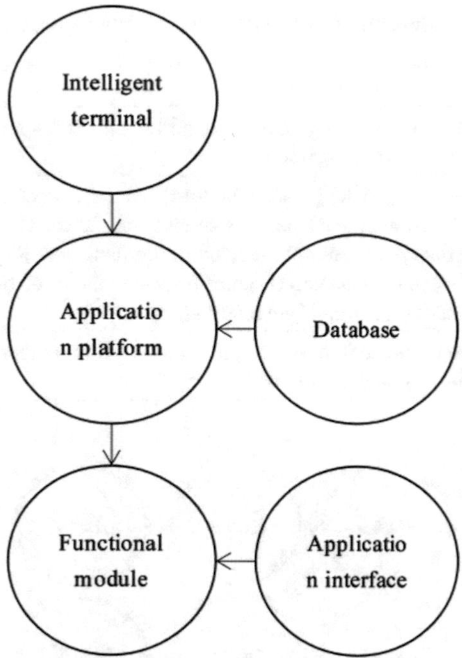

Fig. 1. Basic template of intelligent teaching management system.

error rate is higher, other subject so you understand the reason of the student's wrong topic, so the development of students in the future, there will probably be knowledge of fault, the difficulty of learning increases and the quality of learning is affected. It can be seen that data fusion technology will integrate the collected data and then analyze them to give relevant results [1–3]. This result has good application value for teaching work at least, which can give teachers a clear direction of teaching guidance and greatly improve the teaching quality. It is worth noting that the data fusion technology itself does not have the ability to direct the output results, data fusion, can only in accordance with the rules will be common rules can be seen in Table 1, after the fusion data is equivalent to a hidden as a result, the human is unable to read, so you also need to use intelligent technology for further analysis, so as to make teachers and understand the results, therefore, this technology should be combined with other technologies in teaching in colleges and universities [4–6].

According to Table 1, although there are three or even more fusion rules in data fusion technology, there are only two kinds of relationship between data in each rule, namely point-to-point relationship and point-to-many relationship. For convenience, Fig. 2 and Fig. 3 introduce these two relationships.

Table 1. Common rules of data fusion in data fusion technology

The rules	The connotation of the
All the rules	All data in the range are fused, and the point-to-point relationship between each data is maintained
Filtering rules	According to the preset data items, all the data conforming to the data items within the scope are screened and fused. The data not conforming to the data items are not included in the fusion scope, and the relationship between the data can be point-to-point single relationship or point-to-many multiple relationship
Point-to-many rule	Based on one item of data, it is fused with all other data to form a point-to-many relationship

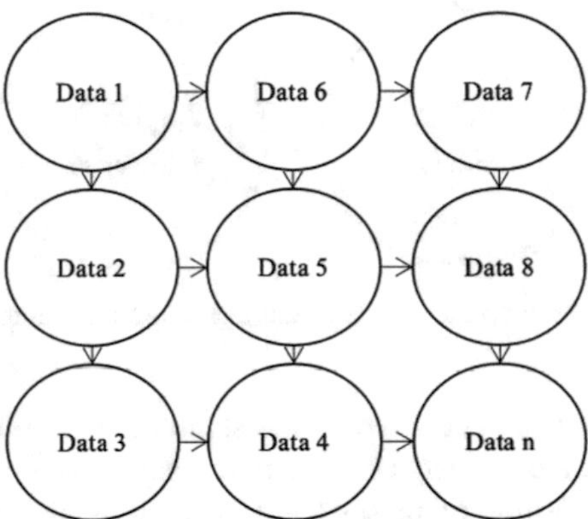

Fig. 2. Point-to-point relationship

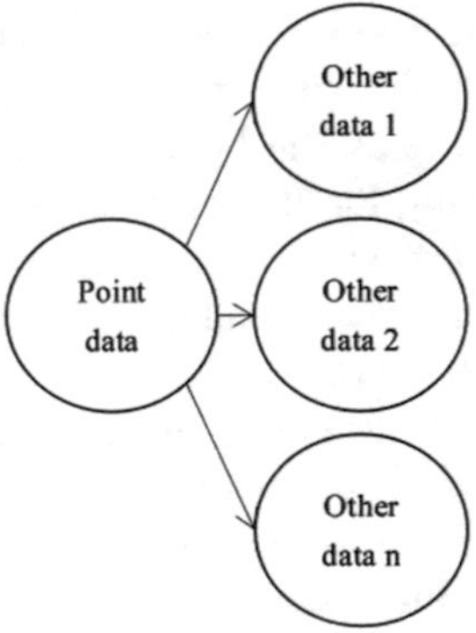

Fig. 3. Point-to-many relationship

3 Significance of Construction of Intelligent Teaching Management System in Colleges and Universities

3.1 Improve the Pertinence of Teaching

The characteristics of the early teaching of colleges and universities in our country are widespread unification, the teacher in the face of all the students equally, will adopt a unified schedule and content of teaching, so students get the basic education services is the same, the individual differences and students themselves, their learning situation under the same education services will still appear difference, part of the student union of teaching has not adapt to the feeling, It has an adverse effect on students' study. This phenomenon has lasted for a long time in the field of higher education in China. It is not until modern times that people realize that this kind of education method is actually unfair and will limit students' success. Therefore, modern people have changed their educational ideas, hoping to improve the pertinence of teaching and requiring teachers to teach according to each student's learning situation [7–9].

Focus on this requirement, the intelligent management system can help teachers to the sustainability of each student learning related data, and then after the data integration is analyzed, the results of the analysis can let the teacher know students learn, so you can targeted teaching, reasonably adjust the teaching according to each student's learning progress and depth of content, it can also timely solve the learning problems that students themselves are not aware of, and comprehensively promote the development of students.

3.2 Help Students Study Independently

Students learning generally divided into two forms, one is the classroom lectures, or continue to learn under the guidance of teachers, the other is autonomous learning, including autonomous learning is beneficial to students' thinking divergence, can not only make students better study problem, also can develop the students' ability of independent thinking, but has the certain difficulty for students autonomous learning, many students in the process of autonomous learning will meet with difficulties, such as students in the autonomous learning has met a themselves unable to understand the concept of knowledge, but no others help to interpret again, can only rely on their own brood, this undoubtedly has the very high difficulty, some students even after a period of independent study found itself has a problem, but I don't know where the problem is, they can't learn on their own. These phenomena are the main reasons leading to the lack of enthusiasm of modern college students in autonomous learning, and for this phenomenon, teachers also hope to provide students with relevant help, but teachers cannot always be around students, so it is impossible to provide students with comprehensive help only by relying on teachers.

Based on this, the teaching of intelligent management system can replace teachers to help students, namely the system analysis will not only show the result of the teacher, also can show students, students just need to combine the results can know themselves what kind of problems, and then search the information related to autonomous learning, problem such as students will be on your own input into the system, through systematic analysis, we can know the types of wrong questions and related knowledge points in the homework, which shows that the system can help students study independently and make students actively carry out independent learning activities [10].

4 Construction Scheme of Intelligent Teaching Management System in Colleges and Universities

4.1 Overall Framework

Based on data fusion technology, the overall framework of intelligent teaching management system in colleges and universities is shown in Fig. 4.

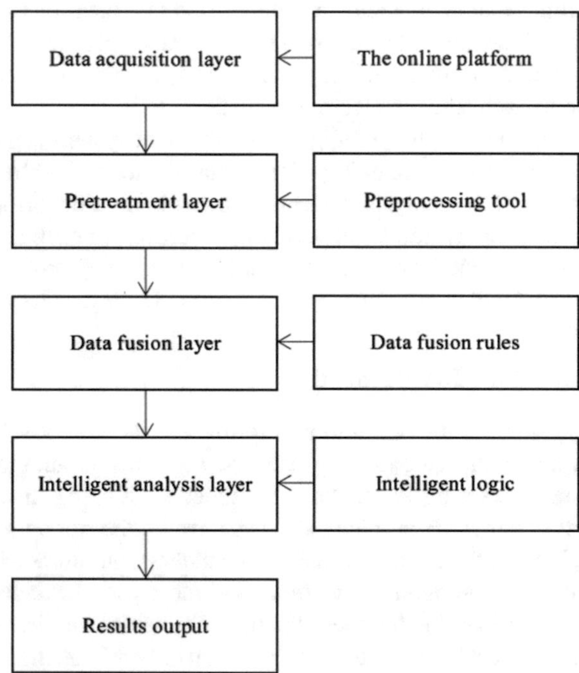

Fig. 4. Overall framework of intelligent teaching management system in colleges and universities

Combined with Fig. 3, the system is mainly composed of four parts, one is the data collection layer, the equivalent of a communication channel, connected with the teaching online platform, platform in various teaching activities under the legacy of data will be stored in the database platform, and then from the database to be accepted by the data collection layer, eventually transferred to the next layer, thus realize the information acquisition, Secondly, pretreatment layer, because get the initialization data possible quality problems, such as data repeat, do not accord with the standard format, such as pretreatment, so will the use of a variety of tools to ensure data quality, data fusion layer, a third is mainly according to the various data fusion rules will deal with the data fusion, the concrete shall be carried out in accordance with the rules of what, depends on the user to select, fourth is the intelligent analysis layer, the layer is mainly will be merged data as a training objects, and then digging through intelligent logic analysis, understand the development trend of the data fusion body, status, etc., the related implied result of fused data can be read from a human, the final output. Figure 5 is an intelligent logic analysis mining mechanism.

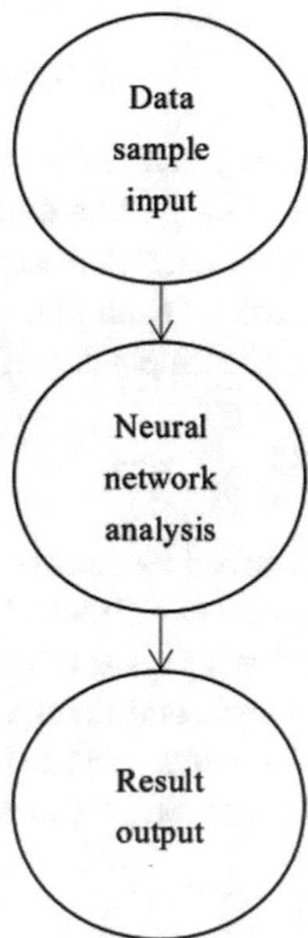

Fig. 5. Intelligent logic analysis mining mechanism.

4.2 Implementation Method

Looking at Fig. 4, the system is implemented as follows.

Information Collection Layer

As a communication channel, the information collection layer itself does not need to have data storage capacity, but needs to have a large communication capacity, and in order to let the data into this channel, also need to design the collection rules. First of all, in terms of communication capacity, it is recommended to choose communication protocols with large communication capacity, such as DAS and TCP. These protocols not only have large communication capacity, but also have low cost and high stability. Their only shortcoming is that communication management is cumbersome, but they do not need frequent management because of their high stability. It meets the communication capacity requirements and other requirements of the information collection layer. Secondly, in terms of information collection rules: First, the online platform is regarded as a data source and temporary data storage is set up internally; Secondly, design the temporary storage time limit in the temporary data repository, such as the storage time limit of each item of data in the database is 7d, the data will be sent automatically when the storage time exceeds 7d; Thirdly, the communication interface is constructed in the temporary database, and the communication protocol of the interface is the same as that of the communication channel, so that the data to be stored and realized will be accepted by the communication channel after being sent, and the information collection will be realized.

Pretreatment Layer
The realization method of preprocessing layer is relatively simple, mainly according to the common quality problems in the initialization data to choose the preprocessing tool, and then develop the preprocessing module, and finally integrate each module. However, in order to ensure the full play of the role of each pretreatment tool, a temporary database should also be established in the pretreatment layer, so that the pretreatment tool can process the data in the stored state.

Data Fusion Layer
The relevant data fusion rules are realized by programming, and then the key design is completed by VB software, so that users can choose fusion rules according to the actual needs, so that the data can be fused with each other to form a fusion data body, and ensure that there are related implicit logic and results in the fusion data body. As for the programming language of data fusion rules, It is recommended to choose Java, which has good implement ability and can meet the programming needs of various data fusion rules.

Intelligent Analysis Layer
Type with the help of a feedforward neural network to realize intelligent logic, the logic of data fusion in every item of data is viewed as the foundation of data nodes, then each basic data import hidden layer nodes, implied mining connection, so you can let the implicit logic and the results of fusion data into intuitive logic and as a result, the end can be exported.

5 Conclusion

To sum up, the intelligent teaching management system in colleges and universities can effectively improve the quality of teaching. Therefore, colleges and universities and other educational organizations should start to construct this system. In the construction of this system, data fusion technology must be mastered, with which the system can operate and achieve various functions. With the help of data fusion technology, intelligent management system can improve the pertinence of teaching and realize the transformation from traditional teaching to modern teaching.

References

1. Chair, Z., Varshney, P.K.: Optimal Data Fusion in Multiple Sensor Detection Systems. IEEE Trans. Aerosp. Electron. Syst. **AES-22**(1), 98–101 (1986)
2. Bloch, I.: Information combination operators for data fusion: a comparative review with classification. IEEE Trans. Syst. Man. Cybern. **26**(1), 52–67 (1996)
3. Rahman, M.J., Morshed, B.I., Preza, C.: A smart health (sHealth)-centric method toward estimation of sleep deficiency severity from wearable sensor data fusion. BioMedInformatics **1**(3), 106–126 (2021)
4. Wang, K., Hossain, S., Habib, K.N.: A hybrid data fusion methodology for household travel surveys to reduce proxy biases and under-representation of specific sub-group of population. Transportation **49**, 1801–1836 (2021)
5. Beć, K.B., Grabska, J., Plewka, N., Huck, C.W.: Insect protein content analysis in handcrafted fitness bars by NIR spectroscopy. Gaussian process regression and data fusion for performance enhancement of miniaturized cost-effective consumer-grade sensors. Molecules **26**(21), 6390 (2021)
6. Romor, F., Tezzele, M., Rozza, G.: Multi-fidelity data fusion for the approximation of scalar functions with low intrinsic dimensionality through active subspaces. PAMM **20**(S1), e202000349 (2021)
7. Chaurasiya, R.B., Shrestha, R.: A New hardware-efficient spectrum-sensor VLSI architecture for data-fusion-based cooperative cognitive-radio network. IEEE Trans. VLSI (VLSI) Syst. **29**(4), 760–773 (2021)
8. Sun, A.Y., Scanlon, B.R., Save, H., Rateb, A.: Combining physics-based modeling and machine learning for GRACE satellite data fusion and reconstruction. In: AGU Fall Meeting Abstracts. AGUFM, pp. 151–152 (2020)
9. Zoghlami, F., Kaden, M., Villmann, T., Schneider, G., Heinrich, H.: Sensors data fusion for smart decisions making: a novel bi-functional system for the evaluation of sensors contribution in classification problems. In: 2021 22nd IEEE International Conference on Industrial Technology (ICIT), pp. 1417–1423. Valencia, Spain (2021)
10. Gomathi, B., Sujatha, R.: Prediction of breast cancer using data fusion of SVM with optimization technique. J. Inform. Comput. Sci. **9**(12), 504–511 (2021)

Location Algorithm of Wireless Sensor Network Based on SDN Network Structure

Shuai Gong[1(✉)], Hang Cheng[1], Xiaoyu Yin[1], Xiaoling Dong[1], and Jingan Wu[2]

[1] State Grid Anhui Electric Power Co., Ltd., Information and Communication Branch, Hefei 230061, China
635926659@qq.com

[2] Anhui Jiyuan Software Co., Ltd., Hefei 230088, China

Abstract. In order to realize wireless sensor network localization based on SDN network structure, this paper will carry out research on relevant algorithms. This paper firstly introduces the basic architecture and key technologies of SDN, then analyzes the characteristics and topology of WIRELESS sensor network based on SDN structure, and finally proposes relevant localization algorithm. Through this study, the positioning algorithm can accurately locate wireless sensors based on the SDN network structure.

Keywords: SDN network structure · Wireless sensor · Network location algorithm

1 Introduction

Wireless sensor has a very wide range of applications, so it involves many application scenarios, some of which require people to locate the sensor, such as underground work, in order to ensure the safety of personnel, or in order to confirm the work place. But such scenes tend to have strong signal interference, lead to people not directly through the positioning technology of sensor signal, so the field put forward the corresponding solutions, to establish SDN network is at the scene, use the network to construct the communication channels are not affected by external interference, in this way can be based on the network can sensor location, And people in the field has encountered new problems, namely SDN network structures, only provides wireless sensor location condition, does not provide specific positioning methods, therefore, how to implement positioning under SDN network structure is still a problem to be solved, draws on relevant algorithm for the problem, in order to help more people master the related algorithms, Relevant research is needed.

2 Basic Architecture and Key Technologies of SDN

2.1 Concept Introduction

SDN network structure is a new network structure, which separates the network control function and forwarding function in the traditional network structure, so it has good

J. H. Abawajy et al. (Eds.): ICATCI 2022, LNDECT 170, pp. 488–495, 2023.
https://doi.org/10.1007/978-3-031-29097-8_58

controllability and can be programmed to control it. This architecture allows people to directly send information to external computing devices from the control layer and network devices. Therefore, its underlying infrastructure is completely transparent to applications and network services, and it is also abstract. Therefore, the network can be interpreted as a virtual logical entity.

The most prominent feature of SDN is the centralized control plane and the distributed forwarding plane. The two planes are independent of each other. The former mainly communicates with the forwarding plane network equipment by means of control and forwarding communication interfaces, centralizes the control of all network equipment, and provides flexible programmable space. The latter is mainly used for forwarding and executing control layer commands. The network structures with these two characteristics are SDN structures in a broad sense.

The control principle of the SDN centralized control plane is not complex, but this control method allows the flow to be generated between the controller and network devices, independent of the data flow generated by the communication between terminals. Therefore, the network devices can directly accept control instructions, and generate a forwarding table to execute forwarding commands. Compared with traditional networks, this method does not require complex distributed network protocols as a support, so it is more convenient.

SND is not a specific network protocol. Theoretically, SND is more inclined to be defined as a network framework. It has multiple interface protocols, including the south interface protocol of OpenFlow. The interface can be summarized to realize the communication between SDN controller and SDN switch, or the north API, which can realize the communication between business applications and SDN controller. The existence of these structures makes the SDN network structure systematic and convenient for people to manage and apply.

2.2 Basic Architecture

Derived from Open Flow, SDN is a software-defined network that can control network traffic by separating the control plane of network equipment and data forwarding. The control process is very flexible and programmable. Figure 1 shows the architecture of SDN.

Combined with Fig. 1, you can see, SDN network architecture is divided into three layers, respectively, the application layer, control layer and infrastructure, including the application layer is composed of several application modules, in the whole network structure of the module is known as the upper application, they mainly by means of superposition integration in SDN in the network, control layer is mainly composed of SDN control software and network equipment, In addition, the combination of the two forms the control data plane in the control layer. Finally, the infrastructure layer is composed of several network facilities, whose main function is to carry out network communication and interaction, which is an important foundation for the establishment of SDN. In addition, a necessary prerequisite for SDN to control the data control plane and network facilities is to confirm the control range. Generally, the control range includes hardware data channel and software program [1–3]. The specific control objectives of the two blocks are shown in Table 1. In SDN under the control of the network, can be in

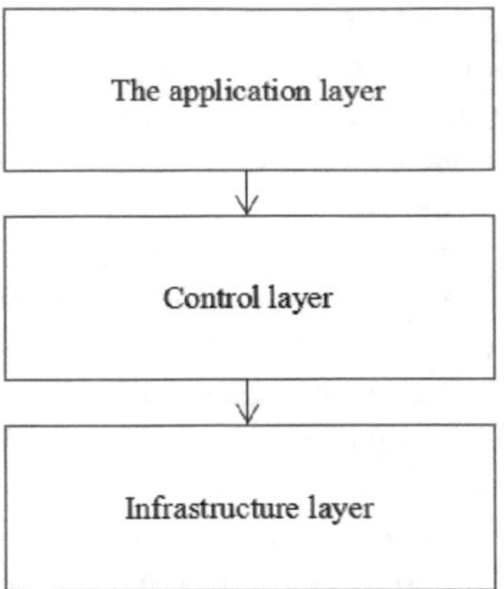

Fig. 1. Architecture of SDN

the network in series with the various network elements and variety series rules, make people in the construction of different types of network element structure of the network environment, such as the correlation rules routing connections, explains the SDN has a strong flexibility in the middle of the practical application [4–6].

Table 1. Specific control objectives of the two control blocks within the control scope of SDN network

Control plate	Control objectives
Hardware data channel	Data processing tool
A software program	Routing

2.3 Key Technologies

SDN is the key technology of network SDN controller, which is equivalent to the entire SDN network operating system, also is the core part in the whole SDN network architecture, its operation principle is, first of all, through the centralized management on the occurrence of the interface protocols, let all control targets in the network and terminal connection, form a distributed concentration relationship, then can for centralized control, Secondly, the underlying network of the monitoring switch is used to monitor the whole network. At the same time, the controller is also linked with the North bridge interface, the main function of the structure is to achieve the programmability of the upper application, the reason is that the north bridge interface is open, the administrator can log in the administrator system and then open the north bridge, so that the program can be carried out in the interface. SDN, moreover, the network data exchange equipment can realize high-speed data packet forwarding plane, in the process, the controller can under the occurrence of the interface protocol of data packet forwarding decision management, which makes the network control management efficiency, also reduces the complexity of the equipment in a network environment can reduce the associated costs, and improve the convenience of management [7–10].

3 Features and Topology of Wireless Sensor Network Based on SDN Network Structure

Under normal circumstances, the network structure environment SDN wireless sensor network is composed of three kinds of fixed nodes, one is the cable connection node, the node is usually used to provide relevant support to the other nodes, such as power supply nodes is wired connection nodes, the sensor is installed on the node, so although node is wired connection, but its integrated with sensor, therefore, it can be regarded as a node in wireless sensor network. The second is the wireless connection node, which usually represents some independent operating facilities, such as batteries, on-site monitoring equipment, etc. Like the wired node, the wireless connection node is also fixed in the application environment. The third is wireless mobile node. Different from the first two types of nodes, this kind of node is not only wireless connected, but also constantly moving. Therefore, it mainly refers to the equipment carried by the staff, such as handheld survey equipment. Combined with these three nodes, the topological structure of SDN network structure is shown in Fig. 2.

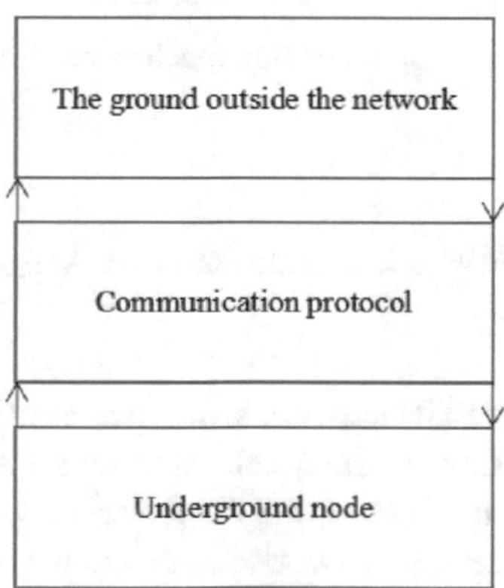

Fig. 2. SDN topology structure

Figure 2 topology structure is built in underground work environment, so the top is called the ground outside the network, its main is composed of communication interface and terminal station, communication interface and the middle layer protocol interface docking, to be able to communicate with the bottom, communication income information is transmitted to the terminal station, artificial can read the information on the terminal equipment, and then make a decision, The decision is set as instruction to be sent externally, and the sent instruction will be received by the middle layer and sent to the specified node through the middle layer. The middle layer is the communication protocol layer, the communication protocol TCP only commonly, the reason is that although there are a lot of network communication protocols, but suitable for SDN network, and can in underground work environment or other complicated work environment of stable operation of only the TCP protocol, the protocol is a typical two-way, able to undertake the top and bottom layers. The lowest layer is the node layer, which is composed of the above three types of nodes. In the SDN environment, the middle layer is directly connected with the middle layer through communication interfaces, and the middle layer is indirectly connected with the uppermost layer. Figure 3 shows the communication mechanism under the topology.

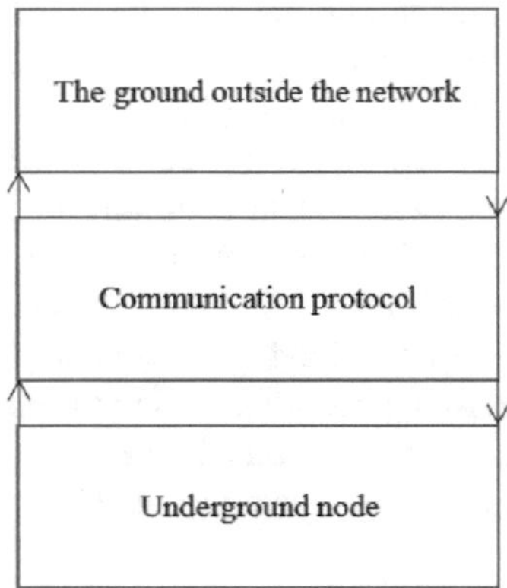

Fig. 3. Shows the communication mechanism under the topology

4 Wireless Sensor Network Localization Algorithm Based on SDN Network Structure

Combined with the feature of SDN network structure based on wireless sensor network, topology structure, the network with the wireless sensor network localization, but did not provide specific positioning method, so the need to choose other methods on the basis of this, and because of the sensor network localization goals more involved, internal structure is changeable, so it can be seen as a math problem, This kind of algorithm is the specific method of wireless sensor location.

4.1 Fixed Node Deployment Algorithm

For underground environmental conditions and is very complicated, fixing the wireless sensor network (WSN), you need to the environment in the relative fixed nodes deployment, and carries on the localization, generally is wired connection here refers to the fixed node, the reason is that the cable connection node is to provide the relevant support node, and these nodes are likely to be affected by environmental factors change, lead to underground foundation support instability, so these nodes are positioning, synchronous monitoring, can understand the environment parameter changes according to the node, or if a cable nodes affected by the environmental factors, then will have corresponding change, if you can even find the change, can not only to manage the nodes in the first place, they also have an understanding of the environment so that they can make accurate decisions. From this point of view, people can install wireless sensors on fixed nodes, and then locate the sensors according to the clustering algorithm. This algorithm is suitable for the network environment with hierarchical characteristics, and SDN is in line with this feature. Under the action of this algorithm, SDN network is divided into several clusters, and one primary cluster is selected from each cluster, and then other clusters conforming to the rules are combined with the primary cluster according to the cluster class rules, so that several cluster classes can be formed. In each cluster class, the relationship between the primary cluster and other clusters is as follows: Other clusters forward their own data to the primary cluster, and then forward data to the ground workstation through the primary cluster. To achieve this, people need to deploy all the wired nodes through the clustering algorithm, which is shown in Formula (1).

$$E_i(t) = e_i(t)/e_i(0) \tag{1}$$

In the formula, e_i (t) and e_i (0) are respectively the geographical location and energy change of E_i (t). Location positioning can be realized according to geographical location, and monitoring positioning can be realized according to energy change.

4.2 Wireless Node Location Algorithm

Wireless nodes in SDN generally refer to independently operated facilities. These facilities cannot be located by clustering algorithm because there is no line to rely on. In this case, we need to change our thinking and adopt other algorithms. According to this condition, location can be considered by ranging ideas, that is, ranging is to measure the distance between two nodes, if the distance is in a reasonable category, and there is no change, we can know that both ends of the line represented by the distance are respectively the location of wireless nodes, so as to achieve positioning. Under the way of thinking, first using the sensor itself connect wireless nodes in sensor node resources, realize the distance between nodes, and then through the correlation algorithm to calculate can, here mainly used algorithm for RSS algorithm, this algorithm is simpler, also do not need to add additional hardware to support algorithm in practice operation, Therefore, it has been widely used. The operation principle of RSS algorithm is as follows: first, deploy a large number of wireless positioning nodes in the environment, and then connect each sensor together according to the three-way positioning principle; Second, after the sensor is connected, the distance data between several non-collinear fixed nodes can be obtained. Third, calculate according to formula (2) to obtain the specific distance data, and then convert it into a straight line of thinking to achieve positioning. In addition, although RSS algorithm has been widely applied, but not on behalf of the algorithm is perfect, in fact the algorithm in some special conditions can not guarantee the accuracy, the reason is that the accuracy of the algorithm stability is low, if the environment cannot be deployed in a large number of infinite node, the precision of the result will drop, at the same time, because the location is a continuous process, Period to obtain distance data, confirm whether the distance between the node changes, so if an infinite node in the process of sensor malfunction, unable to transfer distance data, the algorithm will be failure, in the face of this situation has not been very good method to solve the current, so the location may need to be assisted.

$$S_i(t) = 1 - \sum_{k=1}^{n} \frac{RSS_k(t) - RSS_k(t-1)}{n * RSS_k(t-1)} \tag{2}$$

where, $RSS_k(t)$ represents t moment, n is the distance between two nodes, and k is the k node.

4.3 Mobile Node Location Algorithm

Mobile node is the most difficult node to locate in SDN, because it has the characteristics of random movement, and positioning in different environments will be interfered by realistic factors, so mobile node positioning also needs to select a new algorithm. From this point of view, it is necessary to recognize the environment in which mobile nodes exist. Here, the specific environment of mine operation is selected by referring to the macro concept of underground operation environment. Then, a special SDN network topology is established based on the characteristics of underground manual movement in the mine operation environment, as shown in Fig. 4.

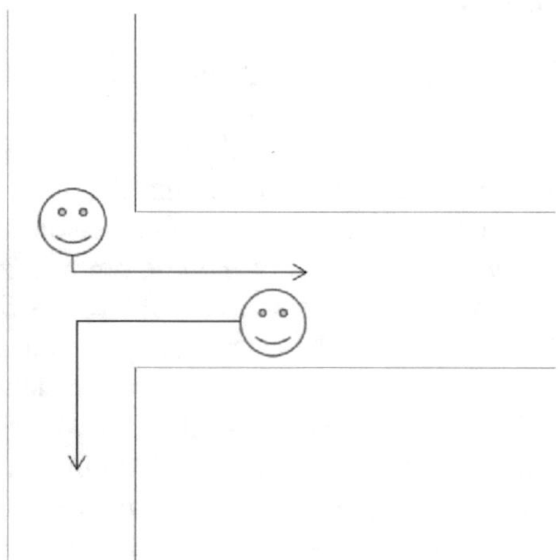

Fig. 4. Characteristics of manual movement in mine operation

Combined with Fig. 4, because the mobile node generally by the artificial hand, so the movement of the artificial features represent the node movement characteristics, from the perspective of the mobile node localization need support from signal transit stations, namely in the work environment to install several signal transit stations, each station signal coverage can not overlap, also cannot have the too big intervals, General advice to control within 1 m, such as artificial is primarily out of work need to move, so mobile artificial mainly to and from each base station is located in the process of the work area, so when the artificial into a work area, the handheld mobile node can communicate with the work area of the base station, at this time of base station signal is read, mobile node localization can be realized, the algorithm used during this period is as follows.

$$E_R = \sqrt{(x - \widehat{x})^2 + (y - \widehat{y})^2} \tag{3}$$

where x and y are the manually moved horizontal and horizontal axes.

5 Conclusion

In conclusion, the wireless sensor in the actual application needs to be in the position, so as to play a role of sensor to control the practical work, work smoothly, avoid the related security problem, but in some complex environment through the sensor location and signal technology, this is where the SDN network structure and the related algorithm to achieve purpose.

References

1. Sundaram, S.: Application of MSVPC-5G multicast SDN network eminence video transmission in drone thermal imaging for solar farm monitoring. Energies **14**(24), 8255 (2021)
2. Cohen, A., et al.: Bringing network coding into SDN: architectural study for meshed heterogeneous communications. IEEE Commun. Mag. **59**(4), 37–43 (2021)
3. Ran, J., Wang, X.: Virtual SDN network embedding algorithm based on load balance. J. Phys. Conf. Ser. **1646**, 012071 (2020)
4. Polverini, M., Galán-Jiménez, J., Lavacca, F.G., Cianfrani, A., Eramo, V.: A scalable and offloading-based traffic classification solution in NFV/SDN network architectures. IEEE Trans. Netw. Serv. Manage. **18**, 1445–1460 (2021)
5. Zhao, B., Liu, Y., Li, X., Li, J., Zou, J.: TrustBlock: An adaptive trust evaluation of SDN network nodes based on double-layer blockchain. PLoS ONE **15**(3), e0228844 (2020)
6. Burdonov, I., Vinarsky, E., Yevtushenko, N., Kossatchev, A.S.: Perfect sets of paths in the full graph of SDN network switches. Proc. Instit. Syst. Programm. RAS **32**(4), 245–260 (2020)
7. Peter, B., Stephen, M., Nelson, B.J., Ron, A., Mak, S.: A flow analysis and preemption framework for periodic traffic in an SDN network. Concurrency Comput. Pract. Experience **32**, e4531 (2020)
8. Alghamdi, K., Braun, R.: Software defined network(SDN) and openflow protocol in 5G network. Commun. Netw. **12**(1), 28–40 (2020)
9. Gautam, Y., Sato, K., Gautam, B.P.: Experimental security analysis of SDN network by using packet sniffing and spoofing technique on POX and Ryu controller. In: NaNA 2020 2020 International Conference on Networking and Network Applications, 11–14 Dec 2020
10. Xie, X., Ouyang, W., Xia, Y., Cai, W., Yang, H.: Identity authentication system for mobile terminal equipment based on SDN network. Int. J. Inf. Commun. Technol. **17**(3), 257 (2020)

Access Control Method of SDN Network Based on Zero Trust

Ming Li[1], Hang Cheng[1(✉)], Wanwan Cao[1], Songbo Yu[1], and Jie Song[2]

[1] State Grid Anhui Electric Power Co., Ltd., Information and Communication Branch, Hefei 230061, China
chenglinluzhu@126.com
[2] Anhui Jiyuan Software Co., Ltd., Hefei 230088, China

Abstract. In order to use relevant methods to control SDN network access based on zero trust, this paper will carry out relevant research. This paper first introduces the basic concepts of zero trust and SDN, then constructs the access control framework of SDN, and puts forward the relevant control principles. Through this study, SDN network access can be effectively controlled on the basis of zero trust, which can greatly guarantee access security.

Keywords: Zero trust · SDN network · Network access control

1 Introduction

Since the high popularity around the world rely on computer networks, network attack problems have existed, the specific construction form of the problem has a lot of, which is leading is "malicious access", namely the attacker will be used in a variety of technical means will own disguised as a normal user, and then to submit an application for access to the network environment, because of the existence of camouflage, So the network management terminal may be blinded by its access request, in which case the attacker is inside the computer network and can wreak havoc inside, or achieve other purposes., under the condition of the related areas have long been in the struggle with the network attack and thus will continue to analyze the problems existing in the network structure, therefore, in recent years has been found on the network access control system in the performance of the previous defects, the reason is that the modern network attack is no longer restricted by factors such as geographical, a network boundary, gradually has the characteristics of globalization, scale, At this time, the original access control system has exposed the defects of rigid mechanism and insufficient flexibility. In view of this, relevant fields have proposed SDN network, which is different from the original network and has strong security. Therefore, the emergence of SDN has brought new enlightenment to people, but also brought some new problems, such as how to play the role of SDN security and reasonably control network access. The latest idea in this field is to build SDN on the basis of zero trust and set up network access control mechanism. In order to achieve this goal, it is necessary to carry out relevant research.

J. H. Abawajy et al. (Eds.): ICATCI 2022, LNDECT 170, pp. 496–504, 2023.
https://doi.org/10.1007/978-3-031-29097-8_59

The application of SDN in the field of network security is very prominent. Its main feature is that the centralized control plane and the distributed forwarding plane are independent but closely related. The former mainly controls the latter, and the latter is responsible for executing the former's commands. In principle, the control plane mainly communicates with the forwarding plane network equipment through the control and forwarding communication interface, centrally controls all network equipment, and provides flexible programmable space. The latter is mainly used for forwarding and executing control layer commands. The network structure with these two characteristics is a generalized SDN structure, and this structure has no narrow definition.

In the early days, people believed that SDN was a specific network protocol with the same nature as HTTP and TCP, but this view was soon overturned. Modern theories defined SDN as a non-specific network protocol. On the contrary, they believed that SDN was more inclined to be defined as a network framework, because SDN had a variety of interface protocols, including the OpenFlow South interface protocol, The interface protocol can be used to realize the communication between SDN controller and SDN switch; The northbound API can be used to realize communication between business applications and SDN controllers. The existence of these structures makes the SDN network structure systematic and convenient for people to manage and apply.

SDN network structure is a new network structure, which separates the network control function and forwarding function in the traditional network structure, so it has good controllability and can be controlled by programming. This architecture allows people to send information directly from the control layer and network devices to external computing devices. Therefore, its underlying infrastructure is completely transparent to applications and network services, and it is also abstract. Therefore, a network can be interpreted as a virtual logical entity.

The control principle of the SDN centralized control plane is not complex, but this control mode allows the flow between the controller and the network equipment, independent of the data flow generated by the communication between terminals. Therefore, network devices can directly accept control instructions, generate forwarding tables, and execute forwarding table commands. Compared with the traditional network, this method does not need the support of complex distributed network protocols and is more convenient.

2 Zero Trust and Basic Concepts of SDN

2.1 Zero Trust

Zero trust is a concept of network security protection and the basic position of network access control. This concept mainly emphasizes the following connotations: In the network environment, people have their own security zones. If other people want to enter other people's security zones, they need to submit an access application [1–3]. Then the "host" of the security zone needs to verify whether the access application submitter will pose a threat to them. On the basis of the basic access control thought, safety area "master" to how to identify whether the access application submission threatening is the key problem, but the problem without a solution, because of the difference between the network environment in the real environment, safety area "master" is not completely to visit application, submit the actual identity identification, that is, the identity information submitted by the access application submitter may be forged and false, which cannot be distinguished. In this case, there is no objective trust between the "host" of the security area and the access application submitter under the security principle, indicating that the "host" should distrust the access application submitter from the very beginning. This attitude of distrust is very thorough, so it is called zero trust. Zero trust isn't put forward to make a safe zone "master" embraces mistrustful attitude all the time, but to let it never trust attitude, to evaluate each access to apply for the identity of the submitter, improve the credibility of the submitter, until its credibility more than standard, break the zero trust framework, to access other network security area [4–6].

2.2 SDN Network

The application of SDN in the field of network security is very prominent. Its main feature is that the centralized control plane and the distributed forwarding plane are independent but closely related. The former mainly controls the latter, and the latter is responsible for executing the former's commands. In principle, the control plane mainly communicates with the forwarding plane network equipment through the control and forwarding communication interface, centrally controls all network equipment, and provides flexible programmable space. The latter is mainly used for forwarding and executing control layer commands. The network structure with these two characteristics is a generalized SDN structure, and this structure has no narrow definition.

In the early days, people believed that SDN was a specific network protocol with the same nature as HTTP and TCP, but this view was soon overturned. Modern theories defined SDN as a non-specific network protocol. On the contrary, they believed that SDN was more inclined to be defined as a network framework, because SDN had a variety of interface protocols, including the OpenFlow South interface protocol, The interface protocol can be used to realize the communication between SDN controller and SDN switch; The northbound API can be used to realize communication between business applications and SDN controllers. The existence of these structures makes the SDN network structure systematic and convenient for people to manage and apply.

SDN network structure is a new network structure, which separates the network control function and forwarding function in the traditional network structure, so it has good controllability and can be controlled by programming. This architecture allows people to send information directly from the control layer and network devices to external computing devices. Therefore, its underlying infrastructure is completely transparent to applications and network services, and it is also abstract. Therefore, a network can be interpreted as a virtual logical entity.

The control principle of the SDN centralized control plane is not complex, but this control mode allows the flow between the controller and the network equipment, independent of the data flow generated by the communication between terminals. Therefore, network devices can directly accept control instructions, generate forwarding tables, and execute forwarding table commands. Compared with the traditional network, this method does not need the support of complex distributed network protocols and is more convenient.

SDN network is a data forwarding and separation network with logical control ability. It is very different from traditional network in structure. It can not only provide smooth evolution ability for The Internet, but also provide brand-new security guarantee for the network. However, SDN can only provide security guarantee conditions, but not security guarantee methods [1, 7, 8]. Therefore, relevant strategies need to be adopted manually to give full play to the role of the network structure to guarantee security to the maximum. On this basis, the current most common SDN network security guarantee strategy is established on the basis of zero trust. See Fig. 1 for the policy process.

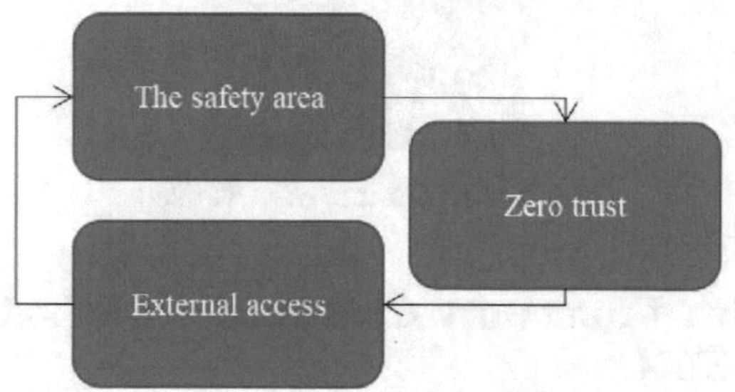

Fig. 1. SDN network security guarantee policy based on zero trust

Combined with Fig. 1, when external access requests to the security zone [3], the security zone has a zero-trust attitude toward external access, and the security zone is divided into two layers, namely the original layer and the access layer. Under the condition of maintaining the zero-trust attitude, external access can only enter the original layer, which has no information. And safety area in the middle of the original layer security protection system to monitor the quality of the behavior of the visitors to the next will, according to its behavior trust assessment, if the trust evaluation standard, then the access will be able to pass the original layer into the access layer, on the contrary will be removed from the original layer, can also according to the trust evaluation results can be used to control access to resources and for the visitor that is, assuming that resource A requires the trust of visitors to reach 100, when the evaluation results show that the trust of visitors is only 99, visitors can not check resource A, but can check it on the contrary. It can be seen that in this case, external access is effectively controlled and the access process is always within the security rules to avoid malicious attacks. In addition, the architecture of SDN itself is shown in Fig. 2.

Combined with Fig. 2, SDN is a typical three-tier network architecture, the first layer for the resource, is refers to the safety area to be in the access to the resource integration center, and security protection object, the second floor for communication and control layer, outer layer of access control in the third layer for data forwarding, if an external access to the trust, Through this layer, the network forwards the resources it is allowed to access to the user.

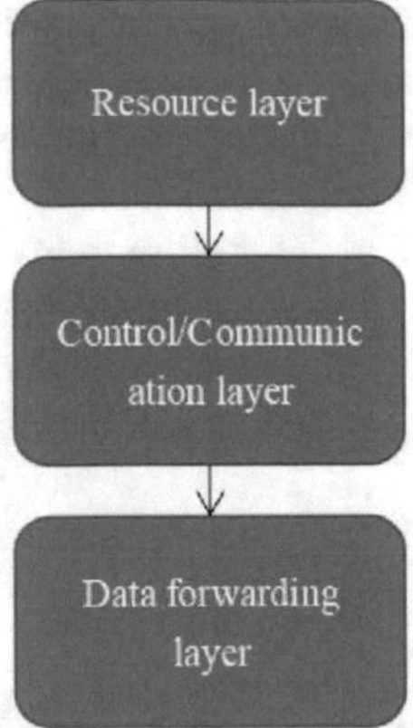

Fig. 2. Architecture of SDN itself

3 Access Control Framework Construction Method and Control Principle of SDN

3.1 General Framework

On the basis of zero trust, the SDN network access control framework is divided into four parts, namely, identity authentication module, access control module, trust evaluation module and IDS module. See Fig. 3 for their relationship.

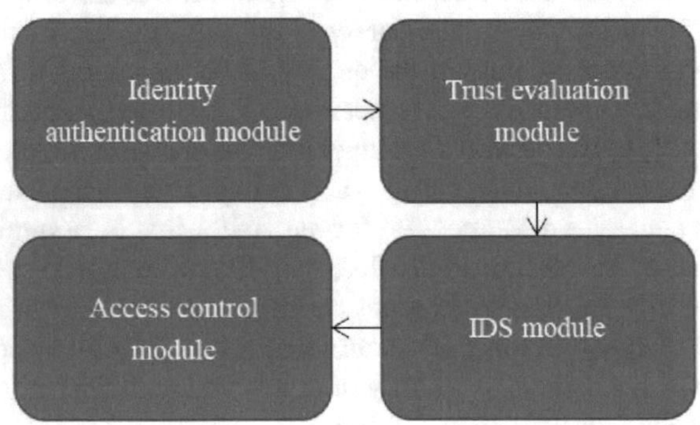

Fig. 3. Relationships among components of SDN Network Access control framework

Combined with Fig. 3, the implementation of the four components of the control framework is as follows.

Identity Authentication Module
The identity authentication module is the first step to realize network access control. Its main function is to identify the identity of visitors. The identification results are generally presented in three forms, as shown in Table 1.

Table 1. Form and logic of identity identification result of identity authentication module

Identification result form	logic
Who has access records	Based on the existing visit records, we can make a preliminary judgment on the trust of visitors. For example, if there are bad behaviors of visitors in previous visit records, it is generally recommended to reject the visit from the perspective of trust. On the contrary, we can consider the next trust evaluation
No visit record	Generally refers to new visitors, because there is no visit record, so it is not possible to make a preliminary judgment of trust, so you can directly enter the trust evaluation step
Inability to recognize visitors	Visitors who cannot confirm whether they have accessed a security zone are generally treated as suspected security threats. In other words, it is recommended to deny access due to the concept of zero trust

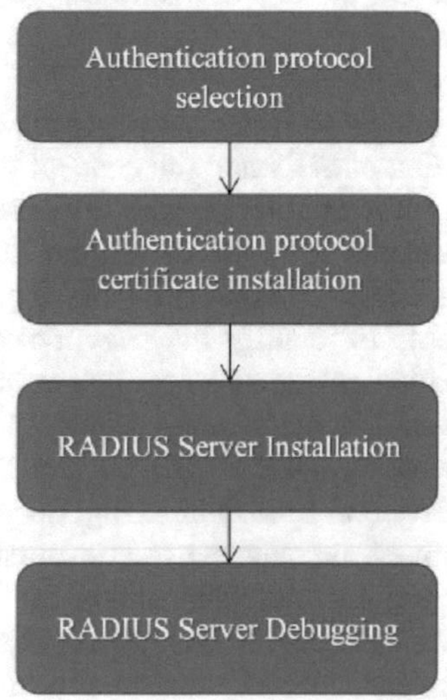

Fig. 4. Identity authentication module implementation steps are introduced

The realization method of identity authentication module is mainly divided into two steps see Fig. 4 for details.

According to the Fig. 4 the device is generally used under 8021X authentication protocol, which is equivalent to the proxy between the visitor and RADIUS authentication server. It can authenticate the account, password and IP address of the login device entered by the defender when logging in to the platform. Second, the RADIUS server is installed. The server is a device that can control user identity and access request permission and complete authentication. It has authentication service middleware and authentication and authorization database.

Access Control Module

The access control module needs to be implemented under the SDN controller and monitoring component, and has the function of centrally controlling the data forwarding subject. Therefore, the first step of this module is to install the SDN controller and monitoring component. After finished the first step can be built on access control module, building work is divided into three steps: first, make grant access middleware, mainly used for access decision received access decision agents give instructions, IDS intrusion detection alarm, these contents will communicate by means of distributed flow table, achieve rapid access authorization or efficient control visitors behavior; Secondly, the access request forwarding middleware is installed. The main function of the middleware is to forward the access request to the trust assessment module after the identity of the visitor is authenticated and passed the authentication, so as to further understand its credibility. Thirdly, the packet parsing middleware is installed, whose main function is to parse decision packets and IDS alarm packets for the next flow standard delivery.

Trust Evaluation Module

The trust evaluation module is used to further evaluate the credibility of visitors. It is a key module for the combination of SDN and the concept of zero trust. There are three steps to realize this module: First, establish credibility assessment agent equipment, it can access for the visitor behavior after the measurement and evaluation, internal is mainly composed of various access middleware behavior measurement, on the basis of the middleware perfect able to identify the network of various types of network attacks, including the incubation period is long, high concealment of network attack behavior can be identified, Note The module has strong security protection capabilities. At the same time, the device of the visitor's behavior of the trust measure evaluation process will continue, to achieve real-time monitoring; Second, set up the behavior trust access decision agent equipment, namely the trust evaluation agency equipment can only give a visitor can trust, and not according to how far they can trust them to make a decision directly, thus to set up behavior trust access decision agents, main effect is proposed according to visitor's dependability numerical corresponding decisions, such as the visitor's dependability is lower than the lowest standard, Then no matter how much the specific value of trustworthiness is, it is regarded as untrustworthy and refused to defend, and when the visitor's trustworthiness is higher than the minimum standard, but does not reach the full value, it will only open the corresponding authority resources according to the specific situation of its trustworthiness, and so on. Third, trust measurement database installation, because visitors dependability assessment is a long process, which involves the visitor behavior history record and the current behavior, so in order to more accurately assess, need to install the credibility measure database, special store every visitor behavior can trust evaluation results, And each trustworthiness assessment will be combined with the results of the previous historical assessment to make a comprehensive consideration.

IDS Module

IDS model is an important network access control modules, namely through the above modules, the system can determine the user can be trusted, and draw the corresponding access control decision, but how to implement the decision or a problem to be solved, and the importance of the IDS module is in this, can perform relevant decision instruction, earnestly to control network access. The implementation method of IDS module is as follows: First, the IDS system is installed, which has the ability to quickly identify various network attacks and has an independent reliability evaluation mechanism. Once the identification results show that the network is under attack, it will send an alarm to the SDN controller. At present, the attack types identified by IDS system include sniffer attack, password attack and flood attack. Secondly, the installation of IDS awareness database is mainly used to obtain and store IDS detection data with the help of SDN switch for better attack identification.

3.2 Control Principles

To control SDN network access to a more reasonable, the staff according to control principle to manage control system, namely, in the whole control process staff shall reasonably set control execution time period and the sliding window size, such ability for visitors to visit behavior can be periodic dynamic trust evaluation, In this way, the principle of continuous monitoring and verification of network access security control under the concept of zero trust is implemented. At the same time, in the control decision-making agent module level, shall, according to network security protection intensity to set the access rules, normally a proposal to visitors as a decision factor, trust levels, access object will defender according to different trust hierarchy, again according to different access level to open different access object, here can be considered the permissions method to achieve the purpose, Such as an organization has A, B, C three kinds of resources, access them value from low to high respectively 1, 2, 3, and generally three level of trust, good, good, so when the visitor's dependability is A regular, just open access to its value of 1 A resource, B and C is the visitors don't have access to resources, and so on can be, This is the rational principle of zero – trust security protection.

4 Conclusion

In conclusion, to comply with the relevant principles, based on zero trust SDN network access control framework can better guarantee the network security, the reason is that the framework can external access request for identification, assessment, its dependability size, again according to the size they can trust them to provide control decisions, execution can avoid most of the security risks, Ensure that every successful visit is based on security.

References

1. Mir, A., Rashid. I., Kumar, K.R.: An augmented smart grid based SCADA security management system (SSMS) based on zero-trust architecture. In: Proceedings of the 2nd International Conference on ICT for Digital, Smart, and Sustainable Development, ICIDSSD 2020, 27–28 Feb 2020, Jamia Hamdard, New Delhi, India (2021)
2. Arabi, A., Ogundijo, A.D., Nyamasvisva, T.E.: A zero-trust model-based framework for managing of academic dishonesty in institutes of higher learning. Turkish J. Comput. Math. Educ. (TURCOMAT) **2**(6), 5381–5389 (2021)
3. Shah, S.W., Syed, N.F., Shaghaghi, A., Anwar, A., Baig, Z., Doss, R.: LCDA: lightweight continuous device-to-device authentication for a zero trust architecture (ZTA). Comput. Secur. **108**, 102351 (2021)
4. Tao, C., Lv, Y., Qi, Z., et al.: An implementation method of zero-trust architecture. J. Phys: Conf. Ser. **1651**, 012010 (2020)
5. Bogner, E.: The zero-trust mandate: never trust, continually verify. Softw. World **50**(4), 9–10 (2019)
6. Lu, L., Han, J., Liu, Y., Hu, L., Huai, J.-P., Ni, L., Ma, J.: Pseudo trust: zero-knowledge authentication in anonymous P2Ps. IEEE Trans. Parallel Distrib. Syst. **19**(10), 1325–1337 (2008)
7. Dunning, D., Anderson, J.E., Schlösser, T., Ehlebracht, D., Fetchenhauer, D.: Trust at zero acquaintance: more a matter of respect than expectation of reward. J. Pers. Soci. Psychol. **107**(1), 122–141 (2014)
8. Beck, E.A.: How zero-trust network security can enable recovery from cyberattacks. ISACA J. **6**, 14–18 (2014)

Application of Permutation Entropy Algorithm in Automatic Mechanical Equipment Running State Monitoring

Xin Lu(✉)

Southwest University, Chongqing 400700, China
1105777632@qq.com

Abstract. In order to realize automatic mechanical equipment operation monitoring function, this paper will carry out related research based on permutation entropy algorithm. This paper first discusses the basic concept of permutation entropy algorithm, and then puts forward the realization idea and method of automatic mechanical equipment running state monitoring function. Through the research of this paper, under the support of permutation entropy algorithm, automatic mechanical equipment running state monitoring work can be achieved, the monitoring effect is good, so the permutation entropy algorithm has certain application value.

Keywords: Permutation entropy algorithm · Automatic mechanical equipment · Operating condition monitoring

1 Introduction

In order to improve work efficiency, ensure the quality of work, China's industry and related fields have long realized the automation of production, its biggest characteristic is to use automatic mechanical equipment to replace manual to complete some production work, can better promote economic development. But with the application of automatic mechanical equipment continues to deepen, people find that production work is more dependent on equipment, and equipment in the continuous application will appear a variety of failures or abnormalities, once this happens, work efficiency and quality will be negatively affected, will limit economic output. In the face of such situation, people realize the equipment failure or abnormal is not to avoid, so the best way to deal with is to monitor equipment, according to the monitoring results to confirm whether the equipment fault, fault specific type, fault location, and then can be the first time, this idea soon got the recognition of people, But how to make the idea into reality is a problem to be solved, and related fields have been researched, and found that the problem is complicated, the reason is that the fault monitoring ranged over a variety of fault types, each fault type and contains a large number of fault information, for the specific situation of the fault will have to deal with this information, and unusual methods cannot do this, So for a long period of time the idea has been stuck in the theoretical stage.

J. H. Abawajy et al. (Eds.): ICATCI 2022, LNDECT 170, pp. 505–513, 2023.
https://doi.org/10.1007/978-3-031-29097-8_60

But in recent years, through continuous exploration of people found some methods can be used in automated machinery and equipment running status monitoring, these methods are known collectively as machinery and equipment running dynamic mutation monitoring method, are essentially mathematical algorithm, the principle is the mechanical equipment in the fault state of operation as a kind of dynamic anomaly, which will contain a signal mutation points, Through the related system and the algorithm can detect the signal mutation points, if any means equipment failure, the current existing and some more concrete form to analyze the fault type, fault location, gives a complete and accurate report, permutation entropy algorithm is such a kind of monitoring method, but the method proposed time is not long, a lot of people didn't know this, therefore, in order to popularize permutation entropy algorithm and meet the requirements of automatic mechanical equipment operating state monitoring, it is necessary to carry out relevant research.

2 Research Background

Automatic detection of mechanical equipment is a concept of detection technology put forward in recent years. This concept has attracted wide attention. Therefore, many researchers have rapidly conducted research on the method of concept realization, and encountered three problems in the process: one is how to realize the automatic detection function of mechanical equipment, that is, to automatically detect mechanical equipment in theory, it is necessary to build an automatic system that can carry the automatic detection function, The system must at least have two modes of regular detection operation and controlled temporary detection operation, and also be able to collect all the information required for mechanical equipment detection. Therefore, in the entire mechanical equipment automatic detection system, the automatic detection function can be regarded as a subsystem. The main task of the system is to automatically execute the detection command and collect information. The second is how to identify the detection information of mechanical equipment, that is, through the automatic detection function, a large amount of detection information will be obtained, which is an important basis for judging the status and fault of mechanical equipment. Because of the huge amount of data, although manual can identify each piece of information, it cannot identify it efficiently, so it needs to be done by other technical means, otherwise, the automatic detection of mechanical equipment will also be impossible, This has also become one of the key problems to be solved in the implementation of the system; The third is how to realize the communication between the automatic detection function and the intelligent system, that is, the automatic detection function belongs to the executive subsystem, while the intelligent system belongs to the terminal system, and there is no direct connection between the two. Therefore, a channel should be established to connect the two together, so that the information collected by the automatic detection function can be transmitted to the intelligent system to promote the intelligent system to identify, How to build such a connection channel has become a problem worth thinking about.

Focusing on the above three issues, modern research has put forward many methods from different angles and established a standard system model. The model consists of three levels. First, the execution unit layer, which mainly refers to the automatic detection function subsystem of mechanical equipment, is responsible for executing the detection work, sending information to the intelligent terminal at the same time, and second, the communication layer, which is responsible for forwarding the information transmitted from the execution unit layer to the intelligent terminal, And protect the information from loss and damage in the transmission process. Finally, the intelligent terminal is responsible for receiving the information, identifying the information, and outputting the detection results. After this model is put forward, the automatic detection system of mechanical equipment is basically realized. The remaining part to be solved is to enable the intelligent terminal to have the ability of information recognition. To give this ability, we need to use mathematical methods, and the replacement entropy algorithm is a good choice.

3 Basic Concepts of Permutation Entropy Algorithm

Permutation entropy algorithm is a metric algorithm whose main function is to measure the complexity of time series. Its main process is shown in Fig. 1.

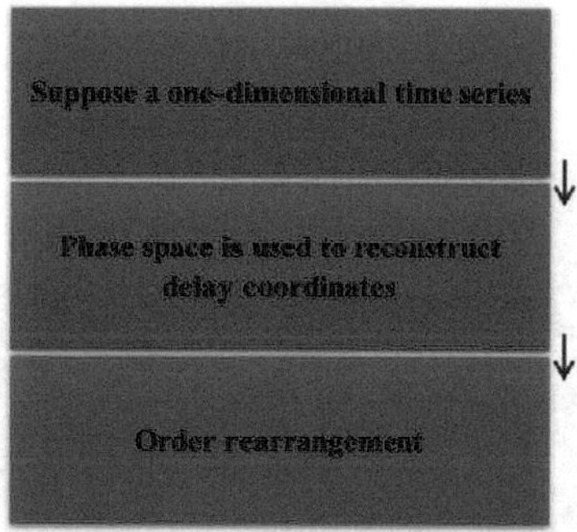

Fig. 1. Main flow of permutation entropy algorithm

According to Fig. 1, permutation entropy algorithm firstly establishes a one-dimensional time series, the expression of which is shown in Formula (1). Then, phase space reconstruction delay coordinate method is adopted to reconstruct phase space delay coordinate, and phase space reconstruction of any element in one-dimensional time series is carried out based on the coordinate, and several sampling points can be obtained. Several samples can be obtained by continuous sampling at each sampling point, and vector reconstruction of one-dimensional time series can be completed. After the reconstruction, to the time sequence of all time to reorder, sorting carried out in accordance with the rules of ascending, at any time after the completion of the statistical line sort see the index of the order, can get another set of symbol sequence, thus the original time symbol sequence and calculation of the time symbol sequence comparison, if there is difference, so differences is the point of signal mutation, Failure occurs at this time, also explain the operation of the equipment is fault location at this time, according to the characters of peak signal mutation point mutations can to preliminary judgment of fault type, such as peak parallel mutation for a long time (generally more than 1 s count too long) means fault belongs to a long period of time, the opposite for the instantaneity of failure [1–3].

$$X = \{x(1), x(2), \ldots x(n)\} \tag{1}$$

where x is the time line in one-dimensional time series.

4 Ideas and Methods for the Implementation of Automatic Mechanical Equipment Running State Monitoring Function

4.1 Implementation Roadmap

In fact, it is not difficult to monitor the running state of automatic mechanical equipment, just need to build a monitoring system, but the difficulty lies in how to obtain the signal mutation point contained in the monitoring information, which needs to use the algorithm [4–6]. According to this requirement, the realization idea of automatic mechanical equipment running state monitoring function in this paper is mainly divided into two parts, as shown in Fig. 2.

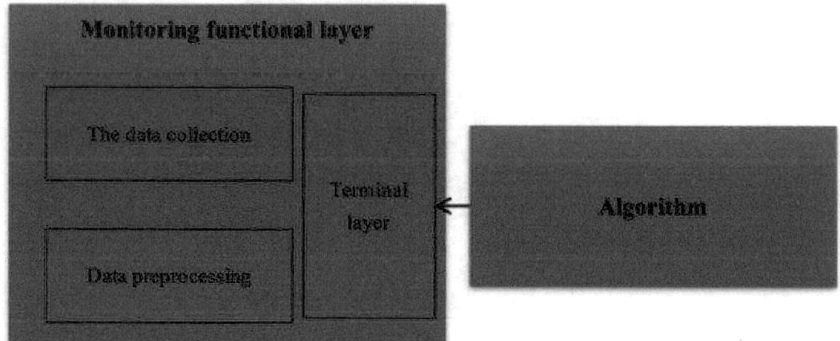

Fig. 2. Realization idea of automatic mechanical equipment running state monitoring function

As can be seen from Fig. 2, the operation status monitoring function of automatic mechanical equipment in this paper is mainly composed of two parts: monitoring function layer and algorithm. The internal architecture of the monitoring function layer is divided into three parts: The first data acquisition function, the function is mainly responsible for the automatic mechanical equipment collection, so as to obtain information, and then send the information in signal format, is received by the data preprocessing function; Because the second function of data pretreatment, data acquisition function signals from the initial format for electrical signals, which may exist noise at the same time, the signal cannot be read from a terminal layer interpretation, or led to the decrease of the final monitoring results accuracy, so through the function of data preprocessing signal format into a digital format, And all the signals converted to digital format are de-noised to ensure signal quality, and then the signal will be sent to the terminal layer; The third terminal, terminal layer is mainly responsible for the signal to read interpretation, access to the information, and on the depth of information analysis, the characteristics of every piece of information, such as the generation of information time, the source device, and then classifying information, to get into formal calculation stage, after the completion of output will eventually results. And although algorithm is automatic mechanical equipment running status monitoring function ideas of another component, but the algorithm itself is on the terminal layer operation, can will be calculated after classification information into yuan, and then by calculating the output as a result, the calculation results will be accepted by the terminal layer, through the data visualization function to export again, Promote artificial understanding of the running state of automatic mechanical equipment, grasp the signal mutation point [7–9].

It is important to note that although automation machinery and equipment running status monitoring function need algorithm as the support, but after a long time of research and exploration, to provide support to the function of the algorithm has a lot of, in addition to the permutation entropy algorithm, Fourier transform, short-time Fourier transform, wavelet analysis, etc., and this article chose permutation entropy algorithm, mainly because other algorithms have some shortcomings: first, although the Fourier transform is good it is concluded that signal the global mutation, can clear tell people signal mutation point exists, but it can't locate the signal mutation point of time, so the result is not completely, using the results obtained with this algorithm cannot allow people to directly handle exceptions; Second, the main characteristic of short time Fourier transform is able to output within a given time interval, frequency interval comprehensive and accurate results, but the actual situation of different equipment in different fault there can be no fixed for a given time interval or frequency interval, first the short-time Fourier transform (STFT) of the instructions in the applicability of the automated mechanical equipment running condition monitoring is bad. Secondly, the algorithm adopts a unified window function in any situation, and the resolution does not change with the change of the situation. Therefore, once the window function is fixed, the STFT change cannot refine the result to an arbitrarily small local time, indicating that it is deficient in the sensitivity of signal mutation feedback. Third, the wavelet analysis in the applicability of the automated mechanical equipment running condition monitoring is very good, also can meet the monitoring requirements of different conditions, can also handle under a state of isolation or other special mutations in the signal, but is depend on the wavelet transform coefficient of wavelet analysis, wavelet function, the selection of these parameters could lead to the result is not accurate, meanwhile, wavelet analysis is easily disturbed by noise. Focusing on the defects of other algorithms, the advantage of permutation entropy algorithm is very obvious, can say that the algorithm is designed to realize automatic mechanical equipment running status monitoring function, may be applied to other fields permutation entropy algorithm is better than other algorithm, but in automatic mechanical equipment running condition monitoring of the optimal permutation entropy algorithm [10].

4.2 Implementation Method

According to the implementation ideas, the following will analyze the specific implementation method of automatic mechanical equipment running state monitoring function.

Implementation of Data Collection

Data acquisition is the automatic mechanical equipment running status monitoring the necessary premise, so the first step to achieve, the method of its implementation is mainly constructed sensing the physical layer, namely sensing the physical layer is mainly composed of several sensors, each sensor will be installed on the automation machinery and equipment, dynamics of equipment of the related indexes data collection, and generate the corresponding information, for example, torque information can be obtained by collecting the torque of the equipment, and then these information will be stored in the electronic unit of the sensor itself, and sent to the data preprocessor through the electronic unit. The realization of the data collection method is simple, but also should pay attention to a point in the process, which must be combined with the actual needs to choose the type of sensor, the different power principle of automation machinery and equipment, which not only represent different mechanical equipment will appear all sorts of different fault operation state, also represent different mechanical equipment to collect different data to obtain information, therefore, sensor selection is very important, we must first understand the specific data to be collected and then select

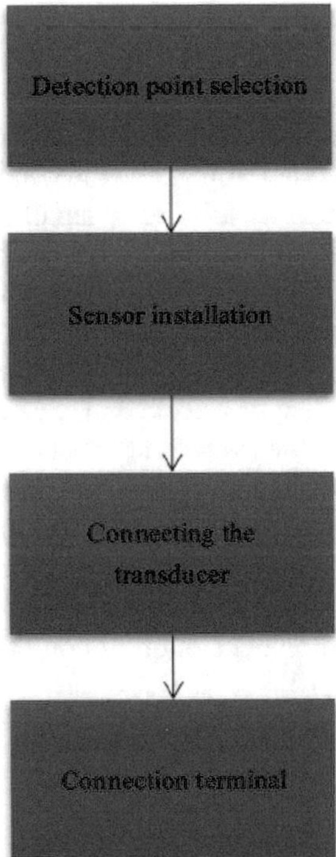

Fig. 3. Installation and working principle of the sensor

the corresponding sensor. Table 1 shows the more common sensors. Figure 3 shows the installation and working principle of the sensor.

Table 1. Shows the more common sensors

Common sensor	Function
Torque sensor	Collect torque data
Friction transducer	Data of friction force are collected
Amplitude sensor	Collect amplitude data

Data Preprocessing Implementation

The data preprocessing end of automatic mechanical equipment operation state monitoring function is mainly responsible for two tasks of data information format conversion and noise reduction. Therefore, this paper adopts two methods on this port: First, because the original information of signal suitable for electrical signals, needs to be converted to digital signals, so you can directly use transducer to achieve the purpose, namely the function of the transducer only is energy conversion, so will send transducer, sensor signal transducer can be related to energy parameters, then the energy parameters of the signal can be converted to the corresponding digital format. The transducer is generally installed in the communication channel between the data acquisition end and the terminal. Second, noise reduction processing, this paper chose the programming technology to develop the matching degree of processing software, it can match the original information and standard information, if the original information with noise, then there will be a match point, when the matching point will not be deleted, and then according to the standard information to fill in as well as to achieve noise reduction goal, the specific process as shown in Fig. 4.

Terminal Layer and Algorithm Implementation

Since the algorithm operates on the basis of the terminal layer, it needs to be realized together with the terminal layer. Thus, the realization method can be divided into two steps: First, terminal layer in the position of the flow in the whole monitoring system, so it must have the corresponding basis function can better processing the signal, to prop up the subsequent algorithm, there must have the function of the database, the information characteristics analysis, the signal to accept/interpretation, and other functions, the main functions can be realized through programming, equipment installation, no further elaboration here; Secondly, based on the terminal layer, the program flow is developed according to the permutation entropy algorithm flow, so that the terminal layer can execute the algorithm and monitor the equipment in real time. It should be noted that although the writing process of the algorithm process is not complicated, the operator must set a termination condition. The reason is that some abnormal phenomena of the equipment in the monitoring process are not faults. If the termination condition is not set, it may lead to misjudgment, which will cause a lot of trouble to the manual work.

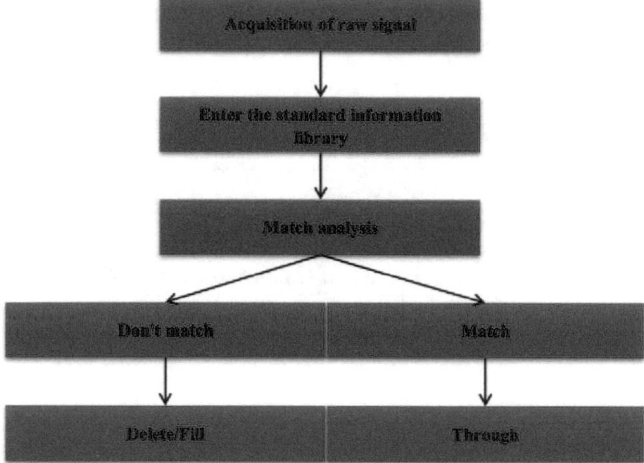

Fig. 4. Basic flow of noise reduction

5 Conclusion

To sum up, the realization of automatic mechanical equipment operation state monitoring function is of great significance, so people should carefully choose related algorithms, and permutation entropy algorithm has advantages compared with other similar algorithms, so this algorithm is one of the best choices, worth promoting. From this point of view, with the support of permutation entropy algorithm, people can understand the faults and anomalies of the running state of automatic mechanical equipment through the monitoring function, so that people can quickly carry out maintenance and processing, and ensure economic benefits.

References

1. Tomescu, D., Heiman, A., Badescu, A.: An automatic remote monitoring system for large networks. In: 2019 IEEE International Conference on Computational Science and Engineering (CSE) and IEEE International Conference on Embedded and Ubiquitous Computing (EUC). IEEE (2019)
2. Manakov, A., Abramov, A., Ilinykh, A., et al.: Comprehensive system of automatic control of track machines operating parameters. Transport. Res. Proc. **54**, 404–410 (2021)
3. Ungar L.Y., Jacobson, N.G., Mak, T.M.: High-speed I/O capabilities added to military automatic test equipment (ATE) using synthetic instruments. In: 2019 IEEE AUTOTESTCON. IEEE (2019)
4. Lackner, C., Chow, J.H., Wilches-Bernal, F.: Performance evaluation of STATCOM equipment using ambient and disturbance data. In: IEEE Milan Power Tech. Rensselaer Polytechnic Institute Troy, NY, US; Rensselaer Polytechnic Institute Troy, NY, US; Sandia National Laboratories, Albuquerque, NM, US (2019)
5. Ab, A., Aa, A., Ab, A., et al.: Automatic control of pneumatic networks of railway train. Transport. Res. Proc. **54**, 274–282 (2021)

6. Hadi, H.H., Sallom, M.Y.: Pneumatic control system of automatic production line using SCADA implement PLC. In: 2019 4th Scientific International Conference Najaf (SICN) (2019)
7. Effert, S.: Automatic monitoring equipment and indication for implantation of electrical pacemakers. Thorax **11**(2), 158–166 (1963)
8. Stephens, K., Sommers, J.: Automatic equipment downtime monitoring in the cement industry. In: Cement Industry Technical Conference. IEEE (2004)
9. Takakura, S.: Development of high-power multi-channel automatic electrical survey equipment for resistivity and self-potential monitoring. Geophys. Explor. **56**, 229–237 (2003)
10. Orlet, J.: System level health and environmental monitoring for automatic test equipment. In: 2018 IEEE AUTOTESTCON. IEEE (2018)
11. Miyake, H., Ashida, E., Kokura, S., et al.: Study on automatic welding of pipe (report 3) – monitoring equipment by remote control welding. Q. J. Jpn. Weld. Soc. **4**(3), 615–621 (1986)

Research on 3D Technology in Animation Design of Digital Media Industry

Sisi Qiu(✉)

School of Comics, Jilin Animation Institute, Changchun 130000, China
jldhqiusisi@sina.com

Abstract. In order to understand the application of 3D technology in animation design of digital media industry and improve the quality of animation design, this paper will carry out relevant research. This paper firstly discusses the basic concepts and main requirements of animation design in digital media industry, then introduces the application advantages of 3D technology, and finally puts forward the application process and application mode of 3D technology in animation design. Through this study, 3D technology can give full play to the role of animation design, with the help of this technology, animation design quality will be significantly improved, and the design work will reduce the burden of artificial.

Keywords: Digital media · Animation design · 3 D technology

1 Introduction

Because modern quality of life has increased, people's spirit of "entertainment" got the booming development, makes the modern people have a lot of entertainment needs, including a large part of the people for the anime is very interested in, this is their entertainment needs, so the development of Chinese animation design field in the rush hour, especially in the era of digital media animation design field quickly completed the digital transformation, Have begun in digital media animation design. However, according to the current situation, although the formation of digital media industry helps the field of animation design to open up the market, it does not change the way of animation design. A large number of designers are still using the traditional two-dimensional perspective design method to work, which has many problems. Such as low efficiency, design difficulty is high, easy to appear problem, which can realize the design of the low quality level, so the animation design field is very difficult to produce high-quality work in colleges and universities, and design staff workload is very big, does not satisfy the quick update in the digital media industry, the development of user requirements, the digital media industry to provide users with a large amount of digital information, and the efficiency of information transmission is very fast, so that user needs will change rapidly, so animation design has yet to be developed.

J. H. Abawajy et al. (Eds.): ICATCI 2022, LNDECT 170, pp. 514–522, 2023.
https://doi.org/10.1007/978-3-031-29097-8_61

From this perspective, through continuous research, relevant fields have realized that the main cause of existing problems in animation design is two-dimensional design. Therefore, the key to solve these problems and promote the development of animation design is to get rid of two-dimensional perspective and adopt a more convenient and intuitive perspective for animation design. By the enlightenment, and some researchers from the angle of the physical environment, put forward the 3 d technology, namely the early physics research is pointed out that the real world has three dimensions, so also known as the three dimensional world, so from the angle of technology that three dimension together, to form the 3 d technology, the first to provide users with a three-dimensional view, Secondly, it can enable users to complete a series of operations from this perspective, and give full play to the advantages of 3D perspective to simplify the previous problems of related operations in 2d perspective. Animation design will also benefit from this, so in order to promote 3D technology in animation design, it is necessary to carry out relevant research.

The advantages of 3D technology lie in three aspects:

The first is the three-dimensional surrounding perspective, that is, when designing animation, designers need to observe the design results from all angles, so as to find potential problems, and then make adjustments, so that the design results can be continuously optimized and finally output high-quality works. However, based on the traditional two-dimensional technology, designers must decompose each plane of the design works if they want to observe the design results from different angles. Then observe the floor plan, and sometimes combine multiple floor plans for comparative analysis. This process will lead to the whole process of design adjustment of works becoming very cumbersome, and at the same time, it is easy to be interfered by human factors, leaving out some problems. However, in contrast to 3D technology, this technology can provide designers with a three-dimensional surrounding perspective. Under this perspective, designers do not need to split the work plane. Instead, they can conduct a surrounding observation of three-dimensional models without dead angles, which can find problems more intuitively, promote the improvement of design efficiency, and better ensure the quality of the final work.

The second is adjustability. 3D technology is usually carried in software programs, while all modern 3D software has good adjustability. This feature represents that designers can directly adjust the design results on the 3D model without changing the actual 2D works repeatedly. Therefore, designers can make many attempts and make a lot of adjustments by relying on 3D technology in animation design, and these operations will not affect the work itself. It is worth mentioning that some two-dimensional design software also has good adjustability, but due to the lack of algorithm support, it can not bring real feedback after each adjustment, which requires designers to conceive, so the adjustability of two-dimensional software is far lower than that of three-dimensional software.

The third is the stereoscopic nature of the works, that is, the animation design works born under the two-dimensional technology are all two-dimensional forms, that is, the traditional plane animation images, which are still used in China and even the world. Most people have long suffered from visual fatigue, but the three-dimensional technology has completely changed this situation. The animation works designed on the basis of this technology are all three-dimensional forms, which can bring people fresh and real experience visually. This will further promote the development of animation design industry.

It is worth mentioning that although 3D technology has high application advantages compared with 2D technology, this technology has also changed the traditional animation design method, and people need to re conceive the design method around this technology, which is also the significance of this study.

2 3D Technology Advantages

The advantages of 3D technology lie in three aspects:

The first is the three-dimensional surrounding perspective, that is, when designing animation, designers need to observe the design results from all angles, so as to find potential problems, and then make adjustments, so that the design results can be continuously optimized and finally output high-quality works. However, based on the traditional two-dimensional technology, designers must decompose each plane of the design works if they want to observe the design results from different angles, Then observe the floor plan, and sometimes combine multiple floor plans for comparative analysis. This process will lead to the whole process of design adjustment of works becoming very cumbersome, and at the same time, it is easy to be interfered by human factors, leaving out some problems. However, in contrast to 3D technology, this technology can provide designers with a three-dimensional surrounding perspective. Under this perspective, designers do not need to split the work plane. Instead, they can conduct a surrounding observation of three-dimensional models without dead angles, which can find problems more intuitively, promote the improvement of design efficiency, and better ensure the quality of the final work.

The second is adjustability. 3D technology is usually carried in software programs, while all modern 3D software has good adjustability. This feature represents that designers can directly adjust the design results on the 3D model without changing the actual 2D works repeatedly. Therefore, designers can make many attempts and make a lot of adjustments by relying on 3D technology in animation design, and these operations will not affect the work itself. It is worth mentioning that some two-dimensional design software also has good adjustability, but due to the lack of algorithm support, it can not bring real feedback after each adjustment, which requires designers to conceive, so the adjustability of two-dimensional software is far lower than that of three-dimensional software.

The third is the stereoscopic nature of the works, that is, the animation design works born under the two-dimensional technology are all two-dimensional forms, that is, the traditional plane animation images, which are still used in China and even the world. Most people have long suffered from visual fatigue, but the three-dimensional technology has completely changed this situation. The animation works designed on the basis of this technology are all three-dimensional forms, which can bring people fresh and real experience visually, This will further promote the development of animation design industry.

It is worth mentioning that although 3D technology has high application advantages compared with 2D technology, this technology has also changed the traditional animation design method, and people need to re conceive the design method around this technology, which is also the significance of this study.

Figure 1 shows the comparison between 3D and 2D images.

Fig. 1. Comparison between 3D and 2D images

3 Basic Concepts and Main Requirements of Animation Design in Digital Media Industry

3.1 Basic Concepts

Cartoon design, digital media industry is to use digital technology to design the cartoon image, create animation works a job, its very test designers not only design professional, digital technology, master degree, also for the designers of cartoon art cognition, thought, etc. have higher requirements, therefore not to be confused with other design work. At the same time, digital media in essence is similar to a network platform, compatible with digital information, the rapid development in recent years, by the vast number of people attention, people began to habits from digital media platforms for all kinds of information, including information such as the animation works, so digital media for animation design not only provides the means of digital technology design, also help animation design opens up the market, the designers can understand through digital media, user requirements, then design the related works, and will work through the digital media platform to release, works with the platform can be seen by a large number of users at the same time, shows that digital media makes the animation design of transmission rate, transmission area, such as improved, It improves the economic value of animation design works and promotes the development of animation design [1–3].

3.2 Main Requirements

Because animation design has been carried out from a two-dimensional perspective for a long time, many designers have felt the existence of problems in their work and put forward their own requirements, which are mainly shown in Table 1.

Table 1. Main requirements of animation design

The main requirements	Demand for purpose
Solve the problem of difficult docking plane perspective	Improve the directness of design perspective
Simplify the tedious design process	Reduce design workload
Enrich the form of design works	On the basis of plane work form, add three-dimensional work form

In combination with the three requirements, in theory, if can fully meet these requirements, animation design will significantly benefit, such as solving the problem of planar social difficult to docking, prompted the design from the point of view of the visual enhancements, can let design professionals better understand design flaws, then even if the adjustment, it is able to guarantee the quality for animation design work, this benefit can be found in other requirements as well, which will not be described here. But the demand has also brought a new problem, that is, how to put these stay in the theoretical stage of the demand into practice, facing this problem 3D technology can help, so the majority of designers to maintain attention to this technology [4–6].

4 Application Advantages of 3D Technology

3D technology does not refer to a particular technology, but a series of 3D perspective characteristics of the technology collectively, different 3D technology is suitable for different occasions, so it should be carefully chosen [7, 8]. From this perspective, there are generally two 3D technologies suitable for animation design, namely Maya and 3Dmax. Figures 2 and 3 are the main flow frames of these two technologies.

According to the Fig. 2 can see, Maya is a typical 3 d modeling software, its all operation started with modeling in the design of cartoon design personnel can choose according to the requirements of work definition pixel parameters, in order to ensure work resolution, can be in the interface of input all kinds of design elements, form model, and then watch for each model, after the adjustment, the corresponding animation works can be generated by inserting all the adjusted models together and the time sequence of frame number. The main advantage of Maya software is that it has clarity debugging

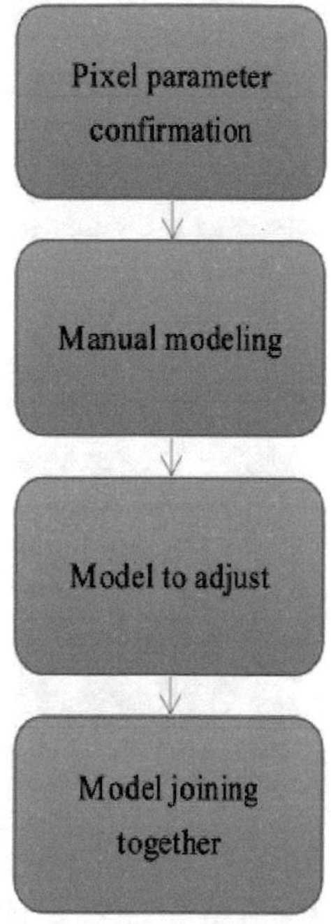

Fig. 2. Maya main flow framework

function, which is not available in other software and requires external technical means, so the process is relatively complicated. Maya also has its own disadvantages, that is, it cannot directly generate 3D models, which requires designers to manually model, which is quite complicated. Higher professional requirements for designers [9, 10].

According to Fig. 3, Dmax is also a typical 3D modeling software, but its process is different from Maya. That is, in 3Dmax, designers can directly import the drawn plane works into the software. According to the import order of the works, the software will automatically conduct modeling, and the initialized model may have problems. But only through simple debugging can get a perfect model, and then the model can be spliced together, combined with the frame sequence to generate animation. Compared to Maya, 3Dmax does not have pixel adjustment, but it has the advantage that it does not require manual modeling by designers, so designers can make choices based on their own needs.

It is worth noting that both Maya and 3 on3dmax, or other 3 d software, compared to traditional two-dimensional design perspective, any 3 d software has obvious advantage, at least with the aid of the software, animation designers put forward by the current three requirements can be fully satisfy, to animation design efficiency provide strong support, the quality of writing, etc. And reduce the work burden of animation designers, so the animation design field should actively introduce 3D technology software.

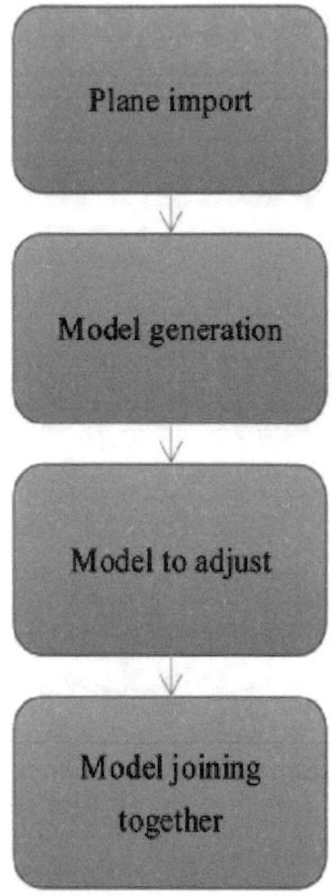

Fig. 3. 3Dmax main process framework

5 Application Process and Method of 3D Technology in Animation Design

5.1 Application Process

In animation design, the specific application process of 3D technology depends on the selected 3D technology software. Therefore, this paper takes 3Dmax software as an example to analyze the application process of technology. In general, animation design and application of on3dmax process is divided into four steps: first, the designers need to according to your own idea, the three-dimensional model is decomposed into a number of plane, character image design, for example, generally can be divided into before and after, left, right, up, down, six plane, and then draw the graphic on the graphics first; Secondly, the drawn plane graphics are imported into 3Dmax software system by digital scanning technology. In the process, attention must be paid to the import sequence of plane graphics, which is usually clockwise. The import sequence is front → right → back → left → up → down; Third, click on the "model generation" button production softwares 3 dmax software, the software will be carried out in accordance with the plane figure of the import order, automatic modeling form six model initialization, the six possible flaw on the initialization model, and does not depend on each other, so need to design personnel of each model first for debugging, and then carried out in accordance with the order together; Fourthly, arrange all the assembled models according to the action sequence of the works, insert the number of frames, establish the sequence of frames, and then export digital animation works after completion.

5.2 Application Modes

There are two main application ways of 3D technology in animation design, which are the combination of animation elements and the switch of model form, as follows.

Combination of Animation Elements. Animation design is the design of the characteristics of a need to use to a lot of anime elements, in order to produce good works, between each kind of anime elements are clever union, to show the user a unique cartoon art aesthetic feeling, but in the traditional two-dimensional perspective as it is difficult to do this which is early cartoon industry development in our country has always been the main cause of rare is silly. On the basis of 3D technology, designers can import all kinds of animation elements into the software system to get the corresponding animation element model, and then combine each animation element model according to their own ideas, or make new attempts to find the best way to combine, so that good animation works can be obtained. Animation element combination is of the most common theme elements combine with scene elements, with the risk to re-turn the book as an example, from the perspective of the designer to see the characters in the works with a variety of scenarios is two 3 d models, the software interface of two model combined with each other, and to manipulate the model, which improve the characters of the model resolution, The scene model is blurred, so that the primary and secondary relationship can be clearly presented in the perspective, but the role of the scene model will not completely disappear.

Switching Model Form. Although all kinds of 3 d technology to 3 d modeling, but does not mean cannot be compatible with the 2 d model, 3 d technology instead of the current mainstream of 3 d technology on the market are compatible with the 2 d model, the representative cartoon designer has two kinds of forms of different model material, the 3 d model and 2 d model is not the same, present the visual perception has great difference, However, animation works have a large space for creation. Sometimes, switching between two model forms to display pictures can bring visual enjoyment or impact to users and meet people's entertainment needs. In a cartoon works as an example, the work is the most of the scenes in the 3 d model, but in some special scenes of switch is unique in the design of the model, such as the designer wants to show the characters surprise of emotions, on the right frames by inserting a 2 d figure surprised expression, when work to the time frames node is, The facial expressions of the characters will suddenly switch from 3D to 2D, which is refreshing and makes the works have the advantages of both 3D and 2D models.

6 Conclusion

To sum up, the establishment of digital media industry has given animation design an opportunity for development. Therefore, animation design field should seize the opportunity and actively introduce 3D technology to seek development. 3D technology can improve the efficiency and quality of animation design, increase the output and quality of animation design on behalf of the strong market benefits of digital media, promote animation works to more users, gain more economic benefits, promote animation culture and so on.

Acknowledgements. The Humanities and Social Sciences Research Project of the Jilin Provincial Department of Education "Research on the Countermeasures of Cultivating Applied Talents in the Whole Industry Chain of Comics" (No: JJKH20211360SK).

References

1. Brehmer, M., Lee, B., Isenberg, P., et al.: A comparative evaluation of animation and small multiples for trend visualization on mobile phones. IEEE Trans. Visual Comput. Graph. **26**(1), 364–374 (2019)
2. Aprianto, H., Saputro, A.: Animation design as an educational media of adolescents' social behavior deviation. J. Phys. Conf. Ser. **1500**, 012109 (2020)
3. Fitriasari, N.S., Suzanti, L., Widjayatri, R., et al.: Interactive animation media of sea biota design for young learners. J. Phys. Conf. Ser. **1811**(1), 012098 (2021)
4. Natalia, D., Fathoni, A.: "Tigerheart" short animation visual communication design. IOP Conf. Ser. Earth Environ. Sci. **729**(1), 012051 (2021)
5. Simarmata, J., Sibarani, C., Silalahi, T.: Development of hybrid learning-based animation media to improve the learning outcomes of multimedia learning. J. Phys. Conf. Ser. **1477**(4), 042067 (2020)
6. Nathania, T.T.: Educational animation: environmentally friendly palm oil based products. IOP Conf. Ser. Earth Environ. Sci. **729**(1), 012108 (2021)
7. Aysolmaz, B., Reijers, H.A.: Animation as a dynamic visualization technique for improving process model comprehension. Inf. Manage. **58**(5), 103478 (2021)
8. Widjanarko, K.C., Kartika, R., Sasongko, H.: Visual communication design of "Voorspel" short film animation. IOP Conf. Ser. Earth Environ. Sci. **729**(1), 012053 (2021)
9. Priyoatmoko, W., Arifah, F.N.: Using 3D animation combined with augmented reality for promotion media: case study of STMIK Bina Patria. J. Phys. Conf. Ser. **1517**(1), 012069 (2020)
10. Webster, J.G., Ksiazek, T.B.: The dynamics of audience fragmentation: public attention in an age of digital media. J. Commun. **1**, 39–56 (2012)

Application of Hospital Computer Wireless Network Bedside Nursing Management System

Juan Lin[1(✉)], Kun Wu[2], and Fenfang Lei[1]

[1] Nursing School of Shaoyang University, Shaoyang 422000, China
270953561@qq.com
[2] The First Affiliated Hospital of Shaoyang University, Shaoyang 422000, China

Abstract. In order to design hospital computer wireless network bedside nursing management system, this paper will carry out related research. This paper first introduces the significance of the system construction, then analyzes the requirements of the system construction, and finally puts forward the system construction scheme according to the requirements. Through this study, the successful design of the hospital computer wireless network bedside nursing management system, the system can better provide call services for inpatients, convenient communication between doctors and patients.

Keywords: The hospital · Computer wireless network · Bedside nursing management system

1 Introduction

Hospital will be responsible for to provide nursing care for hospitalized patients in the hospital management services, and from the perspective of service quality, the need to guarantee service efficiency, service, accurate, but because of the large number of hospitalized patients, so many hospitals are not to do this in manual mode, this leads to the hospital to use technical means to achieve the purpose, which has been the head of a bed nursing management system. Early nursing management system of the head of a bed belongs to cable voice system, the system will be installed in the position of the head of a bed of hospitalized patients with users connected to the nurse station, in line with support patients can notify the nurse came, and then put forward its own demand, the resulting service, and gradually in the long-term use of this system does not meet the needs of patients with large, The modern hospitalized patients in the hospital has a lot of demand, sometimes to express oneself also not clear, lead to the hospital nursing management service does not reach the designated position, at the same time, a lot of hospital human resource is not abundant, once meet day time will appear the phenomenon of no service, slow service, continue to fall so the application value of the system, to realize to the hospital on the system is updated, The emergence and popularization of wireless network technology gives hospitals an opportunity to update

J. H. Abawajy et al. (Eds.): ICATCI 2022, LNDECT 170, pp. 523–531, 2023.
https://doi.org/10.1007/978-3-031-29097-8_62

their systems. Hospitals can use this technology to establish a wireless network bedside care management system. This system can solve previous problems, improve service quality and efficiency, and the system itself is more stable than the traditional system. From this point of view, it is necessary to carry out relevant research on how to construct and apply the traditional nursing system of wireless network.

2 Key Points of Wireless Network Nursing Application

The application of wireless network has attracted the attention of relevant fields for a long time. Therefore, researchers in the field have conducted research and proposed X key points for the application of wireless network in hospital nursing. One is the stability of wireless network, that is, wireless network is different from wired network. Communication nodes are connected and communicated only through signals, while it is known that signals are vulnerable to interference and loss of signals in the transmission process, This phenomenon is easy to cause serious impact in hospital nursing. For example, when a patient suddenly gets sick, the nursing staff will call the nursing staff. If there is signal interference at this time, the nursing staff cannot receive the notice, which may miss the best opportunity for treatment, and there is a certain probability that medical accidents will occur. Therefore, the network must be stable enough in wireless network applications, otherwise it cannot be put into practice; The second is the security of wireless network, that is, while providing people with communication convenience, wireless network also brings some security risks. For example, some criminals may use illegal means to monitor wireless network signal transmission channels to obtain information. This phenomenon is likely to lead to the disclosure of patients' private information or part of the confidential information of the hospital in the hospital nursing work, which has an impact on both sides. Therefore, how to ensure the security of wireless network is also the key point in wireless network nursing application; The third is the smoothness of the wireless network, that is, although the wireless network has a higher communication capacity, the use of capacity is limited. In some special cases, communication congestion may still occur. This phenomenon has a very high incidence in hospital nursing work, because hospital nursing work involves a large number of patients. If a large number of patients happen at the same time, they will use the wireless network to call nurses at the same time, If the communication capacity is insufficient, it will cause communication congestion. Faced with this situation, blindly increasing the communication capacity will bring large costs to the hospital. This method is not desirable, so the capacity of the wireless network needs to be at a reasonable level, which is very worthy of research. If we can meet the requirements of the above points, wireless network can provide strong support for hospital nursing, and its value can also be brought into full play. Figure 1 and Fig. 2 are two basic forms of wireless network nursing applications.

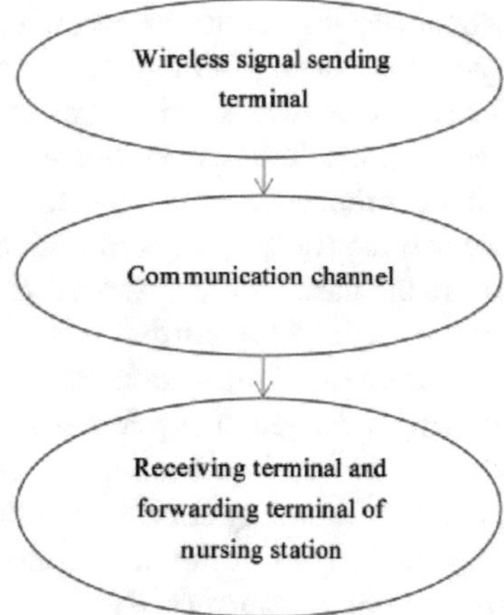

Fig. 1. Wireless network nursing application form 1

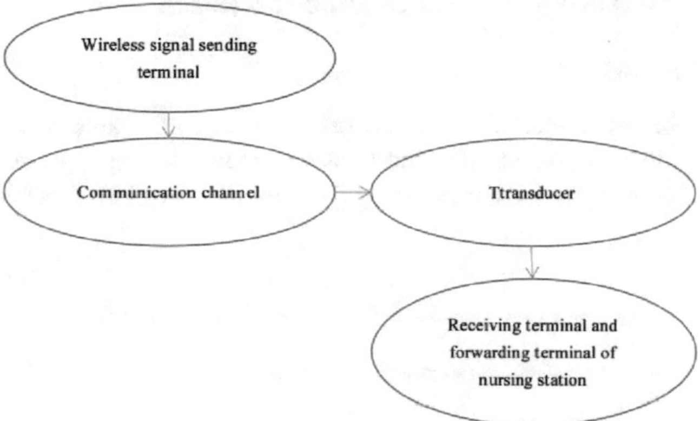

Fig. 2. Wireless network nursing application form 2

3 The Construction Significance of Hospital Computer Wireless Network Bedside Nursing Management System

Compared with the traditional wired voice system, the construction significance of hospital computer wireless network bedside nursing management system is mainly reflected in three aspects: First, the wireless system depend on signal is connected, different from wired connection mode, the difference make hospital bedside care computer wireless network management system can avoid many problems in the cable connection, lines in a wired connection is exposed to the environment, for example [1, 2], is likely to be affected by environmental factors and failure, leading to poor connection, In this case, it will not only lead to inpatients can not get services, but also increase the maintenance burden and cost of hospital line. As a wireless system, hospital computer wireless network bedside care management system does not have these problems, so the system is of great significance for hospital service quality and efficiency. Second, hospital bedside care computer wireless network management system is not phonetic system, on the other hand is a kind of self-help diversified service system, namely the difference in cable system voice input device [3], wireless in hospitalized patients bedside equipment installed in the system is a small computer, under the setting of artificial computer can replace manual provide some service for hospitalized patients, And the computer can carry a variety of services, so on the basis of the system can avoid the inpatient can not express their own needs, but also reduce the burden of hospital nursing management services, indicating that the construction of the system is imperative; Third, the hospital computer wireless network nursing management system of the head of a bed has a strong can expand sex, nursing management services such as hospitals want to launch a new project, then we can design the corresponding service function on the computer, this is the cable system cannot do, that the hospital computer wireless network bedside nursing management system has important significance [4–7].

4 Construction Requirements of Hospital Computer Wireless Network Bedside Care Management System

4.1 Demand Model

Since the hospital computer wireless network bedside care management system is a service system, its construction demand model is shown in Fig. 3 by referring to the demand model of general service system and combining the characteristics of hospital industry.

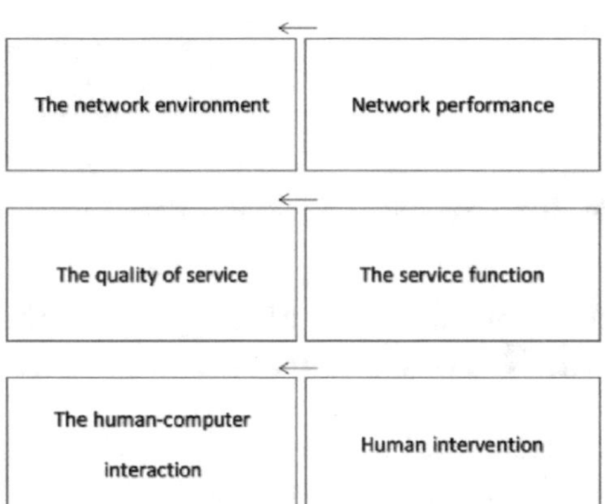

Fig. 3. Demand model of hospital computer wireless network bedside care management system

Based on Fig. 3, it can be seen that the construction requirements of hospital computer wireless network bedside care management system can be divided into three dimensions, namely network environment, service quality and human-computer interaction. The main requirements of each dimension are network performance, service function and human intervention. Details of each requirement are shown below.

4.2 Requirement Analysis

In view of the requirement contents in each dimension of the system construction demand model, the following paragraphs will analyze them and point out the direction of system construction.

(1) Network Performance. Computer wireless network for hospital nursing management system for the laying of the head of a bed and a wireless network environment, and so on wireless network environment is dependent on, it also suggests that if performance problems existing in the network environment, it is bound to affect the hospital bedside care computer wireless network management system of operation, indirectly led to the decrease of the nursing management service quality and efficiency, Therefore, the system construction must focus on the network environment dimension, as far as possible to improve the performance of the network environment. Network environmental performance can be evaluated from three perspectives: The first is the network communication capacity, namely the system in the actual application of may face a large number of hospitalized patients at the same time the phenomenon of communication requests, at this time if the network communication capacity is insufficient, will cause the traffic phenomenon, affect the communication efficiency and service efficiency, so the system of wireless network environment must have enough capacity of communication, and communication capacity should be ready to expand; Is the second signal stability, namely the system of hospitalized patients bedside client and workstation is depend on the signal to connect terminals, the signal is transmitted back and forth in the communication channels, this kind of circumstance is likely to happen communication signals unstable phenomenon, such as distance because both ends, may lead to signal strength after passing a distance attenuation phenomenon, The communication is not smooth, or because the communication channel is interfered, so the communication signal becomes unstable, which is also not conducive to the quality of communication. Therefore, in the system construction at least from these two levels to ensure communication signal stability; And the third is the network communications security, namely the hospital bedside care computer wireless network management system of the wireless network environment belongs to the private network, inside there are a lot of privacy and confidential information, cannot be easily leaked out, and some illegal personnel will use technology to attack private network, or supply private network communication channels, once successful attack will lead to the information leakage, Leakage can have an impact on the hospital itself and its patients. Therefore, network and communication security should be emphasized in system construction [8–10].
(2) Service Functions. Hospital bedside care computer wireless network management system's main function is to provide service to hospitalized patients, and as a service system, the service functions of natural to be consistent with the diverse needs of hospitalized patients, so the system construction should attach great importance to the development of the service function, and in theory should ensure that the service function is all ready, considering the patients demand in all its aspects, In order to better deliver the service, to do that the system uses real value. In addition,

the hospital computer wireless network service function of nursing management system of the head of a bed is divided into three categories, respectively is a self-help, automatic, manual, including artificial class service functions related to human-computer interaction, thus developing should pay attention to distinguish, not only from the perspective of the patients demand for development, but the hospital nurse staff, such as work demand, So that they can better human intervention in the service.

(3) Manual intervention. Manual intervention is a special kind of service, and its content generally includes some work that must be performed by medical personnel, such as dressing change for inpatients, which cannot be performed by computer or other technical means. Therefore, manual service function should be set up in the service function requirements.

5 Construction Scheme and Implementation Method of Bedside Nursing Management System

5.1 Construction Plan

Figure 4 shows the deployment scheme of bedside care management system of hospital computer wireless network.

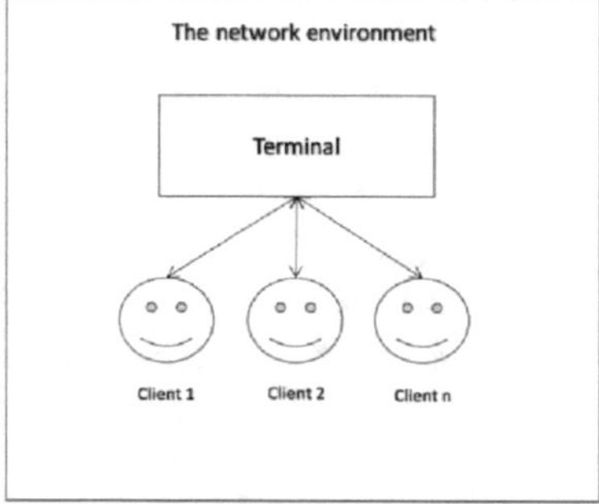

Fig. 4. Deployment scheme of hospital computer wireless network bedside care management system

Figure 2 shows that this system is according to the distribution of centralized deployment, the whole system is divided into two parts in the network environment, terminal and the client, the main terminal installed in nurse station and so on have computer, hospital nursing management departments and client is installed on the computer equipment of the head of a bed of hospitalized patients with, for use in patients. At the same time, each client and terminal maintain a separate two-way connection relationship, so as to meet the two-way communication needs of patients → medical staff, medical staff → users. In addition, the terminal mainly bears the working end of the manual service function, and the client mainly bears the application end of the manual service function and the application end of the other two kinds of service function. According to the scheme, the system implementation method will be analyzed below.

5.2 System Implementation Method

Combined with the construction plan and according to the system construction requirements, the implementation method of hospital computer wireless network bedside nursing management system is divided into three parts, as follows.

Network Environment Construction. In order to meet the requirements of network performance, network protocols should be carefully selected in network environment construction. Sufficient communication capacity, stable communication signals and high security should be guaranteed for network protocols, and then network environment construction should be completed through networking. On this basis, current mainstream network protocols need to be screened. The main reference objects of this paper are shown in Table 1.

Table 1. Network protocols referred to in this paper

The name	The characteristics
The FTP protocol	It is a part of the TCP/IP protocol group and consists of an FTP server and an FTP client. The FTP server is used to store files, and the FTP client allows users to access the server through the client for searching stored files
The HTTPS protocol	Derived from the HTTP protocol, it is equivalent to a secure communication channel, which can support the exchange of information between client computers and servers. The secure Socket layer in this protocol has information exchange capabilities and can encrypt information exchange traffic, which can be regarded as a secure version of HTTP
UDP protocol	It is a user datagram protocol, suitable for transmission of data with low reliability requirements and small data volume

Compared with the above three protocols, UDP protocol has a small communication capacity, so it may be able to meet the daily doctor-patient communication. However, if an emergency occurs, the communication will be blocked, so this protocol is not considered. The HTTPS protocol has all the advantages of the traditional HTTP protocol. However, the traditional protocol itself is often used in the Internet environment, so it has a large communication capacity, which is also reflected in the HTTPS protocol. Meanwhile, THE HTTPS protocol supports the transmission of various forms of information and provides security guarantee. FTP is comparable with HTTPS in terms of communication capacity, but the former has higher communication stability. This is due to the simple structure of the server and client in the FTP protocol, which enables direct communication without interference and guarantees signal strength. However, FTP is not as secure as HTTPS. Therefore, since information security is the most important issue, HTTPS protocol is selected in this paper, and router bridge mode is adopted for construction, which is shown in Fig. 5.

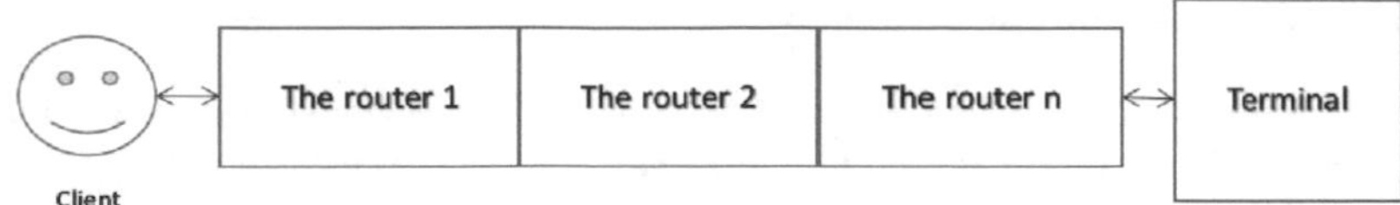

Fig. 5. Shows the router bridge modeanagement system

Service Function Development. Focusing on self-help, automatic and manual functions, the hospital should combine the actual situation to develop: first, autonomous functions generally include nursing service call, nursing knowledge query, etc.; Second, automatic functions include periodic reminder, drug grade, etc. The third artificial function includes notification of dressing change, grade of illness, etc. These functions can be developed in the Java language.

Manual Intervention Design. First design information data acquisition terminal, is used as much as possible of hospitalized patients in the hospital to collect the information, secondly from the information passed to the nurse station and other departments of computer terminal equipment, so that staff can very good to understand the clinical situation, again according to the conditions set manual intervention strategies, for example, according to the hospitalized patients garter st length, the function of plate automatically remind nurses to change medicine, Information is also given on changing drugs so that nurses can perform dressing changes directly without having to go to the field to check.

6 Conclusion

To sum up, the hospital computer wireless network bedside nursing management system has important construction significance, which is of great help to the improvement of hospital nursing management service quality and efficiency, and also indirectly contributes to the rehabilitation of inpatients. Therefore, the hospital should actively build this system. The establishment of the system will replace the traditional wired voice system, can avoid all kinds of problems of the traditional system, ensure the hospital service in place.

References

1. Raja, B., Sandesh, G., Milind, K.: Coexistence System and Method for Wireless Network Devices:US, US8094631 B2. (2019)
2. Saputra, H., Suhandi, A., Setiawan, A., et al.: Real-time data acquisition system with wireless sensor network on damped harmonic motion. Phys. Educ. **56**(5), 053004(5pp) (2021)
3. Naziha, A., Li, F., Elamine, G.M., et al.: a method to construct an indoor air pollution monitoring system based on a wireless sensor network. Sensors **19**(4), 967 (2019)
4. Islam, M.S., Islam, M.T., Almutairi, A.F., et al.: Monitoring of the human body signal through the Internet of Things (IoT) based LoRa wireless network system. Appl. Sci. **9**(9), 1884 (2019)
5. Rusdan, M.: Design of wireless network system for digital village using wireless distribution system. J. Inform. Inform. Syst. Softw. Eng. Appl. **1**(2), 51–59 (2019)

6. Verma, V., Badholia, A., Kashyap, S.K.: Wireless network system based on discrete probability: (Wireless Network System). In: 2019 International Conference on Computing, Communication, and Intelligent Systems(ICCCIS) (2019)
7. Santoso, B., Sari, M.W.: Developing parking queue monitoring system using Wireless Sensor Network and RFID technology. J. Phys. Conf. Ser. **1823**(1), 012056(10pp) (2021)
8. Short, J., Bagrodia, R., Kleinrock, L.: Mobile wireless network system simulation. Wireless Netw. **1**(4), 451–467 (1995)
9. Kawakubo, S., Chansavang, A., Tanaka, S., et al.: Wireless network system for indoor human positioning. In: International Symposium on Wireless Pervasive Computing. IEEE (2006)
10. Subramanian, C., Lapilli, G., Kreit, F., et al.: Experimental and computational performance analysis of a multi-sensor wireless network system for hurricane monitoring. Sens. Transd. **10**(Special Issue), 206–244 (2011)

Cyber Intelligence for AI, VR, Blockchain Applications and Innovations

Analysis on the Development of Visual Communication Design Under the Innovation of Information Technology

Bingnan Pang[1,2], Tiantian Chen[1(✉)], and Minhong Dai[1]

[1] School of Animation, Hainan College of Software Technology, Qionghai, Hainan, China
chentiantian@hncst.edu.cn
[2] Faculty of Art and Design, Universiti Teknologi MARA, Kota Bharu, Kelantan, Malaysia

Abstract. The realization of modern information technology innovation in globalization has had a profound impact on human production, life and even the way of thinking. The huge amount of information and the speed of information processing and dissemination are growing exponentially, and its development has posed new challenges to human learning, mastering and applying knowledge, and has also brought new opportunities for the development of visual communication design.

Keywords: Information Technology · Visual Communication Design · Graphic Design · Art Design

1 Introduction

The rapid development of information technology has brought great changes to today's world. With the popularization and application of computers and the Internet, information technology has played a vital role in social progress. In the realization of the globalization of modern information technology innovation, to the human production, life and even the way of thinking has had a more far-reaching influence, large amount of information and the speed of information processing and transmission in the growth of the double speed, human learning speed has been accelerating, modern information technology has been completely osmosis in human life, Its development brings new challenges to human learning, mastering and applying knowledge, and also brings new opportunities for the development of visual communication design.

2 Overview of Visual Communication Design

Visual communication design is the use of human emotional awareness in order to convey something specific, using visual expression to present a visual message that resonates with the reader emotionally. Visual communication design consists of two main points: visual symbols and communication methods [1]. The so-called "visual symbols", as the name suggests, are the symbols that the eye can see to represent the properties of things,

J. H. Abawajy et al. (Eds.): ICATCI 2022, LNDECT 170, pp. 535–543, 2023.
https://doi.org/10.1007/978-3-031-29097-8_63

such as television, film, photography, graphics, text, etc.; the so-called "communication" refers to the process of information transfer, as show in Fig. 1, which includes "Who", "Says What", "In Which Channel", "To Whom ", "With What Effect" five procedures [2].

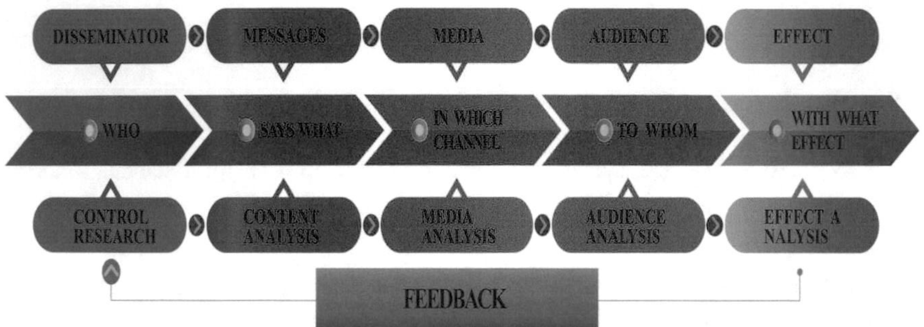

Fig. 1. The Structure and Function of Communication, 1948, by Harold Lasswell

The main content of the visual communication design include the font design, logo design, poster design, pictorial design, packaging design, display design of the two-dimensional space such as image rendering [3], but prefer to visual communication design, interactive design, pay attention to the interaction experience and feelings of the audience, focus on the functional, more have the graphic design of visual aesthetics, It has been widely used in various fields of modern life and has played a good role in information transmission and communication.

3 Transformation of Visual Communication Design under the Innovation of Information Technology

Throughout the development of visual communication design, it originated from the expansion and extension of "Graphic Design" in the 19th century. In the ever-changing information age, the influence of information communication media is becoming more and more important, and European and American countries have combined visual communication design in the culture and art industry, bringing a subversive visual experience for human beings and taking the lead in the world. Visual communication design in mainland China started late, developed slowly, and lacked innovative thinking, but the development of information technology has provided the necessary conditions for changes in the culture and art market and has brought unlimited business opportunities to the economic market. As the information age continues to advance, traditional visual communication design is no longer able to meet the needs of the current cultural market, and therefore needs to learn from the world's forefront of advanced ideas and technologies.

3.1 Shift in Thinking

For a long time, visual communication design as a tertiary sector of the economy has played a vital role in the growth of the gross national product. On this basis, the rapid arrival of the information age, represented by the knowledge economy and 5G network technology, has brought new changes to the mode of thinking and composition of visual communication design, as well as promoting the new development of visual communication design in the cultural industry, and influencing the traditional mode of thinking in a subtle way, as show in Fig. 2. As a cultural form that responds to the changes of the times, the fashion and foresight contained in visual communication design requires designers to have a sense of keeping up with the times and a keen sense of smell. Therefore, in order to respond to the new concepts and high aesthetics of the new era, the concept, methods and approaches of visual communication design must be reconstructed.

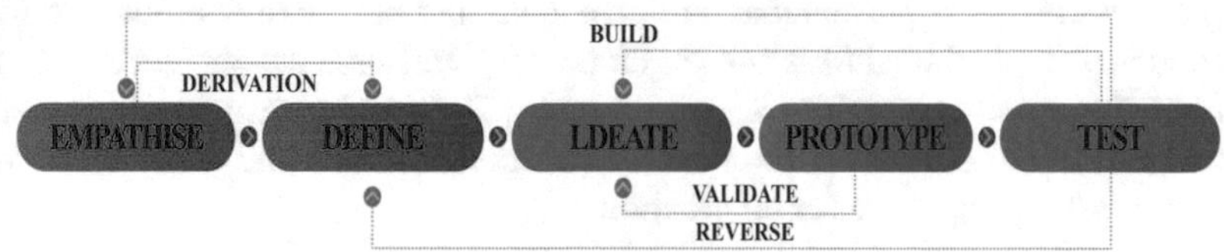

Fig. 2. Design Thinking: Understand - Improve–Apply, 2019, by D.School

With the development of information technology, the integration of interactive thinking in visual communication design makes the limitation of single thinking stronger and stronger, and modern visual communication design has undergone a great change in concept. Through modern information network, computer design works can be spread and displayed on multiple platforms. In the process of communication and interaction, more extensive improvement suggestions can be obtained, so as to achieve better information dissemination purposes. With the intervention of interactive thinking, designers can better perceive the modern design trend, construct a more comprehensive way of thinking, dig out a more unique design perspective, and enhance the sense of participation of readers, as show in Fig. 3. The idea of interactive design in the information age enhances readers' sense of participation, satisfies readers' psychology better, and speeds up the growth of designers. The biggest advantage of this interactive thinking is that it takes into account the different needs of designers and readers [4]. It not only shows the personal thoughts of designers, but also provides a platform for group expression, which further enhances the diversity of visual communication design works and styles.

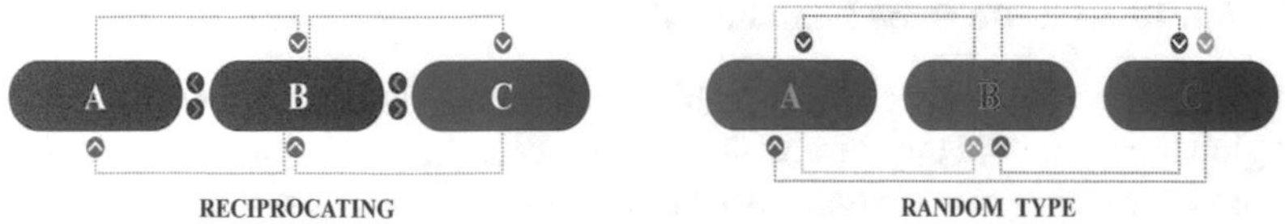

Fig. 3. Interaction diagram

3.2 Technological Shift

Traditional visual communication design relies on printing technology support and exists, designers need to follow the design concept, graphics, text, color and other elements of the combination and processing, and along with the innovation of information technology, a variety of computer design software practicality continues to improve, and can assist designers to greatly enhance the design efficiency, reduce the intensity of work, improve the design effect, more convenient operation. In addition, powerful computer design software is better at completing a variety of complex graphics and exaggerated effects, which can help designers to create more freely, as show in Fig. 4. In particular, the influx of 3D technology makes the flat space present a multi-dimensional dynamic sense, which provides unlimited possibilities for human to free their minds and develop their imagination. However, the excessive application of information technology will make visual communication design tend to be commercialized and process-oriented, making designers ignore personalized creative ideas and pursue high efficiency and fast production in a single thoughtless and patterned way, and such design works are lifeless and will be discarded by the times sooner or later [5]. Visual communication designers should adhere to the bottom of the industry's responsibilities, and really let computer information technology serve for art creation.

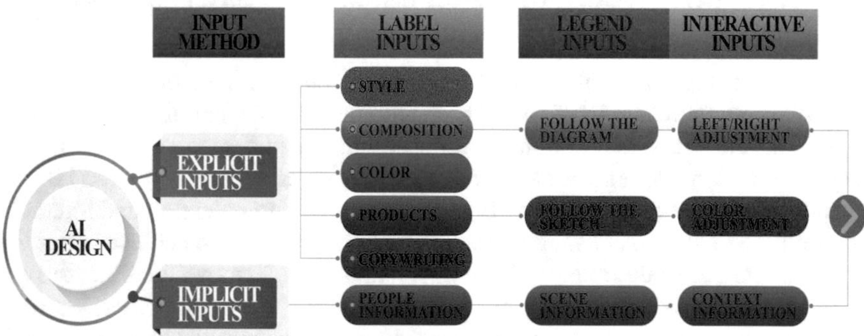

Fig. 4. AI Design-LUBAN Visual Generation Schematic

4 The Characteristics of Visual Communication Design in Information Technology Innovation

4.1 Design Ideas Tend to be Humanized

The continuous breakthrough of information technology has further promoted the process of globalization, and the communication between human beings has become faster and more convenient. 21st century visual communication design works can be quickly delivered to any corner of the world through the Internet, which also leads to the increasingly fierce competition in the culture and art market, mediocre design works can no longer meet the public's thought and aesthetic needs, and the concept of user first has been fully The concept of user first has been fully confirmed by the market. In the information age, visual communication design must achieve respect for people, while requiring designers to have a deep understanding of the design itself, respect the user's own characteristics and needs, define design solutions according to actual needs, and organize visual elements more rationally in design works, so as to show the humanistic features of modern visual communication design, as show in Fig. 5, reflect humanistic care, give more vitality to design contents, and better obtain the recognition of users and the market [6].

Fig. 5. Visual Identity System Design & Visual Design for ICH Events, 2021, by Bingnan Pang

4.2 Design Style Tends to be Personalized

As readers have easier access to information in an information-based society, it has greatly broadened users' horizons, thus placing higher demands on cultural and artistic products. Visual communication design not only needs to accurately reflect the characteristics and attributes of the product, but also needs to show the cultural and spiritual connotation in the design work, highlight the design object's own advantages, emphasize the differences between the design object and other objects, provide a distinctive and personalized design solution, and fully meet the spiritual and psychological needs of users, as show in Fig. 5. This is also a test for the designer himself, not only to understand the content of the ideas required by the user, but also to understand the strengths and weaknesses of the design object, to build on its strengths and avoid its weaknesses, to better reflect the functional advantages, to show a personalized design solution, to make the style and connotation of visual communication design richer, and to fully meet the higher requirements of the information society for visual communication design.

4.3 Design Forms Tend to Diversify

With the innovation of information technology, it has opened up more application functions for visual communication design, and the design state it presents is no longer simply using graphics and words in two-dimensional space. Information technology applications provide more efficient and convenient design tools for visual communication design, creating the necessary conditions for the integration of culture and ideas, and constantly influencing the transformation of visual communication design concepts and expanding more forms of expression, as show in Fig. 6. In the face of different national cultures around the world, visual communication design needs to combine modern information technology, have more inclusiveness, and use a variety of expression techniques and technical means combined in order to fully demonstrate different cultural customs and vividly and effectively express various different ideological connotations in the design [7]. At the same time, the design resources of visual communication design have also been greatly expanded by the application of information technology, creating more favorable conditions for the diversified design forms of visual communication design. In addition, modern designers have applied the advanced design concepts and methods to their practical work through continuous learning and reference, which makes the design methods of visual communication design present a diversified development trend.

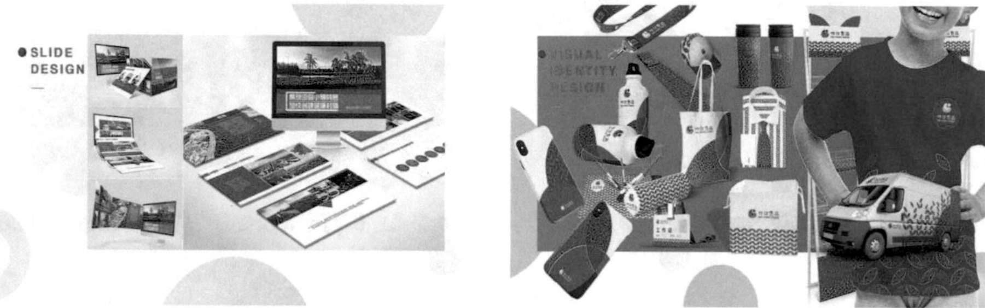

Fig. 6. Qionghai City Government Slide Design & Visual Identity System Design, 2020, by Bingnan Pang

5 The Development of Visual Communication Design in the Context of Information Technology Innovation

5.1 New Media Applications

With the implementation of 5G network, the modern information society has entered the new media era, digital media has just become one with people's lives, and the information conveyed through digital media is getting richer and richer, and the form of conveying these information is inseparable from the participation of visual communication design, so visual communication design has become an important tool to convert the form of information expression and get more audience. The unique means of visual communication design strengthens the effect of information transmission, makes digital media deeply loved by users, and can invariably enhance the competitiveness of brands. In the relevant information released by digital media, user interface, UI design, interaction design and other contents are important components, and most of these contents need to be completed with the help of visual communication design, which requires visual communication designers to master modern advanced design means, so that readers can give feedback while receiving information, thus increasing the interactivity of new media communication, giving full play to the visual communication design in digital media, so that the combination of visual communication design and new media is closer and the scope of application of visual communication design can be better expanded [8].

5.2 Design Ideas Tend to Be Humanized

The expression of art works can fully demonstrate the degree of integration with modern technology. By strengthening the cooperation with modern information technology, visual communication design has reflected the difference from traditional graphic design, gradually getting rid of the limitation of two-dimensional space, no longer limited by traditional technical means, presenting diversified techniques, and beginning to explore the change of multi-dimensionality. The visual construction method using computers as well as software as tools makes visual communication design no longer just a flat and single expression, but also multidimensional and dynamic in time and space [9]. In addition to the rich materials provided by the convenient network resources, designers can also make up for the shortcomings of traditional printing technology by using various information technology design methods to continuously enhance the expressiveness of their works, present the design theme more comprehensively and precisely, and present a rich design concept and a more The designers can also make use of various information-based design methods to make up for the shortcomings of traditional printing technology, to enhance the expression of their works, to present a more comprehensive and accurate design theme, and to present a rich design concept and a more diverse sensory experience, as show in Fig. 6.

From the designer's perspective, multidimensional design refers more to the multidimensionality of the designer's thinking. The application of multi-dimensionality in visual communication design breaks people's traditional way of thinking and better caters to modern people's living habits, while bringing a strong sense of surprise to the reader. The designer's mind is liberated and can freely use his imagination in any extended space, allowing the reader to interpret the meaning conveyed by the design in multiple directions, thus leaving a deep impression on the reader [10]. The concept and method of multi-dimensional design is very much in line with the trend and tendency of people pursuing something new in the information age.

5.3 Multi-sensory Use

The development of computer technology makes the mechanical rigidity blurred, and the computer language gradually realizes the symbiosis between human and object in a more vital form, and the visual communication design no longer focuses on the exploration of the relationship between real human and object, but on the relationship between human and virtual world, and the works become more and more intelligent [11].

In the information age, visual communication design has changed the way of introducing the content by visuals only, breaking the state of over-emphasizing the presentation of images and presenting it to the readers in a more innovative design form, which is no longer a single visual experience but tries to mobilize the multiple senses of the readers, making the work more appreciable. Visual communication design in the information age is more interactive design, and the focus of research is also cognitive psychology, which really achieves the purpose of convenience and practicality.

The information age will develop in the direction of dematerialization, and through the image symbol system, design will become a common language shared by human beings. Due to the integration of many new technological elements, the colors of visual communication design will become richer, accompanied by dynamic elements, which can stimulate the audience's brain at the same time, stimulate other senses to make common associations, and awaken the relevant senses of human beings themselves, so that the communication effect of the work can be effectively improved [12]. People's perception of the work is no longer limited to the visual, but all senses can be experienced, and the joint expression of visual and psychological space makes the reader strongly feel the difference between virtual and real, and enhances the sensory experience comprehensively, and is no longer limited by any linguistic symbols.

6 Prospects of Visual Communication Design Under Information Technology Innovation

How visual communication design can meet the growing high-level needs of human beings in the information age will be the focus of attention in the future, and therefore, visual communication design must form a new standard. The innovation of Internet information technology will definitely bring disruptive changes to the development of visual communication design, and will also allow designers to make groundbreaking innovations in their thinking and concepts. Visual communication design under the Internet and information technology will certainly make breakthrough progress, which is also a manifestation of the progress of the times. Visual communication design will face new opportunities and challenges under the influence of information technology and the Internet, and design expressions will be more flexible and diversified. Visual communication design needs to take advantage of the development of information technology and the Internet to better serve society; designers should also make full use of various advanced modern information technology tools to design better design works and push visual communication design to a higher level.

7 Conclusion

In the context of rapid development of information technology, visual communication design is undergoing qualitative changes, whether in terms of the diversity of software, the medium of transmission, or the concept of thinking, design style, characteristics of the times and the source channels have been richly changed. The innovation of information technology will bring continuous vitality to the development of visual communication design, and only with better awareness of this can we create richer design works.

References

1. Russmann, U., Svensson, J.: Introduction to visual communication in the age of social media: Conceptual, theoretical and methodological challenges. Media Commun. **5**(4), 1–5 (2017)
2. Wenxiu, P.: Analysis of new media communication based on Lasswell's "5W" model. J. Educ. Soc. Res. **5**(3), 245 (2015)
3. Evelyn, R., Aiello, G., Parry, K.: Visual communication: understanding images in media culture. MEDIENwissenschaft: Rezensionen **38**(2–3), 174–175 (2020)
4. Agrawala, M., Li, W., Berthouzoz, F.: Design principles for visual communication. Commun. ACM **54**(4), 60–69 (2011)
5. Chen, T., Pang, B., Dai, M.: Exploring the application of photography in graphic design. J. Phys. Conf. Ser. **1881**(2) (2021). (IOP Publishing)
6. Adami, E., Jewitt, C.: Social media and visual communication. Vis. Commun. **15**(3), 263–270 (2016)
7. Abdillah, R., Azhar, R.S.: Visual communication design and graphic design. Borneo Post Online, p. 1 (2020)
8. David, M. B.: Visual communication and heuristics: challenges and directions from across the disciplines (2019)
9. Maria, D.A., Pettersson, R.: Visual literacy theory: moving forward. Handbook of Visual Communication. Routledge, pp. 433–464 (2020)
10. Kheir, A-K.: Public Participation and Methods of Visual Communicationm. Tall Buildings and the City, pp. 83–119. Springer, Singapore (2020)
11. Song, B.: Visual communication design based on artificial intelligence under the digital media art environment. Solid State Technol. **64**(2), 4882–4890 (2021)
12. Minhong, D., et al.: Application of 3D digital modeling technology in the display of marine fish science in the South China Sea. J. Phys. Conf. Ser. **1881**(3) (2021). (IOP Publishing)

Application of 3D Printing Technology in Product Model Making

Zhongle Wang[1(✉)] and Junmin Kim[2]

[1] Xi'an Academy of Fine Arts, Xi'an, Shaanxi, China
wangzhongle2@126.com
[2] Tech University of Korea, Cheonan-si, Republic of Korea

Abstract. With the continuous updating of digital technology, young industrial designers increasingly rely on digital equipment for product design and model making. The concept of 3D printing technology is defined, the advantages and disadvantages of 3D printing technology and the application of 3D printing technology in industrial design are analyzed, the generation mode of 3D printing digital model is discussed, and the development status of 3D printing technology is studied. The results show that the application and popularization of 3D printing technology have shown an upward trend in the past few years, which has a significant impact on traditional industries.

Keywords: 3D Printing Technology · Product Model · Industrial Design · 3D Printing Digital Model

1 Introduction

Digital manufacturing technology, also known as 3D printing or additional manufacturing, creates physical objects from geometric representations by continuously adding materials. 3D printing technology is a new technology. Today, 3D printing has been widely used all over the world.

With the continuous development of science and technology, many experts have studied the 3D printing model. For example, Mohammadi R, Siavashpour Z, Aghdam S evaluated the feasibility and effectiveness of preoperative planning and used three-dimensional printing model to treat the whole staghorn stone, especially to select the best puncture calyx. 3D models were reconstructed using digital imaging and 3D printers. Three identical models were printed for each patient [1]. Based on reverse engineering, product innovative design and 3D printing technology, Hu Y., Mohammadi R, Siavashpour Z, Aghdam Sconstructed the technical route of rapid design and development of helmet products, optimized the product production process, shortened the product development cycle, improved the production efficiency, completed the personalized design of helmet and improved the comfort of personnel [2]. Shahrubudin n, Chuan l t, ramlan r summarized the types of 3D printing technology and the application of 3D printing technology. Finally, the materials used in manufacturing are introduced [3]. Although

J. H. Abawajy et al. (Eds.): ICATCI 2022, LNDECT 170, pp. 544–551, 2023.
https://doi.org/10.1007/978-3-031-29097-8_64

the research results of 3D printing model are quite abundant, there are still deficiencies in the research of 3D printing technology in product model making.

In order to study the application of 3D printing technology in product model making, this paper studies 3D printing technology and product model making, and establishes an evaluation model. The results show that 3D printing technology is conducive to the production of product model.

2 Method

2.1 3d Printing Technology

2.1.1 Definition of 3D Printing Technology

3D printing technology is a rapid prototyping technology which takes digital CAD model as the basic file and uses sticky materials such as powder metal or plastic to print and build objects layer by layer. 3D printing technology can rapidly manufacture and design product prototypes and effectively shorten the manufacturing time. However, in large-scale manufacturing, the speed of 3D printing is far lower than that of traditional manufacturing methods, which is also the bottleneck restricting the industrialization of 3D printing technology [4]. Now, the material reduction method is extended to cut and polish a whole piece of material to finally form the desired object shape. Relatively speaking, the physical surface of printed matter manufactured by adding materials is rough and needs polishing. The printing processing method of reduced materials has high object fineness and large material selection range [5].

2.1.2 Advantages and Disadvantages of 3D Printing Technology

In terms of traditional manufacturing technology, the manufacturing cost of products increases with the increase of its shape and structural complexity. The most obvious advantage of 3D printing technology is that it will not increase the cost of manufacturing complex items. In addition, when manufacturing finished products, there are few parts, simple assembly, or even no assembly [6]. On the other hand, in terms of unit production space, compared with traditional manufacturing machines, 3D printers occupy less space and can manufacture more and larger items in unit space, so they have stronger manufacturing capacity. However, under the current technical conditions, 3D printing technology also has its own shortcomings. Firstly, in the selection of materials, due to its technical principle, only specific materials that meet the 3D printing and forming mode can be selected. These materials have their own limitations in performance. For example, PLA material is suitable for fdm3d printing and molding, but it has the disadvantages of low strength, poor stability and easy decomposition. The price of achievable metal and ceramic printing materials is relatively expensive [7].

2.1.3 Application of 3D Printing Technology in Industrial Design

In the application of industrial design, the application of 3D printing technology has an impact on the traditional way of industrial design. Under the influence of 3D printing technology, the concept of industrial design has changed. Therefore, the industrial design mode and the quality of industrial products have been improved, which has played an important role in promoting the development of industrial design. In the development and design of industrial products, designers determine the shape of products through repeated drawing and model making processes [8]. Before the application of 3D printing technology, the model making process is relatively complex and requires professional equipment and technicians. After 3D printing technology is applied to industrial design, digital model can be quickly transformed into physical model, simplifying drawing and model making process, reducing model making threshold and optimizing industrial design process. The application of 3D printing technology in industrial design simplifies the model making process, can quickly verify the product design scheme, and effectively shorten the design and development time [9]. Therefore, the application of 3D printing technology plays an important role in promoting the development of industrial design. At the same time, the prototype manufactured by 3D printing technology has higher precision, more realistic appearance effect and closer to the effect of mass production products, which is convenient for early detection of product modeling, function and structure problems.

2.2 Product Model

2.2.1 Product Model Concept

The word "model" can be traced back to "modulus" in Latin, which means sample, proportion or standard. The relationship between them can be simply described as: model is prototype material or conceptual similarity; "Product design model" refers to the product model in which the product function, appearance and structural form conceived in the process of product design can be expressed in two-dimensional or three-dimensional ways and depicted on the plane or space (virtual space) with the help of paper, pen or computer. "Virtual model" is a model for simulation or simulation based on computer technology. It can be easily stored and modified [10]. It is a new forming technology. However, in our actual industrial design, we should not only limit and rely on computer modeling, but also combine and interact with other models.

2.2.2 Product Modeling Design

The application of 3D forming technology makes it easier to process special-shaped surfaces and complex physical structures. Through the method of layer by layer processing, designers can realize the product design that was considered unconstrained or even difficult to realize in the past, and form an independent product form. Digital forming mode makes batch processing different from the original manual mode through the data file of digital model, and realizes the possibility of standardized batch production. Once these products are assembled and combined with other materials, they will produce more rich and diverse product forms [11].

2.2.3 Generation Method of 3D Printing Digital Model

3D printing digital model can be generated in the following ways: first, professionals convert design files in different file formats designed by customers. The 3D printing digital model file after modifying the parameters is a 3D printing digital model that can be converted to 3D printing. Secondly, the existing object is scanned to obtain a 3D printed digital model. By using a 3D scanner to scan the entities we want to print, we can quickly and easily generate a 3D printing digital model [12]. Obtain the 3D print digital model from other channels and modify the drawing style according to the model to obtain a new 3D print digital model. For example, people can obtain some 3D printing digital models on some online communities, forums and other platforms of 3D printing digital models, and then make small modifications or find others to modify them. There will be a modified 3D printed digital model. The manufacturing method of 3D printing technology is also significantly different from the previous value. For designers, the early design modeling process can clearly see the production effect, and there will be no errors in the production of complex structures. Like the previous values, the production process and cost will be different due to different structures. However, the cost of today's 3D printing technology will not be very different whether it is manufacturing one, 100 or even 1000.

2.3 Evaluation Model

Since the early design part does not participate in the later software use differentiation, the evaluation model is mainly aimed at the middle and later stages of the process. Considering the final display effect and process cost efficiency, the score consists of the following:; Physical performance score B is as follows (1):

$$B = Pe(b) + An(b) - B(1) - B(e) \tag{1}$$

Where, B is the physical performance score of the process, PE (b) is the perspective effect, an (E) is the angle effect, B (L) is the software learning complexity of physical performance, and B (E) is the score reduction of physical performance production complexity.

The material light fraction m is as follows (2):

$$M = \mathrm{Re}(m) + Ar(m) - M(1) - M(e) \tag{2}$$

Where m is the material light score of the process, re (m) is the realism of the material light, and AR (m) is the artistic sense of high-quality light.

The moving lens score C is shown in Eq. (3):

$$C = Ra(c) + Ar(c) - c(1) - c(e) \tag{3}$$

Where C is the moving lens score of the process, RA (c) is the degree of freedom of the moving lens, and AR (c) is the artistic sense of the moving lens.

3 Experience

3.1 Object Extraction

First, the initial shape of these products is drawn on paper, and then the shapes of paper, foam and other materials are quickly generated. This can not only shape our short bubble concept, but also timely feel the product shape we conceived, and make some investigation and modification to it as the basis of product modeling design. Therefore, the grass model is more conceptual and simple. It can enable designers to make relatively correct judgments on their designed space and structure in a short time. Because the production of grass model is relatively simple and fast, designers can combine it with design in modeling design, so as to realize the high combination of design and design feedback, so as to better grasp the form of products. Because the material of the grass model is usually paper or foam or mud, these materials are easy to modify. Therefore, designers can constantly modify the grass model, which requires good planning and new inspiration, so that products and designers can grow together. Making 3D product demonstration animation through modeling software is the most common requirement. The designer or engineer has finished creating the 3D model. Generally speaking, they are most familiar with 3D software. In this case, we can directly use the built-in animation function of the software to make simple product demonstration animation. If we only need to achieve the basic animation effect, we can achieve the goal.

3.2 Experimental Analysis

In the first step of the process, the overall animation design is carried out with reference to the 3D animation module. Animation design plays a role in unifying style and clarifying ideas in the whole process, and the cost is not high. In the process of sending set, it is not necessary to carry out the original painting and style design like the preliminary design of 3D animation, but only consider the image segmentation and picture style of the final animation, and determine which animation modules will appear in the final video, so as to select the whole process. The overall animation design first analyzes the purpose of 3D animation, that is, to determine which functions of animation mainly highlight products. Then determine the main color of the animation, and select different background and video styles according to the color of the product. Then, according to the key points to be highlighted, determine the overall rhythm of the animation, that is, speed, urgency and time arrangement. At the same time, there is an overall plan for how to arrange the lens to express the product, or draw the sub lens.

4 Discussion

4.1 3d Printing Technology Development Status

3D printing technology first appeared in the mid-1980s and 1980s. It was invented by MIT and later transferred to 3M company. The first 3D printing device was developed in 1986. At that time, the application and popularization of this technology in the industry was slow, but it had begun to take shape. Compared with later commercial applications, it is inseparable from its origin. After entering the new century, 3D printing technology has gradually matured and developed rapidly, as shown in Table 1.

It can be seen from the above that the increase rate in 2019 is 2.84, with a percentage of 21.4%; The growth rate in 2020 is 3.32, with a percentage of 23.5%; The increase rate in 2021 is 3.74, with a percentage of 27.5%. The specific results are shown in Fig. 1.

Table 1. Growth rate of global 3D printing market

Particular year	2019	2020	2021
Growth rate	2.84	3.32	3.74
Percentage	21.4%	23.5%	27.5%

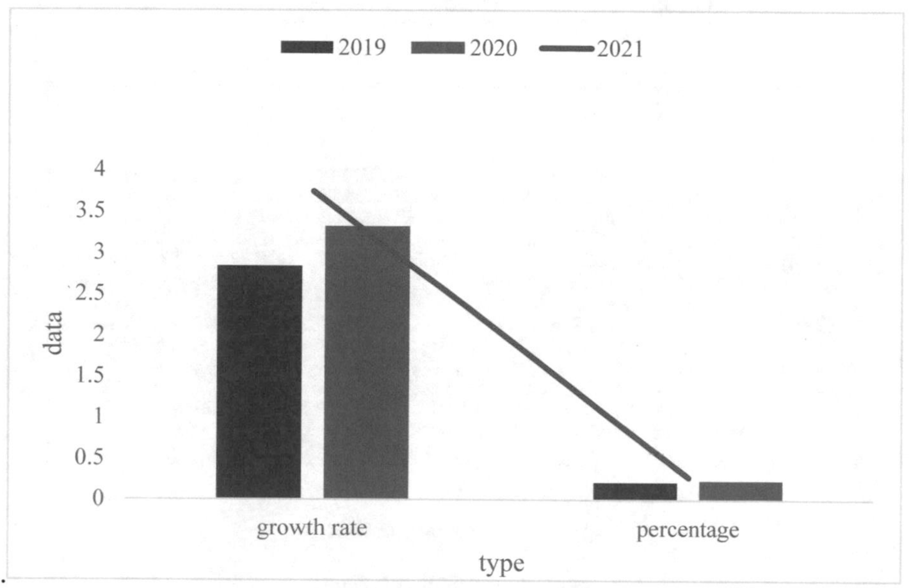

Fig. 1. Growth rate of global 3D printing market

In terms of technical equipment, in 2014, the International Space Organization (ISS) deployed the first 3D printer, and the staff printed daily necessities and machine parts as needed. In recent years, the application and popularization of 3D printing technology have generally shown an upward trend, which has had a significant impact on traditional industries.

4.2 Parameter Optimization Selection

The parameters of 3D printer are tested by orthogonal test method to select reasonable parameters to optimize the values, and the model data are used. Each parameter has a standard value, and the test analysis results are obtained by changing the parameter data up and down the standard value. It is mainly analyzed according to the printing size and printing time of 3D printer, as shown in Table 2.

It can be seen from the above that the layer thickness of group A is 0.28 mm, that of group B is 0.17 mm, that of group C is 0.02 mm, and that of group D is 0.26 mm. The specific presentation results are shown in Fig. 2.

Table 2. Layer thickness

Serial number	Layer thickness (mm)
Group A	0.28
Group B	0.17
Group C	0.02
Group D	0.26

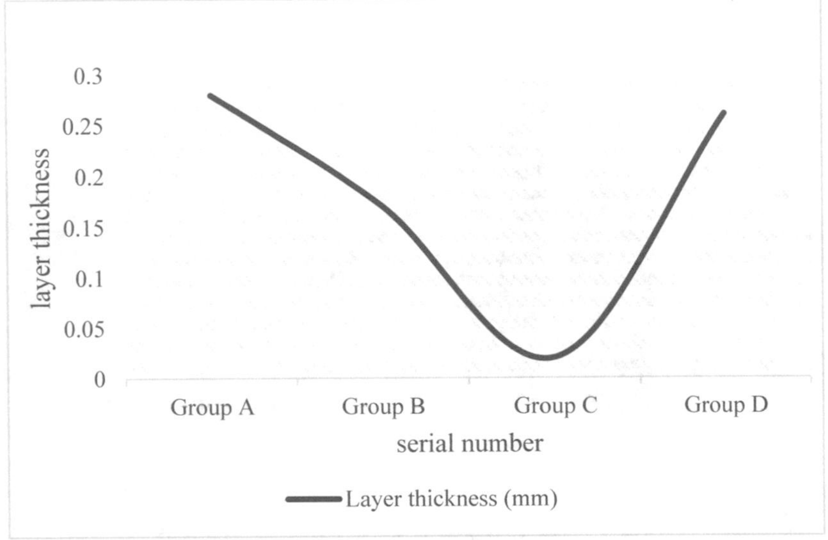

Fig. 2. Layer thickness

As shown in the above figure, four groups of data are used for comparative analysis of layered thickness. Group A has high layering accuracy, the quality of the model is relatively high, and the dimensional accuracy is relatively close, but the printing time is increased to 5 min. Due to the large layer thickness, the printing time of group D is reduced to 2 min, the surface smoothness of the model is significantly lower than that of group I, and the accuracy deviation will also increase. The analysis shows that the layer thickness will affect the printing time, surface quality and accuracy of the model.

5 Conclusion

With the continuous development of 3D printing technology, the manufacturing mode has also changed to a certain extent, which is very different from the traditional manufacturing mode. The product shape and structure are more accurate, and no additional cost will be added in the product manufacturing process. For complex processes completed using 3D printing technology, the time cost will be greatly reduced. In this paper, the parameters of 3D printer are tested. The results show that the layer thickness will affect the printing time, surface quality and accuracy of the model.

Acknowledgment. Scientific Research Program Funded by Shaanxi Provincial Education Department (Program No.21JK0249).

References

1. Xu, Y., Yuan, Y., Cai, Y., Li, X., Wan, S., Xu, G.: Use 3D printing technology to enhance stone free rate in single tract percutaneous nephrolithotomy for the treatment of staghorn stones. Urolithiasis **48**(6), 509–516 (2019). https://doi.org/10.1007/s00240-019-01164-8
2. Hu, Y., Mohammadi, R., Siavashpour, Z., Aghdam, S., et al.: Innovative design of a helmet based on reverse engineering and 3D printing. AEJ **60**(3), 3445–3453 (2021)
3. Shahrubudin, N., Chuan, L.T., Ramlan, R.: An overview on 3D printing technology: technological, materials, and applications. Procedia Manuf. **35**(2019), 1286–1296 (2019)
4. Haleem, A., Javaid, M.: Four-dimensional printing applications in dentistry. Curr. Med. Res. Pract. **9**(1), 41–42 (2019)
5. Ahmed, D., Wang, L.: Optimal area-product model (oapm) based non-iterative analytical design methodology for Litz-wired High-Frequency Gapped-Transformer (LHFGT) in LLC converters. IEEE Access **8**(1), 18134–18148 (2020)
6. Katagiri, K., Chiba, G.: Spatially-dependent nuclear reactor kinetic calculations with the explicit fission product model. Ann. Nuclear Energy **133**(Nov.), 202–208 (2019)
7. Meric, B.N., et al.: Maximum-likelihood parameter estimation of the product model for multilook polarimetric SAR data. IEEE Trans. Geosci. Remote Sens. **57**(3), 1596–1611 (2019)
8. Soloviova, O.: 3d printing technology. Appl. Geometry Eng. Graph. **97**, 136–148 (2020)
9. Melnikov, Y., Zholudev, S., Vladimirova, E., et al.: Precision of production crown manufacturing using 3d-printing technology. immediate temporary restoration after dental implantation. Actual Probl. Dentist. **16**(4), 109–114 (2021)
10. Batrakov, V.V., Krylov, A.I., Saev, V.N., et al.: Upgrade and repair of the instrumentation equipment on the ISS RS simulators using FDM 3D printing technology. Manned Spaceflight **2**(39), 97–110 (2021)
11. Posmyk, A., Marzec, P.: Influence of 3d printing technology of automotive parts made of plastics on their tribological properties. Tribologia **294**(6), 65–70 (2021)
12. Grigore, L.S., Stefan, A.G., Orban, O., et al.: Considerations on the plastic structure of a UAV payload made by 3D printing technology. Materiale Plastice **57**(4), 21–33 (2021)

Intelligent Speech Continuous Recognition System Based on NOSE Algorithm

Liu Yang, Tao Wang, Jiangtao Guo, Hong Li, and Qiakai Hailati(✉)

State Grid Xinjiang Information and Telecommunication Company, Urumchi, Xinjiang, China
702573567@qq.com

Abstract. With the development of speech recognition technology, speech has become more and more popular as an important means of human interaction. People hope to use it to communicate between humans and machines, make the communication between the two more convenient, and improve the human-computer interaction experience. Nowadays, speech recognition technology is relatively mature and developed, and many algorithms and applications have been widely used. However, the processing power of computer technology is limited, and it is difficult to ensure real-time and accuracy in low-capacity systems. Therefore, the implementation of speech recognition technology research work has practical significance and broad application prospects. This paper studies the intelligent speech continuous recognition system based on the NOSE algorithm, and understands the relevant knowledge and theories of intelligent speech continuous recognition on the basis of literature. Then the intelligent speech continuous recognition system based on the NOSE algorithm is designed, and the system is tested. The test result shows that the correct rate of the system in this paper is 85.7%.

Keywords: NOSE Algorithm · Speech Recognition · Intelligent Speech · Recognition System

1 Inductions

Intelligent speech recognition technology has attracted the attention of the industry because of its simple recognition method, convenient use, low cost, and strong versatility. However, intelligent speech recognition technology is not popular and has limitations [1, 2]. With the rapid development of science and technology, not only people need to exchange information, but also between people and machines [3, 4]. Let the machine understand human language, this is the dream of people's dream. With the expansion of human thinking, intelligent speech recognition, as a breakthrough technology, has gradually penetrated into our daily lives in human-computer interaction applications [5, 6]. The research of intelligent speech recognition technology began to focus on speech signal processing, which is an important field in intelligent speech recognition of speech signal processing technology. With the continuous development of the information society, audio signal processing technology has become a subject of in-depth research and development [7, 8]. The ultimate goal of intelligent speech recognition technology is to

J. H. Abawajy et al. (Eds.): ICATCI 2022, LNDECT 170, pp. 552–559, 2023.
https://doi.org/10.1007/978-3-031-29097-8_65

complete speech-based natural language communication between computers and people, so that the computer can respond appropriately to human voice commands [9, 10].

In order to study intelligent speech recognition technology, some researchers apply DNN to a large number of continuous speech recognition vocabulary, and propose a new frame-sensitive large vocabulary speech recognition model. The acoustic model used is based on deep neural network and hidden Marl The Kofu model is a hybrid structure and uses a deep neural network to directly model the output probability of the HMM state. Compared with the traditional speech recognition system, the speech recognition system has made great progress, and the WER word error rate has been reduced by 30%. Since then, speech recognition has also entered the deep learning stage [11]. Some researchers have proposed neural network language models. With the continuous improvement of computing power, neural network language model technology has been fully developed, and gradually showed surpassing performance. Among many neural networks, RNNs are one of the commonly used models for speech recognition and natural language editing, especially suitable for Scenes that require sequence modeling, such as automatic translation, because historical information can be recorded in long text sequences [12]. In summary, there are many research results of intelligent speech recognition, but it needs to be studied in depth in its recognition accuracy.

This paper studies the intelligent speech continuous recognition system based on the NOSE algorithm, and analyzes the current deficiencies of intelligent speech recognition and the factors affecting intelligent speech recognition on the basis of literature data, and then the intelligent speech continuous recognition system based on the NOSE algorithm is designed and tested, and draw relevant conclusions through the test results.

2 Research on Intelligent Speech Continuous Recognition

2.1 Current Shortcomings of Speech Recognition

The introduction of technologies such as deep learning in speech recognition has caused the error rate to drop rapidly, and the recognition accuracy and recognition speed have been significantly improved, but there are still many areas that need to be improved.

(1) Inadequate handling of environmental pressure and noise. The signal-to-noise ratio directly affects the collection of valid model data. Most of the training data when training the model is for denoising, but the key mode of the training set is relatively simple, so a more effective method of collecting enough data samples for all situations still has a long way to go.
(2) The semantic error rate is high. Although the current word error rate is very low, there are still some words that are easily misunderstood. Language is the crystallization of human civilization, and people express emotions through language. There are now some recognition methods that will relatively simple mechanical recognition of various cultural phenomena such as language transitions and puns of human culture.
(3) Unable to successfully place a multi-person chat script to another person. Most of the products currently on the market are mainly for recording and communicating with callers. Subsequent monitoring and analysis processing such as voice segmentation

is very important for how to determine the speaker based on characteristics such as timbre and distance.

(4) The current language recognition is usually isolated, unable to conduct historical connection or sentiment analysis. Interpersonal relationships in life often use clues to help someone understand the other's language. For example, the other's facial expressions and lip movements, the words in conversation, or the prior knowledge of the parties involved. If you allow the voice recognition assistant of your smartphone to read your contact list, the voice assistant can easily determine the name in your address book. At the same time, if the language map search system of the product allows searching, the navigation range of the destination may be restricted to a certain extent. After the introduction of these auxiliary signals, the speech recognition rate will indeed be greatly improved. However, it is necessary to conduct more detailed research on how to filter useless types of environments and how to use them efficiently.

(5) Neural network algorithms applied to language often require larger-scale operations. Therefore, since most of the current language recognition models are used in the cloud, the speed of cloud computing usually lags behind. When some trained models can be applied on mobile devices, but must be applied in a networked situation, this will increase the delay caused by the algorithm calculation and directly affect the customer experience. Therefore, effectively reducing the computing power of the network and improving the system performance is also an urgent problem to be solved.

2.2 Factors Affecting Speech Recognition

The recognition effect of the speech recognition system is greatly influenced by environmental factors. In many practical application scenarios, voice interference outside of certain scenarios, as an unfavorable factor, will result in a completely zero recognition rate, or a worse experience such as misrecognition. The main factors affecting the recognition result are echo, echo, noise, distance, angle and signal-to-noise ratio.

As the distance between the speech source and the recognition system increases, the complexity of speech recognition becomes more complicated. For remote speech recognition, if the user is too far away from the microphone, the recognition system needs to determine the position of the speaker when sending a command between multiple speakers or the issuer of the command. In addition, the room has environmental noise characteristics, which increases the complexity of long-distance speech recognition. Long-distance speech recognition needs to be more robust to adapt to different environments. The angle between the noise and the speaker affects the accuracy and recognition rate of locating the sound source. Due to the inconsistent processing logic of fixed and unstable speech recognition systems, noise can be divided into static noise and non-stationary noise.

The signal-to-noise ratio is defined as the ratio of signal energy to noise energy. The measurement unit is dB, and the calculation method is 10 lg (PS/PN). Here, PS and PN stand for signal strength and noise (using the same energy) respectively. However, when noise and speech are completely mixed, it is difficult to calculate the signal-to-noise ratio. In the case of noise prediction, the noise (clear speech + noise) is subtracted from the actual signal to obtain an almost clear speech signal. In this case, you can use the audition tool to view the waveform of the sound, and measure the level and the level of the low noise area respectively the level of the audio zone. The difference between the two can be used as the value of the signal-to-noise ratio. The higher the signal-to-noise ratio of speech and noise, the clearer the speech signal and the higher the recognition rate.

3 Design of Intelligent Speech Continuous Recognition System Based on NOSE Algorithm

3.1 System Composition

Speech recognition is essentially pattern recognition. Although speech recognition is more complicated than ordinary pattern recognition, it has more complex language information and richer connotations. The speech recognition system is mainly divided into modules such as language signal preprocessing, language end check, feature parameter acquisition and reference. Figure 1 shows the basic schematic diagram of the management system, as well as the library and module mapping.

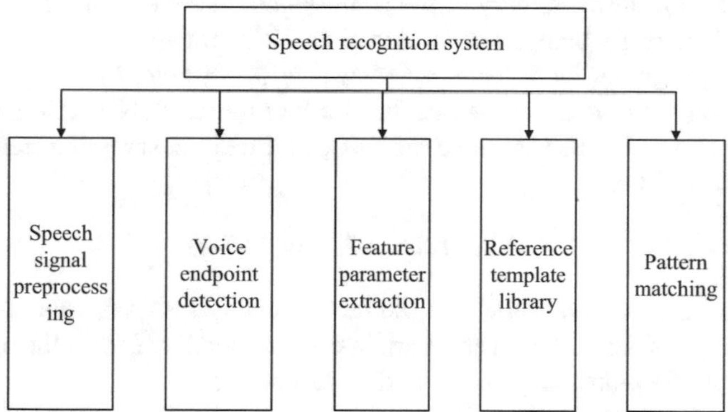

Fig. 1. Composition of intelligent speech continuous recognition system based on NOSE algorithm

The basic structure of the speech recognition system is mainly divided into the following points: (1) Preprocessing includes audio signal collection, enhancement, window, frame and so on. (2) Endpoint detection is to separate effective audio signals from the collected audio. (3) Characteristic parameters. Acquisition refers to acquiring basic characteristic parameters that reflect the characteristics of sound information. (4) The training phase refers to the audio signal input by the user, and the audio signal is usually preset. After processing the final detection points and output attributes, the attribute vector parameters can be obtained. The attribute parameters of each sound information are the template results that can form a reference template library. (5) The identification process is to compare the vector characteristics of the identified sound information with the similarity of the modules in the reference template library, and finally output the closest model result as the identification conclusion.

3.2 Voice Collection

There are generally two ways to collect audio. One is to use a hardware circuit system to collect audio signals, and the other is to directly use a multimedia sound card to collect audio signals, that is, use a computer sound card to collect audio signals, and the frequency of the collected audio signals is between 40–4000 Hz. According to the Nyquist sampling theorem, the sampling rate is more than twice the frequency of the original audio signal for sampling, and the sampling frequency of the receiving unit is $11,025$ kHz.

3.3 Speech Processing

The pre-emphasis mode improves the high frequency part of the audio signal, filters out power frequency interference, and makes the audio signal clearer and more accurate. Suppose that the voice sampling signal at time n is x(n), and the signal received after the preposition is $\hat{x}_n$.

$$\hat{x}(n) = x(n) - 0.9375x(n-1) \tag{1}$$

Due to the short-term stability of the audio signal, the audio signal is divided into several equal periods for analysis and processing. This process can be accomplished by weighting using moving finite length windows. In this article, by comparing window types, Hamming window is selected as the window function. N = 256 audio samples are one frame. It can be seen from the receiving unit that the sampling frequency is 11, 025 kHz. 1 frame time:

$$S = 1/11025 \times 256 = 23(ms) \tag{2}$$

Calculation and analysis show that the frame period is 23 ms, which is just within the defined range of 20–30 ms. Therefore, in order to facilitate FFT, the system uses a frame length of 256 points and a code shift of 128 points.

3.4 Voice Endpoint Detection

Voice endpoint detection is to determine the start and end of the voice by including the part of the signal of the voice. The accuracy of endpoint detection directly affects the noise level of the system, thereby affecting the overall performance of the system. Appropriate endpoint detection has a major impact on the cognitive stability of the entire system. Before extracting features, first remove the silent or noisy parts. An effective endpoint detection algorithm not only eliminates noise interference in the silent part, but also removes most of the noise, reduces system processing time and improves system recognition performance: However, in the real world, the presence of various noises will not only degrade the sound quality, but also It is difficult to detect the end point.

The dual-threshold detection algorithm with short-term energy and short-term zero-pass rate (hereinafter referred to as the dual-threshold method) detects silent zero-pass rate and short-term energy, and detects composite or dual-voice sounds. One is to set the upper and lower limits of short-term energy and short-term zero pass rate. Because its lower limit is small, it is sensitive to changes in audio signals and is easy to overcome, so it can be used as a preliminary screening. The dual-threshold detection process can be divided into three modes: mute, transition and audio. In the silent mode, the frame signal is analyzed frame by frame. If the short-term action or the short-term zero-pass rate exceeds the lower limit, the frame signal is marked as the starting frame and enters the transition mode. At this time, if the two parameter values are lower than the lower limit threshold, the signal returns to the mute state, if one of the values exceeds the upper limit of the transition state, it is considered to be in the voice mode, if the value is lower than the upper limit and remains within the predefined time limit Below, it will be identified as noise and continue to detect subsequent audio frames. Among them, the low threshold is generally determined based on environmental noise (the first frame of the audio signal), and the high threshold is determined based on the overall information of the audio signal.

3.5 Feature Extraction

In this paper, MFCC is selected as the functional parameter of the speech recognition system, the speech input signal is divided into individual syllables, and then the attribute parameters of each frame are derived according to the MFCC export process. In this task, 48 triangular filters will be used as filters to obtain the characteristic parameters of the 24-dimensional MFCC.

3.6 Application of the NOSE Algorithm

The system has deployed two database systems, a professional database system and a reference database system. The professional knowledge database is specially designed for common English pronunciation errors, which can check and correct common language errors for users. The reference database is able to evaluate the user's language through standard vocalization training. The database also uses the NOSE algorithm for evaluation. This is a spoken language evaluation technology that can more effectively evaluate the pronunciation of non-native spoken language. The expert database contains information about ordinary users. The software system establishes a reference database by inputting international language data standards. After the software system receives the user's voice input, it outputs the language function of the application, and then evaluates the application language according to the NOSE algorithm and the standard English language data of the reference database, and performs language error verification according to the evaluation results, and then compares the experts. The most frequently used language errors in the library, correct the applied language.

4 System Detection

Test process: Use manual segmentation data to model each Chinese character separately, count the number of words in the article to create a dictionary, use a multi-level segmentation method to segment the test speech into individual syllables, and then perform translation. It measures the three most likely recognition results for each syllable,

Table 1. System test results

	Correct rate	Error rate	Recognition accuracy
This article speech recognition system	85.7%	14.4%	84.6%
Kaldi: GMM-HMM	86.1%	13.1%	83.1%
HTK: HMM	85.1%	14.1%	83.0%

and uses a dictionary collation algorithm to find the final result. This article judges the designed continuous Chinese speech recognition system from three indicators of correctness rate, error rate and recognition accuracy. The experimental results are shown in Table 1.

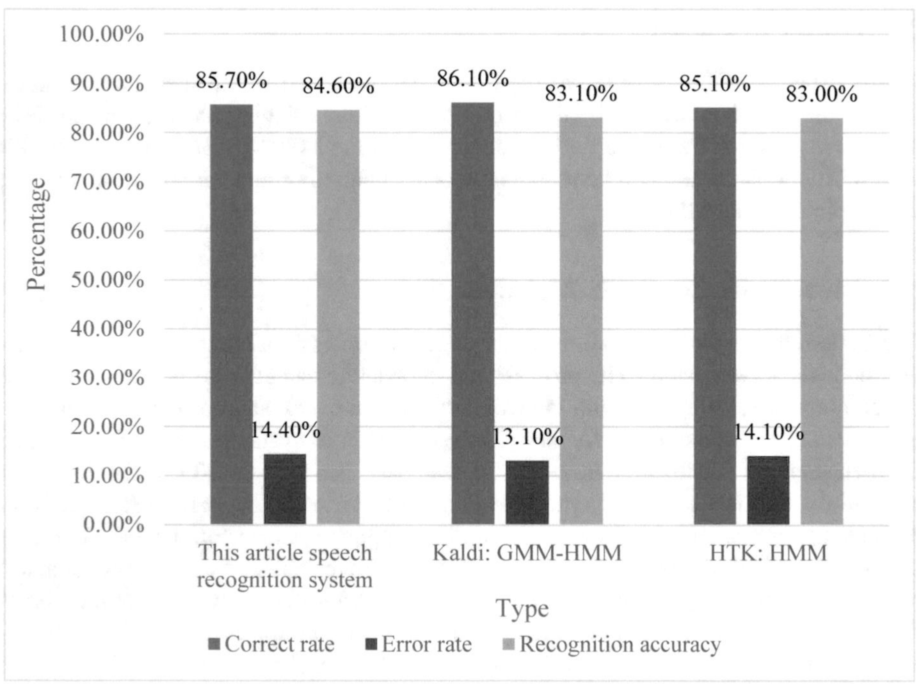

Fig. 2. System test results

It can be seen from Fig. 2 that the recognition accuracy of the continuous speech recognition system designed in this paper is 85.7%, which basically realizes the continuous speech recognition of Chinese with medium vocabulary. At the same time, in this article, two sets of comparative experiments were conducted on the Kaldi and HTK speech recognition platforms, and it was found that the authentication accuracy rate based on the Kaldi platform was higher than that of the system designed in this article, and the recognition accuracy was lower than that of the system designed in this article, and the two sets of experiments were based on the recognition accuracy of the existing platform will be lower than the recognition accuracy of the system designed in this paper.

5 Conclusions

In this paper, the intelligent speech continuous recognition system based on the NOSE algorithm is researched. After analyzing the relevant theoretical knowledge, the intelligent speech continuous recognition system based on the NOSE algorithm is designed, and the system is tested. The test results show that the continuous the recognition accuracy of the speech recognition system reached 85.7%, basically realizing continuous speech recognition of Chinese with medium vocabulary.

References

1. Park, K.I.: A study on the type of user's subjectivity perception for symbolic sculpture. J. Cult. Product Des. **50**(1), 211–220 (2017)
2. Kumar, A., Aggarwal, R.K.: Discriminatively trained continuous hindi speech recognition using integrated acoustic features and recurrent neural network language modeling. J. Intell. Syst. **30**(1), 165–179 (2020)
3. Watanabe, S., Hori, T., Kim, S., et al.: Hybrid CTC/attention architecture for end-to-end speech recognition. IEEE J. elect. Top. Sign. Process **11**(8), 1240–1253 (2017)
4. Ravanelli, M., Brakel, P., Omologo, M., et al.: Light gated recurrent units for speech recognition. IEEE Trans. Emerg. Top. Comput. Intell. **2**(2), 92–102 (2018)
5. Stnpek, E.: Formation of the collection of the foreign artists during the establishment of Istanbul painting and sculpture collection. Turkish Online J. Des. Art Commun. **9**(2), 292–302 (2019)
6. Benitez-Garcia, G., Haris, M., Tsuda, Y., et al.: Continuous finger gesture spotting and recognition based on similarities between start and end frames. IEEE Trans. Intell. Transport. Syst. **99,** 1–12 (2020)
7. Hoople, G., Parde, N., Pratt, Q., et al.: unfolding humanity: cross-disciplinary sculpture design. STEAM **3**(2), 1–7 (2018)
8. Gross, T., Gulliksen, J., Kotzé, P., et al.: Human-computer interaction-INTERACT 2009. Lect. Notes Comput. Sci. **5726**(2), 131–141 (2017)
9. Dinesh, K., Rao, P.V.: Implementing and analysing Far and Frr for face and voice recognition (Multimodal) using Knn classifier. Int. J. Intell. Unman. Syst. **8**(1), 55–67 (2020)
10. You, C., Li, J.: Application of landscape sculpture in interior design-taking wood carving as an example. Open Access Lib. J. **07**(8), 1–5 (2020)
11. Greenberg, S., Honbaek, K., Quigley, A., et al.: Proxemics in human-computer interaction. Dagstuhl Rep. **3**(11), 29–57 (2018)
12. Sarma, C.B.K., Bhuyan, M.K., et al.: Review of Constraints on Vision-Based Gesture Recognition for Human–Computer Interaction, vol. 12, Issue 1, pp. 3–15 (2018)

Power Integrated Energy Collaborative Optimization System Based on PSO-PFCM Clustering Algorithm and Block Chain Technology

Xingchen Heng[1], Wenqi Huang[2], Rongsheng Huang[2](✉), Yao Guo[2], and Wei Yang[2]

[1] China Southern Power Grid Co., Ltd., Guangzhou 510670, Guangdong, China
[2] Digital Grid Research Institute of China Southern Power Grid Co., Ltd., Guangzhou 510663, Guangdong, China
huang_rs2022@163.com

Abstract. With the rapid development of human society, the traditional fossil energy, which is necessary for development, is facing the threat of exhaustion and environmental pollution. As a clean and efficient secondary energy, electric energy is mainly converted from fossil energy to cope with the threat of resource depletion and environmental pollution brought by fossil energy. Faced with the problem of power integrated energy cooperative optimization system, we decide to use quantitative recursive analysis and robustness test of power statistical method. The experimental results show that the number of iterations of this clustering is 11, and the objective function is reduced from the initial 22346.3 to 6061.4. The results show that the comprehensive evaluation of hybrid electric energy utilization effect by using this method is globally optimal with good accuracy and high confidence, and the use efficiency of electric energy is improved.

Keywords: PSO-PFCM Clustering algorithm · Block Chain Technology · Power Integrated Energy · Collaborative Optimization System

1 Introduction

Clustering is based on the simple concept of "mutual clustering" to make the distance between individuals of the same category as small as possible and the distance between individuals of different categories as large as possible. Clustering is also called unsupervised classification. The difference lies in that, through learning, the classification of learning objects or class instances is not marked automatically when combining instances, and the parallel combination must be determined in the organizational learning algorithm, as well as unknown samples, and the clustering combination of all samples. The basic difference between the classification problem and the aggregation problem is that in the classification problem, the classification of the training sample of attribute value is known, while in the aggregation problem, the classification of attribute value must be found in the training sample. Different clustering studies were carried out in

J. H. Abawajy et al. (Eds.): ICATCI 2022, LNDECT 170, pp. 560–567, 2023.
https://doi.org/10.1007/978-3-031-29097-8_66

the early years, resulting in many different methods and techniques. Clustering algorithms can be generally divided into the following categories: classification method, level method, density-based method, grid-based method and model-based method.

Many scholars at home and abroad have carried out researches on the collaborative optimization system of power integrated energy based on PSO-PFCM clustering algorithm and block chain technology. In order to identify the parameters of the Takagi-Sugeno fuzzy model, Fuzzy C mean (FCM) algorithm, possible C mean (PCM) algorithm and possible fuzzy C mean (PFCM) algorithm are proposed. A new clustering algorithm for takagi-Sugeno fuzzy model recognition is proposed [1]. Sendner proposed that software module clustering is to divide complex software systems into multiple subsystems to improve the comprehensibility and maintainability of software systems. In order to improve the clustering speed and optimize the clustering scheme, a software module clustering algorithm based on density PSO (Density PSO) was proposed. Firstly, the software system is transformed into a complex network graph, and then the particle swarm optimization algorithm is improved. The shortest path method was used to initialize the group, and the probabilistic selection method was used to update the particle position [2]. Although a lot of meaningful research results have been achieved in the research on PSO-PFCM clustering algorithm and block chain technology, comprehensive and in-depth research on PSO-PFCM clustering algorithm is still rare, especially the research results on block chain technology of power integrated energy collaborative optimization system. The research of this paper also wants to make a contribution in this aspect.

Based on the chaotic particle swarm optimization (PSO-PFCM) algorithm, this paper realizes the collaborative optimization of integrated energy, electric energy, heat energy and natural gas. The proposed optimization model can significantly improve the performance of information resource search, improve the quality and stability of the algorithm, and effectively avoid the defect that particle swarm optimization algorithm is easy to fall into the local optimal solution.

2 Research on Power Integrated Energy Collaborative Optimization System Based on PSO-PFCM Clustering Algorithm and Block Chain Technology

2.1 Overview of PSO-PFCM Clustering Algorithm

2.1.1 Pedigree Clustering Algorithm

Pedigree clustering algorithm, also known as hierarchical clustering, hierarchical clustering method [3]. It performs a hierarchical decomposition of a given data set. According to different directions of the clustering process, pedigree clustering algorithm can be divided into two categories, namely decomposition method and clustering method. Decomposition divides the entire group of unstructured processes into groups, and then gradually into groups until each data element belongs to a category or satisfies some termination requirement. Instead, clustering rules treat each data element as a separate category, and then combine similar data into a new category until all data elements are grouped into a category or certain termination conditions are met. Most polymer algorithms are the same. The biggest problem with purely synthetic algorithms is that any merge or split cannot be adjusted by implementation. The focus of current research is to combine the level of localization and standardization methods.

2.1.2 Graph Theory Clustering Algorithm

The graph clustering method was first proposed by Zahn, also known as the maximum (small) support clustering algorithm. Graph theory clustering needs to establish a graph suitable for the problem. The nodes of the graph correspond to the smallest unit of the analyzed data, and the edges or arcs of the graph correspond to the similarity measure between the smallest data. Therefore [4, 5], there is a metric representation between each minimum processing unit, which ensures that data locality is easier to handle. The graph theory clustering method uses the local link feature of the sample data as the main information source of clustering, so its advantage is that it is easy to deal with the characteristics of local data.

2.1.3 Clustering Algorithm Based on Density

The method of sampling interval distance collection can be used to determine spherical groups, and the density method can be used to filter data such as "noise" at isolated points and determine random groups. The main idea is that the cluster should remain the same as long as the density (number of samples) of adjacent areas exceeds a certain threshold [6]. That is, all samples from a given group should have at least a corresponding number of samples within a specified interval. Density-based system clustering algorithms mainly include: DBSCAN system clustering method: includes the connection of high-density areas; Optical classification system clustering method, in which the object clustering structure is defined; Also, DENCLUE clustering method based on density distribution function [7, 8].

2.2 Defects of PSO-PFCM Clustering Algorithm

2.2.1 The Number of Clustering C of FCM Algorithm Should be Given in Advance

The aggregation result depends largely on the reasonable choice of user parameters, but the choice of this value C is difficult to evaluate. In particular, ignorance of the form of data allocation creates considerable difficulties for users [9, 10]. The same problem occurs when determining the k value of the K_means algorithm. Some algorithms get better C values by decomposing and merging.

2.2.2 The Algorithm Has Great Dependence on the Selection of Initial Value, and Often Falls into the Local Minimum Solution

Different base values often lead to different results. Site C was randomly selected as the first homologous seed, and repeated localization techniques were used until the algorithm was compiled. Therefore, different initial values lead to polymer instability in the algorithm. In addition, the FCM objective function is a non - explicit function. According to the basic principles of FCM described above, memorization markers usually contain many small local values, only one of which has an attractive region around each part of the objective function. If the selected starting point is close to the attractive point within its region, the optimization process will quickly reach this highest point. Otherwise, the rate of aggregation is very slow.

2.2.3 Modification of Membership Matrix

The clustering process is actually an interactive optimization process of membership matrix graph and clustering analysis center. The synthesis process of clustering analysis is actually an interactive optimization process of gDA and clustering analysis center. However, it cannot be confirmed that the FCM algorithm fully conforms to the optimal state through step-by-step process, nor can it be ensured that every step of each generation is close to the optimal state [11]. Therefore, for all algorithmic problem sets, the most critical thing is to ensure the repetition after the determination of V points, so as to combine the best advantages of FCM algorithm and V points as much as possible.

2.3 PSO-PFCM Clustering Algorithm Improvement

2.3.1 Optimization of Primary Value

If the initial value is outside the gravitational range, the optimization process may accumulate to other parts at a very small time. Thus, by repeating the process, the objective function tends to reach the local lower limit rather than the general lower limit. There are two ways to solve this problem. The first is to choose better priming functions to avoid slipping into optimal levels. Another approach is to combine other algorithms, such as making maximum use of particle swarm and research is prohibited, thus enabling algorithms to jump out of local research to optimize global scope.

2.3.2 Revise Membership Matrix

By optimizing the sequence matrix, the convergence center of the computation can be rationalized, resulting in faster logarithmic reduction. Therefore, the revised summary table will speed up and speed up the consolidation process. Soft polymer algorithm is a kind of fuzzy middle C grade semi-algorithm, is how to adjust the sequence matrix, it may exceed the repetition value of the aggregation center calculated below, thus speeding up the convergence of FCM algorithm. In digital applications, CPOS algorithm is used to replace the original text to replace particle information is a set of distinction, in these applications, each other between the particles are all the best value: the process of seeking information, different species of particles are evolved by similarity calculation, until the evolution process, until the best compensation is the best value.

2.3.3 Particle Interaction of PSO-PFCM Clustering Algorithm

CPOS calculation to the strong interaction between particles was used to optimize optimize resource optimal allocation, and use the best analysis of resources optimization allocation, this analysis method is often used as integrated power combination, not only can achieve the highest rate of resource optimization, at the same time can also solve the user in the allocation of resources is not easy to find problems in the process [12]. In the specific application, the big data algorithm subdivides the original information particles into different subsets, and the information granularity of each subset seeks its own optimal value, so as to share the granularity information of each subset through the common operation of evolutionary matching, and thus achieves the optimal evolutionary compensation.

3 Research and Investigation on Collaborative Optimization System of Power Integrated Energy Based on PSO-PFCM Clustering Algorithm and Block Chain Technology

The basic idea of PSO-PFCM algorithm: on the basis of FCM algorithm, membership function matrix (also called fuzzy partition matrix) and fuzzy coefficient M of samples belonging to different categories are introduced. Compared with FCM algorithm, the classification of data, the calculation of class center and the calculation of objective function are improved. Similar to FCM algorithm, the algorithm gives the steps of FCM algorithm as follows: initialization: given the number of clustering categories C, C 2 $\leq$ N $\leq$, where N is the number of data, set the iteration stop threshold ε, initialize the clustering prototype mode 0 P, and set the iteration counter b = 0.

Step 1: Calculate or update clustering centers by the following formula:

$$u_{ik}^{(b)} = \{ {}^{1 d_{ik}^{(b)} = \min\{d_{ik}^{(b)}, 1 \leq r \leq C\}}_{0}$$

Step 2: Update the clustering center of clustering prototype mode with the following formula:

$$P_i^{(b+1)} = \frac{\sum_{k=1}^{N} u_{ik}^{(b+1)} \cdot x_k}{\sum_{k=1}^{N} u_{ik}^{(b+1)}}$$

This method is relatively simple to implement. Moreover, the candidate initial cluster centers are far apart. In addition, the number of isosphere patterns is taken as the measurement standard, and the points with more numbers, that is, the points in the more dense regions, can be selected as the initial cluster center.

4 Research and Investigation on Collaborative Optimization System of Power Integrated Energy Based on PSO-PFCM Clustering Algorithm and Block Chain Technology

Run the standard PSO-PFCM algorithm once and observe the convergence of its objective function as shown in Table 1. The number of iterations of this clustering is 11, and the objective function is reduced from the initial 22346.3 to 6061.4. The result is globally optimal.

Table 1. Clustering validity function

Iterationcount	1	2	3	4	5	6	7	8	9	10	11
fcn	22346.3	19580.2	11329.4	7087.6	6445.8	6230.3	6136.1	6093.2	6073.7	6065.1	6061.4

The relationship between the objective function value and the number of iterations is shown in Fig. 1.

For the standard PSO-PFCM clustering algorithm, the clustering centers are randomly selected. We can observe its comprehensive clustering effect by clustering several times. Here we run it 10 times to see how it chooses. Experimental data are shown in Table 2. By continuously updating yao iterative calculation of different subpopulations and further setting the iterative step number of certain data yuan can quickly and conveniently obtain the global optimal solution remote.

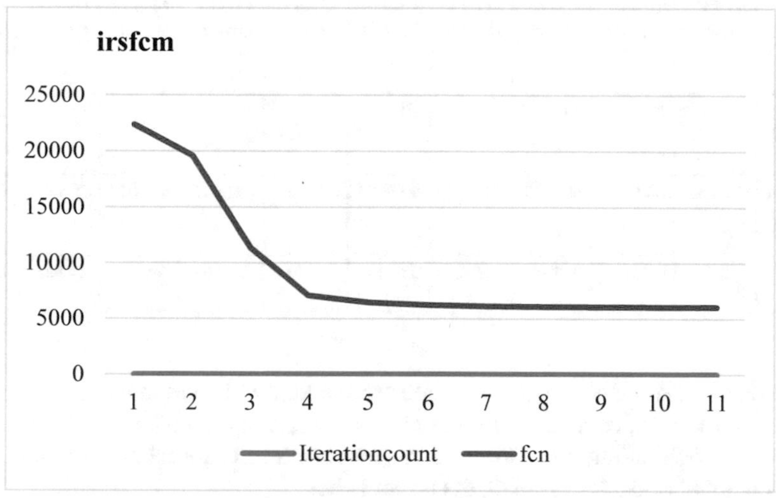

Fig. 1. Objective function and the relationship between the number of iterations

Table 2. Clustering results of standard FCM algorithm

The serial number	1	2	3	4	5	6	7	8	9	10
The initial center	141507	712321	801096	544313	944314	58472	48884	114133	35625	56456
The iteration	28	42	34	30	34	44	54	33	43	38
The objective function	60.5056	60.5056	60.5057	60.5057	60.5057	60.5057	60.5057	60.5057	60.5057	60.5057
The elapsed time	0.4070	0.6250	0.5310	0.4210	0.4690	0.5780	0.7040	0.4530	0.5460	0.5150

The data in Table 1 verify that the accuracy of clustering results obtained by using the standard FCM algorithm is not stable. There are times when the subsets of data to which they belong cannot be found correctly, resulting in local minima, such as times 3 and 6.

If the standard PSO-PFCM clustering algorithm is improved as follows: the data is randomly divided into C classes, and the center of each class is calculated as the initial clustering center. We run the improved algorithm randomly for 20 times, and the clustering results are shown in Table 3.

Table 3. The mean values of the modes randomly divided into class C are the clustering results of the initial center of class

The serial number	1	2	3	4	5	6	7	8	9	10
Randomly divided into C class results	1–79 80–142 143–150	1–26 27–148 149–150	1–41 42–69 70–150	1–132 133–146 147–150	1–21 22–23 24–150	1–135 136–138 139–150	1–45 46–115 116–150	1–43 44–94 95–150	1–10 11–149 150–150	1–66 67–108 109–150
The number of iterations	28	29	33	27	39	30	26	31	30	26
Objective function value	60.5057	60.5057	60.5057	60.5057	60.5057	60.5057	60.5057	60.5057	60.5057	60.5057
Running time (s)	0.3750	0.4840	0.4380	0.3910	0.5940	0.4540	0.3440	0.4540	0.4380	0.5000

The data in Table 3 shows that the data are randomly divided into class C, and then the mean value is calculated as the initial clustering center, which largely avoids the objective function falling into a local minimum. The results are relatively stable. When the total number of data N is small, the effect is best.

5 Conclusions

At present, due to the existence of distributed power and regional power supply model in China still belongs to some pure project style, in order to adapt to the new development needs of electricity market. Therefore, it is extremely urgent to explore a more efficient mode in line with the development needs of the electric power industry. In this study, the improved PSO-PFCM algorithm can realize the collaborative optimization of the comprehensive energy in the power grid. This algorithm has the advantages of strong global search ability, fast convergence rate and high accuracy. This can not only solve the technical problems in distributed power and regional power supply, but also for the research and development of integrated energy management mode in the later period.

References

1. Mohamed, B., Lassad, H., Abdelkader, C.: Robust kernel clustering algorithm for nonlinear system identification. Math. Probl. Eng. **2017**, 1–11 (2017)
2. Sendner, F.M.: Towards energy-autonomous unmanned aerial system: a bionic approach on solar-electric, multi-vehicle aerial platforms for waterborne operations. IOP Conf. Ser. Mater. Sci. Eng. **2020**(1024), 012055 (2021)

3. Sundaram, B.: Performance analysis of collaborative cognitive radio system on generalized and fading channels. Int. J. Adv. Sci. Technol. **29**(8), 5797 (2020)
4. Kilcioglu, E., Mirghasemi, H., Stupia, I., et al.: An energy-efficient fine-grained deep neural network partitioning scheme for wireless collaborative fog computing. IEEE Access **99**, 1 (2021)
5. Kondakci, S.: Routing efficiency of infrastructureless networks: a comparative analysis. IEEE Open J. Commun. Soc. **1**(1), 1171–1184 (2020)
6. Hajebrahimi, A., Kamwa, I., Abdelaziz, M., et al.: Scenario-wise distributionally robust optimization for collaborative intermittent resources and electric vehicle aggregator bidding strategy. IEEE Trans. Power Syst. **99,** 1 (2020)
7. Jooshaki, M., Farzin, H., Abbaspour, A., et al.: A Model for stochastic planning of distribution network and autonomous DG units. IEEE Trans. Industr. Inf. **16**(6), 3685–3696 (2020)
8. Sherrer, M., Simmons, J.W., Dobyns, J.B.: Preoperative risk stratification: identifying modifiable risks for optimization. Curr. Anesthesiol. Rep. **12**(1), 10–25 (2022). https://doi.org/10.1007/s40140-022-00519-z
9. Yoon, S., Do, H., Kim, J.: Collaborative mission and route planning of multi-vehicle systems for autonomous search in marine environment. Int. J. Control Autom. Syst. **18**(3), 546–555 (2020). https://doi.org/10.1007/s12555-019-0666-4
10. Aaa, B., Bsc, A.: PSO-based dynamic distributed algorithm for automatic task clustering in a robotic swarm. Procedia Comput. Sci. **159**, 1103–1112 (2019)
11. Kishore R, Gurugopinath S, Muhaidat S, et al. Energy Efficiency Analysis of Collaborative Compressive Sensing Scheme in Cognitive Radio Networks. IEEE Transactions on Cognitive Communications and Networking, 2020, PP(99):1–1
12. Mohapatra, A., Mishra, S., Das, G., et al.: Application of PSO and K-means clustering algorithm for CBIR. Int. J. Comput. Trends Technol. **67**(5), 141–145 (2019)
13. Soruri, M., Sadri, J., Zahiri, S.H.: Gene clustering with hidden Markov model optimized by PSO algorithm. Pattern Anal. Appl. **21**(4), 1121–1126 (2018). https://doi.org/10.1007/s10044-018-0680-9

Application of Medical Image Segmentation Algorithm Based on Genetic Algorithm in Intelligent Medical Nursing System

Haiyan Tan[1(✉)] and Donghu Gu[2]

[1] School of Health, Yunnan Technology and Business University, Kunming 651701, Yunnan, China
tanhaiyan0620@126.com

[2] School of Intelligent Science and Engineering, Yunnan Technology and Business University, Kunming 651701, Yunnan, China

Abstract. With the progress of society, the level of medical care in our country is constantly improving. In the information age, the application of computers is becoming more and more extensive, and medical image processing technology has become a popular subject. In this paper, genetic algorithm is used to conduct in-depth research on the application of medical image segmentation algorithm in intelligent medical care system. Its purpose is to improve the level of modern medicine and bring better experience to patients. This article mainly uses the experimental method and the comparative method to analyze the application of medical image segmentation algorithm in the intelligent medical care system and the performance of the system. Experimental results show that the coverage and accuracy of the system can reach about 90%, and the time-consuming is about 2 s.

Keywords: Genetic Algorithm · Medical Image Segmentation · Intelligent Medical Care · Nursing System

1 Introduction

With the development of medical technology, people have higher and higher requirements for disease diagnosis and treatment. Traditional medical care can no longer meet the needs of modern society. With the continuous advancement of technology, higher-level requirements are put forward for the quality and technical level of medical staff. In order to meet the intelligent needs of hospitals, it is necessary to design an intelligent medical care system.

There are many research theories on the application of medical image segmentation algorithms based on genetic algorithms in intelligent medical care systems. For example, in order to overcome the shortcomings of standard genetic algorithm, Some scholars proposed an improved Otsu image segmentation algorithm based on dual adaptive genetic algorithm [1]. According to the criterion of maximum variance between clusters and

J. H. Abawajy et al. (Eds.): ICATCI 2022, LNDECT 170, pp. 568–575, 2023.
https://doi.org/10.1007/978-3-031-29097-8_67

poor real-time performance, Some scholars proposed a method that combines the maximum variance between entropy clusters to optimize the swarm intelligence algorithm for the insufficient amount of original image information carried by image segmentation [2]. In order to prevent the local optimization of the bird swarm algorithm, Some scholars proposed a bird swarm image segmentation algorithm based on gene mutation [3]. So this article studies genetic algorithm, medical image segmentation algorithm and intelligent medical care system.

This article first studies image segmentation and elaborates its basic theory. Secondly, the genetic algorithm is analyzed. Then design the system in detail and analyze its principles. Finally, the system is tested through experiments and relevant conclusions are drawn.

2 Application of Medical Image Segmentation Algorithm Based on Genetic Algorithm in Intelligent Medical Nursing System

2.1 Image Segmentation

Generally, the algorithm for segmenting an image is based on two basic attributes of its brightness. First, the segmentation is based on the discontinuity of the brightness value of the points that make up the image, such as the edge detection technology of the image, which detects the edge pixel value of the image. Second, perform image segmentation based on a certain preset rule and the brightness similarity of the joint points. The selection should be made according to the structural characteristics of the segmented digital image, and sometimes these methods must be combined to achieve a good segmentation effect [4, 5].

Compared with the other two types of segmentation methods, threshold segmentation technology is the simplest due to its intuitiveness and simplicity. It greatly simplifies the steps of image analysis and processing and compresses the amount of data required for calculation. It has high computing efficiency and has been widely used [6, 7].

Through the gray histogram of the digital image, the image is divided into two different parts, namely the target part and the background part. This method is called global threshold segmentation. Global threshold segmentation is suitable for digital images with the characteristics of simple image structure and concentrated distribution of gray values. However, the gray histogram of some images has no obvious peaks and valleys. For such complex structures, different images can be compared everywhere. The entire digital image is divided into several different regions, and the threshold is determined according to the gray distribution characteristics and spatial characteristics of each divided sub-region. And each sub-region is divided according to its threshold, and then the results are integrated into a complete segmentation result. This kind of method is called local threshold segmentation method [8, 9].

In fact, medical images need to be pre-processed due to interference from various reasons before segmentation. This is mainly due to the technical parameters of the equipment used to shoot medical images, image quality parameters, and mechanical calibration and debugging. The main visual impact on medical staff is image noise, so before medical image segmentation, noise interference needs to be eliminated [10].

There are many methods for denoising in image processing, including four filters: wavelet, morphology, median and mean. Assuming that the image $p_1, p_2, \ldots, p_{n-1}$ is divided into n parts by a threshold, the between-class variance is expressed as:

$$\lambda(p_1, p_2, \ldots, p_{n-1}) = \sum_{i=1}^{n-1} \max[\lambda(p_i)] \tag{1}$$

The corresponding optimal threshold is:

$$(p_1, p_2, \ldots, p_{n-1}) = \text{agr} \sum_{i=1}^{n-1} \max[\lambda(p_i)] \tag{2}$$

In order to achieve the best image segmentation effect, the optimal threshold is obtained by searching in the target area, and it is selected to be suitable for unconstrained optimization problems. Here, formula (1) is selected as the optimization function, combined with the IAFS-PSO algorithm to optimize the function to obtain the best threshold.

In the genetic algorithm, there are some more important parameters, which may have a great impact on the performance of the algorithm. In the initialization phase and evolution process of the algorithm, these parameters must be reasonably selected and controlled. Genetic algorithm is an optimized search method that can be used in complex systems. It is very robust and easy to parallelize. The algorithm of genetic algorithm is a kind of strong global search ability, high efficiency and self-adaptive environment. However, in practical applications, due to its large computational complexity, it does not have good versatility. Genetic algorithm is a kind of random search algorithm, which is based on the principle of natural evolution, and finds the optimal solution through the processes of encoding, crossover and mutation of natural biological genes in biology. In practical terms, it is to treat multiple individuals as a whole to deal with the problem [11, 12].

2.2 System Detailed Design and Realization

(1) Design Principles. Based on existing construction, integrate existing resources. On the basis of fully protecting the achievements of past informatization construction, realize resource integration and information intercommunication, so that different users can receive and use public resources, communicate and communicate with each other within their specific jurisdiction, and can use information resources. Maximize value.

Highlight needs and implement step-by-step construction. The demand for information interaction is constantly evolving and changing, and business needs cannot be met all at once. The operation and maintenance system of the corresponding system and the monitoring of personnel capabilities will also be gradually expanded. The first is to solve the urgent information exchange and information analysis needs, gradually accumulate experience and continuously improve.

Unify standards and regulations to ensure information security. Through the development of standardized technical guidelines, the integration, management and access control of information resources are realized.

(2) Overall architecture design. The framework of this system is based on the single-chip microcomputer, which is composed of data acquisition module, control management module and communication interface circuit. This system is mainly to realize the automation and intelligence of nursing work, so it is necessary to have a certain understanding of the framework. First of all, the needs of different levels of people need to be considered in the architecture. The second is the software function module aspect. Finally, the required requirements for data transmission between various parts and interface connection issues should meet the requirements of the architecture.

The entire system is managed on a user basis and supports the connection and use of multiple users. The details are shown in Fig. 1:

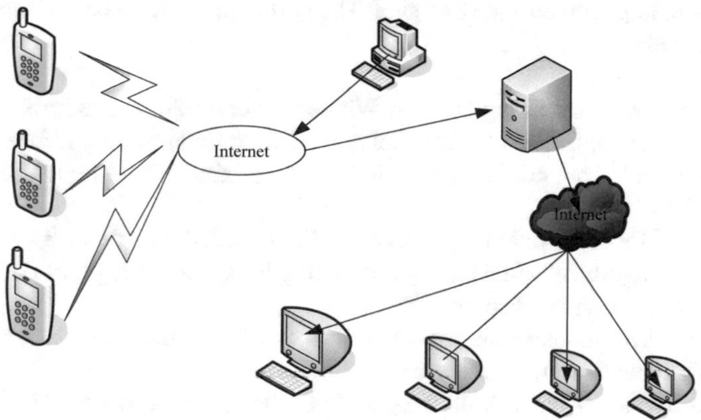

Fig. 1. User-Managed Medical Care System

The entire system architecture adopts an open design. Each device type has a custom and extensible communication format. The first module is the user registration function, that is, users can modify their personal information on the website, and can make corresponding queries on their registered account passwords and identity information. After the user logs in, the personal information management and adding, deleting and saving can be realized.

As the main means for medical staff to enter the village, serve the masses in the community, conduct health inspections, and collect test data, smart terminals will be able to transmit text, information and information, images and data files in a timely manner. End the service platform to ensure the speed and quality of information transmission. In the intelligent medical management and service terminal, there are functions such as "tracking information records at any time", "health file query", "treatment suggestions at any time", and "pushing activity programs". Therefore, communication efficiency and terminal network access quality have become a key component of system design. The best results can be obtained by using wireless access modules of smart terminals such as smart phones and tablet computers.

The intelligent medical management and service terminal system combines modern medical service technology with the latest information technology and Internet technology.

3 Intelligent Medical Monitoring System Test

3.1 Intelligent Medical System

The intelligent medical monitoring system developed in this paper is divided into three parts: parameter acquisition module, database and client software. For the execution of each function, the system test checks whether each part of the function can operate normally. Finally, conduct an overall test and analyze the results to see if the expected function can be achieved.

The hospital's intelligent nursing display system adopts C/S (client/server) structure on the PC side, large screen and PDA side. The software platform configuration of each module is as follows:

(1) Server. Software operating system: Windows Server 2018 Enterprise x84; Using database system: ORACLE 12G; Server software middleware: WebService + IIS 9; System hardware configuration: IBM System x3960 X5, Xeon X8560 2 GHz, 6CPU, 256 GB.
(2) PC client. Terminal operating system: Windows XP, Windows 8, Windows 13; Terminal program: hospital intelligent nursing large-screen display system; hospital intelligent nursing PC display system;
(3) Software environment: Microsoft.Net Framework 4.5;Terminal hardware environment: PC with Core Duo or higher.
(4) PDA handheld terminal. Mobile operating system: Windows Mobile 8.5 system; PDA device model: Motorola MC42N0;Terminal program: Hospital intelligent nursing PDA display system.

3.2 System Test

The smart medical monitoring system test mainly captures the working effect and performance of the entire system, mainly around the following aspects:

Check whether the parameter acquisition module can successfully acquire the human body parameters and transmit them to the coordinator.

Whether the coordinator can transmit data to the PC.

Whether the PC can successfully remotely connect to the database server, and whether the collected data can be saved in the database in time.

Whether the database can meet the requirements of customers, change the database data in time.

Whether the customer's monitoring software can operate normally, and whether each part of the function can be realized.

3.3 Test Preparation

Before the test, first prepare for the test, connect the sensor to the Zigbee node module to obtain human body parameters, connect the Zigbee coordinator to the PC through the RS232 serial line, set the database to remote connection mode and connect to the database server in the client monitoring software.

Click on the smart medical.exe computer desktop to open the smart medical system login interface. Register the user first, click the register button to enter the registration interface. The data query is divided into two parts, one is to display your own data, and the other is to display the master data of all doctors in the community hospital. Enter the real-time display interface, the corresponding TextBox control on the interface will display the corresponding acquisition parameters. When the human body parameter value exceeds the upper and lower limits set by the system, it will be displayed in red. Click the health record button to enter the query parameter record interface. Click the video view button to enter the online user interface and display the current online users.

4 Analysis of Test System Results

4.1 Analysis of Performance Test Accuracy

This experiment analyzes the functions of user login information, data query information, basic situation information, health records, and nursing mode query. The specific data results are shown in Table 1:

Table 1. Performance test accuracy analysis

	Coverage	Accuracy	Time
User login information	89%	91%	2
Data query information	91%	92%	1.8
Basic information	90%	94%	2.2
Health record	96%	95%	2.1
Nursing method query	93%	92%	2

As shown in Fig. 2, we can see that in the five aspects of functional testing, the medical care system has the highest coverage of patients' health records, the largest number of health records testing, and the highest data detection accuracy. Among them, the testing time for data query information is the least, and the accuracy can reach 92%.

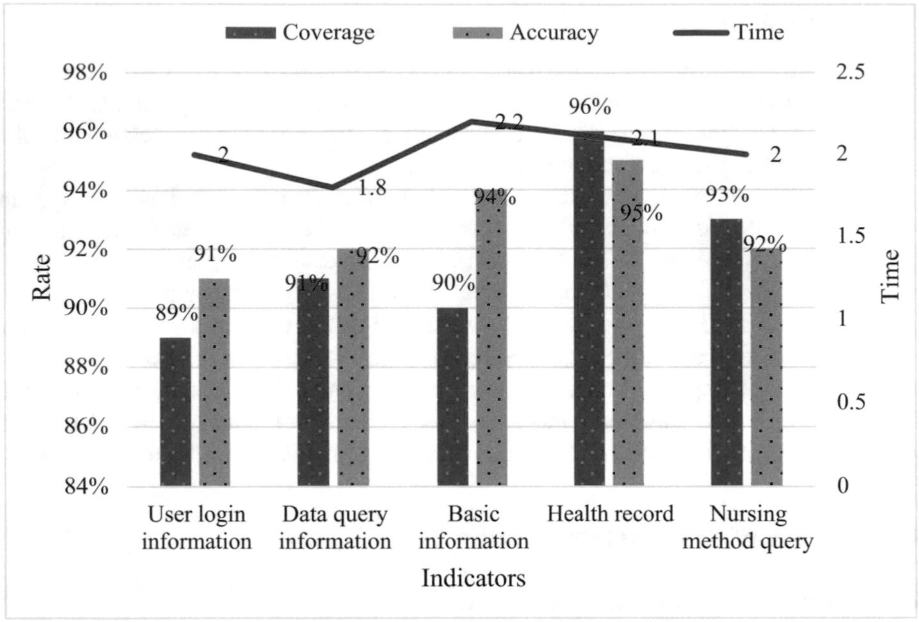

Fig. 2. Performance Test Accuracy Analysis

5 Conclusion

As an important part of biotechnology, medical image segmentation is based on the basic principle of statistical analysis of relevant data extracted from the healthcare system based on different algorithms. Genetic algorithm is a design based on the theory of natural evolution to process information in parallel and simulate the system of the biological world. It has a strong advantage in solving complex problems. Through the research and application of genetic algorithm in intelligent nursing bed, this article mainly aims to improve the comfort of patients' life and reduce the burden on medical treatment.

References

1. Shrivastava, N., Bharti, J.: Automatic seeded region growing image segmentation for medical image segmentation: a brief review. Int. J. Image Graph. **20**(3), 2050018 (2020)
2. Riaz, F., Rehman, S., Ajmal, M., et al.: Gaussian mixture model based probabilistic modeling of images for medical image segmentation. IEEE Access **PP**(99), 1 (2020)
3. El-, S.A., Skobtsov, Y.A., Rodzin, S.I.: Comparison of Hybrid ACO-k-means algorithm and Grub cut for MRI images segmentation. Procedia Comput. Sci. **186**(11), 316–322 (2021)
4. Sirisha, P., Haritha, D.: Hybrid shuffled frog leaping algorithm with probability dispersal method for tumor detection in 3d MRI braintumor images. IOP Conf. Ser. Mater. Sci. Eng. **1074**(1), 012001 (2021)
5. Kalyani, R., Sathya, P., Sakthivel, V.: Multilevel thresholding for medical image segmentation using teaching learning based optimization algorithm. Int. J. Intell. Eng. Syst. **14**(2), 11–21 (2021)

6. Baranov, M., Velichko, E., Shariaty, F.: Determination of geometrical parameters in blood serum films using an image segmentation algorithm. Opt. Memory Neural Netw. **29**(4), 330–335 (2020). https://doi.org/10.3103/S1060992X20040037
7. Hannah, I.H., Azar, A.T., Jothi, G.: Leukemia image segmentation using a hybrid histogram-based soft covering rough k-means clustering algorithm. Electronics **9**(1), 188 (2020)
8. Monteiro, A.: Hematology and digital image processing: watershed transform-based methodology for blood cell counting using the WT-MO algorithm. Med. Technol. J. **4**(3), 576 (2020)
9. Isah, R.O., Usman, A.D., Tekanyi, A.: Medical image segmentation through bat-active contour algorithm. Int. J. Intell. Syst. Appl. **9**(1), 30–36 (2017)
10. Kollem, S., Reddy, K.R., Rao, D.S.: An optimized SVM based possibilistic fuzzy c-means clustering algorithm for tumor segmentation. Multimedia Tools Appl. **80**, 1–29 (2020). https://doi.org/10.1007/s11042-020-09675-y
11. Kishorjit, N., Kanan, K.W., Ranita, K., et al.: An unsupervised cluster-wise color segmentation of medical and camera images using genetically improved fuzzy-Markovian decision relational model. J. Intell. Fuzzy Syst. **35**(1), 1147–1160 (2018)
12. Ali, T., Afzal, M., Yu, H.W., et al.: The intelligent medical platform: a novel dialogue-based platform for health-care services. Computer **53**(2), 35–45 (2020)

Design of Automobile Engine Timing Chain System Under Artificial Intelligence Technology

Shengming Tang[1], Qi Guo[2](✉), Yuandong Sun[1], Jingjie Kang[1], Lijun Zhang[1], Ruolan Wang[1], and Ji Zhang[1]

[1] Ningbo Branch of Chinese Academy of Ordnance Science, Ningbo, Zhejiang, China
[2] BorgWarner Automotive Components (Ningbo) Co., Ltd., Ningbo, Zhejiang, China
qiguo@126.com

Abstract. With the continuous development of automobile technology, the performance requirements of automobile engines are getting higher and higher. Low-noise, low-vibration gear systems are increasingly used in engine timing valve mechanisms and other power transmission mechanisms. However, the technical research, development and application of the domestic timing chain system in the engine timing chain system is still in its infancy, and its independent design and research and development capabilities are insufficient. The basic time chain system design techniques are all foreign chain systems that have existed for a long time. This article focuses on the design of automobile engine timing chain system and the wear experiment. First, use the literature research method to explain the layout position of the timing chain system in the whole machine and the type of wear, and then design the automobile engine timing chain system. In the wear test of the designed chain, the result obtained is that the maximum elongation of the center distance is 0.770 mm, and the total center distance wear elongation is $\varepsilon = 0.33\%$.

Keywords: Automobile Engine · Timing Chain System · Wear Test · System Design

1 Introductions

Major domestic engine factories or technology R&D centers have mastered the development of electronic control systems [1, 2], main engine performance calibration and component structure optimization design and other core technologies in the design and development of independent gasoline engines, but the key subsystems (including timing system and gearbox) also depend on foreign [3, 4], and the company's basic technical capabilities still need to be further improved. Regarding timing system technology, the basic copyright and timing chain design patents are mainly controlled by foreign manufacturers [5, 6]. Therefore, the development of basic engine subsystems with independent intellectual property rights can technically reduce the supply to foreign countries [7, 8].

In the research of automobile engine timing chain system design, many experts and scholars have conducted in-depth research on it, and have achieved good practical

J. H. Abawajy et al. (Eds.): ICATCI 2022, LNDECT 170, pp. 576–583, 2023.
https://doi.org/10.1007/978-3-031-29097-8_68

application results. For example, some researchers start from the direction of reducing the friction and wear of the chain drive system. Through theoretical analysis and practical research and other means, the transmission system layout structure of the toothed chain at a small pitch has been deeply studied for the effect of reducing engine friction, chain wear and engine noise. The article also uses formulas to calculate. The impact energy size between the roller chain and the toothed chain is compared, and the impact of the system's impact excitation on the noise of the automobile engine is mentioned [9]. In addition, some researchers use simulation analysis technology to study the motion force of the timing chain system and the characteristics of the object in its motion process. The focus is on using simulation analysis technology to explore how to achieve low noise and less friction [10]. In addition, some researchers use simulation analysis methods to study the dynamic driving force and kinematic characteristics of the timing chain system. The focus is on using simulation analysis methods to study a timing chain system with low noise, no friction, and good durability, reducing the routine actual test time and research and development costs, thereby improving the engine design workflow [11]. In addition, on the basis of in-depth discussion of the toothed chain meshing mechanism, friction and wear characteristics and noise characteristics, etc., they have made some unique related theoretical and practical researches respectively, and combined with domestic mainframe factories to conduct industry-university-research docking, completed the transition from technology to product [12].

This article focuses on the design of the automobile engine timing chain system and the wear experiment. First, the literature research method is used to explain the layout position of the timing chain system in the complete machine and the type of wear, and then the automobile engine timing chain system is designed, in the wear test of the designed chain.

2 Research on Timing Chain of Automobile Engine

2.1 The Layout Position of the Timing Chain System in the Whole Machine

According to the position of the timing system in the engine, it can be roughly divided into three forms. The first is that the timing system is arranged at the front end of the engine, the second is that the timing system is arranged at the rear end of the engine, and the third is to adopt front and rear phases. The layout form of the combination; and which layout form to adopt not only needs to consider the layout space requirements of the engine body design, but also needs to combine the installation method and layout space of the engine in the whole vehicle.

2.2 Types of Wear

(1) Adhesive wear. When two solids in contact with each other move in a sliding manner or move away from a close contact state, small particles attached to one surface are transferred to the other contact surface. Small particles of material may also adhere to each other on the two contact surfaces. This situation usually allows the two materials in contact to fall freely in the form of a wear debris, and the wear debris changes in the reciprocating sliding direction along the contact surface different.

(2) Abrasive wear. When friction occurs, the material will be exposed to external hard materials or protrusions, causing the surface to fall off. This phenomenon that causes material transfer is called abrasive wear. The abrasive wear range is very wide, especially in the friction and wear process of most mines, agriculture and machinery. Approximately 80% of machine and equipment failures are caused by abrasive wear. When the machine is in direct contact with minerals, mud, etc., there will be hard particles or substances on their contact surfaces, which will cause abrasive wear.
(3) Erosion and wear. The fluid contains solid particles. When the liquid hits the surface of the material or sample, the surface material will be damaged and lost. This phenomenon is called corrosive wear. For example, when the material is transferred to a pneumatic conductor, the impact force on the conductor is corroded and causes wear; when the dust is discharged through the pipe, the impact force of the dust on the pipe causes corrosion and wear.
(4) Contact fatigue wear. Alternating pressure is continuously applied to the sample, causing the material to fall off the surface under fatigue. This kind of wear phenomenon is called contact fatigue wear or surface fatigue wear, and it usually occurs in gears. In the friction and wear of mechanical parts (such as cams and bearings), even if the parts are lubricated in good conditions, contact fatigue and wear cannot be avoided.

2.3 Design of Automobile Engine Timing Chain System

The establishment of the timing chain system and the oil pump chain transmission system is inseparable from the design parameters of the engine. First, determine the system layout coordinate data and its working environment, such as maximum power, maximum torque, maximum crankshaft speed, limit load of chain plates, chain wear elongation, the layout of the transmission system and the spatial layout of the engine box, etc.

(1) Determination of gear data. This paper pre-selects the inner-outer compound meshing round pin toothed chain with a pitch of 8 mm, the thickness of the chain plate is 1.3 mm, and the thickness of the guide plate is 1mm. In the automobile timing chain transmission system, try to make the sprocket center distance 30–50 times the chain plate pitch, so that the number of chain links in the unsupported chain span stage is less than 20, and the number of chain links between the sprocket and the guide rail is average In 3–4.
(2) Determination of gear meshing form. All sprockets in this system adopt involute profile sprockets. Compared with the linear tooth profile, when the involute sprocket tooth profile meshes with the inner tooth profile of the chain plate, anti-rolling occurs, which can significantly reduce the meshing impact, enhance the stability of the system, and reduce stress fatigue. In order to prevent the occurrence of singular links, the chain is usually selected as an even-numbered link. In order to avoid the contact point of the chain and the sprocket from being repeated at a fixed position during the transmission process, that is, to reduce the wear of the chain and the sprocket, the drive sprocket is usually selected. The number of teeth is the base

tooth. The number of teeth of crankshaft, oil pump shaft and intake and exhaust camshaft sprocket is selected as 19, 24 and 38. Due to the limitation of the engine structure, the chain plates of the main system are arranged in a double row 2 × 1 form, and the oil pump chain is arranged in a single row 3 × 2 form.

(3) Chain plate design. The main parameters of the inner-outer compound meshing round pin toothed chain plate include: chain plate pitch p, the unilateral clearance between the chain plate hole and the pin shaft, the hole center distance a, the side center distance f, the tooth profile angle, and the protrusion of the outer convex inner tooth profile of the plate 1 relative to the linear outer tooth profile of the chain plate 2, the radius of curvature r of the inner arc curve and the arc center coordinate o1, and the thickness of the chain plate B. The design parameters of the chain plate are as follows: p = 8 mm, $\Delta = 0.04$ mm, a = 7.92 mm, f = 2.848 mm, a = 30°,$\delta = 0.138$ mm, r = 16.08 mm, B = 1.3 mm.

(4) Sprocket parameter design. 1) Number of teeth z. The selection of the number of teeth is based on the needs of transmission. Usually, the number of teeth is not less than 17 under medium and high speed conditions. At the same time, in order to avoid repeated meshing between the chain plate and the sprocket, the number of teeth of the crankshaft sprocket is selected as the base number. Select the number of teeth of the oil pump sprocket and the sprocket of the intake and exhaust camshafts according to the transmission ratio.

2) Sprocket pitch p, and modulus m. The toothed chain plate and the sprocket mesh are conjugate meshing, so it needs to meet

$$m \cos \alpha = m_1 \cos \alpha_1 \quad (1)$$

where m = p/π, m_1 = p_1/π is the modulus of the chain plate and sprocket, so it satisfies

$$p_1 \cos \alpha = p_2 \cos \alpha_1 \quad (2)$$

3 Automobile Engine Wear Test

3.1 Test Sample Chain

According to the design parameters of the toothed chain drive system, the actual product is manufactured and processed. The number of sprocket teeth is Z1 = 21 teeth and Z2 = 25 teeth, respectively, and the pitch of the chain plate is P = 9.525 mm. The chain consists of a certain number of internal meshing chain links and a certain number of external meshing chain links alternately connected first, and then the head and tail links are riveted, and the processed chain surrounds the motor shaft sprocket and the gearbox shaft chain between the wheels, the normal toothed chain transmission system used in hybrid electric vehicles is formed. In this paper, two chains are selected as the sample chain for this experiment.

3.2 Wear Test Method

The test method selected in this document is the simulation bench test method, which uses a special test bench that can run at high speed. Its structure: the main test gear and the accompanying test gear are respectively connected to the drive shaft, the main gear is connected to the drive shaft, and the test gear is connected to the drive shaft through the automatic drive shaft. The automatic loader can perform the strength of the main test gear and the controlled rotating gear chain related to the main chain test and subsequent tests. At this time, the top of the main test chain is the expansion side, and the bottom of the test chain is also the expansion side. On the test bench, the power consumption of the motor is only used to overcome frictional resistance and is set to act on the main test drive chain. The torque of the wheels is M, the gear speed of the main test drive is n, and the power of the entire transmission system is n.

3.3 Experimental Data

During the test, set the rotation speed of the main test gear to n = 5000 rpm, the motor power P = 8.37 kW, the unilateral voltage of the main test chain and the accompanying test chain F = 500 N, and its lubrication status is the wear test through oil injection lubrication. During the gear wear test, the distance between the center points of the two sprockets in the drive system was measured by a center distance meter, and the test was measured 13.5 h after the start of the wear test. The chain first performs a center distance measurement and then every 23.5 h. Record the center distance test value of each measurement. The total test time is 300 h.

4 Experimental Data Analysis

(1) When measuring the wear test data, the distance between the center of the main test gear and the center of the main test gear measured each time is determined as the actual center distance, denoted by a. The distance between the centers takes the center of the driving gear and the main test gear as the initial center distance, denoted by a0, and takes the change in the distance between the center of the main test gear and the center of the gear. The extension of the center distance of the main test wheel is represented by Δ. The ratio of the extension of the center distance to the initial distance of the center is defined as the extension of the wear of the center distance, expressed as ε. This method of calculating center distance wear elongation is called "center chain distance" calculation method. The results obtained by the friction test are shown in Table 1.

From the above data, it can be seen that after 300 h of high-speed wear test of the toothed chain system, the maximum elongation of the center distance is 0.770 mm, and the total center distance wear elongation is ε = 0.33%.

(2) According to the linear relationship between the wear test time, center distance elongation and center distance wear elongation, plot the center distance elongation Δ and wear time t of the hybrid sprocket in the normal wear process. Curve

Table 1. Results obtained by friction test

Test time	Center distance extension	Abrasion elongation %
0.0	0.0	0.000
13.6	0.45	0.190
38.1	0.56	0.246
61.2	0.58	0.257
85.6	0.63	0.270
108.9	0.64	0.274
134.0	0.64	0.274
156.4	0.64	0.274
179.0	0.71	0.310
203.4	0.73	0.314
226.7	0.74	0.317
249.8	0.74	0.317
273.5	0.74	0.317
300.0	0.76	0.325

relationship diagram (shown in Fig. 1) and the relationship curve between center distance elongation ε and working time t (shown in Fig. 2).

It can be seen from the above two figures that the wear time of 0–20 h belongs to the execution stage, that is, the initial wear stage. In a relatively short period of time, the amount of wear of the gear system is relatively large, the center distance and the wear elongation curve both show a sharp upward trend, and the linear slope is relatively large. This is mainly because the various parts of the gear drive system will produce a large amount of grease during the machining process, and will produce assembly clearance. After processing the toothed chain, there is always a certain degree of roughness on the processed surface. In the initial stage of wear, the contact surface of the component has only a small number of contact tips, the contact area is small, the pressure is high, and the surface of the component is deformed and plastically hardened.

20 ~ 300 h must belong to the permanent wear stage. At present, the wear rate of the gear chain is only 27%, and the amount of wear in the initial wear stage from 0–20 h accounts for 73%. Therefore, in order to reduce the initial wear in the wear stage, the percentage of wear is increased by improving the machining and structure of the gear chain. The precision and strict control of the assembly precision of the gear chain can increase the service life of the gear chain, thereby extending the service life of the gear system.

It can be seen from the figure that in the two time intervals of 100–150 h and 225–275 h, the curve is approximately in a straight state, and there is no change in the amount

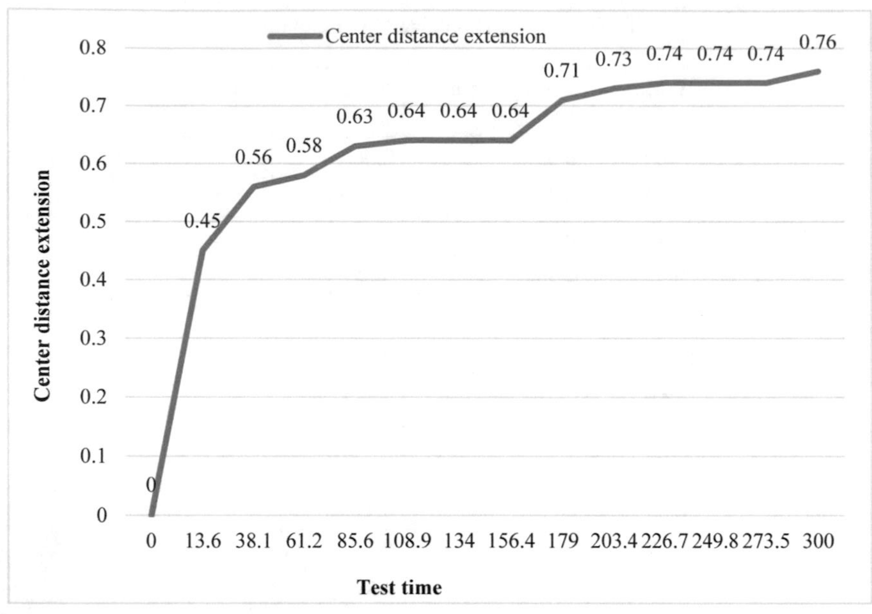

Fig. 1. Toothed chain center distance elongation curve

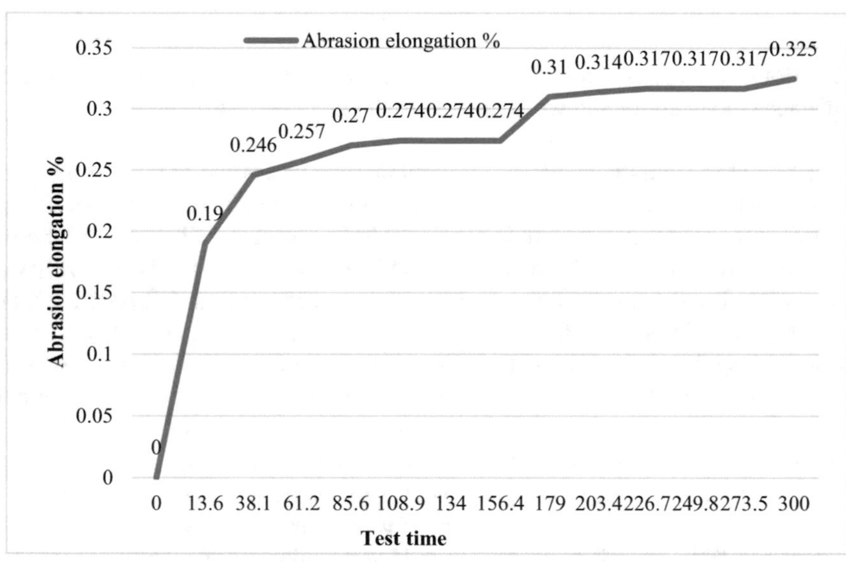

Fig. 2. Abrasion elongation curve of toothed chain center distance

of wear. The toothed chain does not seem to be worn, but it is just an artifact of wear. The components in the system are being repeatedly affected by the alternating stress, the contact surface is undergoing plastic deformation, fatigue cracks are gradually produced, and the wear debris has not fallen off.

5 Conclusions

The chain drive is a flexible transmission mechanism, in which the chain is connected to the drive through a gear to complete the meshing transmission. It is one of the important mechanical transmission systems. Its main advantages are small transmission error and low transmission loss, which is suitable for long-distance transmission. In the actual application of the timing system, it has special advantages such as the same life, high durability, and good wear resistance. It can adapt to constantly changing high-speed engines and engine operating conditions in the timing system. Today, elevated camshafts are well lubricated in high-speed and large-scale heavy-duty operations. More and more engine timing brake systems are used in our country. Due to the reliable economic durability and low maintenance cost of the gearbox, the gear chain transmission system can not only ensure lower noise and vibration within the allowable vibration range at a higher speed, but also it can also ensure the reliability of the high impact rate fatigue system. Its adoption rate in engine timing systems is getting higher and higher, and it is increasingly favored by various automobile engine manufacturers.

References

1. Fan, C.Y., Shu-Xin, X.U., Wang, H.J., et al.: Wear and limiting condition of three-axial timing chain system. J. Jilin Univ. (Eng. Technol. Ed.) **47**(1), 122–128 (2017)
2. Nayak, N., Rane, S., Anarghya, A., et al.: Wear study and EHD lubrication analysis on connecting rod big end bearings of off-highway application engine. Lubr. Sci. **32**(5), 218–229 (2020)
3. Becker, A., Meffert, D., Sauer, B.: Friction and wear investigations on single chain joints. Forsch. Ingenieurwes. **83**(1), 53–63 (2019). https://doi.org/10.1007/s10010-019-00297-x
4. Li, Q.,Guo, J., Hu, J., et al.: Design and test analysis of bionic subsoiler tip. Paper Asia (7), 42–46 (2018)
5. Wang, M., Ge, Q., Jiang, H., Yao, G.: Wear fault diagnosis of aeroengines based on broad learning system and ensemble learning. Energies **12**(24), 4750 (2019)
6. Dong, C., Meng, F., Feng, Z., et al.: 58799 Dynamic simulation and analysis of automotive engine's timing silent chain system (multibody system analysis). Opt. Eng. **45**(10), 514–518 (2017)
7. Kováiková, P., Dubec, A., Kotialiková, D., et al.: Examination of surface wear on the timing chain tensioner depending on the engine oil contamination. Manuf. Technol. **20**(4), 463–467 (2020)
8. Jung, T., Park, Y., Kim, Y.J., Cho, C.: Integrated approach to an optimal automotive timing chain system design. Int. J. Automot. Technol. **18**(6), 1037–1045 (2017). https://doi.org/10.1007/s12239-017-0101-x
9. Kumar, J.S., Giridharan, S., Kannan, T., et al.: Study and overview about process failure mode and effects analysis on automotive timing chain. Int. J. Autom. Mech. Eng. **2**(5), 17–21 (2019)
10. Lee, S., Son, C., Kim, S.-Y., Lee, C.-H.: Timing system development for RAON operation. J. Korean Phys. Soc. **76**(7), 551–559 (2020). https://doi.org/10.3938/jkps.76.551
11. Bazhenov, Y.V., Kalenov, V.P.: The forecasting of the residual resource of electronic engine control system. Russ. Automob. Highw. Ind. J. **2**(54), 52–59 (2017)
12. Nam, K.H., Kim, Y.J., Ye, S.M., et al.: Salinity affects metabolomic profiles of different trophic levels in a food chain. Sci. Total Environ. **s 599–600**(dec.1), 198–206 (2017)

Xuexin Communication APP Based on Blockchain Technology

Fengxiang Xue[1], Hongli Tian[2], Jiawei Luo[3](✉), and Ali Rizwan Jalali[4]

[1] Academic Affairs Office, Guilin Vocational College of Information Engineering, Guilin, Guangxi, China
[2] The Electrical Engineering, Dalian University of Science and Technology, Dalian, Liaoning, China
[3] School of Digital Technology, Dalian University of Science and Technology, Dalian, Liaoning, China
glxxgcxyxfx@sina.com
[4] American University of Afghanistan, Kabul, Afghanistan

Abstract. Blockchain application scenarios have gradually expanded from the initial digital currency to file storage and other fields. The purpose of this paper is to study the design and implementation of ICT APP based on blockchain technology research. The insufficiency of current student educational information systems is elaborated, and the importance of applying blockchain technology to educational information around the world is expounded. Introduced the concept, principle, characteristics and types of blockchain, conducted a simple feasibility study on the design of Xuexintong APP, and then introduced various requirements of the APP in detail, including functional and non-functional requirements, and gave The use case is presented, the user diagram of the system is given, and the system overview is described in detail at the end. This paper introduces the development and testing environment and construction process of the APP, and evaluates the design results of the platform. The index scores are all above 70 points.

Keywords: Blockchain Technology · Xuexintong APP · APP Design · Educational Information

1 Introduction

In the age of cybernetics and globalization, people lack traditional understanding of university reform. Students' technical ability and practice strategy are relatively lagging behind in student behavior and activity certification [1]. These problems will be effectively solved through cluster chain technology characterized by fragmentation and lack of trust. Some scholars proposed the specific application scope and method of zoning technology in education informatization [2]. The industrial chain has practical application in the field of foreign education informatization [3].

The contradiction between online learning and the central management mechanism is becoming more and more obvious, and there are many deficiencies in the actual implementation of domestic decentralized blockchain platforms [4]. Guustaaf E analyzes the

J. H. Abawajy et al. (Eds.): ICATCI 2022, LNDECT 170, pp. 584–591, 2023.
https://doi.org/10.1007/978-3-031-29097-8_69

blockchain capabilities currently in use and the services that will be provided by existing educational projects to leverage blockchain capabilities to improve technology implementation in the educational field [5]. Jagadhesan B discusses leveraging the different components of the blockchain era framework in the context of gadget learning. Areas of machine learning applications where blockchain implementations are beneficial have also been explored [6]. Compared with the pursuit and emphasis on blockchain technology, the practical educational application of information and the development of blockchain technology have not received enough attention [7].

This paper focuses on the development of the blockchain platform, combining theory and practice, provides development cases and practical ideas for the application of blockchain in the field of educational informatization, and explores new ways and methods of human-computer integration. Interactions in the build process System The structure of the system. Starting from the actual needs, through the framework design, model design, communication network design, code application and other computer science research methods, blockchain-based learning and information communication applications are realized. The application of this research is developed based on the browser software system development model. It is an innovation of application mode in the field of blockchain technology, and a pioneering combined application of teaching management, front-end technology and educational software development.

2 Research on the Design and Implementation of Xuexin Communication APP Based on Blockchain Technology

2.1 Blockchain Technology

(1) P2P network

P2P network, namely peer-to-peer network, is a client/server to server (C/S) information exchange system [8]. In the C/S system, the data is stored on the server, and the user receives the data needed to access the server, which has the advantages of simple data consistency management and flexible system management [9]. But in this case, the number of servers is limited, which can easily lead to the failure of the meeting, and the number of clients connecting to the server is also very limited, resulting in poor scale [10]. Each node can provide and receive services for itself in order to use resources purposefully [11]. Unlike traditional distributed systems, blockchain architecture cannot improve performance by adding segments in useful application scenarios, but requires maintaining a glossy network and large storage capacity in the assembly area [12].

(2) Asymmetric encryption

The data exchange process using an encryption algorithm is: the user creates a key pair and introduces everyone to an external key. If user B needs to send confidential information to A, he replaces the information with the public key A and sends it to A. After user A receives the message, he encrypts the confidential information with his private key. When A responds to B, it encrypts the data with B's public key using the same principle, and when B receives the message, B decrypts it with its own private key and displays the information.

2.2 APP Demand Analysis

Considering the APP communication requirements and user functional requirements, the corresponding requirements design strategy is specified according to the system target function. The system and users have their own independent operation functions, and there are functional intersections between them. Some functions need to be triggered by user operations. Therefore, the system needs to be fully supported.

(1) User-related resources: basic information of the user, whether the node is a new node, whether the block storage meets the security requirements, whether the user node is legal and other logical information, the current network status, etc.;
(2) The processing flow of the consensus algorithm, and the evaluation standard of random number difficulty difference;
(3) Resource management during network runtime: dynamic script running mechanism, static dependency file transfer mechanism, management of large static files, channel link management, etc.;
(4) Form content: learning record information generated by the user, whether there is suspicious operation, system interaction error log and other information.

3 Investigation and Research on the Design and Implementation of Xuexintong APP Based on Blockchain Technology

3.1 Overall Architecture Design

The overall architecture design of APP mainly refers to the detailed analysis and design of the content and the technology involved in each part of the APP, which is an indispensable link in the APP development process. The overall architecture of APP is divided into five layers, from the top to the bottom, they are the presentation layer, the business layer, the interface layer, the chain code layer and the storage layer.

Presentation layer: As the entrance of the student education information system, the front-end pages of this system are mainly implemented through languages such as HTML, CSS and JavaScript, as well as the Vue framework, aiming to bring a good visual experience to users.

Business layer: The business layer is the core layer in the overall architecture of the blockchain-based student education information system, and is mainly responsible for the logical realization of various business functions of the entire system. The business functions of this layer mainly include user information management, on-chain application management, application review management, on-chain certificate management and on-chain certificate query, which are implemented through the Java web service built by the back-end framework SpringBoot.

Interface layer: The interface layer is the hub between the business layer, the chain code layer and the storage layer, and is mainly responsible for data interaction with the above three layers. The interface layer includes the calling chain code interface, the IPFS system interaction interface and the MySQL database interaction interface.

Chain code layer: This system designs the basic information chain code of student users, the basic information chain code of college users, the basic information chain code of enterprise users, the chain code of degree information, the chain code of academic achievement information, the chain code of award and honor information, and the chain code of academic achievement information. Code, English level information chain code, bad record information chain code and work experience information chain code.

Storage layer: The storage layer is the bottom layer of the entire system, and is mainly responsible for persisting student education information. The storage layer consists of three parts, namely IPFS system, MySQL database and blockchain. The MySQL database is used in the on-chain application stage, and is responsible for temporarily storing some educational information data, such as award and honor information, academic achievement information and English proficiency information. The IPFS system is mainly used to store file data with large memory. Content-based addressing makes IPFS well integrated with blockchain technology.

3.2 Data Storage Design

According to the characteristics and storage requirements of student education information data, the system stores different student education information data in the Hyperledger Fabric account book, IPFS system and MySQL database, among which MySQL is only used as a temporary database, and the final destination storage of student education information data The location is the Fabric ledger and IPFS system.

3.3 Evaluation Scheme Design

Usability testing is a process of evaluating products through a series of methods, which can find problems in product design and propose improvement directions, so as to achieve usability goals in a targeted manner. Usability testing is a common method for user-centered product evaluation. This paper selects usability testing to evaluate the design scheme of smart education information platform. The main evaluators are platform builders (designers, technicians, etc.) and a total of 20 recruited target users. This questionnaire mainly refers to the system usability scale, according to the usability evaluation indicators selected in this paper, and based on six dimensions of functionality, interactivity, ease of learning, fault tolerance, efficiency and satisfaction, and following the usability principle, the questionnaire design is carried out.

This paper uses SPSS 22.0 software to count the results. The t-test formula as follows:

$$t = \frac{\overline{X} - \mu}{\frac{\sigma X}{\sqrt{n}}} \tag{1}$$

$$t = \frac{\overline{X_1} - \overline{X_2}}{\sqrt{\frac{(n_1-1)S_1^2+(n_2-1)S_2^2}{n_1+n_2-2}\left(\frac{1}{n_1} + \frac{1}{n_2}\right)}} \tag{2}$$

Among them, formula (1) is a single population test, and n is the number of samples. Formula (2) is a dual population test, and n1 and n2 are the sample sizes.

4 Analysis and Research on the Design and Implementation of Xuexintong APP Based on Blockchain Technology

4.1 Function Design and Implementation of Xuexintong APP

The Personal Center page mainly contains user account information, including images, personal signatures, my posts, questions, and settings, as shown in Fig. 1. Click on the title to change personal information, including username, personal information, location, etc. Feedback on questions mainly includes unfinished features or other features set during the use of the Platform. These features can be provided as answers to questions. These settings are mainly used to change basic personal account information, such as changing the password, canceling and updating the application version.

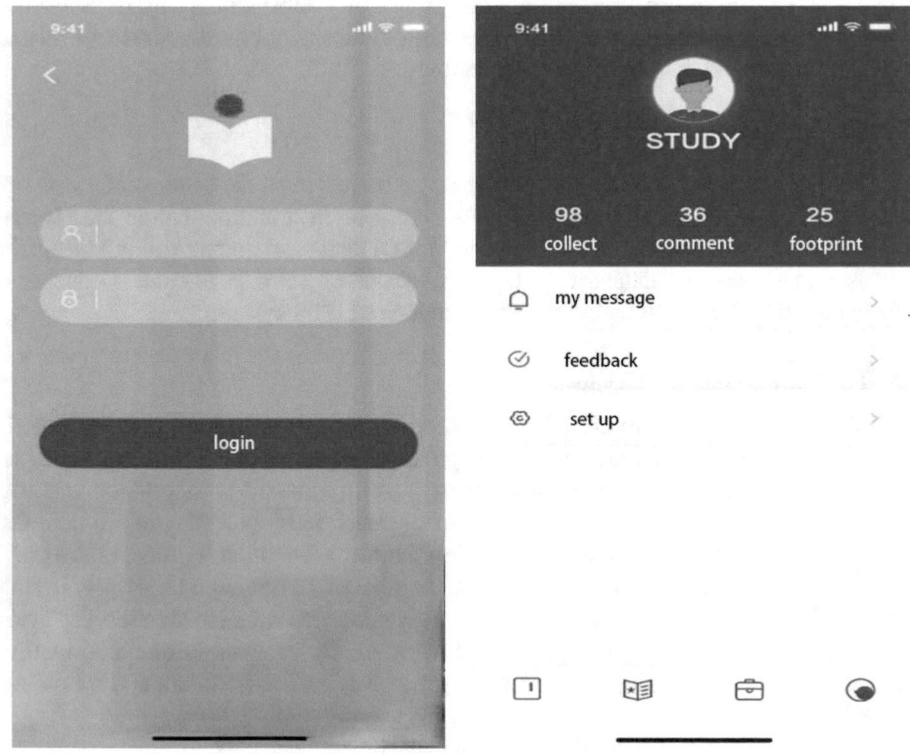

Fig. 1. Personal center interface design diagram

The application review management module includes on-chain application review management of award and honor information, academic achievement information and English proficiency information, involving application approval, application rejection, application list viewing and application details viewing, etc. It is mainly used for university educational affairs. The staff will review the on-chain application submitted by the student. The college clerk first accesses the award and honor information page through the college port, and obtains the application list of award and honor information queried from the database. The application list contains the application list of all the students' award and honor information on the chain. Click on any application in the list to view the detailed content of the application and verify the authenticity of the content.

4.2 Evaluation of Platform Design Effects

The functional index scores are all above 80 points, which is usable, as shown in Fig. 2. Among them, the functional practicability index has the highest score, indicating that the functional design of the smart education information platform can solve practical problems for users, as shown in Table 1. The function comprehensiveness index is the second, indicating that the functional design can meet the user's basic needs, the functional design is not complicated, and the user can operate the product without thinking for a long time, and it has a certain understandability.

Table 1. Indicator Scores

Indicator name	Indicator score	Availability score
Functionality	75	70
Functional compatibility	73	70
functional complexity	72	70
Functionality	76	70
Reasonable color matching	73	65
Reasonable font size	70	65
Willingness to share	72	65
Willingness to continue to use	71	65

Satisfaction index scores are all above 70 points, indicating that a higher evaluation is given on the satisfaction index. In general, blue and white are selected as the keynote, the interface colors are unified and concise, and the layout is more reasonable, which reduces the visual burden of users and provides users with a more comfortable visual experience. Among them, the reasonable score of font size is lower than other indicators. It can continue to optimize and improve the font line spacing, thickness, etc., combined with the overall layout, to further increase the readability of the interface.

The ease of operation and interface similarity scores are within the usability range, indicating that the prototype design of the platform is more in line with the original cognition of the public. The ease-of-use index score is within the critical value range, indicating that the user experience is average in terms of ease of use, and the consideration of ease of use in the design of the scheme is not complete. The operation process should be further simplified, the unnecessary operation buttons should be deleted, the ease of use and the core competitiveness of the product should be improved while ensuring that the normal functional requirements are met.

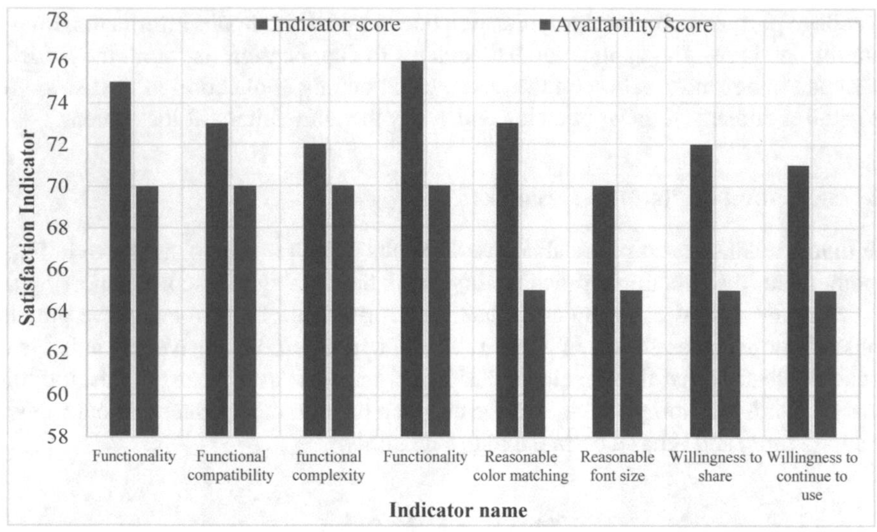

Fig. 2. Metric scores

The three fault tolerance indexes are all within the critical range, indicating that the performance of fault tolerance design is insufficient. You must add a valid input form for the user action. Effective tips and suggestions help reduce the possibility of user error. The form of restricted operation is very poor, and the user's behavior is limited by some obstructive operation requirements to reduce the probability of error. In general, fault-tolerant design reduces the user experience, so fault-tolerant design should be improved to improve availability.

5 Conclusions

Mobile learning is popular because it is not limited by time and place. With the popularization of smart phones, mobile terminals have gradually become mainstream learning tools from the previous auxiliary learning tools, and even replaced computers. This paper attempts to apply decentralized technologies such as blockchain and IPFS to the field of Xuexintong APP to solve some problems in the sharing of educational information in colleges and universities. By uploading and sharing digital educational information, college students can also obtain educational information from other colleges and universities, so that college students can choose other educational information to supplement themselves in their spare time while completing the basic courses in the classroom. It is of great theoretical and practical significance to respond to the country's innovation-driven development strategy.

References

1. Kim, D.K.: Design of lab framework for effective blockchain education. J. Ind. Converg. **18**(6), 147–154 (2020)
2. Guo, J., Li, C., Zhang, G., Sun, Y., Bie, R.: Blockchain-enabled digital rights management for multimedia resources of online education. Multimedia Tools Appl. **79**(15–16), 9735–9755 (2019). https://doi.org/10.1007/s11042-019-08059-1
3. Pence, H.E.: Blockchain: will better data security change chemical education. J. Chem. Educ. **97**(7), 1815–1818 (2019)
4. Vidal, F.R., Gouveia, F., Soares, C.: Blockchain application in higher education diploma management and results analysis. Adv. Sci. Technol. Eng. Syst. J. **5**(6), 871–882 (2020)
5. Guustaaf, E., Rahardja, U., Aini, Q., et al.: Blockchain-based education project. Aptisi Trans. Manag. (ATM) **5**(1), 46–61 (2021)
6. Jagadhesan, B., Saravanan, M.P., Saravanan, M.J.: An impact of blockchain technology on digital education in machine learning. High Technol. Lett. **27**(3), 350–361 (2021)
7. Melnyk, Y.B., Pypenko, I.S.: How will blockchain technology change education future?! Int. J. Sci. Ann. **3**(1), 5–6 (2020)
8. Rivera-Vargas, P., Lindin, C.: Blockchain in education. The search for security: digital world and technological determinism. Rev. de la Educac. Super. (RESU) **49**(194), 159–162 (2020)
9. Arndt, T., Guercio, A.: Blockchain-based transcripts for mobile higher-education. Int. J. Inf. Educ. Technol. **10**(2), 84–89 (2020)
10. Li, T., Duan, B., Liu, D., et al.: Design of outcome-based education blockchain. Int. J. Perform. Eng. **14**(10), 2403–2413 (2018)
11. Lee, D., Kim, S., Park, N.: The blockchain-based online learning platform for the untact education environment in the post-COVID-19 era. J. Korean Inst. Inf. Technol. **18**(11), 109–121 (2020)
12. Bálint, K., Cvetkovi, D., Takács, M., et al.: Connecting bitcoin blockchain with digital learning chain structure in education. Acta Polytech. Hung. **16**(1), 2019–2077 (2019)

Investigate the Application of Virtual Reality Technology into the Expo Building Exhibition Space

Yi Fu and Zihan Zhou(✉)

School of Design and Art, Shenyang Jianzhu University, Shenyang, Liaoning, China
2874737114@qq.com

Abstract. Due to the rapid development of modern society, people's pursuit of life quality gradually improved, the demand for living environment increases, In the continuous progress of science and technology, advanced computer technology gradually becomes mature, also slowly penetrated into people's lives, and virtual reality technology (VR) is as a kind of computer technology. Apply virtual reality technology in expo building exhibition space, not only can be intuitive observation and understanding of design in designing, increase the quality and the workpiece ratio of design work, and can be in the process of viewing expo building exhibition space to realize human-computer interaction, increase interest, and rely on the update of computer technology, expo building exhibition space performance from the science and technology, provides more possibilities for the space. This paper mainly studies the specific practical application of virtual reality technology in the exhibition space and its application characteristics, combined with the design of the exhibition exhibition space and the related theory of computer technology, so as preferable accelerate the growing of virtual reality technology.

Keywords: Computer Technology · Virtual Reality Technology · Expo Building Exhibition Space · Practical Application

1 The Concept of VR Technology

Virtual reality technology refers to the comprehensive utilization of computer graphics systems and various reality and control interface devices [1]. For this technology, its main role is to build relevant scene models based on the real world, and can also be used to conduct 3D model in the virtual world. Virtual reality technology is an artificial simulation environment under the support of computer technology [2, 3]. It is a computer technology related to space, the integration of science and technology can be boring two-dimensional image into a more vivid form of three-dimensional space, and show in front of people, make people can have immersive experience, more practical details of the design, so the use of virtual reality technique can open up a new world shaping the environment.

J. H. Abawajy et al. (Eds.): ICATCI 2022, LNDECT 170, pp. 592–599, 2023.
https://doi.org/10.1007/978-3-031-29097-8_70

In today's era of rapid social and economic development, people's pursuit of the quality of life has also gradually improved, and the demand for the living environment has also gradually increased. Although designers can improve their design level and skills in the field of design, they cannot lack the help of modern science and technology. Indoor environmental quality (IEQ) and its resident's well-being is an significance area that affects the comfort level. The quality of the indoor environment must be evaluated in the context of the impact on the overall performance of the building and the overall indoor environment [4]. And VR technology as a new practical computer technology, it is the application of expo building exhibition space can not only follow the pace of tight information age, let people can watch and experience the human design details, also can satisfy the children visitors more interesting way to obtain knowledge, make people can more intuitive understand the originality of exhibition building exhibition space design (Fig. 1).

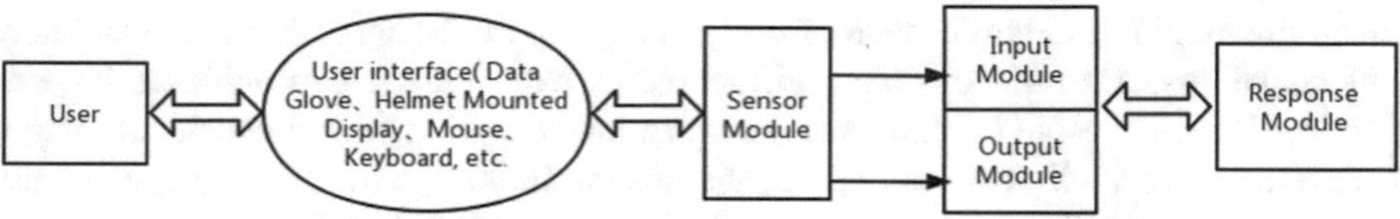

Fig. 1. Composition of a VR system

There is no doubt that the interest in VR technologies is growing and VR technologies are being implemented in new ways across myriad domains [5]. It can not only convert two-dimensional images into virtual three-dimensional space, but also combine the virtual space with the real environment, and use computer technology to make the combination of the two more realistic, and can make people immersive. In both 2D and 3D visualisations presented, both interaction and non-interaction types were used [6].VR technology by using the basic data provided by the user, and upload these data to the database, according to the database data, these data and technology can provide sensitive sensory information for hearing, vision and touch, and then convey these sensory information to people, so that people can interact with the virtual three-dimensional space and produce sensory cognition. VR technology is developed on the basis of computer technology and human-computer interaction technology progress, and has a broad application space and scene advantages. In this context, VR technology should be able to provide various possibilities for conventional museum exhibits [7]. Because the exhibition building exhibition space has a long history, visitors will inevitably encounter obstacles to understanding, and through the application of this technology, not only greatly shorten the understanding of the gap between visitors and the exhibits, but also for the interaction in the display space to play. Virtual reality technology has also been paid attention and recognized by the society, which makes its prospects even broader.

2 The VR Technology Application of the Expo Building Exhibition Space

2.1 The Functionality Presented of VR Technology in the Application of the Expo Building Exhibition Space

VR has been proven successful in achieving learning outcomes in other contexts due to its capacity to provide innovative, immersive and safe learning experiences [8]. The application process of virtual reality technology in the exhibition space is different from the traditional design techniques. It can not only transform the two-dimensional images into a virtual three-dimensional space, but also present all the images and objects in the three-dimensional space, and provides a new way for communication and interaction. Its application characteristics are mainly manifested in the following aspects:

First of all, the powerful functions of virtual reality technology can analyze the complex expo building display space models, Expo exhibition of building space visualization virtual in the initial stage of design can help the designer to more accurately understand the design scheme; next, Virtual reality techniques can be achieved by considering dynamic design thinking and transforming designs in two-dimensional images to designs on three-dimensional space, the design thinking will be more specifically expressed in the display space. So as to help designers to the overall scheme for a more comprehensive and accurate deliberation and plan; Virtual reality technology can combine and integrate all the relevant information, images, text, audio, and other materials displayed in the showcase space, Forming an independent and comprehensive systematic and comprehensive database of the expo building exhibition space, Easy to use in the design process. You can also change the size, shape, or other aspects of virtual world elements by changing the data in the database [9] (Fig. 2).

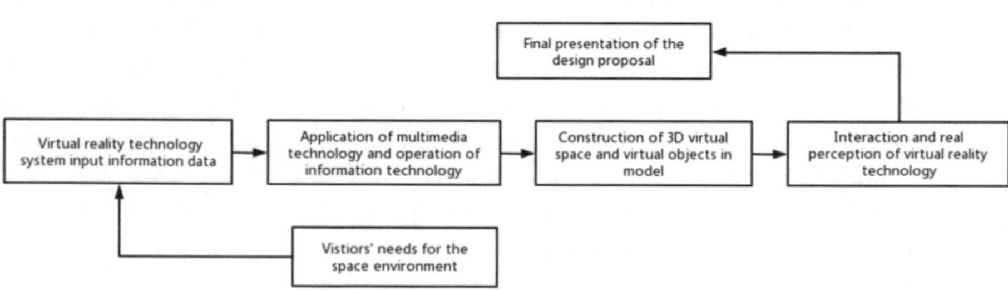

Fig. 2. The course of VR technology actualized in the design

2.2 Performance Method of VR Technology Appliance

In the design of the expo building exhibition space, VR technology generally constructs and the construction of color and light elements to carry out the virtual expression on the two-dimensional images. In the expo building display space design, in order to be able to convey effective information to visitors, effective expression is extra important, and VR technology appliance on three-dimensional virtual space, a good display of the exhibits and the structure of space and modeling, etc., create the atmosphere of the space at the same time, also bring visitors psychological comfort. The creation of color and light and other elements is also very important, with the plane effect expressed through virtual reality technology. According to different colors and light atmosphere according to different space, the visitor's feelings will also change, in their environment, and its specific technical operation is determined according to the specific space environment changes.

Through raytracing, images are visually generated to show people a complete and realistic image. Assuming that people project light from the viewpoint V to the scene through the screen pixel e, and the light intersects the scene in the scene at P1, P2..., and so on, the P1 closest to the viewpoint is the visible point of the picture in the pixel point e, so the brightness of the pixel e is determined by the brightness of the light from the P1 point to the P1V direction. Thus, assuming that I is the luminosity at the visible point P, IC is the local illumination brightness, tsIs is the environmental mirror reflection brightness, and ttIt is the regular transmission brightness: (Fig. 3).

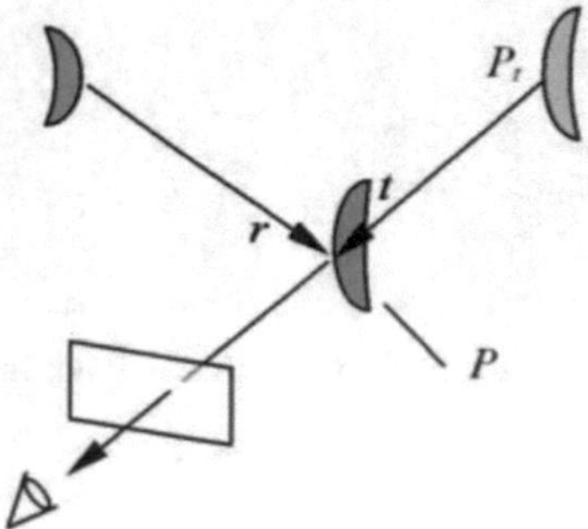

Fig. 3. Ray-tracing technique

$$I = I_C + t_s I_s + t_t I_t$$

2.3 Spatial Characteristics of VR Technology Appliance

VR technology appliance can feel the authenticity in the design process, innovation and conception: Firstly, VR technology appliance can increase the sense of reality, which created by computer technology virtual environment and actual environment with equivalent relationship, make people can with the aid of science and technology equipment practical experience virtual environment, and let people immersive, feel two-dimensional image cannot experience the experience. Secondly, the application of VR technology can bring an interactive experience to people in the virtual environment, and through some forms of science and technology, people can interact with images or objects in the virtual world. This allows a fully immersion in the observed virtual environment and the interaction with its graphic objects and systems without any restriction [10].

When arranging the exhibition, through these human-computer exhibitions that can carry out interaction, visitors can not only experience with the real experience, but also get the feedback from visitors. In environmental design, the appliance of VR technology can be realized in the process. Conceived also known as creativity, technology creates virtual space or virtual scene, can copy the real space, can be on the basis of real space further design and creation, VR technology in the exhibition building display space design, can in visitors easy to accept knowledge and information, can also through imagination and completely virtual space to create creativity, create a better learning atmosphere (Fig. 4).

Fig. 4. Interact with the expo building exhibition space through VR technology

3 Concrete Appliance Process of VR Technology in the Expo Building Exhibition Space

3.1 Use VR Technology to Remedy the Defects in the Design

The building exhibition space will inevitably be restricted by the actual conditions in the expo building in the design process, such as the construction site factors, weather conditions and financial factors, and these factors will become a stumbling block to the implementation in the design process. The appliance of this technology can not only ignore these external factors, but also quickly implement the design, and make out virtual construction for the space and scenes needed in the display space design, so as to make up for some defects in the design process.

3.2 Analysis to Design Feasibility with VR Technology

The application of this technology breaks the limitations of the traditional expo building exhibition space design. In the design scheme comparison, the application of technology can not only cross all the constraints of the actual space, but also quickly switch the design scheme. In addition to the analysis of different design schemes, it can also facilitate the convenience to adjust the deficiencies in the design scheme in time. In terms of design details, designers can use the 3D model to feel the difference between the building appearance and internal structure from the same perspective, so as to compare their own advantages and disadvantages more efficiently and faster, and choose the best design scheme.

Comparing the design scheme with the expo building space design, the application of technology can not only exceed all the limitations in the real space, but also can quickly change the design scheme. In addition to the analysis of different design schemes, it is easy to correct defects in time.

3.3 Display the Design Model Using VR Technique

VR technique can be used in animation modeling, but in the practical application process, a large number of human-computer interaction and animation character imitation technology are added to make the animation characters more flexible [11]. Due to the appliance of virtual reality technique in expo building exhibition space design, through it to make full use of modern technology to create a 3D model, can be in human-computer interaction shows different space design details, as a new era relative to the traditional space display alternative application, let visitors feel not exist objects present at present, let people can better perceive and experience the design originality, also greatly reduce the use of space and property spending. The feeling of being at the center of a real virtual reality environment creates a self-related feeling [12]. Adjustment of the surrounding conditions in an immersive virtual environment allows users to evaluate their designs under more realistic conditions. In addition, VRML LOD (multi-detail level) node can be used for multi-level detailed modelling to achieve dynamic model scheduling management [13].

In addition, designers can think about the full range of information displayed by the 3D model built by VR technology, and timely analyze and modify the design, so as to maximize the design efficiency (Fig. 5).

Fig. 5. The application and presentation of VR technology in Design

4 The Impact of VR Technique on the Exhibition Space of the Expo Buildings

With the rapid progress of VR technique in today's market, it has had a significant impact. Its technology can not only make up for many deficiencies in the design process, but also break the boundary of the rigid spatial structure of the display space design of traditional exhibition buildings. Expo building exhibition space with time is constantly follow up update, its space also follow the update, space transformation more flexible, the demand for characteristic space is increasing, computer technology for the expo building display space is not only in terms of technology, but also promote the designer for the form of space, only, color, light, material and even design grasp of design details.

VR technology can also be used to test the feasibility and rationality of the design results. In the process of expo building exhibition space design results acceptance, the intervention of virtual reality technology enables the designer to fully perceive and observe the design results, and timely for the design error to modify and adjust, reduce the design mistakes, can also be adjust in the virtual space, reduce the cost of the small mistakes in the design. If a design error occurs again during the modification process, rescreen and modify it until all problems can be resolved. Therefore, VR technology can not only improve the efficiency and quality of the design, but also test and optimize the scheme. Besides, the use of virtual reality requires habituation or exercises to be able to explore comfortably and following its objectives [14].

The integration of this technology application improves the spatial sensing leakage of traditional design in the design process, so that people can not only perceive the reality of virtual space, but also bring people an immersive experience of rich three-dimensional space.

5 Conclusions

The progress of digital technology promotes the progress of social development, also promotes the diversification of social space design, virtual reality technology as a comprehensive information technology, let people can perceive and immersed in virtualization space, because the technology has authenticity, innovation, interactive characteristics, so it can open the fields of design research and creation of the new world, will also become an indispensable design tool, help shape the environment, promote the development of design.

For the past few years, virtual reality technology is progressing rapidly, more traditional design techniques, the application of information skill can more break the constraints brought by conventional ideas, can more comprehensive use of design and expression, bring us more free expression space, so the transmission channel quality will not have a greater impact, therefore, the design information generation and transmission quality is not affected, virtual reality technology will gradually become a popular in many industries, and a dominant information technology.

Virtual reality technology is still in the initial stage, but with the progress of society, virtual reality technique will constantly develop and improve, gradually form a mature information technology, there will be more and more people for the application of the technology in different fields of research and deep, and in the combination of expo building display space design is more and more popular.

References

1. Zan, M., Tang, W.: Design and implementation of 3D visualization intelligent home decoration system based on VR technology. Comput. Sci. Appl. **11**(04), 814–820 (2021)
2. Wooldridge, M.: Agent-based computing. Interoper. Commun. Netw. **1**(1), 71–97 (1998)
3. Hagras, H., Doctor, F., Callaghan, V., Lopez, A.: An incremental adaptive life long learning approach for type-2 fuzzy embedded agent in ambient intelligent environments. IEEE Trans. Fuzzy Syst. **15**(1), 41–55 (2007)
4. Shan, N.: Research on indoor environment design based on VR technology and wireless sensor network. Microprocess. Microsyst. **83**, 103999 (2021)
5. Ghobadi, M., Sepasgoza, S.M.E.: An investigation of virtual reality technology adoption in the construction industry. Smart Cities Constr. Technol. (2020)
6. Zammit, D.: Leveraging virtual reality technology to effectively explore 3D graphs. M.Sc. Computer Games Technology, August 2020
7. Hirose, M.: Virtual reality technology and museum exhibit. Int. J. Virtual Reality **5**(2), 31–36 (2006)
8. Akdere, M., Acheson, K., Jiang, Y.: An examination of the effectiveness of virtual reality technology for intercultural competence development. Int. J. Intercult. Relat. **82**, 109–120 (2021)
9. Li, F., Cao, Y.: Indoor space art design based on immersive VR and embedded system. Microprocess. Microsyst. **83**, 104001 (2021)
10. Fahmi, B.: Industrial case studies for digital transformation of engineering processes using the virtual reality technology. Procedia CIRP **90**, 636–641 (2020)
11. Sun, Q., Tsai, S.-B., Huang, C.: An empirical study on application of virtual reality animation technology by big data model. Math. Probl. Eng. **2021**, 1–9 (2021)
12. Schöne, B., Kisker, J., Sylvester, R.S., Radtke, E.L., Gruber, T.: Library for universal virtual reality experiments (luVRe): a standardized immersive 3D/360° picture and video database for VR based research. Curr. Psychol. 1–19 (2021)
13. Phommaly, K., Yu, L.: Research of virtual reality technology in home design. J. Phys.: Conf. Ser. **1624**, 062017 (2020). 2nd International Conference on Computer Modeling, Simulation and Algorithm
14. Wizaka, W., Suharjanto, G., Wangidjaja, W.: The new teaching method using virtual reality technology in building technology subject. Earth Environ. Sci. **426**, 012080 (2020)

Construction of College Students Online Education Ecological Chain Based on Block Chain Technology

Bin Ye[1(✉)], Jieping Liu[1], Jingya Xu[1], and Anli Teekaraman[2]

[1] Chengdu Neusoft University, Chengdu 611844, Sichuan, China
1253958330@qq.com
[2] Vrije Universiteit Brussel, Brussels, Belgium

Abstract. This paper discusses the current situation and trend of online teaching in Colleges and enterprise network education, the challenges of online education quality in colleges and universities, the relationship between "five new" and university entities, and the problems existing in the current education evaluation under the background of epidemic. Combined with the current application scenarios of blockchain technology, the necessity of building an online education ecological chain based on blockchain technology is discussed, and an online education evaluation system based on blockchain technology is constructed, as well as the solutions to the problems existing in the construction mechanism by applying blockchain technology. This paper introduces the online education ecological chain model based on block chain and its operation principle.

Keywords: Blockchain · Universities · Online Education · Ecological Chain

1 Introduction

In January 2021, the research Report on China's Online Education Industry in 2020 released by IResearch Data Company showed that the market size of online education industry in 2020 increased by 35.5% year-on-year to 257.3 billion yuan, with the overall online rate approaching 25%. Both universities and online education-related enterprises, as the proportion of online education is gradually increasing, it will become the normal state of college education and form a new form of education by combining it with traditional offline education [1].

2 The Revolution of Online Education in Universities

Fitst, online learning awareness enhancement; According to China's Ministry of Education, according to official survey data of students' learning autonomy and interaction between teachers and students satisfaction is high, even more than the traditional face-to-face classroom teaching, students after 90, 00 after referred to as "indigenous" online,

J. H. Abawajy et al. (Eds.): ICATCI 2022, LNDECT 170, pp. 600–608, 2023.
https://doi.org/10.1007/978-3-031-29097-8_71

they "Internet +" new era "smart +" study the new characteristics of love, easy to accept, easy to adapt [2].

Second, the willingness of school informatization construction is further improved; Due to the uncertainty of the epidemic, colleges and universities are forced to carry out teaching work in the form of online, and information construction is the foundation of online teaching, including network hardware equipment construction, software platform construction, teaching resources construction, etc. Before the outbreak of the epidemic, online teaching had been promoted to a certain extent in colleges, such as the construction of various high-quality resource sharing courses and open courses.

Third, the informatization construction level of colleges and universities is further improved; With the launch of large-scale online courses, Colleges participate in online education. Through practical exploration and experience summary, colleges and universities will gain more and more experience in online education, and lay a good foundation for the overall informatization construction level, and the informatization construction level will be further improved.

Fouth, The demand for online teaching quality monitoring is further improved;

Before the outbreak of the epidemic, colleges and universities adopted the traditional field teaching method, and online teaching was mainly assisted by MOOCs. According to data from the Higher Education Department of the Ministry of Education of China, the number of online MOOCs in China by 2020 was about 12,500, meaning that before the epidemic, the number of online teaching was less than 12,500. By April 2020, the number of online courses has reached 942,000, a sharp increase of 75 times.

With the sharp increase in the number of online teaching participants and courses, the quality control of online teaching is facing greater pressure, and it is facing serious challenges in terms of manpower, funds and technical means.

Fifth, online teaching is both an opportunity and a challenge for colleges and universities; Under the premise that globalization has been going on for many years, the control of the epidemic is no longer a matter that can be decided by any one country. As countries around the world cannot reach a consensus on the action of the epidemic, it is likely that the epidemic will last for a long time and have a lasting impact on education. Because of its persistence, we will have to think seriously about all kinds of problems in online teaching, and learn from the existing experience, improve and perfect, in order to adapt to the new form of college education, which is both an opportunity and a challenge for colleges and universities.

Sixth, the subjects of education and the new environment should be integrated as soon as possible; As for the quality of online teaching, the Ministry of Education has set a goal of "ensuring that the quality of online learning is substantially equivalent to that of offline classroom teaching". In the context of the epidemic, it is an important challenge to achieve substantial equivalence in teaching activities in the face of new tools, new platforms, new teaching forms, new teaching methods and new learning methods. Of course, if we can realize the integration of "five new" and the main body of each university, we can also promote teaching well. As shown in Fig. 1 below:

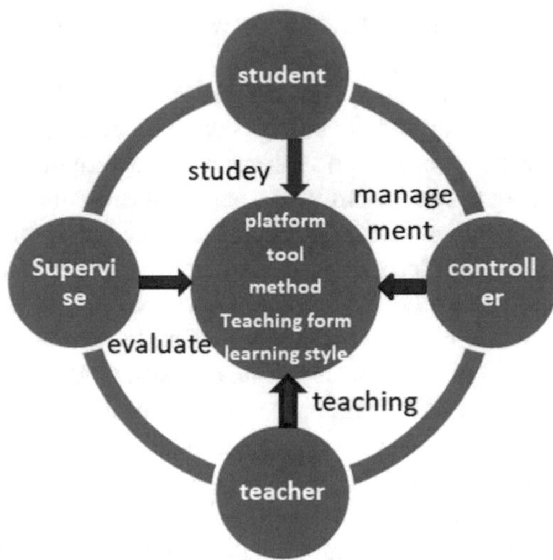

Fig. 1. The integration between "Five New" and the main body of colleges

Subject of colleges and universities through the "five new" as a "bridge" between each other, and achieve their own purposes, such as the supervision evaluation, the management of the managers, students' learning, and teachers' teaching, the teaching activity also because of the "five new", make effect also had the very big promotion, realize "five new" and fusion between the subject of colleges and universities.

3 Application of Blockchain Technology in Online Education

3.1 Key Technologies and Application Scenarios of Blockchain

Blockchain is a technology combining distributed database, data encryption and peer-to-peer network. Due to its high application value in "decentralization", traceability, and data security, imtamability, etc., it has been widely concerned at present.

The blockchain system consists of a data layer, a network layer, a consensus layer, an incentive layer, a contract layer and an application layer [3]. Its core technologies include: consensus mechanism, data storage, encryption algorithm, P2P network protocol, side chain, smart contract. It is because of these core technologies that its technical advantages are obvious, including the immutable data and high data availability, especially the decentralized feature, which makes data management more transparent.

In general, the application of blockchain technology in various industries is gradually becoming mature; In particular, in the field of education, The University of Berkeley has applied blockchain technology in MOOCs to achieve the certification of courses. The "Holburton" school in San Francisco is already using blockchain to store and publish certificates.

3.2 The Technological Advantages and Opportunities of Blockchain Technology in Online Education

First, distributed storage is convenient for information storage. At present, in addition to teaching professional knowledge, teachers also need to record and save all kinds of learning records of students, which are also the records that schools need to keep. And blockchain technology itself also includes distributed storage technology, which provides convenience for data storage and management. On the one hand, teachers can be saved from the onerous data recording and preservation, or even processing activities; On the other hand, due to its data storage and consensus mechanism, the school realizes the tamper-proof of teaching information, making the school administrators get more scientific and effective data.

Secondly, the scientific evaluation mechanism of multi-party consensus is introduced. In education in colleges and universities, colleges and universities of teaching and learning evaluation for many years has been continuously, but its impartiality and accordingly, and the introduction of chain block for multiple evaluation provides a more convenient and effective solution, its characteristic of "decentralization" to students, teachers, supervision, is introduced into the evaluation system of enterprises and other relevant individuals and organizations, To realize "multi-party" and "multi-party" evaluation of teachers, students, schools and enterprises, and form a more complete and scientific evaluation system.

In addition, the realization of authentication information sharing; In the learning stage, students need to accept teachers' assessment and certification to get credits, while many students leave school to participate in vocational qualification certification. As they cannot understand the content and effectiveness of their learning, they need to organize certification exams again. When entering enterprises, they have to face questions from enterprises, or they need to take relevant certification exams again. On the one hand, for students who have truly mastered the knowledge and skills required in all aspects and have relevant qualities, these repetitive certifications or examinations are actually a waste of energy and unnecessary behavior; The parties in the organization, on the other hand, the appraisal of students, there are also some limitations, such as students' status at some stage, or a part of knowledge and ability, not comprehensive, understanding of the student's comprehensive quality, for a long time and waste more inspection cost of manpower and material resources needed for the organization, even unable to really get the people in need. The immutable nature and continuous distributed storage of blockchain make critical information and authentication information permanent and provide resource sharing for all parties.

Finally, the blockchain smart contract mechanism also provides convenience for students' academic early warning and automatic talent selection. On the one hand, early warning trigger conditions can be set for students' learning behaviors through the blockchain smart contract mechanism, which can timely remind students, teachers and schools to further help students complete their learning and achieve the purpose of talent cultivation. On the other hand, smart contract conditions can also be set for some outstanding students or students who meet the needs of enterprises, so that enterprises can quickly get the talents they need.

4 Construction of University Online Education Ecological Chain Based on Block Chain

4.1 The Necessity of Building an Online Teaching Ecological Chain Based on Block Chain

First, the construction needs of the new online education system in colleges and universities;

It is clearly pointed out in the "White Paper on China's Blockchain Technology and Application Development" issued by the Ministry of Industry and Information Technology that "the characteristics of the blockchain system, such as transparency and data cannot be tampered with, are fully applicable to students' credit management, employment, academic, qualification certificate, industry-university cooperation and other aspects, which are of great value to the healthy development of education and employment" [4].

The emergence of new crown outbreak in 2020, when offline education confronted with unprecedented difficulties, the importance of online learning is more apparent, and under the outbreak of normality, besides the function of teaching and learning of basic education, including teaching supervision, learning other early warning, credit certification, qualification certification, companies with the talent demand also gradually manifested in aspects of intelligent matching, The technological advantages of blockchain are also fully reflected, but also more highlights the urgency and necessity of the construction of online education system based on blockchain, big data and other technologies.

Second, the need to establish a credible talent certification system;

A scientific and effective credible talent certification system plays a vital role in the development of higher education, enterprises and the country, and is also the common concern and pursuit of the goal.

First of all, as academic education institutions, colleges and universities have natural requirements and obligations to establish a scientific and credible talent certification system, including credit certification, academic certification, etc.

Third, the reasonable talent certificate system also plays a role of talent incentive, learning in the short term is more purposeful, and will be reflected through credit bank, course certification and other forms;

Finally, it also provides support for enterprise talent strategy; How to evaluate talents efficiently and match the talents needed by enterprises in universities is also an important mission of enterprise talent strategy.

In online education and will be long with offline education appeared at the same time, evaluation on talents, matching of choose and employ persons and enterprises will become more complex, the certification system proposes a bigger challenge, at the same time because of block chain technology of "decentralization" and so on the technical characteristics, makes this challenge into a block chain technology application opportunities, It provides technical support for the establishment of scientific and effective talent certification system.

4.2 Construction Mechanism of Online Education System Based on Block Chain

In the whole construction mechanism, the basic idea is: learning-centered, teaching, management, school-enterprise cooperation to provide external services for learners, while the deficiencies in learning, teaching, management, collaborative education and other processes can be optimized and improved through blockchain technology. The construction mechanism of online education system based on blockchain is shown in the Fig. 2 below:

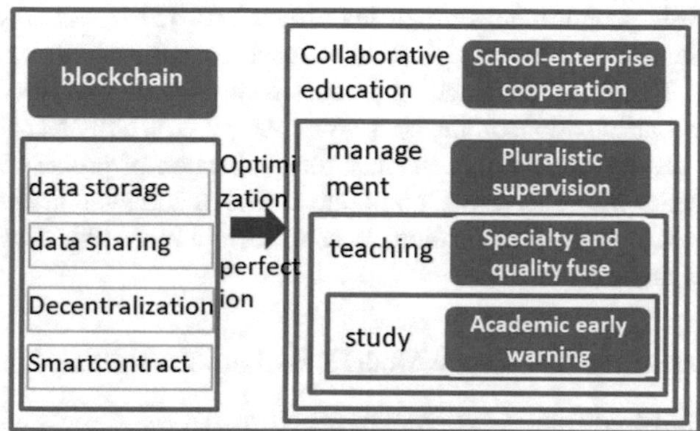

Fig. 2. Construction mechanism of online education system based on block chain

First, School-enterprise cooperation mechanism;School-enterprise cooperation can effectively integrate the talents, technology and resources of enterprises, such as capital and equipment, so that colleges and universities can serve the regional economy and enterprises. At the same time, enterprises can help colleges and universities solve the shortage of funds, hardware equipment and practice places in the process of technological research and development, and ultimately provide support for the cultivation of talents in colleges and universities.

Second, Integration mechanism of professiona and quality education; Integration mechanism of professiona and quality education refers to the integration of students' abilities in all aspects, humanistic qualities, ideological trends and moral cultivation in the process of cultivating students, so as to achieve all-round education and make students fully develop in all aspects.

The talent training mechanism of specialization and element integration combines professional teachers, parents and enterprise mentors who act as "quality teachers" together to cultivate students' all-round ability and improve their comprehensive quality.

Third, Academic early warning mechanism; In the process of talent training, evaluation of students is a very important link, and in order to achieve the aim of talent training, process evaluation is more important for students, especially under the outbreak of normality, unable to understand the student's status, face to face is more easy to ignore the students' learning process, the teacher evaluation, and the school management, It is easy to ignore students' attention to their learning process and warn them of bad learning performance because they cannot see students on the spot. As a result, students cannot achieve ideal results in learning and do not get the opportunity to make up in time during the whole learning process. Therefore, academic warning is especially important during the pandemic, or in online courses.

Fourth, Multiple supervision and evaluation mechanisms;

In summative evaluation, multiple evaluation mechanism should be formed to strengthen classroom participation. In zhang Huafeng's article "The status quo of Chinese College students' Classroom cumulative expression behavior and its influence on learning harvest", through the investigation and analysis of nearly 111852 students from 40 universities for more than two years, it is found that students' classroom participation is positively proportional to their learning effect. [5] If teaching activities are carried out on the Internet, the participation of teachers and students on the Internet is more important. Therefore, the teaching process records, such as video learning time, online classroom participation, online homework completion, daily test and other objective data as the basis of evaluation, enhance the proportion of process assessment; At the same time, teachers, supervisors, corporate mentors, and even learners themselves should join the main body of evaluation, so as to form a more objective and scientific multi-evaluation mechanism.

4.3 Online Education State Chain Model Based on Block Chain

In traditional online education system, education management personnel are mainly relied on to process data in the system. As the competent unit, they can input and process teaching information, and have independent authority to modify the data. They can even evaluate teaching without considering the teaching data itself.

Therefore, there may be a dilemma: on the one hand, the evaluation conclusion of the teaching effect is based on the influence of its own reputation, which may be unfair, but there is no effective monitoring by a third party; If, on the other hand, fully uses the third-party evaluation, teaching valid data may not be able to obtain effectively, lead to insufficient data, not objective evaluation, at the same time, there may be data source is provided by the teaching unit, and the self-evaluation difference is not big, even there is no guarantee that evaluation can be open and transparent [6].

So, if the data can be stored, processed in time, will not be tampered with private, finally can effectively evaluate, reasonably application system is used, is we should focus on, when in even block chain of online teaching model, [7] due to the distributed storage block chain, and the technical features of the decentralized, can well solve these problems, [8]. The online teaching ecological model based on block chain is shown in Fig. 3 below.

In the model, the online teaching system is connected with the data processing interface and data pretreatment module, and the data can be saved and processed in time [9]. The main function of the pre-processing module is to carry out standardized processing according to actual needs. For example, for the evaluation index system, the data requirements of the evaluation index system are collected and processed, and uploaded to the blockchain network to prepare for the final evaluation. Finally, through the user interface, [10] in different decision-making systems, according to different decision-making needs, corresponding data can be extracted on the blockchain to serve different needs such as learning evaluation, teaching evaluation and talent matching.

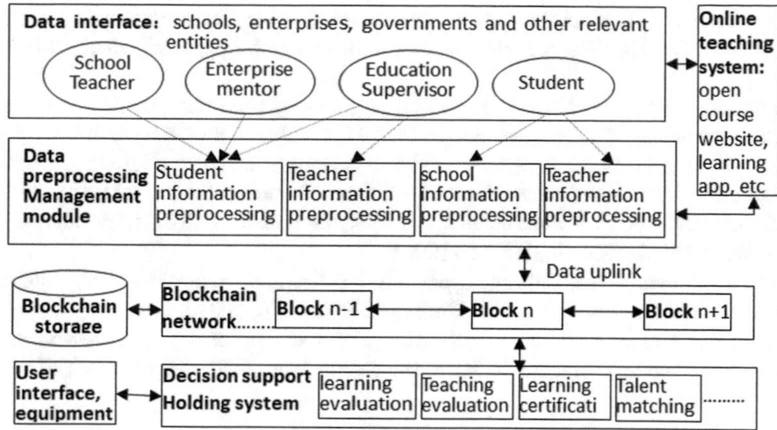

Fig. 3. Online education ecological chain model based on blockchain

Thus realized the data from the generation, collection, processing, chain, application of the entire ecosystem of effective, scientific use.

5 Conclusions

Block chain technology in the application of the online education can technically improve the student's study enthusiasm, improve the education management and evaluation of trust, better talent matching services for the enterprise, and continuously enhance the effectiveness of the teaching activities, for learners, teaching implementation and teaching management, enterprise and society has a positive effect of personnel training, We will foster a healthier education environment.

Acknowledgements. This work was supported by "Key research base of Humanities and Social Sciences in Colleges and Universities in Sichuan Province", "Research Center for application and development of educational informatization in Sichuan Province", Subject: Application of blockchain technology in online teaching evaluation of colleges and Universities, (Project No: JYXX20-015).

References

1. Richard, B.: Why shorten undergraduate medical education over college education? Acad. Med. **96**(2), 165 (2021)
2. Meredian, A.: Mental health impact of online learning: a look into university students in Brunei Darussalam. Asian J. Psychiatr. **67**, 102933 (2021)
3. Frith, K.H.: Is nursing ready for education blockchain technology? Nurs. Educ. Perspect. **43**(2), 139 (2022)
4. Barks, C.: Blockchain technology in the energy sector. Electr. Appar. **74**(11), 21 (2021)

5. Shyr, W.-J., et al.: Development of assessment indicators for measuring the student learning effects of artificial intelligence-based robot design. Comput. Appl. Eng. Educ. **27**(4), 863–868 (2019)
6. Matthew, G., Bailey, H., Denise, H.: P75 Disaggregating nutrition education evaluation data to assess racial equity in program outcomes. J. Nutr. Educ. Behav. **53**(7S), S59–S59 (2021)
7. Broeckelman Post, M.A., Mazer, J.P.: Editors' introduction: online teaching: challenge or opportunity for communication education scholars. Commun. Educ. **71**(2), 145–145 (2022)
8. Shafarenko, A.: A PLS blockchain for IoT applications: protocols and architecture. J. Cybersecur. Sci. Technol. **4**(1), 37–53 (2021)
9. Helles, R., Ørmen, J.: Big data and explanation: reflections on the uses of big data in media and communication research. Eur. J. Commun. **35**(3), 290–300 (2020)
10. Babu, E.S., et al.: Verifiable authentication and issuance of academic certificates using permissioned blockchain network. Int. J. Inf. Secur. Priv. (IJISP) **16**(1), 1–24 (2021)

Application of Video Surveillance Intelligent Analysis System Based on KNN Algorithm

Laihong Zhu[1] (✉) and Sira Yongchareon[2]

[1] Haojing College of Shaanxi University of Science and Technology, Xi'an, Shaanxi, China
zhulaihong750710@163.com

[2] Auckland University of Technology, Auckland, New Zealand

Abstract. Video surveillance intelligent analysis is to use routers or access technology to realize real-time information collection, processing and control of video storage streams and front-end equipment under the condition of limited network bandwidth resources. Therefore, this paper studies the application of video surveillance intelligent analysis system based on KNN algorithm. The research on intelligent analysis of video surveillance is to solve the current problems in network security, multimedia information and so on. This paper mainly uses the experimental method and the comparison method to describe the video surveillance intelligent analysis system based on the KNN algorithm in detail. The experimental data shows that when the bit rate is controlled at 18, the decoding rate changes with the accelerator.

Keywords: KNN Algorithm · Video Surveillance · Intelligent Analysis · System Application

1 Introduction

The main purpose of intelligent analysis of video surveillance is to summarize and classify the image information captured by the current camera, and realize various functions according to different types. In practical applications, the combination of video sensor technology and computer vision system can be used for remote real-time monitoring. At the same time, the communication network technology is combined with the PC to comprehensively control the operation of the whole system.

There are many theoretical achievements in the application of video surveillance intelligent analysis system based on KNN algorithm. For example, Kayabasi A stated that the Multilayer Perceptron (MLP) and K-Nearest Neighbor (KNN) algorithm models are used to determine the operating frequency of the A-Patch Antenna (APA) in UHF band applications. The performance of MLP and KNN models was compared during training and testing [1]. Hu B proposed a classification method that combines correlation-based feature selection (CFS) and k-nearest neighbors (ANN) data mining algorithms [2]. Al-Nawashi M believes that the detection of anomalous activities is crucial in surveillance applications, and there is an urgent need for surveillance systems that can be robust in

J. H. Abawajy et al. (Eds.): ICATCI 2022, LNDECT 170, pp. 609–616, 2023.
https://doi.org/10.1007/978-3-031-29097-8_72

academic settings [3]. So he provides a new video-based framework for automatic real-time surveillance systems. Therefore, from the perspective of KNN algorithm, this paper analyzes video surveillance intelligently to improve the performance of the system.

The KNN algorithm is first studied in this paper, and its algorithm steps are analyzed and described. Secondly, the design and construction of the monitoring system platform are described in detail. Then the moving target detection algorithm is classified and discussed. Then the method of moving target tracking is proposed. Finally, the system is tested by the experimental method, and the data results are obtained.

2 Video Surveillance Intelligent Analysis System Based on KNN Algorithm

2.1 KNN Algorithm

2.1.1 The Process of the KNN Algorithm

Input: In general applications, the training dataset is usually in the form of text, such as TXT format. Creating an input dataset facilitates data entry and does not require manual or batch processing. It is easy to operate without additional data, each entry needs to be reprocessed, and also makes unified data management easier [4, 5].

$$B = (\mathrm{a}_1, \mathrm{b}_1), (\mathrm{a}_2, \mathrm{b}_2), \ldots, (\mathrm{a}_m, \mathrm{b}_m) \tag{1}$$

where $A_1 \in A \in R$ is the feature vector of the instance.

Output: class b to which a belongs.

By calculating the distance between the given points, sorting from small to large, find the K points closest to a in the training set B, and choose the category of a in the points that is the best according to the majority rule.

$$\mathrm{b} = (\arg MAX_{d_k} \sum_{a_i \in Ml(a)} P(b_i = d_k), i = 1, 2, \ldots, M) \tag{2}$$

where P is an indicator function corresponding to a Boolean value.

In the ANN algorithm, the above three distances (Manhattan, Euclidean and Minkowski) can be used as sorting parameters according to the following points.

2.2 Advantages and Disadvantages of ANN Algorithm

Advantages: good classification effect, strong robustness.

Disadvantages: complex calculation, large amount of calculation, large calculation space.

Range: Numerical and nominal.

KNN neural network is a new intelligent learning algorithm. KNN algorithm (also known as neural network) is a distributed processing tool, which is a linear transformation from one node to another node, and obtains the maximum similarity within two or more discrete time intervals. A class of calculation methods for ability. All training examples in this algorithm can be used for classification. It can achieve different types of optimization by intersecting and searching the data, and also has parallel processing capabilities [6, 7].

2.3 Design and Construction of Monitoring System Platform

In the whole system, it is composed of front-end equipment, transmission module and application subsystem. Among them, the application is mainly to collect, organize, and classify data information. After obtaining the camera motion state parameters (scale), the control terminal sends an instruction to the remote terminal to automatically generate the required action execution algorithm to adjust the environmental parameters to ensure work efficiency. The overall functional design of the video surveillance intelligent analysis system is as follows [8, 9]:

Real-time monitoring, including the current state of the equipment and recent inspections. Through continuous tracking and monitoring of the user's location 24 h a day. When a special situation occurs, abnormal information is recorded and uploaded to the background database. The basic functions such as positioning tracking and short-distance data acquisition and processing realize the core part of establishing a video surveillance intelligent analysis system based on the KN algorithm. The system of intelligent analysis of video surveillance can be divided into three parts: information acquisition module, data processing and transmission module and other functions. Among them, the system includes front-end equipment. Its main implementation is to cover and search the space. The server side contains multiple clients. The first is to divide the site area that needs to be controlled into different areas, and determine the communication methods and respective permissions between each grouping group. On this basis, various data types, information acquisition modules, processing and sending modules and other functions in the video surveillance intelligent point cloud storage structure are analyzed. The image acquisition module converts the analog picture information obtained by the camera into recognizable digital signals. The human-computer interaction screen is the main platform for information exchange and emotional expression between the operator and the machine. The core function in the whole process is to realize its function through the B/S structure. The main control board of the monitoring system software is responsible for the functional interfaces of each part, the acquisition terminal program and the operation modules of each subsystem to complete the preprocessing and analysis process control of the video image [10, 11].

The monitoring system plays an important role in security with its intuitive and efficient features, and is widely used in communities, urban subways, logistics halls and other occasions, especially in some important departments and major events, CCTV provides an important security guarantee. Faced with this huge demand, the development of video surveillance technology is becoming more and more mature. Considering the passive monitoring, difficult management, poor security, and high labor cost of traditional video surveillance systems, integrated intelligent video surveillance systems have received extensive attention. Compared with PC-based video surveillance systems, built-in surveillance systems are easy to install, flexible in design, and inexpensive, making built-in surveillance systems more widely used in practice [12].

This system adopts the integrated Cortex-A9 quad-core ARM board as the main control board, which can collect the video information of each alarm point in real time, track the target in the video, and realize the target. At the same time, the entire monitoring system can collect the environmental data of each alarm point in real time, and the monitoring personnel can view the real-time environmental information through the mobile terminal and the PC terminal. The entire intelligent monitoring system consists of three parts: monitoring area terminal, ARM card main control terminal, and mobile Internet client. The overall functional block diagram of the monitoring system is shown in Fig. 1.

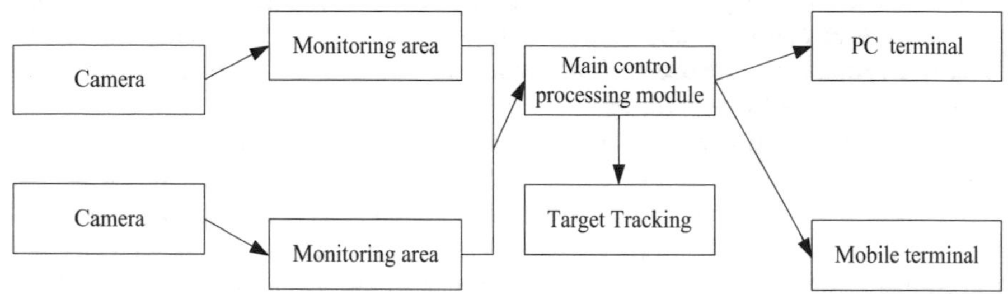

Fig. 1. The overall framework of the monitoring system

The system uses the Cortex-A9 quad-core ARM development board with Exynos4412 chip as the main board. The ARM development board is connected to a USB camera to collect video data, and is connected to a USB WIFI and a monitoring terminal for wireless data transmission. At the same time, the Arduino development board collects the data of each sensor and transmits the data to the ARM development board through the serial port. The ARM development board can also send an alarm command to the Arduino to realize the alarm, including the indicator light and the buzzer alarm request.

Customers include mobile phone monitoring and PC monitoring, the monitor can be monitored on PC, tablet and mobile phone. The monitoring personnel can connect to the IP address of the device by connecting to the local network or the remote network to realize the functions of environmental information display, multi-video monitoring, snapshot, monitoring, and image display.

2.4 Moving Target Tracking

Moving object tracking is essentially obtaining the position of a moving object of interest from a series of images. The amount of video data is huge, and finding the target of interest quickly and accurately is the essence of target tracking. The time complexity of the target tracking algorithm should not be too high, otherwise it will affect real-time monitoring. The motion features extracted from target tracking are the basis for understanding behavior, and can also be used to predict the motion area in the next instant, which is the basis for finding moving targets.

Target tracking is generally based on the properties of the target, using an appropriate adaptive algorithm to locate the target in the image sequence. The quality of target tracking algorithm depends on the choice of target features and the choice of search algorithm. There are many characteristics of moving targets, and various characteristics can be selected according to needs. These include visual features such as image edges and shapes, statistical features such as histograms and moment features, and algebraic features.

These target features must be valid and form the basis of target search comparisons. Depending on the characteristics of the target, the matching algorithm is critical throughout the process.

In order to distinguish various moving objects, an efficient object representation method is required, which can use feature information such as geometry and appearance. This information is a guide for tracking targets and an important basis for target matching. In general, the representation of the target adopts a small bounding rectangle for ease of processing and representation. The rectangle only circumscribes the moving target. The number of pixels the target occupies is defined in terms of area, sometimes the center of the target image is required.

The feature comparison is used to calculate the similarity between the target in the current image and the searched target feature, and the target with the largest similarity is the searched target. The characteristics of the target are generally diverse.

Predicting properties is a very important step in monitoring properties. Real-time target tracking performance requires finding the target to be tracked quickly and accurately. Target prediction can predict the moving range of the tracked target in the next frame, thereby reducing many target feature comparison operations for finding targets and greatly improving the tracking speed. In addition, if the background of the target changes or the influence of other factors changes, it will not be able to find and track the target within one frame, nor can it predict the tracking of the target, and the target will definitely be lost. Using the target tracking algorithm for target prediction, even if no target is detected this time, target tracking can be continued the next time the target appears at the predicted position without losing target tracking.

3 System Test

3.1 System Test Environment

Software environment: Win8 64-bit operating system, VS2018, Eclipse.

Hardware environment: Intel i3-2330M (2.20GHz) processor, AMD Radeon HD769 graphics card, 4G storage, TP-Link router, network card, HTC smart Android phone.

The network environment is Wifi wireless local area network and 5G network in the laboratory.

3.2 System Test Content

This paper mainly tests from two aspects: First, whether the information interaction function of the EMS system between the mobile client and the GAE platform is intact. Second, the communication between the mobile client and the server in the local network includes real-time testing of CCTV images and testing of video decoding rates.

3.3 Test Steps

The basic function test steps of the monitoring system are as follows: First, start the monitoring system. The mobile client enters the multi-function selection interface and selects the video recording button. Then, on the server side, select the monitored video path. Finally, select the record button to save the video locally.

4 Analysis of Test Results

4.1 Multi-channel Video Decoding Rate Data

After testing, the intelligent mobile terminal interacts well with the EMS system used by the GAE platform. There is a delay of about half a second between the monitoring screen of the intelligent mobile terminal and the server display terminal on the local network. The details are shown in Table 1:

Table 1. Multiexed video decode rate data

	Decoding rate	Frame rate	Code rate
First way	54	29	16
Second way	61	29	16
Third way	50	29	16
Fourth way	57	29	16

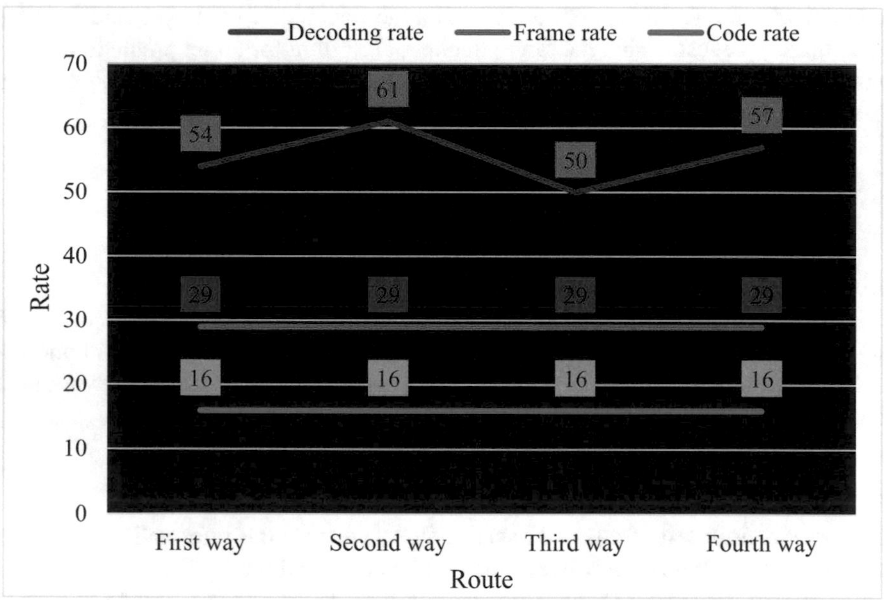

Fig. 2. Multiexed video decode rate data

As shown in Fig. 2, we can find that after using DXVA decoding acceleration, the multi-channel video decoding rate is greatly reduced. There may be other issues with outdoor 5G networks, and this system still needs to be improved and refined. In general, the intelligent mobile video surveillance system of this site mainly meets the basic requirements of the video surveillance system.

5 Conclusion

The system is based on the intelligent analysis of video surveillance and uses its communication interface to divide the on-site equipment control and management into three modules, namely the main control terminal, the mobile terminal and the server. Connect the management center through the KNN network. The system adopts B/S structure, and its main function is to realize intelligent analysis of video surveillance management. The video surveillance intelligent analysis module is used to collect and process the image information from the network, and preprocess the data to obtain the real-time frame format. The character type or feature mark collected at different positions will be collected as a template to determine the specified content area in the current video screen. Smart CCTV means that the computer analyzes and processes the information transmitted by the monitor without human intervention. The intelligent monitoring system is a very complex system. The system adopts artificial neural network algorithm and has achieved good results in target tracking, but there are still many improved methods.

Acknowledgment. Key R & D project in Xianyang City: R & D of mobile communication and navigation equipment for the blind, project (No. 2019k02-20).

References

1. Kayabasi, A.: MLP and KNN algorithm model applications for determining the operating frequency of A-shaped patch antennas. Int. J. Intell. Syst. Appl. Eng. **3**(5), 154–157 (2017)
2. Hu, B., Li, X., Sun, S., et al.: Attention recognition in EEG-based affective learning research using CFS+KNN algorithm. IEEE/ACM Trans. Comput. Biol. Bioinform. **PP**(2), 1 (2018)
3. Al-Nawashi, M., Al-Hazaimeh, O.M., Saraee, M.: A novel framework for intelligent surveillance system based on abnormal human activity detection in academic environments. Neural Comput. Appl. **28**(1), 565–572 (2016). https://doi.org/10.1007/s00521-016-2363-z
4. He, F.: Intelligent video surveillance technology in intelligent transportation. J. Adv. Transp. **2020**(4), 1–10 (2020)
5. Lotfi, M., Motamedi, S.A., Sharifian, S.: Time-based feedback-control framework for real-time video surveillance systems with utilization control. J. Real-Time Image Proc. **16**(4), 1301–1316 (2016). https://doi.org/10.1007/s11554-016-0637-4
6. Arai, K., Bhatia, R., Kapoor, S.: Smart Video Surveillance, pp. 118–126 (2019). https://doi.org/10.1007/978-3-030-02683-7. (Chapter 10)
7. Filho, D.S.M.P., de Macedo, D.D.J.: A model for automated technological surveillance of web portals and social networks. J. Intell. Inf. Syst. **56**(3), 561–579 (2021). https://doi.org/10.1007/s10844-021-00641-0
8. Yogameena, B., Menaka, K., Perumaal, S.S.: Deep learning-based helmet wear analysis of a motorcycle rider for intelligent surveillance system. IET Intell. Transp. Syst. **13**(7), 1190–1198 (2019)
9. Ahmed, A.A., Hawk-Eye, E.M.: An AI-powered threat detector for intelligent surveillance cameras. IEEE Access **PP**(99), 1 (2021)
10. Sri Preethaa, K.R., Sabari, A.: Intelligent video analysis for enhanced pedestrian detection by hybrid metaheuristic approach. Soft. Comput. **24**(16), 12303–12311 (2020). https://doi.org/10.1007/s00500-020-04674-5

11. Sasikala, G., Lekha, R.S.: Intelligent surveillance system for smart security. Indian J. Sci. Technol. **12**(29), 1–9 (2019)
12. Lotfi, M., Motamedi, S.A., Sharifian, S.: Time-based feedback-control framework for real-time video surveillance systems with utilization control. J. Real-Time Image Proc. **16**(4), 1301–1316 (2019)

Research and Implementation of QR Code and Maze Conversion Based on VR

Juan Xiao, Weichen Yang, and Faying Li(✉)

School of Computer and Artificial Intelligence, XiangNan University, Chenzhou, Hunan, China
lifayingvip@163.com

Abstract. This paper combines two-dimensional code with maze, discusses the recognition, generation and transformation of two-dimensional code, and shows it in the form of VR game. It strictly follows the national standard of the people's Republic of China – quick response matrix code QR code: GB/T 18284–2000, NEQ ISO/IEC 18004:2000, and puts forward an effective algorithm for the transformation of two-dimensional code and labyrinth, the determination of whether there is a solution to the labyrinth and the search for the sufficient conditions for whether there is a solution to the labyrinth. Finally, VR game is presented to the players, supplemented by the interaction mode of wireless Bluetooth handle operation, trying to give the players a new immersive game experience.

Keywords: QR Code · Maze Algorithm · VR · OpenGL ES 2.0

1 Introduction

Norman Joseph woodland, the co-inventor of bar code, invented one-dimensional code (bar code) to solve the problems of increasingly frequent commercial activities and low efficiency of manual transactions. This pattern composed of several black-and-white strips with different widths almost subverted the global business model in the first few decades. After decades of changes, the bar code can only carry about 30 characters of data capacity, can only contain letters and numbers, and the size is relatively large, resulting in low space utilization, which calls for the emergence of emerging technologies.

In 1994, Nippon Denso officially announced the disclosure of QR code (quick response code), the world's first two-dimensional code. Compared with the bar code, this square, black-and-white staggered QR code has the advantages of larger data capacity, smaller bar code relative size, surpassing the alphanumeric limit and anti damage ability, which is bound to be popularized rapidly.

Aiming at the visible two-dimensional code, according to its surface characteristics – crisscross black-and-white squares, this paper attempts to explore the conversion between two-dimensional code and its corresponding maze – how to effectively organize the black-and-white square data of two-dimensional code? How to ensure that the generated maze has a solution?

Since its inception, maze game has been greatly loved by people for its unique game mode and different customs clearance paths. Recently, the maze lacks a sense

J. H. Abawajy et al. (Eds.): ICATCI 2022, LNDECT 170, pp. 617–625, 2023.
https://doi.org/10.1007/978-3-031-29097-8_73

of substitution for being tiled on paper or electronic device screens, or because it is displayed in 3D but unfolded from a third person perspective and lacks realism, the playing method is becoming more and more simple [1–4].

The increasing maturity of VR technology almost perfectly solves the above shortcomings. This paper may provide an effective scheme for enriching maze play by virtue of the advantages of immersive experience.

2 Related Technologies and Development Tools

2.1 QR Code

Two dimensional code is actually a "two-dimensional bar code", which is relative to one-dimensional bar code (bar code). It is designed to solve the defects of limited barcode data capacity, limited supported data types and no anti damage ability.

The popular coding systems in the world today mainly include QR code (quick response, QR, quick response matrix code, which is a coding method widely used on mobile devices), PDF417, etc. QR code is the focus of this section and also the code system [5] used in this application.

From the name of "quick response matrix code", we can intuitively feel that high-speed reading is the characteristic of QR code different from other two-dimensional codes. It is this feature that makes it widely used in the fields of automatic production line management, commercial trade transactions and so on.

The position detection pattern, also known as the positioning pattern, is the middle of three identical sizes and shapes in the upper right corner, upper left corner and lower left corner of QR code. It can be regarded as three overlapping concentric squares to enhance the ability to distinguish from other patterns. For each QR code, the position of its three position detection graphics is unchanged [5].

Like other two-dimensional codes, QR codes provide error correction capabilities that bar codes do not have. Error correction refers to the ability of scanning the QR code to decode correctly when the surface of the QR code is dirty, unreadable or damaged.

The four levels of error correction capability of QR code are l, m, Q and h respectively. Their recoverable codeword proportion is given in the form shown in Table 1.

Table 1. Correspondence between error correction level and recoverable codeword ratio

Error correction level	Recoverable codeword ratio
L	7%
M	15%
Q	25%
H	30%

Each QR code symbol is composed of an array of square modules. It consists of coding area and functional graphics including image seeking graphics, separator, positioning graphics and correction graphics. Function graphics are not used for data encoding. The symbol is surrounded by a blank area. Figure 1 shows the basic structure of QR code.

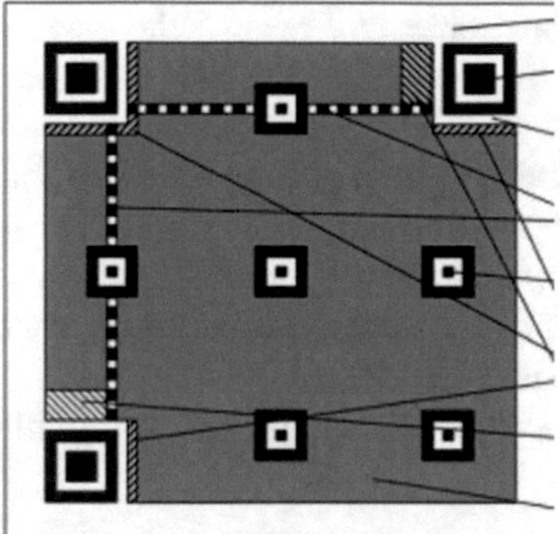

Fig. 1. Basic structure of QR code overview of QR code coding process

QR code coding process:

A. Participate in coding data analysis.
B. Select the desired error correction level.
C. Encoding: for the mode determined in the above stage, convert the data to be encoded into a bit stream.
D. Error correction coding.
E. Construct the final information and "place" black and white modules in the matrix.
F. Mask, generation format information and version information.

Overview of QR code decoding process.

A. Binary image.
B. Find the position detection graphics that determine the position of the QR code.
C. Read format information and version information. Determine the distance d between the center coordinates of the upper left corner position detection graphics and the center coordinates of the upper right corner position detection graphics, and the widths widthleft and widthright of the two detection graphics [6]. First calculate the nominal module width dimension X of the symbol:

$X = (\text{widthLeft} + \text{widthRight})/14.$

Then preliminarily determine the symbol version:

$\text{Version} = ((D/X)-10)/4.$

If the calculated version ≤ 6, the value is the symbol version of the QR code. If version > 6, further calculation is required.
D. Identify the graphic depth symbols and determine the array of 0 and 1.
E. Error correction levels and mask patterns for symbols are obtained.
F. Perform subsequent decoding operations according to rules.

2.2 OpenGL ES

OpenGL (open graphic library) is the software interface of hardware graphics. It contains a series of program functions that allow developers to formulate high-quality graphics, images and three-dimensional objects. Because the development environment of this paper is Android mobile device, it mainly introduces the cropped version of open GL – OpenGL es.

OpenGL es is an application programming interface for advanced 3D. It is a series of APIs (Application Programming Interface) [6, 7] developed by Khronos group for the limitations of CPU computing power, memory and power consumption of handheld and embedded devices, low data access speed and lack of floating-point hardware.

Basic steps of OpenGL es program.

The basic steps of an open GL es program can be roughly divided into the following stages:

(1) Create a rendered surface on the screen with EGL;
(2) Load vertex and clip shaders;
(3) Create a program object, connect vertex and fragment shaders, and link the program object;
(4) Set viewport and clear color buffer;
(5) Rendering simple primitives;
(6) Make the contents of the color buffer visible in the EGL window surface.

The graphics rendering process of programmable rendering pipeline is shown in the following Fig. 2.

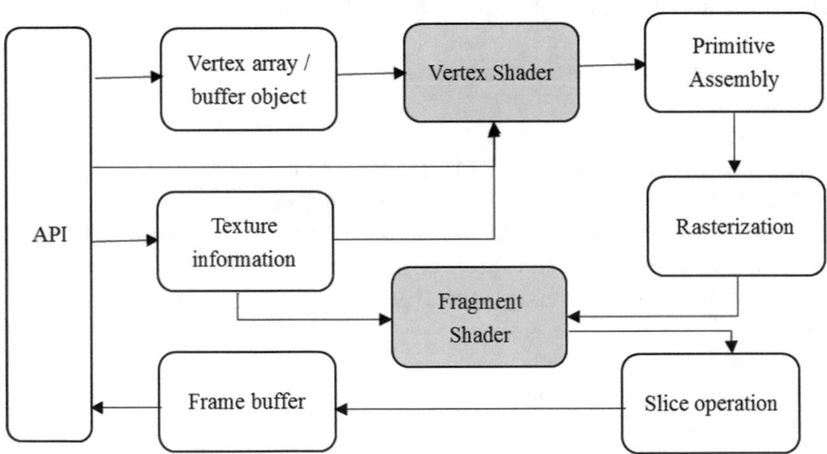

Fig. 2. Graphics rendering process of programmable rendering pipeline

3 Extraction of Effective Data of QR Code and Transformation of Corresponding Maze

(1) Basic idea of algorithm

The extraction process of effective data of QR code picture is largely similar to its decoding (decoding) process. Namely:

A. Binarization: load a static picture containing two-dimensional code, analyze and determine the discrimination threshold between dark pixels and light pixels, and read it;
B. Find the positioning graphics, determine the specific position of the QR code in the picture, and shield the interference of irrelevant pixels.
C. Dark pixels are determined as 0 to represent the non passable walls in the maze, and light pixels are determined as 1 to represent the passable road blocks in the maze.

Finally, according to the actual needs and the version information of the scanned QR code, get through the position detection graphics in the lower left corner and part of the position detection graphics in the upper right corner of the wall as the starting point and end point of the whole maze.

(2) The judgment of maze with or without solution

The corresponding follow-up processing is carried out for the scanned and transformed maze: after opening up a part of the position detection graphics in the lower left corner and the upper right corner of the wall as the starting point and end point of the whole maze, the judgment of whether the maze has a solution is started [8].

Using the data structure of the stack with the characteristics of first in and last out, and using the idea of depth first traversal, all the maze nodes that can be reached from the starting point are exhausted.

The structure of the storage node is defined as:

```
struct node
{
   int row;
   int column;
   int direction;
   struct node *next;
};
```

Row and column store the row and column information of the node respectively. Direction stores the traversal direction when currently passing through the node. Next stores the first address of the next node reached.

In the process of depth first traversal, if the stack returns to the starting point, it indicates that there is no solution to the maze; If the stack element is the end point, it indicates that the maze has a solution.

When the top node of the stack can only retreat, it indicates that the node is the farthest node on the path. When the node is withdrawn from the stack, it is inserted into the impassenode linked list in reverse order from the starting point. The node structure of impassenode is defined as:

```
struct impasseNode
{
    int row;
    int column;
    int direction;
    int distance;
    struct impasseNode *next;
};
```

Similar to the node structure definition, row and column store the row and column information of the node respectively. Direction stores the traversal direction when currently passing through the node. Distance stores the distance between the current node and the starting node (actually the square of the distance, which is consistent with the logic of real distance) [9]. Next links the first address of the next node.

(3) Search of sufficient conditions for the solution of maze

When it is determined that there is no solution to the maze, the search for the sufficient condition for the solution of the maze will be started automatically.

According to the node information stored in the importnode linked list during the process of determining whether there is a solution, select the chain head node (that is, the unprocessed node farthest from the starting point) into the importnode linked list, select the node that is the wall in the next step according to its stored direction information and turn it into a road block, and start a new round of depth first traversal with this node as the starting node. This process continues until the final solution. The importnode node structure is defined as:

```
struct importNode
{
int row, column, initRow,int initColumn;
struct importNode *next;
};
```

The logic of row, column and next is consistent with the previous node or impassenode definitions. Initrow and initcolumn are the nodes that enable the current node to store in the importnode linked list.

4 Realization of Two-Dimensional Code and Maze Conversion Based on VR

When the user clicks to enter the application, the system first obtains the relevant settings of the system preferences such as the opening or closing of background music and sound effects, and performs the corresponding behavior. Load the main interface while obtaining relevant information, and then wait for user operation.

The user clicks to store the game: the system applies for calling the camera permission to obtain several QR code images, and calls the relevant operations of QR code decoding to obtain the information of the QR code specified by the user. The user can also click the "local" button in this interface to select the previously stored QR code from the local album for code scanning and subsequent related operations.

After the two-dimensional code information is obtained and converted into a maze, the system background automatically calls the search algorithm for whether the maze has a solution and when there is no solution, it is the sufficient condition for the maze to have a solution. Automatically jump to the storage interface after completion [10].

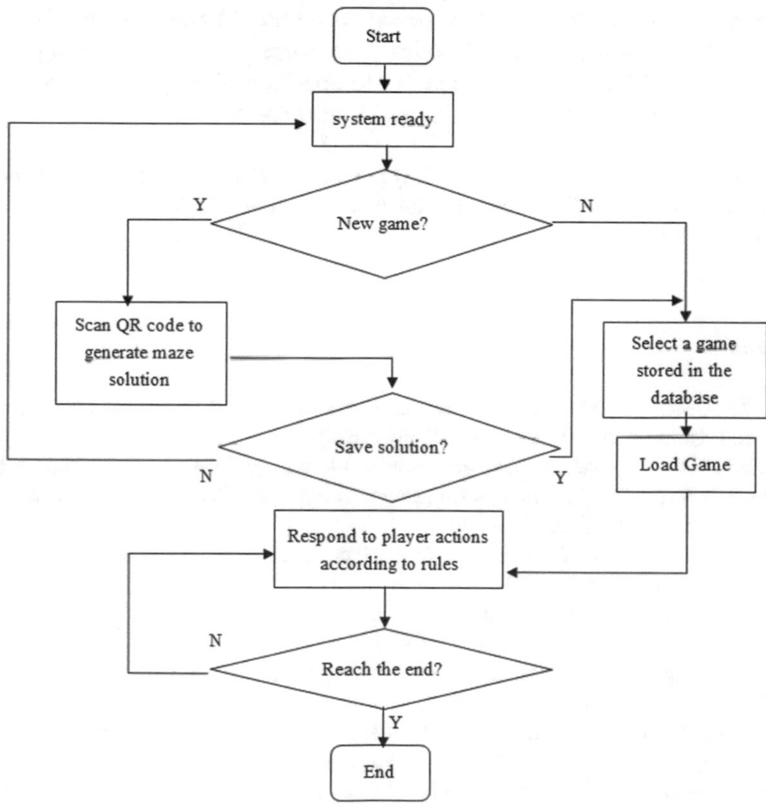

Fig. 3. Overall process of the game

The storage interface will automatically load and read the picture after the QR code is converted into a maze and display it in the left box. The user can independently evaluate the difficulty of the maze and choose to store or give up storing the maze.

When the user clicks save, a dialog box will pop up to prompt the user to name the maze. The system will automatically take the current system time as the default name of the maze. Click OK to complete the storage of the maze.

Users click to read the game: the system reads all mazes stored locally and displays them in the form of two records per page. Users can click the light bulb button to view the solution, and click the joystick button to enter the maze corresponding to the item.

Click system settings by the user: load relevant system settings and respond to user operations.

Click to exit the game: stop playing music and exit the application (see Fig. 3).

5 Conclusion

The exploration of the sufficient conditions to make the unsolved maze solvable is one of the cores of this research and development. According to hundreds of test results, this paper can ensure the realization of the necessary conditions for the program. The solution of discarding a few two-dimensional codes is not surprising. The search algorithm for the sufficient condition for the solution of the unsolved maze proposed in this paper shows that it can give a satisfactory solution in most cases.

Acknowledgments. Open project of Hunan Key Laboratory of embedded and network computing (NO. 20210103, NO. 20210104); Scientific Research Project of Hunan Provincial Department of Education (20C1694).

References

1. Papp, G., Hoffmann, M., Papp, I.: Improved embedding of QR codes onto surfaces to be 3D printed. J. Comput. Aided Des. **131**, 102961 (2021)
2. Al-Saadi, H.S., Elhadad, A., Ghareeb, A.: Rapid and blind watermarking approach of the 3D objects using QR code images for securing copyright. Comput. Intell. Neurosci. **2021**, 2236866:1-2236866:11 (2021)
3. Madsen, S.S., Santos, A., Jørgensen, B.N.: A QR code based framework for auto-configuration of IoT sensor networks in buildings. Energy Inf. **4**(S2), 1–19 (2021)
4. Wahsheh, H.A.M., Al-Zahrani, M.S.: Secure real-time computational intelligence system against malicious QR code links. Int. J. Comput. Commun. Control **16**(3) (2021)
5. Cai, X., Huang, D.: Research on anti-counterfeiting technology based on encrypted QR code. Sci. Technol. Wind **4**, 118–119 (2020)
6. Hill, G.N., Whitty, M.A.: Embedding metadata in images at time of capture using physical Quick Response (QR) codes. Inf. Process. Manage. **58**(3), 102504 (2021)
7. Stark, E., Kucera, E., Haffner, O.: Proposal of IoT devices control using mixed reality and QR codes. J. Autom. Mob. Rob. Intell. Syst. **14**(3), 36–41 (2021)
8. Bi, C., Song, L.: UI design technology in app development. Sci. Technol. Innov. **2**, 61–62 (2020)

9. Darabi, A., Salehi, M.R., Abiri, E.: Newly energy-efficient SRAM bit-cell using GAA CNT-GDI method with asymmetrical write and built-in read-assist schemes for QR code-based multimedia applications. Microelectron. J. **114**, 105117 (2021)
10. Rodriguez Borbon, J.M., Huang, J., Wong, B.M., Najjar, W.A.: Acceleration of parallel-blocked QR decomposition of tall-and-skinny matrices on FPGAs. ACM Trans. Archit. Code Optim. **18**(3), 27:1-27:25 (2021)

Strengthening Research of SYSTRAN in the Field of Artificial Intelligence Automatic Translation

Zhaohui Li[1], Shuai Gao[2](✉), Xin Li[3], and Hooman Bavarsad Asghari[4]

[1] Changchun University of Finance and Economics, Changchun, Jilin, China
[2] Changchun University, Changchun, Jilin, China
gaos83@ccu.edu.cn
[3] College of Humanities and Information, Changchun University of Technology, Changchun, Jilin, China
[4] Islamic Azad University, Tehran, Iran

Abstract. In the field of artificial intelligence, automatic translation is an important technology, which can improve the processing and understanding of information by intelligent machines. This paper studies the application of machine translation and strengthens its application in artificial intelligence automatic translation. Firstly, this paper expounds the principle, function and characteristics of artificial intelligence automatic translation, then introduces the principle of machine translation, then studies the application process of machine translation system, and tests the translation performance of machine translation, so as to improve the ability of intelligent machine to process and understand information to a certain extent. Finally, the test results show that machine translation performs well in artificial intelligence automatic translation, and achieves more than 80% results in word translation accuracy, fluency and accuracy.

Keywords: SYSTRAN Technology · Artificial Intelligence · Automatic Translation · Domain Enhancement

1 Introduction

With the continuous improvement of scientific and technological level and automation and the arrival of the industrial era, the artificial intelligence industry has risen rapidly. Artificial intelligence is a new subject, which is very different from traditional computer technology, automatic control theory and computer science [1, 2]. In the future, intelligent robot will become an important and essential product with great development potential in human society. Therefore, in order to adapt to social needs and better serve human social life, translation has emerged as one of the carriers of information transmission in artificial intelligence [3, 4].

In the field of artificial intelligence translation, researchers at home and abroad have conducted a lot of in-depth discussion and analysis on machine language. Among them, domestic scholars mainly explore the characteristic factors such as semantics

J. H. Abawajy et al. (Eds.): ICATCI 2022, LNDECT 170, pp. 626–633, 2023.
https://doi.org/10.1007/978-3-031-29097-8_74

and semantics. According to the relevant literature at home and abroad, the research and development of artificial intelligence automatic translation system abroad has been quite mature and has a certain market prospect [5, 6]. However, China focuses on the exploratory research on the algorithm implementation and function driving technology involved in the autonomous operating system of intelligent robot. When some scholars combine the "neural network" with other part of speech structures as a new automatic printing technology (RBF) to study the translation of words, they find that they have different degrees of similarity in semantic representation, However, there are still many problems to be solved [7, 8]. Therefore, this paper studies the strengthening effect of SYSTRAN technology on the field of artificial intelligence automatic translation.

With the rapid development of artificial intelligence, people pay more and more attention to intelligent robot technology and machine language. In the field of translation, automatic translation has become a hot topic. This paper will analyze and elaborate from the direction of SYSTRAN. Firstly, this paper introduces the relevant knowledge theory, research background and current situation of autonomous speech recognition system. Then experiments show that this method can improve the quality of the translation and the effect of the original text. Then it introduces the influencing factors of syntax based semantic matching algorithm and nonparametric optimization technology on the accuracy of translated text, provides a reference basis for subsequent chapters, and finally summarizes the main contents and research results of this paper.

2 Discussion on SYSTRAN in the Field of Artificial Intelligence Automatic Translation

2.1 Artificial Intelligence Automatic Translation

1) Principle

   The translation process is determined by the translation information, the content of the original text, the translator's language environment and other factors. The automatic translation rule of artificial intelligence is based on computer technology, which realizes the intelligent processing of text data. As automatic translation is a new translation method, it can not only help translators understand the translation quickly and accurately, but also provide readers with high quality, high reliability and valuable information. The principle of automatic translation is based on language conversion theory. This method processes the information input in the computer and expresses the required information in some form in the output process [9, 10]. Because language itself has the characteristics of complexity, variety, instability and nonlinearity to a certain extent. Therefore, in order to meet the requirements of readers' reading comprehension level and achieve the purpose of accurate transmission of the text, it is necessary to complete the translation with the help of some specific formats, and take into account the possible errors in the translator's translation, which provides a relatively stable and easy space for modification and adjustment.

2) Function

   In the automatic translation system of artificial intelligence, it mainly has the following functions:

(1) It can provide users with an interactive function, dialogue according to different language texts, and feedback error information quickly and timely. As the intelligent automatic translation system is an open platform, it transmits the suggestions and ideas put forward by experts and scholars from different fields and regions to those with the same characteristics or similar views in the neural network through the network, and transmits them in real time. At the same time, it can also provide real-time evaluation results and other service functions to the translation, Thus, the accuracy of translators' understanding of the original meaning and the degree of post translation consistency are improved [11, 12].

(2) It can also modify and interpret the translation information. When text data is entered, sentences can be rearranged through the network. In this way, we can avoid the problems of wrong translation of the original text and poor expression effect due to the mismatch of semantic relations between words. In addition, it can reduce the errors caused by different corpus and grammatical habits, such as high error rate, low accuracy and high repetition.

(3) It can provide a convenient way for translators. In the field of computer, people translate machine language into English, and then convert it into Chinese symbols by decoding. In addition, it can also realize the information exchange and expression ability requirements of different morpheme expression methods in the text, effectively improve the translator's understanding of the original meaning, and accurately translate the translation in the computer language environment.

3) Features

(1) High accuracy. In the process of translation, the translator must make a reasonable interpretation of the translation to ensure that the original text conforms to the reader's understanding level. Therefore, translation should pay attention to the accuracy and coherence of language structure and expression. In addition, we should also pay attention to whether there is a logical relationship between the text content and whether the semantics is accurate, and give corresponding instructions or prompts to help the reading and decoding work be completed smoothly.

(2) Easy to use. The operation interface of artificial intelligence automatic translation system is simple and easy to understand. Users can use it at will with a high degree of visualization. After reading the translation, subsequent processing can be carried out. Intelligent language translation system has the characteristics of good human-computer interaction, universality and easy maintenance and upgrading. It has powerful function, strong compatibility and strong practical value. At the same time, it also has strong fault tolerance and anti-interference ability. This is also one of the unique functions of artificial intelligence automatic decoder, which can quickly and effectively identify and judge all kinds of complex text information.

(3) Real time. Due to the high level of automation of artificial intelligence. Therefore, it is very important to identify, process and judge the translation quickly and accurately. At the same time, computer technology can also be used to

realize the functions of semantic information extraction and storage. It can also carry out speech synthesis and automatic compilation in intelligent robot language programming, so as to achieve the purpose of translation and improve the quality of translation.

2.2 SYSTRAN Technology

1) Principle
   With the rapid development of multimedia information technology such as computer technology and network communication, and the increasingly frequent communication between the field of artificial intelligence and related disciplines such as digital technology and intelligent manufacturing, traditional language translation cannot meet people's requirements for information transmission speed and accuracy. Therefore, machine translation based on artificial intelligence can improve the speed of information transmission. In the process of machine translation, when the translation conforms to the prompt of the original text, the machine will decode the explanatory words according to the original text. Because it is automatic translation, the translation will not be read directly to the original text. Therefore, the translation results can be verified by modifying or adjusting the data and information on the computer, and can be guaranteed to be accurate, so as to realize the functions of automatic recognition and reading of the content and discourse features of the target language.
2) Relevant Technical Support
   Artificial neural network is an interdisciplinary field integrating brain science, neuropsychology and informatics. The goal is to make the computer simulate the thinking mode of the human brain through learning, so that the machine has a certain artificial intelligence. It extends a variety of neural network models according to different neuron connection topologies or learning methods. BP (back propagation) neural network is a hierarchical back propagation neural network with three or more layers. It is composed of input layer, hidden layer and output layer. The layers are fully connected and there is no connection in the layer. BP neural network algorithm is the most widely used neural network at present. Its structure consists of forward propagation and reverse error correction. Then the output value of the network hidden layer unit is:

$$h_j = f(\sum_{i=0}^{i-1} v_{ij} x_i - \Theta)_j) \quad (1)$$

The output value of each unit in the output layer is:

$$y_k = f(\sum_{i=0}^{i-1} w_{jk} h_i - \Theta)_k) \quad (2)$$

BP algorithm transmits information through the forward propagation process, updates the weight through the back propagation process, and carries on again and again until the error is less than the threshold or the iteration reaches the preset number of times.

3 Experiment of SYSTRAN in the Field of Artificial Intelligence Automatic Translation

3.1 SYSTRAN Strengthens the Process in the Field of Artificial Intelligence Automatic Translation

Due to the design and implementation of computer and human-computer interaction interface, users can use the input device at the PC end to obtain corresponding information for language conversion, text decoding, etc. at the same time, they can also transmit the text content to the terminal server through the network to complete the process of document processing and file transmission, So that the translation system can automatically recognize the meaning to be expressed. Therefore, machine plays an important role in the field of artificial intelligence. This paper mainly studies the specific implementation process of SYSTRAN automatic translation in the field of artificial intelligence, and extracts and analyzes the text, grammar and other information in the process of robot translation through machine language. As can be seen from Fig. 1, the translation process mainly includes text translation, grammar and semantics. In the field of computer language processing, machine translation is an effective way to convert English information into understandable symbolic form for expression, while artificial neural network has strong adaptability and robustness, This technology is used to complete the automatic translation of text content and the interaction between texts, so as to improve the quality and accuracy of translation.

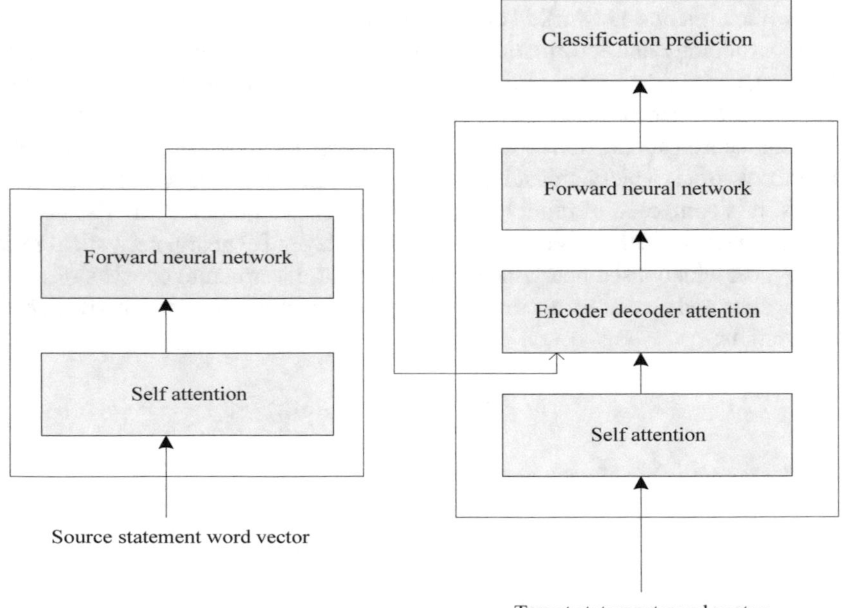

Fig. 1. SYSTRAN strengthens processes in the field of automatic AI translation

3.2 SYSTRAN Strengthens the Performance Test Process in the Field of Artificial Intelligence Automatic Translation

Machine translation realizes the automatic process in the field of artificial intelligence automatic translation, which needs to analyze and process the robot language. Therefore, there is a very important performance index for the robot language: accuracy, fluency and accuracy. The process of machine translation is a dynamic process of continuous adjustment and correction, gradual improvement and finally achieving the desired effect. In the process of translation, we need to test the structural features and semantic expression of machine language. For machine statements used by different users in different fields, their syntax, vocabulary and other parameters may be quite different. Therefore, in order to better help the translation understand and translate the information contained in the standard text that meets the requirements, the amount is small and easy to remember and input, so as to provide certain convenience for the description of the translation. At the same time, we should also consider the problems of language quality degradation and improper semantic expression due to the limited development technology and hardware level of computer translation software. As a result, the machine learning function cannot be realized in the process of automatic translation.

4 Experimental Analysis of SYSTRAN in the Field of Artificial Intelligence Automatic Translation

4.1 Performance Test and Analysis of SYSTRAN in the Field of Artificial Intelligence Automatic Translation

Table 1 shows the SYSTRAN enhanced performance test data.

Table 1. SYSTRAN enhanced performance test data

Word quantity	Accuracy	Fluency	Precision
100	95%	85%	84%
500	89%	94%	86%
1000	90%	90%	82%
1500	87%	80%	90%
2000	85%	89%	83%

In the process of translation, due to the complexity of language and various ways of expression, the translation needs a lot of accurate understanding and accurate description. Automatic translation is not only the simple interpretation, analysis, transformation and modification of the original information. Therefore, in order to improve the decoding quality and achieve better results, it is necessary to carry out machine semantics research to verify whether the algorithm can successfully achieve the expected functional objectives. As can be seen from Fig. 2, machine translation performs well in artificial intelligence automatic translation, and achieves more than 80% results in terms of word translation accuracy, fluency and accuracy.

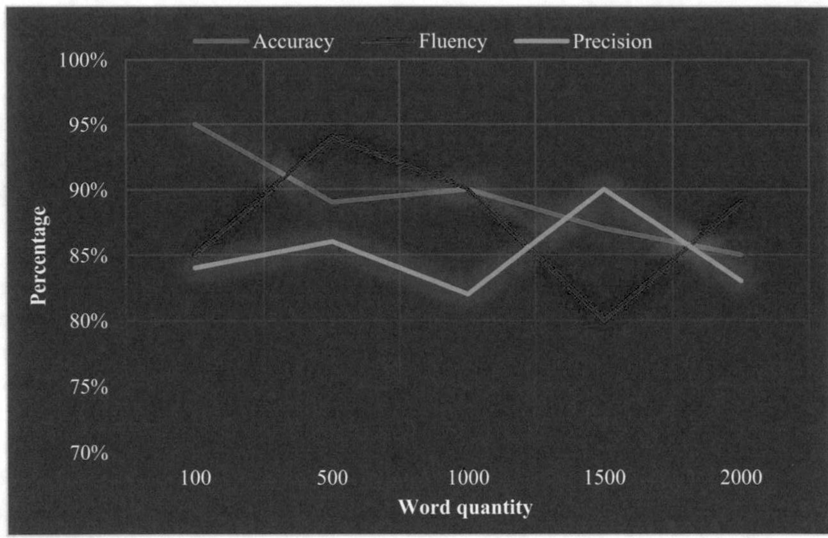

Fig. 2. SYSTRAN enhanced performance test data

5 Conclusion

In the field of artificial intelligence, translation is an important link, which can also be said to be an intelligent machine language processing system. This paper mainly studies the semantic translation method and discourse analysis method of automatic translation robot. Firstly, it introduces the SYSTRAN theoretical model, grammatical rules and syntactic structure characteristics. Secondly, the automatic word recognition algorithm is described in detail, and its functions and characteristics are explained. Then, the artificial intelligence vocabulary set is constructed based on the artificial neural network, and the words extracted from the corpus are transmitted to the computer system in the neural network to realize the automation of language translation process and complete the tasks of semantic information matching and classification.

Acknowledgment. Social Science Research Project of Jilin Provincial Education Department: Anti-War Research in Japanese Post-War Literature (General Project), Project (No.: JJKH20221256SK). 2021 Translation Education Research Project of Jilin Translators Association: Research on Multimodal Teaching Mode of Foreign Language Interpretation in the New Media Environment, Project (No.: JLFY2021YB102). National Foreign Language Teaching and Research Project in Colleges and Universities: Research on the Transformation and Development of Foreign Language Teachers in Colleges and Universities under the Background of New Liberal Arts, Project (No.: 2021JL0126).

References

1. Grover, P., Kar, A.K., Dwivedi, Y.K.: Understanding artificial intelligence adoption in operations management: insights from the review of academic literature and social media discussions. Ann. Oper. Res. **308**(1–2), 177–213 (2020). https://doi.org/10.1007/s10479-020-03683-9
2. Gupta, S., Modgil, S., Bhattacharyya, S., Bose, I.: Artificial intelligence for decision support systems in the field of operations research: review and future scope of research. Ann. Oper. Res. **308**(1–2), 215–274 (2021). https://doi.org/10.1007/s10479-020-03856-6
3. Johnson, M., Albizri, A., Simsek, S.: Artificial intelligence in healthcare operations to enhance treatment outcomes: a framework to predict lung cancer prognosis. Ann. Oper. Res. **308**(1–2), 275–305 (2020). https://doi.org/10.1007/s10479-020-03872-6
4. Lui, A.K.H., Lee, M.C.M., Ngai, E.W.T.: Impact of artificial intelligence investment on firm value. Ann. Oper. Res. **308**(1), 373–388 (2022)
5. Preil, D., Krapp, M.: Artificial intelligence-based inventory management: a Monte Carlo tree search approach. Ann. Oper. Res. **308**(1–2), 415–439 (2021). https://doi.org/10.1007/s10479-021-03935-2
6. Sullivan, Y.W., de Bourmont, M., Dunaway, M.M.: Appraisals of harms and injustice trigger an eerie feeling that decreases trust in artificial intelligence systems. Ann. Oper. Res. **308**(1), 525–548 (2022)
7. Wamba, S.F., Queiroz, M.M., Braganza, A.: Preface: artificial intelligence in operations management. Ann. Oper. Res. **308**(1), 1–6 (2022)
8. Fink, O., Netland, T.H., Feuerriegel, S.: Artificial intelligence across company borders. Commun. ACM **65**(1), 34–36 (2022)
9. Baz, M., Khatri, S., Baz, A., Alhakami, H., Agrawal, A., Khan, R.A.: Blockchain and artificial intelligence applications to defeat COVID-19 pandemic. Comput. Syst. Sci. Eng. **40**(2), 691–702 (2022)
10. Hassan, M.R., et al.: Prostate cancer classification from ultrasound and MRI images using deep learning based explainable artificial intelligence. Future Gener. Comput. Syst. **127**, 462–472 (2022)
11. Oneto, L., Bunte, K., Navarin, N.: Advances in artificial neural networks, machine learning and computational intelligence. Neurocomputing **470**, 300–303 (2022)
12. Holzinger, A., et al.: Natalia Díaz Rodríguez: Information fusion as an integrative cross-cutting enabler to achieve robust, explainable, and trustworthy medical artificial intelligence. Inf. Fusion **79**, 263–278 (2022)

Real-Time Surveillance System of Health Big Data Based on Artificial Intelligence

Xueqing Shi(✉) and Xiaomu Yu

School of Health Care Technology, Dalian Neusoft University of Information, Dalian, Liaoning, China

sonia_neusoft@outlook.com

Abstract. Big data has been defined as techniques and technologies which deal with data at an extreme scale economically. It could be used in varies area. The application of health big data in the fields of prevention, medical treatment and health risk assessment is becoming more and more extensive. It can help improve the efficiency and accuracy of diagnosis, assist in medical decision-making, and optimize the allocation of medical resources. It plays an unconventional role in the identification of high risk of many chronic diseases. It also plays an important role in the prevention and control of infectious diseases. This article introduces the application of health big data in the field of precision health risk assessment and its future development prospects. It also raises issues that need to be considered in applications.

Keywords: Real-time Surveillance System · Artificial Intelligence · Big Data · Precision Health Risk Assessment

1 Introduction

In the past decade, big data has become a common noun, which means a kind of research method involved in complex large-scale data sets. Big data refers to data collection that can't be perceived, accessed, managed, processed and serviced within a limited time by traditional computer technology and software/hardware tools, usually with character of volume, velocity, variety, veracity and value. Theoretically, all medical and health-related data should be called health big data, with no limitation in size and type. Initially health big data was known as case reporting data (e.g., large epidemiological survey), management data (e.g., medical insurance payments data), clinical registration data, electronic health records (EHRs) and so on. Afterwards, other type of resource was gradually incorporated into the study of health big data, including physiological measurement data (e.g., human physiology captured by implanting medical devices and wearable devices), biological markers (e.g., various types of omics data), medical imaging data, network information (e.g., social media), and so on. Health big data has shown big potential in promoting biomedical research, improving the efficiency of medical and developing precision health management. This article focusses on the application of health big data in the precision health risk assessment.

J. H. Abawajy et al. (Eds.): ICATCI 2022, LNDECT 170, pp. 634–641, 2023.
https://doi.org/10.1007/978-3-031-29097-8_75

2 Identification of Individual Health Risks

2.1 Early Recognition and Diagnosis of Disease

Health big data can be used to identify high-risk individuals in the crowd, to whom early intervention can be given, aimed to improve the medical efficiency, and improving outcomes. Upadhyaya and his colleagues have developed the algorithm to identify diabetes prevalence before operation with the EHRs data, expecting by identifying and intervening pre-operative diabetes patients, intra- and post-operative care plan would be improved, and the post-operative complications would be reduced. The algorithm had shown a more than 90% in both sensitivity and specificity (see Table 1) [1]. In 2019 Nature published a study on the use of electronic data in the United States Veterans Medical System, by assessment of dangerous factor via depth learning [2], indicating that health big data has shown a huge application potential in the early recognition in patients with high risk, prevention of potential adverse event, and improving health outcomes, within the therapeutic time window.

Table 1. Validity of Upadhyaya Proposed Method Compared With Other Methods [1]

Measure	Upadhyaya proposed	CCW	DDC	SUPREME-DM	Mayo Clinic manual
Sensitivity	0.99	0.67	0.92	0.64	0.84
Specificity	0.99	1.00	0.94	0.97	0.98
Accuracy	0.99	0.94	0.94	0.91	0.96
P value (sensitivity)	NA	<.01	<.01	<.01	<.01
McNemar's test (sensitivity)	NA	223.00	43.31	239.00	348.01
P value (specificity)	NA	.32	<.01	<.01	<.01
McNemar's test (specificity)	NA	1.00	176.02	98.04	16.20
Total patients, n	684	460	815	543	626

CCW = Chronic Condition Data Warehouse;
DDC = Durham Diabetes Coalition;
SUPREME-DM = Surveillance, Prevention, and Management of Diabetes Mellitus
NA = not applicable;

At present, artificial intelligence (AI) technology has made qualitative progress in automatic identification of medical images and disease diagnosis, which can significantly improve the accuracy of disease diagnosis, in use of health big data. For example, the use of computer-aided testing system to read CT scan results can screen out 70% of lung cancer cases missed by traditional manual reading methods, which can be used as a "second reader" other than radiologists, significantly improving the accuracy of lung cancer diagnosis [3]. The algorithm developed by using deep convolutional neural network can read fundus pictures to identify diabetic retinopathy and diabetic macular edema, showing that the reading results were highly consistent with the judgment of professional fundus physicians. Skin cancer is a common malignancy in humans, which clinical diagnostic procedure included visual observation, clinical screening, and eventually confirmation by dermoscopic analysis or histopathological examination. The use of automatic image classification technology to identify fine-grained changes in the appearance of skin lesions can accurately distinguish skin cancer from benign lesions, and the accuracy was similar to the judgment made by professional doctors [4]. In the daily care of heart disease, deep learning algorithms can read cardiac magnetic resonance imaging (MRI) images and generate indicators including valve damage, hemodynamics, and heart function, which has a great role in improving MRI reading accuracy, shortening reading time, and improving clinical MRI utilization [5]. In the treatment of patients with coronavirus disease (COVID-19), AI medical imaging and CT equipment have been deployed and used in many hospitals in Hubei Province. Intelligent devices can screen lung diseases in seconds, helping doctors identify cases of COVID-19 and alleviating the strain on regional medical resources.

2.2 Precise Definition of Disease Phenotypes

Human disease phenotypes, especially complex diseases, were pervasively heterogeneous. Multiple symptoms might be clinically visible in the same disease (e.g., heart failure), and even patients with the same symptoms might have different physiological mechanisms. In addition, different patients may have other different types of disease. Health big data can be used to establish a disease phenotype atlas (phenomapping), and identify relatively homogeneous patients to form a subgroup, which through the refinement of disease phenotypes, can not only improve the pertinence and accuracy of treatment, reduce the occurrence of unexpected outcomes, safely and efficiently carry out personalized diagnosis and treatment in the clinical practice, but also accurately identify research objects, carry out targeted research, and greatly improve research efficiency in the study work. Current methods of classifying patient phenotypes include traditional methods based on clinical diagnosis and treatment guidelines, natural language mining processes based on clinical records, and machine learning analysis methods based on bioinformatics. With the accumulation of health big data resources, the latter two methods will gradually become the main ways to identify population phenotype subgroups.

2.3 Support of Clinical Decision

The use of various types of health risk scores to assess future disease risk can effectively identify high-risk individuals, help clinicians and patients make clinical decisions, and take corresponding drug and non-drug treatment (see Fig. 1) [6]. Using real-world data, health assessment algorithms developed by using machine learning and AI technologies can identify patterns those cannot be discovered by traditional models, leading to a higher accuracy of risk warning and further promoting the standardization of the clinical diagnosis and treatment process. For example, the in-hospital mortality rate of heart disease patients can be predicted by deep learning algorithm for image data, and the accuracy is significantly better than that of GRACE, TI-MI, MAGGIC, GWTG-HF and other traditional risk assessment models [7]. Evaluation models using EHR data in patients with acute myocardial infarction can accurately predict the risk of future hemorrhagic adverse events and help clinically determine the timing of dual antiplatelet therapy [8]. People with subclinical high genetic risk of atherosclerosis identified by the Global Genetic Risk Score for Coronary Heart Disease could derive greater benefit from statin therapy [9]. The use of health big data to assess individual risks is gradually integrated into the process of personalized medicine. For example, based on EHR data, a research team from Stanford University used machine learning methods to help clinicians quickly query previous cases with similar characteristics to current patients, present previous treatment methods and patient outcomes, and assist clinicians to make decisions in the treatment plan of current patients [10]. With the rapid development and application of various types of omics data, it has shown broad prospects in understanding the pathogenesis of diseases from the perspective of systems biology and guiding personalized preventive treatment in tumor, cardiovascular and other fields [11].

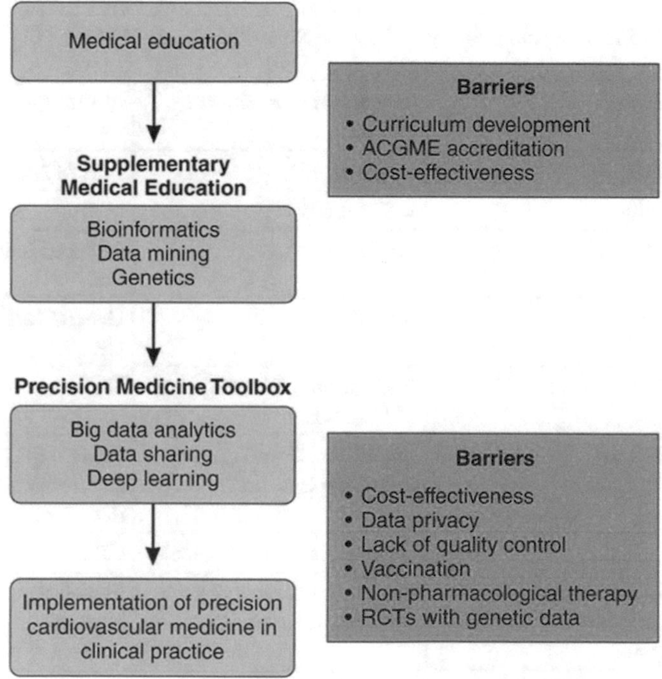

Fig. 1. Implementation of precision medicine in the practice of cardiovascular medicine

3 Group Health Risk Assessment

3.1 Real-Time Surveillance of Infectious Diseases

The use of surveillance systems to identify the occurrence, transmission patterns, epidemic processes and causes of the disease is key to control infectious disease. Traditional disease surveillance systems relied on the data analysis reported and summary by monitoring site, and the entire system was expensive, inefficient and not real-time (see Table 2). Modern real-time surveillance systems for infectious diseases can quickly identify the occurrence, development and transmission patterns of diseases, and provide valuable information for public health interventions by using biological health big data. Researchers have paid great attention to Internet and social media data, hoping to use these data for real-time surveillance of infectious diseases, but the correlation between such data and disease was weak, and various algorithms and external verification were needed to screen the true signals of diseases from a large amount of noise [12]. Compared with data from Internet and social media, others that indicated actual health behaviors and medical practices were more specific and more suitable for real-time monitoring of diseases [11].

Table 2. Presious Study on Disease Surveillance and Big Data [11]

Year	Disease(s)	Researcher/Surveillance System	Institution
1662	Plague	John Graunt	Bills of Mortality, UK
1817	Smallpox	J. C. Moore	
1847	Influenza	William Farr	General Registrar Office, UK
1854	Cholera	John Snow	General Registrar Office, UK
~1900	All	International Classification of Diseases	WHO
1918	Influenza	121 Cities Mortality Reporting System	Centers for Disease Control and Prevention (CDC)
1976	Influenza	Weekly viral surveillance	CDC
1984	Influenza, viral hepatitis, acute urethritis, measles, mumps	First computerized disease surveillance network	French Sentinelles Network
~2000	All	Medical claims forms are first used	Private companies and government
2008	Influenza	Google Flu Trends launched	Google
2015	Influenza	Birth of hybrid systems	Public/private partnership

Disease transmission models, including covariate variables such as demographics, environment, and geography, can further suggest the structural patterns of disease epidemics, identify high-risk groups, and guide disease prevention and control. In the process of fighting the COVID-19 epidemic, China has launched a first-level response to major public health emergencies and implemented control measures throughout the country. Modeling based on daily epidemic data can quickly carry out epidemic prediction, and provide scientific reference for epidemic prevention and control decision-making and measures evaluation. The prediction results of Academician Zhong Nanshan's team showed that if the implementation of control measures was postponed for 5 days, the scale of the epidemic would increase to 3 times, suggesting that the public health interventions implemented in China have effectively controlled the development of the epidemic [13]. Using big data and AI joint laboratories, via linkage with offline medical institutions, can efficiently carry out screening of high-risk groups, assessment of the severity of disease epidemics early, identification of epidemic characteristics, and early warning of epidemic prevention and control, thus provide a large amount of useful information for disease intervention strategies in various countries and regions all over the world. For example, in the process of COVID-19 epidemic prevention and control, if the data of multiple departments such as transportation, communications, and medical care were recorded and shared in real time, possible sources of infection and close contacts would be more accurately identified, regional risk classification management would be realized, and more accurate prevention and control measures would be carried out.

3.2 Group Health Behavior Identification

In the field of chronic diseases, the use of health big data to assess the burden of disease at the group level, through deeply investigating the health behavior, social and environmental determinants of the group, can promote the implementation to carry out targeted control of the modifiable risk factors, such as physical activity, diet, smoking and pollution exposure, which leaded to an increased burden of chronic diseases, and precise disease prevention work. On the other hand, health big data can identify various social and environmental health determinants, such as weather patterns, pollution levels, allergen monitoring, land use, traffic conditions and community criminal records, etc., understanding the health level of groups in a specific social and environmental context, and promoting more comprehensive disease prevention and health promotion. In the field of infectious disease prevention and control, such as the epidemiological investigation of the COVID-19 epidemic, using mobile phones and network data sources can quickly obtain the trajectory of individual activities, trace potential close contacts and virus transmitter, grasp accurate population flow information in the region, and provide important information for the precise prevention and control of the epidemic. In the future, health records will follow individuals for life, and health big data will provide an individual-centered, healthy social based model, which can effectively identify target individuals and subpopulations, and carry out more targeted and personalized public health measures (such as weight control, healthy diet, etc.).

4 Consideration in the Application of Big Data in Health Risk Assessment

4.1 Shared Utilization of Multi-source Data

With the rapid accumulation of EHR, bionomics, imaging, sensing equipment, media, social and public data of millions or more of the population, not only directly health-related data can provide information for disease prevention and control, but a large number of data that seemed to be indirectly related to health can also be integrated and applied to this process. In the future, the full sharing and utilization of multi-source heterogeneous data is the development trend of big data in accurate disease risk assessment.

4.2 Data Evaluation

The availability, applicability and authenticity of all types of health big data must be evaluated before they are used appropriately. First of all, the availability should be evaluated by understanding what data resources were available and the legal norms and ethical constraints that should be adhered to when using them [11]. Second, big data contains a large sample of information covering a variety of medical scenarios. So before using it for health risk assessment, it is necessary to assess the applicability of the data for specific purposes. Third, the most critical element of any study is to ensure the authenticity [12]. In traditional research, researchers always spent a lot of energy to ensure the authenticity of data from various aspects like study design and implementation process, discuss the factors affecting the authenticity of the results, and how to make a correct interpretation of the results. Large biobanks currently established, such as the UK Biobank (UKB), also devoted a great deal of effort to quality control to ensure the authenticity of the data [15, 16]. Since health big data contains a large amount of second-hand data, and it is not collected for research purposes at the beginning, it should be analyzed the characteristics of its data source or cross-verified to understand its authenticity as much as possible before application [17].

4.3 The Application of AI

With the rapid accumulation of data, it has been becoming increasingly difficult to process and analyze large, structured, and rapidly updated health big data in traditional methods. AI is a computer science technique that can mimic human thinking process, learning ability, and knowledge storage, and is capable of dealing with a variety of complex problems. AI has a great potential to leverage health big data for risk assessment [5, 10, 18]. Virtually all diseases are caused by genetic, environmental (e.g., water pollution), and behavioral factors (e.g., exercise), and inherently complex and heterogeneous, thus accurate risk prediction and assessment, rather than simply stratifying in traditional risk scores are needed. Over the past 10 years, AI has been used in disease diagnosis and risk prediction to reduce mortality and improve the quality and efficiency of health care by identifying the exact phenotype of diseases and carrying out targeted treatments. In the future, AI will go beyond traditional analytical methods to promote the transformation of modern medical paradigms to precision medicine through accurate risk assessment of complex and heterogeneous diseases.

References

1. Upadhyaya, S.G., et al.: Automated diabetes case identification using electronic health record data at a tertiary care facility. Mayo Clinic Proc. Innov. Qual. Outcomes **1**(1), 100–110 (2017)
2. Tomašev, N., et al.: A clinically applicable approach to continuous prediction of future acute kidney injury. Nature **572**(7767), 116–119 (2019)
3. Schoenhagen, P., Mehta, N.: Big data, smart computer systems, and doctor-patient relationship. Eur. Heart J. **38**(7), 508–510 (2017)
4. Esteva, A., et al.: Dermatologist-level classification of skin cancer with deep neural networks. Nature **542**(7639), 115–118 (2017)
5. Antes, A.L., Burrous, S., Sisk, B.A., Schuelke, M.J., Keune, J.D., DuBois, J.M.: Exploring perceptions of healthcare technologies enabled by artificial intelligence: an online, scenario-based survey. BMC Med. Inform. Decis. Mak. **21**(1), 221 (2021)
6. Krittanawong, C.: Future physicians in the era of precision cardiovascular medicine. Circulation **136**(17), 1572–1574 (2017)
7. Kwon, J.M., Kim, K.H., Jeon, K.H., Park, J.: Deep learning for predicting in-hospital mortality among heart disease patients based on echocardiography. Echocardiogr. (Mount Kisco N.Y.) **36**(2), 213–218 (2019)
8. Tan, J., et al.: 2021 Asian Pacific Society of cardiology consensus recommendations on the use of P2Y12 receptor antagonists in the Asia-Pacific region: special populations. European Cardiol. **16**, e43 (2021)
9. Natarajan, P., et al.: Polygenic risk score identifies subgroup with higher burden of atherosclerosis and greater relative benefit from statin therapy in the primary prevention setting. Circulation **135**(22), 2091–2101 (2017)
10. Rodriguez, F., Scheinker, D., Harrington, R.A.: Promise and perils of big data and artificial intelligence in clinical medicine and biomedical research. Circ. Res. **123**(12), 1282–1284 (2018)
11. Simonsen, L., Gog, J. R., Olson, D., Viboud, C.: Infectious disease surveillance in the big data era: towards faster and locally relevant systems. J. Infect. Dis. **214**(suppl_4), S380–S385 (2016)
12. Saracci, R.: Epidemiology in wonderland: Big Data and precision medicine. Eur. J. Epidemiol. **33**(3), 245–257 (2018). https://doi.org/10.1007/s10654-018-0385-9
13. Yang, Z., et al.: Modified SEIR and AI prediction of the epidemics trend of COVID-19 in China under public health interventions. J. Thorac. Dis. **12**(3), 165–174 (2020)
14. Hemingway, H., et al.: Big Data for Better Outcomes, BigData@Heart Consortium of 20 academic and industry partners including ESC (2018). Big data from electronic health records for early and late translational cardiovascular research: challenges and potential. European Heart J. **39**(16), 1481–1495
15. Cox, N.: UK Biobank shares the promise of big data. Nature **562**(7726), 194–195 (2018)
16. Bycroft, C., et al.: The UK Biobank resource with deep phenotyping and genomic data. Nature **562**(7726), 203–209 (2018)
17. Reimer, A.P., Madigan, E.A.: Veracity in big data: how good is good enough. Health Informatics J. **25**(4), 1290–1298 (2019)
18. Adedinsewo, D.A., et al.: Cardiovascular disease screening in women: leveraging artificial intelligence and digital tools. Circ. Res. **130**(4), 673–690 (2022)

The Application of Artificial Intelligence Technology in Computer-Aided Translation

Dapeng Li[1](✉), Yuzhe Wang[1], Xuming Yang[1], and Malik Alassery[2]

[1] Changchun University of Finance and Economics, Changchun, Jilin, China
gavinli2021@163.com
[2] Amman Arab University, Amman, Jordan

Abstract. The application of artificial intelligence technology (AIT) in computer-aided translation (CAT) can effectively improve work efficiency, promote the accuracy of information data, and improve the whole social and economic benefits. Firstly, this paper briefly introduces and analyzes AI language. Secondly, it expounds its concept, characteristics, history of development and current situation. Then, combined with specific examples, this paper discusses in detail the existing problems and relevant improvement methods in the process of CAT. Finally, based on AIT, the effect of applying AIT to CAT is tested. The final test results show that CAT takes different time for different semantic materials in translation time, but it can be concluded that the more corpus, the longer time it takes. But the overall time consumption is within 10 s. At the same time, the accuracy and integrity rates were maintained above 80%.

Keywords: Artificial Intelligence Technology · Computer-aided Translation · Applied Translation

1 Introduction

The application of AIT in CAT can not only make CAT more accurate, but also improve work efficiency. At present, the degree of intelligence in China is relatively high [1, 2]. However, due to its short development time, less relevant knowledge and lack of practical experience, there is a large demand for talents in this field, and the personnel structure is complex. In addition, due to the relatively late domestic research on this kind of professional technology and the lack of a certain number of scientific research achievements and innovation ability, it is difficult to achieve the expected effect in practical application. Therefore, it is necessary to strengthen the theoretical study of relevant aspects and apply the research results to CAT [3, 4].

Many scholars have done relevant researches on AIT. In this process, what needs to be solved is how to effectively combine AI technology with traditional language and apply it to practical work [5, 6]. At present, many experts and scholars have conducted relevant research and exploration in China. According to the characteristics and development trend of AIT, some scholars put forward a problem faced in the application of CAT, and give solutions and suggestions according to the specific situation [7, 8]. Some scholars

J. H. Abawajy et al. (Eds.): ICATCI 2022, LNDECT 170, pp. 642–648, 2023.
https://doi.org/10.1007/978-3-031-29097-8_76

combine it with neural network and apply it to computer-aided decoding system. Other scholars use the application of neural network technology in the translation process to extract and analyze the data, so as to draw conclusions and put forward relevant suggestions [9, 10]. The above research has laid the foundation for this paper.

This paper mainly presents problems and solutions with the use of AIT in computerised terms. Firstly, this paper interprets and analyzes the concepts and characteristics of robot and AI in various countries in the world, and then summarizes the deficiencies and deficiencies of China's work in this field at the present stage. Finally, some feasible suggestions are put forward to deal with the challenges and opportunities of AIT.

2 Discussion on the Application of AIT in CAT

2.1 Artificial Intelligence Technology

2.1.1 Development

AIT uses the intelligence of computer to realize automatic translation, recognition and processing of machine language. The development of AIT is a process of continuous improvement and gradual improvement. From the initial research on AIT to now, some achievements have been made, but it is difficult to play a role in practice due to some defects and unstable factors of the computer itself. At present, many robot companies in the world have begun to apply it to practical applications. For example, Toshiba Industry Co., Ltd. of Japan has developed a virtualization system based on fuzzy logic neural network, which can control the robot motion by analyzing and predicting the relationship between input information and output value. Jaguar Automobile Manufacturing Co., Ltd. Has developed an intelligent vehicle chassis controller, which can realize multiple functions such as vehicle automatic driving, driver and control system interaction [11, 12].

2.1.2 Method

The core of AIT is to simulate people. Data processing is carried out by computer, rather than directly calculating the amount of information. A large number of machine languages need to be used in practical work. Therefore, to achieve this, we must have a certain degree of proficiency.

AIT mainly includes the following methods: (1) fuzzy logic reasoning method. This algorithm is based on a certain mathematical model to describe and judge things, and store these information in the computer. It can realize the problems of accuracy, clarity and accuracy, and can also be used for the problems requiring high computing power. It can also be applied to some complex systems or uncertain data, such as engineering quantity control and so on. (2) Neural network technology is a new algorithm based on fuzzy mathematics theory. It uses human brain power and memory ability to process information by simulating human neural system and automatically complete data analysis in computer.

Artificial neural network is an intelligent information processing system proposed by human beings for the composition of human brain nervous system and the way of information transmission. Through the reverse propagation error, you continuously learn the sample data, modify the network weight layer by layer, and calculate the output value. The error will spread back to the network input layer, and the weight of each layer will be corrected until the output result matches the target value set by the network. Initialization of relevant parameters of the algorithm: it mainly initializes the approach distance s and step a of the algorithm. If we have no prior knowledge, we can initialize all to 0, initialize the step size to 1, and adjust it when optimizing the fitting. The specific sequence of algorithms is as follows:

Determine the gradient of the cost function of the current position. For, the gradient expression of the cost function is:

$$\frac{\ell}{\ell\theta_{\mathrm{i}}}k(\theta_0, \theta_1, \ldots, \theta_n) \tag{1}$$

The gradient of the cost function is multiplied by the step size to obtain the falling distance of the current position, that is:

$$\beta\frac{\ell}{\ell\theta_i}\mathrm{k}(\theta_0, \theta_1, \ldots, \theta_n) \tag{2}$$

Equations 1 and 2 show that the network error go hand in hand to the weight of each layer. Therefore, in order to reduce the network estimation error, the weight of the network must be adjusted.

2.2 CAT

2.2.1 Introduction

CAT technology is a new method, which can effectively solve translation problems in computer applications. CAT refers to the use of the function of machine language to realize AIT and information processing and analysis. In the traditional mode, the development process of computer translation is affected by some factors such as certain difficulty and high cost. It can convert traditional language into digital information and improve work efficiency. (1) Artificial intelligence system has the characteristics of strong data processing ability and rapid response. (2) High degree of intelligence and automation. The demand for knowledge is increasing. AIT has become the most important part of CAT application, and with the wide application and in-depth research of science and technology in various fields.

The application of AIT in CAT is mainly to solve the contradiction between traditional language and modern technology, so that machines and people can communicate with each other, and then improve the efficiency of information transmission. At present, it is commonly used to use machine vision to translate and process the required text content. This method can help people understand the meaning of the article quickly and accurately. It can also convert text into computer format file form by image or sound.

2.2.2 Characteristics

The main features of CAT system include:

(1) Accuracy. That is, after the completion of relevant information and data processing, it is necessary to ensure that it has accurate and correct expression of instruction meaning and required form and content. At the same time, it can effectively and reasonably explain these information and finally achieve the expected goal. When designing the CAT system, we must take into account the different requirements of different user groups for software function and interface operation. In the actual application process, it needs to be analyzed combined with the specific situation. For example, intelligent language can be used to edit and process voice, video and other information content.

(2) The principle of coexistence of integrity and practicability. AIT is a system integrating the ability of computer intelligent analysis. It needs to be applied to different fields, which requires that the accuracy of data processing must be guaranteed. Therefore, for different types of documents, only when there is no deviation and error can the corresponding software be used to complete the relevant translation work. For non-structural documents, professional terms or standard specifications need to be used to judge their accuracy and give corresponding opinions.

(3) It has strong rapidity, controllability and real-time tracking ability. Artificial intelligence algorithm can respond to and deal with problems or abnormal situations in the computer, which ensures the high validity of data transmission and large amount of information storage. Due to the differences between different users, the required information needs to be processed. Machine language is an expression tool with high generalization and abstract thinking ability. At the same time, it can also be adjusted according to the specific situation. Therefore, AIT can enable users to operate simply and clearly and complete the task requirements. This is fully reflected and applied in the advantages of CAT.

2.3 Relationship Between CAT and AIT

In the process of CAT, we can complete the rapid and accurate transmission of information through AIT. However, due to the limitations and fuzziness of machine language to a certain extent and people's limited ability to understand it, this situation occurs. This requires the organic combination of the two, so as to realize the effective integration and development innovation of the two. In addition, it can also promote translators to further study and master CAT technology, and apply it to more aspects to solve problems in practical work.

The application of CAT technology is AIT to realize information processing and automation in computer-aided, rather than language conversion in the traditional sense, that is, semantic expression of words, parts of speech or grammar. Both of them are machine translation methods under a new model, and they have their own different characteristics. Firstly, they all use images or text to complete the semantic transmission process. Secondly, they also have the combination of visual perception and auditory perception to a certain extent, and simulate the thinking of human brain, and can convert image and sound information into corresponding text language in computer.

3 Experiment

3.1 Application Process of AIT

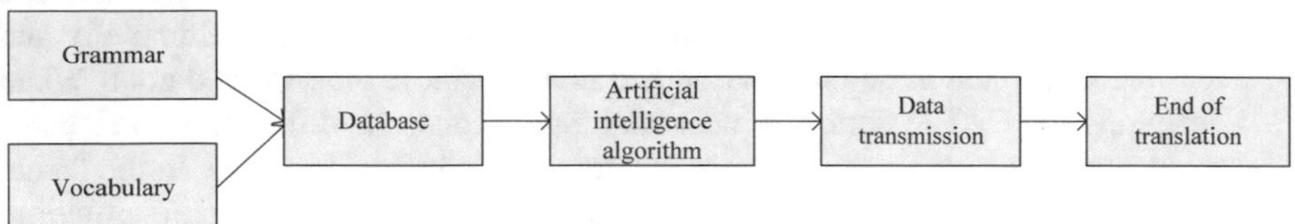

Fig. 1. Translation process

The translation of AIT is based on computer network and uses computer language as a bridge to transfer the required information to machines. In this process, we need the help of the system's database, speech recognition and other auxiliary functions. As can be seen from Fig. 1, the translation of AIT mainly includes the following procedures: (1) Information processing. The intelligent machine needs to have a preliminary understanding of the content to be carried out, and complete the relevant data collection, transmission and analysis through the computer network. At the same time, it can also use the speech recognition system to extract and screen the required information. (2) Semantic expression and logical judgment. In computer-aided language, it plays a very important role and irreplaceable characteristics. When classifying, storing and transmitting different types of data, text files, audio and video or images can be used to process them. In addition, you can also use manual input equipment or directly modify and delete text files. Finally, you should use computer language as a bridge to transfer the required information to the machine, and finally achieve the goal.

3.2 CAT Technology Test Process

In the process of CAT, because the language used is Chinese, we need to test it. First, human translator has to check for errors. For the expressions with errors, we can start from the following aspects: (1) Make grammatical correctness judgment. (2) There will be some deviations in the judgment criteria, methods and program design of logical analysis ability. In addition, it should be noted that all possible error probabilities should be input into the computer as much as possible during the test, otherwise it is easy to cause unnecessary losses or even trouble to users. Secondly, the translation process needs to be comprehensively tested, and the data sets and files need to be tested. The first is to ensure that files, documents and software have been compiled. The second is to automatically generate and upload all program codes to the server through the corresponding system, and then relevant personnel complete the final task according to the program code. Finally, the database or other network resource download equipment can be used to realize the possible problems and troubleshooting in the data set and translation process, and the computer software development needs to be tested according to the preset objectives.

4 Discussion

4.1 Effect Analysis of CAT

Table 1 shows the translation effect data tested in this experiment.

Table 1. Translation effect

	Time of translation (s)	Precision (%)	Full rate (%)
Words and expressions	2	100	100
Short sentence	3	92	98
Screed	5	89	92
Paragraph	7	85	89
Essay	10	82	84

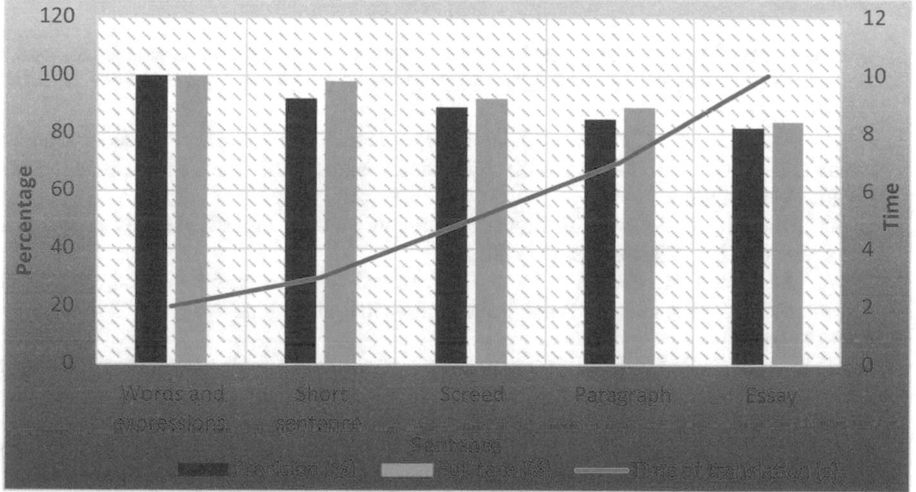

Fig. 2. Comparison of translation effect

As can be seen from Fig. 2, CAT takes different time for different semantic materials in terms of translation time, but it can be found that the more corpus, the longer it takes. But the overall time consumption is within 10 s. At the same time, the accuracy and integrity rates were maintained above 80%. The application of AIT can effectively solve the problems in the application of traditional computer translation. Through the analysis and extraction of a large number of data, intelligent system can be used to realize automatic control in this process, which provides the basis for subsequent processing.

5 Conclusion

The application of AIT has been widely used in various industries. As a new high-tech means, it will have a broader and in-depth development space in the future. This paper mainly introduces the implementation process and related contents of computer intelligent algorithm in computer-aided decoder. Then it analyzes the problems and solutions of traditional technology, and puts forward solutions. Finally, it summarizes and prospects the research in this field. AIT can bring us a lot of development space.

Acknowledgment. Supported by Higher Education Teaching Reform Project of Education Department of Jilin Province, 2021 "Research and Practice of Translation Technology Teaching in the Era of Artificial Intelligence" (NO. JLJY202160065269), Vocational Education Research Project of Jilin Province, 2021 "Research on the Teaching Model of Propaganda and Translation of Folk Culture in Jilin Province from the Perspective of Eco-translation" (NO. 2021XHY211), and Social Science Research Project of Education Department of Jilin Province, 2021 "Research on the English Translation of the Characteristic Vocabulary of Ice and Snow Economy in Cold Regions in Jilin Province from the Perspective of Communication Studies" (NO. JJKH20211391SK).

References

1. Gao, Y., Geras, K.J., Lewin, A.A., et al.: New frontiers: an update on computer-aided diagnosis for breast imaging in the age of artificial intelligence. Am. J. Roentgenol. **212**(2), 300–307 (2019)
2. Ks, A., Wcsia, B., Jjn, B., et al.: Artificial intelligence methods in computer-aided diagnostic tools and decision support analytics for clinical informatics. Artif. Intell. Prec. Health, 31–59 (2020). ScienceDirect
3. Gao, F., Zhu, Y., et al.: Artificial intelligence in computer-aided diagnosis of abdomen diseases. Sci. China Life Sci. **62**(10), 128–131 (2019)
4. Frank, M.S., Dmd, K., Zhang, B.Z., et al.: Artificial intelligence for the computer-aided detection of periapical lesions in cone-beam computed tomographic images. J. Endod. **46**(7), 987–993 (2020)
5. Mori, Y., Kudo, S.E., Morp, K.: Implementation of artificial intelligence into colonoscopy: Experience of research and development of computer-aided diagnostic system for endocytoscopy. Nippon Shokakibyo Gakkai Zasshi **115**(12), 1030–1036 (2018)
6. Kovaerchuk, B., Triantaphyllou, E., Ruiz, J.F., et al.: Fuzzy logic in computer aided breast cancer diagnosis: analysis of proliferation. Artistic Intell. Med. **11**(1), 75–85 (1997)
7. Satiya, J., Dammeyer, K., Ahmad, O., et al.: Is artificial intelligence in colonoscopy ready for prime-time: a systematic review and meta-analysis of randomized controlled trials. Am. J. Gastroenterol. **115**(1), S165–S165 (2020)
8. Alagappan, M., Brown, J., Mori, Y., et al.: Artificial intelligence in gastrointestinal endoscopy: the future is almost here. World J. Gastrointest. Endosc. **10**(10), 19–29 (2018)
9. Kim, E.: Artificial intelligence and computer-aided diagnosis in medicine. Curr. Med. Imaging Rev. **16**(1), 1–1 (2020)
10. Bhargavi, V.R., Rajesh, V.: Computer aided bright lesion classification in fundus image based on feature extraction. Int. J. Pattern Recogn. Artif. Intell. **32**(11), 1850034.1–1850034.15 (2018)
11. Sharma, P., Sarma, K.K., Mastorakis, N.E.: Artificial intelligence aided electronic warfare systems- recent trends and evolving applications. IEEE Access **8**(99), 1 (2020)
12. Kumar, S.: State of the art-intense review on artificial intelligence systems application in process planning and manufacturing. Eng. Appl. Artif. Intell. **65**, 294–329 (2017)

The Development of Computer Translation System Based on Artificial Intelligence Technology

Shihan Hou[1(✉)], Yaxuan Han[1], and Rajit Ragab[2]

[1] Haojing College of Shaanxi University of Science and Technology, Xi'an, Shaanxi, China
houshihan202201@163.com

[2] Vellore Institute of Technology, Vellore, India

Abstract. Computer translation is the process of converting one language into another language by using the programming and computing power of computers. If the corpus is borrowed, more possible automatic translation effects can be achieved by processing grammatical structures, word identification, and vocabulary combinations. Due to the initial underdevelopment of the computer industry, this new technology has not made substantial progress. Later, with the development of computer technology and the improvement of various artificial intelligence technologies (AIT), computer translation developed slowly, and some practical systems were formed. This paper uses artificial intelligence technology to build the overall framework of the computer translation system (CTS), and tests the sensitivity of the system to translate sentence length. Compared with the manual translation method, the computer translation system is more efficient and accurate when translating.

Keywords: Artificial Intelligence Technology · Computer Translation System · Human Translation Method · Translation Efficiency

1 Introduction

In today's society, my country and other countries in the world have carried out deeper economic, cultural, political, military, educational and other exchanges. The translation market continues to expand. However, the number of translators is limited, and the efficiency of manual translation is low and the cost is high. Pure manual translation cannot meet the growing translation market at all. With the vigorous development of AIT, CTS have emerged as the times require. Computer translation can help translators complete translation work efficiently and quickly.

At present, many scholars at home and abroad have carried out in-depth research on the development of computer translation systems based on artificial intelligence technology, and have achieved certain research results. For example, the birth of big data has developed a variety of new technologies for people, artificial intelligence technology is one of them is that it improves the work efficiency of computing devices with its distributed performance and fast computing. It uses a computing database to execute the

J. H. Abawajy et al. (Eds.): ICATCI 2022, LNDECT 170, pp. 649–656, 2023.
https://doi.org/10.1007/978-3-031-29097-8_77

translation model of neural networks, and finally enables researchers to develop practical computer translation software [1]. A scholar proposed an example-based computer translation model. The difference is that it uses the database to derive translation rules indirectly from a statistical point of view, and updates the translation model by integrating different languages in the database, so that computer translation gradually enters a relatively advanced stage development trend [2]. A scholar proposed a heuristic translation method for smoothing different translation versions. The most important thing for this method is to collect a large number of vocabulary, and all the interpretive meanings of various words are included. When translation is required, the matching mode is activated to concatenate words into a complete sentence [3]. Although the research of computer translation system based on AIT has made good progress, if the translation efficiency is higher, AIT should be used in the system design to improve the performance of the CTS.

This paper introduces the translation process of computer translation technology, exemplifies the characteristics of computer translation system based on artificial intelligence technology, and introduces sentence similarity algorithm when designing computer translation system, so that the system can complete translation work according to semantic similarity during translation. Finally, the sentence length sensitivity, translation time and translation accuracy of the manual translation method and the computer translation system are compared, which proves that the computer translation system is more suitable for translation work.

2 Computer Translation System and Similarity Algorithm

2.1 Computer Translation Technology

The process of computer translation can be divided into two steps: one is to interpret the meaning of the sentence in the source language, and the other is to recompile the meaning of the sentence to the target language according to the corresponding grammar rules. Computer translation is to design a program to make it understand like a real person. The meaning of the sentence, and then the result of the sentence processing output is close to the real effect [4]. There are currently two computer translation methods, one is to translate based on the original knowledge, and the other is to translate based on the text data package [5]. The method of translation based on the original knowledge refers to distinguishing various lexical information according to the traditional language rules, analyzing the different combinations and syntax of the original vocabulary, so as to translate the natural language. However, in the face of large-scale real sentences, this method has great limitations and cannot achieve the desired effect. The method of translating on the text data package refers to sorting the word order by counting the text content and matching the vocabulary on the data package. Therefore, this method is very widely used in non-limited fields [6].

Generally speaking, computer translation requires a large number of rules, and it is impossible for these rules to be written entirely by manpower, and it is difficult to maintain the consistency of the rules. Although statistical methods solve the problem of rules, it is difficult to express profound statistical information. Hierarchical, highly general information [7]. There are various methods of computer translation, but no breakthrough has been made in the quality of translation by any translation method.

2.2 Characteristics of Computer Translation System Based on Artificial Intelligence

Compared with traditional machine translation (MT) technology, CTS has many characteristics of its own, mainly in the following aspects:

(1) In the light of language normativeness, CTS can be used in daily communication and dialogue, so the spoken language is strong and contains not standard languages, and differences in personal sense of language make each person comprehend the meaning of the sentence differently. Intermittent, inverted sentences, or emotionally charged expressions in people's speech cannot be translated by speech recognition, and it is difficult to achieve 100% accuracy. Compared with traditional MT, artificial intelligence CTS focuses on correct pronunciation of vocabulary words, remove unnecessary repetition in sentences, and understand key information in translations [8].
(2) In the process of communication, people usually have verbal emotions, which can change the speed. When speaking, some words are masked or combined with words with multiple meanings, so that the system cannot understand the true meaning of the speaker during the recognition process. In the process of talking, the influence of noise in the environment cannot be ruled out will hinder the sentence translation, such as the sound of opening and closing doors, car horns, coughs, all of which lead to imprecise SR results. Therefore, the CTS needs to create a language recognition model with good clarity, as well as the MT technology that can handling various translation order results generated by the AIT and has EC capabilities [9].
(3) Oral communication have no sign to separate sentences like endorsement to express the normal meaning. Oral expressions are relatively short, omitting a lot of basics or conceptual statements, the enunciation is intermittent, and there may even be many topics in a sentence. In this way, the combined results are puzzling even when the enunciation of sentences is connected at the level of translation theory. Therefore, CTS should be based on understanding the intention of the interlocutor, rather than pure sentence-to-sentence translation [10].
(4) For the CTS, people are more inclined to TS that can communicate with the interlocutor, or even imitate the speaking style of TS, which is more difficult for the semantic match(SM) [11].

In short, AI CTS is a comprehensive technology that integrates speech recognition, MT, and SM. Only by understanding sound and text, the best translation results can be obtained through CTS.

2.3 Sentence Similarity (SS) Calculation

In the process of natural language processing, such as English-Chinese dictionary input, test paper analysis, data mining, etc., the calculation of the similarity between two strings is very important. The translation system can judge or cluster the original strings, sentences, paragraphs, texts, etc. according to the calculated results. Similarity calculation is an important parameter to measure translation systems [12].

Although a lot of research on similarity has been carried out, many of these methods often use distance-based calculation methods to calculate the actual similarity. Some translation systems use a relatively simple method, that is, the same method from the surface level of words, and does not consider information such as word semantics. The general calculation formula is:

$$SIM = \frac{S + 2}{M + N} \times 100\% \tag{1}$$

Among them, SIM represents the similarity, S represents the same words in the two sentences, M, N represent the total number of words in the sentence to be translated and the example sentence (ES), respectively. As long as the value of SIM is greater than a certain threshold, it can be considered that the sentence to be translated and the ES have a certain similarity.

When calculating the similarity between two strings a and b, the similarity is calculated by using the cosine theorem commonly used in mathematics.

$$sim(p, q) = \cos(\vec{c}, \vec{d}) \tag{2}$$

Among them, p and q are different characters in strings a and b, and the sizes $\vec{c}$ and $\vec{d}$ of their corresponding vector spaces are calculated as their similarity. The larger the cosine value, the smaller the angle between the two character vector spaces and the higher the similarity between the two characters p and q.

3 Design of Computer Translation System Based on AIT

3.1 Overall Framework of Computer Translation System

Figure 1 shows the overall framework of a CTS based on artificial intelligence technology. The system mainly includes a translation generation module and a similarity calculation module. Input the sentence or article to be translated into the system, the system will segment the content to be translated, and calculate the SS with the sentence to be translated. If the instance exactly matches the sentence to be translated, the sentence to be translated will directly reuse the instance's If the translation is partially matched, that is, the SS to be translated and the instance is greater than or equal to the similarity threshold set by the system, the instance translation is provided to the translator, so that the translator can make corresponding improvements to obtain the translation of the sentence to be translated. The final generated translation can also be stored on system memory.

3.2 System Translation Process Design

Figure 2 shows the translation process of the computer translation system. The translation process can be regarded as the process of information transmission, and a certain source-channel model is used to explain this process. Translate the source text (that is, the text

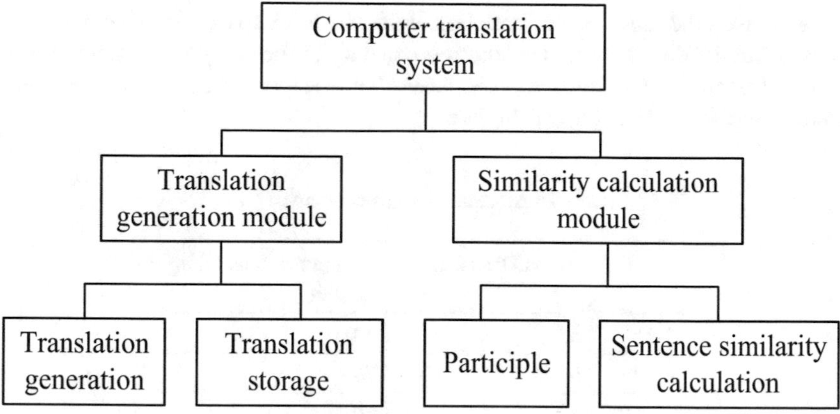

Fig. 1. Overall framework of the system

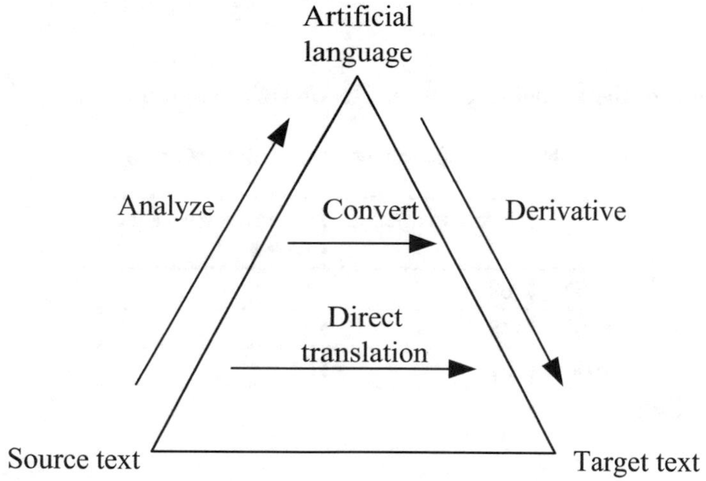

Fig. 2. System translation process

to be translated) into the target text through systematic artificial language analysis and sentence meaning derivation.

4 Application of Translation System

4.1 System Test

In order to verify the influence of sentence length on the quality of translation, the sentence length sensitivity of computer translation system was tested according to different sentence length distributions, and the sentence length sensitivity of human translation was compared. As shown in Table 1, at 0–15, the sensitivity of human translation is 67%, and the sensitivity of computer translation system is 72%, and as the sentence length increases, the sensitivity of both translation methods to translated sentence length decreases, but the drop in sentence length sensitivity of the computer system is smaller, indicating that the system translation is better than the manual translation method, and the quality of the translated text is higher.

Table 1. Sentence length sensitivity (%)

	Human translation	Computer translation system
0–15	67	72
15–35	61	70
35–60	52	69
60–100	40	65

4.2 Analysis of the Actual Application Results of the System

Table 2. Comparison of translation duration (s)

	Human translation	Computer translation system
3000	68	12
8000	115	27
10000	134	32

As shown in Table 2, the time used to translate Chinese with different numbers of words into English using human translation and computer translation system translation. When the number of Chinese characters (CC) is 3000, the manual translation and system translation take 68s and 12s respectively. When the number of CC is 8000, the two methods take 115 s and 27 s respectively. When the number of CC is 10000, the translation time of the two methods is 134 s and 32 s respectively. According to the comparison results of translation time, the translation efficiency of computer systems based on artificial intelligence technology is higher than that of human translation methods.

Figure 3 shows the comparison of the translation accuracy of the two translation methods. According to the data comparison results, the translation accuracy of the CTS is higher than that of the human translation method. The main reason is that because the computer translation system integrates various languages and the semantic library is powerful, it can convert a language into sentence meaning during translation, and finally output the translation that matches the most translation. However, human translation will inevitably lead to confusion of word meanings and grammatical errors.

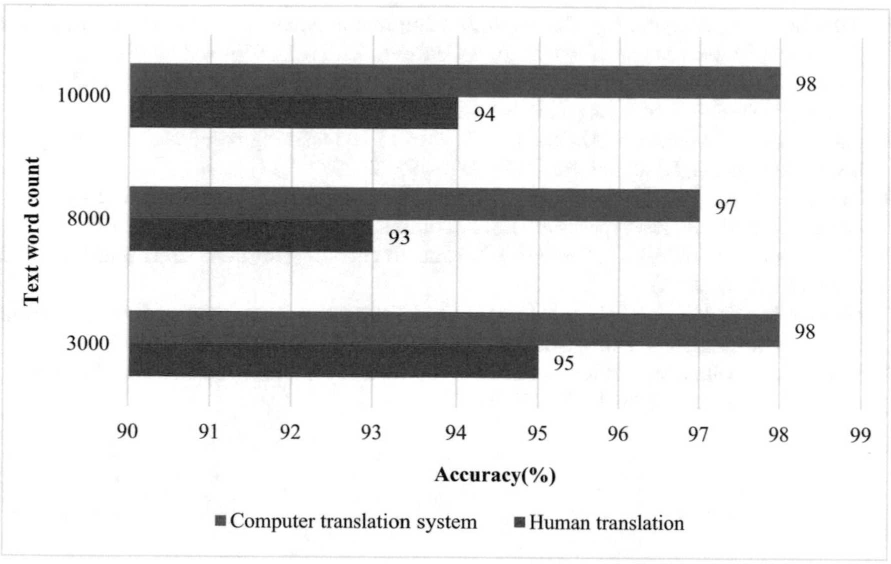

Fig. 3. Comparison of translation accuracy (%)

5 Conclusion

This paper studies the application of computer translation system based on artificial intelligence technology in Chinese-English translation. After designing the computer translation system, the sensitivity of the system is tested to ensure the translation quality. The CTS spends less time in translation and has a higher accuracy rate, indicating that the computer translation system is more adaptable to a large number of translation work, thereby improving translation efficiency.

Acknowledgment. Shaanxi Provincial Department of Education 2020 Scientific Research Plan Project: Research on International Publicity Translation of Guanzhong Folk Culture from the Perspective of Cultural Confidence (No. 20JK0079).

References

1. Puasa, G.: Data Glove for American sign language alphabet and numbers (1–9) translation system. Int. J. Adv. Trends Comput. Sci. Eng. **8**(4), 1128–1133 (2019)
2. Folasade, D., Safiriyu, E.: Development of a machine translation system for date referencing in yoruba language. Int. J. Comput. Appl. **177**(20), 32–38 (2019)
3. Vavruk, E., Kushnir, D.: Mobile system for text recognition and translation with using microsoft cognitive OCR. Comput. Syst. Network **1**(905), 33–42 (2018)
4. Pandey, V., Padmavati, D., Kumar, R.: Hindi chhattisgarhi machine translation system using statistical approach. Webology **18**(Special Issue 02), 208–222 (2021)
5. Bhadwal, N., Agrawal, P., Madaan, V.: A machine translation system from Hindi to Sanskrit Language using rule based approach. Scalable Comput. **21**(3), 543–554 (2020)

6. Albashir, T.M., Alzoubi, H., Albatainih, M.: Improving Arabic instant machine translation: the case of Arabic triangle of language. J. Comput. Sci. **16**(7), 956–965 (2020)
7. Uko, V.S., Dubukumah, G.B., Ayosubomi, I.K.: Microcontroller based speech to text translation system. Eur. J. Eng. Res. Sci. **4**(12), 149–154 (2019)
8. Fahad, A.A., Hassan, H.J., Abdullah, S.H., et al.: Deep learning-based deaf & mute gesture translation system. Int. J. Sci. Res. **9**(5), 288–292 (2020)
9. Asjea, S., Ismail, O., Khawatmi, S.: Support of Arabic sign language machine translation based on morphological processing. Int. J. Comput.r Appl. **177**(15), 28–36 (2019)
10. Majumdar, B., Sarode, S.C., Sarode, G.S., et al.: Technology: artificial intelligence. Br. Dent. J. **224**(12), 916 (2018)
11. Mumin, M., Seddiqui, M.H., Iqbal, M.Z., et al.: shu-torjoma: an English Bangla statistical machine translation system. J. Comput. Sci. **15**(7), 1022–1039 (2019)
12. Duner, I.: Machine translation system for the industry domain and Croatian language. J. Inform. Organ. Sci. **44**(1), 33–50 (2020)

Application of Railway Tunnel Antenna Signal Cover and Artificial Intelligence Technology

Ping Yang(✉), Fei Peng, and Hong Zhang

Beijing Polytechnic, Beijing, China
yangpingeda914@126.com

Abstract. In the recent years, the quick development of railway proposed more higher requirements at railway communication. The tunnel as one form part of the railway, need consider the signal cover of it. The antenna is the important form parts to realize the signal transmission in the radio communication system. We need design one antenna signal cover scheme which able to working in the tunnel environment based on the LTE frequency range to meeting the requirements of future railway development and improve the stability of communication in tunnel.

Artificial intelligence is one new technology science of research and develop the theory, method, technology and application system to imitate, extend and expand human intelligent. This article focus on research one railway tunnel signal cover scheme which combine with artificial intelligence technology to improve the intelligent and valid of railway tunnel communication.

Keywords: Tunnel Signal · LTE · Antenna Cover · Artificial Intelligence

1 Foreword

The railway acting important role in the national economy development as the traffic tool. GSM-R been widely adopted as one communication standard, this technical standard has more obvious advantage used in the railway, more and more wide application range of GSM-R system under the background of national high speed economy development. Thus, it occur many new business at railway along with increasing user quantity, example process video monitor the train, this further more increase the requirements at railway mobile communication system, example the requirements at working band width and delay, but these requirements not completely achieved by GSM-R.

GSM-R is the generation II outputs of mobile communication technology, more narrow communication frequency band and no higher utilize ratio of frequency spectrum, and more slower transmit speed ratio, Mainly applied in voice and some data transmit communication business.

Additionally, along with the continued progress of mobile communication network, the next generation mobile communication system of railway process upgrade and change generation based in GSM-R system is one necessary trend. GSM-R technology is based on 2G standard, so oppositely fall behind compare to the communication capacity, transmission speed rate and other technical index, unable to meet the future

J. H. Abawajy et al. (Eds.): ICATCI 2022, LNDECT 170, pp. 657–664, 2023.
https://doi.org/10.1007/978-3-031-29097-8_78

railway high standard communication speed requirements. International railway league also start to step to step draw up the next one generation communication standard of railway. The international railway league already clearly point out that, directly transmit the GSM-R to exact 4G LTE-R, and applied in the radio communication requirements of railway [1]. More few GSM-R system frequency source and difficult to achieved the current railway communication business requirements, need quicken the speed of develop LTE-R system, confirm the LTE-R frequency as soon as possible, push future development of LTE-R. Compare to GSM-R, LTE-R able to use more lower cost to provide high quality communication, and able to more bigger improve the applicable of data transmission.

The advantages when LTE-R technology applied in the railway communication as below:

The first is the anti-decline characteristics of OFDM technology, able to overcome the Doppler frequency transmission expansion which caused by railway high speed moving, and suitably weaken the multiply pathway transmission effect in the complex environment;

The second is LTE-R adopted same frequency networking, the train able to soft shift applying when required in over the region, validly reduced the interrupt happen ratio during the traditional shift process, improve the reliability of railway communication system;

The third is valid frequency spectrum source, the system dynamic source distribution validly utilize the frequency, greatly improved the efficiency, and able to expand the working band width of railway mobile communication system, make the mobile service which provide broadband to passengers come to be possible;

The fourth is to the current train dispatch and train control [2], LTE-R more stronger safety and reliably to GSM-R, not only provided double way certification of network and users, but also able to encrypt the wire link route, so LTE-R able to improve the transmit efficiency of communication data when high speed railway trains safety running.

LTE (Long-Term Evolution) long term evolution technology is the generation IV communication technology which based on generation III communication technology. Compare to the early communication standards, it has more higher transmit speed and has more bigger capacity to contain more complex system, and reduce the operation cost in a certain degree, make the communication place and time limit of users more smaller, and quicken the communication speed, the communication quality also get more bigger promotion. LTE is one extension of 3G technology, pass through continue development and improvement from 2004, purpose to support more range environment communication, support low and high speed moving user equipment to use [6].

AI artificial intelligence is one new technology science of research and develop the theory, method, technology and application system to imitate, extend and expand human intelligent. This is one branch of computer science to know well about the substance of intelligence, and produce one new intelligent machine which able to do the action similar to artificial intelligence, the main target is make the machine able to undertake the complex works which need be finished by artificial intelligence generally. The advantages as below:

(1) Quick calculate, search, statistic and analysis speed.

(2) High running degree, storage and accuracy.
(3) Simple and clear digit, strong reliability and modelling.

No matter in core technology, or in the typical application, the artificial intelligence technology all already occur the burst type development. Along with the continue updating and break though of platform, algorithm and interactive method, these meaning change of application modes and shown incomparable bright development prospects.

2 Text

Tunnel is one semi close environment, it has more stronger electromagnetism shielding characteristics, so it existing a certain degree difficulties when process valid signal covering in tunnel. So, the cover blind area in the railway tunnel make users radio communication affected after enter into tunnel, seriously affect the user's radio communication experience. More bigger radio communication difference between the radio wave transmission mode under tunnel environment, document [3] proposed one new scheme which used in tunnel communication cover then can cover the tunnel environment through double way antenna.

Micro honeycomb system mainly solve the communication problems in buildings, subway station and in tunnel. The working environment in the tunnel require the antenna has the characteristics such as double way radiation wave petal and low section structure, the antenna height must less than one tenth of wave length, double way direction represent shape 8 at section E. Because the cover area represent one very long distance at longitudinal direction, view from the electric wave transmission, hanging type install antenna more favourable to signal transmission [5]. To the straight line type short distance tunnel, able to install the logarithm period antenna or fix direction plate type antenna at tunnel entrance at both side of tunnel, this able to realize the better cover in tunnel, the install method shown as Fig. 1.

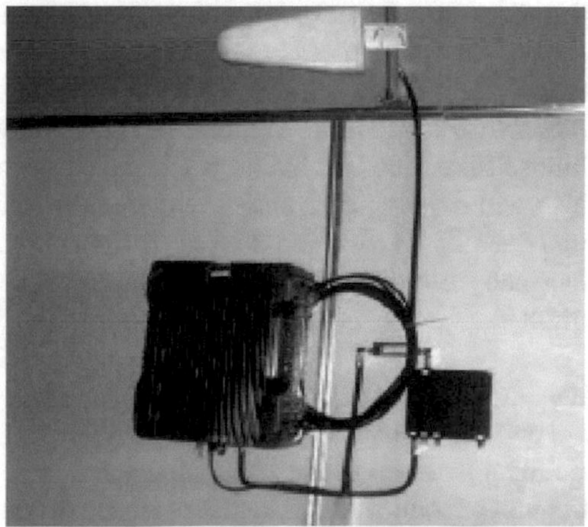

Fig. 1. Antenna installation effect figure

This article emphasized research one antenna system scheme which used in tunnel cover LTE working frequency range (1710 MHz–2690 MHz) to improve the reliability of communication data transmission when train running in the tunnel [10].

Firstly introduce the problems faced by that under tunnel environment communication, and simply state the broadband antenna design principle, explained the necessary of that design LTE double way antenna in the tunnel cover.

Then start from the current tunnel cover scheme, select to use the fix position antenna and double way radiation antenna to process signal cover at tunnel through analyse the advantages and shortcomings of three schemes and the antenna technical requirements of cover scheme.

2.1 Scheme Summary

In the recent years, the tunnel is the important form parts in the railway system because rapidly quick development of national traffic transporting and more mountain area in our geographical conditions. Most limit internal space of railway tunnel, so the train body will caused more bigger affection at signal transmission because the train already occupy most of the space when train pass through the tunnel.

We can divided the tunnel into the below three types according to the length: short tunnel (less than 100 m), middle length tunnel (100–600 m), long tunnel (bigger than 600 m).

Tunnel as one special environment, the radio covers mainly has the below characteristics:

(1) Tunnel cover has strong shield property. The mountain environment around the tunnel generates attenuation to space signal able to achieved above 40 dB, so, to the signal of outdoor base antenna, it bring more bigger affection to user's communication quality because it already passed through more bigger attenuation when transmitted to tunnel.
(2) There has obvious filling effect in the tunnel. The rest space in tunnel are very small when train passing through, this make communication signal transmission in the tunnel been bigger impeded, filling effect should be especially obvious when train high speed running.
(3) Tunnel cover effect able to cause more bigger affection to network index. Because the tunnel structure self has a certain shield performance at the electromagnetism wave, the user also will existing difficulties when communication in the tunnel, so, it existing large quantity cover blind area in the tunnel and able to caused the users unable to normally use the radio communication in the tunnel, make users communication limited.

To the railway tunnel, it able fully fill the whole tunnel space [8] when train passing through. The distance between the out layer of train body and internal wall of tunnel able to make signal rapidly attenuated in the space after train entered into tunnel, at the same time, the consumption of train shell make the users received radio signal strength weaken and affect the communication quality. The antenna in the communication system charge for send and retrieve signal, one pair suitable antenna able to improve the whole set system performance at a certain degree. So, to the communication cover of railway tunnel, we need design one set antenna cover solve scheme.

2.2 Scheme Design

Early we have adopted schemes such as logarithm period antenna or fix direction plate type antenna to cover, this method flexibly design, more easy to realize in the practice, more better suitable to short distance railway tunnel communication. Additionally, able to adopt the scheme that install antenna in the tunnel port and in the tunnel at the same time, and use the coaxial cable to process feed. More smaller attenuation at this method, able to select the suitable antenna gain according to the installation condition, thus confirm the most suitable antenna types. The antenna which adopted high gain able to expanding the cover range. For the commonly adopted leakage cable scheme, the leakage cable like the continue transverse antenna, process radiation through the gap on the coaxial cable, thus cover. This scheme has the advantages such as evenly cover, small floor area, high stability and easily maintain, but very high cost leakage cable. Document [9] proposed one model which suitable to tunnel transmission, this formula is Lpath = 20lgf + 30lgd + Lf (n) – 28 dB.

Among, f is frequency (MHz), d is distance (metre), Lf is storey penetrate consumption factors (dB), n is the storey quantity between terminal and antenna. Aim to the tunnel cover environment which discussed in this article, able to not consider Lf(n). So, able to use formula (1) in the tunnel to process radio transmission estimation, means,

$$\text{Lpath} = 20\text{lgf} + 30\text{lgd} - 28\text{dB} \tag{1}$$

So, adopt coaxial feed distribute type antenna system is one more economy method when process tunnel covering. The design scheme has a certain flexibility, low cost and easily installed. More important is that more smaller attenuation of coaxial cable, adopt high gain double way antenna then can improve the cover distance of antenna, this point most practice under the cover situation in long tunnel. Able to directly place the single piece double way antenna in the tunnel when applying in short tunnel, convenient and practicable [4]. Through comparison we can get that, Table 1 shown the comparison of three types cover schemes. Considering from the practical economy application, this article mainly discuss the solve scheme through coaxial feed non source type antenna.

(1) Short tunnel solve scheme

We can know from the early mentioned tunnel length division that, if more small interval among multiply short tunnel then can place double way antenna at middle of both tunnels when process signal cover the short tunnel which length under 100 m, realize signal cover of tunnel through the micro base station. Also can place the antenna at tunnel port when antenna gain above 5 dBi.

(2) Middle length tunnel solve scheme

The lowest connect in signal electric level required by the above schemes is –85 dBm when process signal cover the middle length tunnel from 100 m to 600 m, because the it will affect the signal transmission after train enter into tunnel, add 10 B protection when place antenna at tunnel port, place 8 dBi gain single way

Table 1. The advantage and shortcoming comparison of three types cover schemes

Cover schemes	Advantages	Shortcomings
Coaxial feed non source distribute type antenna	Cheap price High installation reliability	Antenna gain selection based on limit of installation condition
Leakage cable	Smaller signal wave range Able to cover multiply signal service	A little higher cost
Fibre-optical feed source distribute type antenna	Suitable to complex environment Able to reduce electromagnetism disturb	Smaller single load frequency output power Cover range not vast as non-source antenna

antenna at tunnel port, at the same time, place 5 dBi gain double way antenna at middle of tunnel, then the sum of the cover distance of these two antenna bigger than the whole length of tunnel, able to realize the signal cover of the whole tunnel. Check Table 2.

Table 2. The distance budget when single antenna process tunnel cover

Cover distance	Place antenna at entrance (gain 8 dBi)	Place double way antenna at middle (gain 5 dBi)
Adopt micro base station	400 m	480 m
Adopt repeater station	250 m	360 m

In summary, able to select the suitable antenna place scheme according to the actual situation to the tunnel which length within 600 m. Additionally, select more high gain antenna able to improve the cover distance of antenna when antenna cover distance not enough to meet the whole tunnel length. Able to adopt that place double way antenna at middle of tunnel to cover to the tunnel which has a certain bending ratio.

(3) Long tunnel solve scheme

Aim to the actual application, common tunnel with length above 600 m, able to place the double way antenna per interval a certain distance in the tunnel, use coaxial cable to series multiply radiation units, reasonably distribute the antenna, adopt wall hanging or installed at top of tunnel for the antennas, thus realize the validly cover of space. The detail scheme shown as Fig. 2.

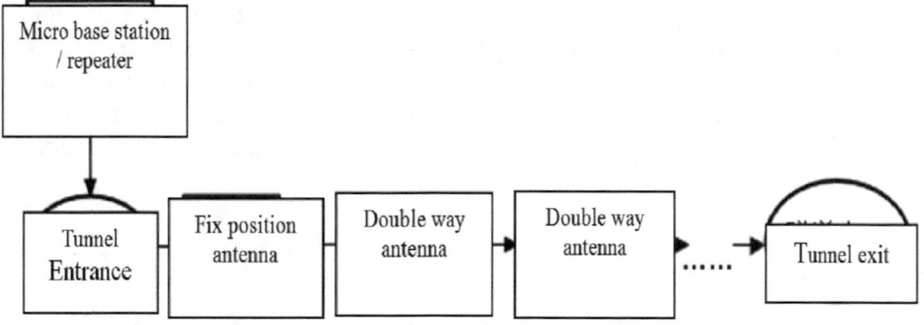

Fig. 2. Long distance tunnel cover scheme figure

2.3 The Related Theory of Double Way Radiation Antenna

Because the characteristic of radio transmission environment in tunnel, so it existing a certain degree difficulties when process signal covering in tunnel. First, the retrieve machine retrieved signal will pass through a series complex reflex and scattering which caused by semi seal environment, and internal wall of tunnel able to absorb and scattering electromagnetism wave. Under the common situation, the frequency range of mobile communication far higher than tens MHz to the tunnel which only has tens MHz stop frequency, tunnel environment easily make signal radiation close field are multiply pathway transmission [4]. Adopt high gain double way radiation antenna when process radio covering, this able to reduce the pathway consumption affection and multiply pathway effect at a certain degree [3]. Generally, the methods used to realize the double way radiation antenna are: back to back assemble two fix position antenna to integrate, process same phase position excitation at the a period of array and through the gap antenna [7], etc.

3 Conclusions

Because the characteristics of tunnel structure, so process signal cover on it also has the characteristics. This article described several main factors which need be considered when process tunnel signal covering through analyse advantages and shortcomings of three types covering schemes, select the coaxial feed non source distribute type antenna system to cover, and research to get the antenna signal cover scheme of three types tunnel [11].

Now artificial intelligence technology already applied everywhere, our world quickly happening changes. Future will be intelligence times, artificial intelligence will merge our work and live together and deeply change everything of human society.

Artificial intelligence technology able to greatly save human power cost, make the complex big data get high efficiency analyse and reasonably applied, this necessarily come to be one main tendency of artificial intelligence technology in the future technology application area. Along with quick development of artificial intelligence technology, we can process data analyse of three types tunnels through tunnel working status, environment conditions, humidity, temperature, train speed, train length and based on artificial intelligence technology (equipment) to improve the intelligence application of railway tunnel signal antenna cover scheme, corresponding to three tunnel cover solve scheme, done the corresponding and in time adjustment of retrieve and lunch antenna working status according to different data analyse result, guarantee the train in time and validly retrieve the usable signal, to realize the better transmission of communication signal in the tunnel [12].

References

1. UIC.LTE/SAE-The Future Railway Mobile Radio System Long-Term Versions on Railway Mobile Radio Technologies. UIC, Paris (2009)
2. Han, Z., Song, W., Sheng, X.: In-band RCS reduction and gain enhancement for a patch antenna array by using a 1-D periodic metasurface reflector. IEEE Trans. Antennas Propag. **67**(6), 4269–4274 (2019)
3. Rajabalipanah, H., Abdolali, A.: Ultrabroadband monostatic/bistatic RCS reduction via high-entropy phase-encoded polarization conversion metasurfaces. IEEE Antennas Wirel. Propag. Lett. **18**(6), 1233–1237 (2019)
4. Yin, L., Yang, P., Gan, Y.-Y., Yang, F., Yang, S., Nie, Z.: A low cost, low in-band RCS microstrip phased-array antenna with integrated 2-bit phase shifter. IEEE Trans. Antennas Propagat. (2020)
5. Chew, W.C., Jin, J.M., Michielssen, S. J.M.: Fast and Efficient Algorithms in Computational Electromagnetics. Artech House, Boston, MA (2001)
6. Carpentieri, B., Duff, I.S., Giraud, L.: Sparse pattern selection strategies for robust Frobenius-norm minimization preconditioners in electromagnetism. Numer. Linear Algebra Appl. **7**, 667–685 (2000)
7. Cui, Y.H., Li, R.L., Wang, P.: A novel broadband planar antenna for 2G/3G/LTE base stations. IEEE Trans. Antennas Propagat. **61**(5), 2767–2774 (2013). https://doi.org/10.1109/TAP.2013.2244837
8. Liu, L.S., Zhang, Z.J., Tian, Z.J., et al.: A bidirectional Endfire array with compact antenna elements for coal/mine tunner communication. IEEE Antennas Radio Propag. Lett. **11**, 342–345 (2012)
9. Joozdani, M.Z., Amirhosseini, M.K., Abdolali, A.: Wideband radar cross-section reduction of patch array antenna with miniaturized hexagonal loop frequency selective surface. Electron. Lett. **52**(9), 767–768 (2016)
10. Paquay, M., Iriarte, J.C., Ederra, I., et al.: Thin AMC structure for radar cross-section reduction. IEEE Trans. Antennas Propag. **55**(12), 3630–3638 (2007)
11. Cheng, Y.-F., Liao, C., Gao, G.-F., Peng, L., Ding, X.: Performance enhancement of a planar slot phased array by using dual-mode SIW cavity and coding metasurface. IEEE Trans. Antennas Propag. **69**(9), 6022–6027 (2021). https://doi.org/10.1109/TAP.2021.3061583
12. Zarbakhsh, S., Akbari, M., Samadi, F., Sebak, A.: Broadband and high-gain circularly-polarized antenna with low RCS. IEEE Trans. Antennas Propag. **67**(1), 16–23 (2019)

Fast Extraction Algorithm of Big Data Information Based on Artificial Intelligence

Gang Chen[1], Rujuan Wang[1](✉), and Hooman Bavarsad Asghari[2]

[1] Changchun Humanities and Sciences College, Changchun 130117, Jilin, China
wangrujuan@ccrw.edu.cn
[2] Islamic Azad University, Tehran, Iran

Abstract. Under the influence of factors such as the explosive growth of information, the rapid rise of the knowledge economy, and new technologies, a large number of diverse massive unstructured data are generated. How to quickly and effectively obtain a large amount of complex and changeable information has become a problem for people to think about. How to quickly and effectively obtain massive amounts of information and accurately extract and analyze them with suitable algorithms has become one of the current research hotspots. Because traditional algorithms can no longer be satisfied with extracting useful information quickly and efficiently. Therefore, people hope to find a better and more reliable method to solve these problems. This article adopts experimental analysis method and data analysis method to better understand the effect of the fast extraction algorithm of big data information. The algorithm has high extraction accuracy, recognition accuracy and recall rate of Sample 4, which are 0.941, 0.933, and 0.908 respectively. In addition, the recall rate of the algorithm has been greatly improved, and the extraction accuracy rate has also been improved, which can meet the needs of rapid information extraction.

Keywords: Artificial Intelligence · Big Data · Information Extraction · Recall Rate

1 Introduction

In recent years, big data has been widely used in various fields and has applications in many aspects. These include industries such as medical diagnosis, smart home, and automobiles. Because the traditional method has the disadvantages of long processing time, low efficiency and high cost, the processing speed is relatively slow and cannot meet the needs of users. Therefore, this article proposes a fast extraction algorithm, which is of great significance to the research of big data information extraction.

At present, Some scholars proposed a method for extracting breast lesion information based on an electrostatic scattering model. After experimental verification, the model can effectively extract the parameter information of breast cancer lesions [1]. Some scholars proposed a point cloud feature information extraction algorithm with two thresholds, which can quickly and accurately extract feature information from scattered and noisy point cloud models, and has strong robustness [2]. Some scholars pointed

J. H. Abawajy et al. (Eds.): ICATCI 2022, LNDECT 170, pp. 665–671, 2023.
https://doi.org/10.1007/978-3-031-29097-8_79

out that automated extraction is an important part of the computerization of procurement. The automatic extraction algorithm takes the multi-item parallel mining as the business premise, and optimizes the rule model by reconstructing the mining process [3]. Therefore, it is of great practical importance and certain research value to conduct research from a new perspective and quickly obtain big data information in combination with AI and related research.

This article focuses on these aspects. First, it describes artificial intelligence and its related research. Next, it introduces the fast algorithm for extracting big data and related research. Finally, the fast big data extraction algorithm is studied experimentally, and relevant experimental and analytical results are obtained.

2 Related Theoretical Overview and Research

2.1 Artificial Intelligence and Related Research

Artificial intelligence uses computers, related algorithms and logic to complete various activities required for human automatic control processes.

The main feature of artificial intelligence is that it has strong intelligent recognition capabilities, can classify according to different objects, and analyze and process these data information. It achieves human goals by processing and analyzing information through the human brain, computer processing capabilities and other related subject knowledge.

Artificial intelligence requires strong analysis, reasoning, and judgment capabilities, while a computer as a machine device needs to meet many aspects. First of all, when performing calculations on hardware, there must be a certain number or types of simple and stable structure, powerful functions and convenient use. The second is that the software program, that is, the algorithm should be simple and fast, and can quickly process complex data information [4, 5].

AI plays a vital role in the life of human beings. It can help people solve some problems, such as real-time detection of the environment, identification and analysis of data and information. These functions are mainly realized by computers. However, due to current computer hardware performance and software design defects. Therefore, if you want to better apply it to your work, you must increase the computer system's operating speed and anti-interference ability to ensure its normal use and efficient operation. It is necessary to strengthen the development and research of artificial intelligence technology to make it more practical and reliable.

Human beings have accumulated a lot of experience and knowledge in the development of biology. Artificial intelligence studies how to use human intelligence to complete complex tasks an automated technology. It has particular corresponding applications in many fields, for example, intelligent robots. Intelligent robots can collect, analyze and process data, and control decision-making, so that human-computer interaction can be realized, and the behavior of the robot can be simulated through computer software programs and various functions and commands required in the process of executing actions can be executed.

Later, artificial neural networks also appeared. It can simplify a large number of complex data processing, information extraction and analysis work into a simple vector composed of nonlinear neurons as a set of simple function approximation process. It also involves intelligent algorithms, including genetic algorithms and particle coding methods, which are of great significance to the research of human behavior simulation [6, 7].

2.2 Big Data Information Fast Extraction Algorithm and Related Research

Various data and information on the Internet are developing rapidly and appearing in front of people. It has become a difficult problem to efficiently construct these data and extract them quickly for people to query. The research of fast information extraction algorithm is carried out under this background.

Many domestic and foreign researchers have conducted research on the extraction of network information. In the new step, some scholars have proposed that content and structural features can be captured in a unified framework to ensure the accuracy of runtime reasoning and extract structured data from pages. Some researchers have proposed a new network information extraction algorithm based on the combination of DOM (document) tree and DBSCAN clustering algorithm. This algorithm has high versatility and can achieve efficient data collection and efficient information acquisition.

Because the big data information extraction algorithm has a great demand for information collection, while collecting information, it is necessary to preprocess the data at the beginning of the algorithm application to improve the information state for better information collection and realize the rapid acquisition of information. In addition, it is necessary to find the internal connection of data in various operational data and establish data connection [8, 9]. The process of big data information mining is shown in Fig. 1.

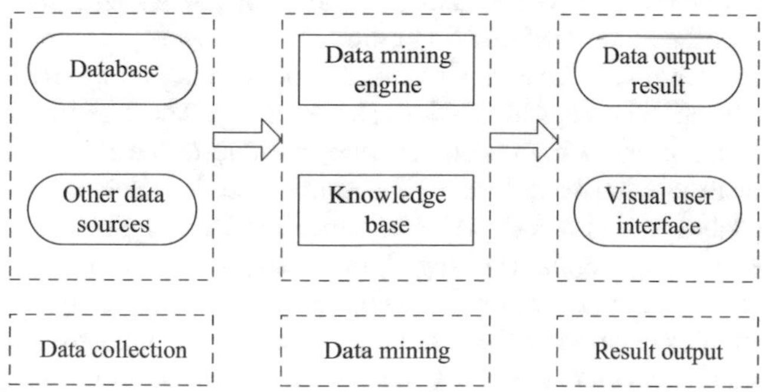

Fig. 1. The process of big data information mining

The process of quickly extracting big data information usually includes the following steps.

(1) Preprocessing the big data information. Its purpose is to remove semi-structured data information and prepare for further information extraction.

(2) Use multiple information models to describe the information to be extracted. Generally, multiple information templates can be predefined according to the information characteristics of specific fields and stored in the template library for users to choose.
(3) Perform template recognition on the preprocessed big data information.
(4) Use a pattern matching program to identify different parts of the specified information pattern.

These are the main methods used when retrieving web page information. One is network information retrieval technology based on DOM tree. All tags of a website form a tree structure, and information can be extracted by acting on the tree structure. It overcomes the limitations of Web data sources and is easy to implement. And the other one is model-based network information retrieval technology. The main purpose of this technology is to extract information from web pages automatically generated based on a unified model.

The technical evaluation of information retrieval initially used the classic information retrieval evaluation indicators, namely recall rate and precision rate, but its definition changed slightly. The revised scoring index may induce breeding phenomenon. There is no data in the entry, but the system may generate it by mistake. The recall rate can be roughly regarded as a measure of the amount of information retrieved correctly, and the precision rate is used to measure the amount of information retrieved correctly [10, 11]. The calculation method of recall rate and precision rate is shown in formula (1) (2).

$$R = \frac{C}{N} \tag{1}$$

$$A = \frac{C}{T} \tag{2}$$

Among them, R is the recall rate, C is the number of extracted correct information points, N is the number of all correct information points, A is the precision rate, and T is the number of all extracted information points.

However, from the two indicators of recall rate and accuracy rate, the existing network information retrieval technology has obvious shortcomings. The extraction method based on package induction has a high accuracy rate, but due to the limitations of manual labeling, its recall rate has been lost. At the same time, it is necessary to maintain a set of learning label pages for each website, which results in high labor costs and poor flexibility. Once the page format is changed, the wrapper will become invalid.

The extraction method based on the webpage structure has a high recall rate, but because it is completely based on the similarity of the performance structure, the system cannot understand the content that the user is interested in, so the extraction result may contain a large amount of data that the user does not need. The extraction method based on natural language processing completely ignores the structural features of the website and focuses on the semantic analysis of the website content. There is very little information on the Internet. It can handle any website. Although theoretically high recognition rate and accuracy can be guaranteed, it must be supported by a huge and extensive knowledge base. Since human research on natural language processing is imminent and the knowledge base is difficult to update, the extraction is only based on natural language processing.

Information extraction has a specific format, which can also be understood as a template. Patterns are terms created from the message understanding lecture series. We call the final result of information retrieval a model, the domain in the model is called a location, and the matching rules used in the information retrieval process are called a model.

The purpose of information extraction is to extract certain information. It breaks through the restriction that humans must read, understand and extract information when retrieving information, and realizes the automatic retrieval, understanding and extraction of information. Information retrieval can further refine the results of information search. Any information retrieval module can be used as an output application [12].

It can be applied to traditional information retrieval systems. After information search, relevant texts are extracted from the information provided, and the process of pure information search is further transformed into a process of information matching (understanding), and the traditional information search system is transformed into an intelligent system that provides information to make users more satisfied. For example, large-scale online search systems, libraries and information retrieval systems, Internet search engines, etc. can be improved by introducing information retrieval technology, or information retrieval can be used as an auxiliary function of user selection.

3 Experiment and Research

3.1 Experimental Background

Aiming at the problems of low information extraction accuracy, recognition accuracy and recall rate in traditional big data information extraction algorithms, this paper proposes a new algorithm for rapid extraction of big data information based on artificial intelligence. By mining the basic information of big data, it can detect the data storage location at all times, estimate the external conditions of the data operation, obtain high-precision initial data, and use the information rule analysis method to improve the data information structure, improve the extraction algorithm, and complete the rapid extraction of big data information.

3.2 Experimental Process

In order to understand the fast extraction algorithm for big data information, this article tested the sample data set, where the test items include extraction accuracy, recognition accuracy, and recall.

4 Analysis and Discussion

The test items include extraction accuracy rate, recognition accuracy rate, and recall rate. The data information extraction results are shown in Table 1.

It can be seen from Fig. 2 that the extraction accuracy, recognition accuracy and recall rate of Sample 4 of this algorithm are relatively high, which are 0.941, 0.933, and 0.908 respectively. It can be seen that the recall rate of the algorithm has been greatly improved, and the extraction accuracy rate has also been improved, which can meet the needs of rapid information extraction.

Table 1. Data Information Extraction Results

Evaluation standard	Sample 1	Sample 2	Sample 3	Sample 4
Extraction accuracy	0.812	0.903	0.867	0.941
Recognition accuracy	0.865	0.897	0.893	0.933
Recall rate	0.793	0.865	0.841	0.908

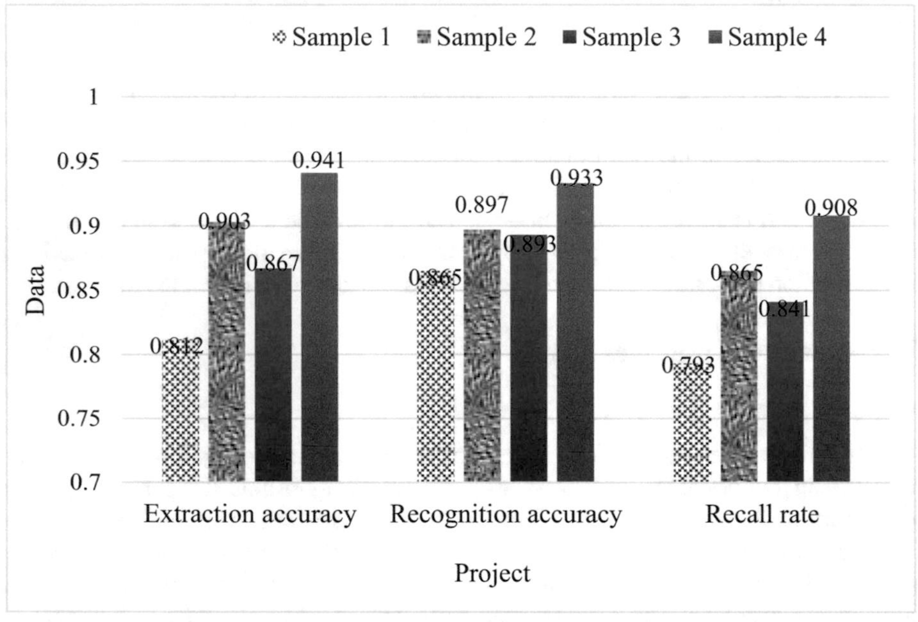

Fig. 2. Data Information Extraction Results

5 Conclusion

With the advent of the era of big data and the advancement of artificial intelligence and other technologies, humans have made great progress in processing massive amounts of information. At the same time, as people's various needs continue to grow, it is necessary to solve complex and tedious problems more efficiently and effectively. People need to obtain a lot of useful information related to their needs. Some of these massive amounts of data can be mined. Therefore, how to improve decision-making efficiency and reduce workload has become a topic of inquiry for researchers. This paper starts from a new perspective, combined with artificial intelligence and related research, to carry out research on the rapid extraction of big data information.

Acknowledgment. 2021 Jilin Provincial Higher Education Research Project: Research and Practice of the Wisdom Classroom Teaching Mode Based on Big Data (No. JGJX2021B27).

References

1. Omar, A., Krittanawong, C., Narula, S., et al.: Echocardiographic data in artificial intelligence research: primer on concepts of big data and latent states – science direct. JACC: Cardiovasc. Imag. **13**(1), 170–172 (2020)
2. Manogaran, G., Chilamkurti, N., Hsu, C.H.: Special issue on advancements in artificial intelligence and machine learning algorithms for Internet of Things. Cloud Computing and Big Data. Int. J. Software Innov. **7**(2), iv (2019)
3. Kedra, J., Radstake, T., Pandit, A., et al.: Current status of use of big data and artificial intelligence in RMDs: a systematic literature review informing EULAR recommendations. RMD Open **5**(2), e001004 (2019)
4. Oh, K., Kim, H.: An analysis of the influence big data analysis-based AI education on affective attitude towards artificial intelligence. J. Korean Assoc. Inform. Educ. **24**(5), 463–471 (2020)
5. Arief, N.N., Gustomo, A.: Analyzing the impact of big data and artificial intelligence on the communications profession: a case study on Public Relations (PR) practitioners in Indonesia. Int. J. Adv. Sci. Eng. Inform. Technol. **10**(3), 1066 (2020)
6. Pachippan, M., Dhandayudam, P., Kannan, B.: Artificial intelligence databases: turn on big data of the SMBs. Int. J. Bus. Inform. Syst. **1**(1), 1 (2020)
7. Yudan, M., Yi, Z., Jing, J., et al.: Relation extraction method combining entity co-occurrence information and sentence semantic features. Sci. China Ser. F **048**(011), 1533–1545 (2018)
8. Kulakli, A., Osmanaj, V.: Global research on big data in relation with artificial intelligence (A Bibliometric Study: 2008–2019). Int. J. Online Biomed. Eng. **16**(2), 31 (2020)
9. Kersting, K., Meyer, U.: From big data to big artificial intelligence?: algorithmic challenges and opportunities of big data. KI - Künstliche Intelligenz **32**(1), 3–8 (2018)
10. Ak, A., Mh, B.: Digital transformation and disruption: on big data, blockchain, artificial intelligence, and other things - sciencedirect. Bus. Horiz. **62**(6), 679–681 (2019)
11. Rahmani, A.M., Azhir, E., Ali, S., et al.: Artificial intelligence approaches and mechanisms for big data analytics: a systematic study. PeerJ Comput. Sci. **7**(2), e488 (2021)
12. Cremer, S., Loebbecke, C.: Artificial intelligence imagery analysis fostering big data analytics. Future Internet **11**(8), 178 (2019)

IoT Big Data Mining Algorithm Based on Artificial Intelligence Technology

Shaohua Fu(✉)

Big Data and Internet of Things School, Chongqing Vocational Institute of Engineering, Chongqing 402260, China
fsh1108@cqvie.edu.cn

Abstract. Artificial intelligence technology has powerful learning ability, its application makes data processing more intelligent, and brings great convenience to people's data analysis work. This paper establishes a data mining system based on Hadoop platform through artificial intelligence technology, and realizes a data mining system based on SpringMVC framework by studying Hadoop platform technology and common data mining algorithms. The system stores data files in the distributed system HDFS by analyzing the RFID data in the Internet of Things, and adds support for the data source format. After integrating the FUP algorithm and the AFUP algorithm of the association rule data mining algorithm, according to the characteristics of the two algorithms, the data discovery amount and the algorithm execution efficiency are compared to test the data mining power of the algorithm.

Keywords: Artificial Intelligence Technology · Data Mining Algorithm · Hadoop Platform · Association Rules

1 Introduction

The data mining system designed by artificial intelligence technology is based on the big data platform. According to the scale and data structure of different data sources, data analysis tasks of different business departments are performed. Users can log in to the system and perform simple operations to realize data analysis and storage, which greatly improves the work efficiency of data processing. Therefore, it is of great significance to establish a data mining system in this paper to help people solve tedious data analysis problems.

At present, many researchers are devoted to the research on the big data mining algorithm of the Internet of Things with artificial intelligence technology, and have achieved good research results. For example, a scholar found that the data mining system under the big data platform can implement a variety of serial and parallel algorithms in the Web-based big data analysis environment. The serial algorithm relies on the open source library of the Weka algorithm to design the algorithm process and related logic processing, providing the latest network-related technologies and plug-ins, which can provide users with a better visual experience [1]. In the process of system application, a

J. H. Abawajy et al. (Eds.): ICATCI 2022, LNDECT 170, pp. 672–679, 2023.
https://doi.org/10.1007/978-3-031-29097-8_80

scholar found that the traditional Kmeans algorithm randomly selects the center point for clustering, which has a great impact on the results, while the existing Kmeans improved algorithm based on the maximum and minimum principle will repeatedly traverse the data when selecting the center point, which brings time Therefore, he proposed an improved M+Kmeans algorithm according to the "maximum principle" to optimize the selection of the center point, and at the same time realized the parallelization of the M+Kmeans algorithm, and integrated the algorithm into the data mining system under the big data platform. It improves the reliability of the algorithm [2]. Although the research results of IoT big data mining algorithms with artificial intelligence technology are good, to improve the performance of the algorithm, it is necessary to use artificial intelligence technology to optimize the data mining system.

This paper expounds the association research, classification prediction, clustering research and other functions of data mining algorithms, and exemplifies several forms of association rule mining algorithms. The number of frequent items and the execution efficiency of the FUP algorithm and the AFUP algorithm under the same support degree indicate that the performance of the AFUP algorithm is better.

2 Data Mining and Its Algorithms

2.1 The Function of Data Mining

With the help of related technologies and functions of data mining, various types of knowledge for decision-making can be obtained. In fact, in many cases, users cannot know the value of data, and even users themselves do not understand the knowledge and models they need [3]. Therefore, a complete data mining system must have the knowledge to search and discover several types of models to meet the needs or applications of various users. The data mining system must provide sections for interactive functions, allowing users to provide guidance, prompts or select preferred patterns to search. Because some patterns cannot match all the data in the database, generally each identified pattern is accompanied by a single measure of credibility [4].

The function of data mining mainly refers to the application in determining the task of data mining and finding the type of patterns required for positioning. Its tasks mainly include two kinds, namely description and prediction. Among them, the former describes the general characteristics of the data in the database. The latter is inferred and predicted based on current data [5]. In addition, there are other functions as follows.

(1) Association research

It mainly applies mining association rules because it describes the association of different items within a given dataset. Generally widely used in shopping basket or transaction data research. That is to find the internal relationship from the massive business transaction data records, so as to make business-related decisions [6].

(2) Classification and prediction

In some applications, the operator expects to find some missing or unknown data values instead of class labels. When the predicted value belongs to a numeric data object, it is often called a forecast. While prediction may be related to prediction of class labels of data, prediction is usually limited to prediction of numerical values and thus is different from classification [7].

(3) Clustering research

Clustering research and classification differ from prediction in that clustering mainly studies the data and does not focus on known class labels [8]. Often class labels are not included in the training data because it is not known where to start. Clustering can be applied to objects that produce such labels. Clustering or grouping is carried out according to the similarity characteristics of the maximizing class and the principle of minimizing the similarity within the class, that is, the object cluster (or cluster) is generated, so that the objects in each cluster are assigned a relatively high value. High similarity, while there is dissimilarity with other objects in the cluster. The resulting single cluster can be viewed as a single object class from which rules can be derived. Clustering is also easy to classify and compile, organize the observed content into a class and hierarchical structure, and arrange similar events in one place [9, 10].

2.2 Association Rules Data Mining Algorithms

(1) FPGrowth algorithm

FPGrowth algorithm is a common material extraction algorithm based on compressed data structure, which overcomes the shortcomings of Apriori algorithm to create a large number of candidate sets. Its basic idea is to create a new data structure by traversing each transaction record, referred to as a frequent pattern tree [11].

The FPGrowth algorithm transforms frequent k-element mining into an item mining process, and long-term mining becomes short-term mining. The FPGrowth algorithm significantly reduces the computational cost, mainly because a new data compression method is adopted in the preprocessing stage, which reduces the memory consumption of the processing process. However, FPGrowth also has some disadvantages. Due to the complex structure of FP-Tree tree, FPGrowth can only have an ideal mining speed when it supports the limit and the algorithm is not suitable for discrete data sets. In the end, in the repeated process, a large number of normal conditional trees will be created, resulting in memory overflow, and the time and cost of building trees are relatively high, which cannot meet people's requirements, but the more data that needs to be processed, the faster the efficiency [12, 13].

(2) FUP algorithm

For transaction $X \in (L_A - L_B)$, if $X \in L_A$ but $X \notin L_B$, , scan B to get the support number X.countB of X in B, and then calculate the support degree of X in (A + B) according to the support number X.count of X in A.

$$X.Support_{A+B} = \frac{X.countB + X.countdA}{|A| + |B|} \tag{1}$$

Among them, A is the transaction data set, B is the new data set, L_A is the frequent itemset in A, L_B is the frequent itemset in B, |A| is the number of transactions in A, and |B| is the number of transactions in B number of transactions.

(3) AFUP algorithm

The calculation process of a frequent itemset obtained on the basis of the FUP algorithm is as follows:

$$item.\sup port = \frac{item.count}{|A| + |B|} \tag{2}$$

where min_sup is the minimum support and remains unchanged.

3 Design of IoT Big Data Mining System Based on Artificial Intelligence

3.1 Overall Architecture of Big Data Mining System

This system is based on the association rule mining system under the Hadoop platform. The main programming language used is JAVA, and the C/S development architecture is adopted. The database has an independent server, that is, the client management terminal. The system can control the quality of data collected by monitoring equipment by eliminating noise. At the same time, to process a large amount of monitoring data, the system must make full use of the hardware environment, combined with cloud computing technology, to improve the ability to analyze and solve problems as much as possible to improve data processing efficiency. The functions of the data mining system are data source selection module, data preprocessing module, data transformation module, and the output and visualization of association rules in Fig. 1.

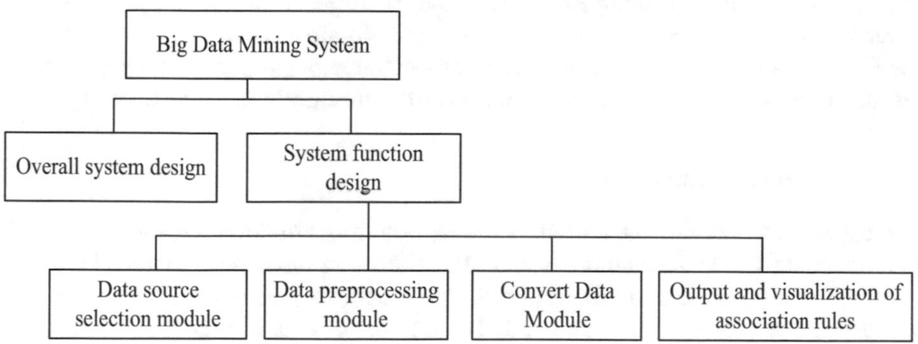

Fig. 1. Association rule mining system

3.2 Architecture Design of HDFS

Since the data type analyzed by this system is big data, the real-time heterogeneous RFID data processing in the IoT big data mining system should be based on artificial intelligence technology and related data mining technology. The Hadoop platform is based on a distributed and reliable file storage system and is the basis for introducing parallel algorithms. The core layer must make algorithm calls according to user requirements. The background processing logic based on SpringMVC uses the Hadoop platform layer to provide an environment for parallel algorithms, supports grouping algorithm tasks, file management, etc., and assists in completing data mining tasks. The processing tasks of data mining are completed with the help of the relevant patterns of Map/Reduce. Generally speaking, the operation process of the big data mining system is as follows: First, the RFID data in the Internet of Things is converted, filtered, and merged, and saved to the distributed HDFS system through PML files. For efficient maintenance and troubleshooting of node failures, a replication strategy can be used to store 2–3 copies of the PML file on each node of the same organization or a node of another organization. Secondly, the master control process Master is responsible for the key creation and management tasks when implementing tasks. When idle, workers can obtain relevant assignment tasks, and at the same time cooperate with Map/Reduce to carry out control and disposal. The results are fed back to the user in a timely manner. The interactive layer provides a graphical operation interface, where users submit algorithm parameters and algorithm requests. The web interface uses the mainstream JQuery technology, which incorporates third-party plug-ins such as Easyui, Bootstrap, Highchart, giving users a friendly interface and richly displaying the algorithm processing results, which is also convenient for future expansion. The SpringMVC framework is a standard MVC framework that provides web layer solutions and is currently the most mainstream MVC framework.

4 Algorithm Experimental Analysis

In order to verify the execution efficiency of the AFUP algorithm, the proposed method is verified by experiments, and the experimental results are analyzed and compared with the FUP algorithm. The experimental environment (including experimental platform and test database), experimental method and results of this algorithm are as follows.

4.1 Experimental Environment

The experimental environment used by the algorithm is Windows XP, the hardware is Intel Pentium(R), and its memory is 256 MB. The algorithm is implemented by Visual Studio. The test data used in the experiment is a data set generated in a random way, and the number of random items is 50. The number of data bars is 500, 1000, 1500, 2000, 2500, a total of five groups of data. Simulate the AFUP algorithm and the FUP algorithm respectively, and compare the number of frequent itemsets found and the execution efficiency of the two algorithms under the same conditions.

4.2 Experimental Results

In this experiment, five sets of data were used to compare the FUP algorithm and the AFUP algorithm. The same data, the same minimum support level and minimum confidence level were used in the test. In order to better reflect the mining ability of the AFUP algorithm, we used the same data every time Both re-evaluate the original data and then add the data to it. Each time the B dataset is added, it accounts for about 5% of the A dataset. In the mining test of mining frequent items, we set a support limit of 15%. Due to the slightly higher threshold, the mining effect of the AFUP algorithm is better than that of the FUP algorithm. The advantages of the AFUP algorithm are not only reflected in the mining ability, but also in the mining efficiency, indicating that the AFUP algorithm has some improvements on the basis of the FUP algorithm. We tested the performance of these two algorithms with the minimum support threshold of 5% and 20% respectively, so as to reflect the efficiency of the algorithm more concretely and intuitively. As shown in Table 1, when the minimum support is 15%, the number of frequent item sets mined by the FUP algorithm and the AFUP algorithm. According to the results, the AFUP algorithm can find frequent data sets better than the FUP algorithm.

Table 1. Comparison of the number of frequent items mined by the two algorithms

	FUP algorithm finds frequent numbers	AFUP algorithm finds frequent
500	87	107
1000	96	129
1500	113	135
2000	124	168
2500	158	167

Figure 2 is a comparison of the execution time of the FUP algorithm and the AFUP algorithm under the condition that the minimum support threshold is 5% and 15%. From the results in Fig. 2, it can be seen that regardless of whether the minimum support is 5% or 20%, the data execution time of the AFUP algorithm is shorter, which proves that the performance of the AFUP algorithm is better than that of the FUP algorithm, and it can not only better discover the relationship, but also And the mining efficiency is also greatly improved.

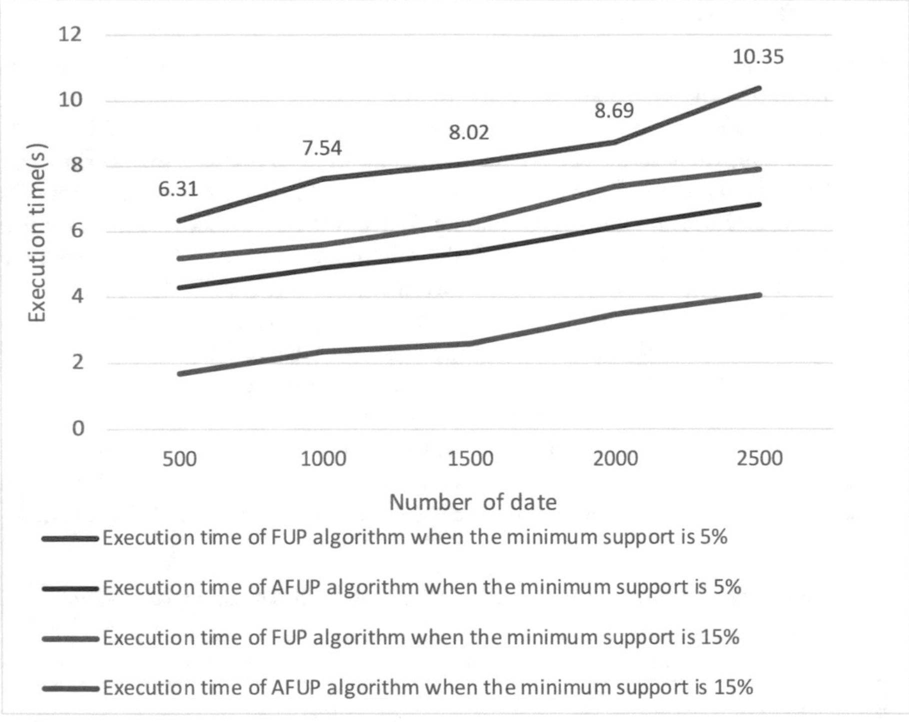

Fig. 2. Comparison of execution time of two algorithms under different support degrees

5 Conclusion

This paper uses artificial intelligence technology to design an association rule data mining system based on the Hadoop platform, and tests the performance of the FUP algorithm and the AFUP algorithm in the association rule algorithm. The number is higher than that of the FUP algorithm, and when processing the same data, the execution time of the AFUP algorithm is less, indicating that the AFUP algorithm is more efficient than the FUP algorithm.

Acknowledgements. This research was supported by the project fund of Chongqing Education Commission Group Innovation: Self-Driving Vehicles Driving Together, (NO. CXQT21032); Scientific research project of Chongqing Vocational Institute of Engineering: Research on cable instantaneous break detection instrument based on embedded chip, (No.kjb202005).

References

1. Kachaoui, J., Belangour, A.: Enhanced data lake clustering design based on K-means algorithm. Int. J. Adv. Comput. Sci. Appl. **11**(4), 547–554 (2020)
2. Tchapga, C.T., Mih, T.A., Kouanou, A.T., et al.: Biomedical image classification in a big data architecture using machine learning algorithms. J. Healthcare Eng. **2021**(2), 1–11 (2021)

3. Cravero, A., Lagos, D., Espinosa, R.: Big Data/IoT use in wine production: a systematic mapping study. IEEE Lat. Am. Trans. **16**(5), 1476–1484 (2018)
4. Sakr, S., Zomaya, A.: Encyclopedia of big data technologies. In: Big Data Stream Security Classification for IoT Applications, pp. 1–5 (2018). https://doi.org/10.1007/978-3-319-63962-8(Chapter 236-1)
5. Lammatha, K.K.: Data mining on 5G technology IOT. Int. J. Sci. Res. Manage. Stud. **8**(5), 24655–24660 (2019)
6. Afzali, G.A., Mohammadi, S.: Privacy preserving big data mining: association rule hiding using fuzzy logic approach. IET Inf. Secur. **12**(1), 15–24 (2018)
7. Manshahia, M.S.: Swarm intelligence-based energy-efficient data delivery in WSAN to virtualize IoT in smart cities. IET Wireless Sensor Syst. **8**(6), 256–259 (2018)
8. Njah, H., Jamoussi, S., Mahdi, W.: Deep Bayesian network architecture for Big Data mining. Concurr. Comput. Pract. Exper. **31**(2), e4418.1–e4418.17 (2019)
9. Alreshidi, E.: Smart Sustainable Agriculture (SSA) Solution underpinned by Internet of Things (IoT) and Artificial Intelligence (AI). Int. J. Adv. Comput. Sci. Appl. **10**(5), 93–102 (2019)
10. Makkar, A., Ghosh, U., Sharma, P.K.: Artificial intelligence and edge computing-enabled web spam detection for next generation IoT applications. IEEE Sensors J. **99**, 1 (2021)
11. Vangala, A., Das, A.K., Kumar, N., et al.: Smart secure sensing for IoT-based agriculture: blockchain perspective. IEEE Sens. J. **99**, 1 (2020)
12. Smith, B.C.: Big data and us: human-data interactions. Euro. Rev. **27**(3), 357–377 (2019)
13. Reshmy, A.K., Paulraj, D.: Data mining of unstructured big data in cloud computing. Int. J. Bus. Intell. Data Mining **13**(3), 147–149 (2018)

Design of Intelligent Calligraphy and Painting Software System Based on Computer Aided Technology

Shuang Gao(✉)

Art Department, Changchun Humanities and Sciences College, Changchun, Jilin, China
g16316300 2022@163.com

Abstract. The advancement of computer technology has brought great flexibility to the development of the industry and has become an indispensable tool for production and survival in today's society. The continuous development of computer software services has made it more and more used in various fields and has become the main driving force for the development of new vision. On the basis of traditional calligraphy and painting theory, the computer-aided design is used to innovate the means of calligraphy and painting art, improve the intelligent calligraphy and painting software system, and inject new vitality into the development of traditional calligraphy and painting. The purpose of this paper is to design the intelligent calligraphy and painting software system based on computer-aided technology. This paper will analyze the advantages of computer-aided technology, deeply study the inheritance of traditional painting and calligraphy, and explore the design and development path of intelligent painting and calligraphy software system based on computer software-assisted technology. It mainly uses the non-normalized method and the Canny optimal boundary detection operator to extract the image element features. Experiments show that the recognition rate of the improved SURF algorithm is improved under different variances, and the larger the variance, the better the anti-noise performance, ranging from 3.8% to 5.8% respectively.

Keywords: Computer Aid · Computer Technology · Intelligent Painting and Calligraphy · Software System

1 Introduction

With the rapid development of computer technology, the technology assisted by intelligent software has been widely used, and the spread of digital technology has brought great changes to the calligraphy and painting system [1, 2]. Intelligent calligraphy and painting software system. In the process of traditional calligraphy and painting, painting, designing or reading is mainly done by freehand in the form of pen and paper, which makes the learning content often easy to mechanize, often simply copying and retelling, losing the fun of creation, and there is no way to break through. Modern calligraphy and painting art needs innovation and absorbs the breath of science and technology in

J. H. Abawajy et al. (Eds.): ICATCI 2022, LNDECT 170, pp. 680–687, 2023.
https://doi.org/10.1007/978-3-031-29097-8_81

the new era. Therefore, with the help of modern computer technology, a new model of computer-aided intelligent calligraphy and painting software system that is perfectly combined with it should be designed and studied.

Computer-Assisted Learning is a complete tool for production, humanities, contextual learning and other research-based skills. Foreign scholars have a very deep understanding of research-based learning, Kurbad A believes: "Research-based learning is a learning method that seeks information and truth, or seeks relevant information for problems. When you have knowledge, you have the ability and attitude to find problem-solving" [3]. For text skeleton extraction, many years ago, Marlowe J proposed the method of central axis transformation [4]. It can be seen that the computer-aided technology is widely used, and the design of the intelligent calligraphy and painting software system cannot be ignored.

Based on computer-aided technology, this paper designs the intelligent calligraphy and painting software system. By analyzing the advantages of computer-aided technology, it deeply studies the inheritance of traditional art, and explores the design and development path of intelligent calligraphy and painting software system based on computer software-aided technology. It mainly uses the non-normalized method and the Canny optimal boundary detection operator to extract the image element features.

2 Research on the Design of Intelligent Calligraphy and Painting Software System Based on Computer Aided Technology

The development of computer-aided technology does not mean the decline of traditional painting and calligraphy, because traditional painting and calligraphy is always the foundation and premise of the development of intelligent painting and calligraphy software systems. In various fields, we should focus on traditional calligraphy and painting theories, and actively explore the innovative design of intelligent calligraphy and painting software systems [5, 6]. In order to follow the law of development, respect the theoretical basis, and make the calligraphy and painting software system more humanized, computer technology can only be used as an auxiliary.

2.1 Advantages of Computer Aided Design

(1) To promote the production of new works. On the basis of the traditional calligraphy and painting system, the auxiliary functions of computer technology are fully exerted to the greatest extent, so as to derive new works, and at the same time promote the continuous development of the intelligent calligraphy and painting software system. New works based on computer-aided technology are different from traditional works in that they have high practicability and strong pertinence. In teaching and business, computer-aided technology is frequently used in calligraphy and painting systems, and many outstanding designers stand out, making the forms of calligraphy and painting systems more and more abundant.

(2) Change the traditional way of calligraphy and painting. The creation and dissemination of traditional calligraphy and painting systems are deeply influenced by computer technology. The intelligent calligraphy and painting system is constantly

improving, and the content of calligraphy and painting software has attracted much attention. Computer-aided techniques are also critical in image rendering. For example, the application of AutoCAD software makes the drawing faster and more accurate and avoids many problems. With the improvement of computer technology, more and more intelligent calligraphy and painting software are developed and applied, the calligraphy and painting system is more free, and the modification of works is more convenient. Since the computer software can also be undone, the error tolerance in the creation process is also higher and the workload is reduced [4, 7].

2.2 The Theoretical Basis of Humanistic Learning in Computer Aided Design

As mentioned in the teaching process, learners can develop complete self-efficacy, behavior and self-esteem through a combination of understanding with purpose, mind and body, which is consistent with the teaching skills of contemporary art education. The computer-assisted learning model integrates the scientific perspectives of self-determination, self-knowledge, and autonomous learning through the humanities to create a focused learning environment that defines the ideal learning environment for everyone.

Humanism has four characteristics: first, the learner's personal full participation, that is, cognitive participation, emotional participation, etc.; second, self-initiated, active learning; third, penetrating learning, learning process, behavior, attitude, behavior and other factors determine the learning effect; fourth, self-evaluation, summarizing the learning content, and supervising the learning. It can be said that humanism is to create a specific learning environment, to help learners establish teaching and content guidance, learners to acquire knowledge independently, and to promote a learning atmosphere [8, 9].

The characteristics of the four elements of humanistic theory are reflected in the multi-learning mode of computer-aided art teaching:

(1) In the computer digital technology environment, the learner is the main body and is not limited by the real space. The learning process is mainly based on the full participation of the individual, including cognitive participation and emotional participation.
(2) Rich digital network resources stimulate learners' intrinsic motivation for learning, which plays an important decisive role in promoting art learning activities.
(3) The digital interactive environment of computer-aided art teaching provides convenient conditions for learners' permeable learning, and promotes learners to form excellent learning behavior, learning attitude and learning character. These factors affect the final learning effect of learners.
(4) In a computer-aided environment, learners must evaluate themselves independently and objectively, and educators only urge them. Computer-aided technology provides learners with means of learning, and the specific learning method is still determined by themselves.

Humanism has great inspiration to let computer technology assist intelligent calligraphy and painting software system. The essence of research-based learning is to allow learners to actively explore knowledge and view the content of professional courses dynamically, rather than statically step-by-step learning, which is to constantly solve unsolved problems according to changes in new situations. It can be said that humanistic theory participates in the construction of intelligent calligraphy and painting software design by computer-aided technology, creates a specific modern art learning environment, and also applies today's latest digital network technology and modern educational concepts [10, 11].

2.3 Research on the Algorithm of Intelligent Calligraphy and Painting Software System Based on Computer-Aided Technology

(1) Image normalization

Normalization methods are divided into linear and nonlinear. Using non-normalization, the image features can be better preserved. The operator expression looks like this:

$$m = \sum_{i=1}^{m} H(\mathrm{i}) \times \frac{M_1}{\sum_{i=1}^{M_0} H(\mathrm{i})}$$
$$n = \sum_{j=1}^{\mathrm{m}} H(j) \times \frac{N_1}{\sum_{j=1}^{N_0} H(j)} \tag{1}$$

$f(x, y)$ before normalization $M_0 \times N_0$, where x = 2,1 0.... M_0, y = 1,2.... N_0, $g(m, n)$ for the normalized $M_1 \times N_1$.

(2) Contour feature extraction

Detecting contours is to detect the gradient and direction of image nodes to maximize the contours. The Canny optimal boundary detection algorithm is used here, and the formula is as follows:

$$G_n = \frac{\partial G}{\partial n} = n\nabla G \tag{2}$$

Among them, $G(x, y) = \frac{1}{2\pi\sigma^2} \exp\left(-\frac{x^2+y^2}{2\sigma^2}\right)$, $n = \begin{bmatrix} \cos\theta \\ \sin\theta \end{bmatrix}$, $\nabla G = \begin{bmatrix} \partial G/\partial x \\ \partial G/\partial y \end{bmatrix}$ n is the direction vector, and ∇G is the gradient function.

3 Experiment Research on Intelligent Calligraphy and Painting Software System Based on Computer-Aided Technology

3.1 Extraction of Image Element Features

(1) Extraction of point features. It is mainly the distant point, the corner point of things and the line intersection generated by the intersection of multiple lines. Mainly

through the interest operator method, namely Moravec operator and Harris operator, etc.; secondly, it extracts the edge points of things, performs wavelet transformation on the image, and takes the maximum value of the obtained transformation modulus value as the edge point of the image after calculation; there are also corner points. Detection, based on grayscale detection, through the grayscale change of the pixel point, the curvature and gradient of the point are obtained by operation for detection.

(2) Extraction of line features. When two adjacent borders in the image tend to be one-pixel wide linear regions, and there are grayscales with almost the same amplitude characteristics in the current region, we call this linear region a line. In terms of line feature extraction, the first step is to extract the line segment information in the image, and then filter out the required line segments by continuously adding constraints to the line features. Among the line extraction operators that are used more frequently, there are operators for differentiating pixels, including gradient and difference operators.

(3) Surface feature extraction. The surface feature is an information feature that is different from other obvious areas in the image, and refers to a closed high-contrast projection area. First extract the features, that is, the weight of this area or the circle of this circle, because the weight or circle will not change relative to the enlargement, reduction and rotation of the image, and grayscale changes and random noise also Stablize. Extract with segmentation to make it more precise.

(4) Color feature extraction. Describe the surface of an image or a certain area of it through color moments, color aggregation vectors and color histograms.

3.2 Calligraphy Character Outline Skeleton Extraction

(1) The maximum circle algorithm is to take the maximum inscribed circle of the stroke as the radius, and gather the bifurcation points in the middle into one point, which is formed. The skeleton is extracted by feature points to obtain skeleton features.

(2) The two-step skeleton optimization algorithm, based on the traditional skeleton refinement theory, adopts single pixelization for the text, and detects the pixels in the vertical and horizontal directions of the text. Because the skeleton has connectivity, an unconnected graph is used to connect the center pixel of the text stroke with the eight neighboring pixels.

(3) A-W thinning algorithm, iteratively analyzes the eight neighboring points on the edge of the image, and then deletes them gradually until the thinning template is satisfied, and the center pixel with a width of one pixel is obtained.

4 Experiment Analysis of Intelligent Calligraphy and Painting Software System Based on Computer-Aided Technology

4.1 Contour Feature Extraction

Contour feature extraction is edge detection, which extracts many closed curves that intersect the contours. It is a method often used for image processing. Generally, contour edge features are extracted first, and redundant information is eliminated to help obtain later feature points. In the study of contour feature extraction, many other algorithms have also been proposed while improving the traditional extraction algorithms. As shown in the table below (Table 1):

Table 1. Comparison of the average running time of the algorithms

image size	maximum circle algorithm	Two-step Skeleton Algorithm	A-W Thinning Algorithm	Algorithm
128 × 128	0.93	0.92	0.89	0.84
256 × 256	3.21	3.02	2.98	2.83
374 × 374	3.40	3.15	2.99	2.97
512 × 512	6.42	6.21	6.01	5.92

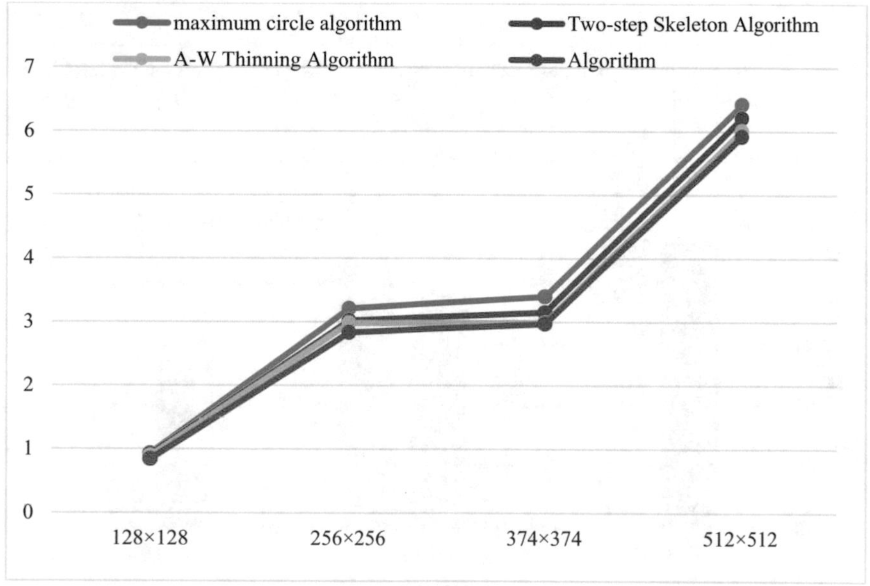

Fig. 1. Comparison of the average running time of the algorithms

Specifically, as shown in Fig. 1, the running time of each algorithm is compared with images of 128 × 128, 256 × 256, 374 × 374, and 512 × 512 pixels, and it is obvious that the algorithm in this paper is better than other algorithms.

4.2 Comparison of Anti-noise Performance

Add white noise with different variances such as 0.02, 0.04, 0.06 to different works, and compare the anti-noise performance of the two. As shown in Table 2:

Table 2. Comparison of anti-noise performance

variance (VAR)	Literature Algorithms (Recognition rate%)	Algorithm (Recognition rate%)
0.02	87.41	91.21
0.04	85.72	90.82
0.06	83.51	89.31

Specifically, as shown in Fig. 2, under different variances, the improved SURF algorithm improves the recognition accuracy and has better anti-noise performance than the feature matching algorithm as a whole. Experiments show that the improved SURF algorithm is very effective in feature extraction and recognition, and the image matching is more accurate.

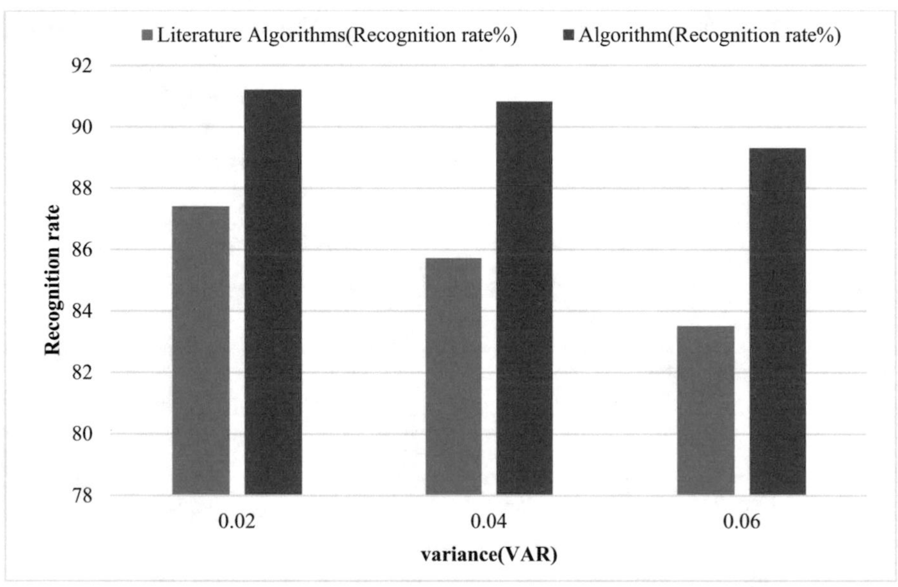

Fig. 2. Comparison of recognition rates under different variances

5 Conclusions

Computer technology-assisted intelligent calligraphy and painting software system is the top priority of the current development of the calligraphy and painting industry. It can not only innovate the form of the calligraphy and painting system and the types of calligraphy and painting works, but also is very suitable for today's high-efficiency and fast-paced life. However, in the process of computer-aided intelligent calligraphy and painting technology, it is still necessary to draw inspiration from the traditional calligraphy and painting system based on traditional calligraphy and painting. In the future development, the performance of computer hardware and software will continue to improve, and the intelligent calligraphy and painting software system based on computer-aided technology will also develop in the direction of humanization. The research on the subject in this paper is relatively preliminary, and there is still a gap for the system to achieve a very ideal state. Therefore, the work in many places needs more in-depth and specific research, in order to achieve the ideal interface effect.

References

1. Nash, M., Kadavigere, R., Andrade, J., et al.: Deep learning, computer-aided radiography reading for tuberculosis: a diagnostic accuracy study from a tertiary hospital in India. Sci. Rep. **10**(1), 210 (2020)
2. Dandıl, E.: A computer-aided pipeline for automatic lung cancer classification on computed tomography scans. J. Healthcare Eng. **2018**, 1–12 (2018)
3. Kurbad, A.: A new milling machine for computer-aided, in-office restorations. Technology, integration, and workflow. Int. J. Comput. Dentist. **20**(2), 201 (2017)
4. Marlowe, J., Tsilomelekis, G.: Accessible and interactive learning of spectroscopic parameterization through computer-aided training. J. Chem. Educ. **97**(12), 4527–4532 (2020)
5. Ajf, A., Eo, B.: Intelligent agents for feature modelling in computer aided design. J. Comput. Des. Eng. **5**(1), 19–40 (2018)
6. Xu, J., Ling, C., Xu, G., et al.: Dynamic spline bas-relief modeling with isogeometric collocation method. Comput. Aided Geometric Des. **81**, 101913 (2020)
7. Sun, H., Zeng, X., Xu, T., et al.: Computer-aided diagnosis in histopathological images of the endometrium using a convolutional neural network and attention mechanisms. IEEE J. Biomed. Health Inform. **26**(6), 1664–1676 (2020)
8. Fricke, A., Schöneberger, J.C.: Know-how protection and software architectures in industry 4. 0 - ScienceDirect. Comput. Aided Chem. Eng. **40**, 2287–2292 (2017)
9. Valentin Plesu, A., et al.: Software tool for computing and visualization of enhanced residue curve maps. Comput. Aided Chem. Eng. **40**, 199–204 (2017)
10. Luu, M.T., Milani, D., Nomvar, M., et al.: Computer-aided design for high efficiency latent heat storage – a case study of a novel domestic solar hot water process. Comput. Aided Chem. Eng. **40**, 1153–1158 (2017)
11. Serfidan, A.C., Turkay, M.: Octane optimization with a combined machine learning and optimization approach. Comput. Aided Chem. Eng. **50**, 221–226 (2021)

Research on Artificial Intelligence Ethical Risk of Threat Intelligence Analysis and Electronic Forensics

Fajian Xu[1], Xu Zheng[2(✉)], Lizhi Lin[3], Shuzhen Chen[1], and Chadi Altrjman[4]

[1] Department of Computer and Information Security Management, Fujian Police College, Fuzhou 350007, Fujian, China

[2] Fujian Ruiyan Judicial Expertise Institute, Fuzhou 350011, Fujian, China
woneichuo75746@163.com

[3] School of Big Data, Fuzhou University of International Studies and Trade, Fuzhou 350202, Fujian, China

[4] Artificial Intelligence Engineering Department, AI and Robotics Institute, Near East University, 10 Mersin, Turkey

Abstract. The research on artificial intelligence ethics is partial to theory, and most of them have no specific practical foothold. We can analyze the ethical principles of artificial intelligence in intelligent policing according to the applications of electronic forensics and threat intelligence analysis. Taking text analysis, image forgery identification, and artificial intelligence research and judgment as examples, we can find and clarify the security ethical boundary of artificial intelligence application. It has great practical significance for social and public security.

Keywords: Artificial Intelligence · Ethical Risk · Electronic Data Forensics · Smart Policing · Threat Intelligence

In recent years, the research on the ethics and limits of artificial intelligence application abroad has a wide range and broad ideas. In 2019, the United Nations Educational, scientific and Cultural Organization (UNESCO) issued the Beijing consensus - artificial intelligence and education, which affirmed that "the design of artificial intelligence should be ethical, avoid discrimination, fair, transparent and auditable". On April 8, 2019, the European Commission of artificial intelligence experts officially issued the "ethical guidelines for trusted artificial intelligence" (AI) [1], This paper puts forward seven key aspects of establishing "trusted artificial intelligence" (trustworthy AI), "human initiative and supervision", "technical robustness and security", "privacy and data management", "social and environmental well-being", "diversity, non-discrimination and fairness", "transparency" and "accountability system". In 2019, the communique of the G20 ministerial meeting on Trade and digital economy issued the G20 principles of artificial intelligence [2]. In 2021, China issued the code of ethics for a new generation of artificial intelligence [3], it puts forward six basic principles of artificial intelligence ethics: improving human well-being, fairness and justice, protecting privacy and security, Controllable and credible, strengthening responsibility and improving ethical literacy.

J. H. Abawajy et al. (Eds.): ICATCI 2022, LNDECT 170, pp. 688–696, 2023.
https://doi.org/10.1007/978-3-031-29097-8_82

There are a lot of research on the application of artificial intelligence in electronic data forensics and identification, but there are few relevant studies on the ethical issues of artificial intelligence in electronic data forensics and identification. Threat Intelligence refers to the knowledge of existing or emerging threats faced by it or information assets, including situations, mechanisms, indicators, inferences and feasible suggestions [4]. Threat intelligence analysis belongs to the interdisciplinary field of network security, big data, informatics and other disciplines, and needs a lot of artificial intelligence technology support like electronic forensics, However, there is little research on artificial intelligence ethics for network threat intelligence analysis. The ethics and limits of artificial intelligence in threat intelligence analysis and electronic data forensics and identification have a lot in common.

The following focuses on the application scenarios of electronic data forensics. At present, intelligent policing is an important direction of artificial intelligence application, and is closely related to social public security. Therefore, the research on the ethics and limits of artificial intelligence in the application scenarios such as electronic data forensics and identification in intelligent policing has great practical significance and application and promotion value.

1 Application of Artificial Intelligence in Electronic Data Forensics and Identification in Intelligent Policing

Through the application research and analysis of artificial intelligence in electronic data forensics and identification, such as forensics tools, process and data analysis, forensics result storage, anti-forensics technology and other related links, to explore the internal relationship between artificial intelligence and electronic data forensics and judicial identification activities.

1.1 Artificial Intelligence Technology Provides the Efficiency of Electronic Data Forensics Software

Electronic data forensics software is to complete the inspection and identification of electronic data such as electronic data collection and processing, data inspection, investigation and analysis, display report are all according to the goal of forensics [5]. Among them, we have to face the common and individual problems of various dimensions, such as massive data, processing efficiency, heterogeneous data, emerging technologies, data encryption, reasoning and induction, analysis depth, anti forensics technology, application breadth and so on [6]. It is urgent to find and analyze these large amounts of heterogeneous electronic data with the help of artificial intelligence technology to improve work efficiency. The intellectualization of forensics tools may also lead to errors in the design or implementation of intelligent analysis algorithms, which may produce inconsistent with expectations or even harmful results, misjudgments and omissions, and have an impact on the existing social equity and ethical system, which is worthy of our attention (as shown in Fig. 1).

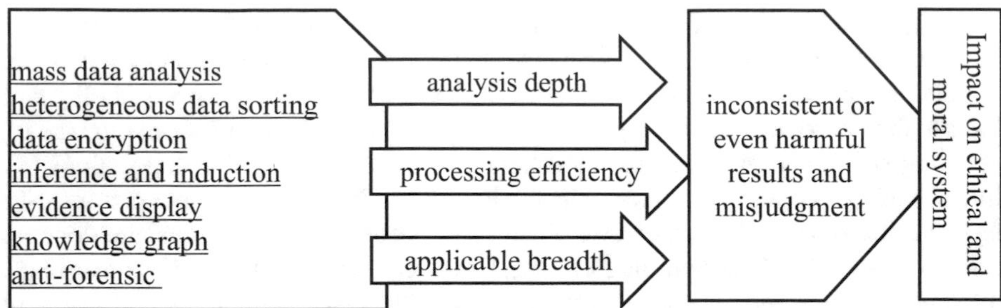

Fig. 1. Research on the ethics and limits of artificial intelligence in intelligent policing application scenarios

1.2 Artificial Intelligence Helps to Mine Network Association and Suspicious Clues to Form the Path of Network Attack, Defense and Forensics

The network system may often have weaknesses. By scanning and analyzing whether some ports are open and seeing if there are common vulnerabilities, a conventional evidence collection channel can be formed. Here, we can also complete the detection, find some correlation or suspicious clues, and conduct logical analysis and derivation through artificial intelligence machine learning, deep learning, computer vision and natural language processing. Combined with the judgment of high-level network security experts, a real network attack point is formed. Moreover, the machine learning model can detect and respond to threats faster. We can more effectively analyze user behavior, deduce patterns and identify various abnormal or abnormal situations in the network by artificial intelligence. Use these data, we can quickly and easily identify network vulnerabilities and prepare for further network forensics.

We can further identify the abnormal behavior of users in the network by analyzing the abnormal behavior of various programs, such as the vast sea of artificial intelligence. However, the security tools based on machine learning may be programmed incorrectly, resulting in the omission of unexpected and even some obviously unreasonable results.

1.3 The Certificate Cloud Based on Blockchain Solves the Notarization Problem of Judicial Expertise

The blockchain electronic data storage system is built on the court or the third-party electronic evidence platform, and is composed of node management, blockchain service layer and judicial application layer. It realizes the closed-loop of electronic evidence collection, fixation and application. In addition to effectively ensuring that the electronic evidence of the cloud platform is legal, true and objective, it will comprehensively refresh the convenience of evidence collection, the professionalism of certificate storage technology and the judicial acceptance rate of the electronic data service industry. The electronic data confirmation letter provided by it is equivalent to the "birth certificate" of electronic data (as shown in Fig. 2).

If it can prove its authenticity through electronic signature, trusted timestamp, hash value verification, blockchain and other evidence collection, fixed and tamper proof

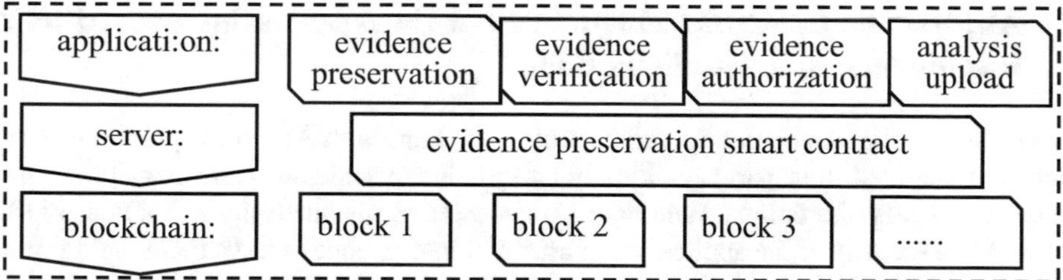

Fig. 2. Deposit cloud of legal norms

technical means, or through the authentication of electronic forensics and certificate storage platform, the Internet court shall confirm it. All of there can ensure the integrity, reliability and standardization of the hardware and software environment of the computer system on which all processes and links of electronic data generation, storage, transmission, extraction and acceptance depend [7].

1.4 Application and Problem Analysis of Artificial Intelligence in Judicial Work

Artificial intelligence has changed some identification processes, identification methods, and even identification ideas. On the other hand, based on electronic data, artificial intelligence analyzes thousands of transaction dispute cases with the help of in-depth learning and migration learning technology, learns the legal provisions in various transaction fields, and arranges into a computer understandable model for each cause of action; At the same time, for each element of the case, it automatically provides relevant legal provisions and other judgment basis, so as to establish a set of judicial knowledge map, which deeply depicts the relationship between the trader, transaction behavior and legal facts; And this relevance is integrated into the artificial intelligence judgment system to give mediation and judgment suggestions for judges. These reflect the application ability of artificial intelligence in judicial work, such as intelligent recommendation, intelligent arrangement, intelligent guide, knowledge transfer and so on. Some people also question the accuracy and interpretability of artificial intelligence algorithm, data security, privacy protection and so on.

1.5 Artificial Intelligence Crime and Its Challenge to Judicial Work Such as Electronic Forensics

Artificial intelligence technology can produce false information content to carry out illegal activities such as fraud. There are also AI related criminal cases in China, involving the use of AI technology for image forgery, video tampering, voice impersonation, handwriting fraud and so on. These problems urgently require us to study and analyze the impact of the application of artificial intelligence on human society, and how to make artificial intelligence more integrated into the diversity, difference and depth of human intelligence.

2 Analysis on the Ethical Principles of the Application of Artificial Intelligence in the Judicial Field

According to the application in various fields, 11 important AI ethical principles have been summarized: transparency, fairness and justice, harmlessness, responsibility, privacy, well-being, freedom and autonomy, trust, dignity, sustainability and solidarity [8]. The ethical security of the application of artificial intelligence in the judicial field is also very important. We should study and analyze these basic ethical principles, and then further test and improve them in the practice of subdivided fields.

When using artificial intelligence system for electronic forensics and other activities, there is an impact of the recognition accuracy of artificial intelligence on the application of forensics analysis; Whether the "skill gap" caused by the skill difference of operators will affect the fairness of electronic forensics and identification; How to standardize the design risk of artificial intelligence in the application of electronic forensics, avoid governance risk, technical risk and ethical risk, and protect the legitimate rights and interests of all parties and private data; Ethical conflicts such as privacy and security issues caused by big data and the regulatory mechanism of artificial intelligence application process need to follow the principles of respect for data collection, understanding of data processing and boundary of data utilization; The application of machine personalized learning algorithm leads to the fact that human beings do not need more innovation and ability, which makes people easily lose the motivation to enter, resulting in the lack of value ethics. Therefore, it is necessary to carry out research on technical means such as inclusive testing, the establishment of multi-layer responsibility distribution system, the improvement of certification system and other artificial intelligence ethical risk prevention, establish the people-oriented principle that human intelligence cannot be replaced by artificial intelligence, the fairness principle of artificial intelligence identifying the conflict between electronic evidence and human cognition, and clarify the boundary and scale of technological power intervention How to embody the principle of interest litigation in the judicial application of artificial intelligence.

3 In the Practice of Electronic Data Forensics and Authentication, We Should Constantly Clarify the Ethical Boundary of Artificial Intelligence Security

Starting from the application of intelligent policing, combined with the research and application of technologies, regulations and standards of intelligent systems and equipment such as police big data, network attack and defense, electronic data forensics and identification, personal information protection, this paper discusses the ethical limits of artificial intelligence application, studies and designs the technical, legal and normative elements that should be studied and integrated in the application of intelligent policing, Further specify the ethical risks and limits of AI in the application context of electronic forensics and identification, and balance the inherent technical bias in the ethical risks of AI application.

The research on the ethics and limits of artificial intelligence in the application scenario of intelligent policing can be carried out from the subdivided fields of intelligent

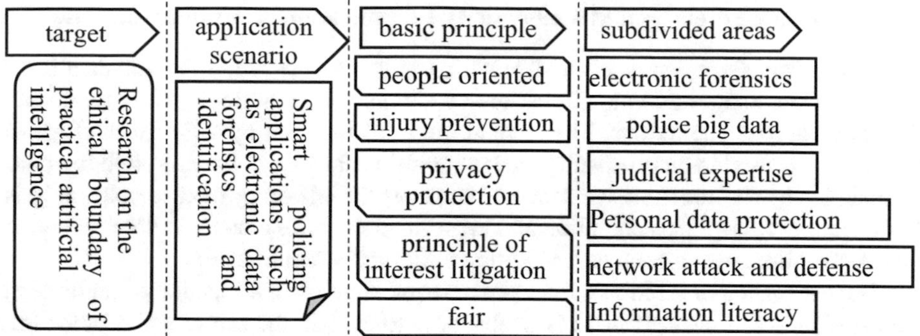

Fig. 3. Research on the ethics and limits of artificial intelligence in intelligent policing application scenarios

applications such as policing big data, network attack and defense, electronic data forensics and identification, and personal information protection, combined with some basic principles of the above artificial intelligence ethics, to study the practice fields with various characteristics such as electronic data forensics and identification, policing big data analysis and so on, And form the corresponding safety ethics boundary of practice and application (as Fig. 3).

We should constantly explore the boundary of electronic data security and artificial intelligence. It is necessary to combine the application ideas and requirements of artificial intelligence in electronic data forensics and the problems encountered in the analysis of electronic data forensics. In the application of electronic data forensics, take text analysis, image forgery identification and analysis, and artificial intelligence comprehensive research and judgment in electronic data forensics as examples to find the safety ethical boundary of artificial intelligence application.

3.1 Application of Text Analysis in Electronic Data Forensics

Text semantic recognition and automatic clue mining can be directly applied to the analysis and discovery of electronic data and clues in the case [9], but there are different search, recognition and research ideas for different case types so it is necessary to analyze and summarize the safety ethics boundary of their AI applications. For example, in pornographic, drug and other cases, similar text keywords, codes, chat session modes and website information will appear in the electronic data text content of the same case type. Through the data extraction, annotation and preprocessing of similar cases [10], build the text analysis model of training artificial intelligence. After machine learning, Intelligent model can quickly find clues like experienced police handling cases.

Text analysis can also use the trained text emotion analysis model to identify those who have serious depression tendency or suicidal tendency, and further identify and find other clues of some cases through the analysis of text emotion views. However, artificial intelligence has no human emotion, so it is more necessary to determine the ethical limits of protecting data and preventing injury; We should also constantly accumulate the boundaries and scales of privacy protection and technology application.

3.2 Electronic Forensics and Authenticity Identification of Multimedia Data

Electronic forensics often needs to deal with a large number of multimedia data in the form of voice, image and video. It is time-consuming and laborious to rely on manual dictation analysis one by one. The recognition model trained by artificial intelligence technology is used to automatically analyze multimedia data in batches, find the clues of specific objects from massive data, and improve the efficiency of case data analysis. For example, voiceprint recognition and identification, voice content and voice source, face recognition, image content recognition and forgery identification.

The conclusion of electronic forensics and authenticity identification of multimedia data is simple and clear. If there is doubt, how to explain the fairness and rationality of the algorithm conclusion is also caused by the opacity of artificial intelligence system. If artificial intelligence has a systematic and practical ethical boundary of artificial intelligence, research on people-oriented, electronic data forensics and other intelligent policing applications, injury prevention, privacy protection, fair interest litigation principle, electronic forensics, policing big data, judicial expertise, personal data protection, network attack and defense, basic principles of information literacy target application scenarios, domain segmentation errors, and even external attacks, Leading to wrong or opposite conclusions, it is difficult to ensure the fairness principle of artificial intelligence application. Therefore, it is necessary to constantly clarify the safety ethical boundary of artificial intelligence transparency in the practice of electronic data forensics and identification, as well as the ethical balance between the existing human cognitive level and the application of intelligence technology.

3.3 Application of Artificial Intelligence Comprehensive Research and Judgment Technology in Electronic Data Forensics

We can use all kinds of artificial intelligence technologies and research and judgment analysis methods mentioned above, summarize the characteristics and laws of corresponding cases according to the research and judgment analysis ideas of cases, sort out and mark the characteristic information of a large number of different cases, sort out and summarize the research and judgment ideas of different cases, form more and better case detection methods and research and judgment models, and form an artificial intelligence assisted case detection model. We also need to constantly adjust and improve the ethical standards of people-oriented and technology use, and also run in to form the ethical tolerance limit of artificial intelligence identifying the conflict between electronic evidence and human cognition.

The complexity of legal reasoning is not only the reasoning and calculation of right and wrong, but also the openness of legal rules. Legal reasoning needs dialectical thinking and continuous renewal of cognition, including grasping the principle of interest litigation on the basis of law and justice, but it cannot be digitized or calculated by algorithm, Therefore, AI cannot realize the cognition matching with people in this technology. The ethical tolerance of the inevitable conflict between artificial intelligence and human cognition also needs to be refined and integrated in practice, reasonably clarify the boundary and scale of technical power intervention, ensure that electronic forensics can truly reflect the fairness of law and judicial power.

4 Summary and Prospect

When big data and intelligent research and judgment are widely used in policing, more and more artificial intelligence is applied to intelligent policing work such as Threat intelligence analysis and electronic forensics. The cognitive recognition ability of artificial intelligence is not invariable, but also needs to be gradually improved through technological development, continuous learning of knowledge and accumulation of experience. It needs to constantly clarify and improve the security ethical boundary of artificial intelligence in the practice of threat intelligence, electronic data forensics and identification. At present, most of the research on the ethics and limits of the application of artificial intelligence in many fields still stays in the stage of general theoretical research and exploration, and there is no specific practical foothold and verifiable way. The research on the ethics and limits of artificial intelligence in the application of intelligent policing for Threat Intelligence and electronic data forensics can have specific application fields and practice objects, and can continuously verify and improve these ethical limits, which is of great practical significance to social public security.

Acknowledgments. 1. "Development and industrialization of threat intelligence platform based on big data intelligent analysis" (2020h6024), a science and technology project of university industry cooperation in Fujian Province in 2020.

2. 2021 Fujian young and middle-aged teacher education and scientific research project (Science and Technology): Research on information security of artificial intelligence equipment (JAT210523).

3. 2021 annual cooperative education project of the Ministry of Education: Innovation of system security training course based on domestic OS (202102373014).

4. 2021 college students' innovation and entrepreneurship training program (provincial level): UAV monitoring river and its environmental security threat intelligence analysis.

References

1. European Commission's High-Level Expert Group on Artificial Intelligence. Ethics guidelines for trustworthy AI. European Commission (2018). Retrieved from https://ec.europa.eu/digital-single-market/en/news/ethics-guidelines-trustworthy-ai
2. IEEE Global Initiative: ethically aligned design: a vision for prioritizing human well-being with autonomous and intelligent systems. IEEE Standards v1 (2016). Retrieved from https://standards.ieee.org/content/dam/ieee-standards/standards/web/documents/other/ead1e.pdf
3. John, L., Nicholas, O.: Global value chains and digital platforms: Implications for strategy. Strateg. Chang. **31**(1), 161–177 (2022)
4. Halima Ibrahim, K., Shareeful, I., Haralambos, M.: An integrated cyber security risk management framework and risk predication for the critical infrastructure protection. Neural Comput. Appl. 1–31 (2022)
5. Singh, J.: Implementation of memristor towards better hardware/software security design. Trans. Electr. Electron. Mater. **22**(1), 10–22 (2021). https://doi.org/10.1007/s42341-020-00269-x
6. Khan, A.A., Shaikh, A.A., Laghari, A.A., Dootio, M.A., Malook Rind, M., Awan, S.A.: Digital forensics and cyber forensics investigation: security challenges, limitations, open issues, and future direction. Int. J. Electron. Secur. Digital Foren. **14**(2), 124 (2022). https://doi.org/10.1504/IJESDF.2022.121174

7. Kusber, T., Schwalm, S., Shamburger, K., et al.: Criteria for trustworthy digital transactions - Blockchain/ DLT between eIDAS, GDPR, data and evidence preservation. Gi-Edition. Proceedings, **P-305**, 49–60 (2020)
8. Jobin, A., Ienca, M., Vayena, E.: The global landscape of AI ethics guidelines. Nat. Mach. Intell. **1**, 389–399 (2019)
9. Aguilera-Prado, M., Salcedo, O., Fernández, E.A.: Citation and similarity in academic texts: colombian engineering case. Cybern. Inform. Technol. **22**(1), 95–103 (2022). https://doi.org/10.2478/cait-2022-0006
10. Kuppa, K., Dayal, A., Gupta, S., Dua, A., Chaudhary, P., Rathore, S.: ConvXSS: a deep learning-based smart ICT framework against code injection attacks for HTML5 web applications in sustainable smart city infrastructure. Sustain. Cities Soc. **80**, 103765 (2022). https://doi.org/10.1016/j.scs.2022.103765

Application of VR Technology in 3D Image Interior Design System

Yihui Du and Yingjuan Ye(✉)

Chongqing College of Architecture and Technology, Chongqing, China
keyanyingjuan@163.com

Abstract. Today the computer technology is updated, and VR technology is widely used in many fields. As an important part of people's livelihood security, housing construction has also widely used VR technology in recent years. At present, the application of VR technology in interior design has been relatively mature. This paper analyzes the advantages of the combination of interior design and VR technology, discusses the main VR technology application characteristics in interior design, and discusses the path of combining the two. It aims to make better VR technology usage to promote innovation and reform of interior design, making interior design more interactive, immersive and perceptive.

Keywords: VR · 3D Graphics · Interior Design · Design Systems

1 Introduction

VR technology is the joint application of various technologies, with the network technology help, multimedia technology, simulated reality, multi-sensing and natural skills; the computer graphics and simulation technology usage to simulate the real environment; the use of sensor technology to connect the simulated environment with people. The perception is connected to realize the simulation of perception including visual perception, touch, hearing, force, taste and smell. It enables users to use natural skills including the user's own eye movements, hand movements or other behavioral actions, recognize these actions through the computer and feed back to the human facial features in real time to complete the user's interaction with the virtual space [1]. That needs to be studied in VR technology is to use computer graphics technology to establish real-time three-dimensional visual effects, to use VR to enhance research and in the field of science and technology development, and to establish an observation interface for the virtual world.

Interior design has been dominated by 2D civilian design. The maturity of 3D modeling software and technology has led many interior designers to try three-dimensional design. As an advanced digital information technology, VR technology has been introduced into the field of digital media art [1]. The VR technology can fully present the three-dimensional and spatial sense of interior design. The specific design is based on the target space and combined with three-dimensional animation. Technology not only allows designers to immerse themselves in the design space, but also recombines people, space and computers to show the design space scene.

J. H. Abawajy et al. (Eds.): ICATCI 2022, LNDECT 170, pp. 697–704, 2023.
https://doi.org/10.1007/978-3-031-29097-8_83

2 Overview of VR Technology

2.1 Development of VR Technology

Current VR technologies most often use VR headsets or multi-projection environments to generate realistic images. A person using a VR device is able to "look around" the artificial world, move around in the artificial world. This effect is usually produced by VR headsets, including head-mounted displays and small screens in front of your eyes, but it is also possible to create multiple large screens through specially designed rooms [2].

2.2 Hardware System

VR technology mainly includes four hardware systems: modeling equipment for building models; 3D visual display equipment that enables images to be displayed in 3D; sound equipment that can emit sound in the VR world; interactive equipment that can interact in the VR world [2]. For example, head-mounted displays (HMDs) can fully immerse users in virtual worlds. Head-mounted VR devices typically to render stereoscopic graphics of a 3D virtual world; a bin-aural audio system for six motion angles Positioning; motion controls with haptic feedback are also included in the options for 3D capabilities for intuitive physical interaction in the virtual world, e.g. an omnidirectional treadmill with great freedom of body movement that can allow users to Exercise in any direction.

2.3 Software System

VR technology includes three softwares: basic modeling software, model-driven software and functional programming software. The basic modeling software first saves various data, such as historical data, measurement data and image data, and then builds a basic model database according to specific requirements; model-driven software can drive virtual models according to specific requirements, and users can manipulate the experience [3]. Functional programming software can realize various user experience requirements such as scene conversion, automatic animation operation, color conversion and so on.

3 Main Application Technologies of VR

3.1 Dynamic Environment Modeling Technology

One of the purposes of VR technology is to create a virtual environment. In order to meet the realistic requirements, this environment needs to be dynamic. The realization of dynamic modeling technology must first obtain various data in the actual environment, and establish a virtual environment that is the same as the actual environment, as shown in Fig. 1. CAD technology is usually used in the process of data acquisition, but this technology can only acquire part of the environment, and the rest must be acquired by Visual Modeling technology [4]. Through these two technologies, the integrity of the actual environment can be guaranteed.

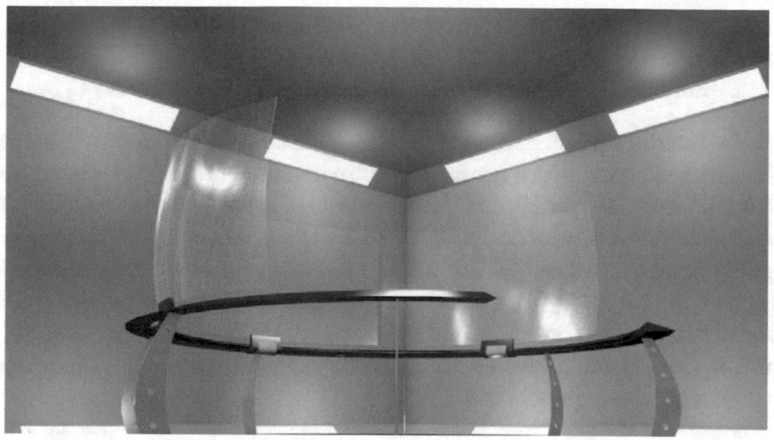

Fig. 1. Dynamic environment simulation

3.2 Real-Time 3D Graphics Generation Technology

In order for a 3D environment to generate images in real time when a human is operating, it is necessary to focus on how to generate it "in real time". In order to achieve the real-time purpose, the designer must ensure that the graphics have a high frame number, and at the same time, the resolution of the graphics cannot be reduced, so as to ensure the quality of the generated graphics [2].

3.3 Stereoscopic Display and Sensor Technology

The development of VR technology is inseparable from the development of stereoscopic displays and sensory devices. There are still many problems in the current VR technology, such as low resolution and small display range [5]. In order to make VR technology an important technology for future development, it is necessary to overcome the existing shortcomings and continuously develop new technologies.

3.4 Application System Development Tools

The VR technology application is based on occasions and objects. It is necessary to use imagination in the virtual world to imaginary and then create some occasions and objects. In order to use technology to generate products, it is necessary to continuously develop new tools to ensure product quality and production efficiency, such as VR System Development Platform, Distributed VR Technology, etc.

3.5 System Integration Technology

In using VR technology process, to create a virtual environment that makes users feel real, it is necessary to apply sensory systems and some corresponding models. Because these sensory models require designers to obtain information and use it in technology, the development of systems integration technology is very important [6].

4 Problems Existing in Traditional Interior Design

4.1 It is Difficult for Designers to Communicate with Householders

In traditional interior design, designers need to consider many factors when designing, such as material selection, space layout, furniture matching, etc. At this time, it is necessary to communicate with the head of the household to ensure that the work is carried out normally. In traditional interior design, designers use paper or computer for drawing. The drawings designed by this method are two-dimensional, and at most, the design scheme can only be expressed through animation. Although this method allows the householder to observe directly, it does not allow him to have a comprehensive experience [6]. At the same time, if the designer designs the scheme through animation, it is time-consuming and has high technical requirements.

4.2 Influence of External Environment

In traditional interior design, designers also need to consider the influence of external forces. With the continuous development of construction, many residential buildings are now high-rise buildings. In such a building, it is necessary to consider some external factors, such as considering the light, whether it is necessary to heat the interior or not; considering the wind power, whether it is necessary to block the balcony; Strengthen the sound insulation effect, etc. [7]. However, in traditional design, designers cannot grasp the impact of these factors.

4.3 It is Inconvenient to Modify the Scheme

In interior design, due to the consideration of many factors, it is necessary to modify the scheme at any time. In traditional design, since the scheme is generally drawn on paper, when the scheme needs to be modified, it needs to be redrawn, which is time-consuming and labor-intensive. The time for interior decoration is generally limited. When the specific design plan is not obtained, the construction period cannot be started on time, which leads to the delay of the construction period [7].

5 Advantages of Using VR Technology in Interior Design

5.1 Model Building Is Easy

Building an indoor model through VR technology enables designers to build a model with the same scale as the on-site house, and on this basis, simulate various factors to test whether the design scheme is feasible and modify the design scheme in time, as shown in Fig. 2.

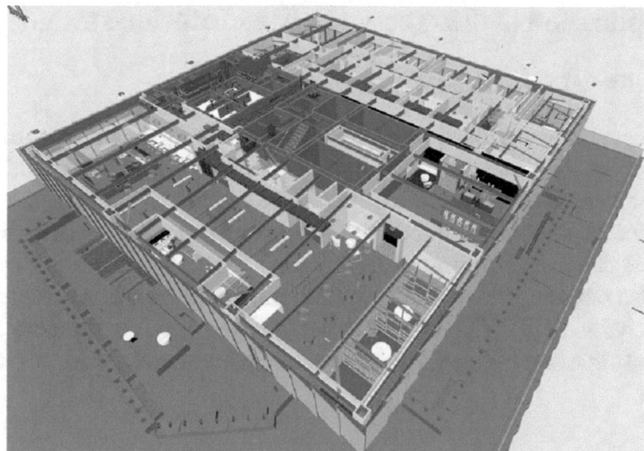

Fig. 2. Model Construction

5.2 Intuitive Display of Design Content

Interior design has a large number of design elements and design contents, and the projects are complex. Not only the requirements of space, color, light and shadow, but also design requirements such as design style, decoration, and furniture are included, as shown in Fig. 3. Combined with VR technology, the interior design is intelligent and humanized, so that it has a more sense of space. By applying VR technology to interior design, designers can not only adjust according to the virtual display of the design, but also stimulate the designer's creative inspiration. And customers, as non-professionals, can also intuitively feel the design effect, and then effectively control the time cost and capital cost for the formation, modification and determination of the design scheme [8]. Finally, the real display of VR technology in interior design solves the incomplete and inaccurate problems of traditional design.

Fig. 3. Intuitive display of interior design content

6 The Application of VR Technology in Interior Design

6.1 Presentation of the Design Scheme

In interior design, there is a lot of work to be done, not only the problem of constructing the drawing model, but also the space layout, material usage, client budget and other issues need to be considered. In order to guarantee the customer's experience, designers need to consider the dynamic nature of the virtual environment [9]. In traditional design work, designers can use animation to display the scheme, as shown in Fig. 4. However, this method also has many disadvantages: first, the display angle is not enough; second, the technical cost is high. With VR technology, customers can simulate the real environment in real time, feel from various aspects, and communicate with designers in a timely manner.

Fig. 4. Display of the design scheme

6.2 Simulation of Various Environmental Factors

When designing interiors, designers need to consider many external factors, and designers cannot simulate the impact of these factors on the final design. For example, the temperature of the upper floors is lower than that of the lower floors. If the designer does not consider this factor, the household owner will be affected by the external wind while living, and the indoor temperature will be too low, resulting in a poor living experience [10]. After the introduction of VR technology, designers can simulate these factors, and then design the perfect solution under the conditions of considering the relevant factors.

6.3 The Designer can Modify the Scheme at Any Time

After designers use VR technology, they can modify the design plan at any time, overcoming the drawbacks of traditional design methods [9]. For example, when customers check the plan, they can have an intuitive experience through VR technology, communicate with the designer in time, and the designer can modify the plan according to the customer's requirements.

6.4 Mobility of Virtual Models

The use of VR technology can enable customers to experience the plan, and can also modify the plan at any time, which greatly saves costs. Moreover, if the customer does not have time to go to the scene for the environment simulation experience, the VR technology can also be loaded into the mobile phone or computer, and the customer can simulate and observe the design scheme anytime and anywhere [10].

7 Conclusion

To sum up, with the leap-forward development of VR technology, digital media art in the art field has also flourished, and interior design bears the brunt of the artistic process of VR technology. The combination of VR and interior design has inherent advantages and has significant characteristics, allowing interior design to develop innovatively from content to form, giving designers more design inspiration and design elements, allowing customers to see and feel before the real thing to the design effect, make the interior design more humanized, intelligent and interactive, make the perfect combination of art and modern technology, and make the interior design move from a dull stylized design to a unique three-dimensional, immersive and experiential style design.

References

1. Zarei, M., DiPaola, S.: VR-designspace: a creativity support environment integrating VR technology into collaborative data-informed evaluation of design alternatives. HCI (38), pp. 367–374 (2021)
2. Palmas, F., Reinelt, R., Cichor, J.E., Plecher, D.A., Klinker, G.: VR public speaking training: experimental evaluation of direct feedback technology acceptance. VR, pp. 463–472 (2021)
3. Yun, Y.: The application of VR technology in residential decoration design. Computer Junkie **11**, 41–45 (2018)
4. Kim, D., Ko, Y.J.: The impact of VR (VR) technology on sport spectators' flow experience and satisfaction. Comput. Hum. Behav. **93**, 346–356 (2019)
5. Budakov, P.: 3D-rendered images and their application in the interior design. Encyclopedia of Computer Graphics and Games (2019)
6. Fangyuan, H.: The impact of VR technology on modern environmental art design. Beauty and Times: Cities **11**(1), 69–72 (2017)
7. Riener, A., et al.: Improving the UX for users of automated shuttle buses in public transport: investigating aspects of exterior communication and interior design. Multimodal Technol. Interact. **5**(10), 61 (2021)
8. Chaytonowicz, J., Maciejko, A.: Interior design and decorations of a house no 83 in Kemer in Turkey. In: Charytonowicz, J., Maciejko, A., Falcão, C.S. (eds.) AHFE 2021. LNNS, vol. 272, pp. 28–35. Springer, Cham (2021). https://doi.org/10.1007/978-3-030-80710-8_4

9. Lin, S.: The application of computer aided technology in interior design. In: Abawajy, J., Xu, Z., Atiquzzaman, M., Zhang, X. (eds.) ATCI 2021. AISC, vol. 1398, pp. 662–667. Springer, Cham (2021). https://doi.org/10.1007/978-3-030-79200-8_98
10. Vazquez, C., Tan, N., Sadalgi, S.: Home studio: a mixed reality staging tool for interior design. CHI Extended Abstracts **431**(1–431), 5 (2021)

Application of Artificial Intelligence Algorithm in the Design of Water Pollution Control System

Yunlong Zhao(✉)

Xi'an Technological University, Xi'an 710021, Shaanxi, China
zyl2275645137@163.com

Abstract. Due to the rapid development of modern industry, the natural resources on which mankind depends for survival have been severely destroyed. How to curb the gradual decline of water resources and take corresponding measures to curb serious water pollution has become an urgent problem for mankind. This paper aims to study the application of artificial intelligence algorithms in the design of water pollution control systems. Based on the analysis of the characteristics of neural networks, the hazards caused by water pollution, the influencing factors of water pollution treatment and the countermeasures of water pollution treatment, aimed at the process of water pollution treatment, starting from the reality of water pollution control, a neural network model of the water pollution control system is established, and simulation experiments are carried out on it. Experimental results show that the control effect of this model can meet the needs of water pollution control.

Keywords: Artificial Intelligence · Water Pollution Control · Neural Network · Nonlinear System

1 Introduction

The problem of water pollution is the main component that affects the development of the national economy and the sustainable development of society. However, due to the rapid development of China's modern manufacturing industry and the continuous expansion of the scale of the city, the increase in population, environmental protection issues, especially the pollution of urban water sources, further increase the problems of natural resource management and water environmental protection, environmental pollution has become a major problem facing the world [1, 2]. Therefore, urban and industrial wastewater treatment has become an important issue in the field of environmental protection. Urban sewage is the most direct factor that produces pollutants in rivers and lakes, and it is also one of the most important factors restricting the sustainable development of many cities in China [3, 4].

Due to its population size, national economic development level and seasonal changes, as well as various factors affecting environmental pollution, water pollution management has the characteristics of non-linearity, time-varying, difference and multi-purpose, and therefore there are many problems and uncertainties. Research on water

J. H. Abawajy et al. (Eds.): ICATCI 2022, LNDECT 170, pp. 705–713, 2023.
https://doi.org/10.1007/978-3-031-29097-8_84

pollution management that uses multiple linear regression, response surface methods, principal component analysis, K-means clustering, etc., often lacks the ability to deal with the uncertainty and complexity of environmental pollution management [5, 6]. The artificial intelligence technologies that emerged in the 1990s mainly include artificial neural networks, vector support machines, genetic algorithms, and particle swarms. It and these methods ensure high accuracy and do not require cumbersome I/O relationships. At the same time, it has gained a strong ability to establish complex formulas and deal with uncertain, interactive and dynamic complexity problems [7, 8]. Moreover, compared with the traditional MILR method, the ecological dissolved oxygen level in the watershed ecosystem evaluated by artificial intelligence technology is closer to the actual test results. In the simulation study of the pollutant treatment process in the watershed, the simulation results of the multi-layer perceptron neural network, radial basis neural network and other methods are compared with the traditional PC, RS and other methods, which are more helpful for post-polluting. The accurate prediction of pollutant removal effect provides a more credible technical means for the reasonable removal of pollutants [9, 10]. By supporting the intelligent pollutant management system formed by the vector machine and automatic pollutant optimization control equipment, it can effectively reduce energy while efficiently solving environmental pollution. The water pollution monitoring system formed by automatic drainage codes and online environmental monitoring equipment can effectively overcome the problem of lagging response to water pollution [11, 12]. Therefore, the use of artificial intelligence technology can provide technical guidance for comprehensively improving the ability to solve water pollution problems.

On the basis of consulting a large number of domestic and foreign references, this paper combines the characteristics of neural networks, the hazards caused by water pollution, the influencing factors of water pollution control and the countermeasures of water pollution control. Set out, establish a neural network model of the water pollution control system, and conduct a simulation experiment on it.

2 Application of Artificial Intelligence Algorithm in the Design of Water Pollution Control System

2.1 Features of Neural Network

The basic characteristics of neural networks are mainly summarized into the following aspects:

(1) Parallel and distributed processing capabilities

Neural network has a highly parallel structure and realization ability, combined with the unique advantages of computers in high-speed data processing, can find the optimal solution as soon as possible.

(2) Non-linear processing capability

When defining things with linearity and nonlinearity, the process of human brain activity is non-linear, so using neural networks to simulate human brain activity is also non-linear. This function is very useful when dealing with nonlinear problems.

(3) Self-learning ability

Training a previously given sample will produce a special neural network, after a certain amount of self-training, it can aggregate all the data.

(4) Can be implemented by hardware

Due to the rapid development of electronic chips, the ability to use very large-scale integration (VLSI) to integrate neurons and connections into chips and create neural networks to solve practical problems has become apparent.

2.2 Harm Caused by Water Pollution

(1) Harm to the environment

First, it has caused great damage to the water quality of the basin. The waters in the basin are polluted by industrial pollutants, heavy metals and various minerals are greatly exceeded; if the amount of pesticides and fertilizers is large and the use rate is low, a large amount of aerobic organic chemicals will remain in the soil and water bodies, causing rapid growth of plankton such as algae, red tides occur frequently in the basin. Secondly, the dissolved oxygen in the water is abnormally reduced and the water quality deteriorates, leading to the death of a large number of aquatic organisms. Finally, when local surface water is contaminated, people collect groundwater. In recent years, groundwater has been collected in excess of the standard in some areas, and the groundwater level has continued to drop, which has had a major adverse impact on the sustainable development of water resources.

(2) Harm to human survival and health

First of all, the most direct consequence of poor water quality is that the population of the basin continues to increase, and there is less and less water that meets human drinking water standards. Faced with huge drinking water pressure, people have to lower the standards. In order to survive, humans can only drink relatively lightly polluted water, which still contains many harmful substances harmful to human health. Studies have shown that copper content in water exceeding 1 PPM can cause red blood cell poisoning; excessive lead content in water can cause diseases such as kidney disease and epilepsy; excessive phosphorus content in water may cause other symptoms such as respiratory distress and poisoning.

(3) Harm to economic development

Surface water pollution caused by improper treatment of industrial wastewater and urban sewage. These polluted water cannot be used as a source of drinking water. Not only that, this wastewater cannot be used as agricultural water, and agricultural irrigation water is insufficient to allow crops to grow normally, resulting in a sharp decline in output. Watershed areas where freshwater aquaculture is the economic center are affected by water pollution. The cost of treatment after pollution continues to increase, and diseases caused by water pollution not only affect the quality of work, but also bring economic burdens to residents. Since water pollution is a problem facing the entire society, the government as a social manager must be responsible for the consequences of water pollution. The government must invest heavily in water pollution control every year.

2.3 Influencing Factors of Water Pollution Control

China's attention to water pollution control projects has continued to increase, and new laws and policies have been continuously promoted to promote the implementation of water pollution control projects. However, there are many factors that affect the smooth implementation of the project.

(1) The results of technical energy saving are not obvious

Industrial structure can improve the economic benefits of water pollution control projects. Technological innovation can not only reduce resource consumption and waste, but also increase capital conversion rate and production capacity. At this stage, China has made considerable progress in water pollution control, and individual regions have reached a leading position. However, the water pollution management capabilities of most regions are still under development. Therefore, strengthening regional governance is a top priority.

(2) Imperfect capital financing system

Funds to manage water pollution are essential, because the completion of a water pollution control project requires a lot of financial support. Since the "13th Five-Year Plan", the Chinese government has increased its support for water pollution prevention and control, and the amount of capital investment has increased again, with nearly 237 million yuan invested during the period. However, only relying on national financial support is not enough to meet the normal function of water pollution control. At this stage, local water pollution control funds are mainly provided through loan banks, but the rear areas cannot receive, collect and use effectively. Various reasons have caused the lack of funds for water pollution control, which affects progress and efficiency.

(3) The performance appraisal mechanism is not sound

The completion of water pollution prevention and control work is mainly measured by administrative means. Unfinished areas will be rejected, and many areas have reached the prescribed standards based on the evaluation of the regulatory authorities. However, since the "Thirteenth Five-Year Plan" period, the water pollution prevention and control challenges in this region have become increasingly difficult, and the time for the completion of the challenges has also been delayed.

2.4 Countermeasures for Water Pollution Control

(1) Strengthen the comprehensive treatment of wastewater pollutants in industrial enterprises

Encourage water conservation, implement the city's planned water management system with large water consumption, and relevant functional departments will focus on strengthening the supervision of major water users. When designing, constructing, and operating the company's main projects, appropriate consideration should be given to supporting water-saving equipment such as papermaking, printing and dyeing, electroplating, and leather. Treatment and reuse, promote the use of recycled water, strictly control the water quota, and make full use of the recycled water processed by the wastewater treatment plant. Carry out industry, urban greening, street cleaning and many other life activities. Environmental protection capacity and environmental protection departments must strictly control the development of high-water and high-pollution industries when approving projects.

(2) Continue to promote infrastructure construction

Pay attention to sewage treatment equipment and projects, transform and treat existing urban sewage treatment plants, and upgrade the pipeline network. When building a new sewage treatment plant in cities and towns, attention must be paid to the improvement of the pipe network, which requires simultaneous participation, construction and commissioning. Rural governance must first purify rivers, innovate governance and maintenance methods, and finally promote the standardization, innovation, and integration of agricultural wastewater management. Promote existing urban sewage treatment equipment and services, and continuously improve the status quo of rural sewage treatment.

(3) Strengthen water environmental protection

Integrate and guarantee the safety of drinking water sources, improve the management methods of drinking water sources of the "four reservoirs and one lake", and build prevention, control and risk management functions; improve the construction of agricultural infrastructure, increase the efficiency of agricultural domestic sewage collection, and prevent groundwater pollution. In order to prevent and control livestock and poultry pollution, corresponding water treatment plants shall be set up in the large-scale livestock and poultry farms maintained to meet the discharge standards.

(4) Improve public participation and supervision mechanism

Promoting the democratization of environmental protection and relaxing the scope of public supervision of water pollution can have a positive impact on environmental water governance. Therefore, the media is a good choice and way to use it. All common radio and television are effective methods, and the current network can provide good transmission effects. Use public opinion to promote water environment governance, increase public participation, and establish smart and effective public participation mechanisms such as environmental protection evaluation, information disclosure system, board of directors, and system letters and visits. In this way, the public ownership of water environment management is continuously strengthened, and the supervision and management of the government and enterprises can promote the accelerated implementation of water environment management.

3 Experiment

The water pollution control process is a typical nonlinear process. At the same time, due to the complex reaction process, it is difficult to determine an accurate mathematical model. First, use a mathematical model to analyze:

$$\begin{cases} \dot{X} = \frac{Q}{V}X_i - \frac{Q_w}{V}CX + \widehat{\mu}\frac{SX}{K_S+S}\frac{O}{K_o+O} - K_dX \\ \dot{S} = -\frac{\widehat{\mu}}{Y}\frac{SX}{K_S+S}\frac{O}{K_o+O} + \frac{Q}{V}(S_i - S) \\ \dot{O} = -\widehat{\mu}\frac{1-ff_xY}{fY}\frac{SX}{K_S+S}\frac{O}{K_o+O} - f_xK_dX + u \end{cases} \quad (1)$$

Among them, S is the concentration of the substrate in the aeration tank, O is the concentration of dissolved oxygen in the aeration tank, X is the concentration of microorganisms in the activated sludge mixture of the aeration tank, K_s is the half-rate constant, Y is the yield coefficient and the self-decay coefficient, K_d is the self-decay coefficient, V is the volume of the biological reaction tank, f_x is the consumption factor, and u is the aeration rate.

Because there are many uncertainties in the mathematical model and many unpredictable effects in the behavior of the control system, here we can use the neural network to learn the direct relationship between the model, modeling and the physical conditions of the control system, and according to the neural network, the network is used for modeling and control, and the adaptive control of neural network is used to carry out control science research. First of all, the water pollution control mode is regarded as a general mode of nonlinear or nonlinear control systems.

$$\dot{O}(t) = f(O(t), X(t), S(t), u(t)) \tag{2}$$

$$\dot{u}(t) = -au(t) + v(t) \tag{3}$$

This is an adaptive neural network test of dissolved oxygen concentration. Here, a low-pass filter is designed for u(t), and the dissolved oxygen treatment process model is a nonlinear triangular structure system, which uses a nonlinear function processed by a neural network, and Design the neural network controller.

4 Discussion

First of all, a simulation experiment is carried out for the iterative control process of the non-linear control system. The selected simulation mode is:

$$\begin{aligned} \dot{x}_1 &= x_2 + x_1 \sin(x_1) \\ \dot{x}_2 &= x_3 + x_1 \sin(x_2) \\ \dot{x}_3 &= u + x_2 \sin(x_3) \end{aligned} \tag{4}$$

The experimental results are shown in Table 1 and Fig. 1.

The ideal substrate concentration S in the aeration tank, the dissolved oxygen concentration O in the ideal aeration tank, the microbial concentration X in the ideal activated sludge mixture, and the ideal sewage treatment process model are non-dimensionally simulated, and a neural network controller is designed for the error system that can be obtained by aeration u can be obtained by Matlab simulation:

It can be seen from Fig. 2 that the neural network controller can gradually limit and stabilize the closed-loop system, control the dissolved oxygen concentration within a reasonable range, and within a limited range in a short time, the control process reduces the exposure. The concentration of substrates and microorganisms in the gas pool is stable in a short time.

Table 1. System state simulation curve

	x_1	x_2	x_3
0	2	−2	1.6
1	1.13	−1.5	0.47
2	0.28	−0.26	0.19
3	0.17	−0.2	0.08
4	0.05	−0.18	0.06
5	0.04	−0.03	0.03
6	0.04	−0.01	0.03
7	0.04	0	0.03
8	0.03	0	0.02
9	0.03	0	0.02
10	0.03	0	0.02

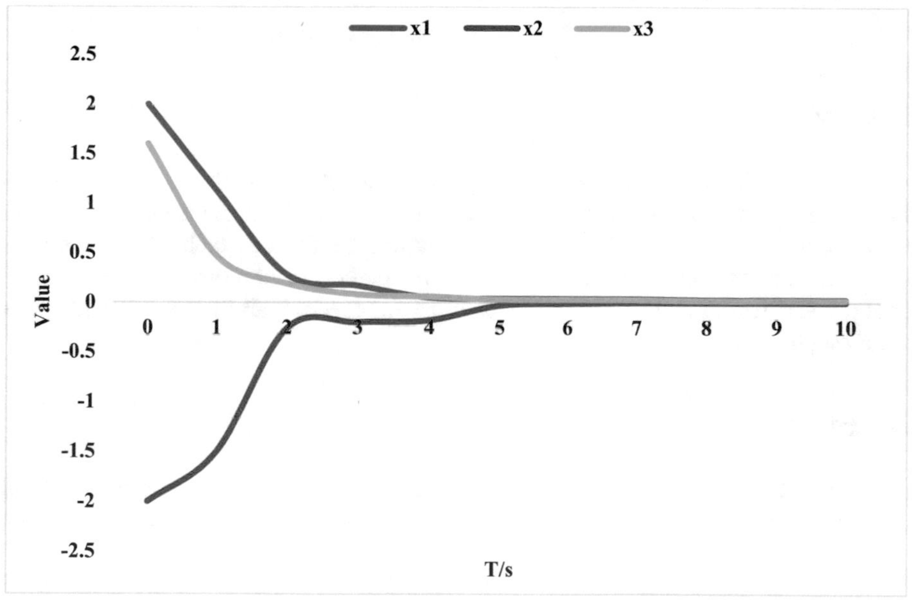

Fig. 1. System state simulation curve

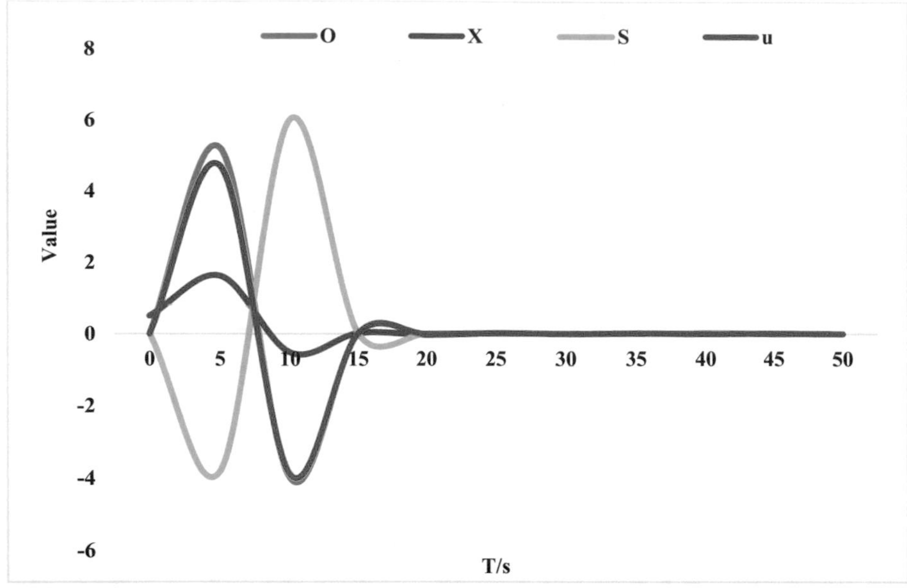

Fig. 2. Sewage treatment system status

5 Conclusions

Although artificial intelligence technology can reasonably reflect the nonlinear relationship between input and output, it still cannot understand the physical mechanism between input and output variables. Therefore, it is a major scientific and technological issue to be studied to integrate artificial intelligence information technology into the pollution mass transfer model and natural mechanism to enhance the ability to deal with water pollution problems.

References

1. Leonard, K.C., Hasan, F., Sneddon, H.F., et al.: Can artificial intelligence and machine learning be used to accelerate sustainable chemistry and engineering? ACS Sustainable Chemistry Eng. **9**(18), 6126–6129 (2021)
2. Joshiba, G.J., Kumar, P.S., Christopher, F.C., et al.: Insights of CMNPs in water pollution control. Nanobiotechnology, IET **13**(6), 553–559 (2019)
3. Hou, L.: Editorial for the special issue on water pollution control. Engineering **5**(5), 813–814 (2019)
4. Pramaningsih, V., Suprayogi, S., Purnama, S.: Strategy of water pollution control base on social economic activitiy, in Karang Mumus River, Samarinda East Kalimantan, Indonesia. E3S Web of Conferences, **31**(1), 04007 (2018)
5. Syafiuddin, A., Boopathy, R.: Effect of algal cells on water pollution control. Current Pollution Reports **7**(2), 213–226 (2021)
6. Licato, J., Zhang, Z.: Correction to: evaluating representational systems in artificial intelligence. Artif. Intell. Rev. **52**(4), 1 (2019)

7. Payedimarri, A.B., Concina, D., Portinale, L., et al.: Prediction models for public health containment measures on COVID-19 using artificial intelligence and machine learning: a systematic review. Int. J. Environ. Res. Public Health **18**(4499), 4499 (2021)
8. Hross, M.: Flow distribution improvements at the stamford water pollution control facility. The NEWEA Journal **53**(2), 26–31 (2019)
9. Campos, B., Motta, J.: Online measuring of robot positions using inertial measurement units, sensor fusion and artificial intelligence. IEEE Access **9**, 5678–5689 (2021)
10. Musulin, J., Egota, S.B., Tifani, D., et al.: Application of artificial intelligence-based regression methods in the problem of COVID-19 spread prediction: a systematic review. Int. J. Environ. Res. Public Health **18**(8), 1–39 (2021)
11. Kim, H., Lee, Y., Lee, E., et al.: Cost-effective valuable data detection based on the reliability of artificial intelligence. IEEE Access **9**, 108959–108974 (2021)
12. Dukhi, N., Sewpaul, R., Sekgala, M.D., et al.: Artificial intelligence approach for analyzing anaemia prevalence in children and adolescents in BRICS countries: a review. Current Research in Nutrition and Food Science Journal **9**(1), 01–10 (2021)

Intelligent Innovative Design of Indoor VR Based on Machine Vision

Jing He(✉)

School of Road Bridge and Architecture, Chongqing Vocational College of Transportation, Chongqing 402247, China
adam_vip_2021@163.com

Abstract. With the development of science and technology, human cognition and perception have changed with the development of science and technology and the progress of culture. For this reason, people try to rely on machine vision and virtual reality technology to develop a more diversified space environment. Interior design and VR technology are also increasingly connected. Based on this, the purpose of this paper is to study the intelligent innovative design of indoor VR based on machine vision. Based on machine vision, this paper uses machine vision to achieve indoor positioning, target detection and positioning tracking and other functions, and then uses virtual reality technology to design, model, and optimize the indoor space. In addition, users can also use virtual reality technology. Technology to personalize interior spaces to your liking. By applying the method in this paper to the interior design, it can make up for the deficiencies in the current design method by taking advantage of its unique advantages. Accurate positioning of interior design, market demand, and customer's wishes have a clear understanding, which has played a good guiding role in interior design and development. In addition, due to the method in this paper, the design can be carried out according to the user's preferences, thus greatly reducing the design cycle. Compared with the past, the design cycle is reduced by 30%, the design cost is reduced by 20%, and the user satisfaction is greatly reduced. Improve.

Keywords: Machine Vision · VR Technology · Interior Design · 3D Modeling

1 Introduction

With the continuous development of our country's scientific and technological strength, people's requirements for interior design are getting higher and higher, especially in first-tier cities, most people tend to have more personalized interior design that suits their preferences, in order to meet the needs of different groups of people. Preference, some scholars propose to realize the interactive interior design function through VR technology, and use VR technology to allow people to actually experience the interior design without leaving home, and propose changes to it according to their own preferences. To this end, this paper proposes an innovative design of indoor VR based on machine vision on the basis of predecessors.

J. H. Abawajy et al. (Eds.): ICATCI 2022, LNDECT 170, pp. 714–722, 2023.
https://doi.org/10.1007/978-3-031-29097-8_85

In the research on the intelligent innovative design of indoor VR based on machine vision, many experts and scholars have conducted in-depth research on it, and have achieved good research results, for example: American scientists Ameethajunaina M designed the immersive VR system for the first time, and produced the first mechanical prototype of the immersive VR device, Sensorama Simulator, which is a multi-channel experience display system with multiple senses. But the concept of virtual reality was not proposed at that time, Ameethajunaina M called it "experience theater" [1]. Nishida I first proposed this new type of graphic display technology in his paper "The Ultimate Display", and in 1968 successfully produced a helmet-mounted stereo display with tracker [2]. This shows that the importance of VR technology is self-evident.

Through literature research, this paper details the basic concepts of VR technology, machine vision technology and interior design, and summarizes and refines its significance in design. Using virtual reality technology, the interior structure and the shape of the house are expressed through virtual technology, so that become visible objects and environments. At the same time, in the early stage of design, designers can simulate their own ideas through virtual reality technology, and can see the actual effect of the interior in the virtual environment in advance, which not only saves time, but also reduces costs.

2 System Design

2.1 Design and Implementation of Indoor Modeling System

(1) Design of system class diagram

At the heart of the modeling system is the document view architecture. As a document manager, an application can open, close and update documents. Each application can only have one currently active document at the same time. A document has one or more corresponding views, and document updates will be automatically updated. Update all its views. The view is the interface interface between the document model and the user. It is responsible for displaying all 3D models and other visual elements in the document, accepting user input and updating the document. Each application can only have A currently active view [3, 4].

(2) Design of application class and main interface

The application class acts as the container of the entire user interface, and manages the drawing and message response of the user interface at the same time, but we encapsulate the interaction between the user and the view and model into commands, and manage and interact with the application class through the command manager. This removes the dependency of each interop command on the user interface.

(3) Implementation of interactive commands

The system encapsulates interactive commands in commands. First, the application receives the command message for creating a room from the menu, control panel or command line interface. The application encapsulates the message and hands it over to the command manager for processing. The command processor creates the corresponding room. An example of an interactive command. Only the interactive command knows how to process the user's input commands and parameters. Once the interactive command is created and started, it takes over the message processing of all interfaces. At this time, the user can input parameters on the command line or directly Move the mouse in the view or use the mouse to input the position of the point, the interactive command will read the input parameters, create a temporary model object and temporarily update the view after verifying the validity of the parameters, enter the next step, repeat the cycle until all parameters are input, and the interactive command sends a message To the command manager, the command manager is responsible for updating the final created room to the document, the document updates all its corresponding views, and finally returns the message processing right to the application, but the executed command object is cached in the command manager's In the stack, it is convenient for repeated execution [4, 5]. Any intermediate steps such as user press or right mouse button to cancel the command, the interactive command will issue a message, the command manager will destroy the current interactive command in the stack, and return the message processing right to the application.

(4) Setting of model elements

Document objects are composed of abstract Element elements. All Element objects are organized into a tree-like structure. Different types of data can be attached to each Element, such as geometric models, lights, cameras, materials, or textures. In the modeling system, the models are divided into two categories, one is the BREP-based model, and the other is the polygon mesh model. The former has curve and surface parametric equations, while the latter does not [5, 6]. The system converts the exact model to an approximate triangular mesh model before rendering, so that the renderer faces a uniform polygon mesh model. The base class of the BREP model is Shape, and all components that can be input and modified by parameter are its subclasses or design patterns.

(5) Realization of automatic wall connection function

Since it is not necessarily end-to-end, and the angle between the two walls is not necessarily a right angle, a more general algorithm needs to be implemented. The solution of the automatic connection problem of any two straight walls at the corners can be solved finally to find the intersection of the inner axis and the outer axis of the two walls, and find these two points, and finally find the local coordinate system of the wall., modify the parameters and parameters of the wall. To find this point, we use the sum of two auxiliary unit vectors as shown in the figure, both of which point to the other end of the wall from the intersection of the median axis, so that the point inside the angle between the two vectors can be obtained, on a defined straight line. By comparing the projection of the points inside the angle on the vector direction, the point with the largest projection can be regarded as the point we are looking for [6, 7].

2.2 Visualization System Design and Implementation

The scene model data is first preprocessed, including BREP model meshing, scene optimization, material and texture assembly, lighting and camera model configuration.

(1) BREP model meshing

For the BREP model, the OpenCascade library itself provides the BrepMesh module. BrepMesh implements adding a triangular mesh to the surface of the BREP model, which is used to hand over the BREP model to Opengl for display.

(2) Material system

In the rendering engine, the material reflects the light scattering properties of the surface of the object, while for the physics engine, the material reflects other physical properties of the object, such as mass, density, etc. Materials must be able to basically accurately reflect the physical properties of materials commonly used in various interior designs in the real world. We first distinguish rigid body, soft body, cloth and fluid, and then include the properties of the corresponding NxMaterial class in the PhyxX engine, so that before rendering, the scene data required by the physics engine can be generated [8, 9].

2.3 Real-Time Photorealistic Rendering

In this paper, photon mapping technology is selected to implement the rendering engine on NVIDIA GPU. First, the scene KD-Tree acceleration structure is constructed. Other data structures, table-based data structures or spatial hash tables can also be used. For static scenes, there is no need to build acceleration structures in real time, and the method of pre-computing in the host can be adopted to save GPU resources [10]. Then enter the first stage, that is, the photon tracing method establishes a photon map, separates the caustics photon map and the global photon map that require high precision, and finally uses distributed ray tracing to process direct illumination. Multiple secondary rays recursively, for non-specular surfaces to cast shadow rays to calculate direct lighting, and for specular surfaces to cast multiple specular and refracted rays to calculate highlight effects. When each first intersection occurs, its indirect illumination must be calculated at the same time. The indirect illumination uses photon mapping technology to improve the calculation speed. In order to improve the image quality, a final collection link is added. The Russian roulette method is used to project the collected rays from the sampling point to a random direction., the indirect illumination along each ray incident sampling point is calculated according to the photon concentration evaluation of the global photon map [11]. Finally, Monte Carlo integration is used to calculate the indirect incident illumination of the sampling point. In the final collection stage, about 200 to 300 rays need to be randomly projected, and the photon concentration around the calculation needs to be evaluated to project the collected rays more efficiently, which narrows the evaluation range by dynamically constructing illuminance segmentation in the GPU.

3 Indoor VR Space Design Experiment Design

3.1 Indoor Positioning Algorithm Based on Machine Vision

The indoor positioning algorithm of machine vision includes three parts: target detection, tracking and positioning. First, the collected target image is subjected to image preprocessing, image noise is reduced through image denoising processing, and the quality of the image is improved. Then, according to the target feature information, identify the target image, and determine the location of the target; when the target deviates from the preset center position, it is necessary to control the rotation of the pan/tilt to track the target; when the target is tracked, the angle sensor information at this time is recorded, and the target position is determined according to the AAE algorithm [12].

(1) Image preprocessing

The target image is small, and the infrared target image has a low noise-to-noise ratio and is easily affected by noise and other sources of interference, so the image should be filtered and denoised. Infrared image preprocessing technology is a technology that improves the signal-to-noise ratio of an image by suppressing noise in the background. In this paper, the mean filter algorithm is used to preprocess the image. The calculation formula of the mean filter calculation is as follows:

$$g(x, y) = \sum\nolimits_{M\,(i,j)\in S}^{1} f(i, j) \tag{1}$$

Among them, f(i, j) is the original image, g(x, y) is the processed image, and M is the number of template pixels.

(2) Image denoising

The signal-to-noise ratio is one of the important criteria to measure the effect of image denoising. The signal-to-noise ratio of an image is the ratio of the signal power to the noise power. Since the image power is difficult to calculate, the ratio of the signal variance to the noise variance is equivalent to the signal-to-noise ratio. The image signal-to-noise ratio formula is as follows:

$$\mathrm{SNR} = 10 \times \lg \frac{\sum (\mathrm{x(i,j)} - \mathrm{x})^2}{\sum (\mathrm{x(i,j)} - \mathrm{X(i,j)})^2} \tag{2}$$

(3) Image segmentation

The image is divided into different regions according to different features such as color, shape and texture of the image. The images in the same region have similar features, but different regions have different image features. The effect of image segmentation results will directly affect the accuracy of target recognition. Because the Roberts edge detection operator has a better effect on the edge extraction of circular objects. Therefore, this paper adopts the Roberts edge detection operator as the image segmentation algorithm.

(4) Improved random Hough transform detection target

First, use the gradient information of the three points to determine whether the three randomly selected edge points are in the same circle, and whether a circle can be determined. If the parameter values of the circle are recalculated in the same circle, the image is subjected to random Hough transform. If the edge points are not in the same circle, the edge points are reselected, and the random Hough transform is not performed until the extracted three edge points are located on the same circle, which can reduce the amount of calculation and the search time of the linked list.

3.2 Experimental Design

(1) First determine the type of experimental project, and use AutoCAD and 3dmax software to make models according to the emotional design requirements of the type, and save the made models as [.osgb] source file format, so that they can be imported into the Vizard system for editing. This paper will take the indoor space of residential buildings, the indoor space of office buildings and the indoor space

of commercial buildings as examples to conduct immersive VR space experiments. Before the model is made, each type of space function should be reasonably divided to meet the needs of the experiment. The model making process will not be repeated in the text;

(2) Establish VR scene roaming, edit and run scripts in Vizard, and associate hardware devices with software models as required to ensure normal experimental operations;
(3) Carry out VR indoor space design experiment.

4 Indoor VR Space Design Experiment

4.1 Collision Detection Experiment of Indoor Objects

In this experiment, the collision detection between a large number of objects is carried out, and the advantages and disadvantages of the two algorithms are compared by comparing the average collision detection time. Array objects are provided, the number of objects and the number of patches are constantly increasing, in order to test whether the real-time performance of the collision detection algorithm has been improved, the experimental results are shown in Table 1.

Table 1. Massive Object Intersection Test Results

Number of noodles (slices)	Average collision detection time in this paper (ms)	Average collision detection time for comparison algorithms (ms)
100	2	4
500	13	15
1000	18	20
1500	21	22
2000	25	27
2500	30	33
3000	34	37
3500	38	41
4000	42	45

It can be seen from the Fig. 1 that with the continuous increase of the number of patches, the method used in this paper has a significant improvement compared with other algorithms, the average collision detection time is shorter, and the system responsiveness is better, which shows that the algorithm in this paper contains a large number of objects and In the case of a large number of patches, the real-time performance of collision detection has been greatly improved, and the purpose of improving the real-time performance of collision detection has been basically achieved.

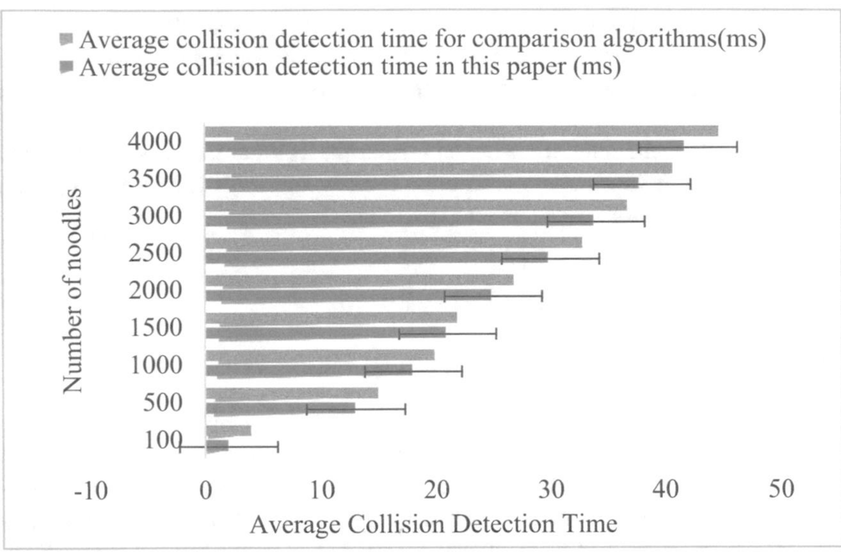

Fig. 1. Average collision detection time for different algorithms

4.2 Comparison with RAPID Algorithm

The RAPID algorithm is a well-known collision detection system. The two objects are rigid body models, so their respective geometric images are unchanged, and one-step preprocessing can be used to generate their own geometric images. The two objects pass through each other at a fixed speed, and the motion of the objects is only translation, not rotation, so there is no need to recreate the hierarchy every frame, the experimental results are shown in Table 2.

Table 2. Massive Object Intersection Test Results

frame number	Collision detection time for this article(ms)	Collision Detection Time of PAPID Algorithm(ms)
200	1	0
400	5.0	4
600	5.0	18
800	5.1	55
1000	5.2	67
1200	5.1	66
1400	5.3	64
1600	5.4	56
1800	5.2	35

It can be seen from the data in the Fig. 2 that when the two objects in the experiment do not collide, that is, when the number of frames is about 400 frames, the speed of this method and RAPID is obviously faster due to its own characteristics. When the two objects actually intersect, the method in this paper begins to show obvious advantages. Therefore, we can conclude that the method in this paper can greatly reduce the rendering time, thereby improving the work efficiency of the designer.

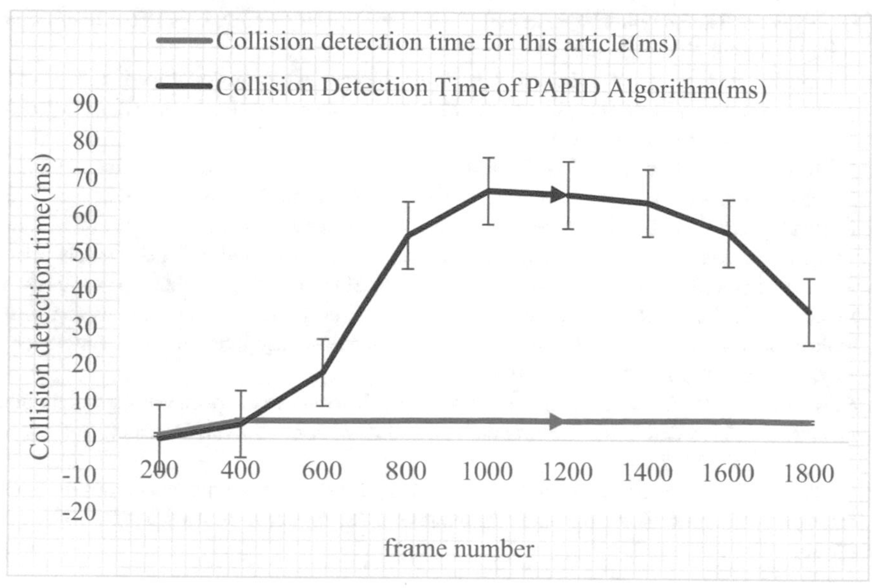

Fig. 2. The collision detection time of the two algorithms varies with the number of frames

5 Conclusions

At present, VR technology is developing rapidly, equipment technology is being updated, and it is gradually being implemented from conceptual design to practical application. The application of new technology will greatly improve the stagnant situation of the traditional interior design industry and promote the prosperity of the interior design industry. Developing.

References

1. Ameethajunaina, M., Abishek, B.E., Rajendren, V., et al.: A survey on fresh produce grading algorithms using machine learning and image processing techniques. IOP Conference Series: Materials Science and Engineering **981**(4), 042084 (2020)
2. Nishida, I., Tsuyama, R., Shirase, K., et al.: Development of innovative intelligent machine tool based on CAM-CNC integration concept - adaptive control based on predicted cutting force -. International J. Automation Technol. **13**(3), 373–381 (2019)

3. Das, S., Sanyal, M.K.: Machine intelligent diagnostic system (MIDs): an instance of medical diagnosis of tuberculosis. Neural Comput. Appl. **32**(19), 15585–15595 (2020)
4. Basha, M., Sibley, M.J., Mather, P.J.: Design and implementation of a long range indoor VLC system using PWM. Annals of Emerging Technologies in Computing **3**(1), 20–27 (2019)
5. Ahmed, O.A., Sayed, H.K., Jalal, K.A., et al.: Design and implementation of an indoor solar emulator based low-cost autonomous data logger for PV system monitoring. Int. J. Power Electron. Drive Syst. **10**(3), 1645 (2019)
6. Natan, O., Gunawan, A.I., Dewantara, B.: Design and implementation of embedded water quality control and monitoring system for indoor shrimp cultivation. EMITTER International J. Eng. Technol. **7**(1), 129–150 (2019)
7. Essahlaoui, F., Elhajrat, N., Abbassi, A.E., et al.: Modeling design and implementation of an embeds system real time over a network of wireless sensors to environmental monitoring. Mod. Appl. Sci. **14**(1), 41 (2019)
8. Salih, M.H., Teng, L.H., Ismail, S.N.: Design and implementation of tracking personnel within building featuring fpga and rfid. J. Eng. Applied Sciences **14**(3), 758–774 (2019)
9. Akilandeswari, J., Jothi, G., Naveenkumar, A., et al.: Design and development of an indoor navigation system using denoising autoencoder based convolutional neural network for visually impaired people. Multimedia Tools and Applications **81**(3), 3483–3514 (2022)
10. Ekstrmer, P., Wever, R., Andersson, P., et al.: Shedding light on game engines and virtual reality for design ideation. Proceedings of the Design Society International Conference on Engineering Design **1**(1), 2003–2010 (2019)
11. Silva, K., Weiller, T.H., Giordani, J., et al.: Oatributo acessode primeiro contatona ateno primáriaà saúdedointeriordo rio grandedo sul: umestudomultinível. Research Society and Dev. **9**(7), 19972865 (2020)
12. Arya, S., Chung, Y.H.: Novel indoor ultraviolet wireless communication: design implementation, channel modeling, and challenges. IEEE Systems Journal **15**(2), 2349–2360 (2020)

Interior Decoration Design System Based on Virtual Reality Technology

Jing He(✉)

School of Road Bridge and Architecture, Chongqing Vocational College of Transportation, Chongqing 402247, China
adam_vip_2021@163.com

Abstract. With the development of my country's economy, the people's living standards have gradually improved, and the housing environment has been continuously improved. At present, most of the house decoration in our country is designed and implemented by the decoration team, and this kind of decoration style is similar, and it is difficult to achieve the expected goals of the residents. Based on this background, the purpose of this paper is to study the interior decoration design system based on virtual reality technology. Based on virtual reality technology, this paper relies on Java language environment, uses Java3D open source three-dimensional graphics modeling engine, and cooperates with SqlServer database and tomcat server software. And Jsp, Servlet and other technologies to achieve system functions. Through the function and performance test of the system in this paper, the test results show that there is no functional failure of the system, and the average response time of the system is less than 3 s. The system works as expected.

Keywords: Virtual Reality Technology · Interior Decoration Design System · Java · Sqlserver Database

1 Introduction

In the interior design industry, traditional floor plans and renderings cannot fully express the designer's design ideas and concepts. Because the construction drawings are too professional, it is difficult to draw a beautiful blueprint for the home environment for consumers. Because the customer cannot see the complete final product form in advance, and the static partial renderings cannot give the customer a comprehensive three-dimensional spatial layout experience, the customer cannot check whether the plan really meets their needs in the design stage, and the design may cause the customer There is a high chance of regret. Therefore, this paper solves the above problems by designing an interior decoration design system based on virtual reality technology.

In the research of interior decoration design system based on virtual reality technology, many scholars have carried out related research on it, and achieved good results. American scholar Martono K T. gave a definition of the essence of virtual reality. He believes that simulation The seven points of sex, interaction, artificiality, immersion,

J. H. Abawajy et al. (Eds.): ICATCI 2022, LNDECT 170, pp. 723–732, 2023.
https://doi.org/10.1007/978-3-031-29097-8_86

telepresence, total immersion, and network communication constitute virtual reality, and represent that the highest level of virtual reality is not in technology but in art. This is exactly in line with the evolution trend of media technology [1]. Baker S divides the medium into cold medium and hot medium by "engagement" in "Understanding the Media", which is consistent with the concept of interaction in virtual reality, virtual reality "doesn't need so many gaps for the recipient to fill or complete" Rather, it provides us with an interactive cyberspace; similarly, the subliminal effect proposed by Baker S has the same characteristics as the "immersion" in Haim's definition of the essence of virtual reality, and to some extent, reflects McLuhan Virtual thinking on media technology research [2]. This shows that the importance of virtual reality technology is self-evident.

By analyzing the actual needs of interior designers and consumer customers, this paper designs and realizes the prototype of interior design software modeling system and visualization system. Design and implementation principles of ray tracing and photon mapping hybrid rendering technology. Finally, a complete and practical interior decoration design system is designed. Therefore, the user can preview the decoration effect at any time, and make timely adjustments, so as to form a decoration effect that truly conforms to his hobbies.

2 Interior Decoration Design System Based on Virtual Reality Technology

2.1 Design Analysis of Interior Decoration Design System Based on Virtual Reality Technology

(1) Rapid 3D modeling

Building a 3D model in interior design starts with the wall, and defines the default wall thickness and storey height. Entering two points can determine a linear wall, and three points can determine an arc wall. Continuous input walls can enclose a room, and adjacent walls are automatically connected at corners. Rectangular rooms can be quickly input through two diagonal points. Continuously input rooms need to detect overlapping walls and merge them to remove unnecessary redundant wall data [3].

(2) Construction drawing output

The software model is linked to a corresponding two-dimensional vector graphics file through the database, which is automatically loaded and placed in the corresponding position when the construction drawing is generated, and the two-dimensional vector graphics are transformed into a matrix according to the different transformations of the model elements. The contour is generated when the model is projected onto a 3D plane through the PolygonalMesh algorithm. This approach relies on model data, and the resulting graph is not easily controllable, but no additional work is produced. In order to be able to revise the output construction drawings, the output files should be able to be imported into commonly used software [4].

(3) Rendering and animation output

Using three-dimensional animation expression, it breaks through the static flat layout of renderings, and can display the spatial structure and design concept of the living environment more comprehensively and dynamically. In order to reduce the complexity, this article does not consider the key frame animation of the spatial position, shape, material, etc. of a single object, and can consider outputting it to professional animation production software, and then carry out animation design. In interior design, most animations are just a combination of consecutive renderings, so it is convenient to set the camera path and then render the output.

(4) Virtual roaming

The process of virtual roaming can show the different effects of the living environment in different time periods and different environments, such as sunny and rainy days, there should be completely different pictures, and there will be corresponding changes during the day and night, and even allow customers to experience The effect of the living space changes dynamically throughout from sunrise to sunset. You can also turn on and off the lights to experience the rationality of the lighting arrangement. If the brightness is not enough, you can increase the number or intensity of the light sources. If the brightness is too strong, you can reduce the energy and quantity of the light sources, so as to provide scientific data reference for the actual installation of lamps. Virtual roaming expands the original single visual perception to a more complete range of perception experience including motion perception, auditory perception, force perception, etc., and brings a three-dimensional virtual environment that makes people feel more immersive. It is bound to greatly improve the customer's consumption experience [5, 6].

2.2 Overall Design of System Modules

(1) Two-dimensional graphic design

Select an existing flat structure and draw a new flat model. In the actual decoration process, the whole house can be divided into different functional units, such as master bedroom, secondary bedroom, bathroom, kitchen, living room and so on. Due to its different functions, each room has its own special decoration method and supporting utensils.

(2) 3D scene modeling

This paper uses Java3D technology to realize the three-dimensional scene modeling of the house structure. At the beginning of the establishment of the house model, it does not include.

Include any decoration. On the 3D model, we call any data objects as elements, such as doors, windows, floors, curtains, furniture, appliances, etc. Through the element embedding function, the display of various elements on the 3D image can be realized. At the beginning of interior decoration design, all elements exist in two-dimensional graphics. In order to realize the transformation from two-dimensional graphics to three-dimensional models, this paper designs a three-dimensional converter, whose function is to convert two-dimensional graphics into three-dimensional shapes that can be recognized by Java3D [6, 7].

(3) Virtual scene data access

This article will realize the access to the 3D scene through object serialization. In the Java language, if you save a class object or class instance as a file, you need to save the object type first, and then save the class attribute [8]. This process of converting an object's state information into a form that can be stored or transmitted is called serialization. Conversely, when data needs to be read, a blank object needs to be created first, and then the corresponding state data is filled on the object.

(4) 3D scene data element maintenance

Data management is the basic function of various management systems. Based on the adopted software and hardware environment, this project uses the Java language to connect to the database through JDBC. For different management operations, construct different Sql statements, and finally return the final result in the form of database query [9].

3 Realization of Interior Decoration Design System Based on Virtual Reality Technology

3.1 Programming Environment and System Design Patterns

This paper adopts the B/S architecture mode, that is, the main function codes are stored on the server side. After an ordinary user logs in to the system, the server-side code is invoked through keyboard input or mouse click to implement specific functions. The processed data results are returned to the user through the browser. In practical applications, the B/S mode can be implemented by various means. In order to cooperate with Java3D engine, this paper adopts JSP technology and Servlet technology to realize the information interaction between user and server [10].

3.2 Implementation of Common User Functions

After logging into the system, ordinary users can complete the design and input of 3D scenes by performing scene creation, 3D modeling, interactive control, model download and information query. First draw a two-dimensional floor plan of the house in the scene. In simple cases, it can be directly saved or output; if the 2D plane needs to be 3Dized, continue to perform the 3D modeling function. In order to increase the decorative effect, three-dimensional furniture, home appliances, etc. can be added to the established three-dimensional house structure. Finally, save or output the designed 3D scene.

3.3 3D Scene Establishment and Realization

Relying on the Java platform, the three-dimensional modeling of the house is realized through the Java3D three-dimensional engine.

(1) The realization of two-dimensional plane graphics rendering

In the interior decoration system, the main graphic elements of two-dimensional drawing include: rectangles and lines. The rectangle corresponds to the ground, and the line corresponds to the wall. First give the grid background in the drawing space. The function of dividing the grid is realized by the drawGrid method, and its realization principle is to divide the entire drawing space into grids according to the specified pixel interval. Each point in the grid can precisely define its two-dimensional coordinate information [11]. In order to realize the correspondence between brush and wall parameters, it is necessary to create a mouse listener object on the grid canvas to monitor different mouse actions.

(2) Two-dimensional floor plan decomposition realization

Before building a three-dimensional model, it is necessary to design the specific house structure and main furniture on a two-dimensional plane. The drawing of two-dimensional graphics is realized through the Swing and AWT toolkits provided by Java. For two-dimensional graphics, this paper constructs a three-dimensional converter to map two-dimensional graphics into three-dimensional space. The conversion process mainly relies on the 3D model interface provided by Java3D. For complex graphics, this paper adopts the decomposition method to split the two-dimensional graphics into several line segments [12]. In this article, all these line segments are an instance of the Line class. All these line segment objects are stored in arrays. When reading these line segment information, the starting point coordinate and end point coordinate information of each line segment can be recorded one by one by traversing the array as the input information of 3D modeling.

(3) Realization of 3D house modeling

It can be seen from the Java3D scene graph that in order to display the three-dimensional effect of a graphic, a content branch with three-dimensional shapes or other contents such as background, sound, and light must be written. In Java3D, this content branch is given by the createSenceGraph method. All content is placed in a BranchGroup object. Under normal circumstances, the shape will be placed at a certain position in the three-dimensional space, therefore, the TransformGroup object needs to be generated in the BranchGroup. In Java3D, the TransformGroup object represents the coordinate system and can be used for coordinate transformation. In addition, in order to display a more friendly three-dimensional effect, it is necessary to place backgrounds and lights in the BranchGroup object.

(4) 3D scene storage implementation

The three-dimensional scene structure based on Java3D is further encapsulated, an independent Java class is constructed, and the call to the Java3D method is executed in the member method of the newly constructed Java class.

(5) Global Illumination Rendering

By using the reverse path tracing method, the light starts from the eyes and passes through each pixel on the screen and is projected into the scene. As long as the light intensity at the point where the light intersects with the object in the scene for the first time is calculated, the corresponding pixel of any point in the image to be generated can be determined. Where does the light that enters the eye from any point in the scene come from? One is that the light is directly projected to this point and then reflected into the eye, which is called direct lighting, and the other is that the light is reflected multiple times by other surfaces and projected. After reaching this point, it is reflected into the eye, which is called indirect lighting. The light reflected by the point into the eye is equal to the sum of the direct and indirect lighting at the point. Assuming that the light intensity of the incident point is known, the material of the point determines the reflected light intensity of the point. The material can be regarded as the object's impact on the incident light

Light response parameters, in the generation of realistic graphics based on physics, the reflection characteristics of materials are generally described as bidirectional reflectance distribution function BRDF, where the BRDF of this point has two parameters, incident light and reflected light direction

$$f(\omega_i, \omega_0) = \frac{dL_0(\omega_0)}{L_i(\omega_i)\cos\theta_i d\omega_i} \tag{1}$$

The BRDF tells us how much of the incident light energy from the direction ω_i is reflected to the ω_0 direction. We denote the light energy radiation of the point in the ω direction as L(p, ω), then the total amount of reflected light in the ω_0 direction can be Expressed as:

$$L(p, \omega_0) = L_e(p, \omega_0) + \int L_i(p, \omega_i) f_r(\omega_i, p, \omega_0)\cos\theta_i d\omega_i \tag{2}$$

4 Experimental Analysis of Interior Decoration Design System Based on Virtual Reality Technology

4.1 Ability Test of System Communication Module Providing Services

Table 1. Experimental Statistics

Send cycle	1 s	0.5 s	0.1 s	0.05 s	0.01 s	0.005 s	0.0001 s
Processing rate (bar/s)	1	4.5	9.5	48.5	90	140.6	750.8
Average demo(s)	0.00104	0.00152	0.00198	0.00224	0.00245	0.00275	0.00306

The ability of the communication module to provide services. This part of the experiment simulates the situation that both the positioning server and the display terminal are one. During the operation of the system, the data sending frequency of each mobile terminal is about 2 s per message. When there is only one mobile terminal, the period for the location server to send the location message is about 2 s, and when the number of mobile terminals increases to N, without considering the delay, the message receiving and sending cycle of the positioning server is 2/N. In order to understand the effect of sending frequency on the message processing capability of the entire communication module, especially the message middleware server, this paper designs a set of experiments with the frequency of message sending as a variable. The experimental statistical results are shown in the Table 1.

The experimental results are shown in Fig. 1. The processing rate in the table is the message processing rate monitored by the server, and the average delay is the data

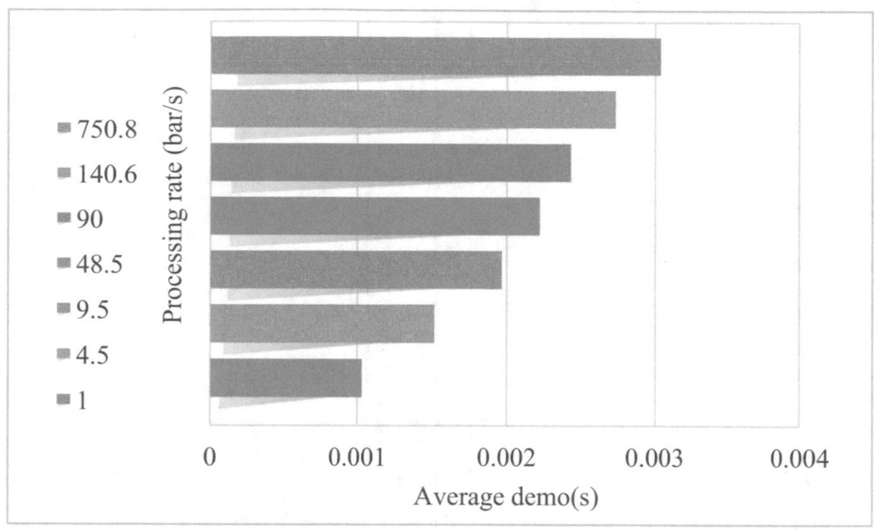

Fig. 1. Relationship between total delay and transmission rate

collected by the client. It can be seen that as the sending period gradually decreases from 1 s to 0.0001 s, the number of messages processed per second on the server side also increases from 1 per second to 750.8 per second, while the average message sending delay has increased, but Has not changed much.

4.2 Analysis of Communication Requirements Between System Communication Module and Mobile Terminal

It mainly tests the ability of the communication module to provide services when the number of data producers increases. This part of the experiment simulates the communication between the mobile terminal and the positioning server when the number of mobile terminals increases. In the experiment, the number of data producers is increased from 10 to 300, the sending rate remains unchanged at 1 per second, and the number of data consumers is 1, the experimental results are shown in Table 2.

From the statistical results in the Fig. 2, we can see that when the client's sending rate is 1 message per second, for the number of concurrency within 300, the change in the number of concurrency has little effect on the ability of the message middleware server to provide services.

Table 2. Experimental Statistics

Number of producers	30	50	60	100	120	140	160	180	200	220	240	260	280	300
Processing rate (bars/s)	30	49	58	98	116	138	157	176	198	216	236	254	276	298
Average delay (s)	0.00125	0.00155	0.00186	0.00215	0.00198	0.00176	0.00186	0.00214	0.00224	0.00189	0.00213	0.00221	0.00227	0.00199

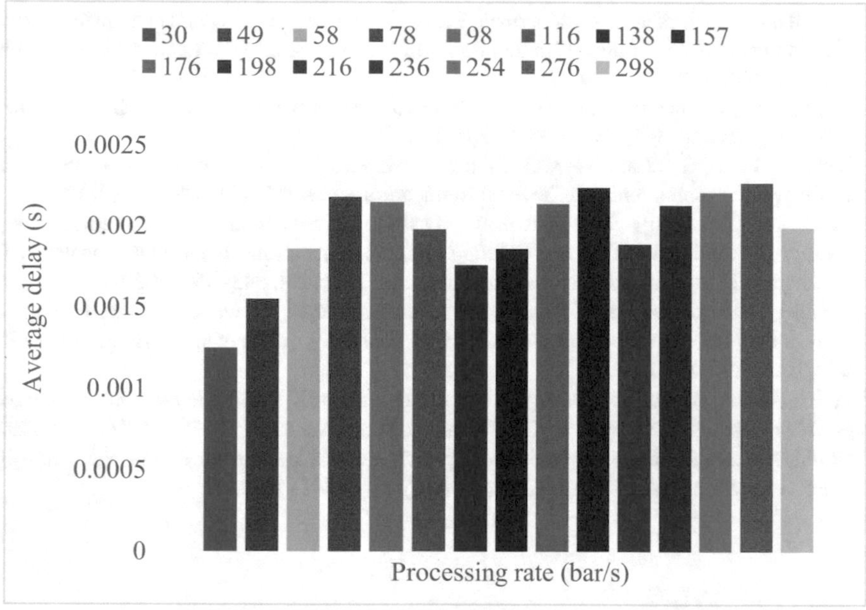

Fig. 2. The effect of changes in concurrency on the ability of messaging middleware servers to provide services

5 Conclusions

Based on Java3D technology, this paper designs and implements a three-dimensional model-based interior decoration system. Starting from the actual needs of users, the system adopts B/S mode, takes virtual reality technology as the core, and integrates functions such as two-dimensional graphics drawing and data maintenance, which is of great significance to the design of interior decoration. By adopting the system to design the interior, the user can greatly reduce the design time and cost, and can meet the actual needs of the occupants. So that the residents feel happy and improve the quality of life.

References

1. Martono, K.T., Eridani, D., Isabella, D., et al.: The design of recognition system of children basic activities based on virtual reality. IOP Conference Series Earth and Environmental Science **704**(1), 012019 (2021)
2. Baker, S., Waycott, J., Robertson, E., et al.: Evaluating the use of interactive virtual reality technology with older adults living in residential aged care. Information Processing & Management **57**(3), 102105.1–102105.13 (2020)
3. Mamaeva, E.A., Suvorova, T.N.: Foreign experience of implementation of 3d modeling and rapid prototyping for forming digital competencies. Informatics in School **1**(7), 18–20 (2020)
4. Goss, M., King, T., Lockstedt, J.: Rapid energy modeling. The Military Engineer **111**(720), 62–63 (2019)
5. Park, S., Zhang, S.: A pilot study of circulation layout based on perceived retail crowding. Journal of Retailing and Consumer Services **49**(JUL.), 305–315 (2019)

6. Haxhibeqiri, J., Karaagac, A., Moerman, I., et al.: Seamless roaming and guaranteed communication using a synchronized single-hop multi-gateway 802.15.4e TSCH network. Ad hoc networks **86**(APR.), 1–14 (2019)
7. Ak, L., Nair, T.: Data storage lock algorithm with cryptographic techniques. Int. J. Electrical and Computer Eng. **9**(5), 3843–3849 (2019)
8. Kim, D., Kime, S., et al.: Network slicing as enablers for 5G services: state of the art and challenges for mobile industry. Telecommunication Systems **71**(3), 517–527 (2019)
9. Muaz, M., Choudhury, S.K.: A realistic 3D finite element model for simulating multiple rotations of modified milling inserts using coupled temperature-displacement analysis. The International J. Advanced Manufacturing Technol. **107**(1–2), 343–354 (2020)
10. Rohmer, S., Gerdessen, J.C., Claassen, G.: Sustainable supply chain design in the food system with dietary considerations: a multi-objective analysis - ScienceDirect. Eur. J. Oper. Res. **273**(3), 1149–1164 (2019)
11. Pelham-Webb, B., Murphy, D., Apostolou, E.: Dynamic 3D chromatin reorganization during establishment and maintenance of pluripotency. Stem Cell Reports **15**(6), 1176–1195 (2020)
12. Mittal, A.K.: A stable time–space Jacobi pseudospectral method for two-dimensional sine-Gordon equation. J. Appl. Math. Comput. **63**(1–2), 239–264 (2020)

Development of Computer Vision Virtual Experiment Based on Virtual Simulation Technology

Xiang Zheng(✉), Min Xian, and Jian Yang

Robotics Engineering Laboratory for Sichuan Equipment Manufacturing Industry, Deyang 618000, Sichuan, China
scetczx@163.com

Abstract. With the rapid development of electronic technology in today's world, computer vision technology has also developed accordingly. In recent years, digital computer has developed rapidly and stably, and has become more and more widely and rapidly popular. This paper provides some help for the development of computer vision virtual by studying all aspects of virtual simulation technology. The main function of virtual network virtual simulation virtual technology will be to imitate and build another semi network real state system by using one of the network simulation virtual systems in the fully real virtual network state. The network virtual simulation technology will also be mainly applied to the virtual simulation and construction in the virtual network protocol system in the future. Firstly, this paper studies the significance and current situation of virtual simulation technology, and expounds the relevant theories of related technologies. Then, based on this theory, the framework of computer vision system is studied, and then the simulation is carried out. Finally, the test results show that the system module can realize the corresponding functions normally and meet the needs of users.

Keywords: Virtual Simulation Technology · Computer Vision · Virtual Reality · Virtual Experiment Development

1 Introduction

Virtual technology simulation is also called virtual reality, which is called virtual VR for short. It is another modern new computer technology, which belongs to a comprehensive application of more advanced new computer technology, which has been gradually developed and applied in China since the end and middle of last century. The whole set of virtual technology scheme takes the basic knowledge of three-dimensional digital computer technology as the main theoretical core, and is combined and applied to more than ten frontier disciplines including computer graphics, computer vision, multimedia technology, simulation control principle, cloud computing and artificial intelligence theory and technology, It has built such a realistic environment of highly real, virtual and realistic three-dimensional digital simulation, which provides an all-round simulation for each computer user's experience in hearing, visual perception experience,

J. H. Abawajy et al. (Eds.): ICATCI 2022, LNDECT 170, pp. 733–740, 2023.
https://doi.org/10.1007/978-3-031-29097-8_87

three-dimensional space tactile experience and so on, Make game users feel as if they are really in a real three-dimensional game environment. Virtual real-time simulation application multimedia technology application refers to the current information technology in modern China, after the comprehensive application research of multimedia technology of traditional computer network multimedia transmission system and multimedia application platform, In several major fields that have become the whole modern information education and information technology field in China, another important application technology is the most promising and widely used technology development, Its requirement is to realize virtual real-time simulation or simulation of the whole three-dimensional real-world scene by using multimedia computer platform technology, So as to finally achieve a rapid and targeted solution to the functional needs of situational reproduction of various knowledge and people's practical application skill requirements for interactive knowledge reproduction of various natural environments in various new media platforms for education, learning and application at the same time, It is another new way with relatively new connotation to provide new ideas for the development of education and the research of auxiliary means of information technology assisted teaching in the future information age. After the sustainable and stable development of our research time in recent ten years, the research on the scientific theory, development ideas and theoretical research methods of the technological frontier in the fields of virtual environment physical modeling and simulation technology application has been and gradually developed, and a number of very rich, practical and effective academic innovation achievements have been obtained [1, 2].

Computer virtual simulation software is not only the main application research and experimental tool of one kind of computer, but also another kind of computer simulation and simulation. Its most basic and main performance feature is to be able to realize the simulation experiment based on the system mathematical model designed by the virtual simulation software system, A series of analog computer simulation system circuits are formed by connecting this series of virtual computer arithmetic unit system models and various analog passive device systems. Then users can directly use the simulated series of simulation circuits to analyze these analog computer system models for some experimental research or research. Through the introduction of mathematical methods to carry out random simulation, although it can improve the work efficiency to a certain extent, the simulation effect has not reached a great breakthrough. At present, in order to solve the difficult problem of establishing simulation model for nonlinear structural system, many scholars in domestic basic research centers have analyzed and studied a large number of input model and experimental output model data of simulation experiment, and established a black box model, which can realize the modeling of complex system when the internal laws of some systems are unknown [3, 4].

This paper studies the simulation of computer visualization based on virtual simulation technology. By constructing a virtual simulation experiment platform, the platform has the characteristics of high efficiency, high precision and low cost [5, 6]. The main work of this paper is to study the virtual simulation technology. Based on the research results of the above technology, the computer vision virtual experiment development based on virtual simulation is developed. Although the combination of virtual simulation technology and sensitivity reliability analysis brings convenience to engineering practice, there are still some difficult problems in this paper due to the limitation of time, energy and personal knowledge.

2 Discussion on the Development of Computer Vision Virtual Experiment Based on Virtual Simulation Technology

2.1 Image Classification

In the design of complex computer vision content analysis and many other graphic analysis application task systems, the purpose of the image classification system of image information content is to make people classify gradually according to the specific graphic content feature information displayed in various images, and reasonably classify all kinds of picture content features, In order to achieve the final substitution, it realizes the classification of complex image information or the analysis of graphic judgment in human vision. Given such a picture, the first question you want to answer for the classification features in the image is likely to be what kind of image content should be described in the two pictures, aircraft or cows? Image classification can be applied to the technical problems of image classification. The first core concept may be the features of image classification. Generally speaking, the automatic classification features of image information can be directly regarded as using an image computer to automatically classify and extract various features from the pixel information database in a given image object, Then extract these image feature data to realize the decision-making process of selecting or matching the object information in line with its specific image category [7, 8]. One of the most important factors that determine the relationship between image feature data classification and data accuracy is two basic factors, that is, how to accurately select image feature types for image feature data classification and how to correctly and reasonably select the extracted image feature methods, One or two of these specific relationship problems can be comprehensively considered or solved by strengthening our own classification learning and training of image features and knowledge comprehensive analysis training or learning training of image data set analysis. Process mode and implementation method of unsupervised image learning unsupervised image learning, as the name suggests, unsupervised image learning process is often that only one image feature extractor can learn and explore some image features in the image by itself, Some supervised image learning process models often guide the image learning process by only giving a mapping relationship between the image features obtained from a learning and training set and the image categories in the learning image set. Generally speaking, some supervised learning and training processes tend to learn more and faster than the training process, and the judgment on the final result is often relatively more accurate [9, 10].

2.2 Target Tracking Algorithm

Automatic target position tracking system has always been said to be another important technical core problem in the current field of computer vision. At present, it is also one of the most important and widely used new technologies in the scope of application in the whole computer field. Another literal meaning of target location, tracking and recognition technology refers to the position detection, recognition, processing and analysis of any still video object or a still image object through algorithm, Finally, the determined object moving in the video image screen of each next frame can obtain the real position information of the found target position. Various types of complex target video tracking algorithm systems can be roughly divided into target video tracking detection and recognition processing and target video tracking detection and recognition. The other one is represented by the extracted image of optical flow method, which is the method of target extraction before and after. Generally, the basic principle of the extracted image is to extract the target feature image through the target position information of the corresponding position in the initial target image slice of the target, Then, using the continuity principle of the target motion process respectively - to determine that the final occurrence time and position of the target information in the initial target image film of a certain frame in the case of the same target should be at least the image near the target that appears in the initial target position of the previous frame of the target - finally, the selected target In the target image slice feature of a frame under the, the initial image position near the target and the target image slice feature corresponding to the image near the target are extracted with the target image feature containing the same meaning information. By analyzing and comparing the moving environment characteristics of the tracked object image film itself and all other relevant characteristics related to tracking the tracked object target environment, it is ensured to correctly determine and calculate the relative accurate position of the tracked object in the new moving environment. However, the use of this research method itself will also have some biggest disadvantages that can not be ignored, that is, when it is found that there are obvious obstacles or covered by buildings in the surrounding environment When the direction of motion change is obviously blurred or the speed of motion change direction after tracking the target object may be faster than before, and so on, the accuracy of the algorithm itself will be greatly affected at the same time. Therefore, an accurate climate algorithm that enables all target areas to be tracked and analyzed in real time quickly, stably and accurately under the above complex and adverse weather conditions has always been the goal of all climate researchers [11, 12].

2.3 Local Matching Algorithm

The local matching algorithm focuses on the methods of matching cost calculation and matching cost collection. The rules of parallax calculation are relatively simple. In the range of maximum parallax, the point with the lowest matching cost is found as the matching point, so as to calculate the parallax. Due to the fast matching speed of local matching algorithm, real-time stereo vision systems mostly use local matching algorithm. However, the local matching algorithm can not guarantee the uniqueness of matching, which can be eliminated by post-processing. The gradient method is to compare the gray gradient values around two points to determine the matching points. Here, it is assumed that the gradient of a point in the image does not change with time. The horizontal transformation of a point from one image to another is as follows:

$$(\nabla_x E)l + E_i = 0 \tag{1}$$

where $\nabla_x E$ the gradient in the horizontal direction, E_i is time factor, l is the conversion between two images.

Since the calculation of gradient transformation direction is more accurate, a constraint condition needs to be added to the calculation: if the parallax in a small range is smooth, the parallax at this point can be solved by the following equation:

$$1 = \left(A^T A\right)^{-1} A^T \beta \tag{2}$$

A^T is gradient matrix, β is time factor matrix, gradient algorithm can achieve half pixel accuracy.

2.4 Module Design

The visual virtual simulation experiment platform can simulate the real scene, and then build a three-dimensional model, so that people can more intuitively understand the surrounding environment and dynamics. The eight functional modules of the virtual simulation experiment platform constitute the platform. The specific modules and the hierarchical relationship between each functional module are shown in Fig. 1.

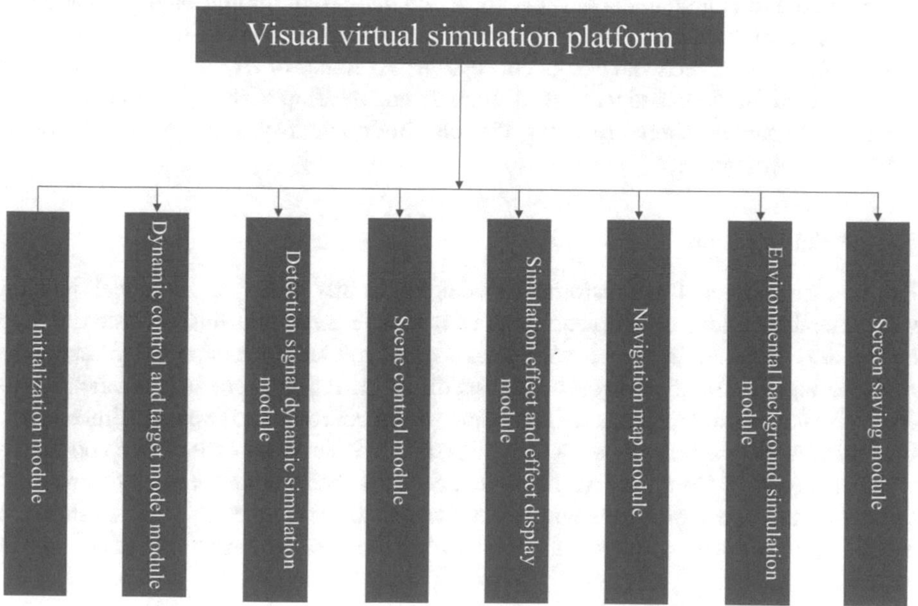

Fig. 1. Module design of visual virtual simulation experiment platform

Through the separate analysis and processing of each module, gradually realize the corresponding functions of each module, and then integrate all modules together to debug the functions. Through experimental simulation, it can be concluded that the corresponding functions can be realized.

3 Experiments

3.1 Common Tools of Virtual Simulation Technology

According to the current research status, there are various development tools for virtual reality software system development. They will have their own unique advantages in one aspect. They can all be used to realize virtual reality works with certain technical requirements, and there are quite a lot of such software. Therefore, We need to choose a suitable software development tool for development, which is also a key place. In this chapter, I will briefly and concisely describe in detail the most popular tool for embedded system design, development and application based on virtual reality technology in our country under the current technical situation, and describe the use of Virtools in detail. Because most of the script editing interfaces used in its scripting are flow chart script editing interfaces, users can generally only select the interactive module (BB) that calls almost all the behaviors they think they need to call in order according to the user's needs, and then drag the mouse to it for various functions and combinations of scripts, It can automatically complete the automatic writing of a script program based on Virtools. It is precisely because Virtools' script system is completely formed by a series of BBS through multiple combinations, even if these BBS can not only be used repeatedly, but also be combined with each other again, resulting in more combinations, In this way, it can be used to complete the design conception and programming implementation of various more complex interaction and functional systems. Even developers without any programming foundation and design conception experience of any system can skillfully use these combinations and use BB to simulate and develop various interactive effects in real-world games, Therefore, using Virtools development system can greatly shorten the development cycle.

3.2 Model Creation

The first is to import the transformed model, but at this time, the 3D model drawing will generally be lost, so we need to convert it separately, and finally paste the map on the 3D model again, which completes the 3D model conversion. At present, the software modeling methods in the market are different, and there are different methods to determine the coordinate system. Three dimensional coordinates have three dimensions. Generally, when establishing a coordinate system, the establishment of two coordinate systems is roughly the same. At this time, only the other one needs to be determined. In order not to cause problems with the relevant setting parameters in the software, it is generally necessary to import the model after the transformation of the coordinate system.

4 Discussion

The background difference method is used to detect the target, and the centroid position of each experimental target is recorded. Before the first landing, 8–10 groups of data can be photographed and 10 groups of data can be processed each time. The following data Table 1 can be obtained through the above steps.

Table 1. 3D coordinates of space points

Times	X	Y	Z
One	446.123	857.209	982.567
Two	441.236	816.036	965.611
Three	428.761	689.123	815.508
Four	423.856	633.031	753.733
Fire	419.783	578.463	693.456
Six	417.352	523.996	621.125
Seven	408.135	411.125	475.466
Eight	397.364	301.922	317.132
Nine	379.531	192.632	143.043
Ten	380.912	125.689	101.153

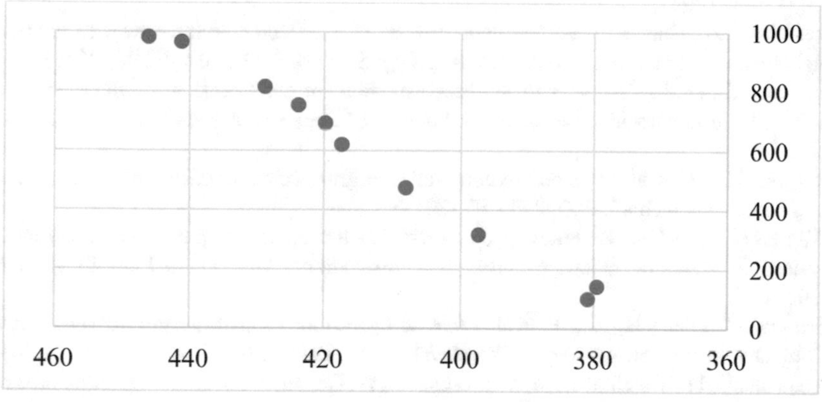

Fig. 2. Coordinates based on raw data

According to the simulation of the original data in the table, the picture shown in Fig. 2 can be obtained. It can be seen that the motion trajectory of the object is probably a parabolic trajectory. Because the object moves in three-dimensional space, it is necessary to fit the x-z and x-yz planes, and the curve fitting is carried out according to the least square method. The connection of the fitted data is a smooth curve, which eliminates the error, indicating that the experiment is relatively successful.

5 Conclusions

No matter what kind of development is inseparable from science and technology, it needs the support of science and technology. Virtual simulation technology has good controllability and repeatability, and has the advantages of safety, economy, energy saving, pollution reduction and less affected by the external environment. With the continuous development of virtual simulation technology, its application fields will be more in-depth and extensive. Through a series of experiments and simulations, this paper can basically achieve the purpose, but it needs to be further improved. In the follow-up work, we need to continue to improve the corresponding data.

Acknowledgements. Supported by Sichuan Science and Technology Program, NO: 22ZDYF2965.

References

1. Hashmi, M.F., Ashish, B., Keskar, A.G., et al.: FashionFit: analysis of mapping 3D pose and neural body fit for custom virtual try-on. IEEE Access **8**(99), 91603–91615 (2020)
2. Doukakis, E., Debattista, K., Harvey, C., et al.: Audiovisual resource allocation for bimodal virtual environments. Computer Graphics Forum **37**(3) (2017)
3. Edsall, D.L., Conrad, K.A.: Virtual team member perspectives on personal development: a sequential explanatory study. New Horizons in Adult Education and Human Resource Dev. **33**(3), 3–27 (2021)
4. Ioannidou, A., Chatzilari, E., Nikolopoulos, S., et al.: Deep learning advances in computer vision with 3D data: a survey. ACM Computing Surveys **50**(2), 20.1–20.38 (2017)
5. Sabanci, K., et al.: Computer vision-based method for classification of wheat grains using artificial neural network. Journal of the Science of Food and Agriculture **97**(8), 2588–2593 (2017)
6. Rathore, M.I.: Computer vision syndrome-an emerging occupational hazard. Research Journal of Science and Technol. **9**(2), 293–297 (2017)
7. Wäldchen, J., Mäder, P.: Plant species identification using computer vision techniques: a systematic literature review. Archives of Computational Methods in Eng. **25**(2), 507–543 (2017)
8. Ertugrul, E., Zhang, H., Zhu, F., et al.: Embedding 3D models in offline physical environments. Computer Animation and Virtual Worlds **31**(4–5), e1959 (2020)
9. Abu Alhaija, H., Mustikovela, S.K., Mescheder, L., Geiger, A., Rother, C.: Augmented reality meets computer vision: efficient data generation for urban driving scenes. Int. J. Comput. Vision **126**(9), 961–972 (2018)
10. Garcia, C.G., Meana-Llorian, D., G-Bustelo, B., et al.: Midgar: Detection of people through computer vision in the internet of things scenarios to improve the security in smart cities, smart towns, and smart homes. Future Generation Computer Systems **76**(nov.), 301–313 (2017)
11. Hamledari, H., Mccabe, B., Davari, S.: Automated computer vision-based detection of components of under-construction indoor partitions. Automation in Construction **74**(feb.), 78–94 (2017)
12. Garcia-Camprubi, M., Bengoechea-Cuadrado, C., Izquierdo, S.: Virtual Sensor Development for Continuous Microfluidic Processes. IEEE Transactions on Industrial Informatics **16**(12), 7774–7781 (2020)

Design and Application of Automatic Evaluation System for Oral English Based on Neural Network

Jing Wang(✉)

Zhengzhou University of Science and Technology, Zhengzhou 450064, China
1296599200@qq.com

Abstract. In order to realize automatic oral English evaluation system based on neural network, help students to evaluate oral English and improve their oral English level, this paper will carry out relevant research. Study firstly introduces common classification of neural network, and analyses the characteristics of all kinds of neural network, then in combination with characteristics of all kinds of neural network, the design of automatic measurement system based on oral English need to selection of neural network, and then based on the neural network has carried on the system design, finally introduced the system application. Through the research, the design goal of automatic test system for oral English is accomplished, and the correct application method can effectively improve students' oral English level.

Keywords: Neural network · Oral English · Oral Evaluation system

1 Introduction

Oral English is an important indicator to measure students' English level, so the development of the students should pay attention to their spoken English, but my spoken English is different from knowledge level development, the students need to continue training, then find out the problem, according to the result of each training so bad accent constantly, this is spoken English peculiar way of development. And during the process of continuous oral training a lot of students will meet problem, namely the students results objectively the existing problems for each of the training, but students themselves unable to identify these problems, mistakenly believe that their own spoken language is no problem, the myth will restrict students' spoken English, students in China is also the main reason for the problems generally exist in oral English. In the face of this kind of phenomenon, although teachers and other education workers have the ability to provide help to the student, but because can't long time with my students, so the actual is very limited, can provide help when people want to be able to have a tool moment for students to provide help for the oral English training, modern related fields will this tool called oral English automatic measurement system, the system is essentially a voice detection system, the internal voice voice print data containing the corresponding standards,

J. H. Abawajy et al. (Eds.): ICATCI 2022, LNDECT 170, pp. 741–750, 2023.
https://doi.org/10.1007/978-3-031-29097-8_88

students through the external input voice equipment, system will be according to the principle of voiceprint matching to determine whether a student input voice voice print with the standard voice voice print enough match, if the match value is lower than the standard value, means that the students oral English pronunciation problems, in addition, students will point out the mismatched pronunciation to help them understand their own problems and then correct them. In theory the system do in training students oral English can give students a lot of help, but how to implement the system or a problem to be solved, one of the most critical problem is how to make the system to identify students enter voice voice print, and matching analysis of the current can be the best way to deal with this problem is neural networks, research is needed to look at this.

2 Common Types and Characteristics of Neural Networks

2.1 Neural Network Concept

Neural network is an intelligent model, which can endow machine system with intelligent logic. The neural network itself is a model created by imitating the neural structure of human brain and the neural conduction process. That is, there are a large number of neural nodes in human beings, and there are direct and indirect connections between each neural node. In this way, a network neural structure is formed. While people are thinking, some neural nodes will start to interact with other neural nodes, forming a neural conduction process. This reticular neural structure and neural conduction process are picked out to form a variety of neural network models. The more common ones are feedforward neural network and convolution neural network. Among them, feedforward neural network is the most commonly used neural network model, which can deal with most problems. The latter is generally applied to some special problems, while the former is generally selected in oral evaluation.

Regardless of the specific type of neural network, neural network is composed of three basic elements, namely input neuron, output neuron and neural link. Input neuron refers to the neuron that specially accepts external input information. It is generally represented by blank value in the neural network model, representing that it can receive all manually input information in digital format. After input, the neuron node will become an input value. The output neuron is mainly responsible for outputting the final result. If it is interpreted by the human brain neuron node, it is equivalent to the neuron node in the human brain that controls speech and behavior. The neural link is special. It is not only a link connecting input neurons and output neurons, but also a link connecting different input neuron nodes, but not a link connecting output neurons. If you need to connect the output neuron links together, you need to convert the information of all output neurons into input information, become an input neuron node, and then re-establish the connection through the neuron link.

2.2 Feedforward Neural Network

Feedforward neural network is composed of input layer, hidden layer and output layer. The number of input layer itself is 1, and the internal layer is composed of basic data nodes [1, 2]. The number of basic data nodes is unlimited, while the number of hidden layer itself is unlimited, and the internal layer is composed of hidden data nodes. Implied data node is derived from the basic data input layer after interaction, namely the basic data of nodes in hidden layer respectively from other basic data nodes interact, forming the relationship between the various nodes, thus every implicit data node contains at least two basic data, according to the result can know the relationship between the data and the correlation, the last output layer is to output the results of data interaction in the hidden layer to achieve the purpose of result visualization [3–5]. Figure 1 shows the model of feedforward neural network.

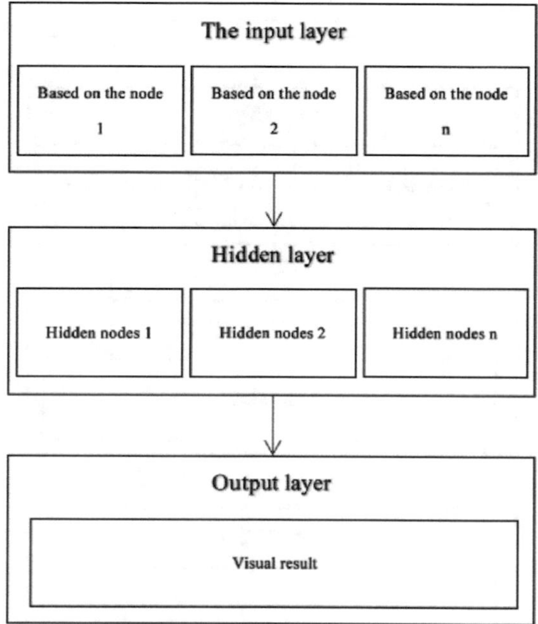

Fig. 1. Model of feedforward neural network

2.3 Feedback Neural Network

Feedback neural network is also known as regression neural network, which is mainly composed of four parts, namely, input layer, hidden layer, receiving layer and output layer. The characteristics of input layer, hidden layer and output layer are the same as those of feedforward neural network, but there is bidirectional relationship between hidden layer and input layer. That is, in feedforward neural network, the data of input layer only enter the hidden layer and then directly obtain the results, while in feedback neural network, after the data of input layer enter the hidden layer, it will continue to return to the input layer for further analysis, and then obtain the visual results through the output layer [6, 7]. At the same time, the output layer of feedback neural network is not only used for output of results, but also plays a linear weighting role, which can make the results clearer. In addition, the continuity layer is used to remember the previous output value of the node of the hidden layer and return it to the input. Generally, it is mainly composed of a one-step delay operator, which is the support of the feedback mechanism of this neural network [8, 9]. Figure 2 shows the model of feedback neural network.

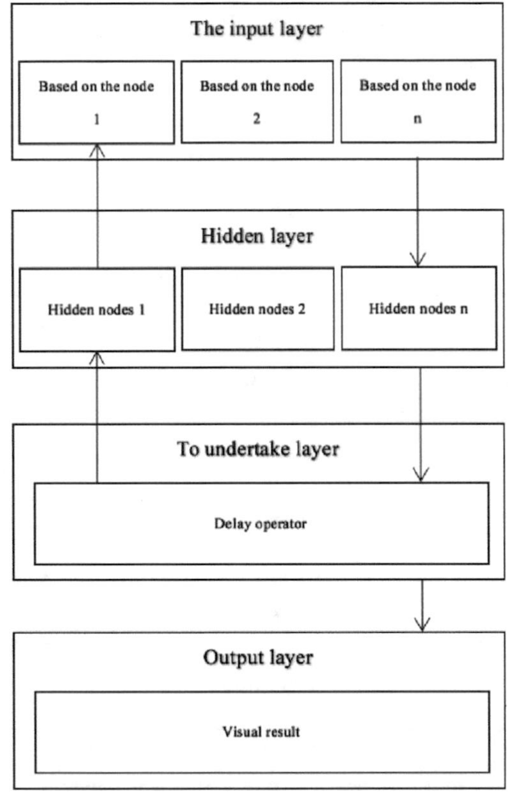

Fig. 2. Model of feedback neural network

3 Neural Network Selection and Oral English Automatic Evaluation System Design

3.1 Selection of Neural Network

Different neural networks overlap in function to a certain extent, but there are still large differences on the whole, so in general, the selection of neural networks should be closely related to the actual problems, and the correct selection can ensure accurate results. Focus on this requirement, the spoken English test is a typical feedforward model problem, so you can choose type feedforward neural networks, the evaluation of students just need to input the voice as the foundation data nodes, through the neural network's input layer to hidden layer, and then let the basic data in the hidden layer nodes and standard voice voice print data matching can get the result, this is in line with the evaluation logic, indicating that the feedforward neural network is the most used in the system. It is worth noting that the applicability of feedforward neural network in the automatic evaluation system of Spoken English is only reflected in the vast majority of cases, some special cases may still need to use feedforward neural network, or other types of neural network, so although the system design chooses feedforward neural network, however, this does not mean that other neural networks have no application value in system design, and the actual design should be selected according to the actual needs, and the possibility of constructing multiple neural network models at the same time cannot be ruled out [10].

3.2 System Design

Figure 3 is the general framework of the automatic test system for Oral English.

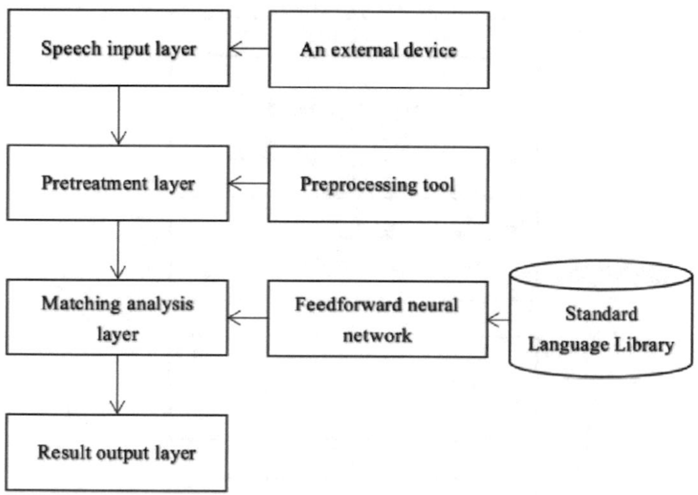

Fig. 3. Automatic Test system for Oral English

3.3 System Implementation Method

Combined with Fig. 3, the design method of automatic oral English evaluation system is as follows.

Voice Input Layer

The voice input layer is supported by external devices. Basically, any device that can accept human voice information can be applied here, such as external microphone, voice input interface of handheld terminal, etc. By using these devices, students can input voice into the system. But it is important to note that the students enter the voice is not in essence a kind of information, but an acoustic energy, the system does not have the function that directly read acoustic energy, therefore, to achieve voice input to the acoustic energy is converted to digital information, look at this point you can refer to the basic principle of transducer, using programming language can create an electron in the program, the basic principle of the program is to judge the audio changes of an input voice with the help of the volume sensing element, that is, there is a fixed relationship between the energy of the input voice sound wave and the level of sound audio, the greater the energy of the sound wave represents the higher the audio, and the audio can be induced by the element, after induction, the audio spectrum is obtained. At this point, recording information in the audio spectrum can complete speech input. In addition, the information within the audio spectrum is the voiceprint data of the input speech. Fig. 4 shows the working principle of the voice input layer.

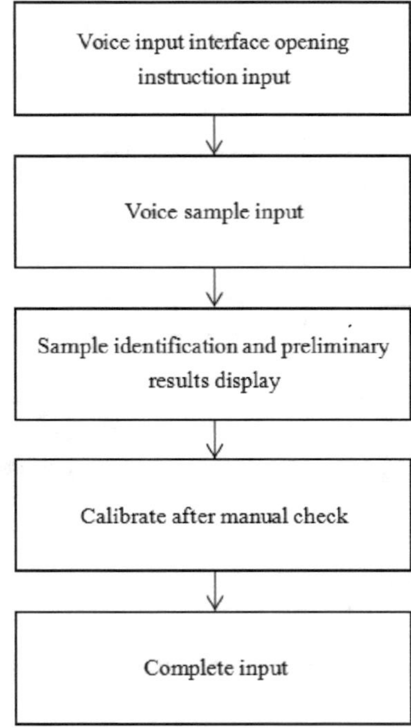

Fig. 4. Working principle of the voice input layer

Pretreatment Layer

Because students is through an external device to input voice, so in the input process is likely to be influenced by external factors lead to voice voice print data quality problems, such as students in the process of the input speech, equipment has been interference of external factors, lead to distortion of audio, so it is concluded that the voice print data would be a phrase, if such voice print data is input into the subsequent matching analysis layer, it will inevitably lead to inaccurate results, which cannot help students to train their oral English. From this point of view, in order to avoid the occurrence of related problems, it is necessary to design a pretreatment layer, which is mainly composed of several pretreatment tools. There are many kinds of such tools and specific selection should be based on the actual situation. Table 1 shows the main pretreatment tools and functions in the system. In addition, for the sake of system availability, the pretreatment of the pretreatment layer mode is specifically designed, or layer normally pretreatment way of pretreatment is to remove all do not conform to the requirements of data, but this may lead to students input speech repeatedly, cause bigger trouble, will give them this system's handling of the pretreatment layer is a priority to repair the input speech, ensure quality of voice, if the input voice quality is still not up to standard after repair, it will be deleted and students will be told to re-input.

Table 1. Main preprocessing tools and functions

Tool	Function
Voice noise reduction	Reduce noise in voice messages
Data to heavy	Remove duplicate audio data

Matching Analysis Layer

Matching analysis layer is where type feedforward neural network, the neural network according to the preset good matching rules matching analysis, rule is: the voice information after pretreatment as the base node, and then input data training within the hidden layer after training is basic nodes and hidden nodes, the gap between the gap size is to match the size. At the same time, the hidden nodes in the feedforward neural network are provided by the standard speech sound database, which contains a large number of standard voice pattern data, which are collected by real teachers, so that matching and comparison can be made. In addition, because everyone there exist certain differences between oral English pronunciation, and usually no mostly difference as long as can be seen as standard, so the match analysis cannot be carried out in accordance with the complete matching rules, on the other hand to set up a compatibility standards, if the analysis results show that the input speech and standard speech voice print data matching degree is more than the standard, it showed that the input voice accurate, less than the standard is not accurate, so the results will be more reasonable. Figure 5 shows how the matching analysis layer works.

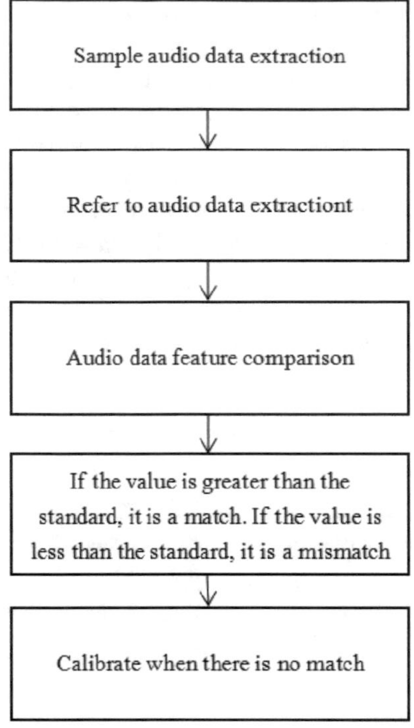

Fig. 5. Matching analysis layer works

4 Application of Automatic Oral English Testing System

As for the specific application of the automatic test system of Spoken English, it can be divided into three steps: First, the teacher can advance in the standard library input standard pronunciation (in the case of oral English teacher itself accurately can let teachers to complete the work, but if the teachers oral English pronunciation is not accurate, should let the professionals complete), and then give each standard voice on number, serial number has no fixed form requirements, all by teacher preferences; Second, after inputting the relevant sounds, the teacher checks the sounds that need to be trained by students according to their numbers, lists them and sends them to students to complete the assignment of oral training; Third, after receiving the speech training homework, students can first play the standard pronunciation and follow the training. After repeated repetitions, if they think they have mastered the relevant spoken pronunciation, they can stop the playing, and then input the voice through the external device. After the input, they wait for the matching analysis results. Fourth, matching the analysis result will first tell the students of the input speech sound and standard phonetic sound lines between matches, the assumption of matching analysis standard is 80%, and the analysis results show that the input speech voice print with standard speech voice print matching degree is 79%, means that students oral English pronunciation is not standard, still need to continue training. Second in addition to students input speech and standard speech voice print exactly match the situation, no matter students enter how much of the matching degree between voice and standard pronunciation and the results will tell the students don't match, play does not match the voice clips at the same time, this way, students can know their own problems, targeted training.

At the same time, if students carry out oral training activities independently, they can also directly check the pronunciation they want to train, and then train according to the following steps. In addition, consider some of the students in the oral training is likely to know their own problems, but can't rely on his master correct English pronunciation points, system has been set up in the "training session" button, the students in such a situation after clicking this button, the whole process of training will be automatically sent to the teachers, so that teachers with specific teaching.

5 Conclusion

In conclusion, the spoken English has been the difficulties of the students' English level in our country, many teachers in the teaching work and therefore, the oral English automatic measurement system can solve this problem, it is a kind of very good spoken English training tool, with the help of the tool teachers oral training teaching work pressure reducing, students can also smooth the oral English training, This is more conducive to the improvement of students' oral English level.

References

1. Sztahó, D., Gábor, K., Gábriel, T.: Deep learning solution for pathological voice detection using LSTM-based Autoencoder hybrid with multi-task learning. In:14th International Conference on Bio-inspired Systems and Signal Processing (2021)
2. Meghraoui, D., Boudraa, B., Gomez, P.: A novel pre-processing technique in pathologic voice detection: application to Parkinson's disease phonation. Biomed. Signal Process. Control **68**, 102604 (2021)
3. Georgopoulos, V.C.: Advanced time-frequency analysis and machine learning for pathological voice detection. In: 2020 12th International Symposium on Communication Systems, Networks and Digital Signal Processing (CSNDSP) (2020)
4. Raj, J.R., Jabez, J., Srinivasulu, S.S., Gowri, S., Vimali, J.S.: Voice pathology detection based on deep neural network approach. IOP Conf. Ser. Mate. Sci. Eng. **1020**(1), 012001 (2021)
5. Mittal, V., Sharma, R.K.: Vocal folds analysis for detection and classification of voice disorder: detection and classification of vocal fold polyps. Int. J. E-Health Med. Commun. (IJEHMC) **12**(4), 23 (2021)
6. Saleem, S., Dilawari, A., Khan, U.G.: Spoofed voice detection using dense features of STFT and MDCT spectrograms. In: 2021 International Conference on Artificial Intelligence (ICAI) (2021)
7. Neelima, M., Santiprabha, I.: Mimicry voice detection using convolutional neural networks. In: 2020 International Conference on Smart Electronics and Communication (ICOSEC) (2020)
8. Narendra, N.P., Alku, P.: Glottal source information for pathological voice detection. IEEE Access **8**, 67745–67755 (2020)
9. Islam, R., Tarique, M., Abdel-Raheem, E.: A survey on signal processing based pathological voice detection techniques. IEEE Access **8**, 66749–66776 (2020)

10. Saleem, S., Dilawari, A., Ghani Khan, M.U., Husnain, M.: Voice conversion and spoofed voice detection from parallel English and Urdu corpus using cyclic GANs. In: 2019 International Conference on Robotics and Automation in Industry (ICRAI), pp. 1–6. Rawalpindi, Pakistan (2019)

Volleyball Training Based on Virtual Reality Technology

Yinfen Song(✉)

Henan Vocational College of Industry and Information Technology, Zhengzhou 454000, China
zyw52718@sina.com

Abstract. In recent years, under the influence of the spirit of the women's volleyball team and the upsurge of winning the women's volleyball team, China has once again set off a wave of promoting and learning the spirit of women's volleyball. At the same time, it has also set off a wave of learning volleyball in the society. However, compared with football, basketball and other ball sports, volleyball focuses more on action skills, especially for beginners. Due to the special sports skills of volleyball, many It is difficult for beginners to get started, and it can even be frustrating. The advancement and development of technology has made us see a lot of impossible possibilities, especially the emergence of virtual reality technology and its wide application in many fields, especially the technical advantages in sports training provide a powerful technology for volleyball training support. The research theme of this paper is to start with virtual reality technology, combine the main characteristics of VR technology, and deeply integrate the technical and tactical movements and sports skills of volleyball to seek an effective integration of the two. Volleyball learning provides more help.

Keywords: VR · Volleyball training · The simulation teaching

With the in-depth application of the industry, virtual reality technology has gradually become synonymous with high-end technology of high-level human-machine interface in the computer field. Virtual reality technology uses the characteristics of network interactivity, conception and immersion, and through the integration of technology Help users achieve a unique multi-scene immersive experience and create a more realistic environmental experience for people. On the other hand, it is well known that most sports have certain risks, such as football, basketball, volleyball, rugby, etc. In the process of daily training, unexpected situations are prone to occur, coupled with the unpredictability caused by some difficult movements risk, which brings certain obstacles to daily training. At the same time, judging from the current situation of volleyball teaching in colleges and universities in my country, the backward teaching mode, unreasonable structure of volleyball teaching teachers, and imperfect infrastructure have all affected the progress of volleyball training, and the combination of virtual reality technology and sports training can be better. Make up for the limiting factors of insufficient material conditions, reduce the risk of difficult movements through the use of network technology, and provide more help for sports training [1, 2].

J. H. Abawajy et al. (Eds.): ICATCI 2022, LNDECT 170, pp. 751–759, 2023.
https://doi.org/10.1007/978-3-031-29097-8_89

1 Research Background

Virtual reality technology is a new technology rising in recent years. The main role of this technology is simulation modeling and simulation body feeling. In terms of simulation modeling, it is mainly to use 3D technology to establish a link that is highly similar to the real environment and has similar texture and physical characteristics. Users can enter the environment visually by wearing devices. Because of the high simulation of the environment, users will get an extremely real visual experience. In terms of simulation somatosensory, it is also the use of specific devices to connect with people's various senses (except for smell), which can provide users with different somatosensory feedback from different senses, and the feedback is also very real. There is a primary and secondary relationship between the two functions of virtual reality technology, in which simulation modeling is the primary and simulation somatosensory is the secondary. The reason is that simulation somatosensory is to calculate the correct somatosensory type and degree according to the user's behavior in the simulation modeling environment, and then give feedback, which means that if the user does not make any action in the simulation modeling environment, the simulation somatosensory function will not operate, For example, if a person strikes a volleyball in the simulation modeling environment, the body feeling feedback of the volleyball on the skin will be simulated through the simulation

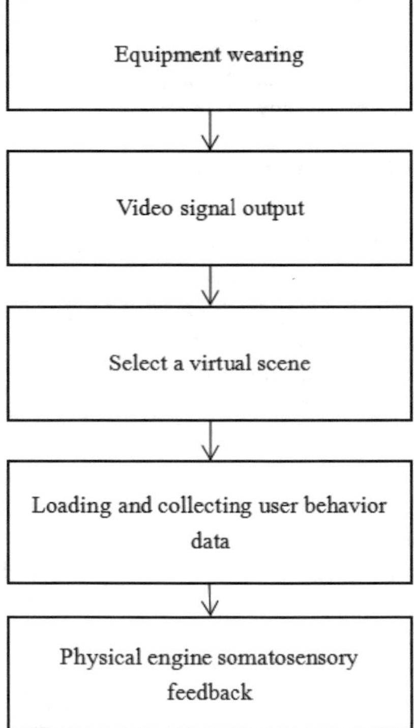

Fig. 1. Operation process of virtual reality technology

body feeling. On the contrary, if a person does not strike a volleyball, he will not get any feedback from the virtual reality technology.

The high simulation characteristics of virtual reality technology have made it widely used in a very short time, that is, the technology did not attract people's attention at the early stage of its introduction, so the technology is generally only used in the entertainment field. However, as time goes by, people find that the technology has a strong universality. With its extremely real simulation modeling environment and simulation somatosensory feedback, people can get a real experience, This kind of experience can make people have a deep feeling about things, so this technology is rapidly popularized and now involves the sports field. The advantage of the application of virtual reality technology in other fields is that it can let people get experience without touching the actual things, so that people can avoid the risks they encounter when contacting the actual things to a certain extent, and also let people better explore. Figure 1 is the operation process of virtual reality technology.

2 Main Features of VR Technology

In fact, with the continuous development of VR technology, this technology has been widely used in various fields of sports training. At present, VR technology is involved in volleyball teaching, aerobics teaching, basketball teaching and other sports training. Therefore, it can be widely used in various fields of sports training, mainly based on its characteristics of Multi-Sensory, Immersion, Interactivity, and Imagination.

(1) Multi-Sensory. The multisensory of VR technology refers to the fact that users can acquire auditory, tactile, force and smell senses in addition to the visual perception provided by general computer technology. The multisensory reflected by the most ideal VR technology includes all the normal perceptual abilities of human beings. Because of this characteristic, Let us experience VR technology when there is a sense of immersive reality [3].

(2) Immersion. The so-called immersion also refers to the sense of presence. VR technology can help users create a relatively real simulation environment, and immersion refers to the real degree that users can feel their presence in the simulation environment. If it is difficult for users to distinguish the true and false simulated environment, this environment is the most ideal state, so that users can be fully invested in the THREE-DIMENSIONAL environment created by VR technology, just as in the real scene [4].

(3) Interactivity. Interactivity is also one of the most prominent features of VR technology. The so-called interactivity means that users can operate and control objects simulated by computers and objects in the environment according to their needs. For example, with the help of VR technology, we can directly touch objects in the virtual environment with our hands, and we can feel the feeling of touch, and even feel the weight of the object. At the same time, the virtual object will move with our touch.

(4) Imagination. So-called idea focuses on VR technology in May be imaginary space, with the aid of VR technology to further expand the cognition of the human, people

have experienced in VR technology under scenario is not only a real environment, and even can according to your own idea, experience and objective scene does not exist or is unlikely to happen in reality [5].

3 The Necessity of VR Technology in Volleyball Training Application

3.1 Better Avoid the Occurrence of Accidents in Training

Volleyball, as a special contact sports, is prone to accidents in the training process, such as general joint injury, knee joint overload, ankle injury, etc., if the training is too heavy, it is easy to lead to injuries. It is because of these factors that many colleges and universities even give up some difficult volleyball training. VR technology can help students or athletes train freely in an unreal simulated environment, and better avoid accidents and other accidents in training through the simulated environment. In the process of training, VR technology can also capture the user's actions, evaluate the training actions, and point out the deficiencies in training in time.

3.2 Insufficient Material Conditions Such as Improving Infrastructure

Volleyball training for infrastructure and other material requirements are very high, but in the reality of volleyball training, especially the volleyball training in colleges and universities often due to the defects of material conditions affect the training effect, and even some training projects due to funds, venues and other material conditions have to cancel. The application of VR technology allows users to experience more comprehensive training conditions and get richer sensory experience. Even the lack of objective material conditions can be remedied by VR technology, and a whole set of training can be completed without a huge training site.

3.3 Avoid Sports Injuries Caused by Difficult Movements

Volleyball players need super explosive power, fast movement ability and super body control, which all require the accumulation of daily training, but some difficult movements can easily cause unpredictable injuries to players. And as a confrontational sport, even in the normal training process, it can be injured due to difficult movements. In the process of sports training, VR technology can combine training requirements to build training scenes that are difficult to achieve in reality, and then help athletes perform virtual action experiments and training according to training requirements. Unnecessary injuries caused by movements or technical tactics, so that athletes have no worries when performing difficult movements.

4 Application Practice of VR Technology in Volleyball Training and Teaching

4.1 Build Virtual Training Venues and Roles

Virtual Training Venue Modeling

First of all, virtual training venue modeling should be carried out with the help of 3D software, which can restore the original architectural design of the training venue according to the original proportion. Meanwhile, attention should also be paid to the proportion relationship between different locations of the training venue and architectural design specifications. Secondly, in the process of stadium modeling, the number of model faces should be reduced as much as possible, and the invisible faces in the stereo model should be deleted. In addition, some models outside the training ground can be simplified as much as possible, or presented by perspective texture, so as to reduce the file size of stadium model. Therefore, in the process of 3D reconstruction, depth value and 3D data are very important. The so-called depth value refers to the gray value of each point in the modeling image, and the real distance between the position of each point in reality and the vertical plane where the camera is located is the specific depth value, as shown in Fig. 2 and Fig. 3:

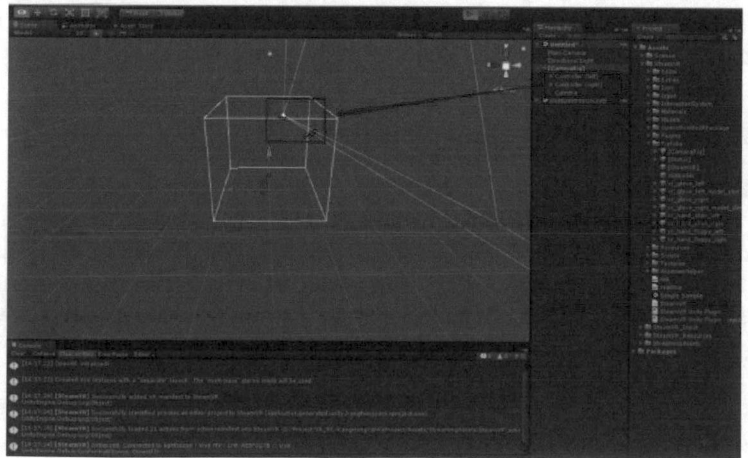

Fig. 2. Setting up a virtual scene

Construct the Role of Virtual Athletes

Virtual athletes need to use 3D graphics and animation software. Because volleyball training requires multiple athletes' roles, professional modeling software should be used as far as possible for the modeling of characters, and relatively tedious 3D graphics and animation software should be avoided. The software involved is character modeling software. The athlete's role animation is realized through the function of human bones

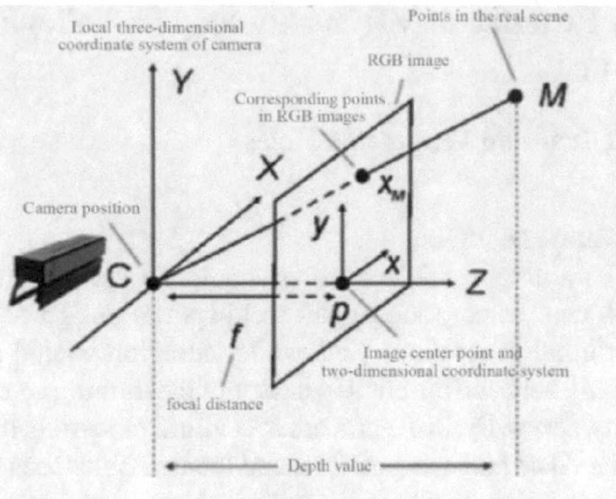

Fig. 3. RGB image and depth values

and joints. Secondly, the setting of the weight of the human skeleton and model will directly impression the effect of the connection between the model and the skeleton, and even affect the deformation of the skeleton of the human model, so it is necessary to edit the weight of the skin to improve the final effect. Virtual scene model optimization is directly related to the file size and interaction environment is smooth, so try to reduce the number of models, while the model and UV texture map should be combined as much as possible, so as to ensure that the model is simple and effective.

4.2 Volleyball Skills and Tactical Simulation

Volleyball technique and tactics as well as the rules of volleyball is changing, whether in volleyball sports popularization, or volleyball teaching of teacher, or national level of volleyball players, their tactics action also is not a complete specifications and reasonable, because they as the growth of the age, is engaged in the work time of growth and falling body quality, Some difficult movements are also difficult to demonstrate in the actual teaching process completely correct, so in volleyball training usually need to be corrected through students or athletes repeated training and continuous action feedback, in order to achieve the correct mastery of skills, tactical action effect. And the combination of VR technology and volleyball, can with the help of virtual reality technology, skill and tactics of volleyball especially difficult simulation, that is commonly used in the volleyball offensive tactics, basic technology and the contents of the textbooks include, by way of virtual reality presented to students, and you can also through the virtual scene practical in-depth analysis of volleyball technology and tactics, Continuous dynamic for students to demonstrate tactics with lines, action essentials and specific operations, so that can be very good to avoid the traditional volleyball teaching and training show inherent defects, volleyball training and volleyball teaching materials in a static way to show the drawbacks of dynamic volleyball skills and tactics. In the process of volleyball training, students can make use of virtual scenes of VR technology to directly and vividly see and hear the key points of technical and tactical movements, which is more conducive to training standard and reasonable technical and tactical movements.

4.3 Help Students Learn Volleyball Theory

Volleyball as a special sport, theoretical knowledge is very broad, contains the history of volleyball, volleyball, volleyball tactics such as basic theory knowledge, but also including volleyball training methods and specific theory and so on, if simply rely on teachers and coaches through traditional books teaching deep understanding and interpretation is very difficult to let the students grasp the volleyball the sport. But with the help of VR technology, volleyball theoretical knowledge and volleyball related content

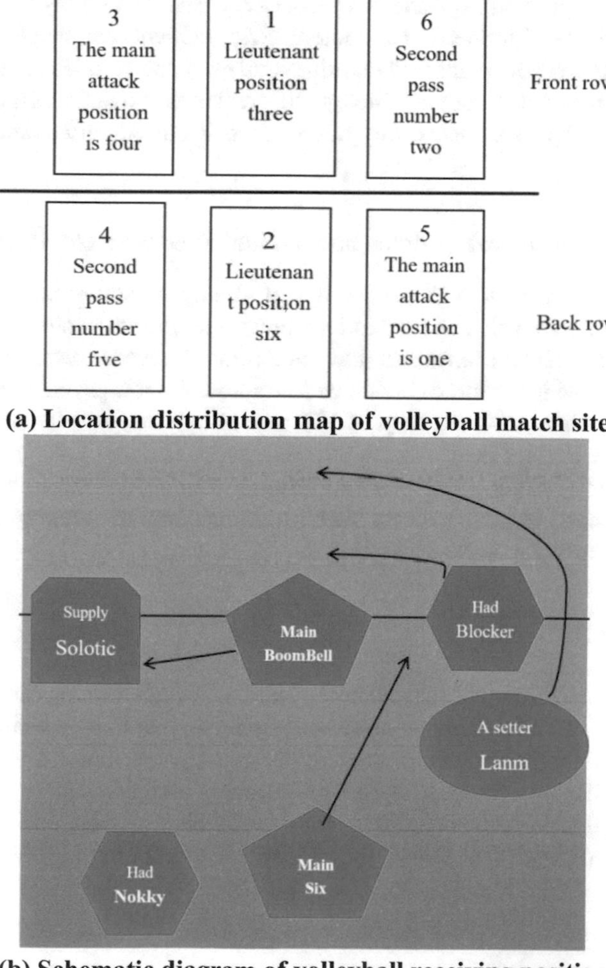

Fig. 4. Schematic diagram of basic volleyball knowledge

through integration into the virtual network space, so that students can experience the development of volleyball through virtual space, innovative theoretical knowledge learning approach, especially to help students better understand some technical and tactical theoretical knowledge, For example, a more realistic virtual role can be built through three-dimensional effect drawings, and a knowledge point can be understood more intuitively through visualized movement. Some basic theoretical knowledge of volleyball matches is shown in Fig. 4.

This can not only increase the interest in learning volleyball, but also through innovative learning ways to improve the effect of volleyball training.

On the other hand, in the process of volleyball training, if the coach's explanation can't let the students really understand the essentials of action and skills, then such training will get twice the result and half the result, and the relatively high level of sports skills of the students is also difficult to find opponents to improve their skills. And based on this problem, if using VR technology, giving full play to the advantages of the characteristics of immersion and interactivity, students can be generalized in the virtual scene into the volleyball skill learning, and even can simulate different levels of volleyball players targeted training, through different scenario Settings meet the training needs of different levels students, Improve volleyball training effect and skill level of students.

4.4 Provide Psychological Training and Virtual Exercise Experiments

Sports events require very high psychological quality of athletes. Before volleyball matches, VR technology can be used to let athletes feel the atmosphere of the competition in advance and feel the tension of the competition in advance, especially for athletes with stage fright, which is more conducive to improving their psychological adaptability. At the same time, with the help of VR technology can also let students through systematic training, combined with the physiological response of students and learning stage requirements, training intensity and training methods for students to carry out a reasonable planning, to help students health training, simulate the movement process, to minimize the injury in volleyball training.

5 Conclusion

To sum up, the combination of VR technology and volleyball training can very good play to the technology, the advantages of more comprehensive auxiliary athletes to participate in volleyball training, through constructing a virtual scene, build a virtual character, improving teaching environment, optimizing training method such as process, lets the student in virtual experience mastery of volleyball technology and tactics in the action and the theory of knowledge, Gradually improve the effect of volleyball training.

References

1. Li, J.: Design of sports training support system based on VR technology. Autom. Technol. Appl. **40**(02), 108–111 (2021)

2. Han, Q.: The application of virtual reality technology in sports volleyball training in higher vocational colleges. Stationery Technol. **24**, 179–181 (2020)
3. Wang, Y.: Feasibility analysis of VR technology in physical education and training. Contemp. Sports Sci. Technol. **10**(26), 37–38+41 (2020)
4. Zhang, P.: The application of computer virtual reality technology in college volleyball teaching – a review of "sports information technology application practice." Sci. Technol. Manag. Res. **40**(11), 263 (2020)
5. Tian. W.: Research on the application of VR technology in physical education and training in colleges and universities. Sports Fashion (02), 32+34 (2020)
6. Ma, Z., Zhang, L., Wang, X.: Theoretical exploration of VR virtual technology in the field of sports training from the perspective of 5G. Sports World (Academic Edition) (02), 111–112 (2020)
7. Yishuo, D.: On the advantages of VR technology in martial arts training. Res. Commun. Power **3**(20), 259 (2019)
8. Zhang, C.: Preliminary exploration of the auxiliary role of "VR" technology in the actual combat training of special police officers. Public Secur. Educ. (06), 23–25 (2019)
9. Zhou, S.: Research on the construction of immersive commander's psychological training environment based on VR. Univ. Educ. (03), 132–134 (2018)

Hidden Markov Model-Driven Speech Recognition for Power Dispatch

Xiaoling Dong[1(✉)], Wanwan Cao[1], Hang Cheng[1], and Tianqi Zhang[2]

[1] Information and Communication Branch, State Grid Anhui Electric Power Co., Ltd., Hefei 230061, China
jy_wxh@163.com
[2] Anhui Jiyuan Software Co., Ltd., Hefei 230088, China

Abstract. In order to use hidden Markov model to realize power dispatch speech recognition and improve the accuracy of recognition results, this paper will carry out relevant research. The paper first introduces the hidden Markov model, then builds a speech recognition system based on the model, and finally tests the system with an example. Through the research, the hidden Markov model has good application value in power dispatching speech recognition, driven by the model, the accuracy of recognition results has been significantly improved, and is very stable.

Keywords: Hidden Markov model · Power dispatching · Speech recognition

1 Introduction

Power dispatching is an important part of power operation and maintenance management. The main work objective is to maintain normal power supply and power use safety, that is, modern power grids are mainly arranged by regions. The total power demand in different regions tends to be stable under normal conditions, but it will change under special circumstances. For example, in hot summer, the power consumption in most regions will increase significantly, and power supply will be carried out according to conventional standards, resulting in insufficient power, At this time, it is necessary to let some idle power facilities in other areas enter the working state through power dispatching, so as to deliver more power to the areas where the power consumption has been greatly increased to meet people's normal power demand. At the same time, the power grid in any area is supported by a large number of power equipment. If a certain equipment fails or is abnormal, the power supply of the power grid will be unstable. At this time, if users continue to use electricity, power consumption accidents may occur. The longer the time, the higher the accident rate. Therefore, when such problems are found, the power organization must carry out power dispatching at the first time, Use other normal equipment to replace the fault or abnormal equipment, and then deal with the fault or abnormal equipment, during which the safety of electricity can be guaranteed.

It can be seen that power dispatching is very important, and China's power organizations have always attached great importance to it. In the early days, because China's

J. H. Abawajy et al. (Eds.): ICATCI 2022, LNDECT 170, pp. 760–768, 2023.
https://doi.org/10.1007/978-3-031-29097-8_90

power grid is not popular, and the internal structure is relatively simple, power dispatching can be carried out only by manual work, and both efficiency and quality can be guaranteed. But as the scale of modern power grid expands greatly and its internal structure becomes more and more complex, the power organization in-depth analysis on the current electric power dispatching work demand, found to make the scheduling work efficiency, and guarantee the quality of work, reduce the pressure of work, you need to develop the corresponding speech recognition system, the system can make scheduling staff anytime and anywhere with terminal system for human-computer interaction, with the help of the voice terminal system to identify the scheduling staff input, and then give feedback, make scheduling work efficiency, quality improvement, also just work a lot of trival matters, the electric power dispatching of speech recognition system application out, but i just have a lot of problems at the beginning of the application of this system, one of the most representative problems is the recognition result is not accurate, the system identification is given after feedback result does not match with the input speech, the reason is that speech recognition system will be input into other voice, this greatly limits the application value of electric power dispatching speech recognition system, but also let electricity group realized that the system needs to be improved, then puts forward many Suggestions to improve related fields, therefore, how to use the model to drive the system and improve the accuracy of recognition results has become a problem worth thinking about, and it is necessary to carry out relevant research on this problem.

2 Significance of Speech Recognition in Power Dispatching

Power dispatching speech recognition is of great significance for improving the efficiency of power dispatching. It can be said that speech recognition function is an essential tool for modern power dispatching and has an indispensable position. Therefore, in the face of various problems of speech recognition function, relevant personnel must start to deal with them to ensure the stability of the function, which is also a topic worthy of discussion.

3 Introduction to Hidden Markov Model

Hidden markov model is a kind of typical statistical model, mainly used to describe an implicit location parameter of markov process, so the model can be regarded as a form of markov chain, able to access parameters identified the implicit parameter of markov process, and then combining with the parameter analysis of relevant problems, that is to say, hidden Markov model can use hidden parameters for further analysis. There are many analysis methods of hidden Markov model combined with hidden parameters, and the most common one is pattern recognition, which is often used in the field of speech recognition [1–3]. Overall, the state of the hidden markov model itself cannot be observed directly, only way to get to know through the observation vector sequence, in which each vector are according to the probability density distribution of the specific deployment, between each state is not the same, the reason is that these vectors are by an independent and existing state of the corresponding probability density distribution sequence, so the analysis process of hidden Markov model is a double stochastic process [4–6].

The hidden Markov model was first applied to speech recognition in the 1980s, and good results have been achieved at that time, which is enough to illustrate the application value of the model in speech recognition, which also makes the model is still the main application means in the field of speech recognition. Hidden markov model, the basic framework as shown in Fig. 1 is a simple dynamic bayesian networks and hidden markov model in the network is the only state transition probability parameter, it also leads to model state cannot be directly observed, but the output of the model was produced under the model state, and the model output can be seen, so you can through the output reverse analysis, have a clear understanding of model state [7–9]. Through the hidden Markov model, people can get the marked sequence, the sequence can provide some state related information for people, which shows that the hidden Markov model has good application value.

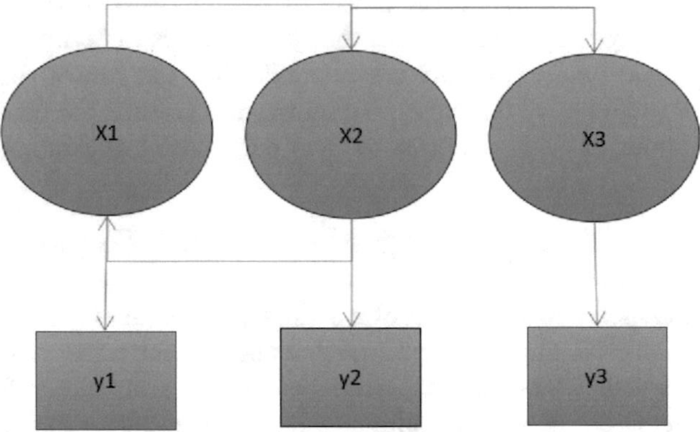

Fig. 1. Basic framework of hidden Markov model (X1, X2… The X3 and y1, y2,…,y3 is the output of various states and states of the model respectively.)

4 Speech Recognition System Construction Based on Hidden Markov Model

4.1 Model Building

In order to make the hidden Markov model become the speech recognition driven basis, the model should be built on the basic framework of the model. The specific process is as follows: Because hidden markov models can use five tuple {N, M, π, A, B}, you can use five yuan group for modeling, N represents the model state in five yuan group number, M is the number of observation symbol may be output in each state,πis the initial state distribution, is the state transition probability matrix, B is the state observation symbol output probability matrix [10]. The model constructed on the basis of quintuples has the characteristics of double randomness, shown in Fig. 2.

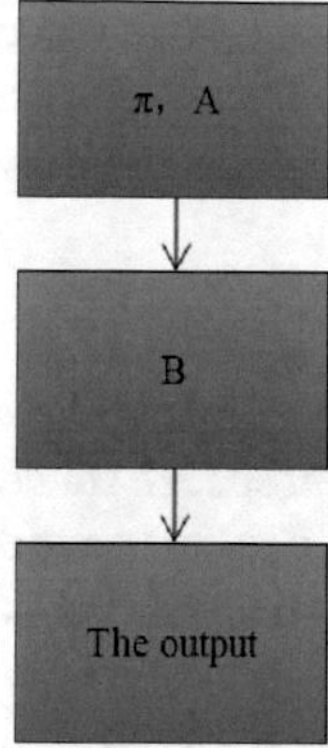

Fig. 2. Hidden Markov model of speech recognition system

4.2 System Design

After the model is built, the power dispatching speech recognition system can be designed. The overall architecture of the system is shown in Fig. 3.

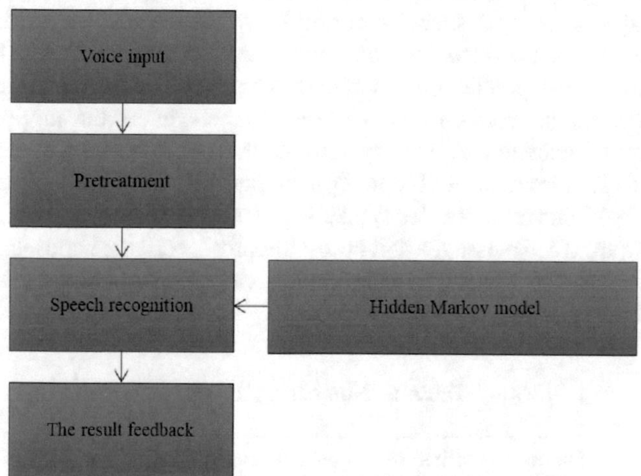

Fig. 3. Power dispatch speech recognition system

Combined with Fig. 3, the operation process of the system is as follows: Firstly, the power dispatching personnel input the voice through the handheld terminal, and the voice will be transmitted to the nearby workstation in the network environment and accepted by the processing terminal of the workstation. Second processing terminal equipment will to the sounds of the input preprocessing, ensure quality of voice information, avoid subsequent caused due to lack of voice information quality identification of misalignment problem occurred as a result, after the pretreatment of speech information can be transmitted to the recognition module, the module is the place of the hidden markov model, with the help of the model can be used to recognize after preprocessing of the voice information analysis. The final identification and analysis results will be sent out as feedback information, which will be received by the hand-held terminal of the staff for manual reference.

Voice Input Design

Although the speech input plate is not directly related to the hidden Markov model, from the perspective of system use, the input speech must be obtained first, so the speech input plate needs to be designed. The design of this section is divided into two steps: The first is to build the network communication channels, namely the voice input dominated by scheduling staff, and because of dispatching work needs, so scheduling staff often far away from the workstation processing terminal of the outdoor environment, in this case in order to ensure the scheduling staff can use the system anytime and anywhere, you need to first set up the network communication channels, The channel can first take the modern popularization of the Internet as the basic environment, and then choose the communication protocol, and then build the communication interface, the interface is located in the terminal of the workstation, under the support of the communication protocol can receive the manual handheld terminal transmission from the voice; The second is to construct communication logic, namely, voice processing terminal will be responsible for not only accept artificial input, and the input speech recognition analysis, but also in charge of the input speech recognition result feedback to artificial handheld terminals, so artificial and system communication logic has the two-way characteristics, in order to achieve this in practice communication logic need to launch logic construction work. Because in the network communication channels, under the support of artificial can one-way communication with the system, so the key to the next question is how to guarantee processing terminal will identify accurate information feedback to artificial handheld terminals, this can consider to take tag technology, namely in the human voice to the system input, voice messages will automatically record the handset number, serial number format shown in Table 1, so in the feedback can be combined with the number of labels to achieve accurate feedback.

Table 1. Numbering format

Numbering structure	paraphrase
N	The name of the terminal
T	Number number

Pretreatment Design

Because during the process of the actual speech input, human input speech may be limited by the interference of external factors, or are influenced by artificial factors in itself and appear quality problem, such as external existence when the input speech strong electrical signal, which could lead to speech noise disturbance, or human input speech without complete expression, make speech incomplete information, as a result, the system cannot accurately identify, so in order to avoid such a situation, pre-processing design is needed. The design method of preprocessing plate is not complicated. It is mainly to choose preprocessing tools, such as noise reduction tools and integrity detection tools, according to the possible quality problems of input voice. These tools are regarded as software programs and can be integrated into the system. It should be noted that, from the perspective of the system structure, the pre-processing plate must be behind the voice input plate, otherwise the process logic of the system operation will be wrong. Figure 4 shows the preprocessing workflow.

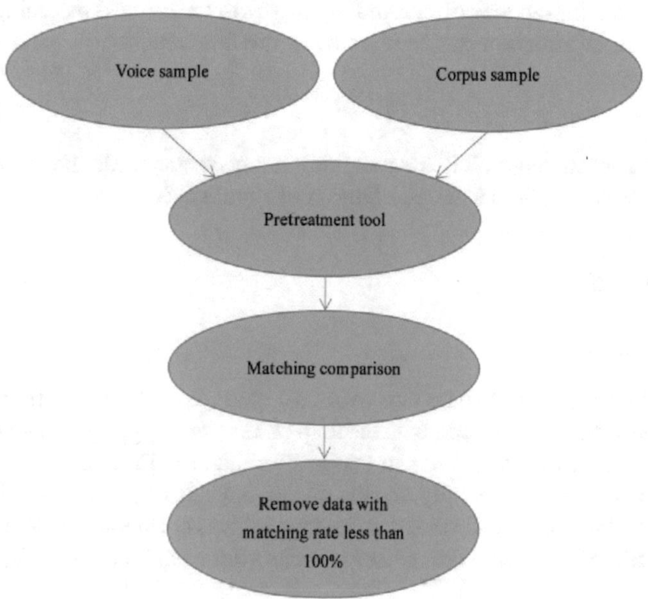

Fig. 4. Power dispatch speech recognition system

Speech Recognition Design

Driven by hidden Markov model, the speech recognition design can be carried out. The design work is divided into three steps: The first is to realize the training of acoustic model. Through training, the system can have the speech recognition function, and the speech recognition ability of the system will be continuously enhanced under continuous training. About acoustic model training method, first of all, based on hidden markov model parameter setting model training, and then set algorithm, it can consider to use Baum Welch algorithm, this algorithm is a typical before and after the bidirectional recursive algorithm, can rapid accumulation of statistics in the training data, makes the model initial parameters constantly updated, at the same time, the algorithm can also enable the system to locally amplify the speech information in speech recognition, so that accurate results can be obtained after convergence. Baum-welch algorithm is shown in Formula (1). The second is the realization of speech recognition logic, here refer to the basic structure of the hidden markov model, voice recognition logic can be divided into two parts, the first part is described the process of model training, the second part is combined with the training results of voice recognition, namely in the training model of the system will accumulate a lot of training, the results can be seen as "knowledge" in the system, the results stored in the database of system can be used as information processing, and then after the system receive the external input voice, can according to the speech vector of MEL frequency cepstrum analysis of the input speech is used to identify the likelihood, and because the likelihood identification analysis results is not entirely accurate, therefore, multiple recognition results with high possibility will be obtained. For these recognition results, matching analysis will be carried out based on the corpus information in the corpus, and the one with the highest matching degree with the input speech information can be selected as the final result.

$$\pi_i^{\mathrm{k}+1} = \gamma_1^{(k)}(i) \tag{1}$$

where i is the speech vector, k is the matching analysis standard, and the error range is $+1$ (not ± 1), γ is the initial state distribution of input speech.

5 System Test

5.1 Test Scheme

In a certain electric power dispatching work, for example, relevant staff working in this day of full use of the voice system, with the help of a system to query the power scheduling information, etc., at the same time to test system has the function of pretreatment of integrity, tests have been specially chosen accent, and there is no similar system using the experience of the staff, make the input voice quality is not stable. Under this condition, firstly, the accuracy of system recognition results after each voice input by the staff will

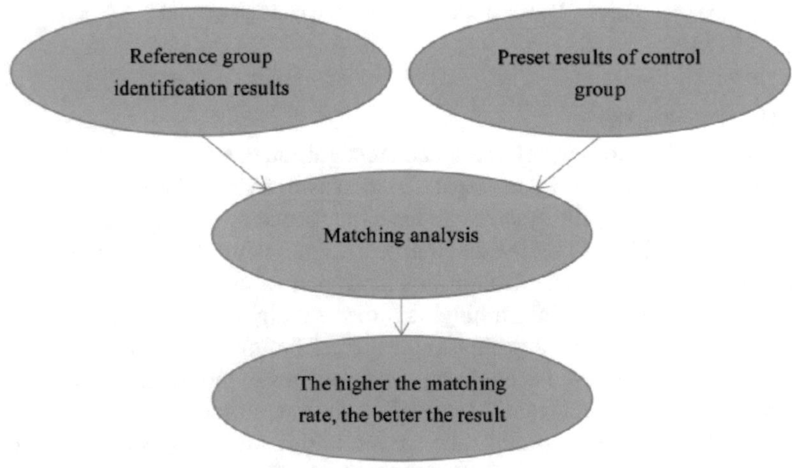

Fig. 5. Scheme flow

be counted. Secondly, the discrimination ability and processing ability of the system for voice quality will be understood. Finally, whether the system can accurately send feedback results to the manual handheld terminal will be confirmed. Figure 5 Scheme flow.

5.2 Test Results

Through the test, the staff used the system to input a total of 16 voices during the day's work. According to the observation of the escort, 12 of the 16 voices had obvious quality problems, while the rest had no problems. According to this result, the system pretreatment was tested. The results show that 12 of the 16 voices are judged by the system to have quality problems and cannot be directly preprocessed. Meanwhile, the system processes all the 12 voices in the way of noise reduction and complete voice input. The voice information processed after noise reduction is very clear and conforms to the quality standard. The actions requiring complete input of speech are also very accurate, so the pre-processing function of the system performs well. At the same time, the result of the system identification of the input speech was observed, the results show that all speech recognition results are accurate, system has made the right feedback action, so the system identification result accuracy is 100%, excellent performance, system will eventually all accurate feedback information are sent to the artificial handheld terminals, so the system has practical value.

6 Conclusion

Voice in conclusion, the electric power dispatching system for power dispatching work is significant to improve the work efficiency and quality, etc., but the current defects identification accuracy of the system, so need to improve, and hidden markov model is a good way to improve choice, related organizations can use this model to improve the system accuracy, Make the system can better serve the dispatching work and staff, but also indirectly improve the quality of power service.

References

1. Jassim, A.K., Al-Bayaty, B.F.Z.: A stochastic approach to identify POS in Iraqi national song using N-iterative HMM using agile approach. IOP Conf. Ser.: Mater. Sci. Eng. **1094**(1), 012019 (2021)
2. Az-Eddine, Z., Abdelaziz, N.: Estimation of steady-state quantities of an HMM with some rarely generated emissions. Monte Carlo Methods Appl. **28**(1), 27–44 (2022)
3. Benhammoud, R., Kacha, A.: Automatic classification of disordered voices based on a hybrid HMM-SVM model. J. Commun. Technol. Electron. **66**(Suppl 2), S139–S148 (2022)
4. Lu, Y., Gu, K., Cai, Y.: Automatic lipreading based on optimized OLSDA and HMM. Soft. Comput. **26**(9), 4141–4150 (2022)
5. Ghimatgar, H., et al.: Neonatal EEG sleep stage classification based on deep learning and HMM. J. Neural Eng. **17**(3), 036031 (2020)

6. Díaz-Valerio, S., Hacohen, A.L., Schppe, R., Liesegamg, H.: IDOPS, a profile HMM-based tool to detect pesticidal sequences and compare their genetic context. Front. Microbiol. **12**, 1471 (2021)
7. Liu, H., Shima, T., Uemura, S.: A fast and fully automated HMM fitting algorithm enables accurate analysis of biophysical data with numerous states. Biophys. J. **120**(3), 266a (2021)
8. Al-Anbagi, L., Al-Dileamy, H., Hamu, A.H.: Evaluation of copeptin hormone function in chronic renal failure stages and dialysis. Plant Archives **202121**(1), 1052–1056 (2021)
9. Ye, L., He, T.: HMM speech recognition study of an improved partic1e SWANN optimization based on se1f adaptive escape AEPSO. IOP Conf. Series: Earth Env. Sci. **634**(1), 012074 (2021)
10. Ubaidi, U., Dewi, N.P.: Penerapan hidden markov model(HMM)dan mel-frequency cesptral coefficients(MFCC)pada e-learning bahasa madura untuk anak usia dini. Jurnal Teknologi Informasi dan Ilmu Komputer **7**(6), 1111 (2020)

Analysis of Machine Translation and Computer Aided Techniques in English Translation

Xiaolin Wang(✉)

Beihua University Teacher's College, Jilin 132013, China
258070188@qq.com

Abstract. In order to understand the application of machine translation and computer-aided technology in English translation, this paper will carry out relevant research. This paper first discusses the current situation and development needs of English translation technology, and then puts forward the application strategies of machine translation in English translation and the main functions of computer aided technology. Through this study, the rational use of machine translation and computer-aided technology can optimize English translation results, better eliminate language differences, and output more accurate translation results.

Keywords: English translation · Machine translation · Computer aided technology

1 Introduction

At present, the increasingly frequent communication between China and the world makes people often troubled by language differences in communication. Therefore, in order to ensure the smooth development of communication, most people hope to translate foreign languages through technical means. As the current universal language, English is the main target of foreign language translation. On this basis, a variety of English translation software has been developed in related fields in China. At the beginning, these software did solve some practical problems and helped people to overcome the language barrier to better communicate to a certain extent. However, with the deepening of application, the initial English translation software gradually exposed many defects. Including the overall translation result of low accuracy, not according to the semantic translation, etc., these defects also make people realize that English translation software needs to be updated iterations, the English translation of reason put forward higher requirements, and in the face of modern requirements, although has made a large number of fields related to try, but for a long time the results are not satisfactory, especially in the aspect of semantic translation, Translation software may produce results that are literally accurate, but do not match the meaning of the sentence, which confuses researchers in related fields.

In view of the above phenomenon, artificial intelligence technology of mature gradually make the English translation software has a broader space, finally find the fields related to solve the previous English translation software translation result is not accurate, not according to be the problem such as route, and on the way to develop a lot of

J. H. Abawajy et al. (Eds.): ICATCI 2022, LNDECT 170, pp. 769–777, 2023.
https://doi.org/10.1007/978-3-031-29097-8_91

new type of translation tools, These tools as a module is installed on the original English translation software can solve the problem, there are two types of these technological tools, respectively referred to as machine translation and computer aided technology, can use these two kinds of technology tools, is the guarantee of English translation result quality, meet the demand of the present people English translation necessary measures, therefore, in order to promote the machine translation and computer aided technology, To solve the problem effectively, it is necessary to carry out relevant research.

The main reason why there were various problems in early machine translation was that the design idea of machine translation software at that time was relatively simple, that is, most of the early machine translation software translated according to corpora. People would collect a large number of corpora before software design, and then generate these corpora into corpora. After the completion of the software design, the software was equivalent to an entrance to receive language information, When the user inputs the information of the language to be translated in the software, the software will search the corresponding corpus in the corpus. If the corresponding corpus is found, it will output the corpus as the final answer to complete the translation. If there are multiple corresponding corpora, it will compare the matching degree and select the highest matching corpus as the final answer to complete the translation. This idea does have some advantages, but it has obvious defects when viewed in depth, that is, in the human language system, the same words can represent different consciousness, such as "I am angry". Sometimes it can really be used to express people's anger, but in some cases it can also be used to "joke". Therefore, according to early logic, machines cannot accurately translate, and various problems naturally occur. The existence of computer aided technology is to make up for the defects of machine translation, but the function of early computer aided technology in this area is not significant. In many cases, even if the computer aided technology module is added, the results obtained are still not accurate, indicating that there is still a lot of research space in this field. This research will produce a certain impetus, so the research has certain value.

2 Research Background

The main reason why there were various problems in early machine translation was that the design idea of machine translation software at that time was relatively simple, that is, most of the early machine translation software translated according to corpora. People would collect a large number of corpora before software design, and then generate these corpora into corpora. After the completion of the software design, the software was equivalent to an entrance to receive language information, When the user inputs the information of the language to be translated in the software, the software will search the corresponding corpus in the corpus. If the corresponding corpus is found, it will output the corpus as the final answer to complete the translation. If there are multiple corresponding corpora, it will compare the matching degree and select the highest matching corpus as the final answer to complete the translation. This idea does have some advantages, but it has obvious defects when viewed in depth, that is, in the human language system, the same words can represent different consciousness, such as "I am angry". Sometimes it can really be used to express people's anger, but in some cases it can also be used to "joke". Therefore, according to early logic, machines cannot accurately translate, and various problems naturally occur. The existence of computer aided technology is to make up for the defects of machine translation, but the function of early computer aided technology in this area is not significant. In many cases, even if the computer aided technology module is added, the results obtained are still not accurate, indicating that there is still a lot of research space in this field. This research will produce a certain impetus, so the research has certain value. Figure 1 Early Machine Translation Ideas.

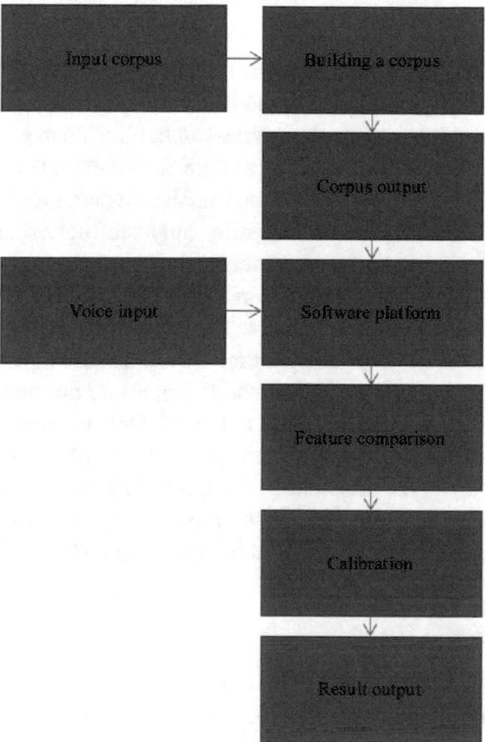

Fig. 1. Early Machine Translation Ideas

3 Current Situation and Development Demand of English Translation Technology

3.1 The Status Quo

The specific flow of current English translation technology is shown in Fig. 2.

In this process, English translation techniques can what kind of results depends on the compatibility of the two corpora in the translation process will results according to two rules: first, according to the preset standard, choice is higher than the standard of the original corpus as a translation as a result, if higher than the standard number

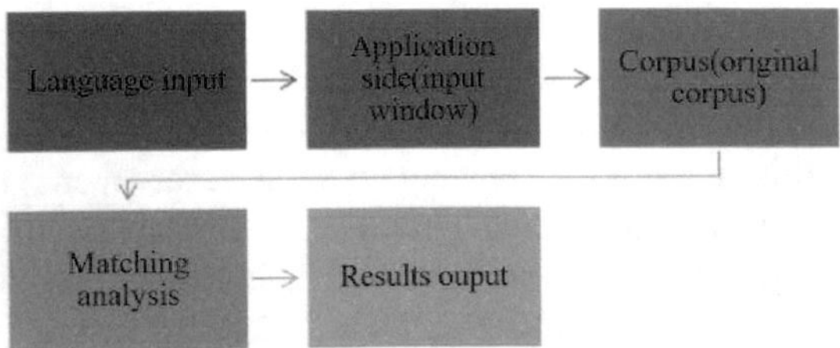

Fig. 2. Main flow and principle of current English translation technology

of primitive data more than 2, then choose the highest matching the primitive data of output; Second, if no original corpus matches the input corpus more than the preset standard, the original corpus with the highest matching degree will be output as the result. According to these two rules, can through the English translation technology to get the results, but will inevitably lead to the translation accuracy is often in a low level, namely, first of all, according to the first rule, only the highest matching degree, and higher than the standard of the primitive data can be translated as a result, but in order to avoid the translation corpora in practical translation standard is too high, The problem of translation results being too limited occurs. The standard preset by people is generally about 85%. Although this standard can guarantee the smooth output of translation results, it also makes the accuracy of translation results cannot be guaranteed. Secondly, in the second rule, among the corpus that does not meet the matching criterion, the corpus with the highest matching degree will be output as the result, which further leads to the decrease of the accuracy of the translation result. It can be seen that the poor performance of English translation technology has played a role, but the quality of the results needs to be improved, and the technology should be updated [1–3].

3.2 Development Needs

Combined with the current situation, there are two requirements for the development of English translation technology, specifically to improve the accuracy of translation results and translate according to semantics, as follows.

Improve the Accuracy of Translation Results

The current situation shows that the translation results of English translation technology are not accurate, and still cannot eliminate the communication barriers under language differences. Therefore, the first requirement of the development of this technology is to improve the accuracy of translation results. There are two main ways to improve the accuracy of translation results: First, improve the preset standard level of translation while giving consideration to the smooth output of translation results; Second, long sentences and texts can be translated in accordance with grammatical rules to avoid all kinds of phenomena unfavorable to the accuracy of translation results. The above requirements have been widely recognized, but on the basis of the original level of technology can not be achieved [4, 5]. Table 1 shows the specific requirements of the two approaches in improving the accuracy of translation results.

Table 1. Specific requirements of the two approaches in improving the accuracy of translation results

Project	Content
Accuracy preset standard level	> 99%
Phenomena unfavorable to the accuracy of translation results should be avoided	Words are confused, sentences are reversed, etc

Translation Based on Semantics
From the perspective of linguistics, when people use language to express by the same language to express different meaning, is particularly common in this kind of expression form in Chinese, so in cross-language communication, it is hoped that English translation techniques can according to their semantic translation, although this point from the Angle of modern technology cannot be fully realized, But at least they can do this under specific rules, such as people enter the two statements, the previous statement shows the person's potential, and then to express, through a statement after so as long as make English translation technology to know the previous statements, can know what the user want to express specific, then according to the rules of semantic translation, it is now possible. On this basis, people hope that English translation technology has the function of translating according to semantics [6–8]. Figure 3 shows the implementation idea of semantic translation function.

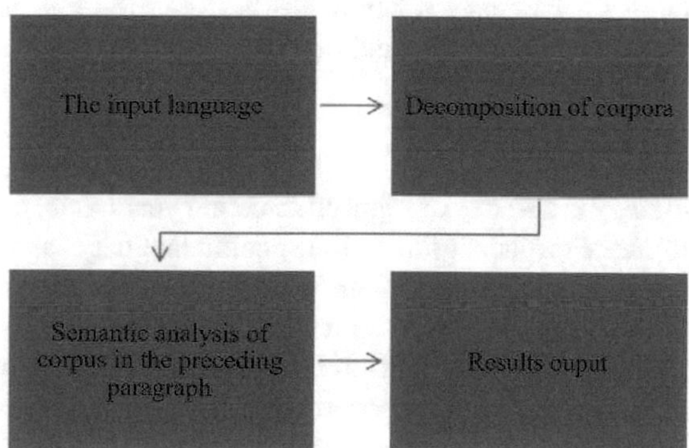

Fig. 3. Implementation ideas of semantic translation function

4 Application Strategies of Machine Translation in English Translation and Main Functions of Computer Aided Technology

4.1 Application Policies

Based on artificial intelligence, and related fields is proposed for a class of translation tools, these tools are collectively referred to as machine translation tools, under the support of artificial intelligence can greatly improve the accuracy of the translation results, and there will be no results can be smoothly, but to do this we must adopt the strategy of different method to use this tool. The main strategies include direct strategy, transfer strategy and central language strategy, which are as follows [9, 10].

Direct Policy

Direct policy is the earliest machine translation tool strategy, and its principle is as follows: According to predefined received the source language into the target language at the same time, the binomial one source all the grammar of the language contains words, words, can be directly connected to the target language in the corpus of the relativity of unit, in this way is established between the source language and target language completely matching relations, the relationship the difference between the previous single matching relationship, Any input language can be guaranteed to have a corresponding answer, so there will be no translation results cannot be output, and the solution of this problem will naturally improve the matching standard to the highest level.

Transfer Policy

Transfer strategy is built on "on behalf of level" thinking a translation strategy, the whole translation process is divided into the analysis phase, phase shift and formation stage, the analysis phase mainly through language and describe the source language, and in accordance with the data found within corpus is consistent with the source corpus of the source language, and is mainly used for the conversion analysis phase transfer results, The goal is to establish a language relationship between the two languages, so that you can enter the final generation stage and directly output the results according to the relationship.

Central Language Strategy

Central language strategy is a strategy proposed in recent years. It is mainly based on any specific language to create text according to the specific language, and then establish the central text according to the language content in the text to get different from the source language and target language, and then the translation results can be output. It can be seen that, compared with other strategies, the principle of central language strategy is simpler, but because it has not been proposed for a long time, people can not confirm the accuracy of the translation results under this strategy, but at least from the current test results, the central language strategy has good reliability.

4.2 Main Functions

The function of artificial intelligence is rich, so they have relevant researchers to convert some of these functions into English translation tools, this kind of tool is referred to as computer aided technology, and through the systematic transformation, these tools are integrated, directly using the system can provide English translation help, one of the most obvious help to realize the logical semantic translation. In general, the main functions of computer-aided technology in English translation include multiple databases, indexing functions and translation records, as follows.

Multivariate Database

Because English translation design language knowledge, translation logic is very diverse, so a single database can not meet the needs of translation, so people in computer aided technology system to establish multiple databases. The specific form of a multivariate database varies, but it contains at least three sections, as shown in Fig. 4.

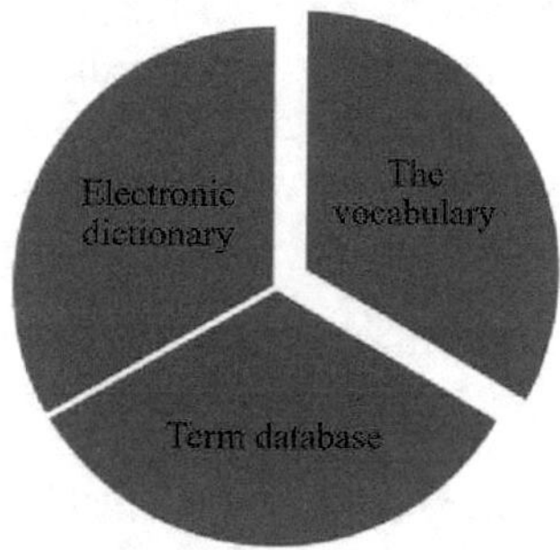

Fig. 4. Basic composition of multivariate database

It can be seen that the multi-component database is mainly composed of electronic dictionary, vocabulary and term database. Among them, electronic dictionary is the dictionary information common in the current network. The content of electronic dictionary is not much different from traditional paper dictionary, but the advantage of electronic dictionary is that it can be well connected with the index function. Make people in a very short period of time from the corresponding data obtained from the electronic dictionary, English translation, the assumption is now a total of 500 words essay, directly through the electronic dictionary index retrieval can get the answer in a few seconds, but if print dictionary is used, the process may be more than one hour. All the words in the glossary come from electronic dictionaries, which can be interpreted by manual or technical means at that time, which supports the semantic translation function. Meanwhile, the glossary database is specially used to store the information of terms in some professional fields, which is similar to the glossary in function, so I will not add more details here.

Index Function

Indexing function is a retrieval function. When users input corpus in English translation, the system will start this function to find corresponding corpus in multiple databases and generate translation results. In essence, indexing is an information processing tool. The basic principles are as follows: According to have a clearly defined corpus, within the corpus retrieval with relevant definitions of translation corpora, and then arrange the serial number of the corpora carried out in accordance with the rules of grammar, in "hello" words in translation, for example, assume that "you" and "good" two words in Chinese grammar in the Numbers 1, 2, respectively, then through the index function to get the serial number of "1–2" format, When all languages are completed and complete numbers are obtained, accurate results can be obtained by compiling languages in accordance with the numbering order.

Translation Records

Translation record is a function based on online bilingual text. It was not used in the field of translation at the earliest, but it attracted the attention of the field of translation in the early 1990s and has been widely used. The basis of translation record is also a database, which is called record database. Its main function is to store the translation result of each time, and then adjust the translation result of the next time according to the translation result of the last time, so as to greatly improve the accuracy of translation result and provide more support for semantic translation. On the basis of artificial intelligence, this function can provide artificial intelligence with translation training samples, that is, every recorded translation result is a training sample, so the translation recording function has high application value.

5 Conclusion

At present, most of the English translation of technical translation results are not accurate enough, and not according to the semantic analysis, but with the help of machine translation and computer aided technology can effectively solve the problem, not only to improve the translation accuracy, and to realize the function of semantic analysis, which makes English translation technology functional enhancements, can provide more high quality translation service to people.

References

1. Ren, J.: Application of computer technology in english translation network teaching. J. Phys.: Conf. Ser. **1648**, 022124 (2020)
2. Chaman Kumar, K.M., Aswale, S., Shetgaonkar, P., Pawar, V., Kale, D., Kamat, S.: A survey of machine translation approaches for Konkani to English. In: 2020 International Conference on Emerging Trends in Information Technology and Engineering (ic-ETITE), pp. 1–6. Vellore, India (2020)
3. Birkenbeuel, J., et al.: Google translate in healthcare: preliminary evaluation of transcription, translation and speech synthesis accuracy. BMJ Innovations **7**(2), 000347 (2021)
4. Pulipaka, S.K., Kasaraneni, C.K., Sandeep Vemulapalli, V.N., Mourya Kosaraju, S.S.: Machine translation of English videos to Indian regional languages using open innovation. In: 2019 IEEE International Symposium on Technology and Society (ISTAS), pp. 1–7. Medford, MA, USA (2019)
5. Ferreira, S., Leitão, G., Silva, I., Martins, A., Ferrari, P.: Evaluating human-machine translation with attention mechanisms for industry 4.0 environment SQL-based systems. In: 2020 IEEE International Workshop on Metrology for Industry 4.0 & IoT, pp. 229–234. Roma, Italy (2020)
6. Tripathi, S., Kansal, V.: Machine translation evaluation: unveiling the role of dense sentence vector embedding for morphologically rich language. Int. J. Pattern Recogni. Artif. Intell. **34**(01), 2059001 (2020)

7. Herbig, N., Pal, S., Vela, M., Krüger, A., van Genabith, J.: Multi-modal indicators for estimating perceived cognitive load in post-editing of machine translation. Mach. Transl. **33**(1–2), 91–115 (2019)
8. Dhanjal, A.S., Singh, W.: An automatic machine translation system for multi-lingual speech to Indian sign language. Multimed. Tools Appl. **81**(3), 4283–4321 (2021)
9. Duygulu, P., Barnard, K., Freitas, J.F.G., Forsyth, D.A.: Object recognition as machine translation: learning a lexicon for a fixed image vocabulary. In: Heyden, A., Sparr, G., Nielsen, M., Johansen, P. (eds.) ECCV 2002. LNCS, vol. 2353, pp. 97–112. Springer, Heidelberg (2002). https://doi.org/10.1007/3-540-47979-1_7
10. Brants, T.: Large language models in machine translation. EMNLP-CoNLL 2007. In: Proceedings of the 2007 Joint Conference on Empirical Methods in Natural Language Processing and Computational Natural Language Learning, 28–30 Jun 2007. Prague, Czech Republic (2007)

Application of 3D Technology in Film and Television Animation System

Xiaoye Zhang[1], Sisi Qiu[1](✉), and Chadi Altrjman[2]

[1] School of Comics, Jilin Animation Institute, Changchun 130000, China
jldhqiusisi@sina.com

[2] Artificial Intelligence Engineering Department, AI and Robotics Institute, Near East University, Mersin 10, Turkey

Abstract. In order to understand the application of 3D technology in film and television animation system, this paper will carry out relevant research. The study first introduces the basic concept and classification of 3D animation technology, then analyzes the key points of 3D technology application in each stage of film and television animation design, and discusses the application performance of 3D technology. Through research, 3D technology plays an important role in modern film and television animation system, and giving full play to the role of this technology can effectively improve the quality of animation and bring visual feast to people.

Keywords: Film and television animation system · 3D technology · Animation design

1 Introduction

Vedio is representative of film and television entertainment industry industry, its main products are all kinds of cartoon work, such work in modern society there is a lot of audience, by people, which makes the economic potential growth of the industry, but also lead to more competitive within the industry, therefore, in order to better cope with competition, practitioners began to constantly seek breakthrough, I hope to design animation works through various advanced technical means and improve the quality of works. This background, 3D technology get the practitioners' attention, make the technology has now begun to film and television animation system related work, but have certain professional 3D technology to the user request, therefore many practitioners using the technology to carry on the design is not very good, this kind of circumstance can lead to limited work competitiveness, it may not be widely available in the market, which is a blow to the industry or other industries. Therefore, in order to enable more practitioners to master 3D technology and give full play to the role of this technology in the design of animation works, it is necessary to carry out relevant research.

J. H. Abawajy et al. (Eds.): ICATCI 2022, LNDECT 170, pp. 778–786, 2023.
https://doi.org/10.1007/978-3-031-29097-8_92

2 Technical Value Analysis and Background Introduction

The reason why 3D technology is highly praised in the field of film, television and animation is that it has many advantages over traditional 2D technology, which are embodied in two aspects: first, performance. 3D technology and 2D technology have many similarities in performance. For example, both can generate pictures quickly, with almost the same efficiency, but the difference is that 3D technology generates picture solids, while 2D technology generates picture planes, the gap between the two is reflected in the three-dimensional and plane. The stereoscopic picture represents that designers can observe the picture from all directions without dead angle, and can see the problems in the picture more intuitively for improvement. There is no blind area in the analysis of problems during the period from the visual point of view, but the plane picture of 2D technology cannot let designers observe from all directions without dead angle. If you want to understand the picture problems, you should separate each plane of the picture, and then observe it separately and in combination, in the process, designers need to rely on themselves to combine different planes for analysis. This method is vulnerable to the influence of personal factors of designers, and blind spots will occur. Some problems that may involve multiple planes will be missed, affecting the quality of results. Therefore, 3D technology has intuitive advantages in performance, which can improve the quality of results; Secondly, in terms of functions, many functions of 3D technology are not available and can not be realized by 2D technology. The most typical function is the 3D physical engine function, that is, from a physical perspective, when people adjust a thing as a whole, other parts of the thing may also change, such as the shape, position and surface of the paper towel may change when a piece of paper towel is pulled, The 3D physics engine function is to attach such changes to the design results. When designers adjust the 3D works, other parts of the works will also change according to the parameters, which is more conducive for designers to design dynamic and real works. However, this function cannot be carried in the 2D technology, because the plane dimension of the 2D technology is not enough to support the physical changes in the three-dimensional dimension. In addition, 3D technology has other functional advantages, which will not be repeated here. However, the above contents should prove that 3D technology has high application value in the field of film and television animation, which is also the significance of this study.

It is worth mentioning that although 3D technology has many advantages and significant application value, when this technology is involved in the design work in the field of film, television and animation, the traditional design mode around 2D technology will be greatly impacted. This impact is likely to cause designers to not adapt, limit the use of 3D technology, and also significantly affect the development of design work in the field of film, television and animation. In the face of this situation, the only way to eliminate the maladjustment is to let more designers understand the specific application methods of 3D technology in the design work. Therefore, it is necessary to carry out relevant research, aiming to show the basic concepts and operation methods of 3D technology, and consolidate the application foundation of 3D technology in the field of film and television animation.

3 Basic Concepts and Classification of 3D Animation Technology

3.1 Basic Concepts

Animation 3D technology is a kind of 3D modeling technology that depends on pre-set parameters. The technology has gradually matured in the continuous development [1, 2]. Now there are a lot of system software, so people can use the technology directly through the system software. In animation design of 3D modeling technology is the main basis of various parameters, common parameters are shown in Table 1, determine the model of these parameters in the design specifications, shape, etc., this is for 3D design attributes, these parameters are adjustable at the same time, the designers in the design good after initial parameters can be inspection on model, if found problems model, or from design concept to the model design results are not satisfied, you can at any time to adjust the parameters of related projects, shows that under the 3D animation design development and realized the parametric and parametric design of the main advantage is high accuracy, the assumption is a cartoon character model body part is too wide, so the designers is to downsize, but to what extent are not stereotypes, need to repeatedly test in actual design process, designers can only over and over again in the face of such situation ever drawing, the whole process is cumbersome, and 3D design personnel you just need to adjust the numbers, this not only makes the design process more convenient, also can ensure every time the progress of the adjustment. In addition, 3D technology also enhances the realization of animation design, that is, although the technology can only be used for modeling, but there are many elements that can be used in modeling, including lighting, texture, color, etc., which can make animation effect more realistic and vivid, indicating that the technology has a high application value in animation design [3, 4].

Table 1. Parameter basis of 3D modeling in animation design

The parameter name	Role
Size	Adjust model specifications
Line	Adjust the model line frame
Brightness	Adjust model light and shade

3.2 Technical Classification

Because of 3D technology has the very high application value, so the related personnel to the years are committed to the development of the technology, makes the technology appeared a variety of types, each type of 3D technology are representative of the system software, there are three kinds of one of the more common, the first is Softimage SXI software, the software is the classic cartoon 3D software, the main function is modeling, but modeling also includes other functional sections, including animation, scene rendering, special effects adding, etc., which make the final model more rich. Under this software, animation design gets rid of the previous linear means, and instead adopts nonlinear animation production means for design. This means is more flexible, which shows that the software has good design performance in animation design, which is also the reason why the software is still the first choice of many animation designers after many years [5, 6]. The software is now widely used in 3DE virtual animation design, the design results of the display effect is very real. Followed by production softwares 3 Dmax software, the software is currently the most widely used in industry of software, the reason is that the 3D animation design is the two most important steps in the scene modeling, modeling and character modelling which workload is far less than the former, but the design difficulty is very high, not only pay attention to the attitude of the characters, but also pay attention to details, With production softwares 3Dmax software is especially good at is used for modelling design, namely the software designers can choose appropriate under operation interface, then on the interface design of the window, the window design method is very convenient, intuitive, can let the designers in the design process found problems in time, coupled with a parameterized debugging, can complete design fast and accurate. Finally, lightwava 3D software, which is relatively complete in functions, but slightly insufficient in performance, is difficult to be applied to some demanding animation design work. The software has high cost performance and easy to understand operation. Basically, as long as the design requirements are not very high, the software can be applied to various animation design [7–9]. Things like games, web sites, movie post production, advertising, etc. In addition, although there are many types of 3D technology, the animation design process on any software is fixed, as shown in Fig. 1, Fig. 2 and Fig. 3 show the comparison of 3D and 2D design results respectively.

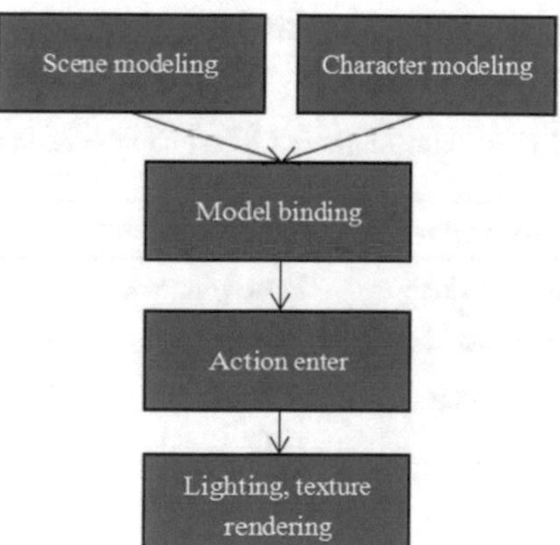

Fig. 1. Animation design process

Fig. 2. 2D

Fig. 3. 3D

4 Application Points and Performance of 3D Technology in Various Stages of Film and Television Animation Design

4.1 Division of Film and Television Animation Design Stage

The stage division of film and television animation design is shown in Table 2.

Table 2. Stage division and content of film and television animation design

Stage division	The main content
Preparation stage	The plane design
The intermediate stage	Modeling and debugging
In the late stage	Action entry, binding, and finally rendering

According to Table 1, graphic design is the main content in the preparatory stage of film and television animation design, which needs to be manually executed by designers. The result is a variety of TWO-DIMENSIONAL plans, but although they are two-dimensional plans, they are closely related to subsequent 3D modeling. Modeling and debugging are the main contents of the middle stage. In this stage, designers are required to complete the initial model construction according to the two-dimensional plan, and check the individual model to find problems and actively debug to ensure that each individual model meets the design requirements. On the basis of meeting the quality requirements of the debugging results in the middle stage, the design enters the later stage. In this stage, human actions are input into the model, and then each individual model is bound to form the overall model. Finally, the final product can be obtained through lighting and animation rendering of the overall model [10].

4.2 Key Points of Technical Application

Each stage of film and television animation design is closely related to 3D technology. To do a good job in each stage, it is necessary to grasp the key points of application of 3D technology in different stages, which will be analyzed below.

Preparation Phase

In the preparation stage, designers do not need to use 3D technology, but they need to take into account some requirements of subsequent 3D technology application in 2d plan design, and ensure that 2D plan can better become the material of 3D model. Two-dimensional layout design is accomplished by manually, the process of drawing it should communicate with comics writers actively, understand the design idea, and then you can better finish drawing, such as the character modelling design, if the 2D layout drawing don't understand the plot, can't guarantee accurate modelling of the characters in drawing expression, this input of the floor plan into the 3D model will result in the modeling not meeting the design requirements, so the preparation phase is very important. In addition, graphic design is only the main content in the preparatory stage of film and television animation design, but not the whole content. Therefore, other staff should also do their own essential work in this stage, as shown in Table 3.

Table 3. Other work related to the application of 3D technology in the preparatory stage

Job content	Purpose
Scene layout	Ensure that subsequent live action input can match the scene
The light	Give designers basic concepts of lighting rendering

Intermediate Stage

After all preparations were made in the preparation stage, film and television animation design design into the middle stage, this stage need to apply directly to the 3D technology, working content only include modeling and debugging, but each job has a lot of internal tasks, and often need to rework, therefore more tedious, need to design personnel to treat seriously. First among modeling, design staff will be ready to stage the floor plan first content input 3D system, and then in 3D software to choose 3D geometry and the corresponding development of geometry, and then the geometry according to the design idea to connect together, such as the design ideas of design of an adult male, then the adult male torso geometry as the foundation, by connecting the extended geometry such as hands and legs on it, the modeling work can be easily completed. However, the model at this time is very rough and in the initial state, so the designer still needs to conduct preliminary parameter debugging to make the model generally meet the requirements. Second in the debugging work (excluding preliminary parameters debugging), designers should actively observe after preliminary debugging model effect, can be considered under the technical support to animate the model rendering, check the movement situation, if found that the problem will be further adjustment of the parameters, the process will often repeatedly, is mainly responsible for the intermediate stage design work tedious, and if the correct choice of 3D technology software, can effectively reduce the tedious degree, promote the mid-term work can be completed as soon as possible.

Later Stage

The later stage of the work content is more, but the operation of the work content is relatively simple, and will not be repeated often, and this does not mean that designers can treat it at will, on the contrary, designers still need to treat the later stage of work seriously on the basis of 3D technology. The first is the action input, that is, the action information has been obtained through the real action actors in the preparation stage, so the designer should input the corresponding action information into the relevant model to make the model move. Here, it is recommended to check each time the input is completed, and if the problem is found, the parameter debugging should be carried out through 3D technology. Secondly, model binding, the main object of this work generally refers to the scene model and character model, the process should be based on the scene model, combined with the design requirements to confirm the character model in the scene location, and then binding, so that the cycle can combine the two models, get the preliminary work. Finally, rendering, which includes three more steps: Texture rendering, the first is the model designers can here in the 3D model to choice of texture, such as choose metal texture, then the model of "rendering" button to get the corresponding texture, the texture at the same time, if the model need more complex, can undertake partition before rendering is done to the model, in the render region separately, complex texture; Second is lighting rendering, here need the reference preparation stage lights change, choose the basic lighting parameters, to adjust the lighting in various parts of the light and shade on the model parameters, and then to adjust the position of the light, so that can give full play to their role as the light in his works, builds a good scene atmosphere, also let more realistic animation; The third is the texture rendering. The texture used in this work is essentially a TWO-DIMENSIONAL plane plan, so it is necessary to use the conversion function in 3D software to apply it to 3D models, let 2d programming 3D, and then carry out texture rendering. In addition, designers also need to complete post-production work such as special effects production, video and audio synthesis, production and editing. Although these work need to use other technical software, the final result will affect the quality of 3D animation works, so designers should pay attention to it.

4.3 Application Performance

In environmental design, film and television animation, for example, using the above methods, and combined with relevant points using 3D design, will be a very good to the environment in the design of the three elements of time, place, characters together each other, and guarantee related to the quality of the model itself, can build a good space stereo feeling, make works from traditional two-dimensional perspective, at the same time, the difference between 3D model and TWO-DIMENSIONAL model is not only reflected in the three-dimensional sense of space, but also reflected in the dynamic, that is, the environmental design results under 3D technology have dynamic characteristics, such as the grass will swing in the wind, which can make the work more realistic, and easier to bring people a feeling of immersive.

5 Conclusion

In conclusion, previous works of film and television animation design output for two-dimensional works, more people for a long time for this kind of work has produced the aesthetic fatigue, thus puts forward new requirements to the industry, and the emergence of 3D technology can help the industry works meet people's needs, bring new experiences to the person, so relevant enterprises it is necessary to fully grasp the 3D technology, or to cause a decline in competitiveness, Eventually eliminated by the market.

Acknowledgement. The Higher Education Teaching Reform Research Project in Jilin Province "Development and Practice of the Construction Standards of the Comics Major for Undergraduate against the Background of New Liberal Arts", Project No: SZ2002.

References

1. Olaru, A.D., Olaru, S.A., Mihai, N.F., et al.: Animation in robotics with LabVIEW instrumentation. Int. J. Model. Optim. **9**(1), 34–40 (2019)
2. Permana, H., Kencana, H.P., Bakri, F., et al.: The development 3-D augmented reality animation on radioactive concept. J. Phys. Conf. Ser. **1402**(6), 066076 (2019)
3. Wald, D.S., Casey-Gillman, O., Comer, K., et al.: Animation-supported consent for urgent angiography and angioplasty: a service improvement initiative. Heart (2020). https://doi.org/10.1136/heartjnl-2019-316227
4. Suprapto, P.K., Suharsono, Chaidir, D.M., et al.: Development of Wimba 3 dimension interactive animation media on plant anatomy. J. Phys. Conf. Ser. **1233**, 012002 (2019)
5. Koehler, M., Usevitch, N.S., Okamura, A.M.: Model-based design of a soft 3-DHaptic shape display. IEEE Trans. Rob. **36**(3), 613–628 (2020)
6. Cabeceiro, D., Esteves, E., Morais, E.P. P romotion of activities of tourist animation in the pages of Facebook of the municipalities of Terras de Trás-os-Montes. In: 14th Iberian Conference on Information Systems and Technologies (CISTI) (2019)
7. Guntara, Y., Saefullah, A., Nulhakim, L., et al.: Development of augmented physics animation (APA) with the integration of crosscutting concepts about the COVID-19 as a supplement to the introductory physics course. In: The 9th National Physics Seminar-UNJ (2021)

8. Milheim, W.D.: How to use animation in computer assisted learning. Br. J. Edu. Technol. **24**(3), 171–178 (2010)
9. Davies, D.M., Fatah, M.F.: Facial animation in Moebius syndrome using the trapezius muscle. Br. J. Plast. Surg. **37**(3), 422 (1984)
10. Wyvill, B., Mcpheeters, C., Garbutt, R.: University of Calgary 3-D computer animation system. SMPTE J. **95**(6), 629–636 (2015)

Optimization and Verification of Artificial Intelligence Image Recognition Algorithm for Transmission Line Defects

Shuang Lin[1], Jianye Huang[1], Yongqian Liu[1], Yan Yang[1], and Xiaofei Liu[2](✉)

[1] Electric Power Research Institute of State Grid Fujian Electric Power Co., Ltd., Fuzhou 350007, China
[2] Anhui Jiyuan Software Co., Ltd., Hefei 230088, China
1357560544@qq.com

Abstract. With the continuous improvement of intelligent operation and inspection requirements of transmission lines in power system, the popularization of online monitoring of transmission lines and uav inspection, a large number of transmission line defect image data are generated. However, traditional identification technology has low efficiency and inaccurate detection in the identification of transmission line defects. Based on transmission line defect source data, automatic transmission line defect image recognition technology based on artificial intelligence is studied. The transmission line defect identification technology is optimized and improved through automatic video image defect identification and defect sample annotation technology of transmission line components. The field scene verification proves that, compared with the traditional identification technology, the proposed adaptive algorithm for the selected area of transmission line defects can effectively improve the identification efficiency of transmission line defects, and the identification accuracy is close to 100%, which provides a guarantee for the stable operation of the power transmission system in the power industry.

Keywords: Image recognition · Transmission lines · Sample labeling · The region to be selected is adaptive

1 Introduction

From a technical point of view, although deep learning has achieved good results in visual recognition tasks at present, it performs well only on the basis of a large amount of label data, and requires a large number of parameters to be trained, consuming a large amount of manpower, which is cumbersome and costly [1, 2]. In practice, we will face the problem of less label data or even no label, so small sample learning occurs [3] and enhanced learning. At present, target detection methods are mainly divided into deep learning model based on regression and deep learning model based on the selected region. The deep learning model based on regression should establish the default box, object box and prediction box according to the idea of regression to train in a certain way

J. H. Abawajy et al. (Eds.): ICATCI 2022, LNDECT 170, pp. 787–795, 2023.
https://doi.org/10.1007/978-3-031-29097-8_93

[4, 5], the most typical model based on the region to be selected is CNN [6–8] series, such as regression based deep learning model and alternative region based deep learning model [9].

From the perspective of application, the transmission line defect detection is an important part of the electric power enterprise inspection work, but the lower level of the current application of artificial intelligence technology in the grid, current inspection detection methods are mainly invest a lot of human resources of artificial checking method is given priority to, low efficiency and operation staff training time, inspection time is not enough [10]. In recent years, with the development of technology, inspection robot, uav inspection began to be applied to transmission line inspection. However, due to the massive data of power transmission defects at present, it is difficult to meet the application effect required by the current transmission line defects detection of electric power enterprises. Therefore, it is urgent to study intelligent and safe image recognition detection methods to improve the inspection accuracy, so as to realize the intelligent transmission inspection upgrade of electric power enterprises. Therefore, the research on automatic image recognition technology of artificial intelligence in transmission scene is carried out and the algorithm is verified.

2 Remote Sensing Image Sample Acquisition

The consistent GLOBAL positioning system coordinate data of transmission lines were obtained through the panoramic inspection image UAV equipment, and remote sensing images were collected, including four single images with different directions and a Mosaic of panoramic images. In the process of image acquisition, the location information of the shooting point is obtained by the global Navigation satellite System (GNSS) equipment, and the location direction information is obtained by the panoramic verification image acquisition system. In order to meet the requirements of the project, the location coordinates collected need to be further transformed. Due to the use of panoramic fisheye lens shooting, image deformity correction is required. Generally, the farther the fisheye lens is from the photo, the better the effect will be. However, the farther the distance is, the image definition will be reduced. Therefore, a moderate distance should be selected to collect as clear and as few deformities as possible. Then the script is written based on CV2 library in OpencV to correct the deformity of the fisheye lens images. The corrected image restores EXIF information lost during the correction process by combining with the original image.

Sample types are interpreted according to the content of panoramic images, such as shock hammer defects, pressure equalizing ring defects, guide wire defects, insulator defects, etc. As the image is a panoramic image taken along the transmission line, unclear data image will be generated in the process of collection. When the clarity does not meet the requirements of sample selection, the unclear image will be marked, and local selection will be carried out according to the original impact clarity of the panorama from the image not taken along the transmission line.

According to the attitude data and position data of the currently selected image, the image azimuth Angle is calculated, and then the interpretation sample classification is preliminarily determined on the satellite remote sensing image based on the texture and color information of the remote sensing image. Then, the location points of the nearby image are selected according to the currently obtained data and the forward direction of the UAV is taken as the reference. Four images with azimuth angles of 60°, 120°, 180° and 240° were opened for evaluation. According to the four separate images generated at the same time, the original image that could be used as the sample was selected. Due to the large field Angle of the adopted panoramic single image reaching 120°, the main body is not obvious in the shooting process and there are many interference ground objects around, the function of dynamically adjusting the selection area of the single image is designed, which requires manual operation on the single image according to the texture, color and ground image of satellite remote sensing image. Adjust the selection box on the single image to determine the range and size of the specific sample field image.

3 Unselected Adaptive Algorithm

Several anchor points were marked in an image in advance, and each anchor point was regression and classified. Finally, N candidate frames were output in the stage of candidate frame generation. Loss function [11]. It can be simplified as:

$$L(p_l, t_l) = L_{cls}(p_l, p_l^*) + \lambda p_l^* L_{reg}(t_l, t_l^*) \tag{1}$$

Among them, t_l and t_l^* is related to the coordinates of the lth anchor point, and the probability that the 1th anchor point is predicted to be the target is P_l, when the lth anchor point is judged as a positive sample, the value of the 1th anchor point is 1.

The input size of the full connection layer of the LAN to be selected is, and N candidate boxes of sizes are generated. The calculated size of the full connection layer is, where K is the output size of the full connection layer. $n \times n \times N$ $n \times n$ $n \times n \times N \times k$. The calculation formula of the full connection layer is:

$$(x_1 x_2 \ldots x_n) \times \begin{pmatrix} w_{11} & \cdots & w_{1n} \\ \vdots & \ddots & \vdots \\ w_{m1} & \cdots & w_{mn} \end{pmatrix} + (b_1 b_2 \ldots b_m) = (y_1 y_2 \ldots y_m) \tag{2}$$

where, the weight matrix and bias are respectively: W and b. Each neuron in the fully connected layer has to be computed to get the final result. If the calculation time consumed by each neuron is T, the total calculation time consumed is:

$$T(N) = 7 \times 7 \times N \times kt \tag{3}$$

The number of candidate boxes is set as 6, 12, 18, 24 respectively in the process of defect detection. Through experiments and formula analysis, it can be seen that in terms of time consumption, the larger the selected area N is, the longer the prediction time will be consumed, but the detection effect of the model is less prone to the phenomenon of missing detection. Therefore, during the experiment, an appropriate region N should be selected to improve the overall effect of model prediction.

Every picture via the multi-value tag icon marking way to generate multiple candidate box, in order to solve the generation of alternatives stage area for different data output candidate box number N is too small to cause the leak and cause additional candidate box number N is too large to predict time consumption, adaptive algorithm is proposed to choose area, in view of the different data, adaptive should choose a reasonable number of candidate area. The ratio of the number of annotation candidate boxes in the image to the image area is defined as the block density of the image, i.e.

$$\rho_i = \frac{N_{proposal}}{S_{img}} \tag{4}$$

That is, the block density refers to the number of candidate boxes per unit area of the picture. The number of detected targets can be calculated based on the block density. Obviously, the higher the block density of the image, the more objects in the image. Generally, images of any size can be accepted in the area network to be selected. Different training sample images have different block densities. In order to detect all targets in the sample images as much as possible, the maximum block density value in the training sample is adopted, i.e. $\rho_{\max}$

$$\rho_{\max} = \max\{\rho_i | i \in [1, N]\} \tag{5}$$

In model prediction, formula (6) is used to calculate the N value of the current test picture in the candidate generation stage according to the length and width of the test picture, i.e.

$$N = \rho_{\max} \times W \times H \tag{6}$$

In order to accelerate the model prediction speed and hardly reduce the model performance, selected to calculate the final region to be selected. The model is sensitive to N value when the number of targets is very small, and the N calculated by the adaptive algorithm may not be enough to satisfy the real target value when the block density is very small. For this reason, N is set to the minimum value M to prevent the calculation of N from being biased when the number of targets is small. Therefore, the number of candidate boxes generated by the adaptive algorithm of the region to be selected is:

$$N = \max\{M, \rho_{\max} \times W \times H\} \tag{7}$$

4 Automatic Recognition of Transmission Line Video Image Defect Target

Using transmission line defect image hazard risk assessment technology, and to evaluate its risks, the establishment of the first to complete the model, the image recognition technology based on artificial intelligence model for training, form the transmission line defect image database, the database directly deployed in unmanned aerial vehicle (uav) embedded computing platform, Automatically detect whether the images captured by the UAV camera contain defective targets and conduct risk assessment. The expression of risk assessment is as follows:

$$\begin{cases} \mathrm{C} = P \times T \times D \\ S = C \times R \end{cases} \tag{8}$$

In the formula, the probability of transmission line defects is represented by P; The exposure time of defect location is represented by T. The amount of current loss caused by defects is represented by D; The transmission line defect risk value is represented by C; The risk value is represented by S; The defect risk adjustment coefficient is represented by R. The probability score and corresponding score interval are divided according to the above calculation results.

According to different molecular intervals, the transmission line defects are classified into grades and automatic identification of transmission line defects is completed. At the same time, in order to ensure the timely processing of defects, the processing process of related defect types can be recorded in the database, and artificial intelligence image recognition technology can be used to automatically process the defects existing in the database. If the defect problem does not meet the treatment conditions for the time being, a new warning should be issued and compared with the last warning. If the defect does not develop, the repeated alarm can be stopped and the defect processing and tracking process can be entered. If a defect develops, algorithmic analysis should be performed immediately to identify the defect.

5 Algorithm Flow

The main flow of the algorithm in this paper is shown in the following table. The overall process is the same as DCF algorithm. Firstly, the feature of multi-channel directional gradient histogram is extracted as the target characterization. Then, the obtained filter template is detected in the subsequent frames. Different from the traditional single detection region, this paper adopts the multi-region detection strategy to obtain the location information of the target in the current frame. Finally, according to the comprehensive

Table 1. Adaptive algorithm of alternative regions for transmission line defects

Input: image sequence I_1, I_2., I_n, the target initial position $P_0 = (x_0, y_0)$ **Output: p for each frame of image**$_t = (x_t, y_t)$
For t = 1, 2,…n, do: (1) Identify target center position (a) Use the target position p of the previous frame$_{t-1}$Determine the ROI position of frame T and extract HOG features (b) Calculation was performed in multiple detection areas to obtain multiple response graphs, and block density was determined according to Formula (4) (c) Extract the maximum value of multiple response graphs as the target center position P_t (2) Model updating (a) Calculate the accuracy and time of the response graph (b) When the comprehensive indexes of accuracy and time are improved, background blocks are extracted to update the model (c) When the comprehensive index of accuracy and time decreased, the corresponding background block was selected as a negative sample to train the model (d) Update the model according to Formula (1)

index of the accuracy and time of the response graph (PT (N)), the specific background region is selected to train and update the model (Table 1).

6 Experimental Demonstration and Analysis

Data sets suitable for transmission scenarios are selected, and data sets for comparative tests are collected from identification items such as bird nest, foreign body, insulator self-explosion, bolt missing pin, smoke volcano, tower vehicle, ground wire broken, pressure equalizing ring damage, identification plate damage, bird protection facilities, shock hammer damage, nut missing and so on. Some samples used in the test are shown in the figure below (Fig. 1):

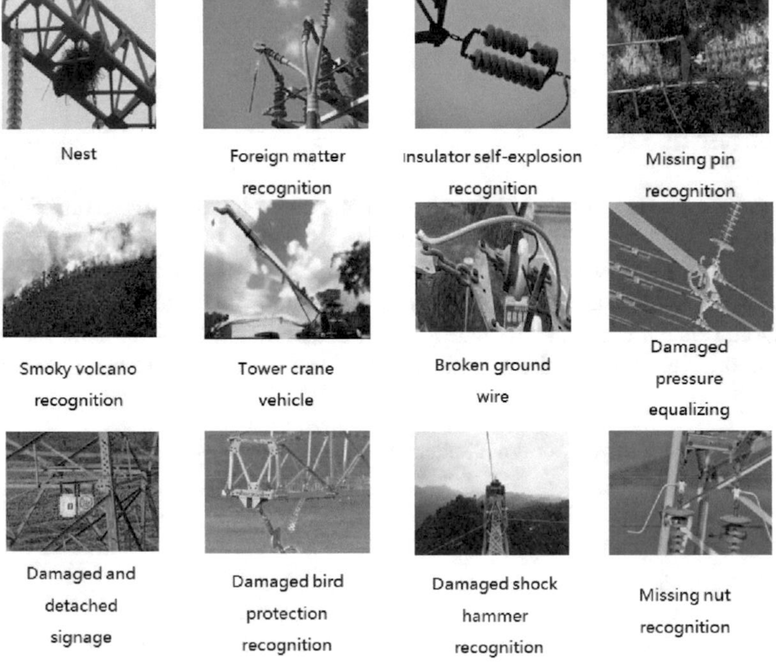

Fig. 1. Experimental sample diagram

The adaptive detection model and the conventional detection model are combined [4, 12, 13]. The prediction accuracy of the same set of data is compared. The result is as follows (Fig. 2):

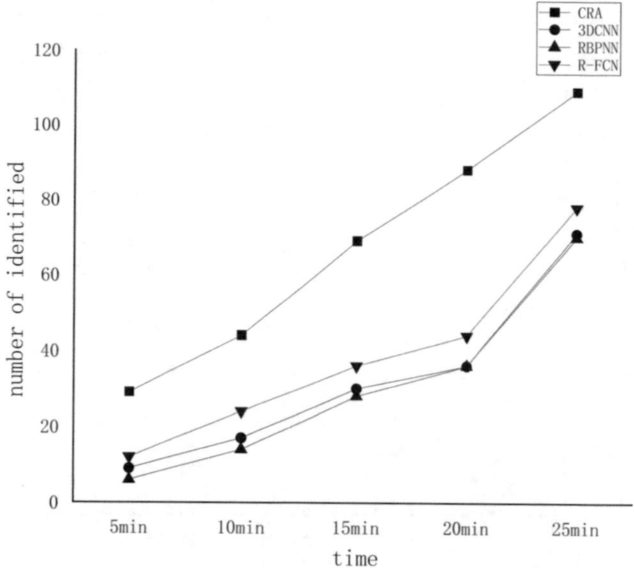

Fig. 2. Time-accuracy identification effect diagram

Transmission line images were collected and integrated into the existing transmission line UAV inspection video image database to complete the response deployment. Transmission line defects were detected based on Python, and the adaptive algorithm for selected areas was verified. The efficiency of the two identification technologies was compared, and the comparative test results were obtained as shown in Fig. 3.

According to Fig. 3, within 25 min, the adaptive algorithm of the area to be selected completed the identification of 106 defective transmission lines and correctly identified all the defects. Only 73 defective transmission lines have been identified by traditional identification technology, and all the permissions have not been correctly identified. Can be seen from Fig. 3, recognition, in 25 min to choose area algorithm of adaptive algorithm to identify the best effect, this is because the paper adopts the improved adaptive algorithm is effective for district of transmission line defect recognition efficiency and accuracy, Fig. 3 card of contrast experiment of the proposed adaptive algorithm recognition rate is higher than the traditional technology, It can find out all the defects of transmission lines efficiently, achieve higher accuracy and have higher application value in the same time.

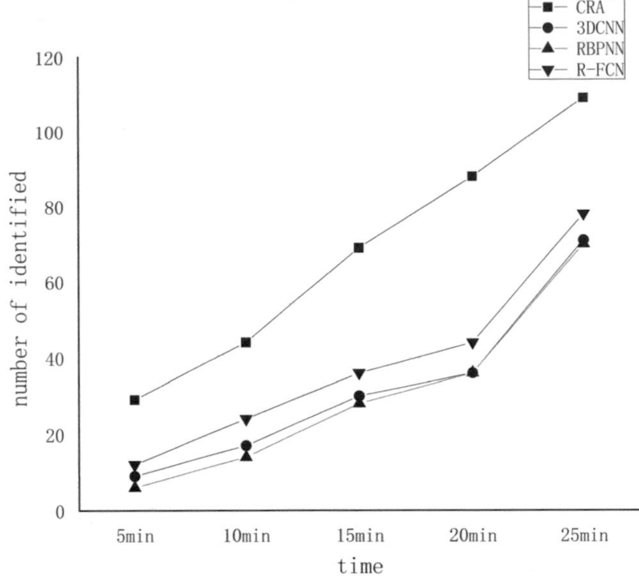

Fig. 3. Effect diagram of iterative identification

7 Conclusion

The automatic identification technology of transmission line defects based on AI image recognition is studied, and the identification of transmission line defects based on defect image region adaptive method is proposed to accelerate model prediction speed without reducing the accuracy of the model. The model can import single picture, multiple pictures and videos. It can provide the recognition function of video, single picture or batch picture for maintenance personnel. The fault location identification of transmission lines based on defect image region adaptive method is proposed, and the effectiveness of this method is verified by identifying the actual transmission line defect images of power grid companies. It is further proved that AI image recognition technology can be introduced into the substation construction scene to improve the efficiency of substation defective equipment identification.

References

1. Wu, S., Zhong, S., Liu, Y.: Deep residual learning for image steganalysis. Multimedia Tools Appl. **77**(9), 10437–10453 (2017). https://doi.org/10.1007/s11042-017-4440-4
2. Srinivasu, P.N., SivaSai, J.G., Ijaz, M.F., et al.: Classification of skin disease using deep learning neural networks with MobileNet V2 and LSTM (2021)
3. Xian, Y., Lampert, C.H., Bernt, S., et al.: Zero-shot learning – a comprehensive evaluation of the good, the bad and the ugly. IEEE Trans. Pattern Anal. Mach. Intell. **99**, 1 (2017)
4. Molchanov, P., Gupta, S., Kim, K., et al.: Hand gesture recognition with 3D convolutional neural networks. In: Computer Vision and Pattern Recognition Workshops. IEEE (2015)

5. Di, W., Pigou, L., Kindermans, P.J., et al.: Deep dynamic neural networks for multimodal gesture segmentation and recognition. IEEE Trans. Pattern Anal. Mach. Intell. **38**(8), 1 (2016)
6. Xie, F., Yu, M.U., Guan, Z., et al.: Oral leukoplakia (OLK) segmentation based on Mask R-CNN with spatial attention mechanism. J. Northwest Univ. Nat. Sci. Ed. (2020)
7. Jiang, Q., Yan, J.: To offload selective search: improving performance of fast R-CNN based on a mobile cloud offloading framework. In: 2019 International Conference on Wireless Communications Signal Processing and Networking (WiSPNET) (2019)
8. Yan, X., Yang, Y., Lu, G.: A target detection algorithm based on faster R-CNN. In: Shi, S., Ye, L., Zhang, Y. (eds.) Artificial Intelligence for Communications and Networks. AICON 2020. Lecture Notes of the Institute for Computer Sciences, Social Informatics and Telecommunications Engineering, vol. 356. Springer, Cham (2021)
9. Yin, X., Yan, L.: Image target detection based on depth convolution neural network. Ind. Control Comput. **30**(4), 3 (2017)
10. Zhou, F., Fang, M., Ma, Y., Pan, H.: Fast detection method of transmission line defects based on YOLO v3. Yunnan Electric Power Technol. **48**(4), 112–116 (2020)
11. Zagoruyko, S., Komodakis, N.: Deep compare: a study on using convolutional neural networks to compare image patches. Comput. Vis. Image Understand. **164**, 38–55 (2017)
12. Huang, X., Zhang, X., Zhang, Y., et al.: Transmission wire defect state recognition based on RADIAL basis probabilistic neural network. Autom. Power Syst. **673**(3), 278–291 (2020)
13. Zou, J., Xu, R., Xi, Y.: Application of deep learning in intelligent defect identification of transmission line engineering acceptance. Jiangxi Electric Power **44**(2), 5–9 (2020)

Cyber Intelligence for Big Data

Dynamic Construction of Pipeline Network Map Data

Qian Li(✉), Yongyi Fang, Baoan Liu, Kun Xing, Lijun Wang, and Hengjian Qian

State Grid Shijiazhuang Power Supply Company, Shijiazhuang, Hebei, China
shmily@sgspsc.com.cn

Abstract. In today's society, more and more cities have begun to develop research on pipeline network map data. Research on city map data can not only speed up the pace of urban informatization, but also allow more cities to have their own urban infrastructure. A comprehensive map information database is used to establish a map information platform based on the database to serve people's lives. In order to maintain the real-time nature of the map information platform, it is necessary to continuously dynamically maintain and update the city pipeline network map database. This paper analyzes the principles and ideas in the field of dynamic information update of map data, uses expert template mapping technology, improves the shortest data association algorithm through Wang's distance and fuzzy-adaptive data association algorithm, and conducts related experimental research. The results show that the improved data association algorithm is more suitable for outdoor conditions with more complex environmental characteristics, and its data association accuracy rate is higher.

Keywords: Map Data · Data Association · Shortest Neighbor Algorithm · Information Platform

1 Introduction

The dynamic construction of urban pipeline network map data has become an important field of informatization construction in many cities. Through the application of computer technology and map information technology, the urban pipeline network map database has developed rapidly, and has served many industries in the city. The construction of the map database is the key core of the dynamic construction of urban pipeline network map data. This database provides a scientific and unified map information sharing platform for the city, sorts out the information resources of the city, and serves the public, which is convenient to solve problems such as map coordinate errors that occur when people travel have made the city's information construction a good and efficient starting point [1]. The urban pipeline network map data system includes basic map information database, map information sharing platform, map information service and other functions. The goal is to have various scales, resolutions, types and other performances on the map information system, and establish a standard and authoritative The pipe network map data system can provide people with high-quality map information application services [2].

J. H. Abawajy et al. (Eds.): ICATCI 2022, LNDECT 170, pp. 799–806, 2023.
https://doi.org/10.1007/978-3-031-29097-8_94

Therefore, the pipeline network map database is an important factor in the construction of an informatized city, and an important data foundation for the establishment of an informatized city [3].

Some scholar Fomichev believes that at the current stage, the development speed of various science and technology in our country is constantly improving, and it has been more maturely applied in all walks of life, and geographic mapping technology is one of them. The relevant personnel can use the integrated geographic information system to effectively collect and store the attributes of various spatial data, which to a large extent also improves the work efficiency and quality, and promotes the development of geographic cartography technology [4]. Doriguzzi-Corin and others believe that map data plays an irreplaceable role in human social and economic activities. Many industries cannot do without the support of map data. It has the characteristics of easy updating, copying and distribution [5]. Ismail and others believe that applying data mining to map data can dig out many aspects of knowledge from map data. Through statistical analysis, we can study the variation and interrelationship of various phenomena in space or within a certain time range, and find out the internal laws of things; through statistical analysis, we can study the spatial distribution of various phenomena, interconnections, dynamic changes, and establish various mathematical models, etc. [6].

So far, many countries have carried out research on map data update technology, and have achieved good research results. Using these research results, it can basically be used in map information database update, real-time map information, accuracy of map information, etc. Nevertheless, the update speed of map information is not fast enough, and it is still a difficult problem to be solved [7]. After exploring the map information update technology at home and abroad, this paper analyzes the problem that the map information update is not fast enough, and proposes a solution to the rapid and real-time update of map information, and the key technology used in the process of research and the process has been carefully studied and thought [8].

2 Research Method of Dynamic Construction of Pipeline Network Map Data

2.1 Principles of Online Update of Map Information

The data update of the city pipeline network map data information sharing platform must be timely and accurate to reflect the real-time nature of map data. Map data is a critical factor for the development of urban informatization. Only real-time map data can be achieved. In order to make full use of the urban pipeline network map data dynamic system. The updated map information must be accurate on the sharing platform and be replaced with the previous map information accuracy, otherwise the system will have problems when the map information data is superimposed [9]. Because the content of the map information sharing platform database is very large, and the data attributes are also very complicated. When updating, not only the map information must be updated, but also the attributes of the data, so that the two kinds of information can be updated at the same time [10]. In the urban pipeline network map data dynamic system, there is a large amount of map information, and the unity of the graphic data can be maintained while updating, so that accurate analysis can be carried out in these data and scientific decisions can be made.

2.2 The Overall Idea of Online Update of Map Information

The characteristics of map information determine that the data must reflect changes in the real world in a timely manner. Only in this way can it meet the needs of all walks of life. The operator determines the spatial range of the elements to be edited on the client by box or click, and transmits the range to the server. The server uses the spatial analysis method to obtain the elements to be edited and returns the elements, and the operator can treat the elements to be edited. For editing, you can change the spatial attribute information by moving, adding points, deleting points, etc., and you can also directly change the field attribute information, or even delete elements. When the operator finishes editing the elements to be edited, they can submit them to the server in batches, and the server will process each element separately, establish a temporary intermediate library, and temporarily store the processing results in the intermediate library. After the operator finishes editing and saves all the elements to be edited, the submit function triggers the operations of updating the data source and saving the history database. The server reads all the temporarily stored operation results of the intermediate library and gradually updates the current data. After the update data is completed, the server will automatically copy the records of the intermediate library to the history database and mark the update of each record Time in order to be used as a historical traceback of the data. Finally, the server clears the records in the intermediate library to avoid wasting unnecessary memory resources.

2.3 Automatic Mapping Technology of Expert Templates

The existence of a map element is mainly expressed by its spatial data and attributes data. According to the standard of the map information public platform data set, the update of spatial data and attribute data only represents the end of the update of the data source. If it is to comply with the map information. The requirements of the public platform data set also need to map the updated elements according to the original map template of the data source. Therefore, it is more appropriate to adopt the automatic map technology based on the expert template. Expert template automatic mapping technology is to make many templates after analyzing the structure and coding of the map data for consistency. As for the data of the same proportion, the mapping work of these data can be completed by relying on a template. Due to the similarity of the data structure and coding of each standard map, including the number of layers, names, and legend content, the difference is not very large. The operation of the next one is a lot of repetitive work with the previous one, in order to avoid a lot of repeated labor, developed multiple map expert templates, built-in cartography technology, to speed up the process of map production.

3 Process of Research on Dynamic Construction of Pipeline Map Data

3.1 Improved Shortest Data Association Algorithm Based on Wang's Distance

In the entire calculation process, the condition for realizing state estimation is data association. The accuracy of association directly affects the accuracy of positioning and the accuracy of map information, which plays a vital role. The shortest algorithm is easy to implement due to its simple principle. However, the correlation condition of the shortest algorithm is Wang's distance. The shortest Wang's distance is selected in the best matching. The accuracy of data association is low and it is easy to cause estimation errors. Regarding the limitations and shortcomings of the shortest algorithm, many researchers have given improved algorithms. However, the method is to combine other correlation algorithms. Although the accuracy of data correlation is improved, the difficulty of calculation and the running time of the system also increase. Compared with the shortest algorithm, the improved algorithm proposed in this paper has the same computational difficulty and running time, but it has a significant improvement in the accuracy of data association.

3.2 Implementation of Improved Wang's Distance Shortest Data Association Algorithm

According to the principle of data association, it can be known that data association calculates the probability that the detected feature is associated with the map feature. Therefore, this article introduces conditional probability $P(n_k|s_k, g_k)$, among them n_k is the data association at time k, s_k is the state vector of the map, g_k is the detected feature value. From the above definition, maximize the conditional probability to get the maximum degree of association n_k. According to the Bayesian rule:

$$P(n_k|s_k, g_k) \propto P(n_k|s_k)P(g_k|s_k, n_k) \tag{1}$$

In the formula, $P(n_k|s_k)$ is the prior probability, which can be regarded as a constant. Set the $h(s_k, n_k)$ observation model, s(n2) is the sum of variance of the observed value and the predicted value, the calculation can be obtained:

$$s(n_k) = p_i^f + q_j^g \tag{2}$$

According to the Gaussian distribution, the formula (1) is:

$$-\log\{p(g_k|s_k, n_k)\} = \frac{1}{2}(d \log 2\pi + d_{min}^m + \log s(n_k)) \tag{3}$$

From the above formula (3), it can be seen that the degree of correlation between the existing features in the map and the observed value is not only related to the Wang's distance, but the variance of the observed value and the predicted value can also affect the accuracy of the data association.

3.3 Fuzzy-Adaptive Data Association Algorithm

In order to solve the problem of low correlation accuracy in a complex environment with greater uncertainty, and to effectively prevent false road signs, this paper uses fuzzy-adaptive algorithms to improve correlation accuracy and positioning accuracy on the basis of fuzzy rules. Applying fuzzy logic rules to the data association between feature observations and estimated values will improve the data association effect; in order to effectively prevent false road signs, the smallest road signs in the environment are tracked according to the observation data, and the two observations with the shortest Euclidean distance One of the road signs is added to the system state as a virtual new road sign, and the other observation is associated with the virtual road sign through fuzzy rules. Simulation experiments show that this method can deal with data association problems, and has excellent performance in dealing with changes in the environment and noise, so that it can deal with false road signs and other problems, and reduce the rate of observation loss. According to the observation data, the minimum distance between road signs in the environment is tracked, virtual association is introduced, and the Zengguo threshold and association threshold are calculated in an adaptive manner.

4 Analysis of the Results of Changes in the Accuracy of Data Association

4.1 Keep the Observation Error Constant, Compare the Change of the Correct Rate of Data Association with the Control Error, and the Results are Shown in Table 1 and Fig. 1

Table 1. Changes in the accuracy of data association with control errors

Control error magnification	Fuzzy algorithm correct rate (%)	Correct rate of adaptive algorithm (%)	Fuzzy-adaptive algorithm correct rate (%)
2	97	92	97
4	95	84	96
6	94	77	94
8	92	70	92

It can be seen from Table 1 and Fig. 1 that when the control error is relatively small, the correlation accuracy of the two algorithms based on fuzzy logic is relatively high; but as the control error continues to increase, the shortest neighbor algorithm based on Wang's distance is correct. The rate drops rapidly. In comparison, the adaptive algorithm based on Wang's distance drops more slowly, but the overall correlation accuracy is not high; only when the control error increases to a certain degree, the accuracy rate decreases. Most of the reason is that as the control error increases, the impact on the uncertainty and ambiguity of the update also increases. The correlation algorithm based on Wang's

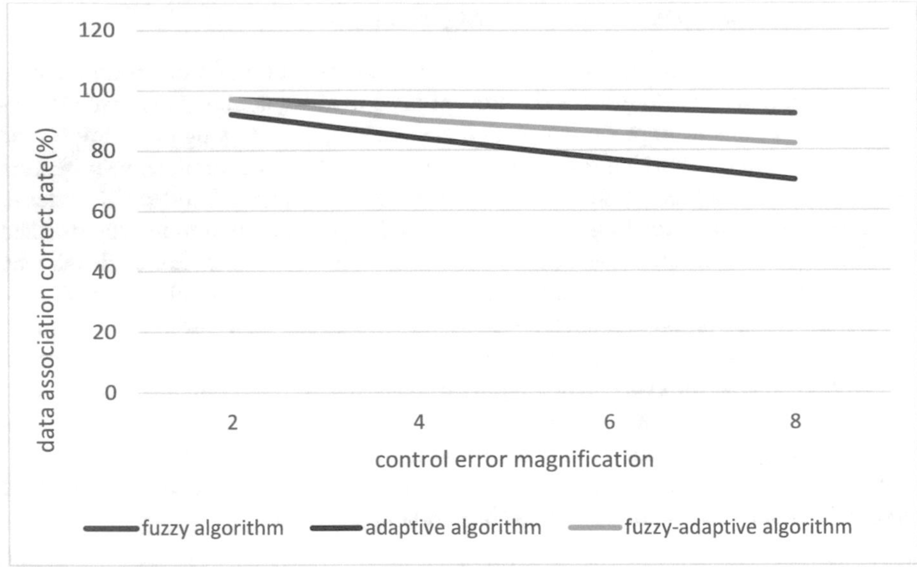

Fig. 1. The correct rate of data association changes with control error

distance does not capture this information, which causes the loss of connection. Failed, and the two algorithms based on fuzzy logic make full use of this information, and the correct rate of association has been improved.

4.2 Keep the Control Error Constant, Compare the Change of the Data Association Accuracy Rate with the Observation Error, and the Results are Shown in Table 2 and Fig. 2

Table 2. Changes in the accuracy of data association with observation errors

Observation error magnification	Fuzzy algorithm correct rate (%)	Correct rate of adaptive algorithm (%)	Fuzzy-adaptive algorithm correct rate (%)
2	96	92	96
4	94	88	93
6	93	86	92
8	91	83	88

From the comparison of Table 2 and Fig. 2, it can be seen that for the accuracy of data association, increasing the observation error has a greater impact than increasing the control error. In terms of the impact on the three algorithms, increasing the observation

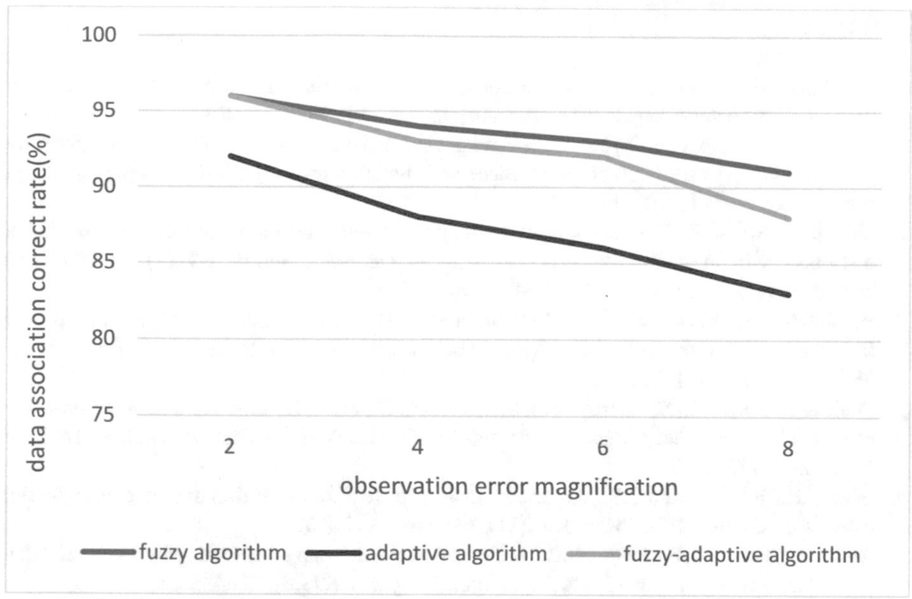

Fig. 2. The correct rate of data association changes with the observation error

error and increasing the control error. it's the same. The above results show that in the case of relatively complex environmental characteristics, the fuzzy-adaptive data association algorithm can perform data association well, and has good anti-interference ability and robustness.

5 Conclusions

In order to solve the problem of low correlation accuracy in a complex environment with large uncertainties and effectively prevent false road signs, this paper adopts fuzzy-adaptive algorithm to improve the correlation accuracy and positioning accuracy. Applying fuzzy logic rules to the data association between feature observations and estimated values will improve the data association effect; in order to effectively prevent false road signs, this method tracks the smallest road signs spacing in the environment based on the observation data, and the two with the shortest Euclidean distance One of the observation road signs is added to the system state as a virtual new road sign, and the other observation is data-associated with the virtual road sign through fuzzy rules. Simulation experiments show that this method can deal with data association problems, and has excellent performance in dealing with changes in the environment and noise, so that it can deal with false road signs and other problems, and reduce the rate of observation loss.

Acknowledgments. This work was financially supported by Research and Application of Underground Pipeline 3D Simulation Technology Based on Cable Pipeline Robot.

References

1. Sandstrom, R., Uwacu, D., Denny, J., et al.: Topology-guided roadmap construction with dynamic region sampling. IEEE Robot. Autom. Lett. PP(99), 1 (2020)
2. Putnins, M., Campagne, O., Mager, D.E., et al.: From data to QSP models: a pipeline for using Boolean networks for hypothesis inference and dynamic model building. J. Pharmacokinet. Pharmacodyn. **49**(1), 101–115 (2022)
3. Ovchinnikova, M.F.: Features of transformation of humic substances in Soddy-Podzolic soil disturbed by main pipeline construction. Moscow Univ. Soil Sci. Bull. **74**(1), 33–39 (2019). https://doi.org/10.3103/S0147687419010095
4. Fomichev, P.A., Fomicheva, E.V.: Estimation of dynamic characteristics of vibration protection systems of a new type for water transport vessels. IOP Conf. Ser. Earth Environ. Sci. **988**(2), 022076 (6p.) (2022)
5. Doriguzzi-Corin, R., Scott-Hayward, S., Siracusa, D., et al.: Dynamic and application-aware provisioning of chained virtual security network functions. IEEE Trans. Netw. Serv. Manage. **17**(1), 294–307 (2019)
6. Ismail, H., Balachandran, B.: Algorithm fusion for feature extraction and map construction from SONAR data. IEEE Sens. J. **15**(11), 6460–6471 (2015)
7. Agyeman, B.T., Bo, S., Sahoo, S.R., et al.: Soil moisture map construction by sequential data assimilation using an extended Kalman filter. J. Hydrol. **598**(6), 126425 (2021)
8. Kim, S., Kim, H., Min, J.K.: An efficient parallel similarity matrix construction on MapReduce for collaborative filtering. J. Supercomput. **75**(1), 123–141 (2019)
9. Ahmed, M., Karagiorgou, S., Pfoser, D., Wenk, C.: A comparison and evaluation of map construction algorithms using vehicle tracking data. GeoInformatica **19**(3), 601–632 (2014). https://doi.org/10.1007/s10707-014-0222-6
10. Monroe, J.G., Allen, Z.A., Tanger, P., et al.: TSPmap: a tool making use of traveling salesperson problem solvers in the efficient and accurate construction of high-density genetic linkage maps. Biodata Mining **10**(1), 38 (2017)

Product Matching Optimization Model Based on Integrated SVM Data Stream Classification Algorithm

Zhiding Zhang[1(✉)] and Khadijah Mansour[2]

[1] School of Economics, Shanghai University, Shanghai, China
zd0662@126.com

[2] University of Garden City, Khartoum, Sudan

Abstract. With the continuous innovation of products, investors are increasingly demanding higher-level products. How to reduce transaction costs has become an urgent problem to be solved at present. This paper optimizes and analyzes the matching of products based on the integrated SVM data stream classification algorithm. And the purpose is to improve the reasonable distribution of products. This article mainly applies the statistical method to the research of the product matching optimization model, and grasps the key points of product matching optimization through comparison. The survey results show that 114 out of 180 people supported the four strategies proposed in this article. These data prove that the recommendations made in this article can be supported by the public.

Keywords: Integrated SVM · Data Stream Classification Algorithm · Matching Optimization

1 Introduction

With the continuous development of the financial industry, the classification of financial products has become more and more extensive. The matching of financial products is very important in the design and optimization of financial service products, and it is an important factor that affects user experience and the overall competitiveness of financial institutions. More and more financial institutions have gradually introduced SVM-based classification models to solve various risk management problems.

There are many research theoretical results on the integrated SVM data stream classification algorithm and product matching optimization. For example, Vinod Desai et al. claim that SVM algorithm is efficacious in improving corporate customer contribution model [1]. When testing the model for predicting stock movement, Kolasani and Assaf found that SVM reduces the amount of data learning, and has higher classification accuracy and practicability than the traditional simple model, in the result, SVM performs the best with the highest accuracy amongst five possible models [2]. According to Meltem Yontar et al., SVM is a small data set created under the classification problem, which can achieve high-latitude nonlinear classification [3]. Therefore, the SVM data stream

J. H. Abawajy et al. (Eds.): ICATCI 2022, LNDECT 170, pp. 807–814, 2023.
https://doi.org/10.1007/978-3-031-29097-8_95

classification algorithm has obvious advantages in data processing, and the design of the matching optimization model of financial products can also be carried out by using this algorithm.

This article first studies the data stream classification algorithm and has a basic understanding of it. Secondly, the support vector machine is researched and elaborated in detail. Then analyze the basic theories and optimal allocation of financial products and resources. Finally, suggestions and strategies are given to the financial product matching optimization model through investigation.

2 Financial Product Matching Optimization Model Based on Integrated SVM Data Stream Classification Algorithm

2.1 Data Stream Classification Algorithm

A data stream is a sequence of a large number of data elements that arrive continuously in real time. It has the following characteristics:

Data comes in very quickly, and if the system cannot complete the processing quickly, a lot of useful information will be lost. The data source from the data stream continues to arrive, as long as the data source is still generating data, the data will not be interrupted. The order of arrival of data is not controlled by humans, and the algorithm can only be processed according to the order of arrival. The data changes dynamically, that is, the implicit concept drift. Commonly used classifier construction methods include decision tree classification, rule-based classification, support vector machines and naive Bayes classification. As an important task form in data stream mining, data stream classification has a very wide range of applications. And we need to classify the data stream. First, the learning algorithm is used to establish a basic classifier for the currently arriving data block, and then it is added to the integrated system, and then the basic classifier is integrated with certain rules. According to the integrated classification result, the classifier is voted by majority or weighted, and the classification error is deleted. The classifier with the smallest weight or the classifier with the smallest weight realizes the update of the integrated classification system [4, 5]. The data stream processing model is shown in Fig. 1.

Integrated learning is the use of a set of learning algorithms to learn the objectives, and in a certain way to integrate the learning results to obtain a better learning effect than single algorithm learning. If the correct class label is entered, the integrated classifier will generate the correct classification result. Assuming that the ensemble classifier K contains individual classifiers, the classification error rate of each individual classifier is $p < 0.5$ [6, 7]. Then for all data, a is the probability that there are l individual classifiers in the set classifier to produce the wrong class label:

$$\frac{K!}{l!(K-l)!}s^{l}(1-s)^{l-l} \tag{1}$$

The probability of the ensemble classifier outputting the wrong label is:

$$\sum_{l=\left\lfloor\frac{K+1}{2}\right\rfloor}^{K}\frac{K!}{l!(K-l)!}s^{l}(1-s)^{l-l} \tag{2}$$

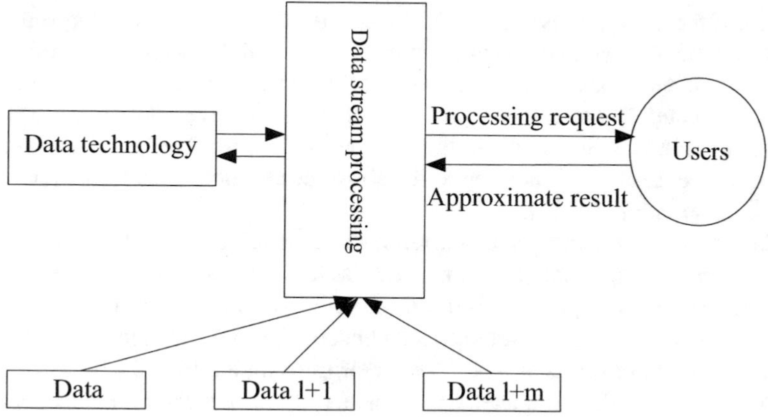

Fig. 1. Data stream processing model

Among them, a is the largest integer not greater than a. The generation of individual classifiers is to establish a single classifier with higher classification accuracy and greater difference through a certain method.

2.2 Support Vector Machine

Support vector machine (SVM) is a new machine learning method that finds the best sample set by analyzing training data. In the process of classification and clustering, statistical theory, statistics and other related principles are used to quantify it. Because of its extensive theoretical foundation and application scenarios and easy implementation, it has been widely used to deal with non-linear programming problems. Support vector machines are mainly trained based on data samples. The data set is mapped into the space and a new individual set is formed by a linear combination of all the attribute values in the feature point marker space within a certain degree. It can classify things in the real world and analyze the correlation between different data categories based on the results obtained. There are two types of support vector machines: linear and nonlinear. Currently the most widely used are linear models and statistical models. The training sample set model, kernel function, and kernel optimal estimation can be used for multi-dimensional data analysis, but its disadvantage is that it has strong dependence on feature space and low classification accuracy. The SVM algorithm has a certain degree of difficulty in theory. First, you need to redistribute the attributes of the sample. Secondly, we must find the appropriate characteristic parameter value and the best characteristic point. Finally, the results obtained after classification are further processed and output information for reference by decision makers [8, 9].

Support vector machine (SVM) is an important method in training and testing classification models. It can concentrate the data to be processed into a set of features so that it can be represented by another location. Through the above analysis of the amount of classified data, we found that in the matching process of financial products, the first thing to do is to ensure that the feature points correspond to the target information. That is, first obtain all the important features of the location area where the sample attribute is located. If there are a large number of categories and categories of the same nature that have the same or similar characteristics or have a high degree of correlation, it is necessary to select a data set with the closest distance, a smaller number and a larger number as the candidate reference object for matching. The characteristics and attributes of financial products will change at different points in time. Therefore, we can regard these data as a collection. According to the above description, each sample has its own unique characteristics [10, 11].

In the financial product matching optimization model, because different algorithms process different sample data types, it is necessary to classify all samples. Since the classification of financial products is to find the most similar features in the known data, we need to select the feature vector that can reflect the attribute information of the sample object as the objective function. The SVM algorithm can improve the processing speed and accuracy by selecting characteristic variables that have a certain independence and spatial distribution type, non-partiality, observable repetitive type, or non-relevant type. In addition, it can extract all the parameters of the classification model and convert them into standard data. By integrating the SVM data model and related clustering algorithms, financial products are matched [12].

2.3 Financial Products and Resources

The constituent elements of financial products include credit line, repayment period, guarantee period, etc. The appropriateness of these elements is an important factor in evaluating whether the product is compatible with the company's strategic development. Products are an important part of corporate strategy. Therefore, slightly chaotic and redundant financial products are already an important factor restricting the company's future development. First of all, the amount of credit line has a great influence on the choice of borrowers. If the funding needs are not met, the ability to purchase the product will decline. The price of the product is the most worrying issue for borrowers and lenders. As long as there are enough interest rates, users who need funds will buy products from financial companies.

Financial resources mainly refer to the collection of a series of objects interacting with the subject and object of financial services in the number, structure, scale, distribution, and effects of the financial services in the financial field. On the whole, understanding the concept of financial resources needs to be developed from three aspects: One is to define financial resources as a basic and core resource. Financial resources are all kinds of monetary capital that we can come into contact with in our daily lives. They are fundamental and are the core resources that can promote the development of production. The second is that financial resources can be integrated among entities to promote the development of the real economy. Among social economic entities, organizations, enterprises, and systems need to use financial resources as the medium and financial organizations and tools to form the environment and conditions for the operation of entities. The third is that the function of financial resources is holistic. Financial resources can organically operate monetary capital borrowing, adjust monetary capital, and adjust industrial structure. Financial resources are mainly savings resources, insurance resources and loan resources. Looking at the allocation function of financial resource allocation from a macro and micro perspective, it can be divided into the macro function and the micro function of financial resource allocation. The internal micro-function mechanism of financial resources is manifested in the fact that financial institutions act as intermediary platforms for financial resources to obtain temporarily unused surplus funds in the hands of savers, pay a certain fee, and loan the collected funds to manufacturers in need of funds. To charge a certain amount of loan interest and obtain certain benefits from it, the maximization of the internal allocation efficiency of financial resources is a state in which in the financial institution system, under the established transaction costs, the capital demanders can obtain the required funds at the lowest cost. From a broader macro perspective, the efficiency of financial resource allocation is obtained by relying on the effective operation of finance. The allocation of financial resources determines efficiency and is closely related to the financial system, financial system, social development, social environment, and regional factors during the same period. Analyzing from the perspective of the optimal allocation of financial resources, in the development process of the new type of financial industrialization, there is a huge demand for financial resources. Therefore, government financial institutions should tap the financial needs of the financial industry and take appropriate measures in due course to promote their own development.

Therefore, government departments must not be confined to traditional concepts, but need to broaden their thinking and horizons to support the development of new-type rural areas. Promote the rapid development of financial industrialization in the region through optimized allocation of financial resources. The redistribution reform of rural financial allocation must first clarify the property rights system. The new financial industrialized financial allocation system must conform to the laws of the market and be able to objectively reflect the institutional requirements of property rights.

3 Design Investigation of Financial Product Matching Optimization Model

3.1 Formulate a Perfect Optimization Plan

First of all, it is necessary to establish a good management, set up an office to manage the development of financial modernization, and the office appoints a person to be responsible for the construction of financial resource allocation. Secondly, the management department takes the lead to draw up the first draft of the optimization plan, and submit it to the regional office for review and revision. Once again, a panel of experts will be composed of review experts. Finally, after the evaluation and demonstration meeting, the revision of the plan is completed and the final draft is completed.

A sound financial support system will play a strong role in promoting the development of regional financial industrialization. First, we must clarify the composition of the financial resource system that can support the development of financial industrialization. Strive to establish a system that can maximize the role of various entities in promoting the development of financial industrialization. Second, we must establish a financial system based on the different characteristics of each subject. To build an effective financing service system, it is necessary for the government to partially enhance the city's overall financing level. As a component of the modern service industry, the financial industry can better play its supporting role in the economy and society on the one hand, and at the same time can further promote the upgrading of the financial industry.

3.2 Questionnaire Design

In order to accelerate the construction of the optimized matching model, this article intends to analyze its model from the perspective of popular sentiment. Therefore, this article adopts a questionnaire survey to study this topic. In this questionnaire design, the core topic is the optimal configuration of financial products. In addition, this article also understands the identity and ability of the person filling in the questionnaire.

3.3 Investigation Process

After setting up questions for this questionnaire, an offline questionnaire was used to analyze the reliability and validity of the questionnaire. On this basis, 200 people were invited to answer the online questionnaire. In the end, 190 questionnaires were collected, of which 180 were validly answered. The questionnaire lasted a total of one week. One week later, sort out and analyze the questionnaire.

4 Analysis of Survey Results

4.1 Suggestions and Countermeasures for Optimizing Financial Products

From the questionnaire survey in this article, it can be analyzed and found that the optimal allocation of financial products can mainly start from the aspects of product line adjustment, price optimization, risk control and marketing reform. The specific support numbers are shown in Table 1:

Table 1. Suggestions and countermeasures for financial product optimization

	Agree	Disagree	General
Product line adjustment	21	10	20
Price optimization	36	3	10
Risk control	30	1	8
Marketing reform	27	4	10

As shown in Fig. 2, we can see that the number of supporters ranks first in price optimization, followed by risk control. This shows that the price and risk of financial products are the two aspects that people are most concerned about. Ranked third is marketing reform, which shows that the attractiveness of financial products is not enough and requires staff to make rectifications.

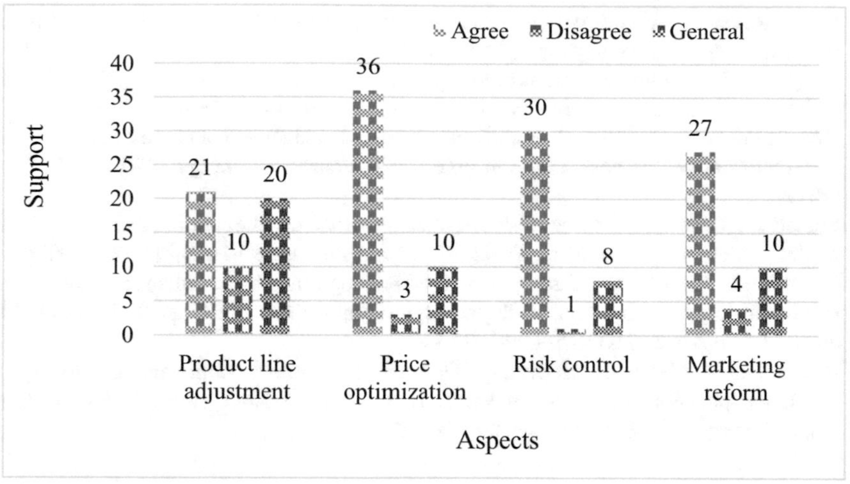

Fig. 2. Suggestions and countermeasures for financial product optimization

5 Conclusion

The research in this paper is based on the integrated SVM algorithm, through matching financial products, to a certain extent, to achieve the goal of reducing risks and increasing returns. Due to the rapid update of Internet technology and computer network information technology, financial products produce massive amounts of complex information. The integrated SVM algorithm shows its talents in the data processing process. Through the investigation, it is found that the design of the allocation optimization model of financial products can start from its price and risk monitoring capabilities. This article also needs to find more excellent data calculation methods to make it more suitable for financial products for precise matching and evaluation.

References

1. Desai, V., Ullagaddi, S.B., Odeyar, V.A.: The key points of measuring the profit contribution of different businesses of bank customerss. J. Glob. Econ. Bus. Finance **3**(11) (2021)
2. Kolasani, S.V., Assaf, R.: Predicting stock movement using sentiment analysis of twitter feed with neural networks. Journal of Data Analysis and Information Processing **08**(04), 309–319 (2020)
3. Yontar, M., Namli, Ö.H., Yanik, S.: Using machine learning techniques to develop prediction models for detecting unpaid credit card customers. J. Intell. Fuzzy Syst. **39**(5), 1–15 (2020)
4. Poongodi, M., et al.: Prediction of the price of ethereum blockchain cryptocurrency in an industrial finance system. Comput. Electr. Eng. **81**(C), 106527 (2020)
5. Zhao, Z., Yin, B.: Financial service quality model optimization and empirical research. Journal of Beijing University of Aeronautics and Astronautics (Social Science Edition) **01**, 102–107 (2017)
6. Abbasi, W.A., et al.: Research on measurement of supply chain finance credit risk based on internet of things. International Journal of Distributed Sensor Networks 15(9), 1550147719874002 (2019)

7. Li, T.: Research on bank product marketing forecast model based on BP neural network method. Modern Marketing **000**(010), 92–93 (2019)
8. Dongning, R., Sihong, H.: Optimization of financial corpus sentiment analysis model based on THUCTC. Journal of Guangdong University of Technology **035**(003), 37–42 (2018)
9. Pavao, L.V., Pozo, C., Costa, C., et al.: Financial risks management of heat exchanger networks under uncertain utility costs via multi-objective optimization. Energy **139**(Nov 15), 98–117 (2017)
10. Frénod, E., Ménard, P., Safa, M.: Two optimization problems of a continuous-in-time financial model. J. Math. Finance **08**(01), 27–42 (2018). https://doi.org/10.4236/jmf.2018.81003
11. Lepinette, E., Tran, T.Q.: Consumption-investment optimization problem in a Lévy financial model with transaction costs and làdlàg strategies. Math. Financ. Econ. **14**(3), 399–431 (2020). https://doi.org/10.1007/s11579-020-00260-3
12. Hekimoğlu, M., Sevim, I., Aksezer, Ç., Durmuş, İ: Assortment optimization with log-linear demand: application at a turkish grocery store. J. Retail. Consum. Serv. **50**, 199–214 (2019). https://doi.org/10.1016/j.jretconser.2019.04.007

3D Printing Floating Body Balance Optimization System Based on Comet Data Analysis

Wei Wang(✉)

Beijing Applied Senior Technical School, Beijing 102600, China
13811985980@163.com

Abstract. 3DP (3D printing) technology has changed the traditional manufacturing methods to a large extent, and has been widely used in many fields such as machinery manufacturing, aerospace, military and so on. The balance of the floating body is the most basic requirement, but if a virtual three-dimensional object is printed, a balance problem may occur. Based on this, this article will start the research of 3DP floating body balance optimization system based on comet data analysis. This paper analyzes the system design method based on Comet data analysis, designs the overall architecture of the 3DP floating body balance optimization system, and designs the system's 3D model acquisition module, 3D floating body model optimization module, printing material recommendation and other functional modules. Through the system's 3DP pipe float (cube) test, the printing errors in the X, Y and Z directions are 0.03 mm, 0.04 mm, and 0.01 mm respectively. This shows that the system has good printing accuracy.

Keywords: Comet Technology · Floating Body Balance · Data Analysis · 3D Printing

1 Introduction

In recent years, due to the development of Ajax technology and the increasing market demand for server technology, comet technology has received more and more attention [1, 2]. At the same time, 3DP technology is a kind of rapid prototyping technology, which is widely used in mold manufacturing, industrial product processing, building construction and other fields, and has become the most popular technology [3, 4]. How to ensure the balance and stability of the 3DP floating body has received extensive attention from the academic community and the industry.

Regarding the research of 3DP technology, many scholars have conducted multi-angle investigations. For example, Eom AH has studied the direct 3D cell printing system of human skin with a functional transwell system [5, 6]; Wei Q studied the selection of functional materials for the surface coating of structural synthetic fibers and the 3DP system [7]; Sabyrov N has studied the metal 3DP technology for the functional integration of catalytic systems [8]. It can be seen that this article uses comet technology in the 3DP system to optimize the balance of the 3DP floating body, which is of innovative significance.

J. H. Abawajy et al. (Eds.): ICATCI 2022, LNDECT 170, pp. 815–822, 2023.
https://doi.org/10.1007/978-3-031-29097-8_96

The purpose of this paper is to design a 3DP floating body balance optimization system based on comet data analysis. This article first analyzes the system design method based on comet data analysis. Then designed the overall architecture of the 3DP floating body balance optimization system, and designed the system's 3D model acquisition module, 3D floating body model optimization module, printing material recommendation and other functional modules. Finally, this article carried out the printing accuracy test of the 3DP system, and analyzed the test results to verify that the system has high printing accuracy.

2 3DP Floating Body Balance Optimization System Based on Comet Data Analysis

2.1 System Design Method Based on Comet Data Analysis

In the requirements modeling stage, it is necessary to design a use case model and analyze the non-functional requirements of the developed 3DP floating body balance optimization system. In the analysis and modeling stage, static and dynamic models of the 3DP floating body balance optimization system need to be designed. In the design modeling, the architecture of the 3DP floating body balance optimization system needs to be designed.

The system modeling method in comet uses the Unified Modeling Language (UML) notation, as shown in Table 1:

Table 1. Comet modeling method UML notation

Modeling stage	Modeling activity	UML notation
Demand modeling	Use case modeling	Use case diagram
	State non-functional requirements	–
	Static modeling	Class diagram
Analytical modeling	Dynamic interaction modeling	Sequence diagram, same sex, activity diagram
	Dynamic state machine modeling	State diagram
Design modeling	Structure building view	Structured class diagram, deployment diagram
	Dynamic concurrent view	Concurrent communication diagram

2.2 Architecture of the 3DP Floating Body Balance Optimization System

(1) 3DP resource layer

The 3DP resource layer is the physical foundation of the entire system and provides support for the operation of various services in the 3DP floating body balance optimization system. Among them, 3DP resources include 3DP hard and soft manufacturing resources. Hard manufacturing resources mainly include various equipment resources (such as printing equipment, scanning equipment, testing equipment, etc.), material resources (such as 3DP materials, parts, etc.); 3DP soft manufacturing resources mainly include various software resources (such as three-dimensional graphics) Software, service billing software, stress analysis software, etc.), human resources (such as model file designers, industry experts, managers, etc.), knowledge resources (such as 3DP examples, industry standards, technical references, etc.) [9, 10].

1) Virtual resource layer

The virtual resource layer mainly realizes the virtualization support and management of 3DP resources distributed in various places. This layer converts 3DP resources into logically existing virtual manufacturing resources through virtualization technology, and realizes management functions such as storage, query, update, and maintenance.

2) Service resource layer

The service resource layer is the core functional layer of the 3DP floating body balance optimization system, and it is the support for users to enjoy various services. This layer mainly manages virtual resource services and provides a consistent service calling interface to the outside, so as to realize free sharing of various services in the comet environment [11, 12]. The functions provided by this layer include: printing material recommendation service, three-dimensional floating model design service, process design service and 3DP service.

(2) Application interface layer

The application interface layer provides a unified interface that is easy for users to access, so that users can freely and efficiently call various system functions. The application interface provided by the 3DP floating body balance optimization system mainly includes the user management of the system, the virtualized access of resources, the packaging of resources as a service, and the intelligent matching of 3DP services.

(3) User layer

The user layer realizes the function of interaction between the user and the system, and the user can access the 3DP floating body balance optimization system through the terminal interactive device.

2.3 Key Functional Modules of 3DP Floating Body Balance Optimization System

(1) Acquisition of 3D model

In the 3DP floating body balance optimization system, there are two main ways to obtain the 3D floating body model file for users to choose freely: the 3D floating body model file library and the personalized design of the 3D floating body model. Among them, the personalized design of the three-dimensional floating body model can be divided into two types, namely: users actively seek out designers to obtain real-time design services, and issue design tenders.

The 3D floating body model of the 3D floating body model file library comes from the uploads of the designers in the system. The designers design 3D floating body models with certain market value according to their own abilities and market needs, and then upload them to the system. The system is responsible for these Model files are centrally optimized and managed so that users can quickly search for suitable three-dimensional floating body model files.

(2) Optimization of 3D floating body model

1) Evaluation of 3D floating body image details

Due to the limited accuracy of the 3D printer, entities smaller than the accuracy threshold in the geometric floating body model cannot be printed, and the hole clogging disappears, causing the loss of the details of the physical results. The system needs to enhance the details that cannot be successfully printed in advance. First of all, after determining the manufacturing size, the system needs to pre-assess the detail area that cannot be successfully printed. Both tones of the binary image in the three-dimensional floating body model need to be considered. Set the detail area of white tones to D_W, and the detail area of black tones to D_B. The image is first corroded, and then expanded, and then the result image of the open operation and the original image are processed to obtain the pre-processed detail area, as shown in formulas (1) and (2):

$$D_W = |I - I \circ S_W| = |I - (I \Theta S_W) \oplus S_W| \tag{1}$$

$$D_B = |I^{-1} - I^{-1} \circ S_B| = |I^{-1} - (I^{-1} \Theta S_B) \oplus S_B| \tag{2}$$

Among them, I^{-1} is the inverted image of the original image, which is used to obtain the detail area of the black tone. S_W and S_B are the structural operators used in the open operation, similar to S_C, their size is related to the resolution of the input image and the physical parameters of the printer. Assuming that the resolution of the input three-dimensional floating body image is v × v and the printing size is u × u, the size ω_w and ω_b of the structure operator for detailed evaluation can be obtained, as shown in formula (3):

$$\omega_\mathrm{w} = \left\lceil \frac{\varphi_s v}{u} \right\rceil, \omega_\mathrm{b} = \left\lceil \frac{\varphi_v v}{u} \right\rceil \tag{3}$$

Considering the effectiveness of the structure operator, ω_w and ω_b must be rounded up to an odd number and greater than or equal to 3.

2) Three-dimensional floating body image detail classification

Since the detailed regions of the three-dimensional floating body image have different geometric and visual features, the performance in the printed results is not the same. In the Comet data analysis, it is necessary to classify the evaluated detailed regions. Assuming that the input floating body image resolution is v × v, and the print size is u × u, then the threshold is defined as shown in formula (4):

$$\omega_t = \frac{\varepsilon v}{u} \tag{4}$$

Among them, the physical meaning of ε is the actual length of the detail feature that needs to be retained in the print result. The system defaults to 3cm, which can be modified. The system will process details with a physical length greater than ε. If it is less than ε, it is considered to be negligible.

(3) Recommended printing materials

The orderly operation of this module needs to rely on the perfect 3DP material database in the system, based on the powerful knowledge gathering ability in Comet data analysis, complete the scientific and reasonable maintenance of the 3DP material database, and provide an important reference for users to make decisions about printing floating materials.

1) Search based on material database

   Searching in this way requires the user to have a comparative understanding of the performance parameters of the product to be printed. After the user enters the desired lower limit of the various performance parameters, the system can screen the material database according to these parameters, and finally select a number of materials that meet the user's performance requirements. Users can read the various performance parameters of the candidate material set in detail, and determine the printing materials independently based on cost factors.
2) Retrieval based on historical print orders

   With the continuous increase in the number of printing service transactions in the 3DP system, historical printing service orders can provide a valuable reference for material decision-making. The 3DP floating body balance optimization system can search the historical printing orders in the system according to the keywords of the products that users need to print, and then sort the materials used in the service orders according to the number of services from high to low, and finally display the sorted material set To the user.

3 System Test Design

3.1 System Accuracy Test Design

Positioning accuracy is a key performance indicator of 3DP. The higher the accuracy, the smaller the gap between the printed object and the model. Therefore, this article will pass the test of absolute positioning accuracy and repeated positioning accuracy to verify the accuracy of the system's printing.

3.2 Absolute Positioning Accuracy

Return the print head, and the system sends a position command to control the print head to move on the 4 vertices of the XY plane in the working space. Record the position of the system command and the actual position of the 3DP in the direction of each vertex.

3.3 Repeat Positioning Accuracy

Repeat the measurement of positioning accuracy, select the four vertices on the oblique plane of the cube with the largest working space to cyclically move, and the command positions input by the system are (360,360,0), (−360,360,0), (−360,−360,180), (360,−360,180). Every time you move a point, you will pause for 15s to measure the vertex position data, and measure the data once every 20 cycles, for a total of 10 measurements.

4 Test Results of 3DP Floating Body Balance Optimization System Based on Comet Data Analysis

4.1 Accuracy of Absolute Positioning and Repeated Positioning

In the accuracy test experiment of absolute positioning and repeated positioning, the difference and standard deviation of the system in each direction of the coordinate system are calculated. The results are shown in Table 2: The variances in the X, Y, and Z directions are 0.0125 mm, 0.0131 mm, and 0.0054 mm, respectively; the standard deviations are 0.114 mm, 0.092 mm, and 0.08 mm, respectively.

Table 2. Measurement data analysis (mm)

Direction	Variance	Standard deviation	Average error
X	0.0125	0.114	0.014
Y	0.0131	0.092	0.012
Z	0.0054	0.08	0.010

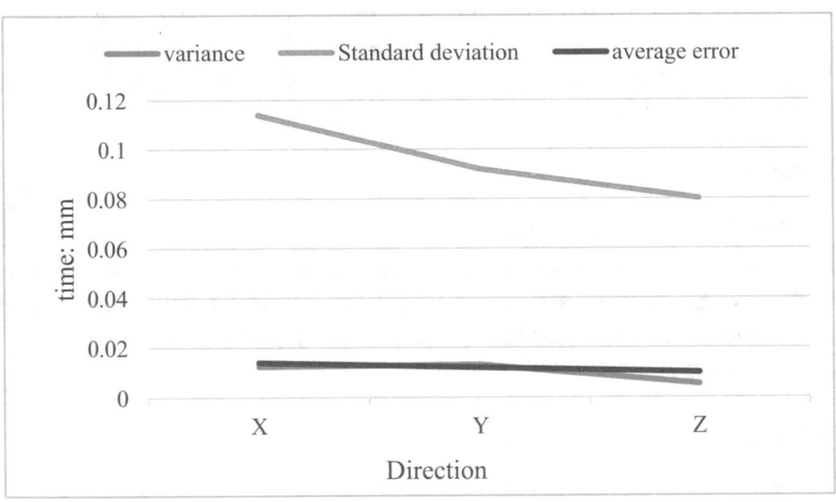

Fig. 1. Measurement data analysis (mm)

It can be seen from the results in Fig. 1 that the positioning accuracy of the 3DP floating body balance optimization system is relatively high, and the error is controlled within 0.02 mm.

4.2 System Printing Verification

After the system starts, it returns to the original position first, keeps the height unchanged, moves the print head in a jog mode and level the printing plane. At this time, the print

Table 3. Printing error (mm)

Floating body	Theoretical value	Actual value	Error
X	60	59.97	0.03
Y	60	59.96	0.04
Z	60	59.99	0.01

nozzle is heated to melt the filaments, and after the temperature stabilizes, the image of the pipeline float model can be downloaded to the controller for printing. After printing, measure the size of the floating body model actually printed, and calculate the size error between the actual and the set floating body model, as shown in Table 3: The printing errors in the X, Y and Z directions are 0.03 mm, 0.04 mm, 0.01 mm, respectively.

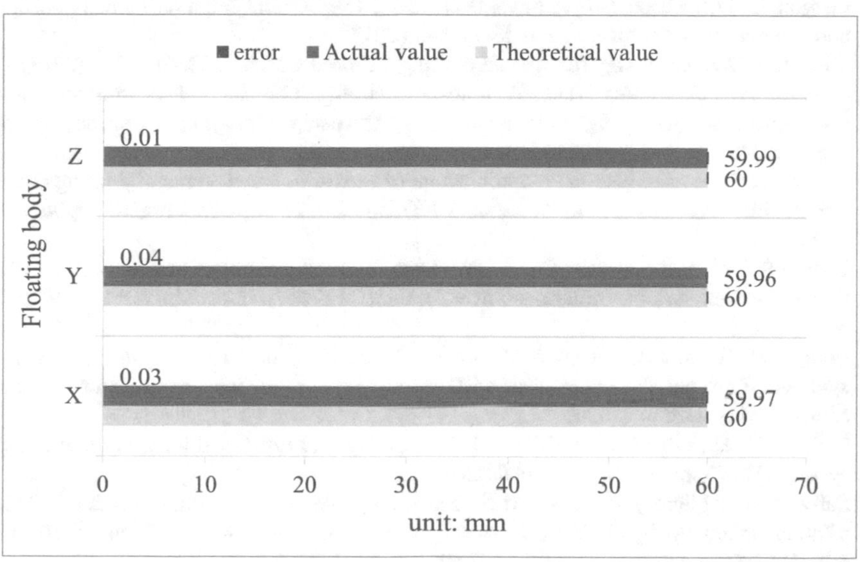

Fig. 2. Printing error (mm)

It can be seen from Fig. 2 that the error of the 3DP system in the Z direction is very small, while the error in the horizontal direction is a bit large. The reason for this problem is that the contour of the surface layer of the floating body is printed first during the printing process, and then the internal filling is performed. At this time, the temperature of the contour filament is difficult to maintain stable, resulting in large shrinkage and deformation, which affects the accuracy of the 3DP floating body in the horizontal direction.

5 Conclusions

With the development of comet technology and 3DP technology, it can be applied in more fields. After research, this article has completed the following work: Analyze the system design method based on comet data analysis, from the 3DP resource layer, application interface the layer and user layer design the system architecture; introduce the key function modules of the 3DP floating body balance optimization system based on comet data analysis, including 3D floating body model acquisition, 3D floating body model optimization, and printing material recommendation function modules; perform system printing accuracy the test shows that the positioning accuracy of the 3DP floating body balance optimization system designed in this paper is high, and the error of the printing floating body is controlled within 0.04 mm.

References

1. Gargon, E., Williamson, P.R., Altman, D.G., et al.: The COMET initiative database: progress and activities update (2015). Trials **18**(1), 1–5 (2017)
2. Turc, T.: AJAX technology for internet of things. Procedia Manuf. **32**, 613–618 (2019)
3. Kontovourkis, O., Phocas, M.C., Tryfonos, G., et al.: Multi-axis 3DP of material reduced shell structures on a reconfigurable supporting system using topology optimization principles. Procedia Manuf. **44**, 379–386 (2020)
4. Ren, L., et al.: Process parameter optimization of extrusion-based 3d metal printing utilizing PW–LDPE–SA binder system. Materials **10**(3), 305 (2017). https://doi.org/10.3390/ma1003 0305
5. Kim, B.S., Lee, J.-S., Gao, G., Cho, D.-W.: Direct 3D cell-printing of human skin with functional transwell system. Biofabrication **9**(2), 025034 (2017). https://doi.org/10.1088/1758-5090/aa71c8
6. Eom, A.H., Won, J.P.: Selection of functional materials for surface-coating of structural synthetic fibres and development of a 3D-printing system for their manufacture. Compos. Struct. **249**(1), 112567 (2020)
7. Wei, Q., Li, H., Liu, G., et al.: Metal 3DP technology for functional integration of catalytic system. Nat. Commun. **11**(1), 4098 (2020)
8. Sabyrov, N., Abilgaziyev, A., Ali, M.H.: Enhancing interlayer bonding strength of FDM 3D printing technology by diode laser-assisted system. Int. J. Adv. Manuf. Technol. **108**(1–2), 603–611 (2020). https://doi.org/10.1007/s00170-020-05455-y
9. Ha, D.-H., et al.: Correction: Indirect fabrication of versatile 3D microfluidic device by a rotating plate combined 3D printing system. RSC Advances **8**(71), 40597–40597 (2018). https://doi.org/10.1039/C8RA90091D
10. Wang, T., Zhou, C., Xu, W.: Online droplet monitoring in inkjet 3dp using catadioptric stereo system. IIE Trans. **51**(2), 153–167 (2019)
11. Wen, C., Zhao, L., Han, T., Li, W., Zhang, G., Li, C.: A versatile dark-field acoustic-resolution photoacoustic microscopy system aided by 3D printing. Rev Sci. Instrum. **90**(8), 083704 (2019). https://doi.org/10.1063/1.5094862
12. Jena, B., Naik, M.K., Panda, R., et al.: Maximum 3D tsallis entropy based multilevel thresholding of brain MR image using attacking manta ray foraging optimization. Eng. Appl. Artif. Intell. **103**(16), 104293 (2021)

The Application of Image Fusion Algorithm in Digital Media Technology Under the Background of Big Data

Junfang Liu(✉)

Dongying Vocational Institute, Dongying 257091, Shandong, China
liu_junfang@163.com

Abstract. In real life, we obtain most of the external information through visual activities all the time. The huge amount of information will increase the difficulty for us to process information, but big data technology can help people solve this problem. With the continuous development of data media technology, the visual information that people are exposed to is more rich and vivid. In order to make the visual image more clearly presented in front of the field of vision, image fusion technology can be used to process the image information to make the characteristics of the image more obvious. This paper takes digital display image fusion as the research object and uses image fusion algorithm to reduce the noise of the display image. It is found that the image resolution of images with different noise standard deviations will be greatly improved after image fusion denoising, and the noise standard deviation will be higher. The larger the image is denoised, the greater the resolution improvement, and for the entire fusion process of the unfolded image, it can be known that the larger the noise standard deviation, the longer the fusion processing time.

Keywords: Big Data · Image Fusion Algorithm · Digital Media Technology · Image Denoising

1 Introduction

With the rapid development of communication technology and the Internet, the digital media industry has gradually entered people's field of vision. On the basis of integrating the characteristics of traditional media technology, digital media can attract people's attention. This is because digital media can bring people an interactive experience, so that people are immersed in a real and interesting virtual world.

Many scholars have conducted in-depth discussions on the application of image fusion algorithms in digital media technology in the context of big data, and have achieved certain results. For example, human vision has a strong ability to perceive images. In real life, people usually represent data features in the form of images, which has caused widespread attention to the development of digital technology. A scholar explored the interactive effects of human behavior through the use of digital technology, explaining The unique importance of digital technology in today's complex information

J. H. Abawajy et al. (Eds.): ICATCI 2022, LNDECT 170, pp. 823–829, 2023.
https://doi.org/10.1007/978-3-031-29097-8_97

processing [1]. A scholar used an image fusion algorithm to denoise the number of digitized film frames. The main purpose is to make the film picture clearer. He intercepted each frame of the film image separately, and used the algorithm to extract the information characteristics of the image. The characteristic images are merged into a new number of movie frames, which greatly improves the quality of the movie picture [2]. Although image fusion algorithms have a wide range of applications in digital media technology, digital media technology has not been widely promoted in daily life, so that people are exposed to relatively few digital media things.

This article explains the concept of image fusion and the three levels of pixel-level fusion, feature-level fusion, and decision-level fusion of fusion algorithms. It lists the dependence, interactivity, virtuality, fashion and fun of digital media technology relative to traditional digital technology. Take the application of digital media technology in digital display image as an example, analyze the denoising application and fusion processing time of image fusion algorithm in the process of display image fusion, so as to illustrate the practicality of the algorithm in digital media technology.

2 Image Fusion and Digital Media Technology Application

2.1 Image Fusion Concept and Fusion Level

Image fusion refers to the use of special fusion algorithms to merge two or more images into a new image to effectively fill in information and remove unnecessary information. The purpose of image fusion is that the new image may be more in line with human perception of the merged object, so that the merged image has more visible features. In recent years, image fusion technology has developed rapidly and has great practical value in many fields. For example, in terms of network information security, some important information can be hidden through fusion technology to improve information security. It is foreseeable that as the research on image fusion continues to deepen, its application range will inevitably become wider and wider [3, 4].

According to the stage of the fusion algorithm in image processing, it can generally be divided into three levels according to the abstract degree of information representation.

(1) Pixel-level fusion. Merging directly at the level of multi-source image data (pixel gray value) is a method of merging low-level images. The advantage of data layer processing is that it stores the largest amount of information, which makes the merged image more accurate. But at the same time, there are some shortcomings, that is, a large amount of processed data makes the real-time performance of the system poor.

(2) Feature level fusion. First, multi-source images need to be processed, and data layer attribute information, such as image texture, image edges, etc., is derived, and then the extracted feature information is fused to obtain the fusion result. Feature layer processing provides effective information compression, thereby improving the real-time performance of the system, but it also reduces the accuracy of image fusion [5, 6].

(3) Decision-making level integration. It is necessary to first develop and derive the features of multi-source images, make decision processing based on the information of the feature level information, and then perform fusion processing on the decision information according to certain fusion rules [7]. The advantage of decision-level processing is to improve the fault tolerance and real-time performance of the system, but there are also some disadvantages, that is, more image information is lost and the accuracy of image fusion is reduced [8].

2.2 Overview of Digital Media Technology

Compared with traditional media technology, digital media technology has its own unique characteristics:

(1) Dependence

The final digital work is to show a certain height and a certain aesthetic. It is a work different from traditional media technology, so it should not be based on a single artistic image. For digital technology, its expression is a complex, diverse, and impure technology, and it contains many topics. Therefore, its dependence and existence are deeper than traditional technology, integrating art and aesthetic qualities. This technology makes people from all walks of life favor it, showing signs of interdisciplinary interdependence [9].

(2) Interactivity

The so-called interactivity has two modes of operation. One is based on its own set of active interaction methods, such as the guidance function of various interactive installation tasks, click, open and scroll gestures, which are all based on computer control. The other is the interactivity of fixed operations, which will change with the participation of participants. This feature uses fixed-point recognition technology. The computer records various sensory channels such as facial expressions, touches, smells, and speech, and enters the non-precise "immersion" virtual environment for the experiencer in time, which can also be achieved through dialogue [10]. In general, one is a single interaction mode in which the computer directly feeds back to the experiencer, and the other is a dual interaction mode in which the experiencer "accidentally" connects to the computer, and the computer then conducts a "dialogue" with the experiencer.

(3) Virtuality

Different from the traditional form of technology, digital media technology is based on numbers. Whether it is digital painting, digital imaging, 3D animation, or digital music, it basically puts people in a digital virtual world. With the help of digital media technology, this world has become a perceptual world, making people live in a ubiquitous "aesthetic fantasy".

(4) Fashion and fun

As one of the most basic needs of people, entertainment begins to seek a breakthrough in form while "creating happiness". Digital media technology is the most advanced art form in technical design and uses technology. Digital technology is forward-looking in the field of modern society. Then, digital media art naturally has a fashionable style.

Digital media must break through self and certain aesthetic inertia. Therefore, directors and producers of digital media works are good at attracting enough attention to current popular phenomena and interesting content and absorbing culture.

2.3 Application of Digital Media Technology

(1) Digital exhibition

Nowadays, exhibitions and exhibitions have become a common social activity. In our daily life, there are all kinds of exhibitions and exhibitions around us almost every day. From small business exhibitions to international exhibitions, it can be described as nothing. Being everywhere allows people to have more dialogues and interactions with exhibits in the virtual space [11]. This has not only changed the traditional forms of exhibitors and visitors, but also expanded the scope of application of digital media interaction technology in exhibitions. Correspondingly, its technology is developing rapidly under the constantly updated demand.

(2) Digital public art

By creating an accurate digital model, we can scientifically calculate the feasibility of the public art work's volume, cost budget, work material durability, and work structure. It is used as an auxiliary means to help designers perform scientific calculations and evaluate the feasibility of the scheme.

(3) Virtual building

A virtual building refers to a virtual building that does not actually exist, or a building that has not yet been completed despite the fact that there are real data information. Virtual building roaming applications are used in many fields. Virtual designers can display the building data, design plans or other content to be designed in a three-dimensional effect, allowing tourists, investors or reviewers to have a full range of "immersive" roaming experience Observe, you can also freely switch multi-angle walking motion, such as: walking, driving, flying, etc., and the route is not controlled by the outside. These are the traditional architectural effects and pre-rendered playback 3D animation cannot be achieved [12].

2.4 Image Fusion Algorithm

The weighted average method is the simplest and most basic fusion algorithm. This method can reduce the mutual interference between multi-source images, but it reduces the image contrast to a certain extent. Its mathematical form is expressed as:

$$I_G = \beta I_A + (1 - \beta) I_B \tag{1}$$

The fusion effect of the weighted average method depends to a certain extent on the choice of weights. The average fusion method is a special form of the weighted average algorithm, that is, at this time, the weighting coefficient $\beta = 0.5$, expressed as:

$$I_G = 0.5(I_A + I_B) \tag{2}$$

In the formula, β is the weighting coefficient, I_A, I_B are the gray value matrix of the image to be fused, and I_G is the gray value matrix of the fused image.

3 Application of Image Fusion Algorithm in Digital Media Technology

3.1 Research Content

This article takes the digital media exhibit image as an example. Projecting it on a large three-dimensional screen allows visitors to interact with the exhibit. The image fusion algorithm can denoise the exhibit image, and the image resolution after denoising will increase. Make digital media exhibition images more clearly presented on the three-dimensional screen, allowing people to have a more realistic viewing experience. In this paper, fusion denoising is performed on digital media display images with noise standard deviations of 15, 20, and 25 respectively, and the effectiveness of the fusion algorithm is proved by comparing the resolution of the images before and after denoising. After that, the processing time of the entire image fusion process is analyzed.

3.2 Research Methods

Contrast method: Compare the resolution of digital media exhibition images with different noise standard deviations after image fusion and denoising to illustrate the effectiveness of the algorithm.

4 Application Analysis of Image Fusion Algorithm in Digital Media Technology

4.1 Image Resolution Comparison Before and After Denoising

Table 1. The image resolution before and after the denoising of the digital media exhibition image

Noise standard deviation	15	20	25
Before denoising	24.72	21.36	18.63
After denoising	46.89	43.57	40.52

From Table 1 and Fig. 1, it can be found that the larger the noise standard deviation of the digital media display image is, the smaller the image resolution before denoising. However, after the image fusion algorithm is denoised, the larger the noise standard deviation is, The greater the resolution of the image after noise is improved, which is reflected in: the resolution of the image with a noise standard deviation of 15 before denoising is 24.72, and after denoising is 46.89, the resolution is increased by about 1.9 times; the noise standard deviation is 20 The resolution of the image before denoising is 21.36, after denoising is 43.57, the resolution is increased by about 2.04 times; the resolution of the image with a noise standard deviation of 25 before denoising is 18.63, and after denoising is 40.53, the image resolution is increased It's about 2.17 times. It can be seen that the smaller the resolution of the image after fusion and denoising, the greater the accuracy of the image resolution, the more effective the fusion algorithm.

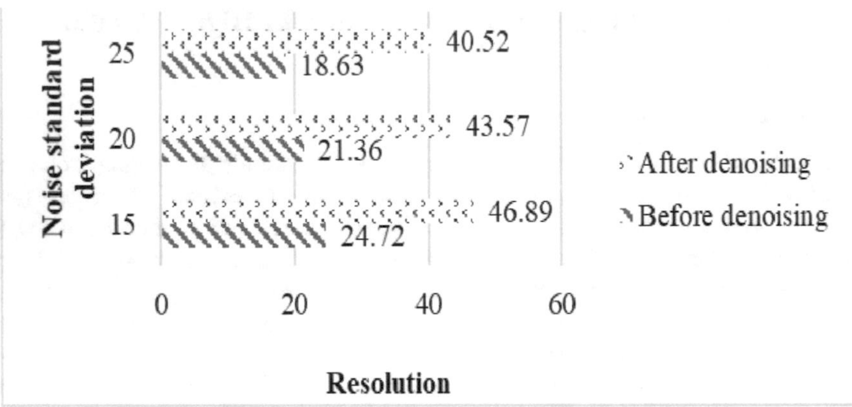

Fig. 1. Image resolution comparison before and after denoising

4.2 Digital Media Exhibition Image Fusion Processing Time

For digital media images to be fully displayed on the display screen, they need to go through image noise reduction, image registration, image fusion, background update, image output and other procedures. From Fig. 2, you can see that the image with the larger the noise standard deviation, each channel The longer the processing time of the process, this is because the more noisy image is more difficult to process, and the processing time of the image output in the whole process is the longest, this is because the image output has high resolution.

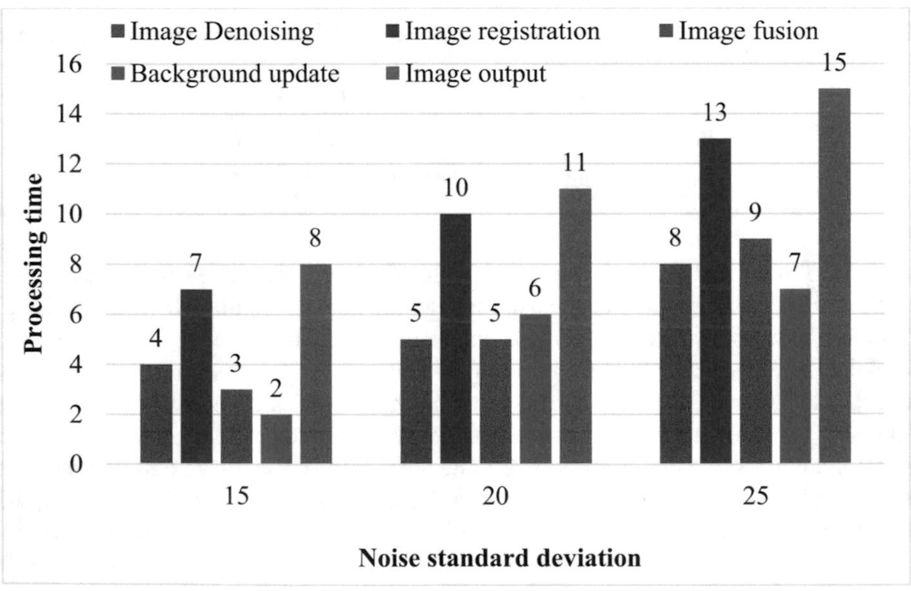

Fig. 2. Processing time (ms) of each program for image fusion

5 Conclusion

This paper analyzes the application of digital media technology under the image fusion algorithm, and uses the fusion algorithm to denoise the image of the exhibited work to increase the resolution of the image for the application of digital exhibition. The resolution of the digital exhibition image has indeed been effectively improved. Through research, it is found that the resolution of the digital exhibition image after algorithm fusion and denoising has indeed been effectively improved. People can interact with the clear image during the exhibition. Visual interaction allows them to experience the immersive exhibition experience brought by digital media.

References

1. Bilal, M., Habib, H.A., Mehmood, Z., et al.: Single and multiple copy-move forgery detection and localization in digital images based on the sparsely encoded distinctive features and DBSCAN clustering. Arab. J. Sci. Eng. **45**(4), 2975–2992 (2020)
2. Testa, D., Carfantan, H., Perrone, L.M.: The SparSpec algorithm and the application to the detection of spatial periodicities in tokamaks: using memory with relaxation*. Plasma Res. Express **3**(2), 025006 (2021). https://doi.org/10.1088/2516-1067/abf947
3. Lubis, A.R., Nasution, M., Sitompul, O.S., et al.: The effect of the TF-IDF algorithm in times series in forecasting word on social media. Indones. J. Electr. Eng. Comput. Sci. **22**(2), 976 (2021)
4. Paramanandham, N., Rajendiran, K.: Multi sensor image fusion for surveillance applications using hybrid image fusion algorithm. Multimed. Tools Appl. **77**(10), 12405–12436 (2018)
5. Goryunov, V.E.: Features of the computational implementation of the algorithm for estimating the lyapunov exponents of systems with delay. Autom. Control. Comput. Sci. **55**(7), 877–884 (2022)
6. Gonopolskiy, M.G., Glonina, A.B.: Research and development of an algorithm for the response time estimation in multiprocessor systems under the interval uncertainty of the tasks execution times. Model. Anal. of Inf. Syst. **27**(2), 218–233 (2020)
7. Medvedev, A.V., Raskina, A.V.: the dual control algorithm of nonlinear dynamic processes in the conditions of incomplete a priori information. IOP Conf. Ser.: Mater. Sci. Eng. **865**(1), 012006 (2020). https://doi.org/10.1088/1757-899X/865/1/012006
8. Arif, M., Wang, G.: Fast curvelet transform through genetic algorithm for multimodal medical image fusion. Soft. Comput. **24**(2), 1–22 (2020)
9. Imen, M.E., Yaln, Y.: A novel hybrid firefly–whale optimization algorithm and its application to optimization of MPC parameters. Soft Computing **26**(4), 1845–1872 (2021)
10. Hagge, J.: Coding to create: a subtext of decisions as early adolescents design digital media. Technol. Knowl. Learn. **23**(2), 247–271 (2018). https://doi.org/10.1007/s10758-018-9359-y
11. Finnemann, N.O.: Hypertext configurations: genres in networked digital media. J. Am. Soc. Inform. Sci. Technol. **68**(4), 845–854 (2017)
12. Colombo, F.: History of digital media: Revolutions and continuity. Technol. Cult. **58**(3), 897–898 (2017)

Monitoring Persistent Change Points in a Heavy-Tailed Sequence with M-Estimation in Big Data Background

Yangyi Cheng(✉), Hao Jin, Gaochang Zhao, and Meiting Liu

College of Science, Xi'an University of Science and Technology, Xi'an, Shaanxi, China
Cyylmh7@126.com

Abstract. In the current era of big data, the financial industry as one of the most important application areas of big data, many financial and economic time series data obtained show the feature of heavy tail. In order to improve the robustness of persistent change detection in heavy-tailed sequences, a moving ratio test method based on M-estimation is posited to monitor the changes between stationary $I(0)$ and non-stationary $I(1)$ sequences. In this paper, the M-estimation method is used to obtain the asymptotic distribution of the coefficients of the autoregressive process β and the mean of the sequence μ in the test problem, and the Moving Ratio statistic is construct. In this paper, we prove the limit distribution of the newly constructed monitoring statistics under the null hypothesis and the consistency under the alternative hypothesis. Studies have shown that the newly statistic is better than the existing monitoring statistics and can effectively solve the problem of thick tail sequence persistent change point monitoring in the era of big data.

Keywords: M-Estimation · Monitor Persistent Changes · Ratio Statistic · Heavy-Tailed

1 Introduction

In today's era of big data, many economic and financial time series are subject to persistent changes. Persistent change is a question of great empirical significance. Especially in the field of big data such as climate change trends, monitoring geological disasters, forecasting financial market fluctuations. The detection of the change point can be divided into two types: post-test and online monitoring. A posteriori test is to detect whether there is a point of change in a sample sequence that has been fully obtained. At present, the academic circle has formed a systematic understanding of this, and some classical methods include Kim [1], Wang [2], Perron and Yamamoto [3, 4], etc.

Online monitoring refers to the continuous test of continuously observed data based on existing models to study whether the new observation values comply with the original model, that is, the continuous test of whether there is a change point. For this problem, scholars have made a series of achievements in recent years, such as HORVÁTH [5], AUE [6], HUŠKOVÁ [7], etc., using the method of least square estimation, online monitoring

J. H. Abawajy et al. (Eds.): ICATCI 2022, LNDECT 170, pp. 830–836, 2023.
https://doi.org/10.1007/978-3-031-29097-8_98

of coefficient change points in linear regression models, and AUE [8] continued to study the monitoring of change points in linear models. Weighted CUSUM based on basic random variables discusses the asymptotic behavior of delay time. Chen et al. [9] proposed two kinds of ratio-type statistics to continuously detect the change points of memory parameters in long memory time series, and proved the limiting distributions of monitoring statistics under the zero hypothesis without change points as well as their consistency under the alternative hypothesis. HSU [10] used the square test of MOSUM to monitor the potential variance change of the new observation value, and obtained its limit distribution. Chen [11] proposed an improved online monitoring method for coefficient change points of linear regression model, which has higher potential and shorter average running length compared with existing methods.

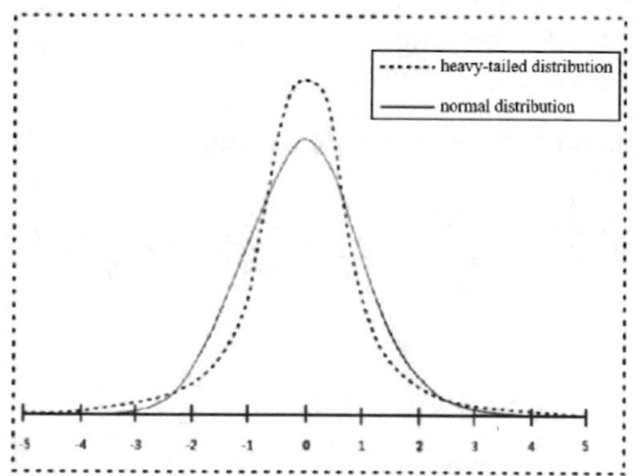

Fig. 1. Heavy-tailed distribution

In practical application, financial market fluctuation is generally a continuous dynamic change process, so it is necessary to use online monitoring method to test the change point. Many data such as return rate of financial assets, stock market analysis and network flow do not conform to normal distribution, but obey the 'peak thick tail' distribution, which make a lot of information trapped in the tail. Figure 1 shows the difference between the heavy-tailed distribution and the normal distribution. The heavy-tail index effectively depicts the tail nature of random variables and reflects the possible risks and losses of assets. Heavy-tail dependent sequences are additive and can describe the characteristics of sharp and thick tails in financial data. However, for thick tail dependent sequences, the test statistics have the defects of over rejecting the null hypothesis and underperforming the potential function. In order to provide more accurate and reliable theoretical basis for decision-makers and avoid the influence of rough tail index in rough tail sequence, and promote the safe, efficient and stable operation of financial system, it is necessary to study the persistence change point of thick-tail dependent sequence.

2 Monitoring Procedure

Let the time series $\{y_t, t = 1, 2, ...\}$ be decomposed as

$$\begin{aligned} y_t &= \mu + \varepsilon_t \\ \varepsilon_t &= \beta\varepsilon_{t-1} + \eta_t \quad t = 1, \cdots, T \end{aligned} \tag{1}$$

where T represents the sample size and $\mu_t = E(y_t) = \delta' d_t$ is a deterministic component, which is modeled as a linear combination of vectors of non random regressors d_t. This article focuses only on $d_t = 1$. Not considering a constant, a temporal trend or dummy variables are typical components of d_t, $\{\eta_t\}$ is an innovation process and meets the assumptions below.

Assumption 2.1. The i.i.d. random variables sequence $\{\eta_t\}$ is in the domain of attraction of a stability law with tail index $E(\eta_t^2) = \infty$ and $\alpha \in (0, 2)$. $\{\eta_t\}$ have a distribution symmetric about 0 when $\alpha = 1$ and if $0 < \alpha < 2$, $E(\eta_t) = 0$.

Lemma 2.1. If Assumption 2.1 is true, it can be obtained.

$$\left(a_T^{-1}\sum_{t=1}^{[Tr]}\eta_t, a_T^{-2}\sum_{t=1}^{[Tr]}\eta_t^2\right) \to^d \left(U_\alpha(r), \int_0^r (dU_\alpha)^2\right) \tag{2}$$

where $r \in (0, 1]$,

$$a_T = \inf\left\{x : P(|\eta_t| > x) \le T^{-1}\right\} \tag{3}$$

and $\xrightarrow{d}$ denotes here convergence in distribution with respect to the Skorohod topology. Moreover, $U_\alpha(r)$ is the Levy α-stable process, $\int_0^r (dU_\alpha)^2 = [U_\alpha]_r$ is the quadratic variation process of $U_\alpha(r)$.

Lemma 2.2. Due to the relevant proof of Solo (1992) [12], the following conclusions can be drawn:

if $y_t \sim I(0)$,

$$\left(a_T^{-1}\sum_{t=1}^{T}\varepsilon_t, a_T^{-2}\sum_{t=1}^{T}\varepsilon_t^2\right) \xrightarrow{d} \left(\sum_{j=0}^{\infty}\varphi_j U_\alpha(1), \sum_{j=0}^{\infty}\varphi_j^2\int_0^1 (dU_\alpha)^2\right) \xrightarrow{d} (Z_1(r), Z_2(r)) \tag{4}$$

$Z_1(\cdot), Z_2(\cdot)$ are the Levy process;

if $y_t \sim I(1)$,

$$\left(a_T^{-1}\varepsilon_{[Tr]}, a_T^{-2}\varepsilon_{[Tr]}^2\right) \xrightarrow{d} \left(U_\alpha(r), \int_0^r (dU_\alpha)^2\right) \tag{5}$$

In this paper, we mainly focus on whether persistent changes occur in heavy-tailed sequences, and monitor whether changes occur between stationary and non-stationary in the above process. Therefore, null hypothesis

$$H_0 : y_t \sim I(0),\ t = 1, 2, ..., T, \tag{6}$$

is opposite to alternative hypothesis

H_1:y_t shows at least one state switch between $I(0)$ and $I(1)$.

where τ^* is the change point that is unkonwn.

The newly monitoring ratio statistic is constructed as follows

$$R_n(\tau) = \frac{\sum_{t=[T\tau]+1}^{[T\tau]+[Th]} (\sum_{i=[T\tau]+1}^{t} \hat{u}_{1,i})^2}{\sum_{t=[T\tau]-[Th]+1}^{[T\tau]} (\sum_{i=[T\tau]-[Th]+1}^{t} \hat{u}_{0,i})^2} \tag{7}$$

where $h < \tau < 1-h$ with some $h \in (0, 1/2)$, τ is the center point of each monitoring moving procedure, if no change point is detected, then $\tau = \tau + h$; if the change point is detected, then $\tau = \tau_j^*$ (the j th change point). $\hat{u}_0$, is an M-estimate of parameter ε_t based on the observations $y_{[T\tau]-[Th]+1}, \ldots, y_{[T\tau]}$, and $\hat{u}_{1,i}$ is an M-estimate of parameter ε_t based on the observations $y_{[T\tau]+1}, \ldots, y_{[T\tau]+[Th]}$. Since the principle of the change of $I(1) \to I(0)$ is the same as that of $I(0) \to I(1)$, it only needs to reverse the statistics, which will not be introduced in detail in this paper.

This paper proposes the stopping time

$$R_T(c) = \min\{N \le n \ge T : R_{T\tau}(n/T) > c\},$$

with conventional min $\Phi = T$, for some critical value c.

Theorem 2.1. We can get a constant under the null hypothesis $c = c(\alpha)$,such that

$$P\left\{\lim_{T\to\infty} \min_{s\in[\tau,1]} R_{T\tau}(s) < c\right\} = \alpha \tag{8}$$

Theorem 2.2. If Assumptions hold, then under the null hypothesis can be obtained

$$R_n(\tau) \Rightarrow \frac{\int_{\tau}^{\tau+h} [Z_1(s) - Z_1(\tau)]^2 ds}{\int_{\tau-h}^{\tau} [Z_1(s) - Z_1(\tau-h)]^2 ds} \tag{9}$$

Theorem 2.3. Under Assumptions, if there occurs a change from $I(0)$ to $I(1)$ or $I(1)$ to $I(0)$ at $[T\tau^*]$, then we have

$$R_{T\tau}(\tau) = O_P(T^2) \tag{10}$$

3 Conclusions

For heavy-tailed time series data in big data, this paper proposes a motion ratio test method based on M-estimation to continuously detect its persistence changes. M-estimation is more robust and stable than OLS estimation. Table 1 provides the empirical reject rate when sample size T = 500 and κ changes at $\tau^* = 0.3, 0.5, 0.75$ with $\beta = 0.3$, and Table 2 with $\beta = 0.8$. This shows that our monitoring process is more sensitive and effective than previous studies.

Table 1. Empirical potential in numerical simulation ($\beta = 0.3$)

		h = 0.2			h = 0.3	
	$\tau^* = 0.3$	$\tau^* = 0.5$	$\tau^* = 0.75$	$\tau^* = 0.3$	$\tau^* = 0.5$	$\tau^* = 0.75$
k = 1.10	1	0.999	0.999	1	0.998	0.995
k = 1.43	1	1	0.999	1	1	0.991
k = 1.80	1	1	0.995	1	1	0.994
k = 2.00	1	1	0.990	1	1	0.990

Table 2. Empirical potential in numerical simulation ($\beta = 0.8$)

		h = 0.2			h = 0.3	
	$\tau^* = 0.3$	$\tau^* = 0.5$	$\tau^* = 0.75$	$\tau^* = 0.3$	$\tau^* = 0.5$	$\tau^* = 0.75$
k = 1.10	0.93	0.852	0.738	0.915	0.874	0.775
k = 1.43	0.955	0.907	0.808	0.953	0.893	0.816
k = 1.80	0.986	0.912	0.897	0.968	0.902	0.842
k = 2.00	0.947	0.906	0.769	0.945	0.908	0.825

More importantly, compared with the literature, this method is more suitable for monitoring the persistent change points of thick tail sequences in big data. Theoretical analysis shows that the longer the start-up time, the longer the delay time of monitoring the change point. Therefore, in practical applications, it is recommended to select a smaller startup time within a larger monitoring range.

4 Mathematical Proofs

Due to the relevant proof of M-estimation by Keith (1991) [13], the following conclusions can be drawn:

Proof of Theorem 2.2
Let $t = [Ts]$,

$$a_T^{-1} \sum_{i=[T\tau]+1}^{t} \hat{u}_{1,i} = a_T^{-1} \sum_{i=[T\tau]+1}^{t} \left(y_i - \hat{\mu}_2\right) = a_T^{-1} \sum_{i=[T\tau]+1}^{t} \left(\varepsilon_i - (\hat{\mu}_2 - \mu)\right)$$

$$= a_T^{-1} \sum_{i=[T\tau]+1}^{[Ts]} \varepsilon_i - a_T^{-1}(s-\tau)T(\hat{\mu}_2 - \mu)$$

$$\Rightarrow Z_1(s) - Z_1(\tau) - a_T^{-1}T^{1/2}(s-\tau)\frac{B(\tau+h) - B(\tau)}{hE(\varphi'(\varepsilon_t))}$$

then the numerator: $T^{-1}a_T^{-2}\sum_{t=[T\tau]+1}^{[T\tau]+[Th]}(\sum_{i=[T\tau]+1}^{t}\hat{u}_{1,i})^2 \to \int_{\tau}^{\tau+h}[Z_1(s)-Z_1(\tau)]^2 ds$

$$a_T^{-1}\sum_{i=[T\tau]-[Th]+1}^{t}\hat{u}_{0,i} = a_T^{-1}\sum_{i=[T\tau]-[Th]+1}^{[Ts]}(y_i-\hat{\mu}_1) = a_T^{-1}\sum_{i=[T\tau]-[Th]+1}^{[Ts]}(\varepsilon_i-(\hat{\mu}_1-\mu))$$
$$= Z_1(s)-Z_1(\tau-h)-a_T^{-1}T^{1/2}(s-\tau+h)\frac{B(\tau)}{hE(\varphi'(\varepsilon_t))}$$

Then the denominator:

$$T^{-1}a_T^{-2}\sum_{t=[T\tau]-[Th]+1}^{[T\tau]}(\sum_{i=[T\tau]-[Th]+1}^{t}\hat{u}_{0,i})^2 \to \int_{\tau-h}^{\tau}[Z_1(s)-Z_1(\tau-h)]^2 ds$$

So

$$R_{Th}(\tau) \Rightarrow \frac{\int_{\tau}^{\tau+h}[Z_1(s)-Z_1(\tau)]^2 ds}{\int_{\tau-h}^{\tau}[Z_1(s)-Z_1(\tau-h)]^2 ds}$$

Proof of Theorem 2.3. The first point of change occurred at $\tau_1 : \tau < \tau_1 < \tau + h$, $(I(0) \to I(1))$.

$$a_T^{-1}\sum_{i=[T\tau]+1}^{t}\hat{u}_{1,i} = a_T^{-1}\sum_{i=[T\tau]+1}^{[T\tau_1]}(y_i-\hat{\mu}_2)+a_T^{-1}\sum_{i=[T\tau_1]+1}^{[Ts]}(y_i-\tilde{\mu}_2)$$
$$= a_T^{-1}\sum_{i=[T\tau]+1}^{[T\tau_1]}(\varepsilon_i-(\hat{\mu}_2-\mu))+a_T^{-1}\sum_{i=[T\tau_1]+1}^{[Ts]}(\varepsilon_i-(\tilde{\mu}_2-\mu))$$
$$= a_T^{-1}\sum_{i=[T\tau]+1}^{[T\tau_1]}\varepsilon_i - a_T^{-1}(s-\tau)T(\hat{\mu}_2-\mu)+a_T^{-1}\sum_{i=[T\tau]+1}^{[Ts]}(\mu+\varepsilon_i-\tilde{\mu}_2)$$
$$\Rightarrow Z_1(\tau+h)-Z_1(\tau)-a_T^{-1}T^{1/2}(\tau_1-\tau)\frac{B(\tau_1)-B(\tau)}{(\tau_1-\tau)E(\varphi'(\varepsilon_t))}$$
$$+a_T^{-1}T(s-\tau_1)\mu - a_T^{-1}\sum_{i=[T\tau_1]+1}^{[Ts]}\tilde{\mu}_2$$

Then the numerator: $T^{-1}a_T^{-2}\sum_{t=[T\tau]+1}^{[T\tau]+[Th]}(\sum_{i=[T\tau]+1}^{t}\hat{u}_{1,i})^2 \to \int_{\tau}^{\tau+h}[Z_1(s)-Z_1(\tau)]^2 ds$.

Then the denominator:

$$T^{-1}a_T^{-2}\sum_{t=[T\tau]-[Th]+1}^{[T\tau]}(\sum_{i=[T\tau]-[Th]+1}^{t}\hat{u}_{0,i})^2 \Rightarrow \int_{\tau-h}^{\tau}[Z_1(s)-Z_1(\tau-h)]^2 ds$$

Hence, y_t goes out of τ_1 with the transformation of $I(0) \to I(1)$,

$$\tau < \tau_1 < \tau + h: \quad \sum_{i=[T\tau]+1}^{t}\hat{u}_{1,i} = O_p(Ta_T), \qquad \sum_{i=[T\tau]-[Th]+1}^{t}\hat{u}_{0,i} = O_p(a_T);$$

According to the continuous mapping theorem,

The numerator of $R_{Th}(\tau)$: $\sum\limits_{t=[T\tau]+1}^{[T\tau]+[Th]}(\sum\limits_{i=[T\tau]+1}^{t}\hat{u}_{1,i})^2 = O_p(T^3a_T^2)$,

The denominator of $R_{Th}(\tau)$: $\sum\limits_{t=[T\tau]-[Th]+1}^{[T\tau]}(\sum\limits_{i=[T\tau]-[Th]+1}^{t}\hat{u}_{0,i})^2 = O_p(Ta_T^2)$.

Hence, $R_{Th}(\tau) = O_p(T^2)$.

Then $\tau = \tau_1$, we can monitor the second change point τ_2.

References

1. Kim, J.Y.: Detection of change in persistence of a linear time series. J. Econometrics **95**(1), 97–116 (2000)
2. Wang, D.: Monitoring persistence change in heavy-tailed observations. Symmetry **13**, 936 (2021)
3. Perron, P., Yamamoto, Y.: Pitfalls of two-step testing for changes in the error variance and coefficients of a linear regression model. Econometrics **7**(2), 22 (2019)
4. Perron, P., Yamamoto, Y.: The great moderation: updated evidence with joint tests for multiple structural changes in variance and persistence. Empirical Econ. **7**, 1–26 (2021)
5. Horváth, L., Hušková, M., Kokoszka, P., Steinebach, J.: Monitoring changes in linear models. J. Stat. Plann. Inference **126**(1), 225–251 (2004)
6. Aue, A., Horváth, L., Hušková, M., Kokoszka, P.: Change-point monitoring in linear models. The Econometrics J. **9**(3), 373–403 (2006)
7. Černíková, A., Hušková, M., Prášková, Z., Steinebach, J.G.: Delay time in monitoring jump changes in linear models. Statistics **47**(1), 1–25 (2013)
8. Aue, A., Horváth, L.: Delay time in sequential detection of change. Stat. Probab. Lett. **67**(3), 221–231 (2004)
9. Chen, Z., Xiao, Y., Li, F.: Monitoring memory parameter change-points in long-memory time series. Empirical Econ. **60**(5), 2365–2389 (2020)
10. Hsu, C.-C.: The MOSUM of squares test for monitoring variance changes. Finance Res. Lett. **4**(4), 254–260 (2007)
11. Chen, Z., Zheng, T., Wei, Y.: Monitoring change in persistence in linear time series. Statist. Probab. Lett. **79**(19–20), 1520–1527 (2010)
12. Solo, P.V.: Asymptotics for linear processes. Ann. Stat. **20**(2), 971–1001 (1992)
13. Keith, K.: Limit theory for m-estimates in an integrated infinite variance. Economet. Theor. **7**(2), 200–212 (1991)

SDP-Based Energy Internet Networking Technology

Hao Feng(✉), Feng Guo, Xianfei Zhang, Zhiyong Zha, Fei Long, Rongtao Liao, Mingyang Yu, Xiong Zhang, Geng Wu, Bo Jin, and Zheng Yu

State Grid Information and Communication Branch of Hubei Electric Power Co., Ltd., Wuhan, Hubei, China
15007194516@139.com

Abstract. In today's increasingly serious environmental pollution and resource depletion, the energy Internet, which supports the large-scale utilization of renewable energy, has attracted much attention. This article aims to study the SDP-based energy Internet networking technology. Based on the analysis of the characteristics of the energy Internet, the privacy threats of the energy Internet and the key technologies of SDP, the SDP-based energy Internet network architecture is designed, and the architecture. The road construction delay and blocking rate performance were tested. The test results show that the SDP-based energy Internet network architecture has good scalability and also has good performance in terms of resource utilization and road construction delay.

Keywords: SDP · Energy Internet · Networking Technology · Privacy Protection

1 Introduction

With the continuous development of information and communication technology, intelligent control technology, network operation technology, and Internet technology, the operation and transmission intelligence level of intelligent networks will continue to improve, and the power business will become more diversified and interactive, and gradually turn to the energy Internet [1, 2]. In this process, the power communication network is facing more and more challenges.

The Internet of Energy includes many smart green concepts such as smart communications, smart grids, smart metaphors, and so on. Compared with the smart grid, the Internet of Energy has the following characteristics: decentralization, wider interconnection, intelligent and advanced, interactive and open. Decentralization is the most important difference between Energy Internet and Smart Grid [3, 4]. The power system is the core of the smart grid, but the energy supply and demand system of the Energy Internet is not. Many distributed power generation and energy storage systems are used. Energy can change based on the relationship between supply and demand. Extensive interconnection refers to regional interconnection and industrial interconnection. Cross-industry refers to the integration of power networks, the Internet, and the Internet of

J. H. Abawajy et al. (Eds.): ICATCI 2022, LNDECT 170, pp. 837–844, 2023.
https://doi.org/10.1007/978-3-031-29097-8_99

Things to create a social service platform. Extensive interconnection makes public service resources including energy resources widely available [5, 6]. Advanced intelligence refers to the realization of deeper intelligence than smart grids through technologies such as telecommunications, the Internet, the Internet of Things, and cloud computing. It has a high degree of intelligence in power generation, transmission, control, and management [7, 8]. In open interaction, openness refers to the opening of the electricity market system. Interaction means that users interact with all kinds of electrical equipment and power grids at the same time as producers and consumers. Energy consumption changes from centralized supply to one-way consumption, and conversion and consumption to a more interactive and flexible production mode [9, 10].

On the basis of consulting a large number of relevant references, this paper combines the characteristics of the Energy Internet, the privacy threats of the Energy Internet and the key SDP technology, and designs an energy Internet network architecture based on SDP. The network architecture includes the device layer, the control layer, and the application layer. Finally, the performance of road construction delay and blocking rate of the architecture is tested to verify whether it meets the requirements of this paper.

2 SDP-Based Energy Internet Networking Technology

2.1 Characteristics of Energy Internet

(1) Long-distance transmission and local consumption of energy

The Energy Internet solves the problem of uneven energy distribution caused by long-distance relationships in terms of structure and technology. Energy, especially the long-distance transmission of clean energy, is an important means to solve the uneven distribution of resources and complementary resource models. In the future, the energy Internet will use UHV technology to solve cross-regional energy problems [11, 12]. Dynamically adjust the transfer and local energy consumption to improve energy, especially clean energy efficiency.

(2) Large-scale network access and distributed consumption of clean energy

Achieving large-scale compatibility of clean energy is one of the goals of the Energy Internet. Large-scale supply-side clean energy is connected to the grid through flexible scheduling, and distributed new high-energy-consuming equipment is locally consumed through autonomous microgrid operation. This is the energy efficiency scale of the clean energy Internet.

(3) Depth perception and effective control

Improving the compatibility and reliability of the clean energy of the Energy Internet depends on a deep understanding and effective control of the information side of the network. The demand for two-way real-time energy information interaction must be based on cutting-edge information technologies such as the Internet of Things, big data, cloud computing, and artificial intelligence, as well as electronic technology and control technology. To achieve deep real-time understanding and ubiquitous effective network control, power information integration is an inevitable feature of the development of the energy Internet.

(4) Hierarchical control and regional autonomy

Compared with the top-down tree control structure of the traditional power system, the Energy Internet provides decentralized and centralized control through "flat supervision and regional autonomy". Internet energy requires two-way information and energy flow within and between layers. The "regional autonomy" control function distributes the control center to all levels of local area networks, conforming to the development direction of information technology such as distributed IT.

2.2 The Privacy Threat of the Energy Internet

(1) Real-time electricity consumption data

Energy Internet uses smart meters to collect users' electricity consumption data in near real-time to optimize energy distribution. Based on these data, the control center can understand the current grid load in real time, and formulate power generation plans and planning strategies accordingly. However, the user's near real-time power consumption data has an increased risk of leaking confidentiality. The intruder can estimate the startup status of each home appliance, analyze the user's energy consumption curve, and further infer the user's behavior and habits. This attack method is also called non-intrusive load detection. The current research on this algorithm is very important. The mature energy decomposition method based on the multi-factor Markov model uses the hidden Markov model to simulate the correspondence between the user's household equipment and the electricity meter data, and trains the hidden Markov model to maintain the parameters. Through sample data: initial state, conversion table. Like the state table, the hidden state of each device is evaluated after the instrument data is replaced. The energy decomposition algorithm based on sparse coding uses data samples to train sparse coding and estimate the key vector of each consumer electronic product. Finally, we obtain the hidden state corresponding to the aggregated data according to the basic vector table.

(2) Distributed energy transaction data

Energy Internet supports the large-scale use of renewable energy, supports the decentralized management of multiple facilities and local energy consumption. On the Energy Internet, users are not only energy consumers, but also energy producers. Users with renewable energy generation capabilities can send information about the energy they can provide across the grid. Users with useful power demand can also sign contracts with decentralized energy generators, depending on the needs of the decentralized power transaction completion method. However, there are many problems with distributed power generation Internet transactions, which are mainly manifested in two aspects: the two parties of the power transaction must send their own power and power demand to achieve the transaction intent, so the data can be used by rejection, malicious users or competitors on the grid. You can use it to infer the power generation and production costs of performance users and other company secrets.

(3) Location of electric vehicle users and battery energy storage data

As the main carrier of the electric transportation network, electric vehicles have brought a heavy burden to the energy Internet. Due to the large volume of electric vehicles, centralized charging of electric vehicles will inevitably cause large fluctuations in the grid load and affect load balance. To achieve the best management of electric vehicles, pay close attention to the energy storage characteristics of electric vehicles, and achieve the effects of peak shaving and valley filling, and load distribution, it is necessary to continuously drive electric vehicles together. Users of electric vehicles send their location and current battery power to the data unit through a wireless signal transmitter. After the receiving unit receives the data of all users in the current signal coverage area, it preprocesses the current data and uploads the processing result to the control center. According to the current network load situation and the processing results of the data processing unit, the control center formulates a wise programming strategy, and signs a charge and discharge contract with the user through remote operation. However, the location and energy storage data of electric vehicles may easily reveal your personal information. The intruder can identify the user in near real-time through the location and energy storage information of the electric vehicle, and further infer the user's outdoor activities and route based on the driving trajectory of the electric vehicle.

2.3 SDP Key Technology

(1) Fine-grained access control

Because the client is the most critical part of the entire SDP structure, it can provide terminal security control that is difficult to achieve in traditional products, and therefore achieve more precise security control. According to the detailed access management strategy based on the principle of least privilege, since SDP can only be accessed directly from the application layer, the intranet cannot be exposed. Therefore, even if a malicious program is installed on the employee's computer, it cannot scan or steal intranet resources. On-demand authentication also helps administrators reasonably restrict the applications that users may log in to the background on the Internet. For example, only employees in the financial department can log in to the accounting system, but they cannot access the personnel management department. This reduces excessive recognition of applications, significantly reduces the attack surface, and reduces the possibility of employee leaks. In addition to the size of the application, you can also control the size of the application's login address, login date, etc.

(2) Access security and identity authentication

The SDP monitoring platform can also remotely clear the account information and device usage traces on the user's device. The company's security is effectively protected in abnormal situations such as the leakage of the company's personnel authentication information or the loss of equipment. Device data can be deleted in real time, so even if the device is connected to the Internet, the user can quickly disconnect the device. The mobile browser can also define more factor authentication methods, such as fingerprints, facial recognition, and mobile text messages to ensure your identity security.

3 Experiment

3.1 Device Layer

The equipment level includes interconnected intelligent power control equipment, charging piles, access switches and switches. The user's electricity consumption information is sent to the remote master station center through the access switch and the local control center under the global control of the flow chart information issued by the controller.

3.2 Control Layer

At the control layer, the device-level data transmission path is managed through the south-facing interface and the OpenFlow protocol stack, and the top-level service drives the application management unit through the north-facing interface. The protocol stack uses the OpenFlow version and OpenFlow messages after the protocol stack analysis is transformed into a high-level data model. This model provides the internal description of resources and strategies for the controller, as well as the programming and control of various services. The conversion process is also a network virtualization process. Virtualization management is achieved by integrating and subtracting various resource information and network state descriptions of the underlying equipment, and provides various basic application network services and controls through their respective APIs and other interfaces. There are three main functions: application management, resource management, basic services and control. The application management module interacts with the upper-level system master station through the north-facing API interface provided by the control layer. Connection detection and global dynamic optimization of network topology, customizable QoS service guarantee mechanism, high-speed access solutions for customized broadband services, EV access information analysis, network status, configuration/topology/network-wide shared information database will be saved. The controller is connected to a group of controllers through the horizontal expansion of the east-west interface to ensure the reliability and stability of the power communication network.

3.3 Application Layer

The application layer includes remote master station applications for various services, such as electricity usage information collection and power management. These applications manage controller applications through the north-facing REST API interface provided at the controller level, and perform logical management of terminal devices at the device level. Efficient calculation and processing of network services and control can receive application level in real time. For electricity consumption information, distributed power supply access information, EV access conditions, information interaction, coordinated management between different business systems, and the distribution of programming information for different business needs, see the corresponding strategies.

3.4 Calculating Trust

The authentication access control model needs to evaluate the reliability of users. The evaluation process mainly considers the following factors:

(1) User historical behavior. Records and ratings of previous operations User operation data is an important parameter for calculating trust, which directly affects system ratings and user trust analysis, and will not disappear with users.
(2) Current user behavior. It can analyze the operation of the current application, record and monitor, analyze the data results of the operation, analyze access and process system resources and data content, and prevent certain malicious functions.

Definition 1. Trust decay function: Estimate the influence of the behavior of the previous action on the current trust. The result is affected by the duration t. The estimation type is shown in (1).

$$\Phi(t) = e^{-pt} \tag{1}$$

where p is the attenuation value of trust.

Definition 2. Visit confidence: Determine and process the past confidence calculated based on the data of the last operation and the confidence of this visit, and calculate the visit confidence by weighting. The calculation is shown in Eq. (2).

$$Cur(a) = History(a) \times \Phi(t) + Entiro(R(a), P(a)) \times \varphi(t) \tag{2}$$

In the formula, History(a) represents the result of historical trust.

4 Discussion

In order to test the ability of the SDP-based Internet energy network architecture to handle a large number of services, the test increased the number of network topology nodes by 200, 600, 800, and 1,000. The network is a single-sector network. As the transaction volume increases, it becomes a network of 1000 performance services. The test data is mainly the delay and blocking rate of road construction. The test results are shown in Table 1, Fig. 1 and Fig. 2:

Table 1. Comparison chart of service intensity blocking rate curves of different network scales

	200	400	600	800	1000
200	0.378	0.347	0.35	0.337	0.336
400	0.521	0.476	0.491	0.452	0.453
600	0.574	0.549	0.558	0.525	0.525
800	0.608	0.573	0.573	0.561	0.562
1000	0.675	0.646	0.655	0.617	0.618

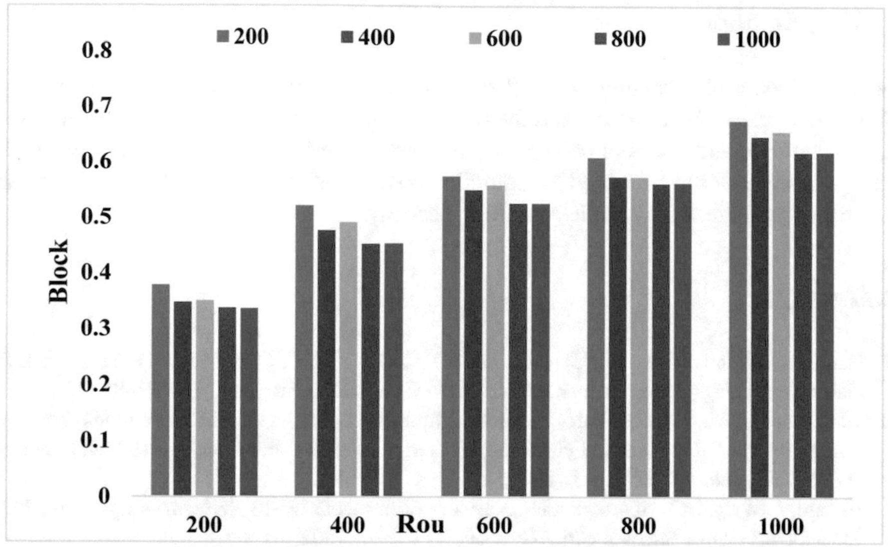

Fig. 1. Comparison chart of service intensity blocking rate curves of different network scales

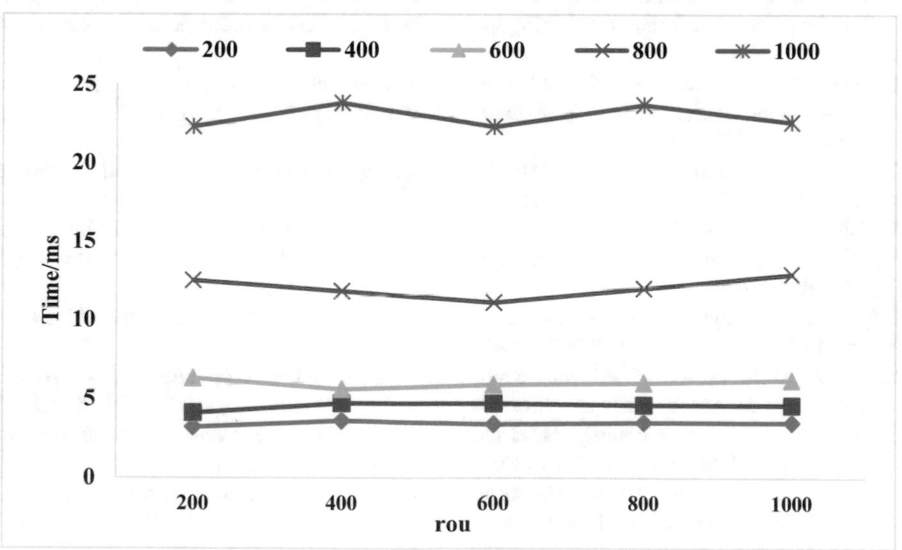

Fig. 2. Comparison chart of the delay curve of business intensity of different network scales

It can be seen from the above chart that the SDP-based energy Internet network architecture has good scalability, and also has good performance in terms of resource utilization and road construction delay.

5 Conclusions

The combined transformation of SDP and Internet energy networking allows end customers to access applications within the scope of authorization, as long as the user passes the verification, and does not need to pay attention to their own rights and interests. IT managers are no longer troubled by complicated authentication rules, and improve ease of use and maintainability while providing security.

References

1. Sani, A.S., Yuan, D., Jin, J., Gao, L., Shui, Yu., Dong, Z.Y.: Cyber security framework for Internet of Things-based Energy Internet. Fut. Gener. Comput. Syst. **93**, 849–859 (2019)
2. Mohammadi, M., Kavousi-Fard, A., Dabbaghjamanesh, M., et al.: Effective management of energy internet in renewable hybrid microgrids: a secured data driven resilient architecture. IEEE Trans. Ind. Inf. PP(99). 1 (2021)
3. Fattahi, J., Wright, D., Schriemer, H.: An energy Internet DERMS platform using a multi-level Stackelberg Game. Sustain. Cities Soc. **60**(1), 102262 (2020)
4. Malik, B.: Networking, transportation education at the Core of SMC~3 logistics technology conference. Inbound Log. **39**(5), 34 (2019)
5. Smerdin, A., Ermachkov, G., Nezevak, V., et al.: Use of sensor networking technology to build a power transformer monitoring system. In: E3S Web of Conferences, vol. 224, no. 4, p. 02021 (2020)
6. Ndubuaku, V., Inim, V., Ndudi, U.C., et al.: Effect of social networking technology addiction on Academic Performance of University Students in Nigeria. Int. J. Recent Technol. Eng. **8**(5), 2277–3878 (2020)
7. Alnaser, A.: Certain contributions in traffic engineering based on software-defined networking technology. J. Comp. Sci. **15**(7), 944–953 (2019)
8. Nishizawa, H., Ishida, W., Sone, Y., et al.: Open whitebox architecture for smart integration of optical networking and data center technology [Invited]. J. Opt. Commun. Netw. **13**(1), A78–A87 (2021)
9. Islam, A., Mantymaki, M., Benbasat, I.: Duality of self-promotion on social networking sites. Inf. Technol. People **32**(2), 269–296 (2019)
10. Rawat, D.B.: Fusion of software defined networking, edge computing, and blockchain technology for wireless network virtualization. IEEE Commun. Mag. **57**(10), 50–55 (2019)
11. Ajitha, C., Jaya, T.: Cognitive radio technology with reduced PAPR and complexity for IoT-based OFDM networks. Peer-to-Peer Netw. Appl. **13**(6), 2065–2077 (2020)
12. Gope, P., Sikdar, B.: An efficient privacy-preserving authentication scheme for energy internet-based vehicle-to-grid communication. IEEE Trans. Smart Grid **10**(6), 6607–6618 (2019)

A Network Camouflage Technology that Randomly Adds False Information to Real Information

Dongling Xiao(✉), Fei Gao, Haohua Meng, Guoru Deng, Xiaoyan Wei, Jundong Huang, Zhaofeng Guo, Cheng Zhang, Rongtao Liao, Zheng Yu, Kai Cheng, and Bo Jin

State Grid Information and Communication Branch of Hubei Electric Power Co., Ltd., Wuhan, Hubei, China
973661626@qq.com

Abstract. With the rapid development of computer networks, while providing people with the convenience of resource sharing, they are also facing the threat of network hacker attacks. It is far from enough to only use network encryption technology to protect network information security. Therefore, the use of network camouflage technology to strengthen network security and prevent information leakage has become a current research hotspot. On this basis, this article randomly adds false information to the real information, uses network camouflage technology and Monte Carlo method to detect the location node of false information in the network, and calculates the detection error to verify whether the network camouflage technology can truly prevent network attacks.

Keywords: True Information · False Information · Network Camouflage Technology · Monte Carlo Method

1 Introduction

Network system attack methods are ever-changing, and it is impossible to prevent them. Incidents of network information data being truncated and modified frequently occur, which seriously hinders the promotion and application of network systems. How to realize the safe transmission of information in the computer network, ensure the correctness and completeness of the information, and prevent the network system from being violated.

Many scholars at home and abroad have conducted research on the network camouflage technology of randomly adding false information to real information, and have obtained effective results. For example, a research team developed a network intrusion detection system, which connects offensive data in network simulation services with network-oriented services, and the prevention system responds to attackers' intrusion methods, thereby protecting the normal operation of network information. The defect of this system is that the defense system does not feedback the attack situation of the intruder to the detection center, nor does it interact with the firewall system, and cannot

J. H. Abawajy et al. (Eds.): ICATCI 2022, LNDECT 170, pp. 845–851, 2023.
https://doi.org/10.1007/978-3-031-29097-8_100

fundamentally solve the problem of network information intrusion [1]. A certain scholar recommends the use of intrusion detection technology to detect network viruses for the intrusion detection mechanism of false data identification. When a node believes that its neighboring node is a virus point, the node sends a signal to all neighboring nodes of the virus point to seek their detection and evaluation of the suspicious virus point, and through these evaluations, comprehensive consideration of whether the suspicious virus point will threaten information security [2]. Although the research results on network anti-counterfeiting technology are good, the network system is still subject to large-scale attacks. In order to ensure network information security, it is necessary to strengthen the research and development of anti-counterfeiting technology.

This article describes the characteristics of wireless sensor networks, which can ensure network information security better than traditional networks. It implements network camouflage technology detection for the spatial random distribution of mixed false information in real information, uses Monte Carlo method to calculate detection errors, and compares them. Detection error when adding different amounts of false information.

2 Overview of the Characteristics of Wireless Sensor Networks and Network Camouflage

2.1 Wireless Sensor Network

Compared with traditional networks, wireless sensor networks have the following characteristics:

(1) Limited energy

Due to the small size and low cost of sensor nodes, exhausted nodes will not be used again. In addition, some sensor networks are highly adaptable, and most of them are distributed in places with harsh environments. It is difficult or even impossible for people to replace the batteries that supply the sensors. Therefore, the battery energy of the sensor is limited, which shortens the life of the network. When designing a sensor network model, we must first consider technical feasibility and power saving. The large-scale distribution of wireless sensor networks refers to the use of a large number of sensor nodes in the monitoring range in order to accurately obtain information [3].

(2) Self-organization

The sensor network places a large number of sensor nodes within the monitoring range, and the deployment location of the nodes cannot be predicted in advance. The wireless sensor network is created through automatic combination and adhesion between nodes, and does not require artificial means or other auxiliary tools to form a network system, the two nodes will coordinate with each other to organize into a huge network system [4].

(3) Robustness

Due to the influence of some factors such as energy exhaustion and environmental interference, some sensor nodes in the network may not work normally or even be damaged. However, the remaining sensor nodes can form a network again through mutual coordination and self-organization, which ensures that the entire sensor network will not be affected by the damage of some nodes [5].

(4) Scalability

When a new sensor node is needed to connect to the network, without other auxiliary tools to identify the security of the node, the wireless sensor network can spontaneously and effectively detect the new node information, so that it can quickly join the network topology, reconstruct the network, and participate in the global work [6].

(5) Transmission through wireless channel

Wired networks can be protected by gateways or other devices, making it difficult for external attackers to attack the network. However, unlike wired networks, wireless sensor networks communicate through wireless channels. Malicious nodes can easily inject, eavesdrop, modify transmission data, or perform malicious attacks by capturing nodes' knowledge of query tasks. This requires wireless sensor networks to have extremely high security [7].

2.2 Network Camouflage

Disguise refers to deliberate performance to cover up the original appearance. The camouflage can be classified according to the nature of the camouflage: one is to merge the hidden information in the network with the material environment in which it is located, so that the hidden information is not easy to be discovered, that is, the material is hidden in another state. For example, in the animal world, stick insects avoid being discovered by natural enemies by simulating their body color as the color of bamboo joints; the other is to conceal the subjective substance, the network information to be disguised combined with its subjective substance characteristics is hidden in In a secret space.

In theory, the adoption of camouflage theory and the introduction of camouflage technology in the field of network security encryption will greatly help improve network security and enhance network node defense. The main purpose of introducing network camouflage in network security is to further improve network security performance and prevent illegal network detection attacks. Network information can be distributed in the scope of security monitoring, by setting false information to disrupt the intruder's attack target direction, reducing the possibility of real information being attacked, and determining appropriate defense measures by analyzing the attack process and technical means. It can also prevent the actual value of the network from being detected by the attacker, and prevent the important or even confidential information in the network from being attacked [8, 9].

Active masquerading prevents detected attacks by actively masquerading specific network values. Active concealment is a unit hidden in the data network to recreate the data stream and modify the corresponding network attributes, such as source IP or destination IP address, ACK sequence number, REQ sequence number, etc. At the same time, the network camouflage unit has a retrieval function, which restores the normal network value of the datagram to the original attribute value of the datagram. In this way, for the application layer, transport layer, and network layer of the attack protocol, whether the active network camouflage system is used in the gateway or the host part, the hidden data information is visible after the data is detected and processed. Network intruders are not aware of the existence of a hidden detection network, so they believe that the exposed network information has been attacked and ignore the security defense of the hidden network [10]. IP address masquerading can also be classified as active masquerading. IP address masquerading is to use idle IP addresses to simulate a large number of virtual hosts with one host through the address translation NAT mechanism. Make the virtual host have the functions of a real host. The real host uses the IP address masquerading module to represent the limited virtual host to communicate with the host in the internal or external network, so that from the internal network or external network host can only see the masqueraded IP address, and cannot see the real internal network host In order to achieve the purpose of hiding the internal network host [11].

Passive camouflage because the simulated datagrams for active detection attacks have certain sequence or header attributes, the target can monitor the active detection attacks through the built-in intrusive detection system, determine the response type, and block active responses through passive camouflage. Identify the purpose of the attack and the actual network value of the attacker. Willing to adopt certain strategies to detect, monitor and slow down active attacks [12].

2.3 Monte Carlo Algorithm

Monte-Carlo algorithm (Monte-Carlo) is based on probability theory and mathematical statistics through independent multiple sampling of random variables or random processes, and the statistical analysis of the sample value is a method to obtain the characteristics of the research object. The Monte Carlo method is a statistical experimental method, which mainly provides solutions to various difficult and miscellaneous problems by implementing statistical sampling experiments in advance, so it is also defined as a random sampling technique. The Monte Carlo method combines the characteristics of a variety of methods. These methods have similarities, that is, the calculation and operation process of the method are similar. This process finds a rough solution to the problem through random number simulation. The theoretical basis is the law of large numbers, given any $n > 1, X_1, X_2, \ldots, X_n$, independent of each other, and have the same mathematical expectation and variance:

$$E(X) = \mu,\ D(X) = \sigma^2 \tag{1}$$

E(X) represents the expectation, D(X) represents the variance, and σ is the standard deviation of the random variable. For any $\lambda_\alpha > 0$, there are:

$$p\left(|\overline{x_N} - u| < \frac{\lambda_\alpha \sigma}{N}\right) \approx \frac{2}{\sqrt{2\pi}} \int_0^{\lambda_\alpha} e^{\frac{1}{2}t^2} dt = 1 - \alpha \tag{2}$$

Among them, α is the significant level and λ_α is the normal difference.

3 Research on Network Camouflage Technology

3.1 Research Purpose

Network security technology is as important as other high-tech technologies, and it will also generate huge benefits and value. At the same time, the development of network security technology is to protect important national information from theft and collusion. Especially in some special fields, information encryption such as satellite communications and electronic warfare plays a very important role. In the current global activities, interactive information on the Internet is the main means of reliance, ranging from simple shopping to war. The issue of information security affects our nerves and affects every corner of our lives. Whether it is from the national or individual level, it is urgent to strengthen the research of network information security technology. The security of the network system can not only stay in the technical algorithm, through the national laws and regulations, confidentiality system and other fields can also ensure information security, each field needs to complement each other to give full play to the greatest advantage.

3.2 Research Methods

This experiment first fixes a network monitoring area, randomly adds false information to the real information in this area, uses network camouflage technology to detect the spatial distribution of false information, and uses Monte Carlo method to calculate the detection error. Verify the effectiveness of network camouflage technology in preventing network intrusion.

4 Analysis of the Effectiveness of Network Camouflage Technology

4.1 Random Distribution of False Information and Real Information in Space

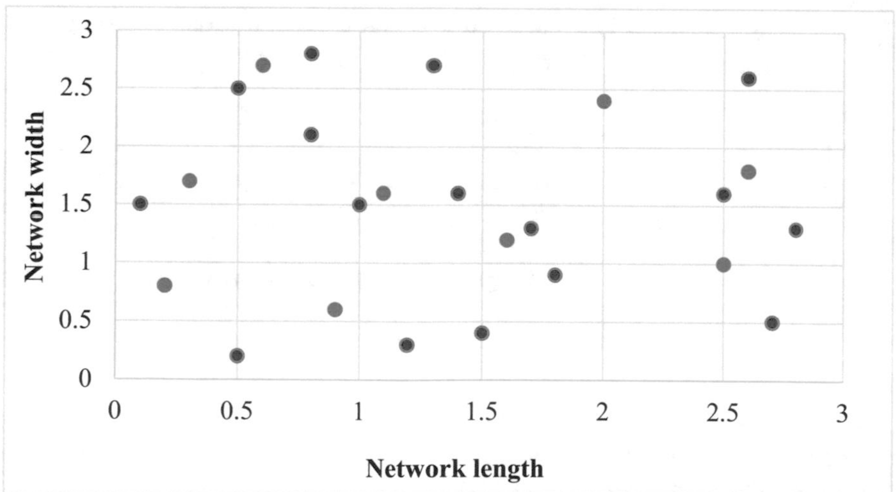

Fig. 1. Schematic diagram of the distribution of false information and real information on network nodes

Figure 1 shows the distribution of real information and false information on some network nodes. The red dots represent real information, and the blue dots represent false information. It can be seen from the figure that after adding false information to the real information, the two kinds of information are randomly distributed. In this disordered information distribution, to find the position of the false information in the real information, you can use the network camouflage technology. Then the error calculation of the detection result is carried out by the Monte Carlo method.

4.2 Network Camouflage Technology Inspection

Table 1. Detection error

	1 s	2 s	3 s	4 s	5 s	6 s	7 s	8 s
30	0.70	0.72	0.73	0.75	0.78	0.80	0.81	0.82
60	0.50	0.51	0.52	0.52	0.53	0.54	0.545	0.55
90	0.4	0.405	0.408	0.412	0.415	0.420	0.421	0.424

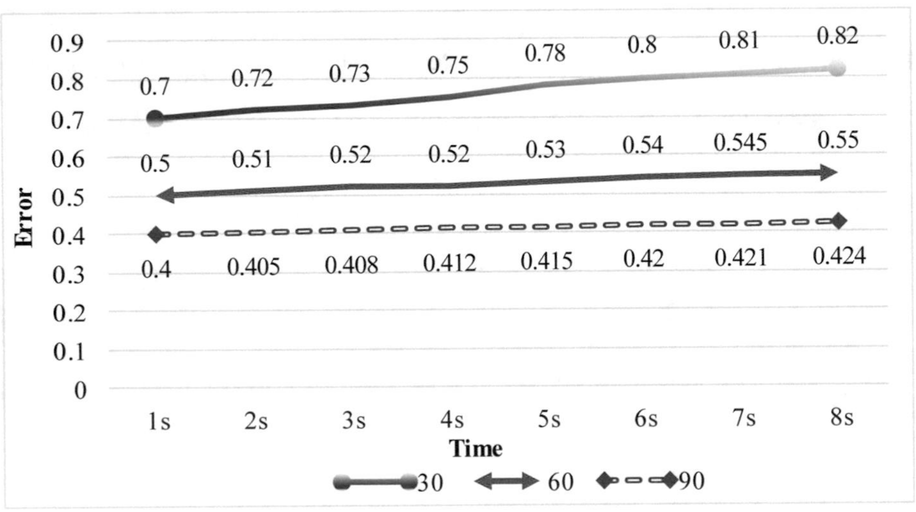

Fig. 2. Detection errors of different amounts of false information

As shown in Table 1 and Fig. 2, the Monte Carlo method is used to calculate the monitoring error when different amounts of false information are added to the real information. The positioning error of 30 false information is significantly larger than the positioning error of 60 and 90 false information, that is It is said that the more false information, the smaller the positioning error. When the more false information is added to the real information, the more false information can be accurately detected. However, as the detection time increases, the detection error will become larger and larger. This is because the degree of fusion of false information and real information increases, which increases the difficulty of detection, thereby reducing detection accuracy and leading to increased errors. On the whole, in the face of more false information, network camouflage technology is more effective.

5 Conclusion

With the continuous development of Internet technology, the opportunities and risks of network intrusion have increased sharply, and network information security has been greatly threatened. Therefore, the design of security measures to prevent unauthorized access to network system resources and information data is an urgent need in the current network security field. The important problem solved. The network camouflage technology studied in this paper is a technology developed in response to network attacks. This technology can effectively distinguish false information when true and false information is mixed. It is hoped that this technology can be promoted to prevent network intrusion and ensure network information security.

References

1. Huang, Q., Gao, H., Yang, S., Ding, D., Lin, Z., Ling, Q.: Ultrastable and colorful afterglow from organic luminophores in amorphous nanocomposites: advanced anti-counterfeiting and in vivo imaging application. Nano Research **13**(4), 1035–1043 (2020). https://doi.org/10.1007/s12274-020-2740-x
2. Darus, E.A.M.Y.: A case study on digital divide and access to Information Communication Technologies (Icts) in Pulau Tuba, Langkawi, Malaysia. Turkish J. Comput. Math. Educ. **12**(5), 1710–1718 (2021)
3. Zheng, Y., Zhang, X., Wang, F., et al.: Detection of people with camouflage pattern via dense deconvolution network. IEEE Signal Process. Lett. **26**(1), 29–33 (2019)
4. Ubaid, S., Shafeeq, M.F., Hussain, M., et al.: SCOUT: a sink camouflage and concealed data delivery paradigm for circumvention of sink-targeted cyber threats in wireless sensor networks. J. Supercomput. **74**(10), 5022–5040 (2018)
5. Chen, Y.-K., Chen, C.-H., et al.: Color adjustment for video-see-through AR and camouflage application. SID Int. Symp. Digest Technol. Papers **49**(2), 1064–1067 (2018)
6. Adelman, R.A.: Security glitches: the failure of the universal camouflage pattern and the fantasy of "identity intelligence." Sci. Technol. Hum. Values **43**(3), 431–463 (2018)
7. Chen, L.P., Yin, H., Yuan, L.G., et al.: A novel color image encryption algorithm based on a fractional-order discrete chaotic neural network and DNA sequence operations. Front. Inform. Technol. Electron. Eng. **21**(6), 866–879 (2020)
8. Elakrat, M.A., Jung, J.C.: Development of field programmable gate array-based encryption module to mitigate man-in-the-middle attack for nuclear power plant data communication network-science direct. Nucl. Eng. Technol. **50**(5), 780–787 (2018)
9. Li, Q., Wang, X., Wang, X., et al.: A novel grayscale image steganography scheme based on chaos encryption and generative adversarial networks. IEEE Access **99**, 1 (2020)
10. Tabash, F.K., Izharuddin, M., Tabash, M.I.: Encryption techniques for H.264/AVC videos: a literature review. J. Inform. Secur. Appl. **45**(Apr), 20–34 (2019)
11. Ratnavelu, K., Kalpana, M., Balasubramaniam, P., et al.: Image encryption method based on chaotic fuzzy cellular neural networks. Signal Process. **140**(nov), 87–96 (2017)
12. Soon, J.M., Manning, L.: Developing anti-counterfeiting measures: the role of smart packaging. Food Res. Int. **123**(Sep), 135–143 (2019)

Intelligent Traffic Flow Prediction Model Based on Deep Neural Network

Yongrong Li(✉)

School of Transportation, Chongqing Vocational College of Transportation, Chongqing 402247, China
13996144078@163.com

Abstract. With the development of my country's economy, people's material development level is getting higher and higher, and people's travel requirements are increasing. However, during peak hours and holidays, traffic congestion is easy to occur, so the importance of traffic flow forecasting is not important. It goes without saying. However, most of the current traffic flow forecasting methods are based on shallow models, which essentially record data through statistical methods. However, this method has the disadvantage of not being able to conduct more in-depth analysis of the data, so its traffic flow forecasting is often inaccurate. Due to its powerful learning ability, deep neural network is widely used in image processing, text recognition and other fields, but it is rarely used for traffic flow prediction. Therefore, the purpose of this paper is to study the intelligent traffic flow prediction model based on deep neural network. This paper is devoted to statistics and analysis of traffic flow data, combined with the improved deep neural network to establish a corresponding traffic model, so as to achieve the purpose of reducing traffic jams and even traffic accidents. The experimental research in this paper shows that the application of this deep neural model to traffic flow prediction, and through our optimized model effect, not only makes the traffic flow prediction close to 9% error rate, but also utilizes the characteristics of the deep neural network on the original basis. Using the function of traffic sign recognition, it can be said that the fitting degree is very good, which fully proves the effectiveness of our improved model.

Keywords: Deep Neural Network · Traffic Flow Prediction · Deep Learning · Traffic Sign Recognition

1 Introduction

With the continuous development of the social economy, the traffic conditions are becoming more and more mature, and the road infrastructure is gradually improving. Especially in the more developed cities, road traffic has taken the lead. However, with the increase in the number of cars, the road congestion, Problems such as traffic accidents are also becoming more and more frequent. In order to solve the traffic problem on the road, people have to turn to smart technology. Therefore, the intelligent traffic flow prediction model of deep neural network emerges as the times require.

J. H. Abawajy et al. (Eds.): ICATCI 2022, LNDECT 170, pp. 852–859, 2023.
https://doi.org/10.1007/978-3-031-29097-8_101

In the research of intelligent traffic flow prediction model based on deep neural network, many scholars have carried out related research on it and achieved good results. For example, in 2006, a famous deep learning article was published by the Bartlett Z team., and proposed a deep belief network (Deep is Belief Nets, DBN) model and an unsupervised learning algorithm, that is, a stacked restricted Boltzmann machine learning model, which laid the foundation for the future learning of deep neural networks [1]. In the ImageNet image classification competition in 2012, Pamula T's research group used the convolutional neural network model to reduce the extremely high error rate to nearly 15%, which aroused high attention and opened the prelude to the research of deep learning. Later The structure of this network is called AlexNet, which shows that deep network learning is very helpful to solve traffic problems [2].

Through the research, this paper adopts the convolutional neural network, combined with the target detection algorithm, and then establishes a machine learning model, and finally designs a complete set of traffic flow early warning scheme, which effectively alleviates various problems caused by traffic. The related research results not only it can bring convenience to traffic and greatly reduce the occurrence of traffic accidents.

2 Research on Two-Channel Convolutional Neural Network Model

2.1 Design of Two-Channel Convolutional Neural Network Model

(1) Overall structure of the model

The structural block of a DCNN network consists of 4 convolutional layers, and the DCNN in the inner layer is realized by the connection of the structural block. The model starts by combining several building blocks and concatenating them with a pooling layer, followed by a fully connected layer, and finally a Softmax detector.

(2) Feature Extraction

The feature extraction part is generally composed of several DCNN structural blocks, each of which is composed of two convolution modules, and the convolution module contains two convolution layers and an activation function layer. Here, first define some notations, denoting the neuron at (s, t) on the i-th input feature map of the convolution layer as, similarly, the coordinates on the j-th input feature map can be at (s, t) The neuron is the weight of the coordinates (u, v) in the convolution kernel, and the convolution kernel connects the ith input feature map and the jth output feature map [3, 4]. Therefore, we define the neurons on the output of a convolutional layer as:

$$z_j, s, t = \sum_{i=1}^{q} \sum_{u=1}^{3} \sum_{v=1}^{3} (w_{ij}, u, v \cdot xi, s + u - 2, t + v - 2) \quad (1)$$

where q is the number of feature maps.

2.2 Feature Classification

Since the convolutional layer adopts almost the same feature classification method as the traditional algorithm, when we design the model of the DCNN model, we must first solve the training problem of the deep neural network. In this paper, we adopt almost the same classification method as ResNet, such as: do not change the settings of parameters such as the number of convolution filters, size and stride [4]. At the same time, because ResNet has many defects in training the neural network, this paper makes corresponding improvements to the network structure of DCNN after finding relevant information. The main differences are as follows:

(1) Since ResNet does not interfere and control the output data, and the Highway controls the final output data through the control gate, it will inevitably increase the number of training times and thus increase the amount of calculation. The threshold for learning this control gate parameter is raised [5, 6]. To this end, this paper summarizes a solution by looking for relevant literature, combining theoretical analysis and doing a lot of experiments: by adjusting the ratio of the convolution channel to the straight channel and introducing the convolution attenuation factor, the above solution is solved.
(2) Due to the inconsistency of dimension size between the through channel and the convolution channel, this paper solves this problem by increasing the feature map.

2.3 Adjustment of Network Weights and Thresholds

(1) Forward propagation stage:

At this stage, the image will change step by step from the input layer and finally to the output layer. This process is also the process performed by the network when it is running normally after training. In this process, the network actually completes the process of dot multiplication of the input image with the convolution kernel (weight) matrix of each layer and pool sampling to obtain the final output result.

(2) Backpropagation stage:

By calculating the difference between the actual output and the corresponding ideal output, calculating the derivative of the weights and thresholds, and then adjusting the corresponding weights and thresholds, the two processes of forward propagation and back propagation are alternated repeatedly until the final until the requirements are met.

3 Training of Convolutional Neural Networks

3.1 Data Preprocessing

The first step is to matrixize the dataset. According to the characteristics of traffic flow data, a one-dimensional array can be expressed as $[z11, z21, x_1^{t1}, x_2^{t1}, x_3^{t1}, \ldots, x_i^{t1}]$, which means the traffic flow distribution from the 1st to the ith day. With weather and holiday factors, each dimension of the array can be represented as $[z11, z21, x_1^{t1}, x_2^{t1}, x_3^{t1}, \ldots, x_i^{t1}]$

[7]. After the training samples are preprocessed, they are mainly processed by convolutional layers and subsampling layers. The convolution operation on the input feature is actually a process of weighting and summing all the values in the local receptive field vector with the convolution kernel each time, and then evaluating it through the activation function. After that, the output is passed to the subsampling layer, and the subsampling layer extracts the features that can best express the original data. This is also the most important reason why the convolutional neural network can perform feature learning. Then, we take the output of the feature through two convolutional layers and two subsampling layers as input. After the training is completed, the prediction can be performed, and the obtained output is the final prediction result.

3.2 Forward Propagation

The convolution operation on the input feature is actually a process of weighting and summing all the values in the local receptive field vector with the convolution kernel each time, and then evaluating it through the activation function. After that, the output is passed to the subsampling layer, and the subsampling layer extracts the features that can best express the original data, which is also the most important reason why the convolutional neural network can perform feature learning. Subsequently, we take the output of the feature after passing through two convolutional layers and two subsampling layers as the input of SVM [8, 9]. For the training of SVM, it mainly includes the selection and adjustment of parameters such as penalty factor and kernel function. When the SVM training is completed, the prediction can be performed, and the obtained output is the final prediction result.

3.3 Error Back Propagation

In the process of error back propagation, the residual ε(i) of the eigenvector node needs to be calculated, and the residual value is the difference between the actual observed value and the fitted value. Calculated as follows:

$$\varepsilon(i) = \sum_{j=1}^{m} \omega_{ij} \varepsilon 0(j) \tag{2}$$

3.4 Calculate Gradient Residuals

By restoring the feature map to be restored to its original form, the residual value of the corresponding previous sub-sampling layer can be obtained. The sub-sampling layer is forwarded to the convolutional layer, and the calculation formula is as follows:

$$\varepsilon_s(i) = \sum_{j=1}^{n} conv(\varepsilon_c(j), rot180(Kij)', l') \tag{3}$$

3.5 Adjustment of Convolutional Neural Network Parameters

The realization of the convolutional neural network passenger flow prediction model mainly includes five modules: data preprocessing, initialization of network structure and parameters, training prediction of CNN-SVM model and evaluation of model performance. The data preprocessing includes the steps of training set division and normalization in the previous chapter. The initialization part of the parameters mainly initializes the relevant parameters of the convolutional layer and the subsampling layer, such as the size of the convolution kernel, the connection weight, the bias, the reduction of the subsampling, etc., as well as the parameters of the SVM. During training, the training task of the model is completed by randomly selecting samples for iteration [9, 10]. In the final evaluation part, we still use the three evaluation criteria of MAE, MRE, and RMSE to evaluate the effectiveness of the model, and visually observe the predicted effect through the test set. For the convolutional network part, the network structure is mainly optimized by adjusting the number of convolutional layers of the convolutional neural network, the size and number of convolution kernels of each convolutional layer, the number of subsampling layers, and the length of samples [10, 11]. The convolution kernel weights are generally assigned between $[-1, 1]$, and the biases are all set to 0. For the size of the convolution kernel, if the size of each convolution kernel is different, the CNN model network will be too complicated, and the training efficiency will be greatly reduced. Therefore, this paper considers the case where the convolution kernels of each convolution layer have the same size but different numbers. In the SVM part, the parameters that need to be adjusted are the penalty factor C and the kernel function. The experiments will discuss the results under different C values and kernel functions respectively [12].

4 Experimental Analysis

4.1 Error Comparison Under Different Learning Rates

For the VAR model, the selection of the learning rate is also very important. If the learning rate is too large, the model will oscillate violently, so that the loss will not converge and decrease with the increase of iterations. If the learning rate is too small, the learning will be ineffective, the descent direction cannot be found quickly. But for now, manually adjusting the learning rate is the most common and better method. We control the number of hidden layers and nodes, and only adjust the learning rate under the same conditions. It is the error comparison of the VAR model under different learning rates. The experimental results are shown in Table 1:

Table 1. Error Comparison Of VAR Models Under Different Learning Rates

Learning Rate	MAE	MRE	RMSE
0.06	11.6	0.1684	4
0.1	9.91	0.1328	1
0.2	11.2	0.1478	3
0.4	10.8	0.1481	3
0.9	11.4	0.1542	5
9	12.8	0.2049	10
90	21.5	0.266	17

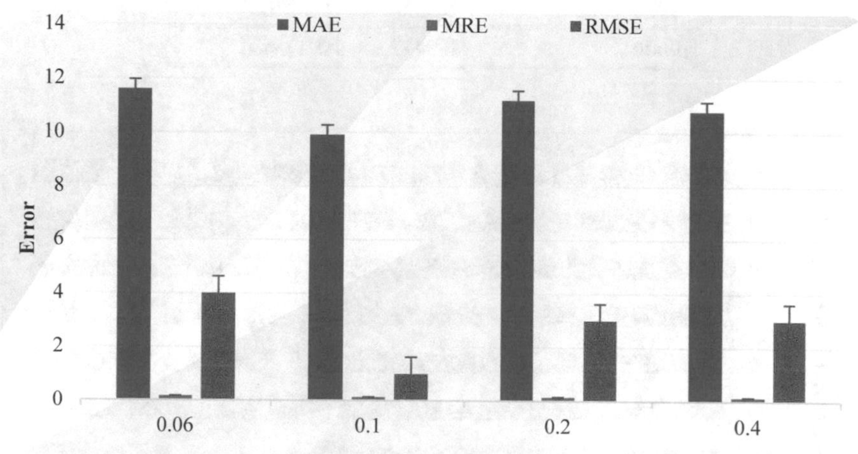

Fig. 1. The Effect Of Learning Rate On The Accuracy Of Model Predictions

As can be seen from Fig. 1, the learning rate has a certain influence on the accuracy of the model prediction. When the learning rate is too small, it does not show better results, and when the learning rate is too large, a large error rate will occur. Therefore, it can be concluded that within a certain range, it is better to choose a relatively small learning rate as much as possible.

4.2 Experimental Analysis of Training Accuracy and Training Loss of Warning Signs

The software Caffe comes with a LeNet-5 model, which uses adaptive gradient descent with a learning rate of 0.0001 for training, the momentum is set to 0.9, the maximum number of iterations is 10000, and the learning strategy is set to inverse proportional descent. The images are single-channel grayscale images with a resolution of 32×32 pixels. First, the warning sign sample set is trained, and the results are shown in Table 2:

Table 2. Warning Mark Training Accuracy and Training loss

Number Of Iterations	Accuracy	Training Loss
1000	0.6321	1.2523
2000	0.7921	0.9821
3000	0.8534	0.7512
4000	0.8943	0.5261
5000	0.9101	0.4531
6000	0.9223	0.4314
7000	0.9145	0.3910
8000	0.9235	0.3543
9000	0.9354	0.2236
10000	0.9472	0.2234

Fig. 2. The Effect Of The Number Of Iterations On The Training Accuracy And Training Loss Of Warning Labels

As can be seen from Fig. 2, as the number of iterations continues to increase, the accuracy rate increases rapidly in the early stage. After reaching 2,000 times, the rate of increase in the accuracy rate slows down. After 3,000 times, it tends to be stable and fluctuates from time to time. The maximum accuracy rate is 94.72%. At the same time, the loss value decreased relatively smoothly, from 1.2523 to the lowest value of 0.2234, and the training loss trended downward.

5 Conclusions

The deep neural network model has powerful feature learning ability, but the effect of time series prediction is often unsatisfactory. In this paper, the deep neural network model is improved, and the improved deep learning model is used to predict the traffic flow. Under the verification of the comparison experiment with the previous shallow structure, it is proved that the deep neural network model can accurately fit the prediction. The trend and the prediction accuracy are higher, and the prediction effect is better.

References

1. Bartlett, Z., Han, L., Nguyen, T.T., Johnson, P.: A Novel online dynamic temporal context neural network framework for the prediction of road traffic flow. IEEE Access **7**, 153533–153541 (2019). https://doi.org/10.1109/ACCESS.2019.2943028
2. Pamula, T.: Impact of data loss for prediction of traffic flow on an urban road using neural networks. IEEE Trans. Intell. Transp. Syst. **20**(3), 1000–1009 (2019)
3. Sivabalaselvamani, D.: Real time traffic flow prediction and intelligent traffic control from remote location for large-scale heterogeneous networking using TensorFlow. Int. J. Future Gener. Commun. Network. **13**(1), 1006–1012 (2020)
4. AbidEeN, Z.U.: The deep 3D convolutional multi-branching spatial-temporal-based unit predicting citywide traffic flow. Appl. Sci. **10**(21), 7778 (2020)
5. Pamua, T.: Impact of data loss for prediction of traffic flow on an urban road using neural networks. IEEE Trans. Intell. Transp. Syst. **20**(3), 1000–1009 (2019)
6. Naik, B., Obaidat, M.S., Nayak, J., et al.: Intelligent secure ecosystem based on metaheuristic and functional link neural network for edge of things. IEEE Trans. Indust. Inform. **99**, 1 (2019)
7. Selvi, K.T., Thamilselvan, R.: An intelligent traffic prediction framework for 5G network using SDN and fusion learning. Peer-to-Peer Network. Appl. **15**(1), 751–767 (2021). https://doi.org/10.1007/s12083-021-01280-6
8. Tedjopurnomo, D.A., Bao, Z., Zheng, B., et al.: A survey on modern deep neural network for traffic prediction: trends, methods and challenges. IEEE Trans. Knowl. Data Eng. **99**, 1 (2020)
9. Ribeiro, M., Samatelo, J., Bazzan, A.: A new microscopic approach to traffic flow classification using a convolutional neural network object detector and a multi-tracker algorithm. IEEE Trans. Intell. Transport. Syst. **99**, 1–5 (2020)
10. Priambodo, B., Ahmad, A., Kadir, R.A.: Spatio-temporal K-NN prediction of traffic state based on statistical features in neighbouring roads. J. Intell. Fuzzy Syst. **40**(6), 1–15 (2021)
11. Engel, V.J.L., Joshua, E., Engel, M.M.: Detection of cyber malware attack based on network traffic features using neural network. Khazanah Informatika : Jurnal Ilmu Komputer dan Informatika **6**(1), 26–32 (2020). https://doi.org/10.23917/khif.v6i1.8869
12. Panimalar, P.: Particle swarm optimization algorithm based artificial neural network for botnet detection. Wireless Pers. Commun. **121**(4), 2655–2666 (2021). https://doi.org/10.1007/s11277-021-08841-1

Data Mining of E-commerce Agricultural Product Reviews Based on Big Data Algorithm

Xiaoning Cui[1] and Haitao Chen[2](✉)

[1] School of Economics and Management, Wuchang Institute of Technology, Wuhan, Hubei, China

[2] School of Marxism, Wuchang Institute of Technology, Wuhan, Hubei, China

cht_wcgxy@163.com

Abstract. This study mainly uses scientific research methods such as questionnaire survey and statistical analysis, uses Octopus data collection tool to obtain relevant corpus, then uses Jieba word segmentation tool to process text word segmentation, and finally analyzes the weight of each comment with the help of TF-IDF algorithm, and finally obtains the most representative comment content, In order to explore the difficulties and solutions of College Students' e-commerce marketing, we hope to provide some reference for the innovation and Entrepreneurship of College Students' campus e-commerce marketing.

Keywords: "Internet+" · Agriculture Products · Electronic Commerce · Big Data · Data Mining

1 Introduction

Since the introduction of the Internet plus plan in 2015, as the country's agricultural base, the practitioners in their field have begun to explore the development path of Internet plus agriculture in the current unique era. In its exploration process, the integration of Internet e-commerce platform and agriculture is undoubtedly the most intuitive and fastest way to improve the value of agricultural products [1]. At present, domestic e-commerce giants have achieved great success in the agricultural field, but for practitioners in the agricultural field, there are the following two deficiencies:

(1) Companies with well-known e-commerce platforms may take advantage of the platform to ask for more benefits from agricultural enterprises as partners, the value created cannot be evenly distributed, and agriculture cannot be fully developed;
(2) From the perspective of competition and development, for future prosperity and development, there should be various forms of agricultural e-commerce platforms in various subdivided fields, rather than just pinning their hopes on the currently mature three or four e-commerce giants [2].

J. H. Abawajy et al. (Eds.): ICATCI 2022, LNDECT 170, pp. 860–867, 2023.
https://doi.org/10.1007/978-3-031-29097-8_102

From the above reasons, it can be concluded that it is indispensable for all localities to develop their own agricultural e-commerce platforms. Although the development technology of e-commerce platform has long been mature, and practitioners everywhere can quickly build a set of e-commerce system, there is a lack of additional big data development system to deal with a large amount of data brought by 5g high-speed network, not to mention mining and analyzing a large amount of user data through artificial intelligence to improve their industrial value. Therefore, this study applies questionnaire survey to obtain relevant e-commerce data of agricultural snacks for college students, and uses Octopus data collector to obtain the snack review data of tmall mall, extract its high-frequency words, and analyze the development status of agricultural snacks by using TF-IDF and other methods for the reviewed data, so as to put forward some targeted countermeasures, To promote college students' e-commerce marketing [3].

2 Research Status at Home and Abroad

The concept of "big data" was put forward by Toffler, a famous futurist, as early as the 1980s. In the context of the second revolution, but the accurate definition of "big data" has not been unified. The National Science Foundation (NSF) defines big data as "a large-scale, diversified, complex and long-term distributed data set generated by a variety of data sources such as sensor devices, Internet transactions, e-mail and personal short messages" [4].

In the era of cloud computing, ubiquitous terminal devices have not stopped producing data, and billions of Internet users have completed clothing, food, housing, transportation and entertainment activities in the online world built by data [5]. The flood of data generated by online transactions has brought a large number of structured business data and semi-structured user behavior data to e-commerce. Based on these data, mathematicians and developers can explore the great value of data by using algorithms full of mathematical logic and data processing programs adapted to algorithms. According to different processing made in different business fields, they can get solutions to optimize the current project. This approach is also the mainstream industrialization scheme in the big data industry [6].

The idea of big data technology abroad has always been ahead of the actual demand. In 2008, when the production of large-scale Internet data was still at the GB level, three pioneering papers on the big data industry, Google filesystem, big table and MapReduce, were published. Since then, visionary foreign Internet companies have been committed to the research and development of big data technology [7]. Facts have proved that they are right. With the update and upgrading of network transmission media and the threshold of networking terminals falling. More and more people participate in data production. Enterprises mine the great value contained in various forms of data through big data technology, and users gain a better service experience. According to the report of the world economic forum in Davos in 2018, the scale of the global big data industry in that year has reached US $45.4 billion, and experts also believe that this figure will exceed US $80 billion in 2021.

In China, the development of big data has also been supported by various national policies. Since the 12th Five Year Plan, the Ministry of industry and information technology and other relevant departments have begun to support a number of big data related project research. In order to support the development of e-commerce logistics, the Ministry of industry and information technology gives policy support and encourages e-commerce, especially big data, to continuously innovate and develop. According to relevant statistics, the output value of China's big data market increased significantly the year before last, close to the level of trillion yuan [8]. According to the prediction of relevant research institutes, by 2022, the scale of China's big data industry will exceed 120 billion yuan. In China, we have a number of e-commerce enterprises led by Taobao, whose audience represents high-quality and a large number of data sources [9].

According to the survey, at present, the use of big data by major e-commerce in China mainly focuses on cloud technology as the core, building user portraits according to behavioral data for marketing activities of various categories, and improving user experience and in-depth development of data processing business at the technical level. As a new technology, big data is developing rapidly. We can use this technology to process and integrate data quickly. Compared with the traditional information integration technology, it is more convenient and fast. It is widely used in the acquisition of product and customer information in all walks of life [10]. Rapidity, magnanimity and diversity are the remarkable characteristics of big data technology. With the help of algorithms, cloud technology and other technologies, people's consumption habits, shopping habits and browsing habits in Internet information can be tracked and recorded, so as to obtain more data about users' personality and provide a technical basis for pushing targeted information. For e-commerce enterprises, obtaining consumer demand data is not only the key information affecting their product R & D and design, but also an effective way to obtain competitive advantage in the industry. After obtaining a large amount of user information, e-commerce enterprises can form a portrait of the target consumer group with the help of data statistics and analysis technology, so as to adjust the marketing strategy, realize the accurate marketing of the target consumer group, and expand the publicity scope and sales of their own products.

3 Research Methods

Firstly, through the method of questionnaire, face the school students and relevant corpus sets obtained through data collection tools such as octopus, then conduct text word segmentation through Jieba word segmentation tool, stop and stop operation through stopping word list, and then extract keywords through word frequency statistics and TF-IDF, TF-IDF algorithm is mainly a common weighting technology for information retrieval and information exploration. The application of this technology can better calculate its weight. Under this background, the weight statistics of keywords in comments are carried out, and the heavier keywords are analyzed according to the marked part of speech to form a multi-dimensional analysis table with brand, taste and satisfaction as attributes, So as to realize text comment analysis. The specific implementation method is shown in Fig. 1.

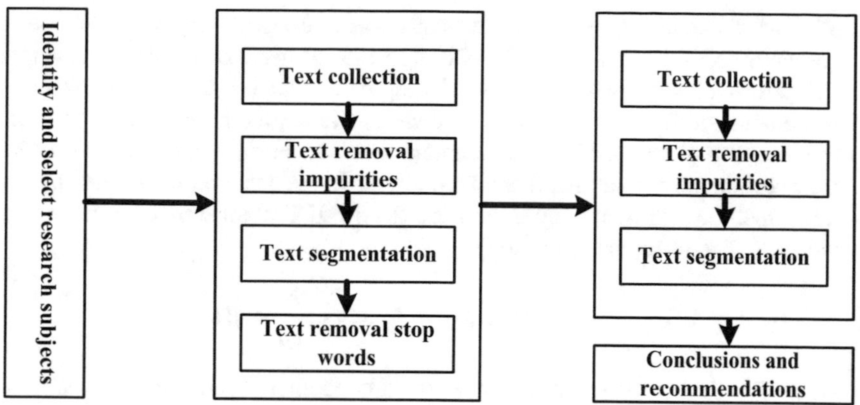

Fig. 1. The review data mining technology roadmap

4 Text Acquisition and Processing

4.1 Data Sources

This paper uses questionnaire survey to obtain relevant e-commerce data of agricultural products and snacks for college students, and uses Octopus data collector to obtain snack review data of tmall mall, so as to realize the collection of text data. There is usually mining value in the reviewed information. Through the analysis of text information, businesses can understand users' preferences and improve product information, so as to obtain the initiative in the market competition. Therefore, the research mainly selects college students for data analysis. Due to layout restrictions, the basic information table of respondents is omitted.

As shown in Table, the people surveyed in this survey are mainly college students, whose ages are mainly 20–25 years old. The service life of tmall mall is mainly 1–3 years, which is relatively long. Therefore, it is representative to select this group.

4.2 Data Processing

The filtered text data is cleaned, mainly including removing duplicate and meaningless text comments, and then the text word segmentation in the comment set needs to be completed. In this paper, Jieba word splitter, which has a good effect on Chinese word segmentation, is used to segment comment articles. After word segmentation, the whole document becomes a word set. For words with low relevance to the text content, the stop words in the comment text are filtered through the "Baidu stop words list".

4.3 Text Statistics and Analysis

(1) word frequency statistics and labeling

After the above calculation, this paper selects words with word frequency greater than 46 for part of speech tagging, and explains the part of speech of the words after word segmentation. They are marked as noun (n), verb (V), adjective (adj) and noun (NS). On this basis, TF-IDF algorithm is used to calculate the weight of the retrieved text keywords. The word weight calculated by TF-IDF represents the importance of the word. The larger the weight value, the higher the importance of the meaning of the word in the comment data set. Further analyze the content of the questionnaire, assuming that all comments in the data are represented by C and a comment is represented by d, and use the TF-IDF algorithm to calculate the de weight i. The formula is as follows:

$$\omega_{ij} = tf \times idf = f_{ij} \times \log(\frac{N}{N_i} + \beta) \tag{1}$$

where f_{ij} refers to the frequency of word i in document d_j, N_i refers to the number of documents in which word i appears in the document, and β is an empirical value, generally 0.01 and 0.1. Through the word system clustering, the weight sum of different dimensions is obtained. Through the graph screening of the weight sum of different dimensions, the selected weight is greater than 0.05 for clustering, then the key words with large weight value are analyzed, and finally the marketing strategy is analyzed.

After filtering and cleaning the captured data, 14 high-frequency comments are obtained, as shown in Table 1.

Table 1. Statistics and footnotes of words in commentary text

Terms	Frequency	Word class	Terms	Frequency	Word class
Eat	153	v	Logistics	87	n
Cheap	118	n	Fast	84	adj
Substantial benefits	106	n	Simple	83	n
Good	97	n	Buy	81	v
Taste	93	n	Attitude	77	n
Original taste	90	n	Buy-back	61	v
Packing	89	n	Happy	54	adj
No add	88	n	Quality	46	n

The results show that the most mentioned words by consumers are "eat", followed by "cheap" and "affordable", which is in line with the business characteristics of agricultural snacks. Snacks of agricultural products are mainly cheap and affordable, which has won the favor of consumers; Secondly, the high-frequency words are "good" and "taste", which shows that the taste of agricultural snacks is in line with the needs of consumers, so as to meet the requirements of consumers.

(2) text center word analysis

Firstly, the comment data set is preliminarily processed. On this basis, TF-IDF algorithm is applied to extract keywords, and the weight and retrograde analysis of the extracted keywords are carried out. At the same time, these words are classified into five levels: commodity and quality, taste, logistics, satisfaction and other value. The weight of keywords in comments mainly reflects the importance of words. The higher the weight value is, the higher the importance is. On the contrary, the lower it is. Through word clustering, the weight sum of different levels is calculated, and then the display diagram is made according to the weight sum of different levels. As shown in Fig. 2.

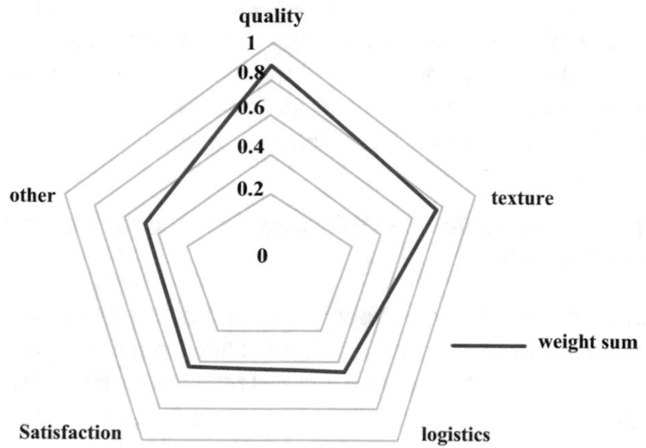

Fig. 2. The radar display diagram of weight sum of each dimension

The quality of products is mainly reflected in the freshness and taste of agricultural products, as well as the packaging of agricultural products. From the weight ratio of quality, with the development of society, people pay more and more attention to the quality of agricultural products. In terms of evaluation, subjective attitudes such as "like" and "satisfaction" account for a relatively high weight, followed by "buy back" and "happy", which means that a large proportion of the commodity consumption comes from the buyback of old consumers. It also shows that the purchase expectation of some customers for products mainly comes from the most primitive demand in their hearts. In terms of logistics distribution, consumers have high demand for the choice of delivery speed and logistics mode, so fast delivery speed can obtain customer satisfaction.

5 Countermeasures and Suggestions

5.1 Strengthen the Standardization System of Agricultural Products and Do a Good Job in Quality Assurance

Quality assurance is the most basic requirement for the sales of agricultural products. Therefore, meeting the needs of consumers, strengthening quality standards and improving quality requirements are the general trend of the development of e-commerce agricultural products. In this context, businesses need to strictly select agricultural products, improve the inspection system of agricultural products, and fundamentally ensure good product quality; Secondly, we should do a good job in the guarantee of quality inspection, strictly control each step of inspection, and fundamentally ensure the quality of products.

5.2 Do a Good Job in Processing and Cold Chain, and Ensure the Taste

The taste guarantee of agricultural products mainly lies in cold chain transportation and product processing. Therefore, in the transportation process, farmers should focus on the upward nature of agriculture, and the government should promote the scientific formulation of supporting technical specifications, government regulations and policies suitable for market expectations, and establish an agricultural production and commodity circulation system suitable for agricultural network goods.

5.3 Promote the Development of Third-party Logistics and Do a Good Job in Logistics Guarantee

From a long-term perspective, the differentiation and diversity of third-party logistics and distribution systems can not only improve the added value and price of agricultural products, realize the scale of economic benefits and reduce logistics costs, but also reduce the necessary costs of operators and producers, so as to increase their interests.

5.4 Do a Good Job in Service Details and Satisfaction Guarantee

In the process of operation and implementation, the company can also establish an automatic customer service management system of the company's network, independently and timely use the Internet to provide high-quality services according to the requirements of customers, so as to form a good impression in the eyes of consumers, so as to encourage consumers to continue shopping and establish a long-term dependence relationship with consumers.

6 Conclusion

This study mainly adopts scientific research methods such as questionnaire survey and statistical analysis, uses Octopus data collection tool to obtain relevant corpus, then uses Jieba word segmentation tool to segment text words, finally analyzes the weight of each comment with the help of TF-IDF algorithm, and finally obtains the most representative comment content, so as to explore the difficulties and solutions of College Students' e-commerce marketing, I hope it can provide some reference for the innovation and Entrepreneurship of College Students' campus e-commerce marketing. At present, e-commerce is developing very rapidly. It is not only the booster of China's economic development, but also provides a steady stream of skills for China's sustainable economic development. Nowadays, with more and more business entities pouring into the e-commerce industry, the competitive pressure among e-commerce enterprises is increasing. How to get more consumers' attention and favor has become the development direction of e-commerce enterprises. The ability of big data technology to process data is very large. On one hand, it can obtain the needs of e-commerce enterprise users, so as to provide the basis for its targeted marketing. In order to further expand product sales, e-commerce enterprises need to improve marketing accuracy with the help of big data technology, so as to impress more consumers. Therefore, e-commerce enterprises can give full play to the important role of big data technology in all aspects of e-commerce from five aspects: changing marketing concepts, paying attention to customer data, using data technology, implementing differentiated marketing and meeting customer needs.

References

1. Sharma, A.: Application research on big data marketing in e-commerce enterprises. Age Wealth **11**, 179–185 (2021)
2. Shrabastee, B.: Research on e-commerce precision marketing application based on big data technology. Econ. Manage. Abst. **22**, 181–187 (2021)
3. Marian, D.R.: Research on marketing management strategy of webcast e-commerce under the background of big data. Eur. J. Oper. Res. **22**, 9–18 (2021)
4. Shrabastee, B.: E-commerce precision marketing strategy under big data environment. J. Mark. Res. **29**, 11–17 (2021)
5. Lim, S.: Application of logistic regression algorithm in e-commerce big data recommendation system. Int. J. Asian Bus. Inform. Manage. **18**, 154–159 (2021)
6. Lucas, M.P.: Construction of e-commerce precision marketing strategy based on big data. Comput. Sci. Rev. **8**, 11–22 (2021)
7. Joris, B.: E-commerce big data classification and mining algorithm based on artificial intelligence. J. Retail. Consum. Serv. **14**, 146–152 (2021)
8. Shreya, B.: Application of business intelligence based on big data in e-commerce data analysis. Electron. Commer. Res. Appl. **6**, 37–49 (2021)
9. Shanthan, K.: Design and application of big data visualization tool – based on e-commerce industry. Decis. Support Syst. **9**, 149–155 (2021)
10. Luohui, Z.: Research on perceived quality evaluation of e-commerce products based on big data analysis. China Bus. Theory **8**, 44–49 (2021)

Design of Equipment Data Analysis Algorithm Based on Convolution and BP Neural Network

Minghu Tang(✉)

Qinghai Minzu University, Xining, Qinghai, China
qh1580971@163.com

Abstract. In view of the increasingly rich types of equipment data, statistical algorithms have been unable to extract various types of feature information and evaluate equipment status. This paper aims to study the design of equipment data analysis algorithm based on convolution and BP neural network. Based on the analysis of the theoretical model of convolution neural network, the structure of BP neural network and the design of BP neural network topology, the equipment data collection and analysis. The system is designed, and the equipment data analysis algorithm is constructed using convolutional neural network and BP neural network. Finally, this paper has tested the system, and the test results show that the performance of the system can fully meet the needs of users.

Keywords: Convolutional Neural Network · BP Neural Network · Equipment Data Analysis Algorithm · Data Acquisition and Analysis System

1 Introduction

With the advent of various types of smart devices and high-precision instruments, every human move will be recorded and protected in the form of digital information [1, 2]. With the vigorous development of big data analysis technology, more industries are looking for ways to discover the value behind these data analyses and extract the meaning behind various data analyses [3, 4].

Domestic scholars have conducted more detailed investigations on the future implementation prospects and development trends of equipment data analysis. In view of the differences in data integration methods and standards between different big data platforms, some researchers have put forward suggestions to promote data standard integration, which allows data to be interoperable between different platforms. By integrating data templates, data can be discovered and embedded, and appropriate guidance for lifestyles can be provided [5, 6]. Some scholars combine the data characteristics of the equipment with the current status of data analysis technology research, explain the source, characteristics and data collection methods, and use data mining, including classification, regression analysis, etc. [7, 8]. The research speed of foreign researchers on equipment data analysis technologies such as association rules and function analysis is faster than that of domestic researchers. Some researchers have summarized the challenges faced by equipment data mining, and summarized these challenges into five

J. H. Abawajy et al. (Eds.): ICATCI 2022, LNDECT 170, pp. 868–876, 2023.
https://doi.org/10.1007/978-3-031-29097-8_103

aspects: data collection, data preprocessing, data conversion, modeling, and model evaluation [9, 10]. Some researchers have proposed a big data analysis system for equipment, and designed a personalized adaptive analysis technology for data conversion and processing. The system updates the knowledge base and provides feedback based on the results of data analysis, and makes personalized recommendations to users. However, the specific sectoral model of data analysis context awareness and data mining has not been resolved [11, 12]. The time series data collected by the equipment is very large, and it is impractical to conduct data mining on the time series directly. Therefore, data discretization technology is usually used, and a specific mapping function is used on time series data to achieve data compression and create a smaller feature data set.

On the basis of consulting a large number of relevant references, this paper combines the theoretical model of convolutional neural network, the structure of BP neural network and the design of BP neural network topology, and designs equipment data collection and analysis based on convolution and BP neural network data analysis algorithms. Analyze the system and test the performance of the system to verify whether it meets the requirements of this article.

2 Design of Equipment Data Analysis Algorithm Based on Convolution and BP Neural Network

2.1 Theoretical Model of Convolutional Neural Network

2.1.1 Partial Receptive Field

Convolutional neural networks are mainly studied based on the definition of receptive fields. The local receptive field is based on the definition from the physiology of the visual cortex. When human beings perceive external things, from the part to the whole. In contrast, the images that are relatively far apart in the graph are relatively weakly related to each other, while the partial images are relatively close, and are relatively closely related to the overall. Therefore, in the process of figure cognition, local cognition can be used to replace global perception. Moreover, a certain method can also be used to combine local cognition at a deeper level to complete global cognition. And this local connection characteristic can also reduce the number of weighting parameters required in the neural network training process.

2.1.2 Parameter Sharing

Although the idea of local receptive fields is adopted, there are still multiple connections inside the neurons in the convolutional neural network. In order to reduce the massive amount of data it generates, a weight distribution strategy is introduced in the convolution mode. Before introducing weight distribution, one must know a principle. That is, the statistical characteristics of some parts of the image must be consistent with the statistical characteristics of other parts. In the local connection, each neuron has one hundred parameters, and these parameters come from the convolution function. When we get the features in the picture by connecting, the string represents a way to get the same attributes regardless of the situation. Attributes derived from one place can also be used elsewhere, so that you can use the same attributes in all images. In other words, that is, each 1M neuron shares the same 100 parameters.

2.1.3 Multiple Convolution Kernels

Through weight sharing, only one feature is learned through a 3x3 convolution window. In this case, the derived function is obviously unreliable and insufficient. To learn as many functions as possible, multiple functions are required. The extraction method is multiple convolution kernels, that is, 100 different convolution kernels, with 100 attributes and 100 feature maps. Different colored windows represent different filters as the core of the convolution. Each Convergence Nucleus creates another new image by navigating the image, which can be viewed as a different channel of the same image.

2.2 BP Neural Network Structure

2.2.1 BP Network Structure Model

The BP network learns and saves a large number of I/O matching relationships, and automatically adjusts network weights and thresholds by using features such as error detection and error redistribution to minimize network errors. The basic structure of BP neural network module is divided into input layer, hidden layer and output layer. The biggest difference between BP network and other computer networks is the use of nonlinear transfer function in the neuron transfer function. The most common are the logsig and tansig functions. The output of the logsig function is:

$$a = \log sig\left(w_p + b\right) \tag{1}$$

2.2.2 BP Network Learning Algorithm

In the forward pass process, the network first accepts input information through the input layer, then reaches the hidden layer, processes the hidden layer, and then reaches the output layer to exit. After the network extracts the results, both the output and the expected value are considered wrong. If the error is large, perform the second step. That is, the error propagates in the opposite direction, and the network sequentially passes the signal error from the output layer to the hidden layer and then to the input layer. Then change the weight and threshold. One for each level of the process. Through the alternate work of the two processes, the network gets better parameters and achieves the expected results.

2.3 BP Neural Network Topology Design

The method to determine the network topology is the basic technology for temperature prediction and early warning using the BP neural network model. The accuracy of sample training and prediction is based on the network topology. BP neural network is a positive neural network, which transforms the weight in the negative gradient direction of network training with the smallest error. Its biggest feature is that it has been trained to minimize the root mean square error between the actual network output and the expected output, and it constantly adjusts the network weight. Therefore, the primary task of designing BP neural network is to design the network topology. This usually includes the following content: It determines the number of layers of the BP neural network model, that is, the number of hidden layers and the number of neurons.

2.3.1 Determination of the Number of Layers of the BP Neural Network Model

A three-level BP neural network can approximate any rational function, and many theories have proved this point. Increasing the number of network layers can reduce errors, but it will also increase the complexity of the network. The current BP network model usually uses a three-layer BP structure. Generally speaking, when dealing with more complex I/O relationships, the prediction effect of the 4-layer network structure is better than that of the 3-layer network.

2.3.2 Determination of the Number of Neurons in the Hidden Layer

Determining the number of neurons in the hidden layer is one of the most important steps in the process of designing a neural network, which directly affects the degree of mapping between the network and complex problems. However, there is still no clear and optimal method to determine the number of neurons in the hidden layer. In other words, there is no universally applicable method. Under normal circumstances, adjustments will be made according to specific problems.

However, in the setup process, if the number of hidden layer neurons is small, the network cannot establish complex mapping relationships, so sample training cannot be completed, and the fault tolerance is low, which affects the convergence speed of the entire network, and the speed will be reduced. Or convergence is impossible. A large number of numbers complicate the network structure, increase the computational complexity of the learning process, and reduce the generalization ability of the network.

Many researchers have tried different methods to determine the number of neurons in the hidden layer, but they have not reached a convincing reason and are prone to errors. Therefore, testing algorithms is the most reliable method. Due to the different problems solved by the BP algorithm and the different conditions of the objects, there is no unified design scheme.

3 Experiment

3.1 Data Acquisition Module

The collected data is divided into historical data and real-time production data. This is because the two systems coexist for a period of time instead of replacing the original system immediately after the system software development is completed. This is also to ensure the smooth operation of the system and avoid losses caused by errors in the new system. Therefore, there are two types of collected data: historical data and real-time updated data. At the same time, the source data is expected to be stored neatly in the library or mysql file, without Excel analyzing the data in the messy storage. In other words, it only focuses on basic functions such as analysis and is easy to apply. The following is an overview of the collection of these two types of data.

3.1.1 Collection of Historical Data

Historical data collection is mainly to transfer data from the old system to the constructed new system, that is, data migration in a sense. There are two common data migration methods. One is to configure the same environment as the source database, and then distribute the data block, the other is to use the direct select and input commands to input line by line. The data transmission method used by this system is based on Nifi. It uses the latter method, but on this basis, it provides configuration possibilities on the Web side and has many powerful processors.

3.1.2 Collection of Real-Time Data

Some equipment data may not have equipment maintenance management data that can be deleted/changed, equipment data that linearly increases in chronological order, or equipment status data that can be deleted/corrected. In the latter case, you can use the Nifi tool to use scheduling methods for intermittent collection. Nifi data is routed to other nodes after the node is closed, but the disadvantage is that the data loaded before the closing time will have to wait for the node to restart and reset. Nifi has many great features, but it is definitely suitable for streaming. This system is not suitable for real-time data collection and data relocation, so here we use Flume + Kafka to collect status data.

3.2 Data Analysis Module

The data analysis module is based on the Spark environment programming and is mainly divided into a real-time computer unit and an offline computer unit. For device data, most of them only need an offline computer to meet the demand.

The data collection unit includes not only the content of data collection, but also the content of how the data is stored. For device management data, the data is directly input into the Mysql database table through the Nifi creation panel, and the device status data is stored in the hdfs file system in chronological order according to kafka. This completes the sending and saving of data.

The offline calculation unit can calculate the service life, cost efficiency, and stability of the equipment based on statistical analysis techniques. The user must use this file to record the device ID and calculation method. Is it cost-effective or stable? It is represented by 1 and 2, respectively, the sending mailbox. There are three parameters, separated by commas. Then open the Nifi website, use the editor and Kafka Name theme configuration file location, you will get the result in the mailbox. The execution process is shown in Fig. 1 below. Use nifi to get the file and immerse in the corresponding Kafka topic. Spark retrieves the data from Kafka, analyzes it, sends the calculation to the corresponding mailbox, and deletes the data in the corresponding Name field after completion.

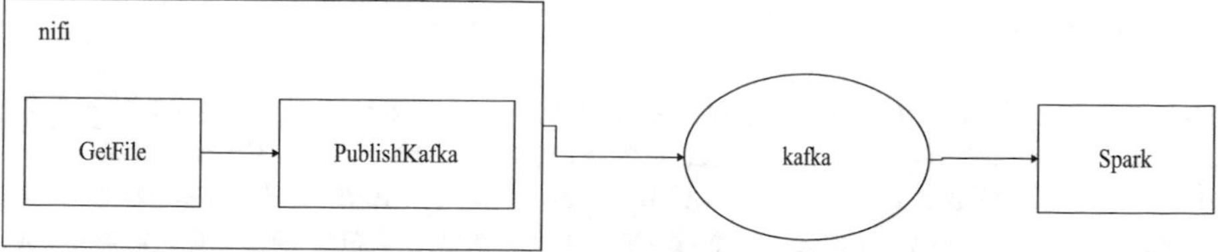

Fig. 1. Cost-effectiveness and stability calculation process

4 Implementation of Equipment Data Analysis Algorithm Based on Convolution and BP Neural Network

When deep learning technology processes big data composed of a large number of natural language words, it must convert these words into text numbers with specific corresponding relationships. First of all, assuming that a Chinese character can be decomposed into words, each word can also be expressed by a vector in the vector character table. Among them, the data in the i-th column of the vector vocabulary represents the vector of the i-th vocabulary in the vocabulary. In this way, we can decompose the same sentence into different vocabulary and find out each corresponding vocabulary vector to obtain the sample vector word list.

This task uses a fully convolutional neural network to represent text-type device data, including five levels: input layer, convolution layer, concentration layer, fully connected layer, and output layer. The text-type device health data is sent to the neural network through the input layer, and the noun vector table is used to measure the data sample. The convolution plane then performs a convolution function on each word vector h from the input level, and sends the result to the concentration level. Among them, the types used for adhesive layers include:

$$h_2 = \tanh(h_1) = \frac{e^{h_1} - e^{-h_1}}{e^{h_1} + e^{-h_1}} \tag{2}$$

Maximize features to select the most important words in the text, because each word in the text has a different meaning. The largest vocabulary h_3 in h_2 is calculated using the following formula:

$$h_3 = \max_{1 \leq n \leq N}(h_2) \tag{3}$$

The neurons in the layer completely connected to the concentration layer communicate with data in a completely continuous method. The result of the interaction is:

$$h_4 = w_4 h_3 + a_4 \tag{4}$$

In the above formula, w_4, a_4 respectively represents the weight and deviation between the two layers of neurons.

5 Discussion

The performance test checks whether users can access the server in real time by applying certain user access rights to the server and letting the server check the performance of the server. Since the system was initially used in small and medium-sized network environments, it was restricted by the network environment and could not be accessed by a large number of users who access the system at the same time.

A total of 5 execution scenarios were designed for this system stress test. After each execution of the test, the server re-applied to avoid affecting the next test result.

Through the execution and analysis of the five scenarios, the users running the simulated load are 50, 100, 150, 200, and 250 respectively. The basic configuration is the same, the thinking time is adjusted when performance problems occur, and the data is the same as in other environments. After the test data is prepared, the steps in each

Table 1. Performance test results

User number	Average transaction response time	Transaction success rate	Throughput per second	CPU utilization	Transactions per second
50	0.045	100%	632935.14	10.745%	1.937
100	0.062	100%	1224709.65	15.133%	3.74
150	0.168	100%	1718944.95	18.408%	5.262
200	0.192	100%	2767703.77	22.171%	8.474
250	3.233	88%	12320000.33	88.787%	10.544

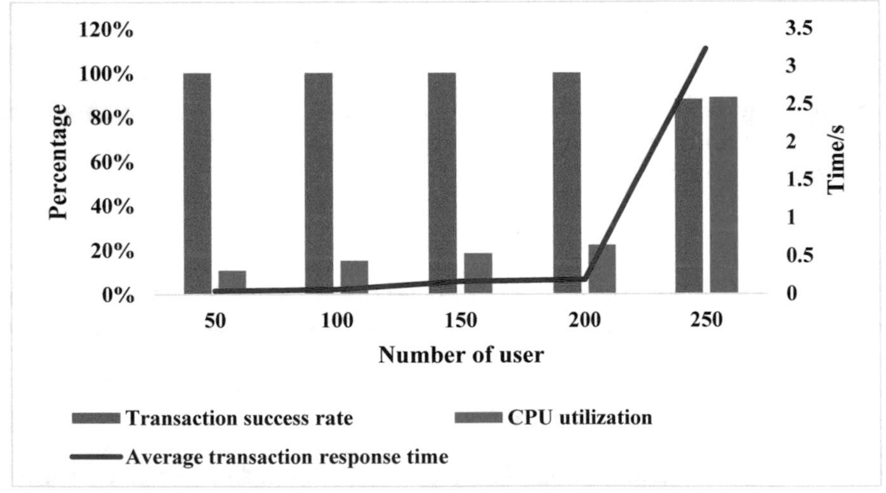

Fig. 2. Performance test results

test are to perform user login tasks and query the data of each transaction (Table 1 and Fig. 2).

No crashing errors were found during the entire test. Other errors focused on the execution of specific operating points and interface performance, all of which were fixed during the test. If the network is in good condition and the load simulation users are less than 200, the average response time of a single transaction is within 1S, and the transaction is completely passed; when the number of loading users is 250, the system transaction response time increases, failed transactions occur, and the success rate decreases. According to the actual business situation of the system, the system can fully meet the needs of users in terms of operation and performance.

6 Conclusions

After studying convolutional neural network and BP neural network algorithms, this paper builds a system with data acquisition, data storage, and data analysis, and realizes a lightweight configuration, which simplifies the learning cost of users, and reduces software development, Time cost of maintenance and testing.

Acknowledgment. The Teaching Reform Research Project of Qinghai Minzu University (2021-JYYB-009).

References

1. Kamalakshi, N., Naganna, H.: Persistent homology residual learning of deep convolution neural network for block match three dimension de-noising algorithm. J. Comput. Theor. Nanosci. **17**(9), 4340–4343 (2020)
2. Baldini, G., Gentile, C., Giuliani, R., et al.: Comparison of techniques for radiometric identification based on deep convolutional neural networks. Electron. Lett. **55**(2), 90–92 (2019)
3. Dabiri, S., Heaslip, K.: Inferring transportation modes from GPS trajectories using a convolutional neural network. Transp. Res. Part C Emerg. Technol. **86**(Jan), 360–371 (2018)
4. Yalçın, O.G.: Convolutional neural networks. In: Applied Neural Networks with TensorFlow 2, pp. 145–160. Apress, Berkeley, CA (2021). https://doi.org/10.1007/978-1-4842-6513-0_7
5. Priyanga, P.S., Krithivasan, K., Pravinraj, S., et al.: Detection of cyberattacks in industrial control systems using enhanced principal component analysis and hypergraph-based convolution neural network (EPCA-HG-CNN). IEEE Trans. Indust. Appl. **56**(4), 4394–4404 (2020)
6. Sahay, A., Amudha, J.: Integration of prophet model and convolution neural network on wikipedia trend data. J. Comput. Theor. Nanosci. **17**(1), 260–266 (2020)
7. Patil, N., Patil, P.N., Rao, P.V.: Convolution neural network and deep-belief network (DBN) based automatic detection and diagnosis of Glaucoma. Multimedia Tools Appl. **80**(19), 29481–29495 (2021). https://doi.org/10.1007/s11042-021-11087-5
8. Kalanidhi, K., Vinodkumar, D., Sridharan, P.: Transmission power line fault detection based on deep leaning methods. Indian J. Power River Valley Develop. **69**(3–4), 35–38 (2019)
9. Babushkina, N., Lyapin, A., Kovaleva, A.: Analysis of neural network results based on experimental data during indentation. E3S Web Conf. **224**(34), 01018 (2020)

10. Xia, T., Zhong, J., Zhang, Y.: Non-invasive continuous blood pressure measurement based on mean impact value method, BP neural network, and genetic algorithm. Technol. Health Care: Official J. Europ. Soc. Eng. Med. **26**(6), 1–15 (2018)
11. He, F., Zhang, L.: Mold breakout prediction in slab continuous casting based on combined method of GA-BP neural network and logic rules. Int. J. Adv. Manufac. Technol. **95**(9–12), 4081–4089 (2018). https://doi.org/10.1007/s00170-017-1517-1
12. Li, D.-J., Li, Y.-Y., Li, J.-X., Fu, Y.: Gesture recognition based on bp neural network improved by chaotic genetic algorithm. Int. J. Autom. Comput. **15**(3), 267–276 (2018). https://doi.org/10.1007/s11633-017-1107-6

Design and Research of Secondary Index of Traffic Big Data Based on Hash

Miao Wang, Quanling Sun(✉), Xiangying Liang, and Li Gao

School of Electronic and Information Engineering, Anhui Jianzhu University, Hefei, Anhui, China

quanling_sun@126.com

Abstract. With the development of smart cities and massive increase in traffic big data, the real-time storage and efficient retrieval of data require urgent attention. Relational databases cannot easily handle the storage and query of massive and multitype data. Compared with relational databases, nonrelational databases, such as HBase, are more flexible. However, because of the way HBase columns are stored and the fact that SQL statements are not supported, HBase does not have an advantage over non-line-key queries. In this study, we propose a secondary index method based on HBase and introduce a hash algorithm into the rowkey design in the secondary index. Consequently, our method not only solves the problem that the multidimensional column value is too long but also significantly improves the retrieval efficiency. It can satisfy the analysis demand of mass data in intelligent transportation.

Keywords: Traffic · Big Data · Hbase · Secondary Index · Hash

1 Introduction

With the development of smart cities, real-time data increase rapidly. Data records applied in various industries can reach more than 100 million. It is challenging for database systems to handle such a large volume of data, whether for storage or retrieval.

After Google published a paper on the BigTable data model in 2006 [1], to effectively handle the storage and query of massive data, NoSQL distributed data storage is used to replace the relational database management systems. These include Apache's HBase, Facebook's (now called Meta) Cassandra, Amazon's Dynamo, and Redis—an in-memory data storage system that can query data efficiently. Unlike the row storage of relational database systems, they employ the column storage technique, among which HBase is the most widely used. The column storage method can support a dynamic increase of columns. Different rows of the same table can have different columns, and all null columns do not occupy storage space. The table design can be very sparse, which considerably saves storage space. In addition, because there is no data type in HBase, it is convenient to store multiple complex types of fields [2].

J. H. Abawajy et al. (Eds.): ICATCI 2022, LNDECT 170, pp. 877–885, 2023.
https://doi.org/10.1007/978-3-031-29097-8_104

There is no doubt that HBase performs well in terms of storage; however, in the face of billion-level massive data, we should consider not only the storage problem but also efficient retrieval, which is the weakness of HBase. One of the reasons HBase does not have an advantage in the efficiency of complex conditional queries is that, unlike relational database systems, it does not support SQL statements and complex aggregation operations, such as join and group by. Its multiconditional queries need to rely on rowkey design. Second, it does not support secondary indexes. It needs to use a third party such as Phoenix to provide secondary indexes [3].

To perform better real-time query and analysis on massive traffic data in intelligent transportation, in this study, we propose a method combining a hash algorithm with a secondary index method to index nonrowkey columns in HBase and store them in spark cluster. According to the size of the column to be indexed and whether the range query is involved, a hash algorithm-based index is developed, and the characteristics of spark parallelization are used to improve the query efficiency.

2 Basic Structure and Index Mechanism of HBase

2.1 Distributed Structure of HBase

HBase distributed design adopts the classic master–slave architecture. The master node is HMaster, and the slave node is HRegionSever. HMaster does not store data; it is only responsible for managing the mapping between data and its HRegionSever. HRegion-Sever is the place where data are stored. The storage structure is HRegion. So, the actual reading and writing of data are based on HRegionSever.

2.2 Query Mechanism of HBase

HBase does not support SQL statements. The following methods can be adopted for the HBase database.

(1) Hive—a batch query engine

Hive has advantages in large-scale batch processing and low real-time requirements. In addition, it has the advantages of simple configuration, low cost, and good compatibility. However, the query speed is slow, the access pressure to the HBase cluster is high, and there are many restrictions on column mapping; hence, the combined primary key cannot be used. So, there is no advantage in the real-time processing of traffic flow big data [4].

(2) Phoenix—an interactive query engine

Phoenix is an SQL middleware based on HBase. Unlike hive, it does not need to be converted to MapReduce tasks. It can convert SQL queries into one or more HBaseScan, and the query efficiency and speed are much better than hive. All data are stored in HBase without other storage methods and media. However, only shell and Java are supported at this stage. Compared with hive, it is suitable for processing traffic big data in terms of execution efficiency.

2.3 Secondary Index Mechanism of HBase

The secondary index of HBase creates a corresponding IndexTable for each DataTable through various methods, ensuring that IndexTable and DataTable Regions correspond one-to-one and are stored on the same RegionServer. The final effect to be achieved is that each IndexTable Region is a local index of the corresponding DataTable Region [5]. When using an index to query, each IndexTable Region will be retrieved to find all DataTable rowkeys that meet the conditions and then read the corresponding DataTable Row from the corresponding DataTable Region according to the DataTable rowkey (Fig. 1).

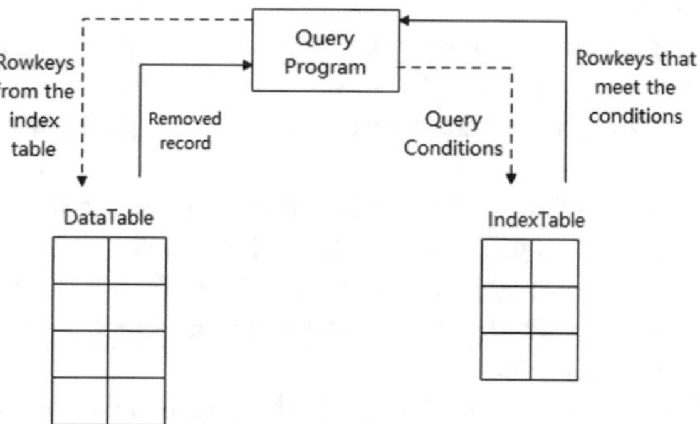

Fig. 1. HBase secondary index reading process

In current open-source secondary indexing schemes, the index rowkey is basically verbose, and a rowkey is stored in each cell. Therefore, the more columns stored in HBase, the more redundant rowkeys are stored, and the more space is wasted. Assuming that an HBase data table has 100 million rows and 10 columns, and the average length of a rowkey is 50 bytes, the redundant storage space occupied by the rowkey is 100000000 × 10 × 50/(1024 × 1024 × 1024), approximately 46.5 GB. So, based on the original secondary index scheme, a hash function is introduced for optimization.

3 Secondary Index Based on Hash Algorithm

3.1 Hash Algorithm

Hash algorithms can map binary values of arbitrary length to shorter fixed-length binary values, which are called hash values. A hash value is a unique and extremely compact numerical representation of a piece of data [6]. After hash mapping, the original continuous value can be hashed (hash values mapped to the same string are the same), and then the hash value is used as the rowkey.

In the org.apache.hadoop.util package provided by Apache, the MD5HASH class is provided. The getMD5AsHex() method in this class can convert a string of arbitrary length to a string of 16-byte length. We employ this algorithm to map index table rowkeys into 16-byte hash values.

3.2 Improved HBase Indexing Process Using Hash Algorithm

Usually, there are three techniques to retrieve data in the HBase table: specifying a single rowkey query, specifying a rowkey range query, and scanning. HBase sorts rowkeys using the dictionary sorting method of the byte array, which can support efficient single-point query and range query of specified rowkeys, but the scanning operation of nonprimary key attributes is expensive [7]. The scale of data records in big data applications is usually in the tens of millions or even hundreds of millions, and the overhead of nonprimary key scans on the entire table is unacceptable.

HBase is retrieved according to rowkeys, so its retrieval efficiency depends on the rowkey design. The system locates the region where the data are located according to rowkeys or rowkey range.

Rowkey design generally has three principles. First, the principle of length; Second, hash Principle; Third, unique Principle.

Based on these principles, we have several rowkey design methods.

(1) Assign a random number in front of a rowkey. According to the random number prefix, it is allocated to different regions. This random number is related to the number of regions and is generally allocated according to the number of clusters. For example, for 200 clusters, the random number is set between 0 and 199 and added to the rowkey header.
(2) The method of prepartitioning. HBase will automatically handle region splitting. When the size of the region reaches a certain threshold, the region will be split into two, and then data can continue to increase in both regions [8].
(3) The above two methods do not really use the hash principle designed by rowkeys. So, we consider hash algorithms. We can use the hash function to calculate a hash value and then modulo the number of regions we need to store; we can store the same input data into the same region [9].

4 Design of Traffic Big Data Rowkey

4.1 Experimental Dataset

The experimental dataset comes from vehicle information recorded using an electronic camera at a certain intersection in Hefei City that Pathon climbed to. The vehicle identification data generated using the high-definition camera can be up to more than 2,000 pieces per second. Among them, the tens of millions of datasets contain data on September 19, 2018, with about 35 million records, and the hundreds of millions of datasets contain data from September 20 to September 23, 2018, with approximately 130 million records. The record contains license plate information, time, etc. Now, some data are extracted and organized into the logical structure of the HBase table model (Table 1).

Table 1. Traffic data information

CarNO	Time	cartype	CarNOtype	property	Speed (KM/H)	carIMG
wanA8232A	2018-09-20 08:01:56	K33	02	A	58	wanA8232A.png
wanA76D45	2018-09-20 08:01:45	K33	02	A	56	wanA76D45.png
wanABC556	2018-09-20 00:01:26	K43	05	A	55	wanABC556.png
wanA23556	2018–09-19 23:59:07	K31	01	B	45	wanA23556.png
wanA675S8	2018-09-19 23:58:59	K33	02	A	57	wanA675S8.png
wanAY1785	2018–09-19 11:01:59	K33	02	A	42	wanAY1785.png
wanA95Z45	2018-09-19 10:58:15	K33	02	A	30	wanA95Z45.png
wanAZ1789	2018-09-19 10:58:12	K33	02	A	67	wanAZ1789.png
wanA8232A	2018-09-19 08:09:45	K33	02	A	55	wanA8232A.png
…	…	…	…	…	…	…

Convert these times into timestamps, such as 2018-09-20 08:01:56 into 1537401716 and 2018-09-20 08:01:45 into 1537401705. The stream data obtained after the data are accessed will be sent to the data queue partition area in a buffer structure. Suppose that there are three queues in each type of buffer waiting to receive data. The hash method is used to divide these timestamps into modulo 3 to obtain the values of 0, 1, and 2. Then, the vehicle data information is stored in the corresponding queues. The data objects in each queue are stored in the memory using a linked list structure (Fig. 2).

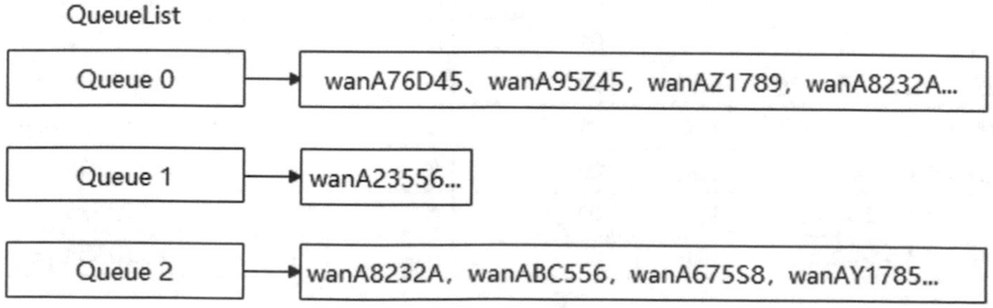

Fig. 2. Vehicle information queue division

After the division is completed, the queue continues to send data to the write buffer according to the window threshold. After the data received by the client write buffer reaches the set threshold, the HBase multithreaded write method is employed to send data to the server for permanent storage.

4.2 Design of Vehicle Data Table Rowkey

If we want to query a vehicle's data table rowkey in real time online, e.g., query traffic flow information for a certain period, it is necessary to support online real-time queries and solve hot issues. To this end, according to the actual needs, a rowkey is designed as a combination of license plate number + monitoring point + detection time. This combination can better resolve the hot issues, but if the time is used as the query condition, it will drastically reduce the efficiency in the process of scanning the cluster database. In addition, through actual monitoring, we found that daily traffic flow from 00:00 to 07:00 and 22:00 to 24:00 significantly reduced compared with other times, and traffic flow from 8:00 to 9:00 and 17:00 to 19:00. During this period, the traffic flow is relatively large, and other times are relatively uniform. In response to the above problems, we consider creating some empty regions in advance to improve the speed of batch writing. As such, when data are written to HBase, they will be balanced within the cluster according to the region partition. We put the timestamp in front and design the rowkey as time + license plate number + monitoring point code. When peak traffic occurs and data are frequently inserted into the region, hot issues will occur. In the rowkey design, there is also a method of adding a random number prefix, this random number is related to the number of regions. If we set up 16 clusters, a random number between 0 and 15 is added to the front of the time as a prefix, and the MD5HASH function is used to take all 16 bytes to obtain a new table (Table 2).

Table 2. Vehicle Data Table

Rowkey(timetamp+ carno+ID)	Cartype	CarNOtype	Property	Speed (KM/H)	CarIMG
50FCBC2A43C8C53C	K33	02	A	58	wanA8232A.png
2E30D72FD6CE4A60	K33	02	A	56	wanA76D45.png
8448EB59EE5BC05A	K43	05	A	55	wanABC556.png
795F19DFC15362F5	K31	01	B	45	wanA23556.png
A9F0062A16F0637E	K33	02	A	57	wanA675S8.png
884FE193B4B02262	K33	02	A	42	wanAY1785.png
EAAFACE2A6C80607	K33	02	A	30	wanA95Z45.png
31FD06CA9B074D96	K33	02	A	67	wanAZ1789.png
0E160342D30B8BEA	K33	02	A	55	wanA8232A.png

5 Experimental Results and Performance Analysis

5.1 Construction of Experimental Environment

In this experiment, a cluster was built on five high-performance computers, and a cluster environment with five nodes was constructed. The hardware and software configuration lists are shown in Table 3.

In this experiment, we compared the response and write times of two query methods with different data volumes of 1, 2, 3, 5, 8, 10, and 50 million.

(1) Comparison of query response time

We compared the response time without hash algorithm and HashHBase on different data volumes (Fig. 3).

Table 3. Configuration table

CPU	IntelXeon E-2278GE
Memory	16GB
DISK	SEAGATE EXOS 7E8 6TB
OS	Red Hat Enterprise Linux Server 6.0
Phoenix Version	4.9.0
Hadoop Version	Hadoop 2.6.0
HBase Version	HBase 1.2.6
Zookeeper version	Zookeeper 3.4.6
JVM version	Java 1.6.0

HBase without a secondary index scans the original table directly, which is significantly affected by the data volume. When the data volume is small, the full table scan performance is good; as the data volume increases, the response time increases. For HBase with the secondary index, because the index table is accessed first before the data table is accessed, the client interacts with the server two times, so the initial.

Time is longer than the direct scan; however, as the data volume increases, its advantages become more obvious.

(2) Comparison of write time

The HBase without a secondary index scheme writes directly, which is basically linearly distributed. However, because HashHBase has an index table, the server needs to maintain the index table information separately, so writing takes a certain amount of time, and the more complex the combined rowkey, the longer the writing time (Fig. 4).

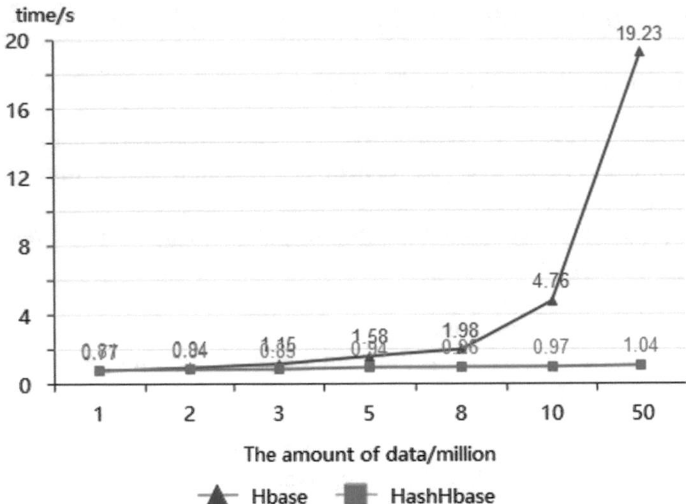

Fig. 3. Response time comparison chart

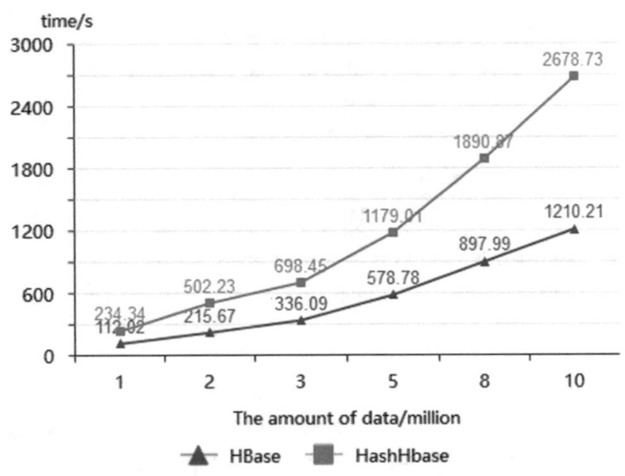

Fig. 4. Comparison of write time

6 Conclusion

To meet the analysis requirements of massive data in smart transportation, efficient processing and query methods are necessary. In this study, we analyze the advantages and disadvantages of the HBase database. To solve the problem of low efficiency of a nonrowkey attribute query, a hash algorithm is introduced into the secondary index method of HBase, and consistent hashing is optimized in the distribution strategy, so.

That even when the number of nodes changes, it can significantly reduce costs. Comparative experiments show that the introduction of the hash algorithm into the index method of HBase can significantly improve the query efficiency.

Acknowledgments. Project funds: 1. Natural Science Research Project of Anhui Provincial Department of Education "Research on key technologies of vector big data management for smart cities"(Project NO:KJ2019JD12).

2. Natural Science Research Project of Anhui Provincial Department of Education "Research and Design of ORB Feature Point Extraction Algorithm Based on FPGA" (Project NO:KJ2021JD17).

References

1. Sharma, T., Shokeen, V., Mathur, S.: Distributed content-based image retrieval of satellite images on Hadoop. Recent Adv. Comput. Sci. Commun. **14**(3), 678–689 (2021). https://doi.org/10.2174/2666255813666191126095114
2. Wei, G., Luo, S., Zhou, W., et al.: HBase: a hierarchical indexing mechanism and system for efficient HBase query. Chinese J. Comput. **39**(1), 140–153 (2016)
3. Hendawi, A., Gupta, J.: Benchmarking large-scale data management for Internet of Things. J. Supercomput. **75**(12), 8207–8230 (2019). https://doi.org/10.1007/s11227-019-02984-6
4. Mary, A., Junaid, R., Wasif, N.M., et al.: Bit-parallelism string matching withmapreduce and Hadoop. Comput. Mater. Continua **68**(3), 3931–3946 (2021)
5. Tsvetelina, M., Yordan, K., Milko, M., et al.: Impact of data compression on the performance of column-oriented data stores. Int. J. Adv. Comput. Sci. Appl. **12**(7), 416–421 (2021)
6. Oussous, A., Benjelloun, F.Z., Lahcen, A.A., et al.: Big data technologies: a survey. J. King Saud Univ. Comput. Inform. Sci. **30**(4), 431–448 (2018)
7. George, L.: HBase: the definition guide. In: Dai, Z., Liu, J., Jiang, J., translated, pp.312–350. People's Post and Telecommunications Press, Beijing (2013)
8. Mershad, K., Hamieh, A.: SDMS: smart database management system for accessing heterogeneous databases. Int. J. Intell. Inform. Database Syst. **14**(2), 115 (2021). https://doi.org/10.1504/IJIIDS.2021.114513
9. Lei, Z., Sun J., Shu, S., et al.: Uncertainty modeling of object-oriented biomedical information in HBase. Acta Geodaetica et Cartographica Sinica **49**(10), 1365–1373 (2020)

User Behavior of Social E-commerce Platform—Based on Big Data Analysis

Yan Li[1](✉), Chunya Li[1], and Muhammad Hamam[2]

[1] Business School of Nantong Institute of Technology, Nantong, Jiangsu, China
muziliyan@163.com

[2] Uni de Moncton, Moncton, Canada

Abstract. Social network is user-centered and has the characteristics of extensive communication and true content, which has a huge impact on the whole e-commerce. Taking Mogujie as an example, this paper analyzes user behavior from the aspects of user group, user time, user motivation, user preference and user experience through data collection, literature collection, questionnaire survey and other methods, analyzes the relationship between social e-commerce platform and users, finds problems and puts forward corresponding measures. Only by constantly introducing active traffic to the platform, releasing high-quality information, improving the platform mechanism, strengthening brand building, encouraging user innovation, improving the management mechanism, improving the practicality and objectivity of the platform, and enhancing user experience, can we gain advantages in the fierce market competition.

Keywords: Social e-commerce · Mogujie · User Behavior

1 Introduction

As the main body of social e-commerce platform, users' consumption behavior affects the development of social e-commerce platform. The analysis of users' preferences, use motivation and use experience has a great impact on the development of domestic e-commerce platform [1]. Search for keywords such as social e-commerce and user behavior on CNKI and other websites, and combined with the research direction of this paper, a total of more than a dozen papers and journals conducive to research have been searched. It is found that most authors will choose mushroom street as the research object, and the research results will finally summarize the problems existing in the development of social e-commerce [2]. At present, many scholars have done a lot of research on social networking sites and e-commerce platforms, but there is little research on social e-commerce platforms. As the social e-commerce platform is an emerging platform, there are still some problems in its communication process [3]. Today, with the rapid changes of the social platform, it is necessary to adjust the communication strategy of the platform. The research of this paper starts with the user behavior of social e-commerce platform, and provides theoretical support for the development of social e-commerce platform from a new perspective [4].

J. H. Abawajy et al. (Eds.): ICATCI 2022, LNDECT 170, pp. 886–894, 2023.
https://doi.org/10.1007/978-3-031-29097-8_105

2 Investigation and Analysis of Mogujie User Behavior

In order to better study the user behavior analysis of social e-commerce platforms, in addition to finding a lot of information from the literature, this paper also obtains reliable data based on questionnaires. A total of 400 questionnaires were distributed and 350 valid questionnaires were collected. The questionnaire has a total of 22 questions, which are designed according to the basic situation of the respondents, time and frequency of use, motivation, preference and experience, and trust.

2.1 Basic Information of the Respondents

From the 350 valid questionnaires collected, it can be seen that most users of mushroom Street are young women. Most of them have the highest education in high school, college and undergraduate. They pay more attention to the topics of beauty, skin care and fashion in their life. Most of them have certain learning ability and judgment ability and always like to share their life state, See Table 1 for details.

Table 1. Statistical table of basic information of user groups

Gender	male		female		
proportion	5%		95%		
Age	Under 16	16–23 years old	23–30 years old	Over 30 years old	
proportion	4%	55%	34%	7%	
occupation	student	Enterprise employees	liberal professions	government-affiliated institutions	other
proportion	57%	20%	16%	6%	1%
Highest education	primary school	junior high school	high school	Diploma or undergraduate	Master degree or above
proportion	1%	11%	38%	48%	2%
monthly income	Below 2000	2000–5000	5000–8000	More than 8000	
proportion	35%	42%	15%	8%	

2.2 User Usage Time and Frequency Analysis

Among the 350 users, only 12% browsed more than once a month in the survey on the browsing time of Mogujie. These people are highly loyal to the platform, and most of them go straight to the topic when using Moguji to find what they want to wear or find. From this, it can be seen that different users have different usage of various functions of the platform. Many of them do not have a clear purpose when using it. Boring time (see Table 2). The social e-commerce platform uses big data to monitor the user's usage time and frequency in the background, conducts specific analysis of the monitored data, and finally incorporates it into the user experience analysis [5].

Table 2. User usage time and frequency statistics

Frequency of use	Every day	Once every three days	once a week	One month or more
proportion	52%	26%	10%	12%
Browse time	Within ten minutes	Within half an hour	Within an hour	More than one hour
proportion	9%	21%	38%	32%

2.3 Analysis of User Motivation

The development of the platform is directly proportional to the degree of its publicity, and users' first impression of the platform comes from its publicity slogan [6, 7]. When the respondents were asked if they remembered Mogujie's promotional slogan, only 17.16% of them remembered its promotional slogan, while the remaining 82.84% did not remember much. The platform is constantly updated and changed, which makes it difficult for users to know the positioning of the platform from its slogan when using it, which invisibly reduces users' loyalty to the platform. In the questionnaire, we did some research on which aspect of the needs of users can be best met by Mogujie, and it was found that users have a high proportion of the four needs, especially the needs for the LOOK function.

1) Release LOOK and get attention

According to the data of the questionnaire, the attention received by the release of LOOK accounted for 20.67% of the valid questionnaires. For many users, it is important to post a Look to gain attention. Social e-commerce platforms like Mogujie can specifically meet their real needs, which is one of the reasons why many users choose to use Mogujie. For example, users can search for the clothes and makeup videos you want on it. At the same time, users can also share their opinions through the platform and get the attention of others. This kind of social mode better satisfies young women who want to discuss fashion and gain attention. Psychological needs. Users can also record the products you have purchased and used by publishing LOOK, and record the price and usage experience in the LOOK for reference when shopping in the future. However, if the LOOK published by the user does not have a certain degree of originality, the platform will not increase the grass-growing power for the user, and it will reduce a certain degree of attention.

2) Profit from platform shopping

The unique "three-sided network" construction has become a major feature of many social e-commerce platforms such as Mogujie. 14.29% of users in this survey chose this option. In the social e-commerce platform, merchants, fashionistas and ordinary users can all obtain their own interests. Fashionistas look for inspiration from published content, and can also buy discounted products from many merchants; merchants get more publicity and exposure, and more importantly, they have a large number of real reviews of free products; while ordinary users can You can search for the information you want on the platform, and you can also interact socially with other people. All three benefit from the platform and are closely integrated to meet the needs of all parties to enter the platform.

3) Socialize to meet social needs

Mogujie has nearby options on the interface, and recommends users on the home page. The principle is that the platform recommends personalized information for each user based on big data.Users can also socialize with other people on the Internet by recommending people and things related in real life, such as regions, etc. Some users are reluctant to communicate too much in real life, and like to seek approval on the Internet. For these people, virtual online socializing is easier and less stressful, but once the user's information is leaked, it will bring unnecessary unnecessary costs to the user. Trouble. In today's widespread network and communication devices, online social networking is more attractive. If the management mechanism is strict, it can provide users with a more comfortable use environment and better integrate social networking into e-commerce [8, 9].

4) Browse LOOK to kill time

In this fast-paced era, many people do not want to spend a lot of time on beauty guides or outfit guides. They prefer to browse the summarized information in a few seconds or minutes, which can save time,but also to avoid "stepping on thunder". In this survey, 28.86% of users chose it, which shows that people do not have a strong purpose to use Mogujie. In Mogujie's LOOK recommendation mechanism, users can see the popular LOOKs on the homepage when they open the software. These LOOKs are all high-quality works selected according to the categories you are interested in, and the number of people who browse is also the largest. Click to browse You can get the information you want, entertain and relax in your leisure time, and gain something while you are relaxing.

2.4 Analysis of User Preference

One of the features of Mogujie is to judge user preferences through big data analysis, which is also a feature of social e-commerce platforms, that is, by investigating the relevant data of users such as browsing time, number of clicks, occupation, and income level, the products are classified according to the user's favorite degree. Arrange and combine. Users can also join various comment areas for discussion, and the wearers in the community can share their own outfits while mobilizing the enthusiasm of other users to interact.

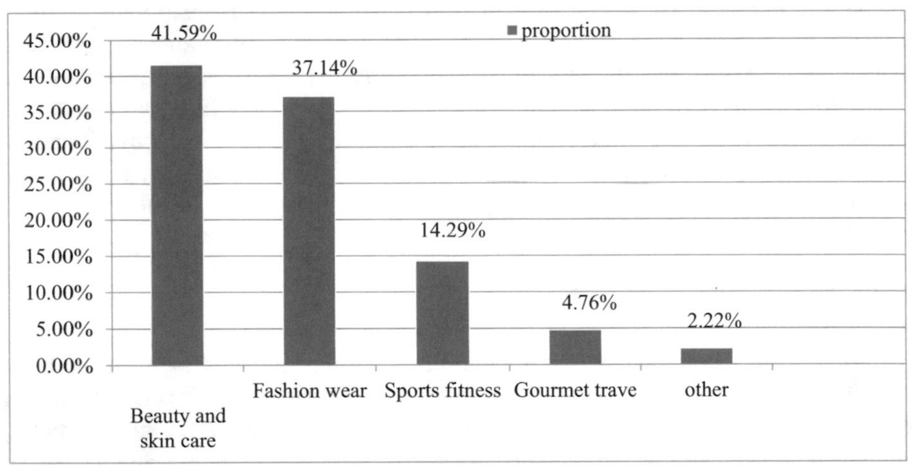

Fig. 1. User preference statistics

For the interface, most of them feel that they cannot quickly find the interface they want. This is because the interface is too confusing and the carousel is used too frequently, but some functions are neatly distributed in "I". There are more people who like to comment and like in LOOK, and these two functions are favored by users. Many users like to interact under other people's posts, which interprets the socialization of the platform very well. From the data, it is found that most of the types of users who prefer to browse LOOK are concentrated in the two types of beauty, skin care and fashion wear, indicating that Mogujie attaches great importance to the creation of beauty, skin care and fashion wear, and puts such look as a key recommendation. Therefore, as a social e-commerce platform, it is necessary to strengthen certain functions according to user preferences, and specify different personalized services according to changes in user preferences (see Fig. 1).

2.5 Analysis of User Experience

73.43% of users said they could not find the corresponding plate problems quickly, which reduced the user's sense of experience on the platform and could not fully attract users to use the platform again. In the user's favorite function look, nearly half of the respondents believe that it is too cumbersome in the publishing process, resulting in users' low interest in publishing look. Most users choose to "dive" to browse other people's posts, which reduces users' sense of achievement in publishing look and users can't integrate into it. When the respondents were asked whether they were satisfied with the functions of the platform, 70% expressed satisfaction. The mushroom street platform can not only shop in the mall, but also participate in discussions to meet social needs, but also use intelligent functions to experience personalized services. However, another 30% of users expressed dissatisfaction with the platform, because the design of some contents is not reasonable, and the distribution of contents of interest to users is messy, so users need to spend a lot of time looking for them. For the question of whether they are willing to continue to use mushroom street, 60% of users said they would continue to use it, while 40% of users said they had low participation and insufficient sense of experience.

2.6 User Trust Analysis

The user's trust in the platform is the premise for users to use the platform, and the loss of trust by the platform's services will lead to the loss of users. This questionnaire asked two questions about users' trust in the Mogujie platform. Do users feel safe in the payment process of Mogujie? 62.57% of the respondents believe that there are certain security problems in the payment process. Mogujie is only limited to cooperation with WeChat, and users can only choose WeChat when choosing a variety of ways to log in. And there is no more secure payment method such as face recognition when choosing payment. 73.73% of users believe that customer service cannot solve problems in a timely and effective manner.

3 Ways to Develop Social E-Commerce Platforms Based on User Behavior Research

3.1 Strictly Manage and Control Information Leakage to Ensure User Social Security

The Mogujie platform should conduct real-name authentication for users when registering. In order to ensure the security background, the user registration screening should be strengthened. For those who already have multiple accounts, the redundant accounts should be cancelled. If someone registers multiple accounts, there should be corresponding penalties. Big data analysis is a feature of the platform, which is to recommend content to users in a targeted manner. If this function leads to data leakage, it will definitely bring unnecessary trouble to users, which is contrary to the original intention of platform development. The platform should Strengthen employee management, punish or even dismiss staff who leak user information to ensure the integrity of the information, prevent this phenomenon from happening in the future, protect the privacy information in the information released by users, and make appropriate use of mosaics, etc. When users are harassed, there must be clear measures to protect user safety, and users can seek help from the platform at any time when they are harassed. At the same time, a complete reporting system is established under the condition that there are sufficient reserve personnel in the background, allowing users to jointly supervise the platform, and setting a reporting button for each user. Once someone reports, the background will verify the situation, and users who have been reported by multiple users we will deal with illegal account bans, and appropriately increase the registration threshold while promoting the implementation of the real-name system for platform users, so as to prevent the phenomenon of trumpets swiping orders and reviews. In the live broadcast room, the user's information should not appear on the public screen. The anchor can watch it through the background, but it is not allowed to say the user's information at will. The platform You should sign a certain confidentiality agreement with the anchors to keep the user information absolutely confidential, and through the joint efforts of the three parties, jointly strengthen the safety of users [10].

3.2 Accurate Platform Positioning to Improve User Loyalty

Compared with other social e-commerce platforms, Mogujie's slogan is "Buy women's clothing, go to Mogujie first". Xiaohongshu prefers beauty makeup and overseas shopping. Mogujie should continue to expand its own advantages. If the business scope is not extensive, it must be proficient in this one, and continue to strengthen the label of "buying women's clothing and matching women's clothing", you can cooperate with traditional social platforms such as Weibo and WeChat, continue to develop the existing live broadcast of celebrities, and introduce more clothing bloggers to publish high-quality clothing information. Mogujie can appropriately expand the user group and expand the promotion of ageing, because the target group is too small to be divided into multiple platforms at the same time, which will inevitably lead to a smaller and smaller user group of the platform. Innovate more sales models on the basis of live broadcast sales and link purchases, fully consider the low consumption level of the user population, and reduce the price of sales products to an appropriate range with small profits but quick turnover. First of all, the platform should strictly control the source of goods for merchants, control the quality of goods, and try to recommend high-quality, low-cost, fashionable and appropriate products to platform users, so that users can trust social e-commerce platforms more, which is the fundamental condition for capturing customers. Secondly, it can also achieve win-win cooperation with some brands, bring price concessions to platform users through discounts, discounts and other activities, enhance user stickiness and increase brand favorability, so as to not only strengthen brand promotion but also increase users' use of the platform,frequency and time. Mogujie should keep up with fashion trends, or actively lead the trend on the basis of big data surveys, develop platform technologies, gain more accurate insights into consumer needs, and provide more refined personalized services.

3.3 Encourage Users to Innovate and Continuously Introduce Traffic

Improve the core technology of the platform, continuously improve the sociality of the platform, attract users to publish high-quality information, and give rich rewards to high-quality bloggers with high popularity and high click through rate [11]. Because many users do not have a strong purpose when browsing the software, but just take it as an entertainment way to kill time, it is necessary to strengthen the entertainment and practicability of recommending some posts, so that users have the interest and desire to browse. Establish a perfect reward and reporting mechanism to reward bloggers who publish excellent appearance for a long time. When the user flow is low, professional activities can be held to encourage users to participate and increase the user activity flow. Mogu street should seize the opportunity, continue to play its cooperation with wechat, actively improve the conversion rate of wechat, and publicize on microblog, Douban and other platforms. Find spokesmen to develop traffic, and increase the number of users through the participation of celebrities, Weibo big V and Internet celebrity phones, so as to improve the popularity and public support of the platform. In addition, we should also seek ways to deepen public memory, find appropriate publicity methods, actively launch TV and offline advertising, realize multi-platform drainage and expand the scale of potential users.

3.4 Improve the Platform Interface and Enhance the User Experience

Improve interface design. In terms of interface design, we should learn from other software, be concise, clear and generous, and retain our own unique style [12]. The slogan of "Mogu Street" is placed on the home page so that everyone who enters the platform can see it, which is virtually publicized and adds memory points to the platform. Accurately push the user's favorite content on the home page, screen the content that the user doesn't like in time, and analyze the user's favorite category after analyzing the background data. Whether it is live broadcast or watching, some suggestions on accountability should be added.For the setting of search box, when users click, they need to set specific subject content, which is similar to the hot search of microblog. When new users enter the platform, they should be asked to select the type of content they are interested in [13].

4 Conclusions

At present, the social e-commerce platform is developing rapidly. Mushroom street, as its typical representative, has its development advantages. It can well combine the social network with e-commerce, use the analysis of background data to locate and promote users' preferences, and use the function of social networking to promote goods and increase communication with users. In today's era of rapid change and continuous development of both e-commerce platforms and social e-commerce platforms, we must make some changes, absorb the advantages of others while retaining our own characteristics, remember our original heart and don't forget our original mission when serving users, so as to highlight the encirclement among a number of social e-commerce platforms.

Acknowledgements. This work was supported by Key Discipline Construction Project of Business Administration in Jiangsu Province During "The 14th Five-Year Plan" (SJYH2022–2/285).

References

1. Iankova, S., Davies, I., Archer-Brown, C., Marder, B., Yau, A.: A comparison of social media marketing between B2B, B2C and mixed business models. Ind. Market. Manag. **81**, 169–179 (2019)
2. Bhat, I.H., Singh, S.: Intention to participate on social commerce platform: a study on e-commerce websites. Allied Bus. Acad. **22**(4), 1–10 (2018)
3. Luo, N., Wang, Y., Zhang, M., Niu, T., Tu, J.: Integrating community and e-commerce to build a trusted online second-hand platform: Based on the perspective of social capital. Technol. Forecast. Social Change **153**, 119913 (2020)

4. Kumar, A., Salo, J., Li, H.: Stages of user engagement on social commerce platforms: analysis with the navigational clickstream data. Int. J. Electron. Comm. **23**(2), 179–211 (2019)
5. Xuedi, J.: Homogenization and heterogeneity of social content e-commerce business models——taking Xiaohongshu and Mogujie as examples. Guangxi Qual. Super. Herald **06**, 122 (2019)
6. Wu, Q., et al.: Dual graph attention networks for deep latent representation of multifaceted social effects in recommender systems. In: Proceedings of the 28th International Conference on World Wide Web, pp. 2091–2102 (2019)
7. Buying behavior: a comparative analysis of luxury sportswear. Elsevier Inc. **69**(12) (2016)
8. Audrey, L.: Influencer fraud is affecting more brands than you think. Glob. Cosmet. Ind. **24**(10), 10 (2018)
9. Lisa, D.: Fast, eco-friendly hair care. Glob. Cosm. Ind. **22**(9), 43–50 (2019)
10. Seth, M.: Catching the wave: product introduction and changing consumer demands breathe life into hair care. Drug Store News **6**, 48–52 (2019)
11. Chen, L., Wu, L., Hong, R., Zhang, K., Wang, M.: Revisiting graph based collaborative filtering: a linear residual graph convolutional network approach. In: Proceedings of the AAAI Conference on Artificial Intelligence, vol. 34, no. 1, pp. 27–34 (2020)
12. Wang, Y., Jia, F., Schoenherr, T., et al.: Cross-border e-commerce firms as supply chain integrators: the management of three tflows. Ind. Mark. Manag. **89**, 72–88 (2019)
13. Sanjit, K.R., Vaibhav, S., Walfriedet, M.L., et al.: Customer engagement behaviors: the role of service convenience, fairness and quality. J. Retail. Cons. Serv. **44**, 293–304 (2018)

Prediction and Analysis of Bank Marketing Data Model Based on K-means Algorithm

Jinfeng Fan(✉)

Nanchang Branch of Jiangxi Radio and TV University, Nanchang, Jiangxi, China
939696850@qq.com

Abstract. Under the background of continuous refinement and depth of computer network algorithm, the model is optimized and improved on the basis of traditional bank marketing data model, and the model algorithm is used for data processing and optimization. At the stage of continuous in-depth development of computer computing power, all walks of life are introducing the most advanced technology, and banks are adopting computer models on a large scale, constantly improving algorithms and optimizing computing power, so that the banking industry has always been in the direction of the most cutting-edge technology development. In the forecast analysis, k-means algorithm is adopted to improve the bank marketing data model, focusing on how to improve the accuracy, forward-looking and fault tolerance of the forecast data. This paper studies the prediction and introduces the definition, concept and model application of K-means algorithm. The model is constantly optimized by imitating the actual scenarios of bank marketing, and the user portrait of bank marketing consumption is depicted by using big data technology, which is finally applied to the banking business. Based on bank marketing data model prediction analysis, computer marketing curve modeling, forecast. Argument according to the experiment, using the K - means algorithm of bank marketing data model prediction analysis of bank business, improve the effect is significant based on repeated tests, compared with the traditional bank marketing data model, this model in bank marketing forecast precision, flexibility and fault tolerance, depth of learning the performance of the composite scores were 80.1%, 85.1%, 92.4%, 97.3% .

Keywords: K-means Algorithm · Bank Marketing · Marketing Data · Forecast Analysis

1 Introduction

With the high development of the computer field, the computer application technology is constantly updated in the algorithm field. Bank marketing business is an important pillar for the survival of banks, how to improve the bank marketing business is the most important. In today's highly developed information technology, it is undoubtedly necessary to develop an excellent software algorithm, so what kind of algorithm depends on how the business scenario. A professional algorithm is specially developed for bank marketing data model to solve problems in bank marketing business. Based on the

J. H. Abawajy et al. (Eds.): ICATCI 2022, LNDECT 170, pp. 895–902, 2023.
https://doi.org/10.1007/978-3-031-29097-8_106

prediction, this paper modeled the bank marketing data, and adjusted and optimized the parameters according to the actual scenarios of the bank, so that the bank business will develop to the perspective of high efficiency and security.

As for the research based on k-means algorithm, Xu Y H, Piol M V, Arejita B, et al. A heterogeneous implementation of industrial use cases based on K-means is proposed and analyzed for symmetric multiprocessing (SMP), GPU and FPGA. We'll show you how to optimize an application from an algorithmic perspective and how this optimization can be performed on two heterogeneous platforms. Programmers can improve performance by explicitly specifying data memory access or replication [1]. Fa Bregas A C, BD Gerardo, Iii B. The k-means algorithm is compared with the improved K-means algorithm, and it is proved that the new initial seed selection method has good effects in both mathematical calculation and reliability [2]. Ozkok F O, Celik M. Proposed a mixed validity index to determine the Kparameter value of the K-means algorithm. The mixed validity indicators proposed in this paper include four internal validity indicators, namely Dunn, Silhouette, C index and Davies-Bouldin index. The method is applied to nine financial and benchmark time series datasets in real life [3]. However, the marketing data model of Chinese banks is still in the initial stage.

To further improve the analysis of the bank marketing data model based on the k-means algorithm, it is necessary to start from the following points: one is to strengthen the research and development based on the k-meas algorithm, the second, to optimize the marketing data model of the bank, and to finally strengthen international cooperation and exchange, and Introducing advanced algorithms.

2 Design and Exploration

2.1 Bank Marketing

Bank marketing refers to the activities and transactions conducted by commercial bank organizations or individuals in accordance with market changes and customer demand-oriented to adapt to the changing environment under certain market environment [4, 5].

(1) Service marketing

Marketing pulls apart the gap in the comparison between competitors, gradually moves away from rivals and wins customers.

(2) Value marketing

A marketing technique in which customers are acquired by making them feel superior value or exceed their expectations.

(3) Rapid marketing

In each link of marketing are like turbo speed rotation, with "quick attack", "early catch" and other ways, one step ahead of competitors, to win customers or other resources.

(4) Quality marketing

Responsible for the quality of the marketing target, so that customers can buy the quality service they are sincerely satisfied with [6, 7].

(5) Word of mouth marketing

In terms of experience, efforts should be made to ensure customer satisfaction, promote the reputation of the bank, publicize the advantages of the bank, and make it impossible for customers to find any opportunity to discover the defects of the bank [8, 9].

(6) Change marketing

Fully aware of the multiple changes in the market, good at based on the current market environment, grasp the law of the market, to change in response to all changes, constantly seeking new, remove old ideas, new ideas, new definition, fully grasp the marketing of the changing market.

(7) Direct marketing

Using multimedia technology, customers can send advertisements for specific customers, do sales, use virtual way to set up business halls or business rooms, all kinds of sales.

2.2 System Function Introduction Based on Bank Marketing Data Model

(1) Connection and management of platforms such as official accounts

Various multimedia platforms have become important means and methods for banks. Bank marketing can expand the interface outward through various means and methods, so that customers and downstream merchants of banks can connect. Through the interface development for various business forms of platform and function, online connection, etc.

(2) Advertiser platform connection

Each big merchant website also borrows the bank's big flow as far as possible in advertising. After the marketing advertisements are put into these platforms, the advertisements provide precise docking for bank marketing. Each AD is assigned to the client with a proprietary ID, and based on these ids, the client is routed and bid in real time, and these parameters are connected to the system.

(3) Content Management

The content of the platform includes pictures, messages, videos, etc. These forms of content contain a large number of customers, and have rich and precise content management. Improper content can be dealt with reasonably, which will not cause harm to public opinion and will not affect the reputation of the platform.

(4) Channel analysis and launch management

Effective analysis of advertising investment and delivery channels, advertising from drainage to the official page, quantitative analysis of advertising, further supervision of customers, purchase behavior. For advertising channels, find channels and methods with high visitor volume and high conversion rate to achieve the best results. Let each investment bring the maximum return.

(5) Sharing mechanism and propagation path analysis

Users will make full use of multimedia platforms to share and forward advertisements, and fully carry out marketing activities to share the rights and interests of product advertisements with customers. The platform can set up relevant rights and interests rules and reward rules, and give rewards when the advertisement reaches a certain amount of forwarding.

(6) Marketing activity management

Marketing campaigns are a very important medium for customers, and there are many kinds, such as Tmall's Singles' Day, JINGdong's 618, and so on. Marketing activities have a variety of concessions, points, coupons, lucky draw, red envelope, collection of five blessings, activities of a variety of content. Each marketing campaign needs careful planning, handling and management, as shown in Fig. 1.

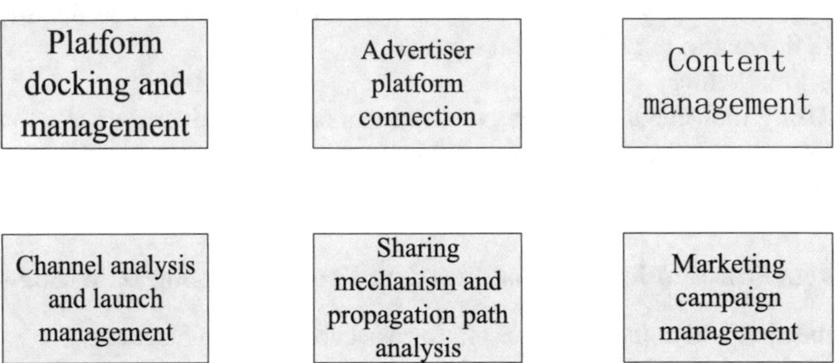

Fig. 1. System function based on bank marketing data model

The most obvious feature of the development and design of the intelligent early education auxiliary system based on the visual algorithm shown in Fig. 1 is to use data mining and other tools to extract [10].

(7) Teacher module

The teacher function module is shown in Fig. 1

2.3 Typical Model Cases of Bank Marketing Data

(1) Forecasting is associated with financial consulting

Sales managers are dissatisfied with current innovation metrics and not doing their job. In a survey conducted by the Boston Consulting Group in the early 2000s, fewer than a third of respondents said they were dissatisfied with innovation measures. In a global McKinsey survey the following year, about a quarter said their firm was doing a good job of measuring innovation.

The predictions made by the model provide some help, mainly in terms of marketing criteria and ROI, while other events can be planned, such as R&D investment, input-output and warehouse inventory. Companies using model forecasting are bound to go beyond a single forecast and confidence interval, and they want to seek multiple different targets through model forecasting.

Prediction: 30 million at best, 35 million at nearly half, 40 million at 10%.

(2) Transfer to interviewee level model

Today's world forecast model is not standard, according to the characteristics of today's consumers, to create a forecast model for consumers, and make necessary research on innovation, for better innovation to do a good job.

(3) Pay more attention to global migration

One way companies control costs is through market forecasts and guesses. For example, Australia is predicted and New Zealand is guessed.

2.4 Analysis of Current Situation of Bank Marketing Data

Take a bank for example, there are a lot of private business customers, these customers have various social people, some huge consumption, some wealth, some ordinary workers, some stock market experts; Some are normal white-collar workers, some are business heroes, and some are people rushing about for survival. Investment ability and preference, consumption ability and preference, reserve absorption potential, service demand, loan demand, repayment ability, deposit and withdrawal habits, transfer frequency and other aspects are closely related to banking business. These feature factors are related to the bank's cost, income, earnings rate of launched products, research and development of new products, service focus of branches, density of branches, capacity of telephone and online banking, planning of market activities and a series of contents inseparable from the bank's input, output and strategic decisions, as shown in Fig. 2.

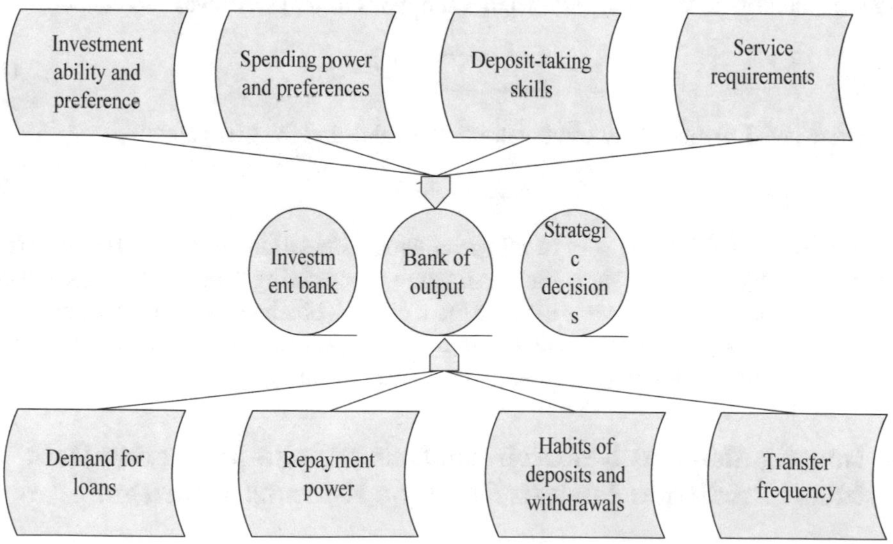

Fig. 2. Influencing factors of bank marketing

In addition, since the bank's marketing staff are used to their own relevant experience as a starting point when conducting marketing campaigns. Therefore, marketing activities are often not based on scientific and objective analysis of market and customer characteristics, on the contrary, it is with a heavy subjective color. In view of the current situation, the bank decided to analyze the massive data it has accumulated through data mining, hoping to get customer segmentation based on the characteristics of customers' consumption behavior rather than subjective judgment, and develop corresponding targeted market strategies on this basis.

3 Research on the Prediction and Analysis Effect

3.1 Quantitative Analysis

In this paper, quantitative analysis is adopted to carry out quantitative analysis and demonstration through the prediction analysis [11, 12].

3.2 Analysis Summary

This paper mainly uses quantitative analysis method to predict and analyze the system model, and records and analyzes the data of the prediction effect.

3.3 Application of Formula

Randomly select K objects from more than 1.5 million data objects as the initial cluster center; For the remaining objects, they are divided into the most similar clusters according to their similarity to these clusters. Then, each new cluster center was calculated.

$$E = \sum\nolimits_{i=1}^{k} \sum\nolimits_{peii} |P - m_i|^2 \tag{1}$$

There is a difference between the predicted value and the true value:

$$m_{i+1} = m_i + \varepsilon \tag{2}$$

E in Formula (1) is the sum of all mean square deviations in the database; P is a point in the object's space; m_i is the mean value of clustering. The clustering standard shown in the above formula aims to make the k clusters obtained as compact as possible and separate as possible. In Formula (2), m_{i+1}, m_i respectively represents the clustering mean values corresponding to each sequence, and ε is parameter.

4 Investigation and Research Analysis of Bank Marketing Data Model Prediction Analysis Based on K-means Algorithm

4.1 Test and Analysis

Firstly, the behavior and results of the bank marketing data model prediction and analysis based on K-means algorithm are analyzed: in the experimental test data, aiming at

Table 1. Model predictive analysis

	Precision	Flexibility	Fault tolerance	Deep learning performance
Test function points	302	205	201	604
Ratio	80.1%	85.1%	92.4%	97.3%
Satisfaction	79	84	85	92

whether the bank marketing data model prediction and analysis based on K-means algorithm can achieve the prediction of bank marketing, the optimization effect of bank marketing parameters is obvious, as shown in Table 1 and Fig. 3.

The bank marketing data model prediction analysis based on k-means algorithm from the experimental test score, compared with the traditional bank marketing data model prediction system is significantly improved. It is necessary to strengthen the research and application of K-means algorithm for a bank marketing data system.

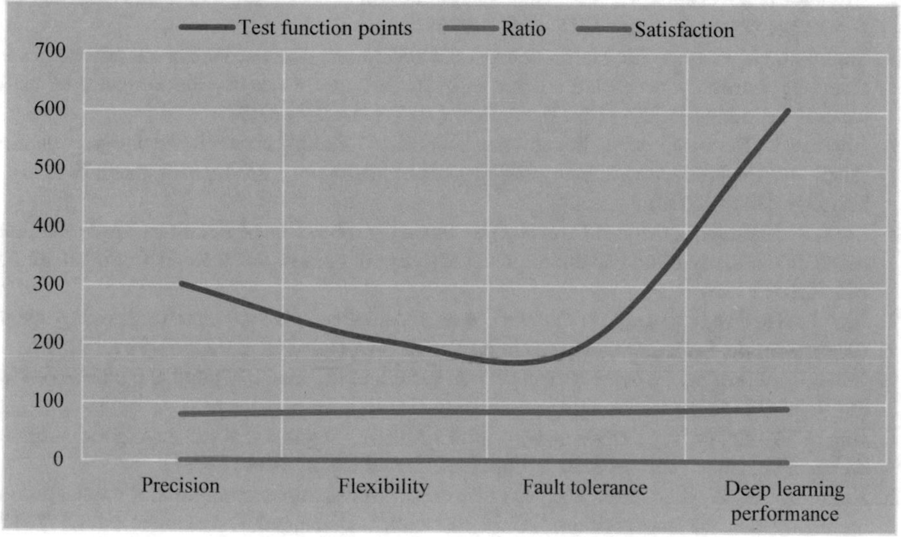

Fig. 3. System function score diagram

5 Conclusions

Based on K - means algorithm in the field of development, thanks to the strengthening of computer technology, marketing business from the bank, China bank marketing data model based on K-means algorithm forecast analysis must be improved, so can in today's international continue to improve, so as to meet the increasing bank marketing data requirements, Only in this way can the future data be predicted more accurately and more fault-tolerant. International background of today, the forecast analysis system also need to blend in, learn foreign advanced technology, enhance and improve the system function, enhance the capacity of deep learning, to make Banks more intelligent marketing model, precision, innovation, so as to promote the bank contribution in the field of marketing.

References

1. Xu, Y.H., Piol, M.V., Arejita, B., et al.: Implementation of the K-Means Algorithm on Heterogeneous Devices: A Use Case Based on an Industrial Dataset. Adv. Parallel Comput. **32**, 642–651 (2018)
2. Bregas, A.C.F., Gerardo, B.D., Iii, B.: Enhanced initial centroids for k-means algorithm. Int. J. Inf. Technol. Comput. Sci. **9**(1), 26–33 (2017)
3. Ozkok, F.O., Celik, M.: A hybrid validity index to determine k parameter value of k-means algorithm for time series clustering. Int. J. Inf. Technol. Decis. Mak. **20**(06), 1615–1636 (2021)
4. Ztürk, A., Güven, F.: Importance of social media as communication channel in bank marketing. Acad. J. Interdisc. Stud. **3**(3), 76–81 (2018)
5. Arun, R.: A novel classifier model efficient prediction of success rate in bank marketing. Int. J. Adv. Sci. Technol. **29**(9), 6955–6961 (2020)
6. Sugiyarti, G., Ardyan, E.: Market sensing capability and product innovation advantages in emerging markets: the case of market entry quality and marketing performance of batik industry in Indonesia. Dlsu Bus. Econ. Rev. **27**(1), 175–189 (2017)
7. Biletska, Y., Pysarevskiy, M., Sokolovska, O., et al.: Marketing research and design of quality function in the production of innovative product of health purpose. Technol. Audit Prod. Res. **3**(4(53)), 41–44 (2020)
8. Cork, B.C., Eddy, T., Cork, B.C., et al.: The retweet as a function of electronic word-of-mouth marketing: a study of athlete endorsement activity on twitter. Int. J. Sport Commun. **10**(1), 1–16 (2017)
9. Jia, Q., Ko, J.M.: A study on the fandom as a voluntary word-of-mouth marketing agent: focusing on the Chinese fandum of N.Flying. J. Korea Cult. Ind. **21**(1), 79–88 (2021)
10. Cabot, J.: WordPress: a content management system to democratize publishing. IEEE Softw. **35**(3), 89–92 (2018)
11. Jing, Y., Wang, J.: Tag clustering algorithm LMMSK: improved K-means algorithm based on latent semantic analysis. J. Syst. Eng. Electron. **28**(2), 374–384 (2017)
12. Xiao-Yu, L.I., Li-Ying, Y.U., Lei, H., et al.: The parallel implementation and application of an improved K-means algorithm. J. Univ. Electron. Sci. Technol. China **46**(1), 61–68 (2017)

Digital Economy Research Based on Spatial Dupin Model Under Big Data Technology

Tiantian Ren(✉)

School of Shanghai University, Shanghai, China
rentian@shu.edu.cn

Abstract. In the context of rapid development of digitalization, the influence of big data and technological progress on economic development cannot be ignored. This paper establishes a spatially fixed durbin model and uses panel data of 30 provinces in China from 2003 to 2017 as a sample to empirically test the effect of big data and technological progress on economic development in the context of digitalization. The empirical results show that: (1) Technological innovation is an important factor to promote economic development, and the improvement of big data can promote economic development. (2) big data, economic development and technological innovation all have significant positive spatial spillover effects. (3) Big data and technological innovation further boost the economy, and benefit the economic development of neighboring areas. This article believes that big data and technological innovation can actively promote the transformation of China's economy to high-quality development.

Keywords: Big Data · Technological Innovation · Spatial Dupin Model

1 Introduction

With the further construction of modernization, breakthroughs and innovations in information and communications technologies are still being made rapidly, the Internet is gaining popularity. Big data and artificial intelligence are gradually being known and applied to various fields [1]. Big data has triggered changes in the economic field, and my country has thus formed a new economic development model. Therefore, the influence of big data on socio-economic development cannot be ignored. The current economic development is operating at a high speed, which means that a new century of digitalization has arrived. Big data is widely used in various industries, science and technology are advanced, information dissemination speeds up, and exchanges in the economic field are getting closer. In the high-tech era, big data affects economic development, social order, national governance, and people's lives. Big data provides the possibility of digital survival for urban development. In the Internet society, data has become the core resource. After industrial transformation and upgrading and the wide application of big data, the ecological scale has expanded rapidly. The integration of the real economy and big data is also deepening. Financial big data, industrial big data, healthcare big data and government affairs big data have promoted the development of digital economy and

J. H. Abawajy et al. (Eds.): ICATCI 2022, LNDECT 170, pp. 903–910, 2023.
https://doi.org/10.1007/978-3-031-29097-8_107

technological iteration. Since the advent of the digital age, the government has issued a number of policies to increase the application of the digital economy, hoping to further improve the speed and quality of my country's economy.

Under the background of the global economic slowdown, whether the application of digital technology can help solve the problem of sluggish economic growth has attracted wide attention. The extensive application of information and communication technology is considered to be one of the key factors for the continuous growth of total factor productivity in advanced economies from the 1990s to the early 2000 [2]. Since 2004, studies have shown that in developed countries, digital technology no longer plays a prominent role in improving labor productivity [3]. Despite the fact that e-commerce, WeChat and other Internet applications have become an indispensable part of Chinese daily life, China's economic growth rate has continued to decline over the past decade, with the per capita GDP growth rate dropping to 6.1 percent in 2016 (Fig. 1). Can the widespread application of big data really drive economic growth today?

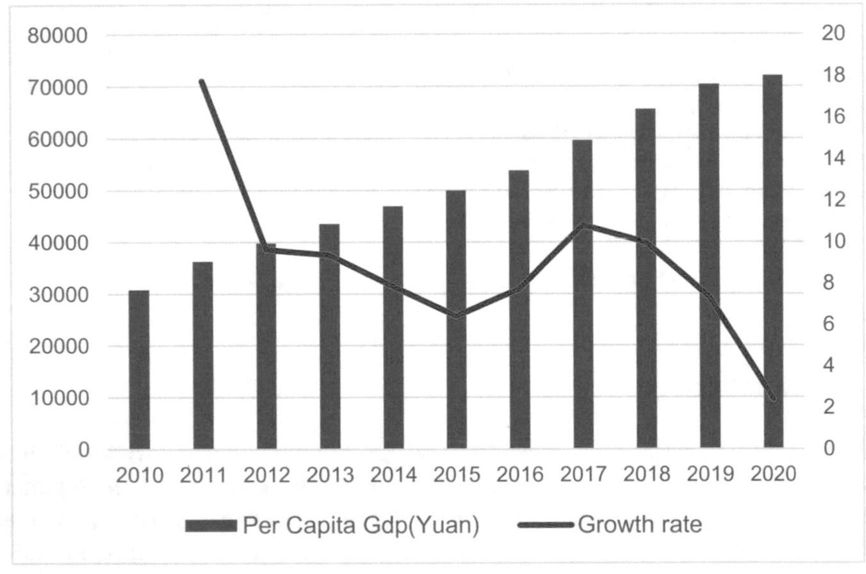

Fig. 1. Economic growth over the years

There is little literature on the economic impact of big data. Existing literature is also quite divided on how big data promotes technological innovation. One type of viewpoint believes that big data can significantly promote technological innovation [4]. For example, Song et al. (2019) [5] believe that big data can significantly promote technological innovation. As one of the basic strategic resources of various countries, data is an important means to promote green innovation of enterprises. Abdul-Nasser et al. (2019) [6] used the PLS-SEM model to verify that big data is one of the important factors that promote enterprise technological innovation and product innovation. Another type of view is that big data is not obvious in promoting technological innovation and even has a certain inhibitory effect. For example, Ghasemaghaei and Calic (2019) [7] used the relevant data of 239 companies in the United States in 2018 and found that big data can improve the innovation level of companies, mainly because of its type and speed attributes. The extensive application of big data has a certain negative impact on improving the innovation efficiency of enterprises.

"Solow surplus" theory holds that total factor productivity is beneficial to increase output. Total factor productivity is an important factor besides labor and capital investment [8]. At present, existing studies have reached consensus on the on the impact of technology on economic development, and all believe that technological innovation can promote economic development. Banerjee and Roy (2014) found technological advancement is one of the important factors contributing to India's long-term economic growth [9]. By studying the factors affecting the economic development of South Asia, Mughal et al. (2022) found that the relationship between technological progress and economic growth is very close [10].

Regarding whether technological innovation and development will be affected by big data, existing research has not yet reached a consensus. Existing studies agree that technological innovation cannot be ignored among the many factors that affect economic development. So, will the development of big data become another important factor driving economic growth?

In summary, existing research cannot determine whether big data can promote economic growth. There are different understandings of the concept of technological innovation in existing research. In the construction of comprehensive evaluation indicators, the indicators selected by existing research are also different. In the research on promoting economic development, the impact of technological innovation cannot be ignored. There are different understandings on the question of how technological innovation can promote economic growth. The analysis of the influence mechanism is mainly theoretical analysis, and there is no empirical test. Many studies also fail to take into account the spatial spillover effects of economic growth and technological innovation. Moreover, few studies have taken into account the association between big data and technological innovation when studying the role of technological innovation in promoting social economy. Big data applications are closely related to technological innovation.

Therefore, this paper adopts the spatial Durbin model to study the impact of big data and technological innovation on the economy in the era of digital economy. Main contributions of this paper are as follows: On the premise of considering the spatial spillover effects, this paper empirically tests the impact of big data industry development and technological innovation on economic development. It takes into account not only the impact of big data level on the economy of the region, but also its impact on surrounding areas. This article more clearly demonstrates the importance of big data development in China's economic development. In addition, the association of big data and innovation has been considered, making empirical results more intuitive.

2 Research Design

2.1 Establishment of Econometric Model

In order to understand the spatial effects of technological innovation and information level on environment, we used the Spatial Dupin Model (SDM) to observe the spatial spillover effects of explanatory variables due to spatial dependence. The SDM model formula is as follows:

$$\mathbf{y} = \beta_0 + \lambda \mathbf{Wy} + \beta \mathbf{WX} + \mu_i + \alpha_t + \varepsilon \tag{1}$$

where y represents the economic development; X is the explanatory variable set. Besides, λ represents the spatial regression coefficient, μ is the cross-section fixed effect, α represents the time fixed effect, and ε represents the error term.

The spatial weight matrix W describes the source and strength of spatial correlation among variables, and generally considers the spatial distance of individuals from the perspective of geography or economic differences. The matrix element $w_{ij} = 1/d_{ij}$, *if* $i = j$, then the w_{ij} is 0, where d_{ij} represents the spatial distance between two places. The larger the distance is, the smaller the weight will be, and the interaction between regions will decrease with the increase of the spatial distance. Therefore, the inverse distance matrix is selected in this paper.

2.2 Variables and Data Description

(1) Independent variable: Economic development (gdp_{it}). Economic development can be reflected in both the increase of gross regional product and the change of growth rate. This article uses the actual per capita GDP of each province to measure economic development.

(2) Technological innovation (tec_{it}). In the existing literature, total factor productivity or R&D expenditure is usually used as the substitute variable of technological progress. As a basic indicator to measure innovation, patent has strong representativeness to measure technological innovation. Therefore, we select the number of patent applications accepted in each province, as the indicator to measure technological innovation.

Table 1. Definition of variables

Variable	Definition
Dependent variable	
gdp	Economic development (GDP per capita)
Independent variable	
data	The level of big data
Tec	Technological innovation (number of patents granted)
Control variable	
Edu	Education level (Education expenditure / government fiscal expenditure)
Open	Opening level (expenditure of direct foreign investment/GDP)
City	Urbanization (Population of municipal districts/ Total population)
Gov	Government intervention (government fiscal expenditure/GDP)

(3) Big data level ($data_{it}$): It is measured by the proportion of the number of employees in the information transmission, computer service and software industries in the total number of employees in the province.

In order to ensure the accuracy of the results, this paper adds control variables such as education level and urbanization (Table 1).

The control variables in this paper are all meaningful, and all data are from China Statistical Yearbook and China over the years. The data in this article has been processed logarithmically, and the price data is based on 2003.Descriptive statistics are used for all variables in this paper (Table 2).

Table 2. Descriptive statistics of variables

Variable	Obs	Mean	Std.Dev	Min	Max
lngdp	450	9.905	0.606	8.190	11.23
lntec	450	9.631	1.648	4.820	13.35
lnopen	450	−4.015	1.061	−8.435	−1.951
lncity	450	−1.008	0.497	−1.815	0
lnedu	450	5.699	1.029	2.536	7.854
lngov	450	−1.539	0.594	−2.593	0.361
lndata	450	−4.288	0.459	−5.240	−2.347

3 Empirical Results and Analysis

The empirical results are as follows (Table 3). Taking into account the spatial effects among variables, we decompose spillover effects into direct and indirect effects. The influence of the local independent variable on the dependent variable is called the direct effect, and the influence of the local independent variable on the dependent variable of the neighboring area is called the indirect effect.

In the main effect, the coefficient of explained variable level are positive. Big data level has a certain impact on economic development. Technological innovation intensity increases by 1%, resulting in economic growth of 0.064%. This shows that technological innovation can actively promote economic development. In the future, government policy makers should still focus on the driving effect of technology in China's economic development, seize the opportunity period of the technological revolution in the new era, strive to create an external environment conducive to technological innovation, build an innovative country, and drive sustained high-quality development. In the control variables, the coefficient the opening is significantly positive. Increased opening will drive economic development. The coefficient of urbanization is positive but not significant. The coefficient of education support intensity is significant, meaning that the increase of education support intensity is beneficial to economic development. The possible reason is that strengthening education is conducive to the cultivation of talents and has an important impact on all aspects of society.

In the decomposition effect, the coefficient of technological innovation under both direct and indirect effects is significantly positive, means that technological innovation can boost the local economy, and benefit the economic development of neighboring areas. The reason is that technological innovation has realized positive spillover effect on neighboring regions through knowledge spillover, technology spillover and synergistic effect. It will promote the economic development of the neighborhood. The spatial lag of the economic development of the explained variable is also significantly positive. It can also promote the development of surrounding areas. The coefficients under the direct and indirect effects of the big data level are both significantly positive. Means that the development of big data shows significant spatial correlation. Big data in this region is beneficial to the economic development of the surrounding areas. The coefficient of opening is significantly positive under the direct effect, while it is not significant under the indirect effect, indicating that the opening has a small impact on adjacent areas. The coefficient of education support intensity under the direct effect is significantly positive, and the coefficient under the indirect effect is significant. That is to say, the improvement of education support intensity is beneficial to the economic development of the local area, and it has an inhibiting effect on the surrounding areas.

Table 3. Analysis of empirical results

variance	main	wx	direct	indirect
lntec	0.064*** (4.2)	0.075* (1.79)	0.067*** (4.25)	0.131** (2.26)
lndata	0.027*** (4.1)	0.047* (1.7)	0.028*** (3.81)	0.078** (2.31)
lnopen	0.015* (1.95)	−0.019 (−0.54)	0.014* (1.89)	−0.023 (−0.44)
lncity	0.005 (0.11)	−0.071 (−0.36)	0.01 (0.2)	−0.077 (−0.27)
lnedu	0.612*** (14.68)	−0.370*** (−5.58)	0.608*** (14.9)	−0.260*** (−4.07)
lngov	−0.398*** (−9.44)	0.2 (1.33)	−0.398*** (−9.80)	0.098 (0.46)

4 Conclusions and Policy Recommendations

The main conclusions of this paper are as follows. Big data and technology can significantly drive the improvement of the quality of comprehensive economic growth. Big data, economic development and technological innovation all have significant positive spatial correlation. Big data and technology play an important role in the development of this region and surrounding areas. Based on the above conclusions, this paper puts forward the following suggestions:

First, we must continue to focus on the digital economy and deeply integrate the real economy with big data. Accelerate digital transformation of the industry. The government should promote resource utilization and increase investment in enterprise innovation. In order to further enhance the level of scientific and technological innovation and promote economic development.

Second, Information sharing systems should be established between regions to achieve knowledge complementarity. Increase the integration of big data and the real economy, which is conducive to economic growth. we need to lead high-quality economic growth through technological innovation. First of all, relevant policies and systems must be introduced to encourage technological innovation. The government should improve supporting policies in fiscal, taxation, finance and intellectual property protection to create necessary external conditions for technological innovation. Let the market play a leading role, break the industrial monopolies and market segmentation that restrict technological innovation, and encourage enterprises to invest in technological innovation. Then establish a good basic scientific research system to support technological innovation. Government decision makers should improve the system mechanism, promote the integrated innovation and development of basic research, establish a good basic scientific research system, and consolidate the foundation of technological innovation. Finally, a good education system should be established to strengthen the introduction and incentive of talents, so as to provide talent guarantee for technological innovation. In light of the development trend of emerging industries such as big data, cloud computing, artificial intelligence and robot manufacturing, we are supposed to establish sound systems of higher education, vocational education and basic education. Cultivate innovative talents who adapt to technological changes and realize industrial upgrading and transformation.

Third, focus on the agglomeration effect of the digital industry, to further enhance the economies of scale. Ensure the concentrated input in regional innovation. Strengthen inter-regional cooperation and exchanges, and actively leverage the beneficial impact of digitalization on economy in underdeveloped regions.

References

1. Teresa Ballestar, M., Diaz-Chao, A., Sainz, J., et al.: Knowledge, robots and productivity in SMEs: Explaining the second digital wave. J. Bus. Res. **108**, 119–131 (2020)
2. Gordon, R.J.: The rise and fall of american growth: the U.S. standard of living since the civil war. Introd. Chapt. **56**(5), 921–924 (2016)
3. Stiroh, K.J.: Information technology and the U.S. productivity revival: what do the industry data say? Am. Econ. Rev. **92**(5), 1559–1576 (2002)
4. Vocke, C., Constantinescu, C., Popescu, D.: Application potentials of artificial intelligence for the design of innovation processes. Procedia CIRP **84**, 810–813 (2019). https://doi.org/10.1016/j.procir.2019.04.230
5. Song, M., Fisher, R., Kwoh, Y.: Technological challenges of green innovation and sustainable resource management with large scale data. Technol. Forecast. Social Change **144**, 361–368 (2019)

6. El-Kassar, A.N., Sanjay, S.K., et al.: Green innovation and organizational performance: the influence of big data and the moderating role of management commitment and HR practices - ScienceDirect. Technol. Forecast. Social Change **144**, 483–498 (2019)
7. Ghasemaghaei, M., Calic, G.: Assessing the impact of big data on firm innovation performance: Big data is not always better data. J. Bus. Res. **108**, 147–162 (2020)
8. Nikolaos, C., Tsaliki, P.: The dynamics of capital accumulation in Marx and Solow. Struct. Chang. Econ. Dyn. **57**, 148–158 (2021)
9. Banerjee, R., Roy, S.S.: Human capital, technological progress and trade: what explains India's long run growth? J. Asian Econ. **30**, 15–31 (2014)
10. Mughal, N., Arif, A., Jain, V., et al.: The role of technological innovation in environmental pollution, energy consumption and sustainable economic growth: evidence from South Asian economies. Energ. Strat. Rev. **39**, 100745 (2022)

Image Segmentation Method Based on Grabcut and Hue-Saturation-Value Color Space Model

Yunhong Zhao[1], Yuhua Xu[2], and Tianbo Wang[3(✉)]

[1] School of Computer Engineering, Jingchu University of Technology, Jingmen, Hubei, China
[2] School of Finance, Nanjing Audit University, Nanjing, Jiangsu, China
[3] School of Mathematics and Statistics, Shanghai University of Engineering Science, Shanghai, China
yh.z@163.com

Abstract. Aiming at the problems of low accuracy and poor effect of image segmentation under complex background, the characteristics and attributes of Hue-Saturation-Value (HSV) color space are analyzed, and a color distance calculation method is defined. Combined with grabcut algorithm, a segmentation method based on grabcut method and HSV color space model is proposed. Experiments show that the image segmentation effect is better when the saturation value and hue value are properly selected to the mask.

Keywords: GrabCut Algorithm · Hue-Saturation-Value Color Space · Image Segmentation

1 Introduction

At present, under the background of artificial intelligence and deep learning, deep convolution is mostly used to extract the features of images, which requires a lot of data to train. But in real life, collecting large amounts of data is not an easy task. For small samples, feature extraction has become the focus of scholars [1]. In practice, small sample extraction involves many factors, which is not a standard data set, and there are a large number of objective factors, such as illumination, complex environment and so on. In this case, image feature extraction and further recognition still cannot achieve accurate extraction and recognition. Therefore, in the case of small samples, it is of great significance to improve the segmentation accuracy, and it is also very important for the subsequent image analysis [2, 3].

In order to solve the problem that it is difficult to accurately segment the target in color image, many scientists have carried out detailed research. People are only interested in image segmentation and image processing [4–6]. In the past, many segmentation methods are for gray images. Even for color images [7–9], Firstly, the color image is converted into gray image, and then it is segmented. In these methods, the color information of the image is not fully utilized, so the segmentation effect of some images is poor [10–13]. Therefore, in the field of color image segmentation, the following two problems must be solved: how is the color information of the image fully utilized? How the accuracy of

J. H. Abawajy et al. (Eds.): ICATCI 2022, LNDECT 170, pp. 911–918, 2023.
https://doi.org/10.1007/978-3-031-29097-8_108

segmentation is greatly improved? For human vision, color information and brightness information are important information of images. For different images, human attention to color information and brightness information is different [14]. Through the saliency map method, the color image segmentation based on visual attention is realized by ITTI, but it is obvious that the segmentation efficiency has not been improved [15–17].

Therefore, a color image segmentation method based on HSV comprehensive significance is proposed in this paper, which makes full use of the color and brightness information of the image to calculate the color and brightness information significance of the image, and gives different weights to the color and brightness information significance according to different scenes, so as to finally form a comprehensive significance map, The threshold segmentation method is used to separate the target and background.

2 Image Segmentation Method Based on Grabcut and HSV

In computer vision, it is key process to image segmentation. Among these technologies, the core method includes subdividing the collected visual input information into a great deal of different regions and removing the noise part, so as to simplify image analysis. A fragment is used to represent a target or part of a target. It consists of a group of digitized pixels or "super pixels", including the corresponding color information or gray information. The segmented image is reorganized according to certain logic algorithm, rather than taking a single pixel as the processing unit. Image analysis is divided into three levels: classification, target detection and segmentation. Classification refers to the classification of the whole image into "human", "animal", "outdoor" and other categories [18]. Target detection is to detect the target in the image, draw a rectangle around it, and then add a label, such as a cat or dog. Segmentation refers to identifying each part of the image, classifying them, and further understanding what kind of object they should belong to. Among them, segmentation is the basis of target detection and image classification.

2.1 Image Segmentation by Grabcut

Ideally, a matching tool can generate continuous alpha values in the Tu area of trimap without hard constraints, and the alpha value can only be 0 or 1. In this way, problems involving complex backgrounds such as smoke, hair and trees can be handled automatically and properly. However, according to in our experience, when there is enough separation between the color distribution of the foreground and the background, the technology designed to solve the general matching problem is effective, but it is ineffective in the pseudo target. In fact, in the case of camouflage, the general matching problem cannot be solved. In a sense, it is difficult for humans to detect the complete shading.

Grabcut is an improved version of graph cut and an iterative graph cut. The algorithm uses the texture and boundary information in the image, that is, color and contrast information in them. As long as a small amount of user interaction, a better segmentation effect can be obtained. The algorithm is described as follows, as shown in Fig. 1:

(1) Enter a rectangle. All areas outside the rectangle must be the background. What's inside the rectangle is unknown. Similarly, any operation of the user to determine the foreground and background will not be changed by the program.
(2) The computer will make an initialization mark on our input image. It marks the foreground and background pixels.
(3) GMM (Gaussian mixture model) is used to model the foreground and background pixels.
(4) Based on user input data, GMM will learn and establish a new pixel distribution. For those pixels whose classification is unknown, that is either foreground pixel or background one; they can be sorted according to their pixel relationship with known classification such as background, just like clustering.
(5) This creates a map based on the distribution of pixels. The nodes in the figure are pixels. Moreover, in addition to pixel nodes, there are still two types of nodes, which are source node and sink node. When the source node is connected, all foreground pixels are the same. And all background ones are connected with sink node.
(6) Connect pixels to source node or end node, the weight of the edges of the node is determined by the probability that they belong to the same class. The weight between two pixels is determined by the edge information or the similarity of two pixels. If the colors of two pixels are very different, the weight of the edges between them will be very small.
(7) Mincut algorithm is used to segment the above graphics. According to the minimum cost equation the node is divided into the source one and the sink one. The cost

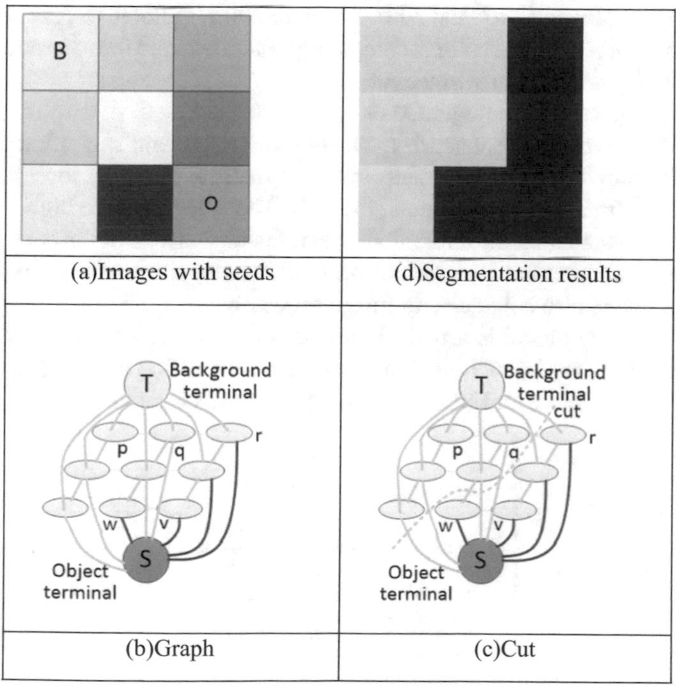

Fig. 1. Image segmentation by grabcut

equation is the sum of the weights of all cutting edges. After clipping, all pixels connected to the source node are regarded as the foreground and all pixels connected to the sink node are regarded as the background.

(8) Continue this process until the classification converges.

2.2 HSV Color Space Model

On the one hand, color space description should conform to the visual perception characteristics of human eyes; on the other hand, it should facilitate image processing. RGB (three primaries) color space is a kind of uneven color one. There is a large gap between the color distance of pixels and the perception of human eyes, which is not suitable for the segmentation of color image; HSV (hue saturation value) color space is a uniform color space, which reflects people's visual perception of color. Firstly, the luminance component is independent of the color information of the image. Secondly, the components of hue and saturation are inseparable from the way people feel color Therefore, the segmentation method in this paper selects HSV color space. In HSV color space as shown in Fig. 2, hue H and saturation S contain color information, while brightness is independent of color information. Hue is related to the main wavelength of light in the mixed spectrum. The hue of color reflects what kind of spectral wavelength the color is closest to. The hue is expressed in terms of angle, without losing generality. It is assumed that the color of 0° is red, 120° is green, 240° is blue, the hue changes from 0° to 240°, covering all the colors of the visible spectrum, and the non-spectral color (purple) visible to the human eye between 240° and 300°. The saturation is related to the brightness of a certain hue. The pure spectral color is completely saturated, and the saturation decreases gradually with the addition of white light.

HSV color space can be regarded as an inverted cone. The long axis represents brightness information, the distance from the long axis represents saturation information, and the angle around the axis represents hue information. The gray tone changes from black to white from bottom to top along the axis. The color with the highest brightness and saturation is located on the circumference of the top surface of the cone. The values of these three components are between 0 and 1. HSV color space reflects the way people observe color and is also conducive to image processing.

HSV color space model is a more intuitive color model, which is widely used in many image editing tools, such as Photoshop and so on. The conversion relationship between HSV color space and RGB color space is shown in Eq. (1)

$$\begin{cases} V = \frac{\max(r,g,b)}{255} \\ S = 1 - \frac{\min(r,g,b)}{V} \\ \theta = \cos^{-1}\left[\frac{\frac{1}{2}[(r-g)+(r-b)]}{\sqrt{(r-g)^2+(r-b)(r-b)}}\right] \\ H = \begin{cases} \theta, g \geq b \\ 2\pi - \theta \end{cases} \end{cases} \tag{1}$$

That is V∈ [0, 1]; S∈ [0, 1]; H∈ [0,2π].

The characteristics of HSV color space are as follows:

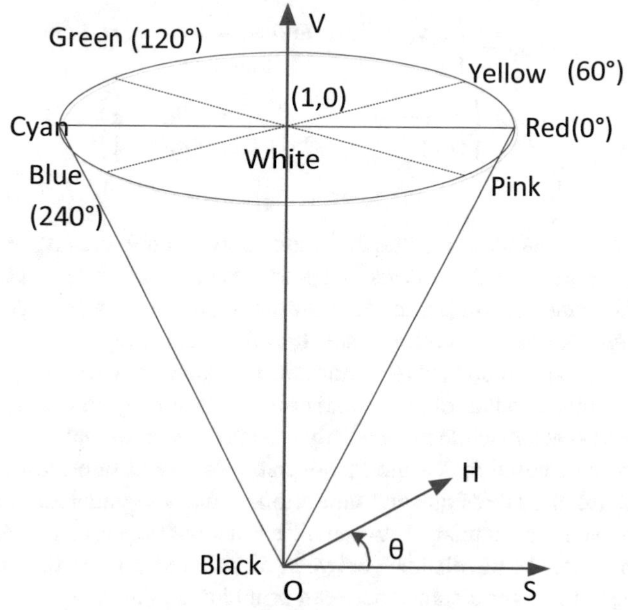

Fig. 2. HSV color space

(1) It has a color description mode consistent with the human visual system;
(2) Allows independent control of hue, saturation, and intensity (values);
(3) In order to facilitate the comparison between colors, HSV color space model can more intuitively express the brightness, hue and brightness of picture colors;
(4) The V component can be isolated from the H and s components, which is required by many color image processing algorithms.

2.3 Image Segmentation Method Based on Grabcut and HSV

Color distance refers to the numerical difference between two colors. Color distance should reflect the difference of color in people's vision. It is the basic basis for image segmentation. However, in HSV color space, color information is only related to H component and S component. The color value in the H component is not linear, but a ring, and people's perception of color is also related to the s component.

The HSV color space is roughly divided by Androutsos et al. through experiments [10]. Areas which brightness is greater than percent 75 and saturation is greater than percent 20 are bright color ones, the ones which brightness is less than percent 25 are black ones, the ones which brightness is greater than percent 75 and saturation is less than percent 20 are white ones, and others are color areas. In the special definition of HSV, white and black are the two colors to be compared. Based on the above conditions, the color distance between two points is defined as:

$$D_{ij} = \begin{cases} k_1 h_{ij} + k_2 s_{ij} & a_i \neq 0 \text{ or } a_j \neq 0 \\ k_1 \left| v_i - v_j \right| & a_i = 0 \text{ and } a_j = 0 \end{cases} \tag{2}$$

$$\text{where } a_i = \begin{cases} 0 & v_i > 75\% \text{ and } s_i < 20\% \; v_i < 25\% \\ 1 & \text{otherwise} \end{cases} \tag{3}$$

$$h_{ij} = \begin{cases} a_i a_{j|h_i - h_j|} & \left(\left|h_i - h_j\right| < \frac{1}{2}\right) \\ a_i a_j\left(1 - \left|h_i - h_j\right|\right) & \left(\left|h_i - h_j\right| > \frac{1}{2}\right) \end{cases} \tag{4}$$

$$s_{ij} = \left|s_{i-} s_j\right| \tag{5}$$

In the above formula, h_{ij} is the hue difference between point i and point j, and a_i is a logical variable. When a_i is 0, it means that point i is black or white. A point is black or white, which is defined according to the research of Androutsos et al. And it is related to brightness and saturation. In other cases, a_i is 1; s_i and s_j represent the saturation of point i and point j, and s_{ij} represents the saturation distance between two points; vi and v_j represent the brightness values of point i and point j respectively, indicating that the color distance of black point and white point is only related to the illumination, and has nothing to do with hue and saturation. k_1 and k_2 are hue factors and saturation factors, which respectively control the role of hue and saturation in image segmentation. Increasing the value of k_1 can make the boundary between different color targets clear, while increasing the value of k_2 can make the distance between different saturation regions in the same color larger. D_{ij} is the color distance between point i and point j.

After defining the color distance, the boundary of color mutation is extracted according to the color mutation and the spatial position relationship of pixels. Generally, there are many operators to extract the boundary, such as Sobel, canny, log, etc., but the color of HSV space is nonlinear, so when the operator is added, the actual mutation region of color cannot be obtained. In this paper, the modified Sobel operator is used, in which only subtraction operation is used. The algorithm calling process is as follows:

(1) Read a picture and mark the foreground with a rectangle.
(2) Call grabcut() to get the segmentation result.
(3) The segmentation result returned by the grabcut function contains four values: the determined background pixel, the possible background pixel, the determined foreground pixel and the possible foreground pixel. Therefore, extract the required value from the returned result as needed.
(4) After extracting the required value (matrix) from the result as needed, the image is segmented through mask.

3 Simulation Experiments

In Python 3.8 environment, Opencv and image segmentation method by grabcut are applied to carry out the simulation experiment of segmented image through the above defined color distance. The image in Fig. 3 is the original input image, Fig. 4 is the mask generated by the method in this paper, and Fig. 5 is the image segmentation result. The simulation results show that compared with the deep learning segmentation networks such as Faster R-CNN and U-Net, which can automatically generate the mask to segment the object (foreground) from the background, although the latter is more efficient and powerful, it will lead to confusion and roughness of the mask, Grabcut can be used to help clean these masks for better results.

Fig. 3. Original input image

Fig. 4. Mask by grabcut and HSV

Fig. 5. Segmented image

4 Conclusion

This paper analyzes the characteristics of HSV color space, defines the color distance calculation method, extracts the boundary with the modified Sobel operator, and then proposes an image segmentation method based on HSV color space and grabcut algorithm. Experiments show that this method can better retain the color information and combine the color information with spatial information. Moreover, by adjusting the saturation factor, it can realize variable color resolution and improve the universality of image segmentation.

Acknowledgements. This work was supported by Research on Application of Big Data in Jingmen Urban Planning, Science and technology planning project of Jingmen science and Technology Bureau (Grant No. 2021YDKY168).

References

1. Kishor, Y.N., Mukesh, S.: A novel fuzzy clustering based method for image segmentation in RGB-D images. Eng. Appl. Artif. Intell. **111**, 104709 (2022)
2. Baraboshkin, E.E., Demidov, A.E., Orlov, D.M., et al.: Core box image recognition and its improvement with a new augmentation technique. Comput. Geosci. **162**, 105099 (2022)
3. Melanie, S., et al.: Semantic segmentation of multispectral photoacoustic images using deep learning. Photoacoustics (2022)
4. Kumar, J.P., et al.: Analysis of RGB plant images to identify root rot disease in korean ginseng plants using deep learning. Appl. Sci. **12**(5), 2489 (2022)
5. Mönchinger, S., Schröder, R., Stark, R.: Methodology for a reverse engineering process chain with focus on customized segmentation and iterative closest point algorithms. MethodsX **9**, 101640 (2022)
6. Valdez-Rodríguez, J.E., Calvo, H., Felipe-Riverón, E., et al.: Improving depth estimation by embedding semantic segmentation: a hybrid CNN model. Sensors **22**(4), 1669 (2022)
7. Jonathan, V., Claire, L., Ruben, C.C.: Flexibly regularized mixture models and application to image segmentation. Neural Netw. **149**, 107–123 (2022)
8. Karmakar, S.: Application of SWSFET in image segmentation. In: Silicon, pp. 1–7 (2022)
9. Bellaj, K., Boujena, S., Guarmah, E.E.L.: An improved approach for image segmentation and three-dimensional reconstruction. Discontin. Nonlinear. Complex. **9**(2), 199–215 (2020)
10. Dhanachandra, N., Chanu, Y.J.: An image segmentation approach based on fuzzy c-means and dynamic particle swarm optimization algorithm. Multimedia Tools Appl. **79**(25–26), 18839–18858 (2020). https://doi.org/10.1007/s11042-020-08699-8
11. Giuliani, D.: Metaheuristic algorithms applied to color image segmentation on HSV space. J. Imaging **8**(1), 6 (2022)
12. Chen, Z., Wang, Y., Tian, W., et al.: Underwater sonar image segmentation combining pixel-level and region-level information. Comput. Electr. Eng. **100**, 107853 (2022)
13. Kang, D., Benipal, S.S., Gopal, D.L., et al.: Hybrid pixel-level concrete crack segmentation and quantification across complex backgrounds using deep learning. Autom. Constr. **118**, 103291 (2020)
14. Liu, S., Zhou, K., Qi, H., et al.: Improved hybrid particle swarm optimisation for image segmentation. Int. J. Parallel Emerg. Distrib. Syst. **36**(1), 44–50 (2021)
15. Gao, J., Wang, B., Wang, Z., et al.: A wavelet transform-based image segmentation method. Optik **208**, 164123 (2020)
16. Wang, Q., Yuan, C., Liu, Y.: Learning deep conditional neural network for image segmentation. IEEE Trans. Multimedia **21**(7), 1839–1852 (2019)
17. Sravani, R., Alex, D.S., Sushma, C.: Lossless image segmentation using adaptive threshold technique. J. Innov. Comput. Sci. Eng. **8**(1), 45–49 (2018)
18. Rahali, M., Loukil, H., Bouhlel, M.S.: Improvement of image compression approach using dynamic quantisation based on HVS. Int. J. Signal Imaging Syst. Eng. **11**(5), 259–269 (2019)

Flight Path Planning of Aircraft Under Multiple Constraints Based on Genetic Algorithm

Huanhuan Guo[1], Wenbin Liu[2,3](✉), and Hany Abdullah[4]

[1] Wuhan Technology and Business University, Wuhan, Hubei, China
[2] Wuchang Institute of Technology, Wuhan, Hubei, China
cq2201@126.com
[3] Department of Public Basic Courses, Wuhan Technology and Business University, Wuhan, Hubei, China
[4] University of Sulaimani, Sulaymaniyah, Iraq

Abstract. Flight path planning has become a significant subject of smart control. Because of system constraints, the location system of such airplane cannot precisely position. When the location error increase to a certain extent, the task could fail. Therefore, it is an key task in flight path to correct the location error. This paper processes the data given on attached annex 1 according to constraints, calculates the direct distance of any two samples and eliminates the path with a distance greater than 30,000 m between samples and a distance greater than 200,000 m from the target point. The 1000 feasible paths were generated by MATLAB as samples, and gradually iterated according to the genetic algorithm. Finally, the shortest route and the minimal correction points was obtained. The aircraft passed nine calibration points during this period, which is the optimal path. That is (A, 417, 210, 148, 155, 142, 254, 403, 594, 501, B), the track distance is 108.7758 km and the algorithm running time is 234.815 s.

Keywords: Route Planning · Genetic Algorithm · Multi-constrained Conditions

1 Introduction

Intelligent aircraft are widely used in agriculture, military, industry, and so on. Flight path programme is the key research content of airplane. [1–3]. It is to seek out the best way from the outset point to the goal point under the premise of satisfying the specific conditions of the system. The goal of the trace planning is to calculate the best or suboptimal flight way at the appropriate time using GPS/INS integrated navigation system based on preset digital maps.Fast trajectory program in intricate environment is a significant subject of smart control [4–6]. Researchers have proposed many methods for track planning [7–9]. There are two types of path planning algorithms: modern intelligent algorithms and classical algorithms. The former mainly includes heuristic optimization search, genetic algorithm, artificial neural network (ANN), etc. The latter mainly includes dynamic programming, optimal control and derivative correlation method. Because of system constraints, the location system of such plane cannot exactly locate. When the

J. H. Abawajy et al. (Eds.): ICATCI 2022, LNDECT 170, pp. 919–926, 2023.
https://doi.org/10.1007/978-3-031-29097-8_109

positioning deviation congest to some extent, the duty could be defeated. Positioning error correction under multiple constraints is an key role of intellectual aircraft route planning. In this paper, the problem of fast flight path planning for intellect aeroplane under the limitation of system positioning precision is studied [10].

Suppose the starting point of the aircraft is pointing A and the finish is point B. Its track constraint condition as follows:

(1) The plane demands real-time registration in spaceflight, and registration errors include vertical deviation and horizontal deviation. Every time the airplane flies 1 m, the horizontal deviation and vertical deviation will be increased by δ dedicated units, hereinafter referred to as the unit. Both vertical and level error should be less than θ units at the endpoint. Suppose when both vertical and horizontal error are less than θ units.
(2) The positioning error of the aircraft during the flight needs to be corrected. The track has some safe locations (called redress points) in flying area.Once the flying machine gets to the point of correction, error correction would be performed through the type of error correction based on location. The site of plumb and level error rectification can be to decide by the topography before orbit programme.After the correction, the distribution position of the flight area varies according to the landform. If the plumb and level error can be rectified in time, the plane can fly in the light of the preset line and eventually get to the end point after several redress points [10].
(3) At point A. the plumb and horizontal deviation of the plane are 0.
(4) Later plumb deviation correction is carried out at the plumb deviation correction point, the perpendicular error of the aircraft would be 0, while the standard error remains constant.
(5) Later the aircraft carries out standard error correction at the level error redress point, its standard error will be zero and the plumb error will hold constant.
(6) Only when the plumb error of the plane does not exceed 1 unit and the level error does not exceed 1 unit, the plumb error redress can be implemented.
(7) Only when the perpendicular error of the plane does not exceed 1 unit and the standard error does not exceed 1 unit, the level error redress can be carried out.

Under the data in Table 1, the flight tracks of the aircraft meeting conditions (1)–(7) are planned respectively, and the following optimization objectives are considered comprehensively:(A) the flight track length should be as small as possible; (B) The corrected area should be corrected as little as possible. The parameters of the data in Annex 1 are:

$$\alpha_1 = 25, \alpha_2 = 15, \beta_1 = 20, \beta_2 = 25, \theta = 30, \delta = 0.001$$

The effectiveness and complexity of the algorithm are discussed.

Table 1. Alternative point

NO.	X coordinates (unit: m)	Y coordinate (unit: m)	Z coordinate (unit: m)	Correction point type
0	0.00	50000.00	5000.00	A point
1	33070.83	2789.48	5163.52	0
2	54832.89	49179.22	1448.30	1
3	77991.55	63982.18	5945.82	0
4	16937.18	84714.34	5360.29	0
5	339.69	14264.46	3857.85	1
6	3941.93	74279.86	9702.92	1
7	45474.01	26849.48	6411.72	1
8	86806.90	5351.31	4409.85	0
9	23602.88	68460.10	88.47	0
10	35987.31	2169.08	2390.30	1
11	58063.15	60162.06	6595.36	0
12	83578.44	52199.36	3003.02	1
13	13777.62	18492.22	4492.09	0
14	85280.70	73958.43	3494.81	0
15	38614.44	50390.52	7020.08	0
16	92783.00	36473.22	6620.87	0
17	68526.50	14799.99	2779.12	1
18	96903.47	58915.30	5468.23	0
19	603.83	71091.65	1769.67	1
20	100000.00	59652.34	5022.00	B point

Note: Column E (correction point type): 1 represents vertical error correction point, 0 represents horizontal error correction point

2 Model Building

This problem needs to plan the optimal track satisfying constraint conditions 1 to 7. The target requires that the way should be as short as possible and the quantity of correction points should be as few as possible. The analysis of the data shows that the X-axis coordinates of the correction point vary the most, so the increasing direction of the X-axis data of the correction point is taken as the forward direction of the aircraft.

Let the target point B be reached after n − 1 correction, and write the correction point as M_1, M_2, M_{n-1}, the outset point A as M_0, the target point B as M_n, and the space position coordinate of the aircraft as $M_i = (x_i, y_i, z_i)$, the track length to target point B is denoted as S(n), and the distance between the (i − 1)-th correction point and the i-th correction point is defined as l_i,

$$l_i = \|M_i - M_{i-1}\| = \sqrt{(x_i - x_{i-1})^2 + (y_i - y_{i-1})^2 + (z_i - z_{i-1})^2} \tag{1}$$

The horizontal error of the i-th rectify point before correction is $\delta_1(i)$, and the vertical error is $\delta_2(i)$, satisfying

$$\begin{cases} \delta_1(i) = \mu_{i-1}\delta_1(i-1) + \delta l_i \\ \delta_2(i) = (1-\mu_{i-1})\delta_2(i-1) + \delta l_i \end{cases} \tag{2}$$

where, μ_i is the controller that performs horizontal correction or vertical correction on the i-th rectify point, meeting the following conditions:

$$\mu_i = \begin{cases} 1\ \delta_1(i) \le \alpha_2, \delta_2(i) \le \alpha_1 \\ 0\ \delta_1(i) \le \beta_2, \delta_2(i) \le \beta_1 \end{cases} \tag{3}$$

When $\mu_i = 1$, vertical correction is executed at the i-th rectify point. When $\mu_i = 0$, level correction is performed at the i-th rectify point.Meanwhile, $\delta_1(0) = \delta_0(0) = 0$, the initial error of the aircraft is 0. At the same time, if the aircraft can reach the destination according to the prescribed flight track, it should also meet the following requirements:

$$\begin{cases} \delta_1(i) \le \theta \\ \delta_2(i) \le \theta \end{cases} \tag{4}$$

Objective function 1: track length should be as short as possible

$$L = \min(\sum_i^n l_i) \tag{5}$$

Objective function 2: the number of correction points as few as possible

$$\min(n-1) \tag{6}$$

This problem uses genetic algorithm.[11–13]. Firstly, MATLAB software is used to preprocess the data given in Annex 1:

(1) Calculate the distance *lij* between any two correction points. According to constraint condition 1, "for every 1 m flight of the aircraft, the plumb error and standard error will raise by 0.001 special units respectively. And when the vertical deviation and level deviation are both less than 30 units, The aircraft can fly normally. When the node distance is smaller than 30/0.001 = 30000 m, there is a connection between the two correction points. Otherwise, no connection [14].
(2) According to constraints 6 and 7, when the distance between correction exceeds $\max\left\{\frac{\alpha_1}{\theta}, \frac{\alpha_2}{\theta}\right\} = 25000$ m or $\max\left\{\frac{\beta_1}{\theta}, \frac{\beta_2}{\theta}\right\} = 25000$ m, the correction points can no longer correct the error. Therefore, 25000 m is regarded as the threshold of the controllable range. For further data processing, except for the correction point

that can directly reach the target point B within a controllable range, the correction points with a distance greater than 25000 m are regarded as no path between the two correction points, otherwise, there is a path. Based on the above data processing, the genetic algorithm is as follows:

1. Generate initial samples (1000)

   (1) If the distance between the two correction nodes before and after exceeds (maximum error/unit distance error), the solution is invalid, and the distance between the points to generate the next correction point must be less than a fixed value (maximum error/unit distance error).
   (2) The error can only be corrected after the airplane arrive to the next redress point, so the current error margin is a constraint on the distance of the next solution, X-axis correction node and Y-axis correction point need to be separately constrained.
   (3) The goal is to go from A to B, so when selecting the next correction point, the probability of the correction point close to B should be magnified.
   (4) Each calibration point can only be used once;
   (5) After reaching the calibration point near B, try to connect to B directly, if successful, then push out.

2. Fitness function

According to the goal, when the school is on time, the shorter the distance is, the better the adaptability is, Take the reciprocal of distance as the fitness function.

3. Selection of the parent sample

Roulette algorithm randomly selects a pair of initial samples as the parent sample, does the crossover.

4. Crossing

A pair of parent samples selected in 3, randomly select a position (the two-parent samples should be in the same position), so the parent samples are cut into four sections and then recombined to form two new paths.

5. Variation

In the new sample generated in 4, the two adjacent correction numbers (probability of variation is 0.01) are randomly swapped, the subsample is updated, and then the subsample is verified whether it meets the constraint condition (i.e., it is not a feasible solution of the track). If not, repeat steps 3, 4, and 5 to obtain 500 new subsamples. The 500 subsamples and the first 500 with a smaller distance in the initial sample form a subsampled set, and then repeat 2–5 and iterate 10 times. Figure 1 shows the algorithm.

Table 2. Track plan

Calibration point No	Vertical error before correction	Horizontal error before correction	Correction point type
0	0	0	Starting point A
417	18.8991	18.8991	01
210	24.0593	5.1602	11
148	18.2132	23.3734	01
155	23.9184	5.7052	11
142	10.2576	15.9628	01
250	18.5744	8.3168	11
403	11.5076	19.8244	01
594	22.5367	11.0291	11
501	11.197	22.2261	01
612	28.062	11.124	endpoint B

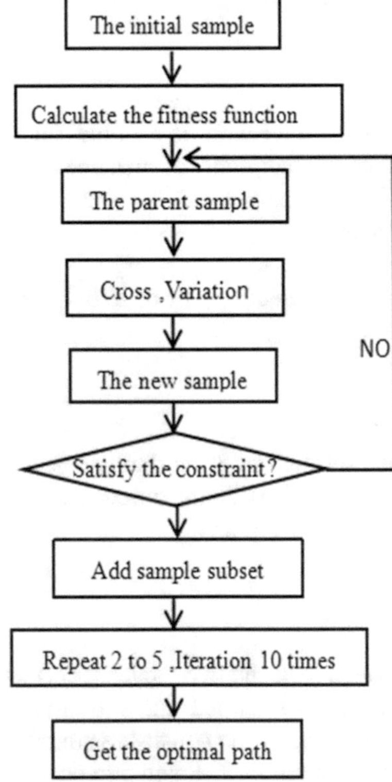

Fig. 1. The algorithm flow chart

The following solutions are obtained: Table 2 and Fig. 2 below represent the optimal track planning table and optimal track path diagram of the data in Table 1 respectively.

Figure 2 shows the good track obtained by the genetic algorithm on the attached page. Starting from the leftmost starting point A, the track proceeds along the increasing direction of the X-axis and finally reaches the rightmost target point B, with 9 correction points in total, Go through the correction point in turn 417 → 210 → 148 → 155 → 142 → 250 → 403 → 594 → 501. The track length is 108.7758 km, and the running time of the algorithm is 234.815 s.

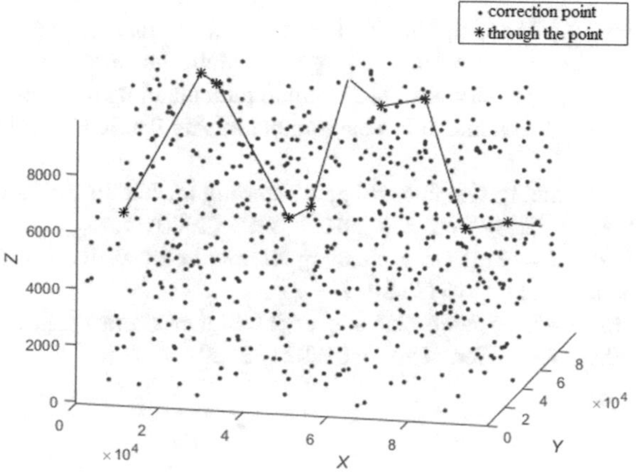

Fig. 2. The optimal path

3 Conclusions

Applicability analysis of this code, this code generates initial solutions through appropriate strategies.It avoids a large number of invalid solutions and starts from the appropriate location, with the ability of large-scale global search and a good advantage in dealing with trajectory planning problems. In what terms complexity, the master complexity of the algorithm lies in this generation stage of the initial solution. If it can be generated in stages, it may have lower complexity.

References

1. Paranjape, A.A., Meier, K.C., Shi, X., Chung, S.J., Hutchinson, S.: Motion primitives and 3-d path planning for fast flight through a forest. Int. J. Robot. Res. **34**(3), 357–377 (2015)
2. Babel, L.: Flightpath planning for unmanned aerial vehicles with landmark-based visual navigation. Robot. Auton. Syst. **62**(2), 142–150 (2014)
3. Al-Jarrah, M.A., Hasan, M.M.: Hils setup of dynamic flight path planning in 3D environment with flexible mission planning using ground station. J. Franklin Inst. **348**(1), 45–65 (2011)
4. Kermani, P., Afzalian, A.A.: Flight path planning using GA and fuzzy logic considering communication constraints. In: 7'th International Symposium on Telecommunications (IST 2014). IEEE (2014)

5. Miwa, H.: Method for flight path planning of unmanned aerial vehicles for manet. IEICE Tech. Rep. Inf. Netw. **113**, 337–342 (2014)
6. Szabolcsi, R.: Flight path planning for small UAV low altitude flights. Land Forces Acad. Rev. **25**(2), 159–167 (2020)
7. Ahmad, Z., Ullah, F., Cong, T., Lee, S.: Efficient energy flight path planning algorithm using 3-d visibility roadmap for small unmanned aerial vehicle. Int. J. Aeros. Eng. **2017**(6), 1–13 (2017)
8. Babel, L.: Flightpath optimization with application to in-flight replanning to changing destinations. Aircraft Eng. Aeros. Technol. Int. J. **90**(8), 1192–1202 (2018)
9. Szabolcsi, R.: 3D flight path planning for multirotor UAV. Rev. Air Force Acad. **18**(1), 5–16 (2020)
10. Zhang, Q., Ding, X., Zhou, J., Nie, Y.: Research on aircraft route planning optimization problem with multi-constraints and dual-targets. J. Math. Ind. **10**(1), 1–15 (2020)
11. Zollars, M.D., Cobb, R.G., Grymin, D.J.: Optimal path planning for unmanned aircraft target observation through constrained urban environments. Air Traffic Control Q. **27**(3), 144–152 (2019)
12. Akshya, J., Priyadarsini, P.: Graph-based path planning for intelligent UAVs in area coverage applications. J. Intell. Fuzzy Syst. **39**(6), 8191–8203 (2020)
13. Heidari, H., Saska, M.: Trajectory planning of quadrotor systems for various objective functions. Robotica **39**(1), 1–16 (2020)
14. Li, F., Li, Z.: Research on rapid planning of intelligent aircraft trajectory under multiple constraints. J. Phys.: Conf. Ser. **1592**(1), 012021 (2020)

Computer Integrated Automation System of Paper Industry Based on Internet of Things Technology

Jinhai Zhang[1](✉) and Ankit Singh[2]

[1] Shandong Jiaotong University, Jinan, Shandong, China
zhangjinhai76@163.com
[2] Jawaharlal Nehru University, New Delhi, India

Abstract. Paper industry is a continuous and semi-continuous production process enterprises. The Internet of things (IOT) is a huge network that combines massive sensor devices with the Internet. In the Internet of things, massive sensing devices constantly collect data and send it to the data center. Computer Integrated Process System (CIPS) for paper industry is a complex integrated automation system. The key problem of CIPS feature is process monitoring and information integration. Its function-oriented structure consists of four subsystems: operation and management, process automation, safe and economic operation, computer network and database support environment. This topic mainly carries on the integration and the development application to its two support environments: the computer network and the database. The production scheduling management system of paper industry realizes seven main functional modules including real-time monitoring module, and meets the requirements of paper production after putting into operation.

Keywords: CIPS · Information Integration · Cloud Platform · Industrial Control Software

1 Introduction

Manufacturing industry is in a pillar position in the world economy. Its development directly affects the national economic foundation and comprehensive national strength. However, there is a considerable gap between China's manufacturing industry and foreign countries in terms of production scale and product quality. After economic globalization and China's entry into WTO, the characteristics of market competition have put forward new requirements and challenges for the survival and development of China's manufacturing industry. Our government fully recognizes this point and puts forward in the communiquy of the Fifth Plenary Session of the 15th Central Committee of the CPC that "Continuing to complete industrialization is an arduous historical task in the process of China's modernization, and vigorously promoting the informatization of national economy and society is overwhelming.“ Strategic measures to cover the overall situation of modernization. To promote industrialization by informationization, give full play to

J. H. Abawajy et al. (Eds.): ICATCI 2022, LNDECT 170, pp. 927–934, 2023.
https://doi.org/10.1007/978-3-031-29097-8_110

the advantages of backwardness, and realize the leap-forward development of social productive forces.

The Internet of Things is a network of "ships of things as data sources." With the continuous development of perceptual devices and network technologies, the Internet of Things is surpassing the Internet as the primary source of information for large data. The traditional Internet is centered on the sea of people. Typical applications include file transmission and electronic mail. Electronic merchants, video on demand, online games and social networks, etc.. The Internet of Things to file & Quot; Or the physical world serves the data source to the society, covering many areas such as urban security, smart cities, target identification and tracking, and location services. In 2013, the Internet of Things data has reached 1.4 ZB, and it is expected that by 2020, the amount of data generated will reach 35 ZB. The World Wireless Research Forum predicts that by 2030, an ubiquitous Internet of Things world will be formed, with more than seven trillion sensors providing ubiquitous services to users.

Pulping and papermaking has always been an important part of China's traditional industries. Since Cai Lun invented papermaking in the Eastern Han Dynasty, paper has played an important role in disseminating knowledge and daily life, and papermaking technology has also evolved from manual papermaking to modern large-scale automated mechanical papermaking [1]. Paper-making methods have also changed with the different raw materials and paper products. At present, the output and consumption of paper and paperboard in China are in the top three in the world, but the advantage of total product does not mean that the pulp and paper industry is in the leading position in the world. There are still gaps in production scale, technical equipment and production efficiency between China's papermaking enterprises and developed countries such as the United States and Finland. The characteristics of China's paper industry are: large number of enterprises, small scale, small variety of products, low technology content and serious pollution. These characteristics and the intensification of domestic and international market competition determine that the traditional industry must be reformed in order to adapt to market competition, survive and develop.

As far as a single pulp and paper enterprise is concerned, it belongs to a small variety and large-scale production mode. It is not only a major consumer of raw materials and energy, such as wood, chemicals, water, electricity and so on, but also one of the industries that cause the most serious environmental pollution. It not only needs to produce high quality and qualified products, but also needs to recycle and treat waste products to reduce pollution, which determines the complexity of its production process. From the point of view of production process, the research and application of how to speed up the time to market, reduce costs, improve product quality, improve customer service, reduce pollution and protect the environment in the pulp and paper industry should not only stay on the surface level of production process. While adopting advanced manufacturing and recycling technology, it is necessary to monitor and control the related production process by taking advantage of the overall advantages of the integrated automation system. On the premise of ensuring that the product meets the quality index, the consumption of chemicals in the production process should be reduced as much as possible, so as to achieve the goal of cleaner production. Therefore, pulp and paper industry can be used as a specific application object of integrated automation system.

CIPS itself is a complex integrated automation system. It integrates control, optimization, scheduling and management. The development of modern management technology, automation technology and computer technology makes the realization of CIPS possible. The information integration technology with comprehensive characteristics is the key to CIPS.

2 The Architecture of CIPS

The characteristics of pulping and papermaking process determine the complexity of production process, and the control equipment of each section is also different. The TDC3000 distributed computer control system (DCS) of Honeywell Company is used in the automatic control system of log processing, wood chip cooking, wood pulp washing and screening, lignin oxidation and decomposition, wood pulp bleaching and hydrochloric acid preparation in pulping process. In the production process of pulp mill, the driving motors of drying section are controlled by high-power thyristor, the cutting, transmission and packing section of cardboard is controlled by programmable logic controller (PLC) of German Siemens Company, the power supply section is monitored by industrial computer, and the other sections are mostly controlled by traditional instruments and instruments. The information between these heterogeneous hardware devices has not been effectively integrated, shared and utilized, and the phenomenon of "automated information island" has been formed in the same manufacturing enterprise. It is necessary to reduce these undesirable phenomena, so as to coordinate and optimize the production equipment in the whole enterprise, improve the overall operational capacity, and maintain the long-term stable and normal operation of production [2].

"Business Management Information Isolation" and "Automation Information Isolation" are common phenomena in domestic chemical process enterprises. These phenomena cannot be solved only by some simple computer control systems. The irregularity of administrative business management and the complexity of chemical process production control system determine that integrated automation system must be adopted to realize the integration and application of enterprise information in order to solve these phenomena and improve the overall efficiency and competitiveness of enterprises. The characteristics and practice of pulp and paper production process and the research content of computer integrated process system (CIPS) determine that the development and application of CIPS in pulp and paper industry is feasible.

Computer Integrated Process System (CIPS) is a complex continuous process production system that integrates modern management technology, manufacturing technology, information technology, automation technology and system engineering technology through computer hardware and software, realizes the integration of enterprise production links, the unified management and integration of human, technology and management, and the organic integration and optimization of energy flow, logistics and information flow. Heterogeneous system.

Because of the unique technical, economic and management characteristics of process industry, CIPS of process industry should have corresponding characteristics. Compared with discrete CIMS, its main difference is:

1) The process industry belongs to the type of continuous production with few varieties and large quantities. The product category is relatively stable, and the logistics and process are single once fixed. The production process is continuous, and the logistics and energy flow are continuous [3]. The process sequence is compact and strict, the process flow is basically unchanged, the buffer space is small, the production is balanced and stable, and accompanied by a series of physical and chemical changes, there are mutations and uncertainties in the production process, which requires high coordination, real-time and reliability of management and control.
2) The basic automation level of process industry is relatively high.
3) Optimizing objectives and means, and integrating contents have their own characteristics. The optimizing objectives are to "grow smoothly and grow smoothly" and reduce pollution. The optimization means are generally the adjustment of process parameters, the optimization of control algorithm and the control of set points. In addition to information integration, the integration includes energy flow, vast integration, human, technology and management integration, multi-mode combination model integration, intelligent control and decision-making theory rental technology integration.
4) The environment is complex and harsh.
5) The loss of malfunction parking is great. Fault handling is a problem that chemical process industry takes great efforts to solve, and the production process is continuous, the process is long, and there are many equipment. Fault occurs in the form of chain, that is, a fault will produce a series of related faults. And the failure parking loss is huge, so fault diagnosis is very important.
6) The parameter measurement is delayed and complex. In the production of paper industry, the quantity and quality of manufactured products and finished products need to be measured by indirect measuring methods and means.
7) The supply chain is close and stable. Therefore, the characteristics of CIPS in chemical process industry are shown in Fig. 1.

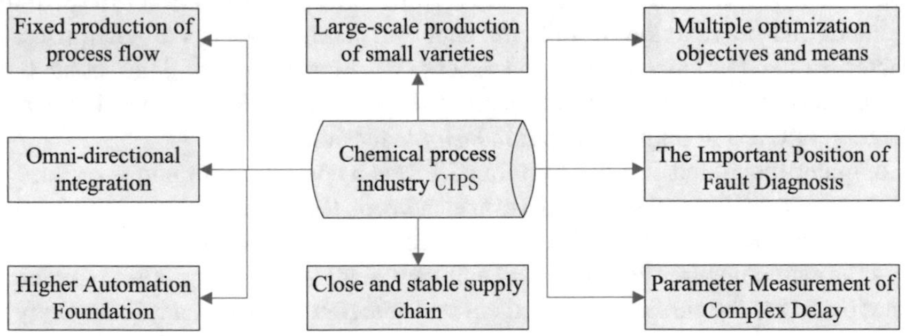

Fig. 1. Characteristics of CIPS in Chemical Process Industry

The structure of CIPS is oriented to system function. It consists of four subsystems: management subsystem, process automation subsystem, safe and economic operation subsystem, computer network and database support subsystem, as shown in Fig. 2.

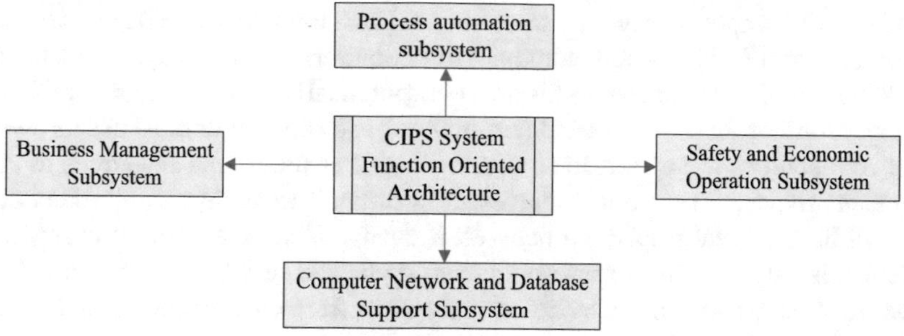

Fig. 2. CIPS System Function Oriented Structural Diagram

CIPS is a complex large-scale system, which can control CIPS by tracking production information flow. Production information flow originates from the top of the whole system. Through market demand, corresponding production objectives and plans are formulated [4]. Then comprehensive production and management plans are formed. According to these plans, optimal production scheduling arrangements are formulated. Finally, the whole production process is completed by starting production settings. Therefore, the control-oriented hierarchical structure of CIPS architecture can be obtained according to the nodes through which information flows. From top to bottom are management layer, management information layer, production scheduling layer, process control layer and equipment control layer [5]. This hierarchical control adapts to information technology and enterprise control mode, and is widely used in process industry.

3 Network Integration and Application

The Ethernet switch detects the MAC (Media Access Layer) address of the source and destination of the packet coming from the Ethernet port, and then compares it with the dynamic lookup table within the system. If the C layer address of the packet is not in the lookup table, the address is added to the lookup table. And send the packet to the corresponding destination port. Ethernet switch is the second layer of OSI reference model [6]. It mainly adopts two working modes: through mode has the advantages of no need of storage, very small delay and very fast switching. The disadvantage is that it can not provide error detection capability. It can not connect input/output ports with different rates directly. The increase of ports makes switching moment. The array becomes complex and difficult to implement. The storage and forwarding method stores the data packets of the input port first, and then carries out cyclic redundancy checking. The destination address of the data packet is extracted by processing the error data packet shield, and the packet is transferred to the output port through the lookup table. This method has a large data processing delay, but it can detect the errors of data packets entering the switch, support the conversion of input and output ports at different speeds, and maintain the cooperation between high-speed ports and low-speed ports.

Computer networks are divided into peer-to-peer networks and server-based networks according to their configuration and information sharing methods. In peer-to-peer networks, there are no dedicated servers and there is no hierarchical difference between computers. All computers have the same status. Each computer can be used as either a client or a server [7]. No system administrator monitors the management of the whole network, but a single user manages his own computer. All users can be arbitrary. Ways to share any resources. In peer-to-peer environments, each computer must use its own large amount of resources to support local users, as well as additional resources to support remote users to access network resources. It is difficult to achieve centralized control, which will have a great impact on network security. When the scale of computer network expands, large parameter networks need dedicated servers, which can only serve as servers and have been optimized to serve requests from clients quickly and ensure the security of files and directories. Advantages of server-based networks: [8].

1) Server-based data sharing can be centralized managed and controlled.
2) Security management is achieved by an administrator who formulates a strategy and applies it to every user on the network.
3) Data backup can be done automatically.
4) By using the data backup method of redundant system, data on any server can be copied and maintained online.
5) Server-based networks can support thousands of users.
6) Client hardware can meet the needs of users.

The network topology structure of production scheduling management system adopts star network topology structure. As shown in Fig. 3. Each node in a certain area is connected to a switch by twisted pair or optical fiber. Each area connects multiple optical terminals and switches by twisted pair or optical fiber [9]. In the General Dispatching room, a 16-port switch is set up as the connection center of the whole network. It connects nine data acquisition and monitoring stations, four dispatching room monitoring stations and servers, and office building switches.

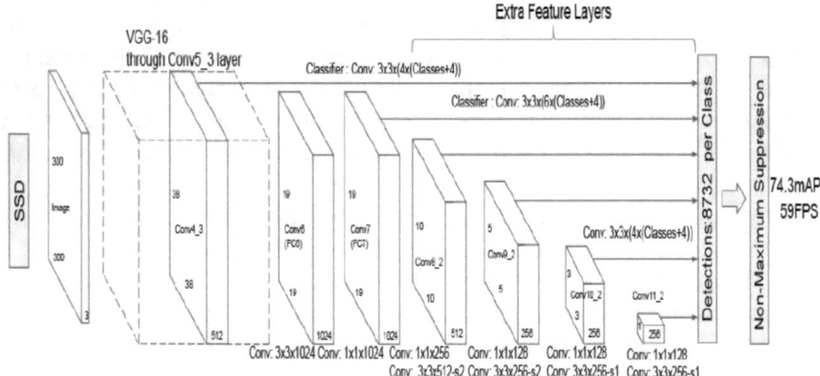

Fig. 3. Industrial Control Application Platform image

4 Data Integration and Application

Data integration is one of the contents of pulp and paper C1PS. The purpose is to organize the data in the system into a whole according to certain rules by certain technical means, so that users can operate the data effectively. It deals with data from various heterogeneous data sources (libraries) in the system.

In the industrial control system, the production status on site is reflected by the values of predefined variables, which are both continuous and discrete. Therefore, industrial control software is required to support a variety of variant types. These variables include t window intermediate variables, intermediate variables, DDE variables, real-time database variables (RDB variables), macro variables, indirect variables and system variables, which determine the scope of variables and data sources [10]. Variable data sources include DDE variable data source and RDB variable data source.

In real-time databases, each variable is a basic data object, also known as a data point (tag). Data point is the basic unit of information preservation and processing in real-time database system. A point is composed of several parameters. The system stores all kinds of information in terms of point parameters. Point parameters are attributes of variables, which are also called variable fields or variable fields. There are five types of standard points in industrial control software: analog I/0 point, digital I/O point, cumulative point, control point and operation point.

In the process control system, real-time data records the real-time information of each control point. The best way to manage real-time data is to use database system. If we use general relational database system to manage real-time data, not only the efficiency is low, but also the data format and storage mode cannot fully meet the requirements of relational data, nor can we complete the processing requirements of real-time data. Therefore, it is necessary to apply real-time database system in process control system.

Real-time database system is also a bridge between control system and other application systems. The real-time database system is established on the upper computer of the control system, and the real-time data in the control system is accessed through the database interface. The advanced control and optimal control of the control system can be realized by designing the algorithm. By connecting real-time data bank system with management information system (MIS), production scheduling management system can be established, and enterprise-level CIWS/CIPS system can also be established. Therefore, industrial control software designs real-time database as a core component [11]. It is a completely independent component with dual interface access mechanism, which enables other functional components to access real-time database (Real-time Data Base RDB) in a more suitable interface mode according to different requirements.

The process of papermaking is very complicated. Paper quantity and moisture are affected by pulp concentration, net pressure, steam pressure, speed and so on. Feedforward control is usually used to control the pulp concentration which has the greatest influence on the quantity, and air pressure cascade control is used to control the moisture content of paper [12]. Smith predictor compensation control is often used to solve the problem of large delay, but the control effect of Smith predictor is sensitive to the change of parameters of the object model. When the model mismatch exceeds a certain range, the control quality of the system using Smith predictor will deteriorate, even less than conventional PID control.

The paper-making process is a slow-changing uncertain system. The model Gp(s) mentioned above is rough. In order to reduce the influence of model errors, an adaptive Smith predictor is designed here. The model parameters of the predictor are identified online, and then the model parameters of the predictor are corrected by the identified parameters.

The tool provided by Genie task editor is used to connect and program to realize Smith predictive compensation control. Basic program is used to judge whether it is automatic or manual, and the manual adjustment value is directly sent to the paper machine under manual condition. In automatic case, the starting PID module is used to calculate the threshold value, and the value is fed to the paper machine. When the deviation exceeds the limit, the result of Basic identification program is sent to Smith compensation algorithm to correct the parameters of the compensator and output the compensator value. The difference of the quantitative setting value minus the output value is taken as the setting value of the PID module (realized by the PRG module), and the paper machine quantity is taken as the feedback input of the PID module.

References

1. Fallah-Tafti, M.: The application of artificial neural networks to anticipate the average journey time of traffic in the vicinity of merges. Knowl.-Based Syst. **33**(5), 6o–72 (2020)
2. Krizhevsky, A., Sutskever, I., Hinton, G.E.: Imagenet classification with deep convolutional neural networks. Adv. Neural Inf. Process. Syst. **59**(4), 216–224 (2020)
3. Chollet, F.: Xception: deep learning with depthwise separable convolutions. In: IEEE Conference on Computer Vision and Pattern Recognition, vol. 46, no. 2, pp. 175–183 (2019)
4. Rastegari, M., Ordonez, V., Redmon, J., Farhadi, A.: XNOR-Net: imagenet classification using binary convolutional neural networks. In: Leibe, B., Matas, J., Sebe, N., Welling, M. (eds.) ECCV 2016. LNCS, vol. 9908, pp. 525–542. Springer, Cham (2016). https://doi.org/10.1007/978-3-319-46493-0_32
5. Dieleman, S., Willett, K.W., Dambre, J.: Rotation-invariant convolutional neural networks for galaxy morphology prediction. Monthly Not. Roy. Astron. Soc. **60**(2), 223–311 (2019)
6. Potena, C., Nardi, D., Pretto, A.: Fast and accurate crop and weed identification with summarized train sets for precision agriculture. In: International Conference on Intelligent Autonomous Systems, vol. 40, no. 3, pp. 106–122 (2020)
7. Walsh, G.C., Hong, Y.E.: Scheduling of networked control systems. IEEE Control Syst. Maga. **32**(3), 61–75 (2019)
8. Mintchelt, G.A.: OPC integrates the factory floor. Controlengineering **38**(5), 111–122 (2019)
9. Bequette, B.W., Ke, B.A.O.: Chemical process control education and practice. IEEE Control Syst. Maga. **33**(4), 243–258 (2020)
10. Walsh, G.C., Hong, Y.E.: Scheduling of networked control systems. IEEE Control Syst. Maga. **32**(3), 99–112 (2019)
11. Huang, S., Sadek, A.W.: A novel forecasting approach inspired by human memory: the example of short-term traffic volume forecasting. Transp. Res. Part C **43**(1), 133–146 (2019)
12. Souza, A.M.D., Yokoyama, R.S., Maia, G., et al.: Real-time path planning to prevent traffic jam through an intelligent transportation system. IEEE Comput. Commun. **32**(2), 198–209 (2019)

Big Data Analysis Technology in Mechanical Intelligent Design and Manufacturing and Its Automation

Xiaojie Hu(✉)

Department of Mechanical Engineering, Dalian University of Science and Technology, Dalian, Liaoning, China
844975062@qq.com

Abstract. With the advancement of science and technology and the improvement of manufacturing level, traditional production methods are often limited to some difficult situations due to the lack of intelligence, adaptability, automation level, and can no longer meet the development needs of the current design manufacturing industry. Faced with this situation, we will introduce new design concepts such as the integration of advanced technologies in various fields and parametric design, automated design, etc., to solve the problems of poor quality and low efficiency in the production process of machinery manufacturing, and at the same time improve the quality of product production and narrow the gap with developed countries, has become a distinctive feature of today's machinery design and manufacturing industry. In this paper, taking the manufacturing of CNC lathe parts as an example, combined with big data analysis technology to analyze the structure of CNC lathes and establish a multi-level example library of CNC lathes, an intelligent machine tool design and manufacturing automation system is designed, and the system is applied to the automatic production of parts and components. Comparing the production speed of manual production and system automated production and the pass rate of product quality inspection, it is found that compared with manual production, the efficiency of using automation technology to produce parts has been significantly improved, and the quality of parts can be guaranteed.

Keywords: Big Data Analysis Technology · Mechanical Design and Manufacturing · Automated Production · Manual Production

1 Introduction

In the new era of manufacturing development characterized by knowledge and information, mechanical product design and mechanical automation applications have also received more and more attention from countries around the world. Product design is a complex activity with human creativity. How to realize the intelligentization of product design, the automation of production process, and the rapid completion of product development and manufacturing has become a major subject of manufacturing production research. In this context, the theory, method and technology of intelligent design are

J. H. Abawajy et al. (Eds.): ICATCI 2022, LNDECT 170, pp. 935–942, 2023.
https://doi.org/10.1007/978-3-031-29097-8_111

widely used. As a new advanced production technology, mechanical intelligent design and manufacturing mainly studies how to apply big data analysis technology to the design of mechanical products, and realize automatic machine production by building a product intelligent production system.

Many scholars have conducted in-depth research on the application of big data analysis technology in mechanical intelligent design and manufacturing and its automation, and obtained good research results. For example, a scholar believes that design and manufacturing automation technology is realized based on the idea of conceptual design. The main goal is to meet the needs of small batches to produce various types of mechanical products on the market, and to reduce the large number of complex and repetitive production by production personnel in the manufacturing process. It can reduce the over-reliance on the work experience of manufacturing personnel, improve production efficiency through automation technology, and shorten the manufacturing period [1, 2]. The advancement of science and technology and the improvement of social needs have forced the competition in the mechanical design and manufacturing industries to become more intense, and the speed of product substitution is getting faster and faster. When many products are updated, their overall topology will not change much, but the model and size of some of their components will change [3]. The big data analysis design is based on the existing design examples and improves them to meet the new design requirements. Since this method makes full use of existing resources, it can quickly launch new products and reduce development costs, and is gradually recognized and used by most enterprises [4]. Based on the concept of data mining, a scholar further researches automated manufacturing technology and develops an automated production system, which transfers most of the decision-making power in the product manufacturing process to the computer, reduces manual intervention steps as much as possible, and significantly improves production efficiency. And quality, and provide new technical support for the rapid development of machinery manufacturing and production [5]. Although the application of big data analysis technology in mechanical intelligent design and manufacturing and its automation is progressing smoothly, it is still necessary to speed up industrial production efficiency to improve economic strength.

The article first lists several big data analysis techniques, including cluster analysis, association analysis, heterogeneous analysis, etc., then summarizes the characteristics of mechanical intelligent design and manufacturing, builds an automated production system for the production of CNC lathe parts, and finally analyzes the automation system. The product quality inspection pass rate of production efficiency is compared with the manual production method to verify the efficiency of automated production.

2 Big Data Analysis Algorithms and Machine Intelligent Production

2.1 Big Data Analysis Technology

(1) Cluster analysis

Unlike classification, clustering belongs to unsupervised learning. Data samples with no classes initially are what it operates on. When clustering, the clustering criterion is to maximize the similarity of data in the same group and minimize the similarity of data in different groups. This standard also provides a basis for the use of cluster analysis methods [6].

(2) Classification and prediction

In a general and literal sense, classification is a combination of different things, and prediction is an analysis of the next evolution of a particular thing. While classification and prediction are combined to obtain a model, the composition structure is closely related to the data, and it is analyzed through the data prediction model, which usually exists in a tree structure [7]. In the field of data mining, classification and prediction are used as general methods in other fields such as weather forecasting, behavioral analysis, and financial risk management.

(3) Concept description

In the popular sense, a conceptual description is an accurate description of a certain type of thing. One refers to the points of difference between different species, and the other refers to the same points that exist within the same species [8].

(4) Association analysis

The most common field of data mining is the analysis of association rules, which analyzes historical transaction data is the most frequently used occasion for association rules [9]. For example, different kinds of commodities may be put together and sold at the same time through a certain relationship, which is obtained by the analysis of association rules. Detecting the database, analyzing the previous transaction volume and using association rules to find information about some hidden values, can obtain the purchasing habits of which kinds of things customers often buy, which can also help sellers how to choose a marketing model and obtain more big profit.

(5) Heterogeneous analysis

Heterogeneous usually means that it does not have some laws that ordinary things have, and it is relative to normal. In the field of data mining, heterogeneous data are considered to be external data obtained from normal data due to some error conditions, and they are relatively small-scale data [10].

2.2 Characteristics of Mechanical Intelligent Design and Manufacturing

Mechanical intelligent design and manufacturing mainly uses computers to replace human experts to process knowledge, data and information, so that computers can undertake complex production and manufacturing tasks at a higher level, and achieve integrated, intelligent and automated production [11]. Modern manufacturing puts forward the following three requirements for intelligent design and manufacturing.

(1) Higher and higher production volumes. As economic growth improves product accuracy, higher requirements are placed on product production quality.
(2) Shorter and shorter design cycles. The competition of products is the competition of efficiency. Intelligent design should reflect the characteristics of automation and intelligence, improve production efficiency and reduce labor intensity.
(3) More and more complex design objects and environmental requirements. Under the conditions of expanding market demand, mechanical products and equipment should have more functions and be more adaptable to complex environments.

2.3 Grey Association Algorithm

The grey relational algorithm counts the similarity of sequences. The closer the geometric curves of two sequences are, the greater the degree of correlation between them. It is an algorithm to compare the similarity from the system category [12]. When using the grey relational algorithm to calculate the similarity, first select the feature sequence that reflects the attributes of the instance, and apply the relevant data of the feature sequence to the grey relational algorithm to determine the similarity between the sequences [13]. The grey correlation coefficient formula is as follows:

$$p(x_0(f), x_i(f)) = \frac{\min_i \min_f |x_0(f) - x_i(f)| + \varepsilon \max_i \max_f |x_0(f) - x_i(f)|}{x_0(f) - x_i(f) + \varepsilon \max\limits_i \max\limits_f x_0(f) - x_i(f)} \quad (1)$$

The grey relational degree formula is as follows:

$$p(x_0, x_i) = \frac{1}{n} \sum\nolimits_{f=1}^{n} w_f p(x_0(f), x_i(f)) \quad (2)$$

Among them, x_0 is the problem sequence to be solved, x_i is the comparison sequence, f is the attribute feature, w_f is the weight of each attribute feature, ε is the resolution coefficient, the value range is between [0, 1], usually 0.5.

3 Construction of Intelligent Design and Manufacturing Platform for CNC Lathes

3.1 Structural Analysis of CNC Lathe

With the advancement of the intelligent manufacturing strategy, domestic machine tool manufacturers are also upgrading their products to meet the new needs brought about by industrial development. The CNC lathe is the most widely used mechanical product in the manufacturing industry. It needs to decompose the product at a reasonable level, and then reorganize the standardized and serialized modules to generate an overall product plan. The reorganization of the genetic algorithm of the CNC lathe is in line with the idea of module design. The CNC lathe is composed of mechanical parts and has the attributes of multi-level characteristics. According to the multi-level characteristics of the CNC lathe, the multi-level structure of the CNC lathe can be decomposed.

3.2 The Establishment of Multi-level Instance Library of CNC Lathe

The decomposed CNC lathe instance is stored in the SQL database, each instance represents an optional design scheme, the storage method is carried out in the form of scheme coding, and the nodes can be uniformly coded. Instances are in different levels in the data lathe, and their locations will also be different. The scheme instance library is used to generate the genetic algorithm scheme. For the mechanical part of the CNC lathe, functional attributes and structural attributes have functional attribute parameters and structural attribute parameters respectively, and the functional attribute database and structural attribute database of lathe parts can be established respectively, which are mainly used in the design of mechanical product parts based on big data analysis. The functional attribute database mainly calls the structural attribute database for the parametric modeling of components.

3.3 CNC Lathe Automation Intelligent Design and Manufacture Frame Structure

The operation process of the CNC lathe intelligent design and manufacturing system mainly includes the system interface, the program background operation and the database data support. The system structure frame can be described as the presentation layer, the function layer and the data layer. The system framework is shown in Fig. 1.

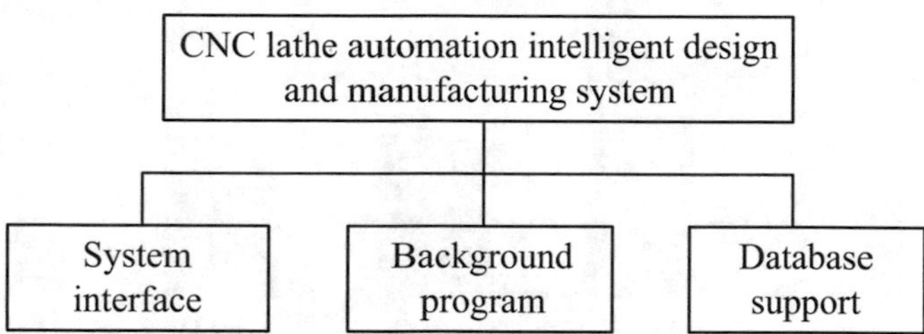

Fig. 1. Framework of CNC lathe manufacturing system

(1) System interface: The system interface is the window for the user to communicate with the design system. The user's design requirements and operation intentions are input through the system interface, and the system operation results are also fed back to the user through the interface. It is an indispensable bridge between the user and the system.
(2) The dialog box included in the background program system interface realizes its function through the running of the background program. The program is written in VC++ language, which mainly includes the generation of genetic algorithm scheme, the retrieval matching based on case-based reasoning and the realization of parametric modeling. It is the core part of the system operation and can realize the connection with the database system.
(3) Database support: The data and instances are stored in the SQL Server database, which mainly includes an instance library of CNC lathe scheme selection, a database of CNC lathe parts parameters and an example library of system operation results. The database provides necessary data support for the operation of the background program, and also stores the results of the operation of the background program, which is the total integrated part of the CNC lathe.

4 Parts Production Application of CNC Lathe Intelligent Design and Manufacturing System

4.1 Comparison of Manual Production and Automated Production Speed

Manual production and CNC lathe automated machines are used to produce lathe parts, and the number of parts per minute produced by the two production methods is compared. As shown in the results in Fig. 2, the number of A-type parts produced by manual

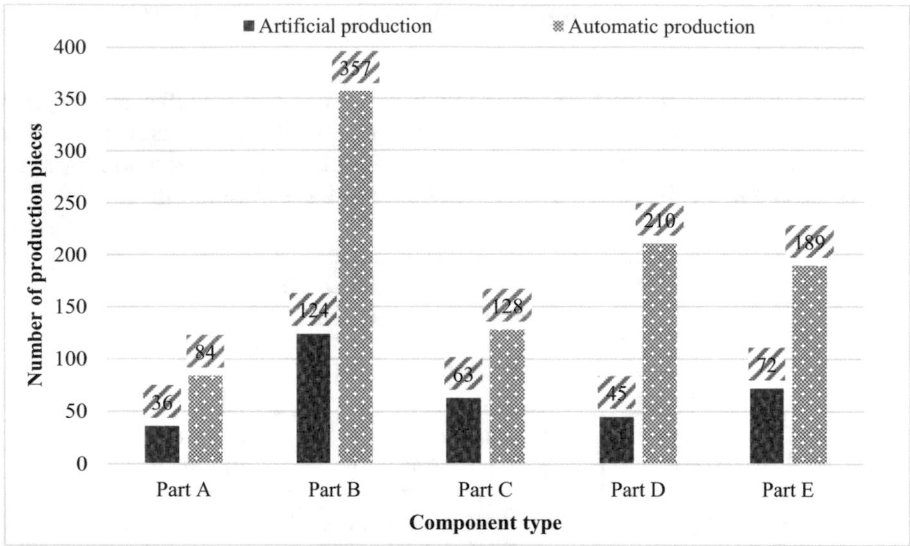

Fig. 2. Number of parts produced in one minute

production and automated machines in one minute is 36 and 84, respectively, and the automated production efficiency is 1.33 times that of manual production; The number of pieces is 124 and 357, and the automated production is 1.88 times more efficient than manual production; other types of parts production are automated and more efficient than manual production. This shows that the intelligent design system of CNC lathe can greatly improve the production efficiency of parts.

4.2 Comparison of Pass Rate of Product Quality Inspection

Table 1. Quality inspection pass rate

	The number of qualified	Number of unqualified	Quality inspection pass rate
Manual detection	868	32	96.44%
Automated detection	843	57	93.67%

The quality inspection of 900 machine tool parts produced was compared, and the qualified rate of product quality inspection by manual inspection and automatic inspection was compared. The results are shown in Table 1. The number of unqualified parts detected by the manual detection method was 32, and the qualified rate of quality inspection was 96.44%, while the number of unqualified parts detected by the automatic system was 57, and the qualified rate of quality inspection was 93.67%, which explained the detection process of the automatic system. More refined, it can screen out unqualified parts that cannot be found by manual inspection methods, and improve the production quality of each batch of parts.

5 Conclusion

In the field of mechanical manufacturing production, if the production plan is modified, the traditional design and manufacturing method has to manually re-formulate the design plan and part drawings before it can be put into production. This process not only wastes time but also increases costs, making it difficult to quickly meet the market need. Driven by this situation, big data analysis technology has emerged to analyze production data and meet efficient production schedules. In this paper, the analysis technology is introduced into actual production, and a mechanical intelligent design and manufacturing system is built to produce mechanical product parts, which improves the production efficiency and reduces the workload of production personnel.

References

1. Farid, A.M.: Measures of reconfigurability and its key characteristics in intelligent manufacturing systems. J. Intell. Manuf. **28**(2), 353–369 (2014). https://doi.org/10.1007/s10845-014-0983-7
2. Liu, Y., Zhao, Y., Li, K., et al.: Design and application research of a digitized intelligent factory in a discrete manufacturing industry. Intell. Autom. Soft Comput. **26**(5), 1081–1096 (2020)
3. Sato, S.: Application of stabilized halogen in deinked pulp manufacturing, and its benefits. Kami-pa-gi-kyō-shi **72**(8), 875–878 (2018)
4. Apalkova, V., Tsyganov, S., Chernytska, T., et al.: Evaluating the economic and ecological effects of investment projects: a new model and its application to smartphone manufacturing in Europe. Invest. Manag. Finan. Innov. **18**(4), 252–265 (2021)
5. Shuai, H.: Development and intelligent design of electrical automation control technology. Revista de la Facultad de Ingenieria **32**(11), 979–983 (2017)
6. Chu, W., Wuniri, Q., Du, X., Xiong, Q., Huang, T., Li, K.: Cloud control system architectures, technologies and applications on intelligent and connected vehicles: a review. Chin. J. Mech. Eng. **34**(1), 1–23 (2021)
7. Katsuda, T., Isshiki, H.: Contribution of medical big data analysis technology to medical innovations. Fujitsu Sci. Tech. J. **54**(2), 59–65 (2018)
8. Oh, M.J., Roh, M.I., Park, S.W., et al.: Estimation of material requirement of piping materials in an offshore structure using big data analysis. J. Soc. Naval Arch. Korea **55**(3), 243–251 (2018)
9. Adamson, G., Havens, J.C., Chatila, R.: Designing a value-driven future for ethical autonomous and intelligent systems. Proc. IEEE **107**, 518–525 (2019)
10. Luo, H., Xie, W.G., Zhang, W., et al.: Design and manufacture of a utility artificial hand for a burned child by three-dimensional printing technology and its application. Zhonghua shao shang za zhi = Zhonghua shaoshang zazhi = Chin. J. Burns **34**(8), 526–528 (2018)
11. Umaras, E., Barari, A., Tsuzuki, M.: Intelligent design tolerance allocation for optimum adaptability to manufacturing using a monte carlo approach. IFAC-PapersOnLine **52**(10), 165–170 (2019)

12. Ostrosi, E., Fougères, A.J.: Intelligent virtual manufacturing cell formation in cloud-based design and manufacturing. Eng. Appl. Artif. Intell. **76**, 80–95 (2018)
13. Teti, R., Masson, P.L., Matsumoto, M., et al.: Special issue on intelligent computation in design and manufacturing. Int. J. Autom. Technol. **12**(3), 273–274 (2018)

Design and Research of Teaching Evaluation System Based on Big Data Technology

Rongxia Wang[1](✉), Fen He[1], Weihuang Yang[1], and Mohammed K. Kumar[2]

[1] Guangzhou Nanyang Polytechnic College, Guangzhou 510925, Guangdong, China
ziming323@126.com
[2] GLA University, Mathura, India

Abstract. At present, the research on the design of teaching evaluation indicators for computer majors is still relatively weak. It is of great significance to establish a scientific, normative and operational evaluation system. The construction of evaluation indicators is the most important part of the teaching quality evaluation of computer majors, and it is also the most difficult part to solve. The purpose of this paper is to study the teaching evaluation system of computer majors in colleges and universities based on educational informatization, mainly including the design of evaluation indicators. First, select the online and offline blended teaching of computer major courses for evaluation, and then collect the original data of evaluation, use big data technology to process and analyze the original data and organize the evaluation results, so as to reflect the students' learning situation more intuitively and accurately, and improve teaching and learning. And learning efficiency. The experimental results show that the surveyors agree with each indicator very well, and the degree of recognition of the learning effect obtained by the achievement index is 84.5% .

Keywords: Computer Courses · Big Data · Evaluation System · Evaluation indicators

1 Introduction

With the implementation of the society, computer-based information technology, multimedia and Internet communications will be applied to education, technological development, information dissemination and educational reform to meet new requirements. The information society is an important choice for education reform and promotion in various countries [1, 2]. Computer courses occupy a very important position in the teaching of computer-related courses. A series of lectures are of great significance for cultivating the knowledge of college students and computer science [3]. In order to successfully manage the teaching and learning process, the education authority must review the current teaching situation and formulate strategies related to teaching revisions based on the evaluation results.

The original version of this chapter was revised: the author's affiliation has been changed to "Guangzhou Nanyang Polytechnic College, Guangzhou, 510925, Guangdong, China". The correction to this chapter is available at https://doi.org/10.1007/978-3-031-29097-8_122.

J. H. Abawajy et al. (Eds.): ICATCI 2022, LNDECT 170, pp. 943–950, 2023.
https://doi.org/10.1007/978-3-031-29097-8_112

Curriculum schedule is the general schedule and macro schedule of educational activities. The course describes the content of the course, students and how to evaluate the course, and finally explains the level of each activity in the course. Given the importance of dental training, Fatimah compared general dental courses with eight of the world's top schools [4, 5]. Materials and methods: In comparative descriptive research, the key components and standards of doctoral programs in many universities around the world are studied and compared with Iranian courses [6]. Tufan A investigated the impact of money, methods, and satisfaction on peer feedback provided by the online system on student performance, self-employment, and technology acceptance. The participants were 32 elementary schools and Turkish teachers, and then registered for two computer science courses at the Turkish National University in the fall of 2013. The assignments and feedback (text or text and video) based on statistical data different from the other two groups of students show that there is no difference between the scores of the students, self, and technical admissions. The feedback provided by students' scores is only related to their technical acceptance scores [7, 8]. It is of great significance to carry out research on the evaluation of computer courses in colleges with the aid of educational information.

This article combines the teaching characteristics, from the perspective of promoting teacher-student interaction and cultivating students' practical application skills, designing a classroom teaching evaluation index system with subject characteristics to implement evaluation. According to modern education theory, classroom teaching is a combination of "teaching" and "learning" [9]. Therefore, we also test the effects, combine the classroom teaching with the effects, and conduct comprehensive evaluations. Reflecting the actual situation of classroom teaching in this subject, diagnosed the problems existing in the teaching of computer professional specialty courses in our school, and proposed improvement measures.

2 Research on Teaching Evaluation System of Computer Specialty Courses in Colleges and Universities

2.1 Educational Information

Environmental information technology is based on traditional teaching methods, through the advanced technology-based teaching network integration environment, to distinguish information from teaching and learning information, and to find ways to display. This is conducive to students' independent learning and collaboration a teaching environment for learning and teaching services for teachers [10, 11]. The concept of deep learning and effective application in multiple fields has accelerated the rapid development of intellectual property rights. With the advent of big data, various fields are focusing on finding valuable things from the data in order to guide the next decision-making. And deep learning is particularly representative, a direction driven by data. With the amazing results of deep learning in many fields in recent years, it has attracted many scientific researchers and companies in the field of computer vision to conduct in-depth research on it, making deep learning develop rapidly, and now it has been applied to many successful products, such as teaching platforms.

2.2 Principles of Constructing Evaluation Index System for Specialty Courses of Computer Application

(1) The principle of goal consistency

The index system needs to fully reflect the specialty needs of curriculum teaching. The various types, indicators and weights in the indicator system need to be suitable to reflect the goals to be achieved; otherwise the evaluation indicators lose their validity.

(2) The principle of operability

The indicator system should be simple and implementable. The expression, design and operation of the evaluation index system should move closer to simplicity. It is convenient for the evaluator to observe, measure and compare. Each indicator should be clear in content, with clear wording and meaning.

(3) The principle of combining comprehensiveness and criticality

The development of the evaluation index system should conform to the comprehensiveness of the index, and essentially reflect and include factors at all levels of the curriculum. Based on the system theory point of view, that is to check the relevance and comprehensiveness of all levels of the curriculum.

(4) The principle of combining independence and relevance

The connotation of indicators at the same level should not have the same relationship, cross relationship, causality, etc., but only one parallel relationship. However, in order to emphasize some very important content, multiple related indicators can also be used to evaluate to improve the reliability of the evaluation. Therefore, when focusing on the independence of indicators, you can choose to design some indicators related to this.

3 Investigation and Research on Teaching Evaluation System of Computer Specialty Courses in Colleges and Universities

3.1 Research Design

This paper uses a questionnaire survey to understand learners' recognition of the evaluation indicators of computer professional courses in MOOC, and then verifies the evaluation system indicators of computer professional courses. The questionnaire mainly consists of two parts: the first part is a survey of personal information and the completion of computer professional courses; the second part is a survey of computer professional course evaluation system indicators. The latitude of the survey is the two latitudes of the course content and the learning effect. Using Likert five-point scale for measurement, they are "strongly disagree", "disagree", "not necessarily", "agree" and "strongly agree".

3.2 Data Collection and Big Data Analysis

The main research objects of this survey are computer majors and learners who have studied computer MOOC courses. There are 215 computer majors, accounting for 94.8%;

the number of students who have studied computer MOOC courses is 223, accounting for 98%. Many courses of computer majors are demanding and the technology is updated quickly. Computer learners also have a higher demand for MOOC learning. The results of the questionnaires we collected were analyzed using the T-test.The T-test formula used in this article is as follows:

$$T = \frac{\overline{Y} - \mu}{\frac{\sigma Y}{\sqrt{N}}} \tag{1}$$

$$T = \frac{\bar{Y}_1 - \bar{Y}_2}{(\frac{1}{N_1} + \frac{1}{N_2})\sqrt{\frac{(N_1-1)P_1^2+(N_2-1)P_2^2}{N_1+N_2-2}}} \tag{2}$$

Among them, formula (1) is the individual test, formula (2) is the overall test, P is the standard sample, $\overline{Y}$ is the sample average and N is the number of samples.

4 Investigation and Analysis of Teaching Evaluation Indexes of Computer Specialty Courses in Colleges and Universities

4.1 Preliminary Establishment of Curriculum Evaluation Indicators

As the provider of MOOC courses, the level of course teachers plays an important role in the evaluation of MOOC computer professional courses. This level is subdivided into two latitudes: teacher skills and attitudes and teacher teaching methods. Under the latitude of teacher skills and attitudes, eight indicators are proposed: knowledge and experience, language expression, teaching media, answering questions, improving teaching, pre-class preparation, reasonable evaluation, and value orientation. Under the latitude of the teacher's teaching method, six indicators of teaching difficulties, teaching methods, teaching goals, meticulous explanation, enlightenment and practicality are proposed.

The content organization, design and structure of MOOC courses are the key to attracting students and improving continuous learning. It is also an important aspect that needs to be considered when evaluating computer professional courses in MOOC. This level is subdivided into two latitudes: course content and course design. Under the latitude of the course content, seven indicators are proposed for complete content, clear division, scientific and cutting-edge, consistent goals, examples and demonstrations, test exercises, and course resources. Under the curriculum design latitude, ten indicators of novelty, representativeness, teaching plan, assessment method, practicality, courseware design, graphic layout, video, time arrangement, and curriculum update are proposed [12, 13].

Learning can be divided into learning behaviors and learning effects, which can reflect learners' learning status and learning expectations, thereby reflecting teaching effects to a large extent. The evaluation of learning is also an important aspect of constructing the evaluation index system of MOOC computer professional courses. This level is subdivided into two latitudes: learning behavior and learning effect. Under the latitude of learning behavior, seven indicators are proposed for careful observation, communication and discussion, progress management, review and review, self-evaluation, sharing and dissemination, and learning needs. Under the latitude of learning effect, six evaluation indicators are put forward for actual operation, achievement, mastery of knowledge, media use, ability improvement, and active learning.

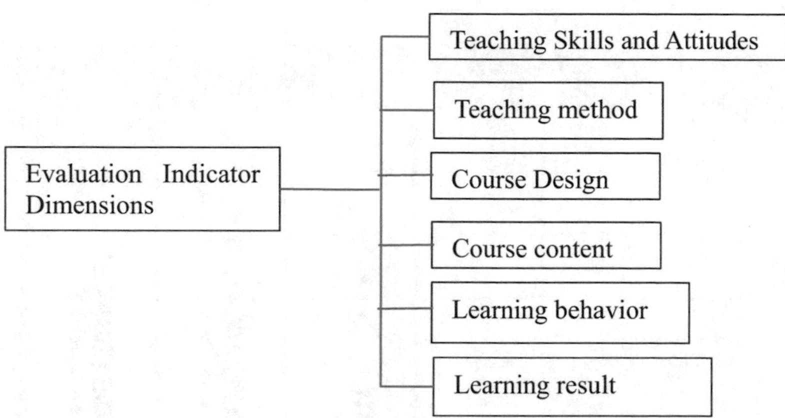

Fig. 1. Evaluation index design

After summarizing and analyzing the evaluation indicators of computer professional courses in MOOC, the six latitudes of the computer professional course rating system, including Teaching skills and attitude, Teaching methods, Course Design, Course content, Learning behavior, Learning result, with a total of 44 evaluation indicators. As shown in Fig. 1.

4.2 Statistical Analysis of Curriculum Evaluation System Indicators

The frequency statistics of the actual operation under the latitude of learning effect and the obtained result indicators are carried out, and the selection of indicators is obtained, as shown in Table 1.

Table 1. Recognition of Learning Effectiveness Indicators

Options	Actual operation	Get results
Strongly disagree	0.8%	2.8%
Disagree	1.6%	4%
Generally	15.6%	8.7%
Agree	43.7%	47.8%
Very much agree	38.3%	36.7%

It can be seen from the recognition table of each indicator of learning effect that the percentages of selecting "agree" and "strongly agree" for the indicators under this latitude are respectively: The actual operation is 82%, and the score is 84.5%, as shown in Fig. 2.

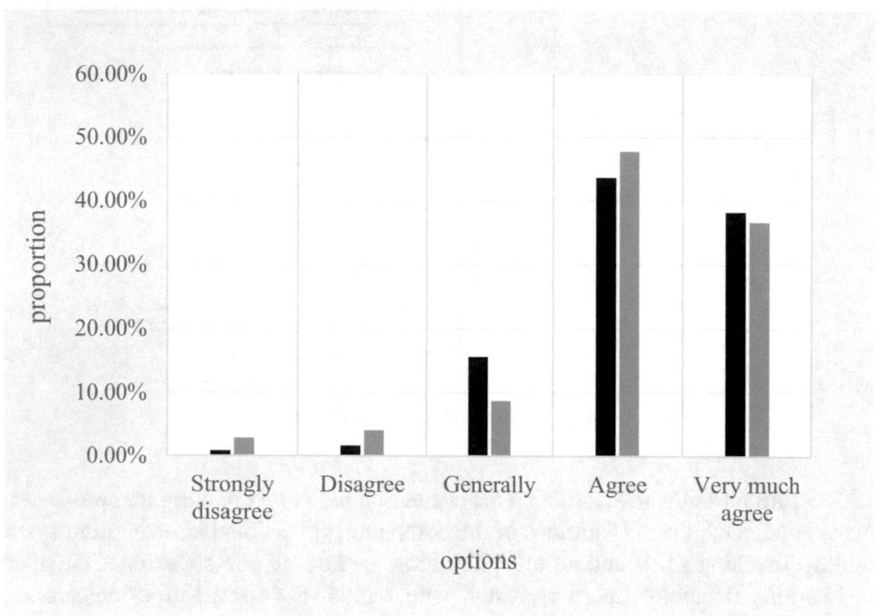

Fig. 2. Practical operation and recognition of score indicators

Mastery of knowledge 86.8%, and media use 80%, the ability is improved by 88.1%, and the active learning is 80.8%, as shown in Table 2. The data shows that under the latitude of learning effect, the surveyors have a high degree of recognition of each index, and the learning effect of learners can reflect the quality of the course to a certain extent, as shown in Fig. 3.

Table 2. Indicator recognition analysis

Options	Grasp knowledge	Media use	Improve ability	Active learning
Strongly disagree	1.4%	3%	2.5%	6.8%
Disagree	5.2%	7%	2.2%	7.9%
Generally	6.6%	10%	7.2%	4.5%
Agree	48.6%	41.5%	38.4%	35.9%
Very much agree	38.2%	38.5%	49.7%	44.9%

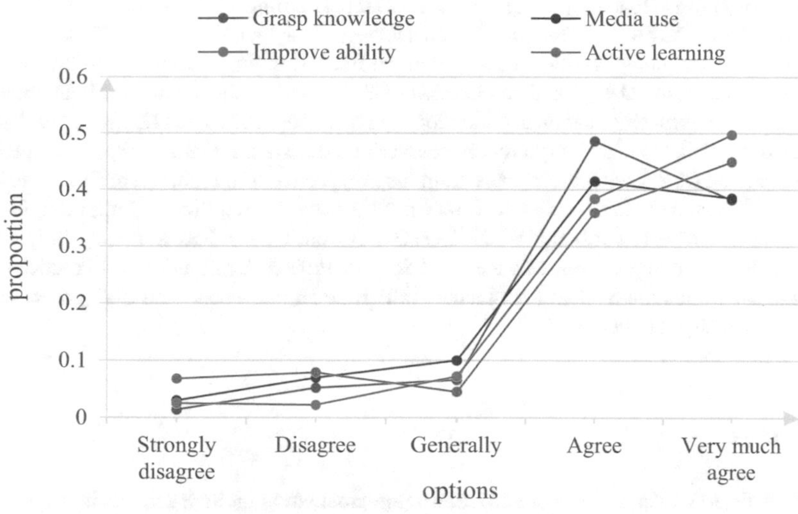

Fig. 3. Learning outcomes reflect the quality of the course

According to the recognition table of each index of the course content, the percentages of selecting the "agree" and "strongly agree" levels for the indicators under this latitude are respectively: 82.6% of the content is complete, 84.5% of the division is clear, 87.8% of scientific and cutting-edge. 82.4% consistent goals, 84% examples and demonstrations, 84.1% test exercises, and 82.6% of external resources. The data shows that in the course content latitude, the investigators pay more attention to the completeness of the course content, the corresponding operation demonstration and test exercises.

5 Conclusions

With the advancement of science and technology and Internet technology, as well as the strengthening of the global economy, specialty computer education in colleges and universities has changed the world. In order to meet the needs of today's information boom and continue to innovate and develop computer technology, specialty computer education in colleges and universities must also continue, and major changes have taken place in their education system. This paper uses big data technology to analyze and study the teaching evaluation system of MOOC computer courses. Based on the education in teaching, good evaluation teaching and the research of specialty computer courses, priority is given to solving the problems existing in the teaching of MOOC specialty computer courses. The factors affecting the teaching of specialty computer courses are analyzed, and on this basis, a quality evaluation system for teachers' specialty computer courses is established.

Acknowledgments. This work was supported by 2021 Educational Science Planning Project of Guangdong Province (Special Project for Higher Education) "Research on the Design and

Application of Hybrid Teaching Mode Based on Deep Learning in the Context of Big Data" (Project No: 2021GXJK493) from Guangzhou Nanyang Polytechnic College; Research Planning Project of Guangdong Vocational and Technical Education Association 2021–2022 Annual Project--Research on the Design and Application of Smart Classroom Teaching Mode under the Background of "Internet+ Vocational Education" (Project No: 202103G122) from Guangzhou Nanyang Polytechnic College; "Exploration Research on the Training Path of Applied Talents in Higher Vocational Computer Professionals from the Perspective of Integration of Production and Education" (Project No: NY-2021CQ-JGYB006); "Big Data and Intelligent Computing Innovation Research Team" (Project No: NY-2019CQTD-02) from Guangzhou Nan yang Polytechnic College; 2020 Guangdong Province Educational Science Project: Exploration and Practice of the Developmental Evaluation System for Teachers in Private Higher Vocational Colleges Based on Big Data (Project No: 2020GXJK532).

References

1. Beta, W.F., et al.: An analysis of curriculum implementation on high schools in Yogyakarta. In: AIP Conference Proceedings, vol. 1911, no. 1, pp. 20002–20002 (2017)
2. Latika, M., et al.: Student feedback in medical teaching evaluation: designing the perfect mechanism. Indian J. Physiol. Pharmacol. **62**(1), 149–155 (2018)
3. Phillips, W.R., Keys, T.: Interprofessional primary care course curriculum and evaluation. Fam. Med. **50**(3), 217 (2018)
4. Fatemeh, D.-T., et al.: Comparison of general dentistry curriculum in iran with eight of the world's top dental faculties. Maedica **14**(2), 104–115 (2019)
5. Chae, H.-W.: A study on development of the job components and competency indicators for apartment community specialist. J. Kor. Hous. Assoc. **28**(3), 1–10 (2017)
6. Dehon, E., Robertson, E., et al.: Development of a clinical teaching evaluation and feedback tool for faculty. West. J. Emerg. Med. **20**(1), 50–57 (2018)
7. Mohamed, S.: An evaluation of culture teaching and learning in a Uniwide Language Program: Teachers and students' perspectives. Lang. Learn. High. Educ. **10**(2), 357–380 (2020)
8. Okoye, K., Arrona-Palacios, A., Camacho-Zuiga, C., et al.: Towards teaching analytics: a contextual model for analysis of students' evaluation of teaching through text mining and machine learning classification. Educ. Inf. Technol. **27**(3), 3891–3933 (2021)
9. Tarraga-Minguez, R., Suarez-Guerrero, C., Sanz-Cervera, P.: Digital teaching competence evaluation of pre-service teachers in spain: a review study. Revista Iberoamericana de Tecnologias del Aprendizaje **16**(1), 70–76 (2021)
10. Furtado, J., Oliveira, S., Chaves, R.O., et al.: An experimental evaluation of a teaching approach for statistical process control in computer courses. Int. J. Inf. Commun. Technol. Educ. Off. Publ. Inf. Res. Manag. Assoc. **17**(1), 154–171 (2021)
11. Lu, X., Zhu, Y., et al.: Learning from multiple dynamic graphs of student and course interactions for student grade predictions. Neurocomputing **431**, 23–33 (2021)
12. Swinnerton, B.J., Morris, N.P., Hotchkiss, S., et al.: The integration of an anatomy massive open online course (MOOC)into a medical anatomy curriculum. Anat. Sci. Educ. **10**(1), 53–67 (2017)
13. Yin, T.: Analysis and research of teaching quality evaluation system of colleges and universities. Int. J. Intell. Inf. Manag. Sci. **8**(3) (2019)

Analysis of Rapid Extraction of Data and Information Based on Deep Learning Technology

Gang Chen[1], Dawei Zhao[1(✉)], and Edris Zeinali[2]

[1] Changchun Humanities and Sciences College, Changchun, Jilin 130117, China
zhaodawei@ccrw.edu.cn
[2] Islamic Azad University, Tehran, Iran

Abstract. The Internet has promoted the explosive growth of network information, providing people with more comprehensive and timely information services, but it also makes it more difficult for users to quickly and accurately find the information they need. Deep learning technology and information extraction can extract effective and concise information from a large amount of data, so as to better realize the needs of users to obtain refined and effective information. Based on this, this article mainly analyzes and researches the rapid extraction of data information based on deep learning technology. This article introduces the data information extraction process based on deep learning technology, including the process of obtaining data information, preprocessing information content, marking target information, multi-target classification, and data information extraction. This paper optimizes the information extraction model from two aspects: "pre-training language model + attention mechanism" and "target task model fine-tuning" to increase the rate of data extraction. In this paper, through data extraction experiments, the results are obtained. The F1 value of the method in this paper is 77.96%, the accuracy rate and the average recall rate are 77.9% and 78.2%, respectively. It can be seen that the data and information extraction scheme proposed in this paper based on deep learning shows good performance in the field of information extraction.

Keywords: Deep Learning · Information Extraction · Text Classification · Data Information

1 Introduction

At present, a thorough understanding of the syntax and semantics of natural language is the most important part of text classification and information extraction [1, 2]. Manually extracting the characteristics of data information is difficult and subjective, and deep learning can mine self-learning characteristics, which improves the possibility of computer natural language understanding [3, 4]. Using the idea of deep learning, it actively learns the syntactic and semantic features of data information, and obtains the deep

J. H. Abawajy et al. (Eds.): ICATCI 2022, LNDECT 170, pp. 951–959, 2023.
https://doi.org/10.1007/978-3-031-29097-8_113

features of the extracted data information, which reduces the difficulty of formulating artificial features, and it has better objectivity [5, 6].

Regarding the research of information extraction, many scholars have conducted multi-angle explorations. For example, Cai HT constructed a soil nutrient information extraction model based on transfer learning and near-infrared spectroscopy [7]; Mishra DK summarized information extraction through investigation the method [8]; Fei-Fei takes plant distribution information as the research object, and studies the plant distribution information extraction method based on remote sensing technology [9]. It can be seen that the research on information extraction is very extensive. Based on deep learning technology, this paper takes web page information extraction as an example, and proposes a method for rapid data extraction.

The purpose of this article is to analyze and discuss the rapid extraction of data and information based on deep learning technology. This article first summarizes the data information extraction process based on deep learning technology, mainly including: obtaining data information, information content preprocessing, marking target information, multi-target classification and other processes. Then, this article optimizes the model for rapid extraction of data and information, and optimizes it from two aspects: "pre-training language model + attention mechanism" and "target task model fine-tuning" to increase the rate of data information extraction.

2 Analysis of Rapid Extraction of Data and Information Based on Deep Learning Technology

2.1 Data Information Extraction Process Based on Deep Learning Technology

2.1.1 Obtain Data Information

The data information extraction model aims to achieve automatic extraction of target information from similar websites with similar structures through neural network training, eliminating the trouble of writing separate extraction rules for each type of website. Suppose that the food portal website is taken as an example. The key information of the food webpage is located in the food introduction area in the center of the page, including food pictures, food names, and ingredients. The key information in each page will be embedded in HTML tags. Therefore, when extracting this type of data information quickly, it need to obtain a certain number of gourmet pages, and mark the key target information specified by the user, and then use the marked sample pages for training, so that the information extraction model can be faster improve the ability to identify target information.

2.1.2 Information Content Preprocessing

After the data information is obtained, the key target information tags located in the HTML code fragments are extracted. Among them, the information in the HTML tags is too complicated, so the information content is preprocessed before training to simplify the information content. The input of the preprocessing function module is the source HTML text document on the Internet, and the output is the document after page cleaning and parsing. In this module, HTML Tidy cleans the poorly formatted HTML of the page and converts it into well-formed XHTML, and then parses the XHTML into a neural network layer through JAXP technology, so that it is convenient to learn the effective data information in the data information extraction task after processing. To a certain extent, reduce the erroneous data in the corpus data, and perform operations such as access control deletion on each node.

2.1.3 Mark the Target Information

As the only link that needs to interact with the user, the step of marking the target information is very critical. Its purpose is to tell the data information extraction system what general information needs to be obtained in the web page structure in order to obtain information during training [10, 11].

The data information extraction system uses text annotation as an important feature of node similarity. Each node has a text label, and has a degree of relevance to the text label. When the similarity is matched, the overall similarity of the text annotation is obtained by weighting according to the relevance of different texts. Correctly labeled data information can be added to the training set as valid samples.

2.1.4 Multi-target Classification-Softmax Classification

The softmax method is often used in the last layer of the neural network to classify the output results in multiple categories [12]. In view of the fact that the number of target information categories in the web page is not fixed, it is the most ideal way to use the softmax function for multi-target classification. The definition of softmax function is shown in formula (1):

$$a_j^L = \frac{e^{z_j^L}}{\sum_k e^{z_k^L}} \tag{1}$$

In the formula, z_j^L represents the output value of the j-th node, and k represents the number of output nodes, that is, the number of categories. After the output of the node sequence RNN is operated by the softmax function, the probability distribution of the node to all target information can be obtained. The number of classifications obtained in the model is the number of all manually marked target information types + 1. In other words, the calculation result of the softmax function can determine whether the text information node of the current input model is target information (non-target information is counted as one type), and if so, the type of target information is output.

2.1.5 Target Data Extraction

Through the previous data information preprocessing, after the location data area extraction rules are obtained, the data extraction stage is easy to implement. It only needs to convert the XHTML data into the target XML data according to the extraction rules, that is, extract the target data information.

2.2 Optimization of Data Information Extraction Model Based on Deep Learning Technology

2.2.1 Pre-training Language Model + Attention Mechanism

The research uses ERNIE + CNN as a model for short text classification in the data information extraction process. Compared with BERT, RoBERTa has made changes in model, data and training methods, so that it can obtain better natural language task processing effects. RoBERTa mainly improved the data generation methods and tasks, that is, canceling the next sentence prediction, using larger and more diverse data, training for longer, using larger batches, adjusting optimizer parameters, and using full word masks.

Taking into account that in each text information, each word contributes to the sentence information to a different degree, so an attention mechanism is added after RoBERTa. Attention is a widely used mechanism in the field of natural language processing. It can intuitively give the contribution of each word to the result. It can also solve the problem of Text CNN's fixed size of the convolution kernel, which makes it impossible to model longer sequences. The Attention mechanism makes the model more intuitive and interpretable, and it can also visually display the importance of each word to the classification category.

The model used in this article is the three-layer weighted model of RoBERTa followed by attention. The specific method is: take the output of the last three layers of the RoBERTa model, and then randomly initialize a weight matrix that can be trained, then multiply the output of the last three layers of the RoBERTa model by this weight matrix, and finally add an attention, which can give the original sentence a higher weight is given to words that are important to distinguish the sentence category, and a lower weight is given to words that are not important to the classification, which improves the speed of short text classification.

2.2.2 Fine-Tuning the Target Task Model

In order to improve the rate of data extraction, this study uses the oblique triangle learning rate. First, the learning rate is increased linearly, and then the learning rate is linearly attenuated over time. The calculation method of the oblique triangle learning rate is shown in formula (2):

$$p = \begin{cases} cut = [T \cdot cut\_\text{frac}] \\ 1 - \dfrac{t/cut}{cut(1/cut\_\text{frac - 1})} otherwise \\ \eta_t = \eta_{\max} \dfrac{1 + p(ratio - 1)}{ratio} \end{cases} \tag{2}$$

In the formula, T is the number of model training times, cut_frac is the number of training times when the learning rate increases, cut is the number of times the learning rate increases to decrease, and t is the learning rate at time t during model training.

3 Data Information Rapid Extraction Experiment Test

3.1 Experimental Data Set

The data set used in this research is the question set of NLPCC Q&A evaluation that can extract answers from search engines.

In this experiment, a total of 26,600 preprocessed data sets (<Q, S, A, y >) are selected, of which the candidate information set (A) and its corresponding candidate information sentence set (S) come from text fragments retrieved by Baidu search engine. If the candidate information is the correct answer, it is marked y as 1; if it is not the correct answer, y is marked as 0. The data set is divided into two parts: training set and test set, and its distribution is shown in Table 1.

Table 1. Data distribution of training corpus and test corpus

Data set	Number of questions (Q)	Candidate information (A)	Candidate information sentence (S)	Total
Training set	576	5420	6580	12000
Test set	121	6320	8280	14600

3.2 Experimental Program

Two experiments are designed to analyze the advantages of this article's deep learning model for data extraction:

Experiment 1: Analyze the contribution to data information extraction from three levels (surface features, syntactic structure features, and depth models combined with deep semantics);

Experiment 2: Compare the data and information extraction accuracy of the optimized deep learning model in this article with other classic models.

3.3 Data Information Extraction Evaluation Index

The indicators used in this experiment to evaluate the performance of the data extraction model include accuracy (P) and recall (R). The calculation method is shown in formulas (3) and (4):

$$P = \frac{A}{(A + B)} \tag{3}$$

$$R = \frac{A}{(A + C)} \tag{4}$$

In the formula, the accuracy rate is the ratio of the total number of extracted correct information to the total number of extracted information. A represents the number of correct information points extracted, B represents the number of incorrect information points extracted, and C represents the number of correct information points extracted. The value range of P and R is [0, 1], which is inversely proportional, that is, P will decrease as R increases, and vice versa. Therefore, when evaluating the performance of the information extraction model, these two values need to be considered at the same time, the F value is calculated. The calculation method is shown in formula (5):

$$F1 = \frac{(\beta^2 + 1) * P * R}{\beta^2 * P + R} \tag{5}$$

In the formula, parameter β determines the degree of support for recall and accuracy, and the value of β is generally 0.5, 1.2.

4 Experimental Results of Rapid Data Extraction Based on Deep Learning Technology

4.1 Results of Information Extraction Experiment

This experiment uses the average accuracy rate, average recall rate and the average F1 three evaluation standards analyze the performance of data information extraction. The results are shown in Table 2: The information extraction accuracy rate of the Su feature is 57.2%, the average recall rate is 69.1%, and the F1 value is 59.24%; Sy and Sy features are integrated. The accuracy of information extraction is 67.7%, the average recall rate is 69.5%, and the F1 value is 68.05%.

Table 2. Results of information extraction experiment (%)

Method	P	R	F1
Su	57.2	69.1	59.24
Su + Sy	67.7	69.5	68.05
Su + Sy + Q + S	77.9	78.2	77.96

It can be seen from Fig. 1 that the performance of the information extraction integrated into the Sy feature is better than that of only using the Su feature to train the classifier, which shows that the Sy feature can better express the understanding of the sentence. The deep learning model that adds sentence semantic related factors shows better advantages. Compared with the surface features, the F1 value is increased by about 18%, and the accuracy and average recall rates are 77.9% and 78.2%, respectively.

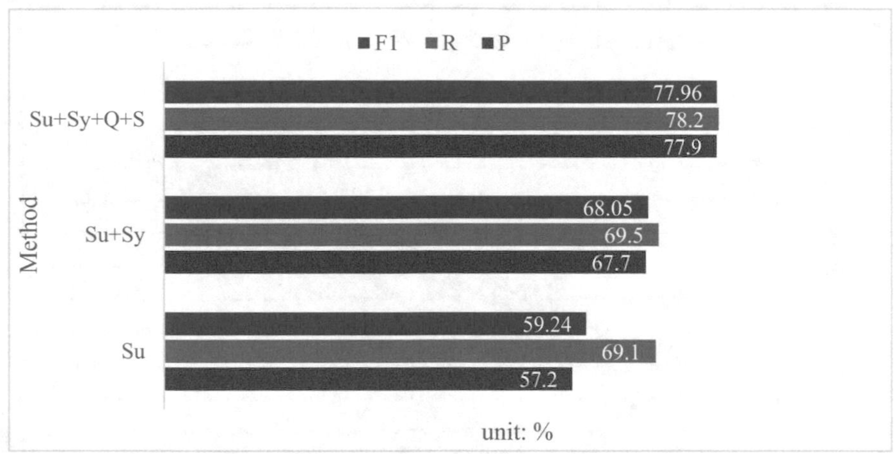

Fig. 1. Results of information extraction experiment (%)

4.2 Comparative Experiment

The data information extraction model in this paper is compared with the evaluation results of other researchers who participated in the NLPCC Q&A evaluation. The results are shown in Table 3. Among them, Team3 is the best score to participate in the evaluation. The value of ACC@N represents the probability that the first N candidate information contains the correct answer.

It can be found from Fig. 2 that the ACC@3 value of the method in this paper reaches 0.864, which means that the number of questions with the correct answer in the first three extracted candidate information accounts for 86.9% of the total number of questions, indicating that the information is based on deep learning ideas. The method of extracting, using the softmax classifier to predict the candidate information and sorting is feasible. It can be seen that the data information extraction scheme proposed in this paper based on deep learning can mine the deep features of the target information and show better performance in the field of information extraction.

Table 3. Comparison of NLPCC Q&A evaluation results

Team	ACC@1	ACC@2	ACC@3	F1
Team1	0.174	0.178	0.195	0.217
Team2	0.241	0.413	0.476	0.499
Team3	0.594	0.592	0.66	0.646
This article	0.791	0.813	0.864	0.841

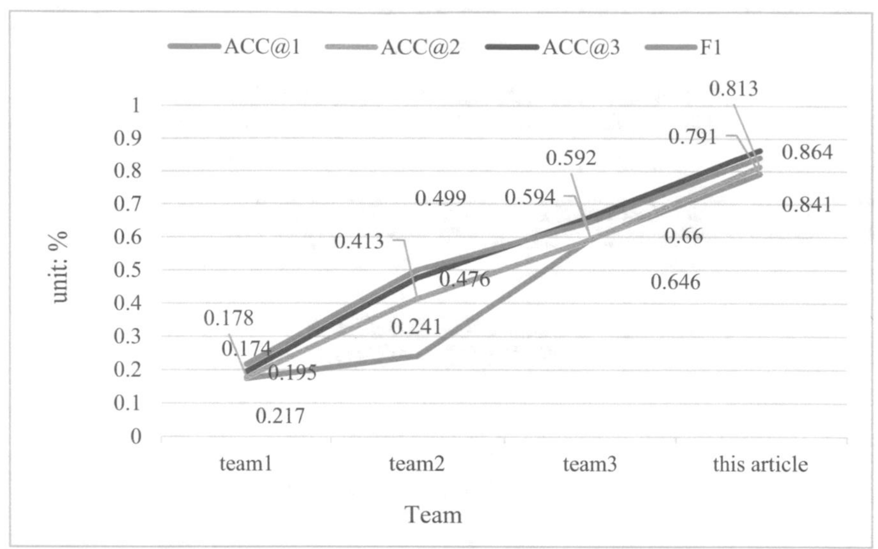

Fig. 2. Comparison of NLPCC Q & A evaluation results

5 Conclusions

With the continuous promulgation of Internet policies and the gradual improvement of infrastructure construction, the current Internet has achieved unprecedented development speed. Information is updated all the time, and massive amounts of data are generated every day, showing massive and diversified characteristics. Data information extraction can allow users to ask questions in natural language, and can return the refined and effective information that users need from a large amount of data information. Therefore, in the future, the rapid extraction technology of data and information based on deep learning will have good application potential in the fields of information filtering, information extraction, and text databases, and will bring us huge social and economic benefits.

Acknowledgment. 2021 Jilin Provincial Higher Education Research Project: Research and Practice of the Wisdom Classroom Teaching Mode Based on Big Data (No. JGJX2021B27).

References

1. Pang, L., Men, S., Yan, L., et al.: Rapid vitality estimation and prediction of corn seeds based on spectra and images using deep learning and hyperspectral imaging techniques. IEEE Access **8**, 123026–1203036 (2020)
2. Elnagar, A., Al-Debsi, R., Einea, O.: Arabic text classification using deep learning models. Inf. Process. Manage. **57**(1), 102121 (2020)
3. Abdi, A., Shamsuddin, S.M., Hasan, S., et al.: Deep learning-based sentiment classification of evaluative text based on Multi-feature fusion. Inf. Process. Manage. **56**(4), 1245–1259 (2019)

4. Nishihara, M., Masaki, N., Nakajima, I., et al.: Ai Navigation system for living donor surgery: seven process classification of laparoscopic nephrectomy by deep learning. Transplantation **104**(S3), S409–S409 (2020)
5. She, X., Long, Y., Kim, D., et al.: ScieNet: deep learning with spike-assisted contextual information extraction. Pattern Recogn. **118**(2), 108002 (2021)
6. Lee, S.J., Yoon, G.Y., Go, T.: Deep learning-based accurate and rapid tracking of 3D positional information of microparticles using digital holographic microscopy. Exp. Fluids **60**(11), 1 (2019). https://doi.org/10.1007/s00348-019-2818-y
7. Cai, H.T., Liu, J., Chen, J.Y., et al.: Soil nutrient information extraction model based on transfer learning and near infrared spectroscopy. AEJ - Alexandria Eng. J. **60**(3), 2741–2746 (2021)
8. Aqib, M., Mehmood, R., Alzahrani, A., et al.: Rapid transit systems: smarter urban planning using big data, in-memory computing, deep learning, and GPUs. Sustainability **11**(10), 2736 (2019)
9. Wei, F.F., Lyu, R., et al.: Extraction of distribution information of Angelicae sinensis plants in Weiyuan county based on remote sensing technology. Zhongguo Zhong yao za zhi = Zhongguo zhongyao zazhi = China J. Chin. Materia Medica **44**(19), 4125–4128 (2019)
10. Wei, Z., Xie, X., Wei, M., et al.: Meteorological information extraction algorithm for ship routing. J. Mar. Sci. Technol. **27**(6), 523–531 (2019)
11. Zhuang, J., Wang, Y., Zhang, S., et al.: A multi-antenna spectrum sensing scheme based on main information extraction and genetic algorithm clustering. IEEE Access **7**, 119620–119630 (2019)
12. Cumbane, S.P., Gidófalvi, G.: Review of big data and processing frameworks for disaster response applications. Int. J. Geo-Inf. **8**(9), 387 (2019)

Application of Data Mining Technology in Microfilm Production

Chulei Zhang[1](✉) and Rasha Almajed[2]

[1] Changchun Humanities and Sciences College, Changchun, Jilin, China
xiaoyazi8880@163.com

[2] American University in the Emirates, Dubai, UAE

Abstract. With the progress of society and the continuous development of computer technology, digital technology has become an indispensable part of the Internet industry. With the continuous development of digital media and the continuous popularization of micro-films, when we change how to use data mining technology for knowledge discovery, we should first have a clear goal, that is, business understanding. Otherwise, even if the mining tools are powerful and the data rich, the conclusions will be drawn. It is also meaningless; furthermore, the purpose of data mining is to dig out valuable information from massive databases, and its purpose is ultimately to be used in actual business operations. Therefore, the data we conduct data mining must be real and Effectively. This article mainly describes the various applications of digital technology in the micro-film industry. How to use digital technology to analyze and explore the production process of movies, take the weight and avoid the light, use AE special effects software, and post special effects production has become an important part of the production of film and television works. Link, it plays an important role both in the visual effect of the work and the success of the film and television works. The digital system is used to make a reasonable analysis of the current development of the micro-film industry. Intelligent and systematic construction for micro-movie audiences, through post-AE software to improve the standardization and accuracy of movies. Finally, the experimental results of digital technology show that the application of digital technology in micro-film production has increased the film transmission rate by 25%, providing technical support for the development of micro-films. The production of film has changed from the past specialization to the current popular micro-film. Moreover, the development of this new platform has lowered the threshold for the release of micro-movies. Only simple operations on common electronic devices such as mobile phones or computers, micro-movies can be uploaded and released to communicate with everyone.

Keywords: Micro-Film · Digital Technology · Ae Special Effects · Data Mining

1 Introduction

Information cannot be equated to knowledge, knowledge should be obtained after sorting and connecting information [1]. This requires communicators to scientifically organize

J. H. Abawajy et al. (Eds.): ICATCI 2022, LNDECT 170, pp. 960–967, 2023.
https://doi.org/10.1007/978-3-031-29097-8_114

the content of popular science micro-videos, following the easy first and then the difficult, the way of connecting. As a form of science and technology communication, the popular video must have authenticity, accuracy and readability [2, 3]. Its authenticity and accuracy are shown as scientific rigor in popular science micro-videos, and visual presentation is an important part of realizing the readability of science and technology communication [4, 5]. For data mining to be finally applied to commercial practice activities, a large amount of real data is required as the premise and foundation. Without real and reliable data, even the most advanced technology is useless [6, 7].

Micro-films have increased the subjective consciousness of film and television advertising audiences who have been in a forced state for a long time, and their individual consciousness has been released [8–10]. In addition, young people as the main consumer are a personalized group, and the story and interactive nature of micro-films are extremely high. It greatly mobilizes the enthusiasm of netizens for participation, and fits the psychology of consumers who pursue individuality and express themselves courageously [11]. Chen L believes that when data mining is used to solve practical problems, it must have a clear goal, that is, business understanding must be accurate, and blindly discussing what data mining methods and software are used is ultimately difficult to succeed [12]. Creativity and the development of scripts are also the top priority of traditional film and television art. However, the results of scientists have timeliness errors, and this timeliness often leads to inaccurate results.

The innovation of this article is to propose the application of digital technology in micro-film production [13, 14]. With the improvement of digital technology and the emergence of digital media platforms, the threshold for the shooting and production of micro-film works has been lowered. A large number of people interested in micro-films have invested in creation. Professionals and non-professionals have gradually increased demand for film and television post-production. For professionals, there are more and better ways to improve professionalism, but for amateurs and beginners, the direction may be unclear. There was a period of confusion. Completed the picture effects that cannot be achieved in reality. As the audience's requirements for the visual effects of film and television works increase, the audience is no longer satisfied with watching movies and only watching the wonderful plot, exquisite picture colors, dreamlike and shocking The visual effect can more attract the audience's attention and capture the audience's heart. These visual effects are part of the production of film and television, and the post-production of film and television plays a very important role for the film.

2 Microfilm Transmission Under Digital Technology

2.1 Network Media Environment

The rapid development of network media has caused a huge impact on the traditional media environment. The low barriers to entry of network media, the strong interaction and the diversified characteristics of the competition subject make the elements of the traditional communication mechanism to be eliminated and reconstructed. Wang C feels that micro-films have inherited the large amount of information carried by film and television media, are more intuitive, and have inherited the narrative style of movies, and have broken the limitations of fixed channels of movie theaters. It is understandable that micro-films are a kind of media by relying on network media for information dissemination [15]. Micro-movie meets all the characteristics of the new new medium defined by Levinson. It should be an audio-visual social medium in the new new medium. Micro-movie has the inherent characteristics of the new new medium, that is, social. The release of micro-movie depends on the video website, and the video website itself It's a social networking site. The commonly used nonlinear functions are shown in the following formula: the formula is as follows.

$$K = (U \times V + 1)^{P1} \tag{1}$$

Throughout more than one hundred years of film history, we will find that all film forms and technologies do not exist outside the field of human history. All film creators have the same qualifications to enter this field. Special historical backgrounds can produce others. A film phenomenon that does not exist in the environment [16]. The development of post-production methods of micro-films in China's digital special effects industry requires high capital investment and high-level professional and technical personnel, but the development of post-production special effects in domestic film and television has not yet formed a sound and commercial benign industry. There are very few professional special effects companies in China. The investment in special effects of film and television works is often not proportional to the return. Such investment and return deter a large number of investors. Domestic special effects companies are still in the early stages of development due to financial and technical reasons stage. Can know the data results and various data formulas of film technology.

$$\mathrm{C}_{\mathrm{Re}sult} = \mathrm{C}_{texture} * L_D \tag{2}$$

Their calculation formula is as follows:

$$X_{\mathrm{t}} = \sum_{j=0}^{\mathrm{q}} \theta_i \varepsilon_{i-j} \tag{3}$$

2.2 An Intelligent System Application

After Effects is a more professional software than Video Studio, Editor and other software. It has more powerful functions and can produce more exciting and detailed special effects. After Effects can be regarded as a post-composite production software for civilians. It is the most commonly used post-composite software for designers and specialized design companies [17]. Such as personal design studios, TV stations, animation companies, etc. A large number of high-quality viewers in the audience have been transferred. In the context of fragmented survival, such as fragmented time and fragmented attention, "micro" products were born in response to the times. The birth of Weibo promotes the prevalence of "micro" marketing. Micro movies are also a means for the advertising industry to respond to crises. Become a new landscape in the advertising landscape [18]. Through the audience's aesthetic requirements and the admiration of special effects movies, it is the general trend for film and television works to add special effects elements at this stage. At this stage, film and television works special effects are produced by professionals from special effects companies, and the ways for students who want to learn special effects to obtain professional knowledge are also compared. Few, most of the methods are online or physical class registration. This method has high tuition and more funds, which has a great impact on the enthusiasm of learners, making learners discouraged.

3 Data Mining System Under Micro-movie

3.1 AE Technical Movie

The micro-film industry is unique. It has some similarities with the film advertising industry chain and TV advertising industry chain, but there are many differences. The academic circles have two different views on the development of the micro-film industry. Some scholars have a positive attitude towards the development of the micro-film industry. They believe that the current development of micro-film is fast, and another view is that micro-film advertising is about to die. Through the research of post-production based on After effects software, I can deeply understand the process and production methods of film and television post-production special effects, and apply them to the graduation design: and be able to write something for students who want to learn about post-production special effects Sexual help, so that the students' works can be completed better. Domestic special effects and micro-films are still in their infancy, especially the theoretical research on special effects micro-films is a lot of blanks waiting for fresh blood to fill.

3.2 The Degree of Media Information Dissemination of Micro-films

In short, value marketing is to pay attention to consumer spiritual pursuits and value trends, and companies strive to respect consumer values and resonate with consumers; the characteristic of the marketing 3.0 era is that consumers have reduced trust in traditional media and turned to social media. They would rather believe in certain product information introduced by unfamiliar netizens than in TV or newspaper advertisements. Enterprises have shifted from tempting consumers to integrating with consumers. It can be seen from this that in the marketing 3.0 era, the audience's use and favor of media has shifted from communication media to social media, and the importance of word-of-mouth communication has greatly increased, and the relationship between enterprises and consumers has become closer, and social responsibility is emphasized. Spread the company's excellent social values. The specific results are shown in Table 1.

Table 1. Annual direct costs of medication for hepatitis B related diseases in environment

Disease name	Inspection fee	Treatment costs	Hospital costs	total
Chronic hepatitis B	577	8176	5550	13726
Liver cirrhosis	1030	13625	10632	252817
Hepatocellular carcinoma	2615	43826	28771	75212

4 Digital Technology Analysis of Micro-films

4.1 Later Development of Micro-films

As a new medium, micro-films are more in line with the communication needs of enterprises in the 30th era of marketing. In the process of forming a unique industry, micro-films are closer to the development characteristics and living conditions of the media industry. Therefore, this article studies the marketing communication value of micro-films from a media perspective. The development of micro-films with digital special effects in China is still in its infancy. Most micro-film creators regard digital special effects as a new thing, but there are also many film lovers, excellent film production teams working in the front line, and colleges. The students of film production have a passion for innovation and have made attempts. They have added digital special effects to their own micro movies through experimental research, step by step attempts and innovation, and the special effects added to their micro movie works are as small as The beam particle halo, and the special effects of fighting between two people, are all innovations in visual effects. The specific results are shown in Fig. 1:

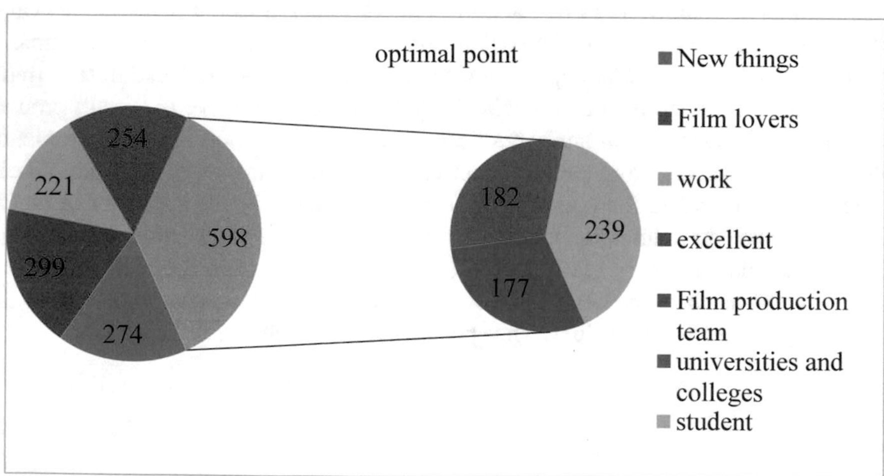

Fig. 1. Incidence number, incidence rate and adjusted incidence rate

4.2 Digital Technology Innovation of Micro Film

An excellent carrier for scientific communication in the new media era. At the same time, on new media platforms, it is easier to produce "aggregation explosion" effects. In traditional media, video playback has to undergo heavy review, focusing on its social significance and educational significance. Many popular science film and television works are designed to be old-fashioned, not close to the current real life, and the audience does not like to see the educational significance, which is naturally difficult to achieve. However, on the new media platform, the design of popular science micro-videos can use more network languages, and the timeliness is greatly enhanced. Case analysis to clarify the role of After effects software in the later stage of the micro-film, and to show the actual effects of After effects software in the later stage of the micro-film. The new media era is an era where wine is fragrant and you are afraid of deep alleys. Various marketing and promotion methods are rich, and publicity channels and forms are diverse. If content producers can make good use of all resources, they can often achieve good results.

Because of the indirectness of spiritual communication, the referential function of media symbols can help the audience understand a certain type of object or a certain concept, thereby communicating the concepts of different individuals and forming a communication. At the same time, with the deepening of understanding, the new functions of the original symbols and new media symbols will continue to be explored to promote the realization of communication. As a new media symbol, popular science micro-film is an important medium for the psychological interaction between the receiver and the receiver. Through the intermediary effect of the symbol, it can promote the interaction and communication between science and technology communication workers and the audience, and better realize the dissemination of science and technology. The specific results are shown in Fig. 2.

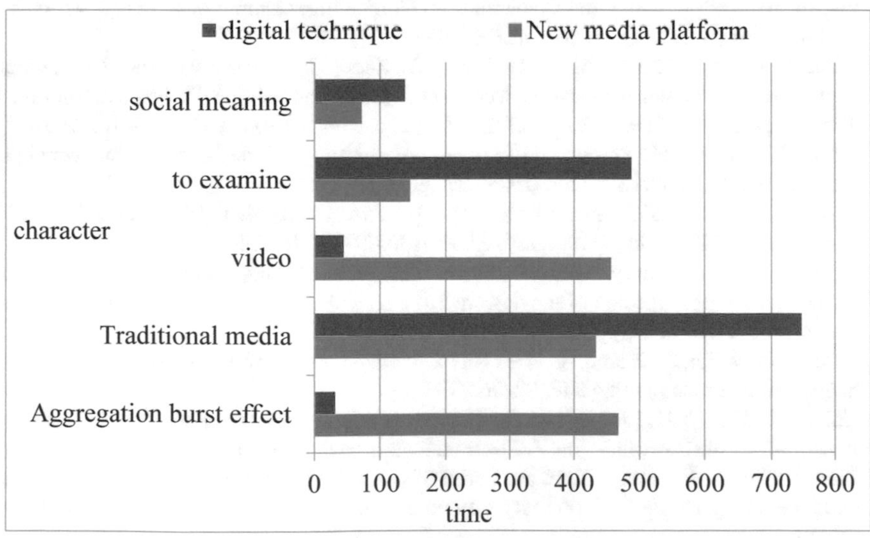

Fig. 2. Advanced science and Technology Communication Workers

5 Conclusions

This paper makes a reasonable analysis on the application of digital technology in micro-film production, and draws various advanced technologies of digital technology in the micro-film industry. In recent years, the field of advertising and marketing has also paid more attention to content marketing. The lack of scientific rigor of scientific micro-videos in my country is mainly manifested in two aspects: one is the lack of systemic video content, and the other is that the production level affects the expression of scientific content. The previous article studied the source of the topic selection of popular science micro-videos. For the current market, the disadvantages of traditional media advertising are becoming increasingly apparent, and the way of advertising in newspapers and magazines is restricted, while TV advertisements and film and television product placements are not only expensive and limited in broadcast time, but also the audience is gradually disgusted. Aggravate. In the development process of special effects technology in film and television, the film special effects team has invented countless special effects technologies and special effects tools after repeated attempts. The software is produced, and the invention, creation and improvement of special effects has been developed from the film special effects team to the film special effects team and nowadays has the ability to research and develop special effects software. In addition, I believe that with the in-depth understanding of micro-films in the future, the production of micro-films based on digital technology will become more and more popular.

References

1. Liao, S.H., Chu, P.H., Hsiao, P.Y.: Data mining techniques and applications–a decade review from 2000 to 2011. Expert Syst. Appl. **39**(12), 11303–11311 (2012)
2. Ahmad, J., Duraisamy, P., Yousef, A., et al.: Movie success prediction using data mining. In: 2017 8th International Conference on Computing, Communication and Networking Technologies (ICCCNT), pp. 1–4. IEEE (2017)
3. Wang, Y., Gong, T., Hassan, M., Li, Q., Huang, S., Zhou, Y.: A feature extraction based support vector machine model for rectal cancer T-stage prediction using MRI images. Multimedia Tools Appl. **80**(20), 30907–30917 (2021). https://doi.org/10.1007/s11042-021-11165-8
4. Latif, M.H., Afzal, H.: Prediction of movies popularity using machine learning techniques. Int. J. Comput. Sci. Netw. Secur. (IJCSNS) **16**(8), 127 (2016)
5. Wang, Y., et al.: McDPC: multi-center density peak clustering. Neural Comput. Appl. **32**(17), 13465–13478 (2020). https://doi.org/10.1007/s00521-020-04754-5
6. Rahim, M.S., Chowdhury, A.Z.M.E., Islam, M.A., et al.: Mining trailers data from youtube for predicting gross income of movies. In: 2017 IEEE Region 10 Humanitarian Technology Conference (R10-HTC), pp. 551–554. IEEE (2017)
7. Parmar, M., Wang, D., Zhang, X., et al.: REDPC: a residual error-based density peak clustering algorithm. Neurocomputing **348**, 82–96 (2019)
8. Chiang, I., Wen, Y.F., Li, M.C., et al.: Using text mining techniques to analyze how movie forums affect the box office. Int. J. Electron. Commer. Stud. **5**(1), 91–96 (2014)
9. Tian, X., Pang, W., Wang, Y., et al.: LatinPSO: An algorithm for simultaneously inferring structure and parameters of ordinary differential equations models. Biosystems **182**, 8–16 (2019)

10. Vijarania, M., Gambhir, A., Sehrawat, D., et al.: Prediction of movie success using sentimental analysis and data mining. In: Applications of Computational Science in Artificial Intelligence, pp. 174–189. IGI Global (2022)
11. Hassan, M., Wang, Y., Wang, D., et al.: Deep learning analysis and age prediction from shoeprints. Forensic Sci. Int. **327**, 110987 (2021)
12. Al-Hashedi, K.G., Magalingam, P.: Financial fraud detection applying data mining techniques: a comprehensive review from 2009 to 2019. Comput. Sci. Rev. **40**, 100402 (2021)
13. Su, Y.S., Wu, S.Y.: Applying data mining techniques to explore user behaviors and watching video patterns in converged IT environments. J. Ambient Intell. Humanized Comput., 1–8 (2021)
14. Ishaq, A., Sadiq, S., Umer, M., et al.: Improving the prediction of heart failure patients' survival using SMOTE and effective data mining techniques. IEEE Access **9**, 39707–39716 (2021)
15. Yang, W., Wang, D., Pang, W., et al.: Goods consumed during transit in split delivery vehicle routing problems: Modeling and solution. IEEE Access **8**, 110336–110350 (2020)
16. Furqon, D.M., Maulana, R.A., Fauzi, A., et al.: Viewer movie predictions based on genres, actors, and directors based on data mining using the eclat algorithm. Gunung Djati Conference Series. **3**, 25–30 (2021)
17. Sahu, M., Saritha, K.: Study on various collabrative filtering techniques to recommend movies. In: 2021 10th IEEE International Conference on Communication Systems and Network Technologies (CSNT), pp. 414–420. IEEE (2021)
18. Alhijawi, Bushra, Awajan, Arafat: Prediction of movie success using twitter temporal mining. In: Yang, X.-S., Sherratt, S., Dey, N., Joshi, A. (eds.) Proceedings of Sixth International Congress on Information and Communication Technology. LNNS, vol. 235, pp. 105–116. Springer, Singapore (2022). https://doi.org/10.1007/978-981-16-2377-6_12

Information Education Evaluation System Under Big Data Technology

Changlin Ke and Baozhi Wang(✉)

Hunan Judicial Police Vocational College, Changsha 410131, Hunan, China
wangbaozhi0418@163.com

Abstract. The ever-changing rules and regulations, pedagogy and teaching methods of countries all over the world ensure that great changes have taken place in the education system. However, the introduction of BD entry technology also has a significant impact on China's education methods and education evaluation. This paper discusses how the data environment of school and school system changes. This paper discusses the typical application of data analysis, expounds the differences of BG, reviews the value of learning analysis in building prediction model, and illustrates the situation of frontier work in CU with examples. The results show that: in the evaluation index system of education informational, the largest proportion of security is 0.637, and the smallest proportion of resources is 0.452.

Keywords: Big Data · Vocational Colleges · Conflicts and Disputes · Information Education

1 Introduction

With the rapid innovation of Internet technology, BD began to spread to various industries and other fields, affecting people's knowledge system and lifestyle.

With the continuous progress of computer technology, many experts have studied the information education mechanism of resolving contradictions and disputes in C and U. For example, some domestic teams have studied the grid governance of HV education management. They use technology describe process selecting obtaining the database, and create a web-based dashboard application, which will communicate with and provide multi-level methods, and effectively improve the evaluation method, so that more participants can participate in the learning process, which helps teachers understand students' learning style. This paper points out the challenges faced by universities in the world, and discusses the potential of BD in dealing with these challenges. Finally, it summarizes the future direction of BD organization project development and implementation. This paper discusses the latest application of BD technology in the field of education, and summarizes the relevant literature of educational DM and learning analysis. The use of, the use of technology to solve problems, solve moral problems in a variety of environments. The approach to analyzing the flow of information from students to other stakeholders includes a hierarchy. All stakeholders can provide opinions and suggestions to enrich students' learning process. This paper lists the structure and

J. H. Abawajy et al. (Eds.): ICATCI 2022, LNDECT 170, pp. 968–975, 2023.
https://doi.org/10.1007/978-3-031-29097-8_115

characteristics of the thinking mode in the era of BD, and puts forward some thoughts on the Enlightenment of the era of BD on Education [1]. Some experts have studied the informatization construction of C and U in order to identify BD and business analysis (BDBA) skills, and further put forward the education and training framework to achieve a successful career in BDBA. Design / method / methods the practical skills of BDBA were summarized through literature review and appreciation survey. This paper puts forward a theoretical framework, which can further help education or training institutions to accept the framework, so as to cultivate young students to acquire BDBA skills. It can also motivate organizations to build courses for BDBA projects. This paper discusses the new methods of curriculum and education data representation, and proposes three visualization and representation methods of education data based on visualization analysis. It shows the possible application of BD concept, but does not indicate what kind of analysis should be carried out to improve the education process. From a technical point of view to design the architecture of this problem, we can deploy some specific software applications to solve this problem. This paper introduces the main concepts of BD: the definition of multiple BD and the framework of BD architecture, which provides a basis for the definition of curriculum structure and public knowledge system in the field of data science and BD technology. Provide background information on existing initiatives and activities related to information exchange and textbook coordination and project development in BD, data science and research data management. This paper introduces status of learning analysis, and summarizes the key technology and analysis model of learning analysis. Taking learning analysis as an example, this paper expounds the application learning analysis in online learning process analysis from the perspectives of managers, tutors and Learners [2]. Some experts have studied the education, and discussed the definition and analysis of BD, the new ideas of learning analysis and education evaluation, as well as the existing problems and challenges. The conclusion is that online course BD will significantly improve the quality of education evaluation, learning analysis and education development. This paper introduces some new opportunities of BD analysis, in order to improve students' learning efficiency and effect, and maximize the retention of knowledge. The concept of BD analysis and DM technology is used to achieve this goal. The method is divided into offline phase (modeling phase) and online phase (Recommendation / decryption phase). In the offline phase, models will be developed for recommendation and conflict resolution respectively. This paper proposes a personalized learning method based on e-learning platform, which provides personalized education network resource recommendation, comparative feedback and personalized bandwidth allocation based on de criticality (reducing the priority of tough guys and tough guys). Taking the design of large-scale open online courses as an example, this paper discusses the nature of design, BD and the interaction between designers and users by using the methods of philosophy and interpretation. An adaptive network English teaching platform based on BD technology has been established to provide the HV students, improve students' ability and reduce differences [3]. Although there are many achievements in the research on the information-based education mechanism of resolving conflicts and disputes in C and U, the research on the information-based education mechanism of resolving conflicts and disputes in HVC is still insufficient under the background of BD.

This paper aims to study the mechanism of resolving contradictions and disputes in HVC under the background of BD. Through the study of BD, it is found that information education, the results show that BD technology is conducive to solving contradictions and disputes in HVC.

2 Method

2.1 Big Data

(1) Big data application

BD processing is roughly divided into five parts: data acquisition and recording, information extraction and annotation, data integration and execution, data analysis and modeling, and data interpretation [4]. Data itself has no value, the important thing is behind, predict on the basis of mining analysis, and guide specific practice [5]. The development of BD has brought challenges and opportunities to higher education [6]. As an organization's strategic resource, data organizational manufacturing data governance can improve the quality of higher education, make the process more transparent, and improve the decision-making efficiency of C and U - manufacturing data governance can be extraction, integration and, interpretation and prediction stages [7].

(2) Characteristics of BD

BD comes from, but in addition to the data itself, it is a technical form combining cloud storage and computing platform, which can help governments, enterprises, organizations and even individuals according to different needs [8]. BD has four characteristics: first; second, wide range; fourth, great value [9]. The application of BD technology has brought great changes to human resource management, making the development of human resource management more scientific and efficient, and improving the intelligent level of human resource management [10]. The first is the large scale of BD, which usually takes TB / Pb as the basic processing unit, while the traditional database data scale is small; the second many types of data in BD, while the traditional database type is relatively single; the third is the relationship between mode and data, which is difficult to determine the mode in advance in many cases; the fourth is that in the era of BD, the value of data processing is far greater than the data itself, and the value of data is far greater The value is also increasing [11]. Data in one field can generate benefits in another.

2.2 Information Education

(1) Information technology

Information technology is another important achievement of Internet development in recent years. This is a concept of interconnection and resource sharing. It can surpass any independent computing and has the advantages of efficient information processing and integration. Informatization is an ideal state of highly applied information technology and highly shared information resources. It constantly uses information to change the traditional economic and social structure.

(2) Characteristics of information education.

Information education has the dual attributes of technology and education. First of all, from the perspective of technical attributes, it is generally believed that the basic characteristics of information education are digitization, networking, intelligence and multimedia. Second, from the perspective of educational attributes, educational informatization is a dynamic development process, a process of comprehensive modern various fields educational attributes and technical attributes, with diversified characteristics. Using information transmission of information is realized on the information sharing platform and modern information network technology. On the established sharing platform, it is conducive to the convenience of information exchange.

(3) Information education mechanism

Information education mechanism refers to the integrate methods, information technology and curriculum knowledge under the guidance of information teaching theory and knowledge. This ability can be reflected in different stages of teachers' information-based teaching process, which is conducive to teachers to maintain good sensitivity, make teachers have based teaching consciousness of autonomous acquisition, application and dissemination of information-based teaching courses, and highly integrate information resources with professional course knowledge and content, so as to give full play to the advantages of information-based teaching. The driving force of the development of information education mechanism lies in how teachers cultivate their own information teaching ideas, consciously improve the information knowledge system of curriculum teaching, so as to better implement information teaching with the help of modern information technology in the information teaching environment, and improve the quality of vocational skills teaching in HVC.

3 Experience

3.1 Extraction of Experimental Objects

Unlike the ordinary teaching tasks, HV education is to transfer the post talents directly to the society, and graduates. At the same time, unlike undergraduate education, HV education tends to export applied talents with technical skills. The quality of HVC directly affects the development of social skills in China. The teaching situation of HVC is very important to the supply and demand of the whole society, which will directly affect whether it can meet the production and management needs of social enterprises and institutions, and affect the improvement of production efficiency in China. This requires teachers in HVC to have strong modern educational teaching ideas and methods skills, in order to cultivate high-quality professional talents to meet the needs of social development.

3.2 Experimental Analysis

First of all, we suggest using supervised learning algorithm, i.e. classification or regression, to try to predict students' academic performance, so as to give early feedback on expected performance before course selection and course selection. Secondly, it is suggested to use these predictions to guide module, course and content recommendation, so as to maximize students' potential, which is reflected in students' learning ability, interest fields, educational objectives and occupation. It is suggested that mechanism of students' learning process, and efforts should be made to determine the best form, style, rhythm and organization of knowledge acquisition process, so as to significantly improve students' academic performance and knowledge retention rate in the long run. Finally, combined with the learning optimization method, through the personalized education curriculum, try to develop a flexible transfer method, so that students are satisfied with faster delivery and longer knowledge retention.

4 Discussion

4.1 Evaluation Index System of Education Informational

The data of judgment matrix are input into the calculation and analysis software for calculation, and the weight of each level index and the consistency test data of judgment matrix are obtained. Generally speaking, when $x \leq 0.1$, the consistency of the judgment matrix is acceptable. As shown in Table 1.

Table 1. Weight chart of education information evaluation index

type	Weight value
achievements	0.532
guarantee	0.637
application	0.521
Informational leadership	0.527
resources	0.452

It can be seen from the above that in the evaluation index system of educational informatization, the weight of achievements is 0.532, that of guarantee is 0.637, that of application is 0.521, that of informatization leadership is 0.527, and that of resources is 0.452. The results are shown in Fig. 1.

It can be seen from the above that in the evaluation index system of educational informatization, the largest proportion of guarantee is 0.637, and the smallest proportion of resources is 0.452.

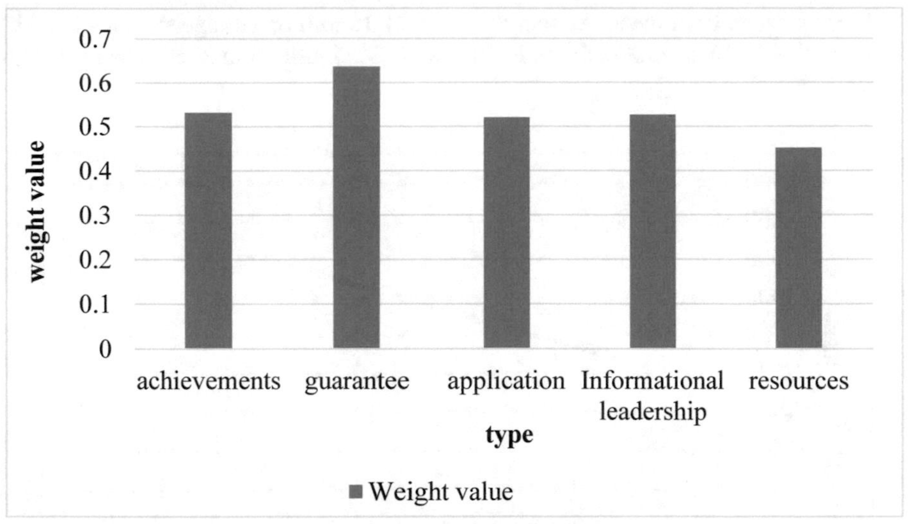

Fig. 1. Weight chart of education information evaluation index

4.2 Statistics of the Number of Students in Higher Vocational Colleges

With the continuous increase of the number of college students and the expansion of the scale of running a school, the problems closely related to the interests of teachers and students are also increasing: on the one hand, the service attitude of individual managers is poor, the teaching level of teachers is low, the teaching conditions can not meet the needs, and the teaching quality is poor. The false propaganda of some C and U may become an important factor leading to the internal instability of C and U; on the other hand, the individual management Poor service attitude, low teaching level of teachers, teaching conditions can not meet the needs, individual differences of students also bring challenges to the safety management of C and U. Due to the immature psychology and physiology of college students, the pressure of study, economy and employment is relatively large, which leads to the increase of psychological problems. If not dredged in time, it may lead to a long-term serious campus crisis. As shown in Table 2.

Table 2. Statistics of college students in 2017–2020(10000 people)

particular year	Number of people
2017	2115
2018	2705
2019	3326
2020	3925

It can be seen from the above that there are 21.15 million college students in 2017, 27.05 million in 2018, 33.26 million in 2019 and 39.25 million in 2020. The results are shown in Fig. 2.

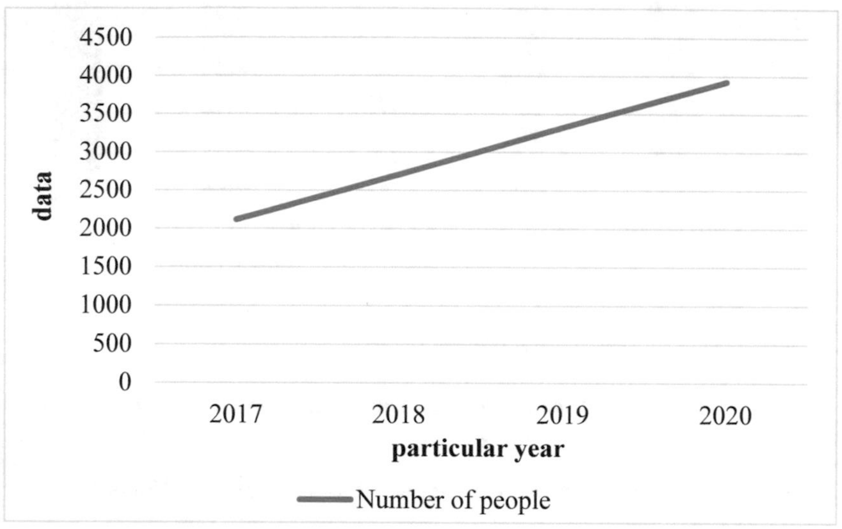

Fig. 2. Statistics of college students in 2017–2020(10000 people)

It can be seen from the above that the number of college students in China every year increases with the growth of the year. In 2020, the number of college students in China will be up to 39.25 million.

5 Conclusion

At present, the era of BD has come, and a large amount of data has been accumulated in the field of education. In the field of education, more and more software systems are deployed to store a large amount of educational resources data. How do you do it? Whether we can make full use of the massive educational data and transform these data into useful information and knowledge, so as to make better use of these data for scientific educational decision-making. This paper summarizes the significance of BD and learning analysis in education from the perspective of data science. This paper holds that in the field of education, through the application of BD technology, various high-quality and high efficiency platforms can be realized. This paper puts forward the main framework of supporting educational research.

References

1. Chen, G.: Research on the integration of information technology and practical writing teaching in higher vocational colleges. Revista de la Facultad de Ingenieria **32**(12), 1090–1094 (2017)

2. Yi, W., Wei, X.: Suggestions and countermeasures of integration of production and education in higher vocational colleges. Int. J. Softw. Eng. Appl. **11**(5), 109–114 (2017)
3. Graebe, J.: Identifying and resolving conflicts of interest for individuals in a position to control educational content. J. Contin. Educ. Nurs. **49**(3), 102–104 (2018)
4. Suzuki, R., Nakamiya, Y., Watanabe, M., et al.: Relationship between stress coping mechanisms and depression in kidney transplant recipients. Transplant. Proc. **51**(3), 761–767 (2019)
5. Nthontho, M.A.: Schools as legal persons: implications for religion in education. S. Afr. J. Educ. **38**(Supplement 2), 1–8 (2018)
6. Lee, E.J., Jeong-Eun, J.O., Yun, M.H.: The effects of education for resolving the information gap on the lifelong learning competence of the middle aged. J. Fish. Marineences Educ. **29**(5), 1313–1330 (2017)
7. Kirchhof, G., Lindner, J.F., Achenbach, S., et al.: Stratified prevention: opportunities and limitations: report on the 1st interdisciplinary cardiovascular workshop in Augsburg. Clin. Res. Cardiol. **107**(3), 193–200 (2018)
8. Hafezi, M.A., Yazdipour, M., Matlabi, D.: What are the aims of managers in developing digital libraries? Iran. J. Inf. Process. Manag. **32**(3), 817–841 (2017)
9. Ruijie, L.I., Weifu, C., Yinye, Y., et al.: First principle calculation of electromagnetic mechanism for Fe2Si bulk material. J. Wuhan Univ. Technol.-Mater. Sci. Ed. **34**(001), 64–68 (2019)
10. Baker, A.: Reviewing big data and learning analytics in higher education: current theory and practice. J. Scholarship Teach. Learn. Christ. High. Educ. **8**(1), 8 (2018)
11. Zeide, E.: The structural consequences of big data-driven education. Big Data **5**(2), 164–172 (2017)

Application of K-Means Clustering Algorithm in Fresh Food Safety Management

Ying Zhou(✉)

Shandong Management, University of Business School, Jinan 250357, China
angel201098@163.com

Abstract. In order to understand the application of k-means clustering algorithm in fresh food safety management, this paper will carry out relevant research. The research mainly discusses the fresh food safety management objectives and supply status, and then introduces the basic concept of K-means clustering algorithm, on this basis for algorithm modeling, index construction work, get the fresh food safety management system. Through this study, the main function of fresh food safety management system is to carry out safety early warning. The early warning function is realized by k-means clustering algorithm, which can guarantee the accuracy of early warning, help relevant organizations to manage in advance, and prevent fresh food safety problems.

Keywords: K-means clustering algorithm · Fresh food supply · Fresh food safety management

1 Introduction

With the continuous development of logistics industry, fresh food market expanding, this also lead to increased mobility of fresh food, and in the constant flow of fresh food safety issues is more serious, the frequency, influence degree that caused the alarm, nature also let the related department to stress the situation to improve. On the basis of this, the relevant departments of the reform work, and actively expand the purpose of the reform from the root place to ensure the safety of fresh food, so you need to establish an early warning system for the fresh food safety management, but whether the system can be used for actual use depends on the accuracy of the early warning system, and the traditional method is not enough to achieve this goal, so slow down the reform progress, the k-means clustering algorithm is proposed to help fresh food safety management to achieve accurate warning with the help of the algorithm. Therefore, how to apply the K-means clustering algorithm in the management has become a problem worth thinking about, and relevant research needs to be carried out.

J. H. Abawajy et al. (Eds.): ICATCI 2022, LNDECT 170, pp. 976–984, 2023.
https://doi.org/10.1007/978-3-031-29097-8_116

2 Fresh Food Safety Management Objectives and Supply Status Issues

2.1 The Target

The objectives of fresh food safety management are divided into many levels, but the ultimate goal is to avoid safety problems as much as possible. The reason is that once most fresh food safety problems occur, they may cause very serious impacts, which may even threaten people's lives and have a very wide range of impacts. At the same time, the impact of the problem is further increased after fresh food joined the logistics industry, so it is the best choice to control the problem before it happens. From this perspective, because fresh food safety issues cannot be totally eliminated, so can only as far as possible to reduce its incidence, according to the goal of fresh food security is put forward the concept of early warning, if can understand its occurred before the safety problem occurred the possibility of size, it can be done to prevent, the problem in the incidence of infinite close to zero [1–3].

Fresh food generally refers to vegetables, meat, seafood and other foods. Such foods have the characteristics of short shelf life and fresh-keeping period. Therefore, the circulation of such foods in China in the early days was not high, and most of them were only circulated within or around cities. Even so, the quality of fresh food varies according to the past, and some occasionally have fresh food safety problems. There are many causes of the problems, including: first, The transportation time of fresh food is too long, so under the characteristics of short shelf life and fresh-keeping period, some fresh food may deteriorate slightly, which is difficult for relevant personnel to find, and then purchase and eat, causing safety problems of fresh food; Secondly, the storage conditions of fresh food during transportation are not good, such as the high temperature of storage space, which will accelerate the deterioration of fresh food, so it may also occur slight deterioration, or severe but not obvious characteristics of deterioration, and food safety problems may occur after being eaten; Third, the fresh food has deteriorated, but has been artificially covered up, causing others to use it, which leads to safety problems. The safety problems of fresh food caused by any reason have serious harm to human health, ranging from diarrhea and vomiting to illness, even to viscera and life threatening. It can be seen that the safety problems of fresh food have a significant impact. However, because such problems were not widespread in the early days, people did not pay attention to the safety of fresh food. Figure 1 Fresh food safety management process and objectives.

K-means is a mean clustering algorithm, which can complete clustering by iterative solution, and then analyze. K-means algorithm is very simple to implement, and has a wide range of applications. Existing research has been used in fresh food safety management related work, which shows that k-means algorithm has good applicability in this study. The main steps of k-means algorithm are not complex. First, a large amount of data needs to be collected and divided into K groups. Then, K random objects are confirmed by random selection, and all random objects are clustered together to form an initialized cluster. According to the cluster, the center value can be obtained, and the distance between each object and the cluster center value, that is, the size of the number difference, is calculated. Secondly, according to the distance between them, each

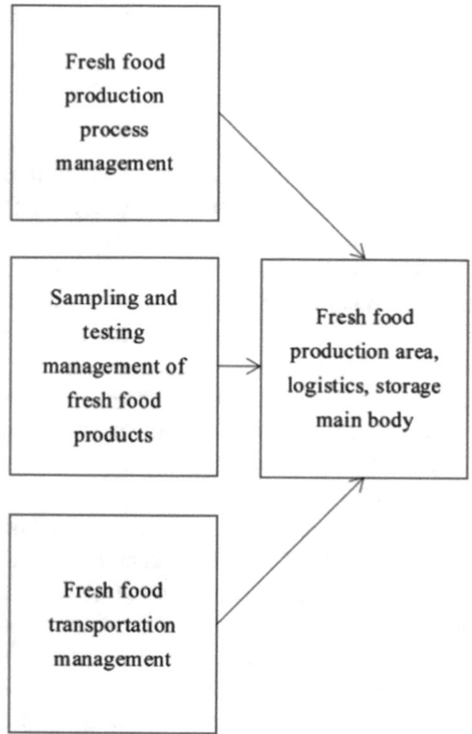

Fig. 1. Fresh food safety management process and objectives

random object is assigned to the nearest clustering unit, so that classification clustering is completed. On this basis, each cluster item and its corresponding center value, and the object assigned to them represent a cluster, and the problem can be analyzed according to the relationship between them. During the operation of k-means algorithm, every time each cluster gets a random object sample, the central value of the cluster will be updated, and each update will be calculated according to all objects in the existing cluster. This process will be repeated until there is no random sample or the preset termination condition is reached. The termination condition has no formula, which is set manually according to their own needs, therefore, k-means algorithm can well meet the personalized needs of people. It is worth mentioning that although there is no definite termination condition, theoretically, k-means algorithm can be divided into three types of termination conditions: no (or the minimum number) objects are reassigned to different clusters; No (or minimum number) cluster centers change again; The sum of error squares is locally minimum, which can be selected manually.

2.2 Current Supply Situation

Although the objective of fresh food safety management is clear, people have been unable to achieve ideal results for a long time. The reason lies in the problems in the current situation of fresh food supply chain under the development background of logistics industry.

First of all, there are two main problems in the fresh food supply chain: First, the supply chain is increasingly complex. Ever fresh food supply chain mainly adopts model is straight for the model, namely the fresh food directly from producer to a sale in the market, so the mode of production and its corresponding market maintained a direct link, this case directly to supervise the production profile of production can be accurate warning, but with the development of the industry, mainly adopt the mode to modern fresh food supply chain contract system centralized transport mode, is a kind of contracted by the enterprise organization for a producer, production personnel in accordance with enterprise requirements for production, then put the fresh food delivered to the enterprise, the enterprise organization is responsible for the hoarding, storage, management, sales, and enterprise organization is divided into several levels, supervisor will be in accordance with the contract will be delivered to the lower part of fresh food, and so on, and finally by the most fresh food delivery to the market to sell at a lower level, or directly from the most to sell at a lower level, it can be seen that the model of fresh food will be unknown for many times, and the production needs to face requirements, therefore, some enterprises in the purpose of the benefit maximization will do anything to make a choice, such as in order to speed up the production efficiency, ensure the production, demand production personnel extensive use of pesticides, fertilizers, etc., may be at a certain phase of transportation storage fresh food spoilage caused by improper management, this will lead to security issues, but in the face of such a complex supply chain environment, related organizations is difficult to conduct a comprehensive monitoring, at least in the manual mode is difficult to achieve early warning; Second, the supply chain transportation technology is not mature and insufficient standardization, that is, fresh food is different from ordinary products, which has the characteristics of short freshness period and fragile, so it cannot be supplied by ordinary transportation mode, instead, it must use special technology, which generally refers to the cold chain transportation technology [4–7]. Cold chain transportation technology can refrigeration principle is used to extend the freshness, and decrease the rate of fresh food is damaged, so the technology has been widely used in the fresh food supply chain, but combined with the actual situation, the technology of the actual effect is not ideal, fresh food under technology protection still exist a lot of bad problems, these problems are hard to find, once entering the market, it may cause serious impacts [8]. Table 1 shows the data of fresh food deterioration in a certain region for three consecutive years, indicating the seriousness of this problem.

Table 1. Data of fresh food deterioration in a region for three consecutive years

Time	Data
The first year,	Spoiled fresh food accounted for 10.3% of the total
In the second year	Spoiled fresh food accounted for 13.1% of the total
In the third year	Spoiled fresh food accounted for 9.2% of the total

Secondly, the two major problems in the current situation of fresh food supply chain affect the safety management of fresh food: First information asymmetry, the fresh food safety management involves a wide range, management organization is not likely to make people actual investigation, and therefore must rely on information to understand the situation of supply chain, but in a complex environment, supply chain management information is asymmetric phenomenon is widespread, such as production technology of production site meet the specification requirements, but when organization for fresh food sampling inspection found problems in chemical excessive amounts of fresh food, production technology of this kind of phenomenon shows that producer may not conform to the requirements of the specification, or because of fresh food in the subsequent transfer of influenced by other factors, caused the management organization will not be able to prevent problems happen, even can't accurate accountability; The second technical management is difficult, that is, because modern technology level is limited, so in theory according to the existing technical level of the supply of fresh food planned transportation route length, avoid the length more than technical scope, but in the face of different regions, different varieties and different types of fresh food in terms of freshness, related organizations cannot undertake unity of technology management, targeted management is not realistic, so the technical management is difficult, the incidence of deterioration is still high [9, 10]. Affected by the above factors, there is the problem of fresh food safety management in every big cities of our country. Figure 2 is the incidence of the fresh food safety management problem for consecutive three years in a city.

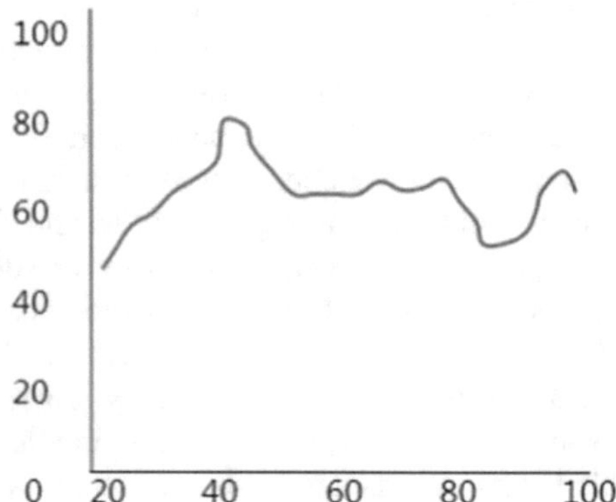

Fig. 2. Is the incidence of the fresh food safety management problem for consecutive three years in a city.

3 Concept Introduction of K-Means Clustering Algorithm and Construction of Security Warning System

3.1 Algorithm Concept

K - means clustering algorithm is a heuristic clustering algorithm, is also a kind of supervised machine learning algorithm, and its basic principle is through iteration to get

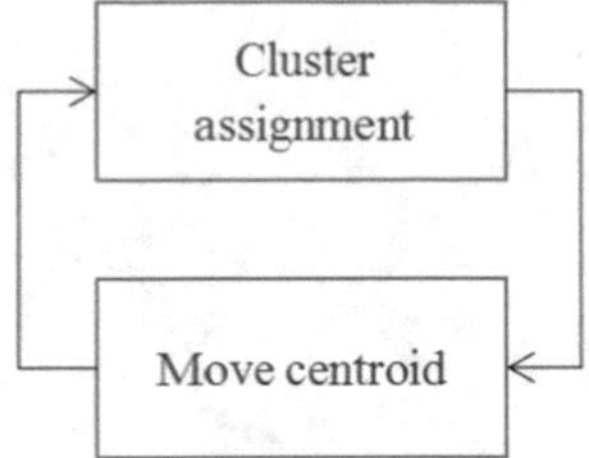

Fig. 3. Two iterative processes of k-means clustering algorithm repeatedly

the optimal solution, and the key points in the process of iteration are: first, the iteration for the first time need to then choose two sample points, the purpose is to construct the clustering center, the establishment of the center determines the direction of the next iteration; Second, because the number of iterations clustering algorithm is unlimited, so if they do not limit leads to a large number of iterations, make limited as a result, the result may not be able to meet the demand of practical application, therefore in the process of iteration algorithm to set up the termination conditions, the conditions when any iteration results meet the requirements of terminate the iteration will stop. The K-means clustering algorithm will repeat two iterations after the initial iteration, as shown in Fig. 3.

Combined with Fig. 2, the first iteration process of the algorithm is to calculate the distance between the sample and the clustering center. In this process, the iterative center is taken as the target, and the sample classification item closest to the center is selected, and other related samples are close to this sample. The second process is mobile clustering center, is the main target in the sample classification, recalculate the classes, the center of the cluster each class the center of the cluster is the optimal solution of the cluster, can help people understand, judgment problem, namely, multiple iterations after the position of the center will not change, on behalf of the algorithm to get the best results at this moment, the results are typically presented in two ways, as shown in Figs. 4 and 5.

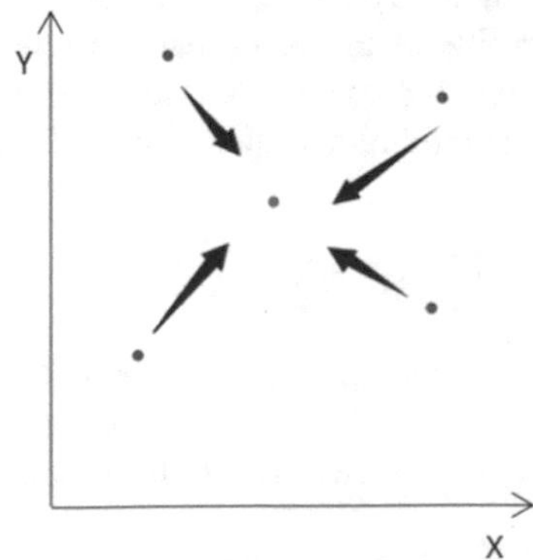

Fig. 4. Result display Method I (red dots are other influencing factors, blue dots are quasi-influencing factors)

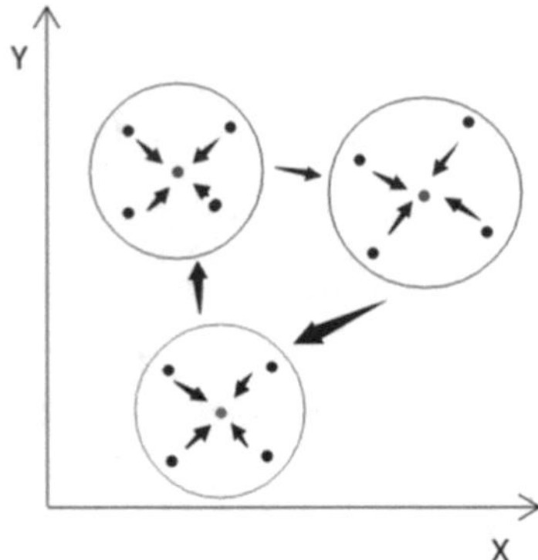

Fig. 5. Result Presentation Method 2 (Each circle represents a classification, the black dot within the circle is other influencing factors of the classification, the non-black dot is the center of the classification, and there is a cyclic relationship between each circle, representing the iterative cycle)

3.2 Safety Early Warning System Construction

The construction of security early warning system on the basis of k-means clustering algorithm can be divided into two steps, namely algorithm modeling and index construction. The specific contents are as follows.

Algorithm Modeling
According to the related theory, the basic principle of K - means clustering algorithm is in accordance with the sample training set, and then coordinate system for the training focus each influence factor is calculated, depend on the coordinate system can be given sample points, to rely on each of a given sample points coordinates calculation of the distance between each sample point size, prompting samples were divided into K type, therefore, the first step of algorithm modeling is to confirm the coordinates of each influencing factor in the coordinate system, and the distance of the center point of class K should be as large as possible in the coordinate selection. The class in the whole process can pass (C1, C2... Ck) represents that the purpose of the algorithm based on this expression is to obtain the minimized squared error E, and the calculation method is shown in Formula (1).

$$E = \sum i = 1k \sum x \in Ci||x - \mu i||_2^2 \tag{1}$$

In the formula, i is the ith influencing factor, k is the KTH type, and x is the coordinate point (the horizontal coordinate point can also be represented by the vertical coordinate point represented by y, but either of them can only be selected in the calculation and cannot coexist).

According to Formula (1), the k-means clustering algorithm model is obtained. The whole calculation steps under this model are as follows: First, the expected value is set according to the actual demand, and several objects are selected as initial points in the expected value; Second, calculate the distance between each influencing factor and the initial point, and classify each influencing factor according to the distance. Third, the standard measure function is calculated for each expectation point, and the result is the termination condition of iteration.

In addition, in order to apply k-means clustering algorithm to fresh food safety management, safety management problems can be realized by using the four-quadrant rule, that is, the four-quadrant rule is a theory that classifies the urgency or importance of events so as to clarify the priority of different events. The levels are from high to low: critical but important, important not urgent, critical not important, and non-critical not important. Combined with the requirements of the rule, security problems can be evaluated after the algorithm results are obtained. If the event level is urgent and important, it means that security problems are about to happen and must be dealt with immediately, and so on, indicating that the four-quadrant rule is helpful for people to interpret the calculation results of k-means clustering algorithm.

Index Establishment
According to the K-means clustering algorithm theory and the four quadrant rule, fresh food safety problems of indexes can be divided into primary, secondary, primary index represents everything in the algorithm could lead to fresh food safety factor sets, such as production factors is set, which contains many factors can lead to the problem, and secondary index contains fresh food safety problems could occur each factor concentration factors, calculated, using algorithm of each factor, fresh food safety issues may be indicated by what happened, also can from the angle of the related factors to determine the incidence of problems, coupled with four quadrant rule to score, and the incidence of judgment problem at any level. As for the specific form of problem incidence scoring, there is no fixed form at present, but the form commonly used in most studies is: urgent and important = 100–90 = 100%-90%, Critical not critical = 80–70 = 80%–70%, Critical not important = 60–50 = 60%-50%, Non-critical not important = less than 50 = less than 50%.

4 Conclusion

In conclusion, the role of K-means clustering algorithm in fresh food safety management is clear, which can help relevant organizations make judgments on safety problems, so as to timely and accurate management and minimize the incidence of problems. Therefore, k-means clustering algorithm is necessary to be promoted in this field. At the same time, in order to ensure that k-means clustering algorithm can play a full role in fresh food safety management, it is suggested that people use it in combination with the four-quadrant rule, which can make people better interpret the calculation results.

References

1. Holilah, D., Bustamam, A., Sarwinda, D.: Detection of Alzheimer's disease with segmentation approach using K-means clustering and watershed method of MRI image. J. Phys: Conf. Ser. **1725**, 012009 (2021)

2. Wang, X., Shen, A., Hou, X., et al.: Research on cluster system distribution of traditional fort-type settlements in Shaanxi based on K-means clustering algorithm. PLoS ONE **17**(3), 1–25 (2022)
3. Elangovan, V.R., Joe, A., Akila, D., et al.: Heart atherosclerosis detection using FCM+kMeans Algorithm. In: 2021 2nd International Conference on Computation, Automation and Knowledge Management (ICCAKM) (2021)
4. Liu, X., Sang, X., Chang, J., et al.: The water supply association analysis method in Shenzhen based on kmeans clustering discretization and apriori algorithm. PLoS ONE **16**(8), e0255684 (2021)
5. Muhajir, M., Ismail, N., Syahreza, S., et al.: Pemutakhiran Zona Musim (ZOM) Provinsi Aceh Menggunakan Data Blending Berbasis Non-Hirarki K-Means Clustering. Jurnal Fisika Flux Jurnal Ilmiah Fisika FMIPA Universitas Lambung Mangkurat **18**(1), 35 (2021)
6. Hidayati, R., Zubair, A., Pratama, A.H., et al.: Analisis silhouette coefficient pada 6 perhitungan jarak k-means clustering. Techno. Com. **20**(2), 186–197 (2021)
7. Arjun, P., Manoj, K.G.: Improved hybrid bag-boost ensemble with K-means-SMOTE–ENN technique for handling noisy class imbalanced data. Comput. J. (1), 1 (2021)
8. Sintiya, S., Laksana, T.G., Tanjung, N.: Kombinasi single linkage dengan K-means clustering untuk pengelompokan wilayah desa kabupaten pemalang. J. Innovat. Inf. Technol. Appl. **3**(1), 17–27 (2021)
9. Hui, C., Tingting, W., Zuoxiao, D., et al.: Power equipment segmentation of 3D point clouds based on geodesic distance with K-means clustering. In: 2021 6th International Conference on Power and Renewable Energy (ICPRE) (2021)
10. Xu, J., Pan, X., Jia, B., et al.: Oil spill detection using LBP feature and K-means clustering in shipborne radar image. J. Marine Sci. Eng. **9**(1), 65 (2021)

Advanced Management System Based on Data Mining Technology

Xianzhu Jin(✉)

College of Education, Jinggangshan University, Jian 343009, China
9920190016@jgsu.edu.cn

Abstract. In order to construct the advanced management system, this paper will focus on data mining technology to carry out relevant research, mainly discusses the basic concept of data mining technology and the characteristics of the advanced management system, and then launch the advanced management system design work, introduce the system design requirements, design scheme, and test the system in the process. The advanced management system designed in this paper can give full play to the role of data mining technology, dig out usable information from the data, and provide accurate management decision-making suggestions for managers.

Keywords: Data mining technology · Advanced management system · Management decisions

1 Introduction

At present, because of the rapid development of social economy, the domestic industry internal appeared a lot of companies, so companies face more pressure of competition, and in order to better cope with competition, and constantly the forward development of the enterprise to pay more attention to their own internal management, the past, internal management and matching system is too rough, simple, The concept of advanced management system is put forward under this background. Higher management system is put forward for enterprises to see the direction of the internal management development, so enterprises are eager to establish advanced management system, and the establishment of the system must have the data mining technology, therefore, enterprises are facing the main problem is how to use the current data mining technology to establish advanced management system, related research is necessary.

2 Basic Concepts of Data Mining Technology and Characteristics of Advanced Management System

2.1 Basic Technical Concepts

Data mining technology is a big data processing tool. Its main function is to dig deeply into the connotation of relevant data in big data, establish the relationship between data

J. H. Abawajy et al. (Eds.): ICATCI 2022, LNDECT 170, pp. 985–992, 2023.
https://doi.org/10.1007/978-3-031-29097-8_117

groups, and output the mining results. The results represent the status quo, development path and possible future development direction of events referred to in data. As for large consumer groups consumption behavior of commodity price data in data mining, by mining can be seen that the consumer's consumption level and consumption level, the development trend of commodity type data, or to the combination of known consumer groups for more interested in what kind of goods, price is reasonable, the judge future sales of goods, if the price is too high or too low, Companies can adjust commodity prices to control the trend and gain maximum benefits. From here you can see, the data mining technology has a high application value, and the application of the technology value, also reflect on the technology of high domestic sex, namely data mining techniques as data processing tool, it is not affected by the data type, data sources, able to handle any data, so the technology can be applied in many fields, This includes the internal management of enterprises. Data mining technology is the core of neural network, the network has feedforward, feedback in two forms, of which the former is based on the original data mining type analysis, then all the mining results summary, form a result output, each mining analysis hierarchy process won't have communication, operation process can only move forward, while the latter is the mining analysis principle of similar, But each layer communicates with each other and has a feedback feature [1–3]. Combined with the internal management needs of enterprises, the data mining technology suitable for the construction of advanced management system is the feedforward neural network mining technology.

2.2 System Features

The so-called senior management system refers to the management system improved on the current popular general management system. Compared with the general management system, it has more rich management functions and can help managers to manage management objectives more deeply. Taking enterprise financial management as an example, the general management system can only provide managers with data collection, storage, search, display and other functions in management work. The actual management analysis, judgment, decision-making and other work need to be carried out by managers. Therefore, the management work still brings great pressure to managers. At the same time, due to the limitations of human capabilities, managers may not be able to find some deep-seated problems in financial management, Such as financial problems hidden under false financial information. On the other hand, the senior management system not only has all the functions of the general management system, but also can, to a certain extent, replace manual work to analyze and judge data, and provide decision suggestions. Manual work only needs to make choices in the decision suggestions according to the actual situation, and then implement them. This reduces the labor burden. At the same time, the system makes management work no longer rely on manual work excessively, Therefore, the limitations of human capacity will not limit the work, so as to facilitate the deep excavation of hidden financial problems. For example, if there is false financial information in the data, the senior management system can carry out information data matching analysis. If it does not match, it represents false information. Then conduct targeted analysis or investigation to find the problem.

It can be seen that the advanced management system has a high application value. Managers of modern entities are very eager to implement the advanced management system, but the implementation of the system is difficult, and needs to be supported by data mining technology, so it is necessary to study.

Data mining technology belongs to data processing technology, but it is different from conventional data processing technology. Its main feature is that it can mine potential information under surface data combination, which is impossible for conventional data processing technology. In principle, data mining technology will regard all surface data as a set, and then list each data in the set separately as a basic data node in a traversal manner, and analyze all data characteristics, adjust different data according to characteristics, so that the overall data set can be divided into several small sets, and then conduct intra set analysis for all basic data nodes in each small set, The process is to directly or indirectly connect the basic data node with other basic data nodes, so that you can know the data relationship in the set, identify the situation of things, and then use the data relationship model of each small set as the basic node to directly or indirectly connect with the data relationship model of other small sets, understand the influence relationship between different data relationship models, and achieve in-depth analysis of problems. This also makes data mining technology have a wide range of applications, which can be used for advanced system implementation.

Higher management system in essence is the enterprise internal management system, but the biggest difference with normal management system lies in the digital management, namely the development of modern enterprises for their own internal management put forward higher accuracy requirement, to meet the requirements must be managed in accordance with the data, but the common management system data capacity is small, the problem such as low, rely on artificial data processing ability, Not according to the internal management of large, and updating of data management, combined with the artificial intervention, the common management system precision is difficult to improve the enterprise internal management, in contrast, advanced management system, the data mining technology, with the help of the powerful data processing ability, logic is similar to human in the process and results shows that it can be concluded that fine data, Only need to use large capacity database to solve the problem of data storage, data mining will be able to get rid of the artificial, artificial is based on the results of mining management decisions can, at this time of management decisions on precision than ever before, at the same time because of the artificial, so mining results tend to be more comprehensive and in-depth, namely human intelligence ability, although able to handle large data, And high precision results, but the person's energy limited, the work efficiency is relatively low, the actual situation of limited, unable to process all the data, so the pure artificial defects that management decisions are widespread, but higher management system does not exist this restriction, natural high quality results, help enterprises to realize digital management, which is the management system of "higher" [4–7].

3 Advanced Management System Design and Testing

3.1 Design Requirements

The general framework of advanced management system in this paper is shown in Fig. 1.

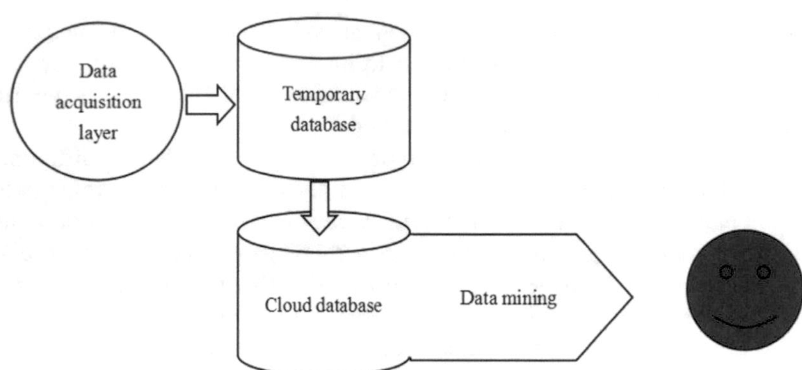

Fig. 1. Overall framework of advanced management system

According to the overall framework of advanced management system, the system design of the main requirements are: first, the data acquisition, the higher is the core of the management system of data mining technology, the technology can't be in the absence of any data to support operation, therefore must implement the data collection, to construct the enterprise internal management of the large data volume to make the system work; Second, because of higher management system will inevitably take up its own network environment and the physical server, so the system will bring a database, the common management system is through the database to store data, and the database capacity is limited, internal management does not meet the requirement of data storage, therefore, how to expand the database capacity is the main problem in the system design, In the face of this problem, this paper will take its own database as a temporary database, ready to connect it with the cloud database to expand the capacity, how to achieve this is a big demand of the system design of this paper; Third, the database layout of the advanced management system makes the data mining technology must be connected with the cloud database, otherwise it cannot schedule all the data, but the technology itself does not belong to the advanced management system, how to integrate it into the system and accurately connect with the cloud database is another big requirement of the system design. In addition, the significance of the existence of advanced management system is to provide favorable help to the internal management of enterprises, so the results of system data mining must be fed back to the artificial, how to achieve this is also a big demand of system design, that is, how to ensure that the system will feed back the mining data to the artificial end, and ensure that the artificial can accurately read.

3.2 System Design Scheme

Combined with the relevant requirements in the design of advanced management system, the system design scheme can be roughly divided into four steps, the specific content of each step is as follows.

Data Acquisition Design

Selection of data collection method: The data collection of advanced management system is special, because the collection object is people, not things, so it cannot use sensors and other technologies to collect relevant data, so we must find another method. From this perspective, this paper chose two methods of data collection, one is the artificial collection report type, the method is mainly dominated by the enterprise internal management personnel, before data collection according to the data acquisition need to design a report, and then send the report to all the management object, let them according to the reality to fill in report, will report after the completion of play, So that you can collect relevant data; Secondly, logging acquisition, the main functions and internal administrative and business system, can automatically record all user behavior records in the two systems, for example, a certain one point when staff login, record the account with the corresponding identity, the staff login access to what information, deal with what business, launched a what kind of work, etc. [8, 9]. The two data acquisition methods have their own defects, but the combination of them tends to be perfect, which can guarantee the integrity of data collection. Figure 2 shows the operation flow of the two data acquisition methods.

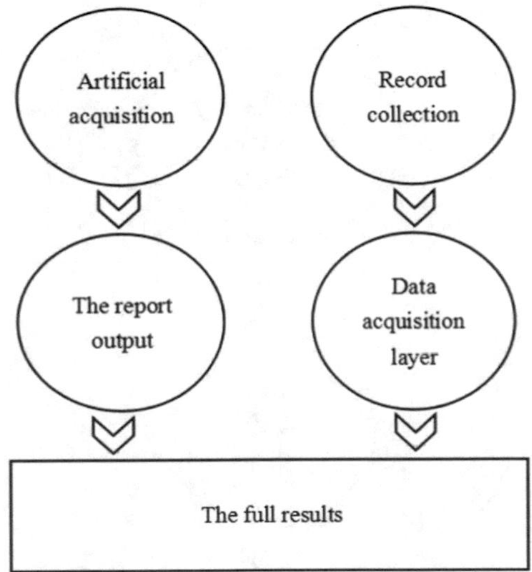

Fig. 2. Operation flow of the two data acquisition methods

Database Design

As mentioned in the general idea, the paper regards the database system at higher management as a temporary database, a cloud of the database and database connections at the same time, the reason lies in cloud data storage space of the database in the network, the network space is infinite, so the cloud database storage capacity is infinite, fully meet the demand of internal management large data storage, so this article idea is feasible. But the core of the database design requirements is to make the temporary database and cloud database connection, this paper carried out design, namely the basic framework of private cloud build a database of cloud, and the statistics of the internal management of big data growth, the results show that the internal management big data about every minute growth articles 1000 to 1200, for the data conversion capacity, which is about 3 gb, On this basis, the temporary database imports data to the cloud database every three days as a cycle. The initial capacity of the cloud database of the private cloud is 3 GB × 3D = 9 GB. Considering the data fluctuation, this paper continues to expand the cloud resource scale on the basis of 9GB, and finally the initial capacity of the private cloud is 15GB, which can meet the requirements of internal management data. And then on a firm's internal LAN (because it is a private cloud, so you can use the LAN, if public clouds are need to use the Internet, but it may increase the risk of a security), will the temporary database to connect to the database of cloud, a database model for temporary data storage time limit set, namely all data priority into the temporary database, storage time limit for three days, All expired data will be stored in the cloud database, or when the temporary database capacity reaches more than 90%, all data will be automatically imported into the cloud database for receiving large data flows. In addition, there are certain security risks in the use of cloud data, so encapsulation design is adopted in this paper. Private cloud is encapsulated and user authentication port is left. All users who

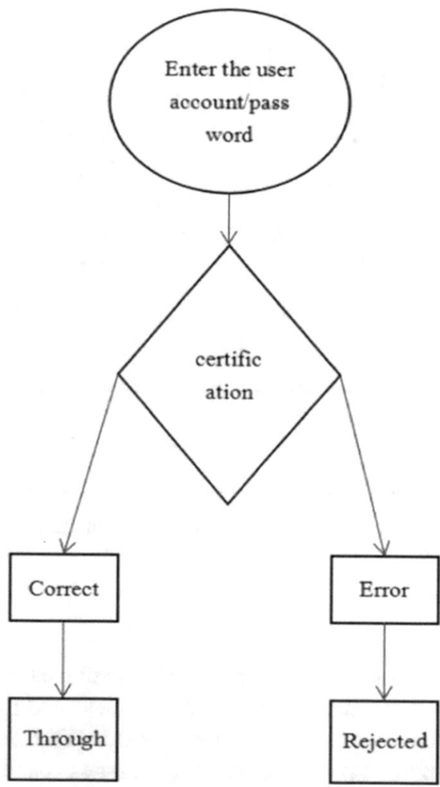

Fig. 3. User authentication process

want to access the cloud database must log in in advance, otherwise they cannot view the data [10]. Figure 3 shows the user authentication process. Figure 4 shows the cloud database security risk protection standard scheme.

Data storage space		
The external network	A firewall	The internal network

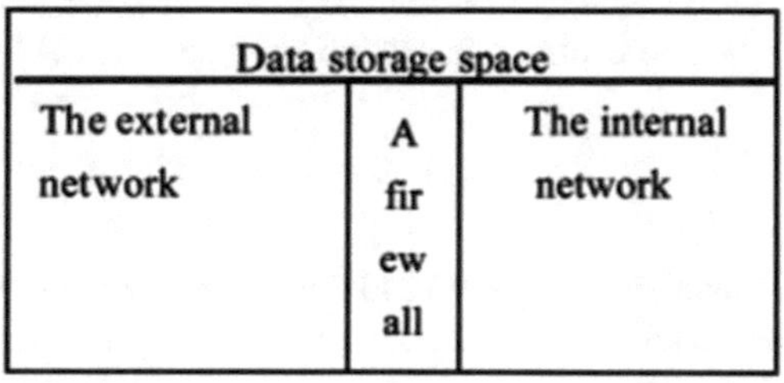

Fig. 4. Cloud database security risk protection standard scheme

Data Mining Module Design

First of all, in order to integrate data mining technology into advanced management system, this paper adopts the modularization method to design the corresponding function module of the technology. The module can be embedded into the advanced management system and become the core function of the system. Second data mining module cannot be arranged in higher management system at random, also must be with the cloud database connection, to the point of this article is not without using external development data mining module, but on the basis of private cloud cloud resources from the independent development of the logic of the data mining technology, and through the programming language packages its success can module, Since this functional module is developed on the private cloud side, it can naturally connect to the cloud database [11, 12].

Human-Computer Interaction Design

The significance of human-computer interaction is to ensure that the system can be controlled by people, otherwise the system will lose its construction significance. From this point of view, this paper carries out human-computer interaction design for modern popular smart phone terminals. In the process, it mainly develops enterprise internal management APP. The network protocol adopted by APP is Internet, so as long as a human can log in to the Internet, he can enter the APP to operate the advanced management system. In addition, considering the APP network protocol is the Internet, the external personnel may use APP invasion of enterprise internal LAN, data security impact, this paper divides the firewall on the border of enterprise internal LAN, any not firewall authentication access request will be rejected, but also to develop the function of APP user login, dual protection of information security.

3.3 System Test

Adopt the above method to design the advanced management system, input 2 million data to it, all data come from the internal management work of an enterprise in previous years, and through the analysis of the internal management status of the enterprise, obtain the best decision-making direction of the development of internal management in advance. On this basis, the advanced management system has carried on the mining to the data, and the output management decision suggestion, compared with the best decision direction, the system output management decision is consistent with the direction, and more in-depth, so the system is effective, design method is feasible.

4 Conclusion

In conclusion, higher management system and the common management system, there's a big difference compared with the former obvious advantages, so it is a new direction of the enterprise internal management, so the companies need to fully understand the data mining technology, using the technology to construct advanced management system, it can improve accuracy of the internal management by way of data mining, It is conducive to the core competitiveness of enterprises and sustainable development.

References

1. Skormin, V.A., Gorodetski, V.I., Popyack, L.J.: Data mining technology for failure prognostic of avionics. Aerosp. Electron. Syst. IEEE Trans. **38**(2), 388–403 (2002)
2. Tavani, H.T.: Genomic research and data-mining technology: Implications for personal privacy and informed consent. Ethics Inf. Technol. **6**(1), 15–28 (2004)
3. Jones, J.K.: The role of data mining technology in the identification of signals of possible adverse drug reactions: value and limitations. Curr. Therap. Res. **62**(9), 664–672 (2001)
4. Delavari, N., Shirazi, M., Beikzadeh, M.R.: A new model for using data mining technology in higher educational systems. In: Proceedings of the Fifth International Conference on Information Technology Based Higher Education and Training, 2004 (ITHET 2004). IEEE (2004)
5. Edwards, M., Rashid, A., Rayson, P.: A systematic survey of online data mining technology intended for law enforcement. ACM Comput. Surv. **48**(1), 1–54 (2015)
6. Bunce, P., Fraser, R., Woodcock, L.: Advanced budgeting: A journey to advanced management systems. Manag. Account. Res. **6**(3), 253–265 (1995)
7. Meliopoulos, A.P.S., Polymeneas, E., Tan, Z., et al.: Advanced distribution management system. IEEE Trans. Smart Grid **4**(4), 2109–2117 (2013)
8. Newing, R.: Advanced budgeting requires an advanced management system. Manag. Account. Magaz. Charter. Manag. Acc. **72**, 28 (1994)
9. Collet, A., Crebier, J.C., Chureau, A.: Multi-cell battery emulator for advanced battery management system benchmarking. In: 2011 IEEE International Symposium on Industrial Electronics, Gdansk, Poland, 2011.
10. Avazov, A.R., Sobinova, L.A., Kuznetsov, G.V., et al.: Advanced distribution management system. EPJ Web Conf. **110**, 01004 (2016)
11. Li, Y., Shi, L., Zhai, P., et al.: Design of distributed temperature and humidity monitoring system based on single-chip microcomputer. In: 2021 IEEE 4th Advanced Information Management, Communicates, Electronic and Automation Control Conference (IMCEC)
12. Chen, M.-C., Cheng, Y.-T., Siang, C.-Y.: Development of inventory management system based on radio frequency identification technology. Sens. Mater. **34**(3), 1163 (2022)

Research on Transaction Mode and Dynamic Pricing Model of Big Data Service Based on Sequential Recommendation Algorithm

Honggang Wang(✉), Wen Liu, and Yirong Wang

Big Data Center, State Grid Corporation of China, Beijing 100052, China
jackdlc@163.com

Abstract. Big data technology has not only become an important driving force of industry digitalization and intelligence, but also a fundamental source of innovation and development. Through the study of pricing model, the application of big data technology in service trading can be more standardized and orderly. As a special data form in the recommendation field, traditional recommendation methods are often difficult to model session sequence data. Although some scholars have improved the traditional recommendation method to solve the problem of session sequence modeling, the effectiveness of the recommendation system is limited by many problems in the model structure. In recent years, with deep learning in image recognition, natural language processing, speech recognition and other breakthroughs in the field of artificial intelligence, will study deep into the recommendation algorithm, can help to effectively solve the traditional recommendation algorithm of multi-source heterogeneous information, data sparse, cold start, data characteristics of the design of artificial rely on such issues, in this case, The results of traditional recommendation methods are often inaccurate, and the results are often lagging and repeatable.

Keywords: Big data services · Sequential recommendation algorithm · Pricing model

1 Introduction

Since the Internet era, the application of information technology has made the circulation and transaction of big data more extensive. Through the analysis, management and optimization of data from different sources, it can not only create more diversified economic and social values, but also become an important information reference for the operation and development of enterprises. Therefore, in order to further strengthen industrial innovation, improve the industrial value created by big data technology, clarify good and effective service mode, and carry out dynamic pricing according to the actual situation have become the key application direction of big data technology at the present stage. With the all-round development of industrial technology, large data service enterprises how to better play the role of supporting data in innovation alliance, how to integrate industry chain development process of big data information generated by the effective

J. H. Abawajy et al. (Eds.): ICATCI 2022, LNDECT 170, pp. 993–1001, 2023.
https://doi.org/10.1007/978-3-031-29097-8_118

integration and management, so as to further control the transaction cost, shorten the trading cycle, promote the accelerated flow of data in the trading environment, Realizing the effective embodiment of the value of big data and promoting the further development of industrial technological innovation has become an important challenge facing the big data industry at this stage [1–3].

Because in information market, influenced by both parties all kinds of objective factors, so the need to implement dynamic pricing strategy, and in the case of network anomaly in a more effective means to control the price, so that the network abnormal phenomenon is more rapid and timely feedback, at the same time so as to further perfect the entire transaction process. However, following the abnormal condition of the network to adjust the price will make the transaction costs continue to rise. Therefore, relevant enterprises and personnel should carry out an effective balance between network anomalies and the formulation of dynamic pricing schemes, and select their main tendencies according to the actual situation, so as to lay a foundation for further strengthening fair pricing and reducing losses caused by network anomalies or fluctuations [4, 5].

Generally speaking, dynamic pricing model refers to the timely adjustment and optimization of pricing model following network fluctuations. In order to solve the problems in the Internet, the current dynamic pricing model should follow the following principles.

(1) Under the influence of transaction type, transaction demand and other factors, the allocation and clarity of network resources also includes differentiated prices.
(2) The operating state of each network node will change with the change of demand, so the resource allocation price should also reflect this state adjustment. Moreover, the demand for network resources in the next stage is restricted to some extent. When the information network load is small, the price of the network should be constantly decreased to stimulate users' demand for the network, while when the information network load is high, the price should be increased according to the demand to control users' demand.
(3) As the information network is composed of nodes in different regions, the price of a regional network should be controlled by the information of the local network usage status, rather than the adjustment of the whole network information.
(4) Simplicity is another principle of dynamic model architecture, which can not only effectively improve the feasibility of pricing, but also control the cost of pricing and make transaction activities more smooth [6, 7].

2 Basic Concepts

Big data is an integration of huge data, and it has good application value, but its value cannot be brought into play actively, and it needs special processing to benefit from it. On this basis, traditional technical means can complete big data processing in colleges and universities, but cannot analyze reliable results from big data, while manual means can analyze reliable results, but cannot process big data efficiently, which leads to big data being difficult to use in pricing. In the face of this situation, relevant fields have introduced a special technology for processing big data through research, known as big data technology.

Big data technology is a technology specially used for deep processing of big data, that is, big data is a huge data integration. Internal data is not only of huge magnitude, but also of various types, and each item of data is representative of real things. Therefore, if you can arrange all the data in order, you can better explain the real things through the data and help people understand the real things more deeply, Or predict the development of things. However, big data is just a data integration body and does not have an active role, so it can only be used after being processed. However, due to the huge quantity and various data types of big data, it can not be processed by manual and conventional technical means. Therefore, a kind of technical tools called big data technology has been born, of which the most representative is AI technology, This technology has machine logic similar to that of human beings. It can process data in a way similar to human thinking, such as extracting features of one-way data, integrating data according to feature definitions, establishing data relationship models, and exposing the correlation between data. Based on this result, people can make accurate judgments or predictions about things, which shows that the depth of data processing of this technology is close to that of human beings, It is different from conventional technical means. At the same time, as a technical tool, this technology does not have the limitations of human capabilities. It can operate at a high speed and stably all day long, and can process a large amount of data in a short time. This shows that this technology has more advantages than humans in terms of data processing efficiency. Artificial intelligence technology is similar to human machine logic, which is derived from artificial neural network. It is a model created by imitating human neural structure. The basic structure can be divided into three parts: input layer, hidden layer and output layer. The input layer and hidden layer also contain several neurons. Each neuron represents a piece of data, and the data entering the input layer is called input neuron, The data entering the hidden layer is called hidden neuron (the neuron can be regarded as the data node of the model). Each neuron can get the result through interaction, and the result will be displayed through the output layer. Figure 1 Big data formation mechanism, Fig. 2 big data application premise.

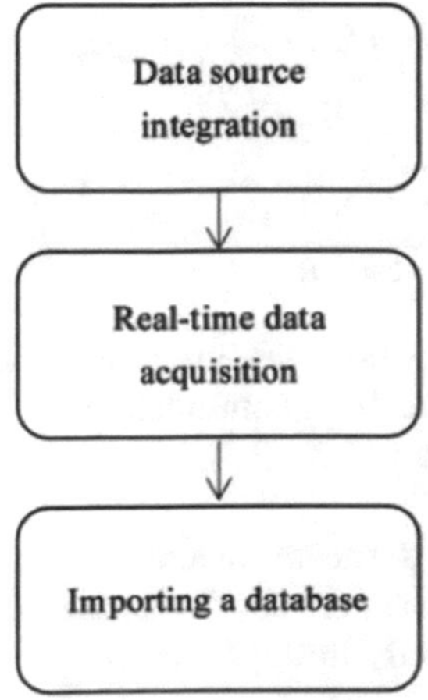

Fig. 1. Big data formation mechanism

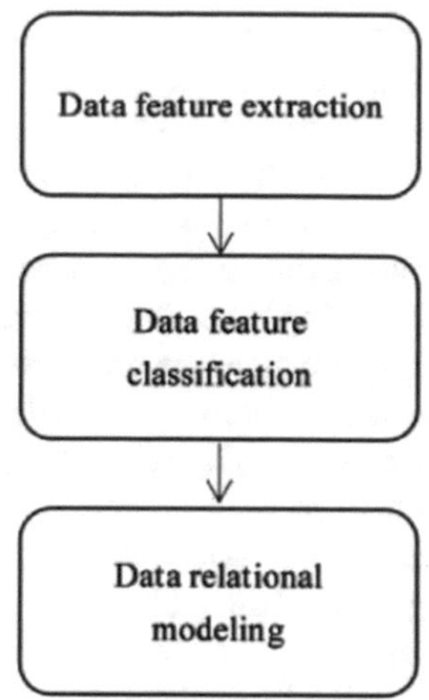

Fig. 2. Big data application premise

3 Related Work

3.1 Sequence Recommendation Algorithm

For many recommendation algorithm problems, a single recommendation method is often difficult to implement because of its own shortcomings. In order to solve these problems, hybrid recommendation method has attracted more and more attention. Through the integration and integration of differentiated recommendation algorithms, users can be provided with more diversified recommendations and choices, so that the recommendation effect is more significant.

$$O_v = \sum_{u \in N[v]} w_{u,v} x_u \tag{1}$$

$$O_{v,j} = h\left(\sum_{i=1}^{f_{k-1}} \sum_{u \in N[v]} w_{i,j,u,v} x_{u,i}\right), (j = 1, \ldots, f_k) \tag{2}$$

In many hybrid recommendation systems, collaborative filtering recommendation and content recommendation are often combined in the following ways to improve the effect of recommendation system M-2:

(1) Use various information retrieval methods and other content-based analysis methods to learn and maintain the user's resume information;
(2) Identify similar users directly through their resume information and recommend them through collaborative filtering.

This means that recommendations can be made to users, whether based on the user's resume or similar users' resume. Furthermore, Shoham argues that the content-based and collaborative approach can be seen as a special case of the hybrid approach. If in the content-based recommendation approach, the content analysis component does not extract any features (such as keywords) from the item, then the hybrid approach is simplified to pure collaborative filtering recommendation. If there is only one user, this hybrid approach simplifies to purely content-based recommendations. Several content-based elements have been added to the collaborative filtering framework. Some elements of collaborative filtering are combined into content-based recommendation framework by using latent semantic indexing technology [8–10].

3.2 SMP Pricing Model

SMP was one of the earliest pricing models for web companies. The idea is to charge for a single packet, depending on how congested the network is, and the price of each packet adjusts over time.

In the process of pricing network resources, differentiated and flexible pricing will generate additional social costs for users. Its social cost includes not only the cost of data packet transmission under normal network conditions, but also the use cost when the information network load is serious. This cost is generated by auction. Each packet of a user is associated with a price, and the price parameters are carried in the packet header. This pricing scheme applies only to the user group that can receive the transmission fee. In short, congestion prices tend to have a fixed value because it is difficult to get effective adjustments in time when networks are congested. Once the user carries out self-transmission, its cost is higher than the current congestion fixed price, the user chooses transmission. Through big data technology, price information of each user can be collected and analyzed, and the threshold can be effectively defined through price information. When the threshold is exceeded, packets are transmitted. Among them, the main influencing factors of threshold include network load and congestion cost of information network. Each packet transmitted is paid on the basis of its marginal congestion cost and not on the basis of its bid fee. The principle of SMP is shown in Fig. 1. The cost of packets transmitted in the information network is mainly the congestion cost during the use of the information network, not determined by the amount of bidding. In summary, the main elements of the SMP pricing model are shown in Fig. 3.

Although SMP pricing model can determine the cost and price of packet transmission according to network congestion and other abnormal conditions, due to its poor connection with current pricing methods, hardware and transport protocol must be effectively optimized to effectively intervene in the current network system. This kind of spending also needs to be taken into account when incorporating SMP pricing models. In addition, pricing information network based on SMP model also requires the real-time information of each user to be transmitted and managed, which also increases transaction costs. During periods of congestion, it may even be necessary to use auction mechanisms. Therefore, it is difficult to provide effective service for network users, and even cannot provide transmission guarantee. The price problem of SMP model makes it difficult to be widely adopted among low-income groups, so it needs universal supervision.

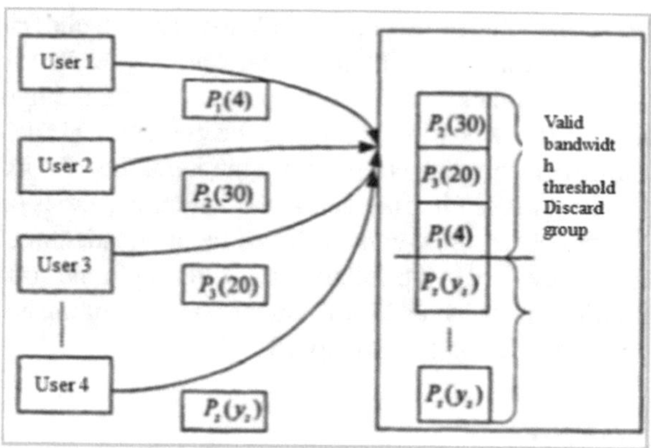

Fig. 3. Schematic diagram of SMP

4 General Framework of Dynamic Pricing Model

Figure 4 illustrates the general structure of the overall framework of the dynamic pricing model. For the sake of description, we call this implementation framework DDPF (Distributed Dynamic Pricing Framework). The Ddpf considers each router as a part of the network node, and requires each router to take responsibility for adjusting the price of network use. In order to further enhance the efficiency of time management, agents can be added between workstations and clients to reduce the response time to network fluctuations and price changes. In the direction of the client, the form of agency by agreement for a long time to negotiate with customers and users to be able to make the network resources on the basis of differentiated service types and their selected business needs of the overall service content, combined with ISP transmission price and all levels of the use of information, will give top priority to the user experience, combined with the actual situation on the network data transmission efficiency optimization. In the current network environment, relevant personnel can realize real-time monitoring of the current network environment through big data technology, and the workstation can analyze and verify the network congestion, and update the resource price within a certain price cycle. Although the price may change multiple times, for the user, it only shows the price when the next protocol cycle arrives.

Actual user requirements may include application directions in different fields, including instant messaging, teleconference, audio and video playback, and so on. In order to strengthen the "cost-performance ratio" of network resource services as much as possible, users can modify related quality parameters according to their predetermined costs, so as to balance the network use cost and their own use requirements. The current user utility maximization problem can be expressed as follows:

$$\max \sum_{I} \left[U^I(X^I) - C^I(X^I) \right] \tag{3}$$

After entering the era of big data, the research scope of information network resources has been expanding, and utility measurement has also been introduced into the field

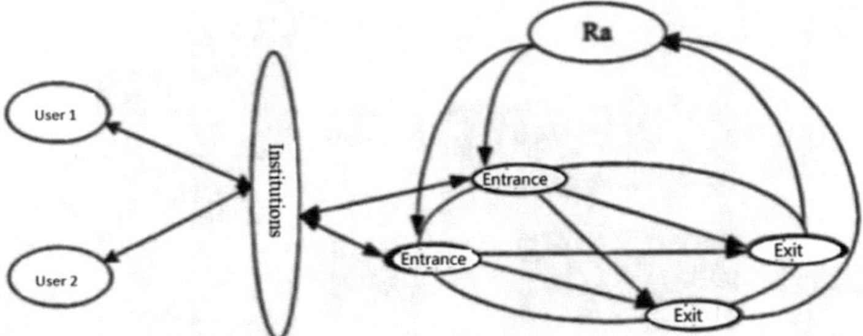

Fig. 4. General framework of dynamic pricing model

of network. This concept covers users' satisfaction with the performance of network resources, and can fundamentally express users' quality of service requirements, so that users' acceptance of network resources can be further quantified. Because network services need to be user-oriented, user satisfaction has an important impact on network services, depending on the type of particular user. In order to further strengthen the monitoring of differentiated network programs, it is very important to study the effectiveness of network.

5 Big Data Pricing Preprocessing of the Recommendation Algorithm

In the deep learning model, data is divided into unified batches for targeted training as required, and the size of data batches plays an important role in the quantity of model training and the final training results. Select the right batch to achieve synchronous improvement of efficiency and capacity.

According to the characteristics of user click sequence, it effectively avoids the phenomenon of modeling difficulty caused by the great difference of sequence. This method decomposes the data in the sequence, taking the last click as the input and the next click as the output, and analyzes the correlation of the data by the relationship between the two. Sets the stop symbol based on the length of the sequence. The data parallel sampling method is shown in Fig. 5.

In order to improve the calculation speed and reduce the calculation cost, negative sampling is carried out on the output results to avoid model calculation and scoring of all commodities. This sampling method not only reduces the size of the positive sample, but also allows the recommendation model to score only a small number of items. In addition to calculating the output of positive samples, the negative sample model also needs to calculate the scores of some negative samples. The negative sample sampling method makes the model have to further modify the weight to improve the output ranking of the positive sample.

User click sequence data is different from other data. There is neither user's characteristic information nor commodity's characteristic information in the data. We can only get the order in which a user clicks items in each shopping event from the click order of the user. Therefore, when using user click order for prediction, we can only rely on the correlation between items for modeling.

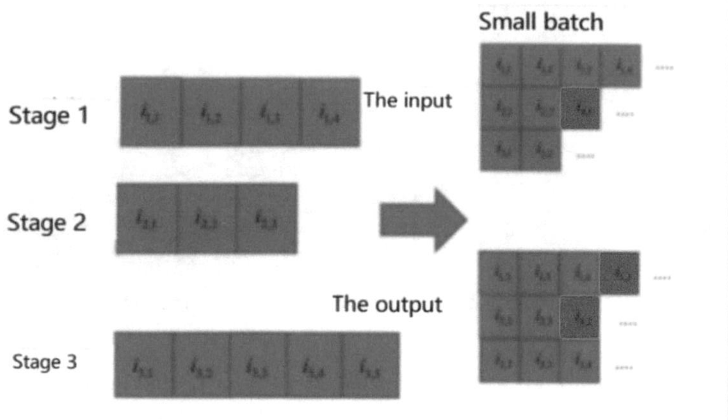

Fig. 5. The data parallel sampling method

6 Conclusion

In the current stage of enterprise business activities, promoting the continuous development of big data technology can realize the effective integration of internal resources and realize the effective embodiment of value. In order to further promote industrial innovation, especially using pricing mechanism to control user behavior becomes an effective method to rationally and efficiently utilize network resources and control network congestion. At present, flat pricing schemes still occupy the main market, but will be phased out due to their low efficiency and inability to provide QoS guarantee. In theory, dynamic pricing scheme can better control congestion and provide QoS guarantee, but there are still implementation problems. The continuous development of network and the continuous improvement of service level and type requirements continue to put forward new challenges to pricing.

References

1. Feng, G., Jian, W.: Research on the innovation support ability of scientific and technological information resources under the background of big data. J. Inf. Sci. Technol. **37**(10), 183–188 (2018)
2. Wu, Y.: Collaborative innovation of Big data in The United States and its implications. **2019**(4), 168–174 (2021)
3. Vasilyev, S., Levichev, I., Acosta, D.P.: Dynamic pricing model for 5G/6G telecommunications. In: ICUMT, pp. 85–90 (2021)
4. Garmabdari, R., Moghimi, M., Lu, J.: Model predictive energy management system in presence of dynamic pricing. In: ISGT Asia, pp. 1–5 (2021)
5. Xiao, X.-C., Zheng, X.-W., Wei, Y., Cui, X.: A virtual network resource allocation model based on dynamic resource pricing. IEEE Access **8**, 160414–160426 (2020)
6. Kandpal, M., Ashesh Patel, K.: Evaluation of the dynamic pricing model in cloud. Int. J. Secur. Priv. Pervasive Comput. **12**(3), 20–29 (2020)

7. Nambiar, M., Simchi-Levi, D., Wang, H.: Dynamic learning and pricing with model misspecification. Manag. Sci. **65**(11), 4980–5000 (2019)
8. Wang, X., Zhu, Z., Yu, J., Zhu, R., Li, D., Guo, Q.: A learning resource recommendation algorithm based on online learning sequential behavior. Int. J. Wavelets Multiresolut. Inf. Process. **17**(2), 1940001:1-1940001:17 (2019)
9. Tsai, C.-Y., Lai, B.-H.: A location-item-time sequential pattern mining algorithm for route recommendation. Knowl. Based Syst. **73**, 97–110 (2015)
10. Fang, H., Zhang, D., Shu, Y., Guo, G.: Deep learning for sequential recommendation: Algorithms, influential factors, and evaluations. ACM Trans. Inf. Syst. **39**(1), 10:1-10:42 (2020)

Design and Application of Automatic Food Packaging Machinery Control System

Xin Lu(✉)

Southwest University, Chongqing 400700, China
1105777632@qq.com

Abstract. In order to control the automatic food packaging machinery reasonably, this paper will carry out relevant research on its control system. Firstly, the characteristics and control requirements of automatic food packaging machinery are introduced, and then the control system is designed according to the control requirements. Through this study, using the correct application strategy can give full play to the role of the control system, and the control system can fully meet the control requirements of automatic food packaging machinery.

Keywords: Automation · Food packaging machinery · Control system

1 Introduction

With the industrial field into the era of automation, there are a variety of automatic machinery and equipment on the market, automatic food packaging machinery is one of the most representative, has been very popular in food related industrial production. But with the mechanical equipment investment, people find themselves, undertake the task of mechanical equipment control is more and more heavy, artificial limits on their own capabilities at the same time also more obvious, so the artificial mode of quality control of food packaging machinery automation gradually low, does not meet the current needs, but it also makes people puts forward new demands to the technology field, the hope can through technical means to control the automation of packaging machinery, it can use technical means to avoid artificial limitations of their own capabilities, improve the efficiency of mechanical equipment operation, ensuring the operation stability, and to reduce artificial burden, reduce the enterprise in the labor cost of mechanical equipment in the control input. In this context, the relevant fields in-depth analysis of automatic food packaging machinery control requirements, proposed the corresponding control system, the system can meet people's demands, so it has good application value.

However, it should be noted that there are differences in the control requirements of automatic food packaging machinery in different situations, so there is no standard formula for the control system, which needs to be designed according to the actual situation. Therefore, how to design the control system is still an urgent problem to be solved, and it is necessary to carry out relevant research on this problem.

The design and research work of automatic food packaging machine has been carried out in relevant fields for a long time. At the beginning, people thought that the design

J. H. Abawajy et al. (Eds.): ICATCI 2022, LNDECT 170, pp. 1002–1010, 2023.
https://doi.org/10.1007/978-3-031-29097-8_120

of the mechanical equipment was not complicated, and it was only necessary to ensure that the equipment could complete automatic packaging. However, with the deepening of the research, people found that the design of the mechanical equipment was difficult, mainly reflected in two aspects: first, the most ideal automatic food packaging machine in theory is a kind of automatic food packaging machine that can achieve a certain volume The equipment for automatic packaging of different foods within the shape range, but in fact, even if there are scope restrictions, the automatic food packaging machine cannot package different foods. The reason is that the automatic food packaging needs to transport the food to the packaging point first, and then packaging. During the transportation process, some foods may change in volume and status due to physical factors, For example, the food originally concentrated together will be dispersed under the influence of vibration force, and the volume will become larger, resulting in the inability to package. In the face of this situation, people have to classify all kinds of food separately and design food automatic packaging machines for specific purposes. This phenomenon has continued to this day, which is a difficult problem to be overcome in the design theory of this mechanical equipment. This research will try to challenge this, at least to ensure that the design results can package more than three kinds of food with the volume and shape within the range; Secondly, because it is an automatic mechanical equipment, the automatic food packaging machine must be operated in strict accordance with the automatic control logic specification. For example, the opening of the packaging device must be carried out when there is no food in the packaging station, otherwise the food cannot enter the packaging station and naturally cannot be packaged. At the same time, because it is an assembly line operation, this situation represents that food accumulation may occur on the assembly line, As a result, subsequent work cannot be carried out. In view of this situation, how to make the automatic food packaging machine recognize the situation of the station and then make the correct action has become a major problem in the design, which is also a problem to be solved in this paper.

In view of the above problems, modern research has given some methods on the theoretical level, and this paper will refer to these methods for design.

2 Current Theoretical Situation

The design and research work of automatic food packaging machine has been carried out in relevant fields for a long time. At the beginning, people thought that the design of the mechanical equipment was not complicated, and it was only necessary to ensure that the equipment could complete automatic packaging. However, with the deepening of the research, people found that the design of the mechanical equipment was difficult, mainly reflected in two aspects: first, the most ideal automatic food packaging machine in theory is a kind of automatic food packaging machine that can achieve a certain volume The equipment for automatic packaging of different foods within the shape range, but in fact, even if there are scope restrictions, the automatic food packaging machine cannot package different foods. The reason is that the automatic food packaging needs to transport the food to the packaging point first, and then packaging. During the transportation process, some foods may change in volume and status due to physical factors, For example, the food originally concentrated together will be dispersed under the influence of vibration force, and the volume will become larger, resulting in the inability to package. In the face of this situation, people have to classify all kinds of food separately and design food automatic packaging machines for specific purposes. This phenomenon has continued to this day, which is a difficult problem to be overcome in the design theory of this mechanical equipment. This research will try to challenge this, at least to ensure that the design results can package more than three kinds of food with the volume and shape within the range; Secondly, because it is an automatic mechanical equipment, the automatic food packaging machine must be operated in strict accordance with the automatic control logic specification. For example, the opening of the packaging device must be carried out when there is no food in the packaging station, otherwise the food cannot enter the packaging station and naturally cannot be packaged. At the same time, because it is an assembly line operation, this situation represents that food accumulation may occur on the assembly line, As a result, subsequent work cannot be carried out. In view of this situation, how to make the automatic food packaging machine recognize the situation of the station and then make the correct action has become a major problem in the design, which is also a problem to be solved in this paper.

In view of the above problems, modern research has given some methods on the theoretical level, and this paper will refer to these methods for design.

3 Characteristics and Control Requirements of Automatic Food Packaging Machinery

3.1 The Characteristics of

Automatic food packaging machinery is a combined mechanical equipment, mainly composed of three parts, as shown in Fig. 1.

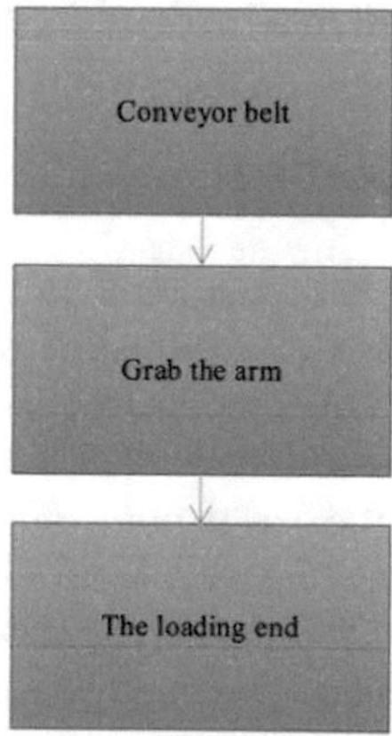

Fig. 1. Composition of automatic food packaging machinery

According to Fig. 1, the automation of packaging machinery by the conveyor belt, grab the arm, loading side of three parts, the conveyor belt is mainly responsible for the transport unpackaged foods to the station, and then wait for grab, grab arm after grasping arms will grab food will be transferred to the loading end, at this time to grab the arm will not loosen, the purpose is to continue to control not packaged foods, So that it can be smoothly into the loading position of the loading end. When the unpackaged food is sent into the loading position, the grab arm will be released, and the loading position will move forward and enter the pre-placed unpackaged packaging bag. After completion, the base of the loading position will exit, and the upper packaging device will start to seal the packaging bag to complete packaging. It is worth mentioning that the conveyor belt of automatic food packaging machinery runs through the whole process, so when the bag is sealed, it will continue to move to the collection point through the conveyor belt, and the single food packaging process is completed at this stage [1–3].

3.2 Control Requirements

Combined with the characteristics of automatic food packaging machinery, the main control requirements in operation are: first, all the components of automatic food packaging machinery are controlled, so that the components of the routine operation; Second, the control of each component, so that the routine operation of each component in line with the requirements of the process (conveying → grasping → loading); Third, the grasping arm is controlled to ensure that it is only released after the unpackaged food enters the loading position, and its movement track is controlled at the same time; Fourth, when any component of automatic food packaging machinery is in an irregular state, special control is carried out on other components, such as conveyor belt stagnation due to failure, so that other components of the same stagnation through control, so as not to cause losses [4, 5].

4 Control System Design

4.1 Control System Framework

According to the ideas of this paper, the basic framework of automatic food packaging machinery control system is shown in Fig. 2.

Combined with Fig. 2, the first by controlling terminal sends a general instructions to the control bus instructions through the bus distributed to the unit, unit number is three, respectively with three part of the automation of packaging machinery connection, so can get correct instruction, let each part of the to perform normal operation, at the same time command sent from the bus to the different unit sequence is different, the ability to move each component according to a standard flow allows a series of routine operations to be performed. The second consider the automation of packaging machinery in continuous operation may malfunction, normal operation was performed on each component of movement, can get the automation of packaging machinery through information collection side of state information, state information will continue to transfer back to the controlling terminal, so control terminal can judge whether the automation of packaging machinery in the normal state, If any abnormal will confirm the fault point according to the information source, and change the instruction of the next cycle, let the other part of

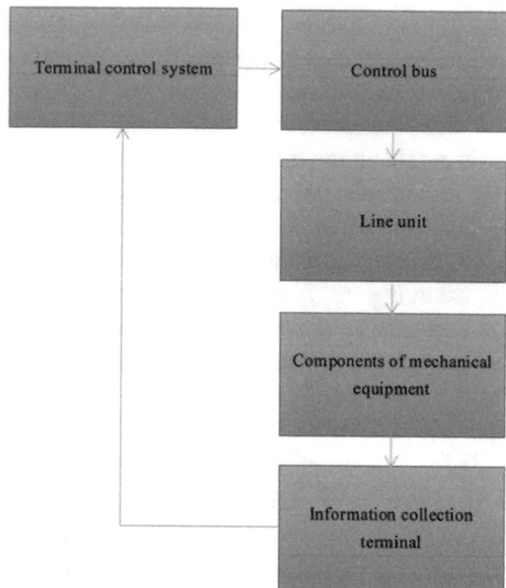

Fig. 2. Basic framework of automatic food packaging machinery control system

stagnation, notify the artificial to deal with at the same time, in the process of information control terminal will constantly monitor, so when after artificial processing fault, terminal can pass the information to confirm the automation of packaging machinery back to normal, then continue to issue the usual operation instructions [6].

4.2 Implementation Method

According to the control framework, the concrete implementation method of the control system will be discussed below.

Implementation Method of Control Terminal

The control terminal is the core of the whole control system, with the function of connecting the preceding and the following, so the first step of the system is to realize the control terminal. Considering the actual control demand, control terminals must not only be responsible for routine operating instructions, and information recognition ability, such ability according to the actual situation to adjust instruction, to control the automation of packaging machinery operation, so the control terminal can be done by artificial intelligence technology, the traditional control terminal intelligent general instructions, not to identify information, the reason is that without the support of artificial intelligence technology, and the control terminal is called automatic control terminal, now does not meet the control requirements, and artificial intelligence techniques neural network can solve this problem, makes the control terminal information recognition ability, this kind of control terminal is called intelligent control terminal, more in line with modern requirements. On this basis, the key step to realize the intelligent control terminal is to construct a neural network, for which feedforward neural network is selected in this paper, and the specific model is shown in Fig. 3 [7–9]. In addition, the feedback network can also be considered, and its structure is shown in Fig. 4.

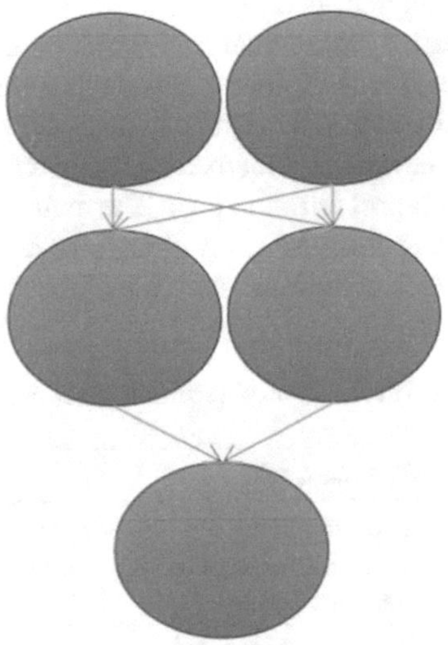

Fig. 3. Feedforward neural network model

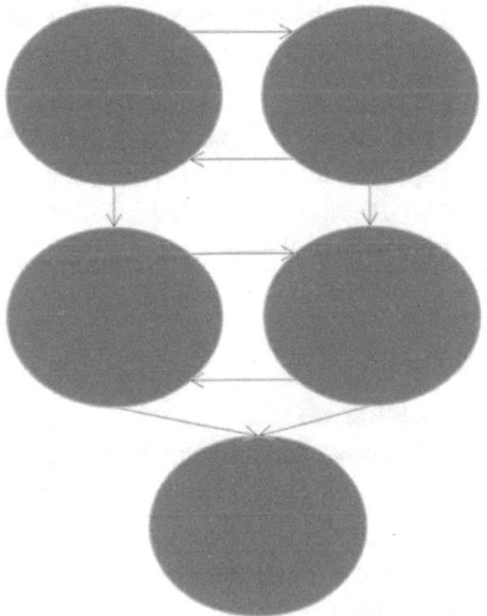

Fig. 4. Feedback neural network structure

Combined with Fig. 3, the feedforward neural network is divided into three layers, namely, input layer, hidden layer and output layer. The input layer is the uppermost layer and contains several neurons. Each neuron represents a kind of information and transmits forward, and then enters the hidden layer. The hidden layer is the middle layer, which is a kind of information interaction rule. In the hidden layer, each neuron in the input layer will interact with each other. After the interaction, the relevant relation between the information can be obtained, such as the correlation degree and matching degree between the information, according to these connotation information, the actual situation can be accurately judged. Finally, the output layer, whose main function is to show the results of information interaction in the hidden layer, so as to make manual judgment. Based on the feedforward neural network model, the control terminal can recognize the information. In addition, in order to make the control terminal conform to the control requirements of automatic food packaging machinery, it is also necessary to develop the function of the control terminal [10]. The main technical means of this work is programming technology. In this paper, Java language is mainly selected to carry out the work and develop related functions, as shown in Table 1.

Table 1. Control terminal functions

The name of the function	The main role
General instruction storage	The routine instructions for the operation of automated food packaging machinery are stored and issued in accordance with standard timing
Sequential standard flow	Responsible for standardizing the timing of each instruction
Signal receiving function	Used to receive signals sent from the signal acquisition terminal
Signal interpretation function	Responsible for interpreting the signal sent by the signal acquisition terminal and obtaining corresponding information from it
Special instruction store	Automatic food packaging machinery control instructions for special situations are stored, and these special instructions will be issued when the terminal finds the machinery in a fault or other special state

Control Bus Implementation Method

To achieve control of the bus, because the actual work of control target is very complicated, and the number has a lot of, so if directly controlled by a single line, can lead to control the difficulty increases, at the same time is also very vulnerable to reality environment factors, there is a big difference to the quality of control, the control bus can solve this problem very well, It integrates all the targets with the same control requirements to form several split-line units, and then transmits the instructions to the corresponding units with specific lines. This control mode is called unit control mode, which has obvious advantages in convenience. From this point of view, this paper chose RS bus, its low cost, good efficiency, although it is not suitable for complex control work, but automatic food packaging machinery control work is not complex, so RS bus has a good applicability in automatic food packaging machinery control. In terms of bus layout, the large-capacity cable is firstly used for field layout, and then the split-line is installed on the cable to connect the related split-line units.

Implementation Method of Split Line Unit

Split line unit is to better docking with the control bus, but also to improve the quality and efficiency of automatic food packaging machinery control and set, if not to achieve the split line unit, the realization of the control bus has no meaning, automatic food packaging machinery control can not continue to carry out, so the work is very important. From this point of view, first of all, it is necessary to divide the components of automatic food packaging machinery according to the requirements. For example, all the conveyor belt is regarded as a unit. Because all the control objectives in the unit are the conveyor belt, the control instructions required are unified, so that the conveyor belt can be reasonably controlled. Other automatic food packaging machinery components of the unit division method is the same, not to be described here. Secondly, communication lines should be used to integrate each unit into THE RS bus. The communication line used in this paper is I/O line, which has high communication efficiency, good stability and good compatibility. Therefore, I/O line is selected in this paper.

Implementation Method of Information Collection End

Information gathering the implementation method is: first, according to the common faults of equipments choice sensor, common sensors such as temperature sensors and power sensors, the amplitude sensor, the sensor can collect food packaging machinery automation operation of all kinds of information, information is stored in the sensor's own electronic unit of waiting. In general, the selection of sensors must ensure that the type is complete, at the same time, when the conditions allow to choose the sensor with good performance as far as possible, otherwise there may be information misalignment problem; Second, to establish a transducer, between the sensor and the terminal is the main function of transducer will signal format is converted to a digital format, namely the sensors to collect the information stored will be the first time to the electronic unit, so the information will be processed into electronic format in electronic cell signal, the format signal cannot be read from a computer terminal, The transducer can convert the signal format into a digital format that can be read by the terminal computer, indicating that the transducer is an important part of the information acquisition terminal.

5 Conclusion

In summary, as an important production equipment in the current related fields, the operation of automatic food packaging machinery is directly related to the economic benefits of the enterprise organization, so it must be controlled reasonably. Based on this, the face of all the defects, artificial mode control design related control system is very important, the system can avoid artificial limitations, and security control efficiency and quality, still can cope with outside situation, makes the automation of packaging machinery can give full play to the role, provide powerful guarantee to the production work.

References

1. Rahman, M.I., Shahriar, S., Bhadra, A.B., et al.: Reduced-order robust nonlinear controller design for speed control of electro-mechanical system. In: IEEE Region 10 Symposium (TENSYMP). IEEE (2020)
2. Nol, J.P., Schoukens, M.: Cross-fertilising research in nonlinear system identification between the mechanical, control and machine learning fields: editorial statement. Mech. Syst. Signal Process. **130**, 213–220 (2019)
3. Sinitsa, M.O., Komshin, A.S.: Development of a control system for large-sized products to improve product quality in mechanical engineering. IOP Conf. Ser. Mater. Sci. Eng. **971**, 022057 (2020)
4. Vanin, V.A., Kolodin, A.N., Rodina, A.A.: Stepper hydraulic motor with pneumatic (jet) control system in machines with hydromechanical forming links. J. Phys: Conf. Ser. **1278**, 012019 (2019)
5. Bachi, I.O., Bahedh, A.S., Kheioon, I.A.: Design of control system for steel strip-rolling mill using NARMA-L2. J. Mech. Sci. Technol. **35**(4), 1429–1436 (2021). https://doi.org/10.1007/s12206-021-0308-7
6. Li, L., Aphale, S.S., Zhu, L.: High-bandwidth nanopositioning via active control of system resonance. Front. Mech. Eng. **16**(2), 331–339 (2021). https://doi.org/10.1007/s11465-020-0619-x
7. Acosta-Cano, J., Sastrón-Báguena, F.: Loose coupling based reference scheme for shop floor-control system/production-equipment integration - ScienceDirect. J. Appl. Res. Technol. **11**(3), 447–469 (2013)
8. Fujimoto, S.: Control system for controlling equipment provided inside a vehicle utilizing a speech recognition apparatus. Trans. Am. Climatol. Clin. Assoc. **37**(4), 142–146 (1994)
9. Asseo, E., Baribaud, G., Boutheon, M., et al.: Integration of existing dissimilar power supply controls into a standardized equipment-orientated control system. IEEE Trans. Nucl. Sci. **26**(3), 3319–3321 (2007)
10. Baek, J.H.: communication consecutive test of train oriented control system for wayside equipment control. Trans. Korean Inst. Electr. Eng. **63**(5), 690–699 (2014)

Instrumental Music Dissemination of Southwest Ethnic Minorities Based on Big Data Technology

Jiaqi Luo(✉)

Southwest University of Science and Technology, Mianyang 621010, China
4403049@qq.com

Abstract. In order to better spread southwest minority instrumental music, this paper will focus on the big data technology to carry out relevant research, mainly discuss the big data of southwest minority instrumental music communication ideas, then put forward the big data technology system framework, synchronous analysis of the system implementation method, and finally verify the system. Through verification and verification, the system gives let the value of big data play out effectively, can help southwest minority instrument Ftal music to find effective transmission path, promote the better transmission of instrumental music culture, at the same time, its role is long-term, can provide communication help in different times background.

Keywords: Big Data Technology · Southwest Minority Instrumental Music · Cultural transmission

1 Introduction

Under the influence of modern music aesthetics, the instrumental music of southwest ethnic minorities inevitably falls into the dilemma of cultural inheritance. From this point of view, because the musical art and cultural system represented by the southwest minority instrumental music is currently in the past, more and more young people no longer pay attention to and love the southwest minority instrumental music, so the continuous development may lead to the loss of the southwest minority instrumental music, which is an unacceptable loss for China. At the same time, previous studies have pointed out that the southwest minority instrumental music lacks the basis for cultural transmission and inheritance. The lack of this aspect not only involves the transmission and inheritance path, but also involves the cultural propaganda among the local people, which will lead to the failure of Hainan folk song resources to be well integrated into some high-quality transmission and inheritance paths. Faced with this situation, the academia believes that big data technology can be used to solve this problem. Therefore, in order to verify this point and better spread the instrumental music culture of southwest ethnic minorities, it is necessary to study it. In fact, people did not attach importance to technology in the early dissemination of southwest minority instrumental music works, and the combination of big data technology has not been formed, so the works mainly rely on manual work, such as manual search of media and platforms, and responsible

J. H. Abawajy et al. (Eds.): ICATCI 2022, LNDECT 170, pp. 1011–1019, 2023.
https://doi.org/10.1007/978-3-031-29097-8_121

for publishing information or other operations. However, until the defects of the manual mode are found, they began to try to use various technical means. Therefore, after the formation and popularization of big data technology, People began to think about "how to use big data to spread southwest ethnic instrumental music" and formed corresponding ideas. According to the current results, the main idea of using big data to spread the southwest minority instrumental music is to collect the preference data of user groups in the current mainstream information dissemination path. The data comes from the topics most often discussed by each user in the path platform, self selected tags and community organizations. Then, these data are classified according to the transmission path, and the user's personal preference is used for secondary classification to build a big data body. After completing the use of private data tools to analyze large amounts of data, mining and understanding the key elements of user groups such as national culture, instrumental music and other relevant user groups, if the path is very suitable for the dissemination of southwest national instrumental music, so as to find an effective path [1–3].

2 Inheritance Value and Ideas

Southwest China is a multi-ethnic region. Different ethnic minorities get along well with each other locally, and have merged with each other through the precipitation of history. This makes the local instrumental music culture very rich and colorful, that is, different nationalities have their own unique music culture, and the multi-ethnic characteristics naturally bring different ethnic instrumental music. At the same time, in the long history of cultural exchanges, the instrumental music of different nationalities, like a hundred flowers blooming, competed for excellence, also gave birth to some instrumental music with different national cultural characteristics, and gradually formed a modern cultural system of southwest instrumental music. There are rich instrumental music resources in the system, which are closely related to each other and have strong national cultural characteristics. Therefore, the southwest national instrumental music is an important part of Chinese traditional music art and culture, which cannot be lost.

Southwest instrumental music resources were born in the long history of human activities in Southwest China. It contains unique local cultural characteristics, is a symbol of local culture, and is also an important treasure of Chinese traditional culture. At the same time, southwest instrumental culture has a long history and unique characteristics. Therefore, from the perspective of cultural value, it has good conditions for sustainable development and research. At the same time, the research potential of southwest instrumental music resources has not been fully developed. In fact, there is still a lot of space. Therefore, its research potential is huge. If we can give full play to its value, it will help the development and inheritance of local culture, and will also bring good economic benefits. Figure 1 The inheritance path of southwest minority instrumental music.

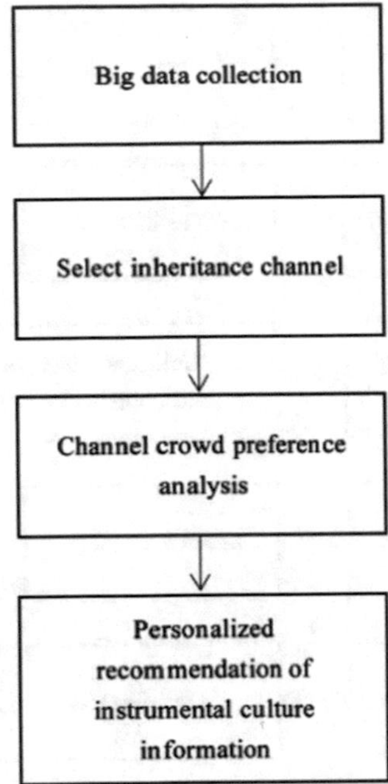

Fig. 1. The inheritance path of southwest minority instrumental music

3 Big Data Technology System Framework and Implementation Method of Southwest Minority Instrumental Music Transmission

3.1 System Framework

Figure 2 is the basic framework of the system, which consists of five parts: data acquisition layer, data storage layer, data pretreatment layer, data analysis layer and result output layer.

(1) Data acquisition layer. Data collection layer is mainly responsible for collecting original data, and the original data from the network data, namely the modern basic can communicate in a network, it has become a social normal phenomenon, and formed a network of community culture under this phenomenon, make network users gather together, gathering place is in the nature of the platform, which leads to the platform of channel is information transmission path, It generally refers to various network ports, etc. Therefore, as long as these ports are mastered, data acquisition tools can be used to obtain the relevant information of users in the corresponding platform, which will be imported into the data storage layer [2]. Considering the quality of data collection results, data collection objects should be set before collection, as shown in Table 1 [4, 5].

(2) Data storage Layer. Although object by setting the data acquisition way to limit the number of data collection, but too much because the number of users in the platform, data growth too fast, so the resulting data information is still huge, at

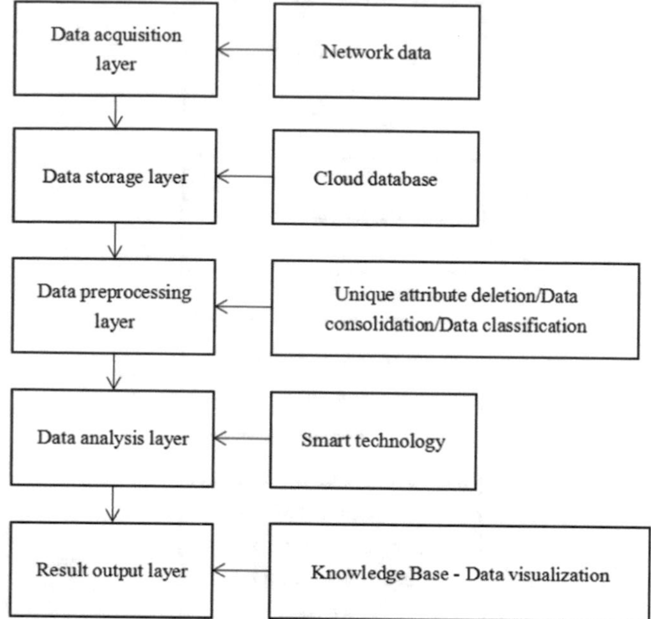

Fig. 2. Big data technology system framework

Table 1. Data collection objects in the data collection layer of the system

Data acquisition object	Describe
Communicate information	Understand the main topics of conversation
Browse information	Know what information users like to view
The tag information	Understand the interest labels that users fill in

this time in order to facilitate subsequent operations, huge data must be stored in the database, so the need for data storage layer design. This paper consider the acquisition data from large scale, common database may not be able to meet the demand of storage capacity, if use circumscribed or other ways to expand capacity, can lead to large cost, so the system database is mainly choose cloud database, the database is infinite capacity theory, good controllability, and storage resources to meet the demand of data storage, Data storage security can also be guaranteed through design [6].

(3) Data preprocessing layer. Because the actual collected initial data may have quality problems, direct application in the follow-up process will lead to insufficient accuracy of the results and mislead the final analysis, so the data pretreatment layer should be set according to common quality problems, which is mainly composed of data reprocessing and noise reduction.

(4) Data analysis layer. Data analysis layer is the core of the system, mainly through the analysis of pre-processed data through relevant technical means, its analysis steps are: first, define the data, then classify according to the definition, and do a good job of data symmetry processing, that is, user-path-preference; Second, through data generation analysis results, show the preferences of users within the platform; Third, the platform on the basis of user preferences and southwest ethnic instrumental music related keywords and docking, if user preferences and a keyword docking success, is suitable for use in southwest ethnic instrumental music transmission platform, effectiveness of grade 1, if the user preferences and more than one keyword docking success, effectiveness is >1 grade, the higher the grade, It shows that the platform is more suitable for southwest minority instrumental music transmission [7, 8].

(5) Result output layer. Since the later dissemination of instrumental music of southwest ethnic minorities still relies on manual work, the results of system analysis must be clearly displayed to manual work, so it is necessary to design the output layer of results. The output layer design is mainly to show the analysis results of the data analysis layer to the manual, and the results are presented in the form of reports to ensure the visualization of results. In addition, in order to let the data analysis results into the result output layer, the system introduces knowledge base in the design of the result output layer, which is specially used to store the analysis results.

3.2 System Implementation Method

The system framework of the five parts of the implementation method and related design content as follows.

Data Acquisition Layer Implementation

The key to realize the data acquisition layer lies in the selection of data acquisition tools and the design of data acquisition schemes. The specific implementation methods of the two points are as follows: First, because the data collection objects are scattered in the network environment of the platform, a widely distributed, so is not suitable for using in centralized data collection acquisition, this paper mainly chose network worm tool, it can crawl in the network environment, according to the preset good data collection requirements, first identify path label is in accordance with the requirements, If it meets the requirements, the path will be brought into the system for collection. For example, if a web platform meets the requirements, the network worm tool will record the domain name of the web page, and then transmit the domain name to the system for collection. Second, although the network worm tool with functions of data acquisition, but the collection efficiency is slow, and there may be gathering omission phenomenon, therefore this system design a network worm+artificial data acquisition scheme, namely the side through the acquisition of network worms, on the other side of the artificial synchronized acquisition (network worms as main collection tools, manual only provides auxiliary, collection is).

Data Storage Layer Implementation

Mentioned above, the system data storage layer mainly include database implementation of cloud, the cloud data cannot be directly applied in this system, otherwise it will cause data storage security issues, namely the cloud database capacity is infinite, because the storage space is to open the network, use can make data directly exposed in the public eye, very easy to data security problems. To this end, the first in a data storage needs consideration, based on cloud database unlimited storage resources part divides the storage resources, create a basic data storage space, space meet the demand of a data storage capacity, then the closed part of resources, leaving only a account audit interface, anyone needs to be approved through the interface to reach the internal data, In this way, data security is ensured on the basis of meeting data storage requirements [9].

Data Preprocessing Layer Implementation
In the data preprocessing layer, this paper mainly selects data matching algorithm and data standard comparison method for preprocessing: First, in order to remove duplicate data, single-word similarity algorithm (a type of data matching algorithm) is selected in this paper to achieve pre-processing. The specific principle of this algorithm is shown in Formula (1). According to the calculation results, if the matching degree between data $= 100\%$, it means that data is repeated and can be deleted. Second, according to different data acquisition program in advance to set up data collection specification, user information, for example, the specification for the user name, user output content, user content output paths, and establish the related forms, will collect data from after break up, fill out the form, if the form, it found that there is no data to fill in a column, means that the data is not complete, Delete the entire data.

$$d = \frac{1}{2}P + \frac{1}{2}s \tag{1}$$

In the formula, *P* represents the similarity of phonetic code, *S* represents the similarity of graphic code, and each of them accounts for 50% of the similarity of the whole word.

Data Analysis Layer Implementation
Data analysis layer of this system is mainly composed of intelligent technology, the technology to the data mining, but the successful application of this technology to realize data mining analysis, must design the logic analysis, this also is a key link in the process of data analysis layer, its core is how to implement the data classification, Fig. 3 for the classification principle of smart technology.

According to Fig. 2, the data classification is realized to know the preferences of platform users. Then, the classification results can be connected with the key words of southwest minority instrumental music, and the results can be generated according to the docking relationship.

Result Output Layer Implementation
Using Java language programming, the results that cannot be manually identified in the system are converted into Chinese, and the result report is generated according to the standard template, and then the report is transmitted to the knowledge base. In view of the knowledge base, this paper mainly selects the common SQL database, which can be accessed manually to obtain the report and consult the results.

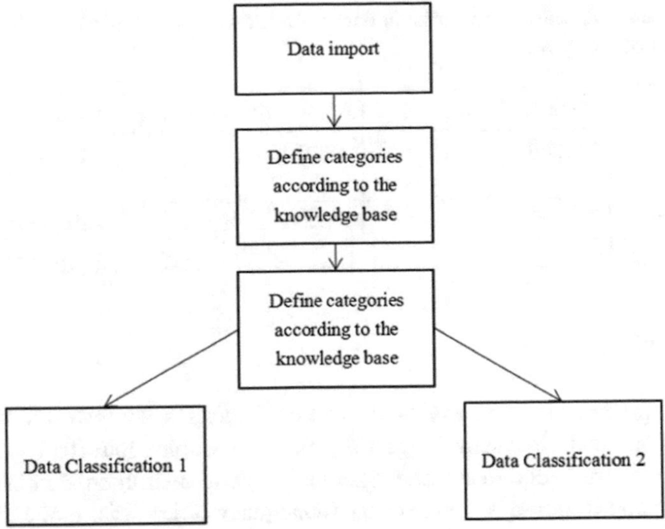

Fig. 3. Classification principle of intelligent technology

4 System Verification

4.1 Verification Scheme

Travels in a department southwest ethnic instrumental music based on the data of a year in a row, looking for a travel path, after the system's year-long southwest ethnic instrumental music transmission, synchronization of statistical data, so the system a year before and after the application of the spread of the southwest ethnic instrumental music two group data, compare two groups of data, If the data in the year after the application of the system is better than that in the year before application, it shows that the application of the system is effective and conducive to the spread of southwest minority instrumental music [10–12].

4.2 Results and Discussion

Table 2 shows the instrumental music transmission data of southwest ethnic minorities before and after the application of the system.

According to Table 2, the instrumental music transmission data of southwest ethnic minorities in the year after the application of the system far exceeded that of the year before, so the application of this system is effective.

Table 2. Instrumental music transmission data of southwest ethnic minorities before and after the application of the system

Time	Click on the quantity	Amount of forwarding	Keyword reference
Year prior to application	128000	90000	113000
One year after application	271000	132000	192000

5 Conclusion

As one of the precious national excellent traditional cultures, southwest minority instrumental music cannot be lost in the long river of time, so people should try to protect southwest minority instrumental music and vigorously spread instrumental music culture. In this context, people should be aware of the inadequacy of artificial mode, draw lessons from the past, and choose big data technology to help southwest minority instrumental music spread.

Acknowledgement. Investigation and Research on the ecological status of Qiang Village Music (No.:xnyy2020009), a key project of Southwest Music Center, a key research base of philosophy and Social Sciences in Sichuan Province in 2020. The 2021 youth project of Southwest Music Center, the key research base of philosophy and Social Sciences in Sichuan Province, "Research on the making skills of Yi folk plucked musical instruments from the perspective of oral history" (No.:xnyy2021036). The 2021 general project of Yi culture research center, the key research base of philosophy and Social Sciences in Sichuan Province, oral history of the making skills of Yi folk.

References

1. Zhou, Z., Gao, C., Chen, X., Zhang, Y., Mumtaz, S., Rodriguez, J.: Social big-data-based content dissemination in Internet of vehicles. IEEE Trans. Industr. Inform. **14**(2), 768–777 (2018)
2. Xu, X.: Cultural communication in double-layer coupling social network based on association rules in big data. Pers. Ubiquit. Comput. **24**(1), 57–74 (2019). https://doi.org/10.1007/s00779-019-01308-y
3. Jing, X.: Research on the method of children's body shape recognition based on big data technology of dressed human body. J. Phys. Conf. Ser. **1790**(1), 012015 (2021)
4. He, X., Wang, Y.: Impact of the application of big data technology on industrial agglomeration and high-quality economic development. In: Atiquzzaman, M., Yen, N., Xu, Z. (eds.) BDCPS 2021. LNDECT, vol. 102, pp. 353–360. Springer, Singapore (2022). https://doi.org/10.1007/978-981-16-7466-2_39
5. Dai, D., Chen, J.: Research on mathematical model of box office forecast through BP neural network and big data technology. J. Phys: Conf. Ser. **1952**(4), 042118 (2021)
6. Youngblood, M.: Conformity bias in the cultural transmission of music sampling traditions. R. Soc. Open Sci. **6**(9), 191149 (2019)

7. Narayanan, S.R., Yang, S.: Airborne transmission of virus-laden aerosols inside a music classroom: effects of portable purifiers and aerosol injection rates. Phys. Fluids **33**(3), 033307 (2021)
8. Mondal, S., Sharma, R.K.: Low power hardware based real time music system and digital data transmission using FPGA. In: 2019 International Conference on Communication and Electronics Systems (ICCES) (2019)
9. Rehfeldt, R.A., Tyndall, I., Belisle, J.: Music as a cultural inheritance system: a contextual-behavioral model of symbolism, meaning, and the value of music. Behav. Soc. Issues **30**, 749–773 (2021). https://doi.org/10.1007/s42822-021-00084-w
10. Liu, X., Yang, C.: Remote music teaching classroom based on embedded system and cloud platform. Microprocessors and Microsystems **82**, 103844 (2021)
11. Bignardi, G., Smit, D.J.A., Ticini, L.F., et al.: On the etiology of the intensity of visual aesthetic appraisal and preferences. Behav. Genet. Int. J. Devoted Res. Inheritance Behav. Animals Man **2020**(6), 50 (2020)
12. Vermeulen, J.R., Schlag, F., Verhoef, E., et al.: Exploring genetic relationships between musicality and social behavior. Behav. Genet. Int. J. Devoted Res. Inheritance Behav. Animals Man **2020**(6), 50 (2020)

Correction to: Tenth International Conference on Applications and Techniques in Cyber Intelligence (ICATCI 2022)

Jemal H. Abawajy, Zheng Xu, Mohammed Atiquzzaman, and Xiaolu Zhang

Correction to:
J. H. Abawajy et al. (Eds.): ***Tenth International Conference on Applications and Techniques in Cyber Intelligence (ICATCI 2022)*****, LNDECT 170,**
https://doi.org/10.1007/978-3-031-29097-8

In the original version of the book, the following belated corrections has been incorporated:

In chapter "Design of Management Cloud Information System for Residential Buildings Based on High-Performance Computing and Face Recognition Technology", the authors' affiliation has been changed from "School of Jilin Engineering, Jilin Engineering Normal University, Changchun, Jilin, China" to "Jilin Engineering Normal University, Changchun, Jilin, China".

In chapter "Design and Research of Teaching Evaluation System Based on Big Data Technology", the authors' affiliation has been changed from "School of Guangzhou, Guangzhou Nanyang Polytechnic College, Guangzhou, 510925, Guangdong, China" to "Guangzhou Nanyang Polytechnic College, Guangzhou, 510925, Guangdong, China".

The correction book has been updated with the changes.

The updated original version of these chapters can be found at
https://doi.org/10.1007/978-3-031-29097-8_16
https://doi.org/10.1007/978-3-031-29097-8_112

J. H. Abawajy et al. (Eds.): ICATCI 2022, LNDECT 170, p. C1, 2023.
https://doi.org/10.1007/978-3-031-29097-8_122

Author Index

J. H. Abawajy et al. (Eds.): ICATCI 2022, LNDECT 170, pp. 1021–1024, 2023.
https://doi.org/10.1007/978-3-031-29097-8

Printed in the United States
by Baker & Taylor Publisher Services